Handbook of Applied Thermal Design

Other McGraw-Hill Reference Books of Interest

Handbooks

Avallone and Baumeister • MARKS' STANDARD HANDBOOK FOR MECHANICAL ENGINEERS
Brady • MATERIALS HANDBOOK
Carrubba and Gordon • PRODUCT ASSURANCE PRINCIPLES
Considine • ENCYCLOPEDIA OF INSTRUMENTATION AND CONTROL
Considine • PROCESS INSTRUMENTS AND CONTROLS HANDBOOK
Crosby • QUALITY IS FREE
Crosby • QUALITY WITHOUT TEARS
Feigenbaum • TOTAL QUALITY CONTROL
Fink and Beaty • STANDARD HANDBOOK FOR ELECTRICAL ENGINEERS
Fink and Christiansen • ELECTRONICS ENGINEER'S HANDBOOK
Gerelle and Stark • INTEGRATED MANUFACTURING
Harrington • THE IMPROVEMENT PROCESS
Higgins • MAINTENANCE ENGINEERING HANDBOOK
Hradesky • PRODUCTIVITY AND QUALITY IMPROVEMENT
Ireson and Coombs • HANDBOOK OF RELIABILITY ENGINEERING AND MANAGEMENT
Juran • MANAGERIAL BREAKTHROUGH
Juran and Gryna • QUALITY PLANNING AND ANALYSIS
Lubben • JUST-IN-TIME MANUFACTURING
Maynard • INDUSTRIAL ENGINEERING HANDBOOK
Ott • PROCESS QUALITY CONTROL
Perry and Green • PERRY'S CHEMICAL ENGINEERS' HANDBOOK
Ross • TAGUCHI TECHNIQUES FOR QUALITY ENGINEERING
Shinskey • PROCESS CONTROL SYSTEMS
Taylor • QUALITY CONTROL SYSTEMS
Woodson • HANDBOOK OF HUMAN FACTORS ENGINEERING DATA

Encyclopedias

CONCISE ENCYCLOPEDIA OF SCIENCE AND TECHNOLOGY
ENCYCLOPEDIA OF ELECTRONICS AND COMPUTERS
ENCYCLOPEDIA OF ENERGY
ENCYCLOPEDIA OF ENGINEERING
ENCYCLOPEDIA OF PHYSICS

Dictionaries

DICTIONARY OF SCIENTIFIC AND TECHNICAL TERMS
DICTIONARY OF MECHANICAL AND DESIGN ENGINEERING
DICTIONARY OF COMPUTERS

For more information about other McGraw-Hill materials, call 1-800-2-MCGRAW in the United States. In other countries, call your nearest McGraw-Hill office.

Handbook of Applied Thermal Design

Eric C. Guyer, Sc.D. *Editor in Chief*
Yankee Scientific, Inc.

David L. Brownell, P.E. *Associate Editor*
Yankee Scientific, Inc.

McGraw-Hill Book Company

New York St. Louis San Francisco Auckland
Bogotá Hamburg London Madrid Mexico
Milan Montreal New Delhi Panama
Paris São Paulo Singapore
Sydney Tokyo Toronto

Library of Congress Cataloging-in-Publication Data

Handbook of applied thermal design.

Bibliography: p.
Includes index.
1. Heat engineering. 2. Engineering design.
I. Guyer, Eric C. II. Brownell, David L.
TJ260.H35 1988 621.402 87-32428
ISBN 0-07-025353-6

1234567890 DOC/DOC 89321098

ISBN 0-07-025353-6

The editors for this book were Betty Sun and David E. Fogarty, the designer was Naomi Auerbach, and the production supervisor was Dianne Walber. It was set in Times Roman by Techna Type, Inc.

It was printed and bound by R. R. Donnelley & Sons Company.

Contents

Part 3 Thermal Insulation

Part 4 Structural Materials in Thermal Design

Part 5 Heat-Transfer Fluids

Part 12 Thermal Sensors

Part 13 Automatic Temperature Controls

About the Editor in Chief

ERIC C. GUYER, SC.D., is president and technical director of Yankee Scientific, Inc., a company that provides engineering and design services for the power, electronics, process, and HVAC industries, among others. Dr. Guyer is a graduate of Lowell University and Massachusetts Institute of Technology and resides in Dover, Massachusetts.

Contributors

Paul David AuCoin, *Product Specialist, Solar-Weston, Houston, Tex.* (PART 12, CHAP. 4)
W. C. Bachman, *Dylon Ceramic Technologies Co., Berea, Ohio* (PART 6, CHAP. 7)
David F. Bartz, *Tecogen Inc., Waltham, Mass.* (PART 5, CHAP. 3)
Walter B. Bienert, *President, Dynatherm Corporation, Corkeysville, Md.* (PART 7, CHAP. 4)
Bennett P. Boffardi, *Senior Research Fellow, Calgon Corporation, Pittsburg, Pa.* (PART 5, CHAP. 2)
David L. Brownell (ASSOCIATE EDITOR), *Yankee Scientific, Inc., Dover, Mass.* (PART 2, CHAP. 1)
Robert E. Buckley, *Mohawk Associates, Norwalk, Conn.* (PART 12, CHAP. 7)
S. L. Camacho, *Plasma Energy Corporation, Raleigh, N.C.* (PART 8, CHAP. 5)
D. Mitchell Carr, *Manager of Applications Engineering, Eurotherm Corporation, Reston, Va.* (PART 13, CHAP. 2)
Majid Charmchi, *Department of Mechanical and Energy Engineering, University of Lowell, Lowell, Mass.* (PART 1, CHAP. 3 AND CHAP. 4)
Wilson A. Clayton, *Director of Engineering, Hy-Cal Engineering, El Monte, Calif.* (PART 12, CHAP. 11)
Kim T. Gilbert, *Mirror Insulation Unit, Diamond Power Specialty Company, Lancaster, Ohio* (PART 3, CHAP. 5)
Michael W. Golay, *Massachusetts Institute of Technology, Cambridge, Mass.* (PART 1, CHAP. 1)
Martin Gollin, *ARCO Chemical Co., Newtown Square, Pa.* (PART 5, CHAP. 1; PART 7, CHAP. 2)
Eric C. Guyer (EDITOR IN CHIEF), *Yankee Scientific, Inc., Dover, Mass.* (PART 1, CHAP. 1 AND CHAP. 2; PART 9, CHAP. 4)
William B. Hampshire, *Assistant Manager, Tin Research Institute, Inc., Columbus, Ohio* (PART 6, CHAP. 2)
Ronald H. Howell, *Professor of Mechanical Engineering, University of Missouri, Rolla* (PART 9, CHAP. 1)
J. P. Hutchinson, *General Products Manager, Harbison-Walker Refractories, Division of Dresser Industries, Inc., Pittsburg, Pa.* (PART 6, CHAP. 4)
Anthony J. Izbicki, *Senior Vice President, Hood & Co., Inc., Division of Dresser Industries, Inc., Hamburg, Pa.* (PART 6, CHAP. 6)
Brian Jones, *Minco Products, Inc., Minneapolis, Minn.* (PART 12, CHAP. 3)
William H. Kielmeyer, *Senior Research Engineer, Manville Corporation, Denver, Colo.* (PART 3, CHAP. 2)
E. Kimmel, *Director of Research and Development, Tempil Division, Big Three Industries, Inc., South Plainfield, N.J.* (PART 12, CHAP. 1)
Jay A. Kohler, *York International Corporation, York, Pa.* (PART 7, CHAP. 3)
Thomas Kopacz, *Marketing Manager, Thermo Electric Company, Inc., Saddle Brook, N.J.* (PART 12, CHAP. 2)
Yair Lahav, *APD Cryogenics, Inc., Allentown, Pa.* (PART 11, CHAP. 1)
George A. Lane, *The Dow Chemical Company, Central Research Laboratories, Inorganic Materials and Catalysis Laboratory, Midland, Mich.* (PART 6, CHAP. 1)
G. Robert Leavitt, *formerly Manager, Quality Assurance, Taylor Instrument Co., Rochester, N.Y.* (PART 12, CHAP. 6)
Steven A. Llewellyn, *Llewellyn Associates, Inc., Feeding Hills, Mass.* (PART 8, CHAP. 6)
Ralph C. Longsworth, *APD Cryogenics, Inc., Allentown, Pa.* (PART 11, CHAP. 1)
Victor Lorenzetti, *Product Manager—Active Cooling, EG&G Wakefield Engineering, Inc., Wakefield, Mass.* (PART 7, CHAP. 5)
Robert P. Madding, *Inframetrics Inc., Bedford, Mass.* (PART 12, CHAP. 8)
George Markus, *Tooling and Plastic Engineer, Advanced Plastics Design, Inc., Fort Wayne, Ind.* (PART 4, CHAP. 3)
Raymond Marlow, *Marlow Industries, Inc., Dallas, Tex.* (PART 9, CHAP. 2)
Jose G. Martin, *Department of Mechanical and Energy Engineering, University of Lowell, Lowell, Mass.* (PART 1, CHAP. 3 AND CHAP. 6)
Harry I. McHenry, *Deputy Chief, Fracture and Deformation Division, National Bureau of Standards, Boulder, Colo.* (PART 4, CHAP. 1)
Jeffrey P. Melly, *Watlow Electric Manufacturing Company, St. Louis, Mo.* (PART 8, CHAP. 2)
Raymond E. Mesloh, *Battelle Memorial Institute, Columbus, Ohio* (PART 6, CHAP. 5)
Donald K. Miller, *Refrigeration Engineering Specialist, York International Corporation, York, Pa.* (PART 9, CHAP. 3)
Ali Reza Moazed, *Research Engineering Associate, Construction Products Division, W. R. Grace & Company, Cambridge, Mass.* (PART 3, CHAP. 3)
Paul Moffitt, *Chief Engineer, J. A. Callanan Company, Chicago, Ill.* (PART 8, CHAP. 3)
Manuel J. Blanco Muriel, *Department of Mechanical and Energy Engineering, University of Lowell, Lowell, Mass.* (PART 1, CHAP. 6)
Hossein P. Nourbakhsh, *Department of Nuclear Energy, Brookhaven National Laboratory, Upton, N.Y.* (PART 1, CHAP. 5)
Gene D. Nutter, *Assistant Director, Instrumentation Systems Center, University of Wisconsin, Madison* (PART 12, CHAP. 7)
E. L. Piper, *Dylon Industries Inc., Berea, Ohio* (PART 6, CHAP. 7)
Thomas A. Pizzitola, *Senior Control Engineer, Fluor Engineers, Houston, Tex.* (PART 12, CHAP. 4)
Richard J. Reed, *North American Manufacturing Company, Cleveland, Ohio* (PART 10, CHAP. 1)
K. P. R. Reddy, *Corning Glass Works, Corning, N.Y.* (PART 4, CHAP. 4)
Jack D. Rickly, *Olin Research Center, Cheshire, Conn.* (PART 5, CHAP. 5)
N. V. Ross, *Ajax Magnethermic, Warren, Ohio* (PART 8, CHAP. 4)
Rocco A. Sapienza, *Liquid Crystal Division, Hallcrest, Glenview, Ill.* (PART 12, CHAP. 10)
Harry J. Sauer, Jr., *Professor of Mechanical Engineering, University of Missouri, Rolla* (PART 9, CHAP. 1)
Walter F. Seifert, *Associate Development & Research Manager for High Temperature Heat Transfer Fluids, The Dow Chemical Company, Midland, Mich.* (PART 5, CHAP. 4)
Larry S. Shu, *Associate Director of Research, Construction Products Division, W. R. Grace & Company, Cambridge, Mass.* (PART 3, CHAP. 3)
Ashok K. Singhal, *President and Technical Director, CFD Research Corporation, Huntsville, Ala.* (PART 1, CHAP. 7)
D. Joseph Stadtmiller, *The Partlow Corporation, New Hartford, N.Y.* (PART 13, CHAP. 1)
Keith E. Starner, *York International Corporation, York, Pa.* (PART 7, CHAP. 3)
William R. Strzepek, *Dow Chemical U.S.A., Granville, Ohio* (PART 3, CHAP. 4)
Robert W. Swindeman, *Oak Ridge National Laboratory, Oak Ridge, Tenn.* (PART 4, CHAP. 2)

R. D. Thompson, *Deceased, formerly Manager, Long Range Planning, Taylor Instruments Co., Rochester, N.Y.* (PART 12, CHAP. 5)

Richard L. Troyer, *Section Manager, Research, Manville Corporation, Denver, Colo.* (PART 3, CHAP. 2)

Ronald P. Tye, *Dynatech Scientific, Inc., Cambridge, Mass.* (PART 3, CHAP. 1)

James H. Van Sant, *Lawrence Livermore National Laboratory, Livermore, Calif.* (PART 1, CHAP. 2)

Graham Walker, *Professor of Mechanical Engineering, University of Calgary, Alberta, Canada* (PART 7, CHAP. 1)

F. H. Walther, *Director, Garber Research Center, Harbison-Walker Refractories, Division of Dresser Industries, Inc., Pittsburg, Pa.* (PART 6, CHAP. 4)

Robert Watson, *The Kanthal Corporation, Bethel, Conn.* (PART 8, CHAP. 1)

Pieter R. Wiederhold, *General Eastern Instruments Company, Watertown, Mass.* (PART 12, CHAP. 9)

Ernest G. Wolff, *Mechanical Engineering Department, Oregon State University, Corvallis, Oreg.* (PART 6, CHAP. 3)

Preface

"Blessings on him who invented . . . , the fire that warms the cold, the cold that moderates heat, . . . , and the simple with the wise."

Don Quixote
[MIGUEL DE CERVANTES (1547–1616)]

This new Handbook will be of value to anyone who must solve heat and temperature problems in engineering design. It provides a general perspective and usable information on the four principal facets of thermal design—heat transfer analysis, materials performance, heating and cooling technology, and instrumentation and controls. We have composed this Handbook to be the basis for creating complete solutions to thermal design problems.

Problems of thermal design are encountered in materials processing systems, manufacturing equipment, consumer products, power machinery, electrical equipment, appliances, energy systems, and scientific apparatus. Indeed, thermal design is a common concern with any device that converts energy or is exposed to a dynamic thermal environment.

This Handbook is comprehensive in that we have attempted to describe the full range of available techniques, technologies, and materials that have application in the solution of thermal design problems. We have avoided focusing on the particular thermal design requirements of specific industries or technologies. Rather, our approach has been to describe the general utility of the methods and things that comprise the modern foundations of thermal design. Thus, this Handbook should be of use in reevaluating solutions to traditional problems in thermal design, solving current problems, and constructing answers to unknown future problems.

The contributing authors represent a wide variety of backgrounds and expertise. A common characteristic among them is their direct involvement in the applied aspects of their respective fields. The reader can be assured that the information and advice provided by the contributors is relevant to the practical solution of thermal design problems.

Part 1 of the Handbook is a general overview of the field of thermal design, a compilation of readily applicable solutions to commonly encountered heat transfer problems, and a perspective on computer-based numerical methods useful in thermal design. Part 2 is a collection of the thermophysical properties of engineering materials. Thermal insulations are covered in Part 3 from the viewpoint of the basic material properties and application considerations. Part 4 provides guidance on the selection of structural materials for elevated- and low-temperature service. The characteristics and application of various fluid media used as heat transfer fluids are described in Part 5. Part 6 covers a wide variety of materials having special application in thermal design including phase-change thermal storage materials, refractories, conductive plastics, heat-transfer compounds, fusible alloys, bimetallic materials, and low-expansion materials. Heat exchangers and related devices are characterized in Part 7 with an emphasis on describing the range of technology options available to the user. The various types of electric power heating are presented in Part 8, and commonly used refrigeration systems are described in Part 9. The design options for combustion heating with clean fuels are presented in Part 10 along with techniques for making estimates of performance. The properties of cryogens and the particular techniques of low-temperature design are detailed in Part 11. The many commonly applied methods of temperature sensing are described in Part 12. To complete the volume, Part 13 explains the technology and use of electrical and self-operated temperature controls.

ERIC C. GUYER
DAVID L. BROWNELL

Acknowledgments

The creation of this Handbook has been possible only through the cooperative and careful efforts of the many contributing authors. The significant effort of Elsie A. Guyer in reviewing the style and composition of the many manuscripts is greatly appreciated and gives the editors confidence that this volume will be a credit to the English language. The continuing and active interest of our colleague Martin K. Gollin has been of great value in keeping the production of the Handbook on schedule. Technical reviews by George J. Champagne and Karl G. Coumou in the areas of instrumentation and controls have also been an important asset. The efforts of Cindy K. Bennett and Rose Heredia to fit the typing of manuscripts and letters into their busy publications activities have also been crucial to our success in maintaining schedule. We wish to acknowledge the direct and indirect assistance of our wives, Deborah J. Guyer and Lorie L. Brownell, in the many tasks involved, over several years, in the editing of the Handbook. Finally, we are appreciative of the support that Betty Sun, the sponsoring editor at McGraw-Hill, has provided from concept to completion.

ERIC C. GUYER
DAVID L. BROWNELL

Part 1

Fundamentals of Thermal Design and Analysis

Chapter **1**

Elements of Thermal Design

ERIC C. GUYER
Editor in Chief

MICHAEL W. GOLAY
Massachusetts Institute of Technology

1.1 THE NATURE OF THERMAL DESIGN

Definition

Thermal design is a creative endeavor that involves the rational application of physical principles, materials, and devices for the purpose of controlling or managing heat flow and temperature. Thermal phenomena are a factor in the design of nearly every product, process, or machine that converts energy or is exposed to a dynamic thermal environment. The importance of thermal design can range from incidental to paramount.

Typically only one aspect of the total activity of an engineering design project, thermal design must often be integrated with mechanical, structural, electrical, chemical, or aesthetic design. The interrelationship with other design objectives and requirements will often have the consequence of constraining achievable thermal behavior.

Both local and distributed heat flow and temperature problems are experienced in the field of thermal design. The problem at hand may be as focused as the temperature of a fastener or as expansive as the regulation of thermal energy flow in a heating network. A distributed and interconnected group of thermal and fluid components, such as heat exchangers, heaters, and chillers, is referred to as a thermal system.[1]

The Concept of Thermal Control

Thermal control, the objective of thermal design, generally involves sustaining a desired temperature or heat flow behavior in a circumstance that acts to alter the desired behavior. Strategies for steady heat flow and temperature control can be related in a general fashion to the basic equation[2]

$$Q = \frac{\Delta T}{R} \tag{1.1}$$

where Q = total rate of heat transfer between two regions
ΔT = effective temperature difference between two regions
R = resistance to heat flow

The resistance to heat flow may be strongly dependent on both the magnitude of the temperature and the effective temperature difference. In circumstances where control of heat flow is required, a general rule would be that the designer may work with either the governing temperature difference or thermal resistance. Conversely, where control of temperature is required, heat flow and thermal resistance are possible design variables. Often, in practical problems, one of the two possible design variables is fixed by the circumstances of the situation. In problems involving heat generation, control of the heat generation rate may be the most direct method of control.

When time-varying heat flow or temperature control is required, the thermal energy capacitance characteristics of the situation become important. In some situations, the thermal capacitance may also be a design variable; an example is the design of an electric overload fuse.

These basic concepts of thermal control apply to either passive or active thermal control. Passive thermal control utilizes the intrinsic thermal properties of materials in a largely static manner. Physical motions may occur in passive control but would be limited to those arising from the extant temperature gradients. Active thermal control utilizes applied thermal power or directed fluid or mechanical movement to affect heat flow or temperature.

Physical Principles

For many practical problems of thermal design, the basic physical principles involved are those of heat transfer and thermodynamics. Some thermal design problems require careful consideration of the relationship between temperature and mechanical stresses and, thus, involve the principles of solid mechanics. Also, temperature and heat flow are often dependent on fluid flow; thus the principles of fluid mechanics become an important element in design.

One or all three modes of heat transfer—conduction, convection, and radiation—may come into play in a particular thermal design problem. Within the context of the definition of thermal design as presented above, thermodynamic considerations are usually limited to those arising from the first law of thermodynamics. The first law of thermodynamics expresses the principle of energy conservation. The definition of thermal design as presented excludes, by intent, the thermodynamic considerations involved in the conversion of thermal energy to mechanical work within the limitations of the second law of thermodynamics. However, the many heat flow and temperature control problems inherent to energy conversion systems are among the most challenging problems in thermal design.

Technology Basis

The opportunities for efficient and optimum thermal design are importantly limited by the actual thermophysical and thermomechanical properties of materials. Materials conduct heat at finite rates, soften or develop high vapor pressures at high temperatures, and have limited ability to retain heat. A knowledge of the performance and limitation of available materials is as important to the thermal design process as is the understanding of the basic physical principles of heat transfer and thermodynamics. Often the accuracy and the extent of available physical properties data are the major factors in design uncertainty.

Knowledge of the technology of thermal devices such as heat exchangers, heaters, and refrigeration units is also key to practical thermal design. A wide range of such thermal power and transport devices is available. Again, limitations exist as such devices operate within the constraints of heat-transfer principles and thermodynamic laws.

Temperature sensing and automatic control comprise the final major area of knowledge and technology within the scope of thermal design. Temperature sensing may be important at any of several stages of the design—from problem diagnosis to performance evaluation to automatic control. As such, design tradeoffs between accuracy, economics, and practicality are routinely encountered. Design options for automatic temperature control range from simple, self-powered mechanical devices to computer-based systems providing multipoint control of a distributed system. Fortunately, many practical automatic control problems can be resolved without a detailed understanding of the dynamics of the thermal behavior.

The Thermal Design Process

The thermal design process has, in general, a logical progression of activity in moving from problem to solution. The steps in this progression can be characterized as follows:

1. Definition of thermal control objective
2. Formulation of analytical model
3. Evaluation of heat-transfer rates, temperature distributions, and thermodynamic limitations
4. Determination of material requirements and material selection
5. Determination of thermal power requirements and selection of thermal power system
6. Formulation of control strategy and selection of control system

The iterative application of some or all of these steps is usually required to achieve a good solution to a thermal design problem.

1.2 THE PRINCIPLES OF TEMPERATURE, ENERGY CONSERVATION, AND HEAT FLOW

Temperature is a measure of the hotness or coldness of a substance. It is a macroscopic manifestation of the physical behavior of substances at a microscopic, molecular level. Heat is defined as the energy transferred by virtue of a temperature gradient. Heat can only be identified as it is transferred by a temperature gradient.

The principle of energy conservation involves the concepts of heat, work, and other energy forms. The first law of thermodynamics is the law of conservation of energy. This law can be stated in a variety of ways, all having essentially the same meaning: energy can neither be created nor destroyed by ordinary chemical means and the sum total of all energy remains constant. Depending on the nature of the system or process under consideration, the statement of the first law can take different forms.

The First Law of Thermodynamics

Heat and work are different forms of energy. Energy is always conserved. Thermodynamic systems, which are simply selected and defined parts of the universe, are a convenient means of accounting for energy. Energy which enters a thermodynamic system as work may leave as heat or energy. The first law of thermodynamics is a statement of the relationship between heat, work, and energy stored in a system. A common expression of the law is

$$Q = \Delta E + W \tag{1.2}$$

where Q = heat transferred to system during a process
W = work transferred from system during a process
ΔE = change in internal energy of system during the process

This equation does not contain an expression for the internal energy E explicitly, but contains the change in internal energy. It can be shown that E, the internal energy, is a property of a system and varies with the condition or state of the system. If an arbitrary or reference value of E is assigned to a particular state, it is possible to determine the values at all other states by the application of the above equation. Knowledge of the absolute value of E is not required to apply the principle of energy conservation. Work, as used in the above equation, is defined as the product of a force and the displacement through which it acts.

Internal energy may be present in any of a number of kinetic and potential energy forms at macroscopic, molecular, and atomic levels. For a system composed of a substance of uniform chemical composition and excluding kinetic and gravitational potential energy, the first law can be expressed as

$$Q = \Delta U + W \tag{1.3}$$

where U is, for a pure substance, an expression of internal energy that excludes kinetic and gravitational potential energy as well as other energies associated with surface tension on electrical or magnetic effects. Changes in internal energy U are associated with changes in temperature and changes of phase. Internal energy U is an extensive property in that, at any given state, it is proportional to the mass of the system. An intensive property, the specific internal energy u is given by

$$u = \frac{U}{m}$$

where m is the mass of the system. The specific internal energy, like internal energy, is a state property, meaning that it is dependent only on the state in which the system exists. The specific internal energy of a pure substance can be characterized as a function of two observable state properties, such as temperature and pressure.

The enthalpy of a substance is a property that is defined as the sum of the internal energy and the pressure-volume product, or

$$h = u + Pv \tag{1.4}$$

where P is pressure and v is specific volume. Enthalpy has convenient and practical applications in the first law analysis of steady flow problems and has the same units as specific internal energy.

Specific Heat Capacity

An important concept related to the application of the first law of thermodynamics is that of specific heat capacity. Two types of specific heat capacity of common interest are the specific heat capacity at constant volume given by

$$C_v = \left(\frac{\partial u}{\partial T}\right)_v \tag{1.5}$$

and the heat capacity at constant pressure given by

$$C_p = \left(\frac{\partial h}{\partial T}\right)_p \tag{1.6}$$

Specific heats C_v and C_p have practical use in the calculation of heat transfer for a change in system temperature. The general form is

$$Q = \int_{T_1}^{T_2} mC\, dT \tag{1.7}$$

where Q = total heat transferred between a closed system and its surroundings
m = system mass
C = C_v or C_p, depending on the kind of process

For small temperature differences where the change in specific heat capacity is not large, the equation

$$Q = mC(T_2 - T_1) \tag{1.8}$$

is usually adequate.

For solids and liquids, values for C_v and C_p are very similar, but generally are functions of temperature. Because of this fact, specific heats for solids and liquids presented in the literature normally are not designated as to process but are presented only as functions of temperature. The difference between C_v and C_p for gases is important. For ideal gases,

$$C_p - C_v = R \tag{1.9}$$

where R is the gas constant in the equation of state for an ideal gas,

$$Pv = RT \tag{1.10}$$

Latent Heat

Another concept important to the application of the first law is that of latent heat. Latent heat is the enthalpy change associated with a change in phase or physical structure of a substance. Latent heat related to the vaporization of a liquid is the heat of vaporization, and latent heat related to the melting of a solid is called heat of fusion. Latent heat is also associated with the process of sublimation and certain solid-phase transitions. Latent heat is considered to be a change in enthalpy of a substance at constant pressure for a given phase change.

Control Volumes

The first law of thermodynamics may be applied to either a thermodynamic system or a thermodynamic control volume. As previously stated, a thermodynamic system is a defined and constant quantity of matter. Heat and work external to the system may interact with the system across its boundary. A control volume is a fixed region in space having a boundary through which matter, heat, and work may flow. By virtue of the first law, the net flow of energy through the boundary of the control volume in a certain time period must equal the change, during the same time, of the energy stored within the control volume. Mass in a control volume must, of course, also be conserved. The first step in the application of the first law is, thus, to define a system when dealing with a fixed mass or a control volume when dealing with a flow of mass.

In many thermal design problems involving stationary systems of solid substances, mechanical work is not a factor and the first law simply becomes, in general,

$$Q = \Delta U \tag{1.11}$$

For the case of steady-state conditions of the system,

$$\sum_{i=1}^{i=n} Q_i = 0 \tag{1.12}$$

meaning that the sum of all heat flows into the system is zero. This simple equation has frequent application.

In Fig. 1.1 the concept of a control volume is applied to a fluid-cooled furnace penetration. Figure 1.1*b* is the simplest application of the control volume concept. More accurate analysis, requiring fewer assumptions, can be achieved by segmenting a component or system into a number of control volumes, as depicted in Fig. 1.1*c*.

Heat Transfer

For engineering and design purposes, heat is considered to be transferred in three different fashions—by conduction, convection, and radiation. Conduction heat transfer results when bodies of different temperatures are in physical contact. Convection heat transfer results from the motion of fluids in the presence of temperature differences. Radiation heat transfer or, more properly, thermal radiation heat transfer is an electromagnetic phenomenon that results in energy flow between bodies at different temperatures, even though they may not be in physical contact. Different combinations of these three modes of heat transfer may be important in a particular situation.

Considerable theoretical and experimental information has been developed for analyzing heat-transfer problems. The theories of heat transfer have practical applications in the quantitative prediction and analysis of temperature distributions and heat flows.

Conduction Heat Transfer Conduction heat transfer is the process by which energy is transported through a mass continuum due to the transfer of kinetic energy at a molecular level. Energy flow occurs in the direction from areas of higher temperature and higher molecular energy level to areas of lower temperature and lower molecular energy level.

Macroscopically, the rate of heat conduction is observed to follow the empirical expression

$$q = -kA\,\frac{dT}{dx} \tag{1.13}$$

where q = heat transferred in direction x
dT/dx = temperature gradient in direction x
A = area normal to temperature gradient
k = thermal conductivity

This equation is called Fourier's law. The thermal conductivity is determined by experiment for specific material compositions and is, itself, generally dependent on temperature.

The heat flow and temperature effects of heat conduction in solids are a common concern in thermal design. General approaches to solving heat conduction problems include

1. Application of an existing analytical solution
2. Formulation and application of a simple numerical model tailored to the specific problem requirement
3. Application of a commercial computer program having the capability for a precise computation of multidimensional temperature distribution and heat flow

Analytical solutions exist for many geometries and conditions of practical interest (Chap. 2). The formulation and application of finite-difference numerical models of conduction heat transfer[3] using a limited number of grid points (say 10 to 20) have become increasingly convenient through the use of personal computers and software capable of solving the resulting simultaneous equations. This approach is often a good compromise between the geometry and boundary condition limitations of available analytical solutions and the expense of more powerful commercial thermal analysis programs, such as those discussed in Chap. 7.

Convection Heat Transfer Convection heat transfer is the transport of energy to or from a surface by the combined action of conduction heat transfer and fluid motion. Convection heat transfer at solid-fluid surfaces is of most common interest, but fluid-fluid interfaces are also encountered. Forced convective heat transfer involves fluids that are moved by some mechanical means. Free or natural convection heat transfer refers to buoyant convection, which results from the influence of gravity upon fluid density differences induced by the heat-transfer process itself.

Convective heat-transfer processes are usually characterized in terms of a convective heat-transfer coefficient h, which is employed in the equation

$$q = hA(T_1 - T_2) \tag{1.14}$$

where q = rate of heat transfer
A = surface area
$T_1 - T_2$ = temperature difference between surface and bulk fluid

This relationship is known as Newton's law of cooling. The convective heat-transfer coefficient h is a characteristic of the particular fluid, flow field, and surface situation.

The flow regime, whether laminar or turbulent, has a major influence on the convection heat-transfer coefficient. Except for certain well-defined laminar flow situations, exact analytical solutions for heat-transfer coefficients do not exist.

Thus the concept of the convective heat-transfer coefficient is practically applied through the aid of empirical convective heat-transfer correlations. Convective heat-transfer correlations, which are based largely on experimental results, relate the convective heat-transfer coefficient to readily definable fluid, flow, and geometric parameters. Generally, application of convective heat-transfer correlations requires a knowledge of the thermophysical properties of the fluid, the fluid velocity field relative to the surface, and the surface geometry.

Heat-transfer coefficients calculated from convection heat-transfer correlations will be uncertain. The level of uncer-

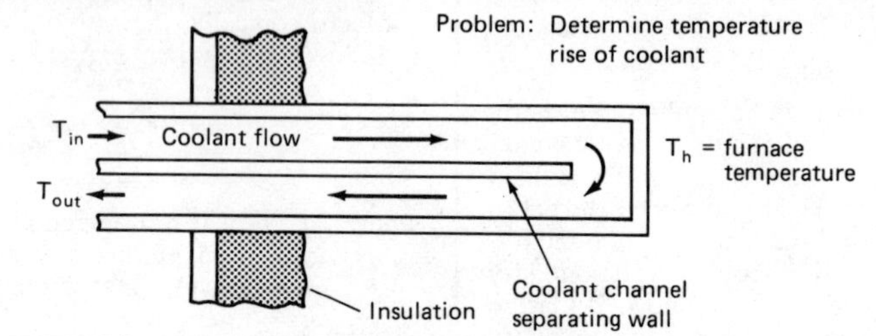

(*a*) Fluid-cooled penetration through insulated furnace wall—problem statement.

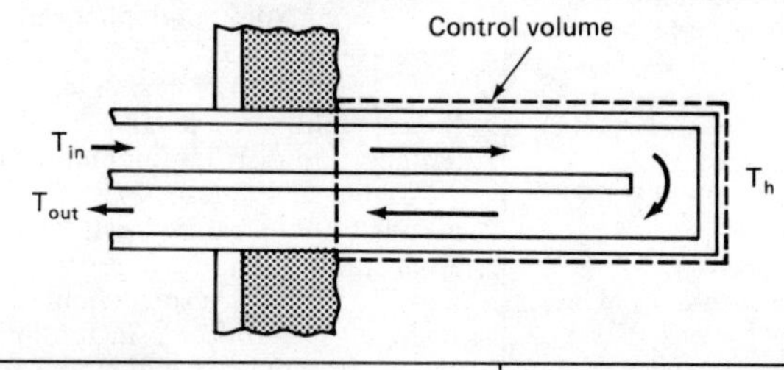

Heat transfer to coolant in control volume

$$Q_c = UA\,\Delta T$$

where U = overall heat transfer coefficient, furnace gas to coolant,
A = exposed area of penetration,
ΔT = mean temperature difference between furnace gas and coolant

For $(T_{out} - T_{in}) \ll (T_h - T_{in})$,

$$Q_c \approx UA\left(T_h - \frac{T_{in} + T_{out}}{2}\right)$$

Change in coolant internal energy in control volume

$$Q = \dot{m}C_p(T_{out} - T_{in})$$

where $\dot{m}$ = coolant mass flow
C_p = coolant specific heat capacity

SOLUTION: $Q = Q_c$; solve for T_{out}.

MAJOR ASSUMPTIONS:
U is uniform over surface.
No heat transfer across coolant channel separating wall.
Physical properties not dependent on temperature.
No heat conduction along channel walls.
No heat conduction through insulation.
Steady-state conditions.
Coolant temperature rise is small.

(*b*) Heat balance and solution using single control volume.

Fig. 1.1 Application of control volume concept to first-law analysis of fluid-cooled furnace penetration.

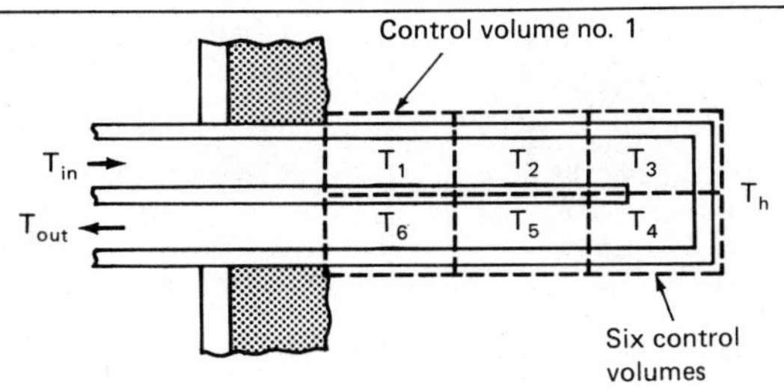

Heat transfer to coolant in control volume 1,

$$Q_{c1} = U_a A_a (T_h - T_1) + U_x A_x (T_6 - T_1)$$

for

$$2(T_1 - T_{in}) \ll (T_h - T_1) \text{ and } \ll (T_6 - T_1)$$

where A_a = local area for heat transfer from furnace gas to coolant
A_x = local area for heat transfer through channel wall
U_a = local overall heat-transfer coefficient, furnace gas to coolant
U_x = local overall heat-transfer coefficient, adjoining channel to coolant

Change in coolant internal energy in control volume 1,

$$Q_1 = 2\dot{m}C_p(T_1 - T_{in})$$

SOLUTION:
Write equations for Q_c and Q for each control volume.
Set $Q_{cn} = Q_n$ for $n = 1, \ldots, 6$.
Solve resulting six simultaneous equations for six unknown temperatures T_1 through T_6.
Adjust temperature-dependent physical properties based on initial solution.
Again solve six simultaneous equations.

MAJOR ASSUMPTIONS:
No heat conduction along channel walls.
No heat conduction through insulation.
Steady-state conditions.
Temperature change in each control volume is small.

(c) Heat balance and solution using multiple control volumes.

Fig. 1.1 (*Continued*)

tainty will depend on how close the situation of interest matches the original experiments, as well as on other factors. For design purposes, a minimum design factor of 15 percent should be applied to heat-transfer coefficients calculated from generalized correlations. Often much larger design factors are warranted, especially in situations of complex fluid flow and surface geometries.

Some general applicable correlations for convective heat-transfer coefficients are contained in Chaps. 3, 4, and 5. Convective heat transfer with so-called high performance or enhanced surfaces is discussed in Part 7, Chap. 3. Typical convective heat-transfer coefficients are as follows.

Fluid	Heat-transfer coefficient h	
	Btu/h · ft² · °F	W/m² · K
Gas		
Free convection	1–5	5–25
Flowing	1–50	5–250
Jet impingement	10–500	50–2500
Liquid		
Nonmetallic, flowing	10–5000	50–25,000
Nonmetallic, jet impingement	500–25,000	2500–125,000
Metallic, flowing	1000–50,000	5000–250,000
Boiling	200–50,000	1000–250,000
Condensing vapors	500–5000	2500–25,000

Radiation Heat Transfer Thermal radiation is a form of electromagnetic energy. It arises by virtue of the molecular and atomic agitation associated with the internal energy level of a substance. Thermal radiation travels at the speed of light in straight lines through homogeneous media. It is characterized by a frequency range of the electromagnetic spectrum that includes the visible portion and extends through the infrared. At any particular temperature, thermal radiant energy is emitted by a body over a wide band of frequencies in this general range.

This form of heat transfer differs from the other fundamental heat-transfer mechanism, conduction, in two important ways. First, the rate of thermal radiant exchange between bodies depends on the difference in the individual absolute temperatures of the bodies, each raised to a power. Second, no medium needs to be present.

While thermal radiation may be the common dominant mode of heat transfer in high-temperature situations such as exist in furnaces and combustion chambers, it can be of subtle importance in situations of moderate and low temperatures. Often the thermal radiation acts in parallel with conduction and convection heat transfer at a significant level.

Generally, analyzing thermal radiation heat transfer is more complex than analyzing conduction heat transfer. Mathematical difficulties arise in doing this because bodies may interact with others beyond those in immediate physical contact. Second, the interaction of thermal radiation with the surfaces of physical bodies depends on many variables, including temperature, wavelength of radiation, angle of interaction, surface smoothness, and actual chemical composition of the surface, including any coatings. Indeed, a number of coatings have been developed specifically for control of the thermal radiation transport and interaction. Analysis of radiation heat transfer is discussed in Chap. 6.

1.3 ANALYTICAL MODELS IN THERMAL DESIGN

The application of heat transfer and energy conservation principles in thermal design can be effective over a wide range of levels of physical detail and mathematical sophistication, depending on the needs of the problems and the abilities of the analyst. While many thermal design problems are still addressed solely by a trial-and-error experimental approach, this is hardly ever justified.

Clear definition of the problem at hand is as essential to thermal design as it is in any engineering activity. Both the desired thermal effect as well as the constraining factors need to be identified and characterized. Some commonly encountered problems are

1. Maintenance of a constant, uniform temperature
2. Attainment of a specific temperature cycle
3. Minimization of heat gain

Some governing constraints that might, for example, apply to these problems would be

1. Existence of a nonuniform temperature in the immediate environment
2. Thermal power capacity of the heating system
3. Limit of the available space
4. Temperature limits of the system materials

Analytical models that can be applied for the purpose of evaluating a thermal design problem can range from simple "lumped"-model analyses to complex numerical models. Often simple control volume expressions of the first law of thermodynamics combined with one-dimensional descriptions of heat transfer processes, as depicted in Fig. 1.1, can provide quick and meaningful insight into a problem.

Modern finite-difference and finite-element numerical methods for evaluating heat transfer have become increasingly convenient to apply, with a number of commercial software packages now being available (Chap. 7). These methods allow the designer to address complex situations with a realistic representation of varying thermophysical properties, irregular geometry, and complicated boundary conditions.

In the design of thermal systems which involve several or more fluid-interconnected components, system simulation and optimization methods may be useful.[1]

1.4 THE THERMAL PERFORMANCE AND LIMITATIONS OF MATERIALS

The thermal properties of materials vary, in magnitude, over a wide range. Considerable opportunity exists for a designer to achieve a required thermal behavior through an exploitation of the diversity of the thermal properties of materials. Nevertheless, the ranges of magnitude of the thermal properties are limited, and feasible thermal designs must obviously be achieved within these limitations.

The material properties which commonly form the basis of thermal design are thermal conductivity, specific heat capacity, latent heat, coefficient of thermal expansion, and surface properties related to radiation heat transfer. Other material characteristics and physical properties that may be of direct importance in thermal design are useful temperature range, density, thermal diffusivity, and transparency to thermal radiation. While nearly all physical properties of matter are influenced by temperature, it is the efficient utilization of these stated properties and characteristics that is the major concern in thermal design.

Thermal Conductivity

Homogeneous substances and nonhomogeneous materials may be classified according to thermal conductivity.

Classification	Range of thermal conductivity	
	Btu/h · ft · °F	W/m · °C
High-performance insulation	<0.015	<0.026
Insulation	~0.015–~0.1	~0.026–~0.17
Low conductors	~0.1–~1	~0.17–~1.7
Conductors	~1–25	~1.7–43
High conductors	>25	>43

The lowest apparent thermal conductivity encountered in practical applications [$\sim 2 \times 10^{-5}$ Btu/h · ft · °F (3.4×10^{-5} W/m · °C)] is attained with evacuated multilayer insulations. Diamond has the highest thermal conductivity [~860 Btu/h · ft · °F (1488 W/m · °C)] of common materials with copper exhibiting the highest thermal conductivity of common engineering materials [~230 Btu/h · ft · °F (398 W/m · °C)]. The ranges of thermal conductivity exhibited by specific classes of engineering materials are more limited.

Material type	Approximate range of thermal conductivity at room temperature and pressure	
	Btu/h · ft · °F	W/m · °C
Common gases	0.002–0.1	0.0035–0.17
Organic liquids and solids	0.01–0.30	0.017–0.52
Metals	5–240	8.6–415
Minerals, glasses, ceramics (solids)	0.1–3.0	0.17–5.2

In general, crystalline materials conduct heat better than amorphous materials, while chemical impurities in crystalline substances lower the thermal conductivity. Pure metals, for example, have much higher thermal conductivities than alloys. Also, the solid phase of a substance has, typically, a higher conductivity than the liquid phase which, in turn, will have higher thermal conductivity than the gas phase. Thermal conductivity is dependent upon temperature. The thermal conductivity of gas increases with temperature, while that of liquids may increase or decrease with temperature, that is, most organic materials exhibit a decrease while water and aqueous solutions exhibit an increase with increasing temperature at atmospheric pressure. The thermal conductivity of amorphous solids typically increases with temperature. The thermal conductivity of most crystalline solids (metallic and dielectric) is greatest at very low temperatures. In general, thermal conductivity increases with pressure.[4]

Specific Heat Capacity

The magnitude of the specific heat capacity varies less than do other thermal properties. Metals have the lowest specific heat capacity, with liquid water and gaseous hydrogen and helium exhibiting the highest specific capacities of common engineering materials.

Material type	Approximate range of specific heat capacity at room temperature and pressure	
	Btu/lbm · °F	J/kg · °C
Gases	0.03–3.4	125–14,200
Organics	0.2–0.7	837–2930
Minerals	0.1–0.3	418–1250
Metals (structural)	0.03–0.43	125–1800

Latent Heat

Latent heats are associated with phase transitions, including solid-to-solid phase transitions. The values of latent heats of vaporization are generally larger than those of latent heats of fusion.

Of common engineering importance are the latent heats of vaporization of heat-transfer fluids and refrigerants.

Material type	Heat of vaporization at atmospheric pressure	
	Btu/lbm	J/kg
Common cryogenic liquids	10–220	2.3×10^4–51.2×10^4
Halogenated hydrocarbon refrigerants	50–100	11.6×10^4–23.2×10^4
Ammonia	592	13.7×10^5
Water	970	22.5×10^5
Alcohols	300–500	6.9×10^5–11.6×10^5

Latent heats of fusion have engineering importance with regard to the performance of phase-change thermal storage materials and fusible alloys (Part 6, Chaps. 1 and 2).

Coefficient of Thermal Expansion

Solids and liquids, in general, expand upon heating and contract upon cooling. Accommodation of this behavior within the function of a temperature-affected device or system is an important aspect of thermal design.

Material type	Approximate range of coefficient of linear expansion	
	in/in · °F	m/m · °C
Metals	$2–20 \times 10^{-6}$	$3.6–36 \times 10^{-6}$
Thermoplastics	$30–100 \times 10^{-6}$	$54–180 \times 10^{-6}$
Thermoset plastics	$9–30 \times 10^{-6}$	$16.2–54 \times 10^{-6}$
Ceramics and glasses	$0.5–7 \times 10^{-6}$	$0.9–12.6 \times 10^{-6}$

Material type	Approximate coefficient of volume expansion at room temperature	
	$°F^{-1}$	$°C^{-1}$
Organic liquids	0.003–0.0010	0.0054–0.0018
Liquid water	0.00011	0.00019
Mercury	0.0001	0.00018

Solids having a linear coefficient of expansion of less than about 1.8×10^{-6} in/in · °F are termed low expansion materials and have special application (Part 6, Chap. 3). Differential rates of thermal expansion are employed to produce mechanical movements as a result of varying temperature (Part 6, Chap. 6).

Thermal Radiation Surface Properties

The radiative surface properties of materials in actual application are dependent not only on the type of material but also on the specific surface chemical composition and physical condition. Generalizations about the radiation surface properties of materials are best limited to a few simple statements:

- Polished metals are relatively good reflectors of infrared radiation.
- Rough surfaces are relatively good absorbers and emitters of thermal radiation.
- The variability among materials in the infrared reflective, absorptive, and emittive characteristics generally is within two orders of magnitude.

Maximum Useful Temperature

The maximum useful temperature of a material is constrained by both mechanical and chemical effects. The maximum useful temperature of a given material may be determined by a degradation in mechanical properties, a degradation due to chemical changes, or deterioration due to environmental factors such as those involved in corrosion. The minimum useful temperature is typically determined by mechanical behavior. Table 1.1 lists the typical maximum temperatures of use for a variety of engineering materials. Metals and alloys, plastics, and ceramics suitable for elevated temperature applications are described in Part 4, Chaps. 2 to 4. Reference 5 provides a perspective on high-temperature composite materials.

Interface Thermal Resistance

At the interface between two solid materials held together in unbonded contact, there is a resistance to heat flow. This interface or thermal contact resistance is a function of the

Table 1.1. Approximate Maximum Temperatures of Various Materials
This information is not for design use; specific materials may have lower limits.

Material	Approximate maximum useful temperature, °F	Limitation
		I. Metals
Carbon steels	450–1000†	
Low alloy steels	650–1200†	
High alloy steels (including stainless)	Up to 1500†	
Aluminum and aluminum alloys	150–400†	
Copper	400	
Brasses, bronzes	150–700†	Mechanical strength per ASME code‡
Copper-nickel (70/30)	600–800†	
Nickel	600	
High nickel alloys	400–1650†	
Titanium and titanium alloys	500–600†	
Zinc	785	
Aluminum alloys	1000–1200	
Brasses and bronzes	1700–1900	
Copper	1980	
Cast gray iron	2150	
Hastelloy C	2350	
Type 304 stainless steel	2600	
Carbon steel (1020)	2760	Melting point
Titanium	3300	
Niobium (columbium)	4470	
Molybdenum	4750	
Tantalum	5400	
Tungsten	6150	
Niobium carbide	6800	
Carbon steel	1050	
Grey cast iron	850	
Type 304 stainless steel	1700	
Type 446 stainless steel	2000	
Nickel chrome alloys	2000–2195	
Iron chrome alloys	1920–2550	
Heat-resistant cast irons	1100–2000	Oxidation in air§
Titanium alloys	Up to 1400 (short duration)	
Molybdenum	850–1100	
Tantalum	930	
Tungsten	930–1300	
Molybdenum	3270	
Tantalum	4000	Reliability as heating element in vacuum of 10^{-4} torr
Tungsten	4350	
		II. Carbon
Graphite	750	Oxidation in air
Graphite	4000	Reduction in mechanical properties
Graphite	6500	Sublimation
		III. Elastomers
Natural and synthetic natural rubber	160–190	
Butadiene	160–190	
Acrylonitrile butadiene (Buna-N)	230	
Styrene butadiene (Buna-S)	212	
Chloroprene (neoprene)	212	
Isobutylene-isoprene (butyl)	248	Degradation and/or loss of mechanical properties in air
Ethylene propylene diene monomer (EPDM)	300	
Fluorocarbon	500	
Polysiloxane (silicone rubber)	400–460	
Perfluoroelastomer	550	

NOTE: Footnotes appear on page 1-11.

Table 1.1. Approximate Maximum Temperatures of Various Materials (*Continued*)

This information is not for design use; specific materials may have lower limits.

Material	Approximate maximum useful temperature, °F	Limitation
	IV. Plastics†¶	
Polystyrene	150–260	Mechanical properties based on deflection temperature at 66 lb/in² (ASTM D648)
Polybutylene	215–235	
Polypropylene	160–330	
FEP	150	
Polyimide	530–680	
Polyamide	155–500	
Polycarbonate	270–325	
Polystyrene	122	Underwriters Laboratories elevated temperature index
Polyvinylchloride	122–194	
Polypropylene	149–257	
FEP	302	
Polyimide (film)	392–464	
Polyamide	149–284	
Polycarbonate	149–257	
	V. Ceramics and Glass	
Silicon carbide (sintered)	2730 (at 40,000 lb/in²)	Creep rate 10^{-5}/h
Alumina	2730 (at 4000 lb/in²)	
Soda-lime glass	1285	Softening point
Borosilicate glass	1510	
Alumino-silicate glass	1670	
Fused silica	2870	

†The actual maximum temperature depends on the specific material composition and other factors, such as heat treatment and use.
‡The temperature indicated is the maximum temperature for which the Boiler and Pressure Vessel Code of the American Society of Mechanical Engineers provides design stress values.
§Based on an oxidation rate of 10 mg/cm² in 1000 h.
¶Includes various formulations and filled and unfilled grades.
CONVERSION FACTOR: °C = (°F − 32)/1.8.

materials in contact (thermal conductivity, surface finish, flatness, and hardness), the contact pressure, the mean temperature, the heat flux, the nature of the interstitial fluid (liquid, gas, vacuum), and the presence of surface oxide or other films.[6] Interface thermal resistance may be an important part of the overall thermal resistance of an assembly. This is most likely when the contacting materials have high thermal conductivity (such as metals) or when the thicknesses of the contacting materials in the direction of heat flow are small. Interface thermal resistance generally decreases with increasing surface smoothness and contact pressure. High-conductivity greases and interstitial gases can reduce the interface thermal resistance.

Often interface thermal resistance data are presented in the form of an interface thermal conductance,

$$h_i = \frac{q''}{\Delta T} \tag{1.15}$$

where h_i = interface thermal conductance
q'' = steady-state heat flux across interface
ΔT = apparent temperature drop across interface

Reference 7 provides a procedure for estimating the interface conductance of metallic surfaces in contact. Reference 8 provides an extensive compilation of thermal contact conductance data. Some representative values of contact thermal conductance at various contact pressures are provided in Table 1.2.

Thermophysical Properties Data Resources

The material physical properties data useful in thermal design are generally obtainable from:

1. Engineering reference books
2. Product literature
3. Engineering/scientific journals and conference proceedings
4. Critically evaluated data compilations
5. Measurements by accredited, commercial laboratories using standardized test methods

Data found in engineering reference books, including the present volume, should be regarded as typical. Such typical data are best used in the initial evaluation and selection of materials.

Product literature of material suppliers is an important resource for physical properties data. Modern design materials of proprietary form, composition, and processing can

Table 1.2. Typical Contact Thermal Conductance Data for Unbonded Metallic Surfaces

Material	Roughness, rms μin, both sides	Material in gap	Temperature, °F	Interface thermal conductance Btu/h · ft² · °F
Carbon steel	1000	Air	200	85 (@100 lb/in²) 110 (@300 lb/in²)
Carbon steel	63	Air	200	500 (@100 lb/in²) 660 (@300 lb/in²)
416 stainless steel	30	Air	200	2,000 (@100 lb/in²)
416 stainless steel	100	Air and 0.0010 in brass shim	200	900 (@425 lb/in²)
416 stainless steel	100	Air and 0.010 in asbestos sheet	200	142 (@425 lb/in²)
Aluminum	120	Air	200	900 (@100 lb/in²) 1,660 (@300 lb/in²)
Aluminum	65	Air	200	1,300 (@100 lb/in²) 2,100 (@300 lb/in²)
Aluminum	53	Hydrogen (1.0 atm)	73	100,000 (@4000 lb/in²)
Copper	174	Air	73	3,000 (@1000 lb/in²) 6,600 (@3000 lb/in²)

CONVERSION FACTORS: 1.0 μin = 0.0254 μm; °C = (°F − 32)/1.8; 1.0 Btu/h · ft² · °F = 5.678 W/m² · K; 1.0 lb/in² = 6.894 kPa.

only be confidently applied if material properties are characterized using standardized or otherwise accepted test procedures. The user of commercial data should be alert to differences between test conditions and the conditions of the intended application.

While being the source of much of the publicly available thermophysical properties data, the general body of engineering and scientific periodicals is at best an inconvenient and at worst an unreliable source of accurate data. Access is facilitated by modern computer-based literature searching services. Proceedings of thermophysical properties conference series (such as the ASME Thermophysical Properties Symposia, the European Conferences on Thermophysical Properties, the International Thermal Conductivity Conferences, and the International Thermal Expansion Symposia) include presentations of new data as well as descriptions of advanced test methods.

For substances with well-known and easily characterized chemical and physical forms, critically evaluated physical properties data compilations can provide an accurate basis for thermal design. The *McGraw-Hill/CINDAS Data Series on Material Properties,* produced by the Center for Information for Numerical Data Analysis and Synthesis at Purdue University, is one such resource for thermophysical properties. This data compilation is an updated and expanded successor of the earlier compilations of the Thermophysical Properties Research Center, entitled *The Thermophysical Properties of Matter*. Such centers can provide assistance in accessing the literature.

Independent technical organizations that undertake materials properties evaluations and characterizations provide the opportunity to confirm directly the thermal performance of materials. Tests can often be performed under conditions (temperature, pressure, environment) specific to the engineering problem of interest. Organizations providing materials testing services are accredited by government agencies and voluntary organizations at national and international levels.

1.5 TECHNOLOGY OF THERMAL POWER SUPPLY, EXCHANGE, AND TRANSPORT

A large number of devices for the supply, exchange, and transport of thermal energy are available for use in the practical control of temperature and heat effects. Proper application is aided through an understanding of the principles of operation and the performance characteristics and limits of commercially available thermal equipment.

Table 1.3 presents a tabulation of the typical efficiency and thermal power range of a number of devices and systems that find practical application in the supply of heating or cooling energy. In each case, a basis for the statement of an efficiency is presented. All the devices listed are capable of continuous thermal energy supply and are powered by thermal, mechanical, or electrical energy. Heating and cooling can also be supplied in a batch manner through the use of thermal or chemical energy storage. Such storage devices allow the separation of thermal power supply capability from the available energy source.

Devices and systems for the exchange and transport of thermal energy exist in many forms. Heat exchangers are available for nearly all types of substances and circumstances and have been fabricated in all types of engineering materials (Part 7, Chap. 1). Generally, heat exchangers involve the transfer of heat between two separate and different fluid or fluidized solid streams. The term heat sink is generally used for devices that exchange heat from solid objects to fluids (Part 7, Chap. 5). Heat exchangers and heat sinks are available as both standard and custom designs. While many use simple geometric shapes for the actual heat-transfer surface (such as round tubes and parallel plates), there are a number of surfaces designed specially for maximizing the efficiency

Table 1.3. Thermal Power Supplies

Type	Efficiency definition	Typical efficiency*	Approximate commercial size range, Btu/h†
	I. Heating		
Electric resistance (Part 8, Chap. 2)	Delivered heat / Electric energy input		Any
Clamped or bonded		0.60–0.90	
Immersion or direct		Near 100	
Radiant		0.45–0.90	
Combustion (Part 10)	Delivered heat / Heat content of fuel	0.50–0.80 (@200°F‡) 0.25–0.40 (@1600°F‡)	500 and up
Dielectric and microwave (Part 8, Chap. 3)	Delivered heat / Electric energy input	0.30–0.50	$5–2 \times 10^6$
Induction (Part 8, Chap. 4)	Delivered heat / Electric energy input	0.60–0.70	Up to 2×10^8
Plasma (Part 8, Chap. 5)	Delivered heat / Electric energy input	0.60–0.65	3×10^4–2×10^7
Laser (Part 8, Chap. 6)	Delivered heat / Electric energy input	0.01–0.15	75–75,000
	II. Cooling		
Single-stage vapor compression (Part 9, Chap. 1)	Cooling power / Electric power input	3.0–5.0 (@40°F)	10^3–10^8 (to −30°F)
Compound or cascade multiple-stage vapor compression (Part. 9, Chap. 1)	Cooling power / Electric power input	0.6–0.8 (@−100°F)	(to −150°F)
Gifford-McMahon cryogenic refrigeration (Part 11)	Cooling power / Electric power input	0.003–0.01 (@−320°F)	0.1–5000 (to −440°F)
Absorption cycle (Part 9, Chap. 3)	Cooling power / Heat input rate	0.3–1.0	30,000 and up (to −75°F)
Air cycle[9]	Cooling power / Air compressor power	Less than 1.0	5000–150,000
Steam jet[10]	Cooling power / Steam energy input	0.2–0.8	10^5–10^7 (to 35°F)
Joule Thompson (Part 11)	Cooling power / Compression power	0.005 (@−320°F)	Up to 10 (to −320°F)
Ranque-Hilsch vortex tube[11]	Cooling power / Compression power	0.10–0.20	Up to 5000 (to −40°F)
Evaporative water cooling; cooling tower[12]	Cooling power / Electric power input	5–50	Up to 10^{10} (to 32°F)
Thermoelectric (Part 9, Chap. 2)	Cooling power / Electric power input	0.1–1.0	0.01–100, with limited application to larger sizes (to −150°F)

*Efficiency of refrigeration equipment highly dependent on temperature at which cooling is supplied.
†1 watt = 3.413 Btu/h.
‡Temperature of material being heated, 60°F combustion air.
°C = 5⁄9(°F − 32).

of heat transfer with respect to volume, weight, and other relevant efficiency parameters (Part 7, Chap. 3).

Heat transport devices and systems generally involve moving thermal energy between two different locations. Heat pipes (Part 7, Chap. 4) and organic-fluid heat-transfer fluid systems (Part 5, Chap. 4) are two well-developed technologies for the specific purpose of heat transport. Many fluids find practical application in heat transport.

- Water and steam (Part 5, Chaps. 1 and 2)
- Brines (Part 5, Chap. 1)
- Nitrogen, hydrogen, and carbon dioxide (Part 5, Chap. 3)
- Organic fluids (Part 5, Chap. 4)
- Inert gases (Part 5, Chap. 3)
- Molten salts (Part 5, Chap. 5)
- Molten metals (Part 2, Chap. 1)

A number of the above types of fluids and other materials have application as expendable coolants (Part 9, Chap. 4).

1.6 THERMAL SENSOR PRINCIPLES AND TECHNOLOGY

Temperature Scales and Standards

The two most commonly used temperature scales are the Celsius and Fahrenheit scales. Both have arbitrary zero points. The Celsius scale was referred to previously as the centigrade scale.

The Celsius scale is currently described by the International Practical Temperature Scale of 1968 (IPTS-68) as adopted by the Comité International des Poids et Mesures.[13] The IPTS-68 was established in relation to the basic thermodynamic temperature having the symbol T and the units kelvin, K. The kelvin is the fraction 1/273.15 of the thermodynamic temperature of the triple point of water. The Celsius temperature, symbol t, is defined as

$$t = T - 273.15 \text{ K} \tag{1.16}$$

The Celsius temperature is expressed in degrees Celsius, °C.

The IPTS-68 has been chosen so that the temperature measured on it closely approximates the thermodynamic temperature. Thermodynamic temperature is the fundamental measure of hotness and coldness and is based, in practice, on the behavior of perfect gases. The IPTS overcomes the practical difficulties associated with the direct realization of thermodynamic temperature by gas thermometry. It is based on the assigned values of the temperatures of a number of reproducible equilibrium states (defining fixed points) and on standard instruments calibrated at those temperatures.

The IPTS-68 also provides for interpolation between the fixed-point temperatures by establishing the relations between the indications of the standard instruments and values of the IPTS. The IPTS has 13 defining fixed points that range from the triple point of equilibrium hydrogen (13.81 K) to the freezing point of gold (1337.58 K). A large number of secondary reference points are also given. With two of the defining fixed points of the IPTS being the triple point of water (0.01°C) and the boiling point of water (100°C), the scale closely agrees with the historical definition of the Celsius scale that the freezing point of water is 0°C and the boiling point at atmospheric pressure is 100°C.

The Fahrenheit scale is also historically established by the freezing point of water (32°F) and the boiling point of water (212°F). The conversions between the Celsius and Fahrenheit scales is thus

$$t\,(°\text{F}) = \frac{9}{5}t\,(°\text{C}) + 32 \tag{1.17}$$

The absolute temperature scale associated with the Fahrenheit scale is the Rankine scale. The relationship is

$$t\,(°\text{F}) = T\,(°\text{R}) - 459.67 \tag{1.18}$$

Temperatures in the Rankine scale are related to the Kelvin scale by

$$T\,(°\text{R}) = 1.8T\,(\text{K}) \tag{1.19}$$

As indicated, the IPTS-68 provides specifications for the instruments to be used in realizing the IPTS at other than the fixed points. Depending on the temperature range, these include a platinum resistance thermometer, a platinum versus platinum-10% rhodium thermocouple, or an optical pyrometer. These are referred to as primary standard thermometers. Primary standard thermometers are calibrated by using fixed points. This service is available from standardizing laboratories such as the National Bureau of Standards. Because primary standard thermometers are expensive instruments, secondary standard thermometers are used for reference testing as a "working" standard. These secondary standards must be periodically checked against the primary standard. Secondary standards may be of the same design as primary standards or they may be of different design, including alternative resistance thermometers, thermocouples, or liquid-in-glass thermometers.

Reference 14 gives four steps for assurance of temperature measurement accuracy:

1. Obtain instruments with operating features, including precision of indication, that are adequate for the task.
2. Obtain certified secondary reference thermometers covering the temperature ranges needed to check operating instruments and to serve as accuracy standards.
3. Using adequate temperature transfer equipment (controlled liquid bath or similar facility), periodically check operating instruments against the secondary standard and keep a record of the tests to provide verification of accuracy.
4. Periodically check the calibration of the secondary reference standards against a primary temperature standard, the ice point of ice, and keep certificates of calibration updated as of each test.

Practical equipment and procedures for the calibration of temperature measuring instruments at three levels of accuracy—±1°C, ±0.1°C, and ±0.01°C—are presented in Ref. 15.

Temperature Sensors

A number of different physical effects and phenomena have been used to produce the great variety in temperature measurement devices and systems in current use. Some of the available devices are characterized in Table 1.4.

Temperature Sensor Application

The accuracy of temperature measurements is dependent upon the characteristics of the measuring system and the method and situation of its application. Factors to be considered in achieving reliable temperature measurements are described below.[24,25]

Accuracy: Difference between the temperature indicated by the measurement system and the actual temperature as defined by the International Practical Temperature Scale. Accuracy is achieved by calibration.

Precision: Confidence level in successive measurements with the same instrument under the same conditions and ability to detect small changes. The precision (repeatability) of most instruments is, in practice, an order of magnitude or more greater than the accuracy.

Stability: Ability of a measurement system to provide the same indicated temperature under constant conditions over a period of time. The change of a measured temperature that is due to the measurement system itself is called drift.

Response time: Measure of the ability of a sensor to detect changes in temperature. The response of sensors is often quantitatively characterized by the thermal time constant. The thermal time constant is the time required for the output of a sensor to achieve 63.2 percent of the final value when directly exposed to a step change in temperature. The functional response characteristics of a sensor depend on the physical features of the sensor as well as on the design of any protection assembly and the heat-transfer characteristics of the sensed medium.

Thermal contact resistance: Measure of the potential for thermal equilibrium between a sensor and a solid body. The lower the thermal contact resistance (the better the thermal contact), the lower will be the potential for local heat-transfer effects (such as convective heat transfer from the sensor) to affect the measurement accuracy. Often specific actions are taken to ensure good thermal contact, such as the use of heat-conducting pastes, spring-loaded sensors, and soldering or welding of the sensor to the surface.

Conduction in connecting wires: Conduction of heat along the electric wires or other connections between the temperature sensor and the readout device. This heat transfer can cause the sensed temperature to deviate in the direction of the ambient temperature. Good design practice calls for the sensor wires to be in good thermal contact with an isothermal portion of a solid surface for a length of about 20 wire diameters.[26]

Radiation shielding: Shielding required when there is a pathway for thermal radiation exchange between a temperature sensor and a remotely located surface or fluid. Radiation heat exchange can affect fluid and surface temperature measurements.

Recovery temperature: Temperature, measured in a flow stream, that is above the temperature that would be measured if the fluid were at rest. The difference between the recovery temperature and the static temperature represents the conversion of fluid flow kinetic energy into internal energy in the vicinity of the probe as the fluid is brought to rest. This temperature difference can be significant in gaseous flow, but is usually small in liquid flow.

Signal transmission: Signal transmission from the sensor to the readout device is accomplished by different means, depending on the type of sensor and the distance from the sensor and the readout device. Transmitters which convert the sensor output locally to standardized measurement and control outputs are used to ensure reliable signal transmission. Certain types of sensors, especially thermocouples, require that close attention be paid to the thermal, electrical, and physical features of the measuring loop.

Sensor protection: Protection routinely achieved with probe-type temperature sensors by using thermal wells and other types of protection tubes.[27] Protection tubes are used to guard a temperature sensor from the effects of pressure, corrosive attack, contamination, flow impingement, and electrical phenomena. A thermal well, the most commonly used type of protecting tube, provides for the insertion of sensors into the thermal environment inside the vessel wall without exposing the sensor to the vessel pressure or fluids. Industrial standards for thermal wells, which are available in many materials, allow for their use with several different common types of sensors. Methods for evaluating the effect of the physical structure of a thermal well upon measurement accuracy have been developed.[28]

1.7 AUTOMATIC TEMPERATURE CONTROL

Automatic temperature control is a specific application of the broad field of automatic process control. Achieving stable temperature control is a primary objective in the implementation of most automatic temperature control systems. Strategies and techniques for achieving stable process control are well developed.[29] The technology of automatic temperature control has been greatly improved and expanded by modern electronic technology.

An automatic temperature controller is a device that can (1) detect deviations in the measured output temperature of a system or process relative to a desired value, and (2) provide a correction to a system or process energy input so that the desired temperature is maintained within an acceptable tolerance.

Types of Temperature Controls

Automatic temperature controllers of many different types are available such as, for example,

- Mechanical thermostats and thermal switches
- Mechanical fluid-filled temperature controllers
- Pneumatic controllers
- Analog electrical and electronic temperature controllers
- Digital electronic temperature controllers
- Programmable process controllers
- Microcomputer-based controllers
- Minicomputer or main-frame computer-based controls

All of these types of automatic temperature control devices and systems have the basic capability of evaluating a measured temperature relative to a desired value and initiating a corrective action. Mechanical thermostats and thermal switches provide simple and inexpensive, although limited, control of temperature through on-off regulation of process or system energy inputs. Mechanical fluid-filled temperature controllers are relatively simple and rugged devices that can provide responsive control without an external power supply (Part 13, Chap. 1).

Pneumatic controllers and control systems can provide full temperature control capability and are especially useful where instrument and control systems must be explosion-proof, where control must be maintained during electrical power outages, and where the control elements are air-operated.

Table 1.4. Temperature Measurement Devices

Type	Approximate temperature range of application	Important characteristics
Thermocouples (Part 12, Chap. 2)	−328–5100°F (−200–2800°C)	Most commonly used sensor High-temperature capability Many types Requires reference junction compensation and matching connecting wires
Platinum resistance temperature detectors (Part 12, Chap. 3)	−436–1292°F (−250–700°C)	High accuracy and stability achievable over wide spans Not as rugged as other common sensors Subject to self-heating errors Requires connecting wire compensation
Thermistors (Part 12, Chap. 3)	−319–572°F (−195–300°C)	Small sensor Highest sensitivity Narrow span Subject to self-heating errors
Transistors[16,17]	−58–392°F (−50–200°C)	Good accuracy, stability, and linearity possible Low maximum temperature Low relative cost and available in several forms Narrow span
Quartz crystals[18]	−112–482°F (−80–250°C)	High accuracy and resolution to ±0.001 K Available with microprocessor-based calibration and control Narrow span
Diodes (Part 11)	Liquid helium to 480°F (248°C)	Silicon and gallium arsenide most common cryogenic sensor Individual calibration required
Johnson noise detectors[19]	Up to 2732°F (1500°C)	Fundamentally linear, absolute method Affected by extraneous noise pickup Relatively new development Very expensive
Filled-system thermometers (Part 12, Chap. 6)	−319–1400°F (−195–760°C)	High stability Limited accuracy Liquid, vapor, gas, and mercury types Simple, rugged, low cost
Liquid-in-glass thermometers (Part 12, Chap. 5)	−300–1100°F (−184–593°C)	Most commonly used thermometer Simple, low cost, no power needed Not easily adaptable to remote reading Not for automatic control uses
Bimetallic thermometers (Part 12, Chap. 4)	−300–1200°F (−184–648°C)	More rugged than liquid-in-glass but also simple and no power needed Wide hysteresis loop
Germanium, carbon, and carbon-in-glass resistance thermometers (Part 11)	Below liquid nitrogen temperature	Increasing resistance with decreasing temperature For cryogenic applications
High-temperature radiation thermometers (Part 12, Chap. 7)		
Optical	1300–7200°F (700–4000°C)	Human eye used to match colors Simple to use
Amplitude-sensing	500–6000°F (260–3300°C)	Radiant energy converted to electrical signal for readout Hand-held and permanent installation types Emissivity adjustment
Ratio	1200–4500°F (650–2500°C)	Measures at two wavelengths Difficulties of filling the field of view, emissivity adjustment, and atmospheric attenuation avoided
Low-temperature radiation thermometer (Part 12, Chap. 7)		
Spectral band	−75–575°F (−60–300°C)	Sensitive in spectral regions where atmospheric absorption is minimized Emissivity compensation required

Table 1.4. Temperature Measurement Devices (*Continued*)

Type	Approximate temperature range of application	Important characteristics
Narrow band	100–950°F (38–510°C)	Sensitive to narrow spectral region Used to measure plastic and glass
Fiber-optic radiation thermometer (Part 12, Chap. 7)	200–4500°F (93–2500°C)	Allows for remote sensing in hostile environments Measurement capabilities similar to those of conventional radiation thermometer Sensing end of fiber can include blackbody cavity
Fluoroptic thermometer (Part 12, Chap. 7)	−325–750°F (−200–400°C)	Sensor stimulated by xenon light source through optical fiber; decay time measured Signal from sensor not affected by electromagnetic radiation
Thermal imaging radiometer (Part 12, Chap. 8)		Displays color or black and white 2-D images Can provide absolute temperatures Video recording and computer processing of data possible
Pyrometer cones (Part 12, Chap. 1)	1100–3600°F (600–2000°C)	Softening point dependent on rate and duration of heating Used mainly in ceramic industry
Fusible materials indicators (Part 12, Chap. 1)	100–2500°F (38–1370°C)	Marking sticks, lacquers, pellets available
Color change indicators (Part 12, Chap. 1)	175–1475°F (80–800°C)	Available as paints and inks Not as accurate as fusible indicators
Liquid crystals (Part 12, Chap. 10)	−4–482°F (−28–250°C)	Produce spectrum of colors over span of temperature Span set by composition
Nuclear quadrupole resonance thermometer[20]	−292–212°F (−180–100°C)	High-accuracy method suitable as practical temperature standard
Ultrasonic thermometer[21]	80–3140°F (30–1727°C)	Relatively new development Most applications at high temperature
Convective null-heat-balance method[22]	Up to 900°F (480°C)	Noncontact method for process control of continuous sheet-type products Based on convective/conduction heat flow between moving product and sensor
Fluidic sensors[23]	Up to 5000°F (2750°C)	Very high temperature contact measurements possible Based on change of properties of gas passing through capillary Relatively new development

Electrical and electronic automatic temperature controllers (Part 13, Chap. 2) encompass a wide range of devices which are designed to accept a temperature signal directly from a temperature sensor, such as a thermocouple or a resistance temperature detector. These controllers provide either direct control of an electric power supply or a control signal for the operation of a valve or other control element. Control algorithms, such as one based on proportional control, are implemented using analog circuitry. Digital electronic temperature controllers provide similar functions as analog controllers but do so through the use of microprocessors. Digital electronic temperature controllers are useful in implementing distributed control systems since the link to a supervising computer is more convenient. Digital electronic temperature controllers are superior to analog controllers with regard to accuracy, stability, and the complexity of control functions which they can provide. However, digital controllers are generally more expensive than the analog types.

Programmable process controllers are digital electronic devices that use a programmable memory to store instructions and to implement specific control functions for processes. While providing any of the features of computer control, programmable process controllers require no special input-output systems, can be located in industrial environments, and are easily programmed.

Microcomputers can be used for temperature control with appropriate analog-to-digital input signal conversion devices. Electric circuit boards for wiring temperature sensor signals directly into personal computers are available. Microcomputer control systems based on personal computers may be most appropriate for test and laboratory temperature control situations. Minicomputers and main-frame computers are used for both direct and supervisory temperature control and can provide advantages in speed, number of inputs and outputs, and data display and analysis.

1.8 THERMAL DESIGN CHECKLIST

Experience reveals that thermal design problems can be resolved successfully if all important factors are considered. Likewise, experience also reveals that initial solutions to thermal design problems are often deficient due to a failure to consider some important aspects of the problem. The following checklist can provide some measure of confidence in ensuring that important design factors are considered. Review of the following list at the beginning and at the end of a design effort is recommended.

1. *General*
   a. *Realism of thermal specifications:* It is not uncommon for thermal performance specifications to be established by those unknowledgeable of the limitations of practical thermal design. This can lead to unrealistic and unnecessary requirements. Difficult specifications should be probed for elasticity and actual requirements determined.
2. *Material factors*
   a. *Variation of thermophysical, transport, and mechanical properties with temperature:* Physical properties, in general, vary with temperature. The effective thermal conductivity of insulation is especially sensitive to temperature as is the viscosity of many liquids, including water. Most materials lose strength as the temperature rises.
   b. *Variation of physical properties with age:* Material composition changes due to aging, especially at elevated temperature, can significantly affect physical properties, including effective thermal conductivity, surface properties, and mechanical strength.
   c. *Compatibility of materials:* Materials commonly used together at one temperature may be chemically incompatible at another, higher temperature.
   d. *Certainty of thermophysical and other properties:* Actual physical properties may differ substantially from handbook values. Radiation surface properties can be particularly troublesome in this regard. Include design margins or have properties measured if performance is crucial.
   e. *Temperature limits of all materials:* Are the maximum service temperatures of gaskets, insulations, instrument components, and sensors as well as those of supporting or connecting members adequate?
3. *Heat transfer and heat balance factors*
   a. *Parasitic heat losses and gains:* Heat gains or losses from supporting components and systems can be significant and can dominate the heat balance. Consider heat transfer to piping systems and through thermal "shorts."
   b. *Overall thermal balance:* A control volume should be defined and the first law of thermodynamics applied.
   c. *Capacity of ultimate heat sink/source:* Will temperature of heat sink/source change in time due to the thermal load?
   d. *Phase-change and chemical reaction heat effects:* The evaporation or condensation of small amounts of water can significantly affect thermal requirements.
   e. *Frictional and structural heat generation:* Effects of heat gain from power and instrument measurement circuits can be important. Fans and compressors add significant heat to gas streams. Bearings generate heat.[30]
   f. *Startup and shutdown requirements:* Whether heating or cooling, the rate of thermal energy exchange to achieve desired startup and shutdown times may significantly exceed normal operating requirements. Thermal power supply must be sized accordingly.
   g. *Three modes of heat transfer:* Have all modes of heat transfer been accounted for? Radiation may be important even at near ambient temperatures. Two-dimensional conduction effects (so-called fin effects) can be overlooked easily.
   h. *Efficiency of convective heat transfer:* Are surfaces for convective heat transfer designed for efficiency? There are a number of techniques for improved performance (Part 7, Chap. 3).
4. *Configuration and operation*
   a. *Thermal expansion:* Thermal expansion can lead to mechanical failure and distortion. Thermal expansion effects may be most pronounced during operating transients.
   b. *Fouling of convective heat-transfer surfaces:* Fouling, by deposition, precipitation, or corrosion can significantly degrade heat-transfer performance and is a potential problem in many situations.
   c. *Operational impact of oversizing:* Conservatism in thermal design leads to oversized systems. Will the oversized system give proper control in part-load situations or if the load is less than expected?
   d. *Safety in the event of loss of thermal control:* Will the failure of the thermal control system result in operator or equipment hazard?
   e. *Effect of varying ambient environment:* Is thermal performance sensitive to changes in the ambient thermal environment, including such factors as humidity and sunlight?
   f. *Temperature capability of system components and fluids:* Components and fluids of a system must perform over the full range of expected temperatures, including startup, shutdown, and other transients.
   g. *Frost and condensation:* Susceptibility to frost accumulation and water damage due to condensation must usually be considered at operating temperatures below ambient. This includes both internal and external surfaces exposed to air.
   h. *Weight and volume restrictions:* Will the thermal components meet weight and volume restrictions?
   i. *Hot surface hazard:* The burn potential of surfaces is dependent on the thermophysical properties and the temperature of the surface material. Personnel should be protected in some manner from burn hazards.
   j. *Performance sensitivity:* Is the design performance sensitive to any of the assumptions about the thermal behavior? For example, laminar-turbulent flow tran-

sitions result in a large change in heat-transfer performance, and small amounts of noncondensable gas can significantly affect a two-phase heat-transfer process.

k. *Fire or explosion hazard:* Have provisions been made to vent fluid vapors if thermal control fails? (See Ref. 31 for guidance on fire safety.)

l. *Auxiliary thermal control:* Do process or instrument lines or vessels need auxiliary heat (heat tracing) for startup, standby, or other purposes?

m. *Temperature stratification and buoyancy effects:* Buoyant flows due to the effect of temperature on fluid density can aid or inhibit thermal processes, depending on the situation.

5. *Measurement and control*

a. *Capability of temperature sensors with regard to control objectives:* Are the accuracy, precision, and time response of temperature measurement loops sufficient to meet the temperature control objective?

b. *Interaction between control loops:* Are different temperature control loops thermally linked, allowing possible control instability?

c. *Location of temperature sensors:* Is the feedback sensor of an automatic temperature control system in the best location to give stable and responsive control?

d. *Sensor heat-transfer effects:* Will connecting wire conduction, poor solid contact, insufficient sensor immersion, radiation, or high gas velocities influence the measured temperature?

REFERENCES

1. W. F. Stoecker, *Design of Thermal Systems,* 2d ed., McGraw-Hill, New York, 1980.
2. Myer Kutz, *Temperature Control,* Wiley, New York, 1968.
3. J. P. Holman, *Heat Transfer,* 4th ed., McGraw-Hill, New York, 1976.
4. L. S. Kowalczyk, "Thermal Conductivity and Its Variability with Temperature and Pressure," ASME Paper 54-A-90.
5. "High Temperature Composites," in *Encyclopedia of Composite Materials and Components,* M. Grayson (ed.), Wiley, New York, 1983.
6. P. J. Schneider, "Conduction," in *Handbook of Heat Transfer,* W. M. Rohsenow and J. P. Hartnett (eds.), McGraw-Hill, New York, 1973.
7. H. Fenech, and W. H. Rohsenow, "Prediction of Thermal Conductance of Metallic Surfaces in Contact," *Trans. ASME, J. Heat Transfer,* **85,** 15–25 (1963).
8. *Heat Transfer Data Book,* Genium Publishing Corp., 1145 Catalyn St., Schenectady, N.Y., sec. 502.5.
9. "Air Cycle Equipment," in *ASHRAE Handbook, Equipment Volume,* American Society of Heating, Refrigerating, and Air Conditioning Engineers, Atlanta, Ga., 1983.
10. "Steam Jet Refrigeration Equipment," in *ASHRAE Handbook, Equipment Volume,* American Society of Heating, Refrigerating, and Air Conditioning Engineers, Atlanta, Ga., 1983.
11. *1986 Product Catalogue,* Vortec Corp.
12. "Evaporative Air Cooling Equipment," in *ASHRAE Handbook, Equipment Volume,* American Society of Heating, Refrigerating, and Air Conditioning Engineers, Atlanta, Ga., 1983.
13. Comité International des Poids et Mesures, "The International Practical Temperature Scale of 1968," *Metrologia,* **5** (2), 35–44 (1969).
14. R. P. Collier, "Temperature Measurement Accuracy," presented at ISA Conf. (Oct. 15–19, 1978), Instrument Society of America, Research Triangle Park, N.C.
15. R. P. Collier, "Calibration with Confidence: The Assurance of Temperature Accuracy," in *Temperature, Its Measurement and Control in Science and Industry,* vol. 5, pt. 2, J. F. Schooley (ed.), American Institute of Physics, New York, 1982, pp. 1311–1315.
16. A. Ohte, M. Yamagata, and K. Akiyama, "Precision Silicon Transistor Thermometer," in *Temperature, Its Measurement and Control in Science and Industry,* vol. 5, J. F. Schooley (ed.), American Institute of Physics, New York, 1982.
17. M. P. Timko, "A Two-Terminal I.C. Temperature Transducer," *IEEE. J. Solid-State Phys.,* **SG-11** (6), (1976).
18. *Measurement, Computation, System Catalogue,* Hewlett Packard, Palo Alto, Calif., 1986.
19. T. V. Blalock and R. L. Shepard, "A Decade of Progress in High Temperature Johnson Noise Thermometry," in *Temperature, Its Measurement and Control in Science and Industry,* vol. 5, pt. 2, J. F. Schooley (ed.), American Institute of Physics, New York, 1982, pp. 1219–1223.
20. A. Ohte, and H. Iwauka, "A New Nuclear Quadruple Resonance Standard Thermometer," in *Temperature, Its Measurement and Control in Science and Industry,* vol. 5 pt. 2, J. F. Schooley (ed.), American Institute of Physics, New York, 1982, pp. 1173–1180.
21. L. C. Lynnworth, "Temperature Profiling Using Multizone Ultrasonic Waveguide," in *Temperature, Its Measurement and Control in Science and Industry,* vol. 5, pt. 2, J. F. Schooley (ed.), American Institute of Physics, New York, 1982, pp. 1181–1190.
22. TRANS-MET Engineering Inc., Bull. 229, LaHabra, Calif., 1977.
23. R. M. Phillippi, T. M. Drzewiecki, T. Negas, and H. S. Parker, "Design of a Fluidic Capillary Pyrometer for Contact Duty at Temperatures to 2750°C," in *Temperature, Its Measurement and Control in Science and Industry,* vol. 5, pt. 2, J. F. Schooley (ed.), American Institute of Physics, New York, 1982, pp. 1345–1352.
24. R. H. Kennedy, "Selecting Temperature Sensors," *Chem. Eng.,* 54–71 (Aug. 8, 1983).
25. R. P. Benedict, *Fundamentals of Temperature, Pressure, and Flow Measurements,* 3d ed., Wiley, New York, 1984.
26. R. J. Goldstein and H. D. Chiang, "Measurement of Temperature and Heat Transfer," in *Handbook of Heat Transfer Applications,* 2d ed., W. M. Rohsenow, J. P. Hartnett, and E. N. Ganic (eds.), McGraw-Hill, New York, 1985, chap. 12.
27. G. R. Leavitt, "Wells and Protecting Tubes," in *Process Instruments and Controls Handbook,* 2d ed., D. M. Considine (ed.), McGraw-Hill, New York, 1974, pp. 2–68 and 2–70.
28. R. P. Benedict, "Heat Transfer Effects in Thermal Measuring Systems," in *Process Instruments and Controls Handbook,* 2d ed., D. M. Considine (ed.), McGraw-Hill, New York, 1974, pp. 2–7 to 2–13.
29. F. Shinskey, *Process Control Systems,* 2d ed., McGraw-Hill, New York, 1979.
30. "Equilibrium Temperatures in Self-Contained Bearing Assemblies," ESDU Engineering Sciences Data Item 78026, 1979.
31. *Fire Protection Handbook,* 15th ed., National Fire Protection Association, Quincy, Mass., 1981.

Chapter 2

Conduction Heat-Transfer Solutions

JAMES H. VAN SANT
Lawrence Livermore National Laboratory

ERIC C. GUYER
Editor in Chief

2.1 INTRODUCTION

Solutions to a selection of common heat conduction problems are presented in this chapter. Most of the solutions are for steady-state conditions. Reference 1 is the source of the solutions and may be consulted for other geometries, boundary conditions, and initial conditions.

2.2 PLATES

1. *Convectively heated and cooled plate* (Fig. 2.1*a*)

$$q = \frac{h_1(t_1 - t_2)}{\mathrm{Bi}_1 + 1 + h_1/h_2}$$

$$\frac{t - t_1}{t_2 - t_1} = \frac{h_1 x/k + 1}{h_1 l/k + 1 + h_1/h_2}$$

2. *Composite plate* (Fig. 2.1*b*)

$$q = \frac{t_0 - t_n}{\sum_{i=1}^{n} (l_i/k_i + 1/h_i) + 1/h_0}$$

$$\frac{t_j - t_0}{t_n - t_0} = \frac{\sum_{i=1}^{j-1} (l_i/k_i + 1/h_i) + x_j/k_j + 1/h_0}{\sum_{i=1}^{n} (l_i/k_i + 1/h_i) + 1/h_0} \qquad j > 1$$

where t_j is the temperature in the jth layer.

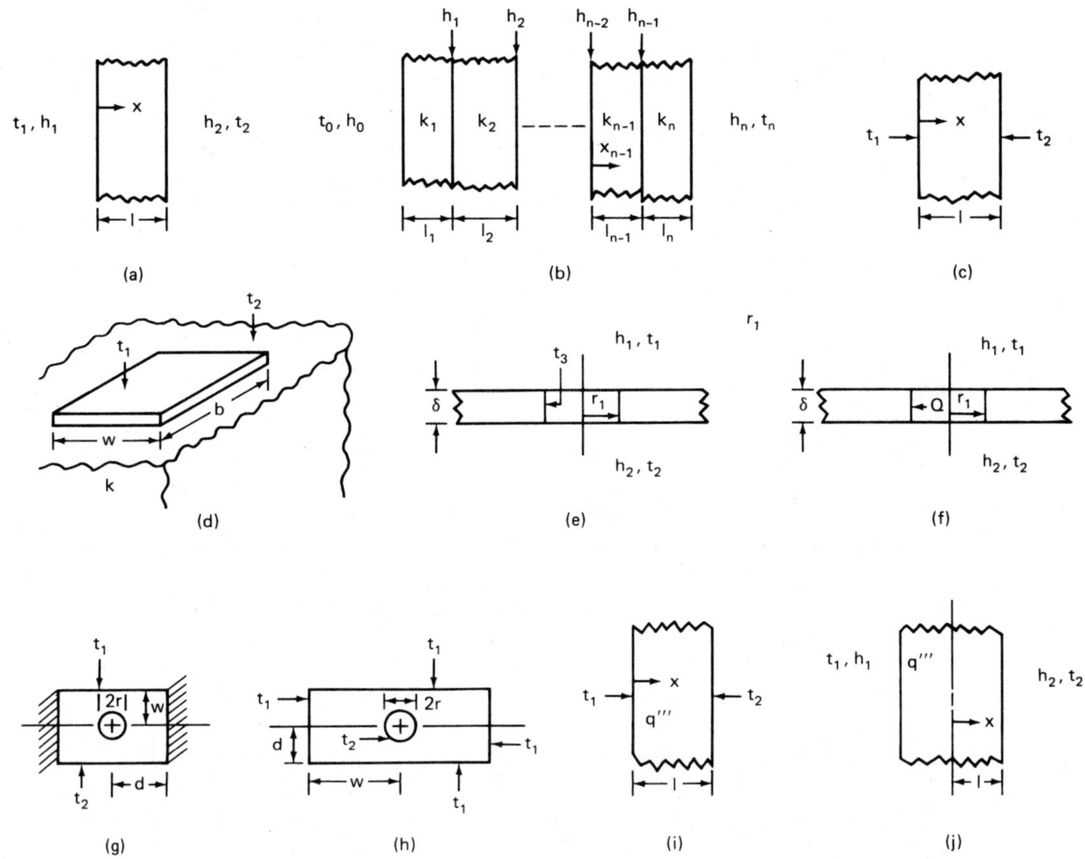

Fig. 2.1 Steady-state plates.

3. *Plate with temperature-dependent thermal conductivity*[2] (Fig. 2.1*c*): With $k = k_1 + \beta(t - t_1)$,

$$q = k_m \frac{t_1 - t_2}{l}$$

$$t - t_1 = \frac{1}{\beta}\left[\sqrt{k_1^2 + 2\beta k_m (t_2 - t_1)x/l} - k_1\right]$$

where

$$k_m = \frac{k_1 + k_2}{2}$$

4. *Thin rectangular plate on the surface of a semi-infinite solid*[3] (Fig. 2.1*d*)

$$Q = \frac{kw\pi(t_1 - t_2)}{\ln (4w/b)}$$

5. *Infinite thin plate with heated circular hole:* With $t = t_3$ at $r = r_1$ (Fig. 2.1*e*),

$$\frac{t - t_\infty}{t_3 - t_\infty} = \frac{K_0(Br/\delta)}{K_0(Br_1/\delta)} \qquad r > r_1$$

or with Q at $r = r_1$ (Fig. 2.1*f*),

$$\frac{k\delta(t - t_\infty)}{Q} = \frac{K_0(Br/\delta)}{2\pi(Br_1/\delta)\,K_1(Br_1/\delta)} \qquad r > r_1$$

where

$$B = \sqrt{\mathrm{Bi}_1 + \mathrm{Bi}_2}$$

$$t_\infty = \frac{t_1 + Ht_2}{1 + H}$$

$$H = \frac{\mathrm{Bi}_1}{\mathrm{Bi}_2}$$

6. *Finite plate with centered hole*[9] (Fig. 2.1g)

$$q' = \frac{k(t_1 - t_2)[\pi d/2w + \ln (w/\pi r)]}{2\pi}$$

7. *Tube centered in a finite plate*[9] (Fig. 2.1*h*)

$$q' = \frac{2\pi k(t_1 - t_2)}{\ln (4d/\pi r) - c} \qquad r < \frac{d}{10}$$

w/d	C
1.00	0.1658
1.25	0.0793
1.50	0.0356
2.00	0.0075
2.50	0.0016
3.00	0.0003
4.00	1.4×10^{-5}
∞	0

8. *Infinite plate with internal heat generation*[2] (Fig. 2.1*i*) With

$$t = t_1 \qquad x = 0$$

$$t = t_2 \qquad x = l$$

$$\frac{t - t_1}{t_2 - t_1} = X + \frac{\text{Po}\, X(1 - X)}{2}$$

9. *Infinite plate with convection boundaries and internal heat generation*[3] (Fig. 2.1*j*)

$$\frac{t - t_2}{t_1 - t_2} = \frac{1 - \text{Po}\,(1/\text{Bi}_2 + 1)}{1 + \text{Bi}_2 + H} + \frac{\text{Po}}{\text{Bi}_2} + \frac{\text{Po}}{2}(1 - X^2) + \frac{\text{Bi}_1[1 + \text{Po}(1/\text{Bi}_2 + \tfrac{1}{2})](1 - X)}{1 + \text{Bi}_1 + H}$$

where $H = h_2/h_1$.

2.3 RODS, TUBES, CYLINDERS, DISKS, PIPES, AND WIRES

1. *Infinite hollow cylinder* (Fig. 2.2*a*)

$$q' = \frac{2\pi k(t_o - t_i)}{\ln (r_o/r_i) + 1/\text{Bi}_i + 1/\text{Bi}_o}$$

$$\frac{t - t_o}{t_i - t_o} = \frac{\ln (r_o/r) + 1/\text{Bi}_o}{\ln (r_o/r_i) + 1/\text{Bi}_i + 1/\text{Bi}_o}$$

where

$$\text{Bi}_i = \frac{h_i r_i}{k} \qquad \text{Bi}_o = \frac{h_o r_o}{k}$$

2. *Composite cylinder* (Fig. 2.2*b*)

$$q' = \frac{2\pi(t_n - t_1)}{\sum_{i=1}^{n-1} (1/k_i) \ln (r_{i+1}/r_i) + \sum_{i=1}^{n} 1/r_i h_i}$$

$$\frac{t_j - t_1}{t_n - t_1} = \frac{\sum_{i=1}^{j-1}\left[\frac{1}{k_i}\ln\left(\frac{r_{i+1}}{r_i}\right) + \frac{1}{r_i h_i}\right] + \frac{1}{k_j}\ln\left(\frac{r}{r_j}\right) + \frac{1}{r_j h_j}}{\sum_{i=1}^{n-1}\frac{1}{k_i}\ln\left(\frac{r_{i+1}}{r_i}\right) + \sum_{i=1}^{n}\frac{1}{r_i h_i}} \qquad j > 1$$

where t_j is the temperature in the jth layer.

3. *Insulated tube* (Fig. 2.2*c*)

$$q' = \frac{2\pi k(t_i - t_f)}{\ln (r_o/r_i) + 1/\text{Bi}_o}$$

where $k << k_{\text{tube}}$ and $\text{Bi}_o = hr_o/k$. Maximum heat loss occurs when $r_o = k/h$.

4. *Infinite cylinder with temperature-dependent thermal conductivity*[2] (Fig. 2.2*d*): With

$$k = k_o + \beta(t - t_o)$$

$$k = k_o \quad \text{at } r_o$$

$$k = k_i \quad \text{at } r_i$$

$$q' = \frac{2\pi k_m(t_i - t_o)}{\ln (r_o/r_i)}$$

$$\frac{(t - t_o)\beta}{k_o} = \left[1 + \frac{2\beta k_m}{k_o^2}\frac{\ln (r_o/r)}{\ln (r_o/r_i)}(t_i - t_o)\right]^{1/2} - 1$$

where

$$k_m = \frac{k_o + k_i}{2}$$

5. *Pipe in semi-infinite solid*[3] (Fig. 2.2*e*): With $d > 4r_1$,

$$q' = \frac{2\pi k_1(t_2 - t_1)}{\dfrac{1}{\text{Bi}_1} + \ln (r_2/r_1) + (1/K)\ln (2D + 2KD/\text{Bi}_2)}$$

where

$$\text{Bi}_1 = \frac{h_1 r_1}{k_1} \qquad \text{Bi}_2 = \frac{h_2 d}{k_2}$$

$$K = \frac{k_2}{k_1} \qquad D = \frac{d}{r_2}$$

6. *Row of pipes in semi-infinite solid*[3] (Fig. 2.2*f*): For one pipe,

$$q' = \frac{2\pi k(t_2 - t_1)}{1/\text{Bi}_1 + \ln \{(d/\pi D r_1)\sinh [\pi 2(D + D/\text{Bi}_2)]\}}$$

where $\text{Bi}_1 = \dfrac{h_1 r_1}{k} \qquad \text{Bi}_2 = \dfrac{h_2 d}{k} \qquad D = \dfrac{d}{s}$

7. *Row of pipes in wall*[3] (Fig. 2.2*g*): For each pipe,

$$q' = \frac{4\pi k(t_2 - t_1)}{1/\text{Bi}_1 + \ln \{(d/\pi r_1 D)\sinh [2\pi(D + D/\text{Bi}_2)]\}}$$

where $\text{Bi}_1 = \dfrac{h_1 r_1}{k} \qquad \text{Bi}_2 = \dfrac{h_2 d}{k} \qquad D = \dfrac{d}{s}$

8. Circular disk on the surface of a semi-infinite solid[6] (Fig. 2.2*h*): With $t = t_1$ and $z \to \infty$,

$$Q = 4r_o k(t_2 - t_1)$$

$$\frac{t - t_1}{t_2 - t_1} = \frac{2}{\pi}\sin^{-1}\left\{\frac{2}{[(R - 1)^2 + Z^2]^{1/2} + [(R + 1)^2 + Z^2]^{1/2}}\right\}$$

where $Z = z/r_o$ and $R = r/r_o$.

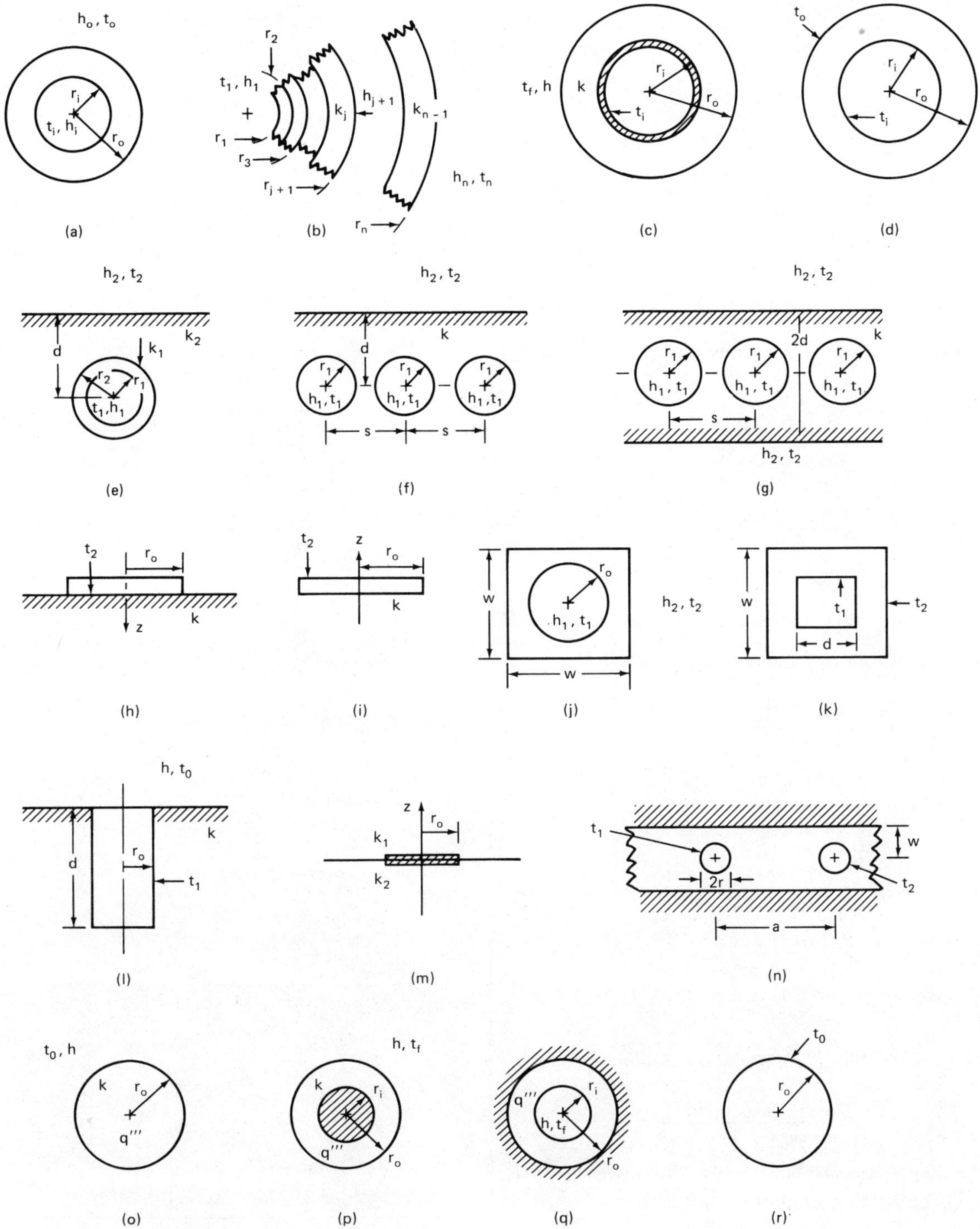

Fig. 2.2 Steady-state rods, tubes, cylinders, disks, pipes, and wires.

9. *Circular disk in an infinite solid*[3] (Fig. 2.2*i*)

$$Q = 8r_o k(t_2 - t_1)$$

10. *Infinite hollow square rod*[3] (Fig. 2.2*j*)

$$q' = \frac{2\pi k(t_2 - t_1)}{k/h_1 r_o + \ln(1.08w/2r_o) + \pi k/2h_2 w}$$

11. *Infinite hollow square pipe*[9] (Fig. 2.2*k*)

$$q' = \begin{cases} \dfrac{2\pi k(t_1 - t_2)}{0.93 \ln(w/d) - 0.0502} & \dfrac{w}{d} > 1.4 \\[2ex] \dfrac{2\pi k(t_1 - t_2)}{0.785 \ln(w/d)} & \dfrac{w}{d} < 1.4 \end{cases}$$

12. *Vertical cylinder in a semi-infinite solid*[3] (Fig. 2.2*l*)

$$Q = \left\{\frac{2D}{\ln[2D(1 + 1/\text{Bi}_d)]} + \frac{\text{Bi}_d}{D}\right\}\pi r_o k(t_1 - t_o)$$

where $\quad \text{Bi}_d = \dfrac{hd}{k} \qquad D = \dfrac{d}{r_o}$

13. *Two semi-infinite regions of different conductivities connected by circular disk*[6] (Fig. 2.2*m*): With

$$t = t_o \qquad z \to +\infty$$

$$t = t_1 \qquad z \to -\infty$$

$$q_z = 0 \qquad r > r_o \qquad z = 0$$

$$\frac{t - t_o}{t_1 - t_o} = \frac{2k_2}{\pi(k_1 + k_2)} \sin^{-1} \left\{\frac{2}{[(R - 1)^2 + Z^2]^{1/2} + [(R + 1)^2 + Z^2]^{1/2}}\right\} \qquad Z < 0$$

$$\frac{t - t_o}{t_1 - t_o} = 1 - \frac{2k_1}{\pi(k_1 + k_2)} \sin^{-1} \left\{\frac{2}{[(R - 1)^2 + Z^2]^{1/2} + [(R + 1)^2 + Z^2]^{1/2}}\right\} \qquad Z > 0$$

$$Q = \left(\frac{4r_o k_1 k_2}{k_1 + k_2}\right)(t_o - t_1)$$

where $\quad Z = \dfrac{z}{r_o} \qquad R = \dfrac{r}{r_o}$

14. *Heat flow between two pipes in an insulated infinite plate*[9] (Fig. 2.2*n*)

$$q' = \frac{2\pi k(t_1 - t_2)}{\pi a/w + \ln(w/\pi r)}$$

15. *Infinite cylinder with convection boundary and internal heat generation*[3] (Fig. 2.2*o*)

$$\frac{(t - t_o)k}{q''' r_o^2} = \frac{1}{4}\left(\frac{2}{\text{Bi}} + 1 - R^2\right)$$

where $\quad \text{Bi} = \dfrac{hr_o}{k} \qquad R = \dfrac{r}{r_o}$

16. *Hollow infinite cylinder with convection boundary on outside surface and internal heat generation*[3] (Fig. 2.2*p*): With $q_r = 0$ at $r = r_i$,

$$\frac{(t - t_f)k}{q''' r_o^2} = \frac{1}{4}\left[\frac{2}{\text{Bi}}(1 - R_i^2) + 1 - R^2 + 2R_i^2 \ln R\right]$$

where $\quad R = \dfrac{r}{r_o} \qquad \text{Bi} = \dfrac{hr_o}{k} \qquad R_i = \dfrac{r_i}{r_o}$

17. *Hollow infinite cylinder with convection-cooled inside surface and internal heat generation*[3] (Fig. 2.2*q*): With $q_r = 0$ at $r = r_o$,

$$\frac{(t - t_f)k}{q''' r_i^2} = \frac{1}{4}\left[\frac{2}{\text{Bi}}(R_o^2 - 1) + 1 - R^2 + 2R_o^2 \ln R\right]$$

where $\quad R = \dfrac{r}{r_i} \qquad \text{Bi} = \dfrac{hr_i}{k} \qquad R_o = \dfrac{r_o}{r_i}$

18. *Electrically heated wire with temperature-dependent thermal and electrical conductivities*[1,4] (Fig. 2.2*r*): With

$$t = t_o \qquad r = r_o$$

$$\frac{k_t}{k_{to}} = 1 + \beta_t(t - t_o)$$

$$\frac{k_e}{k_{eo}} = 1 + \beta_e(t - t_o)$$

$$\frac{t - t_o}{B} = R\left\{1 + B\left[\frac{R}{8} - \frac{\beta_e}{\beta_t}\frac{(2 + R)}{16}\right]\right\}$$

where $\quad B = \dfrac{k_{eo} r_o^2 E^2 \beta_t}{k_{to} L^2} \qquad R = 1 - \dfrac{r^2}{r_o^2}$

and with k_t being thermal conductivity, k_e electrical conductivity, and E voltage drop over length L.

2.4 SPHERES

1. *Spherical shell* (Fig. 2.3*a*)

$$Q = \frac{4\pi r_o k(t_i - t_o)}{r_o/r_i - 1 + (r_o/r_i)\, k/h_i r_i + k/h_o r_o}$$

$$\frac{t - t_o}{t_i - t_o} = \frac{r_o/r - 1 + k/r_o h_o}{r_o/r_i - 1 + (r_o/r_i)\, k/r_i h_i + k/r_o h_o}$$

2. *Composite sphere* (Fig. 2.3*b*)

$$Q = \frac{4\pi(t_1 - t_n)}{\displaystyle\sum_{i=1}^{n-1} \frac{1}{k_i}\left(\frac{1}{r_i} - \frac{1}{r_{i+1}}\right) + \sum_{i=1}^{n} \frac{1}{r_i^2 h_i}}$$

$$\frac{t_j - t_1}{t_n - t_1} = \frac{\displaystyle\sum_{i=1}^{j-1}\left[\frac{1}{k_i}\left(\frac{1}{r_i} - \frac{1}{r_{i+1}}\right) + \frac{1}{r_i^2 h_i}\right] + \frac{1}{k_j}\left(\frac{1}{r_j} - \frac{1}{r}\right) + 1/h_i r_i^2}{\displaystyle\sum_{i=1}^{n-1} \frac{1}{k_i}\left(\frac{1}{r_i} - \frac{1}{r_{i+1}}\right) + \sum_{i=1}^{n} \frac{1}{h_i r_i^2}} \qquad j > 1$$

where t_j is the local temperature in the jth layer.

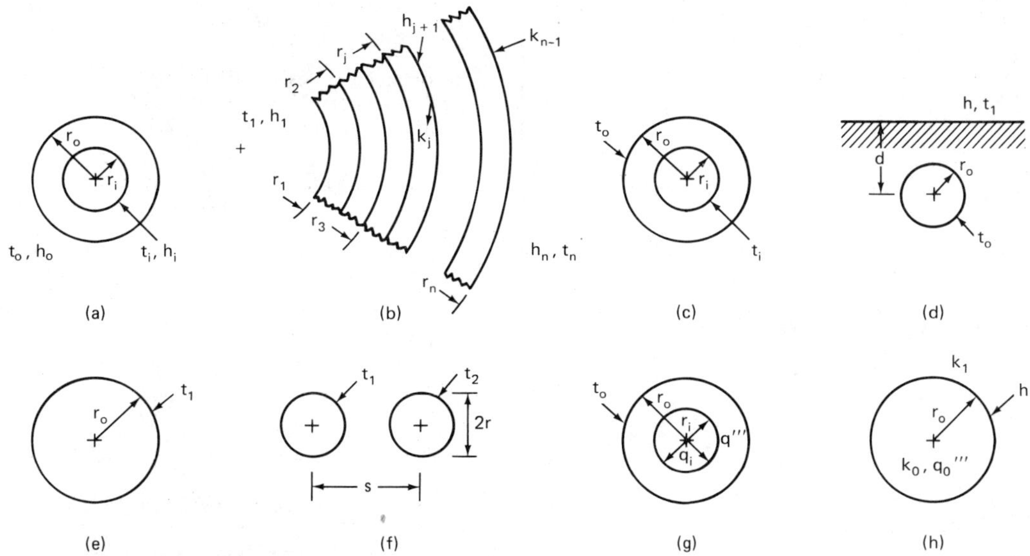

Fig. 2.3 Steady-state spheres.

3. *Sphere with temperature-dependent thermal conductivity*[2] (Fig. 2.3*c*): With

$$t = t_i \qquad r = r_i$$

$$t = t_o \qquad r = r_o$$

$$k = k_o + \beta(t - t_o)$$

$$Q = \frac{4\pi r_o k_m (t_i - t_o)}{r_o/r_i - 1}$$

where $$k_m = \frac{k_o + k_i}{2}$$

With

$$k = k_o \qquad t = t_o$$

$$k = k_i \qquad t = t_i$$

$$\frac{t - t_o}{k_o/\beta} = \sqrt{1 + 2\beta(t_i - t_o)\frac{k_m}{k_o^2}\frac{r_o/r - 1}{r_o/r_i - 1}} - 1$$

4. *Sphere in a semi-infinite solid*[3] (Fig. 2.3*d*)

$$Q = \frac{4\pi r_o k(t_1 - t_0)}{1 + 1/2(d/r_o + 1/\text{Bi})}$$

where $$\text{Bi} = \frac{hr_o}{k}$$

5. *Sphere in an infinite medium*[9] (Fig. 2.3*e*): With $t = t_2$ at $r \to \infty$,

$$Q = 4\pi r_o k(t_1 - t_2)$$

6. *Two spheres separated by a large distance in an infinite medium*[9] (Fig. 2.3*f*)

$$Q = \frac{4\pi rk(t_1 - t_2)}{2(1 - r/s)} \qquad s > 5r$$

For $s > 2r$, error $\simeq 1\%$.

7. *Spherical shell with specified inside surface heat flux and internal heat generation* (Fig. 2.3*g*): With

$$t = t_o \qquad r = r_o$$

$$q_r = q_i \qquad r = r_i$$

$$\frac{(t - t_o)k}{q_i r_i} = \frac{q''' r_i}{6q_i}\left(\frac{2(R - R_o)}{RR_o} + R_o^2 - R^2\right) + \frac{R_o - R}{RR_o}$$

where $$R = \frac{r}{r_i} \qquad R_o = \frac{r_o}{r_i}$$

8. *Solid sphere with internal heat generation in an infinite medium*[6] (Fig. 2.3*h*): With

$$k = k_o \qquad 0 < r < r_0$$

$$k = k_1 \qquad r > r_0$$

$$q''' = q_o''' \qquad 0 < r < r_0$$

$$q''' = 0 \qquad r > r_0$$

$$h = \text{contact coefficient at } r = r_o$$

$$t = t_\infty \qquad r \to \infty$$

$$\frac{(t - t_\infty)k_o}{q_o'''r_o^2} = \frac{1}{6}\left(1 - R^2 + \frac{2}{\text{Bi}} + \frac{2k_o}{k_1}\right) \qquad 0 < R < 1$$

$$\frac{(t - t_\infty)k}{q_o'''r_o^2} = \frac{1}{3R} \qquad r > r_o$$

where $\text{Bi} = \dfrac{hr_o}{k_o} \qquad R = \dfrac{r}{r_o}$

2.5 EXTENDED SURFACES

1. *Infinite rod*[2] (Fig. 2.4*a*): With

$$t = t_0 \qquad x = 0$$

$$t = t_f \qquad x \to \infty$$

$$Q = \sqrt{hkCA}(t_0 - t_f)$$

$$\frac{t - t_f}{t_0 - t_f} = \exp(-mx)$$

where $m = \sqrt{\dfrac{hC}{kA}}$

2. *Finite rod with insulated end*[5] (Fig. 2.4*b*): With

$$t = t_0 \qquad x = 0$$

$$q_x = 0 \qquad x = l$$

$$Q = \sqrt{hCkA}\,\tan(ml)\,(t_0 - t_f)$$

$$\frac{t - t_f}{t_0 - t_f} = \frac{\cosh[m(l - x)]}{\cosh(ml)}$$

3. *Finite rod with end heat loss*[5] (Fig. 2.4*c*): With $t = t_0$ at $x = 0$,

$$Q = \frac{mkA[(h_2/mk) + \tanh(ml)]\,(t_0 - t_f)}{1 + (h_2/mk)\tanh(ml)}$$

$$\frac{t - t_f}{t_0 - t_f} = \frac{\cosh[m(l - x)] + (h_2/mk)\sinh[m(l - x)]}{\cosh(ml) + (h_2/mk)\sinh(ml)}$$

where $m = \sqrt{\dfrac{h_1C}{kA}}$

4. *Straight rectangular fin*[5] (Fig. 2.4*d*): Use the equation for finite rod with end heat loss with

$$m = \sqrt{\frac{2h}{kb}}$$

$$\eta = \frac{\tanh(ml_c)}{ml_c}$$

where $l_c = l + \dfrac{b}{2}$

Recommended design criteria are $2k/hb > 5$; minimum weight criteria are $2l/b = 1.419\sqrt{2k/hb}$.

5. *Straight triangular fin*[5] (Fig. 2.4*e*)

$$q = \sqrt{2hkb}\,\frac{I_1(2\sqrt{\beta l})}{I_0(2\sqrt{\beta l})} \qquad \beta = \frac{2hl}{kb}$$

$$\frac{t - t_f}{t_0 - t_f} = \frac{I_0(2\sqrt{\beta x})}{I_0(2\sqrt{\beta l})} \qquad \eta = \frac{1}{\sqrt{\beta l}}\,\frac{I_1(2\sqrt{2\beta l})}{I_0(2\sqrt{\beta l})}$$

The optimum l/b ratio is

$$\frac{l}{b} = 0.655\sqrt{\frac{2k}{hb}}$$

6. *Cylindrical fin with rectangular profile*[5] (Fig. 2.4*f*): See Fig. 2.5.

7. *Cylindrical fin with triangular profile*[5] (Fig. 2.4*g*): See Fig. 2.5.

8. *Cylindrical pin fin*[7] (Fig. 2.4*h*)

$$Q = \frac{\pi}{4}kd^2m\tanh(ml)\,(t_0 - t_f)$$

$$\eta = \frac{\tanh(ml)}{ml}$$

where $m = \sqrt{\dfrac{4h}{kd}}$

9. *Rectangular pin fin*[7] (Fig. 2.4*h*): Use the equation for cylindrical pin fin with $\pi d^2/4$ replaced with bd, where b is the depth of the fin.

10. *Conical pin fin*[7] (Fig. 2.4*i*)

$$Q = \frac{\pi kb^2M}{8l}\,\frac{I_2(M)}{I_1(M)}\,(t_0 - t_f)$$

$$\frac{t - t_f}{t_0 - t_f} = \sqrt{\frac{l}{x}}\,\frac{I_1(M\sqrt{x/l})}{I_1(M)}$$

$$\eta = \frac{4I_2(M)}{MI_1(M)}$$

where $M = 2\sqrt{2}\,ml \qquad m = \sqrt{\dfrac{2h}{kb}}$

11. *Doubly heated straight rectangular fin*[7] (Fig. 2.4*j*): With

$$t = t_0 \qquad x = 0$$

$$t = t_l \qquad x = l$$

$$\frac{t - t_f}{t_0 - t_f} = \exp(mx) - 2B\sinh(mx)$$

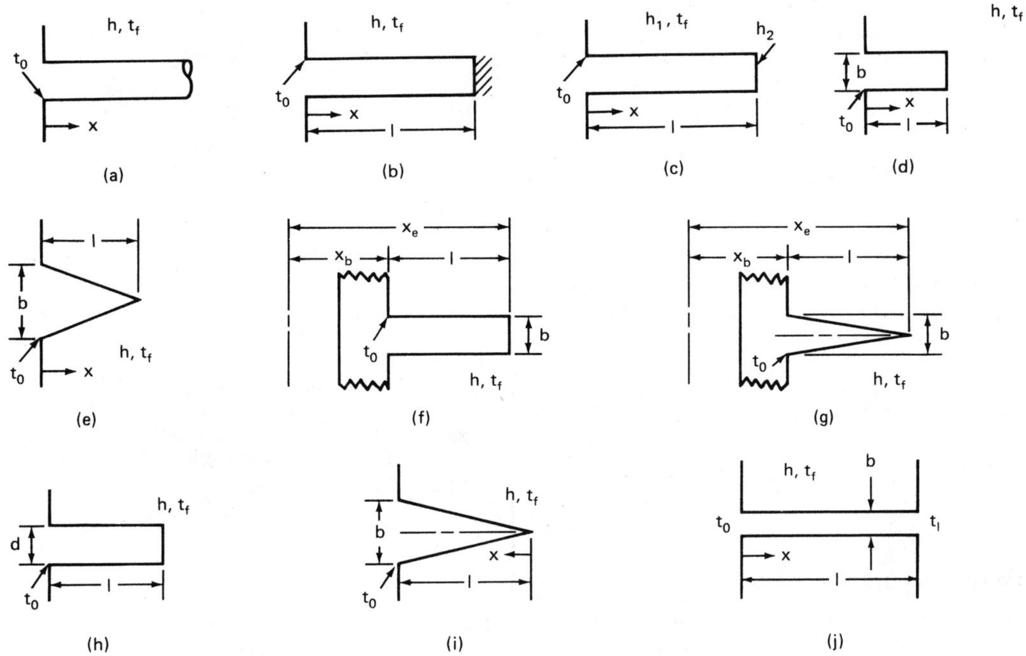

Fig. 2.4 Steady-state extended surfaces.

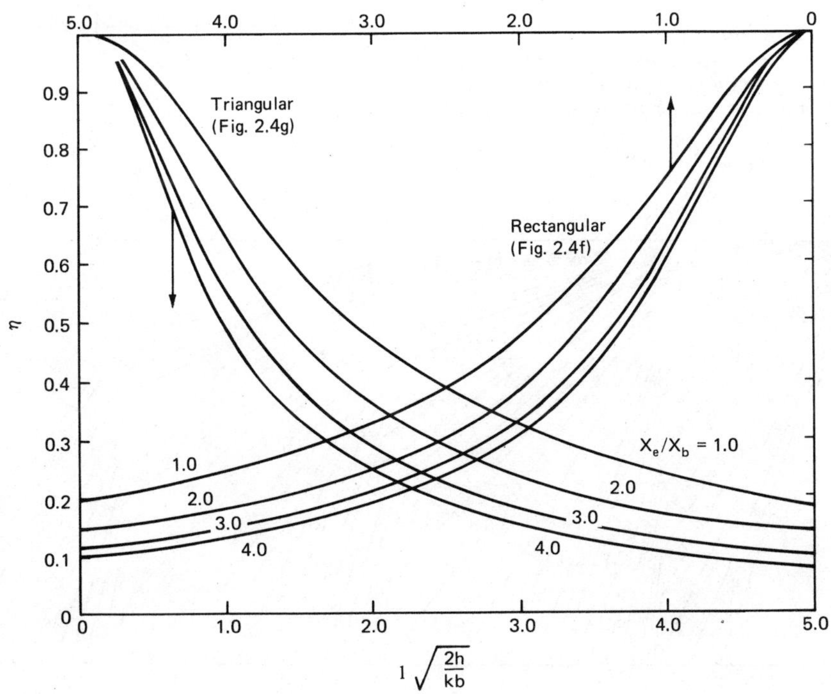

Fig. 2.5 Efficiencies of cylindrical fins.

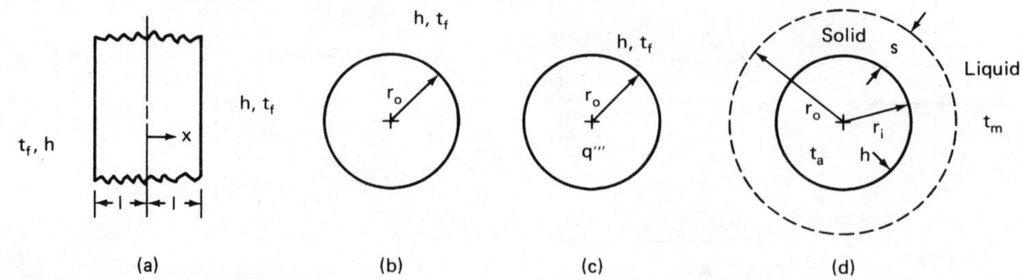

Fig. 2.6 Transient conduction configurations.

$$q_0 = kbm\,(2B - 1)(t_0 - t_f)$$

$$q_l = kbm\,[\exp(ml) - 2B\cosh(ml)](t_0 - t_f)$$

where

$$B = \frac{\exp(ml) - (t_l - t_f)/(t_0 - t_f)}{2\sinh(ml)} \qquad m = \sqrt{\frac{2h}{kb}}$$

2.6 TRANSIENT SOLUTIONS

1. *Infinite plate with constant initial temperature and convection boundary condition* (Fig. 2.6*a*): With t_i being the uniform initial temperature,

$$\mathrm{Fo} = \frac{\alpha\tau}{l^2}$$

See Fig. 2.7 for midplane temperature t.

2. *Infinite cylinder with uniform initial temperature and convection boundary condition* (Fig. 2.6*b*): With t_i being the uniform initial temperature, t_c the axis temperature, and

$$\mathrm{Fo} = \frac{\alpha\tau}{r_o^2}$$

See Fig. 2.8 for axis temperature.

3. *Infinite cylinder with uniform initial temperature, convection boundary condition, and steady and uniform internal heating* (Fig. 2.6*c*): With q_0''' being the steady and uniform heating rate, t_i the uniform initial temperature, t_c the axis temperature, and

$$\mathrm{Fo} = \frac{\alpha\tau}{r_o^2}$$

See Fig. 2.9 for axis temperature.

4. *Freezing on the outside of a tube and convection cooling on the inside*[8] (Fig. 2.6*d*): With t_m being the freezing temperature and the tube resistance being neglected,

$$\frac{\tau\,h^2\,(t_m - t_a)}{\rho\gamma k} = \frac{\mathrm{Bi}^2}{2}\left[\left(\frac{1}{\mathrm{Bi}} - \frac{1}{2}\right)(R^2 - 1) + R^2 \ln R\right] \qquad R \geq 1$$

where

$$\mathrm{Bi} = \frac{hs}{k} \qquad R = \frac{r_o}{r_i}$$

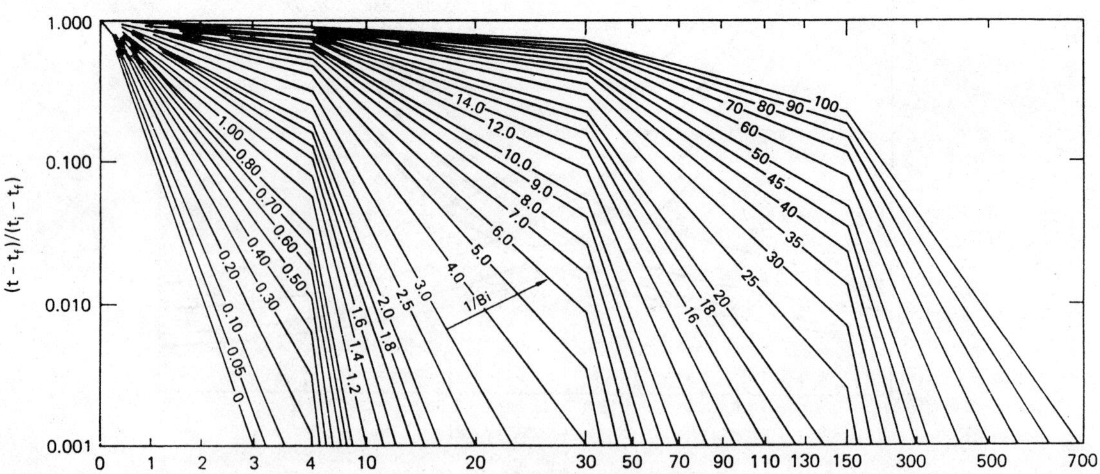

Fig. 2.7 Midplane temperature response of a convectively cooled infinite plate. (*Ref. 10.*)

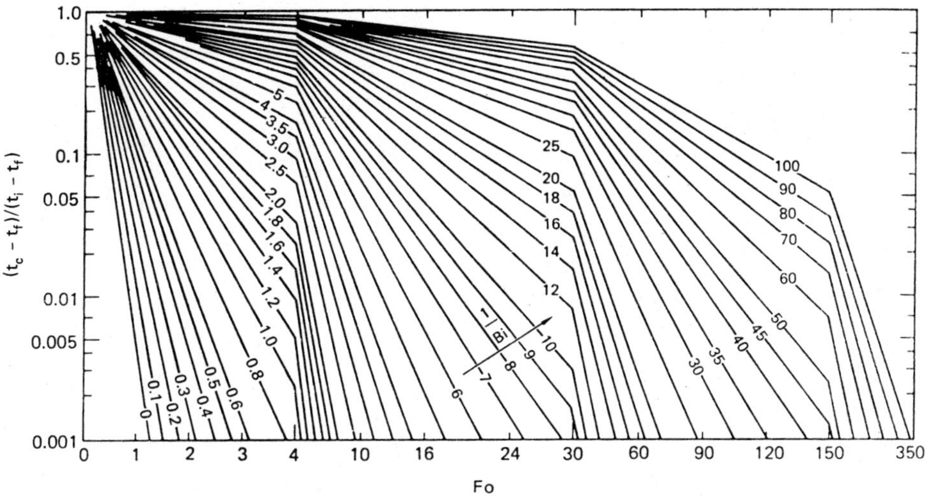

Fig. 2.8 Axis temperature response of a convectively cooled infinite cylinder. (*Ref. 10.*)

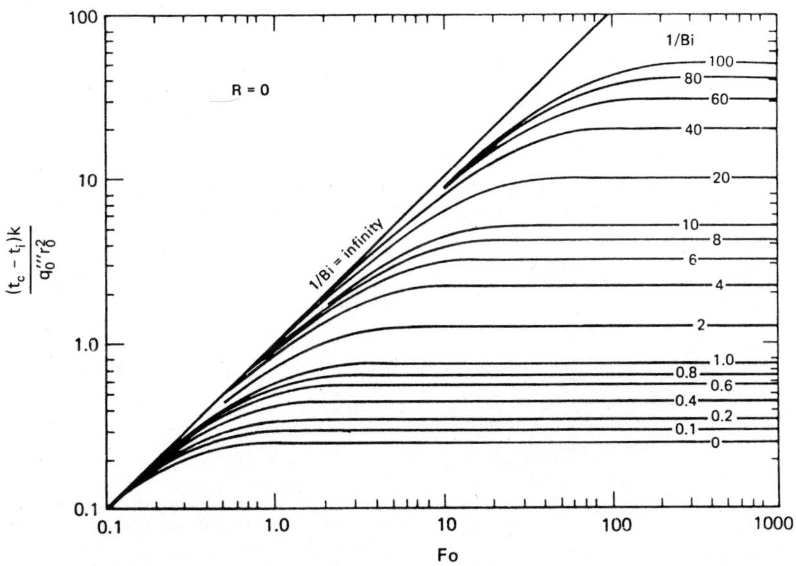

Fig. 2.9 Axis temperature response of a convectively cooled infinite cylinder with uniform internal heating. (*Ref. 11.*)

NOMENCLATURE

A = area, m² (ft²)
b = fin thickness, m (ft)
c = specific heat, J/kg · °C (Btu/lbm · °F)
C = circumference, m (ft)
d = diameter, depth, m (ft)
h = heat-transfer coefficient, W/m² · °C (Btu/h · ft² · °F)
k = thermal conductivity, W/m · °C (Btu/h · ft · °F)
m = $\sqrt{hC/kA}$, m^{-1} (ft^{-1})
l = length, m (ft)
q = heat flux rate, W/m² (Btu/h · ft²)
q' = linear heat flux, W/m (Btu/h · ft)
q''' = volumetric heating rate, W/m³ (Btu/h · ft³)
Q = heat-transfer rate, W (Btu/h)
r = radius, m (ft)

t = temperature, °C, K (°F)
w = width, m (ft)
x, y, z = cartesian coordinates, m (ft)
α = thermal diffusivity = $k/\rho c$, m^2/s (ft^2/s)
β = temperature coefficient, $°C^{-1}$ ($°F^{-1}$)
γ = latent heat of fusion, J/kg (Btu/lbm)
η = fin effectiveness,

$$= \frac{\text{actual heat transferred}}{\text{heat transferred from infinite conductivity fins}}$$

ρ = density, kg/m^3 (lbm/ft^3)
τ = time, s

Dimensionless Expressions

Bi = Biot modulus = hl/k
Fo = Fourier modulus = $\alpha\tau/l^2$
Po = $q'''l^2/k$
X = length ratio = x/l

Mathematical Functions

exp = exponential function
I_n = modified Bessel function of the first kind
J_n = Bessel function of the first kind
K_n = modified Bessel function of the second kind
ln = natural logarithm

REFERENCES

1. J. H. VanSant, "Conduction Heat Transfer Solutions," National Technical Information Service, UCRL-52863 Rev. 1, Aug. 1983.
2. M. Jakob, *Heat Transfer,* vol. 1, Wiley, New York, 1949.
3. S. S. Kutateladze and V. M. Borishanskii, *A Concise Encyclopedia of Heat Transfer,* Pergamon, New York, 1966.
4. R. B. Bird, W. E. Stewart, and E. N. Lightfoot, *Transport Phenomena,* Wiley, New York, 1960.
5. E. R. G. Eckert and R. M. Drake, Jr., *Heat and Mass Transfer,* McGraw-Hill, New York, 1959.
6. H. S. Carslaw and J. C. Jaeger, *Conduction of Heat in Solids,* 2d ed., Oxford Press, London, 1959.
7. D. Q. Kern and A. D. Kraus, *Extended Surface Heat Transfer,* McGraw-Hill, New York, 1972.
8. W. M. Rohsenow and J. P. Hartnett (eds.), *Handbook of Heat Transfer,* McGraw-Hill, New York, 1973.
9. E. Hahne and U. Grigull, "Formfaktor und Formwiderstand der stationären mehrdimensionalen Wärmeleitung," *Int. J. Heat Mass Transfer,* **18,** 751–767 (1975).
10. J. P. Holman, *Heat Transfer,* McGraw-Hill, New York, 1963.
11. F. M. Young and C. R. Savino, "Time-Temperature Charts for One-Dimensional Conduction with Uniform Internal Heat Generation," *J. Heat Transfer,* **92,** 207–210 (1970).

Chapter 3

Natural Convection Heat Transfer

MAJID CHARMCHI
JOSE G. MARTIN
Department of Mechanical and Energy Engineering
University of Lowell

3.1 INTRODUCTION

The heat-transfer process that occurs between a surface and a moving fluid at a different temperature is called convection heat transfer. If the fluid motion is generated by an external source, e.g., by a blower or by a pump, the process is called forced convection (see Chap. 4, "Forced Convection Heat Transfer"). If the flow is driven by the buoyancy force caused by density differences developed due to the heat transfer itself, the process is known as free convection. The buoyancy force is present in almost all earthbound convective heat-transfer problems. Therefore, the process is often called natural convection. When an external force is present, the relative importance of natural versus forced convection must be assessed. If both are important, the process is called mixed convection.[1–14]

It is important to note that other body forces besides gravity can act to move the fluid in a free-convection process, e.g., centrifugal forces, magnetic forces, etc.; hence, a number of variations can exist. This section considers the natural-convection problem where the body force is gravitational and the density gradient is due to a temperature variation.

A summary of analytical and empirical solutions to commonly encountered problems related to natural convection on external surfaces (Fig. 3.1*a*), through open cavities (Fig. 3.1*b*), and in enclosed spaces (Fig. 3.1*c*) is presented.

3.2 BASIC CONCEPTS

Conservation Equations

In the analysis of natural-convection processes, the conservation principles are used to set up the governing equations, i.e., the equation of motion, the equation of energy, and the

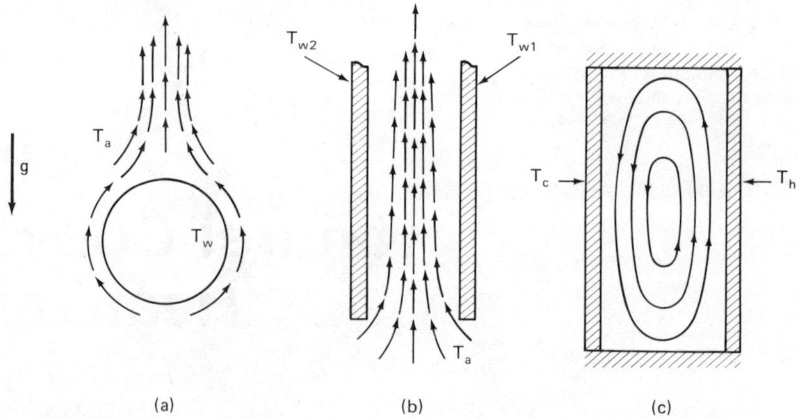

Fig. 3.1 Natural convection flow behavior. (*a*) On external surface; (*b*) through open cavity; (*c*) in enclosed space.

equation of continuity. The solution of these equations determines the change of velocity and temperature with respect to time and position in the system from which the rate of heat transfer can be obtained. In the development of the equation of motion, the term that represents the body force ρg needs special attention.[2] In many analyses, it is assumed that the Boussinesq approximation is valid, i.e., the density is regarded as temperature dependent as it appears in the body force, but elsewhere in the conservation equations, density is a constant. The detailed derivation and the expressions of the conservation equations in different coordinate systems are presented in Refs. 2 and 15.

Heat-Transfer Coefficient

The area-averaged heat-transfer coefficient h is defined as

$$h = \frac{q}{A\,\Delta T} \tag{3.1}$$

where q = overall heat transfer
A = surface area on which heat transfer takes place
ΔT = temperature difference

For external flows (unbounded flow field, e.g., natural convection over a flat plate), the temperature difference ΔT is obtained from

$$\Delta T = T_w - T_a \tag{3.2}$$

where T_w is the surface (wall) temperature and T_a is the temperature of the surrounding fluid. The ambient temperature T_a is assumed to be constant.

For natural convection inside an enclosed space, where heat is exchanged between the surfaces of the enclosure maintained at different temperatures, ΔT is calculated by

$$\Delta T = T_h - T_c \tag{3.3}$$

where T_h and T_c are the temperatures of the hot and cold surfaces, respectively.

Dimensionless Groups

In a given free-convection problem, the heat-transfer coefficient h can vary with many system parameters, i.e., body shape, body size, surface and fluid temperature, buoyant specific weight ($g\,\Delta\rho$), and fluid properties. So,

$$h = f(\text{shape}, L_c, T_w, T_a, g\,\Delta\rho, \rho, \mu, k, c_p) \tag{3.4}$$

where L_c is a characteristic length and the parameters ρ, μ, k, and c_p are the fluid density, viscosity, thermal conductivity, and specific heat capacity, respectively. The principles of dimensional analysis are used to reduce this long list of variables into a more manageable set of dimensionless parameters. In a typical natural-convection problem, the heat-transfer coefficient, in its dimensionless form, can be expressed as

$$\frac{hL_c}{k} = \text{Nu} = f(\text{Gr}, \text{Pr}, \text{generic shape}) \tag{3.5}$$

where the dimensionless variables Nu, Gr, and Pr are called Nusselt, Grashof, and Prandtl numbers, respectively. In free-convection literature another dimensionless parameter appears very often, called the Rayleigh number and expressed by

$$\text{Ra} = \text{Gr Pr} \tag{3.6}$$

The physical interpretation of the dimensionless parameters are as follows:

Nusselt Number Nu: The Nusselt number is the ratio of the convective conductance h to the molecular thermal conductance k/L_c. The area averaged Nusselt number is then expressed by

$$\text{Nu} = \frac{hL_c}{k} \tag{3.7}$$

where L_c is the characteristic length of the body (a geometry parameter) and k is the conductivity of the surrounding fluid.

A Nusselt number of order unity would indicate a sluggish motion little more effective than pure fluid conduction, e.g., laminar natural convection between two closely spaced parallel plates maintained at a small temperature difference ($T_h \approx T_c$). A large Nusselt number means very efficient convection, e.g., turbulent natural convection over a long vertical plate.

Grashof Number Gr: The Grashof number is proportional to the ratio of the buoyancy forces to the viscous forces and is defined as

$$\mathrm{Gr} = \frac{g\,\Delta\rho\, L_c^3}{\rho_r \nu^2} \tag{3.8}$$

where ρ_r is the density at a reference condition and $\Delta\rho$ is the density difference.

Since in natural convection the pressure gradients are negligible, the density of a pure substance varies with temperature only. In moderate temperature ranges and for many common substances it can be assumed that the density varies linearly with temperature as[9]

$$\rho = \rho_r[1 - \beta(T - T_r)] \tag{3.9}$$

The parameter β is the coefficient of thermal expansion; for an ideal gas $\beta \equiv 1/T$, where T is absolute temperature in degrees Rankine or Kelvin. All common gases and many low-boiling-point liquids have a strong density variation nearly comparable to ideal gases. For gases β is evaluated at T_a. For liquids, the mean film temperature $T_f = (T_w + T_a)/2$ or $T_f = (T_h + T_c)/2$ must be used.

Assuming a constant value of β, the Grashof number, in terms of temperature difference, is defined as

$$\mathrm{Gr} = \frac{g\beta\,\Delta T\, L_c^3}{\nu^2} \tag{3.10}$$

In situations where both external and buoyancy forces are present, the relative importance of the natural convection with respect to the forced convection can be assessed by comparing the Grashof number to the Reynolds number squared, $\mathrm{Re}^2 = (UL_c/\nu)^2$, as follows:

$\mathrm{Gr} \gg \mathrm{Re}^2$	dominant natural convection
$\mathrm{Gr} \approx \mathrm{Re}^2$	mixed convection
$\mathrm{Gr} \ll \mathrm{Re}^2$	dominant forced convection

Prandtl Number Pr: The Prandtl number is a fluid property defined by

$$\mathrm{Pr} = \frac{\nu}{\alpha} \tag{3.11}$$

where ν and α are the kinematic viscosity and the thermal diffusivity, respectively. Hence, the Prandtl number is the ratio of the momentum diffusion to the thermal diffusion. This parameter has a moderate to strong effect on convection, especially for single-phase situations.

Rayleigh Number Ra: The natural-convection analyses have shown that for fluids with large Prandtl numbers the effects of the Prandtl and Grashof numbers are similar. This suggests that the product of Grashof and Prandtl numbers is a useful parameter. Called the Rayleigh number, the product is expressed by

$$\mathrm{Ra} = \mathrm{Gr}\,\mathrm{Pr} = \frac{g\beta\,\Delta T\, L_c^3}{\nu\alpha} \tag{3.12}$$

In the heat-transfer literature it is common to propose a Rayleigh-number-type correlation, $\mathrm{Nu} = f(\mathrm{Ra}, \mathrm{Pr})$, rather than the traditional type, $\mathrm{Nu} = f(\mathrm{Gr}, \mathrm{Pr})$.

Constant-Heat-Flux Boundary Condition

In situations where the boundary condition is constant wall heat flux q'' the temperature difference ($\Delta T = T_w - T_a$) is unknown and varies over the surface. Hence, an alternative Grashof (Rayleigh) number is needed that does not contain ΔT. Sparrow and Gregg[16] introduced the modified Grashof (Rayleigh) number as

$$\mathrm{Gr}^* = \mathrm{Gr}\,\mathrm{Nu} = \frac{g\beta q'' L_c^4}{k\nu^2} \tag{3.13a}$$

$$\mathrm{Ra}^* = \mathrm{Ra}\,\mathrm{Nu} = \frac{g\beta q'' L_c^4}{k\nu\alpha} \tag{3.13b}$$

The correlation is then expressed in terms of $\mathrm{Nu} = f(\mathrm{Ra}^*, \mathrm{Pr})$.

With Nu known, the mean temperature difference $\overline{\Delta T}$ can be found from the following relation

$$\overline{\Delta T} = \overline{T}_w - T_a = \frac{q'' L_c}{k\,\mathrm{Nu}} \tag{3.14}$$

where $\overline{T}_w$ is the area-averaged value of the surface temperature expressed by

$$\overline{T}_w = \frac{1}{A}\int_A T_w\, dA \tag{3.15}$$

Laminar and Turbulent Flow

A natural-convection flow field can exhibit transition from a laminar to a turbulent condition. The flow that is caused by a thermal instability, on its pass, may experience small disturbances. When the viscous effects are strong with respect to the buoyancy force (small Grashof number), these disturbances are quickly damped, and hence the flow stays laminar. On the other hand, at large Grashof numbers the disturbances are amplified, and the flow will undergo transition to turbulence. Extensive discussions related to stability and transition effects are given in Refs. 6 and 17.

Transition to turbulence will have an important effect on the heat-transfer rate. Hence, in the evaluation of the Nusselt number, it is essential to determine whether the flow is laminar or turbulent. Due to the complexity of the turbulent flows, most of the correlations for this flow condition are based on experimental results.

3.3 EXTERNAL NATURAL CONVECTION

Empirical correlations have been developed for natural convection in fluids with moderate to large Prandtl number ($\mathrm{Pr} > 0.6$) for common immersed geometries. Typically, they have the form

$$\mathrm{Nu} = \frac{hL_c}{k} = C\,\mathrm{Ra}^n \qquad (3.16)$$

where C and n are constant and Ra is defined by Eq. (3.12). Characteristically, n = ¼ and ⅓ for laminar and turbulent flows, respectively.

The flow may be laminar or turbulent over the entire surface. However, in many practical situations the flow is laminar over one portion of the body and turbulent over another. To blend the laminar and turbulent equations into a single correlation equation that can be used for the entire surface, Churchill and Usagi[18] have proposed the following equation

$$\mathrm{Nu} = [(\mathrm{Nu}_l)^m + (\mathrm{Nu}_t)^m]^{1/m} \qquad (3.17)$$

where Nu_l and Nu_t represent the equations for laminar and turbulent Nusselt numbers, respectively. The parameter m is a constant adjusted so as to fit the experimental data.

Two types of boundary conditions are considered:

1. Constant wall temperature [T_w and ΔT are constant, Eq. (3.2)],
2. Constant wall heat flux (q'' is uniform)

For the second boundary condition, the area-averaged surface temperature $\overline{T}_w$ and temperature difference $\overline{\Delta T}$ are evaluated from Eqs. (3.15) and (3.14), respectively.

All fluid properties are evaluated at the film temperature, $T_f = (T_w + T_a)/2$, except for the coefficient of thermal expansion β. For gases β is evaluated at T_a and for liquids at T_f (absolute temperature for the β calculation).

For some common geometries, empirical correlations are presented in the following sections. For other geometries, the reader may refer to the literature. Alternatively, a new problem may be solved numerically, using the boundary-layer approximation. However, the boundary-layer solution should be considered only as a first step in the determination of the heat transfer; the boundary-layer approximation is based on the assumption that heat conduction along the direction of fluid flow is negligible. This is not generally true in free convection, and the "thin-layer" solution should be corrected. The so-called conduction-layer approximate method (CLAM),[2,19] which has proven quite effective for problems of external natural convection, is based on the premise that thin-layer solutions can be corrected for curvature effects by means of a solution to a problem in conduction heat transfer. This method is developed in Ref. 19, to which the reader is referred for details.

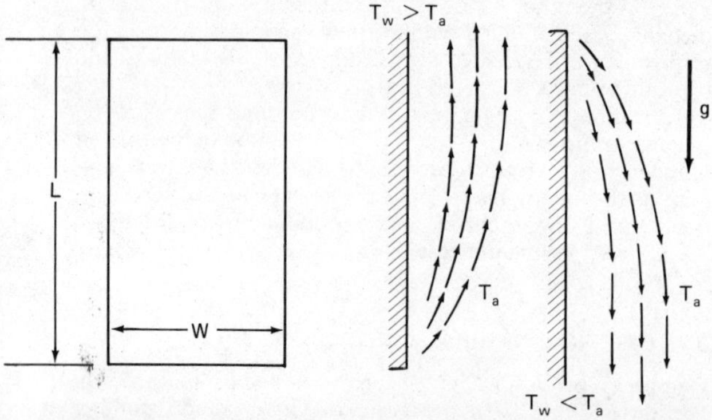

Fig. 3.2 Natural convection on a vertical plate.

The equations listed below are referred to frequently in the correlations that follow.

$$\mathrm{Nu} = \frac{hL_c}{k} = \frac{qL_c}{A\,\Delta T\,k} = \frac{q''L_c}{\Delta T\,k} \qquad (3.18)$$

$$\mathrm{Ra} = \frac{g\beta\,\Delta T\,L_c^3}{\nu\alpha} \qquad (3.19)$$

$$\mathrm{Ra}^* = \mathrm{Ra}\,\mathrm{Nu} = \frac{g\beta q'' L_c^4}{\nu\alpha k} \qquad (3.20)$$

Vertical Flat Plates (Fig. 3.2) Nu, Ra, and Ra* are defined by Eqs. (3.18) to (3.20) with $L_c = L$.

Constant Wall Temperature, $\Delta T = T_w - T_a$. The overall Nusselt number is computed from[2,19,20]

$$\mathrm{Nu} = [(\mathrm{Nu}_l)^6 + (\mathrm{Nu}_t)^6]^{1/6} \qquad 1 < \mathrm{Ra} < 10^{12} \qquad (3.21)$$

where Nu_l and Nu_t are

$$\mathrm{Nu}_l = \frac{2.8}{\ln[1 + 2.8/(C_l\,\mathrm{Ra}^{1/4})]} \qquad (3.22)$$

$$\mathrm{Nu}_t = C_v\,\mathrm{Ra}^{1/3} \qquad (3.23)$$

The parameters C_l and C_v are evaluated from

$$C_l = \frac{0.671}{[1 + (0.492/\mathrm{Pr})^{9/16}]^{4/9}} \qquad (3.24)$$

$$C_v = \frac{0.13\,\mathrm{Pr}^{0.22}}{(1 + 0.61\,\mathrm{Pr}^{0.81})^{0.42}} \qquad (3.25)$$

Some computed values for C_l and C_v are presented in Table 3.1.

Uniform Heat Flux, ΔT from Eq. (3.14). The laminar and turbulent Nusselt equations are[2,21]

$$\mathrm{Nu}_l = \frac{1.83}{\ln[1 + 1.83/(F_l(\mathrm{Ra}^*)^{1/5})]} \qquad (3.26)$$

Table 3.1 Typical Values for Parameters C_l, C_v, and F_l at Various Pr

Pr	C_l	C_v	F_l
0.10	0.387	0.075	0.501
0.71	0.515	0.103	0.624
1.00	0.534	0.106	0.641
2.00	0.568	0.112	0.671
4.00	0.595	0.113	0.694
6.00	0.608	0.113	0.704
10.00	0.622	0.110	0.716
50.00	0.650	0.097	0.738
100.00	0.656	0.091	0.744
1000.00	0.667	0.070	0.753
2000.00	0.668	0.064	0.754

$$\mathrm{Nu}_t = (C_v)^{3/4}(\mathrm{Ra}^*)^{1/4} \tag{3.27}$$

where F_l is evaluated from

$$F_l = 1.2\left(\frac{\mathrm{Pr}}{4 + 9\,\mathrm{Pr}^{1/2} + 10\,\mathrm{Pr}}\right)^{1/5} \tag{3.28}$$

Some computed values for F_l are given in Table 3.1.

Substitution of Eqs. (3.26) and (3.27) into Eq. (3.21) gives Nu that is valid in the range $10 < \mathrm{Ra}^* < 10^{15}$.

Inclined Flat Plates (Fig. 3.3) Equation (3.21) may be used with the following modifications:

1. For laminar flow Ra in Eq. (3.22) and Ra* in Eq. (3.26) are replaced by Ra cos θ and Ra* cos θ, respectively.

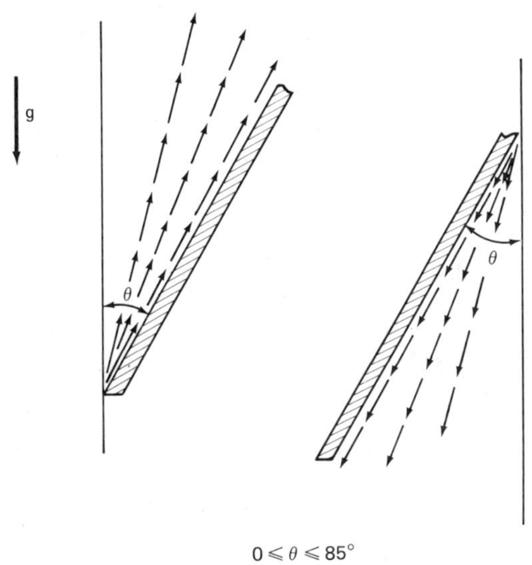

Fig. 3.3 Natural convection on an inclined flat plate.

2. For turbulent flow, C_v in Eqs. (3.23) and (3.27) is replaced by

$$C_v(\cos\theta)^{1/3} \qquad \text{for } 0° \le \theta \le 20° \tag{3.29a}$$

$$0.14(\sin\theta)^{1/3} \qquad \text{for } 20° < \theta \le 90° \tag{3.29b}$$

For further information see Refs. 22–24.

Flat Plates of Various Plan (Fig. 3.4) The overall heat transfer is

$$q = \int_0^W h_x\,\overline{\Delta T}\,L_x\,dx \tag{3.30}$$

For constant wall temperature, $\overline{\Delta T} = T_w - T_a$, Eq. (3.30) becomes

$$\frac{qL^*}{A\,\overline{\Delta T}\,k} = \mathrm{Nu} = \frac{L^*}{A}\int_0^W \mathrm{Nu}_x\,dx \tag{3.31}$$

where L^* is a convenient reference length.

For laminar flow, as an approximation, it may be assumed that $\mathrm{Nu}_l = C_l\,\mathrm{Ra}^{1/4}$, hence Eq. (3.31) can be represented as

$$\mathrm{Nu}_l = \left[\frac{(L^*)^{1/4}\int_0^W L_x^{3/4}\,dx}{A}\right] C_l\mathrm{Ra}^{1/4} \tag{3.32}$$

where $\mathrm{Ra} = g(\cos\theta)\beta\,\overline{\Delta T}\,(L^*)^3/\nu\alpha$.

For constant wall heat flux, $\mathrm{Nu} = F_l(\mathrm{Ra}^*)^{1/5}$, hence

$$\mathrm{Nu}_l = \left[\frac{(L^*)^{1/5}\int_0^W L_x^{4/5}\,dx}{A}\right] F_l(\mathrm{Ra}^*)^{1/5} \tag{3.33}$$

where $\mathrm{Ra}^* = g(\cos\theta)\beta q''(L^*)^4/\nu\alpha k$.

For turbulent flow, C_v in Eqs. (3.23) and (3.27) is replaced by those defined in Eq. (3.29*a* and *b*).

To obtain the Nu valid for both laminar and turbulent flows, the values of Nu_l and Nu_t can be placed in Eq. (3.21).

Horizontal Plates (Fig. 3.5) The values of Nu, Ra, and Ra* are evaluated based on Ref. 25, L_c = plate surface area/plate perimeter = A/P.

Upward Heated Surface or Downward Cooled Surface. The overall Nu can be obtained from[2,25]

$$\mathrm{Nu} = [(\mathrm{Nu}_l)^{10} + (\mathrm{Nu}_t)^{10}]^{1/10} \qquad \mathrm{Ra} > 1 \tag{3.34}$$

where

$$\mathrm{Nu}_l = \frac{1.4}{\ln\,[1 + 1.677/(C_l\,\mathrm{Ra}^{1/4})]} \tag{3.35}$$

and

$$\mathrm{Nu}_t = 0.14\,\mathrm{Ra}^{1/3} \tag{3.36}$$

C_l is defined in Eq. (3.24) and in Table 3.1.

Equation (3.34) to (3.36) are for the condition of constant wall temperature but they may be used for constant wall heat flux by computing $\overline{\Delta T}$, as defined in Eq. (3.14), and Nu simultaneously.

Downward Heated Surface or Upward Cooled Surface. In these situations laminar flow prevails up to very high Rayleigh numbers. Taking $L_c = A/P$,

$$\mathrm{Nu} = \mathrm{Nu}_l = C_n\,\mathrm{Ra}^{1/5} \qquad 10^5 < \mathrm{Ra} < 10^{10} \tag{3.37}$$

where C_h is evaluated by[2,26]

$$C_h = \frac{0.527}{[1 + (1.9/\mathrm{Pr})^{9/10}]^{2/9}} \tag{3.38}$$

Table 3.2 presents some computed values for C_h.

Vertical Cylinders (Fig. 3.6) L_c = cylinder length = L; Nu and Ra are obtained from Eqs. (3.18) and (3.19), respectively. For laminar flow, the vertical flat plate results may be corrected to give Nu_l for vertical cylinders

$$\mathrm{Nu}_{l,\mathrm{cyl}} = \mathrm{Cf}\,\mathrm{Nu}_l \tag{3.39}$$

where Nu_l is from Eq. (3.22), for T_w = constant, or from Eq. (3.26), for q'' = constant, and Cf is the correction factor estimated from[2]

$$\mathrm{Cf} = \frac{1.8\Phi}{\ln\,[1 + 1.8\Phi]} \tag{3.40}$$

$$\text{where } \Phi = \frac{L/D}{(C_l\,\mathrm{Ra}^{1/4})} \qquad \text{constant wall temperature} \tag{3.41}$$

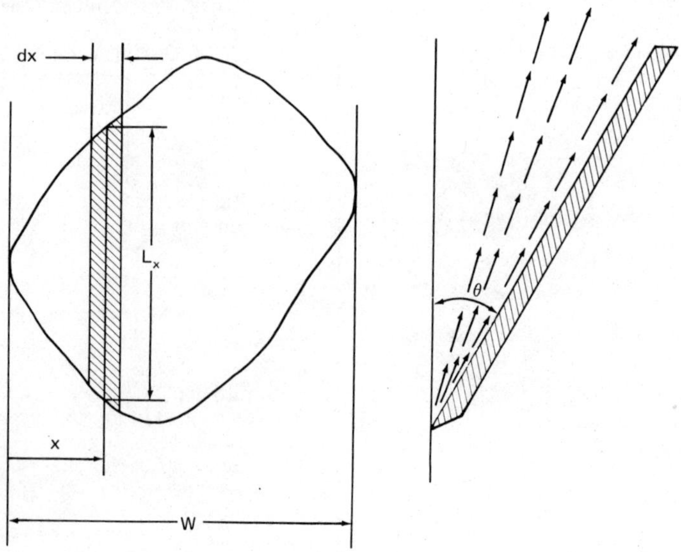

Fig. 3.4 Natural convection on a plate of various plan.

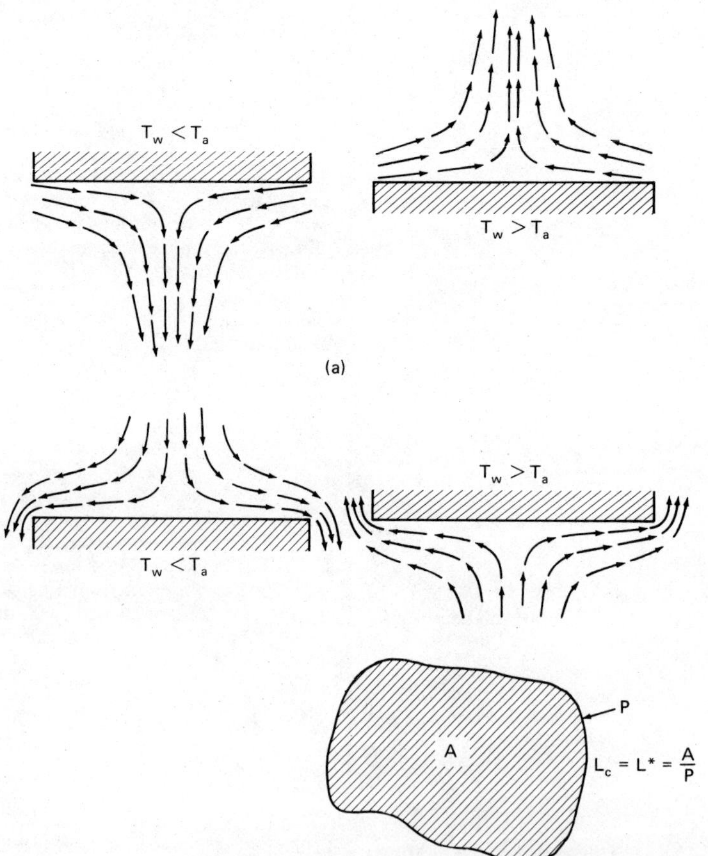

Fig. 3.5 Natural convection on horizontal plates. (*a*) Upward heated or downward cooled surface; (*b*) downward heated or upward cooled surface.

Table 3.2 Typical Values for C_h at Various Pr

Pr	C_h	Pr	C_h
0.10	0.288	5.00	0.488
0.70	0.400	10.00	0.504
1.00	0.420	100.00	0.524
2.00	0.454	1000.00	0.527

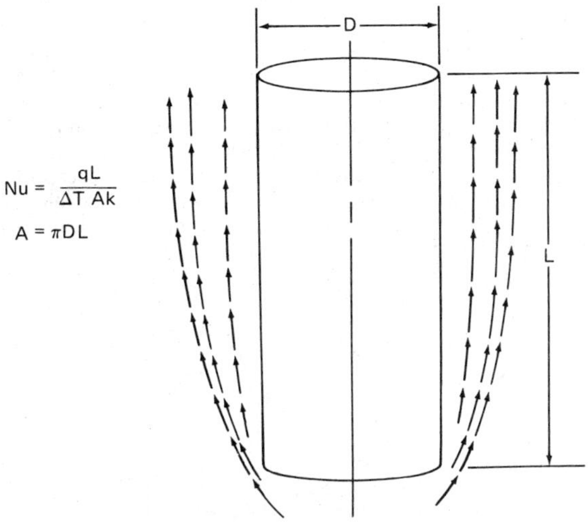

Fig. 3.6 Natural convection on vertical cylinder.

$$\text{and} \qquad \Phi = \frac{L/D}{F_l\,(\text{Ra}^*)^{1/5}} \quad \text{constant heat flux} \qquad (3.42)$$

Nu_t is computed from Eq. (3.23) for T_w = constant, or from Eq. (3.27) for q'' = constant. Introduction of the values of Nu_l and Nu_t into Eq. (3.21) gives a Nu valid for both laminar and turbulent flow.

Horizontal Cylinders Nu and Ra are defined in Eqs. (3.18) and (3.19) with L_c = cylinder diameter = D. The overall Nusselt number Nu can be obtained from[27]

$$\text{Nu} = \left\{0.60 + \frac{0.387\ \text{Ra}^{1/6}}{[1 + (0.559/\text{Pr})^{9/16}]^{8/27}}\right\}^2$$

$$10^{-5} < \text{Ra} < 10^{12} \qquad (3.43)$$

Equation (3.43) is for constant T_w, but for practical purposes Eq. (3.43) can be used for constant q'' with ΔT obtained from Eq. (3.14). Nusselt numbers for horizontal elliptical cylinders can be found in Refs. 2 and 20.

Spheres Taking L_c = sphere diameter = D, the laminar Nusselt number is expressed[28]

$$\text{Nu}_l = 2.0 + 0.878 C_l\, \text{Ra}^{1/4} \qquad (3.44)$$

where C_l is defined by Eq. (3.24) and given in Table 3.1. For turbulent flow

$$\text{Nu}_t = C_t\, \text{Ra}^{1/3} \qquad (3.45)$$

Table 3.3 C_t Coefficients for Eq. (3.45)

Pr	C_t	Pr	C_t
0.01	0.074	6.00	0.111
0.10	0.088	50.00	0.101
0.71	0.104	100.00	0.097
2.00	0.110	2000.00	0.086

where C_t is presented in Table 3.3 for various Pr. Introduction of Eqs. (3.44) and (3.45) into Eq. (3.21) gives a Nu valid for $10 < \text{Ra} < 10^{10}$. Nu for other spheroids are presented in Ref. 28.

Inclined Cylinders (Fig. 3.7) Laminar Nusselt number is expressed by[30]

$$\text{Nu}_l = \frac{1.8}{\ln\,(1 + 1.8/\text{Nu}^*)} \qquad 0° \le \theta \le 90° \qquad (3.46)$$

$$\text{where} \quad \text{Nu}^* = \left\{0.772 + \frac{0.228}{1 + 0.676 p^{1.23}}\right\} \times \left(\cos\theta + \frac{D}{L}\sin\theta\right)^{1/4} C_l\, \text{Ra}^{1/4} \qquad (3.47)$$

and parameter p is defined as

$$p = \left(\frac{2L}{D}\right)\cot\theta \qquad (3.48)$$

Nu and Ra are based on L_c = cylinder diameter = D and C_l is given by Eq. (3.24).

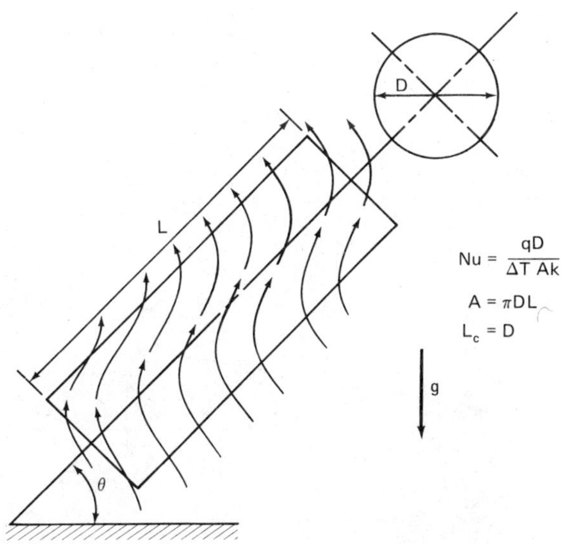

Fig. 3.7 Natural convection on an inclined cylinder.

Horizontal Cylinder with Spherical End Caps (Fig. 3.8) Add the total heat transfer from the horizontal cylinder obtained from Eq. (3.43) and the total heat transfer from the sphere computed using Eqs. (3.44) and (3.45) with Eq. (3.21).[2]

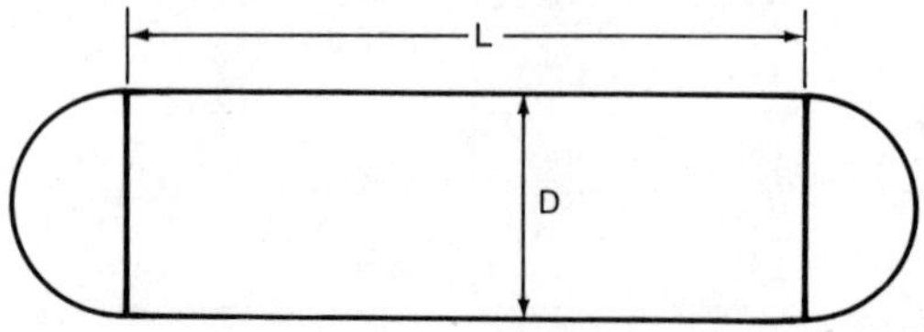

Fig. 3.8 A horizontal cylinder with spherical end caps.

Vertical Cones (Fig. 3.9) For laminar flow[29]

$$\mathrm{Nu}_l = 0.68\ \mathrm{Ra}^{1/4} + \frac{0.9}{\tan\theta} \tag{3.49}$$

Equation (3.49) is tested for $1.5 < \theta < 6$ and $10^7 < \mathrm{Ra} < 4 \times 10^8$. For fully turbulent flow and for $\theta < 20°$, Nu_l may be computed from[2]

$$\mathrm{Nu}_l = C_v\ \mathrm{Ra}^{1/3} \tag{3.50}$$

where C_v is given in Table 3.1.

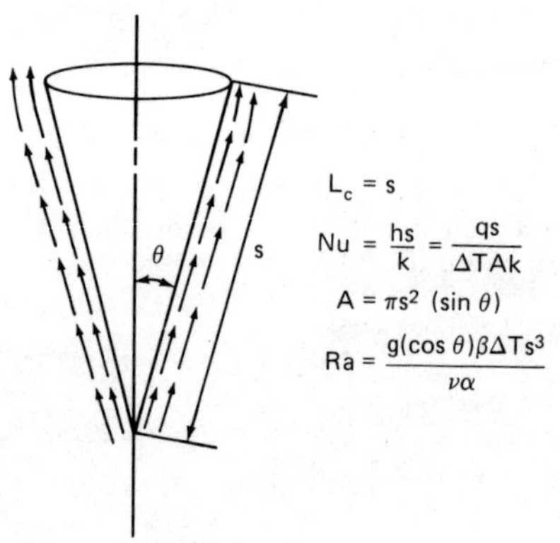

Fig. 3.9 Natural convection on a vertical cone.

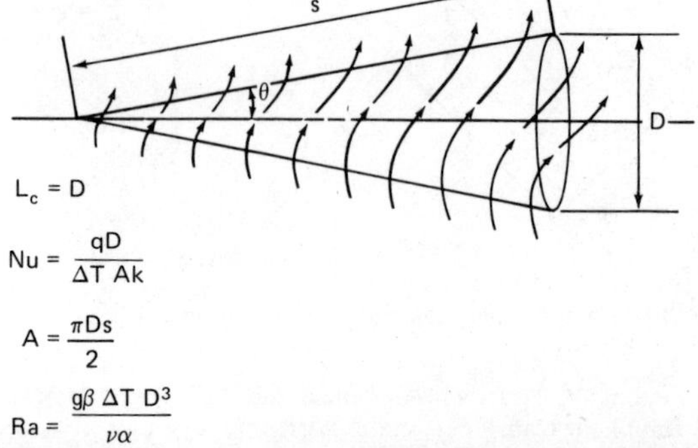

Fig. 3.10 Natural convection on a horizontal cone.

Horizontal Cones (Fig. 3.10) For the range of $7000 < \mathrm{Ra} < 10^6$ and $1.5 < \theta < 6$ the following equation is given[31]

$$\mathrm{Nu} = 0.7 + 0.35\ \mathrm{Ra}^{1/8} + 0.51\ \mathrm{Ra}^{1/4} \tag{3.51}$$

3.4 NATURAL CONVECTION THROUGH OPEN CAVITIES

When the walls of an open cavity are maintained at a different temperature than that of the surrounding fluid, the heat transfer generates buoyancy and drives the fluid through the cavity. The most important areas where open cavity problems arise are natural convection about arrays of fins and heat transfer between the printed circuit boards in electronic equipments.

In very long but narrow open channels (for example, $L/b \gg 1$ shown in Fig. 3.11) the flow and heat transfer become fully developed, i.e., the velocity and temperature profile cease to change. In short channels, however, the developing boundary layers on the walls remain separated and the heat transfers are observed to behave in similar manner to those reviewed in the external flow.

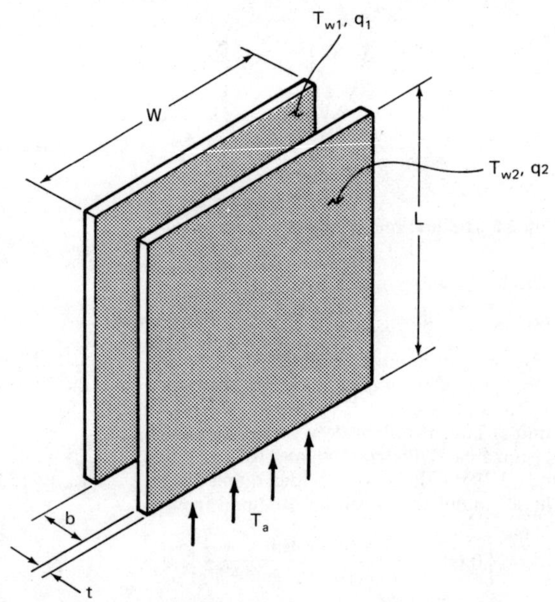

Fig. 3.11 Natural convection between vertical parallel plates.

Parallel Vertical Plates (Fig. 3.11)

Isothermal Plates. It is assumed that $T_{w1} \geq T_{w2} > T_a$. The Nusselt and Rayleigh numbers are defined as

$$\mathrm{Nu} = \frac{qb}{2LW(\overline{T}_w - T_a)k} \tag{3.52}$$

$$\mathrm{Ra} = \frac{g\beta(\overline{T}_w - T_a)b^3}{\nu\alpha}\left(\frac{b}{L}\right) \tag{3.53}$$

where q is the total heat transferred from both plates and $\overline{T}_w$ is defined as

$$\overline{T}_w = \frac{T_{w1} + T_{w2}}{2} \tag{3.54}$$

All of the fluid properties are evaluated at $(\overline{T}_w + T_a)/2$.

Based on the data reported in Refs. 32, 33, and 34 and by using the correlation technique of Ref. 18, Ref. 2 proposes the following equation for $\text{Ra} \le 10^5$

$$\text{Nu} = [(\text{Nu}_{fd})^m + (C_2\,\text{Ra}^{1/4})^m]^{1/m} \qquad m = -1.9 \tag{3.55}$$

where $C_2 = 0.62$ for air;[34] for fluids with $\text{Pr} > 1.0$, C_2 may be assumed, approximately, to be equal to C_l given in Table 3.1. Nu_{fd} is the fully developed Nusselt number defined by[32]

$$\text{Nu}_{fd} = \frac{(4T^{*2} + 7T^* + 4)\,\text{Ra}}{90(1 + T^*)^2} \tag{3.56}$$

where T^* is

$$T^* = \frac{T_{w2} - T_a}{T_{w1} - T_a} \tag{3.57}$$

and where $T_{w2} \le T_{w1}$ then $0 \le T^* \le 1$.

Isoflux Parallel Plates. Nu and Ra are defined as

$$\text{Nu} = \frac{\overline{q''}b}{k(\overline{T}_{w,L/2} - T_a)} \tag{3.58}$$

where $\overline{q''} = (q_1'' + q_2'')/2$, $q_1'' \ge q_2'' \ge 0$, and $\overline{T}_{w,L/2}$ is the average wall temperature at the channel mid-height, and

$$\text{Ra}^* = \frac{g\beta\,\overline{q''}\,b^4}{\nu\alpha k}\left(\frac{b}{L}\right) \tag{3.59}$$

All of the fluid properties are evaluated at $T_f = (\overline{T}_{w,L/2} + T_a)/2$.

Based on the data presented in Refs. 32–35, Ref. 2 suggests the following equation for $\text{Ra}^* \le 10^4$

$$\text{Nu} = \{(\text{Nu}_{fd})^m + [C_3(\text{Ra}^*)^{1/5}]^m\}^{1/m} \qquad m = -3.5 \tag{3.60}$$

where $C_3 = 0.67$ for air.[35] For fluids with $\text{Pr} > 1.0$, C_3 may be assumed to be equal to F_l given in Table 3.1. Nu_{fd} is defined as[32]

$$\text{Nu}_{fd} = 0.29(\text{Ra}^*)^{1/2} \tag{3.61}$$

Reference 36 reports the optimum spacing b_{opt} for negligibly thick plates, at which optimum Nusselt number Nu_{opt} occurs for four special cases that arise very often in practice. They are:

1. Symmetric, isothermal plates, $T_{w1} = T_{w2} = \overline{T}_w = \text{T}_w$

$$\text{Nu} = \frac{qb}{2LW\,(T_w - T_a)k}$$

$$= \left[\frac{576}{(\text{Ra})^2} + \frac{2.873}{\sqrt{\text{Ra}}}\right]^{-0.5} \tag{3.62a}$$

$$q = \left(\frac{\text{Nu}\,k}{b}\right) 2LW\,(T_w - T_a) \tag{3.62b}$$

$$b_{\text{opt}} = 2.714\,P_T^{-1/4} \tag{3.62c}$$

$$\text{Nu}_{\text{opt}} = 1.31 \tag{3.62d}$$

2. Asymmetric, isothermal plates, $T_{w1} = \overline{T}_w = T_w$, with the other wall adiabatic

$$\text{Nu} = \frac{qb}{LW\,(T_{w1} - T_a)k}$$

$$= \left[\frac{144}{(\text{Ra})^2} + \frac{2.873}{\sqrt{\text{Ra}}}\right]^{-1/2} \tag{3.63a}$$

$$q = \left(\frac{\text{Nu}\,k}{b}\right) LW\,(T_{w1} - T_a) \tag{3.63b}$$

$$b_{\text{opt}} = 2.154 P_T^{-1/4} \tag{3.63c}$$

$$\text{Nu}_{\text{opt}} = 1.04 \tag{3.63d}$$

where Ra is defined in Eq. (3.53) and P_T is computed from

$$P_T = \frac{g\beta(T_w - T_a)}{\nu\alpha L} \tag{3.64}$$

3. Symmetric, Isoflux Plates, $q_1'' = q_2'' = \overline{q''} = q''$

$$\text{Nu} = \frac{q''b}{k(\overline{T}_{w,L/2} - T_a)} = \left[\frac{12}{\text{Ra}^*} + \frac{1.88}{(\text{Ra}^*)^{0.4}}\right]^{-1/2} \tag{3.65a}$$

$$q = q''(2LW) = \left(\frac{\text{Nu}\,k}{b}\right) 2LW\,(\overline{T}_{w,L/2} - T_a) \tag{3.65b}$$

$$b_{\text{opt}} = 1.472 P_q^{-1/5} \tag{3.65c}$$

$$\text{Nu}_{\text{opt}} = 0.62 \tag{3.65d}$$

4. Asymmetric, Isoflux Plates, $q_1'' = q''$, $q_2'' = 0$ (adiabatic)

$$\text{Nu} = \frac{q_1''b}{k(\overline{T}_{w1,L/2} - T_a)} = \left[\frac{6}{\text{Ra}^*} + \frac{1.88}{(\text{Ra}^*)^{0.4}}\right]^{-1/2} \tag{3.66a}$$

$$q = q_1''(LW) = \left(\frac{\text{Nu}\,k}{b}\right) LW\,(\overline{T}_{w1,l2} - T_a) \tag{3.66b}$$

$$b_{\text{opt}} = 1.169\,P_q^{-1/5} \tag{3.66c}$$

$$\text{Nu}_{\text{opt}} = 0.49 \tag{3.66d}$$

where Ra^* is defined in Eq. (3.59) and P_q is obtained from

$$P_q = \frac{g\beta q''}{\nu\alpha k L} \tag{3.67}$$

Circular-Cylindrical Channels

Uniform Wall Temperature. The reported heat transfer results of Refs. 37 and 38 for $\text{Pr} \approx 0.7$ (air) are correlated by[2]

$$\text{Nu} = \left[\left(\frac{\text{Ra}}{16}\right)^m + (0.6\,\text{Ra}^{1/4})^m\right]^{1/m} \qquad m = -1.03 \tag{3.68}$$

where Nu and Ra are defined as

$$\mathrm{Nu} = \frac{hr}{k} = \frac{qr}{\pi DL(T_w - T_a)k} \tag{3.69}$$

and

$$\mathrm{Ra} = \frac{g\beta(T_w - T_a)r^3}{\nu\alpha}\left(\frac{r}{L}\right) \tag{3.70}$$

and where r, D, and L are the channel radius, diameter, and height, respectively. The fluid properties are evaluated at $T_f = (T_w + T_a)/2$.

If an unheated entry length is added to the channel, the heat transfer will be reduced.[38]

Uniform Wall Heat Flux. For this case the laminar flow analysis of Ref. 39 can be represented by[2]

$$\mathrm{Nu} = \left\{\left(\frac{\mathrm{Ra}^*}{8}\right)^{m/2} + [0.67\,(\mathrm{Ra}^*)^{1/5}]^m\right\}^{1/m}$$

$$m = -1.7 \tag{3.71}$$

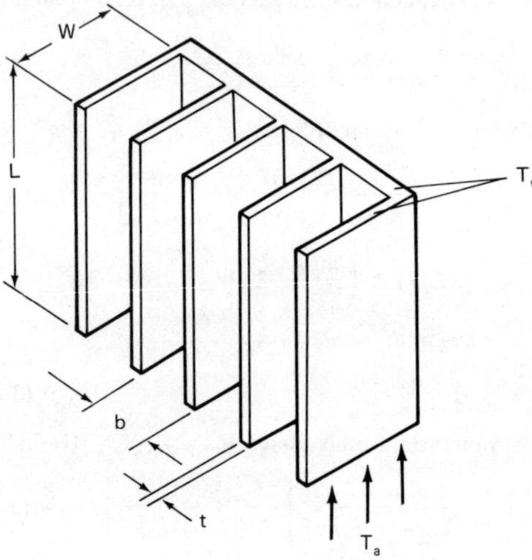

Fig. 3.12 Vertical U-shaped channels.

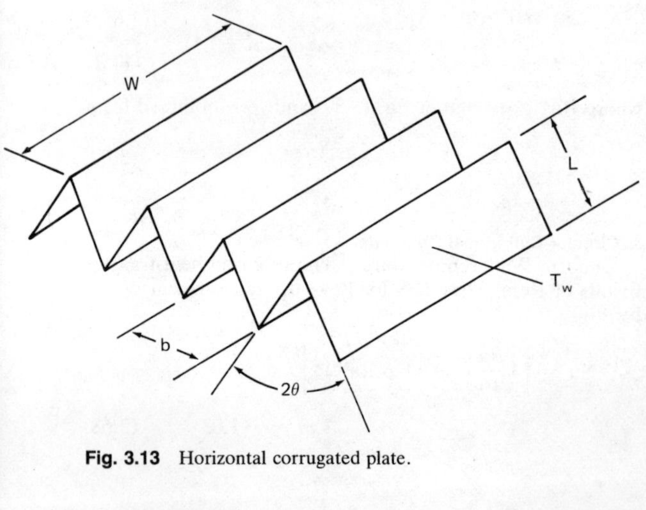

Fig. 3.13 Horizontal corrugated plate.

where Nu and Ra* are obtained from

$$\mathrm{Nu} = \frac{q''r}{(T_w - T_a)k} \tag{3.72}$$

and

$$\mathrm{Ra}^* = \frac{g\beta q'' r^4}{\nu\alpha k}\left(\frac{r}{L}\right) \tag{3.73}$$

Equation (3.71) is for Pr ≈ 0.7 (air) and $0 < \mathrm{Ra}^* < 10^5$.

Vertical, Isothermal U-shaped Channels (Fig. 3.12) For the range $0.3 < W/b < 4.0$, $10 < L/b < 42$, $0.5 < \mathrm{Ra} < 100$, and Pr ≈ 0.7 (air), Ref. 40 proposed the following equation

$$\mathrm{Nu} = \frac{\mathrm{Ra}}{\Psi}\left\{1 - \exp\left[-\Psi\left(\frac{0.5}{\mathrm{Ra}}\right)^{3/4}\right]\right\} \tag{3.74}$$

where

$$\Psi = \frac{24(1 - 0.483e^{-0.17/\alpha^*})}{\{(1 + \alpha^*/2)[1 + (1 - e^{-0.83\alpha^*})(9.14\sqrt{\alpha^*}e^S - 0.61)]\}^3} \tag{3.75}$$

and where the parameters are defined as

$$\mathrm{Nu} = \frac{hL_c}{k} = \frac{q''L_c}{k(T_w - T_a)}$$

$$L_c = \frac{2Wb}{2W + b}$$

$$\mathrm{Ra} = \frac{g\beta(T_w - T_a)L_c^3}{\nu\alpha}\left(\frac{L_c}{L}\right)$$

$$\alpha^* = \frac{b}{W}$$

$$S = \begin{cases} -11.8b & b \text{ in inches} \\ -4.65b & b \text{ in cm} \end{cases}$$

The properties are evaluated at T_w, except β which is computed at T_a.

Horizontal, Isothermal Corrugated Plates (Fig. 3.13) For large W and Pr ≈ 0.7 (air), the laminar and turbulent heat transfer results are correlated by[41]

$$\mathrm{Nu} = \left(\frac{0.46}{\sin\theta} - 0.32\right)\mathrm{Ra}^m$$

$$\text{for } 2 \times 10^4 < \mathrm{Ra} < \mathrm{Ra}_{cr} \tag{3.76}$$

where $\mathrm{Ra}_{cr} = (15.8 - 14\sin\theta) \times 10^5$ and $m = 0.148\sin\theta + 0.187$.

For a turbulent region, $\mathrm{Ra}_{cr} < \mathrm{Ra} < 2 \times 10^7$,

$$\mathrm{Nu} = \left(0.09 + \frac{0.054}{\sin\theta}\right)\mathrm{Ra}^{1/3} \tag{3.77}$$

where Nu and Ra are defined by

$$\mathrm{Nu} = \frac{hL}{k} = \frac{q''L}{(T_w - T_a)k} \quad \text{and} \quad \mathrm{Ra} = \frac{g\beta(T_w - T_a)L^3}{\nu\alpha}$$

Square Isothermal Fins (Fig. 3.14) For the range of $8 < \mathrm{Ra} < 10^4$ and for Pr ≈ 0.7 (air) the experimental data re-

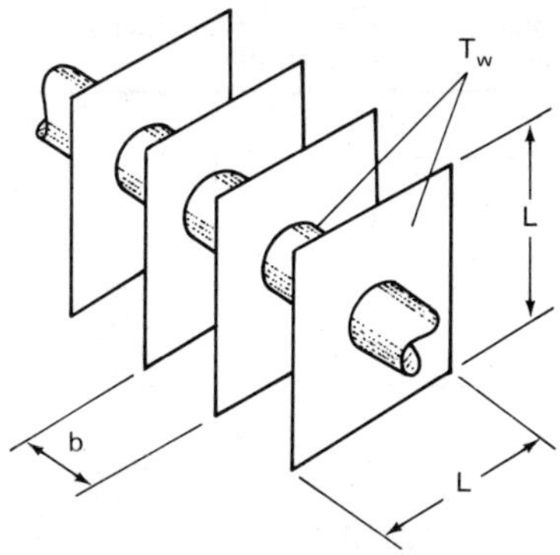

Fig. 3.14 Square fins.

ported in Refs. 34 and 42 are close and can be represented by[2]

$$\mathrm{Nu} = \left[\left(\frac{\mathrm{Ra}^{0.89}}{18}\right)^m + (0.62\ \mathrm{Ra}^{1/4})^m\right]^{1/m} \qquad m = 2.7 \tag{3.78}$$

where Nu and Ra are evaluated from

$$\mathrm{Nu} = \frac{hb}{k} = \frac{q''b}{(T_w - T_a)k}$$

and

$$Ra = \frac{g\beta(T_w - T_a)b^3}{\nu\alpha}\left(\frac{b}{L}\right)$$

Additional information is presented in Ref. 43.

3.5 NATURAL CONVECTION IN ENCLOSURES

When a fluid is confined within an enclosure with its walls maintained at different temperatures, a recirculation motion will be generated due to the gravity force. This recirculation process can occur in countless different configurations, and it tends to enhance the heat transfer between the enclosure walls. For some common geometries empirical correlations are presented in the following sections.

Rectangular Enclosures (Fig. 3.15) This geometry, also called fluid layers, has been studied extensively and comprehensive literature reviews are presented in Refs. 2 and 44. Such layers occur, for example, between window glazings, in wall cavities, and in solar collector cavities.

The heat-transfer results are presented in terms of Nu and Ra, where

$$\mathrm{Nu} = \frac{hL}{k} = \frac{qL}{HW\,(T_h - T_c)k} \tag{3.79}$$

$$\mathrm{Ra} = \frac{g\beta(T_h - T_c)L^3}{\nu\alpha} \tag{3.80}$$

and where q is the total heat transfer and T_h and T_c are the heated and cooled surface temperatures, respectively. All properties are evaluated at $(T_h + T_c)/2$. It is assumed that $W \gg L$ for all of the following cases.

Vertical Rectangular Enclosures, $\theta = 90°$. For air as the fluid ($\mathrm{Pr} \approx 0.7$), Ref. 45 proposed the maximum value of three Nusselt numbers as follows:

$$\mathrm{Nu} = [\mathrm{Nu}_1, \mathrm{Nu}_2, \mathrm{Nu}_3]_{\max} \tag{3.81}$$

where

$$\mathrm{Nu}_1 = 0.0605\ \mathrm{Ra}^{1/3} \tag{3.82}$$

$$\mathrm{Nu}_2 = 0.242\left(\frac{\mathrm{Ra}}{H/L}\right)^{0.272} \tag{3.83}$$

$$\mathrm{Nu}_3 = \left\{1 + \left[\frac{0.104\ \mathrm{Ra}^{0.293}}{1 + (6310/\mathrm{Ra})^{1.36}}\right]^3\right\}^{1/3} \tag{3.84}$$

Equation (3.81) is valid for $5 \le H/L \le 110$ and $\mathrm{Ra} \le 10^8$.

For fluids with $\mathrm{Pr} > 0.7$ and small aspect ratio H/L, Ref. 44 recommends the Berkovsky and Polevikov correlations

$$\mathrm{Nu} = 0.22\left(\frac{\mathrm{Pr\ Ra}}{0.2 + \mathrm{Pr}}\right)^{0.28}\left(\frac{H}{L}\right)^{-1/4} \tag{3.85}$$

valid for $2 < H/L < 10$, $\mathrm{Pr} < 10^5$, and $\mathrm{Ra} < 10^{10}$, and for

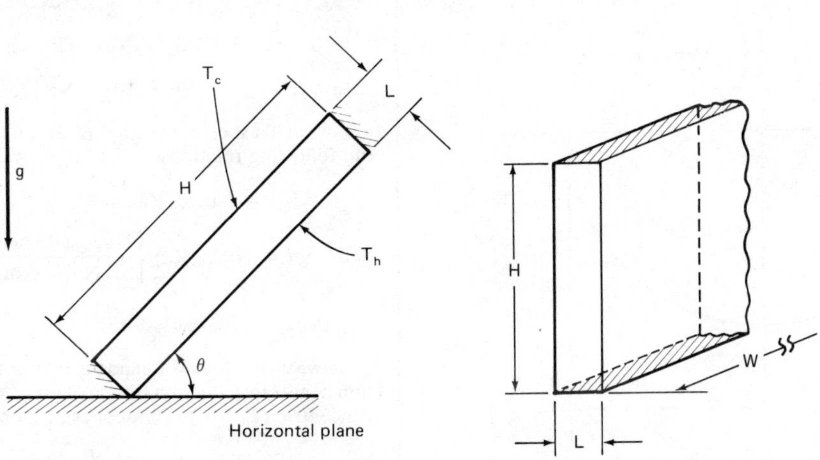

Fig. 3.15 Natural convection in a rectangular enclosure.

even smaller H/L use

$$\mathrm{Nu} = 0.18\left(\frac{\mathrm{Pr\,Ra}}{0.2 + \mathrm{Pr}}\right)^{0.29} \tag{3.86}$$

valid for $1 \le H/L \le 2$, $\mathrm{Pr} < 10^5$, and $1000 < (\mathrm{Pr\,Ra})/(0.2 + \mathrm{Pr})$.

For large values of the aspect ratio, $10 < H/L < 40$, and for $\mathrm{Pr} > 1$ Ref. 46 proposed

$$\mathrm{Nu} = 0.42\left(\frac{H}{L}\right)^{-0.3} Pr^{0.012}\,\mathrm{Ra}^{0.25} \qquad 10^4 < \mathrm{Ra} < 10^7 \tag{3.87}$$

and

$$\mathrm{Nu} = 0.046\,\mathrm{Ra}^{1/3} \qquad 10^6 < \mathrm{Ra} < 10^9 \tag{3.88}$$

Horizontal Rectangular Enclosures, $\theta = 0°$, $H \gg L$. Hollands et al.[47] presented a very accurate correlation for air ($\mathrm{Pr} \approx 0.7$) and water ($\mathrm{Pr} \approx 6$); later in Ref. 2 the correlation was modified to cover a very wide range of Prandtl numbers. That is

$$\mathrm{Nu} = 1 + 1.44\left[1 - \frac{1708}{\mathrm{Ra}}\right]^{\bullet} + \left[\left(\frac{\mathrm{Ra}}{5830}\right)^{1/3} - 1\right]^{\bullet} + \mathrm{Nu}^{+} \tag{3.89}$$

where a bracket: $[\ ]^{\bullet}$ indicates that if the argument inside the bracket is negative, the quantity is to be taken as zero, and Nu^+ is defined as

$$\mathrm{Nu}^+ = 0.0 \qquad \text{for } 0.5 < \mathrm{Pr} < 1.0 \tag{3.90}$$

$$\mathrm{Nu}^+ = 2.0Y^{1-\ln Y} \tag{3.91}$$

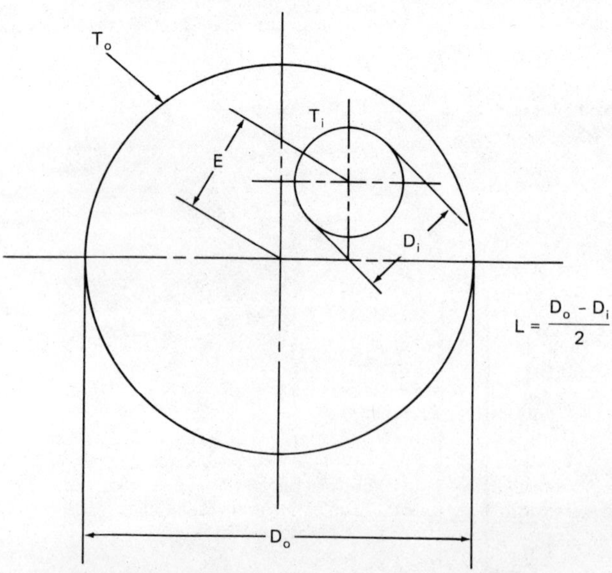

Fig. 3.16 Natural convection between eccentric and concentric horizontal cylinders and spheres.

Table 3.4 Critical Angle for Inclined Rectangular Enclosure

H/L	θ^*	H/L	θ^*
3	53	12	67
5	58	>12	70
6	60		

where Y is computed by

$$Y = \frac{\mathrm{Ra}^{1/3}}{Z} \tag{3.92}$$

and where Z is evaluated from[2]

$$Z = 75\exp(1.5)\,\mathrm{Pr}^{-1/2} \tag{3.93}$$

For water ($\mathrm{Pr} = 6$), $Z = 140$.

Inclined Rectangular Enclosures. For $H/L > 5$, $\mathrm{Pr} \approx 0.7$ (air), and tilt angles less than the critical value θ^* given in Table 3.4, the following correlation is recommended[48]

$$\mathrm{Nu} = 1 + 1.44\left[1 - \frac{1708}{\mathrm{Ra}\cos\theta}\right]^{\bullet}\left[1 - \frac{1708\,(\sin 1.8\theta)^{1.6}}{\mathrm{Ra}\cos\theta}\right] + \left[\left(\frac{\mathrm{Ra}\cos\theta}{5830}\right)^{1/3} - 1\right]^{\bullet} \qquad 0 < \theta \le \theta^* \tag{3.94}$$

where $[\Gamma]^{\bullet}$ implies that, if Γ is negative, set Γ equal to zero. For $\theta^* < \theta < 90°$, linear interpolation between θ^* and 90° may be used:

$$\mathrm{Nu} = \left(\frac{90 - \theta}{90 - \theta^*}\right)\mathrm{Nu}_{\theta^*} + \left(\frac{\theta - \theta^*}{90 - \theta^*}\right)\mathrm{Nu}_{90} \tag{3.95}$$

For fluids with $\mathrm{Pr} > 4$ and for the conditions $H/L \ge 6$ and $0 < \theta < 90°$, the maximum value of the following equations is recommended[2]

$$\mathrm{Nu}_1 = \mathrm{Nu}_{\theta=0}(\mathrm{Ra}\cos\theta) \tag{3.96}$$

$$\mathrm{Nu}_2 = \mathrm{Nu}_{\theta=90}(\mathrm{Ra}\sin\theta) \tag{3.97}$$

then

$$\mathrm{Nu} = [\mathrm{Nu}_1, \mathrm{Nu}_2]_{\max} \tag{3.98}$$

For small aspect ratios, $1 < H/L < 6$, Ref. 44 recommends the following relations.

$$\mathrm{Nu} = \mathrm{Nu}_{\theta=90}(\mathrm{Ra}\sin\theta), \qquad \theta^* \le \theta \le 90° \tag{3.99}$$

and

$$\mathrm{Nu} = \mathrm{Nu}_{\theta=0}(\mathrm{Ra})\left[\frac{\mathrm{Nu}_{\theta=90}(\mathrm{Ra}\sin\theta^*)}{\mathrm{Nu}_{\theta=00}(\mathrm{Ra})}\right]^{\theta/\theta^*} \qquad 0° < \theta < \theta^* \tag{3.100}$$

Between Horizontal Cylinders (Fig. 3.16) Natural-convection heat transfer across the gap between two long, horizontal concentric ($E = 0$) cylinders can be obtained from[19]

$$q' = \frac{2\pi k\,\mathrm{Nu}}{\ln(D_o/D_i)}(T_i - T_o) \tag{3.101}$$

where q' is the heat transfer per unit length of the cylinders and Nu can be evaluated by

$$\text{Nu} = 0.386 \left(\frac{\text{Pr}}{0.861 + \text{Pr}}\right)^{1/4} (\text{Ra}_c)^{1/4} \qquad (3.102)$$

where the modified Rayleigh number Ra_c is

$$\text{Ra}_c = \frac{[\ln (D_o/D_i)]^4}{L^3[(D_i)^{-3/5} + (D_o)^{-3/5}]^5} (\text{Ra}) \qquad (3.103)$$

and where $L = (D_o - D_i)/2$ and the Rayleigh number Ra is defined by

$$\text{Ra} = \frac{g\beta(T_i - T_o)L^3}{\nu\alpha} \qquad (3.104)$$

Equation (3.102) is valid for $\text{Pr} \geq 0.7$ and $100 < \text{Ra}_c < 10^8$.

For an *eccentric geometry* with $E/L \leq \frac{2}{3}$ and for $200 < \text{Ra}_c < 10^8$, Eqs. (3.101) and (3.102) can be used to obtain the overall heat transfer, q' within ± 10 percent.[49]

Results for the large Rayleigh numbers, $\text{Ra}_c > 10^8$, are presented in Refs. 49 and 50.

Between Vertical Cylinders of Different Height (Fig. 3.17) The average heat-transfer coefficient and Nusselt number at the

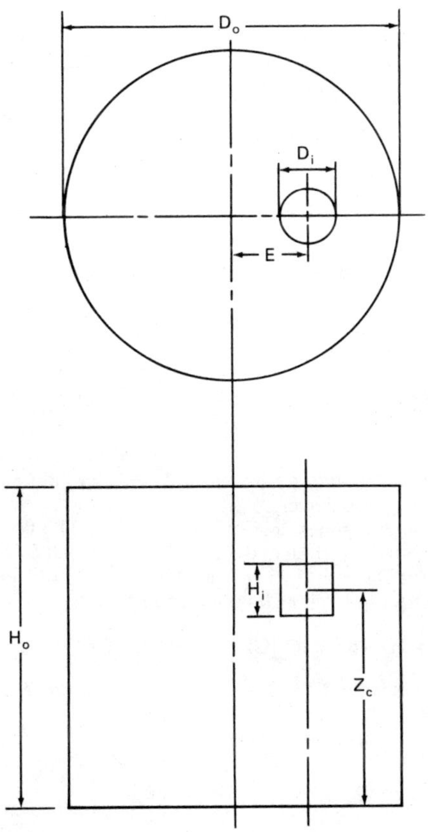

Fig. 3.17 Natural convection between concentric and eccentric vertical cylinders of different height and diameter.

inner cylinder are evaluated by

$$h_i = \frac{q}{A_i(T_i - T_o)} \qquad \text{Nu}_i = \frac{h_i D_i}{k} \qquad (3.105)$$

and at the outer cylinder

$$h_o = \frac{q}{A_o(T_i - T_o)} \qquad \text{Nu}_o = \frac{h_o D_o}{k} \qquad (3.106)$$

where q is the overall convective heat transfer and A_i and A_o are the surface areas of the inner and outer cylinders, respectively.

For the cases of $D_i/H_i = D_o/H_o = 1.0$, the Nusselt number can be obtained from[51,52]

$$\text{Nu}_i = 0.754 \left(\frac{D_o}{D_i}\right)^{0.052} \text{Ra}^{0.204} \qquad (3.107)$$

where

$$\text{Ra} = \frac{g\beta(T_i - T_o)D_i^3}{\nu\alpha} \qquad (3.108)$$

and since $A_i = 3\pi D_i^2/2$ and $A_o = 3\pi D_o^2/2$, it follows that

$$\text{Nu}_o = \left(\frac{D_i}{D_o}\right) \text{Nu}_i \qquad (3.109)$$

Equation (3.107) is valid for the diameter ratio of $D_i/D_o \leq 0.3$, the inner cylinder elevation of $0.15 \leq Z_c/H_o \leq 0.85$, the inner cylinder eccentricity of $0 \leq E/D_o \leq 0.4$, fluids with $\text{Pr} \approx 0.7$, and for $10^3 \leq \text{Ra} \leq 10^6$.

For fluids with $\text{Pr} \geq 1$, Eq. (3.107) may be modified to

$$\text{Nu}_i = 0.914 \left(\frac{\text{Pr}}{0.861 + \text{Pr}}\right)^{0.25} \left(\frac{D_o}{D_i}\right)^{0.052} \text{Ra}^{0.204} \qquad (3.110)$$

Between Spheres (Fig. 3.16) The total heat transfer between concentric ($E = 0$), spheres can be evaluated by[19]

$$q = \text{Nu}\, k\pi \left(\frac{D_i D_o}{L}\right) (T_i - T_o) \qquad (3.111)$$

where

$$\text{Nu} = 0.74 \left(\frac{\text{Pr}}{0.861 + \text{Pr}}\right)^{1/4} (\text{Ra}_s)^{1/4} \qquad (3.112)$$

and the parameter Ra_s is defined by

$$\text{Ra}_s = \frac{L}{(D_i D_o)^4} \left(\frac{1}{D_i^{-7/5} + D_o^{-7/5}}\right)^5 (\text{Ra}) \qquad (3.113)$$

where Ra is evaluated by Eq. (3.104). Equation (3.112) may be used for $\text{Pr} \geq 0.7$ and $100 \leq \text{Ra}_s \leq 10^8$.

For *vertically eccentric spheres,* Eqs. (3.111) to (3.113) are valid within ± 10 percent when $E/L \leq \frac{2}{3}$, where $L = (D_o - D_i)/2$.

Inside Spherical Cavities When the gas inside of a spherical container is at a different temperature than that of the container surface, natural convection will take place. The overall Nusselt number may be obtained by[53]

$$\text{Nu} = \frac{hD}{k} = 0.59\, \text{Ra}^{1/4} \qquad 10^4 \leq \text{Ra} \leq 10^9 \qquad (3.114)$$

$$\text{and Nu} = \frac{hD}{k} = 0.13\ \text{Ra}^{1/3} \qquad 10^9 < \text{Ra} < 10^{12} \tag{3.115}$$

where D is the sphere diameter and Ra is defined by

$$\text{Ra} = \frac{g\beta(T_g - T_w)D^3}{\alpha} \tag{3.116}$$

and where T_g and T_w are the gas and the surface temperatures, respectively.

On a Sphere, Cylinder, or a Cube Placed in a Spherical or a Cubical Enclosure Natural-convection heat transfer between concentrically located isothermal spherical, cylindrical, or cubical inner body and its isothermal spherical or cubical enclosure can be evaluated within ±14 percent using the following correlation equation[54]

$$\text{Nu} = 0.396\ \text{Pr}^{0.0162}\left(\frac{L}{R_i}\right)^{0.496}\text{Ra}^{0.234} \tag{3.117}$$

where L is the hypothetical gap width ($L = R_o - R_i$), and R_o and R_i are the outer and the inner hypothetical radii equal to the radii of spheres having volumes equal to the volumes of the outer body and the inner body, respectively. The Rayleigh number Ra is defined by

$$\text{Ra} = \frac{g\beta(T_i - T_o)L^3}{\nu\alpha} \tag{3.118}$$

The overall heat transfer is then obtained by

$$q = \left(\frac{\text{Nu}\ k}{L}\right)A_i(T_i - T_o) \tag{3.119}$$

where A_i is the inner body surface area.

Equation (3.117) is valid for $0.7 \le \text{Pr} \le 10^4$ and for $2000 \le \text{Ra} \le 10^9$.

3.6 COMBINED NATURAL AND FORCED CONVECTION (MIXED CONVECTION)

If there is simultaneously forced flow past the surface in question, the possible forced-convection heat transfer may have to be accounted for. As noted in Sec. 3.2 under "Dimensionless Groups," the question of mixed convection may arise when the ratio of the Grashof number to Reynolds number squared, Gr/Re^2, is near unity; where the Grashof number is defined by Eq. (3.10) and the Reynolds number is defined by

$$\text{Re} = \frac{UL_c}{\nu} \tag{3.120}$$

and where U is the characteristic flow velocity generated by an external force. The orientation of the velocity U with respect to the direction of the buoyancy force can strongly affect the mixed convection results.

On Vertical Plates When the natural- and forced-convection flows are in the same direction the process is called assisting mixed convection. On the other hand, if the flows are in opposite direction, opposed mixed convection results. For laminar flow and for $0.1 < \text{Pr} < 10$, the presented results in Ref. 55 suggest that the local mixed-convection Nusselt number Nu may be predicted to within 20 percent by using the forced-convection correlation for $\text{Gr}/\text{Re}^2 \le 0.6$ and using the natural-convection correlation for $\text{Gr}/\text{Re}^2 > 0.6$, or in general[2]

$$\text{Nu} = [\text{Nu}_{pn}, \text{Nu}_{pf}]_{\max} \tag{3.121}$$

i.e., the maximum of Nu_{pn} (pure natural convection) and Nu_{pf} (pure forced convection) is used.

For assisting flows the following correlation equation is recommended[56]

$$\text{Nu} = \frac{hL}{k} = [(\text{Nu}_{pn})^3 + (\text{Nu}_{pf})^3]^{1/3} \tag{3.122}$$

where the values of Nu_{pf} may be obtained from Chap. 4 and Nu_{pn} can be evaluated using Eq. (3.21).

On Horizontal Plates Based on the experimental data of Ref. 57, the following correlation is proposed for average heat transfer[2]

$$\text{Nu} = f^{1/2}0.664\ \text{Re}^{1/2}\ \text{Pr}^{1/3} + (1 - f)0.14\ \text{Ra}^{1/3} \tag{3.123}$$

$$\text{where} \qquad f = \left[\frac{29\ \text{Re}}{\text{Gr}^{2/3}}, 1\right]_{\min} \tag{3.124}$$

i.e., f is equal the minimum of the two quantities in the brackets.

On Horizontal Cylinders and Spheres For assisting mixed-convection flow, Eq. (3.122) may be used provided that Nu_{pn} and Nu_{pf} are correctly obtained.

Heat-transfer results for different forced flow orientations are presented in Refs. 58, 59, and 60.

Inside Horizontal Tubes Forced flow in horizontal tubes may experience buoyancy effects when $\text{Gr}/\text{Re}^2 \approx 1$, where Gr and Re are defined as

$$\text{Gr} = \frac{g\beta\ \Delta T\ D^3}{\nu^2} \qquad \text{Ra} = \text{Gr Pr} \tag{3.125}$$

$$\text{Re} = \overline{U}D/\nu \tag{3.126}$$

where D is the tube diameter, $\overline{U}$ is the mean fluid velocity in tube, and ΔT is evaluated by

$$\Delta T = T_w - T_m \tag{3.127}$$

and where T_w and T_m are the wall and the fluid mean temperatures, respectively.

The flow regimes are studied in Ref. 1. The flow may be assumed to be turbulent if $\text{Re} \ge 2000$ when $\text{Ra}(D/L) \le 2 \times 10^4$, or if $\text{Re} \ge 800$ when $\text{Ra}(D/L) \ge 2 \times 10^4$. For mixed convection in laminar flow Ref. 61 suggests

$$\text{Nu} = 1.75(\mu/\mu_w)^{0.14}[\text{Gz} + 0.012(\text{Gz}\ \text{Gr}^{1/3})^{4/3}]^{1/3} \tag{3.128}$$

where Gz is the Graetz number, $\text{Gz} = \text{Re Pr}(D/L)$, and L is the tube length.

For turbulent flow, Ref. 1 recommends

$$\text{Nu} = 4.69\ \text{Re}^{0.27}\ \text{Pr}^{0.21}\ \text{Gr}^{0.07}\left(\frac{D}{L}\right)^{0.36} \tag{3.129}$$

In Eqs. (3.128) and (3.129), all properties are evaluated at fluid mean temperature, except μ_w which is evaluated at T_w.

NOMENCLATURE

A = heat-transfer surface area
A_i = surface area of the inner body
A_o = surface area of the outer body
b = space between heated parallel plates
b_{opt} = optimum spacing between parallel plates
C = constant in Eq. (3.16)
C_2 = constant in Eq. (3.55)
C_3 = constant in Eq. (3.60)
Cf = correction factor, see Eq. (3.39)
C_h = parameter defined in Eq. (3.38) and given in Table 3.2
C_l = parameter defined in Eq. (3.24) and given in Table 3.1
C_p = specific heat of fluid
C_t = parameter in Eq. (3.45) given in Table 3.3
C_v = parameter defined in Eq. (3.25) and given in Table 3.1
D = diameter of a cylinder or a sphere
D_i = diameter of an inner body (cylinder or sphere)
D_o = diameter of an outer body (cylinder or sphere)
E = distance between centers of eccentric cylinders or spheres
F_1 = parameter defined in Eq. (3.28) and given in Table 3.1
f = quantity defined in Eq. (3.124)
Gr = Grashof number, Eqs. (3.8) and (3.10)
Gr* = modified Grashof number, Eq. (3.13*a*)
Gz = Graetz number, Re Pr(D/L), Eq. (3.128)
g = acceleration of gravity
H = height of a body
h = heat-transfer coefficient, Eq. (3.1)
h_i = heat-transfer coefficient on inner body
h_o = heat-transfer coefficient on outer body
h_x = local heat-transfer coefficient
k = thermal conductivity of fluid
L = body length, space between isothermal surfaces, and gap width
L^* = reference length, Eq. (3.31)
L_c = characteristic length
L_x = local plate length, see Fig. 3.4
m = exponent in Eq. (3.17)
Nu = Nusselt number, Eq. (3.7)
Nu^+ = quantity used in Eq. (3.89) and defined in Eqs. (3.90) to (3.93)
Nu^* = quantity used in Eq. (3.46) and defined in Eq. (3.47)
Nu_1 = Nusselt number defined in Eq. (3.82)
Nu_2 = Nusselt number defined in Eq. (3.83)
Nu_3 = Nusselt number defined in Eq. (3.84)
Nu_{fd} = Nusselt number for fully developed channel flow
Nu_l = Nusselt number for laminar flow
Nu_{opt} = optimum Nusselt number for channel flow
Nu_{pf} = Nusselt number for pure forced convection
Nu_{pn} = Nusselt number for pure natural convection
Nu_t = Nusselt number for turbulent flow
Nu_θ = Nusselt number for inclined rectangular enclosure
n = exponent in Eq. (3.16)
P = perimeter of a plate
Pr = Prandtl number of fluid
P_q = quantity defined in Eq. (3.67)
P_T = quantity defined in Eq. (3.64)
p = quantity defined in Eq. (3.48)
q = total heat transfer per unit time
q' = heat transfer per unit time and length of cylinder
q'' = heat transfer per unit time and surface (heat flux)
$\bar{q}''$ = average heat flux over surface
R = radius of cylinder or sphere
R_i = hypothetical radius of inner body equal to radius of sphere having volume equal to inner body, used in Eq. (3.117)
R_o = hypothetical radius of outer body equal to radius of sphere having volume equal to outer body
Ra = Rayleigh number, Gr Pr
Ra* = modified Rayleigh number, Eq. (3.13*b*)
Ra_c = Rayleigh number defined in Eq. (3.103)
Ra_{cr} = critical Rayleigh number, transition from laminar to turbulent
Ra_s = modified Rayleigh number defined in Eq. (3.113)
Re = Reynolds number important for mixed convection, Eq. (3.120)
r = radius of circular channel
s = parameter defined in Eq. (3.75)
T = local temperature
T^* = dimensionless temperature defined in Eq. (3.57)
T_a = ambient temperature
T_c = cold plate temperature in enclosure
T_f = mean film temperature
T_g = gas temperature in enclosure
T_h = hot plate temperature in enclosure
T_i = inner surface temperature
T_m = mean fluid temperature inside tubes
T_o = outer surface temperature
T_r = reference temperature, Eq. (3.9)
T_w = solid surface (wall) temperature
$\bar{T}_w$ = mean value of T_w
$\bar{T}_{w,L/2}$ = mean wall temperature halfway up the channel, Eq. (3.58)
ΔT = temperature difference, Eqs. (3.2) and (3.3)
$\overline{\Delta T}$ = average temperature difference over surface, Eq. (3.14)
U = fluid velocity in forced flow condition
$\bar{U}$ = mean fluid velocity in tubes
W = width of plate or enclosure
Y = parameter defined in Eq. (3.92)
Z = parameter defined in Eq. (3.93)

Greek Symbols

α = thermal diffusivity of fluid
α^* = aspect ratio defined in Eq. (3.75)
β = thermal expansion coefficient
θ = angle of inclination for plates and enclosure
θ^* = critical angle of inclination for enclosure, Table 3.4
μ = dynamic viscosity of fluid
μ_w = dynamic viscosity of fluid evaluated at wall temperature
ν = kinematic viscosity of fluid
ρ = density of fluid
ρ_r = reference fluid density, Eq. (3.8)
Φ = quantity defined in Eqs. (3.41) and (3.42)

REFERENCES

1. B. Metais and E. R. G. Eckert, "Forced, Mixed, and Free Convection Regimes," *J. Heat Transfer,* **86,** 295 (1964).
2. G. D. Raithby and K. G. T. Hollands, "Natural Convection," in W. M. Rohsenow, J. P. Hartnett, and E. N. Ganic (eds.), *Handbook of Heat Transfer Fundamentals,* 2d ed., McGraw-Hill, New York, 1985, Chap. 6.
3. E. R. G. Eckert and R. M. Drake, Jr., *Analysis of Heat and Mass Transfer,* McGraw-Hill, New York, 1972.
4. W. M. Kays and M. E. Crawford, *Convective Heat and Mass Transfer,* 2d ed., McGraw-Hill, New York, 1980.
5. A. Bejan, *Convection Heat Transfer,* Wiley, New York, 1984.
6. B. Gebhart, *Heat Transfer,* 2d ed., McGraw-Hill, New York, 1971.
7. F. P. Incropera and D. P. Dewitt, *Fundamentals of Heat Transfer,* Wiley, New York, 1981.
8. F. Kreith and M. S. Bohn, *Principles of Heat Transfer,* 4th ed., Harper & Row, New York, 1986.
9. F. M. White, *Heat Transfer,* Addison-Wesley, Reading, 1984.
10. J. P. Holman, *Heat Transfer,* 6th ed., McGraw-Hill, New York, 1986.
11. W. M. Rohsenow and H. Y. Choi, *Heat, Mass, and Momentum Transfer,* Prentice-Hall, Englewood Cliffs, N.J., 1961.
12. W. S. Janna, *Engineering Heat Transfer,* PWS Publishers, Boston, 1986.
13. S. Kakas, W. Aung, and R. Viskanta, *Natural Convection: Fundamentals and Applications,* Hemisphere, New York, 1985.
14. Y. Jaluria, *Natural Convection Heat and Mass Transfer,* Pergamon Press, New York, 1980.
15. R. B. Bird, W. E. Stewart, and E. N. Lightfoot, *Transport Phenomena,* Wiley, New York, 1960.
16. E. M. Sparrow and J. L. Gregg, "Laminar Free Convection from a Vertical Plate with Uniform Surface Heat Flux," *ASME Trans.* **78,** 435 (1956).
17. B. Gebhart, "External Natural Convection Flow," *Appl. Mech. Rev.* **22,** 691 (1969).
18. S. W. Churchill and R. Usagi, "A General Expression for the Correlation of Rates of Transfer and Other Phenomena," *AIChE J.,* **18,** 1121 (1972).
19. G. D. Raithby and K. G. T. Hollands, "A General Method of Obtaining Approximate Solutions to Laminar and Turbulent Free Convection Problems," in T. F. Irvine and J. P. Hartnett (eds.), *Advances in Heat Transfer,* vol. 11, Academic Press, New York, 1975, p. 266.
20. G. D. Raithby and K. G. T. Hollands, "Laminar and Turbulent Free Convection from Elliptic Cylinders, with a Vertical Plate and Horizontal Circular Cylinder as Special Cases," *J. Heat Transfer* **98,** 72 (1976).
21. T. Fujii and M. Fujii, "The Dependence of Local Nusselt Number on Prandtl Number in the Case of Free Convection along a Vertical Surface with Uniform Heat Flux," *Int. J. Heat Mass Transfer,* **19,** 121 (1976).
22. J. R. Lloyd, E. M. Sparrow, and E. R. G. Eckert, "Laminar, Transition, and Turbulent Natural Convection Adjacent to Inclined and Vertical Surfaces," *Int. J. Heat Mass Transfer,* **15,** 457 (1972).
23. T. Fujii and H. Imura, "Natural Convection Heat Transfer from a Plate with Arbitrary Inclination," *Int. J. Heat Mass Transfer,* **15,** 755 (1972).
24. G. C. Vliet and D. C. Ross, "Turbulent Natural Convection on Upward and Downward Facing Inclined Constant Heat Flux Surfaces," *J. Heat Transfer,* **97,** 549 (1975).
25. R. J. Goldstein, E. M. Sparrow, and D. C. Jones, "Natural Convection Mass Transfer Adjacent to Horizontal Plates," *Int. J. Heat Mass Transfer,* **16,** 1025 (1973).
26. T. Fujii, H. Honda, and I. Morioka, "A Theoretical Study of Natural Convection Heat Transfer from Downward-facing Horizontal Surfaces with Uniform Heat Flux," *Int. J. Heat Mass Transfer,* **16,** 611 (1973).
27. S. W. Churchill and H. H. S. Chu, "Correlating Equations for Laminar and Turbulent Free Convection from a Horizontal Cylinder," *Int. J. Heat Mass Transfer,* **18,** 1049 (1975).
28. G. D. Raithby, A. Pollard, K. G. T. Hollands, and M. M. Yovanovich, "Free Convection Heat Transfer from Spheroids," *J. Heat Transfer,* **98,** 452 (1976).
29. P. H. Oosthuizen and E. Donaldson, "Free Convective Heat Transfer from Vertical Cones," *J. Heat Transfer,* **94,** 330 (1972).
30. G. D. Raithby and K. G. T. Hollands, "Analysis of Heat Transfer by Natural Convection (or Film Condensation) for Three-Dimensional Flows," *Proc. 6th Int. Heat Transfer Conf.,* vol. 2, 1978, p. 187.
31. P. H. Oosthuizen, "Free Convection Heat Transfer from Horizontal Cones," *J. Heat Transfer,* **95,** 409 (1973).
32. W. Aung, "Fully Developed Laminar Free Convection between Vertical Plates Heated Asymmetrically," *Int. J. Heat Mass Transfer,* **15,** 1577 (1972).
33. W. Aung, L. S. Fletcher, and V. Sernas, "Developing Laminar Free Convection between Vertical Flat Plates with Asymmetric Heating," *Int. J. Heat Mass Transfer,* **15,** 2293 (1972).
34. W. Elenbaas, "Heat Dissipation of Parallel Plates by Free Convection," *Physica,* **IX** (1), 2 (1942).
35. N. Sobel, F. Landis, and W. K. Mueller, "Natural Convection Heat Transfer in Short Vertical Channels Including the Effects of Stagger," *Proc. 3rd Int. Heat Transfer Conf.,* vol. 2, 1966, p. 121.
36. A. Bar-Cohen, W. M. Rohsenow, "Thermally Optimum Spacing of Vertical, Natural Convection Cooled, Parallel Plates," *J. Heat Transfer,* **106,** 116 (1984).
37. W. Elenbass, "The Dissipation of Heat by Free Convection: The Inner Surface of Vertical Tubes of Different Shapes of Cross-Section," *Physica,* **IX,** 865 (1942).
38. J. R. Dyer, "Natural Convective Flow through a Vertical Duct with Restricted Entry," *Int. J. Heat Mass Transfer,* **211,** 1341 (1978).
39. J. R. Dyer, "The Development of Laminar Natural Convection Flow in a Vertical Uniform Heat Flux Duct," *Int. J. Heat Mass Transfer,* **18,** 1455 (1975).
40. D. W. Van De Pol and J. K. Tierney, "Free Convection Nusselt Number for Vertical U-Shaped Channels," *J. Heat Transfer,* **95,** 542 (1973).
41. M. Al-Arabi and M. M. El-Rafaee, "Heat Transfer by Natural Convection from Corrugated Plates to Air," *Int. J. Heat Mass Transfer,* **21,** 357 (1978).
42. E. M. Sparrow and P. A. Bahrami, "Experiments on Natural Convection from Vertical Parallel Plates with Either Open or Closed Edges," *J. Heat Transfer,* **102,** 221 (1980).
43. E. M. Sparrow and P. A. Bahrami, "Experiments on Natural Convection Heat Transfer on the Fins of a Finned Horizontal Tube," *Int. J. Heat Mass Transfer,* **23,** 1555 (1980).

44. I. Catton, "Natural Convection in Enclosures," *Proc. 6th Int. Heat Transfer Conf.* (Toronto, 1978), vol. 6, p. 13.
45. S. M. ElSherbiny, G. D. Raithby, and K. G. T. Hollands, "Heat Transfer by Natural Convection Across Vertical and Inclined Air Layers," *J. Heat Transfer,* **104,** 96 (1982).
46. R. K. MacGregor and A. F. Emery, "Free Convection through Vertical Plane Layers—Moderate and High Prandtl Number Fluids," *J. Heat Transfer,* **91,** 391 (1969).
47. K. G. T. Hollands, G. D. Raithby, and L. Konicek, "Correlation Equations for Free Convection Heat Transfer in Horizontal Layers of Air and Water," *Int. J. Heat Mass Transfer,* **18,** 879 (1975).
48. K. G. T. Hollands, T. E. Unny, G. D. Raithby, and L. Konicek, "Free Convection Heat Transfer across Inclined Air Layers," *J. Heat Transfer,* **98,** 189 (1976).
49. T. H. Kuehn and R. J. Goldstein, "An Experimental Study of Natural Convection Heat Transfer in Concentric and Eccentric Cylindrical Annuli," *J. Heat Transfer,* **100,** 635 (1978).
50. T. H. Kuehn and R. J. Goldstein, "Correlating Equations for Natural Convection Heat Transfer between Horizontal Circular Cylinders," *Int. J. Heat Mass Transfer,* **19,** 1127 (1976).
51. M. Charmchi and E. M. Sparrow, "Analysis of Natural Convection in the Space between Concentric Vertical Cylinders of Different Height and Diameter," *Numer. Heat Transfer,* **5,** 119 (1982).
52. E. M. Sparrow and M. Charmchi, "Natural Convection Experiments in an Enclosure between Eccentric or Concentric Vertical Cylinders of Different Height and Diameter," *Int. J. Heat Mass Transfer,* **26,** 133 (1983).
53. F. Kreith, "Thermal Design of High-Altitude Balloons and Instrument Packages," *J. Heat Transfer,* **92,** 307 (1970).
54. R. O. Warrington, Jr., R. E. Powe, "The Transfer of Heat by Natural Convection between Bodies and Their Enclosures," *Int. J. Heat Mass Transfer,* **28,** 319 (1985).
55. J. R. Lloyd and E. M. Sparrow, "Combined Forced and Free Convection on Vertical Surfaces," *Int. J. Heat Mass Transfer,* **13,** 434 (1970).
56. S. W. Churchill, "A Comprehensive Correlating Equation for Laminar Assisting Forced and Free Convection," *AICHE J.,* **23,** 10 (1977).
57. H. Imura, R. R. Gilpin, and K. C. Cheng, "An Experimental Investigation of Heat Transfer and Buoyancy Induced Transition from Laminar Forced Convection to Turbulent Free Convection over a Horizontal Isothermally Heated Plate," *J. Heat Transfer,* **100,** 429 (1978).
58. R. M. Fand and K. K. Keswani, "Combined Natural and Forced Convection Heat Transfer from Horizontal Cylinders to Water," *Int. J. Heat Mass Transfer,* **16,** 1175 (1973).
59. V. T. Morgan, "Heat Transfer from Cylinders," In T. F. Irvine and J. P. Hartnett (eds.), *Advances in Heat Transfer,* vol. 11, Academic Press, New York, 1975, p. 199.
60. T. Yuge, "Experiments on Heat Transfer from Spheres Including Combined Natural and Forced Convection," *J. Heat Transfer,* **82,** 214 (1960).
61. C. K. Brown and W. H. Gauvin, "Combined Free and Forced Convection," parts I and II, *Can. J. Chem. Eng.,* **43,** 306 (1965).

Chapter **4**

Forced Convection Heat Transfer

MAJID CHARMCHI
Department of Mechanical and Energy Engineering, University of Lowell

4.1 INTRODUCTION

When a surface comes in contact with a moving fluid at a different temperature, heat transfer occurs in such a manner that the heat given off by the surface is carried away by the fluid parcels. This mode of heat transfer is known as heat transfer by convection. Heat diffuses in the fluid primarily by molecular conduction. (In some cases, the thermal radiation heat transfer may also be important.) However, the diffusion process may be altered by relative motion in the fluid. For example, in turbulent flows, the presence of the eddying motion in the fluid has a large effect on diffusion. Therefore convection heat transfer depends on the nature of the fluid motion, and its analysis requires a detailed knowledge of the velocity field.

The convection process is called forced convection when the fluid motion is generated by an external source, such as by a blower or a pump. In many problems encountered in forced convection, the buoyancy force, caused by density differences developed due to the heat transfer itself, can be neglected. However, situations can occur where buoyancy-induced fluid motion can be comparable to that caused by the external forces. Under such conditions, the heat-transfer rate is influenced by both forced and free convection processes, and the combined process is referred to as mixed convection.[1-10] A new experimental work[11] reveals that in cases where flow is driven by buoyancy, under certain conditions, the heat-transfer rate may be predictable from the forced convection relations.

This chapter considers the special aspects, regimes, and problems relevant to forced-convection heat transfer. A summary of analytical and empirical solutions to commonly encountered problems for both internal and external flows is presented. The contents of this chapter are relevant to conditions in which fluid behaves as a continuum. For very low density fluids, the ordinary continuum assumptions are not valid and special techniques are required.[12,13]

4.2 BASIC CONCEPTS

Conservation Equations

In an analysis of convection processes, the conservation principles are used to set up the governing equations. These equations are the conservation of momentum (equation of motion), the conservation of energy, and the conservation of species. (The last is required when mass transfer is involved.) The solution of these equations determines the change of velocity, temperature, and concentration of species with respect to time and position in the system from which the drag forces, the rate of heat transfer, and the rate of mass transfer can be obtained. For solving the momentum equation, the help of one more equation is needed, that is, the conservation of mass (equation of continuity). The detailed derivation and the expressions of the conservation equations in different coordinate systems are presented in Refs. 14 and 15.

Boundary-Layer Concept

When a fluid is flowing over a body, it is observed that the fluid particles immediately adjacent to the body are at rest relative to the surface, except for very low density gases.[4] It is further observed that these particles act to retard the motion of the particles in the adjoining fluid layer, which in turn retard the motion of the next fluid layer (see Fig. 4.1*a*). The region near the surface where the fluid velocity varies from its free-stream value to zero at the surface is called the velocity, or hydrodynamic, boundary layer and was first introduced by Ludwing Prandtl in 1904. The boundary-layer thickness δ is not a precise quantity, since the velocity asymptotically approaches the free-stream value. The boundary-layer thickness is often defined as the distance from the surface where the velocity is 99 percent of the free-stream velocity.

For a newtonian fluid, the shear stress τ occurs between fluid layers. For example, layer *a–a* in Fig. 4.1*a* is given by Newton's law of friction,[17]

$$\tau = \mu \frac{\partial u}{\partial y} \tag{4.1}$$

The shear stress is therefore proportional to the velocity gradient normal to the direction of flow. The proportionality factor μ is called the dynamic viscosity and is a property of the fluid.

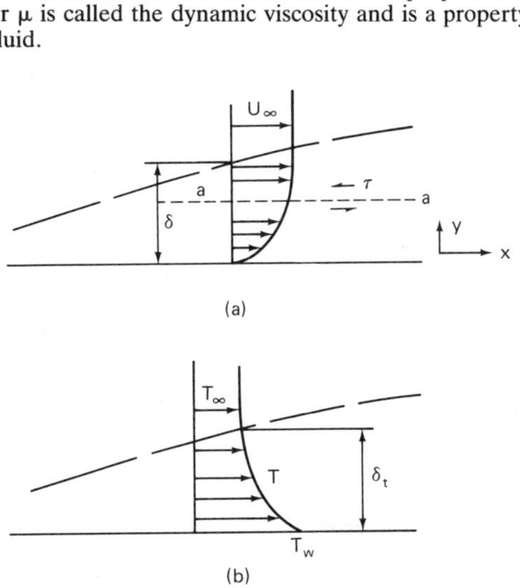

Fig. 4.1 Velocity and thermal boundary layer on solid surface. (a) Velocity boundary layer; (b) thermal boundary layer.

High shear stresses in a flow will occur where large velocity gradients exist. These large gradients are always found to be confined within the boundary layer. Outside of this layer, the velocity variation is much smaller, hence viscous effects become unimportant. Therefore the ideal-fluid analysis is applicable.[16,17] This is the fundamental assumption of the boundary-layer approximation.

Just as it was observed for velocity, when the fluid temperature differs from the surface temperature, the major temperature change occurs in a region close to the surface called thermal boundary layer δ_t (see Fig. 4.1*b*). The heat-transfer rate is obtained by analyzing the intensity of the heat diffusion and convection within this thin layer.

The growth of the thermal boundary layer may be faster than, slower than, or similar to that of the velocity boundary layer, depending on the thermophysical properties of the fluid as characterized by the Prandtl number Pr, expressed as

$$\mathrm{Pr} = \frac{\mu/\rho}{k/\rho C_p} = \frac{\nu}{\alpha} \tag{4.2}$$

where ν is the kinematic viscosity, α is the thermal diffusivity, and both have the same dimensions L^2/t. Hence, the Prandtl number is the ratio of momentum diffusion to thermal diffusion.

Laminar and Turbulent Boundary Layer

The convection heat-transfer rate and the surface drag coefficient depend strongly on the condition of the flow field. Therefore, in the treatment of a convection problem, it is essential to determine whether the boundary layer is laminar or turbulent.

When fluid moves in a streamline manner parallel to the

surface, the flow is called laminar. In a laminar boundary layer, the diffusion of momentum and heat is controlled solely by the molecular transport properties of the fluid, that is, viscosity and thermal conductivity.

If one measures the fluid velocity in a steady-state laminar flow using a fast response instrument, such as a hot-wire anemometer, it would be observed that some small disturbances would appear but they are quickly damped; hence, velocity is essentially independent of time.[9] These disturbances are introduced by the free-stream fluctuations or the surface roughnesses, or both. The damping process is caused by viscosity, whereas the inertia tends to amplify the disturbances. Therefore the Reynolds number, which represents the ratio of inertia to viscous forces, is an important parameter in fluid mechanics. The Reynolds number Re is expressed by

$$\mathrm{Re} = \frac{\rho V_c L_c}{\mu} = \frac{V_c L_c}{\nu} \tag{4.3}$$

where V_c is the characteristic flow velocity and L_c is the characteristic length of the body, a geometry parameter.

In contrast, in a turbulent boundary layer, a short distance away from the surface called the core region, the fluid particles do not travel in a well-ordered fashion. (Next to the surface, a very thin laminar sublayer exists where the fluid velocity is small and the viscous force is in control, damping the turbulent fluctuations.) The core region is filled with random eddying motion, causing components of velocity transverse to the principal direction of the flow, and these components are constantly changing in magnitude. The diffusion of momentum, energy, and mass is controlled predominantly by these macroscopic fluid motions.[17–19] Due to the presence of these eddies, a very efficient mixing takes place, resulting in a thicker boundary layer as well as in more uniform velocity and temperature profiles than those observed in laminar flow. Therefore the gradients of velocity and temperature at the surface are much larger for turbulent than for laminar flow, causing a much larger shear stress and heat-transfer rates.

The unsteady Navier-Stokes equation[14] is able to describe a turbulent flow at every point in space. However, its solution to reveal the complex and fine structure of turbulence is beyond our capabilities even with today's digital computers.[18] In practice, analyses are performed for the time averages of turbulent motion.[19–21]

Heat-Transfer Coefficient

In convection heat transfer it is customary to define a heat-transfer coefficient h as

$$q'' = h(T_w - T_{\mathrm{fl}}) = h\,\Delta T \tag{4.4}$$

where T_w is the surface (wall) temperature and T_{fl} is some reference temperature of the fluid. Equation (4.4) is known as Newton's law of cooling.

As was discussed in connection with the boundary-layer concept, at the surface the fluid velocity is zero and heat transfer takes place by molecular conduction. From Fourier's law of conduction,

$$q'' = -k\left(\frac{\partial T}{\partial y}\right)_{\mathrm{wall}} \tag{4.5}$$

where k is the thermal conductivity of the fluid and q'' is the local heat flux. Combining Eqs. (4.4) and (4.5) yields

$$h = \frac{-k(\partial T/\partial y)_{\mathrm{wall}}}{T_w - T_{\mathrm{fl}}} \tag{4.6}$$

Equation (4.6) suggests that to obtain the heat-transfer coefficient, the fluid temperature gradient at the wall is required. Finding the temperature gradient can be complex; the degree of complexity depends on the geometry and the nature of the flow field. Fortunately, for many practical situations the heat-transfer coefficients have been obtained analytically or experimentally and are reported in the heat-transfer literature in some form of correlation relations.

For forced convection of a single-phase fluid with moderate temperature difference, the heat-transfer coefficient h is independent of the temperature difference ΔT and it may be correlated with stream parameters. There are cases, however, where h is sensitive to ΔT, such as in free convection and in phase-change processes, such as boiling or condensation. For such cases, the use of h as defined by Eq. (4.4) is common, although it may not be as useful.

In general the heat-transfer coefficient h varies over the surface of the body. In practice, it is useful to obtain the area-averaged heat-transfer coefficient as

$$\bar{h} = \frac{1}{A}\int_A h\,dA \tag{4.7}$$

where A is the surface area of the body. The overall heat transfer Q can then be determined by

$$Q = \int_A q''\,dA = \bar{h}A(T_w - T_{\mathrm{fl}}) \tag{4.8}$$

Compressibility

Compressibility is a measure of the change of density of a fluid under the action of external forces. The change in density caused by a pressure increase Δp can be expressed by

$$\Delta p = E\left(\frac{\Delta\rho}{\rho}\right) \tag{4.9}$$

where E is the modulus of elasticity of the fluid. Introduction of Bernoulli's equation into Eq. (4.9) yields[17]

$$\frac{\Delta\rho}{\rho} = \frac{1}{2}\frac{\rho V^2}{E} = \frac{M^2}{2} \tag{4.10}$$

where M is the ratio of the flow velocity to the velocity of sound in the fluid and is known as the Mach number. If $\frac{1}{2}M^2 << 1$, the compressibility can be neglected. If changes in the density of 5 percent can be accepted, the flow can be regarded as incompressible for $M < 0.3$.

4.3 INTERNAL FLOW IN DUCTS

Hydrodynamic Entry Length

Consider a fluid that enters into a duct with a uniform velocity (e.g., flow inside a pipe as shown in Fig. 4.2). As the fluid makes contact with the wall of the duct, viscous effects become important. A boundary layer develops adjacent to the

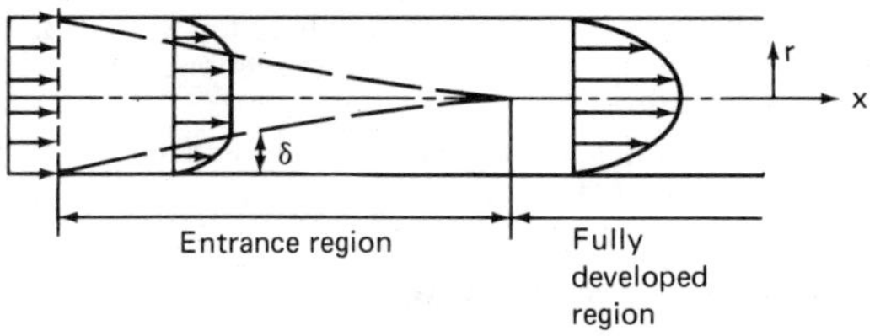

Fig. 4.2 Laminar velocity boundary layer development in circular tube.

wall and its thickness increases along the flow direction. This development occurs at the expense of the shrinking inviscid flow region and concludes when the effect of viscosity is extended over the entire duct cross section where the flow is said to be fully developed. The duct length required to achieve fully developed flow is called the hydrodynamic entry length L_{he}.

Fanning Friction Factor

The mean flow-length-average Fanning friction factor is defined as

$$f = \frac{\tau_w}{\rho u_m^2/2g_c} \tag{4.11}$$

where τ_w is the flow-length-average wall shear stress and u_m is the fluid mean axial velocity and can be determined by

$$u_m = \frac{1}{A_c}\int_{A_c} u \, dA_c \tag{4.12}$$

A_c being the cross-sectional area of the duct.

The pressure drop for the case of fully developed and constant-property flow can be evaluated by

$$\Delta p = f\left(\frac{L}{r_h}\right)\left(\frac{\rho u_m^2}{2g_c}\right) = f_D\left(\frac{L}{D_h}\right)\left(\frac{\rho u_m^2}{2g_c}\right) \tag{4.13}$$

where L = duct length
r_h = hydraulic radius,

$$= \frac{\text{flow cross-sectional area}}{\text{wetted perimeter}} = \frac{A_c}{P} \tag{4.14}$$

D_h = hydraulic diameter = $4r_h = 4A_c/P$
f_D = Darcy friction factor = $4f$

In the hydrodynamic entrance region, pressure drops result from wall shear stresses and a change in the momentum flow rate across the two relevant duct sections. For constant-property flow, the pressure drop from section $x = 0$ to $x = L$ can be obtained from

$$\Delta p = f_{app}\left(\frac{L}{r_h}\right)\left(\frac{\rho u_m^2}{2g_c}\right) = 4f_{app}\left(\frac{L}{D_h}\right)\left(\frac{\rho u_m^2}{2g_c}\right) \tag{4.15}$$

where f_{app} is the apparent Fanning friction factor. For short ducts ($L < L_{he}$), the pressure drop must be obtained using Eq. (4.15).

Heat-Transfer Coefficient

The peripherally average, but axially local heat-transfer coefficient h_x is defined by

$$q_x'' = h_x(T_w - T_b) \tag{4.16}$$

where T_w is the peripheral mean wall temperature obtained by

$$T_w = \frac{1}{P}\int_s T_{w,s}\, ds \tag{4.17}$$

and T_b is the fluid bulk mean temperature evaluated by

$$T_b = \frac{1}{\rho A_c u_m}\int_{A_c} \rho u T \, dA_c \tag{4.18}$$

The flow-length-average heat-transfer coefficient $\bar{h}$ is the integrated average of h_x from $x = 0$ to $x = L$,

$$\bar{h} = \frac{1}{L}\int_0^L h_x \, dx \tag{4.19}$$

In this part, most of the presented heat-transfer data are for $\bar{h}$.

Dimensionless Groups

Reynolds Number Re The Reynolds number is proportional to the ratio of the inertia forces to the viscous forces. For internal flow it is defined as

$$\text{Re} = \frac{\rho u_m D_h}{\mu} = \frac{G D_h}{\mu} \tag{4.20}$$

where G is the fluid mass velocity, $= \rho u_m$.

Nusselt Number Nu The Nusselt number, by its definition, is the ratio of the convective conductance h to the molecular thermal conductance k/D_h. The local Nusselt number Nu_x is then expressed by

$$\text{Nu}_x = \frac{h_x D_h}{k} \tag{4.21}$$

The mean flow-length-average Nusselt number based on $\bar{h}$ is defined as

$$\text{Nu} = \frac{\bar{h} D_h}{k} = \frac{1}{x}\int_0^x \text{Nu}_x \, dx \tag{4.22}$$

where x is the axial length measured from the entrance.

When the effects of axial heat conduction, viscous dissipation, and flow work are neglected within the fluid, it is shown that[22]

$$q'' = \bar{h}(\Delta T)_{av} = \left(\frac{\text{Nu}\, k}{D_h}\right)(\Delta T)_{av} \tag{4.23}$$

where, for constant wall temperature boundary condition,

$$(\Delta T)_{av} = \frac{(T_w - T_{b1}) - (T_w - T_{b2})}{\ln[(T_w - T_{b1})/(T_w - T_{b2})]} \tag{4.24}$$

and for constant wall heat flux,

$$(\Delta T)_{av} = \left(\frac{1}{x}\int_0^x \frac{1}{T_{w,x} - T_{b,x}}\,dx\right)^{-1} \tag{4.25}$$

Peclet Number Pe The Peclet number is defined as

$$\mathrm{Pe} = \frac{\rho C_p u_m D_h}{k} = \frac{u_m D_h}{\alpha} \tag{4.26}$$

Equation (4.26) can be represented as

$$\mathrm{Pe} = D_h \frac{\rho A_c u_m C_p (\partial T_b/\partial x)}{k A_c (\partial T_b/\partial x)}$$

$$= \frac{\text{fluid enthalpy rise}}{\text{axial conduction within fluid}} \tag{4.27}$$

Therefore the inverse of Pe shows the importance of the fluid axial heat conduction. It also can be shown that

$$\mathrm{Pe} = \mathrm{Re}\,\mathrm{Pr} \tag{4.28}$$

Thermal Boundary Conditions

For steady duct flow forced convection problems, typically, three kinds of boundary conditions are encountered:

1. Specified wall temperature, $T_w = T_w(x, y, z)$
2. Specified wall heat flux, $q''_w = q''_w(x, y, z)$
3. Wall heat flux specified via a heat-transfer coefficient and temperature of the surrounding fluid, $q''_w = h_w(T_w - T_a)$

Although the aforementioned thermal boundary conditions present a general three-dimensional situation, in many practical applications a much simpler boundary condition can be considered, that is, mainly uniform wall temperature and constant wall heat flux. The description of several thermal boundary conditions together with their applications is presented in Shah and London.[22]

In this chapter four boundary conditions are considered. The Nusselt numbers that correspond to the different thermal boundary conditions are shown using the following nomenclature:

Nu_T = uniform wall temperature; T boundary condition

Nu_{T1} = uniform axial wall temperature with finite normal wall thermal resistance (see Fig. 4.3); $T1$ boundary condition

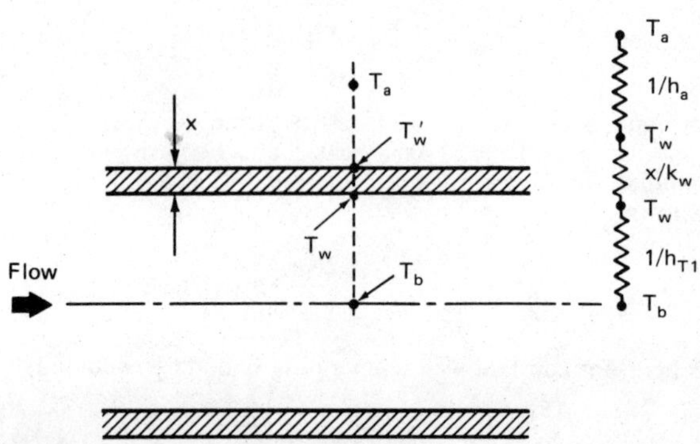

Fig. 4.3 Thermal resistance of $T1$ boundary condition.

Nu_H = constant heat flux in flow direction with uniform wall temperature at any cross section; H boundary condition

Nu_{H1} = constant heat flux both axially and peripherally; $H1$ boundary condition

4.4 FULLY DEVELOPED LAMINAR FLOW IN DUCTS

For hydrodynamically developed laminar flow, it can be shown that[22]

$$f\,\mathrm{Re} = K_f \tag{4.29}$$

where K_f is a constant dependent on the geometry of the duct cross section.

For a long duct where flow is hydrodynamically and thermally developed, the Nusselt number is a constant, dependent on the duct geometry and the thermal boundary condition.

A summary of reported data for the friction factor and heat transfer for fully developed laminar flow of a newtonian fluid through ducts with a variety of cross sections is presented in Table 4.1. With reference to Fig. 4.4, Table 4.1 can be interpreted with the aid of the following remarks.

Circular Ducts

1. *Effects of axial heat conduction:* For uniform wall temperature conditions, axial heat conduction can be important if the flow Peclet number (Pe = Re Pr) is small. Figure 4.5 (curve A) presents Nu_T as a function of Pe for circular ducts.

2. *Overall Nusselt number for T1 boundary condition:* From Fig. 4.3, the heat transfer can be evaluated by

$$q'' = U_o(T_a - T_b) \tag{4.30}$$

where U_o is the overall heat-transfer coefficient obtained by

$$\frac{1}{U_o} = \frac{1}{h_{T1}} + \frac{1}{U_w} \tag{4.31}$$

Also from Fig. 4.3, the wall thermal conductance U_w can be calculated as

$$\frac{1}{U_w} = \frac{1}{h_a} + \frac{t}{k_w} \tag{4.32}$$

where k_w is the thermal conductivity of the wall material. Equation (4.31) can be represented as

$$\frac{1}{U_o D_h/k} = \frac{1}{h_{T1} D_h/k} + \frac{1}{U_w D_h/k} \tag{4.33}$$

or

$$\frac{1}{\mathrm{Nu}_o} = \frac{1}{\mathrm{Nu}_{T1}} + R_w \qquad R_w = \frac{k}{U_w D_h} \tag{4.34}$$

where R_w is the wall thermal resistance.

The following formula can be used to evaluate the fully developed Nusselt number with $T1$ boundary condition Nu_{T1},[24]

$$\mathrm{Nu}_{T1} = \frac{1 + (48/11)R_w}{59/220 + R_w} \tag{4.35}$$

Table 4.1. Developed Laminar Flow Solutions*

Geometry	f Re	Nu_T	Nu_{T1}	Nu_H	Nu_{H1}
Circular duct	16	3.66 (Pe = ∞) Fig. 4.5, curve *A*	Eq. (4.35), Fig. 4.5 $R_w = 0$, curve *A* $R_w = 0.5$, curve *B* $R_w = 0.25$, curve *C* $R_w = \infty$, curve *D*	4.36	4.36
Parallel plates	24	7.54 (Pe = ∞) Fig. 4.5, curve *E*	Eq. (4.36)	8.24	8.24
Parallel plates, one side insulated	24	4.86	....	5.39	
Rectangular duct, Fig. 4.4*a* $\alpha = \frac{2b}{2a}$	Eq. (4.37)	Table 4.2	....	Table 4.3	3.09 (α = 1.0) 3.02 (α = 0.5) 2.95 (α = 0.10) 8.24 (α = ∞)
Equilateral triangular duct, no rounded corners	13.33	....	....	3.11	1.89
Equilateral triangular duct, rounded corners, Fig. 4.4*b*	16.0	....	....	4.21	3.78
Isosceles triangular duct, Fig. 4.4*c*	12.0 (α = ∞, 0.0) 12.6 (α = 4.00) 13.1 (α = 1.87) 13.3 (α = 0.866) 12.7 (α = 0.289) 12.2 (α = 0.134)	0.943 1.81 2.26 2.47 2.00 1.50 and Table 4.4	....	2.06 2.58 2.91 3.11 2.68 2.33 and Table 4.5	0.039 0.173 0.851 1.89 0.620 0.156
Elliptical duct, Fig. 4.4*d*	16 (α = 1.0) Fig. 4.6, curve *A*	3.66 (α = 1.0) Fig. 4.6, curve *B*	....	4.36 (α = 1.0) Fig. 4.6, curve *C*	4.36 (α = 1.0) Fig. 4.6, curve *D*
n-sided polygonal duct	14.7 (n = 5) 15.0 (n = 6) 15.4 (n = 8)	....	....	3.85 4.00 4.15	3.60 3.86 4.10
Annular duct, same heat flux or temperature both walls, Fig. 4.4*e*	20.6 (r^* = 0.02) 21.7 (r^* = 0.05) 22.3 (r^* = 0.10) 23.3 (r^* = 0.25) 23.8 (r^* = 0.50) 24.0 (r^* = 0.90)	5.64 (r^* = 0.02) 6.10 (r^* = 0.05) 6.52 (r^* = 0.10) 7.00 (r^* = 0.25) 7.41 (r^* = 0.50) 7.54 (r^* = 0.90)	....	6.39 (r^* = 0.02) 6.85 (r^* = 0.05) 7.26 (r^* = 0.10) 7.88 (r^* = 0.25) 8.11 (r^* = 0.50) 8.23 (r^* = 0.90)	
Annular duct, inner wall insulated	Same as above	32.3 (r^* = 0.02) 17.5 (r^* = 0.05) 11.6 (r^* = 0.10) 7.37 (r^* = 0.25) 5.74 (r^* = 0.50)	....	....	
Annular duct, outer wall insulated	Same as above	3.99 (r^* = 0.02) 4.05 (r^* = 0.05) 4.11 (r^* = 0.10) 4.23 (r^* = 0.25) 4.43 (r^* = 0.50)			
Annular duct, inner/outer wall heat fluxes different	Same as above	....	....	For each wall, Eqs. (4.38) and (4.39), Table 4.6	
Longitudinal flow over circular cylinders in equilateral triangular pitch, Fig. 4.4*f*	Fig. 4.7	....	....	1.52 (s/d = 1.01) 2.82 (s/d = 1.05) 4.62 (s/d = 1.10) 7.48 (s/d = 1.20) 11.3 (s/d = 1.50) 15.3 (s/d = 2.0)	0.263 1.06 2.94 6.90 11.2 15.3

*For constant fluid properties, evaluated at the mean bulk flow temperature.

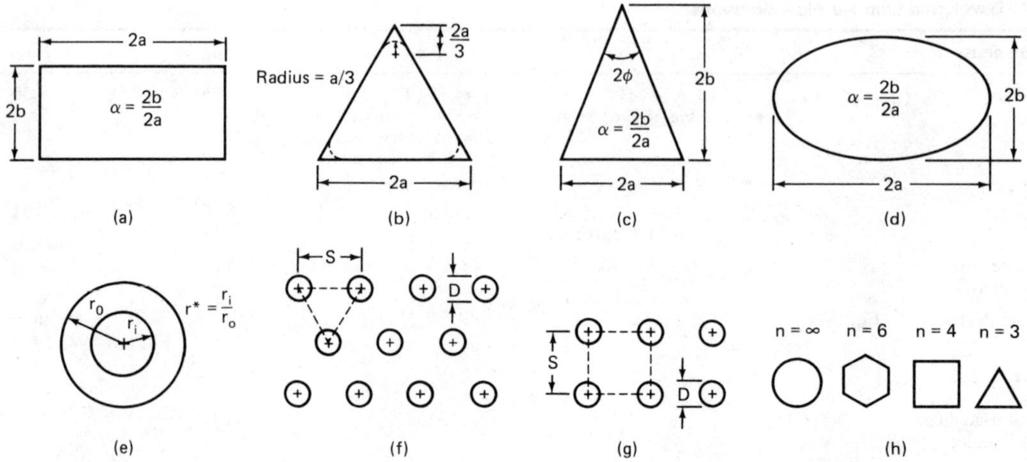

Fig. 4.4 Internal flow geometries. (*a*) Rectangular duct; (*b*) equilateral triangular duct; (*c*) isosceles triangular duct; (*d*) elliptical duct; (*e*) annular duct; (*f*) triangular pitch tubes; (*g*) square pitch tubes; (*h*) regular polygonal ducts.

The overall Nusselt number Nu_o can then be obtained by substituting the result of Eq (4.35) into Eq. (4.34).

The effects of axial heat conduction of Nu_{T1} for different R_w as a function of the Peclet number are presented in Table 4.1 and Fig. 4.5 (curves *A, B, C,* and *D*).

Parallel Plates

1. *Effects of axial heat conduction:* The effects of axial heat conduction on Nu_T as a function of Pe are given in Table 4.1 and Fig. 4.5 (curve *E*).

2. *Overall Nusselt number for T1 boundary condition:* For the *T*1 boundary condition, Nu_{T1} can be evaluated from the formula[24]

$$Nu_{T1} = \frac{1 + (140/17)R_w}{74/561 + R_w} \tag{4.36}$$

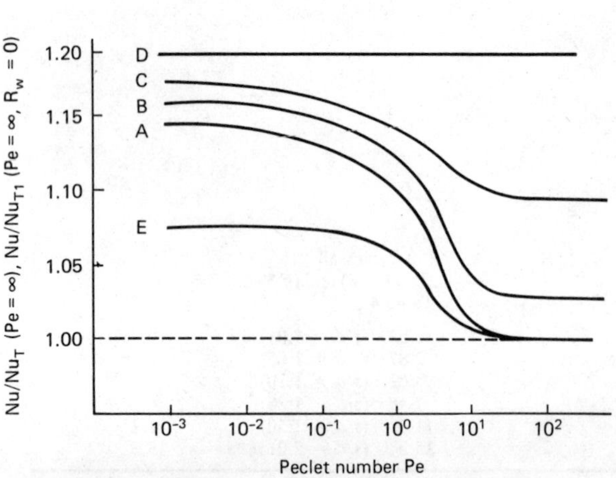

Fig. 4.5 Effect of Peclet number on Nusselt number for fully developed laminar flow (see Table 4.1).

Equation (4.36) together with Eq. (4.34) gives overall Nusselt number Nu_o.

Rectangular Ducts Table 4.1 presents the friction factor and the Nusselt number for the *H*1 boundary condition as a function of aspect ratio α. The factor *f* Re may be approximated by the equation[22]

$$f\,Re = 24(1 - 1.3553\alpha + 1.9467\alpha^2 - 1.7012\alpha^3 + 0.9564\alpha^4 - 0.2537\alpha^5) \tag{4.37}$$

The Nusselt numbers for the *T* and *H* boundary conditions are reported in Tables 4.2 and 4.3, respectively, with one or more walls transferring heat.

Equilateral Triangular Ducts Table 4.1 presents the values of *f* Re, Nu_H, and Nu_{H1} for equilateral triangular ducts with both no rounded corners and three rounded corners, as shown in Fig. 4.4*b*.

Isosceles Triangular Ducts The data for the friction factor and for the heat transfer when all three walls are transferring heat (see Fig. 4.4*c*) are reported in Table 4.1. Tables 4.4 and 4.5 present Nu_T and Nu_H, respectively, when one or two walls are transferring heat.

Elliptical Ducts The friction factors, as well as the Nusselt numbers for the *T, H,* and *H*1 boundary conditions as a function of α (see Fig. 4.4*d*) are given in Table 4.1 and Fig. 4.6.

***n*-Sided Regular Polygonal Ducts** Table 4.1 presents *f* Re, Nu_H, and Nu_{H1} for *n*-sided regular polygonal ducts (see Fig. 4.4*h*).

Concentric Annular Ducts The values of *f* Re, Nu_T, and Nu_H as a function of $r^* = r_i/r_o$ (see Fig. 4.4*e*) are reported in Table 4.1. For the case of constant surface temperature on one wall with the other wall insulated, the Nusselt number for the active surface is also given in Table 4.1.

The Nusselt numbers at the inner wall Nu_i and at the outer wall Nu_o, for the cases of constant equal or nonequal axial heat fluxes specified on both walls, can be evaluated using

Table 4.2. Rectangular Ducts: Nu_T for Fully Developed Laminar Flow, One or More Walls Transferring Heat

$\frac{2b}{2a}$	Nu_T				
	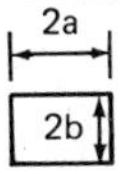	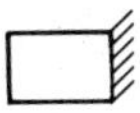	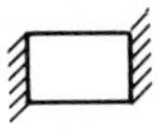	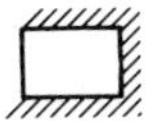	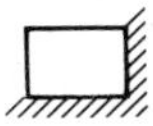
0.0	7.541	7.541	7.541	0	4.861
0.1	5.858	6.095	6.399	0.457	3.823
0.2	4.803	5.195	5.703	0.833	3.330
0.3	4.114	4.579	5.224	1.148	2.996
0.4	3.670	4.153	4.884	1.416	2.768
0.5	3.383	3.842	4.619	1.647	2.613
0.6	3.198	....	....	....	2.509
0.7	3.083	3.408	4.192	2.023	2.442
0.8	3.014	....	....	....	2.401
0.9	2.980	....	....	....	2.381
1.0	2.970	3.018	3.703	2.437	2.375
1.43	3.083	2.734	3.173	2.838	2.442
2.0	3.383	2.602	2.657	3.185	2.613
2.5	3.670	2.603	2.333	3.390	2.768
3.33	4.114	2.703	1.946	3.626	2.996
5.0	4.803	2.982	1.467	3.909	3.330
10.0	5.858	3.590	0.843	4.270	3.823
∞	7.541	4.861	0	4.861	4.861

SOURCE: Ref. 22.

Table 4.3. Rectangular Ducts: Nu_H for Fully Developed Laminar Flow, One or More Walls Transferring Heat

$\frac{2b}{2a}$	Nu_H				
	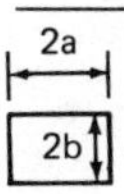	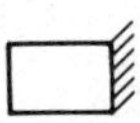	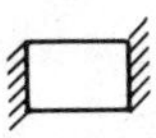	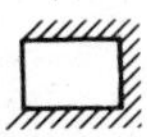	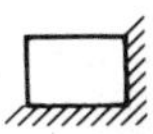
0.0	8.235	8.235	8.235	0	5.385
0.1	6.700	6.939	7.248	0.538	4.410
0.2	5.704	6.072	6.561	0.964	3.914
0.3	4.969	5.393	5.997	1.312	3.538
0.4	4.457	4.885	5.555	1.604	3.279
0.5	4.111	4.505	5.203	1.854	3.104
0.6	3.884	....	....	....	2.987
0.7	3.740	3.991	4.662	2.263	2.911
0.8	3.655	....	....	....	2.866
0.9	3.612	....	....	....	2.843
1.0	3.599	3.556	4.094	2.712	2.836
1.43	3.740	3.195	3.508	3.149	2.911
2.0	4.111	3.146	2.947	3.539	3.104
2.5	4.457	3.169	2.598	3.777	3.279
3.33	4.969	3.306	2.182	4.060	3.538
5.0	5.704	3.636	1.664	4.411	3.914
10.0	6.700	4.252	0.975	4.851	4.410
∞	8.235	5.385	0	5.385	5.385

SOURCE: Ref. 22.

the equations[35]

$$Nu_i = \frac{Nu_{ii}}{1 - (q''_o/q''_i)\,\theta_i^*} \tag{4.38}$$

$$Nu_o = \frac{Nu_{oo}}{1 - (q''_i/q''_o)\,\theta_o^*} \tag{4.39}$$

where θ_i^* and θ_o^* are influence coefficients, Nu_{ii} is the Nusselt number on the inner wall when the inner surface *alone* is heated and the outer wall is insulated, and Nu_{oo} is the Nusselt number on the outer wall when the outer surface *alone* is heated and the inner wall is insulated. The values of Nu_{ii}, Nu_{oo}, θ_i^*, and θ_o^* for constant heat flux are given in Table 4.6.

Longitudinal Flow over Array of Circular Cylinders The hydraulic diameter D_h for triangular and square array arrangements for an infinite rod bundle (see Fig. 4.4*f* and *g*) can be obtained from the following equations. For an equilateral triangular array,

$$D_h = D\left[\frac{3.4641}{\pi}\left(\frac{S}{D}\right)^2 - 1\right] \tag{4.40}$$

for a square array,

$$D_h = D\left[\frac{4}{\pi}\left(\frac{S}{D}\right)^2 - 1\right] \tag{4.41}$$

Figure 4.7 presents the results for the Fanning friction factors versus the ratio S/D.

The heat-transfer results for an equilateral triangular array and for boundary conditions H and $H1$ are reported in Table 4.1.

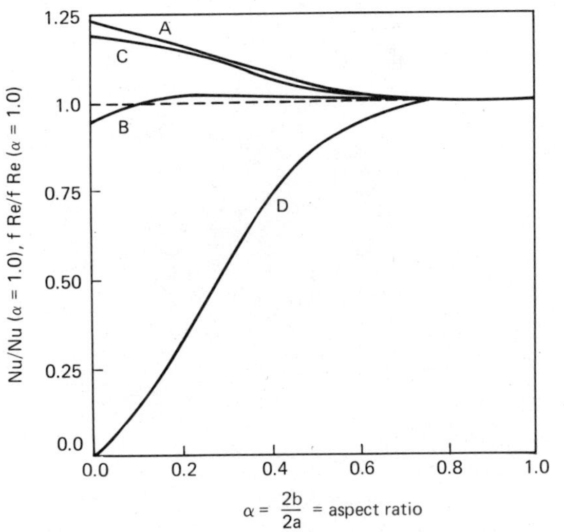

Fig. 4.6 Effect of aspect ratio on Nusselt number for laminar flow in elliptical duct (see Table 4.1).

Table 4.4. Isosceles Triangular Ducts: Nu_T for Fully Developed Laminar Flow, One or More Walls Transferring Heat

$\frac{2b}{2a}$	ϕ, deg	Nu_T			
		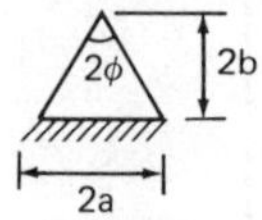	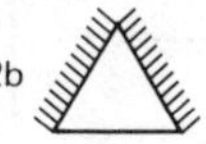	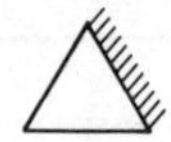	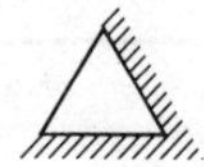
∞	0	1.885	0.000	1.215	1.215
5.000	5.71	.....	0.822	1.416	1.312
2.500	11.31	2.058	1.268	1.849	1.573
1.667	16.70	2.227	1.525	2.099	1.724
1.250	21.80	2.312	1.675	2.237	1.802
1.000	26.56	2.344	1.758	2.301	1.831
0.833	30.96	.....	.....	2.319	1.822
0.714	34.99	2.311	1.812	2.306	1.787
0.625	38.66	.....	.....	2.274	1.735
0.556	41.99	.....	.....	2.232	1.673
0.500	45.00	2.162	1.765	2.183	1.606
0.450	48.01	.....	.....	2.127	1.529
0.400	51.34	.....	.....	2.055	1.433
0.350	55.01	1.923	1.633	1.968	1.315
0.300	59.04	.....	.....	1.861	1.173
0.250	63.43	1.671	1.471	1.733	1.004
0.200	68.20	1.512	1.361	1.581	0.805
0.150	73.30	1.330	1.229	1.401	0.578
0.100	78.69	1.126	1.0710	1.182	0.332
0.050	84.29	0.895	0.878	0.893	0.106
0	90.00	0.6076	0.6076		

SOURCE: Ref. 22.

Table 4.5. Isosceles Triangular Ducts: Nu_H for Fully Developed Laminar Flow, One or More Walls Transferring Heat

$\frac{2b}{2a}$	ϕ, deg	Nu_H			
		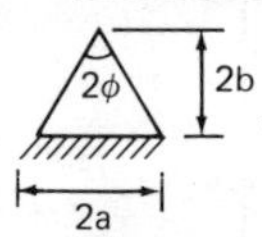	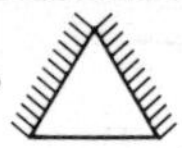	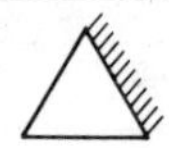	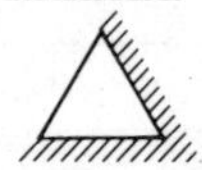
∞	0	2.059	0	1.346	1.346
5.000	5.71	2.465	1.003	1.864	1.739
2.500	11.31	2.683	1.515	2.274	1.946
1.667	16.70	2.796	1.807	2.541	2.074
1.250	21.80	2.845	1.978	2.695	2.141
1.000	26.56	2.849	2.076	2.773	2.161
0.833	30.96	....	....	2.801	2.146
0.714	34.99	2.778	2.146	2.792	2.107
0.625	38.66	....	....	2.774	2.053
0.556	41.99	....	....	2.738	1.989
0.500	45.00	2.594	2.111	2.696	1.921
0.450	48.01	....	....	2.646	1.843
0.400	51.34	....	....	2.583	1.746
0.350	55.01	2.332	1.991	2.505	1.628
0.300	59.04	....	....	2.412	1.486
0.250	63.43	2.073	1.843	2.301	1.316
0.200	68.20	1.917	1.746	2.174	1.114
0.150	73.30	1.748	1.635	2.032	0.874
0.100	78.69	1.576	1.515	1.881	0.587
0.050	84.29	1.418	1.398	1.737	0.244
0	90.00	1.346	1.346		

SOURCE: Ref. 22.

Table 4.6. Concentric Annular Ducts: Fundamental Solutions for Constant Heat Flux and Fully Developed Laminar Flow*

r^*	Nu_{ii}	Nu_{oo}	θ_i^*	θ_o^*
0	∞	4.364	∞	0
0.05	17.81	4.792	2.18	0.0294
0.10	11.91	4.834	1.383	0.0562
0.20	8.499	4.883	0.905	0.1041
0.40	6.583	4.979	0.603	0.1823
0.60	5.912	5.099	0.473	0.2455
0.80	5.58	5.24	0.401	0.299
1.00	5.385	5.385	0.346	0.346

SOURCE: Ref. 22.
*For use in Eqs. (4.38) and (4.39).

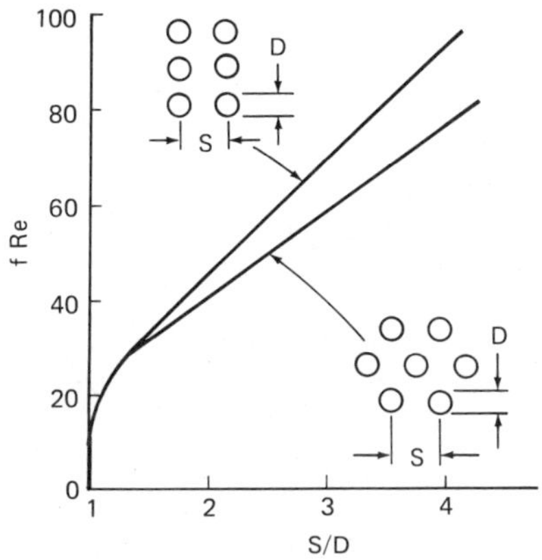

Fig. 4.7 Longitudinal flow over circular cylinder: f Re as a function of spacing ratio S/D for fully developed laminar flow. (*Ref. 36.*)

4.5 DEVELOPING LAMINAR FLOW IN DUCTS

A summary of the data presented in this section is given in Table 4.7.

Circular Ducts The hydrodynamic entry length L_{he} can be evaluated from the equation[40]

$$\frac{L_{\mathrm{he}}}{D_h} = 0.056 \, \mathrm{Re} \tag{4.42}$$

The apparent friction coefficient f_{app} may be obtained from the formula[41]

$$f_{\mathrm{app}} \, \mathrm{Re} = \frac{3.44}{(x^+)^{0.5}} + \frac{16 + 1.25/4x^+ - 3.44/(x^+)^{0.5}}{1 + 0.00021/(x^+)^2} \tag{4.43}$$

where x^+ is defined as

$$x^+ = \frac{x}{D_h \, \mathrm{Re}} \tag{4.44}$$

The pressure loss then can be determined from Eq. (4.15).

The local ($\mathrm{Nu}_{x,T}$ and $\mathrm{Nu}_{x,H}$) and averaged (Nu_T and Nu_H) Nusselt numbers for boundary conditions T and H and for developed velocity profile but developing temperature profile are given in Table 4.7 as a function of $x^* = x/(D \, \mathrm{Re} \, \mathrm{Pr}) = x/(D \, \mathrm{Pe})$.

When the velocity and temperature profiles are developing simultaneously, the resulting Nusselt numbers are higher than those reported in the preceding case, but the differences are only significant for fluids with Prandtl numbers less than about 10.[36] Table 4.7 presents the local and averaged Nusselt numbers for the T boundary condition ($\mathrm{Nu}_{x,T}$ and Nu_T) as well as the local Nusselt numbers for the H boundary condition ($\mathrm{Nu}_{x,H}$).

Parallel Plates Hydrodynamic entry length can be obtained from the formula[40]

$$\frac{L_{\mathrm{he}}}{D_h} = 0.011 \, \mathrm{Re} \tag{4.45}$$

The apparent friction factor can be evaluated by[41]

$$f_{\mathrm{app}} \, \mathrm{Re} = \frac{3.44}{\sqrt{x^+}} + \frac{24 + 0.674/4x^+ - 3.44/\sqrt{x^+}}{1 + 0.000029/(x^+)^2} \tag{4.46}$$

where x^+ is defined by Eq. (4.44).

Table 4.7 presents the local and averaged Nusselt numbers for T and H boundary conditions for thermally developing, but fully developed velocity profiles.

Table 4.7 also gives Nusselt numbers for T and H boundary conditions as a function of x^* and Pr for simultaneously developing laminar flow.

Rectangular Ducts The apparent friction coefficients for different aspect ratios α are presented in Fig. 4.8.

For thermally developing, but fully developed velocity profiles the averaged Nusselt numbers for T and H boundary conditions when all four walls are transferring heat are presented in Table 4.7.

Table 4.7 also provides Nu_T and Nu_H for simultaneously developing flow and for $\mathrm{Pr} = 0.72$ as a function of aspect ratio α and x^*. The effects of the Prandtl number are shown in Fig. 4.9 for $\alpha = 0.5$. For practical purposes, Fig. 4.9 may be used as a guideline to obtain the Nusselt numbers for Prandtl numbers other than 0.72 by extrapolating the data given in Table 4.7.

4.6 TURBULENT FLOW IN DUCTS

Friction and heat-transfer coefficients for turbulent flow through ducts are presented. The reported data are based on the assumption of constant fluid properties. All fluid properties are evaluated at the mean bulk temperature.

Two boundary conditions are considered—constant surface temperature and constant heat flux. The Nusselt number corresponding to the constant heat flow rate is always larger than that related to the constant surface temperature.[4] However, the difference between Nu_H and Nu_T is only significant for fluids with very low Prandtl numbers, $\mathrm{Pr} < 0.03$.

Circular Ducts For turbulent fully developed flow, the Fanning friction factor in general is a function of the Reynolds number and the surface relative roughness e/D. Figure 4.10 presents the friction factor versus the Reynolds number in tubes with various surface roughnesses. The relative roughnesses for commercial pipes are given in Fig. 4.11.

For smooth pipes the applicable equations are[4]

$$f = 0.079 \, \mathrm{Re}^{-0.25} \qquad 4 \times 10^3 < \mathrm{Re} < 2 \times 10^4 \tag{4.47}$$

$$f = 0.046\ \mathrm{Re}^{-0.2} \qquad 2 \times 10^4 < \mathrm{Re} < 10^6 \tag{4.48}$$

and for a wider range of Re,[52]

$$f = [1.581 \ln (\mathrm{Re}) - 3.28]^{-2} \qquad 10^4 < \mathrm{Re} < 10^7 \tag{4.49}$$

The apparent friction coefficients for hydrodynamically developing flow are presented in Fig. 4.12 as a function of the Reynolds number and x/D. As can be seen, the friction factors become constant within relatively short lengths compared to the laminar flow. The hydrodynamic entrance length can be predicted by[36]

$$\frac{L_{\mathrm{he}}}{D_h} = 0.623\ \mathrm{Re}^{0.25} \tag{4.50}$$

The values of Nu_T and Nu_H for fluids with low Prandtl numbers can be evaluated by the following equations.[53] For uniform wall temperature,

$$\mathrm{Nu}_T = 4.8 + 0.0156\ \mathrm{Re}^{0.85}\ \mathrm{Pr}^{0.93} \tag{4.51}$$

Table 4.7. Developing Laminar Flow Solutions

Hydrodynamic entry length solutions	Thermal entry length solutions						
	Developed velocity			Developing velocity			
					Nu_T		
	x^*	Nu_H	Nu_T	x^*	Pr = 0.7	Pr = 2	Pr = 5
Circular Duct							
Eqs. (4.42) and (4.43)	10^{-5}	90	74	5×10^{-4}	30.5	25.2	22.1
	10^{-4}	43	35	10^{-3}	22.2	19.1	16.9
	10^{-3}	20	16	5×10^{-3}	11.3	10.2	9.2
	10^{-2}	9.0	7.0	10^{-2}	8.7	7.8	7.2
	10^{-1}	5.1	4.2	2×10^{-2}	6.8	6.2	5.8
					$\mathrm{Nu}_{x,T}$		
	x^*	$\mathrm{Nu}_{x,H}$	$\mathrm{Nu}_{x,T}$	x^*	Pr = 0.7	Pr = 2	Pr = 5
	10^{-5}	60	48	5×10^{-4}	16.8	14.8	13.6
	10^{-4}	27	22	10^{-3}	12.6	11.4	10.6
	10^{-3}	13	10	5×10^{-3}	6.7	6.2	6.05
	10^{-2}	6.2	4.9	10^{-2}	5.25	4.93	4.92
	10^{-1}	4.4	3.7	2×10^{-2}	4.28	4.17	4.17
					$\mathrm{Nu}_{x,H}$		
				x^*	Pr = 0.7	Pr = 2	Pr = 5
				5×10^{-4}	24	21.3	19.3
				10^{-3}	17.5	16.0	14.5
				5×10^{-3}	9.0	8.2	7.6
				10^{-2}	6.9	6.2	6.1
				2×10^{-2}	5.5	5.2	5.2
Parallel Plate							
					Nu_T		
	x^*	Nu_H	Nu_T	x^*	Pr = 0.1	Pr = 0.72	Pr = 10
Eqs. (4.45) and (4.46)	10^{-5}	105	85	10^{-4}	85	68	52
	10^{-4}	49	40	10^{-3}	28	25	19
	10^{-3}	23	19	10^{-2}	12	10.5	9.7
	10^{-2}	12	10	4×10^{-2}	8.9	8.2	8.1
	10^{-1}	8.7	7.7				
					$\mathrm{Nu}_{x,H}$		
	x^*	$\mathrm{Nu}_{x,H}$	$\mathrm{Nu}_{x,T}$	x^*	Pr = 0.01	Pr = 0.7	Pr = 10
	10^{-5}	70	57	10^{-3}	24	19	15.5
	10^{-4}	32	27	2×10^{-3}	18.5	14.6	13
	10^{-3}	15	13	10^{-2}	10	9.0	8.9
	10^{-2}	8.8	7.9	4×10^{-2}	8.5	8.3	8.3
	10^{-1}	8.3	7.5				

Table 4.7. Developing Laminar Flow Solutions (*Continued*)

Hydrodynamic entry length solutions	Thermal entry length solutions						
	Developed velocity			Developing velocity			
					Nu_T		
	x^*	Nu_H	Nu_T	x^*	Pr = 0.7	Pr = 2	Pr = 5
			Rectangular Duct				
		Nu_T			Nu_T (Pr = 0.72)		
	x^*	$\alpha = 1.0$	$\alpha = 0.5$	$\alpha = 0.25$	x^*	$\alpha = 1.0$	$\alpha = 0.5$	$\alpha = 0.25$
Fig. 4.8	5×10^{-3}	8.15	8.58	9.47	5×10^{-3}	9.30	9.60	10.18
	10^{-2}	6.46	6.84	7.71	10^{-2}	7.10	7.42	7.98
	5×10^{-2}	4.03	4.46	5.44	5×10^{-2}	4.39	4.79	5.56
	10^{-1}	3.50	3.95	5.00	10^{-1}	3.75	4.20	5.11
		Nu_H			Nu_H (Pr = 0.72)		
	x^*	$\alpha = 1.0$	$\alpha = 0.5$	$\alpha = 0.25$	x^*	$\alpha = 1.0$	$\alpha = 0.5$	$\alpha = 0.25$
	5×10^{-3}	10.18	10.64	11.23	5×10^{-3}	14.5	14.9	15.5
	10^{-2}	8.22	8.66	9.39	10^{-2}	11.4	11.7	12.2
	5×10^{-2}	5.19	5.60	6.57	5×10^{-2}	6.60	6.94	7.58
	10^{-1}	4.48	4.94	6.03	10^{-1}	5.43	5.77	6.65

and for uniform wall heat flux,

$$Nu_H = 6.3 + 0.0167\, Re^{0.85}\, Pr^{0.93} \tag{4.52}$$

Equations (4.51) and (4.52) are valid for $Pr < 0.1$ and $10 < Re < 10^6$.

For $Pr > 0.1$, a single Nusselt number is sufficient for both uniform wall temperature and uniform wall heat flux boundary conditions and can be obtained by[53]

$$Nu = 5 + 0.015\, Re^a\, Pr^b \qquad Pr > 0.1, \quad 10^4 < Re < 10^6 \tag{4.53}$$

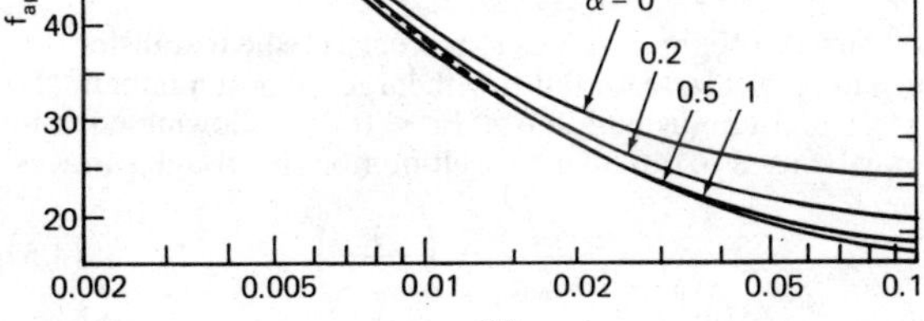

Fig. 4.8 Rectangular ducts: f_{app} Re for developing laminar flow. (*Ref. 22.*)

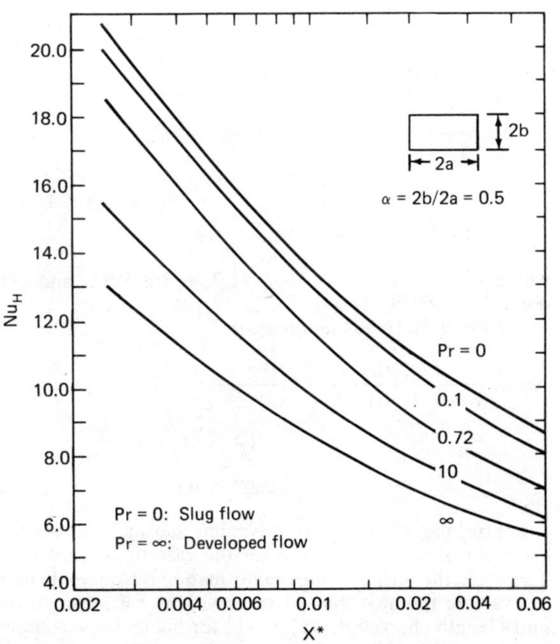

Fig. 4.9 Rectangular ducts ($\alpha = 0.5$): Nu_H as a function of x^* and Pr for simultaneously developing flow. (*Ref. 22.*)

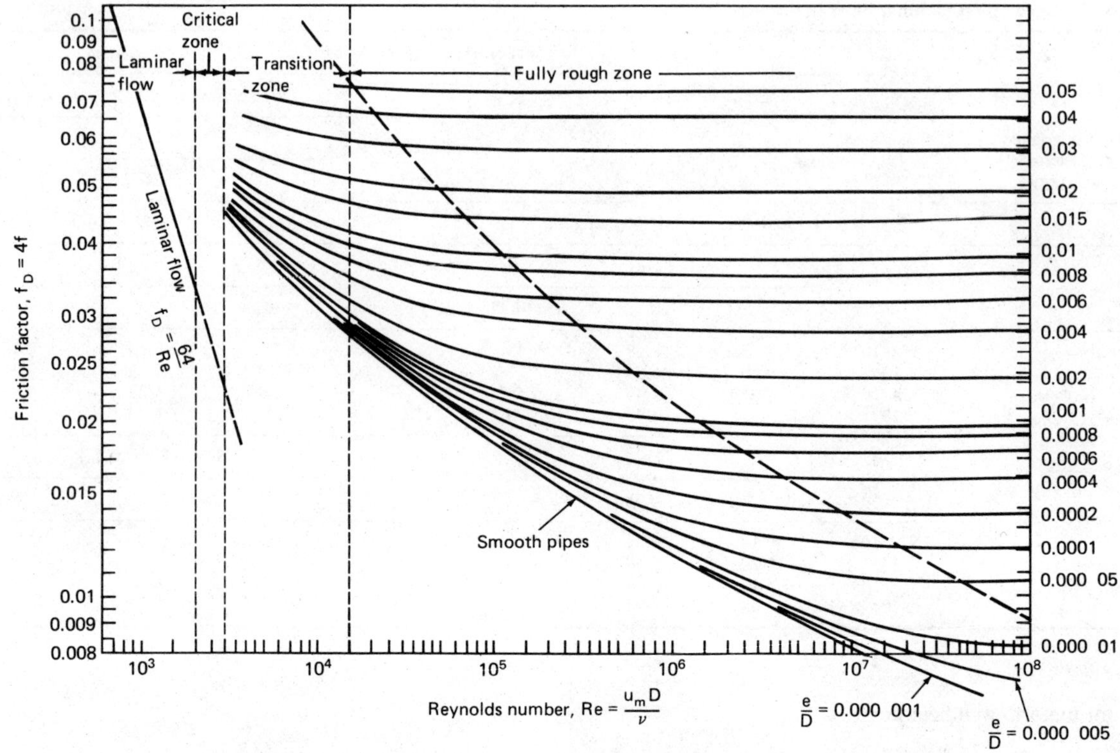

Fig. 4.10 Friction factor for fully developed turbulent flow in circular tube.

where $a = 0.88 - 0.24/(4 + \mathrm{Pr})$ and $b = \frac{1}{3} + 0.5 \exp(-0.6\,\mathrm{Pr})$.

For higher Reynolds numbers,[52]

$$\mathrm{Nu} = \frac{(f/2)\,\mathrm{Re}\,\mathrm{Pr}}{k_1 + k_2(\mathrm{Pr}^{2/3} - 1.0)\sqrt{f/2}} \qquad 0.5 < \mathrm{Pr} < 2000 \qquad 10^4 < \mathrm{Re} < 10^7 \tag{4.54}$$

where $k_1 \doteq 1 + 13.6f$, $k_2 = 11.7 + 1.8/\mathrm{Pr}^{0.33}$, and f is obtained from Eq. (4.49).

For lower Reynolds numbers use[36]

$$\mathrm{Nu} = \frac{(f/2)(\mathrm{Re} - 1000)\mathrm{Pr}}{1.0 + 12.7(\mathrm{Pr}^{1/3} - 1.0)\sqrt{f/2}} \qquad 0.5 < \mathrm{Pr} < 2000 \qquad 2300 < \mathrm{Re} < 2 \times 10^4 \tag{4.55}$$

For turbulent flow the hydrodynamic thermal entry lengths are much shorter than those for the case of laminar flow. Therefore the fully developed information is frequently used to evaluate the heat-transfer rates without reference to the entry length characteristics, except for fluids, such as liquid metals, with low Prandtl numbers, $\mathrm{Pr} < 0.03$, as shown in Fig. 4.13.

Figure 4.14 presents the local Nusselt number Nu_x for air flowing in the entry region of a tube, for various entry configurations. The average Nusselt number can be obtained by integrating the curves. For tubes with $L/D > 20$ the mean Nusselt number Nu can be computed from[4]

$$\mathrm{Nu} = \mathrm{Nu}_\infty\left(1.0 + \frac{C}{x/D}\right) \tag{4.56}$$

where Nu_∞ is the fully developed Nusselt number and the parameter C is as follows:

Fully developed velocity profile	$C = 1.4$
Abrupt-contraction entrance	$C = 6.0$
180° round bend	$C = 6.0$
90° right-angle bend	$C = 7.0$

Equation (4.56) is valid for gases and liquids with $0.6 < \mathrm{Pr} < 10$.

Surface roughness tends to increase the heat-transfer coefficient particularly for fluids with large Prandtl numbers.[4] For gases and liquids with $0.6 < \mathrm{Pr} < 6$ the following equation may be used to obtain a Nusselt number for rough surfaces[4]:

$$\frac{\mathrm{Nu}}{\mathrm{Nu}_{\mathrm{smooth}}} = \left(\frac{f}{f_{\mathrm{smooth}}}\right)^n \tag{4.57}$$

where $n = 0.68\,\mathrm{Pr}^{0.215}$. Equation (4.57) is valid for $f/f_{\mathrm{smooth}} < 4$, beyond which the Nusselt number will no longer

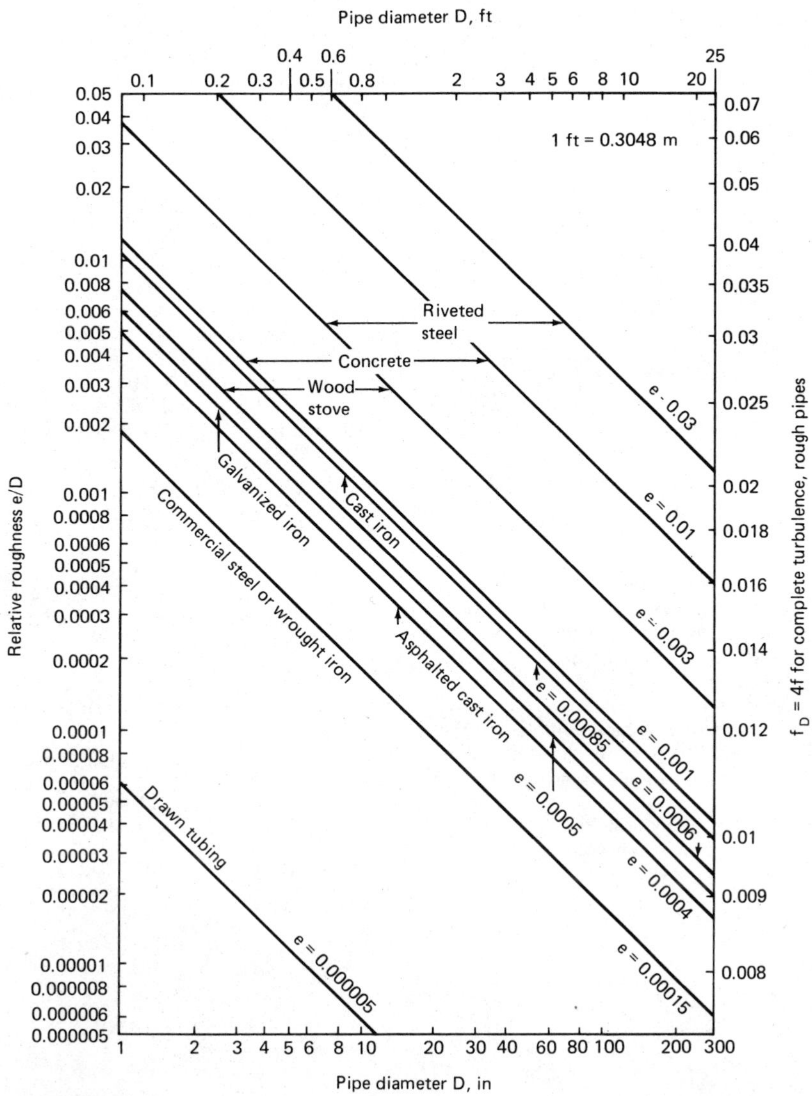

Fig. 4.11 Relative roughness of pipes. (*Ref. 36.*)

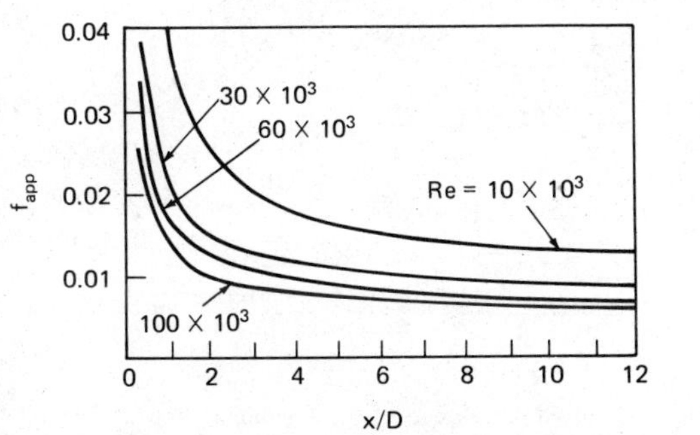

Fig. 4.12 Apparent friction coefficients in entrance region of circular tube with turbulent flow condition. (*Ref. 4.*)

increase. For more information regarding roughness effects see Ref. 55.

Concentric Annular Ducts The friction factor can be evaluated from the equation[36]

$$f = 0.085\,\mathrm{Re}^{-0.25} \qquad 5 \times 10^3 < \mathrm{Re} < 3 \times 10^5 \tag{4.58}$$

where Re is based on the hydraulic diameter $D_h = D_o - D_i$.

The fully developed turbulent Nusselt number at the inner wall, Nu_i, and at the outer wall, Nu_o, can be obtained by invoking Eqs. (4.38) and (4.39) introduced for laminar flow conditions and for similarly prescribed boundary conditions. The values of Nu_{ii} and Nu_{oo} as well as the coefficients of influence θ_i^* and θ_o^* are presented in Table 4.8 for radius

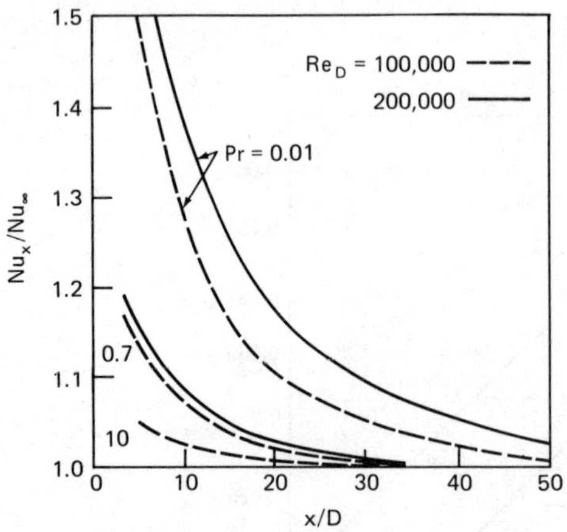

Fig. 4.13 Ratio of local to fully developed Nusselt numbers in entrance of circular tube with turbulent flow condition. (*Ref. 4.*)

ratios r^* of 0.20, 0.50, and 1.0 (parallel plates). For complete data see Refs. 36 and 59.

Longitudinal Flow over Circular Cylinders The friction and heat-transfer coefficients for fully developed turbulent flow parallel to a rod bundle arranged in a triangular array (see Fig. 4.4f) can be evaluated from the circular tube correlations and the curve presented in Fig. 4.15.[4] The rod bundle characteristic length is the hydraulic diameter D_h, computed from Eq (4.40). Nu_{cir} and f_{cir} are the Nusselt number and the Fanning friction factor, respectively, for fully developed flow inside a circular tube having a diameter equal to the rod bundle hydraulic diameter.

Other Noncircular Ducts For many noncircular ducts, the Fanning friction factor for turbulent flow can be obtained, with reasonable accuracy, using the circular tube correlations provided that the hydraulic diameter $D_h = 4r_h$ [see Eq. (4.14)] is employed to evaluate the Reynolds number.[4]

The above procedure somewhat overestimates the friction factor for passages with sharp corners, such as triangular ducts with small angles at one or two corners. For isosceles triangular ducts the following equation is recommended[56]:

$$f_D = 4f = C\,\text{Re}^{-0.25} \tag{4.59}$$

where Re is based on the hydraulic diameter and the parameter C is given in Fig. 4.16.

When the fluid Prandtl number is greater than 0.6, the Nusselt number for turbulent flow may be evaluated using the circular tube correlations and employing the hydraulic diameter in evaluating both Re and Nu.[4] This approximation works very well for rectangular ducts and reasonably well for equilateral triangles. However, as was seen for the friction factor, this procedure overestimates the heat-transfer coefficient for passages with sharp corners.[56–58]

4.7 COILED TUBES

The flow characterization and the associated convection heat-transfer coefficient in coiled tubes are governed by the flow Reynolds number $\text{Re} = u_m d/\nu$ and the ratio of the tube diameter d to the coiled diameter D_c. The product of these two dimensionless parameters is called the Dean number Dn and is defined by

$$\text{Dn} = \text{Re}\left(\frac{d}{D_c}\right)^{1/2} \tag{4.60}$$

The friction factor for laminar flow can be obtained from[60]

$$\frac{f}{f_{st}} = \left\{1.0 - \left[1.0 - \left(\frac{11.6}{\text{Dn}}\right)^{0.45}\right]^{2.22}\right\}^{-1} \qquad 11.6 < \text{Dn} < 2000 \tag{4.61}$$

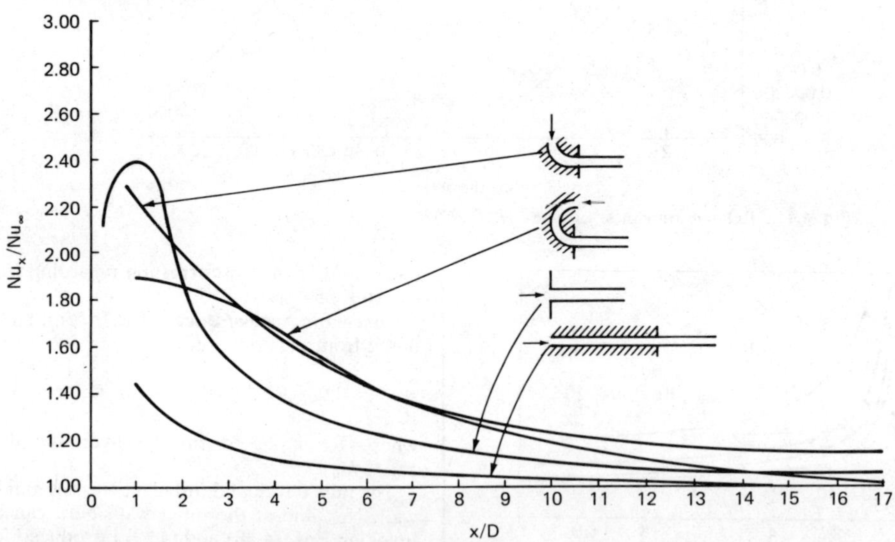

Fig. 4.14 Measured local Nusselt numbers in entry region of circular tube for various entry configurations, air with constant surface temperature. (*Ref. 4.*)

Table 4.8. Concentric Annular Ducts: Fundamental Solutions for Constant Heat Flux and Fully Developed Turbulent Flow*

$r^* = 0.20$, Heating from Outer Tube

	Re = 10^4		Re = 3×10^4		Re = 10^5		Re = 3×10^5		Re = 10^6	
Pr	Nu_{oo}	θ_o^*	Nu_{oo}	θ_o^*	Nu_{oo}	θ_o^*	Nu_{oo}	θ_o^*	Nu_{oo}	θ_o^*
0.001	5.83	0.140	5.92	0.144	6.10	0.151	6.30	0.154	6.92	0.153
0.03	6.22	0.140	7.55	0.140	12.7	0.125	26.3	0.098	80.0	0.074
0.7	29.4	0.063	64.3	0.055	165	0.049	397	0.044	1,070	0.040
1.0	35.5	0.051	80.0	0.046	206	0.042	504	0.039	1,390	0.035
3	60.0	0.026	145	0.026	390	0.024	980	0.024	2,760	0.023
10	98.0	0.013	243	0.013	680	0.012	1,750	0.012	4,980	0.012
100	205	0.003	520	0.003	1,500	0.003	4,000	0.003	12,000	0.003
1,000	380	0.001	980	0.001	2,830	0.001	7,500	0.001	22,500	0.001

$r^* = 0.20$, Heating from Core Tube

	Re = 10^4		Re = 3×10^4		Re = 10^5		Re = 3×10^5		Re = 10^6	
Pr	Nu_{ii}	θ_i^*	Nu_{ii}	θ_i^*	Nu_{ii}	θ_i^*	Nu_{ii}	θ_i^*	Nu_{ii}	θ_i^*
0.001	8.40	1.009	8.40	1.040	8.30	1.020	8.40	1.014	8.90	0.976
0.03	9.00	1.012	10.1	0.943	15.8	0.771	31.7	0.600	81.0	0.374
0.7	38.6	0.412	79.8	0.338	196	0.286	473	0.260	1,270	0.235
1.0	46.8	0.339	99.0	0.284	247	0.248	600	0.229	1,640	0.209
3	77.4	0.172	175	0.151	465	0.143	1,150	0.137	3,250	0.135
10	120	0.120	290	0.074	800	0.072	2,050	0.073	6,000	0.077
100	243	0.014	617	0.014	1,760	0.015	4,630	0.016	13,800	0.016
1,000	448	0.004	1,400	0.002	3,280	0.002	8,800	0.004	26,000	0.003

$r^* = 0.50$, Heating from Outer Tube

	Re = 10^4		Re = 3×10^4		Re = 10^5		Re = 3×10^5		Re = 10^4	
Pr	Nu_{oo}	θ_o^*	Nu_{oo}	θ_o^*	Nu_{oo}	θ_o^*	Nu_{oo}	θ_o^*	Nu_{oo}	θ_o^*
0.001	5.66	0.281	5.78	0.294	5.80	0.296	5.92	0.302	6.40	0.304
0.03	6.03	0.279	7.05	0.284	11.6	0.258	24.4	0.214	64.0	0.163
0.7	28.3	0.137	62.0	0.119	158	0.107	380	0.097	1,040	0.090
1.0	34.8	0.111	78.0	0.101	200	0.092	490	0.085	1,340	0.078
3	60.5	0.059	144	0.058	384	0.055	960	0.054	2,730	0.052
10	100	0.028	246	0.028	680	0.028	1,750	0.028	5,030	0.028
100	207	0.006	530	0.006	1,500	0.006	4,000	0.006	12,000	0.006
1,000	387	0.001	990	0.001	2,830	0.001	7,600	0.001	23,000	0.001

$r^* = 0.50$, Heating from Core Tube

	Re = 10^4		Re = 3×10^4		Re = 10^5		Re = 3×10^5		Re = 10^6	
Pr	Nu_{ii}	θ_i^*	Nu_{ii}	θ_i^*	Nu_{ii}	θ_i^*	Nu_{ii}	θ_i^*	Nu_{ii}	θ_i^*
0.001	6.28	0.620	6.30	0.632	6.30	0.651	6.40	0.659	6.75	0.644
0.03	6.75	0.627	7.53	0.598	12.0	0.533	24.8	0.430	65.5	0.333
0.7	30.9	0.300	66.0	0.258	166	0.225	400	0.206	1,080	0.185
1.0	38.2	0.247	83.5	0.218	212	0.208	520	0.183	1,420	0.170
3	66.8	0.129	152	0.121	402	0.115	1,010	0.114	2,870	0.111
10	106	0.059	260	0.059	715	0.059	1,850	0.059	5,400	0.061
100	220	0.006	558	0.006	1,600	0.006	4,250	0.007	12,600	0.007
1,000	408	0.002	1,040	0.002	3,000	0.002	8,000	0.002	24,000	0.002

Table 4.8. Concentric Annular Ducts: Fundamental Solutions for Constant Heat Flux and Fully Developed Turbulent Flow (*Continued*)

$r^* = 1.00$, Parallel Plates

	$Re = 10^4$		$Re = 3 \times 10^4$		$Re = 10^5$		$Re = 3 \times 10^5$		$Re = 10^4$	
Pr	Nu	θ*	Nu	θ*	Nu	θ*	Nu	θ*	Nu	θ*
0.001	5.70	0.428	5.78	0.445	5.80	0.456	5.88	0.460	6.23	0.460
0.03	6.10	0.428	6.90	0.428	11.0	0.390	23.0	0.330	61.2	0.255
0.7	27.8	0.220	61.2	0.192	155	0.170	378	0.156	1,030	0.142
1.0	35.0	0.182	76.8	0.162	197	0.148	486	0.138	1,340	0.128
3	60.8	0.095	142	0.092	380	0.089	966	0.087	2,700	0.084
10	101	0.045	214	0.045	680	0.045	1,760	0.045	5,080	0.046
100	210	0.009	514	0.009	1,520	0.010	4,030	0.010	12,000	0.011
1,000	390	0.002	997	0.002	2,880	0.002	7,650	0.002	23,000	0.002

*For use in Eqs. (4.38) and (4.39).
SOURCE: Ref. 4.

where f_{st} is the straight-tube friction factor, $f_{st} = 16/\mathrm{Re}$. For turbulent flow the following equation can be used[61]:

$$f\left(\frac{D_c}{d}\right)^{0.5} = 0.00725 + 0.076\left[\mathrm{Re}\left(\frac{d}{D_c}\right)^2\right]^{-0.25} \qquad (4.62)$$

Equation (4.62) is good for $0.034 < \mathrm{Re}\,(d/D_c)^2 < 300$. For $\mathrm{Re}\,(d/D_c)^2 < 0.034$, the friction factor of a curved pipe nearly coincides with that of a straight pipe. For $\mathrm{Re}\,(d/D_c)^2 > 6$, use the equation

$$f = \frac{0.079\,(d/D_c)^{0.5}}{[\mathrm{Re}\,(d/D_c)^2]^{0.2}} \qquad (4.63)$$

The transition Reynolds number can be determined from[61]

$$\mathrm{Re}_{crit} = 2 \times 10^4 \left(\frac{d}{D_c}\right)^{0.32} \qquad (4.64)$$

which is good for the range of $15 < (D_c/d) < 860$. For $(D_c/d) > 860$, Re_{crit} is essentially similar to the critical Reynolds number for straight pipe.

The heat-transfer coefficient for laminar flow, with $50 < \mathrm{Dn} < 1200$, can be determined from[36]

$$\mathrm{Nu} = (0.76 + 0.065\sqrt{\mathrm{Dn}})\,\mathrm{Pr}^{0.175} \qquad 5 < \mathrm{Pr} < 175 \quad (4.65)$$

$$\mathrm{Nu} = 0.913\,\mathrm{Dn}^{0.476}\,\mathrm{Pr}^{0.2} \qquad 0.7 < \mathrm{Pr} < 5 \quad (4.66)$$

For turbulent flow the following equation can be used[36,60]:

$$\mathrm{Nu} = 0.023\,\mathrm{Re}^{0.85}\left(\frac{d}{D_c}\right)^{0.1}\mathrm{Pr}^{0.4} \qquad (4.67)$$

The boundary condition has no appreciable effect on the Nusselt number. The entry length is much shorter than that

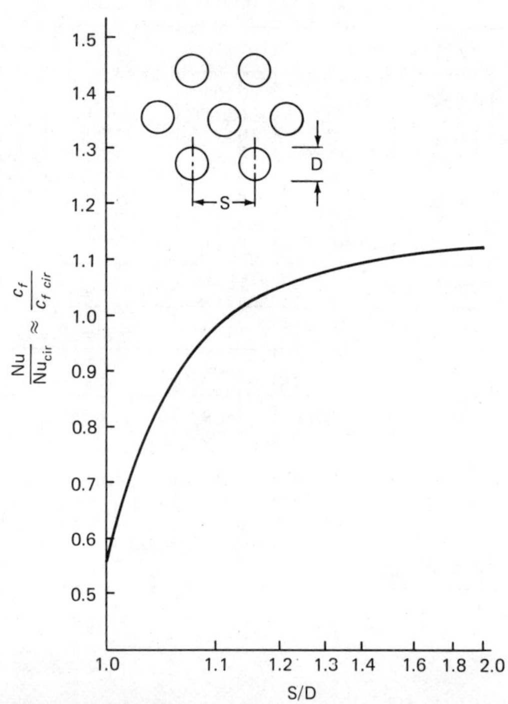

Fig. 4.15 Friction and heat-transfer ratios for circular tube, fully developed turbulent flow parallel to a bank of rods. (*Ref. 4.*)

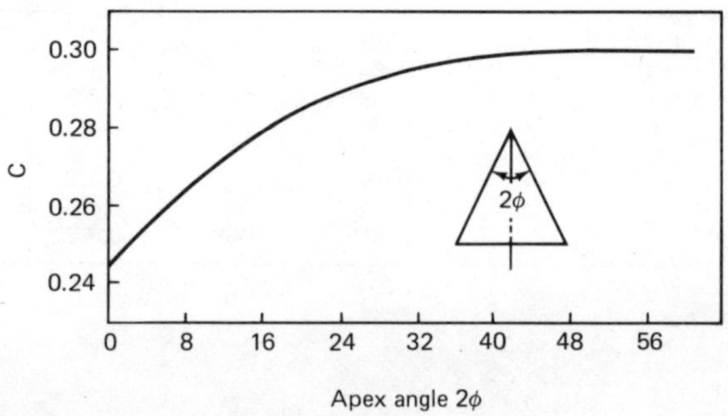

Fig. 4.16 Correlation parameter C for turbulent friction factor, isosceles triangular duct; Eq. (4.59).

for a straight tube. Therefore its effects can be neglected for design purposes.

4.8 EFFECTS OF PROPERTY VARIATIONS

When appreciable heat transfer occurs, large temperature differences may be present and the fluid properties may vary within the flow field. These property variations may distort the velocity and temperature profiles. Thus the pressure loss and the heat-transfer rates may differ from those of near isothermal flow.

For liquids, where the viscosity variation is responsible for most of the effect, the following formulas are recommended to correct the data that are obtained based on the constant-property assumption:

$$\frac{\mathrm{Nu}}{\mathrm{Nu}_{\mathrm{cp}}} = \left(\frac{\mu_w}{\mu_m}\right)^n \tag{4.68}$$

and

$$\frac{f}{f_{\mathrm{cp}}} = \left(\frac{\mu_w}{\mu_m}\right)^m \tag{4.69}$$

For gases, the property variations can be taken into account by the absolute temperature ratio as follows:

$$\frac{\mathrm{Nu}}{\mathrm{Nu}_{\mathrm{cp}}} = \left(\frac{T_w}{T_m}\right)^n \tag{4.70}$$

and

$$\frac{f}{f_{\mathrm{cp}}} = \left(\frac{T_w}{T_m}\right)^m \tag{4.71}$$

where subscripts w and m refer to the properties evaluated at wall temperature and mixed mean temperature, respectively, and cp refers to the constant-property solutions. The values of the exponents m and n are as follows[4].

Liquids

1. Laminar flow in tubes,

$$n = -0.14 \qquad m = 0.5 \qquad (\text{cooling}, T_w < T_b)$$

$$n = -0.14 \qquad m = 0.58 \qquad (\text{heating}, T_w > T_b)$$

2. Turbulent flow in tubes,

$$n = -0.25 \qquad m = 0.25 \qquad (\text{cooling})$$

$$n = -0.11 \qquad m = 0.25 \qquad (\text{heating})$$

Gases

1. Laminar flow in tubes,

$$n = 0.0 \qquad m = 1.0 \qquad (\text{cooling and heating})$$

2. Turbulent flow in tubes,

$$n = 0.0 \qquad m = -0.1 \qquad (\text{cooling})$$

$$n = -0.5 \qquad m = -0.1 \qquad (\text{heating})$$

4.9 EXTERNAL FLOW

When fluid is flowing over a surface, the boundary layers develop freely, without constraints imposed by adjacent surfaces. Therefore there always exists a region of the flow outside the boundary layer in which velocity and temperature variations are negligible—the major velocity and temperature changes occur in the boundary layer close to the surface. The drag force and the heat-transfer rate are obtained by analyzing the intensity of the momentum and the heat diffusion and convection within this layer.

The data presented in this section are related to the low-speed flow, forced convection, where flow is considered to be incompressible (Mach number $M < 0.4$) and friction heating can be neglected [Eckert number $\mathrm{Ec} = u_\infty^2/C_p(T_w - T_\infty) << 1$]. For high-speed forced convection see Ref. 63.

Flow Past a Flat Plate

The local drag coefficient is defined as

$$C_{D,x} = \frac{\tau_x}{\rho u_\infty^2/2g_c} \tag{4.72}$$

where τ_x is the local shear stress and u_∞ is the free-stream fluid velocity. Equation (4.72) can be integrated over a plate of length L to obtain the average drag coefficient

$$C_D = \frac{1}{L}\int_0^L C_{D,x}\, dx \tag{4.73}$$

Both drag coefficients $C_{D,x}$ and C_D can be correlated as a function of the Reynolds number $\mathrm{Re}_x = u_\infty x/\nu$.

The heat-transfer coefficients Nu are found to be correlated as a function of Reynolds and Prandtl numbers. A list of useful correlation equations for a flat plate is presented in Table 4.9.

Flow Past a Single Cylinder

The drag coefficient C_D is defined as

$$C_D = \frac{F_D}{A_f(\rho u_\infty^2/2g_c)} \tag{4.74}$$

where F_D is the drag force per unit length and A_f is the cylinder frontal area for a unit length of cylinder $A_f = D$. The drag coefficient is a function of the Reynolds number and is shown in Fig. 4.17.

The overall heat-transfer coefficient may be obtained from the correlation[65]

$$\mathrm{Nu} = C\,\mathrm{Re}^m\,\mathrm{Pr}^n \left(\frac{\mathrm{Pr}}{\mathrm{Pr}_w}\right)^{0.25} \tag{4.75}$$

where the values of C and m are given in Table 4.10 and $n = 0.37$ for $0.6 < \mathrm{Pr} < 10$ and $n = 0.36$ for $\mathrm{Pr} > 10$.

In the Reynolds number range of $100 < \mathrm{Re} < 10^5$ the following equation can be used[64]:

$$\mathrm{Nu} = (0.4\,\mathrm{Re}^{0.5} + 0.06\,\mathrm{Re}^{2/3})\,\mathrm{Pr}^{0.4}\left(\frac{\mu}{\mu_w}\right)^{0.25} \tag{4.76}$$

Equation (4.76) is valid for $0.65 < \mathrm{Pr} < 300$ and $0.25 < \mu/\mu_w < 5.2$. In Eqs. (4.75) and (4.76), all fluid properties

Table 4.9. Summary of Correlation Equations for Flow Past a Flat Plate*

Coefficient	Correlation	Conditions
	Laminar	
Local drag coefficient	$C_{D,x} = 0.664\ \mathrm{Re}_x^{-0.5}$	$\mathrm{Re}_x < 5 \times 10^5$
Average drag coefficient	$C_D = 1.328\ \mathrm{Re}_L^{-0.5}$	$\mathrm{Re}_L < 5 \times 10^5$
Local Nusselt number	$\mathrm{Nu}_x = 0.332\ \mathrm{Re}_x^{0.5}\ \mathrm{Pr}^{1/3}$	$\mathrm{Pr} \geq 0.6$, $\mathrm{Re}_x < 5 \times 10^5$
Local Nusselt number	$\mathrm{Nu}_x = 0.565(\mathrm{Re}_x\ \mathrm{Pr})^{0.5}$	$\mathrm{Pr} \leq 0.1$, $\mathrm{Re}_x < 5 \times 10^5$
Local Nusselt number	$\mathrm{Nu}_x = 0.453\ \mathrm{Re}_x^{0.5}\ \mathrm{Pr}^{1/3}$	$\mathrm{Pr} > 0.6$, $\mathrm{Re}_x < 5 \times 10^5$ constant heat flux
Average Nusselt number	$\mathrm{Nu}_L = 0.664\ \mathrm{Re}_L^{0.5}\ \mathrm{Pr}^{1/3}$	$\mathrm{Pr} \geq 0.6$, $\mathrm{Re}_L < 5 \times 10^5$
	Turbulent	
Local drag coefficient	$C_{D,x} = 0.0592\ \mathrm{Re}_x^{-0.2}$	$5 \times 10^5 < \mathrm{Re}_x < 10^7$
Local drag coefficient	$C_{D,x} = 0.370(\log \mathrm{Re}_x)^{-2.584}$	$10^7 < \mathrm{Re}_x < 10^9$
Average drag coefficient	$C_D = \dfrac{0.074}{\mathrm{Re}_L^{0.2}} - \dfrac{1700}{\mathrm{Re}_L}$	$5 \times 10^5 < \mathrm{Re}_L < 10^7$
Average drag coefficient	$C_D = \dfrac{0.455}{(\log \mathrm{Re}_L)^{2.584}} - \dfrac{1700}{\mathrm{Re}_L}$	$10^7 < \mathrm{Re}_L < 10^9$
Local Nusselt number	$\mathrm{Nu}_x = 0.029\ \mathrm{Re}_x^{0.8}\ \mathrm{Pr}^{0.43}$	$5 \times 10^5 < \mathrm{Re}_x < 10^7$
Local Nusselt number	$\mathrm{Nu}_x = 0.185(\log \mathrm{Re}_x)^{-2.584}\mathrm{Re}_x\ \mathrm{Pr}^{1/3}$	$5 \times 10^7 < \mathrm{Re}_x < 10^9$, $\mathrm{Pr} > 0.6$
Average Nusselt number	$\mathrm{Nu}_L = 0.036(\mathrm{Re}_L^{0.8} - 9200) \times \mathrm{Pr}^{0.43}\left(\dfrac{\mu_\infty}{\mu_w}\right)^{0.25}$	$5 \times 10^5 < \mathrm{Re}_L < 10^7$, $\mathrm{Pr} > 0.6$

*These correlation equations are taken from Refs. 7, 10, and 64. All properties are evaluated at the film temperature $T_f = (T_\infty + T_w)/2$, except μ_∞ and μ_w, which are obtained at stream and wall temperatures, respectively.

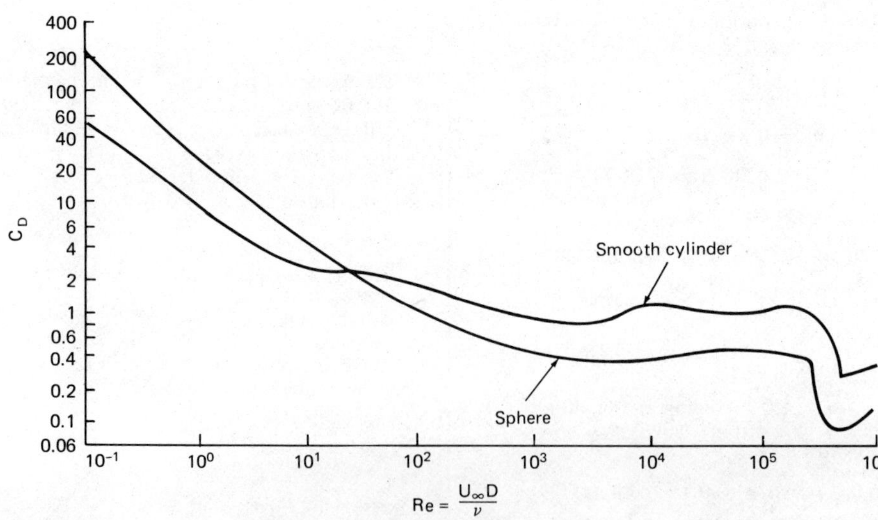

Fig. 4.17 Drag coefficient for smooth circular cylinder in cross flow and for sphere. (*Ref. 17.*)

Table 4.10. Constants for Eq. (4.75), Cylinder in Cross Flow

$\mathrm{Re} = \dfrac{\rho u_\infty D}{\mu}$	C	m
1–40	0.75	0.4
40–10^3	0.51	0.5
10^3–2×10^5	0.26	0.6
2×10^5–10^6	0.076	0.7

are evaluated at T_∞ except Pr_w and μ_w, which are evaluated at wall temperature.

A comprehensive formula that can also be used for liquid metals is[66]

$$\mathrm{Nu} = 0.3 + \frac{0.62\ \mathrm{Re}^{0.5}\ \mathrm{Pr}^{1/3}}{[1 + (0.4/\mathrm{Pr})^{2/3}]^{3/4}} \times \left[1 + \left(\frac{\mathrm{Re}}{282{,}000}\right)^{5/8}\right]^{4/8} \tag{4.77}$$

Equation (4.77) is valid for the range of Pr Re > 0.2, 100 < Re < 10^7. All properties are evaluated at film temperature $T_f = (T_\infty + T_w)/2$.

Flow Past a Single Sphere

The drag coefficient for a sphere is defined as

$$C_D = \frac{F_D}{(\pi D^2/4)\ \rho u_\infty^2/2g_c} \tag{4.78}$$

Figure 4.17 presents the drag coefficient versus the Reynolds number.

The overall Nusselt number can be obtained from[64]

$$\mathrm{Nu} = 2 + (0.4\ \mathrm{Re}^{0.5} + 0.06\ \mathrm{Re}^{2/3})\ \mathrm{Pr}^{0.4} \left(\frac{\mu}{\mu_w}\right)^{0.25} \tag{4.79}$$

where all fluid properties are evaluated at T_∞ except μ_w, which is evaluated at T_w. The correlation is valid for 3.5 < Re < 10^5, 0.7 < Pr < 380, and 1.0 < μ/μ_w < 3.2.

For liquid metals, Pr < 0.03, the following equation may be used[67] for $4 \times 10^4 < \mathrm{Re} < 2 \times 10^5$:

$$\mathrm{Nu} = 2 + 0.386(\mathrm{Re\ Pr})^{0.5} \tag{4.80}$$

Flow Past Noncircular Cylinders

The average Nusselt numbers for the flow of gases over noncircular cylinders are correlated as[68]

$$\mathrm{Nu} = B\ \mathrm{Re}^n\ \mathrm{Pr}^{0.33} \tag{4.81}$$

where the constants B and n are given in Table 4.11. All the fluid properties are evaluated at fluid temperature $T_f = (T_\infty + T_w)/2$.

Table 4.12 presents the drag coefficients for some common geometries.[69]

Flow Past Tube Banks, Cross Flow

A schematic diagram of the tube bank arrangement is shown in Fig. 4.18. The overall mean Nusselt numbers are correlated as[65]

$$\mathrm{Nu} = C\ \mathrm{Re}_{\max}^n\ \mathrm{Pr}^{0.36} \left(\frac{\mathrm{Pr}}{\mathrm{Pr}_w}\right)^{0.25} \tag{4.82}$$

where the fluid properties are evaluated at T_∞ except Pr_w. The Reynolds number $\mathrm{Re}_{\max}$ is defined as

$$\mathrm{Re}_{\max} = \frac{\rho u_{\max} D}{\mu} \tag{4.83}$$

where $u_{\max}$ for the in-line arrangement is

$$u_{\max} = \frac{u_\infty S_T}{S_T - D} \tag{4.84}$$

and for the staggered arrangement,

$$u_{\max} = \frac{u_\infty S_T}{S_T - D} \qquad S_D > \frac{S_T + D}{2} \tag{4.85a}$$

$$u_{\max} = \frac{u_\infty S_T}{2(S_D - D)} \qquad S_D < \frac{S_T + D}{2} \tag{4.85b}$$

Table 4.11. Constants for Use in Eq. (4.81), Flow Past Noncircular Cylinders

Flow direction and profile	Re_D	n	B
	5000–100,000	0.588	0.246
	2500–15,000	0.612	0.248
	2500–7500	0.624	0.289
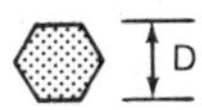	5000–100,000	0.638	0.153
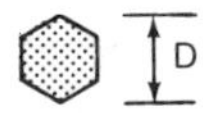	5000–19,500	0.638	0.160
	5000–100,000	0.675	0.102
	2500–8000	0.699	0.177
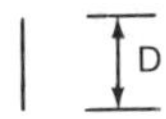	4000–15,000	0.731	0.228
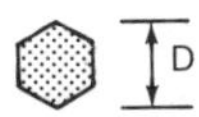	19,500–100,000	0.782	0.0385
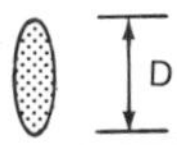	3000–15,000	0.804	0.094

SOURCE: Ref. 68.

Table 4.12. Drag Coefficients

Geometry	C_D	Reynolds number range	Characteristic length	Characteristic area
Flat plate (normal) 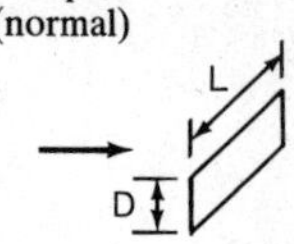	L/D 1 1.18 5 1.2 10 1.3 20 1.5 30 1.6 ∞ 1.95	$Re > 10^3$	D	Plate surface area
Circular disk (normal) 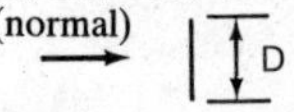	1.17	$Re > 10^3$	D	Disk surface area
Sphere 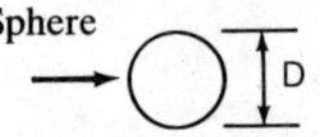	$24/Re_D$ 0.47 0.2	$Re < 1$ $10^3 < Re < 3 \times 10^5$ $Re > 3 \times 10^5$	D	Projected area
Hollow hemisphere 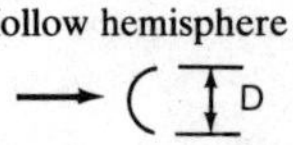 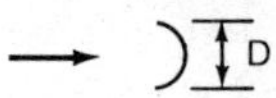	0.34 1.42	$10^4 < Re < 10^6$ $10^4 < Re < 10^6$	D	Projected area
Solid hemisphere 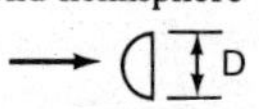 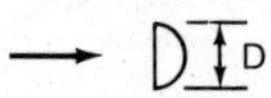	0.42 1.17	$10^4 < Re < 10^6$ $10^4 < Re < 10^6$	D	Projected area
Circular cylinder 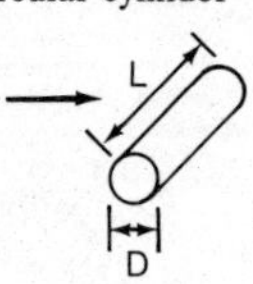	L/D 1 0.63 5 0.8 10 0.83 20 0.93 30 1.0 ∞ 1.2	$10^3 < Re < 10^5$	D	Projected area
Square cylinder 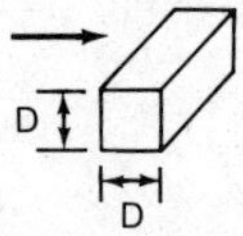	2.0	3.5×10^4	D	Projected area

SOURCE: Ref. 69.

The constants C and n are given in Table 4.13. Equation (4.82) is for a deep bank (i.e., 20 or more rows of tubes are present). For a number of rows N less than 20 use Fig. 4.19 to correct the results of Eq. (4.82).

For calculating the overall heat-transfer rate, the concept of log-mean temperature difference (LMTD) is valid. For example, if the fluid enters the tube bank at T_1 and leaves at T_2, the overall heat-transfer rate is

$$Q = mC_p(T_2 - T_1) = \left(\frac{\mathrm{Nu}\, k}{D}\right) NM\pi DL(\Delta T)_{av} \tag{4.86}$$

where N is the number of rows, M is the number of tubes per row, and $(\Delta T)_{av}$ is computed from

$$(\Delta T)_{av} = \frac{(T_w - T_1) - (T_w - T_2)}{\ln[(T_w - T_1)/(T_w - T_2)]} \tag{4.87}$$

The pressure losses across a tube bank in cross flow are correlated by[70]

$$\Delta p = fN\left(\frac{\rho u_{max}^2}{2}\right)\left(\frac{\mu_w}{\mu}\right)^{0.14} \tag{4.88}$$

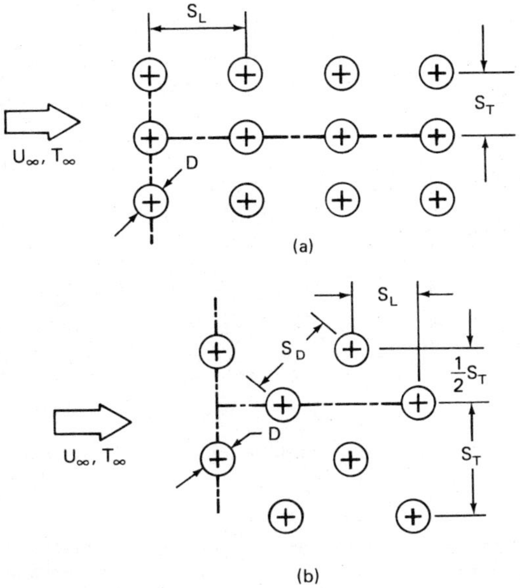

Fig. 4.18 Flow past tube banks. (*a*) In line; (*b*) staggered.

Table 4.13. Constants for Use in Eq. (4.82), Tube Bank in Cross Flow

Reynolds number range Re_{max}	C	n
Staggered		
$100\text{–}2 \times 10^5$, $S_T < 2S_L$	$0.35(S_T/S_L)^{0.2}$	0.6
$100\text{–}2 \times 10^5$, $S_T > 2S_L$	0.4	0.6
$2 \times 10^5\text{–}2 \times 10^6$	0.022	0.84
In line		
$100\text{–}2 \times 10^5$	0.27	0.63
$2 \times 10^5\text{–}2 \times 10^6$	0.021	0.84

where N is the number of rows and f can be computed from the equation

$$f = \left[C_1 + \frac{C_2}{(S_T/D - 1.0)^n}\right] Re_{max}^{-m} \tag{4.89}$$

where the constants are as follows.

1. For staggered arrangement,

$$C_1 = 1.0 \qquad C_2 = 0.472 \qquad n = 1.08 \qquad m = 0.16$$

2. For in-line arrangement,

$$C_1 = 0.176 \qquad C_2 = \frac{0.32 S_L}{D}$$

$$n = 0.43 + \frac{1.13D}{S_T} \qquad m = 0.15$$

Flow Past Pin Fins, Cross Flow

A diagram describing the pin fin arrays in cross flow is presented in Fig. 4.20. This geometry for $1 < h/D < 3$, $0.29 < h/H < 0.875$, and $1000 < Re_{max} < 9000$ has been investigated.[71,72] The reported heat-transfer results are somewhat lower than those found for tube banks, ranging from 6 percent for the tallest pin, $h/H = 0.875$, to 23 percent for the shortest pin, $h/H = 0.29$.

Tube bank Nusselt numbers may be used as a first approximation for engineering calculations. Based on the data furnished in Ref. 71, the tube bank Nusselt numbers Nu_{tb} may be corrected by

$$Nu_{pf} = Nu_{tb}\left[0.67 + 0.33\left(\frac{h}{H}\right)\right] \qquad \text{for } \frac{h}{H} \geq 0.3 \tag{4.90}$$

where Nu_{pf} is the Nusselt number of the pin fin array. For the computation of the tube bank Nusselt number, the required Reynolds number must be evaluated by

$$Re_{max} = \frac{\rho u_{max} D}{\mu} \tag{4.91}$$

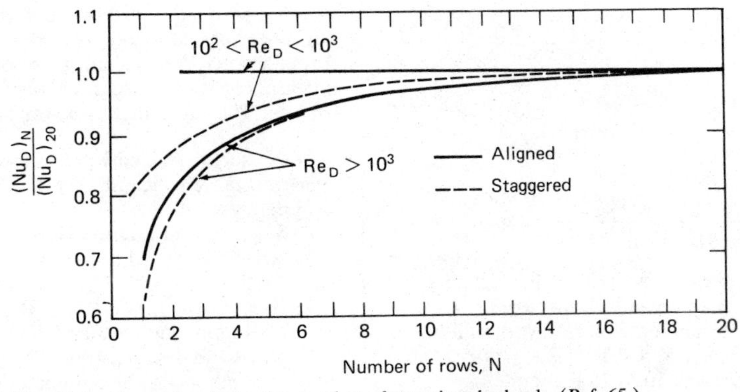

Fig. 4.19 Correction factor for number of rows in tube bank. (*Ref. 65.*)

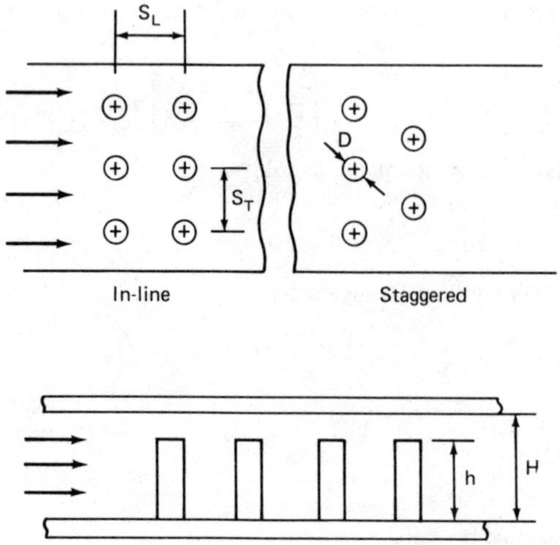

Fig. 4.20 Schematic diagram of pin fin arrays. (*a*) In line; (*b*) staggered.

where u_{max} is obtained from

$$u_{max} = \frac{u_m A_{duct}}{A_{min}} \tag{4.92}$$

where u_m is the mean velocity in the duct and A_{min} is the minimum cross-sectional area. (A_{min} includes both the clearance space above the pins and the intercylinder spaces.) Equation (4.90) should be used within the range of reported data.

Heat-transfer and friction factor data for several pin fin arrangements having $h/H = 1$ are presented in Ref. 73.

Flow Past Finned Tubes

The heat-transfer and pressure loss characteristics of one- and two-row arrays of annular finned tubes, as shown in Fig. 4.21, are reported in Ref. 74. For single-row finned tubes the heat-transfer data were correlated by[74]

$$\mathrm{Nu} = \frac{hD}{k} = 0.0529\,\mathrm{Re}_{max}^{0.704} \qquad \mathrm{Pr} = 0.72 \tag{4.93}$$

where D is the diameter of the base tube and Re_{max} is defined as

$$\mathrm{Re}_{max} = \frac{\rho u_{max} D}{\mu} \tag{4.94}$$

u_{max} being obtained by

$$u_{max} = \left[\frac{(s+t)S_T}{(s+t)\Delta + s(D_f - D)}\right] u_\infty \tag{4.95}$$

Here s is the spacing between fins, t is the fin thickness, D_f is the fin diameter, and Δ and S_T are shown in Fig. 4.21.

Equation (4.93) fits the data for $S_T < 1.52$ very well in the Reynolds number range of $1.5 \times 10^3 < \mathrm{Re}_{max} < 8 \times 10^4$. For $S_T > 2.0$, Nu increases by 10 to 15 percent.

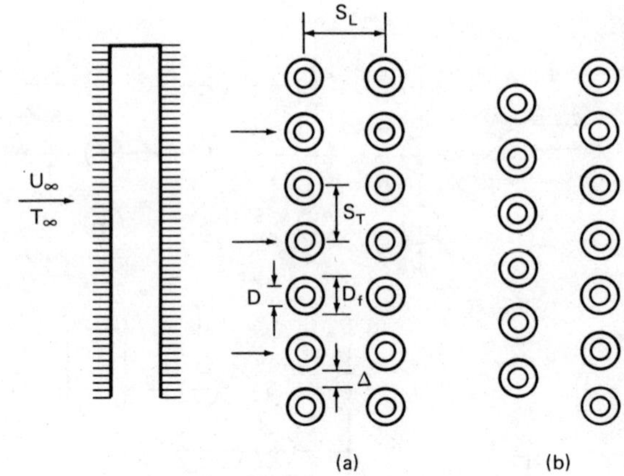

Fig. 4.21 Flow past finned tubes. (*a*) In line; (*b*) staggered.

For two-row in-line arrays, the Nusselt numbers of the first-row finned tubes are insensitive to the presence of the second row and can be obtained from Eq. (4.93). For practical purposes, the Nusselt number of the second row may be obtained from the following equations.

1. For $S_L/D_f > 2$,

$$\mathrm{Nu}_2 = 1.35\,\mathrm{Nu}_1 \qquad 10^4 < \mathrm{Re}_{max} < 6 \times 10^4$$

2. For $S_L/D_f < 2$,

$$\mathrm{Nu}_2 = 1.10\,\mathrm{Nu}_1 \qquad 10^4 < \mathrm{Re}_{max} < 2 \times 10^4$$

$$\mathrm{Nu}_2 = 1.30\,\mathrm{Nu}_1 \qquad 2 \times 10^4 < \mathrm{Re}_{max} < 6 \times 10^4$$

For two-row staggered arrays the Nusselt number of the first row remains essentially unchanged for $S_L/D_f < 1.7$ and for $S_L/D_f > 3$. The Nusselt number of the second row, Nu_2, is about 40 percent higher than Nu_1.

Heat-transfer and friction factors of several finned tube configurations are presented in Ref. 73.

4.10 JET IMPINGEMENT HEAT TRANSFER

Heating or cooling of a large surface area is often carried out in devices consisting of arrays of round or slot nozzles, through which fluid, usually air or other gases, impinges upon the surface. Such processes allow for short flow paths on the surface. Therefore high heat-transfer coefficients are attainable.

In this section the correlations for obtaining integral mean heat-transfer coefficients for the following cases are presented:

1. Single round nozzle (SRN)
2. Arrays of round nozzles (ARN)
3. Single-slot nozzle (SSN)
4. Arrays of slot nozzles (ASN)

A schematic diagram of each case is shown in Fig. 4.22.

Single Round Nozzle (SRN) The integral mean value of the Nusselt numbers can be obtained by the empirical equation[62]

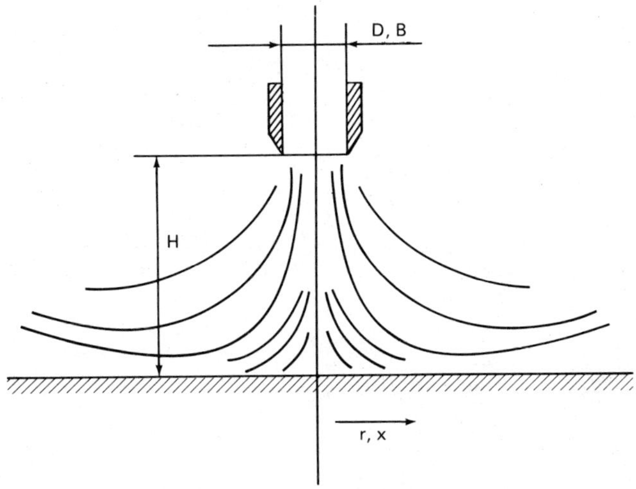

Flow field of impinging flow

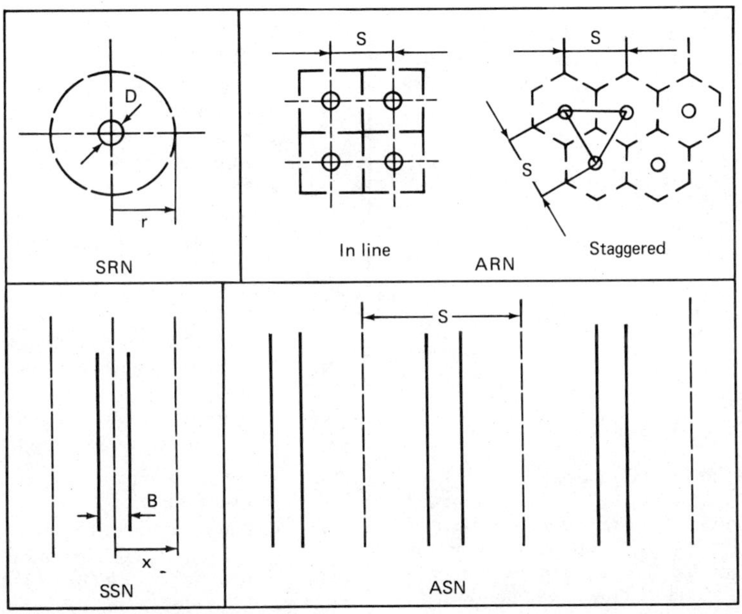

Fig. 4.22 Schematic diagram of nozzle arrangement.

$$\mathrm{Nu_{SRN}} = \frac{D}{r}\left[\frac{1 - 1.1D/r}{1 + 0.1(H/D - 6)D/r}\right] \times \left[2\,\mathrm{Re}^{0.5}\left(1 + \frac{\mathrm{Re}^{0.55}}{200}\right)^{0.5}\right]\mathrm{Pr}^{0.42} \quad (4.96)$$

where, for round nozzles, $\mathrm{Nu} = hD/k$ and $\mathrm{Re} = u_m D/\nu$ (nozzle Reynolds number). The range of validity of Eq. (4.96) is $2000 < \mathrm{Re} < 4 \times 10^5$, $2.5 < r/D < 7.5$, and $2 < H/D < 12$.

Arrays of Round Nozzles (ARN) The average Nusselt number for the case of ARN, and for both in-line and staggered arrangements, can be calculated from[62]

$$\mathrm{Nu_{ARN}} = G\left(\frac{H}{D}, A_r\right) \times \sqrt{A_r}\left[\frac{1 + 2.2\sqrt{A_r}}{1 + 0.2(H/D - 6)\sqrt{A_r}}\right]\mathrm{Re}^{2/3}\,\mathrm{Pr}^{0.42} \quad (4.97)$$

where $A_r = (\pi D^2/4)/A_{\text{square(hexagon)}}$ (see Fig. 4.22), and the function $G(H/D, A_r)$ is defined as

$$G\left(\frac{H}{D}, A_r\right) = \left[1 + \left(\frac{H/D}{0.6/\sqrt{A_r}}\right)^6\right]^{-0.05} \qquad (4.98)$$

Equation (4.97) is valid for the range of $2000 < \text{Re} < 10^5$, $0.004 < A_r < 0.04$, and $2 < H/D < 12$. In the mentioned range, the results of Eq. (4.97) are within 20 percent of experimental data.

Single-Slot Nozzle (SSN) The integral mean Nusselt number can be evaluated from the empirical equation[62]

$$\text{Nu}_{\text{SSN}} = \frac{1.53}{x/2B + H/2B + 1.39} \text{Re}^m \text{Pr}^{0.42} \qquad (4.99)$$

where, for slot nozzles, $\text{Nu} = h(2B)/k$ and $\text{Re} = u_m(2B)/\nu$. The exponent m is calculated from

$$m = 0.695 - \left[\frac{x}{2B} + \left(\frac{H}{2B}\right)^{1.33} + 3.06\right]^{-1} \qquad (4.100)$$

The range of validity of Eq. (4.99) is $3000 < \text{Re} < 9 \times 10^4$, $2 < x/2B < 25$, and $2 < H/2B < 10$. In this range, Eq. (4.99) fits the experimental data within 15 percent.

Arrays of Slot Nozzles (ASN) The following correlation equation may be used[62] to obtain the average Nusselt number $\text{Nu} = h(2B)/k$,

$$\text{Nu}_{\text{ASN}} = \frac{2}{3} C_r^{3/4} \left(\frac{2\,\text{Re}}{A_r/C_r + C_r/A_r}\right)^{2/3} \text{Pr}^{0.42} \qquad (4.101)$$

where, for arrays of slot nozzles, A_r is evaluated by (see Fig. 4.22)

$$A_r = \frac{B}{S} \qquad (4.102)$$

and C_r is computed from

$$C_r = \left[60 + 4\left(\frac{H}{2B} - 2\right)^2\right]^{-0.5} \qquad (4.103)$$

Equation (4.101) is valid for the range of $1500 < \text{Re} < 4 \times 10^4$, $0.008 < A_r < 2.5C_r$, and $1 < H/2B < 40$. Equation (4.101) fits the experimental data within 15 percent.

Equations (4.96), (4.97), (4.99), and (4.101) can be used for Prandtl numbers larger than 0.5 and all of the fluid properties must be evaluated at the fluid mean temperature.

NOMENCLATURE

A = surface area
A_c = cross-sectional area of duct
A_f = frontal area
C_D = drag coefficient
$C_{D,x}$ = local drag coefficient
C_p = specific heat at constant pressure
d = diameter of tube in coil
D = circular duct diameter
D_c = diameter of curvature of coiled tube
D_f = fin diameter
D_h = hydraulic diameter, $= 4r_h$ [Eq. (4.14)]
D_i = inner diameter of concentric annular duct
D_o = outer diameter of concentric annular duct
Dn = Dean number [Eq. (4.60)]
E = modulus of elasticity
Ec = Eckert number, $u_\infty^2/C_p(T_w - T_\infty)$
f = flow-length-average Fanning friction factor
f_{app} = apparent Fanning friction factor
f_D = Darcy friction factor $= 4f$
f_{st} = straight-tube friction factor [Eq. (4.61)]
F_D = drag force
g_c = acceleration of gravity
G = fluid mass velocity
h = heat-transfer coefficient
$\bar{h}$ = area-averaged heat-transfer coefficient [Eq. (4.7)]
h_x = peripherally average, but axially local heat-transfer coefficient
k = thermal conductivity
k_w = thermal conductivity of wall material
K_f = pressure loss constant for fully developed laminar flow in duct [Eq. (4.29)]
L = duct length
L_c = characteristic length
L_{he} = hydrodynamic entry length
M = Mach number, ratio of flow velocity to velocity of sound in fluid
Nu = Nusselt number [Eqs. (4.21) and (4.22)]
Nu_H = Nusselt number for constant heat flux in flow direction with uniform wall temperature at any cross section
Nu_{H1} = Nusselt number for constant heat flux both axially and peripherally
Nu_i = Nusselt number at inner wall of annular duct
Nu_{ii} = Nusselt number at inner wall when outer wall is insulated
Nu_o = Nusselt number at outer wall of annular duct
$\overline{\text{Nu}}_o$ = overall Nusselt number [Eq. (4.34)]
Nu_{oo} = Nusselt number at outer wall when inner wall is insulated
Nu_T = Nusselt number for uniform wall temperature
Nu_{T1} = Nusselt number for uniform axial wall temperature with finite normal wall thermal resistance
Nu_x = local Nusselt number
Nu_∞ = fully developed Nusselt number
p = pressure
P = wetted perimeter
Pe = Peclet number [Eq. (4.26)]
Pr = Prandtl number [Eq. (4.2)]
q'' = heat flux, heat-transfer rate per unit of area
q_i'' = heat flux at inner diameter of concentric annular duct
q_o'' = heat flux at outer diameter of concentric annular duct
Q = overall heat transfer [Eq. (4.8)]
r = radius of circular duct
r_i = radius of inner duct

r_h = hydraulic radius [Eq. (4.14)]
r_o = radius of outer duct
r^* = radius ratio for annular duct, = r_i/r_o
R_w = wall thermal resistance [Eq. (4.34)]
Re = Reynolds number [Eq. (4.3)]
s = spacing between fins [Eq. (4.95)]
S = tube bundle pitch
T = temperature
T_a = ambient fluid temperature
T_b = fluid bulk mean temperature
$T_{b,x}$ = axially local fluid bulk mean temperature
T_f = film temperature
T_{fl} = fluid temperature
T_m = mixed mean temperature
T_w = wall temperature
$T_{w,x}$ = axially wall mean temperature
T_∞ = free-stream fluid temperature
u_m = fluid mean axial velocity in duct [Eq. (4.12)]
u_{max} = fluid velocity through minimum flow area
U_o = overall heat-transfer coefficient [Eqs. (4.30) and (4.31)]
U_w = wall thermal conductance [Eq. (4.32)]
U_∞ = free-stream fluid velocity
V_c = characteristic flow velocity
x = axial (flow direction) coordinate
x^+ = axial coordinate for hydrodynamic entrance region of duct [Eq. (4.44)]
x^* = axial coordinate for thermal entrance region of duct = $x/(D_h^* \text{ Pe})$

Greek Letter Symbols

α = thermal diffusivity
α = aspect ratio (Fig. 4.4)
δ = velocity boundary-layer thickness
δ_T = thermal boundary-layer thickness
Δ = designates difference when used as a prefix
μ = dynamic viscosity
μ_m = dynamic viscosity evaluated at mixed mean temperature
μ_w = dynamic viscosity evaluated at wall temperature
ν = kinematic viscosity
ρ = density
τ = shear stress
τ_x = local shear stress
τ_w = wall shear stress
θ_i^* = influence coefficient [Eq. (4.38)]
θ_o^* = influence coefficient [Eq. (4.39)]

REFERENCES

1. B. Metais and E. R. G. Eckert, "Forced, Mixed, and Free Convection Regimes," *J. Heat Transfer,* **86,** 295 (1964).
2. G. D. Raithby and K. G. T. Hollands, "Natural Convection," in *Handbook of Heat Transfer Fundamentals,* 2d ed., W. M. Rohsenow, J. P. Hartnett, and E. N. Ganić (eds.), McGraw-Hill, New York, 1985, chap. 6.
3. E. R. G. Eckert and R. M. Drake, Jr., *Analysis of Heat and Mass Transfer,* McGraw-Hill, New York, 1972.
4. W. M. Kays and M. E. Crawford, *Convective Heat and Mass Transfer,* 2d ed., McGraw-Hill, New York, 1980.
5. A. Bejan, *Convection Heat Transfer,* Wiley, New York, 1984.
6. B. Gebhart, *Heat Transfer,* 2d ed., McGraw-Hill, New York, 1971.
7. F. P. Incropera and D. P. Dewitt, *Fundamentals of Heat Transfer,* Wiley, New York, 1981.
8. F. Kreith and M. S. Bohn, *Principles of Heat Transfer,* 4th ed., Harper & Row, New York, 1986.
9. F. M. White, *Heat Transfer,* Addison Wesley, Reading, Mass., 1984.
10. J. P. Holman, *Heat Transfer,* 6th ed., McGraw-Hill, New York, 1986.
11. R. J. Moffat and A. Ortega, "Buoyancy Induced Forced Convection," in *Heat Transfer in Electronic Equipment,* A. Bar-Cohen (ed.), ASME Publ. HTD-Vol. 57, p. 135, 1986.
12. G. S. Springer, "Heat Transfer in Rarefied Gases," in *Advances in Heat Transfer,* vol. 7, T. F. Irvine, Jr., and J. P. Hartnett (eds.), Academic Press, New York, 1971, p. 163.
13. W. H. Giedt and D. R. Willis, "Rarefied Gases," in *Handbook of Heat Transfer Fundamentals,* 2d ed., W. M. Rohsenow, J. P. Hartnett, and E. N. Ganić (eds.), McGraw-Hill, New York, 1985, chap. 9.
14. R. B. Bird, W. E. Stewart, and E. N. Lightfoot, *Transport Phenomena,* Wiley, New York, 1960.
15. W. M. Rohsenow and H. Y. Choi, *Heat, Mass, and Momentum Transfer,* Prentice-Hall, Englewood Cliffs, N.J., 1961.
16. E. R. G. Eckert, *Introduction to Heat and Mass Transfer,* McGraw-Hill, New York, 1963.
17. H. Schlichting, *Boundary-Layer Theory,* 6th ed., McGraw-Hill, New York, 1968.
18. F. M. White, *Viscous Fluid Flow,* McGraw-Hill, New York, 1974.
19. J. O. Hinze, *Turbulence,* 2d ed., McGraw-Hill, New York, 1975.
20. B. E. Launder and D. B. Spalding, *Mathematical Models of Turbulence,* Academic Press, London, 1972.
21. A. A. Townsend, *The Structure of Turbulent Shear Flow,* 2d ed., Cambridge University Press, Cambridge, 1972.
22. R. K. Shah and A. L. London, "Laminar Flow Forced Convection in Ducts," in *Advances in Heat Transfer,* T. F. Irvine, Jr., and J. P. Hartnett (eds.), Academic Press, New York, 1978.
23. R. L. Ash and J. H. Heinbockel, "Note on Heat Transfer in Laminar, Fully Developed Pipe Flow with Axial Conduction," *Z. Angew. Math. Phys.,* **21,** 266 (1970).
24. H. J. Hickman, "An Asymptotic Study of the Nusselt–Graetz Problem, Part 1, Large X Behavior," *J. Heat Transfer,* **96,** 354 (1974).
25. C. J. Hsu, "Exact Solution to Entry-Region Laminar Heat Transfer with Axial Conduction and the Boundary Condition of the Third Kind," *Chem. Eng. Sci.,* **23,** 457 (1968).
26. R. K. Shah and A. L. London, "Laminar Flow Forced Convection Heat Transfer and Flow Friction in Straight and Curved Ducts—A Summary of Analytical Solutions," Dept. Mech. Eng., Stanford University, Stanford, Calif., TR 75, 1971.
27. F. W. Schmidt and M. E. Newell, "Heat Transfer in Fully Developed Laminar Flow Through Rectangular and Isosceles Triangular Ducts," *Int. J. Heat Mass Transfer,* **10,** 1121 (1967).
28. R. K. Shah, "Laminar Flow Friction and Forced Convection Heat Transfer in Ducts of Arbitrary Geometry," *Int. J. Heat Mass Transfer,* **18,** 849 (1975).
29. T. S. Lundgren, E. M. Sparrow, and J. B. Starr, "Pressure Drop Due to the Entrance Region in Ducts of Arbitrary Cross Section," *J. Basic Eng.,* **86,** 620 (1964).
30. N. T. Dunwoody, "Thermal Results for Forced Heat Convection Through Elliptical Ducts," *J. Appl. Mech.,* **29,** 165 (1962).
31. V. P. Tyagi, "Laminar Forced Convection of a Dissipative Fluid in a Channel," *J. Heat Transfer,* **88,** 161 (1966).
32. M. Iqbal, A. K. Khatry, and B. D. Aggarwala, "On the Second Fundamental Problem of Combined Free and Forced Convection

Through Vertical Non-Circular Ducts," *Appl. Sci. Res.*, **26,** 183 (1972).

33. K. C. Cheng, "Laminar Forced Convection in Regular Polygonal Ducts with Uniform Peripheral Heat Flux," *J. Heat Transfer,* **91,** 156 (1969).
34. F. S. Shih, "Laminar Flow in Axisymmetric Conduits by a Rational Approach," *Can. J. Chem. Eng.*, **45,** 285 (1967).
35. R. E. Lundberg, P. A. McCuen, and W. C. Reynolds, "Heat Transfer in Annular Passages. Hydrodynamically Developed Laminar Flow with Arbitrarily Prescribed Wall Temperatures or Heat Fluxes," *Int. J. Heat Mass Transfer,* **6,** 495 (1963).
36. W. M. Kays and H. C. Perkins, "Forced Convection—Internal Flow in Ducts," in *Handbook of Heat Transfer Fundamentals,* 2d ed., W. M. Rohsenow, J. P. Hartnett, and E. N. Ganić (eds.), McGraw-Hill, New York, 1985, chap. 7.
37. E. M. Sparrow and A. L. Loeffler, Jr., "Longitudinal Laminar Flow Between Cylinders Arranged in Regular Array," *J. AIChE* **5,** 325 (1959).
38. E. M. Sparrow, A. L. Loeffler, Jr., and H. A. Hubbard, "Heat Transfer to Longitudinal Laminar Flow Between Cylinders," *J. Heat Transfer,* **83,** 415 (1961).
39. O. E. Dwyer and H. C. Berry, "Laminar-Flow Heat Transfer for In-Line Flow Through Unbaffled Rod Bundles," *Nucl. Sci. Eng.*, **42,** 81 (1970).
40. R. Y. Chen, "Flow in the Entrance Region at Low Reynolds Numbers," *J. Fluids Eng.*, **95,** 153 (1973).
41. R. K. Shah, "A Correlation for Laminar Hydrodynamic Entry Length Solutions for Circular and Noncircular Ducts," *J. Fluids Eng.*, **100,** 177 (1978).
42. R. K. Shah, "Thermal Entry Length Solutions for the Circular Tube and Parallel Plates," *Proc. 3d National Heat Mass Transfer Conf.* (Indian Inst. Technol., Bombay, India), vol. I, no. HMT-11-75, 1975.
43. R. W. Hornbeck, "An All-Numerical Method for Heat Transfer in the Inlet of a Tube," ASME Paper 65-WA/HT-36, 1965.
44. C. L. Hwang and L. T. Fan, "Finite Difference Analysis of Forced-Convection Heat Transfer in Entrance Region of a Flat Rectangular Duct," *Appl. Sci. Res.*, sec. A, **13,** 401 (1964).
45. R. M. Curr, D. Sharma, and D. G. Tatchell, "Numerical Predictions of Some Three-Dimensional Boundary Layers in Ducts," *Comput. Meth. Appl. Mech. Eng.*, **1,** 143 (1972).
46. P. Wibulswas, "Laminar-Flow Heat Transfer in Non-Circular Ducts," Ph.D. Thesis, London University, London, 1966.
47. D. P. Fleming and E. M. Sparrow, "Flow in the Hydrodynamic Entrance Region of Ducts of Arbitrary Cross Section," *J. Heat Transfer,* **91,** 345 (1969).
48. M. K. Gangal, "Some Problems in Channel Flow," Ph.D. Thesis, Dept. Math. Stat., University of Calgary, Calgary, Alta., 1974.
49. J. Liu, "Flow of a Bingham Fluid in the Entrance Region of an Annular Tube," M.S. Thesis, University of Wisconsin-Milwaukee, Wis., 1974.
50. C. A. Sleicher and M. Tribus, "Heat Transfer in a Pipe with Turbulent Flow and Arbitrary Wall-Temperature Distribution," *Trans. ASME,* **79,** 789 (1957).
51. L. F. Moody, "Friction Factors for Pipe Flow," *Trans. ASME,* **66,** 671 (1944).
52. B. S. Petukhov, "Heat Transfer and Friction in Turbulent Pipe Flow with Variable Physical Properties," in *Advances in Heat Transfer,* vol. 6, T. F. Irvine and J. P. Hartnett (eds.), Academic Press, New York, 1970, p. 503.
53. C. A. Sleicher and M. W. Rouse, "A Convenient Correlation for Heat Transfer to Constant and Variable Property Fluids in Turbulent Pipe Flow," *Int. J. Heat Mass Transfer,* **18,** 677 (1975).
54. R. G. Deissler, "Turbulent Heat Transfer and Friction in the Entrance Regions of Smooth Passages," *Trans. ASME,* **77,** 1221 (1955).
55. D. F. Dipprey and R. H. Sabersky, "Heat and Momentum Transfer in Smooth and Rough Tubes at Various Prandtl Numbers," *Int. J. Heat Mass Transfer,* **6,** 329 (1963).
56. L. W. Carlson and T. F. Irvine, "Fully Developed Pressure Drop in Triangular Shaped Ducts," *J. Heat Transfer,* **83,** 441 (1961).
57. J. P. Hartnett, J. C. Y. Koh, and S. T. McComas, "A Comparison of Predicted and Measured Friction Factors for Turbulent Flow Through Rectangular Ducts," *J. Heat Transfer,* **84,** 82 (1962).
58. E. R. G. Eckert and T. F. Irvine, Jr., "Pressure Drop and Heat Transfer in a Duct with Triangular Cross Section," *J. Heat Transfer,* **82,** 125 (1960).
59. W. M. Kays and E. Y. Leung, "Heat Transfer in Annular Passages: Hydrodynamically Developed Turbulent Flow with Arbitrarily Prescribed Heat Flux," *Int. J. Heat Mass Transfer,* **6,** 537 (1963).
60. R. A. Seban and E. F. McLaughlin, "Heat Transfer in Tube Coils with Laminar and Turbulent Flow," *Int. J. Heat Mass Transfer,* **6,** 387 (1963).
61. H. Ito, "Friction Factors for Turbulent Flow in Curved Pipes," *J. Basic Eng.*, **81,** 123 (1959).
62. H. Martin, "Heat and Mass Transfer Between Impinging Gas Jets and Solid Surfaces," in *Advances in Heat Transfer,* vol. 13, J. P. Hartnett and T. F. Irvine, Jr. (eds.), Academic Press, New York, 1977.
63. M. W. Rubesin, M. Inouye, and P. G. Parikh, "Forced Convection, External Flows," in *Handbook of Heat Transfer, Fundamentals,* 2d ed., W. M. Rohsenow, J. P. Hartnett, and E. N. Ganić (eds.), McGraw-Hill, New York, 1985.
64. S. Whitaker, "Forced Convection Heat Transfer Correlations for Flow in Pipes, Past Flat Plates, Single Cylinder, Single Spheres, and for Flow in Packed Beds and Tube Bundles," *J. AIChE,* **18,** 361 (1972).
65. A. Zukauskas, "Heat Transfer from Tubes in Crossflow," in *Advances in Heat Transfer,* vol. 8, J. P. Hartnett and T. F. Irvine, Jr. (eds.), Academic Press, New York, 1972, p. 93.
66. S. W. Churchill and M. Bernstein, "A Correlating Equation for Forced Convection from Gases and Liquids to a Circular Cylinder in Crossflow," *J. Heat Transfer,* **99,** 300 (1977).
67. L. C. Witte, "An Experimental Study of Forced-Convection Heat Transfer from a Sphere to Liquid Sodium," *J. Heat Transfer,* **90,** (1968).
68. M. Jakob, *Heat Transfer,* vol. 1, Wiley, New York, 1949.
69. C. P. Kothandaraman and S. Subramanyan, *Heat and Mass Transfer Data Book,* 2d ed., Wiley, New York, 1975.
70. M. Jakob, "Heat Transfer and Flow Resistance in Cross Flow of Gases over Tube Banks," *ASME,* **60,** 384 (1938).
71. E. M. Sparrow and J. W. Ramsey, "Heat Transfer and Pressure Drop for a Staggered Wall-Attached Array of Cylinders with Tip Clearances," *Int. J. Heat Mass Transfer,* **21,** 1369 (1978).
72. E. M. Sparrow, J. W. Ramsey, and C. A. C. Altemani, "Experiments on In-Line Pin Fin Arrays and Performance Comparisons with Staggered Arrays," *J. Heat Transfer,* **102,** 44 (1980).
73. W. M. Kays and A. L. London, *Compact Heat Exchangers,* 3d ed., McGraw-Hill, New York, 1984.
74. E. M. Sparrow and F. Samie, "Heat Transfer and Pressure Drop Results for One- and Two-Row Arrays of Finned Tubes," *Int. J. Heat Mass Transfer,* **28,** 2247 (1985).

Chapter 5

Two-Phase Heat Transfer

HOSSEIN P. NOURBAKHSH

Department of Nuclear Energy, Brookhaven National Laboratory

In this chapter, the general aspects of condensation and boiling heat transfer are discussed. A condensed summary of simply-applied analytical and empirical equations for condensation and boiling heat transfer in commonly encountered conditions and configurations is presented.

This chapter also includes a brief discussion of the factors that can affect these processes significantly. An annotated bibliography of the important reference literature is also included. The objective is to provide an understanding of the potential influencing parameters, the means for making an initial estimate, and direction to the reference literature in this area.

5.1 CONDENSATION

Condensation is defined as the removal of heat from a system in such a manner that vapor is converted into liquid. There are various ways in which condensate may be formed. The different mechanisms are known as the modes of condensation. The modes depicted are described as follows.

1. *Film condensation*: This is the most important mode of condensation as far as industrial condensers are concerned. In this mode, the condensate wets the surface and forms a continuous liquid film.

2. *Dropwise condensation*: This type of condensation has been observed to occur either on highly polished surfaces, on some polymeric material surfaces, or on surfaces contaminated with certain fatty acids. In this mode, the condensate forms individual drops. These drops increase in size and combine with one another until they roll or fall from the surface under the action of gravity or aerodynamic forces, leaving the surface exposed to the formation of new droplets.

3. *Direct contact condensation*: In this mode, condensation occurs directly on a cold liquid interface with no intervening solid surface. Examples of direct contact condensation are condensation of a steam jet in a large pool of cold liquid,[1,2] steam condensation on drops in a spray, and condensation on a coherent liquid jet sparged into a steam environment.[3,4]

Condensation is one of the most common modes of heat transfer. Typical values of the heat-transfer coefficient for the air-free vapors are given in Table 5.1.

This section presents the theoretical and empirical methods of calculating the heat-transfer coefficient for the film

Table 5.1. Orders of Magnitude for Condensation Heat-Transfer Coefficients

Typical values for air-free vapors

	α	
Mode of condensation	Btu/hr·ft²°F	W/m²·K
Film condensation on surfaces		
Pure steam	1000–5000	6000–30,000
Gassy steam	100–1000	600–6000
Hydrocarbons	150–900	1000–5000
Dropwise condensation of steam	10,000–50,000	60,000–300,000
Steam on cold water jet	30,000–100,000	200,000–600,000
Steam jet in cold water	$\approx 1.7 \times 10^5$	$\sim 10^6$

SOURCE: Adapted from G. Hestroni, *Handbook of Multi-Phase System*, McGraw-Hill, New York, 1982, p. 5.3.

and dropwise modes of condensation in commonly encountered conditions and configurations.

Film Condensation

The first attempt to analyze the film condensation on a cooled, constant-temperature vertical surface was that of Nusselt.[5] Nusselt simplified the problem by invoking a number of assumptions. These assumptions were: (1) the film flow is laminar, (2) the vapor is not superheated, (3) subcooling of the condensate may be neglected, (4) the fluid properties are constant, (5) the heat transfer through the film is by conduction only, giving a linear distribution across the film, (6) the vapor is stationary and exerts no drag on the downward motion of the condensate, and (7) the film moves slowly enough to neglect inertia forces within it.

By solving the momentum equation separately and relating the solutions by means of a heat balance, Nusselt obtained the following expression for the average heat-transfer coefficient up to the distance L from the leading edge of a vertical flat plate:

$$\bar{\alpha} = 0.943 \left[\frac{g\rho_L(\rho_L - \rho_G)\lambda_L^3 h_{LG}}{\mu_L L(T_{\text{sat}} - T_W)} \right]^{1/4} \tag{5.1}$$

where ρ_L, ρ_G = densities of liquid and vapor
g = gravitational acceleration
λ_L = thermal conductivity of liquid
h_{LG} = enthalpy change from liquid to saturated vapor
μ_L = dynamic viscosity of liquid
T_{sat} = saturation temperature
T_W = wall temperature

This solution has proved remarkably accurate despite some of the questionable assumptions used in the derivation. Nevertheless, a number of investigations have made significant improvements to the Nusselt theory. The original analysis was extended by Bromley,[6] who considered the effects of condensate subcooling, and by Rohsenow,[7] who also allowed for the nonlinear distribution of temperature through the film due to energy convection. Rohsenow[7] showed that the latent heat of vaporization h_{LG} in Eq. (5.1) should be replaced by

$$h'_{LG} = h_{LG} + \tfrac{3}{8}C_{pL}(T_{\text{sat}} - T_W) \tag{5.2}$$

where C_{pL} is the specific heat of liquid at constant pressure. Figure 5.1 shows the modified heat-transfer coefficient of Eq. (5.1) in the form of the film Reynolds number,

$$\text{Re}_f = \frac{\bar{\alpha}L(T_{\text{sat}} - T_W)}{h_{LG}\mu_L} \tag{5.3}$$

For long vertical surfaces it is possible to obtain large film flow rates such that the film Reynolds number exceeds the critical value at which turbulence begins. Under such circumstances the heat transfer is a complex function of the local film Reynolds number and the Prandtl number.[8] It is recommended that the values for the heat-transfer coefficients under these circumstances be taken from the curves in Fig. 5.1.

A number of studies have been performed to predict film condensation for various geometries and conditions. A summary of some of the analytical and empirical relations developed in such studies is presented in the following paragraphs in a form convenient for quick calculation.

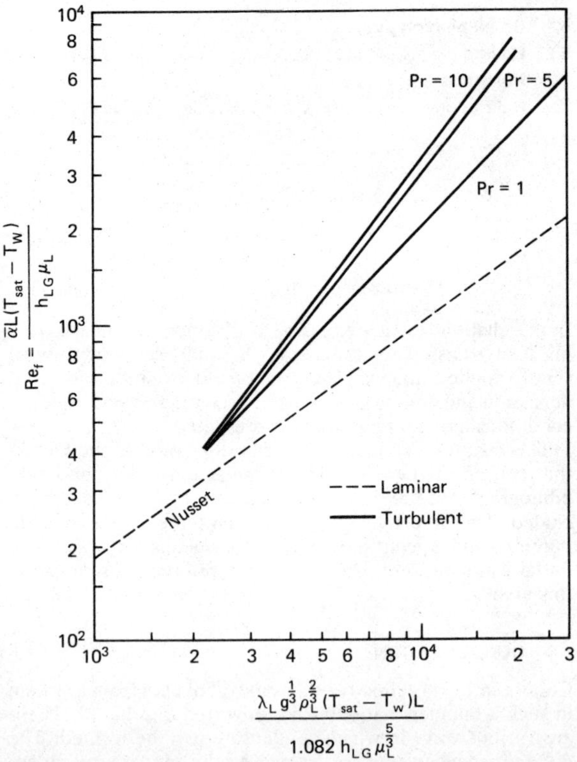

Fig. 5.1 Film condensation heat transfer with laminar and turbulent film.

Inclined Flat Plate Facing up The Nusselt relation holds for film condensation on the upper side of a plate inclined at an angle γ with the horizontal, except that g is replaced by $(g \sin \gamma)$.

Thus the recommended relation for laminar film condensation on a vertical or inclined plate is

$$\bar{\alpha} = 0.943 \left[\frac{g \sin \gamma\, \rho_L(\rho_L - \rho_G)\lambda_L^3 h_{LG}}{\mu_L L(T_{sat} - T_W)} \right]^{1/4} \tag{5.4}$$

Figure 5.1 also applies with this transformation for both laminar and turbulent flows.

Inclined Flat Plate Facing down Gerstman and Griffith[9] investigated this geometry. Several regions are possible. When the angle of inclination is $7.5° < \gamma < 90°$, the procedure given for an inclined flat plate facing up is recommended. When the inclination of the plate is $0 < \gamma < 7.5°$, it is suggested that the values be taken from Figure 5.2.

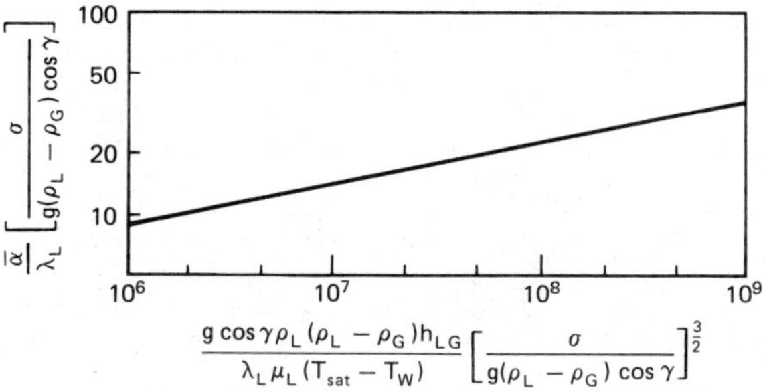

Fig. 5.2 Average condensation heat-transfer rate for inclined flat plates facing down with inclination of up to 7.5° from horizontal. Best line through data is shown. (*Gerstman and Griffith.*[9])

Outside Horizontal Tubes Analyses similar to that of Nusselt for a vertical surface may be carried out for condensation on the outside of a single horizontal tube. The result is

$$\bar{\alpha} = 0.727 \left[\frac{g\rho_L(\rho_L - \rho_G)\lambda_L^3 h'_{LG}}{D\mu_L(T_{sat} - T_W)} \right]^{1/4} \tag{5.5}$$

where D is the diameter of the tube. This relationship yields values that are about 15 percent lower than the existing data.[10]

Most commercial condensers involve a bank of tubes. If the tubes are arranged in a vertical pattern, condensate drops from the top tube to the next and so forth. The film thickness is then greater for the lower tubes and the heat-transfer coefficient is thus reduced. For a vertical stack of N tubes Nusselt showed that the ratio of the heat-transfer coefficient on the Nth tube to that on the first tube is reduced according to

$$\frac{\bar{\alpha}_N}{\bar{\alpha}_{Nu}} = \left(\frac{1}{N}\right)^{1/4} \tag{5.6}$$

Figure 5.3 represents a comparison of theories and data from a variety of sources. For the analyses of complex tube bank arrangements, the reader is referred to Ref. 11.

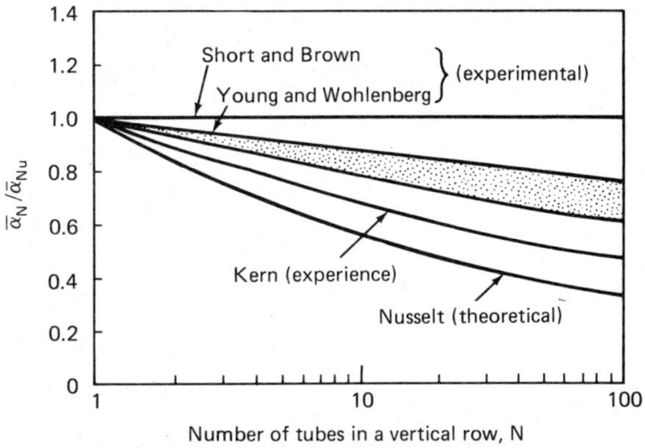

Fig. 5.3 Ratio of condensation heat-transfer coefficient on the Nth tube to that on the first tube for a vertical tube bundle. (*After Bell and Panchal.*[48])

Theories for condensation on the outside of tubes with a variety of orientations have been worked out by Nicol et al.[12]

Inside Tubes For condensation of vapor flowing inside a tube, the relative velocities of vapor and liquid and the resultant interfacial shear alter considerably the hydrodynamic and heat-transfer processes taking place.

For low vapor velocity in a horizontal tube when the inlet Reynolds number of vapor is below 35,000, Chato's recommendation[13] for condensation inside a horizontal tube is

$$\bar{\alpha} = 0.555 \left[\frac{g\rho_L(\rho_L - \rho_G)\lambda_L^3 h'_{LG}}{D\mu_L(T_{sat} - T_W)} \right]^{1/4} \tag{5.7}$$

At low velocity, inside tube condensation is better in down-flowing inclined tubes than in either vertical or horizontal tubes. Chato[13] found maximum heat-transfer coefficient at inclination angles to the horizontal of about 10 to 20°.

For high vapor velocity inside tubes when the flow is annular or slug, the following is recommended for an estimate of the average heat-transfer coefficient inside a tube with 100 percent condensation[14]:

$$\frac{\bar{\alpha} D}{\lambda_L} = 0.024 \left(\frac{\dot{m}D}{\mu_L}\right)^{0.8} \left(\frac{C_{pL}\mu_L}{\lambda_L}\right)^{0.43} \frac{1 + \sqrt{\rho_L/\rho_G}}{2} \tag{5.8}$$

where $\dot{m}$ is the mass flow rate.

For an estimate of local heat-transfer coefficients at high velocity inside tubes of any orientation the reader is referred to Ref. 15.

Effect of Vapor Superheat For calculating the heat transfer from superheated vapor, the enthalpy change on condensation must be augmented for the superheat enthalpy. Thus in Eqs. (5.4), (5.5), and (5.7), h'_{LG} must be replaced by

$$h'_{LG} = h_{LG} + C_{pG}(T_v - T_{sat}) + \tfrac{3}{8} C_{pL}(T_{sat} - T_W) \tag{5.9}$$

where T_v is the temperature of superheated vapor.

Effect of Noncondensable Gases Noncondensable gases seriously reduce the effective condensation heat-transfer coefficient. The vapor that is condensed must diffuse through

Table 5.2. Effect of Air on Condensation Heat Transfer

Weight percent of air in steam	α/α_0
0.5	0.6
1	0.45
2	0.31
3	0.21
4	0.19
>5	0.17

the gas, requiring a decrease in vapor partial pressure toward the liquid water interface.

For air in steam a quick estimate of the effect of air on the condensation heat-transfer coefficient can be obtained from Table 5.2. A more precise answer to this problem requires a lengthy calculation. A procedure developed by Votta[16] is recommended.

For condensation of water vapor (steam) from air-steam mixtures the total heat-transfer coefficient for sensible and latent heat transfer is given by

$$\dot{q} = \alpha_T(T_m - T_W)$$

$$\alpha_T = \alpha_s\left[1 + \frac{h_{LG}}{C_{p,\mathrm{air}}}\left(\frac{dw}{dT}\right)\right]$$

where α_s = single-phase convective heat-transfer coefficient

$\frac{dw}{dT}$ = derivative of humidity ratio w (i.e., $\mathrm{lb_{water}}$ per $\mathrm{lb_{air}}$) of saturated air with respect to temperature evaluated at wall temperature

T_m = sensible temperature of steam-air mixture

T_W = wall temperature

Note that condensation will occur when T_W is below the local water vapor dew point temperature (i.e., the saturation temperature corresponding to the local water vapor partial pressure). However, if the bulk temperature of the mixture is above the dew point, local reevaporation of the condensate may, depending on the situation, partially offset the transfer of heat to the wall due to condensation. However, if the gas mixture is saturated (sensible temperature equals dew point temperature) reevaporation is not possible.

Condensation of Immiscible Mixtures The condensation of a mixture of vapors which form immiscible liquid phases is a complicated topic. The properties of two components are often very different and it often happens that one of these condenses as a film while the other condenses as droplets. An example is the condensation of a water-hydrocarbon mixture in which the hydrocarbon condenses as a film and the water condenses as droplets. Some useful reviews on the subject are those of Boyes and Ponter[17] and Deakin.[18]

Dropwise Condensation

Due to the transient nature of the dropwise condensation process and the continuous clearing of condensate from the surface, heat transfer in this mode of condensation is observed to be five to ten times as great as in the film condensation mode. Considerable work has been directed to the study of this mechanism and in seeking ways to promote its occurrence. For the reviews of the available correlations and models the reader is referred to Merte.[19]

5.2 BOILING

The process of the conversion of liquid into a vapor associated with vapor bubbles is called boiling.

When degasified liquid without impurities is heated in a clean and smooth vessel, it will undergo the liquid-vapor phase change without the appearance of bubbles. However, under normally encountered engineering conditions, the presence of impurities, dissolved gases, and surface irregularities causes the appearance of vapor bubbles on the heating surface. The formation of these bubbles, their subsequent growth and detachment, and so on, profoundly affect the heat-transfer characteristics of the process.

Boiling may occur under various conditions. When the heating surface is immersed in a quiescent pool of liquid, heat is transferred to the liquid by free convection and bubble agitation. This process is called pool boiling. Boiling may also occur simultaneously with fluid motion induced by externally imposed pressure differences, which is referred to as forced convective boiling.

In this section, the various processes occurring during pool and forced convective boiling are described and predictive methods for heat-transfer coefficients occurring during these boiling processes are presented. For more extensive information, the reader is referred to Refs. 20–24.

Pool Boiling

In pool boiling the heat flux through the heating surface depends on both the wall superheat and the bulk subcooling. The wall superheat is the difference $T_W - T_{\mathrm{sat}}$, where T_{sat} is the saturation temperature corresponding to the pressure in the bulk liquid. A positive wall superheat is necessary for boiling to occur. The boiling is termed subcooled (or local surface) boiling if the bulk liquid temperature is below that of saturation at the given pressure. In this instance, the vapor bubbles formed in the superheated wall region condense as they move away into the subcooled liquid. When the bulk liquid is at the saturation temperature or slightly exceeds it, the boiling is referred to as saturated, or bulk, boiling.

The pool boiling characteristics are generally represented on a logarithmic plot of heat flux $\dot{q}$ against wall superheat, as shown in Fig. 5.4. At low wall superheat, the heat is transferred by single-phase natural circulation. As wall superheat increases, the initial natural circulation region *A-B* changes to the nucleate boiling region denoted by *B-C* in Fig. 5.4. The appearance of the first bubble at *B* requires a significant finite superheat.

Continued increase in wall superheat or heat flux causes more and more nucleation sites to become active producers of bubbles. A peak nucleate boiling heat flux, point *C*—called critical heat flux (CHF), first boiling crisis, departure from nucleate boiling (DNB), burnout heat flux (BHF), or dryout—is reached when so many bubble nucleation sites become activated that parts of the surface become covered with vapor, inhibiting liquid flow to the heated surface.

An additional increase of wall superheat beyond that at

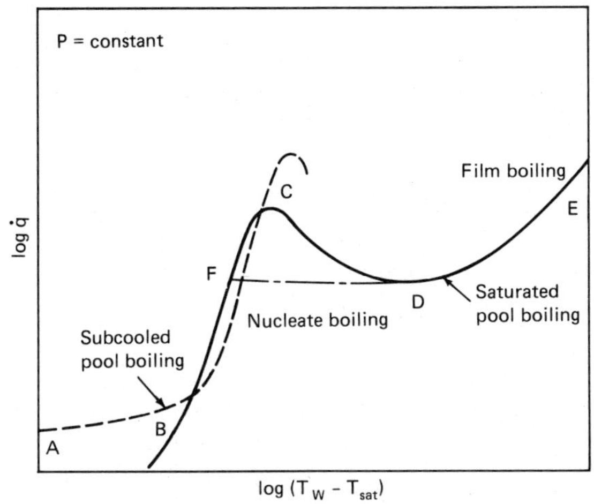

Fig. 5.4 Regimes in pool boiling heat transfer.

the burnout heat flux causes the process to enter the region of transition boiling or partial film boiling, denoted by *C-D* in Fig. 5.4. In this region, the process may be alternating between nucleate and film boiling, in which the surface is blanketed with a film of vapor. Eventually the region of stable film boiling *D-E* is reached, in which the surface is completely covered with a layer of vapor.

In many applications the heat flux is the independent variable (as for electric heating or a boiler with a given rate of heat production from fuel burning) and the surface temperature assumes a value compatible with $\dot{q}$. With a slight increase in $\dot{q}$, the operating condition at point *C* of Fig. 5.4 changes rapidly to point *E*, which is at a much higher temperature.

Reduction of the heat flux below point *D* at the minimum point on the boiling curve causes the operating condition to revert to nucleate boiling at point *F*.

The dashed line shown for subcooled pool boiling lies near the curve for saturated pool boiling. It may lie either to the left or to the right, depending on natural convection effects resulting from different surface geometries.

The mechanism of pool boiling is relatively well understood. However, it is still not possible to predict heat-transfer characteristics for this simplest of boiling systems with the precision associated with single-phase systems. A summary of simply applied working correlations for pool boiling heat transfer is presented in the following paragraphs.

Nucleate Pool Boiling Rohsenow[25] correlated experimental data for nucleate pool boiling with the relation

$$\dot{q} = \mu_L h_{LG} \left[\frac{g(\rho_L - \rho_G)}{g_c \sigma}\right]^{1/2} \left[\frac{C_{pL}(T_W - T_{sat})}{C_{sf} h_{LG} \mathrm{Pr}_L^S}\right]^3 \tag{5.10}$$

where C_{sf} = constant determined from experimental data
σ = surface tension at liquid-vapor interface
Pr_L = Prandtl number of saturated liquid
S = 1.0 for water, but
S = 1.7 for all other fluids

Table 5.3. Values of Coefficient C_{sf} for Various Liquid-Surface Combinations

Fluid–heating-surface combination	C_{sf}
Water–copper[43]	0.013
Water–platinum[44]	0.013
Water–brass[45]	0.0060
Water–emery-polished copper[46]	0.0128
Water—ground and polished stainless steel[46]	0.0080
Water–chemically etched stainless steel[46]	0.0133
Water–mechanically polished stainless steel[46]	0.0132
Water–emery-polished and paraffin-treated copper[46]	0.0147
Water–scored copper[46]	0.0068
Water–teflon-pitted stainless steel[46]	0.0058
Carbon tetrachloride–copper[43]	0.013
Carbon tetrachloride–emery-polished copper[46]	0.0070
Benzene–chromium[47]	0.010
n-butyl alcohol–copper[43]	0.00305
Ethyl alcohol–chromium[47]	0.027
Isopropyl alcohol–copper[43]	0.00225
n-pentane–chromium[47]	0.015
n-pentane–emery-polished copper[46]	0.0154
n-pentane–emery-polished nickel[46]	0.0127
n-pentane–lapped copper[46]	0.0049
n-pentane–emery-rubbed copper[46]	0.0074
35% K_2CO_3–copper[43]	0.0054
50% K_2CO_3–copper[43]	0.0027

SOURCE: Reprinted with permission from J. P. Holeman, *Heat Transfer*, 6th ed., McGraw-Hill, New York, 1986, p. 509.

The value of C_{sf} is a function of the particular fluid-heating-surface combination. Values for different fluid-surface combinations are given in Table 5.3.

It should be emphasized that accurate values of fluid properties are essential in using Eq. (5.10). Also, the relationship applies to a heating surface that is clean. The presence of contamination or deposit on the heating surface can change the exponent of the second term in addition to C_{sf} and S.

Critical Heat Flux in Pool Boiling The critical heat flux (point *C* in Fig. 5.4) is an important parameter in the design of boiling apparatus as well as in connection with the burnout phenomena.

For a very large horizontal flat plate Zuber[26] developed an analytical expression for critical heat flux in nucleate boiling by considering the stability requirements of vapor film and liquid interface. The relation is

$$\dot{q}_c = 0.13 h_{LG} \rho_G \left[\frac{\sigma(\rho_L - \rho_G) g_c g}{\rho_G^2}\right]^{1/4} \left(\frac{\rho_L}{\rho_L + \rho_G}\right)^{1/2} \tag{5.11}$$

However, a coefficient of 0.18 agrees better with the data than the value of 0.13 originally suggested by Zuber. In general, the type of surface material does not affect the critical heat flux, although the surface character can be an influence. Surfaces that are oxidized or coated, which tend to increase wetability, tend to have higher values of critical heat flux.

For a modification of Eq. (5.11) for various geometries the reader is referred to Ref. 27–29.

Subcooling increases the critical heat flux considerably. The ratio of critical heat flux for subcooled and saturated boiling can be expressed as

$$\frac{\dot{q}_{c\ \mathrm{sub}}}{\dot{q}_{c\ \mathrm{sat}}} = 1 + B(T_{\mathrm{sat}} - T_b) \tag{5.12}$$

where T_b is the subcooled bulk temperature.

Kutateladze[30] obtained an expression for B as:

$$B = 0.065 \left(\frac{\rho_L}{\rho_G}\right)^{0.8} \frac{C_{pG}}{h_{LG}} \tag{5.13}$$

Film Pool Boiling Stable film boiling was studied experimentally and analytically by Bromley.[31,32] By considering what is essentially an inversion of the classical Nusselt theory for laminar film condensation, Bromley arrived at the following equation representing the boiling heat-transfer coefficient on a horizontal cylinder:

$$\alpha_b = 0.62 \left\{\frac{\lambda_G^3 \rho_G(\rho_L - \rho_G)g[h_{LG} + 0.4C_{pG}(T_W - T_{\mathrm{sat}})]}{D\mu_G(T_W - T_{\mathrm{sat}})}\right\}^{1/4} \tag{5.14}$$

Since the temperature is generally quite high in film boiling, it is necessary to account for radiation. Radiation contributes to the heat transfer and increases the vapor film thickness. The following equation is recommended for the resultant heat-transfer coefficient:

$$\alpha = \alpha_b \left(\frac{\alpha_b}{\alpha}\right)^{1/3} + \alpha_r \simeq \alpha_b + 0.75\alpha_r$$

where α_r is calculated for radiation between two parallel planes.

Leidenfrost Point (Minimum Heat Flux in DNB Film Boiling) The lower limit of stable film boiling corresponds to the breakdown of the continuous vapor film and the onset of liquid-solid contact.[21] Using the hydrodynamic instability of liquid-vapor interface, the following equation for the minimum heat flux for a flat horizontal surface can be obtained:

$$\dot{q}_{\min} = ch_{LG}\rho_G \left[\frac{\sigma g_c g(\rho_L - \rho_G)}{(\rho_L + \rho_G)^2}\right]^{1/4} \tag{5.15}$$

where the constant c is variously given as 0.177,[33] 0.13,[34] or 0.09.[35]

For the case of small-diameter tubes, the semiempirical equation of Lienhard and Wong[36] is recommended. For larger tubes, it appears that Eq. (5.15) may be applied. However, Kovalev[37] notes that this equation seriously overpredicts the data for water on clean surfaces at pressures above atmospheric.[21]

Forced Convection Boiling

Forced convection boiling is most often encountered in the case of forced flow of a liquid through a channel with boiling occurring on the inner surface. Forced velocity of flow affects bubble growth and the influence of bubble dynamics on the heat-transfer mechanism.

Figure 5.5 depicts the various regions of flow and heat transfer that are observed to develop as a subcooled liquid is forced upward through a vertical heated tube. Figure 5.5 also illustrates how the wall and fluid temperatures might vary along the tube.

As fluid enters the tube, if subcooling is high enough, there is a section of single-phase forced convection (region A). However, boiling is soon initiated, and bubbles begin to appear on the surface as the fluid moves along the tube. As the fluid proceeds further downstream and loses more subcooling, the bubbles grow bigger, depart from the surface, and are taken into the mainstream of liquid. Initially vapor formation takes place in the presence of subcooled liquid. The transition between the subcooled nucleate boiling region (region B) and the saturated nucleate boiling region (region C) is the point at which the liquid reaches the saturation temperature. The location of this point can be predicted by simple heat balance calculations.

The thermodynamic mass "quality" x of the fluid shown in Fig. 5.5 is defined as

$$x = \frac{h - h_L}{h_{LG}} \tag{5.16}$$

The negative and greater than unity values of x have no practical significance other than signifying subcooled liquid and superheated vapor conditions, respectively.

As the quality increases through the saturated nucleate boiling region, the majority of the flow area is occupied by vapor. In this instance, the vapor may flow rapidly in the central portion of the tube while a liquid film is vaporized along the outer surface. Thus the process of boiling is replaced by the process of evaporation. The region beyond this transition is called forced-convective vaporization (regions E and F).

At some critical value of the quality, the complete evaporation of liquid film occurs. This transition is known as dryout and is accompanied by a rise in the wall temperature for channels operating with a controlled surface heat flux.[38] The region between the dryout point and the transition to single-phase vapor (region H) is referred to as the liquid-deficient region (region G), which corresponds to the drop (or mist) flow region.

The preceding comments are restricted to the case when a relatively low heat flux is supplied to the walls of the tube. When the heat flux is high, an abrupt deterioration in the cooling process takes place while bubble nucleation is still occurring. This transition is essentially similar to the critical heat flux phenomena in pool boiling and is referred to as departure from nucleate boiling (DNB). Figure 5.6 shows the regions of two-phase forced convective heat transfer as a function of quality with increasing heat flux.

The literature on two-phase forced convective heat transfer is quite voluminous. Here a summary is given of simple criteria for determining the boundaries of the heat-transfer regions. In addition, methods for calculating heat-transfer rates in each heat-transfer region are also presented.

Incipience of Boiling Bergles and Rohsenow[39] developed a criterion for boiling incipience based on hydrodynamic boundary-layer thickness. They also proposed a convenient numerical expression for the steam-water system in the pressure range of 15 to 2000 lb/in^2 abs (0.1 to 13.6 MPa),

$$\dot{q} = 15.60P^{1.156}(T_W - T_{\mathrm{sat}})^{2.30/P^{0.0234}} \tag{5.17}$$

where P is pressure in lb/in^2 abs, $\dot{q}$ is heat flux in Btu/h·ft^2, and the temperature differences are °F.

Frost and Dzakowic[40] have extended this treatment to cover other liquids.

Point of Net Vapor Generation The net vapor generation

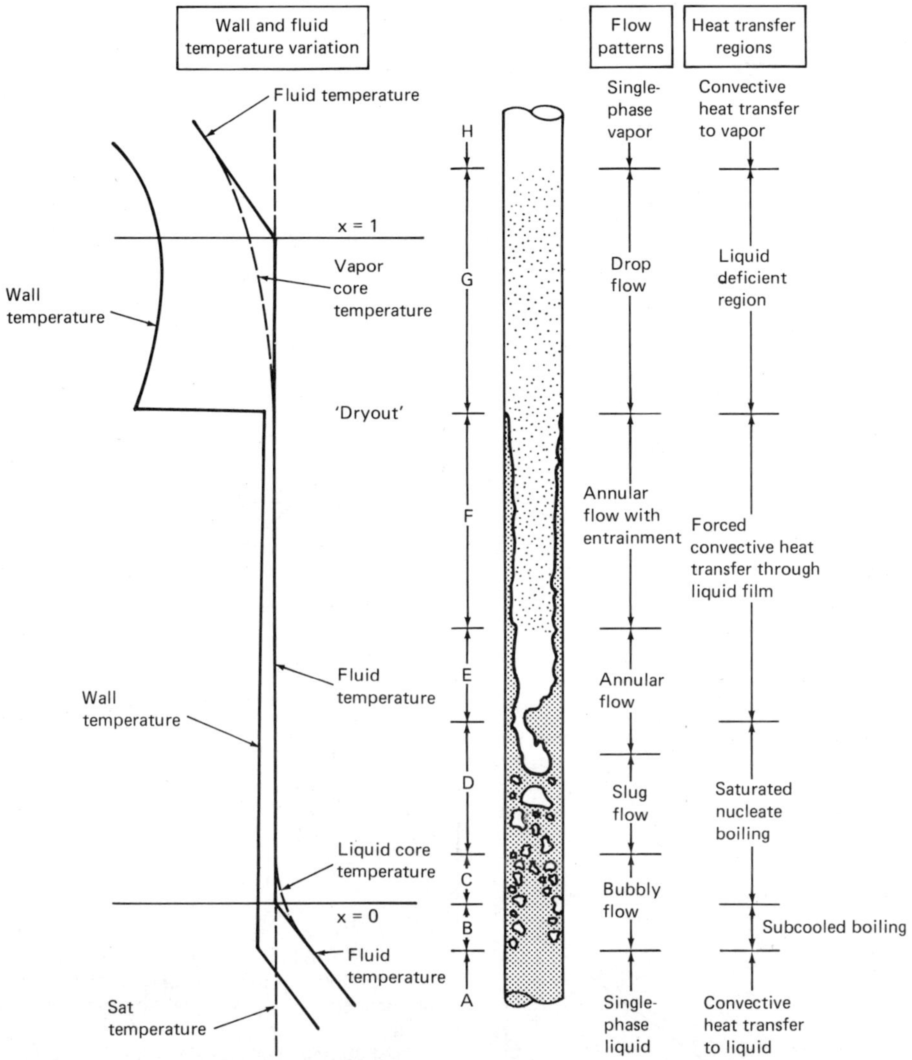

Fig. 5.5 Regions of heat transfer in convective boiling. [*Reprinted with permission from J. G. Collier,* Convective Boiling and Condensation, *McGraw-Hill (UK) 1972.*]

point is located where the subcooling is enough for bubbles to grow and depart into the stream. Saha and Zuber[41] proved a correlation based on the combined hydrodynamic and heat-transfer mechanism. At low flow rates, the bubble detachment is assumed to be thermally controlled, occurring at a fixed Nusselt number. At high flow rates, bubble departure is hydrodynamically induced and occurs at a specific Stanton number. Two regions are distinguished based on the value of the Peclet number,

$$\text{Nu} = \frac{\dot{q}_W D}{\lambda_L (T_{\text{sat}} - T_L)} = 455 \qquad \text{if Pe} < 7 \times 10^4 \tag{5.18a}$$

$$\text{St} = \frac{\dot{q}_W}{\dot{m} C_{pL} (T_{\text{sat}} - T_L)} = 0.0065 \qquad \text{if Pe} > 7 \times 10^4 \tag{5.18b}$$

where the Peclet number Pe is defined as

$$\text{Pe} = \frac{\dot{m} D C_{pL}}{\lambda_L} = \frac{\text{Nu}}{\text{St}} \tag{5.19}$$

Nucleate Boiling and Two-Phase Forced Convective Vaporization Chen[42] has proposed a correlation that has been generally accepted as one of the best available. The correlation covers both the saturated nucleate boiling region and the

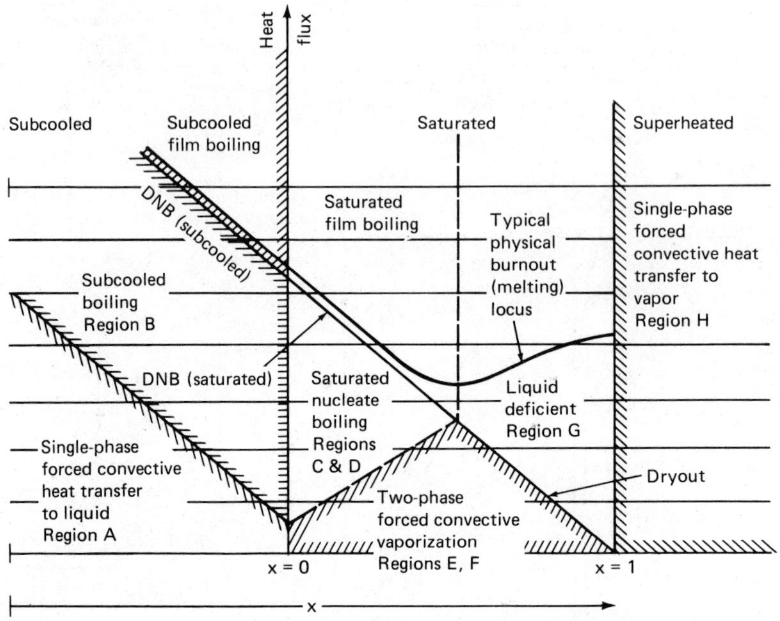

Fig. 5.6 Regions of two-phase forced convective heat transfer as a function of quality with increasing heat flux as ordinate. [*Reprinted with permission from J. G. Collier,* Convective Boiling and Condensation, *McGraw-Hill (UK) 1972.*]

two-phase forced vaporization region. It is assumed that both nucleation and convective mechanisms occur to some degree over the entire range of the correlation and that the contributions made by the two mechanisms are additive. The Chen equation is of the form

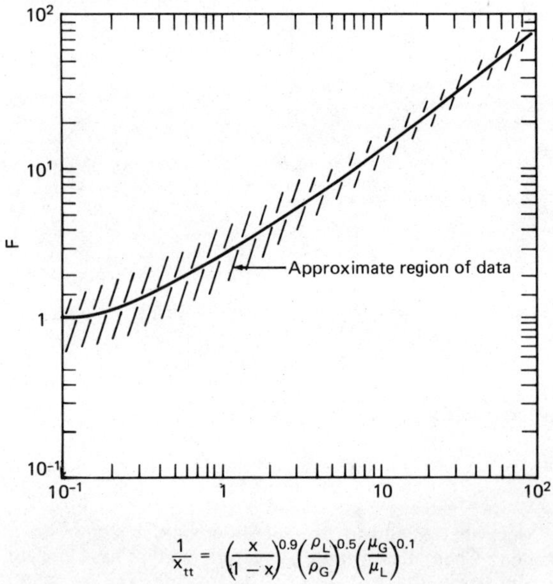

Fig. 5.7 Reynolds number factor F used in Chen's[42] correlation.

$$\alpha = 0.0012 \frac{\lambda_L^{0.79} C_{pL}^{0.45} \rho_L^{0.49} g_c^{0.25}}{\sigma^{0.5} \mu_L^{0.29} h_{LG}^{0.24} \rho_G^{0.24}} (T_W - T_{sat})^{0.24} \times (\Delta P_{sat})^{0.75} S + 0.023 \left[\frac{D\dot{m}(1-x)}{\mu_L}\right]^{0.8} \left(\frac{C_{pL}\mu_L}{\lambda_L}\right)^{0.4} \frac{\lambda_L}{D} F \qquad (5.20)$$

with F given as a function of x_{tt} in Fig. 5.7 and S a function of $Re_L F^{1.25}$ in Fig. 5.8. ΔP_{sat} is the vapor pressure difference corresponding to the wall superheat. This correlation is recommended for use with all single-component nonmetallic fluids.

Post Burnout Heat Transfer A considerable number of empirical equations have been presented for the estimation of

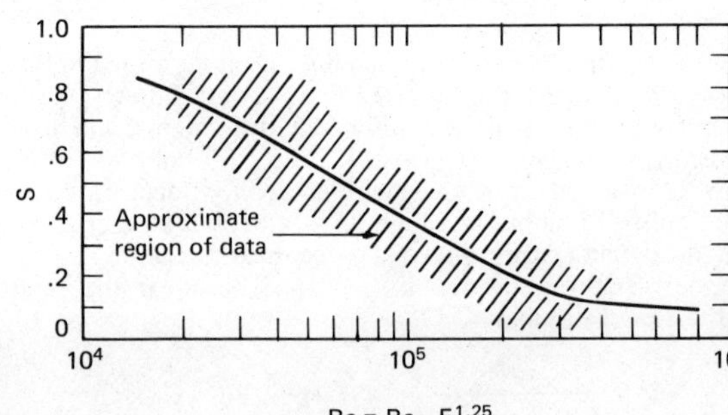

Fig. 5.8 Suppression factor S used in Chen's[42] correlation.

heat-transfer rates in the post burnout region. Almost all of them are the modifications of the well-known Dittus-Boetler relation for single-phase flow.[10] For the reviews of the available correlations and models, the reader is referred to Ref. 10.

NOMENCLATURE

B = parameter defined by Eq. (5.13)
c = constant in Eq. (5.15)
C_p = specific heat capacity at constant pressure
C_{sf} = constant in Eq. (5.10)
D = tube diameter
F = Chen's Reynolds number factor
g = gravitational acceleration
h = specific enthalpy
h'_{LG} = parameter defined by Eq. (5.2)
L = length
$\dot{m}$ = mass flow rate per unit area
N = number of tubes
Nu = Nusselt number
P = pressure
Pe = Peclet number
Pr = Prandtl number
$\dot{q}$ = heat flux density
$\dot{q}_c$ = critical heat flux
Re = Reynolds number
S = Chen's nucleation suppression factor
St = Stanton number
T = temperature
w = humidity ratio
x = quality

Greek Letters

α = heat-transfer coefficient
γ = angle of inclination
λ = thermal conductivity
μ = viscosity
ρ = density
σ = surface tension

Subscripts

b = boiling, bulk
f = film
G = gas phase
L = liquid phase
min = minimum
N = Nth tube
r = radiation
sat = saturation
sub = subcooling
T = total
V = vapor
W = wall

REFERENCES

1. M. Cumo, G. E. Farello, and G. Ferrari, "Direct Heat Transfer in Pressure-Suppression Systems," *Proc. 6th Int. Heat Transfer Conf.* (Toronto, Canada, 1978), vol. 4, pp. 101–106.
2. T. Young, K. C. Yang, and J. Novotney, "The Interaction Between a High Velocity Vapor Jet and an External Co-axial Co-Current Water Flow," University of Notre Dame, Notre Dame, Ind., Tech. Rep. UND-73-6, 1973.
3. T. H. Linehan and M. A. Grolmes, "Condensation of a High Velocity Vapor on a Subcooled Liquid Jet in Stratified Flow," *Proc. 4th Int. Heat Transfer Conf.* (Paris & Versailles, France, 1970), vol. 6, Paper CS 2.6.
4. S. S. Kuateladze, "Heat Transfer in Condensation and Boiling," AEC-tr-3778.
5. W. Nusselt, "Die Oberflöchenkondensation des Wasserdampfes," *Z. Verb. deutsch. Ing.*, **60** (1916).
6. L. A. Bromley, "Effect of Heat Capacity of Condensate," *Ind. Eng. Chem.*, **44,** 2966 (1952).
7. W. M. Rohsenow, "Heat Transfer and Temperature Distribution in Laminar Film Condensation," *Trans. ASME,* **78,** 1637–1643 (1956).
8. A. P. Colburn, "The Calculation of Condensation where a Portion of the Condensate Layer Is in Turbulent Motion," *Trans. AIChE,* **30,** 187 (1933).
9. T. Gerstman and P. Griffith, "Laminar Film Condensation on the Underside of Horizontal and Inclined Surfaces," *Int. J. Heat Mass Transfer,* **10,** 567–580 (1967).
10. G. Hestroni, *Handbook of Multi-Phase Systems,* Hemisphere Publ. Corp., 1982.
11. A. Devor, "How to Design Multiple Condensers," *Pet. Refiner,* **38**(6), 205 (1959).
12. A. A. Nicol, A. Bryce, and A. S. A. Ahmed, "Condensation of a Horizontally Flowing Vapor on a Horizontal Cylinder Normal to the Vapor Stream," *Proc. 6th Int. Heat Transfer Conf.* (Toronto, Canada, 1978, vol. 2, pp. 401–406.
13. J. C. Chato, *J. Am. Soc. Refrig. Air. Cond. Eng.*, p. 52 (1962).
14. L. D. Boyko, and G. N. Kruzhilin, "Heat Transfer and Hydraulic Resistance During Condensation of Steam in a Horizontal Tube and in a Bundle of Tubes," *Int. J. Heat Mass Transfer,* **10,** 361–373 (1967).
15. D. P. Traviss, A. B. Baron, and W. M. Rohsenow, "Forced Convection Condensation Inside Tubes," Dept. of Mech. Eng., M.I.T., Cambridge, Mass., DSR-70591-74, 1971.
16. F. Votta, "Condensation from Vapor-Gas Mixtures," *Chem. Eng. (N.Y.),* **70,** 223–228 (1964).
17. A. P. Boyes and A. B. Ponter, "Condensation of Immiscible Binary Systems," *Chem. Process. Eng.-Heat Transfer Surv.,* 26–30 (1972).
18. A. W. Deakin, "A Literature Survey on the Condensation of Vapors of Immiscible Liquids," UKAEA Dep. AERE-R7453, 1973.
19. H. Merte, "Condensation Heat Transfer," in *Advances in Heat Transfer,* Vol. 9, 1973, pp. 181–272.
20. W. M. Rohsenow (ed.), *Developments in Heat Transfer,* M.I.T. Press, Cambridge, Mass., 1964.
21. A. E. Bergles, J. G. Collier, T. M. Delhaye, and F. Mayinger, *Two-Phase Flow and Heat Transfer in the Power and Process Industries,* Hemisphere Publ. Corp., 1981.
22. D. Butterworth and G. F. Hewitt (eds.), *Two Phase Flow and Heat Transfer,* Oxford University Press, London, 1977.
23. S. Van Stralen and R. Cole (eds.), *Boiling Phenomena,* vols. 1–3, Hemisphere Publ. Corp., 1979.
24. O. E. Dwyer, *Boiling Liquid Metal Heat Transfer,* American Nuclear Society, Hinsdale, Ill., 1976.
25. W. M. Rohsenow, "A Method of Correlating Heat Transfer Data for Surface Boiling Liquids," *Trans. ASME,* **74,** 969 (1952).
26. N. Zuber, "Hydrodynamic Aspects of Boiling Heat Transfer," USAEC Rep. AECU-4439, 1959; Doctoral Dissertation, University of California, Los Angeles, 1959.

27. K. H. Sun nad J. H. Lienhard, "The Peak Pool Boiling Heat Flux on Horizontal Cylinders," *Int. J. Heat Mass Transfer,* **13,** 1425–1439 (1970).
28. J. H. Lienhard and V. K. Dhir, "Hydrodynamic Prediction of Peak Pool Boiling Heat Fluxes from Finite Bodies," *J. Heat Transfer,* **95,** 152–158 (1973).
29. J. H. Lienhard, V. K. Dhir, and D. M. Richard, "Peak Pool Boiling Heat Flux Measurements on Finite Horizontal Flat Plates," *J. Heat Transfer,* **95,** 477–482 (1973).
30. S. S. Kutateladze, "Heat Transfer in Condensation and Boiling," U.S. Atomic Energy Commission, Rep. AEC-tr-3770, 1952.
31. L. A. Bromley, "Heat Transfer in Stable Film Boiling," *Chem. Eng. Prog.,* **46,** 221 (1950).
32. L. A. Bromley, "Heat Transfer in Forced Convection Film Boiling," *Ind. Eng. Chem.,* **45,** 2639–2646 (1953).
33. N. Zuber, M. Tribus, and J. W. Westwater, "The Hydrodynamic Crisis in Pool Boiling of Saturated and Subcooled Liquids," in *International Developments in Heat Transfer,* pt. II, Am. Soc. Med. Eng., New York, 1961, pp. 230–235.
34. N. Zuber, "On Stability of Boiling Heat," *Trans. ASME,* **80,** 711–720 (1958).
35. P. J. Berenson, "Transition Boiling Heat Transfer from a Horizontal Surface," *J. Heat Transfer,* **83,** 351–358 (1961).
36. J. H. Lienhard and P. T. Y. Wong, "The Dominant Instable Wavelength and Minimum Heat Flux During Film Boiling on a Horizontal Cylinder," *J. Heat Transfer,* **86,** 220–226 (1964).
37. S. A. Kovalev, "An Investigation of Minimum Heat Fluxes in Pool Boiling of Water," *Int. J. Heat Mass Transfer,* **9,** 1219–1226 (1966).
38. J. G. Collier, *Convective Boiling and Condensation,* McGraw-Hill (UK), 1972.
39. A. E. Bergles and W. M. Rohsenow, "The Determination of Forced Convection Surface-Boiling Heat Transfer," *J. Heat Transfer,* **86,** 365–372 (1964).
40. W. Frost and G. S. Dzakowic, "An Extension of the Method of Predicting Incipient Boiling on Commercially Finished Surfaces," presented at the ASME, AIChE Heat Transfer Conf. (Seattle, Wash., 1967), Paper 67-HT-61.
41. P. Saha and N. Zuber, "Point of Net Vapor Generation and Vapor Void Fraction in Subcooled Boiling," presented at the 5th Int. Heat Transfer Conf. (Tokyo, Japan, 1974), Paper B4.7.
42. J. Chen, "A Correlation for Boiling Heat Transfer to Saturated Fluids in Convective Flow," ASME Paper 63-HT-34, 1963.
43. E. L. Piret and H. S. Isbin, "Natural Circulation Evaporation Two-Phase Heat Transfer," *Chem. Eng. Prog.,* **50,** 305 (1954).
44. J. N. Addoms, "Heat Transfer at High Rates to Water Boiling Outside Cylinders," Sc. D. Thesis, Massachusettes Institute of Technology, Cambridge, Mass., 1948.
45. D. S. Cryder and A. C. Finalbargo, "Heat Transmission from Metal Surfaces to Boiling Liquids: Effect of Temperature of the Liquid on Film Coefficient," *Trans. AIChE,* **33,** 346 (1937).
46. R. I. Vachon, G. H. Nix, and G. E. Tanger, "Evaluation of Constants for the Rohsenow Pool-Boiling Correlation," *J. Heat Transfer,* **90,** 239 (1968).
47. M. T. Cichelli and C. F. Bonilla, "Heat Transfer to Liquids Boiling Under Pressure," *Trans. AIChE,* **41,** 755 (1945).
48. K. J. Bell and C. B. Panchal, "Condensation," in *Proc. 6th Int. Heat Transfer Conf.,* (Toronto, Canada, 1978), vol. 6, pp. 369–375.

Chapter **6**

Radiation Heat Transfer

JOSE G. MARTIN
MANUEL J. BLANCO MURIEL
Department of Mechanical and Energy Engineering
University of Lowell

6.1 BASIC PRINCIPLES

Thermal radiation is one of the basic mechanisms for energy transfer between regions at different temperatures. It does not depend on the presence of intermediate material to act as an energy carrier. It is associated with electromagnetic waves that result from changes in the energy levels of atoms and molecules.

A participating medium is one that exchanges energy by radiation. When a surface sees either another surface or a participating medium at a different temperature, radiation transfer tends to be important if the difference in the fourth power of their temperatures is high. The last section of this chapter reviews the conditions under which radiation is a predominant means for heat transfer. Radiation lends itself to elegant treatments at several levels. The reader is referred to an excellent literature (for example, Refs. 1–6).

The amount and the characteristics of radiation depend on the atomic structure, the physical form, and the temperature of the source. The rate of emission *from* a source is independent of the surroundings, but the *net* energy transfer does depend on the temperatures and spatial arrangements of the source and the surroundings.

Electromagnetic waves are characterized in terms of frequency ν, with units of inverse time (1/s), and of wavelength λ, with units of distance (such as feet or meters). In any medium, these two properties are related by $c = \lambda\nu$, where c is the velocity of propagation of the wave in the medium. In vacuum, c is a constant—the speed of light in vacuum, equal to 9.836×10^8 ft/s (2.998×10^8 m/s). Other commonly used units for wavelength are the micrometer μm (1 μm = 1.0 × 10^{-6} m) and the angstrom Å (1 Å = 1.0 × 10^{-10} m).

For most problems of interest in heat transfer, thermal radiation lies in the range between 0.1 and 100 μm. Problems involving solar radiation are also of interest to thermal engineers, and most of the solar spectrum lies in the range between the ultraviolet and the infrared, that is, approximately between 0.3 and 25 μm. This includes the visible range. These wavelength bands are shown in Fig. 6.1 in relation to other parts of the electromagnetic spectrum.

When radiant energy is intercepted by matter, some of it is absorbed, that is, it is converted to internal thermal energy. Emission and absorption are, in principle, the results of volumetric phenomena. Emission, for example, is the integrated effect of local emission throughout the volume of a body. The volumetric concept must be used in evaluating radiation phenomena in gases and semitransparent solids. However, for most solids and liquids, radiation is strongly absorbed by adjoining molecules. Thus the emitted radiation originates from molecules that are very close (<1 μm) from the exposed surface. In these cases, emission can be viewed as a surface phenomenon.

Radiant Intensity

In general, the propagation of radiation at any point cannot be represented by a single vector, as in the case of heat flow by conduction. In a homogeneous medium radiation transfer takes place along straight lines. Radiation beams are independent of one another. Thus to specify the radiation incident at a given point, it is necessary to know the radiation from all directions.

A fundamental quantity to describe the amount of radiant power transmitted by a ray in any direction is the monochromatic (spectral) radiation intensity $I_{\lambda\Omega}$. It is defined as the amount of radiative power, per unit wavelength, streaming through a unit area perpendicular to the direction of propagation $\hat{\Omega}$, per unit solid angle around that direction. The integral of this quantity over all wavelengths is intensity I_{Ω}.

The contribution to the (spectral) flux [that is, the (spectral) power emitted or absorbed per unit area by a surface] from a monodirectional beam of (spectral) intensity I is equal to I multiplied by the cosine of the angle between the surface and the direction of propagation. For isotropic radiant distributions the radiant (spectral) power per unit area incident or emitted is equal to π times the (spectral) intensity.

The Blackbody

A black surface, or blackbody, is an idealized surface that absorbs all the radiation that it intercepts. From statistical mechanics or basic thermodynamics, it may be shown that a perfect absorber is also a perfect emitter. The blackbody concept has been a useful concept in the development of physical theory. It is also basic to most practical heat-transfer calculations. At a given temperature, the rate at which energy is emitted per unit area from any body is less than or equal to that from a blackbody. The blackbody concept represents a standard in terms of which all properties that characterize the interaction of radiation with actual surfaces are quantified.

An experimental blackbody may be approximated by a hollow space (Hohlraum) with an opening to the outside which is small compared with the surface of the hollow. Radiation that penetrates into the hollow through the small opening will experience many reflections before reemergence, and will be almost completely absorbed. The hollow space then approximates a blackbody and the radiation in the cavity will be blackbody radiation, whether the walls of the enclosure are good absorbers or not.

Blackbody Spectrum The spectral distribution of blackbody radiation is a continuous function which depends only on the temperature of the blackbody (see Fig. 6.2). Here M_{λ}^{b} is the spectral power distribution of the radiation emitted by the blackbody per unit area. T is the absolute temperature, in degrees Rankine (°R) or in kelvins (K). Note that

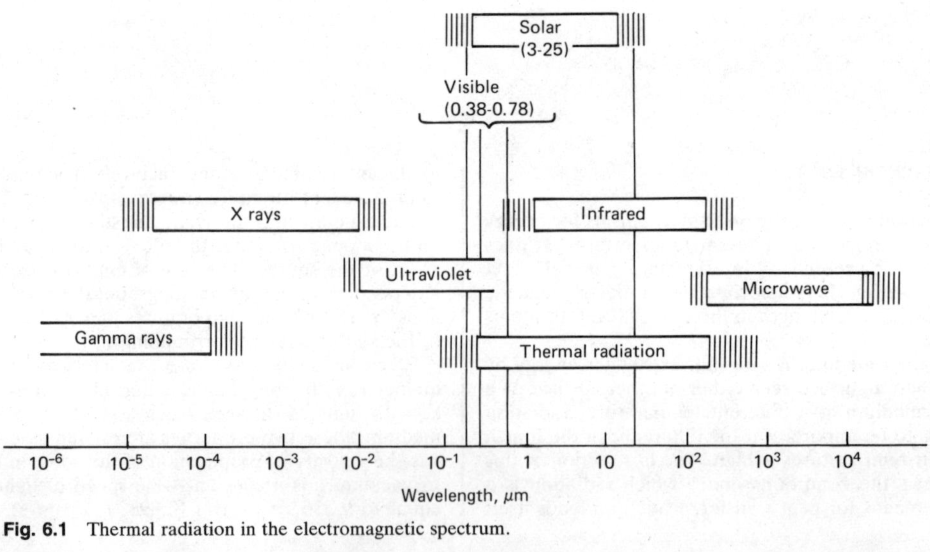

Fig. 6.1 Thermal radiation in the electromagnetic spectrum.

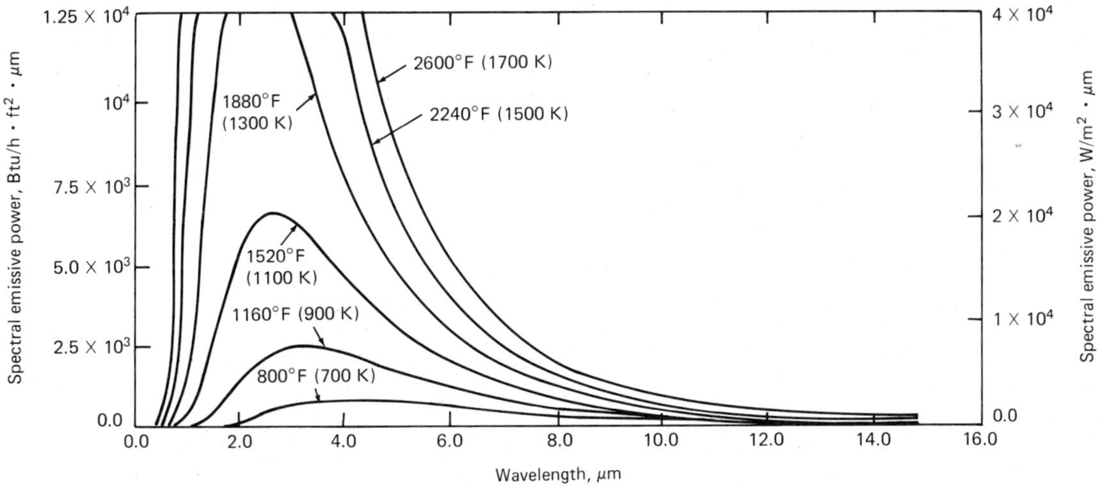

Fig. 6.2 Blackbody spectral emissive power.

the distribution function has the units of power density. If the horizontal axis (wavelength) is given in micrometers, it is convenient to express the distribution function in units of Btu/h · ft² · μm or W/m² · μm.

A mathematical expression for the blackbody distribution as a function of basic physical quantities and constants was given by Max Planck. The spectral emissive power is of the form

$$M_\lambda^b(\lambda, T) = \frac{C_1}{\lambda^5[\exp(C_2/\lambda T) - 1]} \tag{6.1}$$

where the values of the first and second radiation constants C_1 and C_2 are given in Table 6.1. Equation (6.1), known as Planck's distribution, is plotted in Fig. 6.2.

Wien's Law At any wavelength, the magnitude of the emissive power increases with increasing temperature. The spectral region in which the emissive power is maximum depends on temperature, with comparatively more radiation emitted at shorter wavelengths as the temperature increases. Wien showed that the position of the peak of the distribution is given by

$$\lambda = \frac{C_3}{T}$$

where C_3, the third radiation constant is also given in Table 6.1.

Stephan-Boltzmann Law Integrating Eq. (6.1) over the spectrum, one obtains the Stephan-Boltzmann law, which relates the total emissive power of a blackbody M^b to the absolute temperature,

$$M^b = \sigma T^4$$

where σ is the Stefan-Boltzmann constant [$= 1.712 \times 10^{-9}$ Btu/h · ft² · °R⁴ (5.670×10^{-8} W/m² · K⁴)].

For some applications, only the total energy transferred is important; for others, the spectral distribution is important. It is often necessary to know the total emission from a blackbody within a certain wavelength interval or band. This can be calculated by integrating Eq. (6.1) for the appropriate interval. The integral may be written as a function of the product λT. Tables that facilitate the evaluation of this integral can be found in the literature.[6,7]

Table 6.1. Radiation Constants

C_1	C_2	C_3
1.186×10^8 Btu · m⁴/h · ft²	2.590×10^4 m · °R	5.216×10^3 m · °R
3.742×10^8 W · μm⁴/m²	1.439×10^4 μm · K	2.898×10^3 μm · K
3.742×10^{-16} W · m²	1.439×10^{-2} m · K	2.898×10^{-3} m · K

Real Surfaces

The emissive powers given by the Planck and Stefan-Boltzmann laws are maximum values at a given surface temperature. Actual (nonblack) surfaces emit (and absorb) less readily (see Fig. 6.3*a*). The spectral emissive power M_λ of a nonblack surface may be written as

$$M_\lambda(\lambda, T) = \frac{\epsilon_\lambda C_1}{\lambda^5[\exp(C_2/\lambda T) - 1]}$$

where ϵ_λ is the monochromatic hemispherical emittance (Fig. 6.3*b*). Similarly, the total emissive power M may be written as

$$M = \epsilon\sigma T^4$$

where ϵ is called the total hemispherical emittance, or simply the emittance. Representative values of emittance for various surfaces are given in Fig. 6.4. These hemispherical emittances represent fractions which are averaged over all the space above a surface. Actual bodies do not emit or absorb isotropically, and for some problems, directional effects are important.

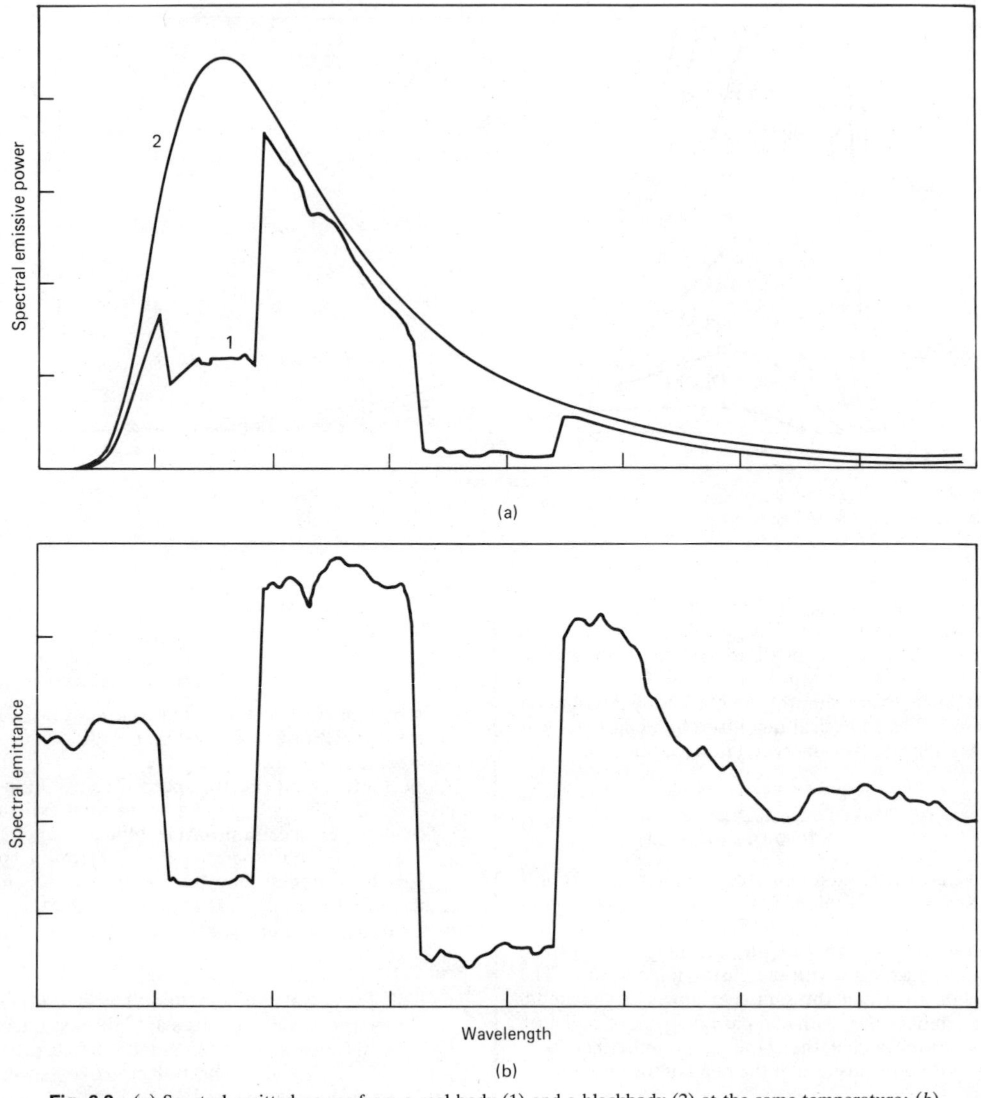

Fig. 6.3 (*a*) Spectral emitted power from a real body (1) and a blackbody (2) at the same temperature; (*b*) monochromatic emittance.

6.2 RADIATIVE CHARACTERISTICS OF OPAQUE SURFACES

Real surfaces do not absorb all the radiation they intercept. When radiant energy falls on a surface, it can be absorbed, reflected, or transmitted through the material. Since energy is conserved, the reflected fraction ρ (reflectance), the absorbed fraction α (absorptance), and the transmitted fraction τ (transmittance) are related by

$$\rho + \alpha + \tau = 1$$

if the fractions are defined consistently. If no energy is transmitted through a material, the material is opaque.

All these fractions, as well as emittance, take modifiers such as *directional* or *hemispherical, spectral* (*monochromatic*) or *total.* Some of them are inherent properties of the surface, in the sense that they are independent of the characteristics of the intercepted radiation; others are not.

The distinction is important for specialized engineering applications, such as selective surfaces for solar applications, heat mirrors for energy conservation in buildings, and the "greenhouse" effect. However, the great majority of common engineering problems find practical solutions using the

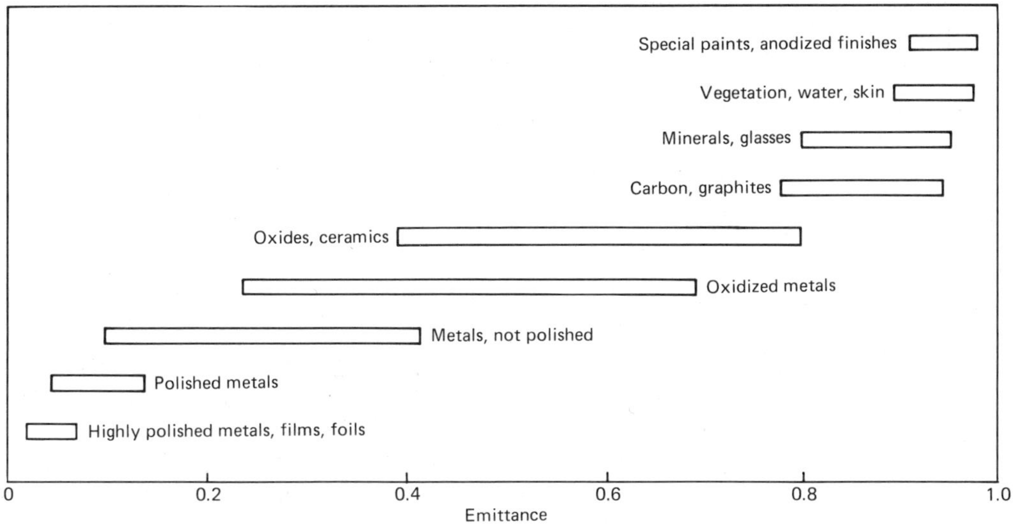

Fig. 6.4 Representative emittance values.

simplest definitions of absorptance and emittance properties. More specific definitions follow.

Emittance

Monochromatic Directional Emittance The ratio of the spectral intensity emitted from a surface in some direction to that emitted by a blackbody at the same temperature is called *monochromatic directional emittance*. This characteristic, also called *directional spectral emissivity,* is clearly an inherent property. If one has complete data for this property for the surface of an opaque body, one knows or can estimate all other radiative properties. Unfortunately, complete data are not available for most materials. This property is measured indirectly, and the measurements are cumbersome.

Directional Emittance The ratio of the intensity emitted from a surface in some direction at all wavelengths to that emitted by a blackbody at the same temperature is called *directional emittance*. This is an inherent property. Since it is in fact the result of an integration over the blackbody spectrum, which depends on the temperature, this property will be temperature-dependent.

Directional emittance or reflectance values are seldom found in the literature. They may be computed for some substances. The directional emissivity of a diffuse emitter is a constant, independent of direction. Although this condition is often a reasonable approximation, all surfaces exhibit some departure from diffuse behavior. Figure 6.5 shows the representative dependence of the directional emittance on direction for conducting and nonconducting materials. Oxide or paint coatings on metal surfaces will cause the surface to emit with characteristics typical of a dielectric.

For metals, the directional emittance is approximately constant over the range of incidence angles θ less than 40°, after which it increases with angle. For dielectrics, it is approximately constant for angles less than 70°, and then it drops sharply to zero. This has the useful consequence that emittance data taken in the direction normal to the surface are a good approximation to hemispherical emittances.

Monochromatic Hemispherical Emittance The ratio of the energy emitted from a surface at some wavelength to that emitted by a blackbody at the same temperature is called *monochromatic hemispherical emittance*. It is also an inherent property.

The manner in which the spectral hemispherical emittance depends on wavelength depends on whether the solid is a conductor or nonconductor; for a metal, it decreases with wavelength. It also depends on the surface treatment.

As noted above, hemispherical values will not differ very much from normal data. Monochromatic normal emittance data for several surfaces are shown in Fig. 6.6.

Hemispherical Emittance The ratio of the energy emitted from a surface to that emitted by a blackbody at the same temperature is called *hemispherical emittance*. It is an inherent property. Hemispherical emittance or, simply, emit-

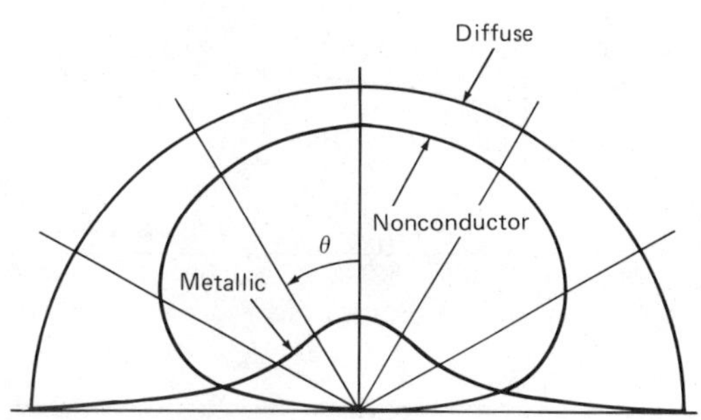

Fig. 6.5 Polar representation of directional emission for different types of surfaces.

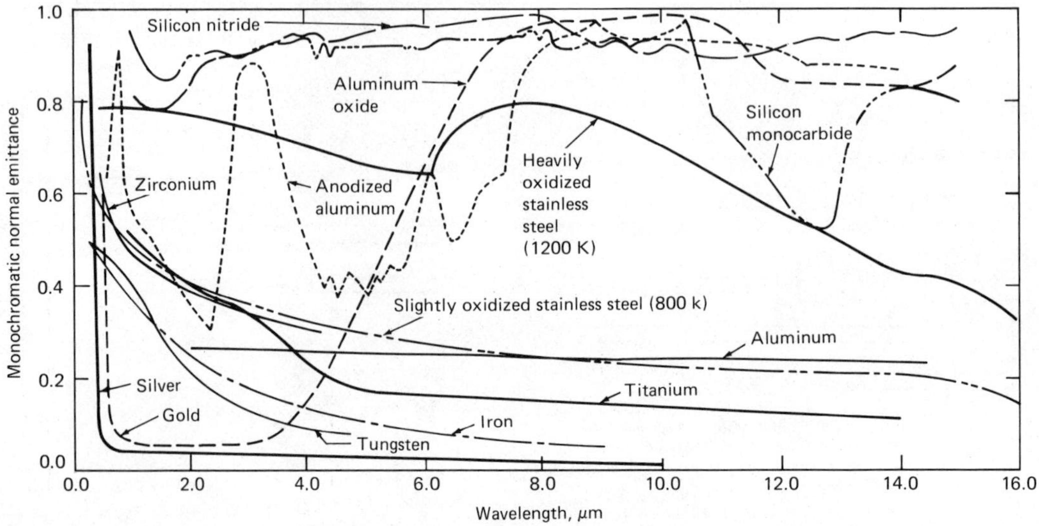

Fig. 6.6 Monochromatic normal emittance for selected surfaces. (*Data taken from Refs. 10 and 12.*)

tance or emissivity is the average value of the hemispherical spectral emittance, weighted by the blackbody spectral power. Thus it is in general a function of temperature.

The total normal emittance ϵ_n is a good approximation to hemispherical emittance ϵ. (The ratio rarely falls outside the range of $0.95 < \epsilon/\epsilon_n < 1.0$ for dielectrics and $1.0 < \epsilon/\epsilon_n < 1.3$ for metals.)

The total normal emittances of selected materials are plotted in Fig. 6.7. Several generalizations may be made.

1. The emissivity of metallic surfaces is generally small. It may be as low as 0.2 for highly polished gold and silver; it does increase with temperature.

2. Oxide layers may increase the emissivity of metallic surfaces significantly.

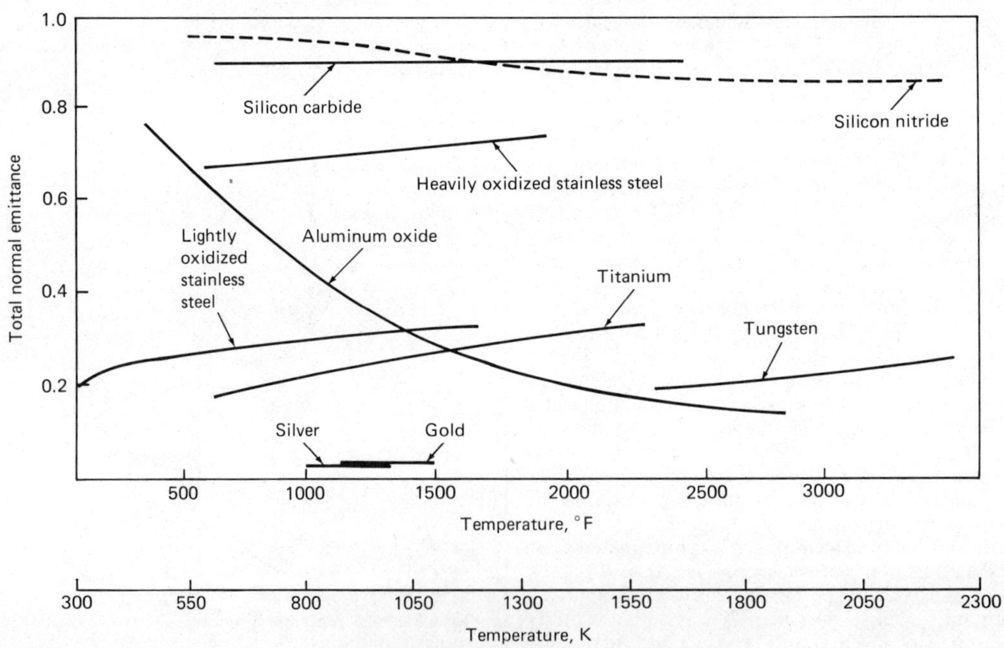

Fig. 6.7 Total normal emittance for selected surfaces (evaluated from the data in Fig. 6.6).

3. The emissivity of nonconductors is relatively large, generally exceeding 0.6. It may increase or decrease with temperature.

The emittance is dependent on the condition of the surface. Experimental data from different sources may vary significantly and much of this variation arises from the way the surface was prepared. For metals the effect of surface oxidation is notable. Figure 6.8 shows the effect of various degrees of oxidation on the hemispherical emissivity of copper.

Absorptance

Monochromatic (or Spectral) Directional Absorptance The fraction of the intercepted monochromatic intensity that is absorbed is called *monochromatic* or *spectral, directional absorptance*. This is an inherent property of the surface.

Directional Absorptance or Directional Total Absorptivity The absorbed fraction of the intercepted intensity is called *directional absorptance* or *directional total absorptance*. Since this characteristic depends on the spectral distribution of the intensity, directional absorptance is *not* an inherent property of the surface.

Monochromatic (or Spectral) Hemispherical Absorptance The fraction of all the energy incident on a surface at some wavelength that is absorbed is called *monochromatic*, or *spectral, hemispherical absorptance*. It is *not* an inherent property.

Hemispherical Absorptance The fraction of all the energy incident on a surface that is absorbed is called *hemispherical absorptance*. Since it involves integrals of the incident radiation over angles and wavelengths, it is *not* an inherent property.

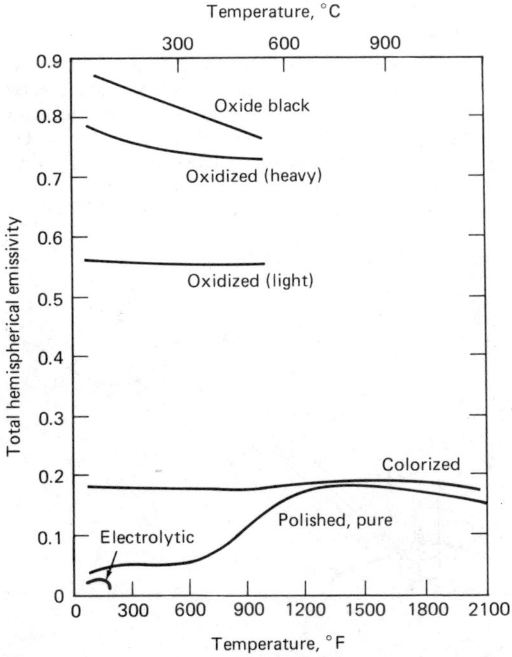

Fig. 6.8 Effect of various degrees of oxidation in the total hemispherical emissivity of copper. (*From Ref. 15; this adaptation is taken from Ref. 5. Copyright © 1973, by John Wiley & Sons, Inc. Reprinted by permission of John Wiley & Sons, Inc.*)

Reflectance

Biangular Reflectance, or Reflectance Function This reflectance is π times the reflected spectral intensity in some direction due to a unit incident spectral flux in some other direction. It is an inherent property.

Spectral Angular to Hemispherical Reflectance The ratio of the reflected flux (in all directions) due to a unit incident flux incident from some direction is called *spectral angular to hemispherical reflectance*. It is an inherent property. Figure 6.9 shows the spectral normal reflectance for selected materials.

Spectral Hemispherical to Angular Reflectance This reflectance is π times the spectral intensity reflected in some direction divided by the hemispherical spectral flux. Since it depends on the directional distribution of the incident radiation, it is *not* an inherent property.

Angular to Hemispherical Reflectance The total reflected flux due to a unit incident flux in some direction is called *angular to hemispherical reflectance*. Since it depends on the spectral distribution of the incident radiation (irradiation), it is *not* an inherent property.

Monochromatic Hemispherical Reflectance The ratio between the reflected spectral flux and the incident spectral flux is called *monochromatic hemispherical reflectance*. It depends on the directional distribution of the incident radiation and is *not* an inherent property.

Hemispherical Reflectance The ratio between the total reflected flux and the total incident flux is called *hemispherical reflectance*. It is *not* an inherent property.

Kirchhoff's Law and Gray Surfaces

Kirchhoff's law, based on thermodynamic considerations, states that the spectral directional emittance from a surface is equal to the spectral directional absorptance. It implies that, whenever the incident radiation is independent of angle or where the surface is diffuse (i.e., emits isotropically), the spectral absorptance is equal to the spectral emittance.

A surface is gray when its properties are not dependent on wavelength. For a diffuse gray surface, the absorptance is equal to the emittance. While this is a common assumption used to simplify analysis, it is not justified for many surfaces and problems. For instance, for many surfaces absorptance for solar radiation is different from emittance at relatively low temperatures.

From energy considerations, the directional monochromatic absorptance is equal to 1 minus the monochromatic directional to hemispherical reflectance. If the incident radiation is independent of angle or the surface is diffuse, the spectral absorptance is equal to 1 minus the monochromatic hemispherical reflectance.

From Kirchhoff's law and the conservation of energy, all radiative surface properties of opaque bodies can be derived if the monochromatic directional absorptance is known.

Practical Assumptions for Surface Characteristics

Simplifying assumptions are usually made when estimating heat-transfer rates between surfaces of different geometries, radiation characteristics, and orientations. Practical applications of the actual directional and spectral surface prop-

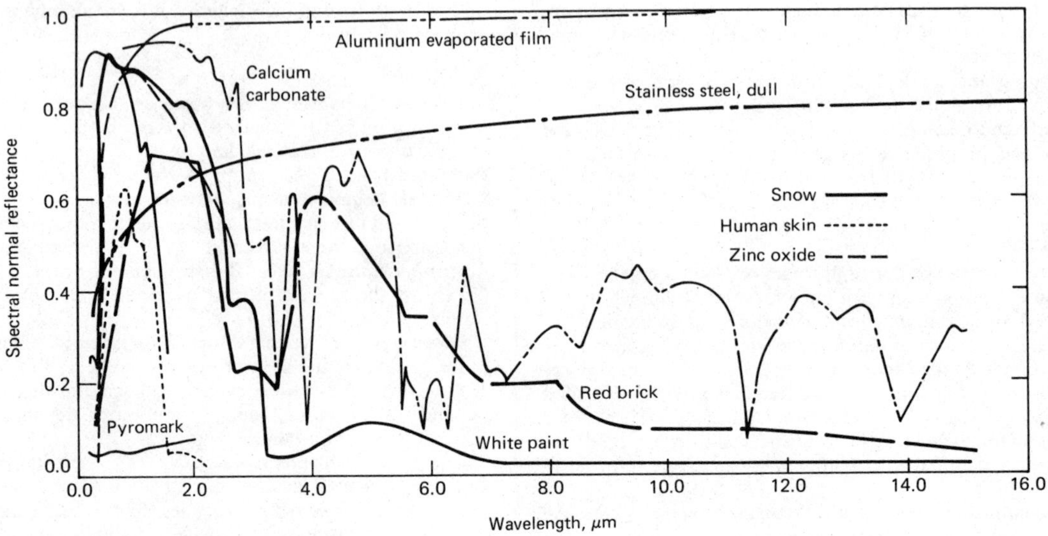

Fig. 6.9 Spectral normal reflectance for selected materials. (*Data taken from Refs. 10 and 12.*)

erties described above are limited by available data and mathematical complexity. Typically, the following is assumed.

- All surfaces are gray or black.
- Radiation and reflection are diffuse.
- Properties are uniform over the surfaces.
- The material in the space between the radiating surfaces does not emit or absorb radiation.

Although it is assumed that the absorptance is equal to the emittance and is independent of the temperature of the source of the incident radiation, it is clear that this assumption is not valid when the surface properties depend on wavelength and the source and the absorbing surfaces are at very different temperatures.

In approaching radiant heat exchange problems, it is important to recognize that accurate prediction of radiant exchange is often an important, but not critical, aspect of thermal design analyses and that there usually will be considerable uncertainty in the physical condition of the radiating surfaces over their lifetime. Thus, it is often appropriate and expedient, for design purposes, to employ the above assumptions with generalized properties data, such as those provided in Fig. 6.4, to make a conservative evaluation of the radiant heat exchange.

6.3 VIEW FACTOR

Definition

The view factor F_{1-2} between surface A_1 and surface A_2 is defined as the fraction of the radiant energy leaving A_1 which reaches A_2. In general, this factor depends on the geometrical relations between the surfaces, their physical properties, and their temperatures, as well as the properties of the medium in the space between them.

Under some conditions, however, the view factor depends only on the geometrical relations between the surfaces under consideration. In such cases, the view factors are called *diffuse view factors*. It is for these cases that the concept is useful since it permits considerable simplification of the analysis of energy transfer by radiation between the surfaces.

General Expressions

Consider two surface elements dA_1 and dA_2, as shown in Fig. 6.10, and assume for simplicity that the space between them is filled with a medium that does not interact with thermal radiation (i.e., a nonparticipating medium). Under these conditions, and using the notation of Fig. 6.10, the rate

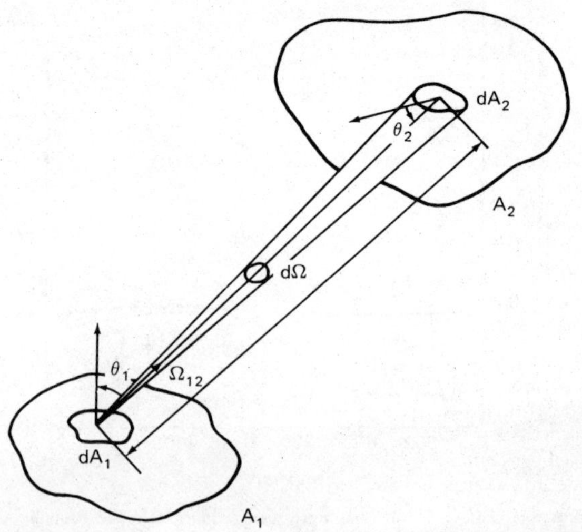

Fig. 6.10 Geometry for radiant interchange between two elementary surfaces.

at which energy reaches dA_2 from dA_1 is given by

$$\underbrace{I(\Omega_{12}) \cos(\theta_1)\, dA_1}_{\substack{\text{rate at which}\\ dA_1 \text{ emits radiant}\\ \text{energy in the}\\ \text{direction of } dA_2}} \underbrace{\cos(\theta_2)\, dA_2/r^2}_{\substack{\text{solid angle subtended}\\ \text{by surface } dA_2 \text{ at}\\ \text{surface } dA_1}}$$

The total radiant energy emitted by the surface element dA_1 is

$$dA_1 \int_{\Omega=0}^{\pi} I(\Omega) \cos(\theta)\, d\Omega$$

By definition the view factor from dA_1 to dA_2 is

$$dF_{1-2} = \frac{I(\Omega_{12}) \cos(\theta_1) \cos(\theta_2)\, dA_2/r^2}{\int_{\Omega=0}^{\pi} I(\Omega) \cos(\theta)\, d\Omega} \tag{6.2}$$

From Eq. (6.2) it is clear that in general the view factor depends not only on the geometrical relation between the surface elements but also on the intensity $I(\Omega)$ of the radiant energy leaving the element dA_1. If $I(\Omega)$ is a constant, that is, if it does not depend on direction, then Eq. (6.2) may be simplified to obtain

$$dF_{1-2} = \frac{\cos(\theta_1) \cos(\theta_2)}{\pi r^2}\, dA_2$$

which depends only on the geometry.

In the case of two finite surfaces A_1 and A_2, the view factor depends only on the geometry if at every point in A_1, the intensity of the radiation leaving A_1 is independent of direction and constant. This diffuse view factor is given by

$$F_{1-2} = \frac{1}{A_1} \int_{A_1} \int_{A_2} \frac{\cos(\theta_1) \cos(\theta_2)}{\pi r^2}\, dA_2\, dA_1 \tag{6.3}$$

Here, for every element dA_1 the integral over A_2 is extended to all the elements of A_2 that dA_1 sees directly.

Diffuse View Factor Properties

Reciprocity Rules These rules relate the view factor from surface 1 to surface 2 with the view factor from surface 2 to surface 1. Three cases are possible: (1) both surfaces are infinitesimal; (2) one surface is infinitesimal and the other is finite; (3) both surfaces are finite. The rules are:

1. $dA_1\, dF_{dA_1-dA_2} = dA_2\, dF_{dA_2-dA_1}$
2. $dA_1\, F_{dA_1-A_2} = A_2\, dF_{A_2-dA_1}$
3. $A_1\, F_{A_1-A_2} = A_2\, F_{A_2-A_1}$

These rules are valid only in those cases in which the intensities $I_1(\Omega)$ and $I_2(\Omega)$ of the radiation leaving A_1 and A_2 are constants (independent of direction and position within each surface), that is, when the view factors depend only on the geometrical relation between the surfaces. Note that the view factor from an elemental surface to a finite surface is finite while the view factor from a finite surface to an elemental surface is infinitesimal.

Rule Based on Energy Conservation The radiant energy leaving a surface which forms part of an enclosure cannot be lost if the medium inside the enclosure does not interact with the radiation.

From this, the following must be true for each surface A_i of an enclosure composed of N surfaces:

$$\sum_{j=1}^{N} F_{i-j} = 1 \tag{6.4}$$

In general, the summation includes the term F_{i-i} because this term is not zero when the surface A_i is concave.

Diffuse View Factor Calculation Methods

The evaluation of the view factors between finite areas requires the determination of two surface integrals [see Eq. (6.3)]. Except for very simple configurations, these integrals are difficult. Some of the more useful methods to estimate view factors are outlined below. These methods apply only to diffuse view factors.

Direct Integration This method consists of solving the quadruple integral in Eq. (6.3) directly. As noted, with the exception of a few configurations, this solution is not simple.

Contour Integrals This method is based on the application of Stokes' theorem to reduce surface integrals to contour integrals. Using this technique, the expression for the view factor between two finite surfaces A_1 and A_2 given by Eq. (6.3) may be reduced to

$$F_{1-2} = \frac{1}{2A_1} \oint_{\substack{\text{contour}\\ A_1}} \oint_{\substack{\text{contour}\\ A_2}} [\ln(r)\, dX_1\, dX_2 + \ln(r)\, dY_1\, dY_2 + \ln(r)\, dZ_1\, dZ_2]$$

where $r = \sqrt{(X_2 - X_1)^2 + (Y_2 - Y_1)^2 + (Z_2 - Z_1)^2}$ and (X_1, Y_1, Z_1) and (X_2, Y_2, Z_2) are the coordinates of the points in surfaces A_1 and A_2, respectively. (For a detailed explanation of this method, see Ref. 6, chap. 7.)

Diffuse View Factor Algebra By using the principle of superposition and the reciprocity rules, the view factor between two complex surfaces may be expressed as a function of known view factors for simpler geometries.

Consider, for example, an enclosure formed by three surfaces which are infinitely long in the direction perpendicular to the figure plane (Fig. 6.11). If the surfaces which make up the enclosure are flat or convex (so that for each surface A_i, $F_{i-i} = 0$) and using the law of conservation of energy and the reciprocity rules, one may write the following system of six equations with six unknowns:

$$\sum_{j=1}^{N} F_{i-j} = 1 \qquad i = 1, 2, 3 \quad \text{(3 equations)}$$

$$A_i F_{i-j} = A_j F_{j-i} \qquad i, j = 1, 2, 3 \quad \text{(3 equations)}$$

Solving this system, one obtains the six unknown F_{i-j}. For example, F_{1-2} is given by

$$F_{1-2} = \frac{A_1 + A_2 - A_3}{2A_1}$$

which in this case may be written as

$$F_{1-2} = \frac{L_1 + L_2 - L_3}{2L_1}$$

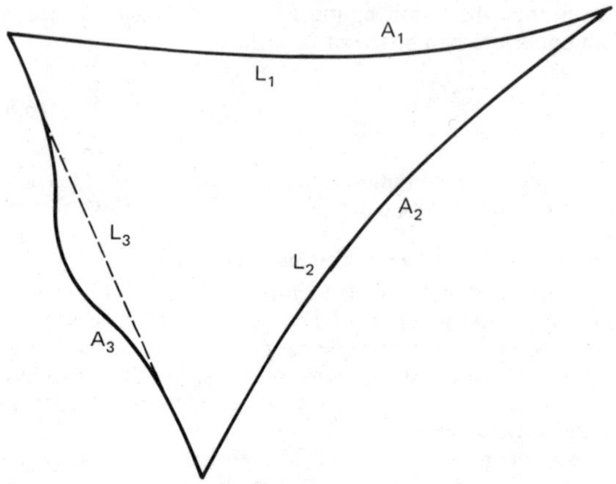

Fig. 6.11 Construction to estimate view factors for a two-dimensional enclosure with three walls.

where L_1, L_2, and L_3 are the lengths of the arcs AB, BC, and CA, respectively.

A generalization of the preceding example is the so-called cross-string method, or Hottel's rule. This rule states that the view factor between two surfaces of an enclosure of the type shown above may be calculated as the sum of the lengths of the cross strings joining the end points of the lines representing the surfaces, minus the sum of the lengths of the noncrossed strings joining the same end points, divided by twice the area of the surface under consideration (see Fig. 6.12),

$$F_{1-2} = \frac{(AE + BHGF) - (AIHGF + BCDE)}{2A_1}$$

View Factor Data

Most texts on heat transfer treat the concept of radiation heat transfer in some detail, but because of space limitations, they reproduce only a few of the hundreds of configuration factors which have been published in the literature.

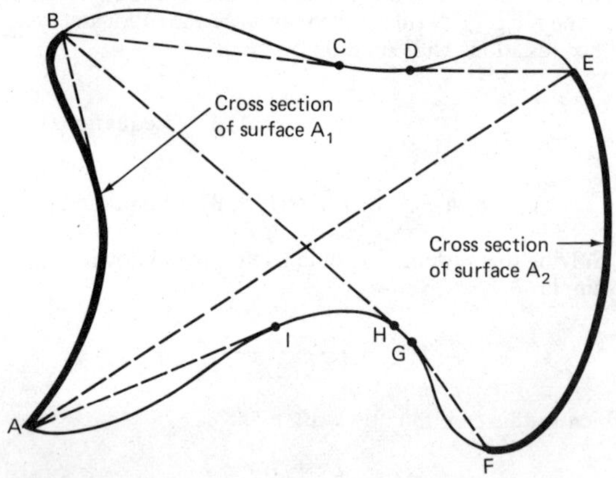

Fig. 6.12 Construction for Hottel's rule.

Reference 8 constitutes a notable attempt to gather many of the most useful published view factors into a single source; it also provides an informative annotated list of references. In this reference, the factors are divided into three types: between differential elements, from a differential element to a finite area, and between two finite areas. Some view factors between finite surfaces for common geometries are shown in Fig. 6.13 and Table 6.2.

6.4 RADIATIVE INTERCHANGE IN NONPARTICIPATING MEDIA

To calculate the rate of radiant energy exchange between surfaces separated by a nonparticipating medium, one must take into consideration all the directions from which radiation may impinge on the surfaces. The concept of an enclosure is useful to ensure that all directions are considered.

To estimate radiation heat transfer between several surfaces, one adds imaginary surfaces to the configuration so as completely to enclose the space in between. Then one assigns radiative properties to the real and imaginary surfaces surrounding the enclosure and specifies their thermal state. The equations which determine the energy exchange are the energy balance equations for the radiation that is incident, reflected, and emitted per each characteristic surface which makes up the enclosure.

In the general case in which the radiative properties of the surfaces vary with radiation wavelength and direction, as well as with position, the governing equations are integral equations whose solution may be difficult. In practice it is often possible to make simplifying assumptions to turn these integral equations into algebraic equations. Some of these simplifications are listed below.

Simplified Zone Analysis

The analysis of radiative heat exchange in an enclosure is greatly simplified if the enclosure envelope is divided into a finite number of zones and one assumes the following conditions to be valid for each zone.

1. The surface properties are uniform and independent of radiation wavelength and direction.
2. Each surface is at a constant temperature or intercepts a constant radiant flux.
3. The radiosity, that is, the radiant energy leaving the surface, is uniform over each zone.
4. The surfaces are opaque.

With these assumptions, the equations that determine the energy exchange within the enclosure simplify to

$$R_i = \epsilon_i \sigma T_i^4 + \rho_i \sum_{j=1}^{N} R_j F_{i-j} \qquad i = 1, \ldots, N \qquad (6.5)$$

$$q_i = R_i - \sum_{j=1}^{N} R_j F_{i-j} \qquad i = 1, \ldots, N \qquad (6.6)$$

where R_i = radiosity from zone i
ϵ_i = emissivity of zone i
ρ_i = reflectivity of zone i
T_i = absolute temperature of zone i
F_{i-j} = view factor from zone i to zone j

Both sets of equations are statements of energy balance. According to Eq. (6.5), the radiosity from surface i is the

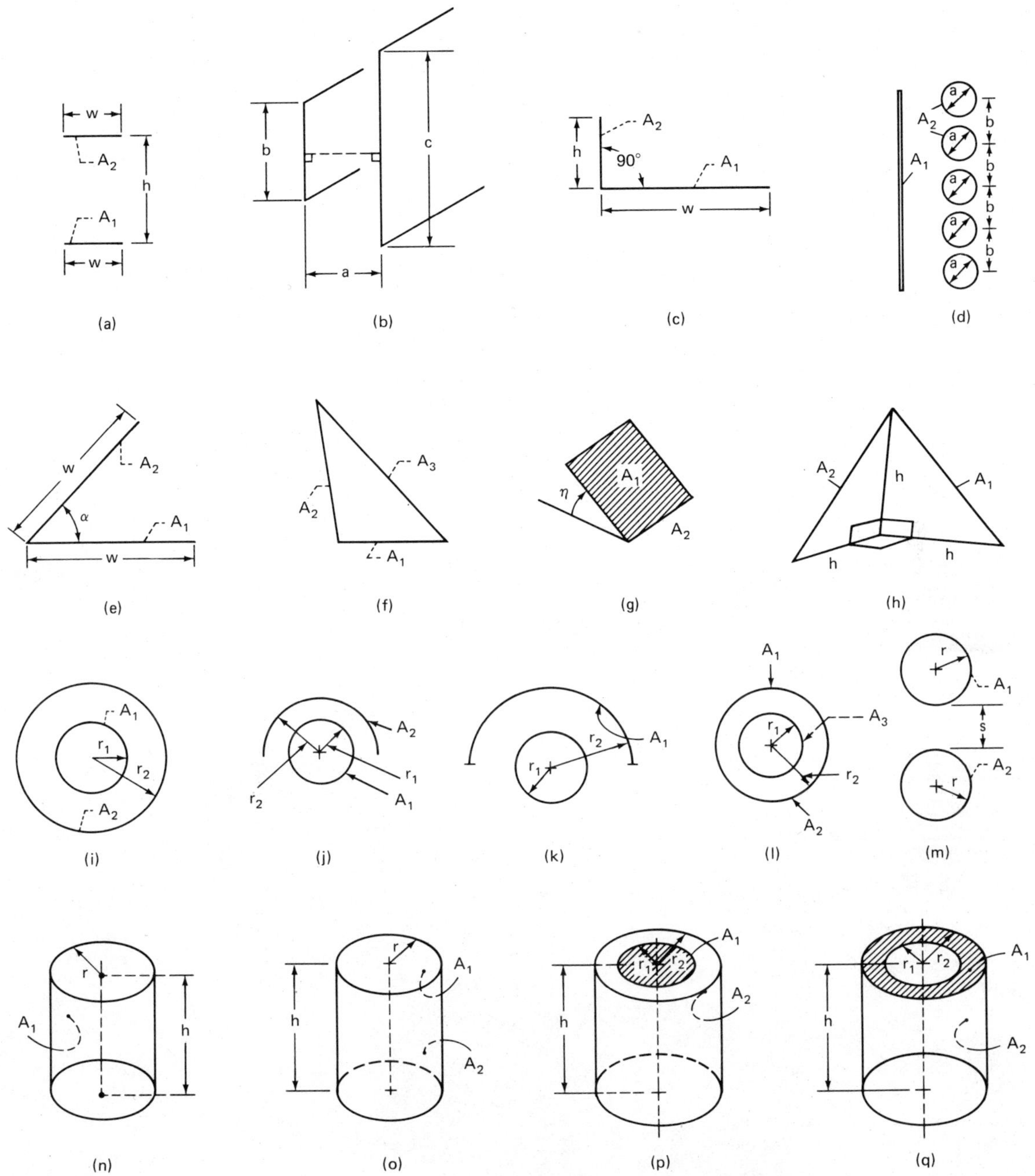

Fig. 6.13 Schematic representation of selected finite surface configurations (Table 6.2). These configurations have been taken from Ref. 8; configuration (*a*) is adapted from Ref. 16; configuration (*b*) is adapted from Ref. 17; and configurations *c, e, i,* and *m* are adapted from Ref. 15.

Table 6.2. View Factors for Selected Finite Surface Configurations (Fig. 6.13)

Configuration	Definition	Governing equation
(*a*) Two infinitely long, directly opposed parallel plates of same finite width.[16]	$H = \frac{h}{w}$	$F_{1-2} = F_{2-1} = \sqrt{1 + H^2} - H$
(*b*) Two infinitely long parallel plates of different widths; centerlines of each plate are connected by perpendicular plane between plates.[17]	$B = \frac{b}{a}, \quad C = \frac{c}{a}$	$F_{1-2} = \frac{1}{2B}[\sqrt{(B + C)^2 + 4} - \sqrt{(C - B)^2 + 4}]$
(*c*) Two infinitely long plates of unequal widths h and w, having one common edge, and at 90° angle to each other.[6]	$H = \frac{h}{w}$	$F_{1-2} = \frac{1}{2}(1 + H - \sqrt{1 + H^2})$
(*d*) Infinite plane to a row of pipes.[16]	$D = \frac{a}{b}$	$F_{1-2} = 1 - (1 - D^2)^{1/2} + D \tan^{-1}\left(\frac{1 - D^2}{D^2}\right)^{1/2}$
For n rows of in-line pipes.		$F_{1-n \text{ rows}} = 1 - (1 - F_{1-2})^n$
(*e*) Two infinitely long plates of equal finite width w, having one common edge, and at included angle α to each other.[6]		$F_{1-2} = F_{2-1} = 1 - \sin \frac{\alpha}{2}$
(*f*) Infinitely long enclosure formed by three plane areas.		$F_{1-2} = \frac{A_1 + A_2 - A_3}{2A_1}$
(*g*) Finite rectangle tilted relative to infinite plane.		$F_{1-2} = \frac{1 - \cos \eta}{2}$
(*h*) Right isosceles triangle to adjacent congruent perpendicular triangle connected along short side.[8]		$F_{1-2} = \frac{1}{2}\left(1 - \frac{1}{\sqrt{3}}\right)$
(*i*) Concentric cylinders of infinite length.[6]		$F_{1-2} = 1, \quad F_{2-2} = 1 - \frac{r_1}{r_2},$ $F_{2-1} = \frac{r_1}{r_2}$
(*j*) Exterior of infinitely long cylinder to interior of concentric semicylinder.		$F_{1-2} = \frac{1}{2}$
(*k*) Interior of infinitely long semicylinder to itself when concentric coaxial cylinder is present.	$R = \frac{r_1}{r_2}$	$F_{1-1} = 1 - \frac{2}{\pi}[(1 - R^2)^{1/2} + R \sin^{-1} R]$
(*l*) Interior of infinitely long semicylinder 1 to interior of semicylinder 2 when concentric parallel cylinder 3 is present.	$R = \frac{r_1}{r_2}$	$F_{1-2} = \frac{2}{\pi}[(1 - R^2)^{1/2} + R \sin^{-1} R] - R$
(*m*) Infinitely long parallel cylinders of same diameter.[6]	$X = 1 + \frac{s}{2r}$	$F_{1-2} = F_{2-1} = \frac{1}{\pi}\left(\sqrt{X^2 - 1} + \sin^{-1}\frac{1}{X} - X\right)$
(*n*) Inside surface of right circular cylinder to itself.	$H = \frac{h}{2r}$	$F_{1-1} = (1 + H) - (1 + H^2)^{1/2}$
(*o*) Base of right circular cylinder to inside surface of cylinder.[8]	$H = \frac{h}{2r}$	$F_{1-2} = 2H[(1 + H^2)^{1/2} - H]$
(*p*) Disk in cylinder base or top to inside of right circular cylinder.[8]	$R = \frac{r_2}{r_1}, \quad H = \frac{h}{r_1}$	$F_{1-2} = \frac{1}{2}\{1 - R^2 - H^2 + [(1 + R^2 + H^2)^2 - 4R^2]^{1/2}\}$
(*q*) Annular ring on cylinder base or top to inside of right circular cylinder.[8]	$R = \frac{r_2}{r_1}, \quad H = \frac{h}{r_1}$	$F_{1-2} = \frac{1}{2}\left(1 + \frac{1}{(R^2 - 1)}\{H(4R^2 + H^2)^{1/2} - [(1 + R^2 + H^2)^2 - 4R^2]^{1/2}\}\right)$

sum of the emitted flux plus the reflected fraction of the total incident flux. Equation (6.6) states that the net flux q_i leaving surface i is equal to the difference between the outgoing and the incoming fluxes. These equations constitute a system of $2N$ algebraic equations, which may be solved if the number of unknowns is also $2N$.

Usually the view factors and emissivities are known. In this case, the number of variables related by the equations is $3N$ (N radiosities, N radiative heat fluxes, and N surface temperatures). Since radiosities are generally unknown, N values among net radiative fluxes and surface temperatures need to be prescribed in order to solve the system of equations.

Suppose that temperatures are prescribed for the first m surfaces of the enclosure and net radiative fluxes are prescribed for the remaining $N - m$ surfaces. In this case, a system of N linear algebraic equations in which the only unknowns are the radiosities can be extracted from the general system of $2N$ equations as follows:

$$\sum_{j=1}^{N} R_i X_{ij} = P_i \qquad 1 \le i \le N$$

where:

$$\left.\begin{aligned} X_{ij} &= \delta_{ij} - \rho_i F_{i-j} \\ P_i &= \epsilon_i \sigma T_i^4 \end{aligned}\right\} \qquad 1 \le i \le m$$

$$\left.\begin{aligned} X_{ij} &= \delta_{ij} - F_{i-j} \\ P_i &= q_i \end{aligned}\right\} \qquad m + 1 \le i \le N$$

δ_{ij} is the Kronecker delta, equal to unity when i is equal to j and 0 otherwise.

This system can be written in matrix notation as

$$\begin{bmatrix} X_{11} & X_{12} & \cdots & X_{1N} \\ X_{21} & X_{22} & \cdots & X_{2N} \\ \vdots & \vdots & & \vdots \\ X_{N1} & X_{N2} & \cdots & X_{NN} \end{bmatrix} \begin{bmatrix} R_1 \\ R_2 \\ \vdots \\ R_N \end{bmatrix} = \begin{bmatrix} P_1 \\ P_2 \\ \vdots \\ P_N \end{bmatrix}$$

or

$$\mathbf{XR} = \mathbf{P}$$

Solving for the radiosities, one gets

$$\mathbf{R} = \mathbf{X}^{-1}\mathbf{P}$$

where $\mathbf{X}^{-1}$ is the inverse matrix of $\mathbf{X}$.

Thus the radiosities can be expressed as a weighted linear combination of the net fluxes or emitted fluxes leaving the surfaces of the enclosure. Once the radiosities are calculated, the determination of the unknown fluxes and temperatures is straightforward, using Eq. (6.5) or Eq. (6.6). This formulation has the advantage that once the elements of the inverse matrix $\mathbf{X}^{-1}$ have been determined, there is no need to solve the system of equations for the enclosure again when different values of T_i ($1 \le i \le m$) and q_i ($m + 1 \le i \le N$) are prescribed.

In general, when the number of zones making up the enclosure is high, it is necessary to use a computer to solve the system of equations.

It is important to emphasize the limitations implied by the assumptions on which the simplified zone analysis is based. The model (to which the literature often refers as the gray-body model) leads to realistic results when the spectral emittances of the different zones are essentially constant within the spectrum range at which the radiant energy exchange takes place. The model may produce erroneous results when the exchanges take place over a wide wavelength spectrum which includes not only the infrared region but also the visible or when there are zones of the enclosure which exhibit strong emission or reflection bands within that spectrum.

In those situations, the electromagnetic spectrum may be divided into L bands within which the surface radiative properties may be considered constant. This leads to a system of $2N$ equations for each band, which may be solved for the band radiosities and fluxes for each zone. The total radiosities and fluxes for each zone may then be evaluated from

$$R_i = \sum_{k=1}^{N} R_{ik} \qquad q_i = \sum_{k=1}^{N} q_{ik}$$

where R_{ik} and q_{ik} are the radiosities and fluxes from zone i within the kth band. Solar central cavity receivers are an interesting example where the surfaces cannot be considered gray (see Ref. 9).

Exchange Between Blackbodies If all the surfaces that compose the enclosure are blackbody surfaces, Eqs. (6.5) and (6.6) reduce to

$$R_i = \sigma T_i^4 \tag{6.7}$$

$$q_i = R_i - \sum_{j=1}^{N} R_j F_{i-j} \tag{6.8}$$

Substituting Eq. (6.7) into Eq. (6.8) and using the energy conservation relation for the view factors [Eq. (6.4)], we get

$$q_i = \sum_{j=1}^{N} \sigma(T_i^4 - T_j^4) F_{i-j}$$

The Two-Surface Enclosure For any two gray surfaces that form an enclosure, Eqs. (6.5) become

$$R_1 = \epsilon_1 \sigma T_1^4 + \rho_1[R_1 F_{1-1} + R_2 F_{1-2}]$$

$$R_2 = \epsilon_2 \sigma T_2^4 + \rho_2[R_1 F_{2-1} + R_2 F_{2-2}]$$

which may be solved for R_1 and R_2. Writing the net radiant exchange $q_{1-2} = q_1 = -q_2 = R_1 - R_2$ and using $\rho_i = 1 - \epsilon_i$,

$$\sum_{j=1}^{2} F_{i-j} = 1$$

and the reciprocity rules, one obtains

$$q_{12} = q_1 = -q_2 = \frac{\sigma(T_1^4 - T_2^4)}{\dfrac{1 - \epsilon_1}{\epsilon_1 A_1} + \dfrac{1}{A_1 F_{1-2}} + \dfrac{1 - \epsilon_2}{\epsilon_2 A_2}} \tag{6.9}$$

The Monte Carlo Method

In general, the Monte Carlo method to analyze a complicated physical process consists of replacing the process by a statistical model which is as simple as possible and is based on chains of elemental events of stochastic nature. It is assumed that the probability of each event in a chain does not depend on prior events; it is independent of history.

Radiative exchange processes inside a system are modeled by following the progress of energy bundles as they traverse the system. Whenever an interaction occurs, the fate of the actual energy bundle is decided by chance (by the value of a random number), but in such a way that the physical constraints of the problem are satisfied. For instance, if the total absorptivity of a surface at a given temperature is 0.75, the decision process that determines whether or not a given energy bundle is absorbed by the surface should operate in such a way that after an infinite number of bundles have struck the surface, 75 percent of them are absorbed.

A detailed description of the Monte Carlo method and its applications to radiative problems is given in Ref. 6.

For very complex problems the Monte Carlo method presents clear advantages over other radiative-exchange calculation methods. It must be noted that, because of the statistical nature of the method, the statistical convergence of the results to their true mean values is not guaranteed.

6.5 SOLUTION TO COMMON PROBLEMS

Radiation Heat-Transfer Coefficient

When the temperature difference ΔT is small compared with the absolute values of the temperatures, one may write

$$\sigma(T_1^4 - T_2^4) = h_r(T_1 - T_2)$$

where the radiative heat transfer coefficient h_r is equal to $4\sigma T_m$ and T_m is the mean temperature at which the radiative exchange takes place.

Special Cases

Important special cases of two diffuse, gray surfaces may be derived from Eq. (6.9) and view factor relations. They are listed in Fig. 6.14 and Table 6.3.

Radiation Shields Thin layers of reflective material may be used to reduce the net radiation transfer between two surfaces. Consider a shield between two large plane surfaces, as shown in Fig. 6.14*b*. The heat transfer between the shield and the two surfaces it sees are given by expressions like the one listed in Table 6.3*a* for case 1. In the steady state, or when the thermal capacity of the shield may be neglected, the heat transfer into the shield must equal the heat transfer out. This condition lowers the overall heat-transfer rate between the two surfaces. The lower rate shown in Table 6.3*b*

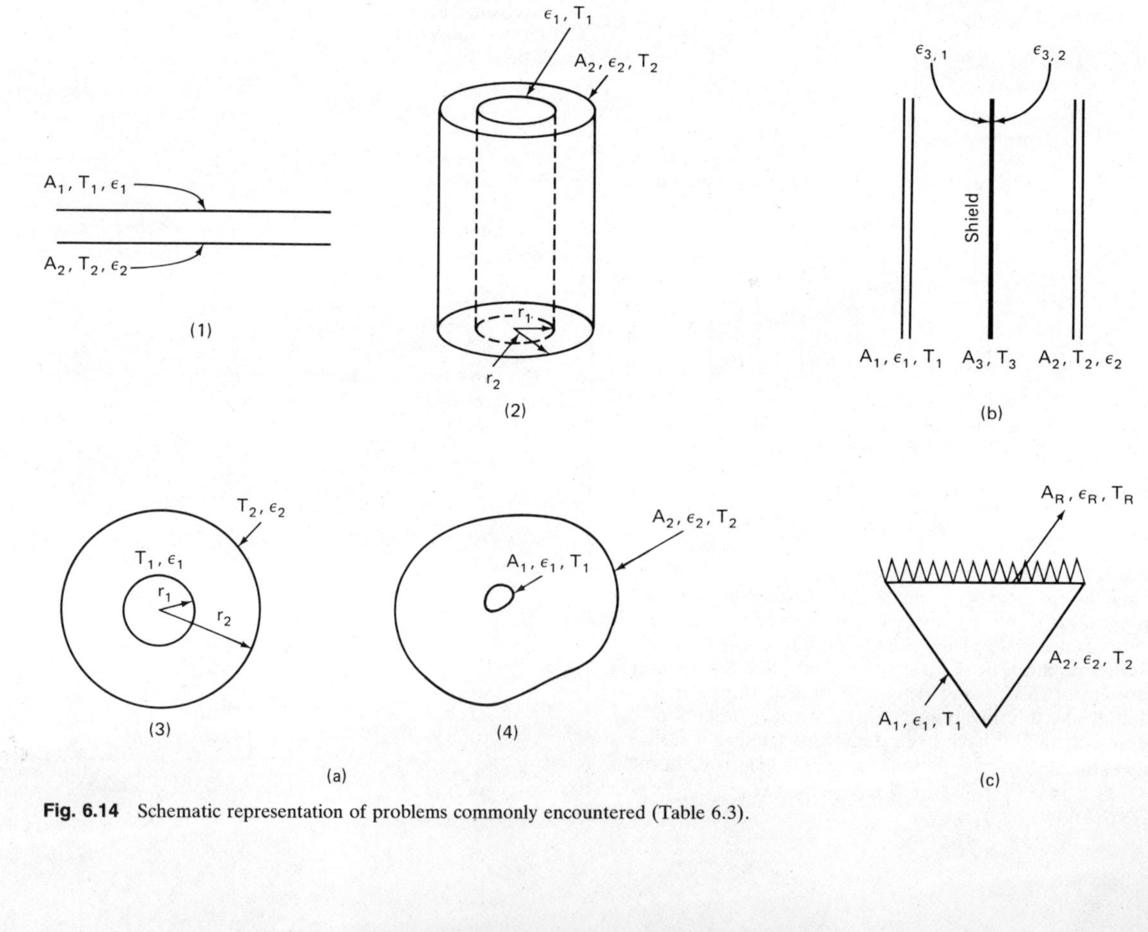

Fig. 6.14 Schematic representation of problems commonly encountered (Table 6.3).

Table 6.3. Solution to Common Problems in Radiation Heat Transfer (Fig. 6.14), All Temperatures Absolute

(*a*) Between Two Surfaces

Case	Solution
1. Large infinite parallel plates, $F_{12} = 1$	$q_{12} = \dfrac{\sigma A_1(T_1^4 - T_2^4)}{\dfrac{1}{\epsilon_1} + \dfrac{1}{\epsilon_2} - 1}$
2. Long (infinite) concentric cylinders, $F_{12} = 1$	$q_{12} = \dfrac{A_1(T_1^4 - T_2^4)}{\dfrac{1}{\epsilon_1} + \dfrac{1 - \epsilon_2}{\epsilon_2}\left(\dfrac{r_1}{r_2}\right)}$
3. Concentric spheres, $F_{12} = 1$	$q_{12} = \dfrac{A_1(T_1^4 - T_2^4)}{\dfrac{1}{\epsilon_1} + \dfrac{1 - \epsilon_2}{\epsilon_2}\left(\dfrac{r_1}{r_2}\right)^2}$
4. Small convex object in large cavity, $F_{12} = 1$	$q_{12} = \sigma A_1 \epsilon_1 (T_1^4 - T_2^4)$

(*b*) Two Large Parallel Plates with Radiation Shield

$$q_{12} = \frac{A_1(T_1^4 - T_2^4)}{\dfrac{1}{\epsilon_1} + \dfrac{1}{\epsilon_2} + \dfrac{1 - \epsilon_{3.1}}{\epsilon_{3.1}} + \dfrac{1 - \epsilon_{3.2}}{\epsilon_{3.2}}}$$

(*c*) Three-Surface Enclosure with One Insulated Reradiating Surface

$$q_1 = -q_2 = \frac{\sigma(T_1^4 - T_2^4)}{\dfrac{1 - \epsilon_1}{\epsilon_1 A_2} + \dfrac{1}{A_1F_{12} + (1/A_2F_{1r}) + (1/A_2F_{2r})} + \dfrac{1 - \epsilon_2}{\epsilon_2 A_2}}$$

From q_1 and q_2, one may determine the radiosities R_1 and R_2. From the radiosities, the radiosity of the reradiating surface R_r is determined,

$$\frac{R_1 - R_r}{1/A_1F_{1r}} - \frac{R_r - R_2}{1/A_2F_{2r}} = 0$$

as well as the temperature T_r of the reradiating surface, $T_r^4 = R_r/\sigma$.

refers to the general case in which the two sides of the shield have different emittances.

For the case of N shields where all the emissivities are equal,

$$q_{1,2} = \frac{q_{1,2,0}}{N + 1}$$

where $q_{1,2,0}$ is the heat-transfer rate with no shields.

Reradiating Surfaces A real surface which is well insulated and for which convection effects are negligible may be approximated by one with zero net radiation transfer. Figure 6.14*c* and Table 6.3*c* refer to the three-surface enclosure with one reradiating surface.

6.6 RADIATIVE INTERCHANGE IN PARTICIPATING MEDIA

Volumetric Absorption

A participating medium is one that interacts with thermal radiation. Spectral radiation absorption in such a medium is described in terms of the monochromatic absorption coefficient K_λ (1/ft or 1/m) and the thickness L of the medium. If a monochromatic beam of intensity I_λ enters the medium, the attenuation may be expressed as

$$\frac{dI_\lambda(X)}{dX} = -K_\lambda I(X)$$

and the spectral transmissivity τ_λ, defined as

$$\tau_\lambda = \frac{I_\lambda(L)}{I_\lambda(0)} = \exp\left(-\int_0^L K_\lambda\, dX\right)$$

The spectral absorptivity is $\alpha_\lambda = 1 - \tau_\lambda$.

Gaseous Emission

From Kirchhoff's law, the spectral emissivity for a gas is given by

$$\epsilon_\lambda = \alpha_\lambda = 1 - \tau_\lambda$$

Gas radiation is the dominant mode of heat transfer in large furnaces. Spectral and directional effects complicate the analysis. A method developed by Hottel[13] is based on determining the emission from a hemispherical gas mass at temperature T_g to a surface element dA, located at the center of the base of the hemisphere.

The emission per unit area of the surface E_g is expressed as

$$E_g = \epsilon_g \sigma T_g^4$$

where ϵ_g is determined from available data. ϵ_g has been correlated in terms of T_g and the total pressure P of the gas, the partial pressure P_g of the radiating species, and the radius L of the sphere.

Figure 6.15 shows the emissivity of water vapor and carbon dioxide at 1-atm pressure as a function of temperature for different values of vapor partial pressure and hemispherical radii. For other pressures, the emissivity from Fig. 6.15 must be multiplied by a correction factor (see Ref. 13).

For geometries other than the sphere, one introduces the mean beam length L, which has been determined for numerous shapes. Representative results are listed in Table 6.4.

With data such as these, the heat-transfer rate to a surface from an adjoining gas may be written as

$$q = \epsilon_g A_g \sigma T_g^4$$

If the surface is black, it will absorb all this radiation. A black surface will also emit, and the net rate at which radiation is exchanged between the surface at T_s and the gas at T_g is

$$q = A_g\sigma(\epsilon_g T_g^4 - T_s^4)$$

ASHRAE[14] suggests the simplified expression

$$q = \sigma A_s \epsilon_g (T_g^4 - T_s^4)$$

6.7 REVIEW OF CONDITIONS UNDER WHICH RADIATIVE HEAT TRANSFER IS IMPORTANT

When two surfaces are at different temperatures and see each other, or when one surface sees a participating medium, the radiative energy transfer tends to be high when the difference

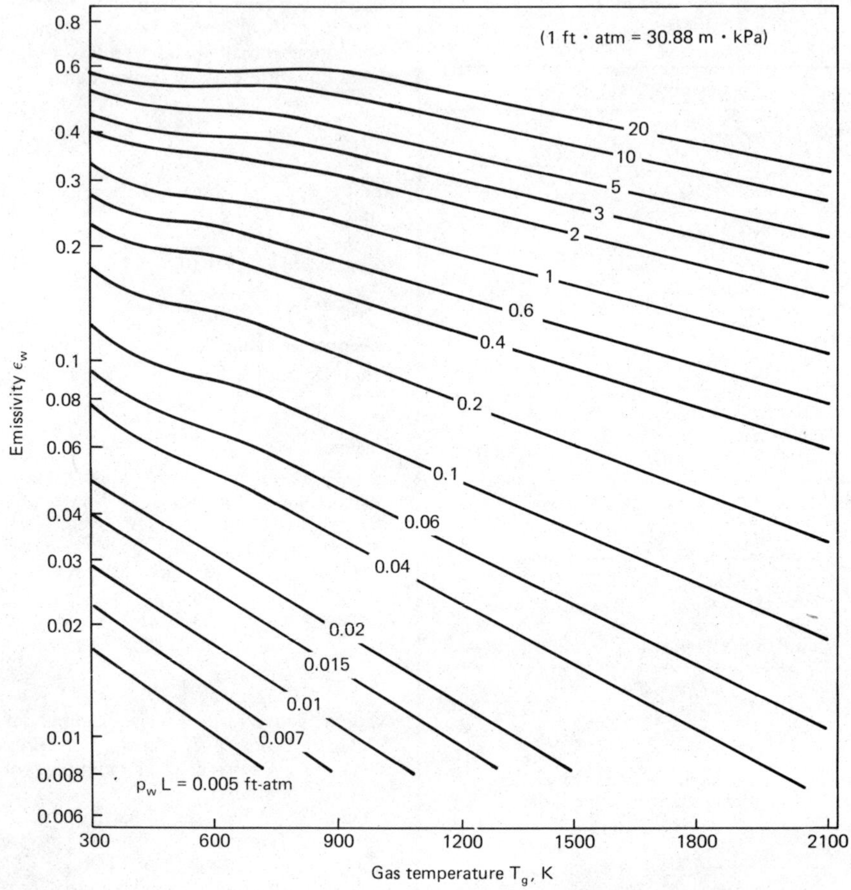

(*a*)

Fig. 6.15 Gas emissivity in a mixture with nonradiating gases at 1-atm total pressure and of hemispherical shape. (*a*) Water vapor; (*b*) carbon dioxide. (*From Ref. 13; these adaptations are taken from Ref. 10. Copyright © 1985, by John Wiley & Sons, Inc. Reprinted by permission of John Wiley & Sons, Inc.*)

Table 6.4. Mean Beam Lengths L_e for Various Gas Geometries

Geometry	Characteristic length	L_e
Sphere (radiation to surface)	Diameter D	$0.65D$
Infinite circular cylinder (radiation to curved surface)	Diameter D	$0.95D$
Semi-infinite circular cylinder (radiation to base)	Diameter D	$0.65D$
Circular cylinder of equal weight and diameter (radiation to entire surface)	Diameter D	$0.60D$
Infinite parallel planes (radiation to planes)	Spacing between planes L	$1.80L$
Cube (radiation to any surface)	Side L	$0.66L$
Arbitrary shape of volume V (radiation to surface of area A)	Volume-to-area ratio V/A	$3.6V/A$

SOURCE: Ref. 10. Copyright © 1985, by John Wiley & Sons, Inc. Reprinted by permission of John Wiley & Sons, Inc.

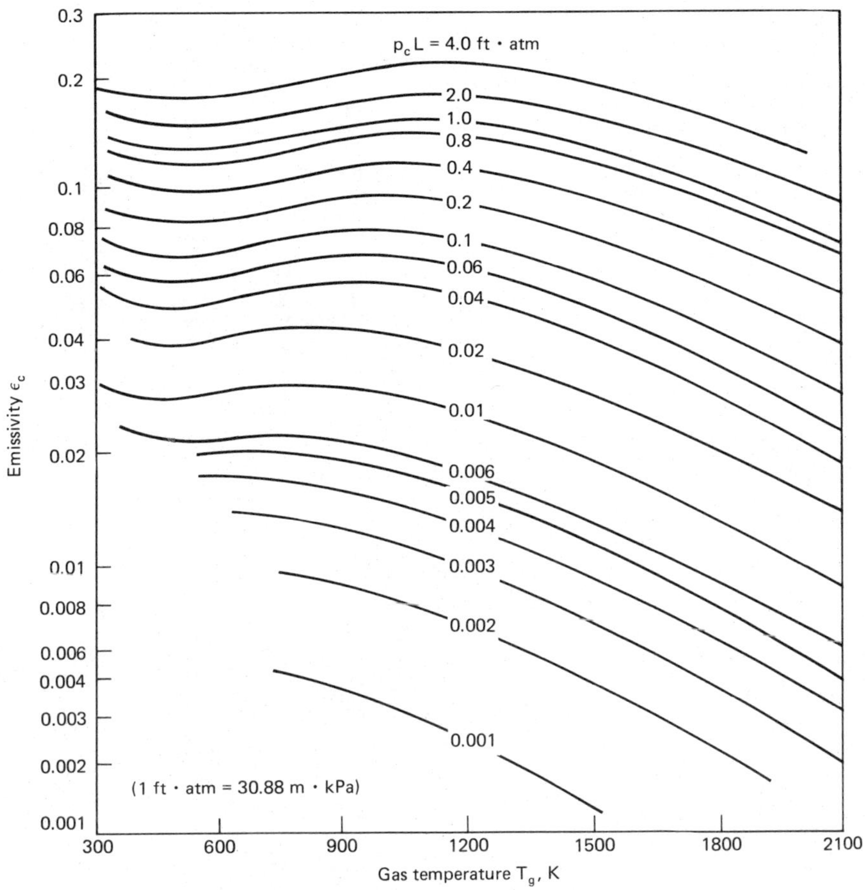

(*b*)

Fig. 6.15 (*Continued*)

in the fourth powers of the absolute temperatures T^4 of the two surfaces is high.

At 300 K the value of $4\sigma T_m^3$ is of the same order as the natural convection coefficient for heat transfer in air. At T_m = 2000 K the radiant heat transfer is 300 times higher (1800 W/m² · K) and is likely to be a dominant transfer mechanism.

Radiative exchange is *not* important when considering transfer *within* a fluid that is *opaque* to radiation in the spectral range which it intercepts. Water is opaque to infrared radiation—radiation heat transfer in the interior of water is only a factor that contributes to energy transfer by conduction.

Similarly, one *does not* consider radiation exchange with fluids that are completely *transparent* in the relevant spectrum. Such fluids are called *diathermal.* Of course, if the walls surrounding such a fluid are at different temperatures from each other, they can exchange energy by radiation.

One does not consider radiative heat transfer in the inside of pipes in a heat exchanger that contains air, oil, or water—in the case of air, because air is diathermal; in the case of oil and water, because these liquids are opaque to infrared radiation.

In short, radiative exchange is important when the temperatures are high and the *distances* between the surfaces are *large* because, in the former, the radiation becomes large, and in the latter, convective exchange diminishes. Radiative exchange is also important in *vacua,* where conduction heat transfer is very low because the fluid density is low.

REFERENCES

1. M. Planck, *The Theory of Heat Radiation,* Dover, New York, 1959.
2. H. C. Hottel and A. F. Sarofim, *Radiative Transfer,* McGraw-Hill, New York, 1967.
3. T. J. Love, *Radiative Heat Transfer,* Merrill, Columbus, Ohio, 1968.
4. E. M. Sparrow and R. D. Cess, *Radiation Heat Transfer,* Brooks/Cole, Belmont, Calif., 1970.
5. M. N. Ozisik, *Radiative Transfer,* Wiley, New York, 1973.

6. R. Siegel and J. R. Howell, *Thermal Radiation Heat Transfer,* 2d ed., Hemisphere, New York, 1981.
7. E. R. G. Eckert, "Radiation—Part 1," in *Handbook of Heat Transfer Fundamentals,* 2d ed., W. M. Rohsenow, J. P. Hartnett, and E. N. Ganic (eds.), McGraw-Hill, New York, 1985.
8. J. R. Howell, *A Catalog of Radiation Configuration Factors,* McGraw-Hill, New York, 1985.
9. A. DeBenedetti and J. G. Martin, "Receiver Thermal Performance: Theory," in *The IEA/SSPS Solar Thermal Power Plants,* vol. 1, P. Hesselring and C. S. Selvage (eds.), Springer, Berlin, 1986, chap. 5.
10. F. P. Incropera and D. P. De Witt, *Fundamentals of Heat and Mass Transfer,* 2d ed., Wiley, New York, 1985.
11. International Organization for Standardization, Norme Internationale ISO 31/6-1980 (F).
12. Y. S. Touloukian et al., *Thermophysical Properties of Matter,* Plenum, New York, 1972.
13. H. C. Hottel, "Radiant-Heat Transmission," in *Heat Transmission,* 3d ed., W. H. McAdams (ed.), McGraw-Hill, New York, 1954.
14. *1985 Fundamentals Handbook,* ASHRAE, Atlanta, Ga., 1985.
15. G. G. Gubareff, J. E. Janssen, and R. H. Torborg, "Thermal Radiation Properties Survey," Honeywell Research Center, Minneapolis-Honeywell Regulator Company, Minneapolis, Minn., 1960.
16. H. C. Hottel, "Radiant Heat Transmission between Surfaces Separated by Non-absorbing Media," *Trans. ASME,* **53,** FSP-53-196, 265–273 (1931).
17. H. Y. Wong, *Handbook of Essential Formulae and Data on Heat Transfer for Engineers,* Longman Group Ltd., London, 1977.

Chapter 7

Numerical Analysis in Thermal Design

ASHOK K. SINGHAL
CFD Research Corporation

This chapter reviews current capabilities and roles of numerical analysis in thermal design. The objective is to introduce the reader to general information on the topic and to provide references for more detailed information.

7.1 INTRODUCTION AND STATUS OF NUMERICAL ANALYSIS

Until recently, thermal design of industrial equipment has been based on simple computations of heat and mass balances and small-scale model tests. Improvements in the design and establishment of safe-operation limits have been based on observations of the actual performance of the equipment under a variety of conditions. Accumulation of such information is slow and, as a result, design modification cycles span over decades. The only way of accelerating this process has been to perform more tests. Either of the two options, either to wait or to perform more tests, is expensive and tedious.

In recent years, numerical analysis has provided a new horizon to thermal designers. In most cases numerical models can be set up to simulate various modes of operation and to predict the corresponding performances. Once the basic model is established, thousands of simulations can be made in a short time. As a result, right from the beginning, the designer has much greater flexibility for optimizing various parameters for both safe and efficient operation of the equipment. Before discussing further advantages, capabilities, and limitations of numerical analysis, let us first establish the context and meaning of numerical analysis.

Heat-transfer problems can be divided into several categories, namely, conduction, convection, radiation, and all combinations of them. Out of these, convection problems have the largest variety, as exemplified by the following categories: natural convection, forced convection, mixed natural and forced convection, laminar and turbulent flows, incompressible and compressible (subsonic, supersonic, and transonic) flows, multicomponent single-phase flows with or without chemical reaction (which itself can be of various levels of complexity), two-phase flows with or without phase change (condensation, evaporation, melting, freezing, etc.), and multiphase, multicomponent flows with or without chemical reaction. In other words, in most of the thermal design problems, fluid flow is an integral part of heat transfer. As a result, the contents of this chapter are heavily biased toward the analysis of convective heat-transfer problems. Conduction and radiation are discussed briefly for situations when either or both of these modes are present in addition to the convective heat-transfer mode.

Some of the problems of fluid flow and heat transfer can be simplified and solved by classical (applied mathematical) methods, analog simulation methods, or integral methods. However, due to inherent flexibility, digital simulation (or discretization) methods have enjoyed faster acceptance in industry.[1,2] This trend has been further accelerated by rapid progress in the capabilities of (1) the digital computers (Fig. 7.1) and (2) the solution techniques of discretization methods. In light of this momentum in industry, the phrase "numerical analysis" is used here to imply an analysis using a discretization method.

There are several types of discretization methods. Their differences will be discussed in the next section. However, first their common features, capabilities, and limitations are briefly described.

All discretization methods are aimed at finding an approximate solution to the equations that govern a process of interest. For fluid flow and heat-transfer problems, basic governing equations express physical laws of mass, momentum, and energy conservation. For ease of understanding, sample equations are given below for a two-dimensional flow in a cartesian coordinate system.

1. *Mass conservation*

$$\frac{\partial \rho}{\partial t} + \frac{\partial}{\partial x}(\rho u) + \frac{\partial}{\partial y}(\rho v) = 0$$

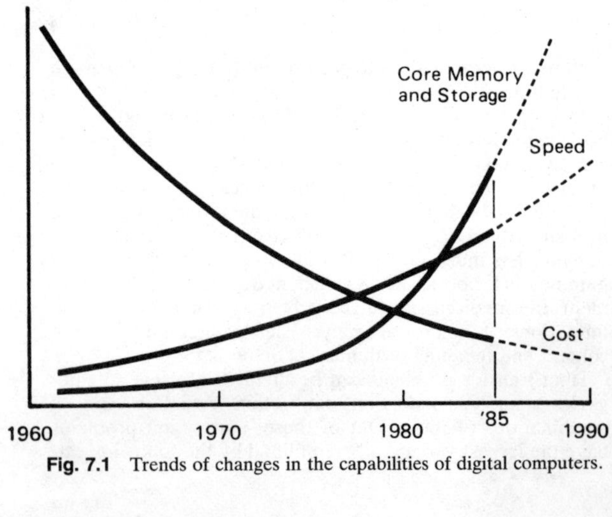

Fig. 7.1 Trends of changes in the capabilities of digital computers.

2. *x-direction momentum conservation*

$$\frac{\partial}{\partial t}(\rho u) + \frac{\partial}{\partial x}(\rho u u) + \frac{\partial}{\partial y}(\rho v u) = \frac{\partial}{\partial x}\left(\mu \frac{\partial u}{\partial x}\right) + \frac{\partial}{\partial y}\left(\mu \frac{\partial u}{\partial y}\right) - \frac{\partial p}{\partial y} + F_x$$

3. *y-direction momentum conservation*

$$\frac{\partial}{\partial t}(\rho v) + \frac{\partial}{\partial x}(\rho u v) + \frac{\partial}{\partial y}(\rho v v) = \frac{\partial}{\partial x}\left(\mu \frac{\partial v}{\partial x}\right) + \frac{\partial}{\partial y}\left(\mu \frac{\partial v}{\partial y}\right) - \frac{\partial p}{\partial y} + F_y$$

4. *Energy conservation*

$$\frac{\partial}{\partial t}(\rho h) + \frac{\partial}{\partial x}(\rho u h) + \frac{\partial}{\partial y}(\rho v h) = \frac{\partial}{\partial x}\left(\frac{k}{c_p}\frac{\partial h}{\partial x}\right) + \frac{\partial}{\partial y}\left(\frac{k}{c_p}\frac{\partial h}{\partial y}\right) + \dot{q}$$

where ρ = fluid density
t = time
u, v = velocity components in x and y coordinate directions
μ = molecular viscosity
p = pressure
F = force per unit volume
h = specific enthalpy
k = thermal conductivity
c_p = specific heat
$\dot{q}$ = heat source per unit volume

The preceding equations take different forms in different coordinate systems (such as cylindrical polar, spherical, orthogonal, and nonorthogonal curvilinear coordinates) and can change for different types of flow. Many text books, such as Ref. 3, provide detailed descriptions of these equations as well as of other transport equations, such as those for chemical species or turbulence parameters.

In a discretization method, the region of interest (calculation domain) is divided into a number of subdomains which are commonly referred to as elements, control cells, or control volumes. Figure 7.2 shows a few sample grid distributions in different coordinate systems. Most methods permit nonuniform grid distributions to allow better resolution of flow and heat-transfer details in selected parts of the calculation domain. For example, finer grid distribution should be used in regions of steeper gradients of flow variables. Discrete, that is, noncontinuous grid or nodal points are located on the intersections of grid lines or at the centers of control volumes.

The required number of spatial coordinates (such as θ, r, and z in a polar coordinate system) determines the dimensionality of the problem. For example, an axisymmetric pipe flow requires a two-dimensional analysis so that the radical (r) and axial (z) variations of velocities (v and w) can be calculated. Now suppose that the flow is swirling, meaning that the circumferential velocity (u) is also to be calculated as a function of r and z. In spite of the fact that three momentum equations for three velocity components are to be

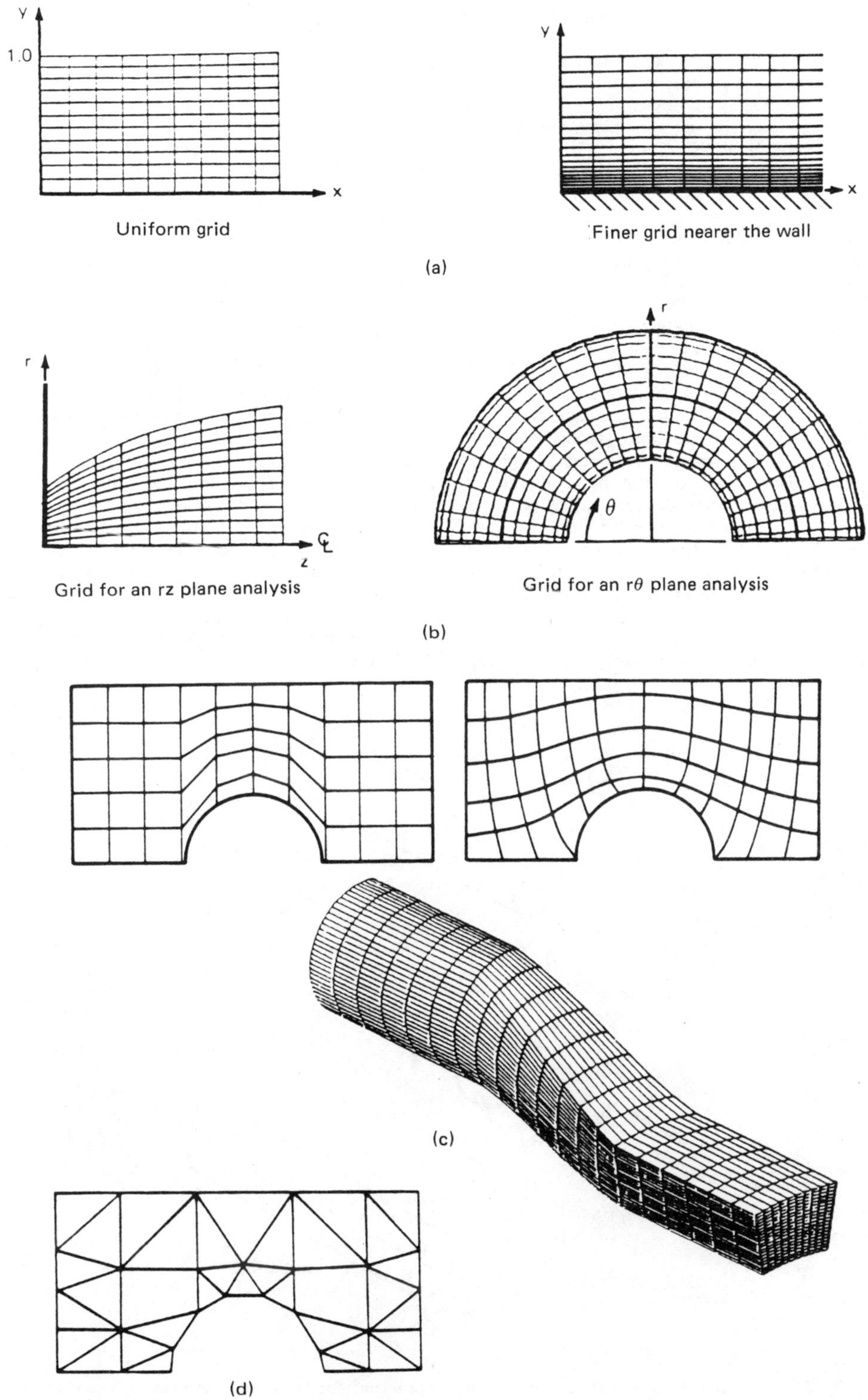

Fig. 7.2 Typical grid distributions. (*a*) Cartesian coordinates; (*b*) cylindrical polar coordinates; (*c*) body-fitted coordinates; (*d*) grid acceptable by finite-element models.

solved, the flow is still two-dimensional because the flow variables are still a function of only two space coordinates. If the pipe has radial baffles or flow injection from ports on the pipe surface, then the flow is three-dimensional, and a three-dimensional analysis is required irrespective of whether or not inlet flow has swirl. Figure 7.3 shows a schematic representation of some two- and three-dimensional flows.

Note that in Fig. 7.3*d*, due to fourfold symmetry in the flow, the calculation domain can be confined to a 90° sector of the pipe. Likewise in Fig. 7.3*e* a 45° sector of pipe should be analyzed.

In all discretization methods, the governing partial differential equations are manipulated to yield separate algebraic equations for each control cell or grid node. Of course, such equations are obtained for each dependent variable such as velocity components, pressure, enthalpy, concentration of selected chemical species, and turbulence parameters. As a result, at each control cell one equation is obtained for each dependent variable. These equations are then solved by an appropriate solution method, and discrete values of all dependent variables are obtained at all grid nodes.

For a transient problem, the time duration of interest is also subdivided into a finite number of time intervals, and discretized (algebraic) equations are obtained for each time interval.

The required computational effort and cost increase with the number of control cells and the number of time steps (intervals). The amount of information available from the solution also increases in the same proportion. This information, on one hand, is far greater than that obtained by a lumped-parameter model (which employs the physical principles over the whole calculation domain rather than over each subdomain). On the other hand, it is far less than that needed for a perfect description of all details of flow and heat-transfer fields (which are continuous rather than made of discrete values as approximated by discretization methods). Also, a numerical solution generally provides far more details than those obtained by well-instrumented experiments.

This detailed information (distributions of velocity, pressure, and temperature) is useful for

1. Determining global performance parameters
2. Determining *local* minimum and maximum quantities, such as velocities, temperatures, and heat fluxes
3. Better understanding of the interactions between geometric, flow, and physical parameters of the problem
4. Providing boundary conditions (inputs) to various other analytical models, such as for vibrations, surface deposition, corrosion, erosion, and thermal stress analysis

The above advantages coupled with the ease and economy of systematic parametric studies make numerical analysis a valuable design tool. The only serious disadvantage of numerical analysis is that if sufficient care is not taken in model setup and verification, the results can be wrong and convincingly misleading. In many problems, particularly when

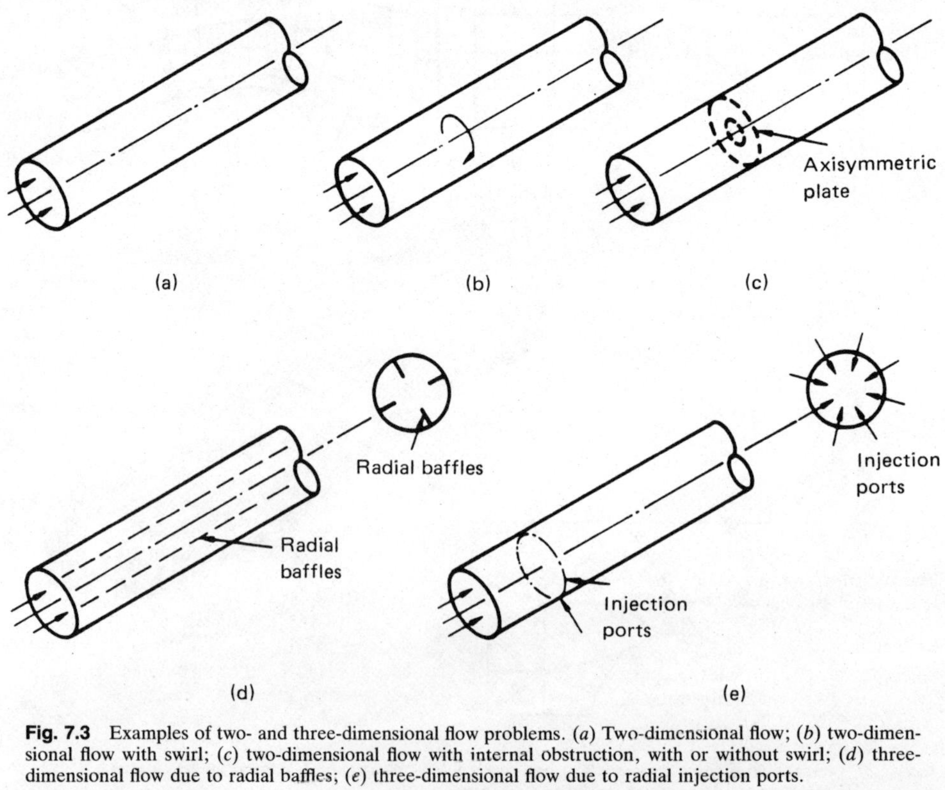

Fig. 7.3 Examples of two- and three-dimensional flow problems. (*a*) Two-dimensional flow; (*b*) two-dimensional flow with swirl; (*c*) two-dimensional flow with internal obstruction, with or without swirl; (*d*) three-dimensional flow due to radial baffles; (*e*) three-dimensional flow due to radial injection ports.

turbulent flow is involved, knowledge and understanding of physical phenomena such as turbulence and chemical reaction are limited, and their gross effects are approximated by physical models. In such cases, the quantitative accuracy of results is primarily controlled by the adequacy of the physical models to the flow situation in question, and hence the contribution of numerical analysis is limited to the prediction of relative trends. Even the capability of predicting relative trends is very useful for

1. Preliminary design optimization
2. Reducing the number of experimental tests for final verification of the designs)
3. Determination of optimum locations for measurement instruments
4. Investigating novel ideas (understanding of flow and heat-transfer fields leads to better designs)

7.2 COMMONLY USED DISCRETIZATION METHODS

There are three broad classes of discretization methods:

1. Finite-element methods
2. Finite-difference methods
3. Finite-volume (control-volume, or finite-domain) methods

These differ from each other mainly with respect to (1) the method of obtaining discretized equations and (2) the method of solving the discretized equations. The former affects the accuracy and stability of a numerical model, while the latter affects the stability and speed (computational efficiency) of solutions. It should be mentioned here that due to the nonlinearity and coupling of governing partial differential equations, the discretized equations need to be solved iteratively until a "converged" solution is obtained. In this regard, the stability of the iterative solution procedure is of paramount importance for practical use of a numerical model. The salient features of each class of numerical models are briefly described here.

Finite-Element Methods

Basic principles and practices of a finite-element approach to fluid flow and heat-transfer problems are discussed in Refs. 4 and 5. In most of the finite-element models, the method of weighted residual is used for obtaining the discretized equations. Generally, an explicit formulation is employed, that is, discretized equations are applicable over a small time interval. As a result, even for a steady-state problem, a solution is obtained in a a time-marching manner, that is, a large number of time steps are taken to progress gradually from the initial field toward the final solution. The time step size is limited by stability considerations such as the Courant condition which, in simple terms, requires that the time step be smaller than the residence time for fluid in the smallest control cell. The major appeal of finite-element methods is their ease of adaptation to irregularly shaped geometries.

Finite-Difference Methods

Reference 6 describes basic principles and numerical properties of both finite-element and finite-difference methods, while Ref. 7 provides similar, but more elaborate and rigorous information about finite-difference methods. In these methods, the discretized equations are obtained by using the Taylor series expansion for each variable around each grid point. Generally only the first few terms are retained and higher order terms are neglected. The resulting methods are referred to as first-, second-, or third-order accurate methods. Formulations such as explicit, semi-implicit, and implicit are used in various methods. Likewise, different methods of solving discretized equations are used. Generally, equations are solved in a point-by-point manner, that is, all variables are first solved for one grid node, then for the next one, and so on. The major appeal of finite-difference methods is their long existence and history.

Finite-Volume Methods

Reference 8 provides a good description of the finite-volume approach. The main idea is to derive the discretized equations by *direct* application of physical principles (conservation of mass, momentum, and energy) to a control volume. As a result, this approach is easy to understand and it does not require a rigorous knowledge of applied mathematics. The numerical methods of this approach employ fully implicit formulations. The discretized equations are solved by various iterative methods with a flexibility for different treatments of different variables. For example, for some problems, different variables are solved in a point-by-point, plane-by-plane, and whole-field manner. The major appeal of finite-volume methods is their stability and computational efficiency.

Out of the above three classes of methods, the finite-volume methods have so far made the greatest penetration in industry for practical problems, such as three-dimensional, elliptic flow with simultaneous heat transfer, turbulence, and often chemical reaction. Of course, all methods are improving and borrowing good features from each other. Some hybrid methodologies are also under development (see, for example, Ref. 9). From the practical engineer's point of view, all these methods are simply means of obtaining solutions to the problems of interest. Therefore it may not always be essential to understand the details of a numerical method. However, there are two additional features, (1) turbulence models and (2) numerical diffusion, which are worth discussing here because they affect the accuracy of solutions in most of the practical flow situations.

Turbulence Models

Most of the flows encountered in industrial equipment are turbulent, and their mean flow (time-averaged) characteristics can be obtained by employing a turbulence model. The commonly used turbulence models are based on the Boussinesq concept of eddy viscosity. The use of turbulent or eddy viscosity accounts for enhanced mixing (diffusion) due to turbulence. Note that the eddy viscosity is generally orders of magnitude larger than the molecular viscosity, and it is a flow property (not a fluid property) which can vary with position and time.

Basic physics of turbulent flows is described in Refs. 10 and 11, and various turbulence models are discussed in Ref. 12. At present the most commonly used model is the two-equation $k \sim \epsilon$ model of turbulence. It requires the solution of two transport equations, one for turbulence kinetic energy k and the other for its dissipation rate ϵ. Local values of turbulent viscosity are determined from the calculated values of k and ϵ. As a result, turbulence length scales do not have to be prescribed a priori (as required in simpler models, such as the Prandtl mixing length model). The $k \sim \epsilon$ model entails five empirical constants, which have been determined by way of comparisons with a large body of experimental data for flows such as in boundary layers, jets, and wakes.

Due to the extensive use and scrutiny of the $k \sim \epsilon$ model, its capabilities, limitations, and deficiencies are well known and documented. Several refinements to this model, for various classes of problems, and several more advanced turbulence modeling approaches are emerging in the literature. The major approaches include

- Reynolds stress modeling (RSM)
- Algebraic stress modeling (ASM)
- Large eddy simulation (LES)
- Two-fluid turbulence modeling

These approaches are also discussed and reviewed in Ref. 13.

Numerical Diffusion

All multidimensional computational methods suffer from some inaccuracy due to numerical diffusion, which is sometimes also referred to as false diffusion. The main source of this inaccuracy is interpolation errors in regions where computational grid lines are skewed to flow streamlines. For illustration, Fig. 7.4 shows a flow of two parallel streams of an arbitrary fluid which has zero viscosity, that is, there should be no diffusion or mixing between the two streams. When such a flow is computed with one set of grid lines aligned with the flow direction and the other set of grid lines orthogonal to the first, the solution is accurate, as shown in Fig. 7.4*a*. On the other hand, if the flow is skewed with respect to the grid lines, the solution shows some mixing which is due to numerical diffusion (Fig. 7.4*b*). The magnitude of numerical diffusion is influenced by flow angle as well as grid size. A finer grid solution has less numerical diffusion (Fig. 7.4*c*).

In practical flows, fluid viscosity is not zero and hence there is always some physical (molecular and turbulent) diffusion as well. Numerical solutions are influenced by both physical and numerical diffusion. Obviously it is desirable that numerical diffusion be as small as possible. There are two ways to reduce numerical diffusion, first to use a finer computational mesh, and second to use a higher order differencing scheme. The first approach is straightforward, but it can be expensive since the mesh sizes may have to be quite small. On the other hand, the second approach is often associated with numerical stability problems and some limitations for complex flows. Further discussion of the basic aspects of numerical diffusion can be found in Ref. 8, and of various schemes of reducing numerical diffusion in Ref. 14.

The effect of numerical diffusion varies from problem to problem. For instance, in the mixing of two streams it is very important since diffusion is the main physical mechanism. However, in many flows diffusion is not the dominant physical mechanism. Other features, such as large pressure and flow variations due to geometric obstructions or heat generation due to electromagnetic stirring or radiation, may dominate. In such cases, numerical diffusion may not have

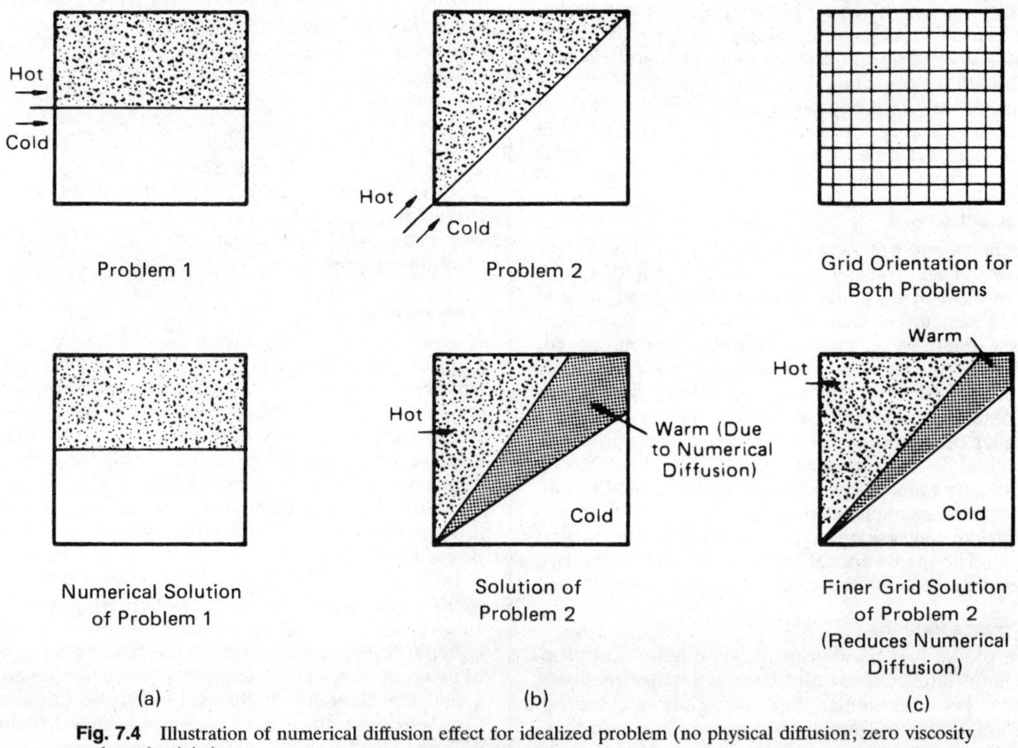

Fig. 7.4 Illustration of numerical diffusion effect for idealized problem (no physical diffusion; zero viscosity and conductivity).

significant influence on the solution. Also, sometimes inaccuracies due to numerical diffusion are confined to small parts of the calculation domain.

7.3 COMMERCIALLY AVAILABLE COMPUTER PROGRAMS

The task of developing a computer code for flow and heat-transfer analysis requires an intimate knowledge of the numerical techniques and a strong discipline to maintain an adequate balance between various desired features of a good code, as depicted in Fig. 7.5.

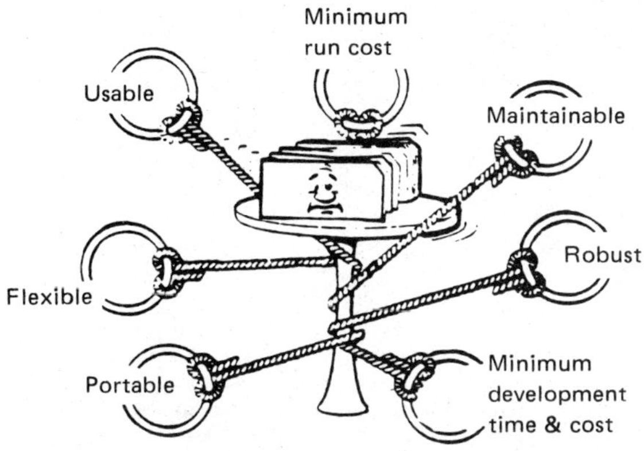

Fig. 7.5 Software development considerations.

As a result, the development of a code can take several-years' effort. Almost equal or greater effort is needed for adequate testing (verification) and documentation. For a practical engineer, the first thing is to determine the availability of suitable computer codes. (The number of commercially available computer codes is growing fairly rapidly.) There are two types of computer codes, (1) specific-purpose codes, designed for a particular equipment or class of problems, and (2) general-purpose codes, designed for much wider use.

Specific-Purpose Codes

In general, specific-purpose computer codes have input-output features tailored to the specific equipment and hence are easier to use. Also it is quite likely that these codes have undergone greater testing for problems and conditions of interest. In the United States, so far, most of the development of specific-purpose codes has been funded by national institutes such as the National Aeronautics and Space Administration (NASA), various laboratories of the Department of Defense (DOD), the Nuclear Regulatory Commission (NRC), and the Electric Power Research Institute (EPRI). Some of the available specific-purpose codes are listed below.

Computer Codes for Rocket Propulsion Problems

BLIMP: JANNAF standard boundary layer integration code for rocket motor's performance calculations (Ref. 15)

REFLAN3D-SPRAY: Code for three-dimensional two-phase flow analysis with combustion and radiation in liquid propellant rocket engines (Ref. 16)

SPF: JANNAF standard plume flow model for two-dimensional analysis of rocket plumes (Ref. 17)

TDK: JANNAF standard two-dimensional inviscid chemical kinetics code using the method of characteristics (Ref. 18)

Computer Codes for Power Plant Components

ATHOS3: Code for three-dimensional steady-state and transient thermal hydraulic analysis of steam generators (Ref. 19)

VIPER-01: Code for subchannel analysis of flow in rod bundles of nuclear reactors (Ref. 20)

RELAP5: Code for simulating system transient responses of light water reactor accidents such as loss of coolant accident (LOCA) (Ref. 21)

RETRAN: Code for thermal-hydraulic analysis of nuclear steam supply systems (NSSS) to LOCA and operational transients (Ref. 22)

TRAC: Code for three-dimensional LOCA analysis (Ref. 23)

VERA2D: Code for two-dimensional analysis of flow, heat, and mass transfer in wet cooling towers of both natural and mechanical draft designs (Ref. 24)

For industrial equipment, such as heat exchangers, air coolers, reboilers, or furnaces, there are no tailor-made programs for detailed flow analysis, but there are several design programs available from organizations such as Heat Transfer Research Inc. (HTRI), Alhambra, Calif., and Heat Transfer and Fluid Flow Service (HTFS), AERE, Harwell, England. At present most of their programs rely quite heavily on an empirical database, and thus provide reliable solutions within only a limited range of conditions.

General-Purpose Codes

One of the basic contributions of numerical analysis is to promote new ideas and new designs. Hence, quite often no specific-purpose code is available for the desired analysis. In recognition of this need, a few general-purpose computational fluid dynamics (CFD) codes have emerged in the last few years. These codes are designed for easy adaptation to different geometries and flow conditions. Three commercially available codes are briefly described below.

PHOENICS: PHOENICS is an acronym for parabolic hyperbolic or elliptic numerical integration code system. It is applicable for one-, two-, and three-dimensional steady-state and transient analysis of single- and two-phase flows with or without turbulence, heat transfer, or combustion. It also applies to compressible (subsonic, supersonic, and transonic) flows and has provisions for orthogonal and nonorthogonal (body-fitted) coordinates, moving grids, and solid objects in flow field. PHOENICS is based on a finite-volume approach and is in wide use worldwide. PHOENICS is available from CHAM of North America, Inc., Huntsville, Ala., and CHAM Ltd., Wimbledon, London, England.

FLUENT: FLUENT is based on a similar methodology as that used in PHOENICS. It is applicable for two- and three-dimensional flows with or without heat transfer and combustion. Currently it does not have provisions for transient analysis, or for the analysis of supersonic or transonic flows. FLUENT is also in wide use and is available from CREARE, Inc., Hanover, N.H.

FIDAP: FIDAP is an acronym for fluid dynamics analysis package. FIDAP employs a finite-element approach. It is applicable for two- and three-dimensional steady-state and transient analysis of single-phase flows with or without heat transfer. It does not have provisions for two-phase flows or for reactive flows. FIDAP is available from Fluid Dynamics International, Evanston, Ill.

One common feature of the preceding three codes is their capability of solving most general (Navier-Stokes) equations for fluid flows. There are some other general-purpose programs which have only limited capability for flow solution (e.g., by using the Darcy law for flow through porous media) and have general capability for heat conduction. Three commercially available computer programs are briefly described below.

ANSYS: ANSYS is a finite-element structural analysis program. It also has the capability of solving linear and nonlinear heat conduction problems. ANSYS is available from Swanson Analysis Systems, Inc., Houston, Pa.

NASTRAN: NASTRAN is also a finite-element structural analysis code with capabilities similar to those of ANSYS. The original code was developed under a NASA contract and is available from NASA's Computer Software Management and Information Center (COSMIC), Athens, Ga. NASTRAN has been modified by several organizations. The most commonly used version MSC/NASTRAN is available from the MacNiel-Schwendler Corporation, Los Angeles, Calif.

SINDA: SINDA is a finite-difference heat-transfer analysis program. It has general capability for the solution of radiation and conduction problems with or without phase changes. SINDA was also originally developed for NASA and is available from NASA's COSMIC Code Library. Commercially, SINDA is available from Network Analysis Associates, Inc., Mountain View, Calif.

The above programs are also available from most of the commercial computer bureaus such as University Computing Company (UCC), Dallas, Tex., Control Data Corporation (CDC), Minneapolis, Minn., and Boeing Computer Service (BCS), Seattle, Wash. The catalogs of engineering applications software provided by various computer bureaus and computer manufacturers are also good sources for identifying suitable thermal-analysis programs.

7.4 STEPS FOR PERFORMING NUMERICAL ANALYSIS

For the effective use of any thermal-analysis computer code, some training and practice are essential. It should also be noted here that these codes are analysis tools which aid the design process but do not produce or optimize the designs automatically. The role of the designer or analyst continues to be crucial for successful designs. Likewise, the success of analysis is directly influenced by the degree of systematic approach in problem formulation and result interpretation. The following step-by-step approach is recommended as a general guide.

STEP 1. List the objectives of the desired analysis, including a list of the questions that need to be analyzed. Group the questions into several categories which require different levels of detail in the analysis. Questions like "What are the dominant physical processes?" must be understood clearly. If several physical processes are present, their relative importance must be established. If different processes are envisioned to be dominant in different parts of the equipment, separate regions should be identified with separate lists of questions. Generally it is neither possible nor advisable to analyze all questions by one analysis. It is far more useful to proceed with analyses of a series of problems, devised with an increasing degree of complexity, than to start directly with the analysis of the whole problem.

STEP 2. For a chosen set of objectives or questions, determine the following:

1. A suitable calculation domain
2. An appropriate coordinate system (which is suitable for the selected calculation domain)
3. The number of independent variables (spatial coordinates and time)
4. The number of dependent variables (such as velocity components, pressure, enthalpy, concentrations of selected species, i.e., variables which are governed by partial differential equations)
5. Auxiliary variables (such as density or temperature, i.e., variables which are determined by algebraic relations with different dependent variables)
6. Appropriate initial and boundary conditions
7. Suitable physical models (for physical phenomena of interest, such as turbulence or combustion)

STEP 3. Select a computer code that is capable of meeting the specific requirements identified in Step 2. Some general considerations for code selection are also indicated after the description of analysis steps.

STEP 4. Set up a numerical model in the following manner.

1. Set up the geometry and grid distributions which are consistent with the selected objectives of the analysis. Most of the codes permit the use of nonuniform grids. This flexibility can be of significant benefit in obtaining the desired resolution of flow details in areas of interest with a moderate number of control cells. Examine, and if necessary modify, the selected grid distribution before proceeding to the next step of model setup.
2. Set up flow and thermal boundary conditions. Care should be taken in determining complete (i.e., not redundant or insufficient) boundary conditions. For example, for a pipe flow, prescribing mass flow rates and pressures at inlet and outlet is an overspecification.
3. Set up physical properties of the fluids in question. Care should be taken in the specification of nonlinear (such as temperature-dependent) physical properties.
4. Select and activate (or set up) appropriate physical models.
5. Set up numerical parameters. For example, initialize all variables (most iterative methods require initialization), underrelaxation parameters, iteration limits, solution-control parameters, convergence criteria, and printout control parameters.

STEP 5. Obtain a numerical solution, and examine its convergence characteristics and physical plausibility by checking the following:

1. Overall mass and heat balances
2. Maximum residuals of *all* equations (a residual is calculated by transferring all terms of an equation to the right-

hand side, a residual is an indication of the local error or imbalance

3. Locations of maximum residuals
4. Trends of iteration-to-iteration reduction of maximum residuals
5. Trends of iteration-to-iteration variations of flow variables at selected monitoring locations
6. Qualitative nature of calculated distributions of velocity, pressure, temperature, and so on

In a converged solution all residuals should be smaller than preassigned limits and all flow variables should be fairly well settled, such as within 1 percent of their nominal values. In addition to examining the convergence characteristics, it is essential to closely examine the plausibility of the calculated flow field. This ensures that the geometry and the boundary conditions have been set up properly. All such checks must be made before proceeding to the next step.

STEP 6. If necessary, depending on the convergence characteristics of the solution, repeat calculations with increased or decreased underrelaxation and/or iteration limits. The solution must be independent of such parameters, since these parameters should influence the rate of convergence only.

STEP 7. If necessary, depending on the confidence and experience with a selected code, repeat the calculation with different initial guesses of flow field. The solution of a steady-state problem must be independent of the initial guess.

STEP 8. Perform calculations with different grid distributions and establish a grid that produces (practically!) grid-independent solutions. Grid independence of solutions is very important. However, the criteria used for judging grid independence should be consistent with the needs and objectives of the analysis. In a mathematical sense, grid-independent solutions can be obtained only with very fine grids and with a large number of iterations. Generally, the required number of iterations (or number of time steps in time-marching methods) increases for finer grids. For many practical purposes, such as for *relative* assessments of design parameters, useful solutions can be obtained with moderate grids. Even otherwise, the use of moderate grids, before the use of fine grids, is a good logical step to ensure that all other aspects of the numerical model are satisfactory.

STEP 9. For a transient analysis, determine the sensitivity of results to time step sizes used. This is done by performing calculations with different time steps. With a fully implicit method, where there is no restriction on time step size, this exercise is more often needed than with an explicit method where sufficiently small time steps have to be used anyway due to the considerations of numerical stability.

STEP 10. Examine the sensitivity of results to the physical models used. Several calculations with different empirical correlations and with the upper and lower bound values of empirical constants should be performed. Establish a suitable set of physical models which should then be used in sensitivity studies of engineering (geometry and flow) parameters.

STEP 11. Perform engineering parametric studies by changing *one* parameter at a time. During these studies, neither grid distributions nor any other numerical parameters should be altered.

STEP 12. Repeat Steps 2 to 11 for the next problem (or set of questions) for thermal analysis of the same equipment.

The described steps clearly indicate that the numerical analysis is an iterative process. For ready reference, the recommended steps are depicted in Figs. 7.6 and 7.7, which also serve as a reminder that even with a given suitable computer code, there are two tasks, (1) the preparation of the numerical model and (2) the use of the model.

Selection of Computer Code

From the preceding description it is clear that the effective use of numerical analysis depends on the ability of making many calculations. Also in most equipment, there are a large number of parameters which need to be analyzed. As a result, the selected analytical tool (computer code) must be robust and economical. Based on these considerations various features of a computer code can be prioritized as follows.

1. Robustness or stability of the iterative solution scheme
2. Accuracy and physical realism of numerical solutions, that is, no wriggles, overshoots, or undershoots of calculated variables
3. Economy of computations, that is, computer time requirements should be low to permit a large number of calculations
4. Flexibility of the computer program so that a large variety of problems can be analyzed with one code
5. User friendliness and cost of the program

7.5 EXAMPLES OF NUMERICAL SIMULATIONS

To illustrate the importance of a step-by-step problem-formulation approach and systematic parametric studies, this section describes one practical example and provides references to some published examples. Figure 7.8 shows a schematic of an incinerator. It has 18 ports, 8 for air and 10 for the supply of gaseous fuel mixed with waste products to be incinerated. The ports are located in the upstream part of the unit, along two axial lines at the cylindrical wall. Generally fuel and air are supplied through alternate ports. Fuel streams enter with a significant tangential velocity component. The incinerator works reasonably well, that is, it burns most of the waste products. However, from air pollution considerations, the quality of exhaust gases is not satisfactory, that is, it has a high level of toxic gases (NOX products). The question is what minimum changes in the operating units of this design will satisfy the requirement.

It is well known that NOX is formed in the regions of high temperature and low oxygen concentration. Therefore one possible approach is to question: "Can the air and fuel supply arrangement be modified with minimum changes in the total flow rates?" Possibilities include: closing certain ports, supplying unequal flow rates through different ports, and adding additional ports. Obviously there are a large number of combinations possible. Trying any fixes on an operating unit (a large-size, high-temperature reactor) is hazardous. The problem was analyzed and a solution was found within a couple of months by performing numerical simulations. Salient features of the numerical analysis are briefly described below.

First the existing design was simulated to study the distributions of velocities, temperature, and concentrations of oxygen and fuel. As a first step, the ports were simulated as circumferential slots (to permit a two-dimensional analysis), and chemical reaction was modeled by a simple turbulent-combustion model.[25] Results (Fig. 7.9) showed that undesirable conditions (low oxygen concentrations in regions of high temperature) were present in the upstream part of the unit.

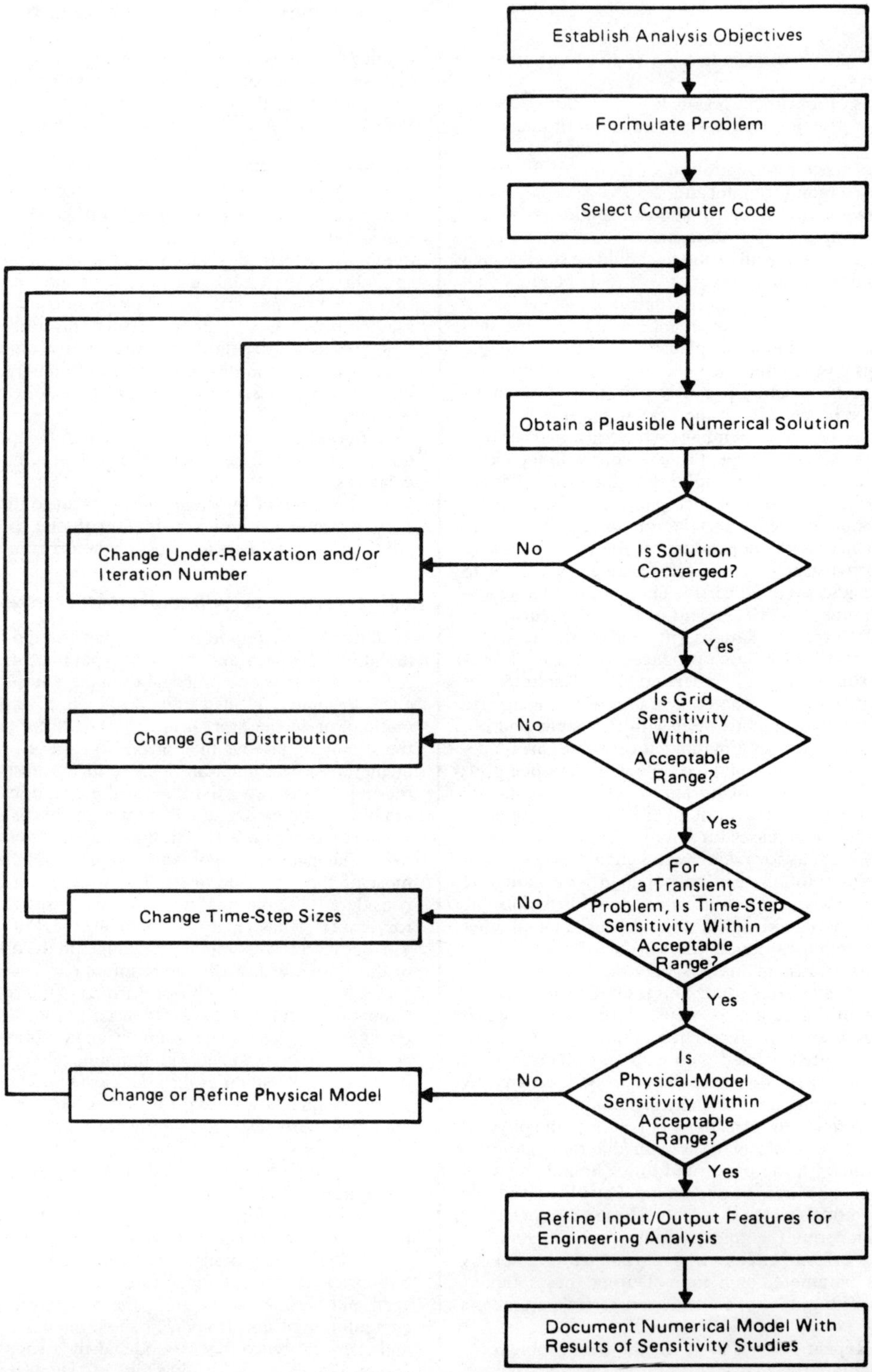

Fig. 7.6 Steps for preparation of numerical model.

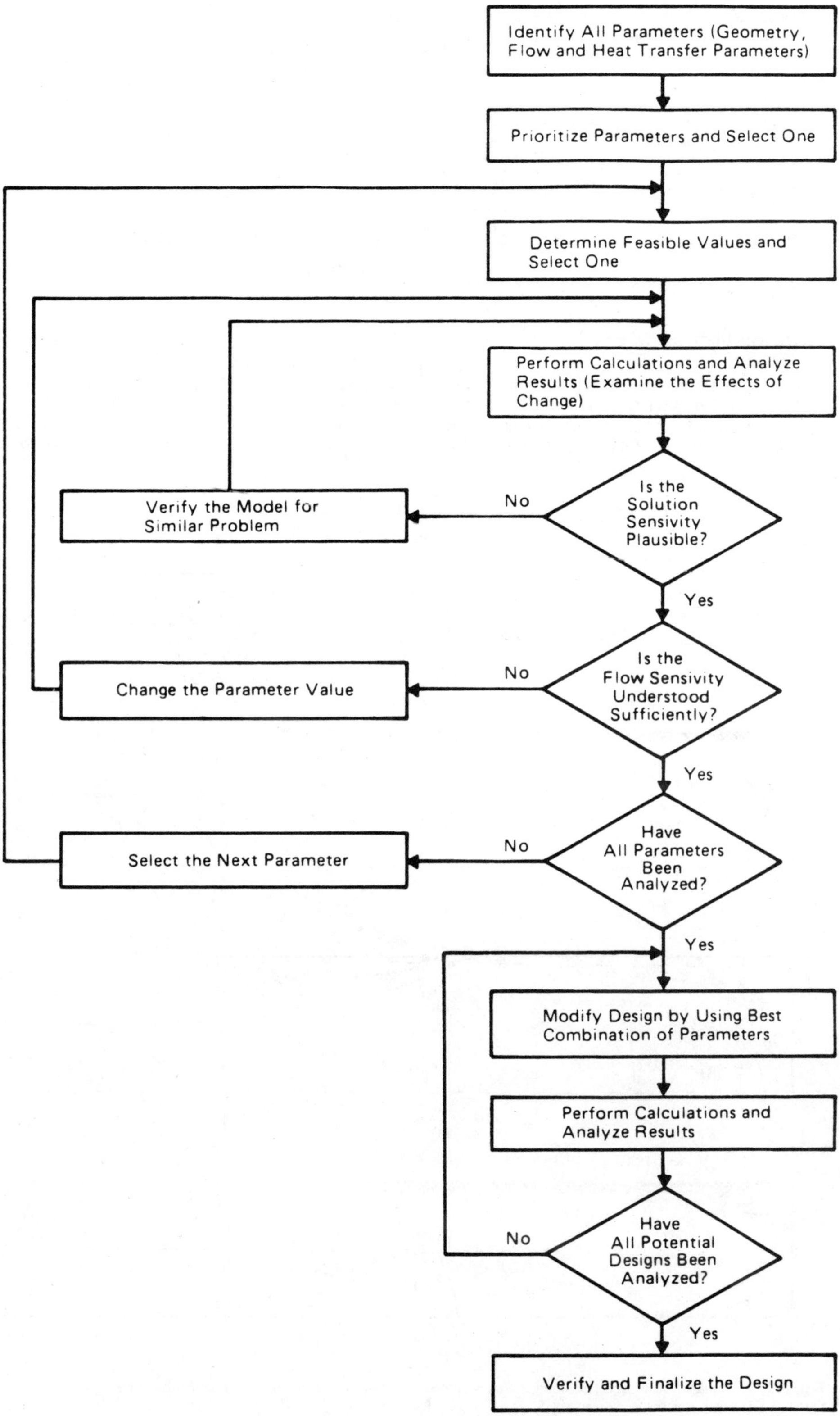

Fig. 7.7 Steps for using numerical model in thermal design.

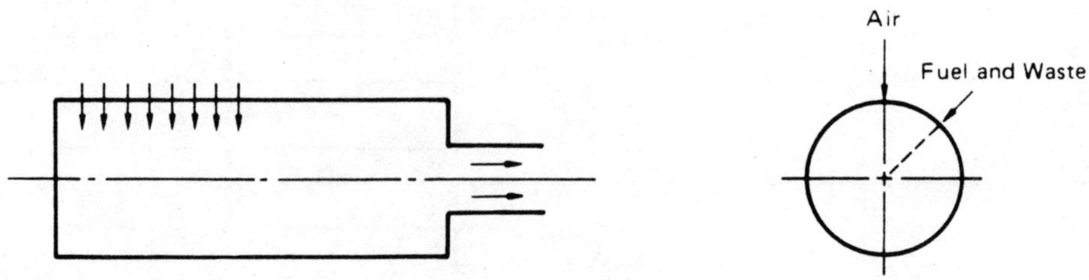

Fig. 7.8 Schematic representation of incinerator.

Then five parametric calculations were performed to study the effects of increasing the oxygen flow rates through upstream ports. Results showed only marginal improvement in conditions. This encouraged further thinking and led to the idea of providing an additional port at the axis of the front cover for supplying a part of the air. Two computations, one with 10 percent and the other with 20 percent oxygen supply through axial nozzle, were performed. Based on the relative analysis of all results, one of the last two cases showed greatest improvement, and one of these (Fig. 7.10) was chosen for final implementation. Several more detailed calculations were performed of the final cases. Some of the basic flow

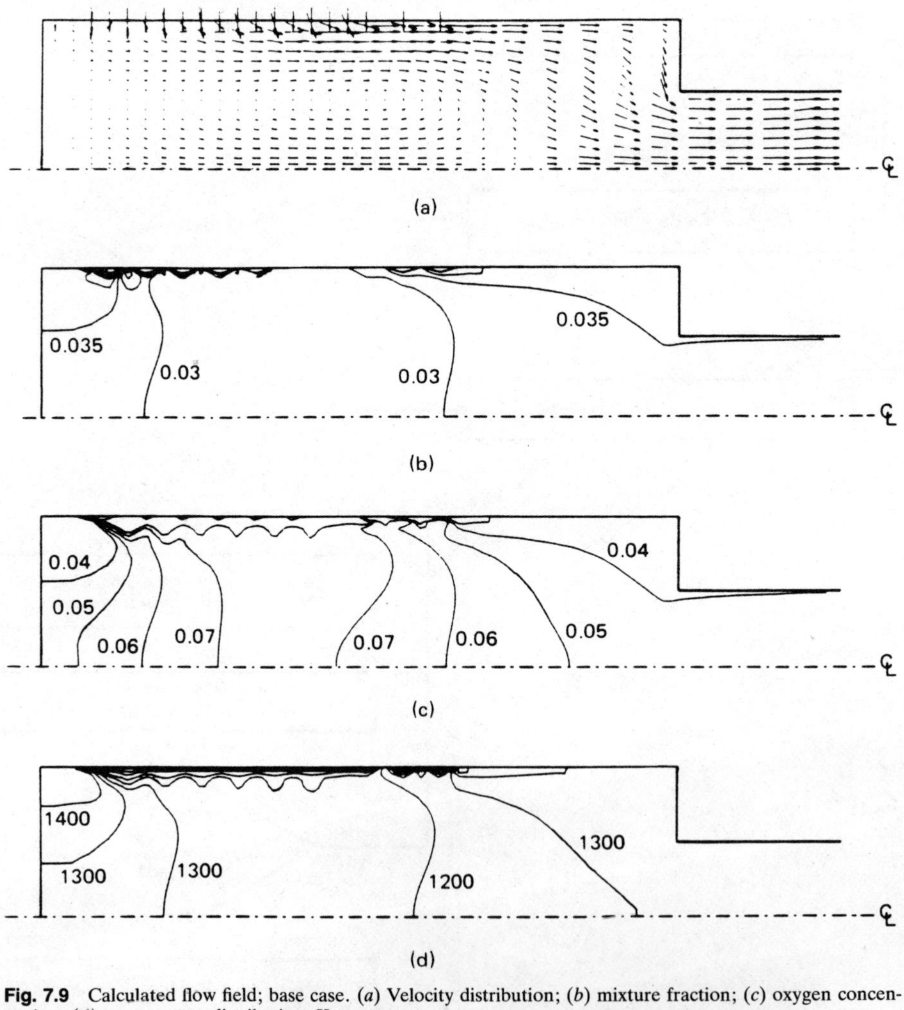

Fig. 7.9 Calculated flow field; base case. (*a*) Velocity distribution; (*b*) mixture fraction; (*c*) oxygen concentration; (*d*) temperature distribution, K.

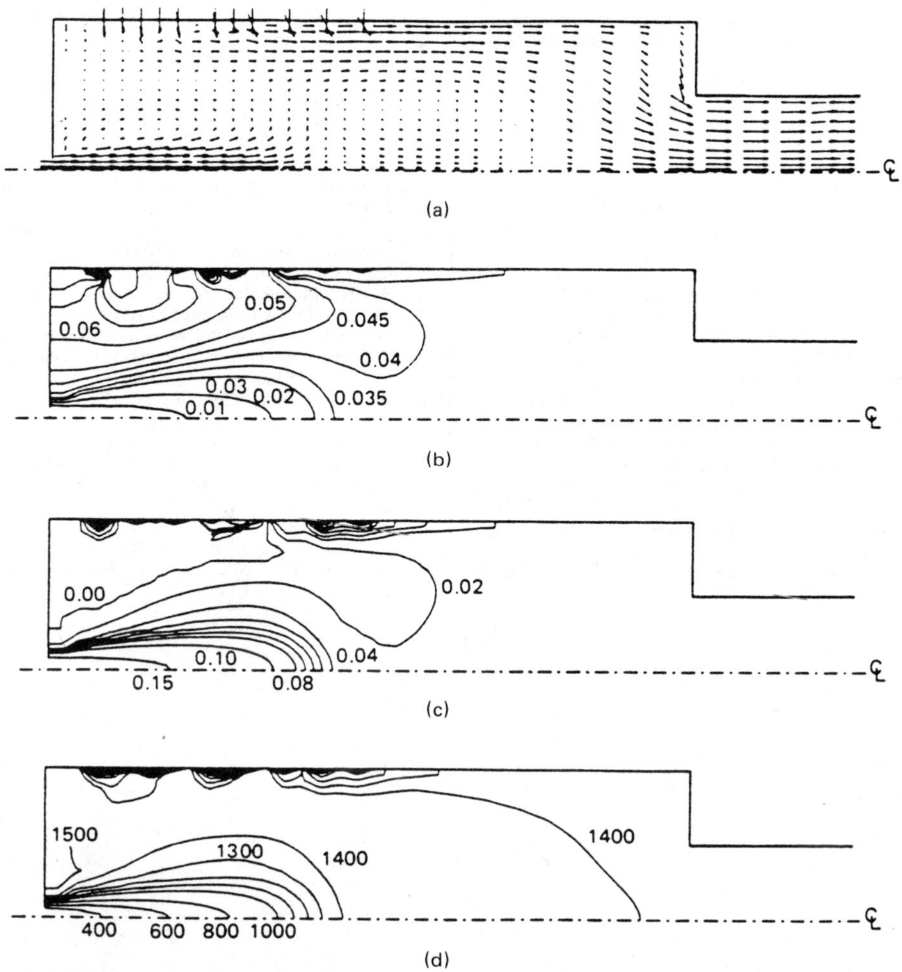

Fig. 7.10 Calculated flow field; modified case. (*a*) Velocity distribution; (*b*) mixture fraction; (*c*) oxygen concentration; (*d*) temperature distribution, K.

features (mixing patterns) were also qualitatively verified by small-scale cold-flow flow-visualization experiments. The final fine-tuned solution was implemented successfully.

References 2, 26, and 27 present a few other applications which illustrate the strength of a systematic approach to problem formulations and analysis.

ACKNOWLEDGMENTS

The author wishes to acknowledge and thank two of his colleagues, Dr. M. Samonds and Mr. C. E. Smith, for constructive comments on the manuscript, and his secretaries, K. King and F. Johnson, for preparing the typescript of this chapter.

REFERENCES

1. D. R. Chapman, "Computational Aerodynamics Development and Outlook," *AIAA J.*, **17,** 1293–1313 (1979).
2. A. K. Singhal, "A Critical Look at the Progress in Numerical Heat Transfer and Some Suggestions for Improvement," *Numer. Heat Transfer,* **8,** 505–517 (1985).
3. F. M. White, *Viscous Fluid Flow,* McGraw-Hill, New York, 1974.
4. O. C. Zienkiewicz and K. Morgan, *Finite Elements and Approximation,* Wiley, New York, 1983.
5. A. J. Baker, *Computation of Fluid Flow by the Finite Element Method,* McGraw-Hill, New York, 1983.
6. T. M. Shih, *Numerical Heat Transfer,* Hemisphere Publ. Corp., New York, 1984.
7. D. A. Anderson, J. C. Tannehill, and R. H. Pletcher, *Computational Fluid Mechanics and Heat Transfer,* Hemisphere Publ. Corp., New York, 1985.
8. S. V. Patankar, *Numerical Heat Transfer and Fluid Flow,* Hemisphere Publ. Corp., New York, 1980.
9. B. R. Baliga and S. V. Patankar, "A Control Volume Finite-Element Method for Two-Dimensional Fluid Flow and Heat Transfer," *Numer. Heat Transfer,* **6,** 245–261 (1983).
10. J. O. Hinze, *Turbulence,* McGraw-Hill, New York, 1975.
11. H. Schlichting, *Boundary Layer Theory,* 6th ed., McGraw-Hill, New York, 1968.

12. B. E. Launder and D. B. Spalding, *Lectures on Mathematical Models of Turbulence,* Academic Press, London and New York, 1972.
13. N. C. Markatos, "Computer Simulation Techniques for Turbulent Flows," in *Encyclopedia of Fluid Mechanics,* vol. 6, N. P. Cheremisinoff (ed.), Gulf Publ. Co., June 1984, and *J. App. Math. Modeling,* **10,** (June 1986).
14. A. O. Nieckele, "Development and Evaluation of Numerical Schemes for the Solution of Convection-Diffusion Problems," Ph.D. Thesis, University of Minnesota, Minneapolis, 1985.
15. R. M. Evans, "Boundary Layer Integral Matrix Procedure," Aerotherm Project for NASA MSFC NAS8-30930, 1975.
16. A. J. Przekwas, A. K. Singhal, and L. T. Tam, *Rocket Injector Anomalies Study,* vols. 1–3, CHAM Project for NASA LeRC, NASA CR-17402 and CR-17403, Aug. 1984, 1986.
17. S. M. Dash and H. S. Pergament, "The JANNAF Standard Plume Flow Field Model (SPF)," U.S. Army Missile Command, Huntsville, Ala., Rep. TR RD-CR 82-9, Apr. 1981.
18. G. Nickerson et al., "JANNAF Rocket Engine Performance Prediction and Evaluation Manual," CPIA Publ. 246, Apr. 1975, and "Two-Dimensional Kinetic (TDK) Reference Computer Program," Ultrasystems, Inc., NASA Contract NAS9-12652, 1973.
19. L. W. Keeton, A. K. Singhal, and G. S. Srikantiah, "ATHOS3—A Computer Program for Thermal-Hydraulic Analysis of Steam Generators," vols. 1–4, Electric Power Research Institute, Rep. NP-4604-CCM, 1986.
20. "VIPER-01: A Thermal Hydraulic Computer Code for Reactor Cores," vols. 1–3, Electric Power Research Institute, Rep. NP-2511-CCM, 1986.
21. K. R. Katsma, et al., "RELAP4/MOD5—A Computer Program for Transient Thermal-Hydraulic Analysis of Nuclear Reactors and Related Systems," Idaho National Engineering Lab., ANCR-NUREG-1335, Sept. 1976.
22. K. V. Moore et al., "RETRAN—A Program for One-Dimensional Transient Thermal Hydraulic Analysis of Complex Fluid Flow Systems," Electric Power Research Institute, Rep. CCM-5, 1978.
23. J. F. Jackson et al., "TRAC-P1: An Advanced Best-Estimate Computer Program for PWR LOCA Analysis," Nuclear Regulatory Commission, Rep. NUREG/CR-0063, LA-727a MS, 1978.
24. A. K. Majumdar, A. K. Singhal, and D. B. Spalding, "VERA2D—A Computer Program for Two-Dimensional Analysis of Flow, Heat and Mass Transfer in Evaporative Cooling Towers," vols. 1–2, Electric Power Research Institute, Rep. CS-2923, Mar. 1983.
25. D. B. Spalding, "Concentration Fluctuations in a Round Turbulent Free-Jet," *Chem. Eng. Sci.,* **26,** 95–107 (1971).
26. T. Mukerjee, A. J. Przekwas, A. K. Singhal, V. K. Duggal, and T. W. Kuo, "Three-Dimensional Modeling of In-Cylinder Processes in DI Diesel Engines," Soc. of Automotive Engineers, Tech. Paper 840227, 1984.
27. A. K. Singhal, L. T. Tam, F. Bachtel, and L. Vaniman, "Thermal Environment Around Space Shuttle with Hot Gas Jets for Ice Suppression," *J. Spacecraft Rockets,* **23**(6), 547–553 (1986).

Part 2

Thermophysical Properties of Selected Engineering Materials

Chapter 1

Thermophysical Properties of Selected Engineering Materials

DAVID L. BROWNELL
Associate Editor

Table 1.1. Thermal Properties of Common Substances

Typical values at or near room temperature unless otherwise noted in parentheses

Material Description	Specific gravity	Thermal conductivity, Btu/h·ft·°F	Specific heat, Btu/lb·°F
Asbestos cement board	1.4	0.35	0.2
Asbestos millboard	1.0	0.08	0.2
Asphalt	2.1	0.43	0.22
Ashes, wood	...	0.04 (122°F)	0.19
Bakelite	1.3	9.8	0.36
Brick			
Common	1.75	0.42	0.22
Hard	2.0	0.75	0.24
Cardboard	...	0.04	
Cellulose, loose fill	0.04	0.024	0.3
Cement			
Portland	1.5	0.17	
Mortar	...	0.67	
Chalk	2.0	0.48	0.22
Charcoal, wood	0.4	0.05	0.24
Clay	1.0	...	0.22
Coal	1.4	0.10 (32°F)	0.24

NOTE: Footnotes appear on the last page of the table.

Table 1.1. Thermal Properties of Common Substances (*Continued*)

Typical values at or near room temperature unless otherwise noted in parentheses

Material Description	Specific gravity	Thermal conductivity, Btu/h·ft·°F	Specific heat, Btu/lb·°F
Coke (petroleum, powdered)	0.99	0.55	0.36 (752°F)
Concrete			
Cinder	...	0.44	
Stone 1-2-4 mix	1.9–2.3	0.79	0.21
Corkboard	0.20	0.025	0.45
Cotton, fiber	1.5	0.024	0.32
Diamond			
Single crystal, high purity (80°F)			
Type I stone (gem quality)	2.4[a]	520	0.12
Type IIa (electric insulator)	2.4[a]	1330	0.12
Type IIb (electric conductor)	2.4[a]	780	0.12
Single crystal, high purity at −334°F (maximum conductivity)			
Type I (gem quality)	2.4[a]	2140[b]	0.0014
Type IIa (electric insulator)	2.4[a]	6930[b]	0.0014
Type IIb (electric conductor)	2.4[a]	3990[b]	0.0014
Diatomaceous earth	0.32	0.035	
Felt			
Hair	0.1	0.03	0.5
Wool	0.3	0.04	0.33
Fiberglass (1.5 lb/ft^3)	0.024	0.022	0.17
Glass			
Soda-lime (window)	2.5	0.6 (200°F)	0.18
Lead	4.3	0.8	0.12
Borosilicate	2.2	0.6 (200°F)	0.20
Gypsum board	0.8	0.1	0.26
Ice			
32°F	0.9	1.3	0.49
−4°F	...	1.4	0.47
Leather, dry	0.9	0.09	0.36
Magnesia (85%)	0.25	0.04	0.2
Magnesite brick	2.5 (212°F)	2.2 (400°F)	0.22
Metals and alloys			
Aluminum, 2024(0)	2.8	110	0.23
Copper, pure	9.0	228	0.092
Gold, pure	19.3	172 (32°F)	0.031
Lead, pure	11.3	21.4	0.031
Silver, pure	10.5	242 (32°F)	0.056
Steel, AISI 1020	7.8	30	0.11
Stainless steel, AISI 304	8.0	9.4	0.12
Tantalum, pure	16.6	31	0.03
Titanium (99% Ti)	4.5	9.1–11.5	0.13
Tungsten, pure	19.3	95	0.033
Mica	2.7	0.4	0.12
Mineral wool blanket	0.1	0.025	0.2
Paper	0.9	0.07	0.33
Paraffin wax	0.9	0.15	0.69
Plaster			
Light	0.7	0.15	0.24
Sand	1.8	0.42	0.22
Plastics, unfilled			
ABS, heat resistant	1.1	0.11–0.19	
Acrylic, cast	1.2	0.10–0.14	0.35
Fluoroplastics, PTFE, FEP, PFA	2.1–2.2	0.14	0.25–0.28
Nylon, type 6, 6/6	1.1–1.2	0.14	0.4
Polycarbonate	1.2	0.12	0.3
Polyethylene			
Low density	0.92–0.93	0.19	0.55
High density	0.95–0.97	0.27–0.29	0.55
Polypropylene	0.9	0.07	0.46
PVC, sheet, tubing	1.3–1.6	0.08–0.13	0.25–0.35
Plastics, foamed			
Polyurethane	0.02–0.05	0.009–0.014[c]	
Polystyrene	0.016–0.08	0.019–0.023	
Porcelain	2.5	0.9	0.22

Table 1.1. Thermal Properties of Common Substances (*Continued*)

Typical values at or near room temperature unless otherwise noted in parentheses

Material Description	Specific gravity	Thermal conductivity, Btu/h·ft·°F	Specific heat, Btu/lb·°F
Quartz, crystal (212°F)			
Parallel to crystal axis	2.6	5.0	0.21
Perpendicular to crystal axis	2.6	3.2	0.21
Rocksalt	2.2	...	0.22
Rockwool (10 lb/ft^3)	0.16	0.023	
Rubber			
Hard	1.2	0.093	
Soft	1.1	0.058	0.48
Sand, coarse	1.5	0.17	0.19
Sapphire, high purity, synthetic			
80°F	3.5[a]	26.6[d]	0.18
−406°F (maximum conductivity)	3.5[a]	12,000[b,d]	
Sawdust	0.15	0.05	0.5
Silica	2.2	0.83 (200°F)	0.32
Snow (32°F)			
Freshly fallen	0.19	0.098	0.5
Compacted	0.40	0.25	0.5
Soils			
Silt			
Hard silt, 10% d.w.	1.9	0.42	0.26[e]
Stiff silt, 20% d.w.	2.1	0.83	0.33[e]
Soft silt, 80% d.w.	1.6	0.54	0.55[e]
Clay			
Loose clay soil, 14% d.w.	1.2	0.22	
Compacted clay soil, 14% d.w.	1.6	0.69	
Stone			
Granite	2.8	1.4	0.20
Limestone	2.3	1.1	0.22
Marble	2.8	1.4	0.19
Sandstone	2.3	1.1	0.17
Slate	2.8	1.1	
Vermiculite, exfoliated	0.13	0.038	0.32
Wood, across grain			
Hardwoods	0.37–1.12	0.065–0.148	0.45–0.65
Ash, white	0.69	0.10	
Elm, American	0.58	0.09	
Mahogany	0.54	0.075	
Maple, sugar	0.72	0.108	
Oak, white	0.75	0.102	0.57
Softwoods	0.14–0.74	0.032–0.093	
Balsa (8.8 lb/ft^3)	0.14	0.032	
Cypress	0.46	0.056	
Fir, white	0.43	0.068	
Pine, white	0.43	0.063	
Pine, yellow	0.64	0.085	0.67
Spruce	0.42	0.065	
Wood shavings	...	0.034	
Wool			
Fabric	0.11–0.33	0.021–0.037	
Fiber	1.3	0.325	

[a]The specific gravity value given is a nominal room temperature value for this material.

[b]The low-temperature thermal conductivity values given are maximum values for this material. See the *TPRC Data Series*, vol. 1, 1970, for further information.

[c]Thermal conductivity is for fluorocarbon blown polyurethane foam. First value is for new foam; second value is for aged foam.

[d]Heat flow measured at 60° from the crystal axis.

[e]Specific heat estimated based on properties and relative amounts of the constituents.

CONVERSION FACTORS: Btu/h·ft·°F × 1.7307 = W/m·K; Btu/lb·°F × 4186.8 = J/kg·K.

SOURCES: Compiled from several sources, including those listed below:

Y. S. Touloukian and C. Y. Ho (eds.), *TPRC Data Series*, Plenum, New York, 1979.

ASHRAE Handbook of Fundamentals, Am. Soc. of Heating, Refrigerating, and Air-Conditioning Engineers, Atlanta, Ga., 1985.

R. Bolz and G. Tuve (eds.), *Handbook of Tables for Applied Engineering Science,* 2d ed. CRC Press, Inc., Boca Raton, Fla., 1976.

Table 1.2. General Properties of Selected Metals and Alloys

Typical values at or near room temperature unless otherwise indicated

				Mean coefficient of linear thermal expansion			
Metal or alloy	Density, lb/in³	Thermal conductivity, Btu/h·ft·°F	Approximate specific heat,[a] Btu/lb·°F	Temperature range, °F	Coefficient, μin/in·°F	Electrical resistivity, μΩ·cm	Approximate melting temperature,[b] °F
Aluminum (99.996%)	0.0975	137	0.215	68–212	13.1	2.65	1220
Aluminum, wrought alloys							
EC(0), 1060	0.098	135–137	0.23 (212)	68–212	13.1	2.8	1215
2024							
(0)	0.100	110	0.23 (212)	68–212	12.7	3.4	1180
(T3, T4)	0.100	69	0.23 (212)	68–212	12.7	5.7	1180
(T6)	0.100	84	0.23 (212)	68–212	12.7	4.6	1180
3003							
(0)	0.099	104	0.23 (212)	68–212	12.9	3.7	1210
(H12)	0.099	94	0.23 (212)	68–212	12.9	4.1	1210
4032							
(0)	0.097	90	0.23 (212)	68–212	10.8	4.3	1060
(T6)	0.097	82	0.23 (212)	68–212	10.8	4.8	1060
5050 (0, H38)	0.097	110	0.23 (212)	68–212	13.2	3.4	1205
6061							
(0)	0.098	104	0.23 (212)	68–212	13.0	3.7	1200
(T4)	0.098	89	0.23 (212)	68–212	13.0	4.3	1200
7075 (T6)	0.101	75	0.23 (212)	68–212	12.9	5.2	1180
Aluminum, casting alloys							
214	0.096	80	0.23 (212)	68–212	13.3	4.9	1185
360	0.095	85	0.23 (212)	68–212	11.7	4.7	1150
750	0.104	106	0.23 (212)	68–212	12.8	3.9	1200
Copper, wrought							
Pure copper	0.324	228	0.092	68	9.2	1.67	1981
Electrolytic tough pitch (ETP)	0.321	226	0.092	68–572	9.8	1.71	1981
Copper, wrought alloys							
Commercial bronze (90Cu-10Zn)	0.318	109	0.09	68–572	10.2	3.9	1910
Red brass (85Cu-15Zn)	0.316	92	0.09	68–572	10.4	4.7	1880
Low brass (80Cu-20Zn)	0.313	80	0.09	68–572	10.6	5.4	1830
Cartridge brass (70Cu-30Zn)	0.308	70	0.09	68–572	11.1	6.2	1750
Yellow brass (65Cu-35Zn)	0.306	68	0.09	68–572	11.3	6.4	1710
Muntz metal (60Cu-40Zn)	0.303	70	0.09	68–572	11.6	6.2	1660
Inhibited admiralty (71Cu-28Zn-1Sn)	0.308	63	0.09	68–572	11.2	6.9	1720
Cupro-nickel 30% (70Cu-30Ni)	0.323	16.9	0.09	68–572	9.0	37	2260
Cupro-nickel 10% (90Cu-10Ni)	0.323	22.9	0.09	68–572	9.5	19	2100
Constantan (55Cu-45Ni)	0.32	12.3	0.09	68–1832	10.4	50	2355
Iron and iron alloys							
Pure iron	0.284	43	...	68	6.5	9.71	2800
Gray cast iron	0.251–0.265	26.6	...	32–212	5.8		
Malleable cast iron	0.260–0.265	...	0.11	68–752	6.7	28–41	2250
Alloy cast irons							
Heat-resistant gray irons	0.246–0.271	12–23	...	70	4.5–10.7	140–170	
Heat-resistant ductile irons	0.257–0.268	7.7	...	70	6.0–10.4	58–102	
Carbon steels, annealed							
AISI 1010	0.28	37[c]	0.10	68–212	6.6	12	
AISI 1020[c]	0.28	30	0.11	68–212	6.5	...	2760
AISI 1078	0.28	27.7	0.11	68–212	6.2	18.0	
AISI 1524	0.28	26.6	0.11	68–212	6.6[c]	20.8	
AISI 4140[d]	0.28	24.8	0.11	68–212	6.8	22.2	
Stainless steels, wrought, annealed							
Austenitic grades							
AISI 202	0.28	9.4 (212)	0.12	32–212	9.7	69	2250–2650
AISI 302, 303	0.29	9.4 (212)	0.12	32–212	9.6	72	2550–2590
AISI 304	0.29	9.4 (212)	0.12	32–212	9.6	72	2550–2650
AISI 310	0.29	8.2 (212)	0.12	32–212	8.8	78	2550–2650
AISI 316	0.29	9.4 (212)	0.12	32–212	8.8	74	2500–2550
AISI 347	0.29	9.3 (212)	0.12	32–212	9.2	73	2550–2600

NOTE: Footnotes appear on the last page of the table.

Table 1.2. General Properties of Selected Metals and Alloys (*Continued*)

Typical values at or near room temperature unless otherwise indicated

Metal or alloy	Density, lb/in³	Thermal conductivity, Btu/h·ft·°F	Approximate specific heat,[a] Btu/lb·°F	Mean coefficient of linear thermal expansion: Temperature range, °F	Mean coefficient of linear thermal expansion: Coefficient, μin/in·°F	Electrical resistivity, μΩ·cm	Approximate melting temperature,[b] °F
Precipitation hardened grades							
15-5 PH	0.28	10.3 (212)	0.10	32–212	6.0	77	2560–2625
17-4 PH	0.28	10.6 (212)	0.11	32–212	6.0	80	2560–2625
17-7 PH	0.28	9.5 (212)	0.11	32–212	6.1	83	2560–2625
PH-13-8Mo	0.28	8.1 (212)	0.11	32–212	5.9	102	2560–2625
Ferritic grades							
AISI 405	0.28	15.6 (212)	0.11	32–212	6.0	60	2700–2790
AISI 409	0.28	...	...	32–212	6.5	...	2700–2790
AISI 430	0.28	15.1 (212)	0.11	32–212	5.8	60	2600–2750
AISI 446	0.27	12.1 (212)	0.12	32–212	5.8	67	2600–2750
Martensitic grades							
AISI 410	0.28	14.4 (212)	0.11	32–212	5.5	57	2700–2790
AISI 416	0.28	14.4 (212)	0.11	32–212	5.5	57	2700–2790
AISI 440 A	0.28	14.0 (212)	0.11	32–212	5.7	60	2500–2700
Lead and lead alloys							
Corroding lead (99.73%Pb)	0.410	20.1	0.031	63–212	16.3	20.6	621
5-95 solder	0.397	20.6	0.032	59–212	15.9	19.5	594
20-80 solder	0.368	21.5	0.036	59–212	14.7	17.5	531
50-50 solder	0.321	26.9	0.041	59–212	13.0	15.6	421
Hard lead (96Pb-4Sb)	0.399	17.7	0.032	68–212	15.4	24.0	570
Hard lead (94Pb-6Sb)	0.393	16.7	0.032	68–212	15.1	25.3	545
Lead-base babbitt (SAE14)	0.352	13.8	0.038	68–212	10.9	28.6	514
Magnesium and magnesium alloys							
Pure magnesium (99.8%Mg)	0.063	89	0.25	68	14.0	4.5	1202
Wrought alloy AZ31B	0.064	56	0.25	68–212	14.4	9.2	1170
Casting alloy AZ63A	0.066	44	0.25	68–212	14.5	11.5	1130
Nickel and nickel alloys							
Pure nickel (99.95%Ni and Co)	0.322	53	0.11 (212)	32–212	7.4	6.8	2651
Grade A nickel	0.321	35	0.13 (32–212)	77–212	7.4	9.5	2635
Grade D nickel	0.317	28 (122)	0.13 (32–212)	...	...	14	2600
Duranickel (94Ni-4.5Al)	0.298	...	0.13 (32–212)	32–212	7.2	43	2635
Hastelloy B	0.334	6.5	0.091	32–212	5.6	135	2460
Hastelloy C-4	0.312	...	0.10 (212)	...	...	125	
Hastelloy D	0.282	12.1	0.109	32–212	6.1	113	2050
Hastelloy X	0.297	5.6	0.116	79–212	7.7	118	2350
Illium G	0.310	7.0	0.105	32–212	6.8	124	2440
Illium R	0.310	7.5	0.110	32–212	6.7	120	2500
Inconel (76Ni-16Cr-8Fe)	0.307	8.7 (32–212)	0.109 (77–212)	68–212	6.4	98	2600
Invar (36%Ni)	0.289	6.3 (68–212)	0.123 (77–212)	68	0–1	...	2600 S
Monel (67Ni-30Cu)	0.319	15 (122)	0.127 (2300)	32–212	7.8	48	2370–2460
Cast Monel	0.312	...	...	77–212	7.2	53	2400–2450
K Monel	0.306	10.8	0.127 (77–750)	77–212	7.8	58	2460
S Monel (cast)	0.302	...	...	70–212	6.8	63	2300–2350
Ni-o-nel	0.294	...	...	81–200	7.2	108	
Tin and tin alloys							
Pure tin (white)	0.264	39 (32)	0.054	32–212	12.8	11.5	449
Soft solder (63Sn-37Pb)	0.304	29 (32–356)	...	59–230	13.7	14.5	361
Tin foil (92Sn-8Zn)	...	34 (32–392)	...	...	...	12.1	390 S
Titanium and titanium alloys							
Commercially pure titanium (99.0%)	0.163	9.1–11.5	0.125	70–212	4.8	49–55	3020
Ti-5Al-2.5Sn	0.162	4.3	0.127	70–212	5.2	157	...
Ti-8Mn	0.171	6.3	0.118	70–212	4.8	91	...
Zinc and zinc alloys							
Pure zinc	0.258	65.4	0.092	68–482	22	5.9	787
AG40A die casting alloy	0.24	65.4	0.10 (68–212)	68–212	15.2	6.4	728
Commercial rolled zinc (0.08Pb)	0.258	60.5[e]	0.094	68–104	18.0/12.8[f]	6.1	786
Rolled zinc alloy (1Cu-0.01Mg)	0.259	60.5[e]	0.096	68–212	19.3/11.7[f]	6.3	792

Table 1.2. General Properties of Selected Metals and Alloys (*Continued*)

Typical values at or near room temperature unless otherwise indicated

Metal or alloy	Density, lb/in³	Thermal conductivity, Btu/h·ft·°F	Approximate specific heat,[a] Btu/lb·°F	Mean coefficient of linear thermal expansion: Temperature range, °F	Mean coefficient of linear thermal expansion: Coefficient, μin/in·°F	Electrical resistivity, μΩ·cm	Approximate melting temperature,[b] °F
Precious metals							
Pure silver	0.379	242 (32)	0.056	32–932	11.4	1.6	1761
Pure gold (99.9%Au)	0.698	172 (32)	0.0312	68	7.9	2.4	1945
Platinum (99.85%Pt)	0.775	40	0.314 (32)	68	4.9	10.6	3217

[a]Numbers in parentheses indicate test temperature in °F.
[b]Normally liquidus temperature is given; an S suffix indicates solidus temperature.
[c]Heat treatment unknown.
[d]Oil hardened, tempered at 1110°F.
[e]Thermal conductivity in the direction of the grain.
[f]First number with grain; second number across grain.

NOTES: (1) For liquid metals see Table 1.3. (2) For thermal conductivity versus temperature for selected AISI carbon steels see Fig. 1.1. (3) For properties of metals and alloys for low-temperature service see Part 4, Chap. 1. (4) For properties of metals and alloys for high-temperature service see Part 4, Chap. 2.

CONVERSION FACTORS: lb/in³ × 27,679 = kg/m³; Btu/h·ft·°F × 1.7307 = W/m·k; Btu/lb·°F × 4186.8 = J/kg·K; μin/in·°F × 1.8 = μm/m·K, °C = (°F-32)/1.8.

SOURCES: Compiled largely from *Metals Handbook* and *ASM Metals Reference Book* (see below):
Metals Handbook, 9th ed., vols. 1–3, Am. Soc. for Metals, Metals Park, Ohio, 1978.
ASM Metals Reference Book, 2d ed., Am. Soc. for Metals, Metals Park, Ohio, 1983.
S. Hoyt (ed.), *ASME Handbook—Metals Properties,* McGraw-Hill, New York, 1954.

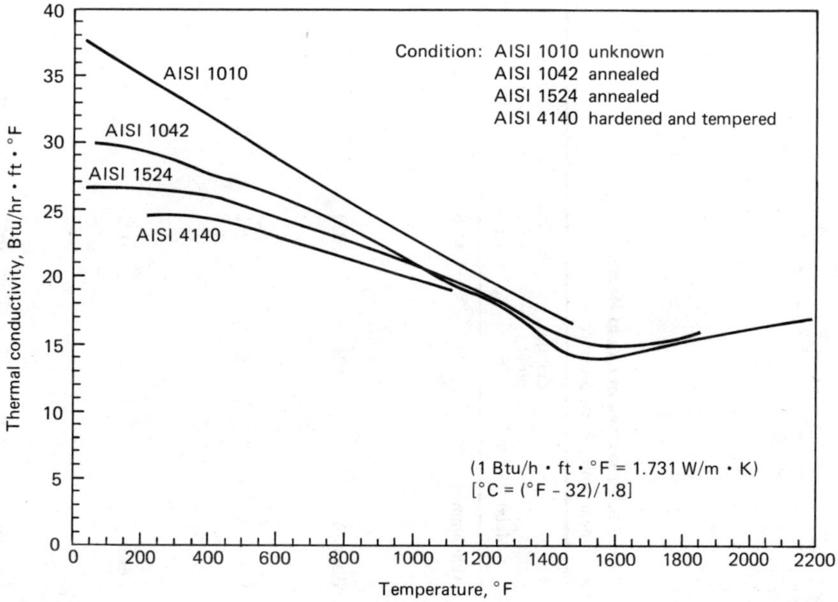

Fig. 1.1 Thermal conductivity as a function of temperature for selected AISI carbon steels. (*Based on data from* The Metals Handbook, *9th ed., vol. 1, ASM, 1978.*)

Table 1.3. Properties of Liquid Metals

At or near atmospheric pressure

Metal	Melting point, °F	Boiling point, °F	Heat of fusion, Btu/lb	Temperature, °F	Specific gravity	Thermal conductivity, Btu/h · ft · °F	Specific heat, Btu/lb · °F	Absolute viscosity, lbm/ft · h
Aluminum	1220	4426	171	1250	2.38		0.259	
				1300	2.37	60.2	0.259	6.77
				1350	2.36	63.4	0.259	5.80
				1400	2.35	64.3	0.259	4.82
				1450	2.34	69.9	0.259	3.89
Bismuth	521	3020	22.3	600	10.0	9.5	0.0345	3.92
				800	9.87	9.0	0.0357	3.2
				1000	9.74	9.0	0.0369	2.7
				1200	9.61	9.0	0.0381	2.2
Cesium	83	1252	6.8	83	1.84	10.6	0.060	
				150				1.38
				250				1.06
				350				0.89
				400				0.83
Lead	621	3182	9.9	700	10.5	9.3	0.038	5.80
				850	10.4	9.0	0.037	4.97
				1000	10.4	8.9	0.037	4.21
				1150	10.2	8.7	0.037	3.67
				1300	10.1	8.6		3.31
Lithium	355	2448	284	400	0.506	24	1.0	1.4
				600	0.497	23	1.0	1.2
				800	0.489	22	1.0	1.3
				1200	0.471			1.0
				1800	0.442			1.0
Magnesium	1203	1994	158	1250	1.55		0.318	
				1301	1.53		0.320	
				1350	1.49		0.322	

Mercury	−38	674	4.9	50	13.6	4.8	0.033	3.85
				200	13.4	6.0	0.033	3.0
				300	13.2	6.7	0.033	2.7
				400	13.1	7.2	0.032	2.4
				600	12.8	8.1	0.032	2.1
Potassium	147	1400	26	800	0.74	22.8	0.183	0.51
				1000	0.71	21.1	0.182	0.41
				1200	0.69	19.5	0.183	0.35
				1400	0.67	18.0	0.187	0.32
56% Sodium/ 44% Potassium	66	1518	…	200	0.89	14.9	0.270	1.36
				400	0.86	15.3	0.260	0.92
				600	0.83	15.7	0.255	0.71
				800	0.81	16.0	0.251	0.50
				1000	0.79	16.0	0.250	0.49
				1200	0.76	16.0	0.251	0.41
Sodium	208	1621	49	200	0.93	49.1	0.331	1.73
				400	0.90	46.7	0.320	1.10
				600	0.88	43.8	0.311	0.80
				800	0.85	40.1	0.305	0.61
				1000	0.82	37.2	0.302	0.56
				1200	0.79	35.0	0.301	0.48
				1400	0.76	32.7	0.303	0.42
Tin	449	4712	25.4	500	6.94	19	0.058	4.39
				700	6.86	19.4	0.060	3.5
				850	6.81	19	0.062	3.1
				1000	6.74	19	0.064	2.7
				1200	6.68	19	0.066	2.4
Zinc	787	1670	49	600	6.97		0.123	
				850	6.90		0.119	7.56
				1000	6.86		0.116	6.19
				1200	6.76		0.113	5.00
				1500	6.74		0.107	3.54

CONVERSION FACTORS: Btu/lb × 2326.0 = J/kg; Btu/h · ft · °F × 1.7307 = W/m · K; Btu/lb · °F × 4186.8 = J/kg · K; lbm/ft · h × 4.1338 × 10^{-4} = Pa · s; °C = (°F − 32)/1.8.

SOURCE: Reprinted with permission from R. Bolz and G. Tuve (eds.). *Handbook of Tables for Applied Engineering Science,* 2d ed., copyright, CRC Press, Inc., Boca Raton, Fl., 1976. Data for sodium and potassium are based on *Liquid Metals Handbook,* 3d ed., Office of Naval Research, U.S. Government Printing Office, Washington, D.C., 1955.

Table 1.4. Typical Properties of Selected Plastics at Room Temperature

Plastic classification	Specific gravity	Thermal conductivity, Btu · in/h · ft² · °F	Approximate specific heat,[a] Btu/lb · °F	Coefficient of linear thermal expansion, μin/in · °F	Tensile yield strength,[b] ksi	Modulus of elasticity in tension,[c] ksi	Deflection temperature under flexural load: 66 lb/in², °F	Deflection temperature under flexural load: 264 lb/in², °F
ASTM Method:	D792	C177	…	D696	D638/D651	D638/D651	D648	D648
ABS (acrylonitrile-butadiene-styrene)								
Injection MC, heat resistant	1.05–1.08	1.3–2.3	…	33–52	4.3–7.0	300–360	230–245A	220–240A
Extrusion grade	1.02–1.06	…	0.3–0.4	33–72	4.3–6.4	130–420	170–235A	170–220A
Acetal, homopolymer	1.42	1.6	0.35	56	9.5–12.0	520	328–338	255–260
Acrylic, sheet, cast	1.17–1.20	1.2–1.7	0.35	28–50	…	350–450	165–235	160–215
Allyl, DAP molding compound, glass filled	1.70–1.98	1.5–4.4	…	5.6–20	…	1400–2200	…	330–550
Cellulose acetate, sheet	1.28–1.32	1.2–2.3	0.3–0.5	56–83				
Epoxy								
Bisphenol MC, glass reinforced	1.6–2.0	1.2–2.9	…	6.1–28	…	3000	…	225–500
Novolak MC, filled, high temperature	1.85–1.94	4.9–7.0	…	19	…	2300–2400	…	500
Casting resins								
Unfilled	1.11–1.40	1.3	0.25	25–36	…	350	…	115–550
Silica filled	1.6–2.0	2.9–5.8	0.2–0.27	11–22	…	…	…	160–550
Aluminum filled	1.4–1.8	4.4–7.3	…	3.1	…	…	…	190–600
Fluoroplastics								
PCTFE	2.08–2.2	1.4–1.5	0.22	20–39	5.3	150–300	258	
PTFE, granular	2.14–2.2	1.7	0.25	55–84	…	58–80	250	132
PFA	2.12–2.17	1.8	…	67	…	…	164	118
FEP	2.12–2.17	1.4	0.28	46–58	…	50	158	124
PVDF MC, EC	1.76–1.78	0.7–0.9	0.33	80–85	6.0–7.0	160	300	195
Melamine formaldehyde								
Cellulose filled	1.47–1.52	1.9–2.9	…	22–25	…	1100–1400	…	350–390
Glass reinforced	1.5–2.0	2.9–3.3	…	8.3–16	…	1600–2400	…	375–400
Nylon								
Type 6								
MC, EC	1.12–1.14	1.7	0.4	44–46	11.7[d]; 7.4[e]	380[d]; 100[e]	365–375[d]	155–185[d]
30–35% glass reinforced	1.13–1.15	…	0.35–0.5	11–17	…	1450[d]; 800[e]	425–430[d]	410–420[d]
Type 6/6								
MC	1.13–1.15	1.7	0.4	44	8.0[d]; 6.5[e]	230[d]; 500[e]	474[d]	167[d]
30–35% glass reinforced	1.37–1.38	1.5–3.4	0.3	8.3–11	…	…	495–500[d]	485–490[d]
Type 6/12								
MC	1.06–1.10	1.5	0.3–0.4	…	5.8[d]; 3.1[e]	218–290[d]; 123–180[e]	330[d]	180[d]
30–35% glass reinforced	1.30–1.38	3.0	…	…	…	1200[d]; 900[e]	400–430[d]	415–425[d]
Phenolic								
MC, high strength, glass reinforced	1.69–2.0	2.3–4.1	…	4–12	…	1900–3300	…	350–600
Casting resin, unfilled	1.24–1.32	1.0	…	38	…	400–700	…	165–175
Polybutylene, EC	0.91–0.93	1.5	0.45	71–83	1.7–2.5	30–40	215–235	130–140
Polycarbonate, unfilled, MC	1.2	1.4	0.28–0.30	38	9.0	345	280	270

Polyester								
Thermoplastic								
PBT, unfilled	1.31–1.38	1.2–2.0	...	33–53	...	280	240–375	122–185
PET, unfilled	1.34–1.39	0.96–1.0	...	36	...	400–600	100–106A	
Liquid crystal, unfilled	1.35	0.6	...	...	...	1900–2400	...	639–671
Polyarylate, injection molding	1.21–1.22	1.2	...	34–35	9.9–10.0	300–330	355	338–345
Thermoset/alkyd								
Cast, rigid	1.10–1.46	...	...	31–56	...	300–640	...	140–400
Glass reinforced BMC, TMC	1.72–2.05	5.2–6.4	...	...	...	1500–2500	...	320–350
Polyetheretherketone, unfilled	1.30–1.32	1.7	...	26	13.3	...	...	320
Polyetherimide	1.27	1.5	...	31	15.2	430	410	392
Polyethylene								
Low to medium density	0.92–0.93	2.3–2.9	0.55	56–122	0.6–2.3	14–55	100–165	90–120
High density	0.94–0.97	3.2–3.6	0.55	61–72	3.1–5.5	60–180	140–190	110–130
Ultrahigh molecular weight	0.94	3.2	...	78	4–6	29–110	155–180	110–120
Cross linked, molding grade	0.95–1.45	...	...	56	...	50–500	130–225	105–145
Polyimide, thermoplastic, unfilled	1.36–1.43	0.67–0.75	...	25–31	12.5	300	...	530–680
Polyphenylene sulfide, unfilled	1.3	2.0	...	27	...	480	...	275
Polypropylene								
Homopolymer	0.90–0.91	0.81	0.46	45–56	4.5–5.4	165–225	225–250	120–140
40% glass reinforced	1.22–1.23	2.4–2.6	...	15–18	...	1100–1500	330	300–330
Polystyrene								
General-purpose polymer	1.04–1.09	2.4–3.3	...	33–44	5–12	400–600	180–230	190–220
Impact-modified polymer	1.03–1.10	1.0–3.0	...	19	1.5–7	140–500	180–220	160–200
Polysulfone, injection molding, EC	1.25	1.8	...	31	10.2	360	358	345
Polyarylsulfone	1.29–1.37	...	...	17–27	0.4–12.0	390	...	400
Polyethersulfone, unfilled	1.37	0.93–1.3	0.26	31	12.2	390	...	397
Polyphenylene oxide and polyphenylene ether, modified	1.06–1.10	1.1–1.5	...	18–43	6.5–9.0	340–380	230–279	180–265
Polyurethane, liquid casting resin	1.1–1.5	1.5	0.42–0.44	56–112	...	10–100	Wide range	Wide range
PVC								
Rigid MC, sheet, rod, tubing	1.30–1.58	1.0–1.5	0.25–0.35	12–56	6.0–8.0	350–1000	135–180	140–170
Chlorinated, MC and EC	1.49–1.58	0.96	0.33	38–42	6.0–8.0	341–475	215–247	202–234
Silicone								
Flexible casting resin (including RTV)	0.99–1.5	1.0–2.2	...	167–444				
Epoxy encapsulating compound	1.84	4.6	...	17				
Vinyl, formal MC and EC	1.2–1.4	1.0	...	36	...	350–600	...	150–170

[a]Values for specific heat are based on *Guide to Plastics—Property and Specification Charts*, McGraw-Hill, New York, 1975.
[b]Note that factors such as creep resistance are often more important than tensile yield strength for selecting design stresses for plastic components.
[c]Compressive modulus can be significantly different from tensile modulus for plastics.
[d]Dry, as molded (approximately 0.2% moisture content).
[e]As conditioned to equilibrium with 50% relative humidity.

NOTES: (1) Use data compilation as a guide for typical property values. Actual property values can vary widely depending on the exact formulation, processing methods, and temperature. Design values should be obtained from manufacturers. (2) For properties of foamed plastics see Table 1.5. (3) For cellular plastic insulations see Part 3, Chap. 4. (4) For information on the use of plastics at high temperatures see Part 4, Chap. 3. (5) For the characteristics of thermally conductive plastics see Part 6, Chap. 4.

CONVERSION FACTORS: Btu · in/h · ft^2 · °F × 0.14422 = W/m · K; Btu/lb · °F × 4186.8 = J/kg · K; μin/in · °F × 1.8 = μm/m · K; ksi × 6.8948 = Pa; °C = (°F − 32)/1.8.

ABBREVIATIONS: MC, molding compound; EC, extrusion compound; A, annealed.

SOURCE: Abstracted with permission from *Modern Plastics Encyclopedia*, vol. 62, no. 10A, McGraw-Hill, New York, 1985.

Table 1.5. Typical Properties of Selected Foamed Plastics

At or near room temperature

Plastic classification	Density, lb/ft^3	Thermal conductivity, Btu · in/h · ft^2 · °F	Coefficient of linear thermal expansion, μin/in · °F	Tensile strength, lb/in^2
ASTM Test Method	D1622	C177	D696	D1623
ABS, injection molding type, pellets	40–56	0.58–2.1	37–95	2000–4000
Acrylic, boards	2.6–6.2	0.22	29–32	
Cellulose acetate, rigid closed cell	6.0–8.0	0.31	25	170
Epoxy				
Precast foam blocks, slabs, sheet	10.0	0.28	...	180
	20.0	0.32	...	650
Syntactic, foam in place, 2-part liquid	2–2.3	0.11–0.13		
Phenolic				
Foam-in-place liquid resin	0.33–1.5	0.21–0.28	...	3–17
	2–5	0.20–0.22	5	20–54
	7–10	0.24–0.28	...	80–130
Syntactic, 2-part liquid and paste	50–60	1.0	100	>1000
Polycarbonate, 5–10% glass fiber reinforced	50–66	0.90–1.05	18–20	6100–6300
Polyethylene				
Low density, sheet	2.1–3.3	0.28–0.34	23	35–100
High density, molded parts	25–50	0.92	42	1200
Cross linked, sheet, rolls	1.6–2.4	0.25–0.28	...	40–70
	3.6–4.4	0.30	...	70–110
	5.5–7	0.32–0.34	...	110–210
Polyimide	0.6–1.0	0.25	...	8
Polypropylene				
Low density, sheet, rolls	0.6	0.27	...	20–40
High density, molded parts	35–42	...	50–65	1600–1800
Polystyrene, shapes molded from PS beads	1	0.26–0.28	30–40	16–35
	2	0.23–0.24	30–40	25–70
	5	0.246	30–40	148–172
Polyurethane, rigid closed cell, molded	1.3–3.0	0.11–0.17*	40–80	15–95
	9–12	0.19–0.25*	40	230–450
	19–25	0.34–0.42*	40	775–1300
PVC, flexible, closed cell, molded shapes	4.0–11	0.24–0.28	...	50–150 (D412)
Silicone				
Open cell, sheet	10.0	0.6	...	45
Closed cell, sheet	25–34	...	...	80–100
Styrene, acrylonitrile, shapes molded from beads	0.5	0.32	...	
	1.0	0.29	...	30
Urea formaldehyde, block, foam in place	0.8–1.2	0.18–0.21	...	Poor

*Thermal conductivity is for fluorocarbon blown polyurethane foam. First value is for new foam; second value is for aged foam.

NOTES: (1) Use data compilation as a guide for typical property values. Actual values can vary widely depending on exact formulation, processing methods, and service temperature. Design values should be obtained from manufacturers. (2) For properties of cellular insulations see Part 3, Chap. 4.

CONVERSION FACTORS: lb/ft^3 × 16.018 = kg/m^3; Btu · in/h · ft^2 · °F × 0.14422 = W/m · K; μin/in · °F × 1.8 = μm/m · K; lb/in^2 × 6894.8 = Pa.

SOURCE: Abstracted with permission from *Modern Plastics Encyclopedia,* vol. 62, no. 10A, McGraw-Hill, New York, 1985.

Part 3

Thermal Insulation

Chapter 1

Characteristics and Application of Thermal Insulation

RONALD P. TYE
Dynatech Scientific, Inc.

This chapter reviews the relationship of the physical properties of thermal insulations to the requirements for thermal insulation system design, installation, and operation.

There are three "mass types" of thermal insulation material.

1. *Fibrous:* Fine-diameter, usually <20-μm, organic or inorganic materials in the form of low-density [<~10 lb/ft^3 (160 kg/m^3)] loose fill, batts, blankets, and felts; sometimes bonded or rigidized into higher-density [>~10 lb/ft^3 (160 kg/m^3)] boards, tapes, ropes, cements, and preformed shapes.

2. *Cellular:* Closed- or open-cell organic and inorganic materials in the form of expanded or extruded flexible or rigid boards, blocks, or shapes and foamed or sprayed-in-place products.

3. *Flake or granule:* Composed of small flakes, nodules, or particles, generally of inorganic materials. In some cases, these can be bonded together to form boards, blocks, preformed shapes, cements, and mastics. If the particle diameter is small enough (<1 μm), both particle and rigidized forms are microporous, having special thermal performance properties due to restricted gas conduction.

The thermal performance of these various types of thermal insulation is based on the ability to divide an air space into very small pockets, thereby minimizing solid and gas conduction and radiation and, for most applications, eliminating convection.

In addition, there is the reflective type of insulation. This is composed of parallel thin foils or sheets of metal, or metallized plastic, having low emittance and spaced so as to reflect radiative heat transfer toward the source and to minimize convective or gas conductive heat transfer. The thermal performance of this type depends on the design and construction of the system rather than on a single material property.

In application, all of these forms are installed separately or in combination together with other materials, such as vapor and air retarders or protective coverings, and they perform as an insulation system. Insulation is used for one or more of the following reasons: to conserve heat, maintain desired temperature conditions, retard changes in temperature, prevent condensation, provide fire protection, and provide safety from burn hazards. It is therefore essential that the system perform as designed.[1] Thus the performance properties, and especially the factors that affect them, are an important consideration to economic installation, lifetime, and thermal efficiency.

The variety of insulation types and forms have properties that vary over significant ranges, and thus the uses of a material can also vary widely. The selection of an insulation is determined by considering a number of properties-related factors. Thermal performance, while being the ultimate requirement, may not necessarily be the major parameter that determines the choice of a particular material.[2] In addition, for some applications a particular property may be important, whereas for another its influence can be minor.

The following sections discuss both the various properties of thermal insulation and the factors that affect them in relation to the choice for a particular application.

1.1 PROPERTIES AND THEIR SIGNIFICANCE

Properties of thermal insulation are listed in Table 1.1 together with a brief discussion of each to illustrate its significance and importance to an application.

In general, most of the listed properties are evaluated according to recognized ASTM test methods at specific test conditions. Note that not all of the properties obtained and listed by manufacturers or other sources are carried out for or under the same temperature, conditions, or other related criteria for all applications. Thus, great care should be exercised in obtaining and comparing such "ideal" information and applying it for a specific application. This is particularly true for the thermal performance characteristics. However, it may be equally applicable to some of the other properties, specifically where information is provided at only one temperature or for one condition. Where doubts or uncertainties exist, measurements under the conditions of application are recommended.

A final consideration is health and safety. While no specific test methods and criteria are available, ASTM publishes a standard classification relating to potential concerns in this area.[25] This recommends that reliance be placed on information from responsible manufacturer and user sources. It

Table 1.1. Significant Properties of Thermal Insulations

Property	Significance of need in application
Physical and mechanical	Structural loading and heating and cooling rate requirements
Density, compressive, flexure, shear, tensile, and breaking strengths	Need to support structural loading, or conversely to take up dimensional changes
Abrasion and vibration resistance	Need to withstand wear due to mechanical and environmental reasons
Dimensional stability, expansion, shrinkage	Need to retain shape, dimensions, spacing of joints, etc.
Incidence of cracking, warpage	Affects both strength and thermal performance
Compaction, settling, recovery	Particularly important for batt, blanket, and loose fill to retain design thicknesses
Air resistance	Reduce convective effects and erosion
Adhesion, dry and wet	Important to ensure that materials will adhere to surfaces
Chemical and fire	
Alkalinity, acidity, resistance to acids, alkalies, and solvents	Important to service life in application
Corrosion resistance and chemical reactivity	Insulation may corrode or be corroded by surface or other materials in contact
Combustibility	Melting point, flame point, burning rate, heat-contributory smoke density, toxicity of combustion products, all affect choice for an application
Moisture, moisture vapor, and environment	
Hygroscopicity, capillarity, and absorption	Ability to absorb and retain liquid affects thermal and mechanical performance
Vapor permeance and migration	Ability to absorb and retain moisture vapor can affect dimensions and performance
Weather resistance	Ability to resist effects of the elements
Thermal	
Temperature limits and thermal shock resistance	Determines use temperature for continuous and/or cyclic operation
Emittance, reflectance	Determines use of insulations and coverings based on radiant heat transfer
Heat capacity, thermal diffusivity	Very important in applications where heat storage, rapid or slow heating and cooling rates are required
Thermal conductivity, resistance, conductance	Determines ultimate insulation peformance

Table 1.2. Considerations in Selecting Insulations for Particular Applications

Application	Parameters
Toxic chemicals	Nonreactive, nonabsorbent, temperature limits
Transportation	Density, vibration resistance, fire resistance
Cold storage	Resistance to moisture absorption, high mass specific heat product
Cyclic processing	Low mass specific heat product, ease of removal and installation
Nuclear	No dust formation, short radioactive half-life, easy and rapid flooding with water
Roofs	Resistance to weathering and moisture and moisture vapor transmission, high strength to weight, ease of maintenance and repair
Petroleum processing	Good fire resistance characteristics, low thermal diffusivity
High-temperature furnaces	High strength, nonreactive, low shrinkage
Appliances	Good structural characteristics, high mass specific heat, high termal resistance per unit thickness
Pipelines	Location and type of exposure—indoors, outdoors, underground, underwater, moisture mechanisms, environment, soil temperature (protection of permafrost, for example)

contains a list of government regulatory agencies, trade associations, and other reliable information sources on different material types.

There are very few applications where only one particular insulation becomes a sole choice. In most cases, various candidates are available. The final choice must be made by considering all aspects of an application. The final selection of the most suitable insulation is based on the available information with regard to the properties relevant to the requirements. Some brief examples of such considerations are given in Table 1.2.

There are four major stages in the life of an insulation: (1) manufacture and fabrication, (2) shipping and storage, (3) installation and operation, and (4) maintenance. For any application, the choice of an insulation depends not only on the previously described operational parameter considerations but also on the lifetime factors. Table 1.3 summarizes the major issues and considerations related to these phases.

There is also the particularly important matter of economics. The total costs involve those for each phase during the lifetime, including initial materials and components; shipping, storing, and handling; fabrication; installation (including overhead for supervision, etc.); maintenance (both regular and abnormal); and depreciation. These have to balance against such factors as savings of energy, use of smaller equipment, added value of products, reduced insurance, and better health and safety of occupant, all due to the use of a well-designed and operated insulation system.

In summary, for the final choice of an insulation, there is a need to:

- Establish the basic functions the insulation must perform for a particular set of criteria determined by the application
- Determine the properties of materials and systems which will fulfill the functions
- Select candidate insulation or insulations
- Determine the thermal resistance required for the design and hence the thicknesses required

The thermal performance characteristics become a dominant factor only in the final stages of selection. The operation of the insulated system depends on the insulation attaining its desired performance levels. Any influencing effect which negates performance attainment must be considered in the final selection. The following section discusses the various factors that affect performance, including illustrations that provide guidance on the quantification of these effects.

1.2 FACTORS INFLUENCING THERMAL PERFORMANCE

The following chapters include properties information for the various types and classes of thermal insulation. Such information, with few exceptions, has been generated on good samples of materials, usually under controlled laboratory conditions, using the appropriate standard test methods and practices. While this information is useful as an initial guide, it must be realized that true performance in situ, under actual operating conditions, is not represented by such values. Consequently, many designers or users build in a large safety factor, often of 2 or 3, to try to ensure that the system will perform adequately. The final systems are often larger than necessary, costly, and inefficient, while those designed solely on data contained in data books or specification sheets will usually provide a poorly performing system.

The major parameters affecting thermal performance are shown in Table 1.4. They fall into two classes, materials variables and external variables, and these can often be interrelated.

Product Variability

The thermal performance is a direct function of the thermal resistance R,

$$R = \frac{\text{thickness } x}{\text{apparent thermal conductivity } \lambda_{\text{app}}} \tag{1.1}$$

In general,

$$\lambda_{\text{app}} = f(\text{density } \rho) \tag{1.2}$$

Thus, any discrepancies in the values of either or both parameters between actual products and their design values will contribute to a difference in performance, since in many cases the effects can be additive. Results for fibrous materials used both for building-envelope application[3,4] and for industrial-process purposes[5,6] illustrate the potential extent of these effects.

Glass and other mineral fiber products are utilized widely for insulating walls or attic spaces of buildings. These 4- to 6-μm-fiber-diameter fiber products are designed to have spe-

Table 1.3. Lifetime Factors Related to Choice of Thermal Insulations

Phase	Requirement	Relevant properties
Manufacture, fabrication	Minimum variability	Uniformity of raw materials and of product dimensions
	Ease of handling	Strength properties, strength-to-weight ratio, dimensional stability, nontoxic safety
Shipping, storage	Ease of handling in minimum space	All strength properties, dimensional stability
	Nontoxic, good resistance to fire	Chemical and fire resistance properties Resistance to all moisture, rodents, vermin, weathering, etc.
Installation	Ease of handling, space limitations	Uniformity of dimensions, ease of cutting and forming, minimum dusting and flaking
	Minimum breakage and waste, accomplished rapidly	All mechanical properties, safety, and toxicity
		In addition, for insulations foamed in place or applied wet, need for quick setting and drying times, adhesive properties, resistance to chemicals, resistance to elements, especially wind and rain
Operation, maintenance	Withstand mechanical damage due to impact, vibration, expansion, and contraction	All mechanical properties, expansion coefficients, chemical resistance, acidity, alkalinity
	Compatibility with other components	Corrosion resistance, all moisture-related properties
	Resistance to elements and environment	Resistance to animals and insects, moisture, and air movement
	Safety	Reactivity, absorptivity, adsorptivity, combustion and fire properties, burn hazards
	Thermal performance	Temperature limits, coefficients of expansion, thermal resistance, specific heat, density, thermal diffusivity

cific thermal resistances for given thicknesses. Due to economic factors, the densities of the products are relatively low, 0.6 lb/ft^3 (10 kg/m^3) for glass and 2 lb/ft^3 (32 kg/m^3) for rock fiber. Two common thermal resistance values are R = 11 h · ft^2 · °F/Btu (1.94 m^2 · K/W) at 3.5-in (89-mm) thickness and R = 19 h · ft^2 · °F/Btu (3.34 m^2 · K/W) at 6.9 in (152 mm). For these materials, the change in λ_{app} versus ρ is significant. For example, λ_{app} changes from approximately 0.030 to 0.023 Btu/h · ft · °F (0.053 to 0.041 W/m · K) for a change in density from 0.5 to 0.87 lb/ft^3 (8 to 14 kg/m^3). Thus, it is critical that the products have a reproducible density since small changes will have a large impact on the limited range in which they are manufactured. However,

Table 1.4. Factors Influencing Thermal Performance of Insulations

Materials variables	External variables
Product variability: type, material content, density, thickness	Temperature: temperature difference, time dependence
Flaws and inhomogeneities: voids, cracks, irregular geometry	Environment: type and pressure of gas, moisture pickup and retention
Anisotropy: preferred orientation of fibrous and layered products	Orientation: direction of heat flow
Shrinkage: dimensional, density, form changes, aging	Mass transfer: natural and forced convection, type and velocity of gas, moisture migration
Retained moisture: loss of moisture, changes in form and density	Installation: incorrect application and maintenance practices

Table 1.5. Results for Fibrous Glass Products of one Manufacturer

Product*	Number of measurements	Property*	Mean	Property range
Fiberglass at T_N = 3.5 in	47	T_R	3.59	2.85–4.22
		ρ_R	0.59	0.48–0.77
	17	ρ_N	0.61	0.41–0.77
		R	10.7	9.4–11.9
Fiberglass at T_N = 6.25 in	33	T_R	5.70	4.29–7.22
		ρ_R	0.68	0.52–0.82
	11	ρ_N	0.63	0.50–0.88
		R	18.5	17.1–21.3

* T_N = design thickness, in; T_R = recovered thickness, in; ρ_R = calculated density at T_R, lb/ft^3; ρ_N = calculated density at T_N, lb/ft^3; R = measured thermal resistance at T_R, ft^2 · °F · h/Btu.

CONVERSION FACTORS: 1 in = 2.54 mm; 1 lb/ft^3 = 16.01 kg/m^3; 1 ft^2 · °F · h/Btu = 0.175 m^2 · K/W.

Table 1.6. Summary of Measurements on Fibrous Building Products

Product type	Number of specimens	Percentage below nominal value	Percentage below 90% nominal value
3.5-in fiberglass	48	85	19
6.25-in fiberglass	30	93	23
3.07-in rock wool	12	100	67
3.5-in rock wool	15	53	0

CONVERSION FACTOR: 1 in = 25.4 mm.

an additional factor is the thickness and its recovery after packaging. This not only affects the density and, hence, the value of λ_{app}, but also the total thermal resistance.

Table 1.5 contains summarized data for randomly sampled products of one manufacturer. Results for products of all other manufacturers of similar products showed essentially the same order of 20 percent variability, as illustrated in Table 1.6. The products are heterogeneous and are somewhat difficult to manufacture economically in a highly reproducible form. It must also be mentioned that recovery characteristics are age-dependent. Thus, while testing is usually carried out soon after manufacture, packages of insulation can be stored, often for long periods, before installation, and thus recovery may be somewhat less, providing further uncertainty in the thermal resistance.

For higher-temperature applications, various types of fibrous products are used, depending usually upon the maximum temperature and environmental conditions. Such products are much higher in density than those for buildings, but since radiation becomes the dominating heat transmission mode, the overall effects at elevated temperatures are similar.

Mineral fiber products in the density range of 4 to 14 lb/ft^3 (64 to 220 kg/m^3) are the most common type of insulation used in applications for temperatures up to 1200°F (650°C). The term mineral fiber covers both glass and rock. Rock fiber diameters are less than 7 μm. In the United States, the term rock wool is applied to material manufactured from both virgin rock and the slag obtained from the steel-making process. The latter contains greater than 20 percent of iron oxide and is different from the true rock products, which contain less than 10 percent of this oxide which has a higher λ. These products vary in both form and performance. The true rock fiber is more homogeneous, with a tendency to form a longer, smaller-diameter fiber. In addition, it contains little or no unfiberized product (shot), whereas the slag wool can often contain up to 50 percent. The effects of density and composition variables on the performance of this type of product are shown in Table 1.7. It can be seen, for example, that a nominal rock wool of 9.4 lb/ft^3 (150 kg/m^3) at 500°F (260°C) can have 20 to 30 percent improved performance over a 12.5-lb/ft^3 (200-kg/m^3) slag wool. Alternatively, for a given performance value it is possible to choose a lower-density product, which could be more economical.

For higher-temperature applications, refractory oxide products with 5- to 7-μm fiber diameters are now widely used, especially in oxidizing environments. These can range from relatively lightweight blankets [3.7 to 12.5 lb/ft^3 (60 to 200 kg/m^3)] to rigidized boards [12.5 to 25 lb/ft^3 (200 to 400 kg/m^3)], where some structural integrity is needed as well as good thermal performance. Table 1.8 illustrates the variability that can be experienced with different fiber-type products at design thicknesses of 1 in (25.4 mm).

Overall, the results indicate that a high variability in the blanket products can have a widely significant effect on performance. The effects are much smaller with the higher-density board products. The designer of a high-temperature system must be aware of the extent of these effects. Newer products having higher-purity fibers and smaller fiber diameters are now available.[6] These can be manufactured in a much more homogeneous form and as a result have significantly improved properties at comparable densities.

For even higher temperatures, and especially for vacuum and inert environment applications, carbon and graphite fiber products can be used.[7,8] Currently they are available as needled felts having long, 7- to 12-μm-diameter organic fibers, which have been heated at temperatures up to 2900°F (1600°C) to yield carbon grades and above 4000°F (2200°C) to provide graphite grades. In addition, rigidized forms are available. These consist of short chopped fibers, 2 to 7 μm in diameter, which are mixed with a resin binder and graphite flakes. The product is vacuum cast and fired at temperatures above 3600°F (2000°C) to provide a rigidized mass having some structural integrity. One form consisting of a rigidized felt material is also available.

The properties of the felts can be affected significantly by density and, in particular, by the degree of graphitization of the fiber. The rigidized forms are affected both by the amount of added graphite flake and binder and by the final density. The former can give a 30 to 50 percent increase in thermal

Table 1.7. Apparent Thermal Conductivity of Mineral Fiber Industrial Insulation

Material	Density lb/ft^3	λ_{app}, Btu/h · ft · °F 194°F	500°F	1022°F	1202°F
Slag wool, 40–50% shot	8.87	0.029	0.049	—	—
Slag wool, 40–50% shot	11.8	0.027	0.050	0.112	—
Slag wool, 20–30% shot	9.12	0.026	0.041	0.101	0.130
Slag wool, 20–30% shot	11.7	0.026	0.040	0.087	0.112
Rock wool, <10% shot	9.37	0.024	0.039	0.086	0.109
Rock wool, <10% shot	8.43	0.023	0.037	—	—

CONVERSION FACTORS: 1 lb/ft^3 = 16.01 kg/m^3; 1 Btu/ft · h · °F = 1.73 W/m · K; °C = (°F − 32)/1.8.

Table 1.8. Variability of Some Typical Refractory Fiber Products

Material	Nominal density, lb/ft³	Variability, %		Average effect on thermal resistance at 1832°F (1000°C), %		
		Thickness	Density	Thickness	Density	Total
Alumina blanket	6	+36 → +24	−60 → −42	+38	−21	+9
Alumina board	15	+16 → 0	−11 → +4	−8	−3	+5
Mullite blanket	6	−51 → −42	61 → 67	−46.5	+5	−41.5
Mullite board	12.5	+5 → −4	−10 → +9	+1	−4	−3
Zirconia blanket	6	+8.5 → +2.5	−56 → −32	−5.5	−18	−23.5
Zirconia board	30	+2	+2.5 → 3.8	+2	−1	+1

CONVERSION FACTOR: 1 lb/ft³ = 16.01 kg/m³.

conductivity for increases in content of less than 15 percent. Finally, the rigidized form is also much more highly anisotropic (up to 2.5 times) than the typical 20 to 30 percent anisotropies seen in most fibrous batt and blanket materials. Table 1.9 illustrates the effects of some of these variables.

Flaws, Nonconformities, and Shrinkage

To attain design performance, it is essential that the thermal insulation fill the cavity or space for which it is intended. The presence of any flaw, void, or crack through which air movement can take place will negate much of the resistive effects.

For buildings, Burns ct al.[9] and Bankvall[10] have illustrated analytically and experimentally the degrading effects of both natural and forced convection in fiber-insulated cavity walls. More recently, Vinieratos and Verschoor[11] showed that for insulated wall and ceiling constructions containing air gaps and voids, reductions in thermal performance of up to 50 percent are possible.

Figure 1.1 contains a summary of the results obtained on a typical cavity wall construction which was insulated successively with the same insulation cut to dimensions and installed in such a way that regular increases in void area and volume were simulated.[12] In addition, the experimental results are compared in Table 1.10*a* with values derived from two widely used analyses[13,14] and the magnitude effects on walls of different thermal resistances are shown in Table 1.10*b*.

Table 1.9. Thermal Conductivity, Btu/h · ft · °F, For Carbon and Graphite Fiber Insulations in an Argon Environment

Temperature, °F	Carbon felt (8.12)*	Graphite felt (6.25)	Rigidized fiber (13)†	(12.25)‡	(8.88)‡	(7.75)‡
392	0.033	0.050	0.149	0.099	0.094	0.064
1112	0.066	0.086	0.222	0.157	0.153	0.116
1832	0.129	0.144	0.291	0.251	0.248	0.212

* Values in parentheses are density, lb/ft³.
† Experimental material containing chopped pitch fibers and large amounts of flaked graphite.
‡ Production material containing chopped PAN and pitch fibers and smaller amounts of graphite flakes.
CONVERSION FACTORS: °C = (°F − 32)/1.8; 1 lb/ft³ = 16.01 kg/m³; 1 Btu/ft · h · °F = 1.73 W/m · K.

Similarly, in industrial insulated pipes, for example, the existence of air gaps between the hot surface and the insulation or at radial or longitudinal joints between adjacent sections will give rise to a significant decrease in thermal performance. The latter can be particularly important for high-density molded insulations where both eccentricity and mismatching of undersize half-sections can cause gaps. Table 1.11 contains results which illustrate the extent of such effects. In each case, the results can be compared with design values allowed in the appropriate ASTM material specifications.[15]

Normally, in testing to provide values for comparison with the specification, great care is taken to ensure that the joints or gaps are sealed or that well-matching uncracked pieces of insulation are tested. This is not possible in the field, since not every piece of material delivered can be checked for uniformity of dimensions and form. Clearly, for such insulation to perform as specified, it must be handled carefully and installed with great care, all cracks and joints being well taped or otherwise closed.

Anisotropy

Many thermal insulations, and especially fibrous materials and multifoils, exhibit anisotropy in performance properties. This is due either to a manufacturing process that causes cells

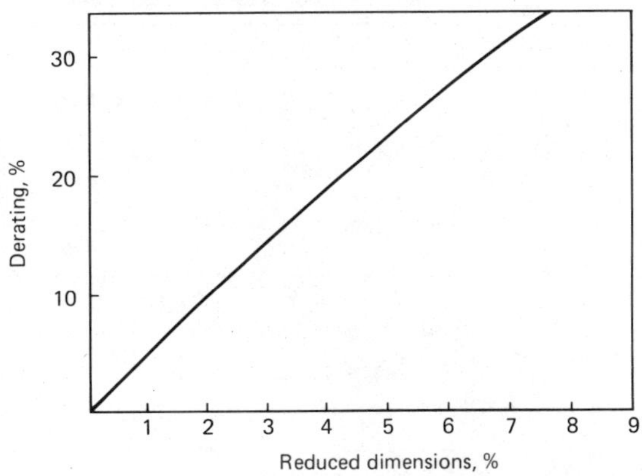

Fig. 1.1 Derating of typical insulated cavity wall for increased air gap dimensions.

Table 1.10*a*. Comparison of Measured and Calculated Effective Thermal Resistances of Cavity Insulation

	Effective thermal resistance, h · ft² · °F/Btu			Reduction in effective thermal resistance, %		
Percent reduction in dimensions	NBS model	Series/ parallel model	Measured	NBS model	Series/ parallel model	Measured
0	18.55	18.55	15.6			
1	17.0	15.45	14.5	8.4	16.7	7.1
2	15.7	13.25		15.4	28.6	
3	14.5	11.6	12.25	21.8	37.5	21.5
4	13.5	10.35		27.2	44.2	
5	12.65	9.35	10.2	31.8	49.6	34.6
7	11.15	7.85	8.85	39.9	57.7	43.3

CONVERSTION FACTOR: 1 h · ft² · °F/Btu = 0.175 m² · K/W.

to be elongated or fibers to be laid down in a preferred (layered) direction rather than at random, or to the fabrication of spaced layers of air gaps separated by metallic foils. When such insulations are applied to surfaces, the preferred direction is such that the heat flow is perpendicular to the fiber or layer direction to take advantage of the resistive barriers. However, in some applications and orientations, especially furnace linings where some structural integrity and resistance to erosion is required, the insulations are utilized such that the heat flow is parallel to the fiber orientation. In this situation, heat transmission is enhanced. Thus, when designing systems for such applications, anisotropy effects need to be determined.

The effects of anisotropy can cover a wide range, depending on the insulation type and form, such as 15 to 30 percent for some cellular plastics and fibrous blankets, factors of 2 to 6 for rigidized fibers, and orders of magnitude for multifoils. Figure 1.2 shows typical effects of anisotropy and also of the gaseous environment on the performance of a microporous fiber material.

Aging

Time and environmental factors can have a significant impact on thermal performance. Recovery characteristics have already been mentioned. Dimensional stability at operating temperature or over a temperature range, moisture gain or loss, excessive shrinkage at elevated temperatures, changes due to weather or wind or other environmental factors, and chemical reaction are but some of the factors that affect materials over time.

While the above are all aging effects, the term aging is most commonly used in the context of cellular plastics manufactured with a halogenated fluorocarbon gas in the cells. Initially the thermal resistance is high due to the presence of the low-conductivity gas. However, as discussed in Chap. 4 on cellular plastics, there is oxygen, nitrogen, and carbon

Table 1.10*b*. Effect of Thermal Resistance of Cavity Insulation and Wall Construction on Reduction of Thermal Performance (NBS Model)

Percent reduction in dimensions	Reduction of effective thermal resistance, as a function of thermal resistance of cavity fill* (R_{wall} = 2.6), %			Reduction in effective thermal resistance as a function of fixed thermal resistance of wall construction* (R_{foam} = 18.55), %	
	R = 11.0	R = 14.7	R = 18.55	R = 2.6	R = 6.6
1	5.3	6.7	8.4	8.4	5.4
2	10.0	12.8	15.4	15.4	10.2
3	14.2	18.0	21.8	21.8	14.6
4	18.3	22.9	27.2	27.2	18.7
5	21.9	27.2	31.8	31.8	22.4
7	28.2	34.5	39.9	39.9	29.3

* All values of R in h · ft² · °F/Btu.
CONVERSION FACTOR: 1 h · ft² · °F/Btu = 0.175 m² · K/W.

Table 1.11. Effect of Cracks and Nonconformities on Thermal Performance of Calcium Silicate Insulation

Type	Density, lb/ft³	Btu/ft · h · °F 300°F	500°F	700°F	Remarks
Block and pipe	15.0 max	0.042	0.050	0.060	ASTM C533 specification
Block	13.8	0.039	0.047	0.055	Typical uncracked
Block	10.0	0.038	0.050	0.065	Cracked
Pipe	13.1	0.039	0.049	0.062	Cracked
Pipe, horizontal	12.5	0.039	0.051	0.066	~1/8-in gap along section of inner surface
Pipe, horizontal	13.1	0.038	0.047	0.055	Well-taped joints
Pipe, horizontal	12.5	0.039	0.051	0.066	Untaped joints
Pipe, vertical	11.8	0.039	0.049	0.061	Uncracked

CONVERSION FACTORS: 1 lb/ft³ = 16.01 kg/m³; 1 Btu/h · ft · °F = 1.73 W/m · K; °C = (°F −32)/1.8.

dioxide diffusion into the cells, and the thermal resistance falls significantly over a period of time (i.e., these materials "age").

This gas diffusion aging process has been modeled and verified extensively.[16,17] While much work has been expended in developing means to minimize or eliminate the phenomenon in materials and products, both field[18,19] and laboratory[20] experience have indicated that claimed thermal performance cannot be sustained unless the cellular plastic is sandwiched between thick metal impermeable skins.

In general, reliable thermal performance of cellular plastics with aged R values in the range of 5.5 to 6 h · ft² · °F/Btu (0.96 to 1.05 m² · K/W) can be attained, provided the material has greater than 95 percent closed-cell content, a small, <0.080-in- (0.20-cm-) diameter, regular-geometry cell size, a density in the range of 1.8 to 2.2 lb/ft³ (29 to 35 kg/m³), and a high initial fluorocarbon gas content. For faced products, R values in the range of 6.5 to 7 h · ft² · °F/Btu (1.14 to 1.22 m² · K/W) can be attained if, in addition to the above material criteria, low-permeance metal skins thicker than 0.001 in (0.0025 cm) are used in conjunction with an adhesive and provided the shear plane layer at the product surfaces is eliminated.

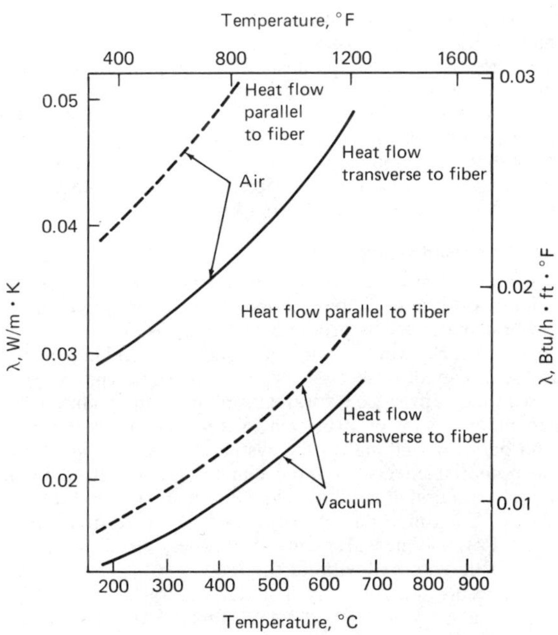

Fig. 1.2 Effect of fiber orientation and gaseous environment on thermal performance of MinK microporous insulation.

Moisture

Moisture in any form can have a deleterious effect on the thermal performance of insulation materials. Some materials are significantly more prone than others to the absorption of liquid water and water vapor. Other materials can both absorb and transmit moisture vapor.

Two major studies[21,22] using high water-vapor driving forces have shown that, under particularly extreme conditions, all insulations can absorb and retain moisture such that the thermal performance is reduced in excess of 50 percent in some cases. Results of the first study for the most extreme conditions are summarized in Table 1.12. The effects were lower when less extreme moisture driving forces were used. However, these examples illustrate what can happen should liquid water come in contact with the insulation. A more recent study on both extruded and expanded polystyrene and a phenolic foam[22] used driving forces more common to those of the northern U.S. winter climates. The results illustrated that overall decreases in thermal resistance were significantly less than 10 percent, even for 128-day continuous exposure to such conditions.

One final point on the effect of moisture concerns higher-temperature inorganic materials, especially those containing cementitious binders. In these, some of the components contain bound water which is driven off at temperatures in the 750 to 950°F (400 to 500°C) region, and the resulting thermal performance can be improved due to loss of the moisture. Such effects are usually less than 20 percent, but they need to be considered in the design, especially where shrinkage

Table 1.12. Effect of Moisture on Thermal Performance of Various Insulations

Material type	Moisture pickup,* wt %	Reduction in R value, %
Fiberboard	150	55
Glass fiber	300	80
Cork board	200	50
Perlite board	300	60
Polyurethane	280	70
Polyisoanurate	300	70
Expanded polystyrene	30	10
Extruded polystyrene	2000	55
Cellular glass	15	15

* Under conditions of 85°F and 100% relative humidity to 40°F and 75% relative humidity, generating an average vapor pressure differential of 1.03 in Hg for 100 days.

may also occur. In such cases, the loss in thermal performance due to thickness reduction may negate most of the improvement due to moisture loss.

Orientation and Mass Transfer

Multifoil insulations are used particularly in nuclear reactor power systems. These usually consist of polished metal foils separated by crumpled foils or low-density inorganic fiber layers such that the foils are some 6 to 10 mm apart. Thus, long convection paths can occur in sections which are usually on the order of 1 m long. These insulations are used for both horizontal and vertical applications, and clearly there is a much different aspect ratio depending on the orientation, with both free and forced convection more likely to occur for the latter case.

Typical results for this type of pipe insulation are shown in Fig. 1.3. The performance in the vertical orientation is some 70 percent lower than that for the horizontal orientation. Similar measurements on a dense block insulation showed a difference on the order of 6 percent, while for a

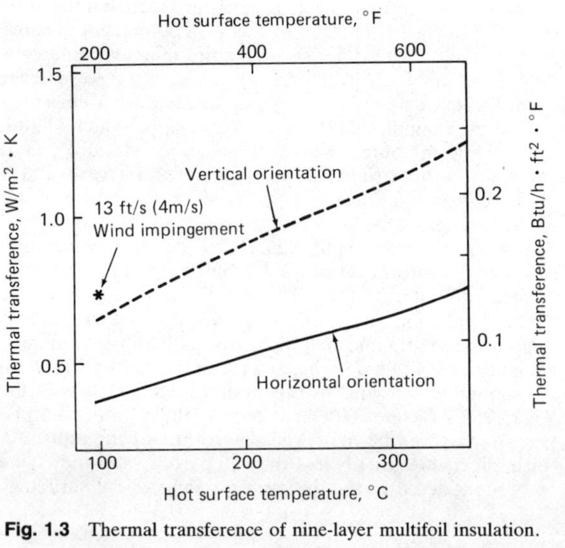

Fig. 1.3 Thermal transference of nine-layer multifoil insulation.

Table 1.13. Effect of Heat Flow Direction on Thermal Performance of Mineral Fiber Insulation

Density, lb/ft³	Temperature, °F Cold	Hot	Mean	Apparent thermal conductivity, Btu/h · ft · °F	Heat flow direction
5.4	−285	66	−108	0.0912	Up
5.4	−295	70	−112	0.0154	Down
10	−306	91	−108	0.0151	Up
10	−299	70	−116	0.0154	Down
15	−308	73	−117	0.0168	Up
15	−297	70	−114	0.0171	Down

CONVERSION FACTORS: 1 lb/ft³ = 16.01 kg/m³; 1 Btu/ft · h · °F = 1.73 W/m · K.

medium-density mineral fiber the difference was found to be on the order of 15 percent. One additional measurement shown in the figure illustrates the effect of a wind of 13 ft/s (4 m/s) impinging on the insulation system. The measured transference is seen to be on the order of 25 percent greater than for the case of no wind (i.e., natural convection). It must also be recognized that installation practices are important since wind impingement on a system containing cracks, poor seals and joints, etc., will cause significant reductions in performance.[6]

Heat flow direction and temperature gradient when combined can cause the onset of convection. This effect is illustrated in Table 1.13, where results for a cryogenic application using different densities of mineral fiber blankets are shown. It is seen that for a nominal 5- to 6-lb/ft³ (86-kg/m³) mineral fiber material convection can cause an approximate 5:1 difference between heat flow up and down. This effect is eliminated with higher density.

Clearly, high-mass-type insulations are preferable for applications where significant forced convection is foreseen. Measurements in the United States[23] and Europe[10] on building insulations having density less than 3 lb/ft³ (48 kg/m³) show that reductions in thermal performance of 20 percent and greater are possible in horizontal attic applications with wind speeds of less than 13 ft/s (4 m/s).

Installation and Maintenance

It cannot be stressed too often that poor installation and maintenance cause the best designed insulation systems using the best materials or products available to perform inadequately. Cracks, voids, poorly fitting sections, and loose joints and seals can all affect the design performance significantly.

A common practice is for the installation and maintenance personnel to subject insulation to compression by the common habit of walking on the system. Table 1.14 illustrates the potential effects on three commonly used insulations, namely, mineral fiber blanket and board and molded calcium silicate. Each material was subjected to a load of 14 lb/in² (0.1 MPa), and measurements of thickness, density, and λ_{app} were made with and without the load.

The results show clearly that the mineral fiber materials in particular were affected significantly in that the overall thermal resistance was reduced by up to 50 percent. However, on removal of the load, these materials do recover, but not completely, which illustrates that permanent change can occur.

Table 1.14. Effect of 14-lb/in² (0.1-MPa) Applied Load on Thermal Performance

Material	Thickness, in		Density, lb/ft³		λ_{app}, Btu/h · ft · °F at 700°F (371°C)		R, h · ft² · °F/Btu at 700°F (371°C)	
	Unloaded	Loaded	Unloaded	Loaded	Unloaded	Loaded	Unloaded	Loaded
Mineral fiber blanket	1.0	0.48	8.7	17.3	0.066	0.052	1.20	0.74
Mineral fiber board	1.0	0.50	9.3	18.6	0.060	0.060	1.31	0.74
Calcium silicate	1.0	0.98	14.1	14.3	0.054	0.057	1.54	1.43

CONVERSION FACTORS: 1 in = 25.4 mm; 1 lb/ft³ = 16.01 kg/m³; 1 Btu/h · ft · °F = 1.73 W/m · K; 1h · ft² · °F/Btu = 0.175 m² · K/W.

REFERENCES

1. R. P. Tye and A. O. Desjarlais, "Factors Influencing the Thermal Performance of Thermal Insulations for Industrial Applications," in *Thermal Insulation Materials and Systems for Energy Conservation in the '80's,* F. A. Govan, D. M. Greason, and J. D. McAllister (eds.), ASTM STP 789, American Society for Testing and Materials, Philadelphia, PA, 1983, pp. 733–748.
2. W. Turner and J. Malloy, *Thermal Insulation Handbook,* McGraw-Hill, New York, 1981.
3. R. P. Tye, A. O. Desjarlais, D. L. McElroy, and D. W. Yarbrough, "An Experimental Study of Thermal Resistance Values (*R*-Values) of Low-Density Mineral-Fiber Building Insulation Batts Commercially Available in 1977," Oak Ridge National Laboratory, Rep. ORNL/TM7266, 1980.
4. R. P. Tye, E. Ashare, E. C. Guyer, and A. C. Sharon, "An Assessment of Thermal Insulation Materials for Building Applications," in *Thermal Insulation Performance,* D. L. McElroy and R. P. Tye (eds.), ASTM STP 718, American Society for Testing and Materials, Philadelphia, PA, 1980, pp. 9–26.
5. R. P. Tye, "Upgrading Thermal Insulation Performance of Industrial Processes," *Chem. Eng. Prog.,* **81**(2), 30 (1985).
6. R. P. Tye, A. O. Desjarlais, and S. E. Smith, "The Thermal Transmission Properties of High Temperature Insulating Materials," in *Thermal Conductivity,* vol. **18,** T. Ashworth and D. R. Smith (eds.), Plenum, New York, 1985, pp. 317–326.
7. Z. L. Ardary and C. D. Reynolds, "Carbon Fiber Thermal Insulation," Union Carbide Corp., Oak Ridge Y12 plant, Rep. Y-1803, Mar. 1972.
8. R. P. Tye, L. L. Lander, and K. C. Meiler, "The Thermal Transport Properties of Fibrous Carbon and Graphite Products," in *Thermal Conductivity,* vol. **19,** D. W. Yarbrough (ed.), Plenum, New York, 1988.
9. P. J. Burns, L. C. Chow, and C. L. Tien, "Convection in a Vertical Slot Filled with Porous Insulation," *Int. J. Heat Mass Transfer,* **20,** 919 (1977).
10. C. G. Bankvall, "Forced Convection: Practical Thermal Conductivity in an Insulated Structure Under the Influence of Workmanship and Wind," in *Thermal Transmission Measurements of Insulation,* R. P. Tye (ed.), ASTM STP 660, American Society for Testing and Materials, Philadelphia, PA, 1978, pp. 409–425.
11. E. R. Vinieratos and J. D. Verschoor, "Influence of Insulation Deficiencies on Heat Loss in Walls and Ceilings," in *Thermal Insulation Performance,* D. L. McElroy and R. P. Tye (eds.), ASTM STP 718, American Society for Testing and Materials, Philadelphia, PA, 1980, pp. 142–159.
12. R. P. Tye, A. O. Desjarlais, S. C. Spinney, and J. G. Bourne, "Effective Thermal Performance of an Insulated Standard Stud Wall Containing Air Gaps," in *Thermal Performance of the Exterior Envelope of Buildings I,* ASHRAE SP 28, American Society of Heating, Refrigeration and Air Conditioning Engineers, Atlanta, GA, 1981, pp. 965–977.
13. *Handbook of Fundamentals,* American Society of Heating, Refrigeration and Air Conditioning Engineers, Atlanta, GA, 1977.
14. B. Peavy, "Formulation of Heat Transfer Coefficients for Air Spaces Comprised of Enclosed Rectangular Slots," National Bureau of Standards, July 1977; see also NBS Tech. Note 946, July 1977.
15. "Specification for Calcium Silicate Block and Pipe Thermal Insulation," ASTM C533, vol. 04.06, American Society for Testing and Materials, Philadelphia, PA, 1985.
16. F. J. Norton, "Thermal Conductivity and Life of Polymer Foams," *J. Cell. Plastics,* **3,** 23 (1967); also **17,** 300 (1982).
17. D. W. Reitz, M. A. Schuetz, and L. R. Glicksman, "A Basic Study of Aging of Foam Insulation," *J. Cell. Plastics,* **20,** 104 (1984).
18. D. A. Zarate and R. L. Alumbaugh, "Thermal Conductivity of Weathered Polyurethane Foam Roofing," U.S. Naval Civil Eng. Rep. TNNO N-1643, Sept. 1982.
19. S. P. Muhlenkamp and S. E. Johnson, "In-Place Thermal Aging of Polyurethane Foam Roof Insulations," *Proc. 7th Conf. on Roofing Technology,* National Bureau of Standards/National Roofing Contractors Assoc., Chicago, IL, Apr. 1983, pp. 49–55.
20. R. P. Tye and A. O. Desjarlais, "Aging Characteristics of Cellular Plastic Insulation Materials and Insulated Building Envelope Components," *Polyurethane World Congress 1987: 50 Years of Polyurethanes,* Fachverband Schaumkunstoffe e.V. Frankfurt FRG, The Society of the Plastics Industry, New York, 1987, pp. 91–98.
21. W. Tobiasson and J. Ricard, "Moisture Gain and Its Thermal Consequence for Common Roof Insulations," *Proc. 5th Conf. on Roofing Technology,* National Bureau of Standards/National Roofing Contractors Assoc., Chicago, IL, Apr. 1979.
22. L. I. Knab, D. R. Jenkins, and R. G. Mathey, "The Effect of Moisture on the Thermal Conductance of Roofing Systems," *NBS Building Science Ser.,* vol. 123, U.S. Dept. Commerce, Apr. 1980.
23. R. P. Tye and C. F. Baker, "Development of Experimental Data on Cellular Plastic Insulations Under Simulated Winter Exposure Conditions," *Thermal Insulation: Materials and Systems.* ASTM STP 922, F. J. Powell and S. L. Matthews (eds.), American Society for Testing and Materials, Philadelphia, PA, 1987, pp. 518–537.
24. A. L. Berlad, N. Tutu, Y-J. Yeh, R. Jaung, R. Krajewski, R. Hoppe, and F. J. Salzano, "Air Intrusion Effects on the Performance of Permeable Insulation Systems," in *Thermal Insulation Performance,* D. L. McElroy and R. P. Tye (eds.), ASTM STP 718, American Society for Testing and Materials, Philadelphia, PA, 1980, pp. 181–194.
25. "Standard Classification of Potential Health and Safety Concerns Associated with Thermal Insulation Materials and Accessories," ASTM C930, vol. 04.06, American Society for Testing and Materials, Philadelphia, PA, 1985.

Chapter 2

Fibrous Insulations

WILLIAM H. KIELMEYER AND RICHARD L. TROYER
Manville Corporation

2.1 OVERVIEW

Thermal insulating materials made of fibers are evident everywhere, in such divergent applications as wearing apparel, the exterior tile of the space shuttle, building insulation batts and blankets, and lightweight liners for high-temperature industrial furnaces. Moreover, the origins, compositions, and characteristics of the fibers themselves are as diverse as their applications. They consist of protein (animal fibers, such as fur, silk, and wool), cellulose (vegetable fibers, including cotton and wood pulp), cellulosic derivatives and synthetic polymers (most synthetic organic fibers), natural minerals (asbestos), and processed minerals (glass, ceramic, and mineral fibers). They are available in a wide variety of forms: as bulk filaments; as woven, knitted, and felted (nonwoven) fabrics; as papers and sheets; as blankets, batting, and felts; and as boards and vacuum-formed shapes. The broad base of fiber composition and properties and the large number of ways these fibers can be combined and processed account for their extensive and varied uses.

To say that any mass or assemblage of fibers has, or could have under the right conditions, some value as a thermal insulator is probably an oversimplification. Nevertheless, history has observed such coarse and unlikely fibrous materials as hemp, sawdust, wood chips, and straw purposefully used as insulators. That these substances have largely given way to other fibrous materials is based on a number of factors, including relative availability, accessibility, processibility, cost, thermal efficiency (thermal resistivity as a function of fiber mass), and side traits such as odor.

Nevertheless, although the number of materials used extensively for insulating purposes has been reduced, it is still very large, so much so that it is preferable to address the materials in groups. Fiberglass, for example, is a generic name for a family of insulation materials, each having a unique set of properties based on many controlled (and a few uncontrolled) process parameters. In this chapter, an attempt is made to provide either a range of significant properties or representative properties for the various material groups, rather than for individual products (unless the group consists of only one or two of these). With the myriad of products in some groups and with new products added regularly, it is entirely possible to have materials within a group that do not conform to the properties or property ranges described for that group.

For purposes of discussion, it is helpful to separate the kinds of fibrous materials into two broad classes: (1) *personal* insulations, those which conform to or envelop the human body, and (2) *structural* insulations, those which surround or protect buildings and equipment. This division is somewhat arbitrary, but *most* (although not all) organic fibers fall into the first category, while most (but not all) inorganic fibers fall into the second. Personal insulations tend to be flexible and removable, that is, they are designed for extensive handling and moderate abuse, so their *physical* properties (especially resilience under repeated deformation) are as important as their thermal performance. Structural insulations, on the other hand, tend to be permanent and relatively immobile once in place, so that properties related to repeated distortion and recovery are not especially significant.

2.2 HEAT TRANSFER IN FIBROUS MATERIALS

Modes of Heat Transfer

Convection, solid fiber conduction, air conduction, and radiation can all be heat-transfer modes that are present in fibrous insulations. These modes of heat transfer can combine in a complex manner to give a total conductivity as shown in Fig. 2.1. The exact shape of the curve in Fig. 2.1 is dependent upon the mean temperature, the type of fibrous insulation, the fiber orientation, the specimen size, and other factors.[1,2]

Solid-Fiber Conduction Solid-fiber conduction occurs when, because of a temperature gradient, heat energy is transmitted down one fiber and passes to another fiber at their point of contact. At very low densities this factor is small; but as the density increases, the fibers are packed closer together and the solid-fiber conduction increases. In fact, as Fig. 2.1 shows, this is the one mode responsible for an increase in the total thermal conductivity beyond its minimum value as density increases.

Convection Convection, the transmission of heat by the mass movement of air (or gas) particles, depends on many factors, including the size of the cavity occupied by the particles, the cavity orientation, and the temperature gradient. Convection can be significant in the absence of insulation, but drops off when insulation impedes the gas currents. Commercially available fibrous insulation is designed at a high enough density to make convection within the insulation insignificant. If fibrous insulation is installed such that it essentially fills a cavity, convection within the cavity will not be significant. If, however, the insulation is installed with one or two boundaries exposed to air, the potential for natural or forced convection exists.[3,4]

Radiation Radiant heat transmission is a very large factor when there is no insulation present, but it is reduced rapidly as the insulation density increases. Radiation is reduced by scattering from or absorption on and reemission by the fibers. At densities beyond the minimum total conductivity point, radiant heat transfer is negligible.

Other factors that influence radiation are the temperature level and the surface emissivity according to the following equation:

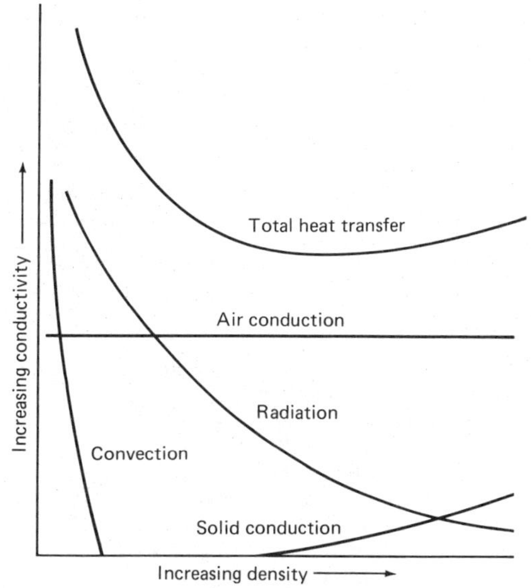

Fig. 2.1 Modes of heat transfer in fibrous insulations.

$$\text{Radiation heat transfer} = \frac{\text{constant}(T_h^4 - T_c^4)}{1/\epsilon_c + 1/\epsilon_h - 1 + N}$$

where T_h = hot-surface absolute temperature
T_c = cold-surface absolute temperature
ϵ_c = emissivity of cold surface
ϵ_h = emissivity of hot surface
N = opacity factor

Since radiation is controlled by the difference of the absolute temperatures to the fourth power, it becomes increasingly important as the temperature increases. For this reason it is common for high-temperature insulations to be higher in density than insulation used for more moderate temperature applications.

Still-Air Conductivity Fibrous insulation even at relatively high densities is, by volume, mostly air. Thus still-air conductivity is an important contributor to the total conductivity. Its influence is nearly constant over a wide range of densities.

The cumulative effect of these four modes of heat transfer is a minimum in apparent thermal conductivity as a function of density (Fig. 2.1). It could be assumed that manufacturers of fibrous insulation would design around this minimum value; but in fact, where possible, they try to produce at a density that is lower and more economical than that corresponding to the thermal conductivity minimum. Where space is available, as in a typical attic, the thermal resistance, or R value, can be achieved at less cost by increasing the insulation thickness rather than by raising the density. Some types of insulation, most notably industrial insulations, may be designed on or about the minimum thermal conductivity. With these kinds, other factors, such as structural strength and fire resistance, may be as important as or more important than thermal cost efficiency.

Effect of Thickness

It is well established that when low-density fibrous insulation is tested at increasing thickness, the tested conductivity increases. This phenomenon has commonly become known as the "effect of thickness." Although many commercially available fibrous insulations are tested at full thickness, residential insulations and other rather thick materials are routinely tested at less than full thickness due to time constraints on the test equipment. The Federal Trade Commission has promulgated a ruling that insulation used in residences must be tested at "representative thickness," which is the minimum thickness at which, with increasing thickness, the conductivity changes no more than 2 percent. The effect of thickness is density-dependent so that with increasing density the effect of thickness decreases, as shown in Fig. 2.2.

Even at low densities, the effect does not appear to be large. The National Bureau of Standards has completed a very thorough study of the effect of thickness on low-density fiberglass used for thermal conductivity standards.[5] For that specific lot of material, the study indicates an "effect of thickness" at a fiberglass density of 0.6 lb/ft^3 (9.0 kg/m^3) to be 3.5 percent when the test thickness is varied from 1 to 6 in (2.54 to 15.2 cm). The insulation in this study was toward the low end of density for commercial insulations. So although the effect of thickness may be somewhat higher than 3.5 percent for some insulations, it is near the maximum that would be expected. As the density of insulation increases, the effect of thickness becomes smaller.

Moisture in Fibrous Insulation

The effects of moisture on heat transfer in fibrous insulation are not completely understood. Much of the work completed

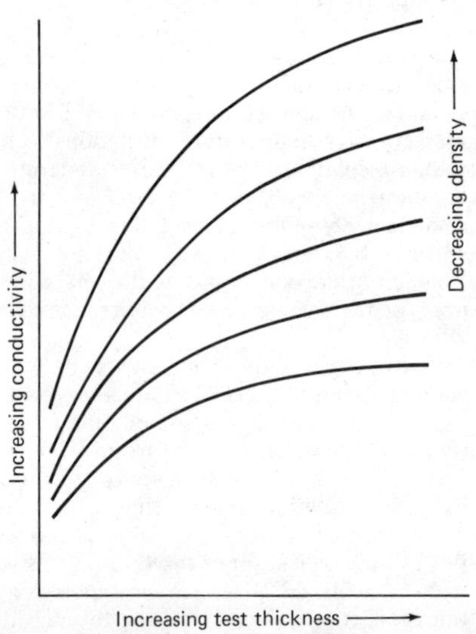

Fig. 2.2 Effect of thickness on thermal conductivity of fibrous insulations.

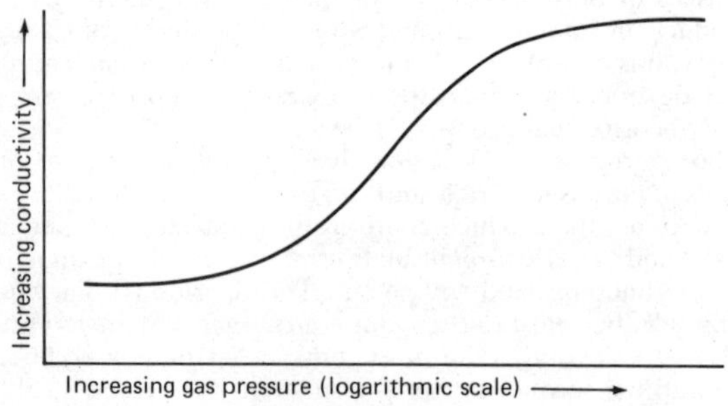

Fig. 2.3 Effect of gas pressure on thermal conductivity of fibrous insulations.

to date involves transient conditions. Since fibrous insulation is very porous, moisture can migrate with little resistance from a warm surface to a cold surface. During this period of migration, the measured conductivity may be several times the conductivity of dry insulation. Most researchers have called this a quasi steady state to indicate that it only appears to be a steady state but does not meet the true requirements of steady state. If the total moisture in a sample of insulation is low (less than 2 vol %), a thermal conductivity test on the sample can be run long enough so that the conductivity returns to near the value for dry insulation. It may take days or even weeks for the moisture to equilibrate. The total heat transfer for these conditions is closely approximated by the heat transfer for dry insulation plus the latent heat load for the migrating moisture.

Other studies with moisture contents up to 25 percent by volume have been conducted. These studies simulate the effects of a leak on a flat roof deck where the moisture is retained in the insulation because of a nearly impervious membrane. For this kind of condition, the effective conductivity is much higher than that for dry insulation and remains high as long as the water is present in that quantity.

Effect of Atmospheric Pressure

As noted earlier, a significant contributor to total heat transfer is the conductivity of still air. As the air pressure is changed in fibrous insulation, the conductivity of still air changes, and thus the total apparent insulation conductivity changes, as represented in Fig. 2.3. Near a pressure of 1 atm the conductivity is, for practical purposes, independent of atmospheric pressure. However, as the pressure falls, the conductivity starts to drop rather sharply. This decline occurs when the mean free path of gas molecules is near the cell size of the insulation. The conductivity continues to fall until approximately 0.1 Pa of pressure. At this point the conductivity of air is insignificant, and further changes in pressure do not change the total conductivity appreciably.

Thermal Conductivity Versus Temperature

Figure 2.4 shows a typical relationship between thermal conductivity, density, and mean temperature. Typically, the thermal conductivity increases with increasing mean temperature. As the mean temperature increases, the minimum conductivity shifts toward higher densities.

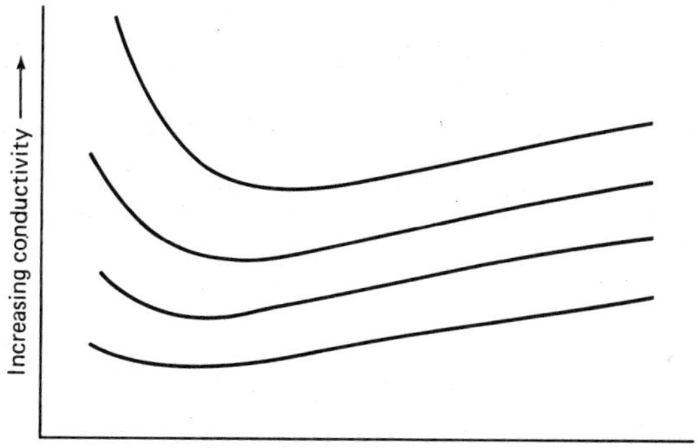

Fig. 2.4 Effect of density and increasing temperature on conductivity of fibrous insulations.

2.3 PERSONAL INSULATION

For a conventional textile material, the two most important properties influencing thermal performance are (1) its thickness, which determines its thermal resistance, and (2) its resilience, which affects thickness over time. Bodily comfort also dictates that the material be light and easy to handle and, if placed next to the skin, soft, flexible, and absorbent. Thus, ideally, a textile product for personal insulation should have a relatively high loft, high recovery from distortion, and friendly feel.

For the range of textile fiber diameters, fabric weights, and thicknesses normally encountered, the ratio of thermal resistivity and thickness is nearly constant—3.6 h · ft² · °F/(Btu · in.) [or 3.6*R* per inch or 0.025 K · m²/(W · mm)].[6] Thus, if a fabric is fluffed, its thermal resistance increases in proportion to its increased thickness, even though its surface weight may not have changed.

Woven Fabrics

Among woven materials, the standard for warmth is wool. Wool's outstanding characteristics stem from its fiber composition and morphology. The molecules in a fiber of wool consist of long, coiled, cross-linked chains, each a copolymer of 18 different amino acids. This arrangement gives wool fiber an asymmetric structure—a kind of permanent, natural crimp. When a tightly knit wool fabric is brushed, some of the fibers are released, effectively becoming like miniature springs that not only increase the thickness and thermal resistivity of the fabric, but also aid recovery. Wool recovers about 99 percent from a 2 percent distortion, ranking it among the highest of the natural fibers. Wool, unlike most synthetic fibers, also has a high moisture absorbency (up to a third of its dry weight); a wool fabric next to moist skin feels dry.[7,8]

The tactile and useful qualities of wool (and for that matter, cotton or any other natural fiber) are somewhat offset by its mediocre abrasion resistance and its susceptibility to attack by moths and other insects. Synthetic or chemical fibers, derived from either cellulose or synthetic polymers, offer more versatility in product properties by virtue of their processing. Although no chemical fiber has yet been made to duplicate the complex and unique structure of wool, fiber synthesists have come very close to duplicating and, for some applications, improving on wool properties.

The general properties of the most significant natural and chemical fibers are presented in Tables 2.1[7,9] and 2.2.[7,10,11] Like the natural fibers, chemical fibers are made up of long-chain molecules of high molecular weight. It is beyond the scope of this work to cite all of the synthetic polymers used for thermal insulation. However, except for polybenzimidazole, novoloid, and aramid fibers, they have all been involved in fabrics, either as single components or mixed with other fibers, to help keep people warm.

Chemical fibers are typically formed by spinning (extruding through fine holes or orifices) a solution, emulsion, or melt of the desired composition. During, or subsequent to, the spinning process, the filaments are normally drawn or stretched to align the molecules in the direction of drawing. The extent to which this is done determines the degree of crystallinity and stiffness of the filaments. Both continuous filaments and staple fibers are formed the same way. However, considerably more filaments are drawn from a spinneret for staple use, and the products of several spinnerets are combined into a very large bundle called a tow. Tows are conventionally stretched and mechanically crimped before being cut into staple fibers.

Crimp, along with a low fiber density and high recovery,

Table 2.1. Properties of Natural Fibers

Fiber type	Melting or decomposition temperature, °F	Physical and chemical properties
Cotton	≈250 (dec)	ρ_f = 1.55 g/cm³; tenacity 3.2 g/den†; elongation at break 5–9%; moisture regain 8.5%; filament diameter 14–21 μm; poor resilience; swells in and absorbs water, acids, bases, and aqueous salt solutions.
Silk	≈340 (dec)	ρ_f = 1.25 g/cm³; tenacity 3.4–4.0 g/den; elongation to break 20%; moisture regain 11%; textile filament diameter 15–19 μm; dissolves in concentrated acids and alkalis; absorbs weak acids and most solutions.
Wool	≈210 (dec)	ρ_f = 1.30 g/cm³ tenacity 1.4–1.8 g/den; elongation to break 30–50% (dry); moisture regain 16%; excellent resilience; swells in water; combines with strong acids; degraded by strong bases and oxidizing agents.

* ρ_f = filament density.
† g/den = grams of tensile force per denier of fiber where denier is the fiber weight in grams per 9000 m of length.
CONVERSION FACTORS: °C = (°F − 32)/1.8; 1 g/den = 0.0882 N/Tex. 1 lb/ft³ = 0.016 g/cm³.

Table 2.2. Properties and Uses of Synthetic Organic Fibers

Generic type or family	Physical and chemical properties	Insulating uses
Acrylic (*A*) and modified acrylic (*M*) ρ_A = 1.14–1.17 g/cm³† ρ_M = 1.31–1.40 g/cm³ $M_p \geq$ 446°F (dec)‡	Filament tenacity 2.2–3.6 g/den; elongation to break 33–64%; (*A*) excellent resistance to sunlight; (*M*) good flame resistance and controlled shrinkage at high temperature; not affected by weak or dilute acids, bases, most solvents, oxidizing agents; dissolves in concentrated acids (e.g., HNO_3 or H_2SO_4); weakened by strong bases.	(*A*) Bulky sweaters, synthetic furs, pile, fleece, blankets, curtains, drapes, socks, coats. (*M*) Flame-resistant clothing and furnishings, linings for outerwear.
Aramid (heat-resistant) ρ = 1.35–1.45 g/cm³ M_p > 750–1022°F (dec)	Filament tenacity 2.7–9.2 g/den; elongation to break 7–41%; excellent flame resistance (chars on exposure); maximum use temperature >575°F; chemical resistance comparable to nylon 6.6.	Military and industrial fire-resistant clothing; aircraft and ship furnishings; fire-protection shields; asbestos substitute.
Cellulose acetate (*A*) and triacetate (*T*) ρ = 1.30–1.32 g/cm³ M_{pA} = 500°F M_{pT} = 572°F	Filament tenacity 1.2–1.4 g/den; elongation to break 25–45%; normally blended; similar to rayon in sunlight degradation; low abrasion resistance; little effect from weak acids or bases; degraded by strong acids and bases.	Apparel, household fabrics, velours.
Novoloid ρ = 1.27 g/cm³ M_p = 480°F (dec)	Filament tenacity 1.4–1.8 g/den; elongation to break 30–60%; excellent thermal insulating properties but prone to punk; good property retention to low temperatures; good light and nuclear radiation resistance; excellent flame resistance (chars with low shrinkage and evolution of nontoxic gases); unaffected by nonoxidizing acids, weak bases, organic solvents; attacked by concentrated or hot H_2SO_4 or HNO_3 and strong bases.	Precursor to carbon fiber; fire-protective clothing; thermal insulating gloves; asbestos substitute.
Nylon ρ = 1.13–1.14 g/cm³ M_p = 420–482°F (M.P.)	Filament tenacity 4.3–9.5 g/den; elongation to break 12–45%; high abrasion resistance; good elasticity and flexural properties; stable in dilute acids and alkalis; hydrolyzed by hot concentrated HCl and attacked by formic acid, phenol and cresol.	Normally blended with other fibers in clothing (e.g., thermal underwear); improves wear life.
Polybenzimidazole ρ = 1.43 g/cm³ M_p = 1025°F (dec)	Filament tenacity 3.1 g/den; elongation to break 30%; high moisture regain (14% at 68°F); good abrasion resistance; fabrics char but remain pliable when exposed to flame; continuous use at 575–650°F; stable in most chemicals except bases at high temperature.	Lightweight, comfortable, replacement for asbestos in fire and heat-protective clothing; possible use in aircraft fabrics and high-temperature specialty insulations.
Polyester ρ = 1.38 g/cm³ M_p = 496–505°F	Filament tenacity 2.3–9.7 g/den; elongation to break 20–65%; high resilience (80–100% from 2% stretch); good sunlight resistance; good resistance to cleaning solvents and mineral acids except concentrated HCl and H_2SO_4; attacked by strong bases.	Normally blended with other fibers (cotton, rayon, wool, etc.) for clothing; carpets; fill for sleeping bags, ski wear.
Polyolefin [polypropylene (*P*) and polyethylene (*E*)] ρ = 0.90–0.96 g/cm³ M_{pE} = 266–280°F M_{pP} = 330–350°F	Filament tenacity 1.1–6.8 g/den; elongation to break 20–120%; very low (<0.1%) moisture regain; generally excellent resilience (90–100% from 5% stretch); poor natural resistance to sunlight and flame (correctable); at room temperature, resistant to inorganic acids and bases and organic solvents; attacked by oxidizing acids, aromatic and chlorinated solvents at high temperatures.	Carpeting and home furnishings; high coverage fill for sleeping bags and ski wear; normally blended with other fibers.
Rayon [viscous (*V*) and polynosic (*P*)] ρ = 1.5–1.53 g/cm³ M_p = <350°F (chars)	Filament tenacity 1.2–10.0 g/den; elongation at break 5–30%; high moisture regain (11–13%); (*V*) swells in water (25% diameter increase); (*P*) improved strength and firmness over (*V*); microfibrillar, properties like those of cotton; unaffected by cold acids, solvents, and weak alkalis; attacked by strong alkalis and acids at high temperatures.	Precursor for inorganic fiber development; clothing; home furnishings.

Table 2.2. Properties and Uses of Synthetic Organic Fibers (*Continued*)

Generic type or family	Physical and chemical properties	Insulating uses
Vinyl ρ = 1.35–1.40 g/cm^3 M_p = ≈300°F	Filament tenacity 0.9–3.8 g/den; elongation to break 12–180%; low softening temperature; nonflammable, but could release HCl in decomposition; negative static charge when fabric next to skin said to be therapeutic; water repellent; good thermal insulation; excellent resistance to strong acids and alkalis; softened by some organic liquids.	Clothing, thermal underwear; protective clothing; home furnishings, normally blended with other fibers.

* Letter designations in parentheses refer to specific fiber types.
† ρ = filament density, subscripted for specific fiber types.
‡ M_p = melting point, subscripted for specific fiber types.
CONVERSION FACTORS: 1 lb/ft^3 = 0.016 g/cm^3; °C = (°F − 32)/1.8; g/den = 0.0882 N/Tex.

accounts for the use of some fibers in sweaters, thermal underwear, and other insulating clothing. A *chemical* crimp, similar to that of wool, can be obtained by extruding, from the same orifice, a single fiber from two polymer solutions with different characteristics. Such bicomponent or conjugate filaments can provide yarns with high bulk and resilience. Lofty acrylic yarns can also be obtained by blending high-shrinkage fibers with regular, nonshrinking fibers and exposing the blend to heat. The high-bulk fibers shrink, distorting the nonshrinking ones. An interesting variation of this occurs when high- and low-shrinkage acrylics are woven together into fabrics and heated: the high-shrinkage fibers contract around the regular fibers. Properly done, the final product looks like natural fur, with the shrunken fibers resembling a dense fur undercoat and the regular fibers resembling long, silky guard hairs.[7]

Bulk and Nonwoven Fibers

Although high-loft woven fabrics generally have good thermal properties, most of the fibers composing them are prevented by the weaving process from contributing significantly to thermal resistance. These are the tightly bound fibers in the yarns that define the fabric structure. If an assemblage of fibers is to be used primarily for its insulation value, then a bulked or nonwoven form, if the fibers are resilient, tends to provide more thermal resistance per unit weight than a comparable woven form. Nowhere is this more evident than in ski clothing and sleeping bags, where not only is thermal resistance important, but low weight and resilience are essential. At present insulations in these applications are represented by only one major family of natural fibers—waterfowl (goose and duck) down—and a handful of unique synthetic-fiber materials.

Electron micrographs reveal that true waterfowl down consists of three-dimensional snowflakelike clusters of very fine (<4-μm) barbed filaments radiating from a single point (quill). Down's ability to insulate stems from a locking action between the filaments of adjacent particles to form a buoyant continuum.[12,13] In actual service, true down is never found; it is always mixed with feathers and other related materials. Nevertheless, even though there are impurities present, high-quality down, in terms of weight and thickness, surpasses every other natural and synthetic fiber for insulation efficiency. A Canadian Defence Research investigation in the early 1980s[14] found that the thermal resistance of a sample of down and feather filled sleeping bag was 47.4R per unit surface mass (lb/ft^2) (1.7 × 10^{-3} m^2 · K/W per g/m^2) and 4.13R per unit thickness (inches) (0.029 m^2 · K/W per mm). The corresponding values for samples filled with 100 percent polyolefin microfibers ranged from 13.7 to 17.1R per lb./ft^2 (0.50 to 0.62 × 10^{-3} m^2 · K/W per g/m^2) and 4.58 to 4.94R per in (0.032 to 0.034 m^2 · K/W per mm) of thickness; for a blend of polyolefin microfiber and polyester staple, 25.4R · ft^2/lb (0.91 × 10^{-3} m^4 · K/W · g) and 4.44R/in (0.031 m^2 · K/W per mm); for continuous-filament polyester, 46.0 to 50.4R · ft^2/lb (1.7 to 2.3 × 10^{-3} m^4 · K/W · g) and 2.64 to 3.14R/in (0.018 to 0.022 m^2 · K/W per mm); and for hollow-filament polyester, 44.9 to 52.2R · ft^2/lb (1.6 to 1.9 × 10^{-3} m^4 · K/W · g) and 3.16 to 3.54R/in (0.022 to 0.025 m^2 · K/W per mm). Thus, although the natural fiber did not achieve the highest resistance either per unit mass or per thickness, it achieved an excellent balance in thermal efficiency between the two.

The chief drawbacks of down are (1) its high relative cost, (2) its loss of insulation value when wet, (3) an allergic reaction frequently associated with it, (4) a characteristic odor, and (5) its susceptibility to microorganisms. The synthetic-fiber alternatives to down—mainly polyesters and polyolefin polyester blends—are less costly and overcome the other deficiencies of down. It is instructive to compare these products, if only to appreciate the ingenuity of the developers and the variety of ways synthetic fibers can be used for insulation purposes. Polarguard, a product of Celanese Corporation, consists of crimped, continuous 23-μm polyester filaments bonded into a light batting with acrylic resin (in much the same way as fiberglass batting). Hollofil™ and Quallofil™, materials made by DuPont, are hollow, staple 23- to 26-μm fibers, also of polyester; the main advantage of hollow fibers over solid fibers is added coverage per unit weight of material (that is, an increase in fiber packing per unit surface weight). Thinsulate™ C, a 3M product, consists of a blend of 60 percent polyolefin microfibers (1 to 3 μm) and 40 percent coarse polyester staple fibers (38 μm); presumably the latter provides stiffness to the blend and aids resilience. Two other unique, recent developments from DuPont are Thermolite™ and Thermoloft™, which consist of 10-μm bicomponent fibers having a heat-softenable polyester "skin" around a more durable polyester core. In its formation, the Thermolite batting is subjected to high enough temperatures to weld contacting fibers together. Thermoloft is a blend of these heat-fusable sheath-core fibers and thicker Quallofil fibers.

In a recent study of the thermal transfer mechanisms in various synthetic-fiber battings for sleeping bags,[15] the U.S. Army observed that the natural convective heat loss in all materials was negligible, conductive heat transfer was a function of thickness, independent of fiber type, but that radiative heat loss dropped with increased bulk density and decreased fiber diameter. The study concluded that, on an equal bulk density and thickness basis, a batting containing fibers under 10 μm could provide as much as 25 percent higher thermal resistance than one containing larger fibers. It recommended that future insulating battings address radiation transfer by either incorporating reflective layers in their construction or blending microfibers with the standard fibers.

Heat-Resistant Personal Insulations

When the temperature of a person's environment greatly *exceeds* the person's body temperature, a number of factors govern the kind of protective clothing the person should wear: the external temperature; the time of exposure; the chemical, physical, and thermal stability of the insulating garment, and the thermal resistance of the insulation, which may itself be variable. Most of the natural and common synthetic fibrous materials covered in Table 2.2 are significantly stable to offer at least limited protection between 200 and 400°F (93 and 204°C). For higher temperatures of exposure, fire-resistant organic synthetics with high melting temperatures—specifically aromatic polyamides (or aramids) and, to a lesser extent, novoloids (phenolic-based) and heterocyclic polymers such as polybenzimidazole—have joined the more traditional inorganic fibers of asbestos, glass, minerals, and treated natural fibers in protective clothing. It is of interest that at high temperatures, such as in a fire, aramids, such as Nomex, and polybenzimidazole decompose, forming a highly insulating char. As in all insulations, the thickness of the fabric is important. A 1983 study[16] of fabrics designed for firefighters' clothing revealed that material thickness, rather than fiber type, influences its resistance to radiant and convective heat transfer and to loss of body heat.

2.4 STRUCTURAL INSULATIONS

A high thermal resistivity and structural stability within its environment are two qualities which a material used for insulating equipment or buildings shares with one used for personal protection. Because structural insulations generally do not experience the recurrent flexing, compression and expansion, and general physical handling applied to textile types, certain physical characteristics related to recovery from distortion are of secondary importance. For this reason, building insulations consist chiefly of silicon-based inorganic fibers, which are inexpensive, but lack the abrasion resistance of the organics and break down with repeated flexing. On the other hand, the only organic fiber used extensively in structural insulation is cellulose derived from waste paper because of its widespread availability and low cost.

Low-Temperature Applications

Effectively maintaining an environment in the cryogenic temperature range [−150°F (−101°C) to absolute zero] requires an insulation *system* that restricts or controls heat flow in each of its transfer modes—radiation, convection, and conduction. At present, several systems using fibrous insulations are employed, all variations or extensions of the Dewar flask, discovered by James Dewar in 1898. Each consists of a double-shelled vessel, with insulation installed within the space between the two shells.

The first, and the one yielding the lowest thermal conductivity of any cryogenic insulation thus far developed, consists of multiple layers of reflective sheets, such as aluminum foil or metal-coated mylar, separated by thin layers of low-conductivity glass microfiber mats or organic fiber netting. Properly installed in an evacuated Dewar space, the insulation has an apparent thermal conductivity of 2.8 to 4.9 × 10^{-4} Btu · in/h · ft^2 · °F (4 to 7 × 10^{-5} W/m · K) between 81°F (27°C) on the warm side and −418°F (−250°C) on the cold side. A second system consists of loose glass or mineral fibers or mats in the evacuated space. The thermal conductivity of an evacuated fiber system depends on fiber type and configuration (loose or mat form), insulation density, fiber diameter, and gas pressure, but values as low as 0.017 Btu · in/h · ft^2 · °F (2.5 × 10^{-3} W/m · K) can typically be achieved with fine-fiber glass mats. The third system consists of fibers in a gas-filled space. Because the mechanism for thermal transfer in this system is predominantly gas conduction, the system thermal conductivity is only slightly above that of the gas used in the interstitial space and is significantly higher than that obtained in evacuated vessels.

The choice of system depends on the application. The evacuated systems predominate for long-term storage of volatile liquefied gases, especially in small vessels. The multilayer system in particular has been utilized for protecting vehicles in space. For the storage of large quantities of fluids, and in the temporary storage and transfer of fluids of low volatility, a nonevacuated system is generally used.[17]

Sales of fibrous insulations for cryogenic applications in the second and third systems have been relatively small in comparison to expanded perlite and other fine particulate materials. Perlite is less expensive than fibers, is less sensitive to vacuum levels in evacuated systems, and provides about the same resistivity value at low temperatures.[18]

Fiberglass and mineral fiber blankets have also been used to insulate cold-storage rooms and refrigerated warehouses, motor trailers, ships, and railroad cars to −22°F (−30°C). A continuous vapor barrier is mandatory next to the exterior shell of the structure to prevent the migration of moisture through the insulation from the warm side. The condensation of such moisture at its dew point inside the insulation could ultimately destroy the effectiveness of the insulation, both thermally and structurally.[19]

Applications at or Near Ambient Temperature

Three types of fibrous insulating materials predominate at or near normal atmospheric temperatures: (1) fiberglass, (2) mineral wool (also called rock or slag wool), and (3) cellulose fiber. The fibers in all three are staple, as opposed to continuous filaments, and in most of the noncellulose materials they are bonded at points of contact with a thermosetting resin to impart shape, resilience, handleability, and strength to the fiber mass. Unlike stable textile unwovens, which originate as continuous filaments that must be drawn, crimped, cut, blended with binders, and laid down as webs, the majority of fiberglass and mineral wool products begin as staple fibers spun or drawn directly from a melt. Shortly after formation, while still suspended in air, the filaments are sprayed with liquid resin; then they are combined on a moving chain to form a continuous pelt or blanket. Finally, the mass of uncured wool is heated under pressure to set the resin and so define the structure of the blanket, which may be further trimmed and cut into shapes suitable for use. Depending on the type and amount of binder (if used), the size and composition of the fiber, the thickness and density of the insulation, and the methods of fiber processing and binder curing, a wide variety of products are obtainable, as shown in Table 2.3, which is based on Ref. 20 and product literature. In this table the separations into product categories under the fiberglass and mineral wool classifications are quite arbitrary. Density is perceived as the primary determinant of product performance, but binder content, thickness, and technique of fiber formation, among others, have significant influences on the way insulation behaves. The categories are best viewed as "fuzzy," with a considerable amount of overlap.

It is of interest that glass and mineral fibers for insulation have diameters that are much finer and cover a narrower dimensional range than synthetic textile fibers. Whereas synthetic filaments generally range from 10 to more than 40 μm for clothing, nonwovens, and upholstery, and as high as 80 μm for carpets, the inorganic insulation fibers typically run on average from 2.5 to 7 μm for most conventional purposes and as low as 0.5 μm for aerospace and specialty applications. That light and resilient products are achievable with such fine fibers can be attributed to the fiberization process to some degree. Equally significant, however, is the chemical structure of the glass—a continuous, three-dimensional, nonordered arrangement of interconnected silicon-oxygen tetrahedral units—that gives the fiber high tenacity, on the order of 0.7 g/Tex, and low extensivity (2 percent to break). Glass is too stiff (and brittle) in the textile fiber range for common apparel uses, but large-diameter fibers serve in a variety of other noninsulating applications (such as reinforcements).

The fact, as shown in the table, that mineral wool products are heavier than comparable glass fiber products is noteworthy. It can be traced to differences in the manufacturing processes. Mineral wool originates as a blend of natural rocks and by-product slags, which is melted in layers of coke in a large vertical furnace. A hot fluid stream flows out of the furnace onto the rims of a series of rapidly spinning vertical wheels, into a large rotating flat horizontal bowl with indentations along the rim, or in front of a high-pressure steam or air jet. It is by these devices or methods that the fibers are formed. By contrast, glass fibers originate from a mixture of pure oxides and clays in a gas of electrically heated furnace, and the molten stream produced therefrom is generally forced through fine holes in a horizontal high-speed spinner to form the fibers. The first process is about one-half the cost of the second due to the nature of the raw materials and the source of fuel for melting. However, because of compositional variations in the raw materials and because the mineral wool fibers are in effect "thrown" from their fiberizers, as opposed to being extruded through holes, they contain a high percentage of unfiberized glass beads, called shot, which adds to the weight of the product but contributes little to its insulation value. The heavier mineral wool batts tend to provide the same thermal resistance or *R* value with less loft than comparably priced glass fiber batts, an important consideration when space is limited. On the other hand, fiberglass batts and blankets are generally more thermally efficient per unit weight. Mineral fibers have melting points well above those of fiberglass and can thus insulate to higher temperatures and serve as fire barriers.

One form of fibrous material gaining in acceptance and therefore worthy of discussion is loose-fill insulation. This form consists of finely divided bundles of fibers that are applied to attics, walls, and other surfaces by mechanical or pneumatic means or by spraying with air-setting adhesives. All three types of insulation fibers are available in this form. Cellulose loose fill is finely ground paper or wood pulp that has been treated with chemicals for flame retardancy. Mineral and glass fiber loose fills are cut or nodulated pieces of wool that is either dedicated to the end product (i.e., "virgin" wool) or a by-product of the manufacture of insulation in other forms (such as trim or quality rejects).

For many applications, loose fill has certain advantages over other forms of fibrous insulations. For a professional applicator with the proper blowing equipment, loose fill installs much more easily and faster in residential attics, especially those with truss construction, than batts or rolls; and it is the only form of fibrous material that can fill confined spaces, such as the concealed walls of finished buildings. Unlike batts or rolls, loose fill is not thickness or thermal-resistance specific, that is, an applicator can vary the thermal resistance or *R* value of the insulation by blowing in more or less material.

Loose fill does have shortcomings, however. For any other commercially available insulation product, the resistance value is a function of installed thickness. Thus, as long as the product is installed at its designated thickness, its designed or labeled *R* value can be assumed. The *R* value of loose fill, on the other hand, is a function not only of thickness, but also of its installed weight per unit surface area. The installed density of a given loose-fill product can vary, depending on the source and the age of the material, the type of blowing machine used, and the blowing machine settings. Thickness is an easy gage for application, but may not be a particularly reliable one.

Loose-fill insulation may be spray-applied with a wet adhesive. Spray-applied insulation is heavier than blown insulation. Spraying insulation is considered a convenient way to insulate irregularly shaped surfaces and structures, but frequently, alternative methods using other forms of insulation prove to be less expensive and give better results because less material is used and vapor barriers can be included.

High-Temperature Applications

When used to retard the transfer of heat from a medium or body significantly hotter than room temperature, resin-bonded fiberglass and mineral wool insulations are effective to around 450°F (230°C) in air, the temperature above which the binder begins to decompose. Ordinary unbonded fiberglass, or fiberglass in which the heat can burn out the binder without destroying the product usefulness, can insulate to 1000°F (535°C). Under certain circumstances specially formulated glass fibers can withstand temperatures near 1200°F (650°C) for extended periods. Conventional mineral wool high-density boards are useful insulators to 1350°F (730°C),

Table 2.3. Properties and Uses of Structural Insulations Near Ambient Temperatures

Fiber type and basic product description	Representative uses
Cellulose	
1. Loose fill ρ_i = 2.0–3.2 lb/ft³* k = 0.27–0.32 Btu · in/h · ft² · °F†	Pneumatically installed insulation for horizontal surfaces and enclosed cavities (e.g., walls).
2. Spray-applied ρ_i = 2.0–6.2 lb/ft³ k = 0.24–0.30 Btu · in/h · ft² · °F	Pneumatically applied insulation and fire-proofing with suitable adhesives onto compatible surfaces.
Fiberglass	
1. Low-density flexible standard fiber batts and rolls ρ_i = 0.30–1.0 lb/ft³ k = 0.26–0.40 Btu · in/h · ft² · °F	Building attic, ceiling, wall, and floor insulation and sound control products; water heater insulation (kits); flexible ducts.
2. Medium-density flexible to rigid standard fiberglass boards, sheets, rolls, and blankets ρ_i = 1.0–3.0 lb/ft³ k = 0.23–0.28 Btu · in/h · ft² · °F	Duct liner; appliance, air-conditioning, and equipment insulation and liners; solar collector insulation; curtain wall insulation; flexible exterior insulation for pipes, ducts, furnaces, boilers, precipitators, heated vessels, and tanks; masonry insulation.
3. High-density rigid standard fiberglass boards, sheets, and molded parts ρ_i = 3.0–12.0 lb/ft³ k = 0.22–0.25 Btu · in/h · ft² · °F	Preformed round and custom plenum ducts; rigid exterior insulations for pipes, ducts, and equipment; exterior building insulation; insulating roofing baseboard or form.
4. Loose fill ρ_i = 0.5–1.0 lb/ft³ k = 0.28–0.50 Btu · in/h · ft² · °F	See cellulose.
5. Spray-applied ρ_i = 3.5 (rep)‡ lb/ft³ k = 0.25 (rep) Btu · in/h · ft² · °F	Pneumatically applied insulation (with suitable adhesive) to compatible surfaces.
6. Low-to-medium-density, 1–2-μm fiberglass sheets and rolls ρ_i = 0.4–3.0 lb/ft³ k = 0.21–0.28 Btu · in/h · ft² · °F	Aerospace and specialty insulation.
Mineral Wool	
1. Low-density flexible batts and rolls ρ_i = 1.2–4.0 lb/ft³ k = 0.25–0.27 Btu · in/h · ft² · °F	Building attic, wall, and floor insulation and sound control products.
2. Medium-density flexible to rigid sheets, boards, and rolls ρ_i = 4.0–8.0 lb/ft³ k = 0.23–0.28 Btu · in/h · ft² · °F	Fire-rated partition, curtain wall, door, and bulkhead (ship) insulation; industrial and commercial equipment insulation; exterior insulation for boilers, precipitators, breechings, ducts, and process tanks; safing; fire-proofing felts.
3. High-density rigid boards and felts ρ_i = 8.0–20.0 lb/ft³ k = 0.23–0.28 Btu · in/h · ft² · °F	Semirefractory insulation for high-temperature industrial applications (to 1900°F); preformed pipe insulation.
4. Loose fill ρ_i = 1.5–2.0 lb/ft³ (typ)‡ k = 0.30–0.37 Btu · in/h · ft² · °F (typ)	See cellulose.
5. Bulk ρ_i = 3.0–12.0 lb/ft³ k = 0.25–0.28 Btu · in/h · ft² · °F	Fill and packing for cavities, especially in industrial equipment.
6. Spray-applied ρ_i = 5.0–15.0 lb/ft³ k = 0.25–0.31 Btu · in/h · ft² · °F	See cellulose.

* ρ_i = installed density.
† k = thermal conductivity at installed density (75°F mean temperature).
‡ rep = representative values; typ = typical values.
CONVERSION FACTORS: °C = (°F − 32)/1.8; 1 lb/ft³ = 0.016 g/cm³; 1 Btu · in/h · ft² · °F = 0.144 W/m · K.

Table 2.4. Properties of Selected Refractory Fibers

Fiber type	Fiber density, g/cm³	Fiber diameter, μm	Liquidus or melting point, °F	Maximum service temperature, °F*
Alumina (95%)	3.4	3	3700	3000
Alumina (99%)	3.9	20	3700	2400
Alumina-silica	2.54–2.73 (rep)	2.0–5.5	3000–3300	1600–2600
Alumina-silica-boria	2.7 and 3.1	11	3300 and 3400	2200 and 2600
Alumina-silica-chromia	2.6	2.5–4.0	3200	2600
Potassium titanate	3.1–3.5	<1	2300–2400	2200
Silica	2.2	1–10	3000–3100	1800–1940
Zirconia (cubic)	4.8	3–6	4700	4000
Carbon	1.5–2.0	5–8	6600	4500†
Boron nitride	2	7–8	5400	3000†
Silicon carbide (whiskers)	3.2	0.5–10	4875	3275
Silicon nitride (whiskers)	3.2	1–10	3450	3275†

* For long-term or extended service. Fibers may be exposed to higher temperatures for short durations.
† In nonoxidizing atmosphere.
CONVERSION FACTORS: 1 lb/ft³ = 0.016 g/cm³; °C = (°F − 32)/1.8.

and specialty "semirefractory" products can face temperatures as high as 1900°F (1035°C).

For thermal protection above these temperatures, there are a group of oxide and nonoxide ceramic fibers collectively called refractory fibers. Table 2.4 lists the common members in this group, along with representative fiber diameter and density ranges and ultimate temperatures for continuous service.[21–24] Among the important properties affecting the performance of an insulation at elevated temperature is its physical stability. Normally, if a fibrous product shrinks less than 5 percent in one dimension during full exposure to a temperature for 24 h, it can be used continuously at that temperature.[21]

Alumina-Silica Fibers In terms of quantity produced and cost, fine fibers composed of alumina (Al_2O_3) and silica (SiO_2) are the most significant of the group. Formed from a melt by spinning or hot-air or steam attenuation, alumina-silica fibers yield a wide range of products for industry and aerospace use. Blankets and felts made of them are folded into modules to line high-temperature furnaces, kilns, reformers, and boilers. In plain sheet form, the fibers are used as asbestos blanket substitutes; insulations for flexible high-temperature pipe, nuclear reactors, and thermal reactors; and fire barriers for pressure and cryogenic vessels. Bulk fiber serves as fill and packing insulation. Boards of the fiber are used as high-temperature baffles and muffles, trough liners for molten metal, heat shields, high-temperature duct linings, and backups for brick and castable cements. The alumina-silica fiber is also woven into cloth, formed into paper, and wet-formed under vacuum into special configurations for specific insulation service.

The alumina content in this class of fibers varies from 35 to 60 percent of the composition by weight, with the maximum use temperature more or less a function of the concentration of this component. Higher-temperature service has also been obtained in alumina-silica fibers by introducing a small amount (<3 wt %) of chromia (Cr_2O_3).

In fibrous insulations at elevated temperatures, heat migrates mainly by radiation and to a smaller extent by gas or air conduction. To block radiation transfer effectively, the alumina-silica fibers tend to be fine, averaging between 2.5 and 5.5 μm in diameter. The densities of insulation products made of these fibers typically range from 3 to 8 lb/ft³ (0.048 to 0.128 g/cm³) for blankets to as high as 30 lb/ft³ (0.48 g/cm³) for boards. Because of the method of fiberization, which is similar to that of mineral wool, about 20 to 50 percent of the product weight is from shot.

Two relatively new members of the alumina-silica family are boria-modified fibers called Nextel 312 and 440, both developed by 3M Company. Unlike other aluminosilicates, Nextel fibers are formed using a modified textile process, by extruding a concentrated solution through a spinneret and heating the resulting filaments to stabilize them. The 11-μm shot-free continuous filaments of Nextel 312, having a composition close to $3Al_2O_3 \cdot B_2O_3 \cdot 2SiO_2$, are adaptable to textile operations and, therefore, are used in high-temperature fabrics for furnace curtains, thermocouple insulation, and protective sleeves around hoses, wires, and electric cables. On the space shuttle, Nextel 312 finds application as a low-shrinkage separator between the thermal protection tiles and as a reinforcing aid within the tiles.

Nextel 440, also about 11 μm in filament diameter, has a composition close to that of mullite, $3Al_2O_3 \cdot 2SiO_2$, and, in fact, contains mullite crystals around 0.04 μm in size. Nextel 440 is expected to find application chiefly as a reinforcing fiber, but it is an excellent thermal insulator and is usable to higher temperatures than Nextel 312.[22]

Silica Fibers Fibers of >99 percent silica have been developed either by attenuating quartz rods heated with oxy-hydrogen flames or by acid-leaching filaments of low-melting glasses to remove nonsilica components. Normally, the fibers produced have diameters upward of 4 μm and find uses as bulk fibers and textiles for reinforcement and insulation in ablative coatings; resin "prepregs"; furnace curtains; heat-protection garments and fabrics; wire, hose, and cable sleeves; gasketing; and insulative packing material. A fine (1- to 2.5-μm) staple leached fiber of high silica content is the primary insulating material in the external thermal protection tile on the space shuttle. Amorphous silica is specified for this use due to its low and relatively constant coefficient of thermal expansion over a wide temperature range.

Alumina Fibers Like Nextel, alumina (Al_2O_3) fibers are generally formed by spinning from a solution or slurry and then heating to remove fiberizing aids and eliminate porosity. In one type, called Saffil™, developed by Imperial Chemical Industries, the final fiber (about 3 μm in diameter) is polycrystalline and contains about 5 percent silica as a phase stabilizer. Like the discontinuous aluminosilicate fibers, this alumina fiber is available in bulk, board, mat, paper, and vacuum-formed shapes for insulating liners, asbestos replacements, fillers, and gaskets. Generally, because of their higher cost, Saffil fibers are blended with alumina-silica fibers in these products.

Another fiber, Fiber FP™, a DuPont development, is a continuous fiber with a composition of >99 percent Al_2O_3. It is heat-treated to provide a structure consisting of microcrystals of α-Al_2O_3. Being about 17 μm in diameter, Fiber FP is used mainly for reinforcement.

Zirconia Fibers Zirconia (ZrO_2) fibers are produced either from solutionlike alumina fibers or by a technique called the precursor process. In the latter, cellulosic fibers, usually rayon, preswollen from exposure to water, are impregnated with a solution of zirconyl chloride containing some yttrium salt (for high-temperature phase stability). On heating in a low-oxygen atmosphere, the organic fiber pyrolyzes, leaving a fibrous residue of polycrystalline zirconia. Usually 3 to 6 μm in diameter, zirconia fibers are sold in a variety of forms for insulating at temperatures above those possible with alumina fibers.

Potassium Titanate Fibers Formed from a molten bath of potassium carbonate and titanium dioxide with potassium molybdenate as a flux, potassium titanate fibers with a theoretical composition $K_2O \cdot 6TiO_2$ are truly monofilament whiskers with a diameter of <1 μm and lengths of only 20 to 30 μm. Their fine dimensions, high refractive index, relatively low cost, high abrasion resistance, good chemical resistance, and ready dispersibility in liquids make them excellent reinforcing and insulating agents in papers, felts, resins, and paints.

Nonoxide and Other Fibers Silicon carbide (SiC), silicon nitride (Si_3N_4), boron nitride (BN), boron, and carbon are available as high-temperature fibers and whiskers used principally where their high strength and modulus values predominate, as in reinforced plastics, metals, and ceramics. Carbon fiber, however, is employed as insulation in nonoxidizing processes. It can withstand temperatures upward to 4500°F (2480°C) for extended periods in inert atmospheres.

REFERENCES

1. G. G. Bankvall, "Mechanisms of Heat Transfer in Permeable Insulation and Their Investigation in a Special Guarded Hot Plate," in *Heat Transmission Measurements in Thermal Insulations,* ASTM STP 544, American Society for Testing and Materials, Philadelphia, PA, 1974, pp. 34–48.
2. C. M. Pelanne, "Does the Insulation Have a Thermal Conductivity? The Revised ASTM Test Standards Require an Answer," in *Thermal Transmission Measurements of Insulation,* R. P. Tye (ed.), ASTM STP 660, American Society for Testing and Materials, Philadelphia, PA, 1978, pp. 60–70.
3. G. O. Hardegord and N. B. Hutcheon, "Thermal Performance of Frame Walls," *Trans. ASHRAE,* **58,** 171–188 (1952).
4. S. Wolf, K. R. Solvason, and A. G. Wilson, "Convective Air Flow Effects with Mineral Wool Insulation in Wool-Frame Walls," *Trans. ASHRAE,* **72**(2), III.3.1–III.3.8 (1966).
5. B. Rennex, "Low Density Thermal Insulation Calibrated Transfer Samples—A Description and a Discussion of the Material Variability," NBS Spec. Publ. NBSIR 82-2538, June 1982.
6. B. Holcombe, "The Thermal Insulation Performance of Textile Fabrics," *Wool Sci. Rev.,* **60,** 12–22 (Apr. 1984).
7. R. W. Moncrieff, *Man-Made Fibres,* 6th ed., Wiley, New York, 1975.
8. P. Mehta, "The Requirements of Moisture Transport in Underwear," *Wool Sci. Rev.,* **60,** 23–46 (Apr. 1984).
9. H. R. Mauersberger (ed.), *Matthews Textile Fibers,* 5th ed., Wiley, New York, 1947.
10. *Kirk-Othmer Encyclopedia of Chemical Technology,* 3d ed., Wiley, New York, 1978–1984.
    (a) W. J. Roberts, "Fibers, Chemical," vol. 10, pp. 148–166.
    (b) P. H. Hobson and A. L. McPeters, "Acrylic and Modacrylic Fibers," vol. 1, pp. 355–386.
    (c) J. Preston, "Aramid Fibers," vol. 3, pp. 213–242.
    (d) G. A. Serad and J. R. Sanders, "Cellulose Acetate and Triacetate Fibers," vol. 5, pp. 89–117.
    (e) J. S. Hayes, Jr., "Novoloid Fibers," vol. 16, pp. 125–138.
    (f) J. H. Saunders, "Polyamide Fibers," vol. 18, pp. 372–405.
    (g) G. W. Davis and E. S. Hill, "Polyester Fibers," vol. 18, pp. 531–549.
    (h) D. R. Buchanan, "Olefin Fibers," vol. 16, pp. 357–385.
    (i) J. Lundberg and A. Turbak, "Rayon," vol. 19, pp. 855–880.
11. G. M. Moelter, R. F. Tetreault, and M. J. Hefferon, "Polybenzimidazole Fiber," *Polymer News,* **9,** 134–138 (1983).
12. J. V. E. Hansen, "Clothing for Cold Climes," *Natural History,* **90,** 90–97 (Oct. 1981).
13. H. McAlpin, "Down," *Country J.* (Nov. 1981); reprinted in *Science Annual 1983,* Grolier, Danbury, CT, pp. 161–166.
14. R. M. Crow, R. W. Nolan, S. W. Cattroll, and M. M. Dewar, "Comparison of Down and Feathers and Synthetic Insulants for Use in Sleeping Bags," Defence Research Establishment Ottawa (DREO), Rep. 866, July 1982.
15. C. K. Lee, "Heat Transfer of Fibrous Insulation Battings," U.S. Army Natick Research and Development Center, Natick, MA, Final Rep. Natick/TR-84/055, Jan. 1984.
16. B. V. Holcombe, "The Heat Related Properties of Protective Clothing Fabrics," *Fire Safety J.,* **6,** 129–141 (1983).
17. "Cryogenics," in *ASHRAE Handbook, 1982 Applications,* chap. 49, American Society of Heating, Refrigerating and Air-Conditioning Engineers, Atlanta, GA, pp. 49.18–49.23.
18. P. E. Glaser, "What You Should Know about Cryogenic Insulations," *Machine Des.,* 146–152 (Aug. 17, 1967).
19. *ASHRAE Handbook, 1982 Applications,* American Society for Heating, Refrigerating and Air-Conditioning Engineers, Atlanta, GA, "Refrigerated Warehouse Design," chap. 40, pp. 40.9–40.13; "Trucks, Trailers, and Containers," chap. 43, pp. 43.2–43.5; "Railroad Refrigerator Cars," chap. 44, pp. 44.7–44.8; "Marine Refrigeration," chap. 45, pp. 45.2–45.4.
20. "Design Heat Transmission Coefficients," in *ASHRAE Handbook, 1985 Fundamentals,* chap. 23, American Society of Heating, Refrigerating and Air-Conditioning Engineers, Atlanta, GA, pp. 23.1–23.22.
21. W. C. Miiller, "Refractory Fibers," in *Encyclopedia of Composite Materials and Components,* M. Grayson (ed.), Wiley, New York, 1983, pp. 1069–1081.
22. H. G. Sowman and D. D. Johnson, "Ceramic Oxide Fibers," *Ceramic Eng. and Sci. Proc.,* **6,** 1221–1230 (Sept.–Oct. 1985).
23. L. R. McCreight, H. W. Rauch, Sr., and W. H. Sutton, *Ceramic and Graphite Fibers and Whiskers, a Survey of the Technology,* vol. 1 of *Refractory Materials,* J. L. Margrave (ed.), Academic Press, New York, 1965.
24. A. P. Levitt (ed.), *Whisker Technology,* Wiley-Interscience, New York, 1970, pp. 292, 302.

Chapter 3

Granular and Powdered Insulations

ALI REZA MOAZED AND LARRY S. SHU

Construction Products Division, W. R. Grace & Co.

3.1 DEFINITION

Granular insulations consist of granular solids in the form of nodules which contain voids, while powdered insulations consist of particulate solids. Both granular and powdered insulations perform an insulating function by filling and finely dividing the volume of a space of any shape.

3.2 SELECTION OF POWDERED AND GRANULAR INSULATIONS

The proper selection of powdered or granular insulating material depends on the method of installation called for by the particular application under consideration as well as on the physical and thermal characteristics that are required.

Installations Requirements

The required method of installation depends on the geometry of the cavity or fill space and its accessibility. For example, if accessibility is a problem, the insulating material will have to be blown into the cavity. As a result, the potential candidates of insulation material are limited to those that can be installed by this method.

Physical and Thermal Characteristics

The following properties are important in the proper selection of granular and powdered insulations. Often properties other than those of a heat-transfer nature dictate the choice of material.

Alkalinity (pH) or acidity: A measure of the tendency of a material to have a basic (alkaline) or acidic reaction. It

Table 3.1 Granular and Powdered Insulations

ID number	Insulation material, generic type	Description and general characteristics	References
1	Alumina-silica fibers, chopped and milled	Used in fill applications, such as packing furnace joints	2
2	Calcium silicate, granular	Used for industrial piping insulation in high-temperature processes	1, 5
3	Cork, granulated	Available in various grades and particle sizes	1, 2, 4
4	Diatomaceous silica, fine powder	Milled to a fine powder for filling cavities	1, 2
5	Diatomaceous silica, calcined powder	Calcined to withstand higher temperatures; used as fill or with concrete to form an insulating concrete	1, 2
6	Diatomaceous silica, coarse powder	Milled to a coarse powder for filling cavities	1, 2
7	Glass, cellular pellets	Various sizes, hermetically sealed; used for pouring into cavity spaces	2
8	Gilsonite, granules	Available in various grades for various service temperatures; used as fill around underground hot piping	1, 2
9	Gypsum, pellets, bulk	Used for pouring into cavity spaces	1, 2
10	Lampblack, powder	Used for pouring into cavity spaces	2
11	Mineral fibers, granulated	Used for pouring or blowing into cavity spaces	1, 2
12	Perlite, expanded	Available in various size granules; used in intermediate- and high-temperature-range applications	1, 2, 5
13	Sand, granules	Available in various grades and sizes; used as inexpensive insulation in construction	4, 6
14	Sawdust, various kinds, loose	Low-cost insulating material used in buildings	4, 8
15	Silica, fine powder	Used for pouring into cavity spaces, in intermediate- and high-temperature-range applications	4
16	Silica-aerogel, granules	Small spheres, very low conductivity	1, 2
17	Soil, dry	Used for insulating underground piping	4,6
18	Titanium oxide, finely ground powder	Used at high temperatures, up to 3200°F	4
19	Vermiculite, expanded	Expanded mica, available in various sizes for filling cavity spaces; used in intermediate- and high-temperature-range applications; also used with concrete to make light insulating concrete	1, 2, 5
20	Zirconia, powder	Ceramic insulation used as high-temperature insulation in metallurgical industry	4
21	Carbon-silica, powder	Used in vacuum cavity as cryogenic insulation	1, 9
22	Glass, beads	Used in vacuum cavity as cryogenic insulation	7
23	Perlite, powder	Used in vacuum cavity as cryogenic insulation	1
24	Silica-aerogel, powder	Used in vacuum cavity as cryogenic insulation	1, 5
25	Silica-aerogel and aluminum, powder	Opacified with aluminum powder in vacuum cavity; used as cryogenic insulation	1
26	Silica-aerogel and copper, powder	Opacified with copper powder in vacuum cavity; used as cryogenic insulation	1

is measured on the pH scale with all readings above 7.0 alkaline and below 7.0 acidic. A reading of 7.0 indicates neutral. In applications where the insulation comes in contact with metals, the pH of the insulation should not be in the range that can be detrimental to the metal.

Capillarity: The property of a material which enables it to draw liquid up into or through itself. An example of the adverse effect of high capillarity is the occurrence of a water leak in an insulated system. High capillarity of the insulation material will spread the water damage. This property can be measured in accordance with ASTM C80.

Chemical reaction: The property of a material which measures its tendency to combine chemically (or react) with other materials which may come into contact with or be absorbed by it. This information is very important since such a reaction may, in certain instances, create a fire hazard. If an insulation material reacts chemically with a combustible liquid or solid, the reaction may create a fire hazard by lowering the flash or self-ignition point of the latter.

Combustibility: A measure of the tendency of a material to burn. Granular and powdered insulation materials are tested for combustibility in accordance with ASTM E906.

Compaction or settling: A measure of the change in density and volume of a granular or powdered material resulting from loading or vibration, thus producing a change of thermal efficiency. Some materials tend to fluff up when subjected to vibration whereas others tend to compact. This factor has a definite relation to the suitability of the insulation for certain applications.

Corrosion to substrates: The property of a material in-

Table 3.2. Thermal Conductivity of Noncryogenic Insulation Materials (Btu · in/ft² · h · °F)

ID number	Generic type	Form	Density, lb/ft³	Mean temperature, °F: −100	0	100	200	300	400	500	600	700	800	900	1000	1100	1200
1	Alumina-silica fibers	Chopped and milled	6	...	...	...	...	...	...	...	0.85	0.97	1.09	1.22	1.35	1.55	
2	Calcium silicate	Granular	13	...	...	0.34 at 75°F											
3	Cork	Granulated	12–14	0.31	0.34												
4	Diatomaceous silica	Fine powder	13–17	...	...	0.42	0.45	0.48	0.51	0.55	0.58	0.62	0.65	0.69	0.74	0.80	
5	Diatomaceous silica	Calcined powder	32 (max.)	...	...	0.92	0.97	1.03	1.08	1.13	1.18	1.25	1.31	1.37	1.43	1.49	
6	Diatomaceous silica	Coarse powder	25–31	...	...	0.63	0.68	0.77	0.84	0.92	0.99	1.06	1.14	1.22	1.32	1.42	
7	Glass	Cellular pellets	6	...	...	0.40											
8	Gilsonite	Granules	44–50	...	...	0.75	0.85	1.00									
9	Gypsum	Pellets	20–30	...	...	0.45–0.95											
10	Lampblack	Powder	4.7–18	...	...	0.20–0.26											
11	Mineral fibers	Granulated	6–12	...	...	0.30	0.35	0.44	0.54	0.65	0.75	0.86	0.97				
12	Perlite	Expanded	8	...	...	0.36	0.40	0.50	0.55	0.65	0.73	0.85	0.98	1.10			
13	Sand	Granules	100	...	...	2.2	...	2.5	...	2.6							
14	Sawdust	Powder	12.5	...	...	0.41 at 86°F											
15	Silica	Fine powder	50	...	...	...	...	...	...	...	0.54	...	...	0.57	...	...	0.62
16	Silica-aerogel	Granules	4–5.5	...	...	0.21	0.23	0.24									
17	Soil, dry		127	...	3.4	3.7											
18	Titanium oxide	Powder	52	...	...	...	...	...	...	...	...	...	...	...	0.49		
19	Vermiculite	Expanded	8–10	...	...	0.47	0.57	0.63	0.74	0.84	0.94	1.05	1.15	1.25	1.36		
20	Zirconia	Powder	113	...	...	...	...	...	...	...	1.35	...	...	1.50	...	...	1.70

CONVERSION FACTORS: 1 lb/ft³ = 16.02 kg/m³; 1 Btu · in/ft² · h · °F = 0.1442 W/m · K.

dicating the chemical effect it has on various metals with which it comes in contact. This is of utmost importance in applications where insulation comes in contact with metals. Any aggressive ions present in insulation material may be corrosive to metals. Therefore the corrosive effect of the insulation on metals must be determined in these types of applications. To eliminate this effect, either another insulation should be used or the substrate metal should be protected by a proper coating.

Dry flow: A measure of the flowability of a powder or granular material. It is important in applications where powdered or granular insulation needs to be installed by pouring or pumping into a cavity. Dry flow for granular and powdered insulation materials is tested in accordance with ASTM D1755.

Hygroscopicity: A measure of the ability of a material to absorb and retain water, in either the liquid or the vapor state, from the ambient air. The retained moisture is in a state of equilibrium with the moisture in the surrounding air. In certain applications and for some insulations, the presence of water or moisture has an adverse effect on thermal efficiency. In such circumstances, some type of barrier to inhibit

Table 3.3. Thermal Conductivity of Cryogenic Insulation Materials

ID number	Generic type	Form	Density, lb/ft³	Conductivity,* Btu · in/ft² · h · °F	Recommended vacuum (absolute pressure in μm Hg)
21	Carbon-silica	Powder	8.0	0.0048	2.0
22	Glass†	Beads	9.9	0.0007–0.0028	0.1
23	Perlite	Powder	8.0	0.0087	2.0
24	Silica-aerogel	Powder	6.0	0.0115	2.0
25	Silica-aerogel and aluminum	Powder	10.0	0.0027	2.0
26	Silica-aerogel and copper	Powder	11.0	0.0028	2.0

* Based on mean temperature of −200°F, under partial vacuum.
† Based on mean temperature of −54°F, under high vacuum (0.1 μm Hg).
CONVERSION FACTORS: 1 lb/ft³ = 16.02 kg/m³; 1 Btu · in/ft² · h · °F = 0.1442 W/m · K.

Table 3.4. Physical Properties

ID number	1	2	3	4	5
Generic type	Alumina-silica fibers	Calcium silicate	Cork	Diatomaceous silica	Diatomaceous silica
Form	Chopped and milled	Granular	Granulated	Fine powder	Calcined powder
Alkalinity, pH	N/A	N/A	6.5–7.5	5–7	6–7
Combustibility	Noncombustible	Noncombustible	Combustible	Noncombustible	Noncombustible
Flash point temperature, °F	. . .	. . .	570		
Flame point temperature, °F	. . .	. . .	620		
Heat release, Btu/lb	. . .	. . .	8000		
Self internal heating	. . .	. . .	N/A		
Smoldering or afterglow	. . .	. . .	N/A		
Density					
As received, lb/ft^3	7.5	13	8–12	10–12	22
Packed, lb/ft^3	7.5	N/A	12–14	13–17	42
Hydroscopicity, wt %	N/A	N/A	N/A	N/A	N/A
Maximum temperature limits, °F					
Short time	2300	1200	200	1600	2000
Continuous	2300	1200	200	1600	2000
Specific heat, Btu/lb · °F	0.20	0.28	0.27	0.25	0.25
Thermal diffusivity, ft^2/h	N/A	0.008	N/A	0.01	0.01
Water absorption, wt %	High	High	Low	High	High
Capillarity	N/A	N/A	N/A	N/A	N/A

ID Number	6	7	8	9	10
Generic type	Diatomaceous silica	Glass	Gilsonite	Gypsum	Lampblack
Form	Coarse powder	Cellular pellets	Granules	Pellets	Powder
Alkalinity, pH	5–7	6.5–7.5	7	N/A	N/A
Combustibility	Noncombustible	Noncombustible	Combustible	Noncombustible	Noncombustible
Flash point temperature, °F	. . .	. . .	600		
Flame point temperature, °F	. . .	. . .	600		
Heat release, Btu/lb	. . .	. . .	17,000		
Self internal heating					
Smoldering or afterglow	. . .	. . .	Will smolder		
Density					
As received, lb/ft^3	25–27	. . .	40	12–20	4.7–18
Packed, lb/ft^3	25–31	. . .	44–50	20–30	N/A
Hydroscopicity, wt %	N/A	0	Very low	N/A	N/A
Maximum temperature limits, °F					
Short time	1600	800	300–460	800	1200
Continuous	1600	800	300–460	800	1200
Specific heat, Btu/lb · °F	0.25	0.20	0.5 at 200°F	N/A	N/A
Thermal diffusivity, ft^2/h	0.25	0.18	0.0023 at 200°F	N/A	N/A
Water absorption, wt %	High	. . .	Low	High	N/A
Capillarity	N/A	N/A	N/A	N/A	N/A

the migration of moisture into the insulation may be required.

Resistance to acids: The property of a material which indicates its ability to resist decomposition by various acids to which it may be subjected. Since insulation is used in all kinds of atmospheres and conditions of chemical contamination, its resistance to the acids to which it may be subjected affects its service life.

Resistance to caustics: The property of a material which indicates its ability to resist decomposition by various caustics to which it may be subjected.

Resistance to solvents: The property of a material which indicates its ability to resist decomposition by various solvents to which it may be subjected.

Specific gravity—apparent: The ratio of the specific weight of a material, including all its voids, to the weight of an equal volume of water. This can be obtained by simply dividing the weight per cubic foot of the material (density) by the

Table 3.4. Physical Properties (*Continued*)

ID number	11	12	13	14	15
Generic type	Mineral fibers	Perlite	Sand	Sawdust	Silica
Form	Granulated	Expanded	Granules	Powder	Fine powder
Alkalinity (pH)	7–9	N/A	N/A	N/A	N/A
Combustibility	Noncombustible	Noncombustible	Noncombustible	Combustible	Noncombustible
Flash point temperature, °F	. . .	. . .	. . .	N/A	
Flame point temperature, °F	. . .	. . .	. . .	N/A	
Heat release, Btu/lb	. . .	. . .	. . .	N/A	
Self internal heating	. . .	. . .	. . .	N/A	
Smoldering or after-glow	. . .	. . .	. . .	will smolder	
Density					
As received, lb/ft^3	7–10	N/A	94–125	9–12.5	50
Packed, lb/ft^3	6–12	3–11	N/A	N/A	N/A
Hydroscopicity, wt %	N/A	N/A	N/A	High	N/A
Maximum temperature, limits, °F					
Short time	1000–1200	1800	N/A	90	1500
Continuous	1000–1200	1800	N/A	90	1500
Specific heat, Btu/lb · °F	0.12	N/A	N/A	32	N/A
Thermal diffusivity, ft^2/h	N/A	N/A	N/A	N/A	N/A
Water absorption, wt %	High	High	N/A	High	N/A
Capillarity	N/A	N/A	N/A	N/A	N/A

ID number	16	17	18	19	20
Generic type	Silica-aerogel	Soil, dry	Titanium oxide	Vermiculite	Zirconia
Form	Granules		Powder	Expanded	Powder
Alkalinity, pH	3.5 to 4.0	N/A	N/A	6–10	N/A
Combustibility	Noncombustible	Noncombustible	Noncombustible	Noncombustible	Noncombustible
Flash point temperature, °F					
Flame point temperature, °F					
Heat release, Btu/lb					
Self internal heating					
Smoldering or after-glow					
Density					
As received, lb/ft^3	3.5–5.5	13–19	52	4–10	113
Packed, lb/ft^3	4–5.5	N/A	N/A	4–10	N/A
Hydroscopicity, wt %	10–15	N/A	N/A	2	N/A
Maximum temperature, limits, °F					
Short time	1300	160	3200	1400	1800
Continuous	1300	160	3200	1400	1800
Specific heat, Btu/lb · °F	0.2 at 150°F	N/A	N/A	0.24 at 300°F	N/A
Thermal diffusivity, ft^2/h	0.02 at 150°F	N/A	N/A	0.025 at 150°F	N/A
Water absorption, wt %	100	High	N/A	300	N/A
Capillarity	Will wick	Will wick	N/A	N/A	N/A

CONVERSION FACTORS: 1 lb/ft^3 = 16.02 kg/m^3; 1 Btu/lb = 2.3260 kJ/kg; 1 Btu/lb · °F = 4.184 kJ/kg · °C; 1 ft^2/h = 0.0929 m^2/h; °C = (°F − 32)/1.8.

weight of a cubic foot of water. This property is used in calculating the installed weight or volume of the insulation.

Specific gravity—real: The ratio of the weight of the solids contained in the insulation material to the weight of water which has a volume equal to that of the solid. This property gives an indication of the weight or volume of the solids in the insulation materials. To illustrate the difference between apparent and real specific gravities, glass fiber insulation has an approximate apparent specific gravity of 0.05 and a real specific gravity of 2.5.

Specific heat: The ratio of the amount of heat required to raise a unit mass of material 1°F to that required to raise a unit mass of water 1°F at some specified temperature. It is essential to know this property where heat storage and temperature-time lag are important.

Temperature limits: The upper and lower limiting temperatures within which a material will experience no essential change in its properties. Within its limits a material should not fail to function as designed. These limits are determined by the effect of the temperature on the properties listed in this section. Furthermore, time exposure to temperature and the number of changes in temperature have an effect on some materials. For this reason, in most instances it is necessary to specify the following:

Maximum temperature—continuous operation
Maximum temperature—intermittent operation
Maximum temperature—cyclic operation
Minimum temperature—cyclic operation
Minimum temperature—continuous cyclic operation

Thermal conductivity: The property of a homogeneous body measured by the ratio of the steady-state heat flux (time rate of heat flow per unit area) to the temperature gradient (temperature difference per unit length of heat flow path) in the direction perpendicular to this area. Since, in general, the thermal conductivity of an insulation varies with the mean temperature, it is essential to know the conductivity as a function of temperature. The reciprocal of thermal conductivity is thermal resistivity. The commonly used units for thermal conductivity are Btu · in/ft^2 · h · °F (W/m · K), with 1 Btu · in/ft^2 · h · °F = 0.1442 W/m · K. Thermal resistance is now being used extensively to include the effect of the thickness of insulation. It is defined as the product of thermal resistivity and thickness. For example, a 4-in-thick insulation with a conductivity of 0.25 Btu · in/ft^2 · h · °F has a thermal resistance of 16°F · ft^2 · h/Btu. In the United States it would be advertised as having an R value of 16.

Thermal diffusivity: The property that measures the time rate of temperature change through a material. It is not a measure of the amount of heat or heat transfer. In many applications where the insulation material has a high specific heat, or a high density, thermal conductivity alone is not indicative of the thermal performance of the insulation. Instead, thermal diffusivity becomes the important parameter.

Temperature rise from internal self-heating: A property of some materials indicating that when heated above a certain temperature, an internal reaction will occur causing an internal temperature rise in excess of the temperature to which they are subjected. This internal temperature not only causes damage to the material itself, but can be very hazardous since fires can result.

Vapor (water) migration: A property of a material which measures the rate at which water vapor penetrates it under a water vapor pressure gradient.

3.3 CLASSIFICATION

The applicability and the properties of thermal insulations change with the application temperature. Hence, granular and powdered insulations can be classified effectively and conveniently on the basis of this application temperature. For convenience, the temperature scale is divided into the following ranges:

1. *Cryogenic:* −450°F (−267°C) $< T <$ −150°F (−101°C)
2. *Low temperature:* −150°F (−101°C) $< T <$ 212°F (100°C)
3. *Intermediate temperature:* 212°F (100°C) $< T <$ 900°F (482°C)
4. *High temperature:* $T >$ 900°F (482°C)

Cryogenic Range, −450°F (−267°C) < *T* < −150°F (−101°C)

Evacuated powdered insulations are used in the cryogenic range. Vacuum cryogenic thermal insulation systems typically consist of highly polished metal walls with a vacuum space between them. Within this vacuum space there may be various inorganic or organic insulation materials. Among these some are coated with metals (so-called opacified powders) and some are uncoated oxide powders, depending upon the application temperature. This group of insulations covers a thermal conductivity range of as low as 0.0007 Btu · in/ft^2 · h · °F (1.0×10^{-4} W/m· K) to about 0.0115 Btu · in/ft^2 · h · °F (1.6×10^{-3} W/m · K). Under conditions of very low pressures (1 to 10 μm Hg) conduction and radiation are the dominant modes of heat transport within the granular and powdered insulation materials.

Low-Temperature Range, −150°F (−101°C) < *T* < 212°F (100°C)

Thermal insulation materials that are used in this temperature range include some of the evacuated systems in the cryogenic range as well as organic and inorganic "massive" insulations such as cork and vermiculite. The conductivity of some insulation materials employed for cryogenic and low-temperature applications increases dramatically in service because of permeation by water vapor or the exchange of the in-place low-conductivity gas with air.

In the temperature range from about −22°F (−30°C) to 212°F (100°C), granulated cork, expanded vermiculite, and perlite are most commonly used. Where the insulation is used at atmospheric pressure, all three heat-transfer modes, that is, convective, conductive, and radiative, are present in the insulation.

Intermediate-Temperature Range, 212°F (100°C) < *T* < 900°F (482°C)

This temperature range is very important for industrial applications since it includes temperatures typically encountered in steam systems and most chemical and petrochemical processes.

The most notable insulation types used in this range are diatomaceous silica and vermiculite. For powdered insulation, conduction and radiation are the predominant modes of heat transfer in this temperature range. Granular insulation materials exhibit convective as well as conductive and radiative heat transfer in the intermediate-temperature range.

High-Temperature Range, *T* > 900°F (482°C)

Thermal insulation materials available for use in this temperature range become limited in number when the temperature exceeds about 1200°F (650°C). Typical insulations used in this temperature range are calcium silicate, perlite, vermiculite, silica and titanium oxide (both finely ground), and zirconia powder. At high temperatures, radiation and conduction modes of heat transfer are most important for the powdered insulation materials, whereas for granular insulation materials convection, conduction, and radiation are all present.

There are some expensive microporous insulation products that have exceptionally high thermal performance in the high-temperature range. Thermal conductivities less than that of still air have been reported for commercially available products consisting of ultrafine powders, reinforcing fibers, and opacifiers. These ingredients are "packed" in the form of sheets, blocks, or panels in which the void size is smaller than the mean free path of air molecules.

These products are not listed in this chapter because the commercially available products are not sold in granular or powdered form.

3.4 GENERAL CHARACTERISTICS OF GRANULAR AND POWDERED INSULATIONS

Commonly used granular and powdered insulation materials and their characteristics are summarized in Table 3.1.

3.5 THERMAL CONDUCTIVITY OF GRANULAR AND POWDERED INSULATIONS

Thermal conductivity for noncryogenic and cryogenic granular and powdered insulation materials of particular densities or density ranges are presented in Tables 3.2 and 3.3, respectively. For some materials the available data do not cover the full temperature range, and in a few instances thermal conductivity is given only for one temperature.

3.6 PHYSICAL PROPERTIES OF GRANULAR AND POWDERED INSULATIONS

Physical properties, other than thermal conductivity, are summarized in Table 3.4. Some properties are not available for a number of materials, and the user is advised to use the manufacturer's literature or perform independent testing to determine the value for a particular property.

REFERENCES

1. W. C. Turner and J. F. Malloy, *Thermal Insulation Handbook,* McGraw-Hill, New York, 1981.
2. J. F. Malloy, *Thermal Insulation,* Van Nostrand Reinhold, New York, 1969.
3. L. Y. Hess (ed.), *Insulation Guide for Buildings and Industrial Processes,* Noyes Data Corp., Park Ridge, NJ, 1979.
4. G. B. Wilkes, *Heat Insulation,* Wiley, New York, 1950.
5. D. E. Croy and D. A. Dougherty, *Handbook of Thermal Insulation Applications,* Noyes Publications, Park Ridge, NJ, 1984.
6. J. R. Barned, *Thermal Conductivity of Building Materials,* Division of Building Research, CSIRO, Rep. R.2, 1970.
7. M. G. Langseth, F. E. Ruccia, and R. E. Wechsler, "Thermal Conductivity of Evacuated Glass Beads: Line Source Measurements in a Large Volume Bead Bed Between 225 and 300 K," in *Heat Transmission Measurements in Thermal Insulations,* ASTM STP 544, American Society for Testing and Materials, Philadelphia, PA, 1974.
8. *The Chartered Institution of Building Services Engineers Guide,* London, UK, 1980, sec. A3, table A3.22.
9. B. L. Knight, "Thermal Properties of Selected Evacuated Powder Insulations at Cryogenic Temperatures," Ph.D thesis, Dept. of Chemical Engineering, University of Colorado, Boulder, 1969.

Chapter **4**

Cellular Insulations

WILLIAM R. STRZEPEK
Dow Chemical U.S.A.

4.1 GENERAL CHARACTERISTICS

The use of cellular materials as thermal insulations has become popular due to the good thermal insulating properties and high strength-to-weight ratio exhibited by these materials. Also, cellular materials generally are reasonable in cost when compared with their energy-saving potential.

There are many organic and inorganic substances which are capable of being processed to form stable foams. Basically the substance must have the capability of being processed as a fluid, foamed or expanded mechanically or by a gas while in that state, and then solidified while maintaining the cellular matrix established during the foaming process. Cellular materials can be produced with varying degrees of rigidity, depending primarily on the chemical composition and rigidity of the substance used to form the cellular matrix. For example, flexible, semiflexible (or semirigid), and rigid materials are available. However, in general only rigid cellular materials are used as thermal insulations. The most common types of organic, cellular thermal insulations are manufactured using polystyrene resins, polyurethane and polyisocyanurate chemistry, urea formaldehyde resins, and phenol formaldehyde resins (which are used in the manufacture of phenolic foams). The most common inorganic cellular insulation is produced from glass. (Perlite and vermiculite are also expanded inorganic products; but since they are commonly available as loose-fill materials, they are addressed in Chapter 3.)

Cellular materials consist of a gas phase, which is made up of a multitude of fine gas bubbles, dispersed in a solid phase, which is the cellular matrix. The physical properties of these materials are determined by both the gas and the solid phases. The gas phase consists of the cell gas atmosphere, which can be air, chlorofluorocarbons, carbon dioxide, or other blowing agents used to foam or expand the solid substance. Some blowing or foaming agents are fugitive in that they readily diffuse through the cellular matrix and remain a component of the cell gas atmosphere for only a short period of time. The solid phase consists of the cellular matrix. It is composed of the organic or inorganic substance of which the matrix is formed and other ingredients added for a variety of reasons, such as foaming agents, fire retardants, and dyes or pigments. In low-density cellular materials, such as those used as thermal insulations, the gas phase accounts for the majority of the volume of the material, whereas the solid phase accounts for virtually all of the weight.

The fine cells or bubbles, which constitute the gas phase, may be interconnected so that gas can pass readily from one cell to another. Materials with this characteristic are called open-cell. If cells are not interconnected and the gas phase of each cell is relatively independent of that of the surrounding cells, the material is closed cell. In practice, no material is entirely open- or closed-cell; the terms imply that the material is predominantly one or the other. Cellular thermal insulations are primarily closed-cell, but some phenolic and urea formaldehyde foams are notable exceptions.

The base resin constituting the solid phase of foamed plastics may be either thermoplastic or thermosetting. Foam insulations made using thermoplastic resins such as polystyrene will soften or melt (if exposed to a sufficiently high temperature) before undergoing pyrolysis. Thermosetting foams, such as polyurethane and polyisocyanurate, urea formaldehyde, and phenolic, will not. These materials generally will form a char as they pyrolyze, depending on the conditions. This and some of the other differences between these two classes of foams can be important considerations in some end uses.

4.2 MATERIALS AND MANUFACTURING PROCESSES

The processes for manufacturing cellular thermal insulations are similar in a general sense regardless of which substance is being foamed. However, there are specific differences in the processes, which are worth noting.

Polystyrene Expanded polystyrene (EPS) can be produced in either a molding process or an extrusion process. In the molding process small polystyrene beads, which have been impregnated with a blowing agent (typically pentane), are expanded in two stages. In the first stage the expandable beads are exposed to pressurized steam and the blowing agent vaporizes, which causes the beads to expand. The second steam expansion is done in a large mold. The partially expanded beads are expanded fully and fused together to form a billet, which can then be cut into boardstock or various shapes.

Polystyrene is also commonly expanded in an extrusion process. A solution of the molten polymer, a pressurized blowing agent in a liquid state, and other additives is extruded at elevated temperature and pressure through a slot-shaped orifice onto a conveyer system at ambient temperature and atmospheric pressure (or in some cases at a partial vacuum). As the polymer solution passes through the orifice, the blowing agent vaporizes because of the reduction in pressure, and the polymer is expanded into a foam. The material is extruded continuously as boards or billets with naturally occurring high-density skins.

Polyurethane and Polyisocyanurate Polyurethane (PUR) and polyisocyanurate (PIR) foams are produced by reacting two main ingredients, a polyol and polyisocyanate (both in a liquid state) in the presence of a catalyst, a blowing agent, and other additives. The physical properties of the resultant foam can vary greatly depending on the specific type and amount of ingredients selected as well as the particular foaming process used. Since the foaming process occurs at the same time as the polyurethane or the polyisocyanurate polymer is being produced, the material can be foamed on site, which may be advantageous in some situations. Polyurethane and polyisocyanurate foams can be produced using a spray-applied technique, a pour-in-place process, a slabstock process, or a free-rise or restrained-rise continuous lamination process.

In the spraying process the ingredients are mixed and dispersed as a spray, and the surface being insulated generally must be clean and dry to ensure good adhesion of the foam. In the pour-in-place process, the chemical components are mixed and the resulting blend is poured into a mold or cavity in which the expansion takes place. Billets are occasionally made using this process as well. To make boardstock and billets, reactants are mixed and deposited on a conveyer in a continuous process. After the foam cures, billets can be cut into boardstock and various shapes. In the free-rise or restrained-rise continuous lamination process, the components are poured or sprayed between two facing materials, and the mixture adheres to the facings as it expands. The restrained-rise process uses top and bottom conveyers to maintain thickness.

Polyisocyanurate foams are closely related to polyurethane products. In general, the same basic ingredients are used, but the specific formulations are different. Different reaction conditions are established which result in the production of the polyisocyanurate polymer. There are differences in some of the physical properties: polyisocyanurate foams typically have improved thermal stability and flammability characteristics compared with polyurethane foams.

Urea Formaldehyde Urea formaldehyde (U-F) foams are typically poured in place on site. To produce these foams, compressed air is mechanically dispersed into an aqueous solution of a catalyst and a foaming agent in a mixing gun, and a foam is generated. A second aqueous solution of partially cured urea formaldehyde resin is then blended mechanically into the foam and the bubbles are coated with the resin. The mixture is then poured under pressure and the foam cures in place over time as the resin reacts with the catalyst.

Phenolic Phenolic foams used as thermal insulations are typically produced in a continuous slabstock or double-band-lamination process similar to that used for the manufacture of polyurethane and polyisocyanurate foams discussed previously. A phenol formaldehyde resin, produced by reacting phenol and formaldehyde, is mixed with a catalyst, a blowing agent, and a foaming agent to generate the foam. Either open- or closed-cell materials can be obtained, depending on the formulation used and the foaming conditions selected.

Cellular Glass Cellular glass insulation is generally manufactured from a mixture of silicates and alkali oxides, which is fused at high temperatures and cellulated as a result of a high-temperature reaction between carbon and oxygen. The product is available in blocks, boards, and various shapes.

4.3 PHYSICAL PROPERTIES

Typical physical properties for cellular thermal insulations are shown in Tables 4.1–4.4 and Figs. 4.1–4.3. The data are taken primarily from general references, such as technical sales literature, material specifications, and published articles and papers; however, some data are included from private communications with insulation manufacturers and proprietary research reports. The data reported are for the purposes of general comparison, and the values may not be representative of every product available in the marketplace. For any given entry in the tables the range in values may be considerable due to the variability between specific products within a given material type and other factors relating to the testing of the products. Also, the data represent typical properties, not minimum values or specification limits. To obtain appropriate values for use in design and information concerning the test methods followed, the insulation manufacturer should be consulted.

Thermal Transmission

Perhaps the most important physical property of cellular thermal insulations is thermal transmission. Typical thermal transmission data for the common cellular insulations are listed in Tables 4.1 and 4.2. In cellular insulations thermal energy is transferred by three different mechanisms: conduction through the solid portion of the foam, conduction through the gaseous portion, and radiation through the cel-

Table 4.1. Typical Physical Properties of Cellular Thermal Insulations[a]

Property	Expanded polystyrene (molded)	Expanded polystyrene (extruded)	Polyurethane and polyisocyanurate	Urea formaldehyde	Phenolic	Cellular glass
Apparent density, lb/ft^3 [b]	0.8–2.0	1.4–4.0	1.7–3.0	0.8–1.2	2.0–3.0	8.5–12.0
Open-cell content, %	5–40	1–7	2–10	30–95	5–90	0–1
Apparent thermal conductivity[c] Btu · in/h · ft^2 · °F[d]	0.23–0.30	0.20	0.13–0.18	0.24	0.12–0.23	0.35–0.57
Heat capacity, Btu/lb · F[e]	0.27–0.31	0.27–0.31	0.20–0.25	0.36	0.38	0.20
Apparent water vapor permeability,[f] perm in[g]	0.6–5.0	0.4–1.4	2.0–4.0	4.5–100	1.4–3.5	0.0
Water absorption,[h] vol%	<2.5	<0.3	2–5	Very high	<10	Nonabsorptive
Compressive strength, lb/in^2 [i]	5–33	15–125	20–60	5	10–33	100–210
Coefficient of linear thermal expansion, in/in · °F[j] ($\times 10^{-6}$)	25–40	25–40	30–60	50	10–20	1.6–4.6
Dimensional stability,[k] % linear change	<2	<2	2–10	30	<2	Very low
Maximum use temperature, °F[l]	165	165	250 PUR 250–300 PIR	120	300	900
Combustibility	Combustible	Combustible	Combustible	Combustible	Combustible	Noncombustible
Flame spread[m]	5[n]	5[n]	25–50 PUR 20–25 PIR	10–25	20–25	5
Smoke developed[m]	45–200[n]	45–200[n]	150–500 PUR 50–200 PIR	5–70	5–15	0

[a] The data reported are for the purposes of general comparison, and the values may not be representative of every product available in the marketplace. The range in values may be considerable due to the variability between specific products within a given material type and other factors relating to the testing of the products.

[b] 1 lb/ft^3 = 16.0 kg/m^3.

[c] Materials are measured at a 75°F (24°C) mean temperature. In general, values are intended to be representative of long-term performance. However, values may not be appropriate for all products in the marketplace due to a lack of long-term data and uncertainties regarding product quality and aging conditions.

[d] 1 Btu · in/h · ft^2 · °F = 0.144 W/m · K.

[e] 1 Btu/lb · F = 4.19 kJ/kg · K.

[f] Values are based on the cellular material only. The effects of facings are not included. Values are determined using the desiccant method as described in ASTM E96.

[g] 1 perm · in = 1.46 ng/Pa · s · m.

[h] Materials are immersed for 24 h in accordance with ASTM C272.

[i] 1 lb/in^2 = 0.006895 MPa.

[j] 1 in/in · °F = 1.8 m/m · °C.

[k] Values are representative of materials conditioned at approximately 158°F (70°C) and 97% relative humidity for 7 days. See ASTM D2126.

[l] °C = (°F − 32)/1.8.

[m] Values are indicative of the response of these materials to heat and flame under controlled laboratory conditions. To describe or appraise the fire hazard or fire risk under actual fire conditions, other factors need to be considered as well. Values may vary with thickness. See ASTM E84.

[n] Ignition of molten residue on the floor of the furnace is not considered.

Table 4.2. Typical Thermal Transmission Properties, 1.0-in (2.54-cm) Thickness*

Mean temperature, °F	Expanded polystyrene: Molded, ρ = 1.0 lb/ft³†	Expanded polystyrene: Molded, ρ = 2.0 lb/ft³	Expanded polystyrene: Extruded, ρ = 1.4–4.0 lb/ft³	Polyurethane and polyisocyanurate: Unfaced or high-permeance facings, ρ = 1.7–3.0 lb/ft³	Polyurethane and polyisocyanurate: Low-permeance facings, ρ = 1.7–3.0 lb/ft³	Urea formaldehyde: ρ = 0.8–1.2 lb/ft³	Phenolic: Open-cell, ρ = 2.0–3.0 lb/ft³	Phenolic: Closed-cell, ρ = 2.0–3.0 lb/ft³	Cellular glass: ρ = 8.5 lb/ft³	Cellular glass: ρ = 12.0 lb/ft³
Apparent thermal conductivity Btu · in/h · ft² · F‡										
25	0.23	0.20	0.18	0.16–0.18	0.13–0.15	0.21	0.20	0.115	0.31	0.54
40	0.24	0.21	0.19	0.15–0.17	0.12–0.14	0.22	0.21	0.115	0.32	0.55
75	0.26	0.23	0.20	0.16–0.18	0.13–0.15	0.24	0.23	0.12	0.35	0.57
110	0.28	0.25	0.22	0.17–0.19	0.14–0.16	0.26	0.25	0.125	0.38	0.59
Thermal resistance, °F · ft² · h/Btu§										
25	4.3	5.0	5.6	5.6–6.2	6.7–7.7	4.8	5.0	8.7	3.2	1.9
40	4.2	4.8	5.4	5.9–6.7	7.1–8.3	4.5	4.8	8.6	3.1	1.8
75	3.8	4.3	5.0	5.6–6.2	6.7–7.7	4.2	4.3	8.3	2.9	1.8
110	3.6	4.0	4.7	5.3–5.9	6.3–7.1	3.8	4.0	8.0	2.6	1.7

* The data reported are for the purposes of general comparison, and the values may not be representative of every product available in the marketplace. The range in values may be considerable due to the variability between specific products within a given material type and other factors relating to the testing of the products. In general, values are intended to be representative of long-term performance. However, values may not be appropriate for all products in the marketplace due to lack of long-term data and uncertainties regarding product quality and aging conditions.

† 1 lb/ft³ = 16.0 kg/m³.

‡ 1 Btu · in/h · ft² · °F = 0.144 W/m · K.

§ 1°F · ft² · h/Btu = 0.176 m² · K/W.

lular matrix from cell wall to cell wall.[1] Convection heat transfer within the cells is generally not considered because the cell size is usually too small to support convective movement.[1]

For many products measures are taken to minimize the heat-transfer contribution from one or more of these mechanisms. For example, chlorofluorocarbon blowing agents are commonly introduced not only to help foam the fluid but also to reduce conduction through the gaseous portion of the foam. The use of facings laminated or bonded in some way to chlorofluorocarbon-blown foams also can help improve thermal performance. These measures are discussed further in the section on aging effects.

To minimize radiation through the cellular matrix, the density can be increased in order to provide more material in the cell walls to absorb infrared radiation. Although an increase in density can reduce the radiation component of heat transfer, it simultaneously increases the conduction through the solid portion of the foam. For example, the optimum balance point for radiation and conduction effects is generally 1.8 to 2.5 lb/ft³ (28.8 to 40.0 kg/m³) for cellular plastics.[1,2,3] However, because other factors generally need to be considered, such as raw material costs and mechanical strength requirements, the density of cellular plastics typically ranges from less than 1.0 to 4.0 lb/ft³ (16.0 to 64.0 kg/m³).

Radiation transfer can also be reduced by decreasing the cell size so that more cell walls are provided to absorb the radiation. However, in some cases the cell size can be reduced only to a certain point before the product becomes difficult or uneconomical to produce. It can also be difficult to control the cell size or produce a uniform cellular matrix.

Aging Effects The thermal resistance, commonly termed *R* value, of some cellular insulations decreases with time, as shown in Fig. 4.1. This phenomenon is termed aging. Aging can occur only with those closed-cell materials expanded with agents other than air, such as chlorofluorocarbons, and reflects a change in composition of the cell gas atmosphere. Products containing only air, such as molded bead polystyrene (MEPS), do not age in this way. However, the *R* value can also decrease as a result of environmental factors relating to a particular end use, such as physical deterioration and water absorption. Aging occurs in addition to these other mechanisms of *R*-value degradation and in some cases can be accelerated by them. Regardless of the mechanism at work, aging effects and *R*-value degradation are important to consider when designing or selecting an insulation system.

The decrease in the *R* value of chlorofluorocarbon-blown foams can be most easily comprehended by recognizing that the phenomenon is the result of two independent mechanisms which take place simultaneously. At the time of manufacture, the cells of these materials contain virtually no air, only the chlorofluorocarbon and perhaps some other fugitive blowing agents, such as carbon dioxide, which readily diffuse out of the product. Because of the difference between the partial pressure of the air in the ambient atmosphere and that in the cell gas atmosphere, air immediately begins to enter the cells. The diffusion of air into the product continues until the partial pressure of the air in the foam equals the ambient air pressure. The chlorofluorocarbons have a lower thermal conductivity than that of air, so the net effect of the entry of air into the foam is the reduction of the *R* value of the product. Air intrusion constitutes one of the two mechanisms involved in aging. It is reflected in the curves in Fig. 4.1. For most products and applications this mechanism of aging occurs early enough in the service life of the product that it is significant and needs to be taken into account. However, the presence of well-bonded low-permeance facings, such as some aluminum foils, can significantly retard this air influx.

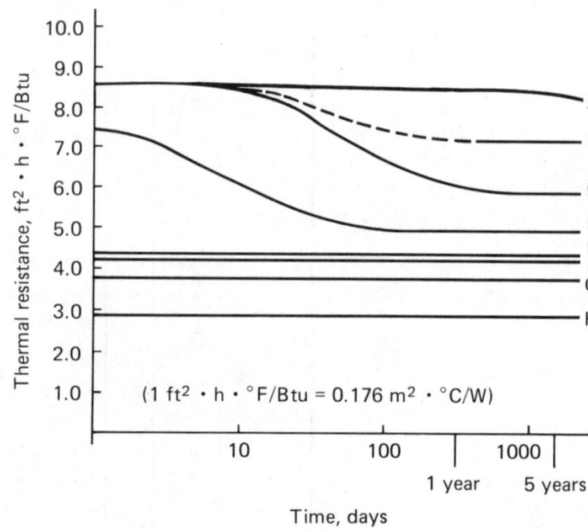

Fig. 4.1 Thermal resistance versus time for 1.0-in (2.54-cm) materials aged at ambient conditions and measured at a 75°F (24°C) mean temperature. *A*: phenolic, 2.0–3.0 lb/ft³ (32.0–48.0 kg/m³), closed-cell. *B*: polyurethane and polyisocyanurate, 1.7–3.0 lb/ft³ (27.2–48.0 kg/m³), low-permeance facings. *C*: polyurethane and polyisocyanurate, 1.7–3.0 lb/ft³ (27.2–48.0 kg/m³), unfaced or high-permeance facings. *D*: expanded polystyrene (extruded), 1.4–4.0 lb/ft³ (22.4–64.0 kg/m³). *E*: phenolic, 2.0–3.0 lb/ft³ (32.0–48.0 kg/m³); open-cell. *F*:urea formaldehyde, 0.8–1.2 lb/ft³ (12.8–19.2 kg/m³). *G*: expanded polystyrene (molded), 1.0 lb/ft³ (16.0 kg/m³). *H*: cellular glass, 8.5 lb/ft³ (136 kg/m³). *Notes:* (1) The data reported are for the purposes of general comparison and the values may not be representative of every product available in the marketplace. The range in values may be considerable due to the variability between specific products within a given material type and other factors relating to the testing of the products. (2) Curve *A* is based on limited measured data from Ref. 14.

The second mechanism of aging involves the diffusion of the chlorofluorocarbon blowing agent out of the foam and occurs because of the difference between the partial pressure of the chlorofluorocarbon gas in the cell gas atmosphere and that in the ambient atmosphere. This mechanism also begins immediately after manufacture but generally proceeds very slowly compared with air intrusion. It is completed when there is virtually no chlorofluorocarbon remaining in the cellular matrix, only air. Diffusion of the chlorofluorocarbon is generally so slow that it is of no consequence in most applications. For example, for an unfaced 1.0-in (2.54-cm) sample of extruded polystyrene (XEPS) the second mechanism takes approximately 100 years to complete.[4] For thicker products and those with well-bonded low-permeance facings, diffusion of the chlorofluorocarbon out of the product can involve a much longer time frame.

The rate at which each mechanism occurs for a given product depends on the rate of diffusion of air and chlorofluorocarbons through the particular cellular matrix. The diffusion rates through the various types of cellular materials are different. For example, in general the diffusion of air and chlorofluorocarbons through polyurethane and polyisocyanurate foams proceeds at a slower rate than diffusion through extruded polystyrene (XEPS) foams, as illustrated in Fig. 4.1, and diffusion through closed-cell phenolic foams is slower than that through polyurethane and polyisocyanurate foams. To retard diffusion effectively, the percentage of open cells in the product must be very low. If the cells are not completely closed, the aging phenomenon can take place much more rapidly.[5]

The timing of each mechanism also depends on the gas-permeation resistance of any facings attached to the foam core, the quality of the bond between the facings and the core, and the thickness of the product. If a chlorofluorocarbon-blown foam is totally encapsulated with an effective gas-diffusion barrier during or immediately following manufacture, then air cannot enter the product and the chlorofluorocarbon cannot exit. The *R* value of such a product could be as great as 9.00 F · ft² · h/Btu per inch (62.5 K · m²/W per meter). However, in practice most products are not totally encapsulated and aging does occur. Even for those products in which the major surfaces are protected by aluminum foil or foil-composite facings, aging can be significant.[6,7]

In most cases the aluminum foils used on such products are too thin to be totally free of pinholes. As a result diffusion of gases is only impeded to some degree depending on the thickness of the foil. Generally a foil must be at least 1.0 mil in thickness before it can be considered pinhole-free and hence a true diffusion barrier.[8] However, some foil or foil-composite facings may act as diffusion barriers at lower thicknesses, and in some cases the 1.0-mil thickness may not provide a pinhole-free barrier. Most foil products used in facings range between 0.3 and 1.0 mil in thickness. Generally facings of this type should be considered retarders, not barriers. Also, in some cases facings are made of materials which are not effective barriers, such as kraft paper and thin polyethylene film. Such facings should not be considered barriers either. The impact of the barrier properties of a facing on the *R* value of polyurethane and polyisocyanurate foams is indicated in Table 4.2.

The quality of the bond between the facing and the core material is also important. If the facing is not properly laminated to the core or the bond is in some way broken, diffusion can take place, which can increase the rate at which aging occurs.[8,9] In addition, product thickness also has an effect on aging: thicker products age more slowly. Diffusion of air into the cellular matrix and chlorofluorocarbons out of the product takes longer for thicker products because there is more material through which the gases must diffuse.

Mean Temperature Effects The thermal resistance of all cellular insulations changes with the mean temperature. Generally speaking, at lower mean temperatures the *R* value is greater, as indicated in Table 4.2 and Fig. 4.2. The thermal performance of the insulations generally improves as the mean temperature decreases because the thermal conductivities of the cell gases decrease with temperature. Like aging effects, mean temperature effects should also be considered in the design or selection of insulation systems.

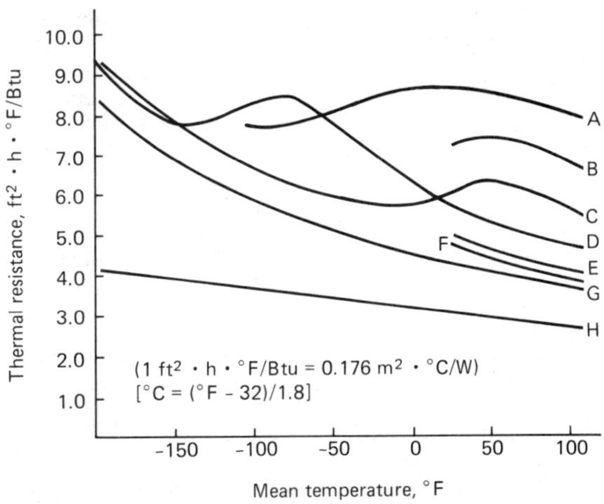

Fig. 4.2 Thermal resistance versus mean temperature for 1.0-in (2.54-cm) materials aged at ambient conditions. *A*: phenolic, 2.0–3.0 lb/ft³ (32.0–48.0 kg/m³), closed-cell; *B*: polyurethane and polyisocyanurate, 1.7–3.0 lb/ft³ (27.2–48.0 kg/m³), low-permeance facings; *C*: polyurethane and polyisocyanurate, 1.7–3.0 lb/ft³ (27.2–48.0 kg/m³), unfaced or high-permeance facings; *D*: expanded polystyrene (extruded), 1.4–4.0 lb/ft³ (22.4–64.0 kg/m³); *E*: phenolic, 2.0–3.0 lb/ft³ (32.0–48.0 kg/m³); open-cell; *F*: urea formaldehyde, 0.8–1.2 lb/ft³ (12.8–19.2 kg/m³); *G*: expanded polystyrene (molded), 1.0 lb/ft³ (16.0 kg/m³); *H*: cellular glass, 8.5 lb/ft (136 kg/m³). *Notes:* (1) The data reported are for the purposes of general comparison and the values may not be representative of every product available in the marketplace. The range in values may be considerable due to the variability between specific products within a given material type and other factors relating to the testing of the products. In general, values are intended to be representative of long-term performance. However, values may not be appropriate for all products in the marketplace due to lack of long-term data and uncertainties regarding product quality and aging conditions. (2) The curve *A* is based on limited measured data from Ref. 14. There is uncertainty regarding the reason for the large temperature range between the inflection points for the phenolic foam product (at approximately 25 and −100°F) compared with the corresponding ranges shown in curves *C* and *D*. The differences in the relative magnitudes of the temperature ranges shown in curve *A* and curves *C* and *D* may be the result of differences in experimental procedure.

As shown in Fig. 4.2, the curves for extruded polystyrene (XEPS), unfaced polyurethane and polyisocyanurate, and phenolic foam products have inflection points which occur around −75 and −150°F, 50 and −25°F, and 25 and −100°F, respectively. The occurrence of inflection points in a curve indicates the presence of a chlorofluorocarbon [which explains why there is no peak in the curve for the molded polystyrene (MEPS) product, for example]. The specific location of the inflection points is characteristic of the particular chlorofluorocarbon that is contained in the foam. For example, refrigerant-11 is generally used with polyurethane and polyisocyanurate foams and refrigerant-12 with extruded polystyrene foams, so it is not surprising that the inflection points occur at different locations for these products.

Over the mean temperature range bounded by the inflection points, the R value decreases as the mean temperature decreases. This phenomenon is observed because the chlorofluorocarbons condense from a vapor to a liquid over this range of mean temperatures. As the condensation continues, the cell gas atmosphere becomes increasingly rich in air, which is a poor insulator relative to the chlorofluorocarbons. Therefore the R value of the product decreases. Once all the chlorofluorocarbon has condensed, then the R value again increases with decreasing mean temperature.

Thickness Effects For low-density cellular insulations the R value per inch of a given material may be lower for products of greater thickness. In other words, the thermal resistance of a 3.0-in (7.62-cm) product, for example, may not be as great as the resistance of a similar 1.0-in (2.54-cm) product multiplied by 3. The effect is primarily caused by the fact that the radiation component of thermal transmission can be different for similar products of different thicknesses. The transfer of heat by radiation through cellular thermal insulations is influenced by several factors including the material that constitutes the solid phase; the product thickness, density, and cell size; and the mean temperature. For cellular insulations the reduction in R value attributable to the thickness effect can be as great as 5 percent of the thermal resistance of the product (or perhaps even greater in some cases).[10,11,12] For more information contact the insulation manufacturer.

Other Considerations When deciding which cellular thermal insulation is most appropriate for a particular application, it is important to compare products on an equal basis. As far as product aging is concerned, the comparison should generally be made with R values that reflect the first mechanism of aging, that is, intrusion of air. Several sample conditioning procedures have been used in the industry in an attempt to represent this aging condition: 180-day aging at 75°F (24°C) and 50 percent relative humidity, 90-day aging at 140°F (60°C), and 5-year aging (or longer) at 75°F (24°C). Whether or not these procedures represent the first mechanism of aging accurately depends on many factors which relate to the rate at which air and chlorofluorocarbons pass into and out of the products. In general, R values based on conditioning procedures involving long time periods or high temperatures are more indicative of long-term performance.

It is also important to compare the thermal resistances of the insulations at the same mean temperature. The particular mean temperature selected should be representative of the particular end use in question. This affects the comparison

Table 4.3. Typical Moisture Transmission and Absorption Properties[a]

	Expanded polystyrene					
Property	Molded, ρ = 1.0 lb/ft^3[b]	Molded, ρ = 2.0 lb/ft^3	Extruded, ρ = 1.4–4.0 lb/ft^3	Polyurethane and polyisocyanurate ρ = 1.7–3.0 lb/ft^3	Phenolic ρ = 2.0–3.0 lb/ft^3	Cellular glass ρ = 8.5 lb/ft^3
Apparent water vapor permeability,[c] perm · in[d]	1.2–3.0	0.6–1.5	0.4–1.4	2.0–4.0	1.4–3.5	0.0
Water absorption, vol %						
ASTM C272 (24-h immersion)	<2.5	<1.0	<0.1	2–5	5–10	Nonabsorptive; only moisture retained is that adhering to surface cells after immersion.
ASTM D2842 (96-h immersion)	1–6	1–4	1	0–2	5–20	
ASTM D2842 (500-h immersion)	4–7	1–5	1	1–4	15–40	
Water vapor diffusion[e] (28-day exposure)	8–25	4–25	1–4	5–40	10–60	
Freeze-thaw cycling[f] (1000-cycle exposure)	15–80	15–50	2–10	4–90[g]	50–70	70[h]

[a] The data reported are for the purposes of general comparison, and the values may not be representative of every product available in the marketplace. The range in values may be considerable due to the variability between specific products within a given material type and other factors relating to the testing of the products.

[b] 1 lb/ft^3 = 16.0 kg/m^3.

[c] Values are based on the cellular material only. The effects of facings are not included. Values are determined using the desiccant method as described in ASTM E96.

[d] 1 perm · in = 1.46 ng/Pa · s · m.

[e] Materials are subjected to both thermal and water vapor pressure gradients. Specimens are placed between a cooling plate at 34°F (1°C) and a water bath in which the temperature is regulated to maintain a temperature gradient of 46°F per inch (10°C per cm) of thickness. See Swiss standard SIA 279.

[f] Materials are subjected to freeze-thaw cycling by freezing in air at 0°F (−18°C) followed by thawing in water at 40°F (4°C). See ASTM C666.

[g] The higher water absorption levels are representative of materials with glass fiber reinforcement, and in many cases the test must be terminated prematurely because of excessive damage to the specimens.

[h] Generally the test must be terminated prematurely because of excessive damage to the specimens.

because the thermal performance of one product with respect to another can change over a given range of mean temperatures. For example, the *R* value per inch of extruded polystyrene at a mean temperature of approximately 10°F (−12°C) is equivalent to that of unfaced polyurethane and polyisocyanurate. This is not the case at a mean temperature of 75°F (24°C), as shown in Fig. 4.2. The particular mean temperature selected also establishes the *R* value per inch that is used to determine the thickness of insulation required. To simplify matters somewhat a mean temperature of 75°F (24°C) is often used for purposes of comparison because it is widely accepted as a standard within the industry. Therefore, most manufacturers report *R* values for their products at this mean temperature, even though it is not representative for many applications.

Moisture Transmission and Absorption

Moisture can have an adverse effect on the thermal transmission (and perhaps mechanical strength) properties of cellular thermal insulations. In most applications, good system design and proper installation of vapor retarders minimize the danger of moisture accumulation in insulating materials. Despite these precautions it is still possible (and in some applications expected) that insulations will be exposed to moisture. Therefore, it is important to consider the moisture transmission and absorption characteristics of insulating materials. These properties are given in Tables 4.1 and 4.3 for cellular thermal insulations. Compared with other common insulating materials, such as fibrous and granular insulations, cellular insulations exhibit good resistance to moisture intrusion.

The values for apparent water vapor permeability shown in the tables apply to unfaced materials. When facings are added, which is the case with many products, they may be rendered essentially impermeable, depending on the barrier qualities of the facings or laminates. In general, products with a water vapor permeance of 0.5 to 1.0 perm (28.7 to 57.4 ng/Pa · s · m^2) or less are considered vapor retarders, so some cellular insulations may be installed in certain applications without any moisture protection. For a given product type, higher-density materials generally have lower permeability and absorption values. However, this is not always true because these properties are also dependent on other factors such as the integrity of the cellular matrix. For example, some closed-cell foams often contain interstitial spaces or voids between the cells. As a result, these products can exhibit some of the characteristics of open-cell foams, such as relatively high water absorption values, depending on the number and size of the voids.

It is important to consider the severity of the moisture exposure expected for a given application. The severity of the exposure to which the insulation is to be subjected is a major factor in determining which material is selected and what protective measures are appropriate, if any. Various moisture transmission and absorption tests are available to help assess the moisture resistance of various insulations. Some of these tests are listed in Table 4.3. Results from such tests should not be used indiscriminately, however. The tests differ greatly with regard to the severity of the moisture exposure involved, so care should be taken to ensure that the method or methods in question relate to the intended end use.

As mentioned previously, the thermal performance of insulations can be adversely affected by the presence of moisture, depending on the amount present and its location. This is because the thermal conductivities of liquid water and especially ice are significantly greater than that of the insulation. Figures 4.3*a* and 4.3*b* show the water absorption and *R*-value retention of several cellular insulations following a 14-day exposure in a conditioning apparatus where specimens were subjected to both water vapor pressure and thermal gradients.

Mechanical Strength

Approximate mechanical strength properties for cellular thermal insulations are presented in Table 4.4. As mentioned previously, the range in the values for a given property may be considerable due to the variability between specific products and differences in test procedures and test specimen dimensions. As a result, the data reported may not be representative of all the materials available. Some materials, extruded polystyrene and polyurethane and polyisocyanurate products, for example, exhibit anisotropic behavior, that

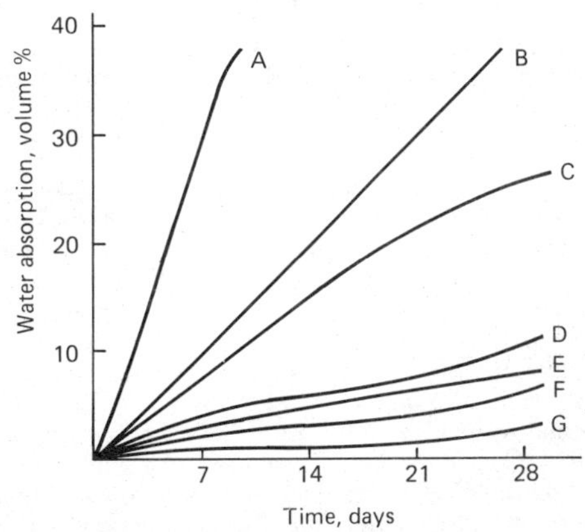

Fig. 4.3a Water absorption versus time for 2.0-in (5.08-cm) materials subjected to thermal and water vapor pressure gradients in a water vapor diffusion test (modified Swiss standard SIA 279). *A*: phenolic, 2.5–3.0 lb/ft^3 (40.0–48.0 kg/m^3), open-cell, perforated and corrugated kraft paper/aluminum foil facings. *B*: polyurethane, 2.4, lb/ft^3 (38.4 kg/m^3), kraft paper facings. *C*: expanded polystyrene (molded), 1.6 lb/ft^3 (25.6 kg/m^3). *D*: phenolic, 3.0 lb/ft^3 (48.0 kg/m^3), closed-cell, corrugated kraft paper/aluminum foil facing on one surface, flat kraft paper facing on the other. *E*: expanded polystyrene (molded), 0.9 lb/ft^3 (14.4 kg/m^3). *F*: polyisocyanurate, 2.0 lb/ft^3 (32.0 kg/m^3), aluminum foil facings. *G*: expanded polystyrene (extruded), 1.4–4.0 lb/ft^3 (22.4–64.0 kg/m^3). *Note:* The data reported are for the purposes of general comparison, and the values may not be representative of every product available in the marketplace. The range in values may be considerable due to the variability between specific products within a given material type and other factors relating to the testing of the products. Water absorption data for materials *B* and *C* are taken from Ref. 15. Other data are taken from internal reports of The Dow Chemical Company.

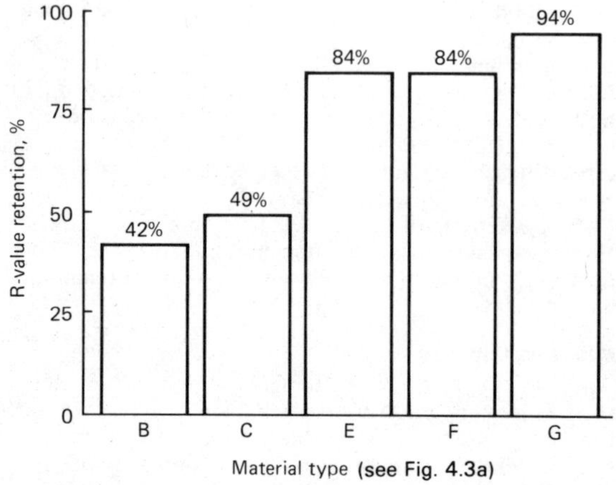

Fig. 4.3b *R*-value retention for 2.0-in (5.08-cm) materials subjected to thermal and water vapor pressure gradients for 14 days in a water vapor diffusion test (modified Swiss standard SIA 279). *Note:* The data reported are for the purposes of general comparison and the values may not be representative of every product available in the marketplace. The range in values may be considerable due to the variability between specific products within a given material type and other factors relating to the testing of the products. *R*-value retention is calculated using equations developed by Dechow and Epstein, which are given in Ref. 15. Data for materials *A* and *D* are not shown because equations for predicting the thermal transmission properties of phenolic foam materials containing moisture are not included in Ref. 15.

is, the mechanical properties differ according to the direction of measurement. Hence, the direction of measurement, either parallel or perpendicular to rise, is indicated in the table. Generally speaking, both the length and width directions are perpendicular to rise. (Rise is the thickness direction, which is the principal direction of expansion.) In some cases there may be differences in the values measured in the length and width directions. Such differences have not been identified to keep the table as simple as possible.

The facings that are laminated or bonded to some materials can increase the strength of the products significantly as well as help prevent the foam core from being damaged. For some properties such as tensile strength, the strength of the composite can be reduced if the bond between the facing and the core is poor. The values shown in the table apply to unfaced materials. For data on faced products consult the insulation manufacturer. In addition to the obvious variation in properties with density and product type, the manufacturing process and related factors can also be important. At similar densities extruded polystyrene (XEPS) products are stronger than molded bead (MEPS) products; and the strength properties of rigid polyurethane and polyisocyanurate boards and buns, which are reflected in Table 4.4, may be different than those of spray-in-place and pour-in-place products. The strengths and moduli also generally increase with decreasing temperature. For data at other temperatures within the ranges shown, consult Ref. 13 and contact the insulation manufacturer.

Miscellaneous

Maximum use temperature and solvent resistance must be considered for many applications. In general, thermosetting foams can be used at higher temperatures and exhibit greater solvent resistance than thermoplastic foams. Cellular glass products can be used at higher temperatures than other cellular insulations, and they also have excellent solvent-resistance properties. Dimensional stability may also be a significant factor, especially if the materials are exposed to high temperatures and humidities. Urea formaldehyde foams generally exhibit considerable shrinkage as they cure, and this should be taken into account when these products are used. Polyisocyanurate foams typically exhibit improved dimensional stability at high temperatures compared with polyurethane foamed plastics. In certain applications corrosivity may also be a concern with some phenolic foams because of the acid catalyst used in the reaction to produce the polymer.

In regard to product handleability, friability should also be taken into consideration. In general, thermosetting foamed plastics are more friable than thermoplastic foams, which is one of the reasons why many of the thermosetting products are offered with facings. Also, prolonged exposure to solar radiation will result in the deterioration of the surface of unfaced foamed plastics. The severity of the degradation will vary depending on the foam type and specific formulation, and the intensity of the solar exposure. The surface degradation will not have a measurable effect on the insulating value of the products unless deterioration is allowed to continue until the thickness of the foam is reduced measurably. All products, faced and unfaced, should be protected during storage and after installation with a covering or coating that is opaque to ultraviolet radiation.

4.4 APPLICATIONS

The largest markets for cellular thermal insulations are building construction applications involving both new and retrofit construction in residential, commercial, agricultural, and industrial structures. Cellular insulations in board form are routinely installed in conventional built-up roofs and beneath single-ply membranes; and some products, namely, extruded polystyrene, are installed above the membrane in protected membrane roof applications. Insulating boards are also used in cathedral ceilings in residential and light commercial construction. They are often used as sheathing on wood-frame, metal-frame, and masonry walls in a variety of systems, which are installed on both the interior and the exterior of the structures. They are also widely used as core materials in a variety of sandwich-panel constructions. In addition, these products are used as foundation insulations in basement, crawl-space, and slab-on-grade applications. Cellular insulating boards are the predominant insulations used in the walls, roofs, and floors of low-temperature storage and processing facilities. In these applications they are installed as boards individually or incorporated in various sandwich-panel systems. These products are also widely used as insulation in metal buildings and agricultural buildings.

Polyurethane, urea formaldehyde, and other urea-based foams are poured into wood-frame, metal-frame, and masonry wall cavities, and they are also used as core-fill insu-

Table 4.4. Typical Mechanical Strength Properties,* lb/in²†

Property	Expanded polystyrene (molded)		Expanded polystyrene (extruded)				Polyurethane and polyisocyanurate				Urea formaldehyde	Phenolic	Cellular glass	
	ρ = 1.0 lb/ft³,‡ @75°F§	ρ = 2.0 lb/ft³, @75°F	ρ = 1.4 lb/ft³, @75°F	ρ = 2.0 lb/ft³ @75°F	ρ = 2.0 lb/ft³ @−325°F	ρ = 4.0 lb/ft³, @75°F	ρ = 2.0 lb/ft³, @75°F	ρ = 2.0 lb/ft³, @−325°F	ρ = 6.0 lb/ft³, @75°F	ρ = 6.0 lb/ft³, @−325°F	ρ = 0.8–1.2 lb/ft³, @75°F	ρ = 2.0–3.0 lb/ft³, @75°F	ρ = 8.5 lb/ft³, @75°F	ρ = 12.0 lb/ft³, @75°F
Compressive strength, parallel to rise (thickness)	10–14	25–33	15–25	25–50	45–65	115–125	25–35	40–50	100	190	5	10–35	100	210
Compressive modulus, parallel to rise, (thickness), (× 10³)	0.2	0.5	—	1.3	2.5	3.0	0.9	2.0	2.6	10.5	—	1.0	150	180
Flexural strength, perpendicular to rise	25–30	50–75	40–70	50–100	—	100–140	50	—	210	—	15	25	80	110
Shear strength, perpendicular to rise	20	35	—	35	30	70	20	20	100	175	—	12	50	—
Tensile strength, parallel to rise (thickness)	18	25	25	50	25	125	45	25	140	200	1	20	50	—
Tensile modulus, parallel to rise (thickness), (× 10³)	0.2	0.5	—	3.2	4.2	4.7	1.6	3.0	4.0	10	—	0.2	—	—

* The data reported are for the purposes of general comparison, and the values may not be representative of every product available in the marketplace. The range in values may be considerable due to the variability between specific products within a given material type and other factors relating to the testing of the products.

† 1 lb/in² = 0.006895 MPa.

‡ 1 lb/ft³ = 16.0 kg/m³.

§ °C = (°F − 32)/1.8.

lation for concrete masonry units. Spray-applied polyurethane and polyisocyanurate foams are used as roof insulation and are often used to insulate irregularly shaped building envelope components, which would be difficult to insulate with rigid boardstock. Polyurethane foam, sprayed or poured in place, is sometimes used as an air infiltration barrier. Rigid-board and spray-applied cellular insulating materials are also widely used in industrial applications as tank, vessel, and pipe insulation in a great variety of systems and processes. In addition, fabricated products for insulating pipes and pipe fittings are routinely prepared from large billets.

Cellular insulating boards have been used to insulate highways, airport runways, and railroad embankments to prevent the earth from freezing and thereby to eliminate damage from frost heaving. In some cases these materials are installed to prevent the earth from thawing in order to avoid large seasonal changes in the load-bearing capacity of the underlying soils. Cellular insulations are also occasionally used to insulate water and sewage utility lines and oil and gas pipelines. These products are often used in appliances such as refrigerators and freezers as well as in vehicles used to transport foods and other goods that require refrigeration. They are also typically incorporated in structural panels which are used in the manufacture of recreational vehicles.

The various cellular insulations present unique combinations of physical properties and cost. They all can be used in many applications, although no one material is recommended for use in every application. The appropriateness of a material for a given application can only be determined after the physical properties of the product and the demands of the end use have been reviewed carefully.

4.5 HEALTH AND SAFETY CONSIDERATIONS

With the exception of cellular glass, the cellular thermal insulations discussed in this chapter are combustible, as indicated in Table 4.1, and may constitute a fire hazard if improperly used or installed. The combustibility characteristics for the various types of foamed plastics are different, and the presence of various facings can influence these characteristics. Phenolic foams are recognized for their superior fire-resistance properties, and polyisocyanurate foam insulations also exhibit improved flammability characteristics. Polystyrene foams typically melt when exposed to a fire, which helps account for the low flame-spread and smoke-developed ratings for these materials. Many of these materials contain flame-retardant additives to inhibit accidental ignition during storage and after installation. Those that do not are inherently fire resistant, such as phenolic foams.

The use of cellular thermal insulations in construction applications is regulated by the various building codes. Various small-scale and large-scale fire tests on products and systems incorporating these materials may be required depending on the application, and in most cases a thermal barrier must be installed to protect the foamed plastics. Maximum limits for the flame-spread and smoke-developed ratings of cellular insulations are also included in the codes. The numerical flame-spread and smoke-developed ratings presented in Table 4.1 are not intended to reflect hazards presented by cellular thermal insulations under actual fire conditions. The values are indicative of the response of these materials to heat and flame under controlled laboratory conditions. To describe or appraise the hazard under actual fire conditions, several other factors should also be considered.

The products of combustion for cellular plastic insulations are also important to consider. Toxicity of combustion products is an extremely complex issue, and at the present time there is no universally accepted test for toxicity. Most experts agree, however, that the foamed plastics used as thermal insulations are no more toxic than other commonly used combustible materials. The production of carbon monoxide and the generation of smoke during fires are the two issues of primary concern. For some cellular plastic insulations relatively small amounts of other toxic gases are generated during the course of combustion, but they are not believed to have a major impact on overall toxicity.

Some products contain additives or unreacted components that may be a concern in applications where direct contact with food is anticipated. Also, formaldehyde vapor may be emitted from urea-formaldehyde foams (as well as other materials) as a result of a variety of factors relating to the formulation, foaming process, and end-use conditions. Above a certain level the vapor is a strong irritant and can cause illness.

REFERENCES

1. R. E. Skochdopole, "Thermal Conductivity of Foamed Plastics," *Chem. Eng. Prog.*, **57,** 55–59 (Oct. 1961).
2. L. D. Booth and W. M. Lee, "Effects of Polymer Structure on K-Factor Aging of Rigid Polyurethane Foam," *J. Cellular Plastics,* Jan.–Feb. 1985, pp. 26–30.
3. F. J. Norton, "Thermal Conductivity and Life of Polymer Foams," *J. Cellular Plastics,* **3,** 23–27 (Jan. 1967).
4. Proprietary computer model developed by The Dow Chemical Company to study thermal transmission characteristics of cellular plastic insulations.
5. S. E. Johnson and S. P. Muhlenkamp, "In-Place Thermal Aging of Polyurethane Foam Roof Insulations," *Proc. 7th Conf. on Roofing Technology,* National Bureau of Standards/National Roofing Contractors Assoc., Chicago, IL, Apr. 1983, pp. 49–55.
6. G. F. Baumann, "*R*-Value of Rigid Polyurethane Foam—An SPI Research Project," *J. Thermal Insulation,* **6,** pp. 39–47 (July 1982).
7. A. O. Desjarlais and R. P. Tye, "Experimental Methods of Determining the Thermal Performance of Cellular Plastic Insulation Materials Used in Roofs," *Proc. 8th Conf. on Roofing Technology,* National Bureau of Standards/National Roofing Contractors Assoc., Chicago, IL. Apr. 1987, pp. 12–22.
8. M. R. Ascough, S. A. Baitinger, and K. T. Dishart, "Barrier Packaging Technology—A New Approach to the Thermal Aging Problem of Rigid Foam Insulation," *Proc. 30th Ann. Polyurethane Technical/Marketing Conf.,* Society of the Plastics Industry, Oct. 1986, pp. 388–392.
9. L. R. Glicksman and A. G. Ostrogorsky, "Laboratory Tests of the Effectiveness of Diffusion Barriers," *J. Cell. Plastics,* **22,** 303–313 (July 1986).
10. W. F. Cammerer, "Thermal Conductivity as a Function of the Thickness of Insulating Materials," *International Institute of Refrigeration Bulletin* (*Annexe*), Issue No. 4, 1973, pp. 189–200.
11. T. T. Jones, "The Effect of Thickness and Temperature on Heat Transfer through Foamed Polymers," *Proc. 7th Conf. on Thermal Conductivity,* National Bureau of Standards, Washington, D.C., Sept. 1968, pp. 737–748.
12. C. J. Shirtliffe, "Effect of Thickness on the Thermal Properties of Thick Specimens of Low Density Thermal Insulation," in *Ther-*

mal Insulation Performance, D. L. McElroy and R. P. Tye (eds.), ASTM STP 718, American Society for Testing and Materials, 1980, pp. 36–50.

13. L. L. Sparks, "Low-Temperature Properties of Expanded Polyurethane and Polystyrene," in *Thermal Insulation Performance,* D. L. McElroy and R. P. Tye (eds.), ASTM STP 718, American Society for Testing and Materials, Philadelphia, PA, 1980, pp. 431–452.
14. Private communication from E. W. Kifer, Koppers Company, Inc., October 5, 1987.
15. F. J. Dechow and K. A. Epstein, "Laboratory and Field Investigations of Moisture Absorption and Its Effect on Thermal Performance of Various Insulations," in *Thermal Transmission Measurements of Insulation,* R. P. Tye (ed.), ASTM STP 660, American Society for Testing and Materials, Philadelphia, PA, 1978, pp. 234–260.

Chapter 5

Reflective Insulation

KIM T. GILBERT
Mirror Insulation Unit,
Diamond Power Specialty Company

5.1 INTRODUCTION

Reflective insulation is conceptualized as a series of thin sheets of metal with low thermal emittance, which are spaced and oriented to "reflect" radiant energy back toward the thermal source. These sheets are usually positioned parallel to each other and may also be arranged, as a system, to minimize the effects of natural convection and conduction within the insulation. One possible arrangement is shown in Fig. 5.1. The thermal resistance attributed to reflective insulation is strongly dependent on the materials used in construction, but clearly it is also dependent on the assembly geometry. The theory of reflective insulation is well developed and thermal performance predictions account for conduction, convection, and radiation heat-transfer modes that are present, in varying proportions, within any insulation product. However, because reflective insulation is not a homogeneous substance, special test methods, performance parameters, and vocabulary have been developed to describe its use. For some specific applications, reflective insulation exhibits thermal performance equivalent to more common mass insulations. The unique features of a metal insulation product can make reflective insulation a desirable alternative.

5.2 TYPICAL APPLICATIONS

Nonevacuated reflective insulation has found wide use in nuclear power containments, where it has been employed to insulate piping of all diameters, valves, and pressure vessels. This type of insulation, constructed entirely of sheet metal with low emittance, has also been supplied to fossil power plants and to chemical and petrochemical as well as food processing industries. Using the same heat-transfer principles in different settings, reflective insulation has been supplied for building insulation to replace or complement mass insulations. Typical applications provide one or more alumi-

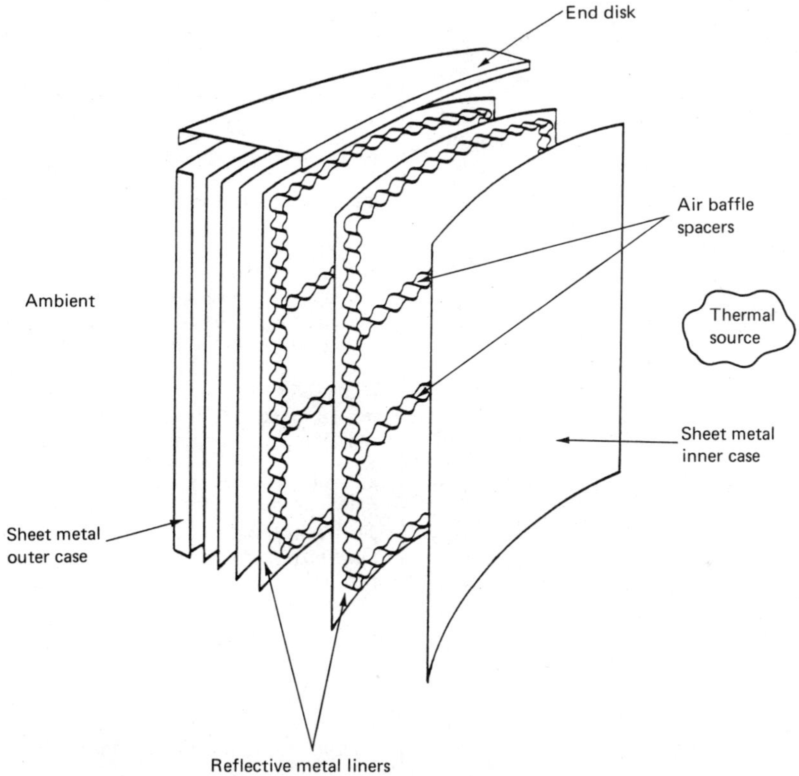

Fig. 5.1 Construction of prefabricated reflective insulation panel.

num foils bonded to paper or polymer sheets to provide mechanical strength. Aluminum foil is also used as facing material for mass insulations to reduce air infiltration and water vapor transport.[1] Typical building applications include installation in wall cavities, between ceiling and floor joists, in roofing systems, and behind paneling mounted on masonry walls. Finally, NASA has developed and used reflective multilayer insulation systems for the space program, which demonstrate further potential for reflective systems.[2]

A logical extension of reflective insulation principles led to the development of insulation for cryogenic service. Such insulation achieves "superinsulation" performance by evacuating the space between the insulation reflective layers. This essentially eliminates conduction and convection heat transfers within the free space of insulation, leaving the low-emittance foils to deal efficiently with radiative heat transfer. (The subject of cryogenic insulation is discussed separately in Part 11 of this handbook.)

5.3 ADVANTAGES AND DISADVANTAGES OF REFLECTIVE INSULATION

The two major classifications of reflective insulation, building insulation and prefabricated all-metal insulation, offer a number of important advantages when compared to mass insulations. Some of the advantages are common to all reflective insulations, while others are unique to one of the classifications. The advantages are complemented by some potentially significant disadvantages. The summary provided in Table 5.1 describes advantages compared to conventional mass insulations. The comparisons indicated in Table 5.1 may or may not be significant in a specific application or when compared to a specific mass insulation alternative.

5.4 IMPORTANT PARAMETERS

Features that determine the thermal resistance of reflective insulation are material selection (determines emittance), uniformity of spacing between reflective sheets, distance between reflective layers, quantity of reflective layers, insulation panel size and curvature, external construction and supports, insulation orientation, internal insulation construction, and insulation durability.

Material Selection

Aluminum (including aluminum foil) and stainless steel have been the predominant choices for use in nonevacuated reflective insulation. Since thermal resistance is a strong function of material emittance (resistance decreases as emittance increases) and because emittance increases due to both elevated temperature and time, it is necessary that tests be conducted on each material to determine the relationship

Table 5.1. Advantages and Disadvantages of Reflective Insulation

Characteristic	Reflective building insulation	Prefabricated all-metal reflective insulation
Advantages		
Resistance to rot and decay	Yes	Yes
Built-in moisture vapor barrier	Yes	N/A
Low weight or bulk for equivalent insulation	Yes	Compare
Dustless and noncontaminating or irritating	Yes	Yes
Nonabsorbent	No	Yes
Self-draining	No	Yes
Nonflammable	Compare	Yes
Nuclear radiation resistant	N/A	Yes
Resistant to damage from external load or vibration	No	Yes
Removable or permanent designs available	No	Yes
Simple installation or self-supporting	Yes	Compare
Resistance to solvents and corrosive atmospheres	N/A	Yes
Disadvantages		
Thermal performance strongly dependent on method of construction	Yes	Yes
Thermal performance orientation dependent	Yes	Yes
Susceptible to increased convective effects	N/A	Yes
Particular attention to be given to penetration sealing	Yes	Yes
Susceptible to damage during handling and installation	Yes	Yes
Relatively high outer (cold) surface temperatures compared to fabric-covered or painted surfaces	N/A	Yes
Custom geometric design required for good fit	N/A	Compare
Thermal performance a strong function of insulation mean temperature	Compare	Compare

between emittance, time, and temperature. ASTM C835 is the standard test method for determining total hemispherical emittance.[3] Some test results conducted in accordance with this ASTM guide are summarized in Table 5.2.

Spacing Between Reflective Layers

The effect of natural convection between and over reflective layers can be important even though the spaces between layers are narrow or the paths for airflow are irregular. Proper design should provide a method for maintaining uniform spacing between liners. The spacing between reflective layers varies substantially depending on the application. Reflective building insulations are usually offered with spaces between reflective surfaces of 0.375 in (10 mm) or 0.75 in (19 mm). Reflective insulations used in nuclear power applications have spacings between reflective sheets ranging from 0.25 in (6 mm) to 0.5 in (13 mm), with the most common spacing being 0.33 in (8 mm).

Number of Reflective Layers

Theory indicates that the radiative component of heat transfer between infinite parallel planes of emittance ϵ, separated by a quantity of k reflective shields of emittance ϵ, is decreased by a factor equal to $1/(k + 1)$ when compared to the case of two infinite parallel planes with no reflective shields between them.[4] This relationship does not include conduction and convection effects. Because of the inverse relationship between the number of reflective layers and the radiation heat flux through the insulation, there is a diminishing return for each additional layer. As a point of reference, for a 3.0-in (76-mm) thickness of pipe insulation, an increase in the number of reflective layers from 10 to 25 results in a 22.0 percent reduction in heat flux for a typical reflective construction.

Insulation Panel Size and External Construction

The contribution of insulation panel joints to the total heat flux through fabricated insulation panels is determined by the design of the joints. These joints routinely provide the structural connection that holds the individual panels together. The material thickness and design details of the structural connections combine with the physical panel size to determine the percentage contribution of the panel joints to the total heat flow through the insulation. Insulation design is a compromise among thermal performance, structural requirements, and insulation installation access.

Insulation Orientation

Thermal radiation and conduction are essentially independent of the insulation orientation, but natural convection be-

Table 5.2. Material Emittance for Materials Routinely Used in Reflective Insulation

Aging time (at temperature), h	Temperature °F	Temperature K	Emittance, dimensionless
304 Stainless, B.A. Finish			
0	500	533	0.147 ± 0.004
8,000	500	533	0.142 ± 0.004
21,000	500	533	0.149 ± 0.004
36,000	500	533	0.154 ± 0.004
0	900	755	0.179 ± 0.005
8,000	900	755	0.243 ± 0.005
21,000	900	755	0.251 ± 0.005
36,000	900	755	0.252 ± 0.005
3003H16 Aluminum, Bright Finish			
0	300	422	0.042 ± 0.002
7,000	300	422	0.048 ± 0.002
15,000	300	422	0.049 ± 0.002
0	750	672	0.056 ± 0.001
6,000	750	672	0.062 ± 0.001
18,000	750	672	0.073 ± 0.002

Table 5.3. Relative Heat Loss as a Function of Insulation Orientation

Insulation orientation	Typical relative heat loss
Building Insulation*	
Horizontal, heated from above	1.0
Horizontal, heated from below	2.8
Vertical	3.75
All Metal, Factory Prefabricated†	
Horizontal, heated from above	1.0
Horizontal, heated from below	1.24
Vertical‡	1.32

* Multiple-layer radiant barrier insulation used in building construction, based on data from *ASHRAE Handbook, 1981 Fundamentals,* used with permission.

† Hot surface temperature 500°F (533 K), ambient temperature 100°F (311 K), 3 in (7.6 cm) thick.

‡ Internal construction designed to reduce natural convection by reducing dimensions of cells where convection occurs.

tween reflective layers is orientation-dependent and can have a significant impact on insulation performance. Relative heat flux data for two fabricated insulation panels are given in Table 5.3.

Internal Construction

Convection stops can be used to reduce natural convection effects (see air baffle spacers in Fig. 5.1). The convection barriers create smaller cells within the insulation and thereby reduce convection between the foils and around the edge of the foils. For heat flow through vertically oriented reflective insulation, tests have shown a reduction in heat flux of approximately 20 percent due to such convection stops.

Insulation Durability

Consistent thermal performance depends on the ability of the reflective insulation system to withstand damage during its installation and service life. For piping and vessel insulation, reflective insulations can be designed to support foot traffic and maintenance activities without sacrificing the insulating value.

5.5 THERMAL PERFORMANCE

One factor that has limited the acceptance of reflective insulation is that thermal resistance is not an intrinsic material property. Uncertainty about the thermal resistance stems from the wide variations in the methods of construction used by the various manufacturers. In addition, the point of application of the insulation (including insulation mean temperature and orientation) determines the relative proportion that each mode of heat transfer will contribute to the total for a given construction. In order to account for all three modes of heat transfer properly, including the effect of joints, service conditions, and installation, a thermal resistance or overall heat-transfer coefficient must be used to predict performance.

Reflective Building Insulation

The steady-state thermal performance of reflective building insulation assemblies is measured according to ASTM C236[5] or ASTM C976.[6] Results are usually reported in terms of thermal resistance R in units of h · ft² · °F/Btu (K · m²/W). The products are labeled as having a specified R factor, which is analogous to the R factors specified for mass insulations (the higher the R value, the higher the thermal resistance). The overall thermal resistance R_u may be obtained from the measured resistance R by adding the outside surface resistance typical of the service conditions. State-of-the-art theoretical prediction methods and representative test results are presented in Ref. 1. Performance data from a variety of sources clearly indicate that reflective building insulation can be an effective thermal barrier. Indications are that combinations of reflective and fibrous insulation will see increased application.

Prefabricated Reflective Panels

The thermal performance of prefabricated metallic insulation is reported in terms of thermal transference Tr. Measurement procedures are defined in ASTM C691.[7] By definition, Tr includes the effect of the cold surface heat-transfer coefficient and joint losses. Therefore, it is an overall heat-transfer coefficient that can be used to calculate heat loss through the insulation according to the simple formula[6]

$$q = \mathrm{Tr}\, A_0(t_0 - t_a)$$

where Tr = thermal transference, Btu/h · ft² · °F (W/m² · K)
q = heat loss through insulation, Btu/h (W)
A_0 = surface area of hot surface, ft² (m²)
t_0 = temperature of hot surface, °F (K)
t_a = ambient temperature, °F (K)

Reflective insulation manufacturers specify product performance in terms of this overall heat-transfer coefficient since the typical job specification requirements are written to limit "through"-insulation heat loss. Since the hot surface and ambient conditions are usually specified, a simple calculation using Tr values yields the overall average through-insulation heat loss. The main disadvantage to this approach of specifying insulation performance is that mass insulation performance values cannot be directly compared to evaluate different insulating options.

For the sake of comparison, it is possible to calculate an artificial value of thermal conductivity for a reflective insulation if thermal transference, insulation surface temperatures, and insulation geometry are known. For reflective pipe insulation, including the effects of joints, orientation, construction, etc., the apparent thermal conductivity may be calculated by[8]

$$k = \mathrm{Tr}\, A_0(t_0 - t_a)\frac{\ln (r_2/r_0)}{2\pi L(t_0 - t_2)}$$

where k = apparent thermal conductivity, Btu/h · ft · °F (W/m · K)

Tr = thermal transference, Btu/h · ft^2 · °F (W/m^2 · K)
A_0 = surface area of hot surface, ft^2 (m^2)
t_0 = temperature of hot surface, °F (K)
t_a = ambient temperature, °F (K)
r_2 = insulation outside radius, ft (m)
r_0 = insulated pipe radius, ft (m)
L = insulated pipe length, ft (m)
t_2 = average insulation cold surface temperature, °F (K)

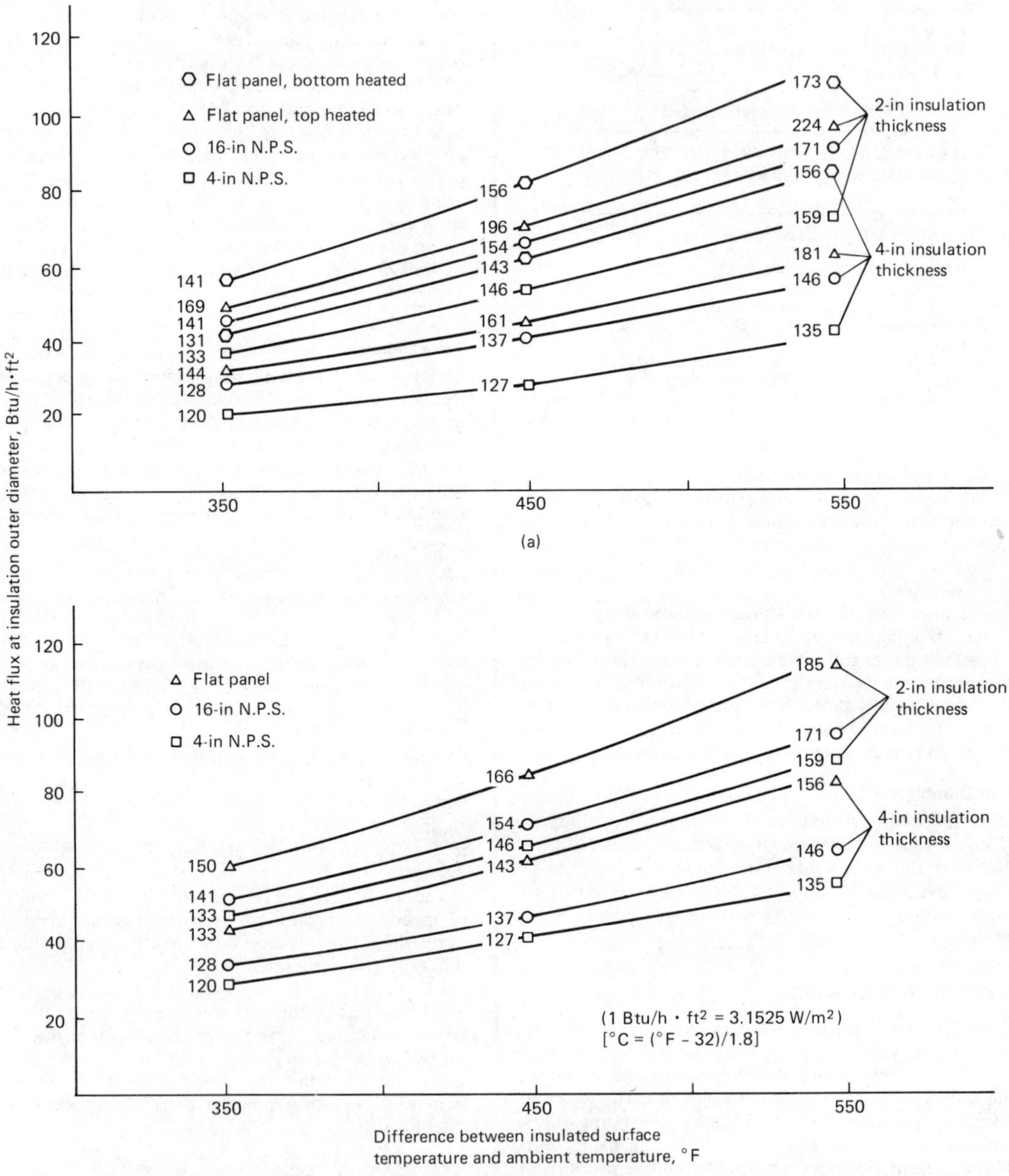

Fig. 5.2 Reflective insulation heat flux versus overall temperature difference. (*a*) Horizontal insulation; (*b*) vertical insulation. Numbers adjacent to data points are insulation cold (outside) surface temperatures in the center of the insulation panel. N.P.S. = nominal pipe size.

Table 5.4. Overall Average Effective Thermal Conductivity of Reflective Insulation, Including Joints, Installation Tolerances, and Insulation Orientation

Mean temperature, °F	Insulation thickness, in	Nominal pipe size, in	Effective conductivity, Btu/h · ft · °F
415 ± 10	2.0	4	0.051–0.065
300 ± 10	2.0	4	0.039–0.053
400 ± 10	4.0	4	0.052–0.073
290 ± 10	4.0	4	0.041–0.061
420 ± 10	2.0	16	0.056–0.063
300 ± 10	2.0	16	0.044–0.051
405 ± 10	4.0	16	0.059–0.069
295 ± 10	4.0	16	0.047–0.057
415 ± 15	2.0	Flat panel	0.046–0.051
290 ± 10	2.0	Flat panel	0.036–0.042
410 ± 10	4.0	Flat panel	0.050–0.067
290 ± 10	4.0	Flat panel	0.040–0.054

CONVERSION FACTORS: K = (°F + 460)/(1.8); 1 in = 2.54 cm; 1 Btu/h · ft · °F = 1.73 W/m · K.

Note that it is necessary to adjust values of mass insulation thermal conductivity tabulated in the literature to take into account the effects of joints, insulation thickness, insulation orientation, insulation jacketing, mean temperature, and ambient conditions before comparing any data with reflective insulation conductivities calculated by the above method.

Published data on reflective insulation thermal performance are scarce, with the majority of work being done for the insulation manufacturers by independent laboratories. One recent independent database is reported[9] where four metallic insulation samples were tested on a 6-in nominal-pipe-size test facility constructed per ASTM C691. Based on these results, a one-dimensional heat-transfer model was constructed, and good agreement between predicted and measured heat fluxes was obtained.

Samples of heat-transfer data for the construction shown in Fig. 5.1 are given in Fig. 5.2. Extrapolations of thermal performance beyond the bounds of the data or for use with other constructions are not recommended. Individual reflective insulation manufacturers should be consulted to predict the performance of their products, and a clear basis for the predictions should be established. For reference, Table 5.4 gives representative thermal conductivity values calculated by the preceding methods, based on data from Fig. 5.2.

5.6 SUMMARY

The main technical advantage of reflective insulation does not lie in a superior thermal insulation effect, but rather in an ability to maintain functionality for indefinite periods in adverse environments without causing or contributing to damage to the insulated component. Thermal insulation can be designed to be thermally efficient and still be removable, reusable, nonabsorbent, self-draining, noncontaminating, weather- and waterproof, durable, solvent and chemical resistant, self-supporting, and radiation and fire resistant.

The theory supporting thermal performance predictions and the special ASTM test methods required to support and confirm those predictions are available. The thermal performance and structural integrity of reflective insulation panels depends to a significant degree on the details of construction. Reflective insulation is susceptible to external convective effects, and attention must be paid to eliminate unnecessary gaps and openings in the insulation system.[10–12]

REFERENCES

1. D. W. Yarbrough, "Assessment of Reflective Insulations for Residential and Commercial Applications," Oak Ridge National Laboratory, Rep. ORNL/TM-8891, Oct. 1983.
2. L. D. Stimpson and W. A. Hagemeyer, "Predicting Spacecraft Multilayer Insulation Performance from Heat Transfer Measurements," ASTM STP544, American Society for Testing and Materials, Philadelphia, PA, 1974, pp. 85–94.
3. "Standard Test Method for Total Hemispherical Emittance of Surfaces from 20 to 1400°C," ASTM C835-76, American Society for Testing and Materials, Philadelphia, PA.
4. J. P. Holman, *Heat Transfer,* McGraw-Hill, New York, 1963.
5. "Standard Test Method for Steady-State Thermal Performance of Building Assemblies by Means of a Guarded Hot Box," ASTM C236-80, American Society for Testing and Materials, Philadelphia, PA.
6. "Standard Test Method for Thermal Performance of Building Assemblies by Means of a Calibrated Hot Box," ASTM C976-82, American Society for Testing and Materials, Philadelphia, PA.
7. "Standard Test Method for Steady-State Thermal Transmission Properties of Nonhomogeneous Pipe Insulation Installed Horizontally," ASTM C691-78, American Society for Testing and Materials, Philadelphia, PA.
8. "Standard Test Method for Steady-State Heat Transfer Properties of Horizontal Pipe Insulation," ASTM C335-79, American Society for Testing and Materials, Philadelphia, PA.
9. E. C. Guyer and D. L. Brownell, "Industrial Reflective Insulations: Performance Measurements and a Modelling Technique," ASME Paper 83-WA/HT-58.
10. K. Gilbert, R. Patel, and K. Grant, "Containment Air Temperatures; Design and Management Issues," Diamond Power Tech. Paper 8310-1, Diamond Power Specialty Co., Lancaster, OH.
11. K. Gilbert, R. Patel, and K. Grant, "Heat Loss from an Insulated Pressurizer During Hot Functional Testing," Diamond Power Tech. Paper TPA47, Diamond Power Specialty Co., Lancaster, OH.
12. K. Gilbert and R. Patel, "Improving Primary System Insulation Performance," Diamond Power Tech. Paper TPC9, Diamond Power Specialty Co., Lancaster, OH.

Part 4

Structural Materials in Thermal Design

Chapter **1**

Metals and Alloys for Low Temperatures

HARRY I. McHENRY

Fracture and Deformation Division, National Bureau of Standards

Many metals and alloys have been used successfully for a wide range of structures and devices that operate at low temperatures. Selection of a specific alloy is based on the thermal, elastic, and mechanical properties of that alloy at the design temperature. These properties and their temperature dependence are discussed first, and then the structural alloys being used for cryogenic applications in commercially significant quantities are described.

The principal source of information used in the preparation of this section is *Materials at Low Temperatures*.[1] This book is recommended to the reader for further information and for use as a guide to the technical literature.

1.1 THERMAL PROPERTIES

Specific Heat

Specific heat relates the total heat per unit mass added to the material to the resultant temperature change of the material. At cryogenic temperatures, specific heat is a strong function of temperature, as shown in Fig. 1.1.

Thermal Expansion

The ratio of the change in volume of a material to the change in temperature is the thermal expansion. Linear thermal expansion is the change of length L as the temperature is changed

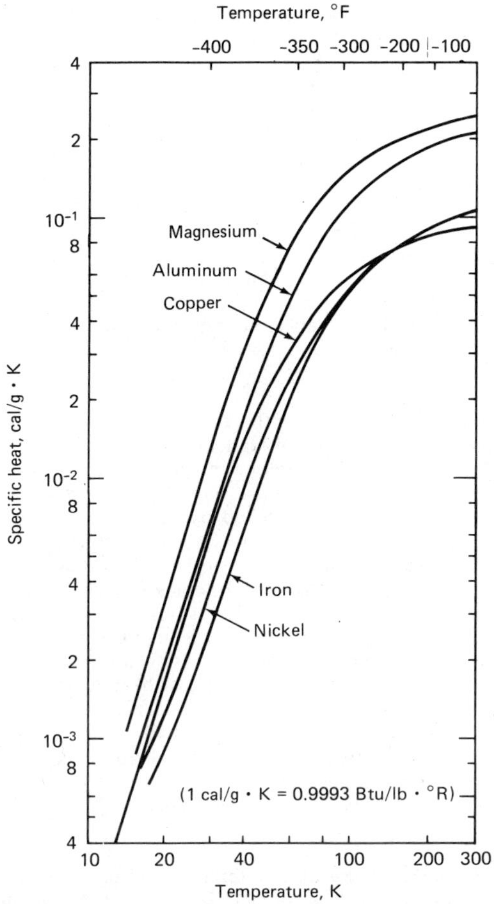

Fig. 1.1 Specific heat as a function of temperature for several metals.[2]

from a reference temperature (usually room temperature). Data for several metals are shown in Fig. 1.2. The coefficient of linear expansion α is the slope of the curve

$$\alpha(T) = \frac{1}{L}\frac{dL(T)}{dT} \cong \frac{1}{L}\frac{\Delta L}{\Delta T}$$

Thermal Conductivity

Thermal conductivity is the ratio of heat flow to temperature gradient. Thermal conductivity data for several materials at cryogenic temperatures are shown in Fig. 1.3. For pure metals at low temperatures, thermal conduction by electrons is extremely sensitive to the "electronic purity" of the metal, which is characterized by the residual resistance ratio RRR (the ratio of electrical resistivities at 273 and 4 K). This is illustrated in Fig. 1.4 by the data for pure aluminum and its alloys.

1.2 ELASTIC PROPERTIES

Elastic constants, being physical properties related to interatomic potentials, are insensitive to small variations in composition and metallurgical structure.

Young's Modulus

Young's modulus, also called the modulus of elasticity, is the ratio of axial stress σ to axial strain ϵ in a uniaxially stressed rod,

$$E = \frac{\sigma}{\epsilon}$$

Young's modulus data on 40 alloys are summarized in Fig. 1.5. The moduli typically increase 5 to 15 percent between room temperature and absolute zero.

Poisson's Ratio

Poisson's ratio is the negative ratio of transverse strain ϵ_x and axial strain ϵ_z in an uniaxially stressed rod,

$$\nu = -\frac{\epsilon_x}{\epsilon_z}$$

As temperature is decreased, Poisson's ratio decreases, as shown in Fig. 1.6 for six metals.

1.3 MECHANICAL PROPERTIES

The mechanical properties of a material relate the stresses acting on the material to the resulting deformation or fracture. These properties can be markedly affected by relatively minor changes in the metallurgical structure, caused by al-

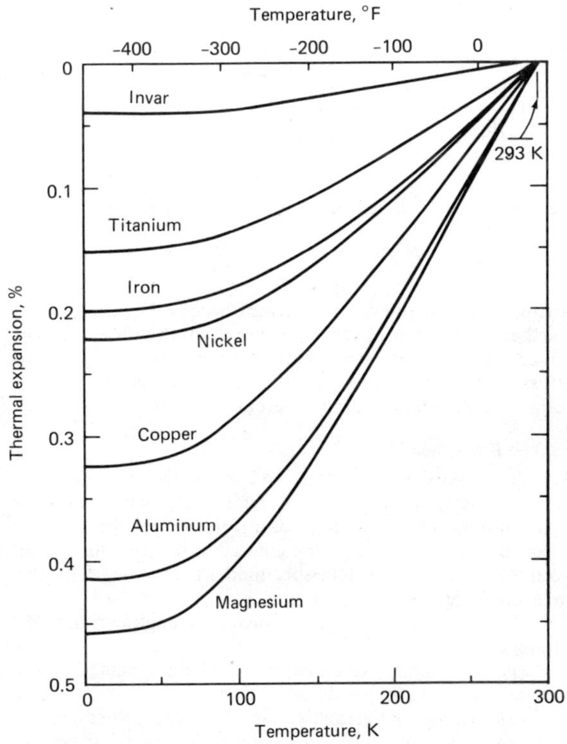

Fig. 1.2 Linear expansion as a function of temperature for several metals.[1]

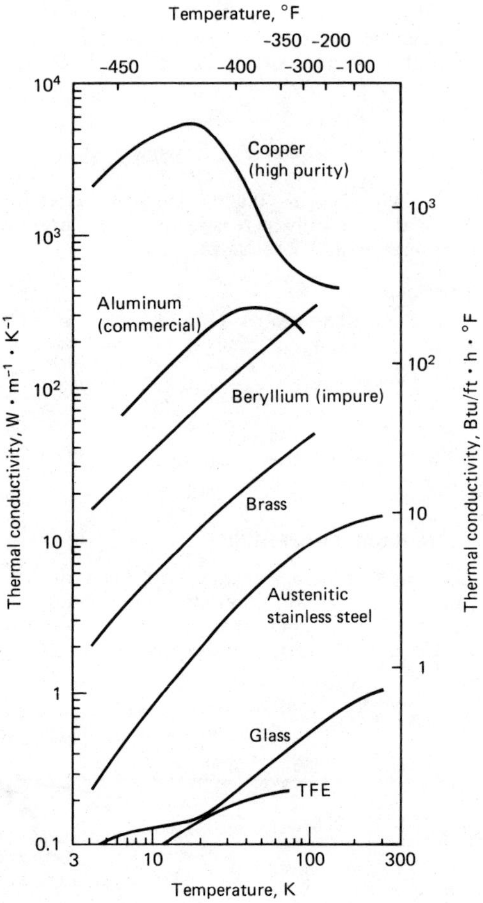

Fig. 1.3 Thermal conductivity as a function of temperature for several materials.[1]

loying, heat treatment, mechanical processing, and so on. Furthermore, the properties depend upon the measurement conditions, such as temperature, strain rate, and state of stress. Thus, for design, specific data with regard to the material and the service temperature are of interest.

Tensile Properties

Tensile properties are determined by axially loading a bar and measuring the engineering stress P/A_0, where P is the load and A_0 the initial area, as a function of the average strain $\Delta L/L$, where ΔL is the change of length over a gage section of length L, until the specimen breaks. The properties measured are:

1. *Ultimate tensile strength:* Maximum engineering stress attained in test
2. *Yield strength:* Engineering stress at a specified plastic strain, usually 0.2 percent
3. *Elongation:* Average strain (in percent) at fracture
4. *Reduction of area:* Percentage change of cross-sectional area at fracture

At cryogenic temperatures, the tensile properties are dependent upon the crystal structure of the material. As the temperature is lowered, the face-centered cubic (fcc) metals, such as aluminum, copper, nickel, and their alloys, have a slight increase in yield strength, but the ultimate tensile strength increases markedly. Thus, there is substantial strain-hardening and the fcc metals retain their ductility. For body-centered cubic (bcc) metals, notably iron and steel, the yield strength, and to a lesser extent the ultimate tensile strength, increase markedly. Thus, the difference between yield and ultimate strengths is reduced and the ductility drops, sometimes to the point where brittle fracture occurs. The hexagonal close-packed (hcp) metals, such as titanium, also have increases in yield and tensile strengths, but to a lesser extent than the bcc metals.

Toughness

Toughness is a measure of the ability of a material to resist fracture in the presence of high stresses in the vicinity of a notch or crack. It is commonly determined by the Charpy V-notch impact test, which measures the energy required to break a standard sized notched bar in bending. More recently, toughness has been is measured by fracture mechanics methods which characterize the fracture resistance of materials in the presence of a sharp crack in terms of various fracture criteria, such as critical stress intensity K_{Ic}, J integral $K_{Ic}(J)$, or crack tip opening displacement CTOD.

At cryogenic temperatures, the crystal structure is a useful guide for a qualitative prediction of the temperature dependence of toughness. The fcc alloys typically exhibit high toughness at low temperatures, bcc alloys exhibit precipitous decreases in toughness at temperatures below a critical transition temperature, and hcp alloys are noted for comparatively low toughness at all temperatures.

Fatigue

Fatigue is the failure of a material due to the repeated application of stress. It is characterized by plotting the amplitude of the repeated stress (or strain) cycle versus the number of cycles to failure. At cryogenic temperatures, the fatigue strength of materials that retain their ductility generally improves, except at the highest strain amplitudes where failure occurs in less than 1000 cycles. The fatigue behavior of AISI 304L stainless steel at room and cryogenic temperatures is plotted in Fig. 1.7. Similar data for 10 alloys commonly used at cryogenic temperatures are given in Ref. 4.

1.4 AUSTENITIC STAINLESS STEELS

The austenitic stainless steels are iron chromium alloys with sufficient nickel and manganese to stabilize the fcc phase, austenite. A wide range of mechanical properties may occur in the wrought austenitic stainless steels, depending on their chemical composition and thermal mechanical processing. Four general classes are useful for cryogenic applications.

1. The annealed AISI 300-series grades (e.g., 304 and 316) have moderate strength, excellent toughness and are readily available in a variety of product forms.
2. The annealed nitrogen-strengthened grades (e.g., 304N and 316N) exhibit a remarkable increase in yield strength at cryogenic temperatures and retain most of the advantages of the 300-series grades.

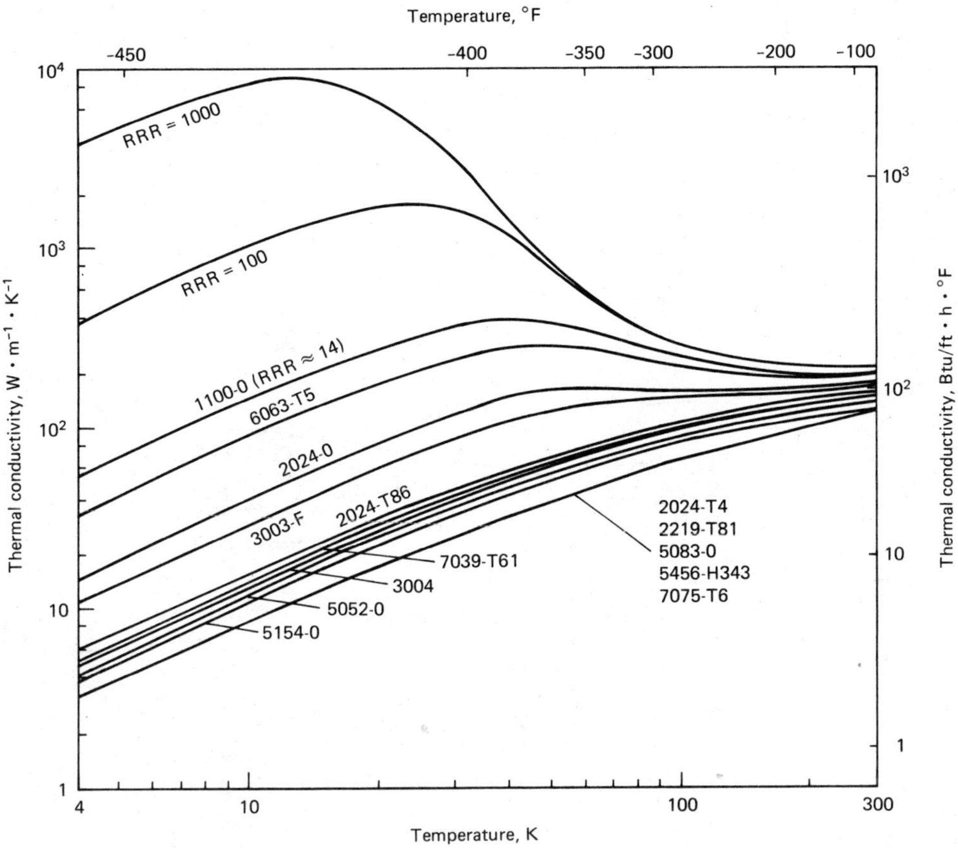

Fig. 1.4 Thermal conductivity as a function of temperature for aluminum and its alloys.[1]

3. Cold-rolled sheets of the 300-series grades have very high strength, but more limited applicability.

4. The precipitation-hardened austenitic grades (e.g., ASTM A453) have high strength at room and cryogenic temperatures.

Properties

The chemical compositions and representative tensile properties of austenitic stainless steels suitable for cryogenic applications are summarized in Table 1.1. The physical properties of AISI grades 304, 310, and 316 are presented in Table 1.2.

Nitrogen has a pronounced effect on the yield strength of austenitic stainless steels, particularly at cryogenic temperatures, as shown in Fig. 1.8. Unfortunately, as the strength increases, the toughness decreases, and the fatigue strength is not necessarily improved. The nitrogen-strengthened stainless steels have essentially the same fabrication characteristics and physical properties as the conventional grades.

Welding

The austenitic stainless steels, with and without nitrogen strengthening, are readily weldable by all of the common welding processes, provided the appropriate consumables and procedures are used. However, the properties of the welds at −452°F (4 K) are strongly influenced by the welding materials (electrodes, fluxes, etc.), processes, and procedures. Toughness is improved by using low-carbon filler metals to avoid sensitization, by balancing weld metal chemistry to avoid ferrite, by using low-nitrogen welding practices, and by using inert-gas-shielded welding processes.

1.5 NICKEL STEELS

The nickel steels have a predominantly bcc crystal structure and consequently undergo a ductile-to-brittle transition as the temperature is reduced. The transition temperature decreases with increasing nickel content, as shown in Fig. 1.9, and may be further reduced by heat treatment. The alloy chosen for a specific application should undergo the ductile-to-brittle transition at temperatures below the minimum service temperature.

Properties

The chemical compositions, minimum service temperatures, and representative tensile properties of the 3.5 to 9 percent

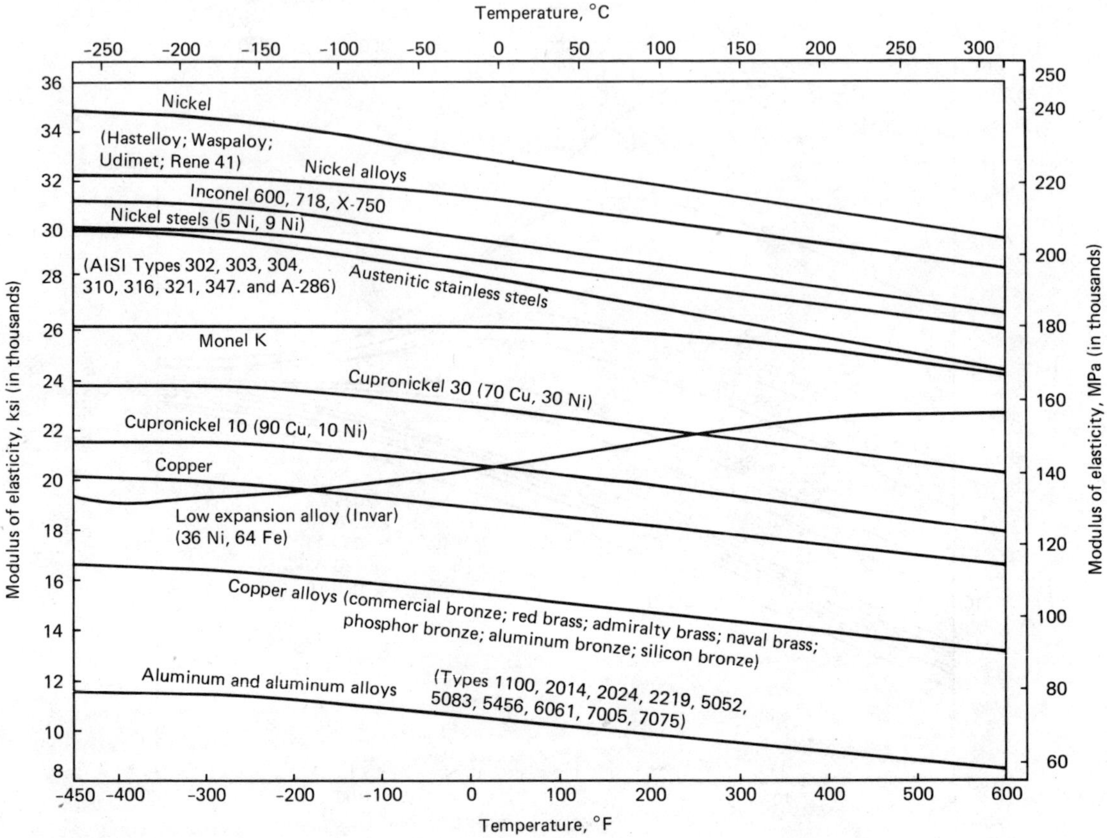

Fig. 1.5 Young's modulus as a function of temperature for 40 alloys.[3]

nickel steels are summarized in Table 1.3. Their physical properties are presented in Table 1.4. The steels are normally supplied as heat-treated plates for welded construction, but they are also available as forgings, sections, bars, and pipes.

The fracture toughness and fatigue properties of the nickel steels have been studied extensively by means of variety of test procedures. The results consistently indicate that the 5 to 9 percent nickel steels retain satisfactory properties at −260°F (111 K) (LNG temperature). As the temperature is reduced from −260 to −320°F (111 to 77 K), the fatigue and fracture resistance may be degraded, particularly for the 5 percent nickel steel. The temperature at which property degradation may occur is a function of microstructure as well as chemistry. Thus, care must be exercised in the production, heat treatment, and fabrication of the nickel steels to ensure the retention of satisfactory properties at cryogenic temperatures.

Welding

The nickel steels are readily weldable, but since their low-temperature properties are enhanced by heat treatment, the welding heat input must be limited to avoid embrittling the heat-affected zone. In addition, the filler metal must have satisfactory toughness in the as-deposited condition because postweld heat treatments are generally impractical. The most common filler metals are nickel-base alloys with 60 to 70 percent nickel and 15 to 20 percent chromium.

1.6 ALUMINUM ALLOYS

Aluminum alloys have an fcc crystal structure and retain strength, ductility, and toughness at cryogenic temperatures. The preferred alloys are (1) Al-4.5Mg (5083), a weldable alloy generally used in the annealed condition, (2) Al-1Mg-0.6Si (6061), a versatile alloy available in a variety of product forms and heat treatments, and (3) Al-6Cu (2219), a high-strength alloy used in the precipitation-hardened condition. Several other alloys have excellent properties at low temperatures and may be used for applications requiring extensive forming (1100 or 3003), brazing (3003), or maximum strength (2014 or 7005).

Properties

The chemical compositions and representative tensile properties of five aluminum alloys are summarized in Table 1.5.

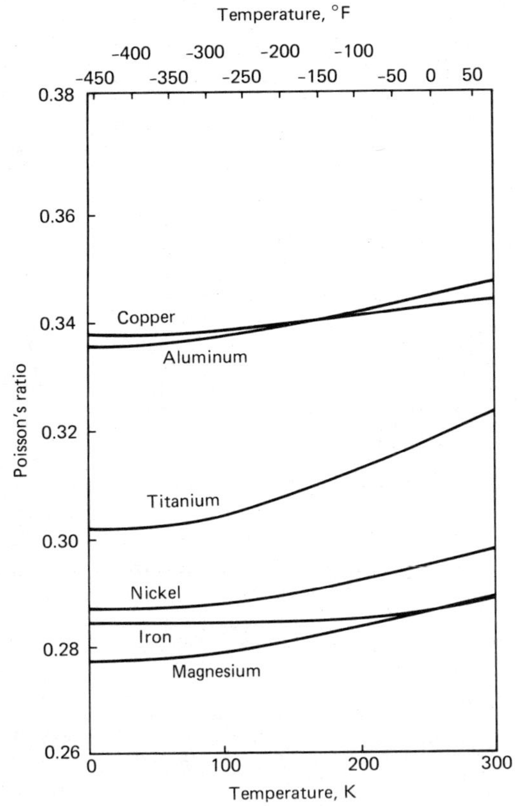

Fig. 1.6 Poisson's ratio as a function of temperature for several metals.[1]

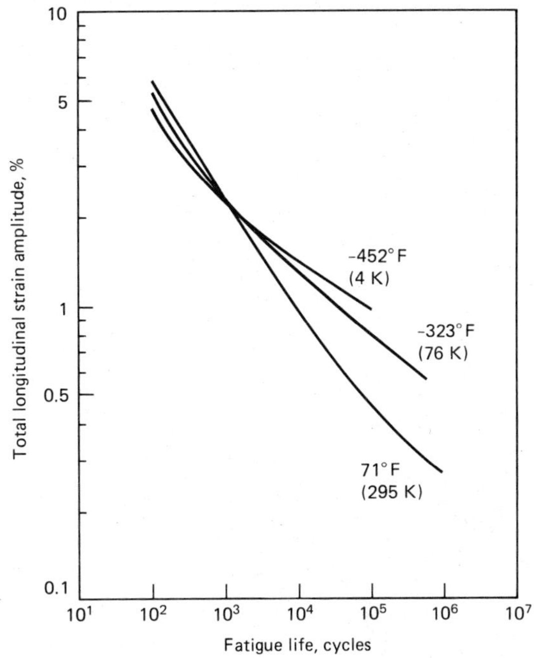

Fig. 1.7 Strain cycling fatigue behavior of AISI 304L stainless steel at three temperatures.[4]

Table 1.1. Tensile Properties of Austenitic Stainless Steels at Room and Cryogenic Temperatures

Alloy: Designation (Reference)	Alloy: Principal elements*	Temperature, °F	Yield strength, 10^3 lb/in^2	Tensile strength, 10^3 lb/in^2	Elongation, %
AISI 304 (6)	20Cr, 12Ni	70	33.9	92.8	76
		−320	40.9	221	45
		−452	58.2	250	30
AISI 304N (5)	20Cr, 12Ni	70	41.4	85.5	56
		−320	98.8	217	52
		−452	118	247	43
AISI 310S (7)	26Cr, 22Ni	70	30	83	58
		−320	75	161	69
		−452	106	192	56
AISI 316 (8)	18Cr, 14Ni, 3Mo	70	34.5	81.9	60
		−320	75.2	180	65
		−452	88.4	212	53
AISI 316LN (8)	18Cr, 14Ni, 3Mo, 0.16N	70	94.1	49.6	50
		−320	177	111	48
		−452	218	143	43
ASTM XM-11 (9)	21.5Cr, 7.5Ni, 10Mn, 0.40N	70	59.2	107	55
		−320	145	218	46
		−452	180	240	23

* Maximum specified.

CONVERSION FACTORS: 1 GPa = 0.145×10^6 lb/in^2; K = (°F + 459.67)/1.8.

Table 1.2. Physical Properties of Austenitic Stainless Steels at Room and Cryogenic Temperatures

Temperature, °F	Density, lb/in³	Young's modulus, 10^6 lb/in²	Shear modulus, 10^6 lb/in²	Poisson's ratio	Thermal conductivity Btu/ft · h · °F	Thermal expansion (mean), 10^6/°F	Specific heat, 10^{-3} Btu/lb · °F	Electrical resistivity, μΩ · in	Magnetic permeability (initial)
				AISI 304					
70	...	29.0	11.2	0.290	8.5	8.8	115	27.7	1.02
−320	0.284	31.1	12.2	0.278	4.6	7.2	45	20.2	
−452	...	30.5	11.9	0.279	0.16	5.7	0.45	19.5	1.09
				AISI 310					
70	...	27.7	10.6	0.305	6.6	8.8	115	34.4	1.003
−320	0.283	29.8	11.5	0.295	3.4	7.2	43	28.5	
−452	...	30.0	11.6	0.292	0.14	5.7	0.53	27.0	1.10
				AISI 316					
70	...	28.3	10.9	0.294	8.5	8.8	115	29.5	1.003
−320	0.288	30.3	11.8	0.283	4.6	7.2	45	22.3	
−452	...	30.2	11.8	0.282	0.16	5.7	0.45	21.2	1.02

CONVERSION FACTORS: GPa = 0.145 × 10^6 lb/in²; 1 W/m · K = 0.578 Btu/ft · h · °F; 10^{-6}/K = 0.556 × 10^{-6}/°F; 1 J/kg · K = 0.239 × 10^{-3} Btu/lb · °F.
SOURCE: Ref. 1.

The physical properties of alloys 5083, 6061, and 2219 are presented in Table 1.6.

Aluminum alloys have a wide range of mechanical properties, depending on chemistry and heat treatment. The yield strengths of several alloys are shown as a function of temperature in Fig. 1.10. Tensile ultimate strengths show a much larger increase with decreasing temperature than yield strengths. For the preferred alloys, ductility, notched-tensile

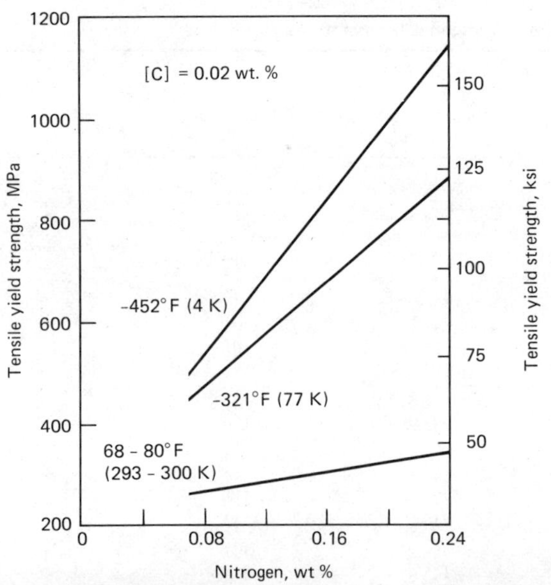

Fig. 1.8 Yield strength as a function of nitrogen for AISI 304 stainless steel at three temperatures.[5]

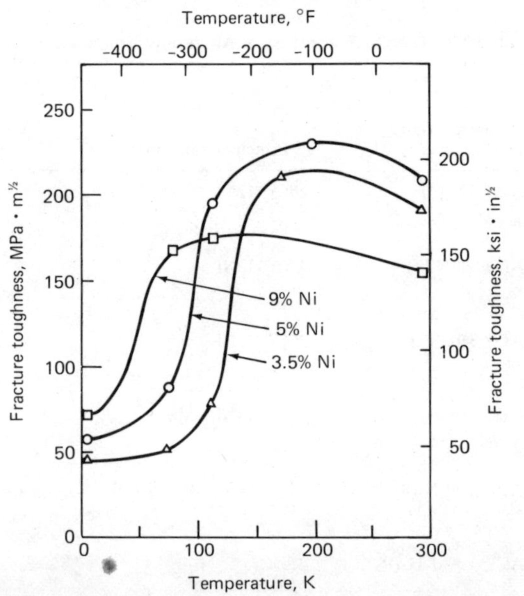

Fig. 1.9 Fracture toughness as a function of temperature for three nickel steels.[1]

Table 1.3. Tensile Properties of Nickel Steels and Invar at Room and Cryogenic Temperatures

Alloy: Designation	Alloy: ASTM specification	Minimum service temperature, °F	Temperature, °F	Yield strength 10^3 lb/in²	Tensile strength, 10^3 lb/in²	Elongation, %
3.5Ni	A203-D	−150	70	54	75	37
			−150	67	90	39
5Ni	A645	−275	70	70	102	32
			−150	75	130	31
			−275	97	157	30
9Ni	A553	−320	70	100	110	28
			−150	112	137	30
			−320	145	167	32
36Ni (Invar)	A658	−452	70	42	75	28
			−320	94	135	30
			−452	122	155	24

CONVERSION FACTORS: 1 GPa = 0.145 × 10^6 lb/in²; K = (°F + 459.67)/1.8.
SOURCE: Ref. 7.

properties, fracture toughness, fatigue strength, and fatigue crack growth resistance at cryogenic temperatures tend to be similar or better than room temperature values.

Welding

Aluminum alloys are readily weldable by the gas-metal-arc and gas-tungsten-arc processes, but of the preferred alloys, only annealed 5083 alloy retains full strength in the as-welded condition. The cold-worked and precipitation-hardened alloys are locally weakened by welding.

1.7 OTHER METALS AND ALLOYS

From a tonnage standpoint, nearly all of the alloys used for cryogenic applications are the austenitic stainless steels, nickel steels, and aluminum alloys. However, certain important ap-

Table 1.4. Physical Properties of Nickel Steels and Invar at Room and Cryogenic Temperatures

Temperature, °F	Density, lb/in³	Young's modulus, 10^6 lb/in²	Shear modulus, 10^6 lb/in²	Poisson's ratio	Thermal conductivity, Btu/ft · h · °F	Thermal expansion (mean), 10^{-6}/°F	Specific heat, 10^{-3} Btu/lb · °F
				3.5Ni			
70	0.284	29.6	11.5	0.282	20	6.6	108
−150	. . .	30.5	11.9	0.281	17	5.7	84
				5Ni			
70	0.282	28.7	11.2	0.283	18	6.6	108
−260	. . .	30.1	11.8	0.277	12	5.2	60
−320	. . .	30.3	11.8	0.277	9.2	4.9	36
				9Ni			
70	0.283	28.3	10.7	0.286	16	6.6	108
−260	. . .	29.6	11.2	0.281	10	5.2	60
−320	. . .	29.8	11.3	0.280	7.5	4.9	36
				36Ni (Invar)			
70	0.282	22.1	8.08	0.284	8.0	0.65	
−320	. . .	20.4	7.37	0.307	3.6	1.00	
−452	. . .	20.6	7.33	0.332	0.18	0.77	

CONVERSION FACTORS: 1 GPa = 0.145 × 10^6 lb/in²; 1 W/m · K = 0.578 Btu/ft · h · °F; 10^{-6}/K = 0.556 × 10^{-6}/°F; 1 J/kg · K = 0.239 × 10^{-3} Btu/lb · °F.
SOURCE: Ref. 7.

Table 1.5. Tensile Properties of Aluminum Alloys at Room and Cryogenic Temperatures

Alloy					
Designation	Principal elements	Temperature, °F	Yield strength, 10^3 lb/in^2	Ultimate strength, 10^3 lb/in^2	Elongation, %
1100-0	99Ac, min.	70	6.9	13.5	46
bar		−320	9.0	27.7	56
		−423	9.4	47.6	54
2219-T851	5.8–6.8Cu	70	53.8	67.6	11
plate		−320	63.9	82.4	14
		−452	70.2	95.6	15
5083-0	4.0–4.9Mg	70	20.5	46.7	19
plate		−320	22.9	63.0	32
		−452	25.8	80.8	32
6061-T651	0.4–0.8Si, 0.8–1.2Mg	70	42.2	44.8	16
plate		−320	48.9	58.3	23
		−452	55.0	70.1	25
7075-T5351	1.0–1.8Mg, 4.0–5.0Zn	70	55.0	62.0	15
extrusion		−320	67.5	83.9	17
		−452	75.6	97.5	17

CONVERSION FACTORS: 1 GPa = 0.145 × 10^6 lb/in^2; K = (°F + 459.67)/1.8.
SOURCE: Ref. 10.

plications require the use of other metals, usually copper, nickel, titanium, and their alloys.

Copper

Copper is used in applications where its excellent electrical and thermal conductivities are needed. It can be alloyed with zinc to form brasses, with tin to form bronzes, with nickel to form cupronickels, and so on. Copper and most of its alloys have an fcc crystal structure, and thus, toughness and ductility are retained at cryogenic temperatures. The yield strengths of Cu-Ni and Cu-Zn alloys are shown in Figure 1.11.

Nickel Alloys

Nickel-base superalloys such as the Inconels (X 750, 718 and 706) are suited for applications where strength and toughness are needed at both room and cryogenic temperatures. For example, Inconel 718 has a yield strength of 170 × 10^3 lb/in^2 (1170 MPa) at room temperature, and the strength, ductility, and toughness all increase 10 to 20 percent at cryogenic temperatures.

Table 1.6. Physical Properties of Aluminum Alloys at Room and Cryogenic Temperatures

Temperature, °F	Density, lb/in^3	Young's modulus, 10^6 lb/in^2	Shear modulus, 10^6 lb/in^2	Poisson's ratio	Thermal conductivity, Btu/ft · h · °F	Thermal expansion (mean), 10^{-6}/°F	Specific heat, 10^{-3} Btu/lb · °F	Electrical resistivity, μΩ · in
				5083-0				
70	0.096	10.4	3.89	0.333	69	12.8	215	2.23
−320	. . .	11.6	4.41	0.320	32	10.1	81	1.31
−452	. . .	11.7	4.45	0.318	1.9	7.8	0.067	1.19
				6061-T651				
70	0.097	10.2	3.83	0.338	. . .	12.8	215	1.55
−320	. . .	11.2	4.22	0.328	. . .	10.1	81	0.65
−452	. . .	11.3	4.23	0.327	. . .	7.8	0.067	0.54
				2219-T851				
70	0.102	11.2	4.22	0.330	69	12.8	215	2.2
−320	. . .	12.3	4.68	0.319	32	10.1	81	
−452	. . .	12.4	4.71	0.318	1.7	7.8	0.067	1.1

CONVERSION FACTORS: 1 GPa = 0.145 × 10^6 lb/in^2; 1 W/m · K = 0.578 Btu/ft · h · °F; 10^{-6}/K = 0.556 × 10^{-6}/°F; 1 J/kg · K = 0.239 × 10^{-3} Btu/lb · °F.
SOURCE: Ref. 1.

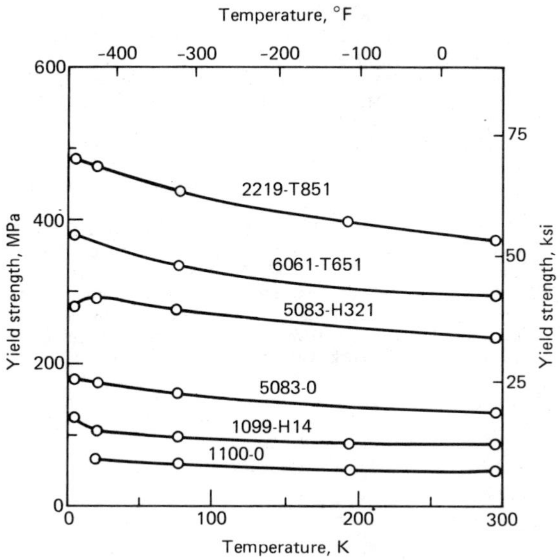

Fig. 1.10 Yield strength as a function of temperature for several aluminum alloys.[1]

The Fe-36Ni alloy known as Invar has a remarkably low thermal expansion at room and cryogenic temperatures. As shown in Fig. 1.2, it contracts about 0.04 percent between room temperature and −320°F (77 K) and no further between −320 and −452°F (77 and 4 K). It is commonly used in applications requiring close dimensional tolerances or to

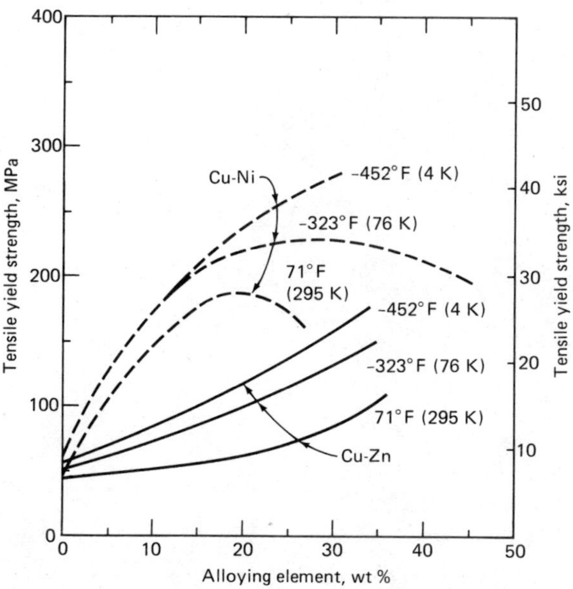

Fig. 1.11 Yield strength as a function of alloy content for cupronickel alloys and brasses at three temperatures.

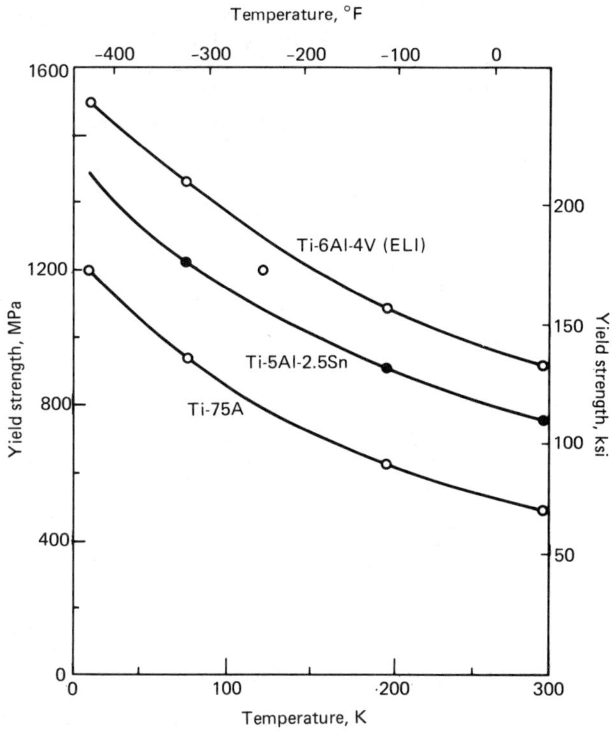

Fig. 1.12 Yield strength as a function of temperature for three titanium alloys.[1]

prevent large thermoelastic stresses. Invar has an fcc crystal structure. But unlike many fcc metals, it has a substantial increase in yield strength at cryogenic temperatures as shown in Table 1.4.

Titanium Alloys

Commercially pure titanium, Ti-2.5Sn, and Ti-6Al-4V, have been used at cryogenic temperatures for aerospace applications where the strength-to-weight ratio is the primary concern. The strengths of these alloys improve as the temperature is reduced. For example, a twofold increase in yield strength occurs between room temperature and −424°F (20 K) as shown in Fig. 1.12. Similar trends occur for tensile strength, but ductility and toughness decrease as the temperature is reduced. To improve toughness, extra-low-interstitial (ELI) grades are specified, and the alloys are used in the annealed condition. Even with these improvements, titanium alloys have very low ratios of fracture toughness to yield strength at cryogenic temperatures and, consequently, are susceptible to brittle fracture.

1.8 SUMMARY

Many commercially available alloys are suitable for use in a wide variety of cryogenic applications. The most commonly used alloys are austenitic stainless steels, nickel steels, and aluminum alloys. There are also numerous alloys, particularly with fcc crystal structures, which can be used at cryogenic temperatures, for example, nickel-base alloys, copper and its alloys, and special-purpose alloys such as Invar.

REFERENCES

1. R. P. Reed and A. F. Clark, (eds.), *Materials at Low Temperatures,* American Society for Metals, Metals Park, OH, 1983.
2. Y. S. Touloukian and E. H. Buyco, "Specific Heat, Metallic Elements and Alloys," in *Thermal Physical Properties of Matter,* vol. 4, Plenum, New York, 1970.
3. H. M. Ledbetter, "Temperature Behavior of Young's Moduli of Forty Engineering Alloys," *Cryogenics,* **22,** 653–656 (1982).
4. A. J. Nachtigall, "Strain-Cycling Fatigue Behavior of Ten Structural Metals Tested in Liquid Helium, Liquid Nitrogen, and Ambient Air," in *Properties of Materials for Liquefied Natural Gas Tankage,* ASTM STP 579, American Society for Testing and Materials, Philadelphia, PA, 1975, pp. 378–396.
5. N. J. Simon and R. P. Reed, "Structural Materials for Superconducting Magnets; III, AISI 304 Stainless Steel," National Bureau of Standards, Boulder, CO, 1985.
6. *Handbook on Materials for Superconducting Machinery,* MCIC-HB-04, Battelle Memorial Institute, Columbus, OH, 1977.
7. D. B. Mann (ed.), *LNG Materials and Fluids,* National Bureau of Standards, Boulder, CO, 1978.
8. N. J. Simon, and R. P. Reed, "Structural Materials for Superconducting Magnets; I, AISI 316 Stainless Steel," National Bureau of Standards, Boulder, CO, 1982.
9. N. J. Simon and R. P. Reed, "Structural Materials for Superconducting Magnets; II, 21Cr-6Ni-9Mn Stainless Steel," National Bureau of Standards, Boulder, CO, 1983.
10. J. G. Kaufman, K. O. Bogardus, and E. T. Wanderer, "Tensile Properties and Notch Toughness of Aluminum Alloys at −452°F in Liquid Helium," in *Advances in Cryogenic Engineering,* vol. 13, Plenum, New York, 1967, pp. 294–308.

Chapter 2

Metals and Alloys for High Temperatures

ROBERT W. SWINDEMAN

*Oak Ridge National Laboratory**

2.1 INTRODUCTION

A large selection of alloys is available for service at high temperature, but these may be divided into different groups according to potential applications. The design criteria and data needs are closely associated with these specific applications. Pressure vessel alloys, for example, are expected to survive many years of service under low-stress creep conditions. Bolting alloys are selected on the basis of high yield strength and resistance to relaxation. Turbine alloys must resist creep and creep-fatigue loading, while hot engine parts could undergo thermal fatigue. Factors entering into alloy selection include physical properties, mechanical properties, fabricability, corrosion resistance, and cost. This chapter focuses on the physical and mechanical properties important to the material selection for an initial design. All of the alloy groups are treated together on the basis of the property of interest, rather than by alloy group. Also, the properties

* Operated by the U.S. Department of Energy, under contract DE-AC05-840R21400 with Martin Marietta Energy Systems, Inc.

Table 2.1. Compositions for Some Typical High-Temperature Alloys

Alloy	Weight percent												Applications	References
	C	Mn	Si	Cr	Ni	Co	Mo	W	Al	Ti	Fe	Other		
Group 1														
1¼Cr-½Mo	0.15	0.5	0.5	1	...	...	0.5	...	...	...	Bal	...	Vessels, piping	1–3
2¼Cr-1Mo	0.15	0.5	0.5	2.2	...	...	1	...	...	...	Bal	...	Vessels, piping	1–9
1¼Cr-½Mo-V	0.3	0.5	0.6	1.2	...	...	0.5	...	...	...	Bal	0.25V	Rotors, bolting	1, 2, 6, 7
9Cr-1Mo-V	0.1	0.5	0.4	9	...	...	1	...	...	...	Bal	0.25V, 0.05Nb	Vessels, piping	1
12Cr-1Mo-V	0.25	1	0.3	12	0.2	...	1	...	...	...	Bal	0.25V	Rotors, bolting	1, 2
410 SS	0.12	0.6	0.4	12	...	...	...	...	...	...	Bal	...	Blades	1, 5, 10, 11
Group 2														
304 SS	0.06	1.5	0.6	19	9	...	...	...	...	...	Bal	...	Vessels, piping	1–4, 6–10
316 SS	0.06	1.5	0.6	17	12	...	2.5	...	...	...	Bal	...	Vessels, piping	1–4, 6–8, 10
347 SS	0.06	1.5	0.6	18	11	...	...	...	...	...	Bal	1.5Nb + Ta	Vessels, piping	1–3, 10, 12, 13
310 SS	0.25	1.5	1.5	25	21	...	...	...	...	...	Bal	...	Vessels, piping	1, 6, 10, 12
Incoloy 800H	0.06	0.8	0.5	21	32	...	...	...	0.35	0.35	Bal	...	Vessels, piping	1, 3, 6, 9, 10
19-9DL	0.3	1.1	0.6	19	9	...	1.2	1.2	...	0.3	Bal	0.4Nb	Rotors, disks	2, 10, 11
N155	0.12	1.5	0.5	21	20	20	3	2.5	...	...	Bal	1Nb	Liners, blades, bolting	2, 10, 11
A286	0.05	1.5	0.4	15	26	...	1.2	...	0.2	2.1	Bal	0.003B	Rotors, blades, bolting	2, 3, 6, 7, 10, 11
Group 3														
Monel K500	0.13	0.8	0.25	...	Bal	...	...	...	2.7	0.6	...	34Cu	Pumps	1, 5, 10
Hastelloy C	0.08	1	1	15	Bal	...	16	4	...	...	5	...	Liners, piping	1, 2, 5, 10, 14, 15
Hastelloy X	0.1	1	1	22	Bal	1	9	0.6	0.3	...	...	0.003B	Liners, piping	1, 2, 5, 8, 10, 14, 15
Inconel 600	0.08	0.5	0.2	15	Bal	...	...	...	...	...	8	...	Vessels, piping	1, 2, 5, 10, 14, 16
Inconel 617	0.1	0.5	0.5	22	Bal	12	9	...	1.2	0.3	1	...	Piping	10, 16
Inconel 718	0.04	0.2	0.2	18	Bal	...	3	...	0.5	0.9	18	5Nb	Bolting, disks	2, 5, 8, 10, 14, 16
Inconel X 750	0.04	0.5	0.2	15	Bal	...	...	...	0.7	2.5	7	1Nb	Bolting, springs, blades	2, 5, 10, 14, 16
Waspaloy	0.07	0.3	0.3	19	Bal	13	4.3	...	1.4	3	...	0.006B,0.1Zr	Blades, disks	2, 5, 10, 11
Group 4														
L605	0.1	1	<1	20	10	Bal	...	15	...	...	...	...	Vessels, liners	2, 5, 10, 14, 15
Haynes 188	0.1	...	...	22	22	Bal	...	14	...	...	...	0.05La	Vessels, liners	5, 10, 15
S816	0.38	1	0.4	20	20	Bal	4	2	...	...	4	2Nb	Bolting, springs, blades	2, 10, 14
Group 5														
Inconel 713	0.05	<0.25	<0.5	12	Bal	...	4.5	...	6	0.5	...	0.01B,2Nb,0.1Zr	Disks, blades	5, 10, 14
IN 100	0.18	...	...	12	Bal	15	3	...	5.5	4.5	...	1V	Blades	5, 10
U-500	0.08	...	...	10	Bal	18	4	...	3	3	...	0.007B	Blades	2, 5, 10, 14
Mar-M200	0.15	...	...	9	Bal	10	...	12	5	2	...	0.015B,2Nb,0.05Zr	Blades	5, 10
B-1900	0.15	...	...	8	Bal	10	6	...	6	1	...	0.015B,0.1Zr,4Ta	Blades	5
Group 6														
Cb-752	0.004	...	...	...	...	...	...	10	...	...	...	2.5Zr,Bal Nb	...	15, 17
Mo-½Ti	0.04	...	...	...	...	...	Bal	...	...	0.5	...	...	...	6,14,17
TZM	0.03	...	...	...	...	...	Bal	...	...	0.5	...	0.08Zr	...	6, 14, 17
T-111	0.002	...	...	...	...	...	...	8	...	...	...	2.5Hf, Bal Ta	...	6, 10, 14, 17

NOTES: Inconel, Incoloy, and Monel are registered trademarks of the International Nickel Corporation; Hastelloy is a registered trademark of Haynes Alloys, Inc.; Waspaloy is a registered trademark of United Technologies Corporation.

presented should be considered as typical, rather than minimum or average. The scatter of the property values about the illustrated trend is not indicated, and the user should recognize that mechanical properties strongly depend on fabrication route, product form, and heat treatment. For precise information on specific alloys more comprehensive codes, standards, and handbooks should be consulted. Some relevant documents are cited in the references. Data from these and other sources are represented in graphic form in this chapter.

2.2 TYPICAL ALLOYS

The chemical compositions for approximately 30 high-temperature alloys are listed in Table 2.1. The first group consists of alloy steels containing chromium and molybdenum. These steels may be heat-treated to produce ferritic, bainitic, or martensitic microstructures with corresponding differences in short-time strength. For pressure vessel and piping service at 1000°F (538°C) and above, the alloys are often used in the annealed condition. For bolting, rotor, and high-pressure service the vanadium bearing steels are often used in the bainitic or martensitic condition. The second group of alloys listed in Table 2.1 are austenitic alloys having iron as the major constituent. These include the 300 series stainless steels, which have a very wide range of applications as pressure vessels, piping, boiler tubing, pumps, and the like in the power and petrochemical industries. In this group the A286 alloy requires heat treatment for optimum strength and ductility. The third group consists of nickel-base wrought alloys. These are used where good elevated temperature strength is needed in a fabricable alloy. Aerospace and petrochemical industries are typical users. Some of these alloys, such as alloy 718, alloy X 750, and Waspaloy, require heat treatment to produce optimum strength and ductility. The fourth group consists of fabricable cobalt-base superalloys often used in very-high-temperature structures. The fifth group consists of cast nickel-base superalloys, while the final group consists of refractory alloys, suitable only for service in vacuum or inert atmospheres. Cast versions of the first four groups can be obtained, with some modification of composition.

2.3 SPECIFIC HEAT

The specific heats for several metals and alloys are plotted against temperature in Fig. 2.1. Trends reveal that the specific heat for ferritic alloys increases appreciably with temperature to at least 1200°F (649°C). A sharp decrease in specific heat occurs when the temperature reaches the ferrite-to-austenite transformation temperature above 1350°F (732°C). Data for most of the austenitic alloys cluster to-

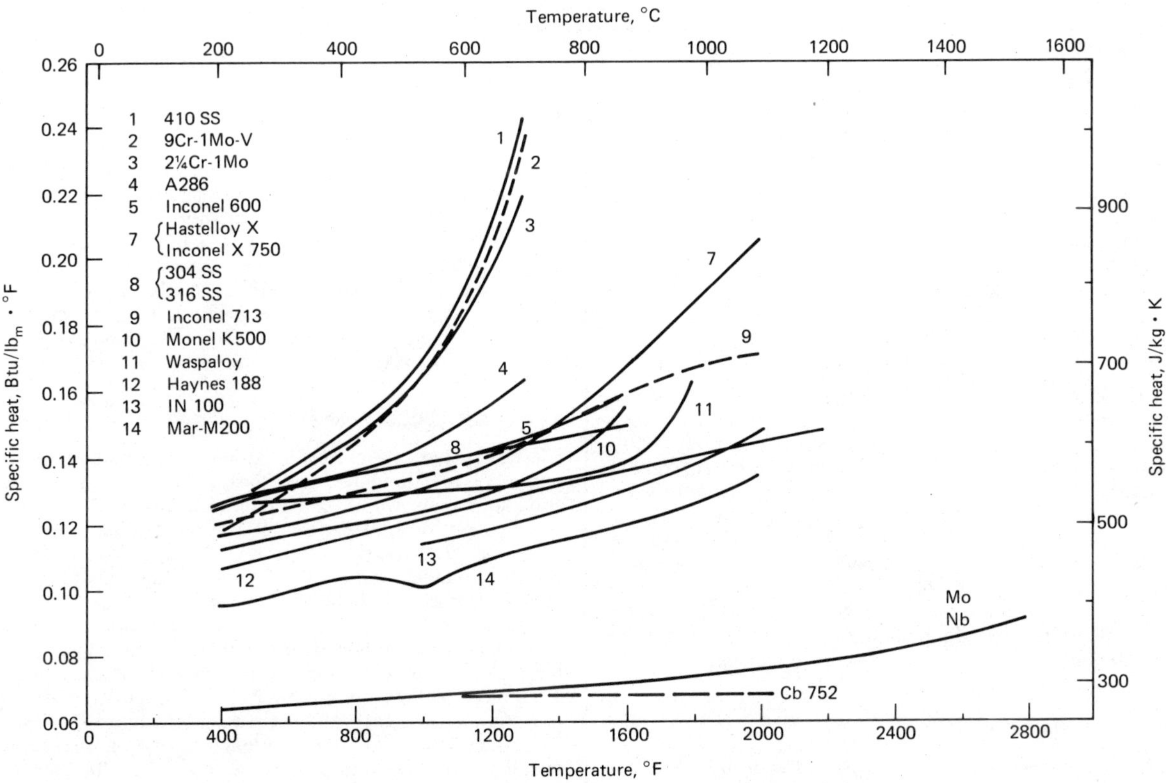

Fig. 2.1 Specific heat versus temperature for high-temperature alloys.[1,5,11,14]

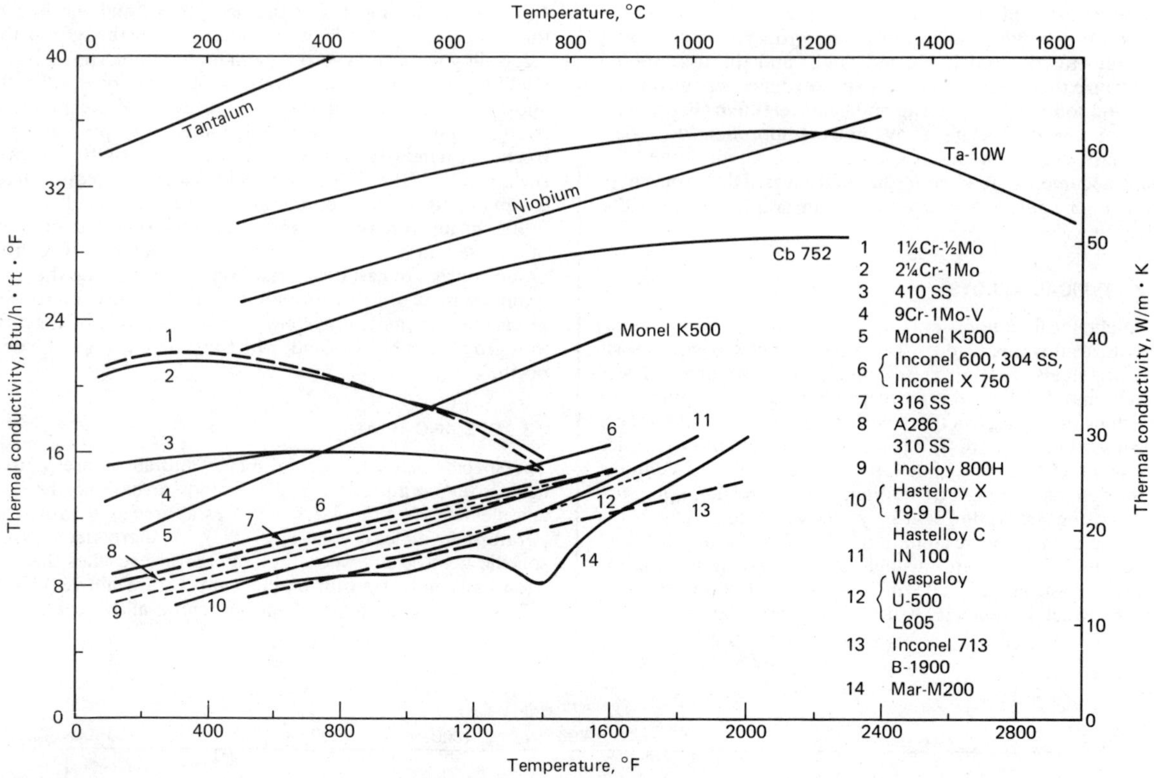

Fig. 2.2 Thermal conductivity versus temperature for high-temperature alloys.[1,5,11,14]

gether and exhibit modest increases in specific heat with increasing temperature. Cobalt-base alloys and nickel-base superalloys exhibit similar trends. The exceptions are the cast nickel-base superalloys such as Mar-M200 and B-1900 that have lower specific heats. The refractory metals and alloys have very low specific heats.

2.4 THERMAL CONDUCTIVITY

The thermal conductivities for 30 alloys are plotted against temperature in Fig. 2.2. The refractory metals have high thermal conductivities that increase with temperature up to at least 1600°F (871°C). The ferritic alloys exhibit a maximum thermal conductivity between room temperature and 1000°F (538°C). Above 1000°F (538°C) the conductivity decreases. The austenitic alloys such as the stainless steels and the nickel-base alloys have low thermal conductivities at room temperature and show only a small variation from one alloy to the next. Conductivity increases with temperature throughout the useful temperature range for the alloys. The cast superalloys have the lowest thermal conductivities.

2.5 THERMAL EXPANSION

The thermal expansion data for metals and alloys are frequently described on the basis of the mean coefficient of thermal expansion between room temperature and the temperature of interest. Alternatively, the expansion curve is reported. In Fig. 2.3 the thermal expansion curves are shown for a wide range of alloys. The austenitic stainless steels and alloy 800H exhibit the greatest expansion, followed by nickel-base alloys, cobalt-base alloys, and then ferritic alloys. The refractory metals and alloys have the least thermal expansion. The thermal expansion coefficient can be estimated from the slope of the curve of interest.

2.6 ELASTIC CONSTANTS

Elastic constants of interest in design include Young's modulus, the shear modulus, and Poisson's ratio. Many metals and alloys are isotropic, and unique values for each constant can be used to represent the elastic behavior as a function of temperature. Some cast alloys, especially the unidirectionally solidified cast alloys and electroslag welds, tend to be anisotropic, and the elastic constants may vary with orientation. The data reported here apply to isotropic materials. A second consideration with regard to elastic constants is the technique used to produce the data. Techniques most often cited include the mechanical straining of test bars at low strain rates that produces a static modulus and high-frequency low-strain-amplitude vibratory or ultrasonic straining that produces a dynamic modulus. The Young's modulus data produced by the two methods generally agree

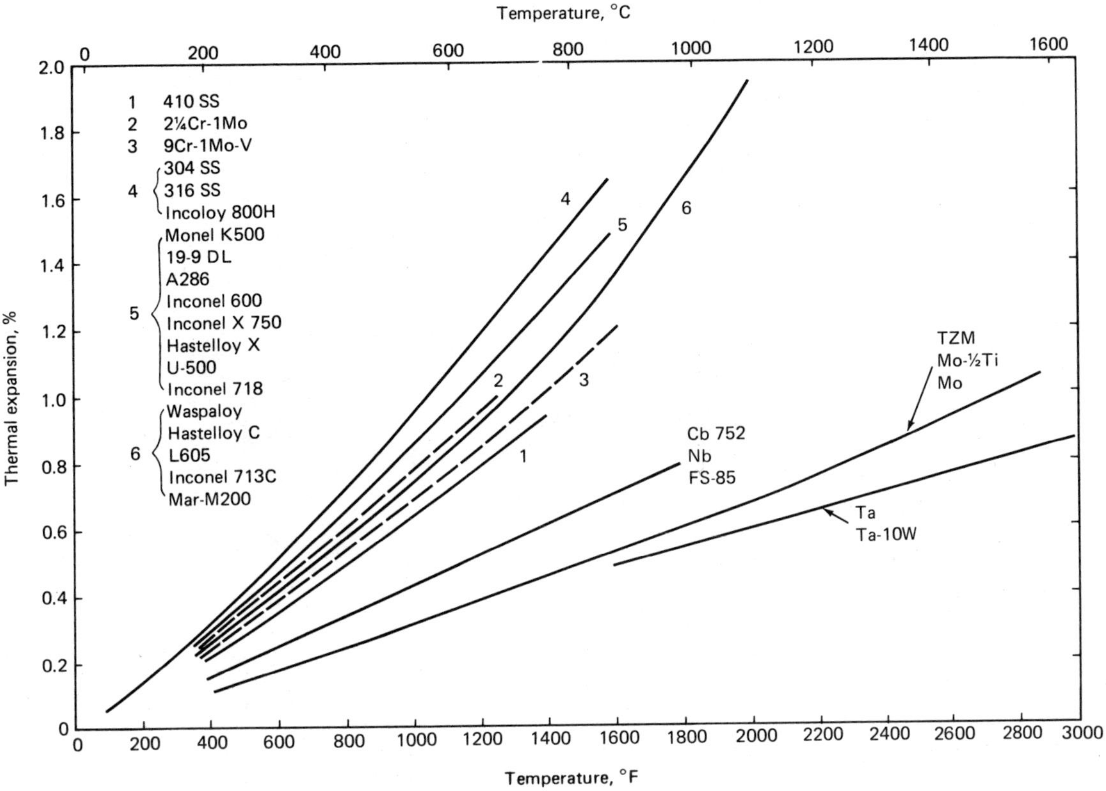

Fig. 2.3 Thermal expansion versus temperature for high-temperature alloys.[1,5,11,14]

within 10 percent up to the temperature where relaxation mechanisms are important. Here the static modulus becomes dependent on the strain rate and may diminish to two-thirds of the dynamic modulus. Most of the data represented in Fig. 2.4 were derived from static tests and are applicable to slow strain rates and large strains. Tungsten and its alloys have the highest moduli, followed by tantalum, cobalt alloys, superalloys, and nickel-base alloys. The ferritic alloys exhibit a greater decrease in modulus with temperature than the austenitic alloys. The stainless steels have relatively low moduli, but the niobium alloys have moduli that are much below any other high-temperature alloy group.

Poisson's ratio data are relatively scarce. Around room temperature the ratio for many alloys is in the range of 0.28 to 0.31 and increases slightly with temperature. Often Poisson's ratio goes through a maximum at intermediate temperatures; then it may decrease toward 0.3.

The shear modulus may be determined from the knowledge of Young's modulus and Poisson's ratio.

2.7 TENSILE PROPERTIES

Tensile data are useful in high-temperature design to minimize the likelihood of excessive distortion, buckling, collapse, and thermal ratcheting. The yield strength is of particular use in developing isochronous stress-strain curves and partitioning strains for fatigue and creep-fatigue analyses.

Strain rates used for tensile testing are typically in the range of 0.005 to 0.05 per minute for yield strength determination and up to 0.2 per minute for the ultimate strength. The designer should bear in mind that the applicable strain rate for a specific problem should be compared to this range of strain rate before making use of the data. The definition of the yield stress is also of importance. In the United States the yield strength most often determined for high-temperature applications is based on the 0.2 percent offset strain. Such data are reported here. Both the yield strength and the ultimate strength vary with the chemistry and the thermal mechanical treatment. Alloys are usually purchased to a specification that requires that the product meet or exceed some minimum strength at room temperature. In most design problems the use of minimum strength is conservative or adequate, and such data may be obtained in appropriate codes and standards. The data provided in Fig. 2.5 for yield strength and in Fig. 2.6 for ultimate strength represent typical, rather than minimum, strengths, and it is up to the designer to determine the appropriate allowances for heat-to-heat variations. Most high-temperature alloys exhibit only small temperature-dependent changes in the yield and ultimate strengths up to 1000°F (538°C). The exceptions are the ferritic alloys. They start to lose strength above 800°F

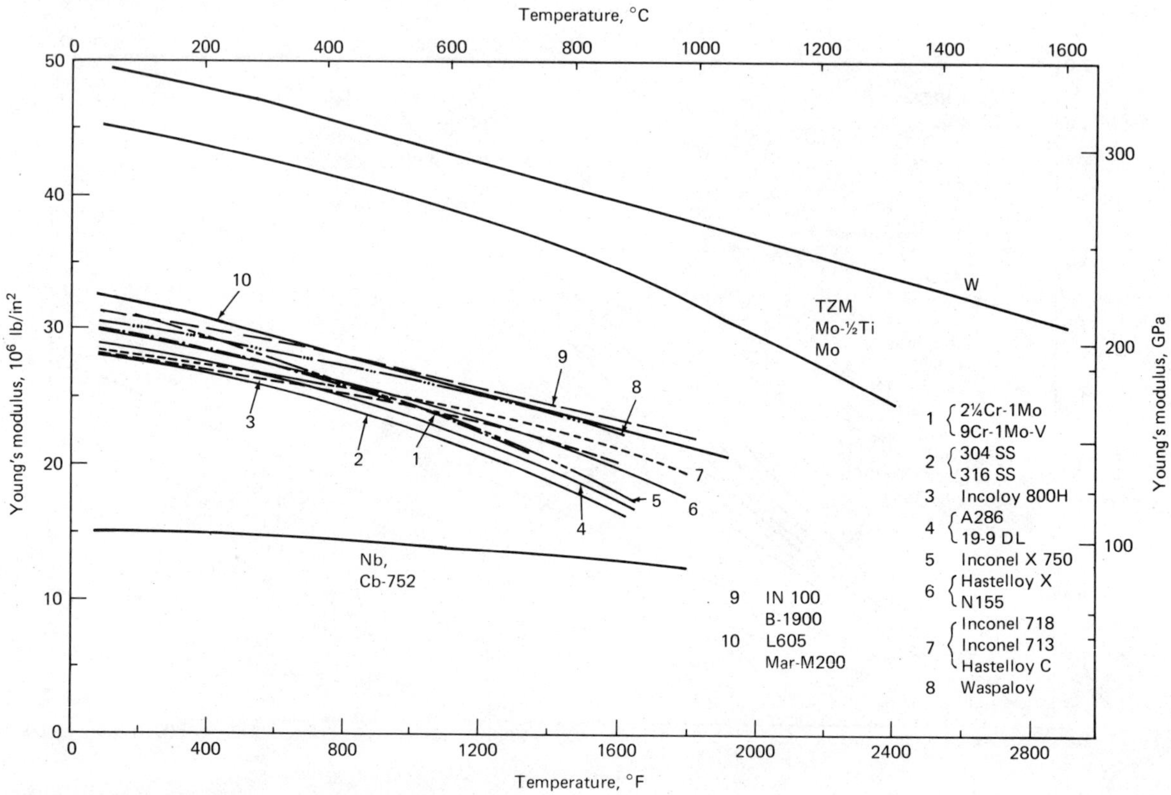

Fig. 2.4 Young's modulus versus temperature for high-temperature alloys.[1,5,11,14]

(427°C). Cold-working and heat-treating effects become less significant at high temperature. Hence, there will be less heat-to-heat variation in the strength properties near the upper limit of a material's use.

2.8 CREEP AND STRESS RUPTURE

The pressure vessel industry is concerned with service for very long times at temperatures in the creep range. Allowable design stresses are most often controlled by the minimum or average stress to produce rupture in some specific time, say 100,000 h. The gas turbine industry, on the other hand, is concerned with deformation. The design of a component is often based on the stress to produce a specific deformation in a specific time, say 1 percent strain in 10,000 h. Creep deformation and stress rupture are related processes, however, and the selection of an alloy with good stress rupture resistance may be adequate for the initial design. In Fig. 2.7 the logarithm of the stress to produce rupture in 100,000 h is plotted against temperature for a number of pressure vessel alloys. To allow for the selection of other design lives, these and other data have been plotted on the basis of the logarithm of stress versus the Larson-Miller parameter in Fig. 2.8. Here the number of alloys has been greatly expanded. The curves for most of the alloys have been based on short-time test data, say 100 to 1000 h. Hence, the use of the curves should be restricted to short-time applications, say 10,000 h or less. The refractory alloy data correspond to tests in vacuum. All other data were obtained in air. It is important to recognize that thermal mechanical treatment and product form influence the relative strength of the alloys. The trends shown are selected to be typical of each alloy.

2.9 RELAXATION

Stress relaxation is an important consideration in many design problems. Typical uses for relaxation data include the estimation of the loss of a load in a bolt or spring as a function of time, the decrease in the residual stress introduced during fabrication or as a result of a thermal transient, and the redistribution of stresses in a component subjected to thermal or stress gradients. Relaxation data are derived from many types of tests and many specimen geometries. These include uniaxial, multiaxial, and bending stress conditions. The data reported in this section were largely produced from tests conforming to ASTM E328. Here the specimen is strained to the desired stress or strain level, then the strain is held constant to within ±20 microstrain, and the load needed to keep the strain constant is recorded as a function of time. Depending on the specific alloy, the initial strain may exceed the proportional limit. In some tests the speci-

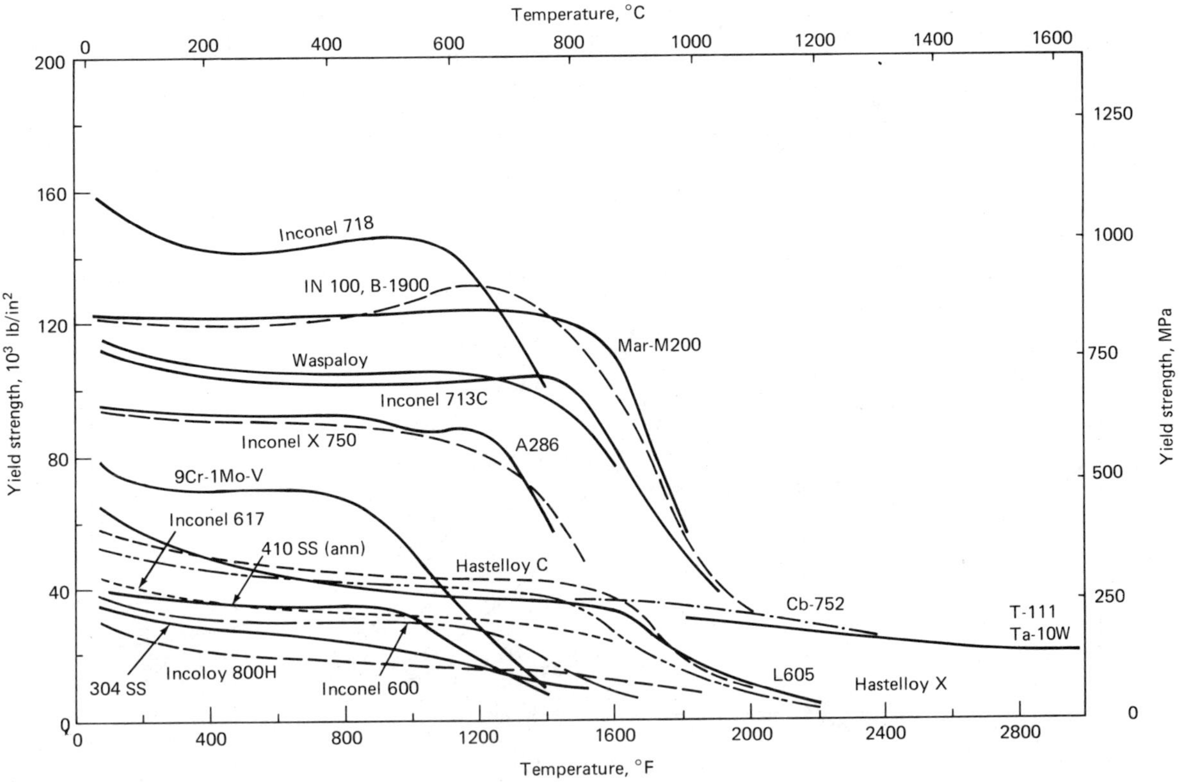

Fig. 2.5 Typical yield strength versus temperature for high-temperature alloys.[4,5,10–14]

men may be reloaded in tension several times, while in other tests the specimen may be cyclically strained from tension to compression before the relaxation period. Most of the data reported here represent the first loading behavior in tension from uniaxially strained test bars. Relaxation tests can last anywhere from 0.1 to 10,000 h. For bolting, long times are of most interest. Structures tend to complete most of their relaxation-induced stress redistributions within 100 h, however, so this time interval was chosen as a basis for reporting the data. A second reason for this choice stems from the fact that long-time relaxation behavior can be estimated reasonably well from a knowledge of the stress dependence of the minimum creep rate. Shorter-time relaxation behavior, however, is dominated by primary creep, which is not very well documented for most engineering alloys. With repeated loadings austenitic steels with high hardening capability become stronger. Bainitic and martensitic steels with low hardening capability tend to weaken.

Relaxation data are provided in Fig. 2.9, which plots the logarithm of the 100-h residual stress against temperature. Only a few alloys are represented, but often similar alloys in the same class exhibit similar relaxation strengths. The strengths depend to a significant extent on the starting strain and the precondition of the alloy and will vary from one heat to another. Taking these factors into account, the trend curves represent the typical behavior of an alloy initially strained in the range of 0.2 to 0.3 percent. The relaxation strengths more or less follow a ranking similar to the rupture strengths reported earlier.

2.10 FATIGUE

The type of fatigue data needed in high-temperature design vary greatly with the application. However, most of the available data on metals and alloys fall into two categories, (1) low-cycle-fatigue data produced by strain-controlled tests and limited to less than 10^6 cycles and (2) high-cycle-fatigue data produced by stress-controlled tests that usually fall into the cycle range from 10^5 to 10^8. It is possible to estimate the low-cycle-fatigue response for continuous cycling from a knowledge of the true fracture strain and the ultimate tensile strength, and such guidance is cited in Ref. 7. Generally, alloys of similar strengths and ductilities exhibit similar low-cycle-fatigue behavior, with only a modest effect of temperature. Typical data are provided in Fig. 2.10, which shows a number of plots for log strain range versus log cyclic life. The ductile pressure vessel alloys exhibit good fatigue resistance at high strains, while the strong cast and precipitation-hardened alloys exhibit the best fatigue resistance in the high cycle regime.

The fatigue curves such as those shown in Fig. 2.10 are representative of behavior under rapidly cycling conditions,

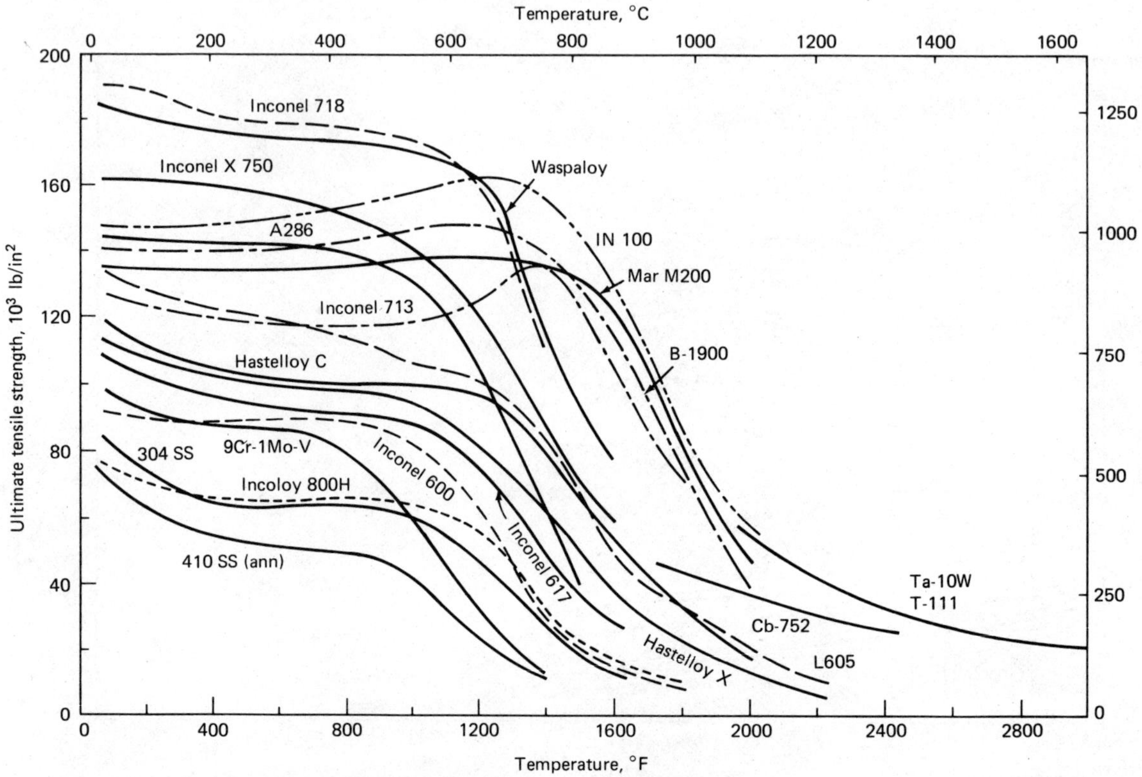

Fig. 2.6 Typical ultimate strength versus temperature for high-temperature alloys.[4,5,10–14]

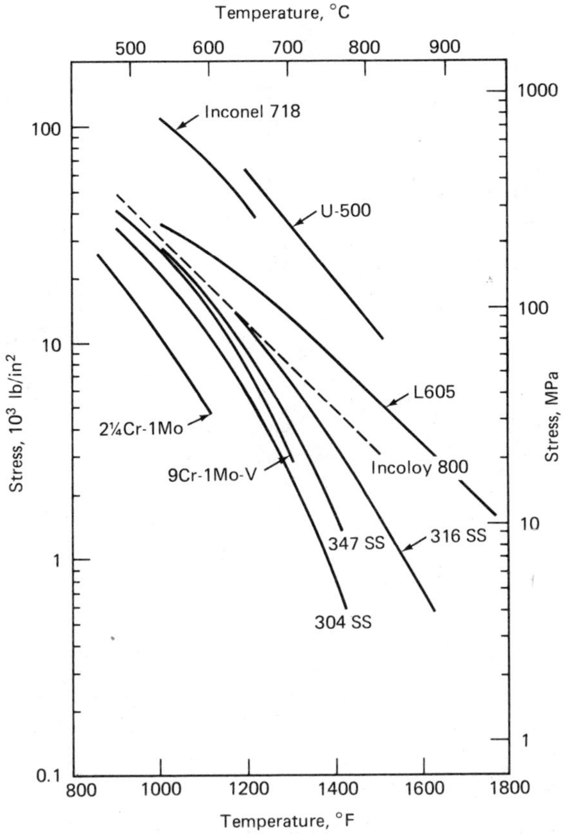

Fig. 2.7 Typical stress to produce rupture in 100,000 h for high-temperature alloys.[3,4,10,12,14]

say 0.1 Hz or more. At lower cycling rates, creep and environmental effects reduce the life. For many of the austenitic alloys it is possible to sum damage using Minor's rule for fatigue and Robinson's rule for creep. When the damage sum adds to unity, or some specified value less than unity, failure is said to be imminent.

The stress fatigue limit in the high cycle regime is often taken as the stress to produce failure in 10^8 cycles. In Fig. 2.11 the stress amplitude corresponding to this definition is plotted against temperature. Data are available on only a few alloys. Generally, the fatigue limit can be taken as a fraction of the ultimate or yield strength, and comparison of the data in Fig. 2.11 with the curves in Figs. 2.5 and 2.6 reveals similar rankings for the alloys. One exception is for the high-strength superalloys containing aluminum, such as alloy 713. These show relatively poor fatigue strength at low temperature and an increase in strength above 1200°F (649°C). The high-cycle-stress fatigue data are often produced by bending and flexure tests, rather than by axial tests, hence the data do not always agree with data produced by the extrapolation of strain-controlled fatigue curves. As before, however, many design problems involve the superposition of cyclic stresses on mean stresses. For many alloys the linear damage summation, mentioned above, is conservative. As with all fatigue problems, greatly reduced life may be encountered when notches are present.

2.11 OTHER CONSIDERATIONS

No data are provided here on the emerging crack growth technology. Indeed, high-temperature fatigue crack growth and creep crack growth are subjects of much concern in high-temperature design. The designer may find such information in Refs. 5, 8, 14, 18.

No data have been provided on environmental effects that are very important in high-temperature design, especially in the area of fatigue and crack growth.

Aging effects have not been addressed. This may be an important consideration in alloys that embrittle during high-temperature service. Information for a few alloys is contained in Refs. 5 and 14.

Cast metal and weld metal behaviors have not been included. Information on the cast metal versions of the fabricable alloys (groups 1 through 4 in Table 2.1) may be found in Refs. 5, 12, 14–16. Often the cast metals have poor short-time strengths but approach wrought metal properties at long times. A list of matching weld metals may be found in Ref. 10. Weldment strength relative to base metal is highly var-

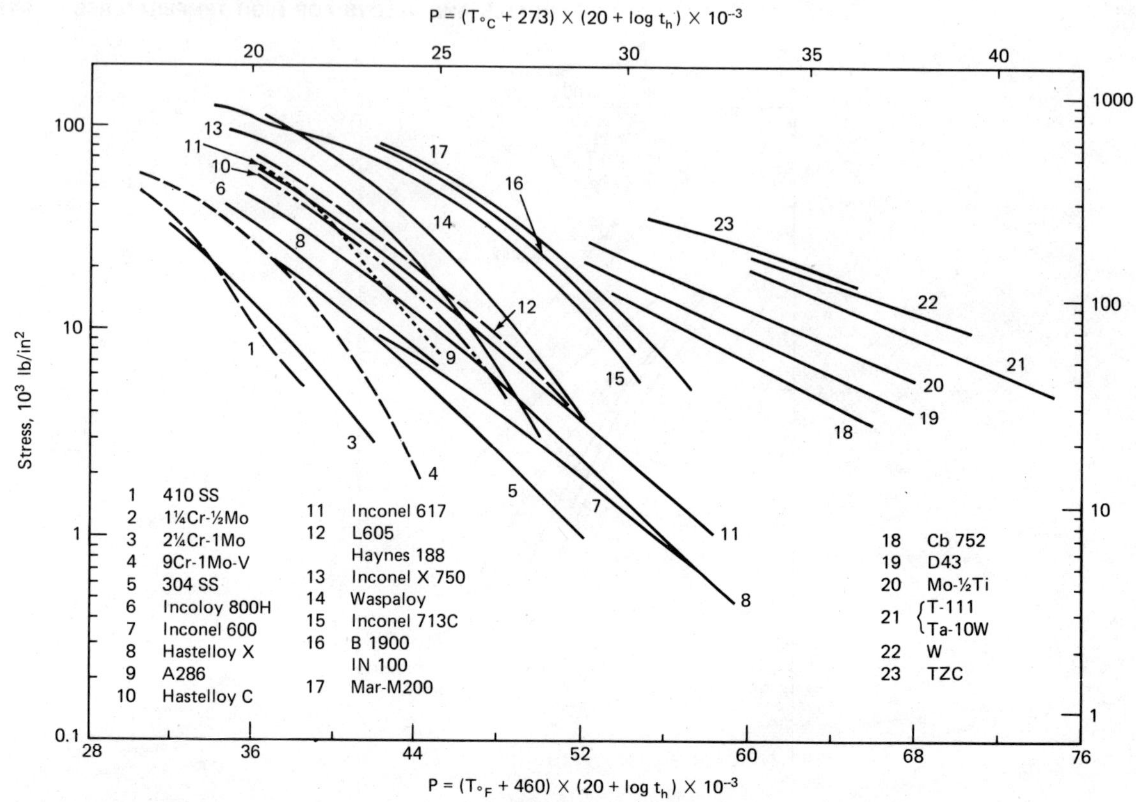

Fig. 2.8 Typical rupture stress versus Larson-Miller parameter for high-temperature alloys.[4–6,10,12–14,17]

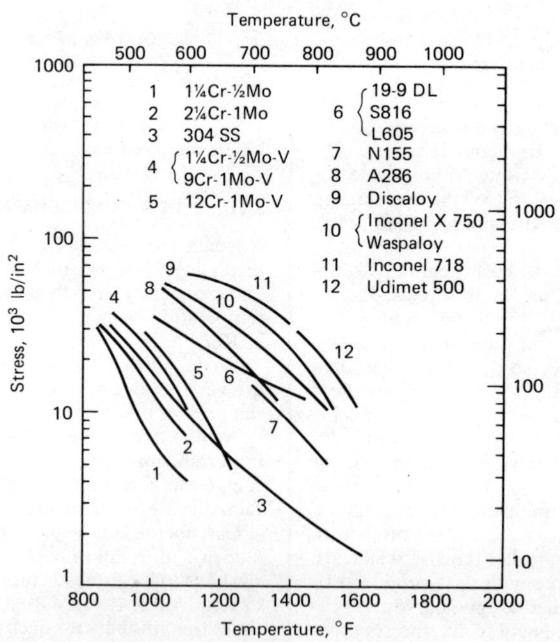

Fig. 2.9 Typical residual stress at 100 h versus temperature for high-temperature alloys.[2,6,14]

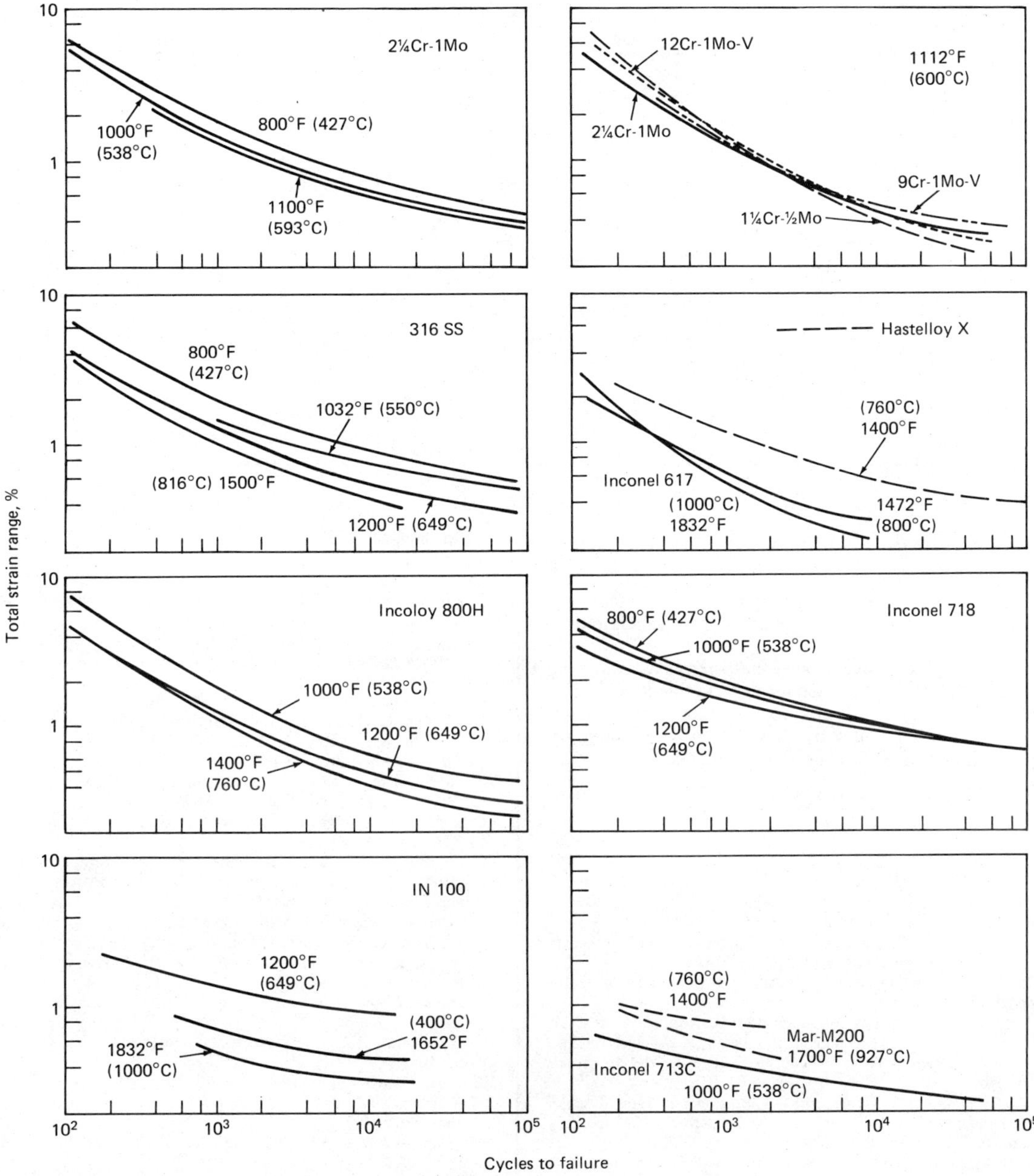

Fig. 2.10 Typical low-cycle-fatigue data for high-temperature alloys.[3,6–8]

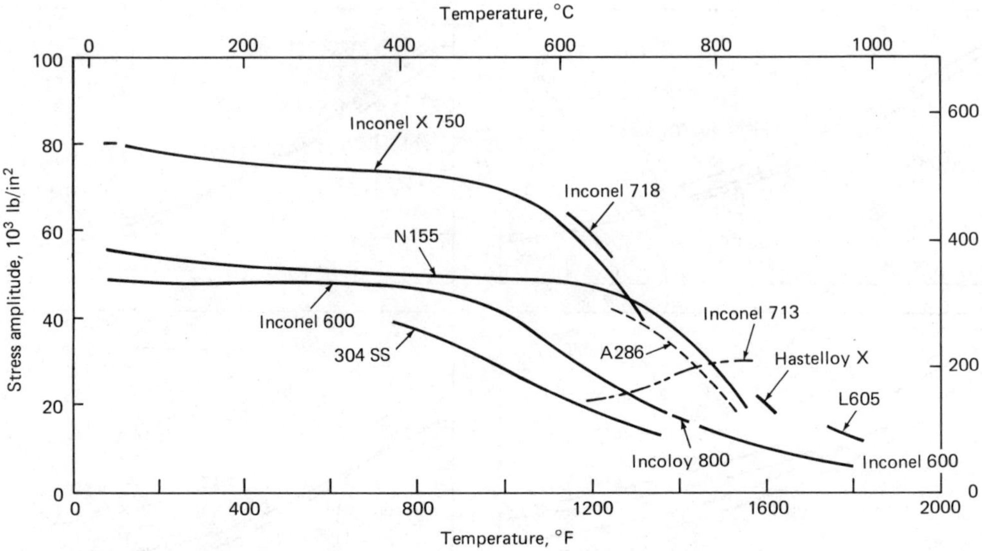

Fig. 2.11 Typical fatigue strength at 10^8 cycles versus temperature for high-temperature alloys.[5,9,14]

iable. The group 1 steels usually require heat treatment after welding, and care must also be taken with the precipitation-hardened alloys.

Finally, it should be mentioned that there exist many excellent alloys that have not been included in this compilation. Specific needs in regard to corrosion resistance, fabricability, or strength could be met by these alternative materials. Again, Refs. 5 and 10 provide some guidance.

REFERENCES

1. "ASME Boiler and Pressure Vessel Code," sec. VIII, div. 2, American Society of Mechanical Engineers, New York, 1984.
2. M. Manjoine and H. R. Voorhees, "Compilation of Stress-Relaxation Data for Engineering Alloys," DS 60, American Society for Testing and Materials, Philadelphia, PA, 1982.
3. "NRIM Creep Data Sheets," National Research Institute for Metals, Tokyo, Japan.
4. "Progress in Analysis of Fatigue and Stress Rupture," MPC-23, American Society of Mechanical Engineers, New York, 1985.
5. *Metals Handbook,* vols. 1–3, American Society of Metals, Metals Park, OH, 1983.
6. J. B. Conway, R. H. Stentz, and J. T. Berling, "Fatigue, Tensile, and Relaxation Behavior of Stainless Steels," TID-26135, U.S. Dept. of Commerce, Springfield, VA, Jan. 1975.
7. G. R. Halford and S. S. Manson, "Application of a Method of Estimating High Temperature Low Cycle Fatigue Behavior of Materials," NASA-TM-X-52357, 1967.
8. C. R. Brinkman, "High Temperature Time-Dependent Fatigue Behavior of Several Engineering Alloys," *Int. Metals Rev.,* **30**, 235–258 (1985).
9. P. Soo and J. G. Y. Chow, "The Effects of Mean Tensile Stresses on High-Cycle Fatigue Life and Strain Accumulation in Some Reactor Materials," Rep. BNL-NUREG-50654, Brookhaven National Laboratory, Brookhaven, NY, May 1977.
10. "Metals Progress 1985 Materials and Processing Databook," vol. 128, no. 1, American Society of Metals, Metals Park, OH, 1985.
11. *Carpenter High Temperature Alloys,* Carpenter Steel Company, Reading, PA, undated.
12. F. Simmons and J. A. Van Echo, "Elevated-Temperature Properties of Stainless Steels," DS 5-S1, American Society for Testing and Materials, Philadelphia, PA, 1965.
13. "Use of Computers in Managing Material Property Data," MPC-14, American Society of Mechanical Engineers, New York, 1984.
14. *Aerospace Structural Materials Handbook,* Syracuse University Press, Syracuse, NY, 1967.
15. *Haynes High-Temperature Alloys,* Haynes Stellite Co., Kokomo, IN.
16. *Huntington Alloys,* Huntington Alloy Products Division, Huntington, WV, 1970.
17. J. B. Conway and P. N. Flagella, *Creep-Rupture Data for the Refractory Metals to High Temperatures,* Gordon and Breach, New York, 1971.
18. *Atlas of Fatigue Curves,* American Society of Metals, Metals Park, OH, 1985.

Chapter 3

Plastics for Elevated Temperatures

GEORGE MARKUS

Advanced Plastics Design, Inc.

3.1 BASIC CONCEPTS AND DEFINITIONS

Introduction

Plastic materials are specified in ever more demanding applications. Their success in technically challenging fields is based invariably on appropriate combinations of performance characteristics or functionalities. Commonly, performance characteristics of interest include heat and chemical resistance, good electrical properties, light weight, high mechanical strength, and productive processibility.

Heat-resistant plastics products have applications on aircraft, on automotive engines, in electrical/electronic equipment, and in almost all aspects of demanding, contemporary life. In order to meet the demand, high-performance plastics, commonly referred to as engineering and specialty plastics, have been developed.

Designers face a complex task in selecting the proper material grade for new applications due to the unique behavior mechanisms of plastics and the complexity of requirements which are presented by multifunctional applications.

Definitions

Creep: Time-dependent dimensional change of a material under load following the initial elastic deformation. It is referred to as "cold flow" when it occurs at room temperature, and it is evaluated using ASTM D2990, "Test Methods for Tensile, Compressive, and Flexural Creep and Creep Rupture of Plastics."

Creep (apparent) modulus: Time-dependent modulus of elasticity, determined by the applied stress over the sum of the initial and the time-dependent creep strain. The creep modulus is often presented as a function of time at specific stress and temperature levels on a log-log coordinate system. Creep-modulus data (typically collected over an approximately 1000-h time span) can, within limits, be extrapolated in a log-log coordinate system. The effect of thermal oxidative or heat aging has to be taken into consideration when determining the working stress from extrapolated values. Normally the creep modulus is applied with a safety factor in actual design.

Creep rupture strength: Load-bearing capability for a specific period of time under specific stress and temperature conditions.

Glass transition temperature: Temperature region in which an amorphous polymer or the amorphous portion of a partially crystalline polymer changes upon an increase in temperature, from a relatively hard and brittle state to a viscous and rubbery state.

Heat aging: Degradation of the physical properties of a plastic due to chemical changes and reactions associated with long-term exposure to elevated temperature. See ASTM D794, "Standard Practice for Determining Permanent Effect of Heat on Plastics," ASTM D1870, "Standard Test Method for Elevated Temperature Aging Using Tubular Oven," and ASTM D3045, "Standard Recommended Practice for Heat Aging of Plastics Without Load."

Heat distortion temperature (*also referred to as deflection temperature under flexural load*): Temperature at which a standard test bar deflects 0.010 in (0.0254 cm) under a stated load of either 66 or 264 lb/in^2 (455 or 1820 kPa). Tests are performed using ASTM D648, "Standard Method of Test for Deflection Temperature of Plastics Under Load." It is not intended to be a direct guide to high-temperature limits for specific applications, but it may be useful in comparing the relative behavior of different plastics.

UL temperature index [*also referred to as relative thermal index* (*RTI*)]: Defines different maximum temperatures for the retention of electrical properties (dielectric strength), for the half-life of tensile strength excluding impact strength, and for the half-life of tensile strength including impact strength. The specimens for the tests are exposed to elevated temperatures in air and under stress-free conditions. The UL temperature index is obtained by extrapolation. The RTI is not applicable if a part is exposed to mechanical load or to chemicals.

Types of Plastics

Plastic materials can be placed into three major categories: thermoplastics, thermosets, and elastomers. The primary emphasis of this chapter is on thermoplastics and thermoset materials.

Among the thermoplastic materials, the ones that possess an amorphous molecular structure generally exhibit good surface appearance, flatness, high impact strength, low shrinkage, and transparency. Crystalline and semicrystalline polymers possess, in general, good lubricity, fatigue strength, dimensional stability, and higher heat resistance than that of amorphous materials. Crystalline structure also results in good chemical resistance, desirable melt flow, and fast, clean molding properties. Blends, alloys, and special filler, reinforcement and additive systems allow the designer to tailor functional properties of thermoplastic materials.

The thermoset materials possess, in general, good mechanical, dimensional, and electrical stability; good heat and chemical resistance; and low material cost.

Effects of Temperature

The effect of elevated temperature on the properties of a plastic material are dependent upon the exposure time and the molecular structure of the polymer. Generally, if a thermoplastic (amorphous or semicrystalline) material is exposed to an elevated temperature for a short period of time, a decrease in load-bearing capability, an increase in impact strength, and an increase in dimensions can be observed. These changes are usually accompanied by changes in electrical properties. Crystalline and semicrystalline thermoplastic materials show a significant loss of mechanical load-bearing capability if the temperature exceeds the glass transition temperature. If a thermosetting material is exposed to an elevated temperature for a short period of time, no significant changes in mechanical properties can normally be observed.

When thermoplastic materials are subjected to a high temperature for a long period of time, additional effects can be observed beyond the short-term changes: creep (if the part or specimen is exposed to load) and thermal oxidative degradation. No significant creep is observed with thermosetting materials because of the existence of strong chemical bonds between the molecules. The main cause of failure with these plastics is thermal aging.

The heat resistance of plastic materials is expressed in terms of their resistance to thermal aging or their short- and long-term load-bearing capability at elevated temperatures. The UL temperature index, the strength versus temperature curves, and the creep resistance data are provided by plastics suppliers to furnish information for these characteristics. However, quite often additional environmental factors (chemicals, ultraviolet radiation, etc.) that are not accounted for in commercial data sheets, but are important to the applications, must be considered. This is because these additional factors tend to degrade performance.

Factors Affecting Material Performance

The performance of a plastic material is greatly dependent upon environmental factors. The most common environmental factors affecting the overall (resultant) performance of a plastic material are

- Temperature
- Time
- Mechanical load
- Exposure to chemicals and water
- Ultraviolet exposure

Other nonenvironmental factors which also affect the performance include

- Molecular or fiber orientation
- Degree of crystallinity
- Molded-in stresses
- Weld-line integrity
- Surface imperfections (gate mark, etc.)

3.2 MATERIAL SELECTION—A SYSTEMATIC APPROACH

Due to the complexity of the behavior of plastics and the need to satisfy the engineering, manufacturing, and marketing objectives of the finished part, the plastic material selection is best executed with a systematic approach. This approach should identify all the performance requirements for the part and consider all the interacting factors of the comprehensive environment of the plastic product. It should also determine the resultant material performance as a consequence of the interplay of environmental factors and compare it to the required performance. The comprehensive environment includes the firm and foreseeable abnormal conditions in the field, in processing, in assembly, and in shipping and handling. The interplay of the different factors of the environment of the product can be extreme; for example, processing can impair chemical resistance through molded-in stresses.

The more detailed design of the part takes into consideration specific (design-dependent) characteristics of the selected material, that is, the design is usually executed around specific properties of the selected materials. This fact puts extra emphasis on the importance of the selection of the appropriate material grade.

The task of material selection begins with the identification of the requirements that the plastic part must satisfy during its service life. These requirements must reflect precisely the performance of the expected part and will form the basis of the communication between the designer and the material suppliers.

Gathering the Requirements

A good way to clarify all aspects of material requirements is to employ a properly designed questionnaire. An appropriate and fully completed requirement specification combined with the knowledge of material limitations are the key factors of successful part performance. A carefully prepared questionnaire is especially useful in soliciting information beyond that which is contained in or explicit in publicized material data sheets.

The first section of information of the questionnaire should describe the required agency approvals. The next section of information should apply to the elements of the product's environment, including concentration, exposure time, and temperature of all ambient fluids and chemicals. In this section the resistance to ultraviolet and outdoor exposure is also to be clarified together with the time of exposure.

The following section of the questionnaire should apply to the aesthetic requirements of the plastic product in terms of surface finish, transparency, color, and allowable surface imperfections.

The next section should describe the mechanical requirements that are to be met by the product, in terms of the types, the magnitude, and the cycle (if applicable) of stresses, along with both operating and peak temperatures. Maximum allowable deflection should also be specified.

Usually, the geometry-dependent requirements of part performance necessitate a translation into standard material characteristics. This can be a difficult task. For example, a correlation between the required drop impact strength of a given part and the standard notched or unnotched impact values of a material may be difficult to achieve. Advanced computer programs are of great help to designers in defining certain aspects of the required part performance. Finite-element analysis programs can be used to determine the mechanical strength criteria or the temperature distribution of a part.

The last section of the questionnaire should apply to the electrical requirements, with a consideration of the operating temperature and the relevant wall thicknesses of the product. The performance specifications should also clarify the service life of the product and the possible processing constraints of the candidate materials.

After establishment of the required performance profile, the following preliminary categories of required, inherent material properties can be defined:

1. Essential and design-independent properties, such as chemical resistance, agency recognition, heat resistance, processibility, moisture absorption, and electrical performance
2. Essential but design-dependent properties, such as creep rupture strength, modulus, impact strength, and fatigue resistance

These categories can be used as screening devices to eliminate those material grades which do not possess the desired, inherent properties, such as heat resistance, agency recognition, and chemical resistance.

Computerized Material Selection

Computer programs for the selection of plastics material that are based on data banks of material suppliers are of great help to designers. One new, computerized material selection program, described in Ref. 1, can be used to accelerate the process of selection of candidate materials by providing logic access to a large database. This program enables the user to interpolate within given ranges of variables to determine the effect of interacting factors. The program also takes into consideration economic aspects of the material selection and it can be tied directly to molding (injection molding) optimization programs. A significant amount of expense and time, needed for the traditional trial-and-error approach to material selection, can be avoided with this and similar techniques.

Selection from Material Supplier Data Sheets

There are nonetheless simple applications for which the final material grade can be selected directly from standard supplier data sheets. This approach should be restricted to those cases having a very limited number of interacting factors. This makes possible the direct translation between required part performance and the usually single-valued data-sheet properties.

To evaluate the resultant material performance for a complex plastics material application involving elevated temperatures, the following procedure can be used. In this procedure two major categories are established: short-term (hours, days) and long-term exposure. This emphasizes the dominant influence of time on the performance of plastics at elevated temperatures.

After definition of the resultant material performance as a function of its environment, the comparison with the required performance profile can be executed and the final choice can be made. To evaluate materials from the economy standpoint, ranking can be carried out based on values of unit strength per material cost.

It is also important that part design be in accordance with the material supplier's recommendations with regard to fillet and corner radii, wall transitions, minimum and maximum wall thicknesses, sidewall tapers, and surface finish. Material suppliers usually also provide design guidelines for assembly methods, such as ultrasonic welding, adhesive bonding, and snap fits.

Procedure for Material Selection

Category I: Short-Term Exposure

CLASS 1: *Elevated Temperature Only*

1. Select candidate materials based on melting point, glass

Table 3.1. Plastics with High UL Temperature Index

Generic name	Molecular structure	Characteristics	Processing techniques	Applications
Allyl	Thermoset	Excellent retention of electrical properties at elevated temperatures and at high humidity; good chemical resistance, high mechanical strength, good dimensional stability	Compression molding, composite technology	Electronic connectors for electrical/electronic, computer, and aerospace systems requiring high reliability under long-term, harsh environmental conditions; in a form of impregnated glass reinforcement, including tubing, junction boxes, aircraft and missile components
Bismaleimide	Thermoset	Excellent resistance to chemicals at elevated temperatures, especially to acids	Compression molding, composite technology	Wiring boards
Fluoroplastics (polytetrafluoroethylene, fluorinated ethylene propylene copolymer, perfluoroalkoxy, polychlorotrifluoroethylene, ethylene-chlorotrifluoroethylene copolymer, ethylene-tetrafluoroethylene copolymer, polyvinylidene fluoride, polyvinyl fluoride)	Thermoplastics	Operating temperature range of −200 to 260°C, outstanding electrical characteristics; excellent flammability properties and outstanding chemical resistance and lubricity	Melt processing, compression molding	Wire jacketing for high-performance wire and cable applications, coatings, and linings for chemical processing equipment, filter housing, pump casing, and impellers for fluid- and gas-handling systems whenever corrosion resistance is crucial; also in connectors, appliances, aircraft, mechanical seals, rings, bearings
Phenolic	Crystalline thermoset	Good electrical insulating characteristics and dimensional stability, heat and chemical resistance, and high rigidity; retention of electrical insulating properties at elevated temperatures; good creep resistance, low shrinkage, and high stress crack resistance	Injection, compression, transfer molding	Metal replacement, automatic transmissions, brake components, rotors, fuse blocks, coil towers in the automotive industry; also in appliances: plugs, receptacles, circuit breakers, light sockets, connectors in the electrical/electronics industries
Polyamide-imide		Excellent mechanical performance, including exceptional creep resistance, high heat deflection temperature, and outstanding thermal stability; good electrical insulating properties, excellent radiation and chemical resistance	Injection molding	Metal replacement in the aerospace industry, jet engine components, compressor and generator parts, and electronic devices; bushings, seals, and flow control parts for fluid- and gas-handling systems; transmission parts, universal joints in the automotive industry
Wholly aromatic copolyester	Crystalline thermoplastic [liquid crystal polymer (LCP) or self-reinforcing polymer]	Extremely high melt temperature, good moldability, excellent mechanical property, outstanding thermal oxidative stability, and superior chemical resistance	Injection molding	Electrical components, electronic devices, chemical process and oil-field equipment, and aerospace, whenever high heat and exceptional chemical resistance are needed
Polybutylene terephthalate (PBT)	Semicrystalline, thermoplastic	Excellent mechanical and weatherability performance, good chemical resistance and thermal and electrical insulation properties	Extrusion, injection molding	Reinforced and unreinforced grades in all major industrial segments, e.g., transportation, automotive, electrical/electronics, household, appliance, medical, power tool, lighting equipment, safety, and protective items

Polycarbonate, polyphthalate carbonate	Amorphous thermoplastic	Excellent clarity and colorability with UV and thermal stability; excellent mechanical properties including high strength and impact strength; good inherent flame retardancy and surface appearance	All thermoplastic processing methods	Business machine and computer housings and internal structural components; lenses: automotive head and tail lamps, ophthalmic, and industrial, automotive interior, trim parts, instrument panels, lamp reflectors, home appliances, medical and surgical equipment, packaging, including bottles
Polyethylene terephthalate (PET)	Semicrystalline thermoplastic	Excellent mechanical, thermal, and electrical properties, good weatherability and chemical resistance	Extrusion, injection molding	Similar to those of PBT; widely used in packaging
Polyester	Thermoset	Good mechanical performance, stable electrical properties in elevated-temperature environment, good molded-in colorability	Injection, compression, transfer molding	Parts for the electrical/electronics industry, appliance components
Polyetheretherketone (PEEK)	Semicrystalline thermoplastic	Excellent chemical resistance, even at elevated temperatures; outstanding radiation resistance; good mechanical properties; inherent UL94 V-O rating and very low smoke emission	Extrusion, injection molding	Military, aerospace, nuclear plant, oil wells, and compressor parts
Polyetherimide	Amorphous thermoplastic	Good chemical resistance; good hydrolytic stability; excellent autoclave resistance; good UV and radiation resistance; excellent mechanical strength and retention of strength under load and temperature; excellent flame and smoke resistance	Extrusion, injection molding	Electrical/electronic, medical, computer, and business equipment, appliances, and military parts; also in transportation and aircraft applications
Thermoplastic polyimide	Semicrystalline thermoplastic	Exceptionally high temperature resistance, toughness, flammability; good electrical properties, dimensional stability, and chemical resistance	Compression molding	Aerospace, aircraft, automotive, transportation, electronics, military, and environmental protection applications
Thermoset polyimide		Excellent mechanical performance and its retention at elevated temperatures; creep very small, even at high tempratures; good electrical properties and chemical resistance	Compression, injection, transfer molding and extrusion	Aircraft engine parts, printed wiring boards, military applications, bushings and bearings, chip carriers, and automotive parts
Polyphenylene sulfide	Crystalline thermoplastic	Good mechanical properties with the exception of elongation; inherent flame resistance and good chemical resistance	Injection molding	Electrical/electronic applications, chemical processing equipment; in automotive, in under-the-hood use
Silicone	Thermoset	Silicone belongs to the family of elastomers; however, because of its outstanding temperature resistance, low compression set, tear strength, and the ease of processing, it is included in this listing	Injection, transfer, compression molding, extrusion, dispersion, foaming	Automotive seals, boots, O-rings, insulators in the electrical/electronic industry; in the form of dispersion-coated fabrics in hot gas-handling systems
Polysulfone	Amorphous thermoplastic	Transparent, good electrical properties and their retention at elevated temperatures; good chemical and hot-water resistance and acceptable mechanical performance, even at elevated temperatures	All thermoplastic forming methods	Medical equipment, appliances, including cookware for microwave ovens, connectors, bobbins, television components, insulator film, circuit boards for the electrical/electronic industry; also in chemical processing equipment, fluid and gas handling, and aerospace components

Table 3.1. Plastics with High UL Temperature Index (*Continued*)

Generic name	Molecular structure	Characteristics	Processing techniques	Applications
Polyarylsulfone	Amorphous thermoplastic	High heat resistance, stable electrical properties, good mechanical performance, acceptable resistance to environmental stress (cracking), transparent	Injection molding, extrusion	Circuit boards, connectors, lamp housings, and electrical motor parts; also in gas- and fluid-handling systems
Polyethersulfone	Amorphous thermoplastic	Good mechanical performance including high ductility and good creep resistance; outstanding thermal ability and inherently low flammability and oxygen index; good chemical resistance; however, certain cleaning solvents attack it if the part is heavily stressed	Injection molding, extrusion, blow molding, compression molding, vacuum or pressure forming	Integrated circuits, boards, connectors, terminal blocks, printed circuit boards for the electrical/electronic industry, pump housings, bearings, bearing cages for fluid-handling systems; also in transparent indicator lights and glasses and many medical parts

transition temperature T_g, deflection temperature under load T_{dul}, and flammability.

2. Check electrical properties (if applicable) as a function of temperature.

3. Evaluate dimensional control: postmolding shrinkage, effect of glass transition, thermal expansion, and the possibility of warpage.

CLASS 2: *Elevated Temperature Plus Load Only*

1. Use apparent modulus and strength values since these decrease with increasing temperature.

2. Check impact strength. It generally improves with increasing temperature.

3. Evaluate abrasion resistance (if applicable). It decreases at higher temperatures. Individual tests are usually required.

4. See also steps 2 and 3 of Class 1 of this category.

CLASS 3: *Elevated Temperature Plus Chemicals Only*

1. Select material based on standard chemical resistance data (usually available). If data are not available, tests should be conducted, for example, by exposing material to the particular chemicals through immersion, at elevated temperatures, under adequate safety measures. The results can be evaluated from the standpoint of surface appearance or dimensional change.

2. Check dimensional control. It can be influenced by absorption of the chemicals.

3. See also steps 2 and 3 of Class 1 of this category.

CLASS 4: *Elevated Temperature Plus Load Plus Chemicals*

1. Carry out tests under precautionary measures if data are not available (ASTM D2990). The specimen is immersed in the chemical at an elevated temperature and placed under tensile stress through the cap of the stainless steel container. The apparatus should maintain a constant stress level despite the elongation of the specimen. The test results can be evaluated based on the onset of yield.

2. Check dimensional control. It can be influenced by the absorption of chemicals.

3. See also steps 2 and 3 of Class 1 of this category.

Category II: Long-Term Exposure

CLASS 1: *Elevated Temperature Only*

1. Select the material based on the UL temperature index.

2. Test color stability (if applicable).

3. See also steps 2 and 3 of Category I, Class 1.

CLASS 2: *Elevated Temperature Plus Load Only*

1. Obtain creep resistance data from databases or tests.

2. Obtain fatigue resistance data (if applicable), because it decreases with increasing temperature.

3. See also steps 2 and 3 of Category I, Class 1, and step 2 of Class 1 of this category.

CLASS 3: *Elevated Temperature Plus Chemicals Only*

1. Conduct tests (as described in step 1 of Category I, Class 3) if chemical resistance data at continuous elevated temperatures are not available. Time to failure can be established at unacceptable surface deterioration or at the loss of 50 percent of original strength values.

2. See also steps 2 and 3 of Category I, Class 1, and step 2 of Class 1 of this category.

CLASS 4: *Elevated Temperature Plus Load Plus Chemicals*

1. Conduct tests with equipment noted in step 1 of Category I, Class 4, if data are not available. The specimen is immersed in the chemical and placed under a constant tensile stress at elevated temperature. Any extrapolation of the data from the stress testing should be carried out with extra caution due to the unpredictable behavior of plastics in this complex environment. A safety margin should be applied.

2. See also steps 2 and 3 of Category I, Class 1, and step 2 of Class 1 of this category.

3.3 CANDIDATE PLASTICS FOR ELEVATED-TEMPERATURE APPLICATIONS

The various types of plastic materials listed in Table 3.1 can have a relative temperature index of 120°C (248°F) or above and are candidate materials for elevated-temperature applications. Reference 2 is a detailed and comprehensive review of the properties of useful heat-resistant polymers.

REFERENCES

1. E. C. Bernhardt and L. T. Manzione, *Application of CAE in Injection Molding,* Hanser/Macmillan, New York, 1987.
2. J. P. Critchley, G. J. Knight, and W. Wright, *Heat-Resistant Polymers,* Plenum, New York, 1983.
3. G. Markus, "A Systematic Approach to Selection of Plastics for High Temperature Performance," presented at the 1985 Conf. on High-Heat-Resistant Plastics, Business Communications Co., Stamford, CT.
4. J. L. O'Toole, "Design Guide," in *Modern Plastics Encyclopedia, 1985–1986,* McGraw-Hill, New York, 1985.

Chapter **4**

Glass and Ceramics for Elevated Temperature Applications

K. P. R. REDDY
Corning Glass Works

Some of the elevated-temperature applications in which glass and ceramic materials are being used are catalyst substrates, heat exchangers, extrusion dies, lamp envelopes, cutting tools, refractories, cooking vessels, furnace tubes, spark plug insulators, nozzles, oxygen sensors, and space shuttle tiles. The characteristics of the material required for satisfactory performance in each of these applications are different. In this chapter a review of ceramic materials will be presented along with a discussion of some of their physical properties.

The quantitative design methodology for the reliable performance of brittle materials in high-stress applications has not been fully developed. However, there are some design guidelines which are useful, and these are described later in this chapter.

4.1 CERAMIC MATERIALS AND APPLICATIONS

Ceramics are essentially composed of inorganic nonmetallic materials, such as oxides, nitrides, borides, and carbides. The word *ceramic* is used to describe a broad range of materials, which we can classify as traditional ceramics and advanced ceramics. Traditional ceramics include products such as porcelain, pottery, cement, bricks, tiles, refractories, abrasives, and glass products. Only glass products from this group are covered in this chapter. Advanced ceramics represent newer materials developed in the last 20 years, which have unique or outstanding properties, such as greater temperature resistance, superior mechanical properties, greater chemical resistivity, or special electrical properties. Some of the advanced ceramics that are useful in high-temperature applications are alumina, mullite, zirconia, silicon nitride, silicon carbide, sialon, whisker composites, and glass-ceramics. It is preferable to classify the materials as glasses, glass-ceramics, and ceramics, basing the classification on their microstructure and refractoriness. Important materials in these three categories are described in the following.

Glasses

ASTM defines glass as an inorganic product of fusion that has been cooled to a rigid condition without crystallizing. Except for the nearest neighbors, the atomic arrangement in glasses is random (amorphous). When a liquid glass is cooled from the molten state, there is a continuous change in density with temperature, making it difficult to assign a single temperature as the melting temperature. Glasses are capable of accommodating many different elements, and therefore their properties can be tailored to a wide variety of applications. Most commercially important glasses are silicates. In particular, soda lime silica glasses (similar to Corning 0080) account for about 90 percent of the total glass tonnage made. This glass is relatively easy to melt from inexpensive raw materials. It is used for containers, flat glass, and lighting products, where exceptional chemical durability and heat resistance are not required. Another significant family of commercial glasses is borosilicate glass (e.g., Corning 7740). These glasses have lower coefficients of thermal expansion and good chemical durability. Some of the products made from these glasses are ovenware and sealed-beam headlamps. Glasses for higher-temperature use can be produced by incorporating alumina in an alkali silicate glass. These aluminosilicate glasses (similar to Corning 1723) are used for aircraft windows, lamp envelopes, and top-of-stove applications. Among single-component glasses, vitreous silica (Corning 7940) is the most important. It has excellent thermal shock resistance and is used for space vehicle windows and crucibles for growing silicon and germanium. However, this glass is expensive to produce due to higher processing temperatures and nonconventional manufacturing methods.

Glass products are manufactured by either molding, blowing, pressing, drawing, or rolling the molten glass. Strengthening methods such as tempering (rapid cooling of the surface of the glass articles), glazing/laminating (surface glass layer with lower thermal expansion coefficient than the core glass), or ion exchange (replacement of smaller ions by larger ions at the glass surface) can be employed to increase glass strength. All these methods strengthen the glass by introducing residual compressive stresses near the surface regions. To balance these compressive stresses, residual tensile stresses are generated in the interior of the glass article. Incorporation of significant tensile stored energy in the glass makes the glass fracture more violent, that is, more fragments are formed when strengthened glass fractures. The strengthening methods used depend therefore on the nature of the application.

Glass-Ceramics

Glass-ceramics contain both crystalline and amorphous (glassy) phases. The processing of these materials is similar to that for glass, except that the articles are heat-treated after forming to cause crystallization. The addition of nucleating agents such as titania, zirconia, platinum, and silver helps the glass crystallize homogeneously. The glass-ceramics are more refractory than their parent glasses. The morphology and the composition of the crystalline phase can be varied by adjusting the composition and the heat treatment schedule. Some important glass-ceramics are given in Table 4.1 along with their applications. The advantage of glass ceramics is that they can be produced economically, using glass-processing methods with tailored physical properties.

Ceramics

Ceramics are more refractory than glasses and glass-ceramics because they are essentially composed of crystalline phases of higher melting temperatures. The processing methods used for manufacturing ceramic materials are different from those used for glass. Normally the ceramic powder is molded to the desired shape by pressing the powder in a mold. Other forming methods such as injection molding and slip casting are also used. The shaped articles are fired at high temperatures to eliminate porosity and to develop strong bonding between the particles. The densification of the powder compact occurs by the migration of atoms from the interior of the particles to the pore surfaces. The size and shape distribution of the powder particles together with the firing schedule determine the porosity in the final product. Typically, the porosity in sintered parts for engineering applications ranges from 1 to 5 percent. For difficult to sinter materials it is necessary to incorporate additives that increase the atomic diffusion rates either by forming a liquid phase at the sintering temperature or by changing the lattice point defect concentration. Liquid-phase sintering additives in most cases limit the maximum use temperature of the materials.

The important ceramics that are useful in high-temperature engineering applications are alumina, zirconia, mullite,

Table 4.1. Commercial Glass-Ceramics

Material designation	Properties	Applications
Schott ZeroDur	Nearly zero expansion	Electric range tops, telescope mirrors
Corning 9608	Low expansion, high chemical durability	Cooking utensils
Corning 9455	Low expansion, high thermal and mechanical stability	Heat exchangers
Corning 9606	Low expansion, high transparency to radar waves	Missile radomes
Corning 9658	Machinable, thermally stable, high dielectric strength	Precision dielectric components, vacuum components

silicon carbide, silicon nitride, and sialon. A brief description of these materials follows.

Al_2O_3 Among the materials mentioned above, alumina has been the most widely used ceramic for applications such as furnace tubes, crucibles, high-pressure sodium vapor lamp envelopes, molten metal filters, metal cutting tools, electronic substrates, pump seals, and grinding balls. Addition of a small amount (0.1 wt %) of MgO, Y_2O_3, or NiO is necessary to produce dense alumina ceramics. In high mechanical stress applications, alumina can be used up to at least 1830°F (1270 K). Recently the mechanical properties of alumina have been improved by making a composite with zirconia or silicon carbide whiskers (about 20 wt %).

ZrO_2 Strong and tough zirconia ceramics have been produced recently by alloying zirconia with MgO, CaO, Y_2O_3, or CeO_2. The strengthening mechanism of zirconia can be understood by considering its crystal phases. The crystal structure of pure zirconia depends on the temperature. The transformation temperatures for the three crystalline modifications are

$$\text{Monoclinic} \underset{(1171^\circ\text{C})}{\overset{2140^\circ\text{F}}{\rightleftarrows}} \text{tetragonal} \underset{(2371^\circ\text{C})}{\overset{4300^\circ\text{F}}{\rightleftarrows}} \text{cubic}$$

The tetragonal-to-monoclinic transformation during cooling is accompanied by a volume increase which introduces strains on the order of 1.5 percent. Because this transformation strain exceeds the fracture strain, pure zirconia ceramics fall apart during cooling from the sintering temperature. Alloying zirconia with oxides such as MgO makes it possible to retain the cubic phase at room temperature. Fully stabilized zirconia (100 percent cubic phase) has a very high thermal expansion, making its use very limited. Best engineering properties are obtained with partially stabilized zirconia (PSZ). The type of crystalline phases required for optimum properties depends on the alloying oxide. In the case of Mg-PSZ, one type of microstructure contains tetragonal precipitates in a cubic matrix. The stress-induced transformation of tetragonal precipitates ahead of the crack tip is responsible for improved room temperature mechanical properties. If the tetragonal precipitates in Mg-PSZ are converted to monoclinic by heat treating at around 2010°F (1370 K), the thermal expansion coefficient decreases, and this material performs better in high-temperature applications. For zirconia stabilized with yttria or ceria the microstructure for optimum mechanical properties contains mostly the tetragonal phase with a grain size of 0.5 to 1 μin (0.012 to 0.025 μm). Depending on the application, the properties of partially stabilized zirconia can be tailored by either changing the concentration of the alloying oxide or changing the firing schedule. Zirconia ceramics are used in applications such as extrusion dies, oxygen sensors, nozzles for casting steel, oil pump valve parts, pen tips, scissors for cutting tough fibers, pump seals, heating elements, and refractories.

$3Al_2O_3 \cdot 2SiO_2$ Mullite is being considered for high-temperature applications due to recent advances made in the processing of this material. Sintering of alumina-zircon ($ZrSiO_4$) mixtures has yielded a composite of mullite-zirconia with good high-temperature mechanical properties. It was demonstrated that zirconia-toughened mullite performs well as a cover plate for the cylinder head of a piston engine. The present uses of mullite include furnace tubes and thermocouple protection tubes.

SiC Among the existing commercial engineering ceramics, silicon carbide is the most refractory material. Two types, namely, reaction-bonded silicon carbide and sintered silicon carbide, are commercially available. The reaction-bonded silicon carbide is produced by firing a powder compact of silicon carbide and carbon that has been infiltrated with silicon. Carbon reacts with silicon and forms silicon carbide at the particle contacts and at the pore surfaces. It usually contains free silicon and therefore is less refractory than sintered silicon carbide. Sintering aids such as boron and carbon, aluminum and carbon, or beryllium and carbon have been used to produce dense silicon carbide by conventional sintering methods. Potential applications in which silicon carbide can be used are thermocouple protection tubes, heat exchanger components, advanced gas turbine engine parts, rocket nozzles, heating elements, and applications in which wear and corrosion resistance are required (seals, bearings, pump sleeves, valve trim).

Si_3N_4 Silicon nitride received considerable attention due to its promising physical properties. However, pure silicon nitride could not be sintered to high density because of the low atomic diffusion rates and significant decomposition of silicon nitride at elevated temperatures. Two different methods are used commercially to manufacture silicon nitride. In the reaction sintering method, the silicon powder is compacted to the desired shape, and later the compact is reacted with nitrogen gas at high temperatures. The reaction between silicon and nitrogen is highly exothermic, and therefore care must be exercised to prevent the melting of silicon. The molar volume increase caused by the formation of silicon nitride reduces the porosity. Typically, the porosity after reaction sintering is around 20 percent. This material is used for thermocouple sheaths, kiln furniture, welding torch nozzles, and for handling molten nonferrous metals.

Dense silicon nitride is produced using sintering aids, such as MgO, Al_2O_3, Y_2O_3, and Mg_3N_2. These additions cause the formation of a grain boundary phase which is less refractory than silicon nitride grains. The potential applications in which sintered silicon nitride could be used are draw dies, cutting tools, bearings, and advanced gas turbine engine components.

Sialon The composition of sialon can be represented by $Si_{6-z}Al_zO_zN_{8-z}$, where z denotes the number of nitrogen atoms substituted by oxygen atoms. Commercially, sialons are produced by sintering a powder mixture of silicon nitride, alumina, and yttria. The high-temperature mechanical properties have been improved by crystallizing the grain boundary phase (yttrium aluminum garnet). Recently it was shown that this material is suitable for high-stress applications up to 2300°F (1525 K). Among the nitride ceramics, sialons are more promising for high-temperature applications since strong, refractory materials can be produced using conventional powder-processing methods. Sialons are presently used for cutting tools and in other wear applications.

4.2 MECHANICAL PROPERTIES

Elastic Properties

The elastic properties of ceramic materials are central in determining the mechanical behavior and thermal shock resistance. The elastic extension of a body under stress corresponds to increased separation between atoms and thus

Table 4.2. Elastic Properties

	Melting temperature, °F	Density, lb/in^3	Poisson's ratio	Young's modulus RT, 10^6 lb/in^2
		Glasses		
0080*	1285†	0.09	0.22	10.2
1723	1666	0.095	0.24	12.5
7740	1510	0.081	0.20	9.1
7940	2885	0.079	0.16	10.5
		Glass-ceramics		
9606*	...	0.094	0.24	17.2
9608	...	0.09	0.25	12.5
9658	...	0.092	...	9.4
		Ceramics		
ZrO_2 (Mg-PSZ)	4530	0.206	0.23	30
Al_2O_3	3760	0.141	0.26	54
Mullite	3490	0.116	0.25	21
Cordierite	2730	0.094	0.30	20
Si_3N_4 (sintered)	3450	0.119	0.24	45
Si_3N_4 (reaction sintered)	...	0.094	0.24	20
SiC (sintered)	4890	0.114	0.14	58
SiC (reaction bonded)	...	0.097	0.22	41
Sialon	3400	0.116	0.29	46

* Corning Glass Works material codes.
† Softening point.
CONVERSION FACTORS: K = (°F − 32)/1.8 + 273; g/cm^3 = 27.68 × lb/in^3; GPa = 6.894 × 10^6 × lb/in^2.

depends on the crystal structure and the strength of interatomic forces. For isotropic materials, the elastic response can be characterized by two elastic constants, namely, Young's modulus E and shear modulus G, which are related through Poisson's ratio ν ($\nu = E/2G - 1$).

Glasses and most engineering ceramics can be treated as isotropic materials on a macroscopic level. At low and moderate temperatures these materials show elastic behavior until the failure stress is reached. The elastic moduli are commonly measured using a sonic reasonance technique described in ASTM C623. In this method the sample is vibrated in the flexural or torsional mode by an audio frequency signal and the resonance frequency of the sample for the desired mode of mechanical vibration is measured. The elastic modulus is computed from the resonance frequency using the density and the sample geometry. The elastic properties of some glass and ceramic materials are given in Table 4.2. The elastic constants of well-bonded two-phase materials are between those of the individual phases. If the second phase is void due to porosity, Young's modulus of the porous body can be estimated from

$$E = E_0 (1 - 1.9P + 0.9P^2) \tag{4.1}$$

where E_0 is the modulus of the fully dense material and P is the fractional porosity.

Strength

When a ceramic material is progressively loaded, it fractures at a certain stress value. The stress at which it fractures is defined as the strength of that sample. At low and moderate temperatures ceramic materials undergo brittle fracture. As a consequence, the material fails with no warning prior to fracture. Another characteristic is that these materials are notch sensitive; the stress concentration at the crack tip and other discontinuities cannot be relieved by plastic flow. Furthermore, fracture of brittle materials occurs at a much lower stress under tensile loading than under compressive loading. Unlike elastic and thermal properties, the strength of the ceramic materials is an extrinsic property. It depends on such variables as the quality of the external surface, internal defects, the test environment, and residual stresses. For example, the strength of annealed glass can range from 5000 to 100,000 lb/in^2 (35 to 690 MPa), depending on the surface condition. The strength of ceramic materials is determined by the size of the largest flaw present in the maximum stress region. The probability of finding a given size flaw increases with increase in the area under stress. Therefore the observed strength is related to the specimen size. When several samples are tested under similar conditions, there is usually a considerable spread in the strength values. The strength S is commonly represented by a two-parameter Weibull distribution,

$$P = 1 - \exp\left(-\frac{S}{S_0}\right)^m \tag{4.2}$$

where P = cumulative failure probability at applied stress S
S_0 = scale factor
m = Weibull modulus (measure of scatter in strength values)

Materials with higher values of m are more reliable because of the narrower distribution in strength values. The value of m ranges from 5 to 20 for most engineering ceramics.

A flexural method, described in ASTM C158, is commonly used to measure the strength of ceramic materials. The area under maximum stress in most flexural tests is in the range of 0.1 to 0.3 in^2 (60 to 200 mm^2). Glass and glass-ceramic materials are normally tested after abrading with high-velocity silica sand; but in the case of ceramics, the strength is measured using carefully machined samples. Typical strength values of several materials are given in Table 4.3. When comparing strength values it is necessary to take into consideration the sample preparation and test details.

The high-temperature strength behavior is determined by the refractoriness and grain size of the material. For less refractory materials the strength drop occurs at a lower temperature. One can select more refractory materials using melting temperatures (the softening point for glasses) as a guide, but in most cases impurities present in the material form secondary phases which are not as refractory as the matrix. The high-temperature strength is determined by the less refractory phase. Figure 4.1 shows strength as a function of temperature for a few materials. Sintered silicon carbide retains its strength to at least 2735°F (1775 K). For other materials the strength drop occurs at lower temperatures.

Table 4.3. Mechanical Properties of Glass and Ceramic Materials

	Strength (room temperature), lb/in^2	Fracture toughness, (lb/in^2) in$^{1/2}$	Fatigue constant n (70–250°F)
	Glasses		
0080* (annealed)	6,900	650	16
1723 (annealed)	6,400	760	30
7740 (annealed)	6,800	690	28
7940 (annealed)	6,800	700	39
	Glass-ceramics		
9606*	25,000	2,180	55
9608	12,000	1,500	
9658	15,000	1,365	
9606 (etched)	38,000		
	Ceramics		
ZrO_2	60,000–100,000	3,640–12,740	45–100
Al_2O_3	70,000	4,095	45
SiC (sintered)	65,000	4,185	High >60
Si_3N_4 (sintered)	90,000	5,095	High >60
Sialon	90,000	4,550	High >60
Si_3N_4 (reaction sintered)	35,000		

* Corning Glass Works material codes.

CONVERSION FACTORS: MPa = 6.894 × 10^{-3} × lb/in^2; (lb/in^2)in$^{1/2}$ = 909.8 × MPa · m$^{1/2}$; K = (°F − 32)/1.8 + 273.

Fracture Toughness

Fracture toughness is a measure of material resistance to crack propagation and is an intrinsic property. Materials with higher toughness values are more resistant to machining and impact damage and hence are more reliable. The fracture toughness and strength are related by

$$K_{IC} = YSC_f^{1/2} \tag{4.3}$$

where K_{IC} = fracture toughness
S = strength
C_f = crack length at instant of catastrophic crack propagation
Y = constant

Equation (4.3) describes one of the criteria for the fracture of ceramic materials. In most cases the fracture toughness is determined by testing notched samples. At present the use of toughness values (Table 4.3) in design considerations is limited. However, the concept is useful in developing new materials.

Fatigue (reduction in strength with time)

The mechanism that governs the time dependence of strength depends on the use temperature. At ambient temperatures the failure occurs by the growth of preexisting flaws via a stress corrosion mechanism. The rate of growth of the crack in this temperature regime depends on the stress intensity factor at the crack tip, relative humidity, pH, and temperature. As illustrated schematically in Fig. 4.2, three regions are observed in crack growth behavior. Below the threshold value of the stress intensity factor K_0 no crack growth is observed. The approximate value of K_0 is $0.2K_{IC}$. In region I the velocity of crack growth V increases with an increase in the stress intensity factor K_I, and is represented by

$$V = AK_I^n \tag{4.4}$$

A and n are the crack growth parameters. The value of n is referred to as fatigue constant. Higher values of n imply better fatigue resistance. The controlling factor in region I is the reaction of the corrosive species (such as water molecules) with the atomic bonds at the crack tip. The entire line of region I shifts to higher crack velocities with increasing test temperature up to at least 440°F (500 K). In region II the velocity of crack growth is constant and water diffusion to the crack tip is the controlling factor. Region III corresponds to high crack velocities and is only of academic interest. It is worth noting that when the material is stressed, the crack initially grows in a stable manner. As the crack grows, the stress intensity factor at the crack tip increases, causing an increase in the crack velocity, that is, crack growth is accelerated as it proceeds. Failure occurs when the stress intensity factor at the most critical flaw reaches the K_{IC} value. Cracks grow more slowly in region I compared to regions II and III. Therefore it is sufficient to consider only region I for estimating failure times.

The relationship between failure time and applied stress can be derived from Eqs. (4.3) and (4.4) for different loading conditions. When a constant tensile stress S is applied instantaneously (static fatigue test), it can be shown that the quantity $S^n t$ is a constant. If the failure time t_1 for a given

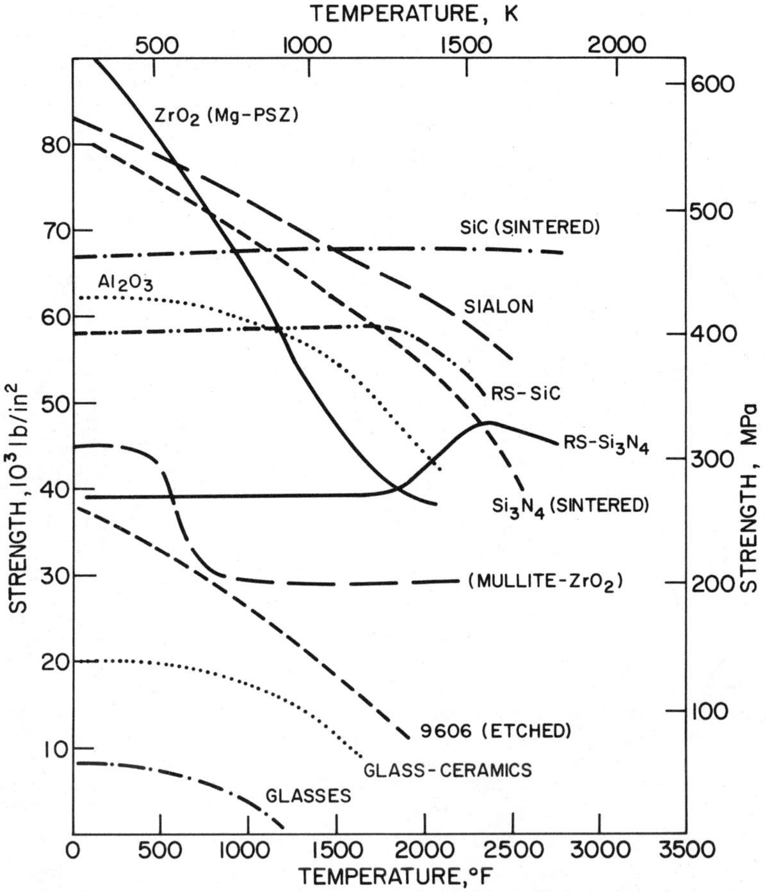

Fig. 4.1 Flexural strength of various ceramic materials as a function of temperature.

applied stress S_1 is known, the failure time t_2 under a different stress S_2 can be computed from the relationship

$$\frac{t_1}{t_2} = \left(\frac{S_2}{S_1}\right)^n \tag{4.5}$$

In strength tests the material is tested at a constant strain rate. To construct strength-probability-time (SPT) diagrams it is necessary to relate the failure time in the static fatigue test t_s to the failure time in the strength test t_e. The two failure times are related by

$$t_e = (1 + n)t_s \tag{4.6}$$

If the fatigue constant is known, the time to failure t_s when the fracture stress in the strength test is applied instantly, can be found from Eq. (4.6).

An SPT diagram, as shown in Fig. 4.3, is constructed to determine the design stress when the product life and the acceptable failure probability level are known. The construction of the diagram starts with a computation of t_s values from the failure times observed in the strength tests. By substituting the strength values with the corresponding t_s values in Eq. (4.5) the failure stresses for 1-s life are computed. These failure stresses are plotted on Weibull axes. The curves for each decade increase in failure time run parallel to the initial line and can be constructed using Eq. (4.5). As the value of n decreases, the spacing between the lines in Fig. 4.3 increases.

Before concluding the discussion of slow crack growth by stress corrosion mechanisms, it is appropriate to mention briefly the techniques used to measure the fatigue constant. Several techniques have been suggested to measure the fatigue constant; among these, dynamic and static fatigue techniques are more common. In the dynamic fatigue method the fracture stress is measured at different stress rates $\dot{S}$. The data are plotted on a graph of log S versus log $\dot{S}$, and the fatigue constant is determined from the slope of the line. In the static fatigue test the specimen is loaded to a constant stress. The failure times for different values of applied stress are measured. Figure 4.4 shows the static fatigue data for

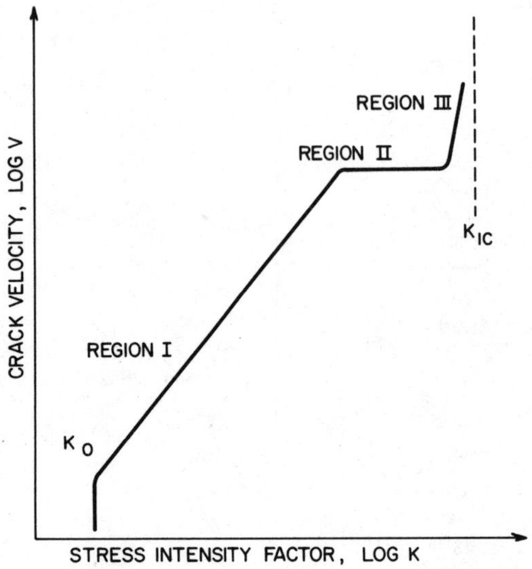

Fig. 4.2 Schematic diagram of crack velocity versus stress intensity factor.

silicon nitride. The slope of the line in the graph of log S versus log t is determined to obtain the fatigue constant.

Elevated-temperature fatigue behavior depends on the stress level and the temperature. For failure times of 100 h or less (high stresses at intermediate temperatures or low stresses at high temperatures), the fracture originates from preexisting flaws and the above analysis can be adopted for predicting the lifetime. The mechanism of slow crack growth at high temperatures is not clear. For failure times greater than 1000 h (low stresses and intermediate temperatures), failure occurs from deformation (creep) induced cracks. The delayed failure analysis in the creep-controlled regime has not been developed, and therefore the creep resistance (deformation rate) is used as the criterion for material selection. Figure 4.5 shows typical creep curves of a glass-ceramic at different temperatures and stresses. The creep rate, that is, the rate of change in plastic strain, is measured from the linear portion of the curve. The creep rate depends on both temperature and applied stress and can be represented by the empirical relation

$$e = A_0 S^w \exp\left(-\frac{Q}{R_0 T}\right) \tag{4.7}$$

where T = temperature
Q = activation energy
w = stress exponent
A_0, R_0 = constants

Lower values of activation energy indicate less sensitivity of the creep rate to temperature change. The value of the stress exponent is typically between 1 and 3. Both porosity and glassy grain boundary phases have a significant effect on the creep resistance. For comparing the creep resistance of different materials, one can measure the creep rate, keeping both temperature and stress constant. Table 4.4 shows the torsional creep rates of different materials at 2370°F (1575 K) under a stress of 1800 lb/in² (12.4 MPa). It is seen that alumina and coarse grained spinel have good creep resistance. Since these types of data are not readily available for

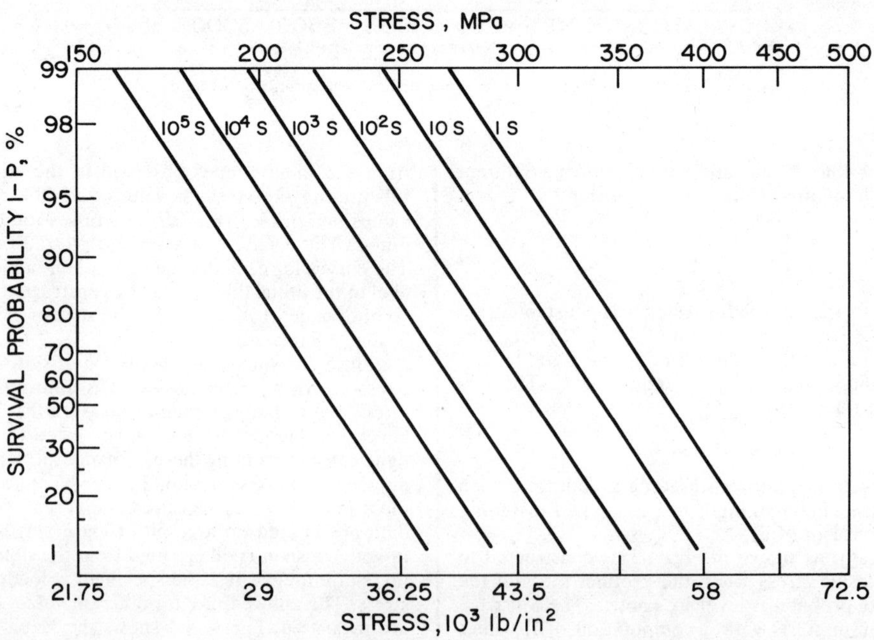

Fig. 4.3 SPT diagram of 95% alumina at room temperature. (*Reproduced with permission from Ref. 2.*)

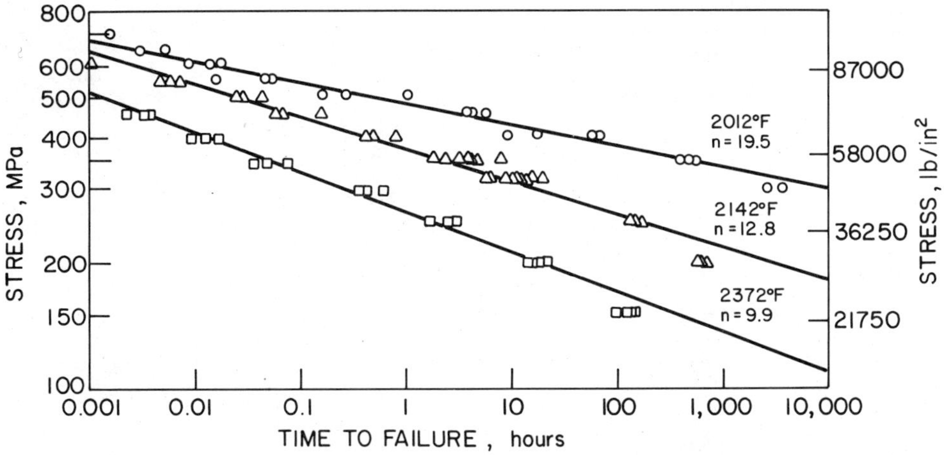

Fig. 4.4 Static fatigue results of silicon nitride (Norton NC 132). (*Reproduced with permission from Ref. 6.*)

other materials, we have given in Table 4.4 the combinations of temperature and stress required to observe a creep rate in the range of 10^{-4} and 10^{-5} per hour. Sintered silicon carbide has the best creep resistance among the current commercial materials.

4.3 THERMAL PROPERTIES

Thermal properties determine the usefulness of ceramic materials in many high-temperature applications. The properties of interest are heat capacity, coefficient of thermal expansion, and thermal conductivity. Heat capacity and thermal conductivity determine the rate of temperature change in a ceramic during heat treatment and therefore are fundamental in calculating the thermal shock resistance.

Heat Capacity

Heat capacity is the quantity of heat required to raise the temperature of a material by 1 degree, usually at constant pressure or constant volume. For most oxides and carbides the heat capacity increases with increasing temperature and

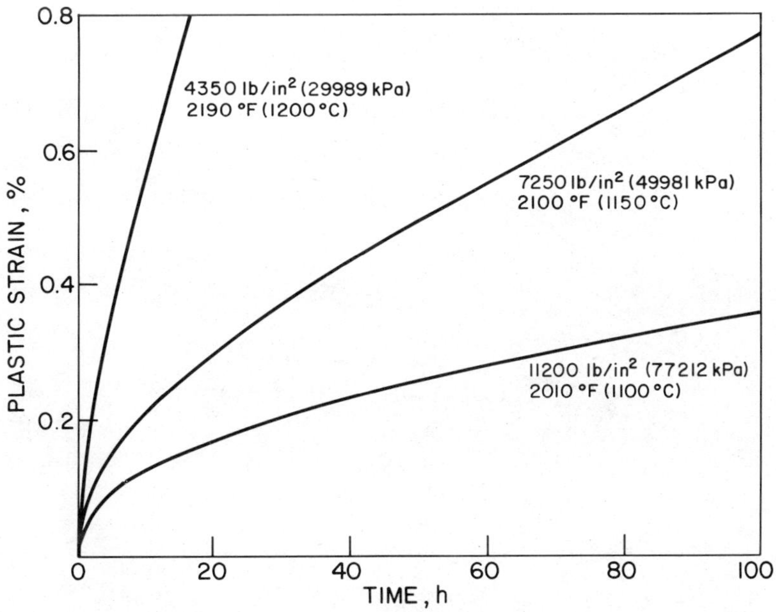

Fig. 4.5 Creep curves for cordierite glass-ceramic. (*Ref. 7.*)

Table 4.4. Creep Results of Ceramics

Material	Stress, lb/in²	Temperature, °F	Creep rate, 10^{-5}/h
Alumina	1,800	2,372	0.13
BeO	1,800	2,372	30.0
MgO (slip cast)	1,800	2,372	33.3
MgO (isopressed)	1,800	2,372	3.3
$MgAl_2O_4$ (fine grain)	1,800	2,372	26.3
$MgAl_2O_4$ (coarse grain)	1,800	2,372	0.1
ThO_2	1,800	2,372	100.0
Silica	1,800	2,372	20,000
SiC (sintered)	40,000	2,730	1
Al_2O_3	4,350	2,730	1
Al_2O_3	17,400	2,730	10
Si_3N_4 (RS)	10,000	2,730	10
ZrO_2 (Y-PSZ)	18,000	1,830	10
ZrO_2 (Mg-PSZ)	18,000	1,830	0.20
Sialon (59D)	10,000	2,550	7
HP-Si_3N_4 (Y_2O_3 doped)	20,000	2,460	7.5

CONVERSION FACTORS: kPa = 6.894 × lb/in²; K = (°F − 32)/1.8 + 273.
SOURCE: Reproduced in part from Ref. 1.

reaches a constant value above 1800°F (1250 K), as can be seen from Fig. 4.6.

Thermal Expansion

The linear coefficient of thermal expansion is the ratio of the change in length per degree of temperature change to the length at room temperature. An increased amplitude of atomic vibrations is responsible for the increase in the sample dimensions with temperature. Similar to heat capacity, the thermal expansion coefficient increases with temperature and reaches a nearly constant value at higher temperatures. Figure 4.7 and Table 4.5 show the thermal expansion coefficients of several glass and ceramic materials. In using thermal expansion coefficients in stress analysis, one should remember that the value used should correspond to the temperature range of application.

In some crystals the thermal expansion coefficient is different in different crystallographic directions. Microcracks formed in these materials cause hysteresis in thermal expansion curves. For microcracked ceramics the overall expansion coefficient of the material is lower than that of the individual grains; therefore they have better thermal shock resistance. The hysteresis in thermal expansion curves is also observed in two-phase materials when the second phase undergoes a phase transformation (ZrO_2–PSZ in Fig. 4.7) or when the thermal expansion coefficient of the second phase is substantially different from that of the matrix phase.

Thermal Conductivity

Thermal conductivity (the amount of heat conducted through a unit area of the material per unit time and per unit temperature gradient) in ceramic materials arises mostly from phonon-phonon interactions. At low temperatures [80 to 800°F (25 to 425°C)] the thermal conductivity is proportional to the inverse temperature. Figure 4.8 shows the temperature dependence of thermal conductivity for several ceramics. The materials with complex crystal structures have lower thermal conductivities. Also the presence of foreign atoms decreases the thermal conductivity. For example, zirconia stabilized with MgO or CaO has low thermal conductivity and is very useful as a high-temperature refractory material. The thermal conductivity can also be lowered by adding second-phase materials which form microcracks.

Above temperatures of about 750°F (400°C) radiative con-

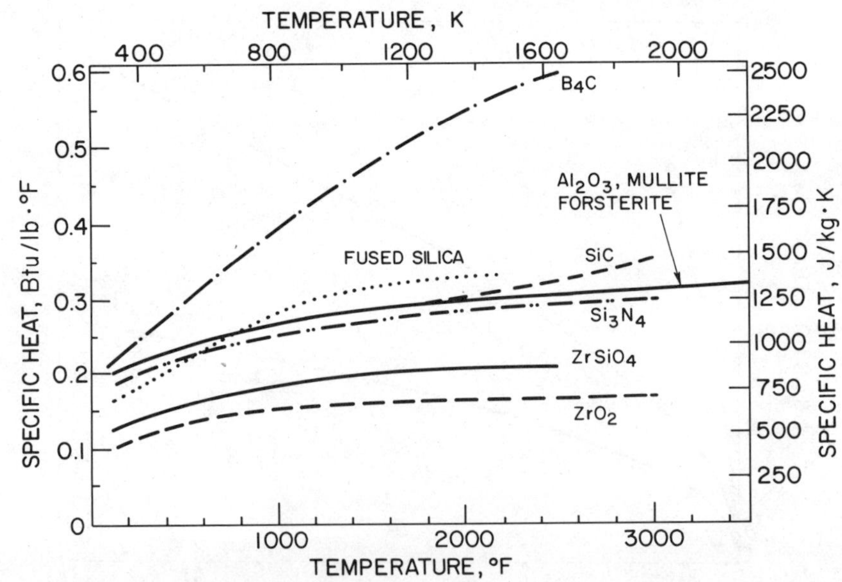

Fig. 4.6 Heat capacity of ceramic materials.

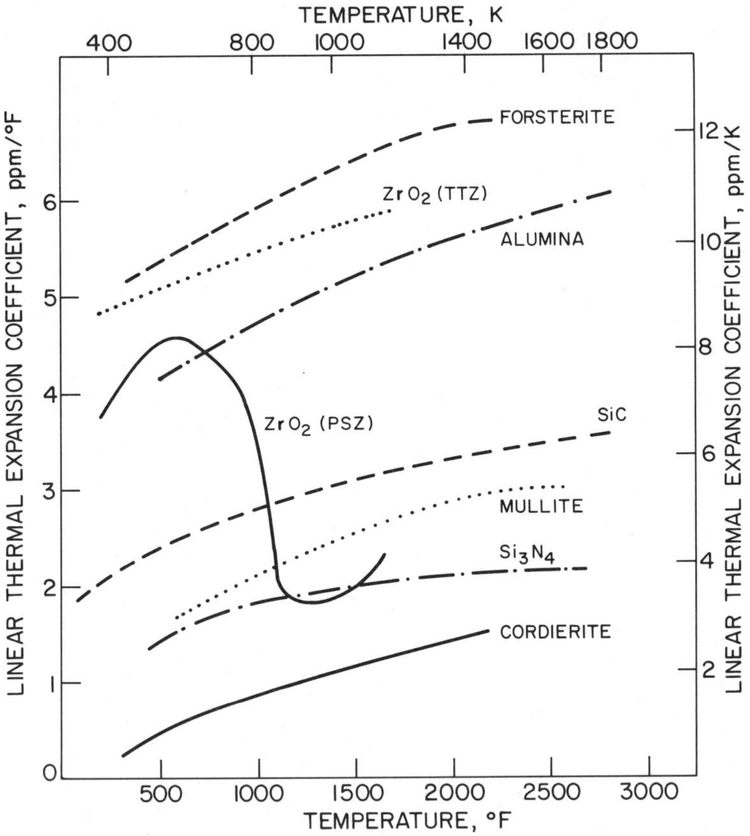

Fig. 4.7 Thermal expansion coefficient as a function of temperature.

duction becomes increasingly important. This is because (1) the amount of energy in the radiation field increases as the fourth power of the absolute temperature (Stefan-Boltzmann law), and (2) the energy spectrum shifts to shorter wavelengths where the materials have lower absorption (Wien displacement law). At high temperatures the details of heat transfer depend on sample configuration, so a material value cannot be assigned that is useful in all situations. A detailed discussion on thermal conductivity is beyond the scope of this chapter; therefore, the reader is referred to Ref. 1.

Thermal Shock Resistance

In high-temperature applications the ceramic components may be subjected to temperature gradients which vary with time and position. Thermal stresses are generated when the free expansion of individual volume elements due to the temperature change is constrained. These stresses can exceed the material strength and cause component failure. The magnitude of thermal stresses depends on the physical properties, heat-transfer conditions, and the geometry of the component. There are two approaches for assessing the thermal shock resistance of the ceramic materials.

The first approach is applicable for strong materials and is based on the avoidance of fracture initiation. The two thermal shock parameters that characterize the material resistance for crack initiation in thermal stress environment are

$$R = \frac{S(1 - \nu)}{E\alpha} \tag{4.8}$$

$$R' = KR \tag{4.9}$$

Where K is the thermal conductivity and α is the coefficient of thermal expansion. Higher values of R and R' signify better thermal shock resistance. Favorable material characteristics for avoiding fracture initiation are high values of strength and thermal conductivity and lower values of Young's modulus and thermal expansion coefficient. The parameter R represents the maximum temperature gradient the material can withstand without fracturing when the surface of the body instantly attains the temperature of the quenching medium. R' has the dimension of heat flux and seems to be applicable in a wider range of thermal stress situations. It should be noted that in computing the values of R and R' the physical properties should correspond to the temperature range of the application.

Table 4.5. Thermal Properties

	Specific heat, Btu/lb-°F	Thermal conductivity (room temperature), Btu/h · ft-°F	Thermal expansion coefficient, ppm/°F
		Glasses	
0080*	0.19	0.61	5.19
1723	0.18	0.59	1.81
7740	0.18	0.63	2.56
7940	0.18	0.85	0.31
		Glass-Ceramics	
9608*	0.19	1.19	0.22–1.11
9606	0.22	1.33	3.16
9658	0.18	...	4.94
		Ceramics	
Al_2O_3	0.19	19	4.45
ZrO_2	0.096	1.4	1.6–6.67
SiC (sintered)	0.24	72.6	2.56
Si_3N_4 (sintered)	0.20	12.6	1.83
Sialon (sintered)	0.17	13.2	1.78

* Corning Glass Works material codes.
CONVERSION FACTORS: J/kg · K = 4184 × Btu/lb · °F; W/m · K = 1.73 × Btu/h · ft ·°F; ppm/K = 1.8 × ppm/°F.

The second approach applies for weak materials such as refractories. It is based on minimizing the distance of crack propagation after the thermal stress fracture has been initiated. The thermal shock damage resistance parameter R_{st} for this situation is given by

$$R_{st} = \left(\frac{W_f}{\alpha^2 E}\right)^{1/2} \tag{4.10}$$

where W_f is the work of fracture and represents the energy consumed to create a unit area of fracture surface. It is usually determined by measuring the area under the stress-strain curve of a notched sample.

The thermal shock resistance is characterized experimentally by measuring the retained strength when the test samples are quenched from different temperatures. Figure 4.9 shows schematically the retained strength as a function of the temperature difference between the specimen and the quenching medium ΔT. For strong materials, a sudden drop in strength is observed at ΔT_c due to unstable crack propagation. The strength retained after thermal stress failure is inversely related to the initial strength. Between ΔT_c and $\Delta T_c'$ the strength remains the same, indicating that the length of the largest flaw is constant. However, the number of cracks formed increases as ΔT increases from ΔT_c to $\Delta T_c'$. Above $\Delta T_c'$ the strength drop is more gradual. In the case of refractories the cracks propagate in a stable manner and the retained strength shows a gradual decrease with increasing ΔT.

4.4 BRITTLE MATERIALS DESIGN

The successful development of ceramic components for a high-stress application is commonly achieved through an iterative process. Figure 4.10 shows a schematic diagram of the design process. The process starts by identifying an application where the use of ceramic material either offers a performance advantage or results in the invention of a new product. Next the critical physical properties required for the application are identified, knowing the conditions in which the component must operate. An evaluation of the available materials is then made, taking into consideration the physical property needs for the intended application. After selecting the candidate material, a detailed analysis is performed to determine the safe design stress for the material.

The stresses that develop in a component depend on its geometry and on the operating conditions. In order to determine the magnitude of the stresses in the component it is necessary to perform a stress analysis. If the geometry of the component is simple and the operating conditions are well defined, one can estimate stresses using formulas documented in the mechanics literature. For complicated shapes and complex loading conditions it is necessary to use numerical methods such as finite-element analysis. Once the magnitude of the operating stresses is determined, a comparison of the design stress with the computed stresses is made. If the computed stresses everywhere in the component are below the design stress, the component is field tested in the real application. Unless the material capability substantially exceeds the material requirements, the preliminary field test usually results in premature component failure. The failed components are examined to determine the failure mode, and depending on the results, either material improvement or component design improvement is implemented for subsequent field tests. When the component performs successfully in the field test, the product is commercialized. The number of iterations required depends on the accuracy of the material design stress and the operating stresses in the component. These two aspects are discussed in more detail below.

Design Stress

A major role of the material scientists in the design process is to provide the design engineers with material property data in a format useful in designing the component. This task is a little more difficult when dealing with ceramics because these materials are brittle. It should be noted that brittle does not mean weak. It means that the material exhibits very small strain to failure, usually less than 0.25 percent. The lack of ductility restricts the ability to distribute the load in the presence of stress concentrations. As a result, ceramics are sensitive to conditions that give rise to stress gradients such as point loads, flaws, and thermal gradients. Under these conditions the peak stress can exceed the material strength and cause component failure, although the average stress is well below the material design stress. Furthermore, variable flaw size due to surface damage and porosity causes a broad distribution in strength values. For these reasons one has to talk in terms of probability of failure when designing ceramic components. The probability of failure depends on the overlap in distributions of the applied stress and the material strength, as shown in Fig. 4.11. The design goal is

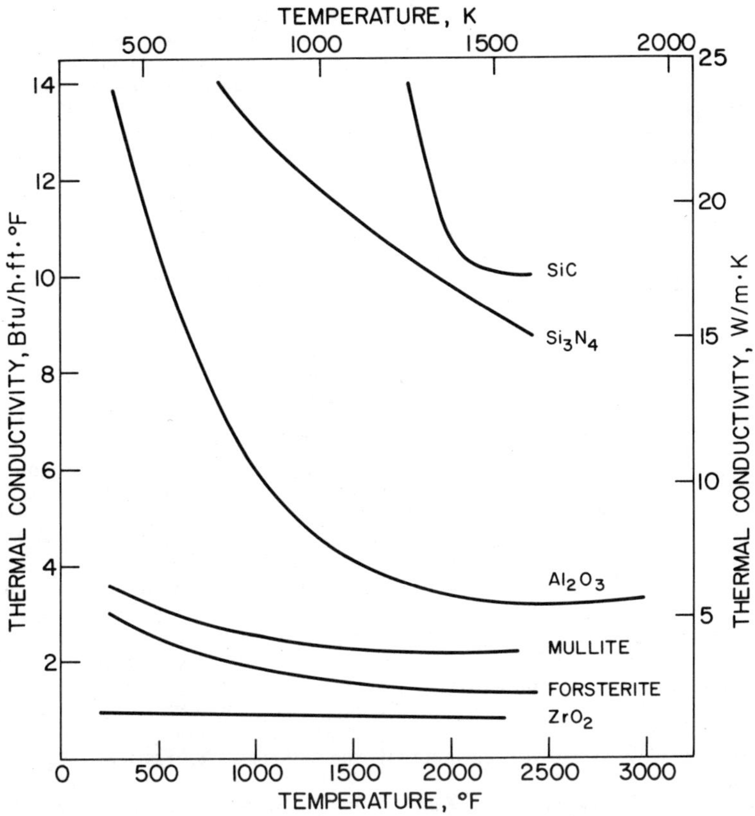

Fig. 4.8 Thermal conductivity of ceramics as a function of temperature.

to reduce the overlap such that the component performs for the duration of the expected life with a low probability of failure.

The probability approach in designing ceramics has been validated through accelerated life tests in which the failure times are in the range of 10 to 1000 h. But for long-term applications the accuracy of life prediction is not known because of the uncertainty in extrapolating data from short-term tests. Owing to the lack of such information, at this time only a simplified procedure can be offered for determining the design stress. It is hoped that the use of this procedure eliminates guesswork. Better quantitative treatment is required before ceramics can be used extensively in high-stress applications. There are several steps involved in determining the design stress.

STEP 1: *Determine the mean value of strength and the Weibull modulus.* It is customary to measure strength in four-point flexure using relatively small-size samples with carefully prepared surfaces. The reported values of the modulus of rupture typically show a coefficient of variation of 5 to 10 percent. For design purposes it is necessary to test at least 20 samples to determine the strength and the Weibull modulus.

STEP 2: *Correct the strength value (mean) for the size difference between the strength specimen and the product.* In most applications the product size is larger than the amount of material under maximum stress in the strength specimen. It has been observed experimentally that the strength decreases with increasing specimen size. This can be demonstrated by breaking a long fiber in tension and later retesting the broken halves.

Before estimating the strength correction due to the size difference, it is necessary to identify whether strength-controlling flaws are at subsurface defects or at surface defects. In most cases the fracture occurs from surface flaws, hence surface areas must be used in the computation instead of the volumes. When two samples with areas A_1 and A_2 are under uniform stress, the ratio of stresses for the same probability of failure can be estimated from the relationship

$$\frac{S_1}{S_2} = \left(\frac{A_2}{A_1}\right)^{1/m} \tag{4.11}$$

where S_1 and S_2 are the stresses in samples 1 and 2, respectively. Let A_1 = 0.15 in^2 (0.97 cm^2), A_2 = 15 in^2 (97 cm^2), S_1 = 50,000 lb/in^2 (345 MPa), and m = 10. Then S_2 = 31,500 lb/in^2 (217 MPa). A hundredfold increase in area causes the strength to drop by about 37 percent. For

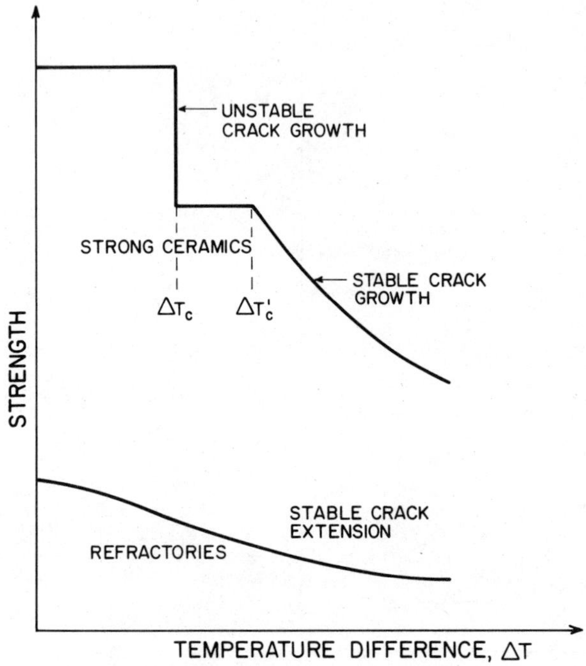

Fig. 4.9 Schematic illustration of retained strength versus ΔT.

lower values of m the strength drop becomes even more significant.

STEP 3: *Allow for the difference in the flaw sizes of the strength specimen and the product.* The product usually sees much more handling damage than do the strength specimens. Furthermore, if the product is manufactured using sintering methods, the pore size distribution could be different from that in the test bars. For silicon nitride and silicon carbide, oxidation in air can change the flaw population. The allowance for these factors depends on the individual situation, and it is therefore recommended that the strength be measured on the bars cut from the product after the product has been put in use for a short time.

For strengthened glass articles the design strength is typically reduced by 50 percent to allow for strength loss due to handling damage.

STEP 4: *Allow for slow crack growth.* The cracks grow slowly until the stress intensity factor at the crack tip reaches a critical value. For a given material the rate at which the crack grows depends on the stress and the environment in which it is stressed. The allowance for slow crack growth depends on the fatigue constant, stress duration, and the expected life of the product. To serve as an illustration, we can consider an application where the stress distribution is uniform. Let $n = 20$ and the failure stress for 10-s life be 30,000 lb/in^2 (207 MPa). Then the design stress for 10-yr life computed from Eq. (4.5) is 12,650 lb/in^2 (87 MPa). To account properly for slow crack growth it is necessary to identify whether the critical flaw responsible for fracture lies in the high-temperature or the low-temperature region of the component. The fatigue constant should be measured at the temperature the critical flaw sees in real application.

STEP 5: *Proof test to improve reliability.* Since the strength of the ceramic materials is represented by a probability function, there seems to be no absolutely safe working stress. For any combination of applied stress and failure times, there is a finite probability of premature failures. Proof testing can be used to eliminate the weak samples from the initial strength distribution (see Fig. 4.11). The procedure for proof testing involves subjecting all the components to a stress between the operating stress and the fracture stress

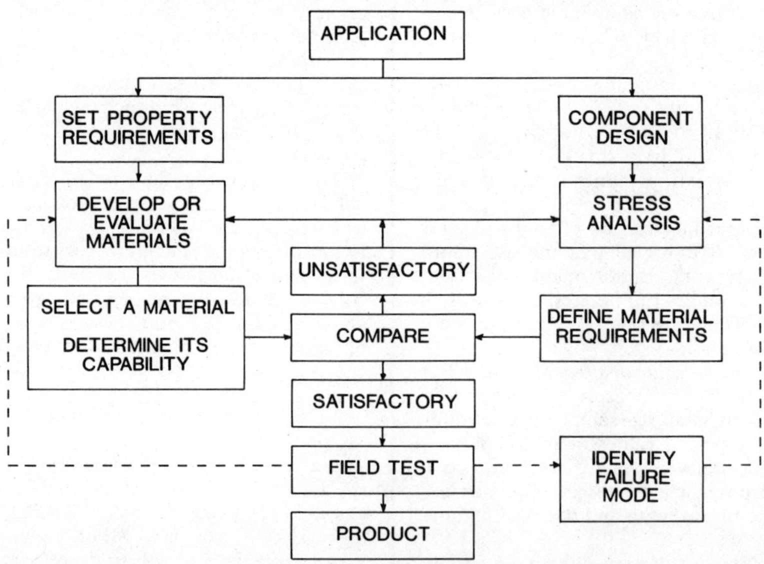

Fig. 4.10 Design process. (*Reproduced with permission from Ref. 2.*)

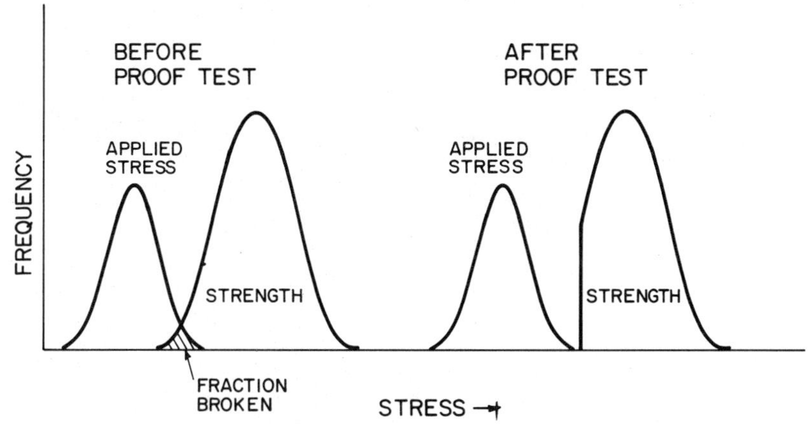

Fig. 4.11 Frequency distribution of strength and applied stress.

such that a certain portion of the components that are weak will fail. The components that survive will have a definite life in which no failures occur in service. The minimum time to failure, t_{min} is given by

$$t_{min} = BS_a^{-2}\left(\frac{S_p}{S_a}\right)^{n-2} \tag{4.12}$$

where B is a constant that depends on the fracture toughness and crack growth parameters A and n, and S_p and S_a are the proof stress and the applied stress, respectively. As seen from the above relationship, the minimum time to failure depends on the ratio of proof stress to applied stress, the fatigue constant, and the applied stress. Equation (4.12) makes no reference to the probability of failure and all terms in the equation can be determined experimentally. For a proof test to be effective, slow crack growth during proof testing should be minimized by testing rapidly in a dry environment. Furthermore, caution should be exercised not to introduce a new flaw population after the proof test.

From the preceding discussion it is evident that the value of design stress depends on several factors. For glass, experience has shown that the safe design stress in many applications is about 1000 to 1200 lb/in^2 (6.9 to 8.3 MPa).

Operating Stresses

In order to eliminate poor designs, it is necessary to perform a stress analysis of the component. In most cases the component geometry and loading conditions are complex; therefore, numerical methods such as finite-element analysis are required. Several finite-element computer programs are readily available for performing the stress analysis. In this method the component is divided into discrete regions, that is, elements, and these elements are interconnected at nodal points. The displacements at the nodal points due to the applied thermal and mechanical loads are computed, and the displacements are converted to stresses using elasticity equations. The accuracy of the computed stresses depends on the accuracy of the stated boundary conditions, the knowledge of thermomechanical properties, and the description of the finite-element grid itself. It is necessary to define the boundary conditions accurately, because they have a direct influence on the magnitude of the stresses. The finite-element grid should be finer in the regions of high-stress gradients. Once the stress states everywhere in the component have been determined, the computed stresses are compared with the material design stress. If the stresses in the component exceed the design stress, the geometry of the component is altered and the stress analysis is repeated. After a satisfactory design has been achieved, the component is field tested. When it is not possible to do a finite-element analysis, there are a few rules of thumb one can follow in designing ceramic components, and these are given in Table 4.6.

Table 4.6. Rules of Thumb for Ceramic Design

1. *Avoid point loads.* To minimize the stress where loads are transferred, it is best to use areal loading (spherical surfaces are particularly good); line loading is next best.

2. *Maintain structural compliance.* Use compliant layers, springs, radius mating parts (to avoid lockup)

3. *Avoid stress concentrators.* Sharp corners, rapid changes in section size, undercuts, holes, etc., are to be avoided or minimized.

4. *Minimize the impact of thermal stresses.* Use the smallest section size possible; the higher the symmetry, the smaller is the impact of thermal stresses (a cylinder will perform better than a prism). Breaking complex components down into simpler subcomponents with higher symmetry will help.

5. *Keep components as small as possible.* The flaw distribution of ceramics makes the strength size dependent. Thus minimizing the component size increases the reliability.

6. *Minimize the severity of impact.* Where impact (i.e., particulate erosion) cannot be avoided, design for low-obliquity impacts (20–30°).

7. *Machine components very carefully.* The most critical strength-reducing flaws are often determined to have been surface or near-surface cracks introduced during a grinding or other machining process.

SOURCE: By permission from Ref. 3.

REFERENCES

1. W. D. Kingery, H. K. Bowen, and R. D. Uhlmann, *Introduction to Ceramics,* Wiley, New York, 1976.
2. R. W. Davidge, *Mechanical Behavior of Ceramics,* Cambridge University Press, London, 1979.
3. R. N. Katz, "Application of High Performance Ceramics in Heat Engine Design," in *Engineering Ceramics, 1984,* M. Ish-shalom and D. G. Brandon (eds.), Elsevier, New York, 1985, p. 227.
4. D. C. Boyd and D. A. Thompson, "Glass," in *Encylcopedia of Chemical Technology,* 3d ed., vol. 11, Wiley, New York, 1980, p. 807.
5. R. J. Roark and W. C. Young, *Formulas for Stress and Strain,* McGraw-Hill, New York, 1975.
6. G. D. Quinn, "Static Fatigue Resistance of Hot Pressed Silicon Nitride," in *Fracture Mechanics of Ceramics,* vol. 8, R. C. Bradt, A. G. Evans, D. P. H. Hasselman, and F. F. Lange (eds.), Plenum, New York, 1986, p. 319.
7. T. I. Barry, L. A. Lay, and R. Morrell, "Refractory Glass-Ceramics," *Proc. Brit. Ceramic Soc.,* no. 22 (1973).

Part 5

Heat-Transfer Fluids

Chapter 1

Water and Brine Heat-Transfer Media

MARTIN GOLLIN,
ARCO Chemical Company

The liquids discussed in this chapter are water and brines, which include alcohol-water mixtures. The use of these fluids as heat-transfer media is widespread.

Each fluid is discussed with respect to industrial use, useful physical properties, safety, cleanliness, and corrosion.

1.1 WATER

Water is the fluid most widely used for heat-transfer systems. Applications include domestic heating systems, industrial cooling, low-pressure steam generation for industrial use, and the generation of high-pressure and even supercritical-pressure steam for the generation of electric power. Steam is also important in the petrochemical industry, where it is used as a reactant, coolant, heating fluid, and power generator working fluid.

Advantages

Water possesses unique properties. Some of these are listed below.

Availability Water is readily available and, in most parts of the world, inexpensive. This allows water-based systems to be widely used, which in turn has resulted in the development of low-cost water handling and treatment systems.

If the availability of water at a specific site is in question or the cost of the water is high, then the economics of a water-based system will require examination. An example of this would be the question of whether water coolers or air coolers should be used in a petrochemical plant to be constructed in an arid region.

The use of water in environments where ecologically damaging materials are being processed must be considered carefully. The ease with which water passes into the environment and its high chemical solubility may have a strong influence upon the type of cooling system selected.

Range of Applicability The range of conditions in which water systems are used is extremely wide. Processes can be cooled down to the limit of the freezing point of 32°F (0°C) with refrigeration. Direct evaporative air cooling of water can be used to produce cooling to near ambient wet-bulb temperatures. Steam is used up to about 480°F (250°C) for process heating and to over 930°F (500°C) when passed through steam turbines for power generation.

Physical Properties Water possesses physical properties which allow both a large amount of heat to be carried (per unit mass of fluid) and high heat-transfer rates to be obtained. These characteristics reduce the size of equipment in comparison to systems utilizing other heat-transfer media. This is true for high-temperature applications as well as for ice storage facilities and ice-water slurries for cooling duties.

The combination of high latent heat, high thermal conductivity, and relatively high specific heat and density together with moderate viscosity makes water a very attractive fluid for use in heat-transfer systems.

Figure 1.1 shows the effect of variations in the temperature, size, and velocity upon the convective coefficient of water flowing inside and outside pipes.

Limitations

Water does possess some characteristics which can cause problems in the operation and maintenance of associated equipment. Some are listed below.

High Pressure Levels for Steam Production In order to generate steam at elevated temperatures, the pressure of the system must be high. (Table 1.1 gives the saturation pressure for a range of temperatures.) This results in both thick-walled vessels and the potential for large energy release should the pressure-containing units fail.

Impurities The degree of impurities which are acceptable in water is dependent upon the application. At one extreme, water supplied to very-high-pressure steam-generation systems must have impurity levels measured in the parts per million range. At the other extreme, the use of water in once-through cooling systems normally requires no water treatment. Table 1.2 summarizes the problems which impurities in water can cause.

For intermediate services a compromise between ideal requirements and the cost of providing the treatment must be made. For example, in heat exchanger design, a substantial fouling factor is used to account for the buildup of deposits on the heat-transfer surfaces over the operating life of the unit. This fouling factor can amount to a significant proportion of the total thermal resistance of the system and results in additional surface area being installed. Care should be taken to ensure that the fouling factors specified are realistic, taking into account the characteristics of the supply water and the water treatment which is being applied. Designing equipment such that the heat-transfer surfaces can be cleaned mechanically and chemically will ensure that the buildup of deposits does not reach a level at which the ability of the equipment to perform its specified duty is destroyed. It is also important to design equipment so that the water velocity is high enough to ensure that sedimentation of any solids contained in the water does not occur. A minimum water velocity of 3 ft/s (0.9 m/s) is considered satisfactory for flow inside tubes. For flow outside of tubes in a shell-and-tube exchanger the problem is complicated by the low flow regions which exist at the baffles close to the outside of the bundles. For such equipment the tubes should be placed on a rectangular pitch and the tube bundle should be removable from the shell so as to allow mechanical cleaning of the surfaces. Table 1.3 gives values for outside fouling factors for various waters.

Water systems are susceptible to fouling due to micro-

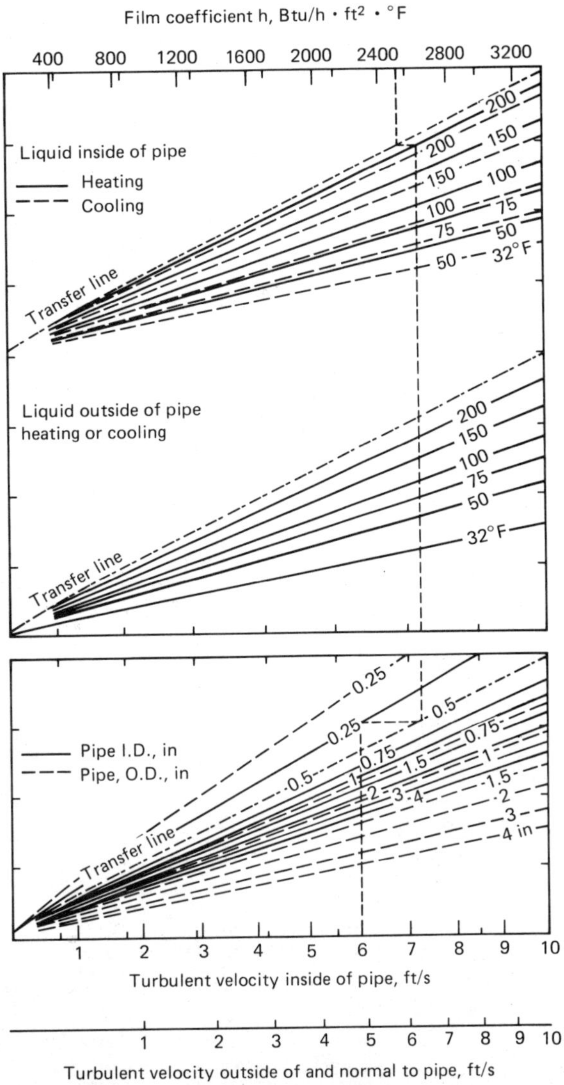

Fig. 1.1 Film coefficient for water in turbulent flow (over 1 ft/s) within a tube and in turbulent flow (over 1 ft/s unless Re > 100) outside and normal to a single tube. (*From Ref. 1. Reproduced by permission.*) (Conversion factors: 1 ft = 0.3048 m; 1 ft/s = 0.3048 m/s; 1 Btu/h · ft² · °F = 5.678 W/m² · °C.)

biological organisms. The use of chlorination, fungicides, and toxicants can inhibit the growth of such organisms.[3,4]

A more important problem caused by water impurities is that of corrosion. Many of the more commonly used austenitic stainless steels are susceptible to stress corrosion cracking, pitting, crevice corrosion, and other phenomena associated with chloride (or other halogen compounds) or caustic impurities in the water or condensate stream. This problem is complex since the degree to which a unit will be susceptible is a function of the impurity content, the stainless steel type, the extent to which cavities are present, and the

Table 1.1. Saturation Temperatures and Pressures for Steam

Temperature, °F	Pressure, lb/in² abs	Volume, ft³/lb		Enthalpy, Btu/lb		Entropy, Btu/lb · °F	
		Liquid	Vapor	Liquid	Vapor	Liquid	Vapor
32.018	0.08865	0.016022	3302.4	0.000	1075.5	0.0000	2.1872
35	0.09991	0.016020	2948.1	3.002	1076.8	0.0061	2.1767
40	0.12163	0.016019	2445.8	8.027	1079.0	0.0162	2.1594
45	0.14744	0.016020	2037.8	13.044	1081.2	0.0262	2.1426
50	0.17796	0.016023	1704.8	18.054	1083.4	0.0361	2.1262
55	0.21392	0.016027	1432.0	23.059	1085.6	0.0458	2.1102
60	0.25611	0.016033	1207.6	28.060	1087.7	0.0555	2.0946
65	0.30545	0.016041	1022.1	33.057	1089.9	0.0651	2.0794
70	0.36292	0.016050	868.4	38.052	1092.1	0.0745	2.0645
75	0.42964	0.016060	740.3	43.045	1094.3	0.0839	2.0500
80	0.50683	0.016072	633.3	48.037	1096.4	0.0932	2.0359
85	0.59583	0.016085	543.6	53.027	1098.6	0.1024	2.0221
90	0.69813	0.016099	468.1	58.018	1100.8	0.1115	2.0086
95	0.81534	0.016114	404.4	63.008	1102.9	0.1206	1.9954
100	0.94294	0.016130	350.4	67.999	1105.1	0.1295	1.9825
110	1.2750	0.016165	265.39	77.98	1109.3	0.1472	1.9577
120	1.6927	0.016204	203.26	87.97	1113.6	0.1646	1.9339
130	2.2230	0.016247	157.33	97.96	1117.8	0.1817	1.9112
140	2.8892	0.016293	122.98	123.00	1122.0	0.1985	1.8895
150	3.7184	0.016343	97.07	117.95	1126.1	0.2150	1.8686
160	4.7414	0.016395	77.27	127.96	1130.2	0.2313	1.8487
170	5.9926	0.016451	62.06	137.97	1134.2	0.2473	1.8295
180	7.5110	0.016510	50.225	148.00	1138.2	0.2631	1.8111
190	9.340	0.016572	40.957	158.04	1142.1	0.2787	1.7934
200	11.526	0.016637	33.639	168.09	1146.0	0.2940	1.7764
210	14.123	0.016705	27.816	178.15	1149.7	0.3091	1.7600
212	14.696	0.016719	26.799	180.17	1150.5	0.3121	1.7568
220	17.186	0.016775	23.148	188.23	1153.4	0.3241	1.7442
230	20.779	0.016849	19.381	198.33	1157.1	0.3388	1.7290
240	24.968	0.016926	16.321	208.45	1160.6	0.3533	1.7142
250	29.825	0.017066	13.819	218.59	1164.0	0.3677	1.7000
260	35.427	0.017089	11.762	228.76	1167.4	0.3819	1.6862
270	41.856	0.017175	10.060	238.95	1170.6	0.3960	1.6729
280	49.200	0.017264	8.644	249.17	1173.8	0.4098	1.6599
290	57.550	0.01736	7.4603	259.4	1167.8	0.4236	1.6473
300	67.005	0.01745	6.4658	269.7	1179.7	0.4372	1.6351
320	89.643	0.01766	4.9138	290.4	1185.2	0.4640	1.6116
340	117.992	0.01787	3.7878	311.3	1190.1	0.4902	1.5892
360	153.01	0.01811	2.9573	332.3	1194.4	0.5161	1.5678
380	195.73	0.01836	2.3353	353.6	1198.0	0.5416	1.5473
400	247.26	0.01864	1.8630	375.1	1201.0	0.5667	1.5274
420	308.78	0.01894	1.4997	396.9	1203.1	0.5915	1.5080
440	381.54	0.01926	1.2169	419.0	1204.4	0.6161	1.4890
460	466.87	0.01961	0.99424	441.5	1204.8	0.6405	1.4704
480	566.15	0.02000	0.81717	464.5	1204.1	0.6648	1.4518
500	680.86	0.02043	0.67492	487.9	1202.2	0.6890	1.4333
520	812.53	0.02091	0.55956	512.0	1199.0	0.7133	1.4146
540	962.79	0.02146	0.46513	536.8	1194.3	0.7378	1.3954
560	1133.38	0.02207	0.38714	562.4	1187.7	0.7625	1.3757
580	1326.17	0.02279	0.32216	589.1	1179.0	0.7876	1.3550
600	1543.2	0.02364	0.26747	617.1	1167.7	0.8134	1.3330
620	1786.9	0.02466	0.22081	646.9	1153.2	0.8403	1.3092
640	2059.9	0.02595	0.18021	679.1	1133.7	0.8686	1.2821
660	2365.7	0.02768	0.14431	714.9	1107.0	0.8995	1.2498
680	2708.6	0.03037	0.11117	758.5	1068.5	0.9365	1.2086
700	3094.3	0.03662	0.07519	825.2	991.7	0.9924	1.1359
702	3135.5	0.03824	0.06997	835.0	979.7	1.0006	1.1210
704	3177.2	0.04108	0.06300	854.2	956.2	1.0169	1.1046
705.47	3208.2	0.05078	0.05078	906.0	906.0	1.0612	1.0612

CONVERSION FACTORS: °C = (°F − 32)/1.8; 1 lb/in² = 0.006895 MPa; 1 ft³/lb = 0.0624 m³/kg; 1 Btu/lb = 2326 J/kg; 1 Btu/lb · °F = 4186.8 J/kg · °C.

SOURCE: Ref. 2. Reproduced by permission.

Table 1.2. General Problems Caused by Impurities in Water

	A	B	C	D	E	F	G	H
Uniform corrosion	×	×	×	×	×	×	×	×
Localized corrosion	×	×	×	×	×	×	×	×
Stress corrosion	. . .	. . .	. . .	. . .	×	×	. . .	. . .
Dezincification of brass	×	. . .	×	×	. . .	. . .	×	×
Ammonia attack of brass	. . .	. . .	×	×	. . .	. . .	. . .	. . .
Dissimilar metal corrosion	×	×	×	×	. . .	. . .	×	×
Chloride attack of stainless steel	×	. . .	×	. . .	. . .	. . .	×	. . .
Caustic attack	. . .	. . .	. . .	. . .	. . .	×	. . .	. . .
Hydrogen damage	. . .	. . .	. . .	. . .	. . .	×	. . .	. . .
Formation of inorganic deposits	×	×	×	×	×	×	×	×
Microbiological fouling	×	. . .	×	×	. . .	. . .	×	. . .
Deterioration of cooling tower lumber	. . .	. . .	×	. . .	. . .	. . .	. . .	. . .
Carryover	. . .	. . .	. . .	. . .	×	×	. . .	. . .

KEY: A = once-through cooling water systems; B = closed circulating cooling water systems; C = open circulating cooling water systems; D = air-conditioning systems; E = low-pressure steam-generating systems; F = high-pressure steam-generating systems; G = process water; H = potable water.
SOURCE: Ref. 3. Reproduced by permission.

metal temperature of the surfaces involved. Use of stainless steels with higher alloy contents or modified grain structure, placing the cooling water on the tube side of a shell-and-tube exchanger (where the water cannot contact the crevices at the rear of the tube to tubesheet joint), and the design of the equipment to give the lowest possible surface metal temperatures are all methods by which stress corrosion problems can be mitigated.[5,6]

Water systems are also susceptible to galvanic corrosion; the use of sacrificial anodes to protect equipment is widespread. Table 1.4 summarizes operating conditions that can affect corrosion.

Freezing A major limitation of water systems is that of freeze-up due to low ambient conditions. This is generally a problem during shutdowns where water, trapped in "dead legs" of equipment and piping, freezes and can cause the fracture of lines. Problems of this nature can often be solved by ensuring that all equipment is self-draining and that drainage points are available at each low point to allow liquid to be removed before freezing can occur. Freezing can also occur during the startup of steam systems. This can often be avoided by first flushing with heated liquid water.

Under extreme conditions, however, freezing of lines can occur even during operation, and for these situations a

Table 1.3. Fouling Factors* (°F · h · ft²/Btu) for Waters of Varying Quality

	Water velocity,† ft/s		Water velocity,‡ ft/s	
	Up to 3	Over 3	Up to 3	Over 3
Seawater	0.0005	0.0005	0.001	0.001
Brackish water	0.002	0.001	0.003	0.002
Cooling tower and artificial spray pond				
Treated makeup	0.001	0.001	0.002	0.002
Untreated	0.003	0.003	0.005	0.004
City or well water, Great Lakes	0.001	0.001	0.002	0.002
River water				
Minimum	0.002	0.001	0.003	0.002
Mississippi, Delaware, Schuylkill, East River, and New York Bay	0.003	0.002	0.004	0.003
Chicago Sanitary Canal	0.008	0.006	0.010	0.008
Muddy or silty	0.003	0.002	0.004	0.003
Hard (over 15 grains/gal)	0.003	0.003	0.005	0.005
Engine jacket	0.001	0.001	0.001	0.001
Distilled	0.0005	0.0005	0.005	0.0005
Treated boiler feedwater	0.001	0.0005	0.001	0.001
Boiler blowdown	0.002	0.002	0.002	0.002

* For inside surface fouling, multiply these values by the outside-to-inside surface ratio.
† At heating medium temperatures up to 240°F (116°C) and water temperatures of 125°F (52°C) or less.
‡ At heating medium temperatures from 240°F (116°C) to 400°F (204°C) and water temperatures over 125°F (52°C). If the heating medium temperature is over 400°F (204°C) and the cooling medium is known to cause scaling, these ratings should be modified accordingly.
CONVERSION FACTORS: 1°F · h · ft²/Btu = 0.176°C · m²/W; 1 ft/s = 0.3048 m/s.
SOURCE: Ref. 1. Reproduced by permission.

Table 1.4. Factors Affecting Corrosion

Chemical	Physical
1. *pH* Acid-soluble metals—oxides more soluble as pH decreases; increased corrosion Amphoteric metals—oxides soluble at low or high pH; protection favored at intermediate pH Noble metals—oxides insoluble at any pH; inert to corrosion	1. *Relative areas* In a galvanic couple, as ratio of cathodic to anodic area increases, corrosion increases
2. *Dissolved salts* Chloride, sulfate—can penetrate passive metal oxide films and promote local attack Calcium, magnesium—alkalinity may precipitate to form protective barrier deposits	2. *Temperature* Increased temperature favors oxygen depolarization, lowers hydrogen overvoltage, increases corrosion Higher-temperature areas become anodic to other areas Higher temperatures change metal potentials (e.g., reverse galvanizing)
3. *Dissolved gases* Carbon dioxide—reduces pH and promotes acid attack Oxygen—depolarizes corrosion reaction at cathode; oxygen-deficient areas become anodic (differential aeration cell) Nitrogen—aggravates cavitation corrosion Ammonia—selectively corrosive to copper-base metals Hydrogen sulfide—promotes acid attack; forms deposits that promote galvanic corrosion Chlorine—promotes acid attack; strips corrosion inhibitor films	3. *Velocity* High velocity promotes erosion corrosion, removes certain passivating corrosion products Low velocity increases sedimentation and differential aeration cell corrosion, decreases amount of corrosion inhibitor reaching and passivating metal surfaces
4. *Suspended solids* Mud, sand, silt, clay, dirt—settle to form deposits promoting differential aeration cell corrosion	4. *Heat transfer* Favors oxygen depolarization by "hot wall effects" Favors differential aeration cell formation by increasing precipitation and sedimentation of solids
5. *Microorganisms* Promote acid attack, differential aeration, cell corrosion, cathodic depolarization, galvanic corrosion	5. *Metallurgy* Surface flaws—cuts, nicks, scratches, etc., favor anodic site formation Stress—internal stresses promote anodic site formation Microstructure—metal inclusions, precipation at grain boundaries, differing adjacent grains, etc., promote galvanic cell formation

SOURCE: Ref. 3. Reproduced by premission.

method of heating the pipelines and equipment must be installed. Such a system would include electric or steam heat tracing, together with additional insulation. The extra expense and the complexity of such systems is a factor to be considered in the choice of heat-transfer fluid.[7]

Vacuum If steam is placed in a closed system and allowed to condense, the pressure in the vessel will decrease. This decrease in pressure must be accounted for either by the installation of vacuum breakers or by designing the vessel or component so that it can withstand the full vacuum internal condition in conjunction with the maximum external pressure.

Physical Properties

Figure 1.2 gives the specific heat, thermal conductivity, viscosity, and density of liquid water for temperatures between 32 and 640°F (0 and 340°C). An interesting point to note is the variation of viscosity over this range of temperatures. The strong variations in the heat-transfer coefficient which will occur as a result of this viscosity variation is shown in Fig. 1.1.

Tables 1.5 to 1.7 give the specific heat, the thermal conductivity, and the viscosity of steam for a range of conditions. It is important to note the variation of the specific heat of steam with temperature. The saturation pressure, the saturation temperature, enthalpy values, and specific volumes for a range of conditions are given in Table 1.1.

There are many detailed tabulations of steam physical properties and saturation conditions and the reader is referred to these if further information is required.[9–15]

Thermodynamic Behavior

Terms used to define the thermodynamic state of water follow.

Unsaturated (subcooled) water: Water below the saturation temperature.

Saturated water: Liquid water which is at the saturation temperature at a given pressure. The addition of any further heat will result in the generation of steam.

Wet steam: A two-phase steam-water mixture which exists during the vaporization of water to steam or the condensation of steam to water.

Saturated steam: Steam that is at saturation temperature for a given pressure and which contains no liquid phase. The extraction of any heat or work will result in the formation of drops of liquid and a wet-steam condition.

Superheated steam: Steam heated above the equilibrium saturation temperature at a given pressure. Another term often used for this condition is dry steam.

Degree of superheat: The amount by which the actual temperature of steam exceeds the saturation temperature at that pressure.

Cooling Water Systems

Three basic types of cooling water systems can be identified, (1) once-through, (2) open recirculating (evaporative), and (3) closed recirculating (nonevaporative). In certain environments, radiant cooling and ground-coupled systems are feasible.[16]

Once-Through System As its name suggests, cooling water from the local environ is used once and then rejected from

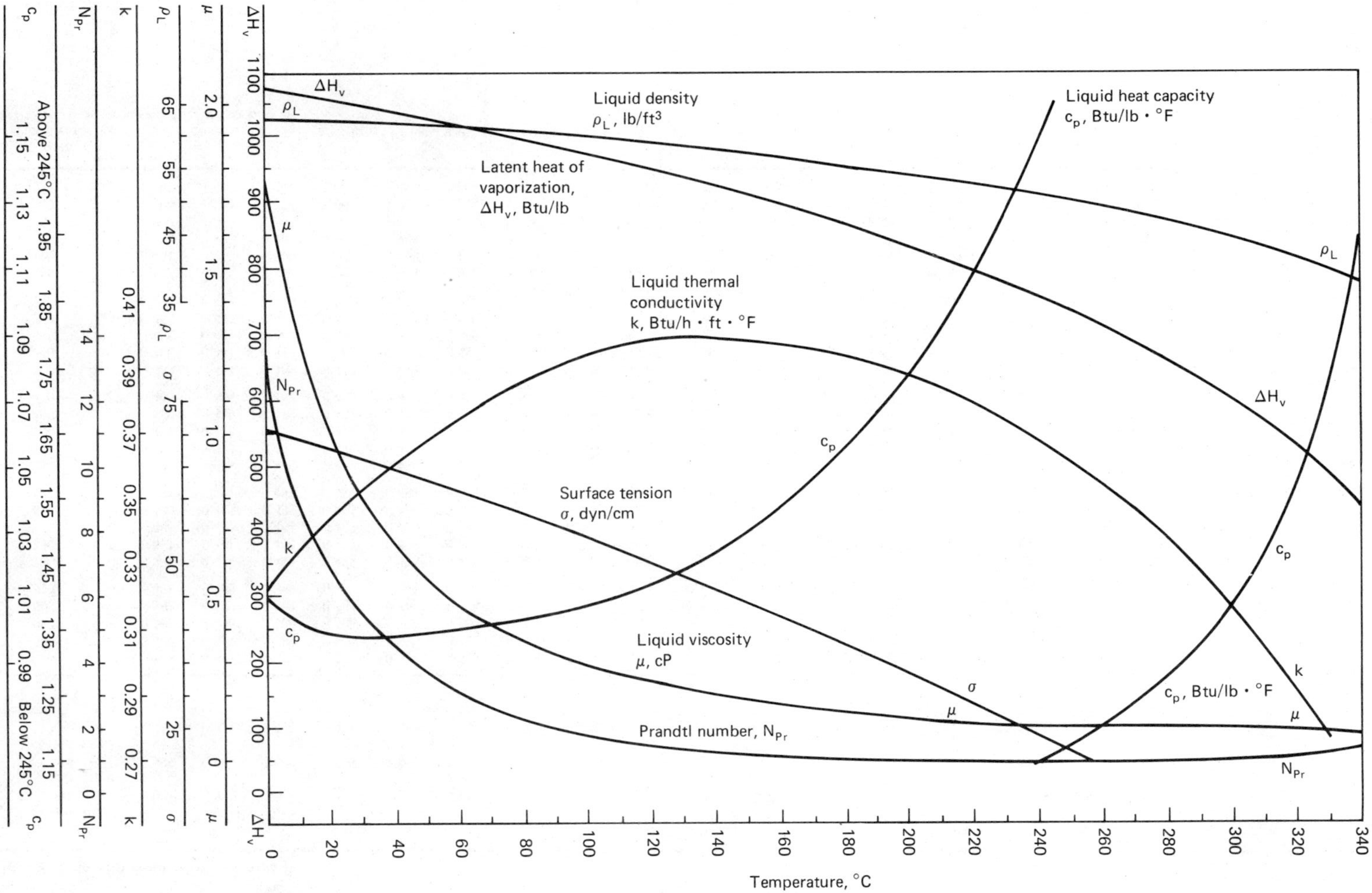

Fig. 1.2 Physical properties of liquid water. (*From Ref. 8. Excerpted by special permission from* Chemical Engineering, *Apr. 6, 1959. © 1959 by McGraw-Hill, Inc., New York.*) (Conversion factors: 1 Btu/lb = 2326 J/kg; 1 cP = 0.001 N · s/m² = 2.42 lb/ft · h; 1 lb/ft³ = 16.02 kg/m³; 1 dyn/cm = 0.001 N/m = 6.85 × 10^{-5} lb_f/ft; 1 Btu/h · ft · °F = 1.73 W/m · °C; 1 Btu/lb · °F = 4186.8 J/kg · °C.)

Table 1.5. Specific Heat of Water and Steam

Temperature, °F \ Pressure, lb/in² abs	C_p, Btu/lb · °F																			
	1	2	4	8	15	30	60	100	150	300	600	1000	1500	2000	3000	4000	6000	8000	10,000	15,000
Saturated water	0.998	0.999	1.000	1.003	1.007	1.014	1.026	1.039	1.054	1.093	1.168	1.286	1.492	1.841	7.646	—	—	—	—	—
Saturated steam	0.450	0.454	0.461	0.471	0.485	0.508	0.543	0.582	0.624	0.729	0.915	1.191	1.667	2.557	13.66	—	—	—	—	—
980	0.513	0.513	0.513	0.513	0.514	0.514	0.516	0.518	0.520	0.528	0.544	0.567	0.601	0.640	0.737	0.858	1.180	1.475	1.531	1.312
940	0.509	0.509	0.509	0.510	0.510	0.511	0.512	0.515	0.517	0.526	0.543	0.570	0.610	0.658	0.778	0.934	1.376	1.708	1.639	1.299
900	0.506	0.506	0.506	0.506	0.506	0.507	0.509	0.512	0.515	0.524	0.544	0.576	0.624	0.683	0.834	1.048	1.716	1.932	1.660	1.259
860	0.502	0.502	0.502	0.503	0.503	0.504	0.506	0.509	0.512	0.523	0.546	0.584	0.644	0.718	0.918	1.240	2.316	2.019	1.593	1.212
820	0.498	0.499	0.499	0.499	0.499	0.501	0.503	0.506	0.510	0.522	0.550	0.597	0.672	0.767	1.054	1.620	2.886	1.888	1.503	1.175
780	0.495	0.495	0.495	0.495	0.496	0.497	0.500	0.503	0.508	0.522	0.557	0.615	0.712	0.840	1.312	2.550	2.547	1.670	1.416	1.142
740	0.491	0.491	0.492	0.492	0.493	0.494	0.497	0.501	0.507	0.524	0.567	0.642	0.770	0.960	1.913	8.119	1.886	1.493	1.332	1.114
700	0.488	0.488	0.488	0.489	0.490	0.491	0.495	0.500	0.506	0.528	0.582	0.681	0.861	1.181	6.145⊙	2.237	1.557	1.358	1.250	1.089
660	0.484	0.485	0.485	0.485	0.486	0.489	0.493	0.499	0.507	0.534	0.604	0.738	1.015	1.639	1.851	1.587	1.369	1.256	1.187	1.071
620	0.481	0.481	0.482	0.482	0.483	0.486	0.491	0.499	0.509	0.543	0.637	0.826	1.308	1.614	1.455	1.362	1.252	1.184	1.136	1.056
580	0.478	0.478	0.478	0.479	0.481	0.484	0.491	0.500	0.513	0.558	0.685	0.969	1.393	1.351	1.289	1.243	1.178	1.134	1.102	1.046
540	0.475	0.475	0.475	0.476	0.478	0.482	0.491	0.503	0.519	0.580	0.756	1.272	1.249	1.229	1.196	1.169	1.128	1.098	1.074	1.031
500	0.472	0.472	0.473	0.474	0.476	0.481	0.492	0.508	0.530	0.611	0.865	1.181	1.169	1.157	1.137	1.120	1.092	1.069	1.051	1.017
460	0.469	0.469	0.470	0.472	0.475	0.481	0.495	0.516	0.545	0.657	1.132	1.125	1.117	1.110	1.096	1.084	1.064	1.047	1.033	1.004
420	0.466	0.466	0.467	0.470	0.473	0.482	0.501	0.528	0.568	0.724	1.091	1.087	1.081	1.076	1.067	1.058	1.042	1.029	1.018	0.994
380	0.463	0.464	0.465	0.468	0.473	0.484	0.509	0.546	0.601	1.065	1.063	1.059	1.056	1.052	1.044	1.038	1.026	1.015	1.006	0.986
340	0.460	0.461	0.463	0.467	0.473	0.488	0.521	0.572	1.045	1.044	1.042	1.039	1.036	1.033	1.028	1.022	1.013	1.004	0.996	0.979
300	0.458	0.459	0.462	0.466	0.475	0.495	0.539	1.029	1.028	1.028	1.026	1.024	1.022	1.019	1.015	1.010	1.002	0.995	0.988	0.973
260	0.456	0.457	0.461	0.467	0.478	0.505	1.017	1.016	1.016	1.016	1.014	1.013	1.011	1.009	1.005	1.001	0.994	0.998	0.982	0.968
220	0.454	0.456	0.460	0.468	0.484	1.008	1.008	1.008	1.008	1.007	1.006	1.005	1.003	1.001	0.998	0.994	0.988	0.982	0.977	0.964
180	0.452	0.455	0.460	1.003	1.003	1.003	1.003	1.002	1.002	1.002	1.001	0.999	0.998	0.996	0.993	0.989	0.983	0.978	0.973	0.961
140	0.451	0.454	1.000	1.000	0.999	0.999	0.999	0.999	0.999	0.998	0.997	0.996	0.994	0.992	0.989	0.986	0.980	0.974	0.969	0.958
100	0.998	0.998	0.998	0.998	0.998	0.998	0.998	0.998	0.997	0.997	0.996	0.994	0.992	0.990	0.986	0.983	0.976	0.970	0.965	0.955
60	1.000	1.000	1.000	1.000	1.000	1.000	0.999	0.999	0.999	0.998	0.996	0.994	0.991	0.989	0.984	0.979	0.970	0.963	0.956	0.942
40	1.004	1.004	1.004	1.004	1.004	1.004	1.004	1.003	1.003	1.002	1.000	0.997	0.993	0.989	0.983	0.976	0.965	0.954	0.945	0.920
32	1.007	1.007	1.007	1.007	1.007	1.007	1.007	1.006	1.006	1.005	1.002	0.999	0.994	0.990	0.983	0.975	0.962	0.949	0.937	0.904

⊙ Critical point.

CONVERSION FACTORS: 1 Btu/lb · F = 4186.8 J/kg · °C; 1 lb/in² = 6894.8 Pa; °C = (°F − 32)/1.8.

SOURCE: Ref. 9. Reproduced by permission.

Table 1.6. Thermal Conductivity of Water and Steam

Temperature, °F \ Pressure, lb/in², abs	$k \times 10^3$ Btu/h · ft · °F											
Pressure, lb/in², abs	1	5	10	20	50	100	200	500	1000	2000	5000	7500
Saturated liquid	362.3	382.1	388.6	393.2	395.6	393.1	385.3	362.5	329.7	272.7	—	—
Saturated vapor	11.2	12.8	13.7	14.9	16.9	19.0	22.0	28.2	36.9	60.3	—	—
1500	63.5	63.5	63.5	63.5	63.6	63.7	63.9	64.5	65.5	67.8	76.5	86.0
1400	58.9	58.9	58.9	58.9	59.0	59.1	59.3	59.3	61.0	63.5	73.2	84.1
1300	54.4	54.4	54.4	54.4	54.5	54.6	54.8	55.5	56.7	59.3	70.3	83.2
1200	49.9	50.0	50.0	50.0	50.1	50.2	50.4	51.1	52.4	55.3	68.0	84.1
1100	45.6	45.6	45.6	45.7	45.7	45.8	46.1	46.9	48.3	51.6	67.0	88.6
1000	41.4	41.4	41.4	41.4	41.5	41.7	41.9	42.7	44.3	48.1	68.6	102.2
900	37.3	37.3	37.3	37.4	37.5	37.6	37.9	38.8	40.6	45.3	77.6	143.5
800	33.4	33.4	33.4	33.5	33.5	33.7	34.0	35.1	37.2	43.7	136.7	221.4
775	32.4	32.5	32.5	32.5	32.6	32.7	33.1	34.2	36.5	43.7	182.2	240.3
750	31.5	31.5	31.5	31.6	31.6	31.8	32.1	33.3	35.8	44.1	215.5	258.2
725	30.6	30.6	30.6	30.6	30.7	30.9	31.2	32.4	35.2	44.9	239.2	275.0
										⊙		
700	29.6	29.6	29.7	29.7	29.8	30.0	30.3	31.6	34.6	46.7	261.3	290.9
675	28.7	28.7	28.7	28.8	28.9	29.1	29.4	30.9	34.3	50.4	281.2	305.7
650	27.8	27.8	27.8	27.9	28.0	28.2	28.6	30.1	34.1	55.3	298.7	319.6
625	26.9	26.9	27.0	27.0	27.1	27.3	27.7	29.4	34.0	282.8	314.5	332.5
600	26.0	26.1	26.1	26.1	26.2	26.4	26.9	28.8	34.2	302.9	328.5	344.2
575	25.2	25.2	25.2	25.2	25.4	25.6	26.1	28.2	35.0	319.6	341.1	355.0
550	24.3	24.3	24.4	24.4	24.5	24.8	25.3	27.8	36.5	334.0	352.4	364.7
525	23.5	23.5	23.5	23.6	23.7	23.9	24.6	27.6	340.2	346.5	362.5	373.6
500	22.7	22.7	22.7	22.7	22.9	23.2	23.9	27.6	351.9	357.3	371.5	381.6
475	21.8	21.8	21.9	21.9	22.1	22.4	23.2	28.0	362.0	366.8	379.5	388.7
450	21.0	21.0	21.1	21.1	21.3	21.7	22.7	368.5	370.7	375.0	386.5	395.0
425	20.2	20.3	20.3	20.3	20.5	21.0	22.3	376.1	378.1	382.0	392.5	400.4
400	19.5	19.5	19.5	19.6	19.8	20.3	22.0	382.5	384.3	387.8	397.5	404.9
375	18.7	18.7	18.7	18.8	19.1	19.7	386.7	387.7	389.3	392.6	401.6	408.5
350	17.9	18.0	18.0	18.1	18.4	19.3	390.7	391.7	393.2	396.2	404.6	411.1
325	17.2	17.2	17.3	17.3	17.8	393.4	393.7	394.5	396.0	398.8	406.7	412.8
300	16.5	16.5	16.5	16.7	17.2	395.1	395.4	396.2	397.6	400.2	407.7	413.5
275	15.8	15.8	15.8	16.0	395.6	395.7	396.0	396.7	398.0	400.5	407.6	413.2
250	15.1	15.1	15.2	15.4	394.9	395.1	395.3	396.0	397.2	399.6	406.4	411.7
225	14.4	14.4	14.5	393.0	393.0	393.1	393.4	394.1	395.2	397.5	404.0	409.2
200	13.7	13.8	13.9	389.7	389.8	389.9	390.1	390.8	391.9	394.1	400.4	405.3
175	13.0	13.1	385.0	385.1	385.1	385.2	385.5	386.1	387.2	389.3	395.4	400.3
150	12.4	378.9	378.9	378.9	379.0	379.1	379.3	379.9	381.0	383.0	389.0	393.8
125	11.8	371.1	371.1	371.1	371.2	371.3	371.5	372.1	373.2	375.2	381.1	385.9
100	361.5	361.6	361.6	361.6	361.7	361.8	362.0	362.6	363.6	365.7	371.6	376.4
75	350.0	350.1	350.1	350.1	350.1	350.3	350.5	351.1	352.1	354.2	360.3	365.3
50	336.3	336.3	336.3	336.3	336.4	336.5	336.7	337.4	338.5	340.7	347.1	352.3
32	324.7	324.7	324.7	324.7	324.8	324.9	325.1	325.8	327.0	329.3	336.1	341.6

⊙ Critical point.
CONVERSION FACTORS: 1 Btu/h · ft · °F = 1.731 W/m · °C; 1 lb/in² = 6894.8 Pa; °C = (°F − 32)/1.8.
SOURCE: Ref. 9. Reproduced by permission.

Table 1.7. Viscosity of Water and Steam

	$\mu \times 10^7$ lb · s/ft²								
Pressure, lb/in², abs	1	5	20	50	200	1000	5000	10,000	15,000
Saturated liquid	139.7	81.6	53.8	41.5	28.8	19.1			
Saturated vapor	2.1	2.4	2.6	2.8	3.3	4.0			
Temperature, °F									
1500	8.6	8.6	8.6	8.6	8.6	8.6	9.1	10.0	11.0
1400	8.1	8.1	8.1	8.1	8.1	8.2	8.7	9.7	10.9
1300	7.7	7.7	7.7	7.7	7.7	7.7	8.3	9.5	10.9
1200	7.2	7.2	7.2	7.2	7.2	7.3	7.9	9.3	11.2
1100	6.8	6.8	6.8	6.8	6.8	6.8	7.5	9.4	11.7
1000	6.3	6.3	6.3	6.3	6.3	6.3	7.1	10.0	12.8
900	5.8	5.8	5.8	5.8	5.8	5.8	6.9	11.7	14.4
800	5.3	5.3	5.3	5.3	5.3	5.3	7.9	14.4	16.6
750	5.1	5.1	5.1	5.1	5.1	5.1	11.6	15.9	17.8
700	4.9	4.9	4.9	4.9	4.8	4.8	⊙ 14.5	17.3	19.0
650	4.6	4.6	4.6	4.6	4.6	4.6	16.6	18.9	20.5
600	4.4	4.4	4.4	4.4	4.4	4.3	18.6	20.5	22.0
550	4.1	4.1	4.1	4.1	4.1	4.0	20.6	22.3	23.8
500	3.9	3.9	3.9	3.9	3.9	21.3	22.9	24.5	25.9
450	3.7	3.7	3.7	3.6	3.6	24.1	25.5	27.1	28.5
400	3.4	3.4	3.4	3.4	3.3	27.5	28.9	30.5	32.0
350	3.2	3.2	3.2	3.2	31.8	32.1	33.5	35.2	36.8
300	3.0	3.0	3.0	2.9	38.3	38.6	40.1	41.9	43.6
250	2.7	2.7	2.7	48.0	48.0	48.4	49.9	51.8	53.6
200	2.5	2.5	63.4	63.4	63.4	63.7	65.3	67.2	69.2
150	2.3	89.9	89.9	89.9	89.9	90.2	91.6	93.4	95.3
100	142.3	142.3	142.3	142.3	142.3	142.4	142.9	144.0	145.5
50	273.1	273.1	273.1	273.0	272.8	271.7	267.3	263.8	262.5
32	374.3	374.2	374.2	374.1	373.6	371.0	359.9	350.3	344.7

⊙ Critical point.
CONVERSION FACTORS: 1 lb · s/ft² = 47.88 Pa · s; 1 lb/in² = 6894.8 Pa; °C = (°F − 32)/1.8.
SOURCE: Ref. 9. Reproduced by permission.

the system. This results in low capital costs but possibly higher operating costs than competing systems. It may be the economical choice for smaller systems, but for larger systems, or for situations where rejection of hot water to the environment is regulated, it may not be the best choice.

Open Recirculating (Evaporative) System In this system the heated cooling water is cooled before being recycled to the plant to repeat its cooling duties. The cooling of the water is accomplished by evaporative means. This may be either a cooling pond or a cooling tower.

A cooling pond consists of impoundments exposed to the open air. Cooling takes place due to the evaporation of water and can be enhanced by the installation of sprays. While simple in concept, the use of cooling ponds requires large areas and gives poor control of the cooling water return temperature.

Cooling towers contact the hot water and air intimately by spraying the water over a packing, which results in large surface areas being available for evaporation. Although natural draft towers can be used, the increase in performance that results from the use of a fan to move the air through the unit can usually be justified on economic grounds. The design of cooling towers is a specialized subject and decisions such as whether forced- or induced-draft fans should be used, whether a cross-flow or a countercurrent design is superior, and what type of packing is to be used should be discussed with suppliers of this type of equipment.

One problem which exists with evaporative cooling systems is that the evaporated water does not carry any significant level of impurities. Makeup water must be added continuously to replace the evaporated water, and this results in the concentration of impurities in the circulating water. In order to combat this, the water must have a proportion continuously removed (blowndown), which is wasteful. Water lost from the system due to both evaporation and blowdown must be replaced. The replacement water carries impurities and the cost of treating the water may be high. Even with good treatment, the water circulating in such a system often contains sufficient impurities to require the use of large fouling factors in heat exchanger design and to cause concern about corrosion.

Closed Recirculating (Nonevaporative) System In this type of system the heated water is cooled without being exposed to air. This may be carried out by a separate cooling system (which is then cooled evaporatively), by air cooling, by a sea or river water stream, or by a refrigeration system.

The advantage of this system is that there is no buildup of solids in the cooling water. This results in the reduction in blowdown rates and an improvement in water quality, which in turn results in a reduction in the fouling tendency and the ability to install smaller, less expensive heat exchangers.[17]

Cooling Tower Specification Specification of the cooling tower duty is important in determining the water temperatures and flow rates required for the heat exchangers in the system.

The most important temperature is the maximum value of the ambient wet-bulb temperature. It gives the theoretical limit to which the water can be cooled by simple evaporative cooling. The wet-bulb temperature is a function of the ambient dry-bulb temperature and the relative humidity. The selection of the design conditions is affected by seasonal climatic variations and by the extent to which a reduction in the performance of the cooling system can be accepted during periods of high wet-bulb temperatures.

Cooling Water Treatment The reader is referred to Chapter 2 for a discussion of water treatment.

Steam-Heating Systems

The use of steam for heating purposes is widespread in industry and ranges from small heating systems to large utility boilers producing superheated steam used for driving steam turbines and generating electricity. The principles involved remain the same for all systems, as do the precautions and care required in selecting the auxiliary equipment. The design of steam systems is a specialized field involving thermal, hydraulic, and material factors and the need to comply with state, local, and ASME codes. An experienced design company should be used to implement the system. A range of self-contained package units are available. This is the most economic way of purchasing steam power if the required steam conditions and duty can be matched by an off-the-shelf unit.

The replacement of steam-heating systems by those using hot water has been proposed. Advantages claimed for such a method include reduced losses, lower makeup water requirements, and smaller piping diameters.[18]

Components The major components of a steam-generation system are shown in Fig. 1.3 and described in the following.

Boiler This is the unit where the water is vaporized to produce steam. The heating fluid can be on the tube side of the unit, in which case the unit is described as being a fire-tube boiler; or on the shell side (outside of the tubes), in which case it is defined as being a water-tube boiler. Both types of units are suitable for certain service conditions, but the water-tube design is more common. The majority of steam raising is achieved by the combustion of fuel to produce flue gas, which then transfers heat to the water in the boiler. The use of a water-tube design places the high-pressure fluid inside the tubes and results in the most economical design.

There may be one or more boilers in a steam system, depending upon its complexity, and care must be taken in these situations that no undesirable interactions occur during startup or transient conditions.

There are two distinct methods by which the water is circulated through the boiler, namely, natural circulation and forced circulation. In natural-circulation systems the difference in density between the cold feedwater and the steam-

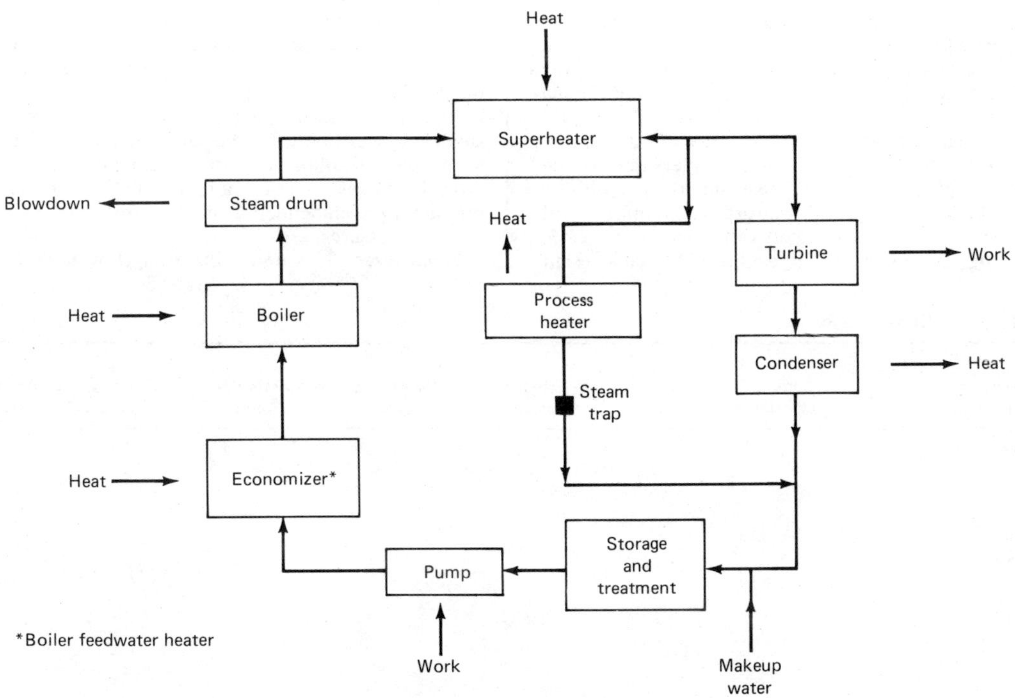

Fig. 1.3 Major components of a steam-generation system.

water mixture produced in the boiler is used to provide the motive force to bring water into the boiler and prevent dryout; such systems utilize a steam drum. Forced-circulation systems, which become a necessity above a certain pressure, utilize a pump to supply the water; in this situation there is no reserve water supply. The majority of industrial systems utilize natural circulation.

Steam Drum When present, the steam drum provides water storage to the boiler and the static head necessary to drive natural-circulation systems. In addition the steam drum carries out the separation of the steam water mixture that enters from the boiler. The efficiency of this separation is important because if water droplets are carried out of the drum with the steam, their impingement upon superheater tubes or turbine blades can be catastrophic.

If saturated steam is required, steam from the steam drum is passed directly to the unit where it is to be used, be it a heater or a process unit. If the steam is to be used in a steam turbine to produce power, it is usually passed to a steam superheater.

Superheater This unit takes the dry steam from the steam drum and heats it above its saturation temperature (i.e., it increases the level of superheat). The purpose of this is to add additional energy to an already high energy stream. This has been determined to be the most thermodynamically efficient manner of operating a steam turbine cycle. The high temperatures and pressures at which superheaters are required to operate require the use of low-alloy steels (such as those containing a nominal 2¼ percent of chromium so as to improve the high-temperature properties).

Although stainless steels are used for these units, it is important to consider the effects that trace chlorides in the steam may have (relating to the possibility of stress corrosion cracking) and the effects of operation for long periods at temperatures which may cause an alteration of the grain structure and a change in the physical properties of the material.

Several designs of superheaters are used. Consideration should be given to the self-draining characteristics of a design. For example, the pendant type of superheater with top headers and U-tubes hanging downward is not naturally self-draining, and problems associated with condensate collection, freezing, and corrosion are possible with such a design.

Table 1.8. American Boiler Manufacturers Association Recommended Maximum Impurity Levels in Boiler Water

Operating pressure, lb/in^2 gage	Total solids, ppm	Total alkalinity, ppm	Suspended solids, ppm
0–300	3500	700	300
301–450	3000	600	250
451–600	2500	500	150
601–750	2000	400	100
751–900	1500	300	60
901–1000	1250	250	40
1001–1500	1000	200	20
1501–2000	750	150	10
2001 and higher	500	100	5

CONVERSION FACTOR: 1 lb/in^2 = 6894.8 Pa.
SOURCE: Ref. 3. Reproduced by permission.

Condenser The low-pressure steam exhaust from a turbine or other process component is condensed using cooling water, or occasionally air, so as to produce liquid condensate. The condensers used for these services exhibit very high rates of heat transfer and contain features to allow the venting of inert gas which accumulates in the steam. The effect of even small quantities of inert gas upon the heat-transfer rates of condensing steam are dramatic. Inert gas contaminants, if not removed, will cause condensers to underperform seriously.

Condensate Treatment Depending upon the pressure level at which the system is operating, there will be a requirement for condensate treatment. The condensate should be deaerated and may be dosed with chemicals before being passed to the boiler feedwater pump, where it is raised to the system operating pressure. The condensate then goes to the economizer or boiler feedwater heater, where it is preheated before being passed to the steam drum. There will be some small amount of blowdown from the system to help maintain a low level of solids in the water; a figure of 1 percent is a typical value assumed in initial calculations. Table 1.8 gives minimum recommended levels of purity for boiler water at various pressures.

Economizer This unit utilizes any low-grade heat avail-

Table 1.9 Typical Brine Applications

Application	Sodium chloride	Calcium chloride	Ethylene glycol	Propylene glycol	Methanol water	Ethanol water	Chlorinated or fluorinated hydrocarbons
Breweries	. . .	. . .	. . .	×			
Chemical plants	×	×	×	. . .	×	×	×
Dairies	×	×	. . .	×			
Food freezing	×	×	. . .	×	. . .	×	
Heat pumps	. . .	×	×	. . .	. . .	. . .	×
Meat packing	×	×					
Preheat coils (air-conditioning systems)	. . .	. . .	×				
Skating rinks	. . .	×	×	. . .	×		
Snow melting	×	×	×				
Low temperature (special)	. . .	×	×	. . .	. . .	. . .	×
Ice cream	. . .	×	. . .	×			

SOURCE: Ref. 21. Reproduced with permission from *Handbook of Air-Conditioning System Design,* Carrier Air Conditioning Co., © 1965, McGraw-Hill, Inc.

Table 1.10. Brine Properties

Temperature, °F	Brine	Solution, wt %	Density, lb/ft³	Specific heat, Btu/lb °F	Thermal conductivity, Btu/h · ft · °F	Viscosity, cP	Freezing point, °F	Boiling point, °F	Gal/min/ton /10°F rise	h_b*	V_b†	Relative cost per gallon of solution
30	Sodium chloride	12	68.2	0.86	0.28	2.2	17.5	215.	2.55	941	1.61	1
	Calcium chloride	12	69.2	0.83	0.32	2.4	19.0	213.	2.62	971	1.78	3
	Methanol water	15	61.5	1.00	0.28	3.2	13.5	187.	2.45	781	2.63	13
	Ethanol water	20	61.0	1.04	0.27	5.5	12.0	189.	2.37	621	4.60	20
	Ethylene glycol	25	64.7	0.92	0.30	3.7	12.9	217.	2.52	775	2.92	42
	Propylene glycol	30	64.5	0.94	0.26	8.0	13.0	216.	2.47	525	6.35	43
15	Sodium chloride	21	72.8	0.80	0.25	4.2	1.0	216.	2.57	693	2.90	1
	Calcium chloride	20	74.8	0.72	0.31	4.8	1.0	214.	2.77	730	3.28	5
	Methanol water	22	60.4	0.97	0.26	5.3	4.5	182.	2.56	599	4.44	19
	Ethanol water	25	61.0	1.02	0.25	8.2	4.5	187.	2.41	504	6.85	25
	Ethylene glycol	35	66.0	0.86	0.28	6.8	0.0	219.	2.65	576	5.25	60
	Propylene glycol	40	65.3	0.89	0.24	20.0	−4.2	218.	2.58	103	‡	58
−5	Calcium chloride	25	78.4	0.67	0.29	10.3	−21.0	215.	2.85	513	6.75	6
	Methanol water	35	60.0	0.89	0.23	9.9	−22.0	176.	2.82	98	8.40	30
	Ethanol water	36	60.6	0.95	0.22	13.5	−16.0	183.	2.62	97	‡	35
	Ethylene glycol	45	67.4	0.79	0.25	17.2	−15.5	223.	2.82	103	‡	78
	Propylene glycol	50	66.5	0.83	0.23	80.0	−29.0	222.	2.72	98	‡	75
−30	Calcium chloride	30	82.1	0.63	0.28	27.8	−47.0	216.	2.90	110	‡	8
	Methanol water	45	60.0	0.80	0.22	18.0	−45.0	171.	3.13	91	‡	39
	Ethanol water	52	59.5	0.81	0.19	20.2	−50.0	179.	3.11	83	‡	50
	Ethylene glycol	55	69.0	0.73	0.22	75.0	−43.0	227.	2.98	93	‡	97
	Propylene glycol	60	67.2	0.77	0.21	700.0	−55.0	227.	2.90	91	‡	90

* h_b = coefficient of heat transfer between brine and surface (Btu/h · ft² · °F), at 7-ft/s velocity for 0.554-in ID tubing.
† V_b = minimum brine velocity (ft/s) at Re = 3500 for 0.554-in ID tubing.
‡ Above 10 ft/s.
CONVERSION FACTORS: 1 lb/ft³ = 16.01 kg/m³; 1 Btu/lb · °F = 4186.8 J/kg · °C; 1 Btu/h · ft · °F = 1.731 W/m · °C; 1 cP = 0.001 N · s/m² = 2.42 lb/ft · h; 1 Btu/h · ft² · °F = 5.678 W/m² · °C; 1 ft/s = 0.3048 m/s; 1 in = 2.54 cm; 1 gal (U.S. liquid) = 0.003785 m³.
SOURCE: Ref. 21. Reproduced with permission from *Handbook of Air-Conditioning System Design,* Carrier Air Conditioning Co., © 1965, McGraw-Hill, Inc.

able in the system to preheat the feedwater and improve the overall thermal efficiency of the process. If the temperature of the incoming boiler feedwater can be lower than the dew point of a gas stream which is heating it under any condition (startup, shutdown, part load, or transient), then consideration should be given to using a corrosion-resistant steel for this unit.

Under certain conditions the feedwater in the economizer may boil. This is not necessarily a problem as long as the potentiality for this to occur is recognized and the unit is designed to accept the relatively small amount of two-phase flow that will occur. The piping and valves between the economizer and the steam drum should also be checked to ensure that the system will cope with two-phase flow.

Steam Traps In order to separate steam and condensate, steam traps are used. The purpose of separating the steam and the condensate is to minimize steam loss, prevent condensate buildup (which may interfere with heat transfer and cause corrosion and water hammer), and vent noncondensable gases from the system. Various types of traps are available, which rely upon liquid, temperature, or energy level to perform their function. The savings in energy obtained by the effective use of steam traps can vary between 10 and 30 percent, assuming good maintenance is carried out on the units.[19,20]

1.2 BRINES

Brines may be any fluid which is used to transfer heat in a refrigerant system, does not change phase in the process, and has a flash point above 150°F (65°C). Originally the term brine referred to aqueous inorganic salt solutions such as calcium chloride or sodium chloride which had freezing points low enough to be used in refrigeration systems. Aqueous solutions of ethylene and propylene glycols are examples of nonsaline solutions which are used as brines. Table 1.9 gives a summary of typical brine applications and Table 1.10 lists the variations in physical properties, heat-transfer coefficients, flow requirements, and relative costs for a range of brines and operating temperatures. For further information on the physical properties of various brines the reader should consult Refs. 10, 11, and 21.

When selecting a brine, a rule of thumb is to choose a solution which has a freezing point approximately 20°F (11°C) lower than the lowest operating temperature expected in the system. This method provides a margin of safety while minimizing the decrease in heat-transfer performance and the increase in pumping requirements associated with the change in physical properties as the concentration of the solution is increased.

Advantages

The major advantage of using a brine is the ability to operate a cooling system below the freezing point of water while still maintaining reasonably high heat-transfer rates. In addition, when water-glycol mixtures are used, the cooling system is able to operate at temperatures higher than the boiling point of water without the requirement for high operating pressures.

Limitations

As the operating temperature is lowered, the thermal performance of brines generally decreases and their cost increases (see Table 1.10). Other disadvantages are listed below.

Corrosion Corrosion of piping and equipment should be considered when specifying brine systems. Since the operating pressures of such systems are not generally high, the additional corrosion allowances which should be considered do not result in large economic penalties. Sodium dichromate is a satisfactory and low-cost corrosion inhibitor which can be used with aqueous solutions of salts. Manufacturers of alcohols and glycols should be contacted for advice as to which corrosion inhibitors are most suitable for a given service condition. The choice of materials for systems utilizing brines must be given careful consideration because operating conditions, brine type, and the type of inhibitor are all important variables.

Safety The toxicity and flammability of brines (and additives) must be considered in any system. Certain brines, such as calcium chloride or methanol, should not be used with foodstuffs; very careful consideration must be given to the use of materials where any contamination of food is possible. Alcohol is flammable, and again care must be taken to ensure that such a brine is not used in systems where sources of ignition are present. For any brine the possibility of leakage into the process stream must be considered and reviewed to ensure that no undesirable effects will result. Table 1.11 lists brine toxicity data and Table 1.12 gives information on flammability.

"Natural" Brines

Due to concern over toxicity and the effect on the environment the use of a brine composed of "natural" material would be beneficial in many instances. Such a material is marketed.[22] It is composed of D-limonene, which is the major component of orange oil, and has a freezing point of −142°F (−96.7°C) and a boiling point of 136°F (57.8°C). Its viscosity

Table 1.11. Brine Toxicity Data

Brine	Threshold limit values, ppm*	Remarks
Sodium chloride	...	Nontoxic; suitable for direct contact with food
Propylene glycol	...	Nontoxic; suitable for direct contact with food
Calcium chloride	...	Essentially nontoxic but not suitable for direct contact with food
Freon-11	1000	Nontoxic
Ethanol water	...	Nontoxic
Methylene chloride	500	Very few toxic effects
Ethylene glycol	...	As toxic as methanol water but considered less dangerous because it is not absorbed as readily
Methanol water	200	Toxic; outdoor use normally recommended
Trichloroethylene	100	Toxic; outdoor use normally recommended

* American Conference of Governmental Industrial Hygienists, 1966.
SOURCE: Ref. 10. Reproduced with permission from *Perry's Chemical Engineer's Handbook*, 6th ed., © 1984, McGraw-Hill, Inc.

Table 1.12. Brine Flammability Data

Brine	Ratings*
Sodium chloride†	Nonflammable
Calcium chloride†	Nonflammable
Freon-11†	Nonflammable
Trichloroethylene†	Nonflammable at ordinary temperatures
Methylene chloride†	Practically nonflammable at ordinary temperatures
Propylene glycol‡	Flammable, moderate fire hazard; flash point 210 to 225°F undiluted
Ethylene glycol‡	Flammable, moderate fire hazard; flash point 232 to 240°F undiluted
Methanol water‡	Highly flammable, fire hazard; flash point 54 to 60°F undiluted; flash point 75°F at 30% solution
Ethanol water‡	Highly flammable, fire hazard; flash point 55°F undiluted

* The higher the flash point, the more safely the liquid can be handled. Liquids with flash points under 70°F should be regarded as highly flammable.

† *ASHRAE Guide and Data Book, Handbook of Fundamentals,* American Society of Heating, Refrigerating and Air-Conditioning Engineers, Atlanta, GA, 1967.

‡ Sax (ed.), *Dangerous Properties of Industrial Materials,* 5th ed., Van Nostrand Reinhold, New York, 1979.

CONVERSION FACTOR: °C = (°F − 32)/1.8.

SOURCE: Ref. 10. Reproduced with permission from *Perry's Chemical Engineers' Handbook,* 6th ed., © 1984, McGraw-Hill, Inc.

is 1 cP at −40°F (−40°C). The physical properties of this material allow low-temperature operation with high heat-transfer rates and low pressure drop.

Low-Temperature Materials

As the brine systems generally operate at low temperatures, the use of materials that are suitable for these conditions must be specified. At temperatures above −20°F (29°C) the problem should not be significant, but below this, impact-tested carbon steels or high-alloy steels should be used.

REFERENCES

1. *Gas Engineers' Handbook,* Industrial Press, New York, 1966.
2. R. H. Perry and C. H. Chilton (eds.), *Chemical Engineers' Handbook,* 5th ed., McGraw-Hill, New York, 1973.
3. *Principles of Industrial Water Treatment,* Drew Chemical Corp., Boonton, NJ, 1979.
4. P. Puckonis, "Controlling Corrosive Microorganisms in Cooling Water Systems," *Chem. Eng.,* **85** (Oct. 23, 1978).
5. O. Jones, "Guidelines Help Designers Protect Against Localized Corrosion," *Power,* **130** (Aug. 1986).
6. R. G. Kunz, A. F. Yen, and T. C. Hess, "Cooling Water Calculations," *Chem. Eng.,* **84** (Aug. 1, 1977).
7. C. Bertram, V. Desai, and E. Interess, "Designing Steam Tracing," *Chem. Eng.,* **79** (Apr. 3, 1972).
8. W. R. Gambill, "Physical Properties of Water," *Chem. Eng.,* **66** (Apr. 6, 1959).
9. ASME Thermodynamic and Transport Properties of Steam," American Society of Mechanical Engineers, 1977.
10. R. H. Perry and D. Green (eds.), *Perry's Chemical Engineers' Handbook,* 6th ed., McGraw-Hill, New York, 1984.
11. *ASHRAE Handbook, 1985 Fundamentals,* American Society of Heating, Refrigeration and Air-Conditioning Engineers, Atlanta, GA, 1985.
12. *Handbook of Heat Transfer Fundamentals* and *Handbook of Heat Transfer Applications,* W. M. Rohsenhow, J. P. Hartnett, and E. N. Ganić (eds.), McGraw-Hill, New York, 1985.
13. E. A. Avallone and T. Baumeister III (eds.), *Marks' Standard Handbook for Mechanical Engineers,* McGraw-Hill, New York, 1987.
14. *Steam—Its Generation and Use,* Babcock and Wilcox, New York, 1975.
15. "Engineering Science Data," Chem. Eng. ser. 67031, 68008, 68009, 68010, Engineering Science Data Unit, London, UK, 1967.
16. B. Givoni, "Options and Applications of Passive Cooling," *Energy and Buildings,* **7,** 297–300 (1984).
17. I. S. Rattan and V. K. Pathak, "Special Problems in the Design of Closed-Loop Cooling Systems," *Chem. Eng.,* **91** (Sept. 17, 1984).
18. W. Teller, W. Diskant, and L. Malfitani, "Conserving Fuel by Heating with Hot Water Instead of Steam," *Chem. Eng.,* **83** (June 21, 1976).
19. E. S. Moonroe, "Select the Right Steam Trap," *Chem. Eng.,* **83** (Jan. 5, 1976).
20. P. M. Goodhall (ed.), *The Efficient Use of Steam,* IPC Science and Technology, London, 1980.
21. Carrier Air Conditioning Co., *Handbook of Air-Conditioning System Design,* McGraw-Hill, New York, 1978.
22. "FC Heat Transfer Fluid Data Sheet," Florida Chemical Co., Lake Alfred, FL, 1987.
23. K. A. Selby, "Efficient Condensate Polishing Hinges on Process Control," *Power,* **126** (Sept. 1982).
24. C. O. Mackey, W. N. Barnard, and F. O. Ellenwood, *Engineering Thermodynamics,* Wiley, New York, 1957.

Chapter 2

Treatment of Water in Heat-Exchanger Systems

BENNETT P. BOFFARDI
Calgon Corporation

2.1 TYPES OF WATER HEAT-TRANSFER SYSTEMS

Water is the most common heat-transfer fluid in heat exchange equipment. However, water is corrosive to most alloys and contributes to the buildup of insulating deposits through water-formed and water-borne foulants.

There are basically three types of cooling systems: once-through, open recirculating, and closed recirculating systems. In once-through systems, the cooling water passes through the heat-transfer equipment only once before it is discharged. The mineral content of the water remains essentially the same. Open recirculating cooling water systems continuously reuse the water which passes through the heat-transfer equipment. Evaporative cooling to the atmosphere dispels the unwanted heat transferred to the cooling water. Makeup water is added to replace the evaporative losses. Additional makeup replaces the intentionally discharged water (blowdown) to maintain an acceptable level of dissolved minerals in the cooling water. Open recirculating cooling systems are oxygen saturated and contain a high level of dissolved solids. Closed recirculating cooling systems have little water loss and continuously recirculate the same water. The heat absorbed from the heat-transfer equipment is dissipated to another heat sink, which is cooled by other methods. Since there are no evaporative losses, the makeup water is minimal, and the mineral content remains essentially constant. However, corrosion by-products can easily accumulate and foul heat-transfer equipment since there are no means to remove them from the system. Figure 2.1 is a schematic of the three types of cooling systems.

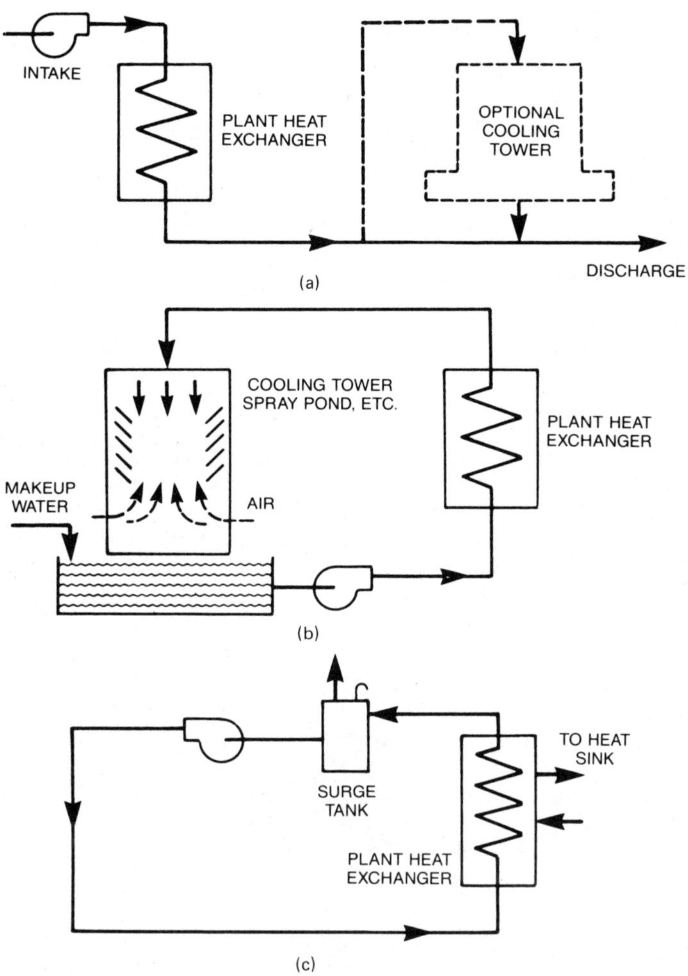

Fig. 2.1 Cooling water systems. (*a*) Once-through; (*b*) open recirculation; (*c*) closed recirculation.

2.2 CORROSION AND FOULING

It is important that both corrosion processes and fouling of heat-transfer equipment be understood if long-term reliability is to be achieved. We can define corrosion in a very practical sense as the deterioration of metals caused by the reaction with their surrounding environment. Since water is one of the most common heat-transfer fluids, it is not surprising that a majority of the problems associated with corrosion and deposits are water-related. Chemically, water is expressed as H_2O. However, dissolved gases (oxygen, carbon dioxide, chlorine, etc.), dissolved salts (calcium, magnesium, chloride, sulfate, bicarbonate ions, etc.), and suspended solids make water something other than pure H_2O.

Corrosion Reactions

For the corrosion reactions to take place between water and the metal surface, a potential difference needs to exist between different areas on the surface. This causes the passage of electric current through the metal from the area of low potential to that of high potential. Thus, corrosion of metals in contact with water is electrochemical in nature. The basic reactions occur at the region of lower potential, which is the anode. The dissolution reaction causes metal ions to form and go into solution. The anodic oxidation reaction can be generally represented by

$$M \longrightarrow M^{n+} + ne^-$$

where M represents the metal that has been oxidized to its ionic form having a valence of n^+ and the release of n electrons. For the more common heat-transfer materials, the individual reactions are as follows:

$$Fe \longrightarrow Fe^{2+} + 2e^-$$

$$Cu \longrightarrow Cu^+ + e^-$$

$$Al \longrightarrow Al^{3+} + 3e^-$$

The liberated electrons which migrate through the metal to areas of higher potential are utilized in the reduction of other ions or oxygen in the water. These reactions occur at the cathodic site on the metal surface.

Interaction between the products of the anodic and the cathodic reactions can occur, forming solid corrosion products on the metal surface. For example, ferrous ions coming from the corrosion of metallic iron will react with hydroxyl ions produced from the reduction of dissolved oxygen,

$$Fe^{2+} + 2OH^- \longrightarrow Fe(OH)_2$$

Ferrous hydroxide is further oxidized to form ferric hydroxide. The latter is unstable and subsequently transformed to hydrated ferric oxide (Fe_2O_3), which is common red rust,

$$4Fe(OH)_2 + O_2 + 2H_2O \longrightarrow 4Fe(OH)_3$$

$$2Fe(OH)_3 \longrightarrow Fe_2O_3 + 3H_2O$$

A buildup of rust occurs at anodic sites and forms mounds known as tubercles. Under these mounds, localized corrosion continues to accelerate. Figure 2.2 illustrates this process.[1] Not all corrosion products are detrimental. The protective oxide films on copper (Cu_2O) and aluminum (Al_2O_3) are the result of corrosion,

$$2Cu^+ + 2OH^- \longrightarrow Cu_2O + H_2O$$

$$Al^{3+} + 3OH^- \longrightarrow Al(OH)_3$$

$$2Al(OH)_3 \longrightarrow Al_2O_3 \cdot 3H_2O$$

These films are self-limiting, inhibiting the corrosion process after they have been fully developed.

The corrosion process between water and metal surfaces can take many forms.[2,3] The more common forms that have been observed to occur in heat-transfer equipment are described below.

General Corrosion, or Uniform Attack It occurs when the anodic areas on the metal surface keep shifting to different sites. This continual shifting results in relatively uniform metal removal. This type of corrosion can often be predicted so that material loss can be taken into account.

Galvanic Corrosion It takes place when dissimilar materials are in contact in a conducting fluid (water). Accelerated corrosion occurs with the least resistant alloy while the more resistant alloy is protected. The resistance of alloys can be found from the galvanic series, which ranks materials according to their chemical reactivity in seawater. The metals close to the active end of the series will behave as the anode and corrode, whereas those at the noble end will behave as the cathode and be protected. Table 2.1 lists an abridged galvanic series.[2] The intensity of attack is related to the relative surface areas of the metals in electrical contact. Large cathodic areas coupled to small anodic areas will aggravate galvanic corrosion and cause severe dissolution of the more active metal. The reverse situation, large anodic areas coupled to small cathodic areas, produces very little galvanic current.

It is obvious that galvanic corrosion must be considered in the design of heat-transfer equipment. Alloys close to one another in the series should be used. Less obvious are the damaging effects that can occur when a dissolved noble metal is transported through the water and is capable of depositing on an active metal. For example, copper ions can plate out onto the surfaces of steel heat-transfer tube, setting up local corrosion cells.

Erosion Corrosion This is normally restricted to copper-base alloys. It occurs in area where the intensity of turbulence at the metal surface is high enough to cause mechanical or

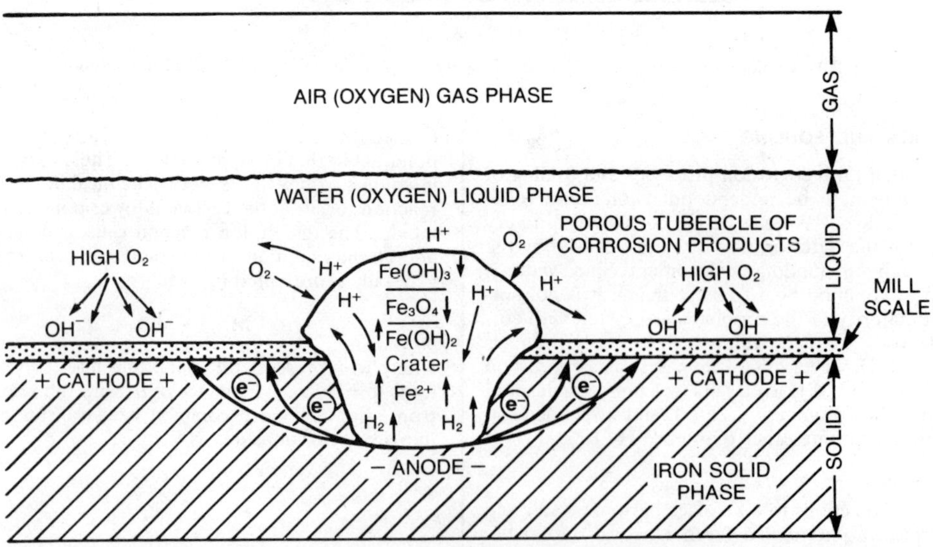

Fig. 2.2 Tuberculation and pit development in typical corrosion cell.

Table 2.1. Galvanic Series of Some Commercial Metals and Alloys in Seawater

	Platinum
	Gold
Noble or cathodic	Graphite
	Titanium
	Silver
	Chlorimet 3 (62Ni, 18Cr, 18Mo)
	Hastelloy C (62Ni, 17Cr, 15Mo)
	18-8 Mo stainless steel (passive)
	18-8 stainless steel (passive)
	Chromium stainless steel 11-30Cr (passive)
	Inconel (passive) (80Ni, 13Cr, 7Fe)
	Nickel (passive)
	Silver solder
	Monel (70Ni, 30Cu)
	Cupronickels (60–90Cu, 40–10Ni)
	Bronzes (Cu-Sn)
	Copper
	Brasses (Cu-Zn)
	Chlorimet 2 (66Ni, 32Mo, 1Fe)
	Hastelloy B (60Ni, 30Mo, 6Fe, 1Mn)
	Inconel (active)
	Nickel (active)
	Tin
	Lead
	Lead-tin solders
	18-8Mo stainless steel (active)
	18-8 stainless steel (active)
	Ni-resist (high Ni cast iron)
	Chromium stainless steel, 13Cr (active)
	Cast iron
	Steel or iron
	2024 aluminum (4.5Cu, 1.5Mg, 0.6Mn)
Active or anodic	Cadmium
	Commercially pure aluminum (1100)
	Zinc
	Magnesium and magnesium alloys

electrochemical disruption of the protective film. The anodic reaction is promoted at these sites and forms horseshoe or crescent-shaped indentations facing upstream of the water flow. The process is usually accelerated when abrasive solid particles (i.e., sand) are entrained in the water. Since turbulence increases with velocity, areas having higher water velocities are prone to attack.

Concentration Cell Corrosion (Crevice or Underdeposit Corrosion) This is an electrochemical attack due to differences in the corrosive environment between a shielded area and its surroundings. Attack usually occurs in areas having a small volume of stagnant solution, such as at tubesheet supports, under deposits or tubercles, and in threaded joints.

Pitting Corrosion This is one of the insidious forms of attack. It takes place at small discrete areas where overall metal loss is negligible. The pit develops at a localized anodic site on the surface and continues to grow due to a large cathodic area surrounding the anode. The reactions within the pit become self-sustaining (autocatalytic) with very little tendency for them to be suppressed, ultimately causing penetration through the base metal. Pitting corrosion has also been associated with both crevice and galvanic corrosion. It should be noted that metal deposition (copper ions plated on steel surface) also can create sites for pitting attacks.

Factors in Corrosion

Water Quality The corrosivity of water is influenced significantly by concentrations of dissolved species including gases, pH, temperature, suspended matter, bacteria, and so on. The effect of the entities on corrosion are assumed to be independent. Although this is not totally correct since interactions do exist, it is easier to visualize their contribution when considered alone.

Dissolved Gases Dissolved *oxygen* is a major factor contributing to the natural corrosion of steel. Steel corrosion is proportional to the oxygen content in the water. Increasing temperature decreases oxygen solubility, while increasing its partial pressure (Henry's law) increases its solubility. Concentrated electrolytes decrease oxygen solubility. (Molar salt concentrations exist in pit cavities and crevices.) For open recirculating cooling water systems, the concentration of dissolved oxygen is approximately 6 mg/L. The presence of oxygen is necessary for the formation of protective oxides on copper (Cu_2O), aluminum (Al_2O_3), and steel (γ-Fe_2O_3). In the absence of dissolved oxygen, the corrosion of steel can be negligible. A uniform protective film of magnetite (Fe_3O_4) is formed in the absence of oxygen by

$$3Fe + 4H_2O \longrightarrow Fe_3O_4 + 4H_2$$

Overall the effect of oxygen can have either a negative or a positive impact on steel corrosion. In aerated systems, an uneven distribution of corrosion products can form on the metal surface, giving rise to localized corrosion. Under deaerated conditions, the magnetite film isolates the base metal from the water, drastically reducing corrosion.

Carbon dioxide is more soluble than oxygen in pure water (1.3 g/L at 30°C) and will convert to carbonic acid, producing a solution having a pH of less than 6 where acid attack can predominate. At higher pH, carbonic acid (H_2CO_3) will dissociate to form bicarbonate ions (HCO_3^-) and subsequently carbonate ions (CO_3^{2-}). Calcium ions in the water will react with bicarbonate species to produce calcium carbonate ($CaCO_3$), which will precipitate on heat-transfer surfaces. (See discussion on calcium carbonate.) The effect of CO_2 is most important in boiler condensate systems. Rapid general thinning of steel can occur. The addition of soluble amine inhibitors neutralizes the carbonic acid to suppress the corrosive attack. Copper alloys are susceptible to increased attack in the presence of carbon dioxide.

Chlorine is not a natural constituent of cooling waters, but is added for biological control. When dissolved in water, chlorine will convert to hypochlorous acid (HOCl), which will suppress the pH,

$$Cl_2 + H_2O \longrightarrow HOCl + HCl$$

Acid attack on steel is a concern when the pH falls below 7. Above this pH, chlorine does not have any deleterious effect on steel. Adequate biological control can be achieved if the pH is maintained in the vicinity of 7.5. However, chlorine will accelerate the corrosion of copper alloys, even at alkaline pH. It is one of the most aggressive species to copper alloys in cooling waters.

Another gas that can have an impact on heat-transfer equipment is *ammonia* (NH_3). Although its effect on ferrous alloys is minimal, it has a drastic impact on brasses and alloys rich in copper.[4] These alloys can experience both rapid general thinning and stress corrosion cracking. Ammonia is originally formed from the thermal degradation of various nitrogen-containing compounds added to a boiler to reduce ferrous corrosion.

Temperature Effect Temperature plays a dual role with respect to oxygen corrosion. Increasing the temperature will reduce oxygen solubility. In open systems, where oxygen can be released from the system, corrosion will increase up to a maximum at 176°F (80°C), where oxygen solubility is 1.3 grains/ft^3 (3 mg/L). Beyond this temperature, the corrosion rate of carbon steel decreases, and at boiling water conditions the temperature effect is similar to the condition at room temperature when a high oxygen content exists. For closed systems, in which oxygen cannot escape, corrosion continues to increase linearly with temperature.

Increasing the temperature will increase the rate of oxygen to the metal surface, thus increasing the corrosion rate since more oxygen is available for the cathodic reduction process. The viscosity of water will decrease with increasing temperature, which will aid oxygen diffusion. Ionic mobility will also increase with temperature, increasing the overall conductivity of water.

An unusual temperature effect can occur with copper alloys.[5] Temperature differences of at least 65°C between the ends of copper conduits will cause the cold end to be cathodic to the hot end. Copper ions will migrate to the cold end and dissolve (corrode) at the hot end. This effect is known as thermogalvanic attack.

One other effect of temperature should be noted, although it is not related to corrosion. The solubility of calcium carbonate and calcium sulfate will decrease with an increase in temperature, causing them to precipitate and form a thick barrier deposit at the hottest areas.

Suspended Solids Suspended matter such as clays, silt, and corrosion products are always present in open recirculating cooling water systems. Particulates scrubbed from the air add to the suspended solids concentration. These materials are usually soft and nonabrasive. They are capable of depositing in low flow areas, forming a physical barrier and preventing oxygen from reaching the metal-solution interface. This buildup will contribute to the formation of differential aeration cells (underdeposit corrosion), promoting localized attacks.

Effect of pH The normal pH range for an open recirculating cooling water system is 6.5 to 9. Closed systems operate at a pH of 8.5 to 9. In boilers, the pH of the water is maintained at 11.

These pH values are the bulk pH of the water, while the actual pH at the metal surface can be different depending upon prevailing surface reactions. Oxygen reduction will produce OH^- ions, raising the pH, whereas underdeposit corrosion products can depress the pH.

When the bulk water pH is moderately acidic, having a pH of 5, uniform corrosion is the predominant form of attack, which increases with decreasing pH. In mineral acid, with a pH of 4 or below, the protective oxide film dissolves, exposing bare metal surfaces. Corrosion is further accelerated when dissolved oxygen is reduced at the metal surface at low pH. As the pH rises above 4, iron oxides precipitate from solution to form deposits. The diffusion of oxygen to the metal surface is impeded by these deposits. As the pH increases, the nature of the iron oxide deposits changes from loosely adherent at pH 6 to hard and tenacious at pH above 8. Although the corrosion of steel in aerated waters decreases within the normal operating pH range of 6.5 to 9, the rates are sufficiently high that chemical treatments must be added to these systems to bring the rates within a manageable level of less than 5 mils (0.127 mm) per year.

Copper alloys are not as sensitive to pH as carbon steel. Acid pH will accelerate general corrosion. As the pH increases, uniform corrosion decreases drastically. However, general thinning is not nearly as great a concern as the formation of cupric ions into the cooling water, which can cathodically deposit on steel and create active sites for pitting attacks.

Apart from the effect of pH on corrosion, increasing the pH of water having moderate levels of calcium and alkaline values can result in the precipitation of calcium carbonate. The deposition of this alkaline scale can impede the diffusion of oxygen to the metal surface, besides forming a heat-transfer barrier.

Effect of Dissolved Salts Dissolved constituents in water can have a variety of effects, both individually and through their interactions. The effects include increased corrosion as well as scale and deposit formation. Increasing the dissolved solids content of water increases its conductivity. Galvanic effects due to the coupling of dissimilar metals are extended in water having a higher salt content compared to water of low conductivity.

Hardness ions (calcium and magnesium), besides bicarbonate ions, are inhibitive and will suppress corrosion, whereas chloride and sulfate ions are antagonistic and will increase the rate of some forms of corrosion.[6] A reduction in the aggressiveness of water can be achieved by increasing the concentration of hardness ions. Most waters used for cooling have a pH in the vicinity of neutral or above. This environment is very conducive to the formation of a protective film of calcium carbonate, even when the water is below the saturation level for this salt.

The overall aggressiveness of water is related to its hardness and alkalinity. Soft water, which is low in calcium, is more corrosive than hard water. Figure 2.3 shows the decrease in corrosion rates with an increase in calcium concentration for three different water compositions.

The bicarbonate ion is the predominant alkaline species in natural water within the pH range of 6.5 to 9. By itself, it is a mild, inhibitive species. Its reaction with calcium ions is quite obvious, producing bulk precipitation of calcium carbonate once saturation is reached. Localized corrosion cells normally develop under the deposits. Increasing the bicarbonate concentration beyond that required for calcium carbonate saturation can be aggressive to steel due to the formation of a nonprotective iron carbonate ($FeCO_3$) film.[7]

Copper alloys are also attacked by high bicarbonate concentrations. The formation of greenish-blue nodules, which consist of basic copper salts, usually creates pit sites.

Chloride and sulfate ions are known to have a deleterious effect upon steel. Pitting is the most common form of attack. The small exposed areas where chloride ions are adsorbed are anodic to the large cathodic passive oxide surface. High current densities are generated at the chloride site. Once corrosion starts, hydrolysis of the metal ions from the anodic reaction causes a drop in pH, which discourages film repair and accelerates attack. The level of chloride ions needed to

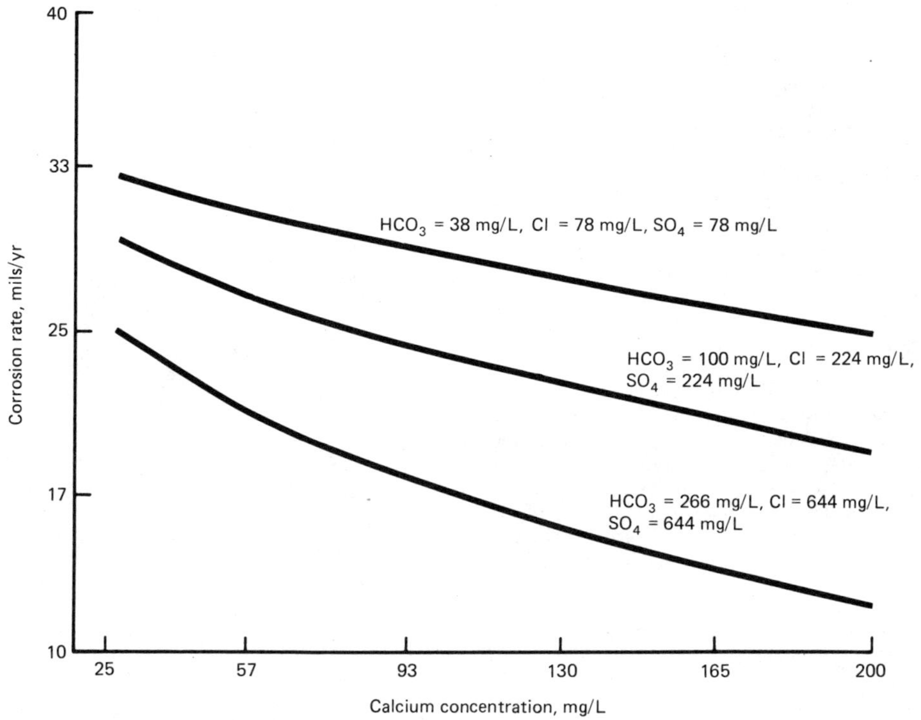

Fig. 2.3 Effect of calcium concentration on general corrosion of carbon steel.

initiate attack can be as low as a few milligrams per liter for some stainless steel in high-purity water. The more susceptible an alloy is to general attack, the less effect there is by chloride ions, carbon steel being a good example. Copper alloys are also subject to degradation by chloride ions, which modify the cuprous oxide structure.

Sulfate ions are somewhat more elusive in their effect upon corrosion. These ions do not appear to have the film-piercing properties that chloride ions display. In fact, there is evidence that sulfate ions can inhibit pitting attacks on some stainless steels.[8]

Other ions found in cooling waters which have an effect on corrosion are manganese, sulfide, phosphate, and nitrate. Sodium and potassium ions are considered neutral species and have no discernible effect. Manganese is found in both well and surface water and can cathodically deposit on stainless steels and copper-base alloys. Attack usually occurs under these deposits. Sulfide ions from putrefaction of organic matter, sulfate-reducing bacteria, or pollution will attack steels and copper-base alloys. The cuprous oxide film becomes nonprotective through a substitution of sulfide ions for the oxygen ion, thus accelerating corrosion.

Phosphate ions can act both as an accelerator or as an inhibitor for steel corrosion, depending upon their concentration. Nitrate ions can be reduced on steel surfaces, producing nitrites, ammonia, and hydroxyl ions. Nitrite ions are inhibitory to copper and steel alloys, whereas ammonia will attack copper.

Scale Deposition

Water-formed deposits commonly referred to as scale can be defined as a crystalline growth of an adherent layer (barrier) of insoluble salt or oxide on a heat exchange surface. The rate of formation is a complicated function of many variables, including temperature, concentration of scale-forming species, pH, water quality, and hydrodynamic conditions. The normal solubilities of scales increase with temperature, but a few have the reverse trend, such as calcium carbonate and calcium sulfate. Unfortunately these scales are commonly found in cooling water systems.

Calcium Carbonate Calcium carbonate ($CaCO_3$) is probably the most common scale found in cooling water systems. Both calcium and bicarbonate ions are needed to form this extremely tenacious scale. An increase in heat or pH will cause calcium bicarbonate to decompose to carbon dioxide and calcium carbonate,

$$Ca(HCO_3)_2 \longrightarrow CaCO_3\downarrow + CO_2\uparrow + H_2O$$

The greatest concentration of $CaCO_3$ will occur at the hottest areas along the heat-transfer surfaces.

Many methods have been proposed to predict the formation of calcium carbonate. However, they are all based upon the thermodynamic equilibria of carbonic acid and alkalinity corrected for temperature and dissolved solids (ionic strength). The more commonly used equations or indices are the Langelier saturation index[9] and the Ryzner stability index.[10]

It should be noted that these indices indicate only the tendency for calcium carbonate to deposit and not the rate or capacity for deposition. Also, these values do not take into account the tendency for calcium carbonate to supersaturate.

Calcium carbonate formation can be controlled by the addition of acids or specific chemicals tailored to inhibit its formation or modify the crystal lattice.

Sulfuric acid, which is inexpensive, is most often used. Other acids, such as hydrochloric acid or sulfamic acid, are also suitable. Acid addition produces more soluble salts compared to $CaCO_3$. These salts can reach saturation and must be controlled to prevent precipitation on heat-transfer surfaces.

Calcium carbonate can also be inhibited by chemical treatment. The most cost-effective materials are the polymeric inorganic phosphates. This class of compounds includes the salts of pyrophosphate, tripolyphosphate, and hexametaphosphate. Other phosphate compounds found to be effective belong to the class known as phosphonates. Two of the more common compounds are AMP [nitrilo tris (methylene phosphonic acid)] and HEDP (1-hydroxyethylidine 1,1 diphosphonic acid).

Organic polymers have also been found to be effective calcium carbonate inhibitors. These include polycarboxylates such as polyacrylates, polymethacrylates, polymaleates, and their copolymers.

The polymers adsorb onto the $CaCO_3$ crystal structure, which will limit their growth and ultimately scale formation. These polymers are more frequently considered dispersants and retard $CaCO_3$ scale by maintaining small particles of distorted crystalline material in suspension.

Bifunctional compounds are a recent development for calcium carbonate control. Phosphinocarboxylic acids contain an organic phosphorous and carboxylic acid groups. Simple molecules combining both the phosphono and the carboxylic groups have also emerged and have been found to be very effective calcium carbonate inhibitors.[11] However, the cost performance of these bifunctional compounds is poor compared to that of the traditional low-molecular-weight polymers, polyphosphates, or phosphonates.

Calcium Sulfate Calcium sulfate can exist in various forms in cooling water systems, the most common being gypsum ($CaSO_4 \cdot 2H_2O$). The hemihydrate and anhydrite forms are much less common. Their solubility as a function of temperature[12] is shown in Fig. 2.4. Since gypsum is at least a factor of 50 more soluble than calcium carbonate, it will precipitate only after the latter scale has been formed, within the normal pH range of 7 to 9. This phenomenon provides the basis for sulfuric acid addition to control $CaCO_3$ in recirculating cooling water systems. The normal upper limit for calcium and sulfate concentrations in the absence of an inhibitor is expressed by the relationship

$$[Ca] \cdot [SO_4] = 500{,}000$$

where the bracketed values are the ionic concentrations expressed in milligrams per liter.

The control of calcium sulfate scale can be achieved most effectively with polyacrylate,[13] its copolymers,[14] and phosphinocarboxylates.[15] As with calcium carbonate, excellent calcium scale control is achieved with polymers having low molecular weights in the range of 1000 to 10,000.

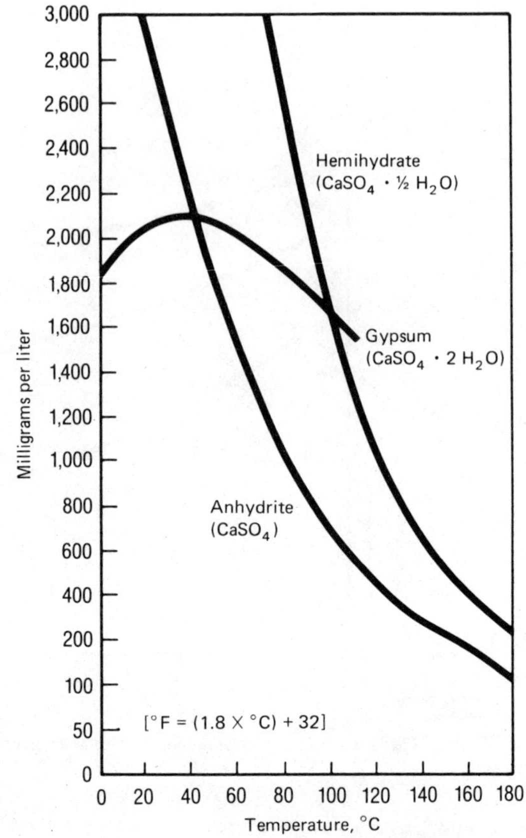

Fig. 2.4 Solubility of calcium sulfate.

Another chemical having efficacy to control calcium sulfate precipitation is the phosphonate AMP.[13] In fact, the family of aminophosphonates has been found to be extremely effective.

Calcium Orthophosphate Calcium orthophosphate scale has become more common in recirculating cooling water systems. Increasing pH and calcium concentration as well as phosphate addition from chemical treatment have provided the potential for this deposit to form on heat-transfer surfaces. Makeup waters obtained from agricultural runoff and partly treated sewage can have high orthophosphate ion levels of at least 4.3 grains/ft^3 (10 mg/L).

Due to the low solubility of calcium orthophosphate ($\approx 10^{-30}$), deposits can easily form in waters containing 2.1 grains/ft^3 (5 mg/L) of PO_4 ions and 131 grains/ft^3 (300 mg/L) of calcium ions at a pH of 7 to 7.5.

Calcium phosphate is an extremely difficult scale to inhibit. Intensive research has brought forth new chemistry to control this tenacious deposit. These new inhibitors[16,17] are acrylic acid–hydroxypropylacrylate copolymer; styrene sulfonic acid–maleic acid copolymer; acrylic acid–sulfonic acid copolymer; and phosphino/acrylic acid–organo sulfonic acid copolymer.

Silicate Scales Calcium and magnesium silicate scales tend to develop under more alkaline cooling water conditions, where the pH is approximately 8.5 or greater. These scales are very tenacious, dense, and difficult to remove from heat-transfer surfaces. Although the solubility of silica increases with pH, the alkaline silicates decrease with increasing pH.

An upper limit for the silica concentration is 65 grains/ft^3 (150 mg/L) in the recirculating water, although other factors impact upon this value. Magnesium silicate can precipitate on heat-transfer surfaces with magnesium concentration as low as 21.6 grains/ft^3 (50 mg/L) and 65 grains/ft^3 (150 mg/L) SiO_2. A rule-of-thumb pseudosolubility product of

$$[Mg] \times [SiO_2] \leq 8400$$

has been developed where the bracketed values are in milligrams per liter. The addition of chemical treatment as a preventive measure is essentially nonexistent. The most effective means of control is to keep the silica concentration in the recirculating cooling water below the 65-grains/ft^3 (150-mg/L) limit.

Fouling Deposits

Water-borne deposits commonly known as foulants are those loose, porous, insoluble materials suspended in water. They include such diverse substances as particulate matter scrubbed from the air, migrated corrosion products, silt, clays, and sand suspended in the makeup water, as well as organic contaminants (oils), biological matter, floc carryover from clarifiers, and extraneous materials including leaves, twigs, and wood fibers from cooling towers.

Fouling interferes with the flow of water and thus may affect heat transfer. High flow rates [5 to 8 ft/s (1.5 to 2.4 m/s)] can sweep away ordinary deposits, whereas with low flows [less than 2 ft/s (0.6 m/s)], the suspended foulants will drop out and deposit.

Mechanical methods can be used to reduce fouling. With once-through cooling systems, coarse filters (bar screens, trash racks, traveling screens) are used to remove large debris. Fine filtration is not practical. This is not the case with open recirculating systems. In situations where the makeup water contains appreciable concentration of suspended matter, it can be most advantageous to use side stream filtration. In general, passing a few percent of the recirculating water through the side stream filter will reduce the suspended solids by 80 to 90 percent at 5 to 8 cycles of concentration.

Another effective mechanical method used to control foulants is to pass scrapers, brushes, or balls through the heat exchanger tubes, wiping them free of deposits. This technique has found greatest use in power utility surface condensers and other critical exchangers.

Chemical treatment also has been found to be an effective control method. Synthetic polymers are the more commonly used chemicals to disperse foulants. Polyacrylates, polymaleates, partially hydrolyzed polyacrylamides, or their copolymers constitute the majority of dispersant chemicals. Natural dispersants such as tannins, lignin sulfonate, and carboxymethyl cellulose are occasionally used, but are not as effective as the synthetic polymers.

Biological Fouling The presence and growth of living organic matter is commonly referred to as biofouling. Its existence is capable of interfering with the flow of water through heat exchangers and other conduits, inhibiting heat transfer while contributing to corrosion and general deterioration of the entire cooling system.

Recirculating cooling water systems are ideal incubators, promoting growth and proliferation of microorganisms. Waters saturated with oxygen, exposed to sunlight, maintained at a temperature of 86 to 140°F (30 to 60°C), and having a normal pH range of 6 to 9 ensure abundant nutrients and an appropriate environment for life-sustaining growth.

Although biofouling interferes with heat transfer, it does so in a manner quite different from inert deposits. The buildup of biofoulants causes losses in heat transfer through the insulating properties of the biofilm. Kinetic energy is absorbed from the flowing water due to the ripple surface.[18] Increased pumping energy is required to overcome its frictional resistance. This characteristic surface is illustrated in Fig. 2.5. Even though the film is 95 to 98 percent water, the pressure drop produced by a 20-mil (500-μm) film is greater than would be expected.

There are three major classes of microorganisms associated with cooling water systems: algae, fungi, and bacteria. Algae can exist as very simple single-cell (unicellular) plants to multicell species. All algae contain colored pigments of which the most important is chlorophyll. Algae usually flourish on wetted surfaces exposed to oxygen and sunlight, such as cooling tower lumber, mist eliminators, screens, and distribution trays. Their impact upon metal corrosion can be severe. Large slime masses contribute to both underdeposit and pitting attacks. Massive growth can inhibit proper water distribution by plugging screens, restricting flow, and interfering with pump suction.

Fungi are similar to algae since they are plants, but fungi contain no chlorophyll. The major forms of interest are yeasts and molds. They require both moisture and air, but

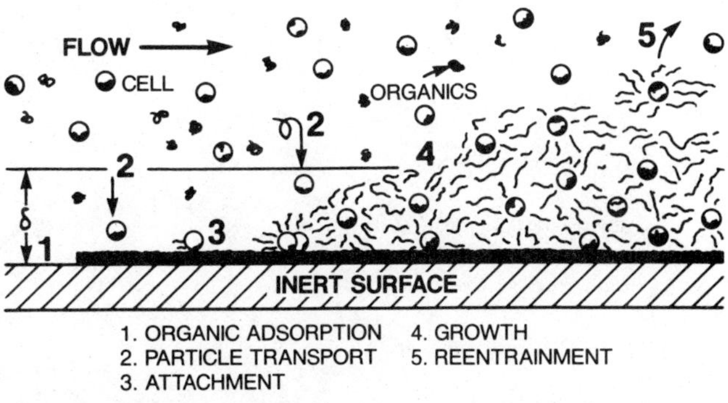

Fig. 2.5 Biofilm processes. Biofilm formation is initiated when small organic molecules become attached to an inert surface (1) and microbiological cells are absorbed on the resulting layer (2). The cells send out hairlike exopolymers to feed on organic matter (3), adding to coating (4). Flowing water detaches matter, producing equilibrium layer δ (5).

no sunlight, and they exist on nutrients found in water or substances to which they are attached (e.g., bacteria, algae). Mold fungi are filamentous in form, whereas most yeasts are unicellular. Fungi consume wood components, causing serious surface deterioration and internal decay (wood rot).

Bacteria are unicellular microscopic plantlike organisms similar to algae, but lacking chlorophyll. They exist in three basic forms: rod-shaped (bacillus), spherical (coccus), and spiral (spirillus). Wet environments high in organic content become most suitable for the proliferation of bacteria slime. Heat exchangers are ideal incubators for this biomass. They are capable of reducing heat-transfer efficiency significantly and of aggravating underdeposit corrosion attacks. Not all species of bacteria require the presence of oxygen. Aerobic bacteria need free oxygen for growth, whereas anaerobic bacteria grow in the absence of oxygen.

Acidic waters produced by some forms of bacteria will attack metal surfaces directly. One quite common species are sulfate-reducing bacteria. These organisms are anaerobic and convert dissolved sulfur compounds (i.e., SO_4^{-2}) to hydrogen sulfide. Carbon steel, stainless steel, and copper-base alloys can be corroded severely by H_2S acid. *Desulfovibrio desulfuricans* is the most prevailing sulfate-reducing species. It exists mainly under deposits that are devoid of oxygen. The corrosive attack on carbon steel by this bacterium is of quite distinctive form, recognizable by the smooth, disk-shaped, concentric rings formed on the metal surface. Corrosion rates as high as 100 mil/yr can occur under optimum growth conditions.

Thiobacillus is an aerobic sulfur bacterium capable of oxidizing sulfur, sulfides, or sulfates to sulfuric acid. Localized pH depression as low as 1 can occur, causing severe general thinning of steels where these organisms are in contact with the metal.

Another species common to cooling water systems are the nitrifying bacteria, which are capable of oxidizing ammonia to nitrate accompanied by a lowering in pH,

$$NH_3 + 2O_2 \longrightarrow HNO_3 + H_2O$$

Rapid general thinning can occur with copper-base alloys as well as steels. Nitrite-base steel corrosion inhibitors can be rendered ineffective by this species of bacteria due to their oxidation to nitrate.

Several other groups of bacteria exist in cooling water environments. Iron-depositing bacteria can oxidize water-soluble ferrous ions to insoluble ferric oxide, which will subsequently deposit on the inside of piping, reduce flow, and aggravate underdeposit corrosion. Slime-forming bacteria form dense sticky biomasses which impede water flow and contribute to fouling by sustaining the growth of other organisms.

The most practical and efficient method used to control microbiological activity in cooling waters is through the use of biocides.[19] Table 2.2 lists some of the microbiocides used in cooling water systems.

2.3 CORROSION CONTROL

Corrosion is an electrochemical process which occurs due to differences in potential at points on a metal surface. The potential difference is caused by many factors such as differences in the size of the crystal lattice or its orientation and boundary, differences in composition within a metal, discon-

Table 2.2. Biocide Selection Guide

Microbiocide	Bacteria*	Fungi*	Algae*	Comments
Chlorine (Cl_2)	E	S	E	Oxidizing, reacts with $-NH_2$ groups, effective at neutral pH, loses effectiveness at high pH. Use concentration: 0.1–0.2 mg/L continuous free residual; 0.5–1.0 mg/L intermittent free residual.
Chlorine dioxide (ClO_2)	E	G	G	Oxidizing, pH insensitive, can be used in presence of $-NH_2$ groups. Use concentration: 0.1–1 mg/L intermittent free residual.
Bromine (HOBr)	E	S	E	Oxidizing, substitute for Cl_2 and ClO_2, effective over broad pH range. Use concentration: 0.05–0.1 mg/L continuous free residual; 0.2–0.4 mg/L intermittent free residual.
Organo-bromide (DBNPA)	E	NA	S	Nonoxidizing, pH range 6–8.5. Use concentration: 0.5–24 mg/L intermittent feed.
Methylene bisthiocyanate (MBT)	E	S	S	Nonoxidizing, hydrolyzes above pH 8. Use concentration: 1.5–8 mg/L intermittent feed.
Isothiazolone	E	G	E	Nonoxidizing, pH insensitive, deactivated by HS^- and $-NH_2$ groups. Use concentration: 0.9–13 mg/L intermittent feed.
Quaternary ammonium salts	E	G	E	Nonoxidizing, tendency to foam, surface active, ineffective in highly oil or organic fouled systems. Use concentration: 8–35 mg/L intermittent feed.
Organo-tin/quaternary ammonium salts	E	G	E	Nonoxidizing, tendency to foam, functions best in alkaline pH. Use concentration: 7–50 mg/L.
Glutaraldehyde	E	E	E	Nonoxidizing, deactivated by $-NH_2$ groups, effective over broad pH range. Use concentration: 10–75 mg/L intermittent feed.

* E = excellent; G = good; S = slight; NA = not applicable.
CONVERSION FACTOR: 1 mg/L = 0.433 grains/ft^3.

Table 2.3. Inhibitors for Corrosion Control

Chemical	Type	Metal	Concentration range, mg/L	pH range	Temperature range, °C	Sensitivity and influence factors	Advantages	Cautions
Chromates	Anodic	Ferrous and nonferrous	Initial, 500–1000; maintenance, 200–250	6–11	30–70	Reducing conditions can deplete treatment concentration.	Extremely effective; rapid formation of protective oxide film.	Can cause pitting if applied below critical concentrations; toxic blowdown disposal restricts its use.
Nitrites	Anodic	Ferrous	500–800	Above 7.5	30–70	High concentrations of sulfate and chloride ions can interfere with protective oxide film.	Used in closed recirculating system when formulated with borax.	Not suitable in open recirculating system; critical concentration required to prevent pitting; nutrient for some species of bacteria.
Molybdates	Anodic	Ferrous and nonferrous	100–1000	5.5–8.5	30–70	Requires oxidizing (O_2) agent to impart protective film; sensitive to chloride and sulfate concentrations.	Essentially nontoxic to aquatic life; can inhibit pitting and differential aeration attack.	Potential for calcium molybdate precipitation in moderate- to high-calcium waters.
Orthophosphate	Anodic	Ferrous	15–20	Above 7.5	30–70	Used in oxygenated system; sensitive to water quality (especially chlorides) and pH; below pH 7.5 not as effective.	Rapid film formation, very tenacious; synergistic with other treatments.	Potential for calcium phosphate precipitation; low concentrations can cause pitting attacks.
Zinc	Precipitating	Ferrous	0.5–2	Below 7.5	30–70	Above pH 7.5 zinc will precipitate unless combined with zinc stabilizers.	Synergistic when combined with other inhibitors.	Not used alone in cooling systems.
Polyphosphates	Cathodic	Ferrous	15–20	6.5–7.5	30–50	Requires calcium ions for effective inhibition.	Relatively insensitive to water quality; synergistic with zinc; controls calcium scales.	Will revert to orthophosphate ions; potential for calcium phosphate precipitation.
Phosphonates	Mixed	Ferrous	15–20	Above 7.5	30–50	Sensitive to water quality and temperature.	Synergistic with zinc; reversion to orthophosphate significantly reduced; controls calcium scales.	Can accelerate corrosion of copper alloys; amine phosphonate degraded by chlorine.

Table 2.3. Inhibitors for Corrosion Control (*Continued*)

Chemical	Type	Metal	Concentration range, mg/L	pH range	Temperature range, °C	Sensitivity and influence factors	Advantages	Cautions
Zinc chromate	Cathodic	Ferrous and nonferrous	5–10 zinc, 5–10 chromate	Below 7.5	30–70	No sensitivity to water quality and temperature.	Very effective; requires very low level of chromates.	Zinc will precipitate above pH 7.5; chromate discharge restriction applies.
Zinc polyphosphate	Cathodic	Ferrous and nonferrous	0.2–1.5 zinc, 10–15 polyphosphate	6.8–7.2	30–70	Low sensitivity to water quality.	Rapid protective film formation.	Reversion to orthophosphate ions.
Zinc phosphonates	Cathodic		2–3 zinc, 8–10 phosphonate	7–9	30–70	Low sensitivity to water quality; zinc reduces attack on copper alloys.	More stable than zinc polyphosphate; rapid film formation; calcium scale control.	Degradation to orthophosphate ions by chlorine may occur.
Polyphosphate/ phosphonate	Cathodic	Ferrous	10–20 polyphosphate, 5–10 phosphonate	6–8	30–60	Some sensitivity to water quality.	Nonheavy metal treatment.	Reversion of polyphosphate to orthophosphate ions; attack on copper alloys is possible.
Polyphosphate/ orthophosphate	Mixed	Ferrous	5–10 polyphosphate, 10–20 orthophosphate	6–8.5	30–70	Moderately sensitive to water quality; chloride ions can promote pitting attacks.	Nonheavy metal treatment.	Potential for calcium phosphate precipitation unless controlled with Ca/PO_4 inhibitor.
Mixed phosphonate (AMP/HEDP)	Mixed	Ferrous	10–20 AMP, 5–10 HEDP	7.5–8.5	30–60	Some sensitivity to water quality.	Synergistic nonheavy metal treatment.	Attack on copper alloys may occur.
Tolyltriazole or benzotriazole	Cathodic	Copper-base	1–2	4–9	30–70	Sensitive to residual chlorine.	Specific copper inhibitor.	May be deactivated in presence of chlorine; activity returns after chlorine dissipates.
Mercaptobenzo-thiazole	Cathodic	Copper-base	1–2	4–9	30–70	Very sensitive to residual chlorine.	Specific copper inhibitor.	Cannot be used with residual chlorine in water.

CONVERSION FACTORS: 1 mg/L = 0.437 grains/ft^3; °F = 1.8 × °C + 32.

tinuous surface oxide films, inclusion of foreign or dissimilar material, applied or residual stresses, differences in electrolyte or oxygen concentration along the metal surface, and contact with dissimilar metals.

When corrosion occurs, a neutral metal atom goes into solution as a charged ion at the anodic area (oxidation reaction). Simultaneously, one or more cathodic reactions occur at cathodic areas (reduction reaction). The electrons lost by the metal dissolution reaction are utilized in cathodic reactions.

For example, corrosion of ferrous alloys in an oxygenated water system at neutral pH can be simply written

$$2Fe \longrightarrow 2Fe^{2+} + 4e^- \qquad \text{(anodic reaction)}$$

$$O_2 + 2H_2O + 4e^- \longrightarrow 4OH^- \qquad \text{(cathodic reaction)}$$

$$2Fe + O_2 + 2H_2O \longrightarrow 2Fe(OH)_2$$

$$2Fe(OH)_2 + \tfrac{1}{2}O_2 + H_2O \longrightarrow 2Fe(OH)_3 \qquad \text{(rust)}$$

The characteristic of the electrolyte can influence the degree of corrosive attack. Some of the more important variables include oxygen concentration, amount of dissolved or suspended solids, solution pH, velocity, and temperature.

Several approaches can be used to reduce corrosion by cooling waters. These include designing systems that use more corrosion-resistant materials, applying protective coatings (paints, epoxy), using sacrificial anodes (cathodic protection), and applying chemical treatments.

For chemical treatment or corrosion inhibitor programs to be effective, they must protect all exposed metal from corrosive attack; be effective at low concentration; not cause deposits on the metal surface; remain effective under a broad range of pH, temperature, water quality, and heat flux; prevent scale formation and disperse deposits; and have a minimal toxicological effect when discharged.

Economics must also be considered when assessing the merits of a chemical treatment program. In once-through cooling systems, protection must be achieved with a few milligrams per liter, otherwise the cost may be prohibitive. For these types of systems, the use of more corrosion-resistant alloys often is a more economical approach to corrosion control. With respect to open recirculating cooling systems, chemical addition to the makeup water can be around 50 mg/L. Cost is not as prohibitive due to the recycling of the water system, which also includes the inhibitor. Closed recirculating cooling water systems require high treatment levels, in the range of several thousand milligrams per liter. These systems have very little water loss.

Inhibitors can be classified broadly according to the rate process being controlled. Anodic inhibitors suppress anodic reactions, that is, the rate of metal ions being transferred into the aqueous environment is reduced. Cathodic inhibitors impede the cathodic reaction, such as the oxygen reduction reaction. Mixed inhibitors hinder both reactions.

Inorganic inhibitors usually affect the anodic process. Rapid suppression of general corrosion occurs. However, when these inhibitors are below a critical concentration, they can stimulate localized pitting attacks and are thus classified as "dangerous" inhibitors.

Cathodic inhibitors reduce corrosion by mainly interfering with the oxygen reduction reaction. This family of inhibitors decreases general corrosion and does not stimulate pitting attacks. They are generally considered "safe" inhibitors, although not as effective as the anodic inhibitors. When both the anodic and the cathodic reactions are affected by the chemical treatment, a mixed mode of inhibition is in effect.

Precipitating inhibitors fall under the domain of cathodic inhibitors. They produce insoluble films on the cathode under conditions of locally high pH and isolate the cathode from solution.

The performance of an inhibitor can be assessed by its effectiveness in reducing the rate of corrosion. Generally, this is expressed as a rate of thinning or penetration in micrometers or mils per year.

Today's approach to cooling water treatment is toward the more alkaline or high pH conditions. This was not always the case. Historically, acid treatment was widely used to prevent scale deposition on heat-transfer surfaces from occurring. The system pH was slightly acidic (pH 6.5 to 7.0), and corrosion was controlled by using highly effective anodic inhibitors. A multitude of different treatment programs have evolved to control corrosion under a variety of conditions. The basic building blocks used in these treatments are summarized in Table 2.3.[20]

REFERENCES

1. M. F. Obrecht and M. Pourbaix, *J. Am. Water Waters Assoc.*, **59**, 977 (1967).
2. M. G. Fontana and D. N. Green, *Corrosion Engineering,* 2d ed., McGraw-Hill, New York, 1978, pp. 28–113.
3. C. P. Dillon (ed.), *Forms of Corrosion, Recognition and Prevention,* National Association of Corrosion Engineers, Houston, TX, 1982.
4. H. Leidheiser, Jr., *The Corrosion of Copper, Tin, and Their Alloys,* Wiley, New York, 1971, pp. 146–150.
5. V. G. Ereneta, *Corrosion Sci.* **19,** 507 (1979).
6. B. P. Boffardi and G. W. Schweitzer, "Water Quality, Corrosion Control and Monitoring," presented at the 9th International Congress on Metallic Corrosion, Toronto, Canada, June 3–7, 1984.
7. G. T. Burnstein and D. H. David, *Corrosion Sci.* **20,** 1143, (1980).
8. P. H. Leckie and H. H. Uhlig, *J. Electrochem. Soc.*, **113,** 1262 (1966).
9. W. F. Langelier, *J. Am. Water Works Assoc.*, **28,** 1500 (1936).
10. J. W. Ryzner, *J. Am. Water Works Assoc.*, **36,** 472 (1944).
11. R. H. Ashcraft, "Scale Inhibition under Harsh Conditions by 2-Phosphonobutane-1,2,4-tricarboxylic Acid (PBTC)," Corrosion/85, National Association of Corrosion Engineers, Boston, MA, March 25–29, 1985, paper 123.
12. E. P. Partridge and A. H. White, *J. Am. Chem. Soc.*, **51,** 360 (1929).
13. P. H. Ralston, *Mater. Perform.*, **11**(6), 39 (1972).
14. L. J. Persinski, P. H. Ralston, and R. C. Gordon, Jr., US Patent 3,928,196.
15. M. J. Smith and P. Miles, US Patent 4,046,707.
16. R. C. May, G. E. Geiger, and D. A. Bauer, *Mater. Perform.*, **20**(4), 34 (1981).
17. B. P. Boffardi and G. W. Schweitzer, "Advances in the Chemistry of Alkaline Cooling Water Treatment," Corrosion/85, National Association of Corrosion Engineers, Boston, MA, March 25–29, 1985, paper 132.
18. J. F. Gary et al. (eds.), *Condenser Biofouling Control,* Ann Arbor Sci. Publ., Ann Arbor, MI, 1980, pp. 49–76.
19. J. W. McCoy, *Microbiology of Cooling Water,* Chemical Publ. Co., New York, 1980.
20. B. P. Boffardi, *Mater. Perform.*, **23**(11), 17 (1983).

Chapter 3

Common Gases as Heat-Transfer Fluids

DAVID F. BARTZ
Tecogen, Inc.

3.1 PROPERTIES OF COMMON GASES AT ATMOSPHERIC PRESSURE

Tables 3.1 to 3.8 list the thermal properties (density ρ, absolute viscosity μ, specific heat at constant pressure C_p, thermal conductivity k, Prandtl number Pr, and buoyancy coefficient $g\beta/\nu^2$) for eight common gases (air, ammonia, argon, carbon dioxide, helium, hydrogen, nitrogen, and refrigerant-12) at atmospheric pressure and at temperatures of −100 to 2000°F (−73.3 to 1093.3°C).

3.2 FUNCTIONAL RELATIONSHIPS OF GAS PROPERTIES WITH TEMPERATURE AT ATMOSPHERIC PRESSURE

The following information is provided to calculate, by interpolation, properties of the common gases at a desired temperature, given two or more values at known temperatures. The procedures presented will yield an error of less than 0.5 percent for the temperature range of Tables 3.1 to 3.8.

Viscosity

Reid[10] recommends the Sutherland equation,

$$\mu = \frac{bT^{3/2}}{S + T} \tag{3.1}$$

where μ = absolute viscosity, $lb_m/ft \cdot h$
T = temperature, °R
b, S = constants particular to the gas

Table 3.1. Thermal Properties of Air at Atmospheric Pressure

T_{gas}, °F	ρ, lb_m/ft^3	μ, lb_m/ft · h	C_p, Btu/lb_m · °F	k, Btu/h · ft · °F	Pr	$g\beta/\nu^2$, 1/°R · ft^3*
−100	0.1106	3.227E−02	2.403E−01	1.044E−02	7.427E−01	1.363E+07
0	0.0863	3.948E−02	2.396E−01	1.310E−02	7.220E−01	4.342E+06
32	0.0807	4.161E−02	2.398E−01	1.391E−02	7.174E−01	3.194E+06
68	0.0752	4.395E−02	2.398E−01	1.481E−02	7.118E−01	2.315E+06
100	0.0709	4.598E−02	2.400E−01	1.556E−02	7.093E−01	1.773E+06
200	0.0602	5.197E−02	2.411E−01	1.780E−02	7.039E−01	8.472E+05
300	0.0522	5.764E−02	2.427E−01	1.993E−02	7.020E−01	4.510E+05
400	0.0462	6.285E−02	2.448E−01	2.200E−02	6.994E−01	2.617E+05
500	0.0413	6.774E−02	2.473E−01	2.401E−02	6.980E−01	1.619E+05
600	0.0374	7.237E−02	2.504E−01	2.595E−02	6.983E−01	1.053E+05
800	0.0315	8.099E−02	2.567E−01	2.963E−02	7.018E−01	5.008E+04
1000	0.0272	8.886E−02	2.631E−01	3.323E−02	7.036E−01	2.673E+04
1200	0.0239	9.617E−02	2.688E−01	3.672E−02	7.041E−01	1.552E+04
1400	0.0213	1.032E−01	2.740E−01	3.997E−02	7.074E−01	9.587E+03
1600	0.0193	1.101E−01	2.788E−01	4.357E−02	7.047E−01	6.197E+03
1800	0.0176	1.167E−01	2.835E−01	4.694E−02	7.048E−01	4.178E+03
2000	0.0161	1.231E−01	2.881E−01	5.043E−02	7.035E−01	2.908E+03

* $g = 32.2$ ft/s². $\beta = 1/T_{gas}$(°R^{-1}); $\nu = \mu/\rho$ (ft²/h).

CONVERSION FACTORS:

Property	Symbol	USCS units	SI units
Density	ρ	1 lb_m/ft^3	= 16.02 kg/m³
Absolute viscosity	μ	1 lb_m/ft · h	= 4.13 × 10^{-4} kg/m · s
Specific heat	C_p	1 Btu/lb_m · °F	= 4.187 kJ/kg · °C
Thermal conductivity	k	1 Btu/h · ft · °F	= 1.731 W/m · °C
Prandtl number	Pr		
Buoyancy coefficient	$g\beta/\nu^2$	1/°R · ft³	= 63.58/K · m³

SOURCES: Refs. 1–5.

This can be rearranged to give

$$\frac{T^{3/2}}{\mu} = \frac{S}{b} + \frac{T}{b} \tag{3.2}$$

This relationship is solved as a linear equation for the gas constants S and b.

Density

For real gases,

$$\rho = \frac{144PM}{RTZ} \tag{3.3}$$

where ρ = density, lb_m/ft^3
Z = compressibility factors = 1.0 for an ideal gas
P = pressure, lb/in² abs

Table 3.2. Thermal Properties of Ammonia at Atmospheric Pressure*

T_{gas}, °F	ρ, lb_m/ft^3	μ, lb_m/ft · h	C_p, Btu/lb_m · °F	k, Btu/h · ft · °F	Pr	$g\beta/\nu^2$ 1/°R · ft^3
0	0.0517	2.108E−02	5.275E−01	1.167E−02	9.533E−01	5.46E+06
32	0.0481	2.259E−02	5.238E−01	1.267E−02	9.344E−01	3.85E+06
68	0.0446	2.426E−02	5.197E−01	1.379E−02	9.145E−01	2.67E+06
100	0.0420	2.579E−02	5.204E−01	1.489E−02	9.013E−01	1.98E+06
200	0.0354	3.064E−02	5.342E−01	1.864E−02	8.781E−01	8.47E+05
300	0.0307	3.560E−02	5.563E−01	2.279E−02	8.693E−01	4.09E+05
400	0.0271	4.062E−02	5.813E−01	2.726E−02	8.661E−01	2.16E+05
500	0.0243	4.566E−02	6.066E−01	3.201E−02	8.652E−01	1.23E+05
600	0.0220	5.069E−02	6.314E−01	3.691E−02	8.673E−01	7.40E+04
800	0.0185	6.066E−02	6.791E−01	4.689E−02	8.787E−01	3.07E+04

* See Table 3.1 for unit conversions.
SOURCES: Refs. 6, 7.

Table 3.3. Thermal Properties of Argon at Atmospheric Pressure*

T_{gas}, °F	ρ, lb_m/ft^3	μ, lb_m/ft · h	C_p, Btu/lb_m · °F	k, Btu/h · ft · °F	Pr	$g\beta/\nu^2$, 1/°R · ft^3
−100	0.1527	3.925E−02	1.251E−01	7.195E−03	6.822E−01	1.754E+07
0	0.1192	4.858E−02	1.247E−01	8.906E−03	6.803E−01	5.463E+06
32	0.1114	5.138E−02	1.247E−01	9.424E−03	6.797E−01	3.989E+06
68	0.1038	5.444E−02	1.246E−01	1.001E−02	6.779E−01	2.873E+06
100	0.0978	5.708E−02	1.246E−01	1.051E−02	6.767E−01	2.190E+06
200	0.0830	6.489E−02	1.245E−01	1.201E−02	6.727E−01	1.034E+06
300	0.0720	7.215E−02	1.244E−01	1.341E−02	6.695E−01	5.472E+05
400	0.0636	7.897E−02	1.244E−01	1.474E−02	6.666E−01	3.151E+05
500	0.0570	8.542E−02	1.244E−01	1.600E−02	6.641E−01	1.936E+05
600	0.0516	9.159E−02	1.244E−01	1.719E−02	6.625E−01	1.251E+05
800	0.0434	1.032E−01	1.243E−01	1.942E−02	6.606E−01	5.866E+04
1000	0.0375	1.140E−01	1.243E−01	2.147E−02	6.600E−01	3.088E+04
1200	0.0330	1.243E−01	1.243E−01	2.338E−02	6.607E−01	1.768E+04
1400	0.0294	1.339E−01	1.243E−01	2.517E−02	6.616E−01	1.082E+04
1600	0.0266	1.433E−01	1.243E−01	2.687E−02	6.632E−01	6.953E+03
1800	0.0242	1.524E−01	1.243E−01	2.847E−02	6.652E−01	4.660E+03
2000	0.0222	1.612E−01	1.243E−01	3.000E−02	6.680E−01	3.228E+03

* See Table 3.1 for unit conversions.
SOURCES: Refs. 1–4.

R = universal gas constant = 1545.3 ft · lb_f/lb · mol · °R
T = temperature, °R
M = molecular weight, lb_m/lb · mol

At constant pressure and small variations in T, $\rho = C_1/T$, where C_1 is a constant.

Thermal Conductivity

Thermal conductivity can be considered to be linear with temperature,[10]

$$k = C_2 T \tag{3.4}$$

Specific Heat at Constant Pressure

For the gases considered, the variation of C_p with temperature is very small, so linearity is an accurate assumption,

$$C_p = C_3 T \tag{3.5}$$

3.3 EFFECT OF PRESSURE ON THERMAL PROPERTIES

While the density of gas is proportional to pressure the effect of pressure on the other properties of gas is small, except near the gas critical point. This section provides estimates of the deviation of the gas thermal properties at high pressure

Table 3.4. Thermal Properties of Carbon Dioxide at Atmospheric Pressure*

T_{gas}, °F	ρ, lb_m/ft^3	μ, lb_m/ft · h	C_p, Btu/lb_m · °F	k, Btu/h · ft · °F	Pr	$g\beta/\nu^2$, 1/°R · ft^3
−100	0.1708	2.348E−02	1.812E−01	5.451E−03	7.806E−01	6.140E+07
0	0.1323	3.116E−02	1.937E−01	7.673E−03	7.866E−01	1.637E+07
32	0.1235	3.322E−02	1.976E−01	8.439E−03	7.777E−01	1.173E+07
68	0.1149	3.548E−02	2.020E−01	9.301E−03	7.705E−01	8.292E+06
100	0.1082	3.747E−02	2.058E−01	1.009E−02	7.646E−01	6.218E+06
200	0.0916	4.349E−02	2.174E−01	1.258E−02	7.516E−01	2.804E+06
300	0.0794	4.921E−02	2.281E−01	1.510E−02	7.435E−01	1.428E+06
400	0.0701	5.466E−02	2.381E−01	1.762E−02	7.385E−01	7.985E+05
500	0.0628	5.985E−02	2.468E−01	2.018E−02	7.321E−01	4.788E+05
600	0.0569	6.480E−02	2.551E−01	2.276E−02	7.264E−01	3.033E+05
800	0.0479	7.399E−02	2.687E−01	2.812E−02	7.069E−01	1.385E+05
1000	0.0413	8.281E−02	2.799E−01	3.274E−02	7.080E−01	7.106E+04
1200	0.0363	9.078E−02	2.891E−01	3.663E−02	7.166E−01	4.022E+04
1400	0.0324	9.823E−02	2.967E−01	4.030E−02	7.232E−01	2.442E+04
1600	0.0293	1.053E−01	3.029E−01	4.355E−02	7.322E−01	1.565E+04
1800	0.0267	1.120E−01	3.081E−01	4.650E−02	7.420E−01	1.047E+04
2000	0.0245	1.184E−01	3.124E−01	4.928E−02	7.505E−01	7.265E+03

* See Table 3.1 for unit conversions.
SOURCES: Refs. 1–4.

Table 3.5. Thermal Properties of Helium at Atmospheric Pressure*

T_{gas}, °F	ρ, lb_m/ft^3	μ, lb_m/ft · h	C_p, Btu/lb_m · °F	k, Btu/h · ft · °F	Pr	$g\beta/\nu^2$ 1/°R · ft^3
−100	0.0153	3.649E−02	1.240	6.810E−02	6.646E−01	2.027E+05
0	0.0119	4.318E−02	1.240	8.024E−02	6.675E−01	6.929E+04
32	0.0112	4.520E−02	1.240	8.394E−02	6.679E−01	5.167E+04
68	0.0104	4.739E−02	1.240	8.810E−02	6.672E−01	3.803E+04
100	0.0098	4.938E−02	1.240	9.168E−02	6.681E−01	2.936E+04
200	0.0083	5.540E−02	1.240	1.026E−01	6.699E−01	1.425E+04
300	0.0072	6.104E−02	1.240	1.133E−01	6.686E−01	7.681E+03
400	0.0064	6.655E−02	1.240	1.235E−01	6.683E−01	4.459E+03
500	0.0057	7.183E−02	1.240	1.334E−01	6.679E−01	2.751E+03
600	0.0052	7.688E−02	1.240	1.430E−01	6.667E−01	1.783E+03
800	0.0044	8.681E−02	1.240	1.617E−01	6.658E−01	8.326E+02
1000	0.0038	9.622E−02	1.240	1.791E−01	6.665E−01	4.357E+02
1200	0.0033	1.054E−01	1.240	1.936E−01	6.659E−01	2.470E+02
1400	0.0029	1.144E−01	1.240	2.128E−01	6.666E−01	1.490E+02
1600	0.0027	1.228E−01	1.240	2.288E−01	6.659E−01	9.505E+01
1800	0.0024	1.312E−01	1.240	2.446E−01	6.653E−01	6.315E+01
2000	0.0022	1.393E−01	1.240	2.596E−01	6.658E−01	4.329E+01

* See Table 3.1 for unit conversions.
SOURCE: Ref. 8.

from values at atmospheric pressure. The pressure and temperature ranges where pressure effects are significant are also given.

Viscosity

A criterion for determining whether the effect of pressure on the viscosity of pure gas is significant is given in Ref. 10. The assumption that the viscosity of a gas at high pressure and temperature is equal to the viscosity of the gas at the same temperature and atmospheric pressure yields an error that is less than 1 percent of the actual gas viscosity if

$$P_{rgc} \leq 0.18T_{rgc} - 0.08 \tag{3.6}$$

where $P_{rgc} = P_{gas}/P_{crit}$
$T_{rgc} = T_{gas}/T_{crit}$
P_{gas} = gas pressure, atm
P_{crit} = gas critical pressure, atm
T_{gas} = gas temperature, °R
T_{crit} = gas critical temperature, °R

Experimental measurements of the variations of the viscosities of gases at 68°F (20°C) at different pressures are given in Ref. 11. These viscosity values at elevated pressures are approximated by

$$\mu = \mu_0(1 + CP_{gas}) \tag{3.7}$$

Table 3.6. Thermal Properties of Hydrogen at Atmospheric Pressure*

T_{gas}, °F	ρ, lb_m/ft^3	μ, lb_m/ft · h	C_p, Btu/lb_m · °F	k, Btu/h · ft · °F	Pr	$g\beta/\nu^2$ 1/°R · ft^3
−100	0.0077	1.638E−02	3.228	7.385E−02	7.161E−01	2.549E+05
0	0.0060	1.938E−02	3.357	9.168E−02	7.099E−01	8.728E+04
32	0.0056	2.030E−02	3.377	9.691E−02	7.074E−01	6.503E+04
68	0.0052	2.129E−02	3.400	1.028E−01	7.041E−01	4.782E+04
100	0.0049	2.216E−02	3.414	1.076E−01	7.034E−01	3.696E+04
200	0.0042	2.483E−02	3.443	1.208E−01	7.075E−01	1.798E+04
300	0.0036	2.739E−02	3.461	1.323E−01	7.167E−01	9.668E+03
400	0.0032	2.985E−02	3.462	1.436E−01	7.198E−01	5.615E+03
500	0.0029	3.227E−02	3.466	1.547E−01	7.227E−01	3.450E+03
600	0.0026	3.461E−02	3.470	1.659E−01	7.240E−01	2.227E+03
800	0.0022	3.893E−02	3.487	1.877E−01	7.233E−01	1.050E+03
1000	0.0019	4.322E−02	3.516	2.101E−01	7.233E−01	5.471E+02
1200	0.0017	4.726E−02	3.550	2.320E−01	7.234E−01	3.117E+02
1400	0.0015	5.129E−02	3.594	2.538E−01	7.265E−01	1.880E+02
1600	0.0013	5.521E−02	3.643	2.754E−01	7.306E−01	1.193E+02
1800	0.0012	5.898E−02	3.696	2.965E−01	7.352E−01	7.932E+01
2000	0.0011	6.257E−02	3.755	3.181E−01	7.387E−01	5.436E+01

* See Table 3.1 for unit conversions.
SOURCES: Refs. 1–4.

Table 3.7. Thermal Properties of Nitrogen at Atmospheric Pressure*

T_{gas}, °F	ρ, lb$_m$/ft^3	μ, lb$_m$/ft · h	C_p, Btu/lb$_m$ · °F	k, Btu/h · ft · °F	Pr	$g\beta/\nu^2$ 1/°R · ft^3
−100	0.1070	3.106E−02	2.491E−01	1.053E−02	7.347E−01	1.376E+07
0	0.0835	3.803E−02	2.482E−01	1.307E−02	7.223E−01	4.378E+06
32	0.0781	4.013E−02	2.482E−01	1.384E−02	7.197E−01	3.212E+06
68	0.0727	4.242E−02	2.482E−01	1.471E−02	7.158E−01	2.325E+06
100	0.0686	4.440E−02	2.482E−01	1.544E−02	7.139E−01	1.779E+06
200	0.0582	5.024E−02	2.487E−01	1.757E−02	7.111E−01	8.482E+05
300	0.0505	5.565E−02	2.499E−01	1.958E−02	7.102E−01	4.526E+05
400	0.0446	6.070E−02	2.515E−01	2.154E−02	7.085E−01	2.625E+05
500	0.0400	6.543E−02	2.537E−01	2.337E−02	7.105E−01	1.624E+05
600	0.0362	6.991E−02	2.562E−01	2.512E−02	7.132E−01	1.056E+05
800	0.0305	7.823E−02	2.620E−01	2.848E−02	7.198E−01	5.022E+04
1000	0.0263	8.586E−02	2.686E−01	3.155E−02	7.311E−01	2.679E+04
1200	0.0231	9.294E−02	2.746E−01	3.447E−02	7.404E−01	1.555E+04
1400	0.0206	9.953E−02	2.802E−01	3.724E−02	7.490E−01	9.637E+03
1600	0.0186	1.058E−01	2.852E−01	3.987E−02	7.566E−01	6.281E+03
1800	0.0170	1.117E−01	2.895E−01	4.251E−02	7.609E−01	4.265E+03
2000	0.0156	1.173E−01	2.932E−01	4.521E−02	7.609E−01	2.998E+03

* See Table 3.1 for unit conversions.
SOURCES: Refs. 1–4.

where
μ = viscosity at 68°F and P_{gas}
μ_0 = viscosity at 68°F and 1 atm
C_{air} = 0.0013 (P_{gas} = 0–60 atm)
C_{argon} = 0.0009 (P_{gas} = 0–30 atm)
$C_{carbon\ dioxide}$ = 0.0010 (P_{gas} = 0–20 atm)
$C_{nitrogen}$ = 0.0011 (P_{gas} = 0–60 atm)
C_{helium} = 0.0 (P_{gas} = 0–60 atm)
$C_{hydrogen}$ = 0.0 (P_{gas} = 0–80 atm)

A more general determination of the viscosity as a function of pressure and temperature is given in Ref. 12.

Density

As previously given,

$$\rho = \frac{144PM}{RTZ} \qquad (3.3)$$

A chart of the compressibility factor Z for pure gases as a function of T_{rgc} and P_{rgc}, is given in Ref. 13. As a rough guide, the value of Z is between 0.98 and 1.02 for the following gases when the gas temperature is greater than T_{min} and the gas pressure is less than P_{max}.

Gas	T_{min}, °F	P_{max}, atm
Air	0	40
Argon	80	70
Carbon dioxide	400	60
Hydrogen	−200	40
Nitrogen	0	60

Ammonia and refrigerant-12 have relatively high critical temperatures and as a result, the compressibility is significant at moderate pressure and temperature. For compressibility data for these two gases and for helium Ref. 10 should be consulted.

Table 3.8. Thermal Properties of Refrigerant-12 at Atmospheric Pressure*

T_{gas}, °F	ρ, lb$_m$/ft^3	μ, lb$_m$/ft · h	C_p, Btu/lb$_m$ · °F	k, Btu/h · ft · °F	Pr	$g\beta/\nu^2$ 1/°R · ft^3
0	0.3676	2.682E−02	1.389E−01	4.544E−03	8.198E−01	1.704E+08
32	0.3414	2.845E−02	1.420E−01	5.030E−03	8.032E−01	1.221E+08
68	0.3161	3.024E−02	1.455E−01	5.576E−03	7.889E−01	8.636E+07
100	0.2969	3.168E−02	1.488E−01	6.072E−03	7.765E−01	6.545E+07
200	0.2502	3.623E−02	1.589E−01	7.565E−03	7.609E−01	3.015E+07
300	0.2165	4.055E−02	1.677E−01	9.098E−03	7.474E−01	1.565E+07
400	0.1909	4.435E−02	1.749E−01	1.064E−02	7.293E−01	8.991E+06

* See Table 3.1 for unit conversions.
SOURCE: Ref. 9.

Thermal Conductivity

For gas pressure between 0.001 and 10 atm thermal conductivity increases by about 1 percent per atmosphere,[10]

$$k = k_0(1 + 0.01P_{gas}) \tag{3.8}$$

where k = thermal conductivity at T_{gas} and P_{gas}
k_0 = thermal conductivity at T_{gas} and 1 atm
P_{gas} = gas pressure, atm

At gas pressures below 0.001 atm the influence of pressure on the thermal conductivity is much more significant and k is nearly proportional to pressure.

Specific Heat at Constant Pressure

The value of C_p at 10 atm or less is within 2 percent of C_p at 1 atm at gas temperatures greater than T_{min}.

Gas	T_{min}, °F
Air	−10
Argon	100
Hydrogen	−240
Carbon dioxide	350
Nitrogen	0

3.4 COMPARISON OF HEAT TRANSFER AND MECHANICAL POWER REQUIREMENT CHARACTERISTICS

For the different gases it is possible to make a general quantititative comparison of the convective heat-transfer coefficient h and the mechanical power requirement E per unit heat transfer area.

Fixed Heat-Capacity Flow Rate

The relative heat transfer and the mechanical power requirement of different gases at a constant heat-capacity flow rate are important applied characteristics of the thermal properties of gases.

A common application of gases in thermal design is in the exchange of heat with solid objects. In these applications, the temperature change of the gas must be within certain temperature limits. The temperature change of the gas may be limited by temperature uniformity requirements or simply by the difference between the gas inlet temperature and the solid object temperature. In situations where the heat-transfer duty and the temperature range are fixed, the required heat-capacity flow rate of the gas wC_p, as given below, is also fixed,

$$q = wC_p(T_{out} - T_{in}) \tag{3.9}$$

where q = heat duty
w = mass flow rate
C_p = gas specific heat at constant pressure
T_{out} = gas outlet temperature
T_{in} = gas inlet temperature

If a fixed gas flow passage geometry is further assumed, such as a tube cross section of a specific shape and dimension, a specific quantitative heat-transfer and power requirement comparison among gases is possible. Figure 3.1 illustrates, for several gases, the relative heat-transfer coefficient and the mechanical power requirement related to wall skin friction for fully developed incompressible turbulent flow at atmospheric pressure through a smooth tube of fixed diameter. Results are shown relative to the performance of air at 68°F (20°C).

The gases having the highest heat-transfer coefficients will require the least amount of heat-transfer area or the smallest temperature difference between the surface and the gas. However, this must be reconciled with the power consumption E which, in Fig. 3.1, is presented as the flow energy expended (or converted from pressure to thermal energy) per unit area of heat-transfer surface per unit time. The substantially superior performance of hydrogen compared to other gases is noteworthy.*

The performance for fluids and pressure conditions other than those shown in Fig. 3.1 can be compared, for turbulent and laminar flow conditions, by using the following equations.

Turbulent Flow

$$\frac{h_1}{h_2} = \left(\frac{\mu_2}{\mu_1}\right)^{0.4}\left(\frac{C_{p2}}{C_{p1}}\right)^{0.4}\left(\frac{k_1}{k_2}\right)^{0.6}$$

$$\frac{E_1}{E_2} = \left(\frac{\mu_1}{\mu_2}\right)^{0.2}\left(\frac{\rho_1}{\rho_2}\right)^{-2.0}\left(\frac{C_{p1}}{C_{p2}}\right)^{-2.8}$$

Laminar Flow

$$\frac{h_1}{h_2} = \frac{k_1}{k_2}$$

$$\frac{E_1}{E_2} = \left(\frac{\mu_1}{\mu_2}\right)\left(\frac{\rho_1}{\rho_2}\right)^{-2.0}\left(\frac{C_{p1}}{C_{p2}}\right)^{-2.0}$$

The above equations apply specifically to the case of fully developed, incompressible flow through smooth tubes, but may be used generally for initial evaluations.

Fixed Mechanical Power Requirement

Another applied comparative characteristic of the thermal properties of gases is the relative magnitude of the heat-transfer coefficient and the relative magnitude of the mass flow that can be achieved in a heat-transfer system at a fixed level of mechanical power consumption for gas circulation. The relative performance of eight gases in this regard is presented in Fig. 3.2 for incompressible fully developed turbulent flow in a smooth tube of fixed diameter at atmospheric pressure. The ratio of mass flows can be multiplied by the ratio of specific heat capacity to determine the ratio of heat-capacity flow rates at the same mechanical power requirement.

* Reference 15 is a classic paper on the subject of hydrogen as a cooling fluid for large electrical machinery.

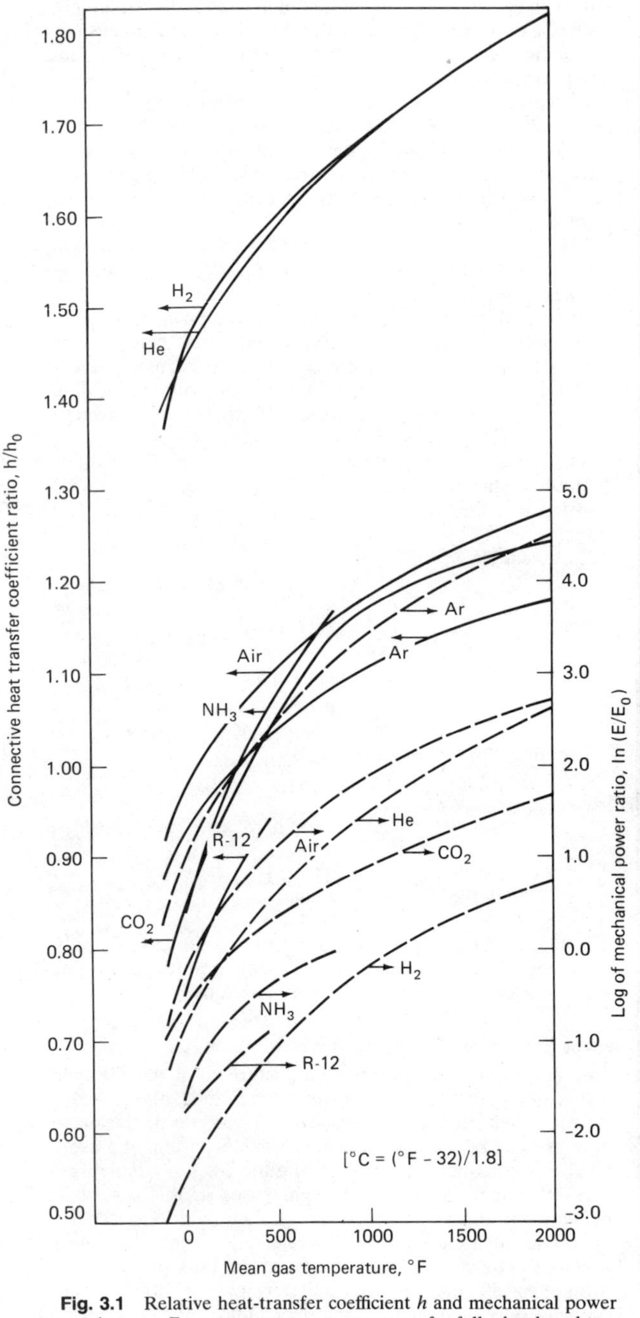

Fig. 3.1 Relative heat-transfer coefficient h and mechanical power requirement E versus mean gas temperature for fully developed turbulent gas flow in smooth tube with constant heat capacity flow rate. All gases at atmospheric pressure; h_0 = convective heat-transfer coefficient of air at 68°F (20°C); E_0 = mechanical power requirement for air at 68°F (20°C) (power per unit area).

The performance for other fluids and pressures can be compared, for turbulent and laminar flow conditions, by using the following equations.

Turbulent Flow

$$\frac{h_1}{h_2} = \left(\frac{k_1}{k_2}\right)^{0.60} \left(\frac{\mu_1}{\mu_2}\right)^{-0.457} \left(\frac{C_{p1}}{C_{p2}}\right)^{0.40} \left(\frac{\rho_1}{\rho_2}\right)^{0.571}$$

$$\frac{w_1}{w_2} = \left(\frac{\mu_1}{\mu_2}\right)^{-0.71} \left(\frac{\rho_1}{\rho_2}\right)^{0.71}$$

Laminar Flow

$$\frac{h_1}{h_2} = \frac{k_1}{k_2}$$

$$\frac{w_1}{w_2} = \left(\frac{\mu_1}{\mu_2}\right)^{-0.5} \left(\frac{\rho_1}{\rho_2}\right)$$

Natural Convection Heat Transfer

For an isothermal vertical plate the average heat-transfer coefficient over the length is found from

$$\overline{\text{Nu}}_f = \frac{\bar{h}_L L}{k} = C(\text{Gr}_f \, \text{Pr}_f)^m$$

where $\overline{\text{Nu}}_f$ = average Nusselt number
$\bar{h}_L$ = natural-convection heat-transfer coefficient averaged over length of plate
C, m = constants
Pr_f = Prandtl number
L = length scale

All property values correspond to the film temperature $T_f = (T_{\text{plate}} + T_{\text{gas}})/2$. The Grashof number Gr_f is defined as

$$\text{Gr}_f = \frac{g\beta(T_{\text{plate}} - T_{\text{gas}})L^3}{\nu^2}$$

where $\beta = 1/T_f$. The Grashof number represents the ratio of buoyant to viscous forces, and is related to the quantity $g\beta/\nu^2$. A large value of $g\beta/\nu^2$ yields a high Grashof number and a relatively large value of h. The constants C and m depend on the quantity Gr_fPr_f. For the gases considered and at the film temperature of 100°F, $C = 0.59$ and $m = 0.25$.[16] Table 3.9 is a comparison of the average natural convection heat-transfer coefficients for different gases.

3.5 SOME GENERAL CHARACTERISTICS OF COMMON GASES

Table 3.10 provides a summary of the key properties of the eight gases listed previously. Note also the following.

1. The adiabatic head requirement (ft · lb_f/lb_m) to compress a gas is inversely proportional to its molecular weight. It requires much more energy per unit mass to pressurize hydrogen than carbon dioxide.

2. For a fixed pressure and temperature, the gas volume flow rate across an aperture (as through a pinhole leak or

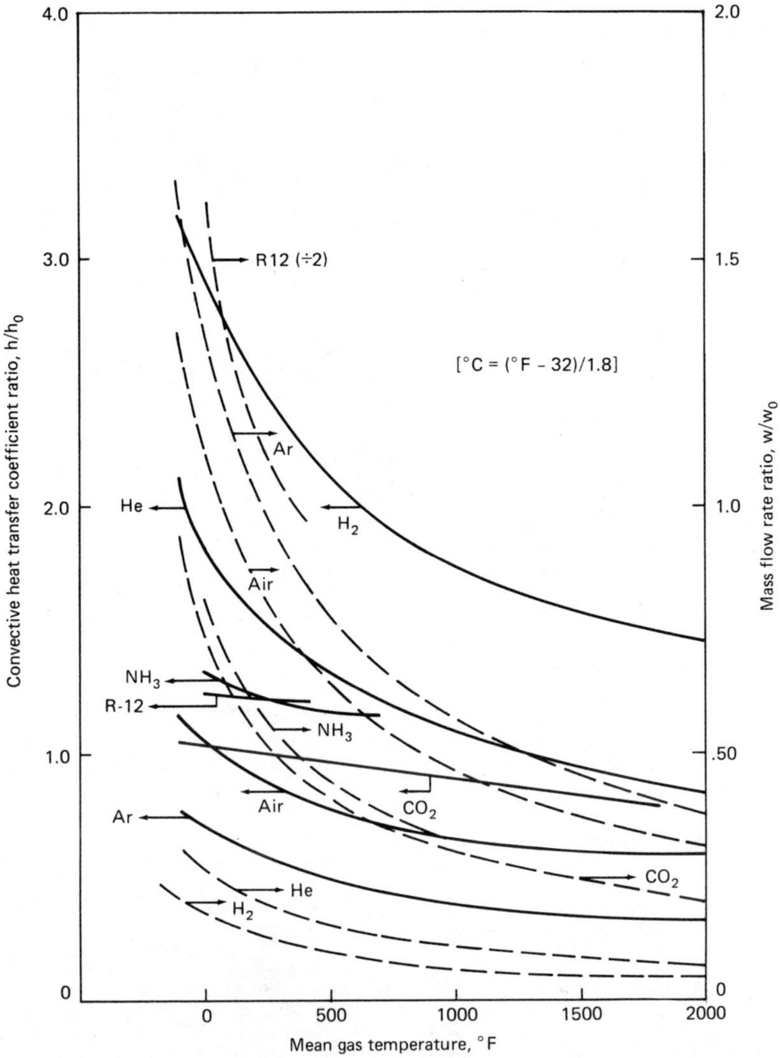

Fig. 3.2 Relative heat-transfer coefficient h and mass flow rate w versus mean gas temperature for fully developed turbulent gas flow in smooth tube with constant mechanical power requirement. All gases at atmospheric pressure; h_0 = convective heat-transfer coefficient of air at 68°F (20°C); w_0 = mass flow rate of air at 68°F (20°C).

seal) is inversely proportional to the square root of the gas molecular weight.[18]

$$Q = C_1 M^{-1/2}$$

where Q = volume flow rate
M = molecular weight
C_1 = constant

This is true for choked flow and subcritical flow. The ratio of the volume flow rate of hydrogen to air for these conditions is 3.8.

3.6 PROPERTIES OF MOIST AIR

Definitions

Moist air: A mixture of dry air and water vapor.

Saturation: A state of thermodynamic equilibrium between moist air and condensed water phase (liquid or solid) at a given temperature.

Table 3.9. Relative Natural-Convection Heat-Transfer Coefficients for Common Gases with Vertical Plate

Gas	$\bar{h}_L/\bar{h}_{L0}$
Air	1.00
Ammonia	1.04
Argon	0.704
Carbon dioxide	0.905
Helium	2.08
Hydrogen	2.62
Nitrogen	0.995
R-12	1.01

ASSUMPTIONS: $\bar{h}_{L0}$ = 0.817 Btu/h · ft² · °F; T_{plate} = 125°F; T_{gas} = 75°F; T_f = 100°F; L_{plate} = 1 ft; P_{gas} = 1 atm.

Absolute humidity: The ratio of the mass of water vapor in moist air to the total volume of moist air.

Humidity ratio W: The ratio of the mass of water vapor to the mass of dry air. W_s is the humidity ratio of moist air at saturation; W_s^* is the humidity ratio of moist air at saturation at the thermodynamic wet-bulb temperature.

Relative humidity ϕ: The ratio of the mole fraction of water vapor x_w in a given sample of moist air to the mole fraction x_{ws} in air saturated at the same temperature.

Degree of saturation μ: The ratio of the air humidity ratio W to the humidity ratio W_s of saturated air at the same temperature.

Specific volume: v_a is the volume of dry air per unit mass; v_s is the volume of moist air at saturation per unit mass of dry air.

Temperature (dry-bulb): T is the absolute temperature (°R or K); t is the dry-bulb temperature (°F or °C).

Dew-point temperature t_d: The temperature of moist air, saturated at the same pressure, with the same humidity ratio as that of the given sample of moist air.

Thermodynamic wet-bulb temperature t^*: The temperature at which water, by evaporating into a mixture of air and water vapor in the absence of external heat transfer, will bring the air to saturation at the same temperature t^*.

Pressure: p_s is the vapor pressure of water in saturated moist air (essentially equal to the saturation vapor pressure of water at the given temperature); p is the total pressure of moist air; p_a is the partial pressure of dry air; p_w is the partial pressure of water vapor in moist air.

Enthalpy: h_a is the enthalpy per unit mass of dry air with an assigned value of zero at 0°F (0°C) and standard atmospheric pressure; h_s is the enthalpy of moist air at saturation per unit mass of dry air; h is enthalpy of moist air per unit mass of dry air.

Perfect gas: $p_aV = n_aRT$ for dry air; $p_wV = n_wRT$ for water vapor. Here n_a and n_w are the number of moles of dry air and water vapor, respectively; V is the total volume of the mixture; R is the universal gas constant. [8.31441 J/g · mol · K (1545.32 ft · lb$_f$/lb · mol · °R)]; and T is the absolute temperature.

Thermodynamic Properties

Perfect gas relationships provide a convenient and a generally sufficiently accurate basis for evaluating the thermody-

Table 3.10. Characteristics of Some Common Gases

Gas	Formula	Molecular weight	Boiling point at NTP,* °F	Critical temperature, °F	Critical pressure, lb/in² abs	Density at NTP, lb$_m$/ft³	C_p at NTP, Btu/lb$_m$ · °F	Safety considerations	Typical cost,‡ $/lb
Nitrogen	N_2	28.01	−320	−232	493	0.073	0.247	Noncombustible, NO_X formation; asphyxiant	0.0057
Argon	Ar	39.95	−303	−188	710	0.103	0.125	Inert, nonreacting gas; asphyxiant	0.0102
Hydrogen	H_2	2.016	−423	−400	188	0.00521	3.42	Flammability limits in air 4.0–7.5 vol %; asphyxiant	0.175
Helium	He	4.003	−452	−450	33.2	0.0103	1.25	Inert, nonreacting gas; asphyxiant	0.122
Carbon dioxide	CO_2	44.01	−109	88.0	1070	0.115	0.202	TLV† = 5000 ppm; asphyxiant	0.367
Ammonia	NH_3	17.03	−27.9	271.4	1657	0.0446	0.520	TLV = 25 ppm; skin/eye irritant	0.847
Dichloro-difluromethane (R-12)	CCl_2F_2	120.93	−21.6	233.6	597	0.316	0.146	TLV = 1000 ppm; narcotic in high concentration	2.34
Air		28.97	−318	−221.3	547	0.075	0.240		0.0084

* NTP = normal temperature and pressure, 70°F (21.1°C) and 14.7 lb/in² abs (10.1 × 10⁴ Pa).
† TLV = threshold limit value, = maximum concentration for repeated exposure with no ill effects.[17]
‡ Typical 1987 prices for 285-ft³ tank at 99% or better purity.
CONVERSION FACTORS: °C = (°F − 32)/1.8; 1 lb$_m$/ft³ = 16.018 kg/m³; 1 Btu/lb$_m$ · °F = 4187 J/kg · K; 1 $/lb = 2.2 $/kg.

Table 3.11. Properties of Moist Air at Standard Atmospheric Pressure, 29.921 in Hg

Temperature, °F	Humidity ratio W_s, lb_w/lb_a	Volume, ft³/lb dry air		Enthalpy, Btu/lb dry air		Vapor press. p_s, in Hg	Temperature, °F	Humidity ratio W_s, lb_w/lb_a	Volume, ft³/lb dry air		Enthalpy, Btu/lb dry air		Vapor press. p_s, in Hg
		v_a	v_s	h_a	h_s				v_a	v_s	h_a	h_s	
−80	0.0000049	9.553	9.553	−19.221	−19.215	0.000236	−44	0.0000615	10.465	10.466	−10.570	−10.505	0.002943
−79	0.0000053	9.579	9.579	−18.980	−18.975	0.000255	−43	0.0000656	10.491	10.492	−10.329	−10.261	0.003137
−78	0.0000057	9.604	9.604	−18.740	−18.734	0.000275	−42	0.0000699	10.516	10.517	−10.089	−10.016	0.003343
−77	0.0000062	9.629	9.629	−18.500	−18.493	0.000296	−41	0.0000744	10.541	10.543	−9.849	−9.771	0.003562
−76	0.0000067	9.655	9.655	−18.259	−18.252	0.000319	−40	0.0000793	10.567	10.568	−9.609	−9.526	0.003793
−75	0.0000072	9.680	9.680	−18.019	−18.011	0.000344	−39	0.0000844	10.592	10.593	−9.368	−9.280	0.004039
−74	0.0000078	9.705	9.705	−17.778	−17.770	0.000371	−38	0.0000898	10.617	10.619	−9.128	−9.034	0.004299
−73	0.0000084	9.731	9.731	−17.538	−17.529	0.000400	−37	0.0000956	10.643	10.644	−8.888	−8.788	0.004575
−72	0.0000090	9.756	9.756	−17.298	−17.288	0.000430	−36	0.0001017	10.668	10.670	−8.648	−8.541	0.004866
−71	0.0000097	9.781	9.782	−17.057	−17.047	0.000463	−35	0.0001081	10.693	10.695	−8.407	−8.294	0.005175
−70	0.0000104	9.807	9.807	−16.817	−16.806	0.000498	−34	0.0001150	10.719	10.721	−8.167	−8.047	0.005502
−69	0.0000112	9.832	9.832	−16.577	−16.565	0.000536	−33	0.0001222	10.744	10.746	−7.927	−7.799	0.005848
−68	0.0000120	9.857	9.858	−16.336	−16.324	0.000576	−32	0.0001298	10.769	10.772	−7.687	−7.551	0.006214
−67	0.0000129	9.883	9.883	−16.096	−16.083	0.000619	−31	0.0001379	10.795	10.797	−7.447	−7.302	0.006601
−66	0.0000139	9.908	9.908	−15.856	−15.841	0.000665	−30	0.0001465	10.820	10.822	−7.206	−7.053	0.007009
−65	0.0000149	9.933	9.934	−15.616	−15.600	0.000714	−29	0.0001555	10.845	10.848	−6.966	−6.803	0.007442
−64	0.0000160	9.959	9.959	−15.375	−15.359	0.000766	−28	0.0001650	10.871	10.873	−6.726	−6.553	0.007898
−63	0.0000172	9.984	9.984	−15.135	−15.117	0.000822	−27	0.0001751	10.896	10.899	−6.486	−6.302	0.008381
−62	0.0000184	10.009	10.010	−14.895	−14.876	0.000882	−26	0.0001858	10.921	10.924	−6.245	−6.051	0.008890
−61	0.0000198	10.035	10.035	−14.654	−14.634	0.000945	−25	0.0001970	10.947	10.950	−6.005	−5.798	0.009428
−60	0.0000212	10.060	10.060	−14.414	−14.392	0.001013	−24	0.0002088	10.972	10.976	−5.765	−5.545	0.009995
−59	0.0000227	10.085	10.086	−14.174	−14.150	0.001086	−23	0.0002214	10.997	11.001	−5.525	−5.292	0.010594
−58	0.0000243	10.111	10.111	−13.933	−13.908	0.001163	−22	0.0002346	11.022	11.027	−5.284	−5.038	0.011226
−57	0.0000260	10.136	10.137	−13.693	−13.666	0.001246	−21	0.0002485	11.048	11.052	−5.044	−4.783	0.011893
−56	0.0000279	10.161	10.162	−13.453	−13.424	0.001333	−20	0.0002632	11.073	11.078	−4.804	−4.527	0.012595
−55	0.0000298	10.187	10.187	−13.213	−13.182	0.001427	−19	0.0002786	11.098	11.103	−4.564	−4.271	0.013336
−54	0.0000319	10.212	10.213	−12.972	−12.939	0.001526	−18	0.0002950	11.124	11.129	−4.324	−4.013	0.014117
−53	0.0000341	10.237	10.238	−12.732	−12.697	0.001632	−17	0.0003121	11.149	11.155	−4.084	−3.754	0.014939
−52	0.0000365	10.263	10.263	−12.492	−12.454	0.001745	−16	0.0003303	11.174	11.180	−3.843	−3.495	0.015806
−51	0.0000390	10.288	10.289	−12.251	−12.211	0.001865	−15	0.0003493	11.200	11.206	−3.603	−3.235	0.016718
−50	0.0000416	10.313	10.314	−12.011	−11.968	0.001992	−14	0.0003694	11.225	11.232	−3.363	−2.973	0.017679
−49	0.0000445	10.339	10.340	−11.771	−11.725	0.002128	−13	0.0003905	11.250	11.257	−3.123	−2.710	0.018690
−48	0.0000475	10.364	10.365	−11.531	−11.481	0.002272	−12	0.0004128	11.276	11.283	−2.882	−2.447	0.019754
−47	0.0000507	10.389	10.390	−11.290	−11.237	0.002425	−11	0.0004362	11.301	11.309	−2.642	−2.182	0.020873
−46	0.0000541	10.415	10.416	−11.050	−10.994	0.002587	−10	0.0004608	11.326	11.335	−2.402	−1.915	0.022050
−45	0.0000577	10.440	10.441	−10.810	10.750	0.002760	−9	0.0004867	11.351	11.360	−2.162	−1.647	0.023289

Table 3.11. Properties of Moist Air at Standard Atmospheric Pressure, 29.921 in Hg (*Continued*)

Temperature, °F	Humidity ratio W_s, lb_w/lb_a	Volume, ft^3/lb dry air		Enthalpy, Btu/lb dry air		Vapor press. p_s, in Hg
		v_a	v_s	h_a	h_s	
−8	0.0005139	11.377	11.386	−1.922	−1.378	0.024591
−7	0.0005425	11.402	11.412	−1.681	−1.108	0.025959
−6	0.0005726	11.427	11.438	−1.441	−0.835	0.027397
−5	0.0006041	11.453	11.464	−1.201	−0.561	0.028907
−4	0.0006373	11.478	11.490	−0.961	−0.286	0.030494
−3	0.0006722	11.503	11.516	−0.721	−0.008	0.032160
−2	0.0007088	11.529	11.542	−0.480	0.271	0.033909
−1	0.0007472	11.554	11.568	−0.240	0.552	0.035744
0	0.0007875	11.579	11.594	0.0	0.835	0.037671
1	0.0008298	11.604	11.620	0.240	1.121	0.039694
2	0.0008742	11.630	11.646	0.480	1.408	0.041814
3	0.0009207	11.655	11.672	0.721	1.699	0.044037
4	0.0009695	11.680	11.699	0.961	1.991	0.046370
5	0.0010207	11.706	11.725	1.201	2.286	0.048814
6	0.0010743	11.731	11.751	1.441	2.584	0.051375
7	0.0011306	11.756	11.778	1.681	2.884	0.054060
8	0.0011895	11.782	11.804	1.922	3.188	0.056872
9	0.0012512	11.807	11.831	2.162	3.494	0.059819
10	0.0013158	11.832	11.857	2.402	3.804	0.062901
11	0.0013835	11.857	11.884	2.642	4.117	0.066131
12	0.0014544	11.883	11.910	2.882	4.433	0.069511
13	0.0015286	11.908	11.937	3.123	4.753	0.073049
14	0.0016062	11.933	11.964	3.363	5.077	0.076751
15	0.0016874	11.959	11.991	3.603	5.404	0.080623
16	0.0017724	11.984	12.018	3.843	5.736	0.084673
17	0.0018613	12.009	12.045	4.084	6.072	0.088907
18	0.0019543	12.035	12.072	4.324	6.412	0.093334
19	0.0020515	12.060	12.099	4.564	6.757	0.097962
20	0.0021531	12.085	12.127	4.804	7.107	0.102798
21	0.0022592	12.110	12.154	5.044	7.462	0.107849
22	0.0023703	12.136	12.182	5.285	7.822	0.113130
23	0.0024863	12.161	12.209	5.525	8.187	0.118645
24	0.0026073	12.186	12.237	5.765	8.558	0.124396
25	0.0027339	12.212	12.265	6.005	8.935	0.130413
26	0.0028660	12.237	12.293	6.246	9.318	0.136684
27	0.0030039	12.262	12.321	6.486	9.708	0.143233

Temperature, °F	Humidity ratio W_s, lb_w/lb_a	Volume, ft^3/lb dry air		Enthalpy, Btu/lb dry air		Vapor press. p_s, in Hg
		v_a	v_s	h_a	h_s	
28	0.0031480	12.287	12.349	6.726	10.104	0.150066
29	0.0032984	12.313	12.378	6.966	10.507	0.157198
30	0.0034552	12.338	12.406	7.206	10.917	0.164631
31	0.0036190	12.363	12.435	7.447	11.335	0.172390
32	0.0037895	12.389	12.464	7.687	11.760	0.180479
32	0.003790	12.389	12.464	7.687	11.760	0.18050
33	0.003947	12.414	12.492	7.927	12.170	0.18791
34	0.004109	12.439	12.521	8.167	12.587	0.19559
35	0.004277	12.464	12.550	8.408	13.010	0.20356
36	0.004452	12.490	12.579	8.648	13.441	0.21181
37	0.004633	12.515	12.608	8.888	13.878	0.22035
38	0.004820	12.540	12.637	9.128	14.322	0.22920
39	0.005014	12.566	12.667	9.369	14.773	0.23835
40	0.005216	12.591	12.696	9.609	15.233	0.24784
41	0.005424	12.616	12.726	9.849	15.700	0.25765
42	0.005640	12.641	12.756	10.089	16.175	0.26781
43	0.005863	12.667	12.786	10.330	16.660	0.27831
44	0.006094	12.692	12.816	10.570	17.152	0.28918
45	0.006334	12.717	12.846	10.810	17.653	0.30042
46	0.006581	12.743	12.877	11.050	18.164	0.31206
47	0.006838	12.768	12.908	11.291	18.685	0.32408
48	0.007103	12.793	12.939	11.531	19.215	0.33651
49	0.007378	12.818	12.970	11.771	19.756	0.34937
50	0.007661	12.844	13.001	12.012	20.306	0.36264
51	0.007955	12.869	13.033	12.252	20.868	0.37636
52	0.008259	12.894	13.065	12.492	21.441	0.39054
53	0.008573	12.920	13.097	12.732	22.025	0.40518
54	0.008897	12.945	13.129	12.973	22.621	0.42030
55	0.009233	12.970	13.162	13.213	23.229	0.43592
56	0.009580	12.995	13.195	13.453	23.850	0.45205
57	0.009938	13.021	13.228	13.694	24.484	0.46870
58	0.010309	13.046	13.262	13.934	25.131	0.48589
59	0.010692	13.071	13.295	14.174	25.792	0.50363
60	0.011087	13.096	13.329	14.415	26.467	0.52193
61	0.011496	13.122	13.364	14.655	27.157	0.54082
62	0.011919	13.147	13.398	14.895	27.862	0.56032
63	0.012355	13.172	13.433	15.135	28.582	0.58041

Table 3.11. Properties of Moist Air at Standard Atmospheric Pressure, 29.921 in Hg (*Continued*)

Temperature, °F	Humidity ratio W_s, lb_w/lb_a	Volume, ft³/lb dry air		Enthalpy, Btu/lb dry air		Vapor press. p_s, in Hg	Temperature, °F	Humidity ratio W_s, lb_w/lb_a	Volume, ft³/lb dry air		Enthalpy, Btu/lb dry air		Vapor press. p_s, in Hg
		v_a	v_s	h_a	h_s				v_a	v_s	h_a	h_s	
64	0.012805	13.198	13.468	15.376	29.318	0.60113	100	0.043219	14.107	15.084	24.031	71.761	1.93492
65	0.013270	13.223	13.504	15.616	30.071	0.62252	101	0.044634	14.133	15.143	24.271	73.583	1.99396
66	0.013750	13.248	13.540	15.856	30.840	0.64454	102	0.046090	14.158	15.203	24.512	75.452	2.05447
67	0.014246	13.273	13.577	16.097	31.626	0.66725	103	0.047592	14.183	15.264	24.752	77.373	2.11661
68	0.014758	13.299	13.613	16.337	32.431	0.69065	104	0.049140	14.208	15.326	24.993	79.346	2.18037
69	0.015286	13.324	13.650	16.577	33.254	0.71479	105	0.050737	14.234	15.390	25.233	81.375	2.24581
70	0.015832	13.349	13.688	16.818	34.097	0.73966	106	0.052383	14.259	15.455	25.474	83.460	2.31297
71	0.016395	13.375	13.726	17.058	34.959	0.76528	107	0.054077	14.284	15.521	25.714	85.599	2.38173
72	0.016976	13.400	13.764	17.299	35.841	0.79167	108	0.055826	14.309	15.588	25.955	87.799	2.45232
73	0.017575	13.425	13.803	17.539	36.743	0.81882	109	0.057628	14.335	15.657	26.195	90.061	2.52473
74	0.018194	13.450	13.843	17.779	37.668	0.84684	110	0.059486	14.360	15.727	26.436	92.386	2.59891
75	0.018833	13.476	13.882	18.020	38.615	0.87567	111	0.061401	14.385	15.799	26.677	94.776	2.67500
76	0.019491	13.501	13.923	18.260	39.583	0.90533	112	0.063378	14.411	15.872	26.917	97.237	2.75310
77	0.020170	13.526	13.963	18.500	40.576	0.93589	113	0.065411	14.436	15.947	27.158	99.760	2.83291
78	0.020871	13.551	14.005	18.741	41.592	0.96733	114	0.067512	14.461	16.023	27.398	102.362	2.91491
79	0.021594	13.577	14.046	18.981	42.633	0.99970	115	0.069676	14.486	16.101	27.639	105.035	2.99883
80	0.022340	13.602	14.089	19.222	43.701	1.03302	116	0.071908	14.512	16.181	27.879	107.786	3.08488
81	0.023109	13.627	14.132	19.462	44.794	1.06728	117	0.074211	14.537	16.263	28.120	110.617	3.17305
82	0.023902	13.653	14.175	19.702	45.913	1.10252	118	0.076586	14.562	16.346	28.361	113.530	3.26335
83	0.024720	13.678	14.220	19.943	47.062	1.13882	119	0.079036	14.587	16.432	28.601	116.528	3.35586
84	0.025563	13.703	14.264	20.183	48.238	1.17608	120	0.081560	14.613	16.519	28.842	119.612	3.45052
85	0.026433	13.728	14.310	20.424	49.445	1.21445	121	0.084169	14.638	16.609	29.083	122.792	3.54764
86	0.027329	13.754	14.356	20.664	50.681	1.25388	122	0.086860	14.663	16.700	29.323	126.065	3.64704
87	0.028254	13.779	14.403	20.905	51.949	1.29443	123	0.089633	14.688	16.794	29.564	129.432	3.74871
88	0.029208	13.804	14.450	21.145	53.250	1.33613	124	0.092500	14.714	16.890	29.805	132.907	3.85298
89	0.030189	13.829	14.498	21.385	54.582	1.37893	125	0.095456	14.739	16.989	30.045	136.482	3.95961
90	0.031203	13.855	14.547	21.626	55.951	1.42298	126	0.098504	14.764	17.090	30.286	140.163	4.06863
91	0.032247	13.880	14.597	21.866	57.355	1.46824	127	0.101657	14.789	17.193	30.527	143.965	4.18046
92	0.033323	13.905	14.647	22.107	58.794	1.51471	128	0.104910	14.815	17.299	30.767	147.878	4.29477
93	0.034433	13.930	14.699	22.347	60.271	1.56248	129	0.108270	14.840	17.409	31.008	151.916	4.41181
94	0.035577	13.956	14.751	22.588	61.787	1.61154	130	0.111738	14.865	17.520	31.249	156.076	4.53148
95	0.036757	13.981	14.804	22.828	63.343	1.66196	131	0.115322	14.891	17.635	31.489	160.370	4.65397
96	0.037972	14.006	14.858	23.069	64.940	1.71372	132	0.119023	14.916	17.753	31.730	164.796	4.77919
97	0.039225	14.032	14.913	23.309	66.578	1.76685	133	0.122855	14.941	17.875	31.971	169.374	4.90755
98	0.040516	14.057	14.969	23.550	68.260	1.82141	134	0.126804	14.966	17.999	32.212	174.084	5.03844
99	0.041848	14.082	15.026	23.790	69.988	1.87745	135	0.130895	14.992	18.127	32.452	178.957	5.17258

Table 3.11. Properties of Moist Air at Standard Atmospheric Pressure, 29.921 in Hg (*Continued*)

Temperature, °F	Humidity ratio W_s, lb_w/lb_a	Volume, ft³/lb dry air		Enthalpy, Btu/lb dry air		Vapor press. p_s, in Hg	Temperature, °F	Humidity ratio W_s, lb_w/lb_a	Volume, ft³/lb dry air		Enthalpy, Btu/lb dry air		Vapor press. p_s, in Hg
		v_a	v_s	h_a	h_s				v_a	v_s	h_a	h_s	
136	0.135124	15.017	18.259	32.693	183.987	5.30973	172	0.46905	15.926	27.820	41.366	573.504	12.7880
137	0.139494	15.042	18.394	32.934	189.179	5.44985	173	0.48829	15.951	28.352	41.607	595.767	13.0823
138	0.144019	15.067	18.534	33.175	194.548	5.59324	174	0.50867	15.976	28.913	41.848	619.337	13.3831
139	0.148696	15.093	18.678	33.415	200.092	5.73970	175	0.53019	16.002	29.505	42.089	644.229	13.6894
140	0.153538	15.118	18.825	33.656	205.824	5.88945	176	0.55294	16.027	30.130	42.331	670.528	14.0010
141	0.158552	15.143	18.978	33.897	211.754	6.04256	177	0.57710	16.052	30.793	42.572	698.448	14.3191
142	0.163748	15.168	19.135	34.138	217.892	6.19918	178	0.60274	16.078	31.496	42.813	728.073	14.6430
143	0.169122	15.194	19.297	34.379	244.233	6.35898	179	0.63002	16.103	32.242	43.054	759.579	14.9731
144	0.174694	15.219	19.464	34.620	230.802	6.52241	180	0.65911	16.128	33.037	43.295	793.166	15.3097
145	0.180467	15.244	19.637	34.860	237.600	6.68932	181	0.69012	16.153	33.883	43.536	828.962	15.6522
146	0.186460	15.269	19.815	35.101	244.651	6.86009	182	0.72331	16.178	34.787	43.778	867.265	16.0014
147	0.192668	15.295	19.999	35.342	251.949	7.03435	183	0.75885	16.204	35.755	44.019	908.278	16.3569
148	0.199110	15.320	20.189	35.583	259.514	7.21239	184	0.79703	16.229	36.793	44.260	952.321	16.7190
149	0.205792	15.345	20.385	35.824	267.356	7.39413	185	0.83817	16.254	37.910	44.501	999.763	17.0880
150	0.212730	15.370	20.589	36.064	275.490	7.57977	186	0.88251	16.280	39.113	44.742	1050.892	17.4634
151	0.219945	15.396	20.799	36.305	283.943	7.76958	187	0.93057	16.305	40.416	44.984	1106.298	17.8462
152	0.227429	15.421	21.017	36.546	292.705	7.96306	188	0.98272	16.330	41.828	45.225	1166.399	18.2357
153	0.235218	15.446	21.243	36.787	301.816	8.16087	189	1.03951	16.355	43.365	45.466	1231.848	18.6323
154	0.243309	15.471	21.477	37.028	311.273	8.36256	190	1.10154	16.381	45.042	45.707	1303.321	19.0358
155	0.251738	15.497	21.720	37.269	321.118	8.56871	191	1.16965	16.406	46.882	45.949	1381.783	19.4468
156	0.260512	15.522	21.972	37.510	331.359	8.77915	192	1.24471	16.431	48.908	46.190	1468.238	19.8652
157	0.269644	15.547	22.233	37.751	342.012	8.99378	193	1.32788	16.456	51.151	46.431	1564.013	20.2913
158	0.279166	15.572	22.505	37.992	353.112	9.21297	194	1.42029	16.481	53.642	46.673	1670.430	20.7244
159	0.289101	15.598	22.788	38.233	364.685	9.43677	195	1.52396	16.507	56.435	46.914	1789.793	21.1661
160	0.29945	15.623	23.082	38.474	376.737	9.6648	196	1.64070	16.532	59.578	47.155	1924.188	21.6152
161	0.31027	15.648	23.388	38.715	389.325	9.8978	197	1.77299	16.557	63.137	47.397	2076.466	22.0714
162	0.32156	15.673	23.707	38.956	402.457	10.1353	198	1.92472	16.583	67.218	47.638	2251.102	22.5367
163	0.33336	15.699	24.040	39.187	416.175	10.3776	199	2.09975	16.608	71.923	47.879	2452.547	23.0092
164	0.34572	15.724	24.388	39.438	430.533	10.6250	200	2.30454	16.633	77.426	48.121	2688.205	23.4906
165	0.35865	15.749	24.750	39.679	445.544	10.8771							
166	0.37220	15.774	25.129	39.920	461.271	11.1343							
167	0.38639	15.800	25.526	40.161	477.739	11.3965							
168	0.40131	15.825	25.942	40.402	495.032	11.6641							
169	0.41698	15.850	26.377	40.643	513.197	11.9370							
170	0.43343	15.875	26.834	40.884	532.256	12.2149							
171	0.45079	15.901	27.315	41.125	552.356	12.4988							

SOURCE: Abstracted by permission from *ASHRAE 1985 Handbook of Fundamentals*. Compiled by J. A. Goff and S. Gratch.

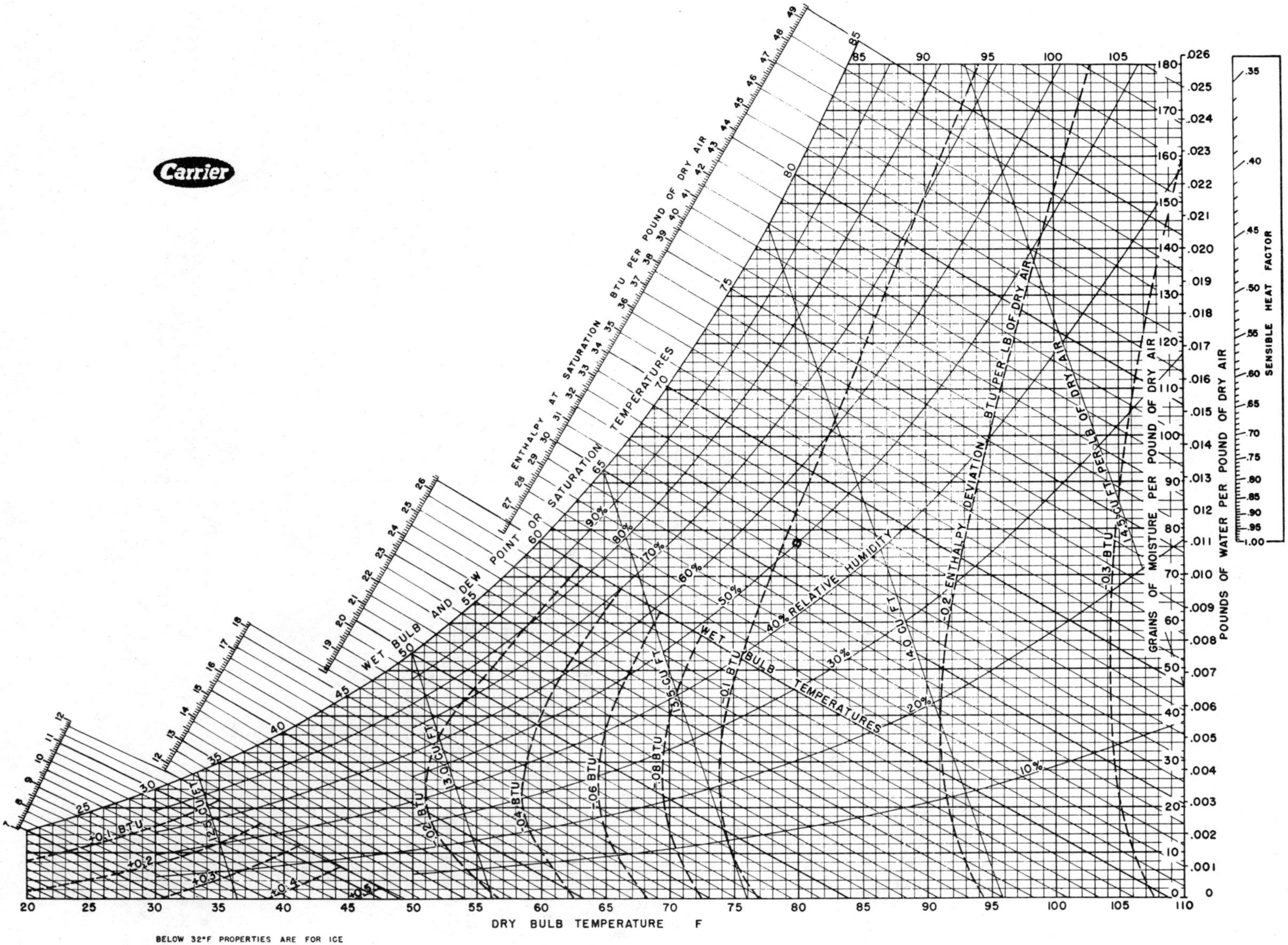

Fig. 3.3 Psychrometric chart at atmospheric pressure. (*Reproduced Courtesy of Carrier Corporation. © 1961 Carrier Corporation.*)

namic properties of moist air in heating and cooling processes.[19] The applied relationships are[20]

$$W = 0.62198 \frac{p_w}{p - p_w}$$

$$W_s = 0.62198 \frac{p_s}{p - p_s}$$

$$\mu = \frac{W}{W_s}$$

$$\phi = \frac{p_w}{p_s} = \frac{\mu}{1 - (1 - \mu) p_s / p}$$

$$v = \frac{RT}{28.9645 p} (1 + 1.6078 W)$$

$$h = 0.240t + W(1061 + 0.444t) \qquad \text{(Btu/lb dry air)}$$

$$= t + W(2501 + 1.805t) \qquad \text{(kJ/kg dry air)}$$

$$W = \frac{(1093 - 0.556t^*)W_s^* - 0.240(t - t^*)}{1093 + 0.444t - t^*} \qquad \text{(USCS units)}$$

$$W = \frac{(2501 - 2.381t^*)W_s^* - (t - t^*)}{2501 + 1.805t - 4.186t^*} \qquad \text{(SI units)}$$

$$p_s(t_d) = p_w = \frac{pW}{0.62198 + W}$$

The above relationships, along with the saturation vapor pressure and the temperature relationship for water, can be used to calculate the thermodynamic properties of moist air from the observable pressure, dry-bulb temperature, and either wet-bulb temperature, dew-point temperature, or relative humidity. Vapor pressure of water is given in Table 3.11 from −80 to 200°F (−62 to 93°C), while an accurate polynomial expression for the same is given in Ref. 19. An iterative trial-and-error solution of the above relationship that is implicit in t^* is necessary if the wet-bulb temperature t^* is unknown.

Alternatively, the thermodynamic properties of moist air as calculated from more accurate relations are given in Table 3.11. With good accuracy up to 158°F (70°C), the following equations apply to the data given in Table 3.11.

$$v = v_a + \mu(v_s - v_a)$$

$$h = h_a + \mu(h_s - h_a)$$

Figure 3.3, a psychrometric chart or graphic representation of the thermodynamic properties of moist air, is of practical value in the graphic evaluation of moist air thermal processes. Charts for higher and lower temperatures and for altitudes other than sea level are available from the American Society of Heating, Refrigerating and Air-Conditioning Engineers.

Transport Properties

Up to a temperature of about 120°F (50°C), the thermal conductivity and the viscosity of moist air are approximately equal to the values for dry air. At higher temperatures and water vapor contents, the transport properties of moist air can be roughly approximated by a simple mole-fraction-weighted addition of the property values for water vapor and dry air.

REFERENCES

1. Y. S. Touloukian, P. E. Liley, and S. C. Saxena, *Thermophysical Properties of Matter;* vol. 3, *Thermal Conductivity, Nonmetallic Liquids and Gases,* IFI/Plenum, New York, 1970.
2. Y. S. Touloukian and T. Makita, *Thermophysical Properties of Matter;* vol. 6, *Specific Heat, Nonmetallic Gases and Liquids,* IFI/Plenum, New York, 1970.
3. Y. S. Touloukian, S. C. Saxena, and P. Hestermans, *Thermophysical Properties of Matter;* vol. 11, *Viscosity,* IFI/Plenum, New York, 1975.
4. J. Hilsenrath et al., "Tables of Thermal Properties of Gases," NBS circ. 564, Washington, DC, Nov. 1955.
5. D. J. Poferl, R. A. Svehla, and K. Lewandowski, "Thermodynamic and Transport Properties of Air and the Combustion Products of Natural Gas and of ASTM-A-1 Fuel with Air," NASA Tech. Note D-5452, Washington, DC, 1969.
6. L. Haar and J. S. Gallagher, "Thermodynamic Properties of Ammonia," *J. Phys. Chem. Ref. Data,* **7,** 635 (1977).
7. *ASHRAE Thermophysical Properties of Refrigerants,* American Society of Heating, Refrigerating and Air-Conditioning Engineers, Atlanta, GA, 1976.
8. R. D. McCarty, "Thermophysical Properties of Helium-4 from 2 to 1500 K with Pressures to 1000 Atmospheres," NBS Tech. Note 631, Washington, DC, Nov. 1972.
9. J. T. R. Watson, "Thermophysical Properties of Refrigerant 12," National Engineering Laboratory, Edinburgh, UK, 1975.
10. R. C. Reid, J. M. Pravsnitz, and T. K. Sherwood, *Properties of Gases and Liquids,* 3d ed., McGraw-Hill, New York, 1977.
11. J. Kestin and W. Leidenfrost, *Physica,* **25,** 1033 (1959).
12. O. A. Uyehara and K. M. Watson, *Nat. Petroleum News,* **36,** R-714 (1944).
13. L. C. Nelson and E. F. Obert, *Trans. ASME,* **76,** 1057 (1954).
14. W. M. Rohsenow, J. P. Hartnett, and E. N. Ganic, *Handbook of Heat Transfer Fundamentals,* 2d ed. McGraw-Hill, New York, 1985.
15. E. Knowlton, C. W. Rice, and E. H. Freiburghouse, "Hydrogen as a Cooling Medium for Electrical Machinery," *Trans. AIEE,* **44,** 922–932 (1925).
16. J. P. Holman, *Heat Transfer,* 4th ed. McGraw-Hill, New York, 1976.
17. N. I. Sax, *Dangerous Properties of Industrial Materials,* 4th ed., Van Nostrand Reinhold, New York, 1975.
18. C. M. Van Atta, *Vacuum Science and Engineering,* McGraw-Hill, New York, 1965.
19. J. L. Threlkeld, *Thermal Environmental Engineering,* 2d ed., Prentice-Hall, Englewood Cliffs, NJ, 1970.
20. *ASHRAE Handbook, 1985 Fundamentals,* American Society of Heating, Refrigerating and Air-Conditioning Engineers, Atlanta, GA, 1985.

Chapter 4

Organic Heat-Transfer Fluids

WALTER F. SEIFERT
The Dow Chemical Company

4.1 COMPARING VARIOUS HEAT-TRANSFER FLUIDS

The selection of a high-temperature heat-transfer fluid for a given application begins with the definition of general parameters for fluid selection and the gathering of information on various available fluids. Unfortunately, there are a number of methods for presenting fluid physical and stability data. Units of measurement and the methods for measuring fluid characteristics may not be directly comparable. Before an informed selection can be made, there must be some assurance that important specifications can be compared and evaluated accurately.

The fluid information presented here is the result of gathering the latest available data from various manufacturers. The data have been converted as necessary to assure like units of measurement. Test methods, on the other hand, may still not be comparable because some of the data were developed before the new industry testing standards were adopted. These exceptions are noted in the data presented here when measurement methods were detailed in the specific manufacturer's literature.

Process Temperature Range

The first consideration in fluid selection is the process temperature range. Before delving into the process of selecting a high-temperature organic heat-transfer fluid, one should be sure that the application conditions warrant the cost of an organic fluid. Some of the alternatives and points to consider are discussed below.

For temperatures between 32 and 350 °F (0 and 177°C), water is the most economical and efficient heat-transfer fluid. Even occasional freezing temperatures can often be tolerated. For example, if system shutdown would be required should ambient conditions fall below the freezing point, freeze protection can be provided by heat tracing piping for equipment exposed to these low temperatures. If the application involves chilling, freezing, or frequent ambient conditions well below freezing, the most common heat-transfer fluids are vapor compression refrigerants, brines, and glycol-water solutions. A citrus derivative, D-limonene, has been found to be a useful low-temperature heat-transfer fluid in the temperature range from −140 to +179°F (−95.5 to 81.6°C).

Vapor compression refrigerants are useful in many chilling and freezing applications. Brines are an inexpensive solution for many low-temperature applications. Unfortunately they are also corrosive to most inexpensive, commonly used metals. To render them less corrosive, toxic inhibitors such as chromates are commonly added. Because of these factors, care must be exercised in the use of brine and in the selection of equipment materials for construction.

Glycol-water solutions are another low-cost option for applications requiring temperatures in the range of −60 to 250°F (−51 to 121°C). Glycol alone can be corrosive in contact with air, so a low-toxicity corrosion inhibitor is generally included in the glycol formulation. Some inhibited glycols are specifically designed for applications where accidental contact with food or potable water must be considered.

Both brines and glycols offer good low-temperature capabilities; however, they are unstable or corrosive at temperatures much above the boiling point of water. Even the use of steam as a heat-transfer fluid has some limitations above 350°F (177°C) because both the vapor pressure of water and the associated process equipment costs increase significantly.

Table 4.1 Physical Properties of Organic Heat-Transfer Fluids[a,b]

Fluid	Dowtherm A	Dowtherm G	Dowtherm LF	Dowtherm J	Dowtherm HT	Syltherm 800	Therminol VP-1	Therminol 66
Composition	Mixture DPO − DP[c]	Mixture of di- and triaryl ether	Aromatic blend	Alkylated aromatic	Hydrogenated terphenyl	Dimethyl siloxane	Mixture DPO − DP[c]	Hydrogenated terphenyl
Temperature use range								
Liquid, °F (min/max)	60/750	20/700	−40/650	−100/600	25/650	−40/750	60/750	15/650
Vapor, °F (min/max)	495/750	....	....	358/600	....	....	495/750	
Minimum pumping temperature, °F[d]	53.6[e]	−4	−50	<−100	31	<−50	53.6[e]	30
Flash Point, °F (COC)[f]	255	285	275	145	355	320	240	350
Fire Point, °F (COC)[f]	275	295	295	155	375	350	260	380
Autoignition temperature, °F ASIM D-2155 (new ASIM method E-669-78)	(1139)	(1083)	(932)	(788)	(662)	725	1150	705
Liquid physical properties at 600°F								
Vapor pressure, lb/in² gage (mm Hg)	30.6	9.5	17.3	159.8	(386)	50.0	31.0	(434)
Thermal conductivity, Btu/h · ft · °F	0.061	0.059	0.054	0.0627	0.062	0.052	0.056	0.0545
Specific heat, Btu/lb · °F	0.579	0.540	0.590	0.721	0.630	0.49	0.564	0.628
Density, lb/ft³	49.29	53.94	48.93	35.46	49.58	42.00	49.78	50.30
Viscosity, cP	0.19	0.30	0.24	0.10	0.38	0.43	0.183	0.40
Price range[h]	C	D	D	C	D	E	C	D

[a]Generic chemical equivalent fluids:

Diphenyl-diphenyloxide	Hydrogenated terphenyls	Manufacture
Dowtherm A	Dowthern HT	Dow Chemical Company
Therminol VP-1	Therminol 66	Monsanto Company
Diphyl		Bayer Leverkusen
Therm S 300	Therms S 900	Nippon Steel
Gilotherm DO	Gilotherm TH	Rhône Poulenc
Thermex		ICI/England
Santotherm VP1	Santotherm 66	Monsanto/Europe

Temperature		Vapor pressure of water	
°F	°C	lb/in² gage	bar
350	177	120	8.3
450	232	405	27.9
550	288	1035	71.4
600	343	1540	106.2

By way of comparison, most *organic* heat-transfer fluids, at 600°F (343°C), have vapor pressures between 0 and 30 lb/in² gage (0 to 2 bar). This low vapor pressure makes these special fluids more attractive for high-temperature applications. In fact, low vapor pressure at high temperatures is the major reason for choosing an organic fluid over steam.

Organic fluids also offer advantages at temperatures below the freezing point of water. Some high-temperature fluids may be practical to pump down to −100°F (−73°C). In cases where both low- and high-temperature capabilities are necessary, high-temperature organic heat-transfer fluids can often be the optimum choice.

The low-temperature capabilities of organic fluids may merit consideration even when the process normally requires only high-temperature capabilities. For instance, in geographic locations where seasonal temperatures can drop well below freezing, low-temperature capabilities become equally important to prevent a system shutdown.

Establishing the fluid temperature range is only the starting point in the process of selecting a fluid. A number of other factors, including fluid vapor pressure, flammability, corro-

		Caloria		Mobiltherm			Marlotherm	Diphyl	Therm S	Gilotherm
60	55	HT43	Texatherm	600	603	605	S	DT	600	ALD
Polyaromatic	Synthetic hydro carbon	Parafinic oil	Parafinic oil	Mineral oil	Mineral oil	Mineral oil	Isomeric dibenzyl toluenes	Isomeric dimethyl diphenyl oxide	Alkyl diphenyl	Alkyl benzene
−60/600	0/600	15/600	5/600	5/600	25/550	5/550	−4/662	−4/626	−39/707	32/590
−50	4	>15	>5	35	>25	20	−18	−46	−22	−49
310	350	390	430	350	380	400	374	275	266	358
320	410	430	ND[g]	ND	ND	ND	ND	ND	ND	367
835	675	670	675	ND	ND	ND	932	1013	ND	ND
6.9	(341)	(110)	(150)	(210)	(160)	(120)	(134)	15.3	11.3	(310)
0.0540	0.0535	0.046	0.064	0.060	0.065	0.065	0.0644	0.0534	0.0569	0.0609
0.622	0.699	0.70	0.70	0.70	0.67	0.670	0.608	0.552	0.625	0.645
49.05	42.71	44.42	42.4	47.30	42.43	42.43	50.85	49.79	47.96	43.70
0.30	0.36	0.40	0.54	0.45	0.43	0.47	0.38	0.273	0.322	0.35
D	A	A	A	A	A	A	ND	ND	ND	ND

[b] All reported physical property data were received from the fluid manufacturers after May 1985. The manufacturers and their respective trademarks are: Dowtherm (Dow Chemical Company), Syltherm (Dow Corning Corporation), Therminol (Monsanto Company), Caloria (Exxon Company), Texatherm (Texaco Incorporated), Mobiltherm (Mobil Oil Corporation), Diphyl (Farbenfabriken Bayer AG, Leverkusen, Germany), Gilotherm (Rhône-Poulenc, Paris, France), Marlotherm (Chemische Werke Hüls AG, Germany), Therm S (Nippon Steel Chemical Company, Tokyo, Japan).

[c] Eutectic mixture of 26.5% diphenyl (DP) and 73.5% diphenyl oxide (DPO).

[d] Temperature where fluid viscosity = 1000 cP.

[e] Freeze point.

[f] (COC)—Cleveland open cup.

[g] ND—No data available from manufacturer.

[h] Price range, \$/gal: A = \$2.00–5.00; B = \$5.00–\$10.00; C = \$10.00–\$15.00; D = \$15.00–\$20.00; E–over \$20.00.

CONVERSION FACTORS: °C = (°F − 32)/1.8; bar = mm Hg/750.2; W/m · °C = 1.7307 Btu/h · ft · °F; kJ/kg · °C = 4.1868 Btu/lb · °F; kg/m³ = 16.021 lb/ft³; mPa · s = 1.00 cP.

sion, toxicity, engineering properties, thermal stability, and cost must also be considered.

Vapor Pressure

For systems already in operation, the pressure capabilities of the system should be taken into account in the selection of a replacement heat-transfer fluid. In the design of new systems, the fluid can dictate the system pressure parameters. Normally the equipment with the lowest pressure rating in a heat-transfer system is the expansion tank or process vessels. The expansion tank should have pressure capabilities sufficient to take into account the pressure created when the fluid thermally degrades or if water or other contaminants enter the system.

Generally an inert gas such as nitrogen is used to blanket the air space above the heat-transfer fluid. The expansion tank should be selected to withstand the inert gas pressure *in addition to* the other possible sources of pressure mentioned.

Flammability

Most organic heat-transfer fluids are flammable when the temperature is sufficiently high and a source of ignition is present. At the flash point, the fluid momentarily ignites on application of a flame or spark. At the fire point, vapor is generated at a rate sufficient to sustain combustion. At the autoignition temperature, no ignition source is required to initiate combustion. The autoignition temperature of most organic fluids is far above their maximum recommended use temperatures. In all cases, air must be present at sufficient concentrations to support combustion.

Corrosion

The heat-transfer fluids listed in Table 4.1 are all noncorrosive to mild steel, so material selection can generally be based on operating temperatures and pressures of the process medium. For example, at low temperatures stainless steel may be required for strength, even though neither the process fluid nor the heat-transfer fluid is corrosive. Specific material compatibility is available from most fluid and heat-transfer equipment manufacturers.

Toxicological and Environmental Concerns

Most high-temperature heat-transfer fluids commercially available are moderately toxic but should present no unreasonable risk to operating personnel if normal operating precautions are taken. The fluid manufacturers provide detailed information on how to handle a spill or leakage from the system.

Pumpability and Engineering Efficiency

The fluid selected should have temperature capabilities that extend beyond those of the process *at both the high and the low ends of the range.* It is desirable to select a fluid with higher temperature capabilities than required in order to safeguard the system against unexpected overheating and degradation of the fluid. (See the following discussion of fluid thermal stability.)

The minimum temperature required for the system's fluid may be dictated by either the process or the ambient conditions.

When low process temperatures are not a concern, the fluid should remain pumpable at the lowest ambient temperature of the geographic location in the event of an extended shutdown. It is possible to steam or to trace heat-transfer lines electrically in order to keep fluid pumpable for cold weather startups, but a fluid that remains pumpable without tracing will generally prove more economical in the long run.

It is important to make the distinction between pour point and minimum pumping temperature when comparing fluids. The pour point is determined by the ASTM D97-66 procedure. In this test procedure, the fluid is cooled to the temperature where the fluid will not flow when the container is tilted on its side. For determining pumpability, the viscosity is a major factor to evaluate.

The minimum pumping temperature is determined by the temperature at which the fluid viscosity reaches 1000 cP. A standard centrifugal pump, designed for normal operating conditions, will experience a significant reduction in flow rate when the viscosity increases above this limit. Pump and heater bypass loops can be designed into the system, but they also become ineffective as the viscosity rises.

Engineering Properties

Table 4.1 lists typical properties for some of the most frequently used heat-transfer fluids. Included are reported use range, minimum pumping temperature, flammability characteristics, thermal properties, and relative price range. This table is based on the conversion of manufacturers' reported data to U.S. customary units with metric conversions as required. In Table 4.2 the relative heat-transfer coefficient and the pressure drop for each fluid, as calculated from the data of Table 4.1, are presented.

4.2 FLUID THERMAL STABILITY

If any single factor in selecting a high-temperature heat-transfer fluid has a dominant effect on long-term performance and operating costs, it is the thermal stability. Low thermal stability will cause premature degradation of the heat-transfer fluid, possibly resulting in loss of fluid, shortening of fluid life, loss of heat-transfer efficiency, increase in viscosity and pumping costs, system fouling, and equipment damage.

Causes of Degradation

Degradation occurs when sufficiently high temperatures cause molecular bonds within the fluid to break down. Some of the new materials produced have lower molecular weights and volatilize at process temperatures. The degradation products are known as "low boilers" or "lights."

As these vapors build up, they can create excessive system pressure, cause pump cavitation, and restrict normal circulation. In normal operation low boilers are vented periodically. The lost fluid is replaced by makeup fluid. High makeup volume is the major result of using low-stability fluids that degrade and produce a high percentage of lights.

Not all degradation products are lights. As degradation occurs and bonds break, some of the molecules can recombine to form high-molecular-weight components. These high-molecular-weight components are good heat-transfer products as long as their concentration does not exceed the manufacturer's recommended level.

The rate at which degradation occurs increases with tem-

Table 4.2 Comparison of Heat-Transfer Fluid Engineering Properties

Fluid	Maximum use temperature,* °F/°C	MPT,† °F/°C	Film coefficient at 600°F, Btu/h · ft² · °F, Velocity‡ 5.0 ft/s	7.0 ft/s	Pressure drop at 600°F, lb/in², Velocity‡ 3.0 ft/s	7.0 ft/s	Film coefficient at 315.5°C, W/m² · °C, Velocity‡ 1.52 m/s	2.13 m/s	Pressure drop at 315.5°C, mbar/m, Velocity‡ 0.91 m/s	2.13 m/s
Dowtherm A	750/399	53.6/12	464	607	1.3	6.9	2630	3450	2.9	15.6
Therminol VP-1	759/399	53.6/12	449	588	1.3	7.0	2550	3340	2.9	15.8
Syltherm 800	750/399	<−50/<−45	239	313	1.2	6.2	1360	1780	2.7	14.0
Dowtherm G	700/371	−4/−20	387	506	1.5	7.6	2200	2870	3.4	17.2
Dowtherm HT	650/343	30/−1	353	462	1.4	7.2	2000	2620	3.2	16.3
Therminol 66	650/343	30/−1	319	417	1.4	7.3	1810	2370	3.2	16.5
Dowtherm LF	650/343	−50/−45	384	503	1.3	6.9	2180	2860	2.9	15.6
Dowtherm J	600/316	<−100/<−73	522	684	0.9	4.9	2960	3880	2.0	11.1
Therminol 60	600/316	−50/−45	355	465	1.4	7.0	2020	2640	3.2	15.8
Therminol 55	600/316	4/−15	302	395	1.2	6.2	1710	2240	2.7	14.0
Caloria HT 43	600/316	>15§/>−9	268	351	1.3	6.5	1520	1990	2.9	14.7
Texatherm	600/316	>5§/>−15	280	367	1.3	6.4	1590	2080	2.9	14.5
Mobiltherm 600	600/316	35/2	319	417	1.4	7.0	1810	2370	3.4	15.8
Mobiltherm 603	600/316	>20§/>−7	310	406	1.3	6.3	1760	2310	2.9	14.3
Mobiltherm 605	600/316	20/−7	298	390	1.3	6.3	1690	2210	2.9	14.3
Marlotherm S	662/350	−18/−28	366	479	1.4	7.4	2080	2720	3.2	16.6
Diphyl DT	626/330	−46/−43	359	469	1.4	7.1	2040	2660	3.2	16.0
Therm S 600	707/375	−22/−30	350	458	1.4	6.9	1990	2600	3.2	15.5
Gilotherm ALD	590/310	−49/−45	330	433	1.3	6.4	1870	2460	2.9	14.4

* Manufacturer's maximum recommended use temperature.
† MPT = minimum pump temperature, where viscosity reaches 1000 cP.
‡ Velocity in 1-in schedule 40 pipe.
§ Reported pour-point temperature.

perature. A general rule for the degradation rate for organic fluids is that degradation doubles for every 18 to 21°F (10 to 12°C) increase in fluid temperature. Figure 4.1 shows the degradation rate versus temperature for Dowtherm* G heat-transfer fluid.

Degradation problems can be avoided by selecting the right fluid and monitoring the fluid condition periodically. The fluid should be replaced when the concentration of degradation product exceeds the manufacturer's recommended limits. Where applicable, fluids can be reprocessed to remove degradation products. In this way, a large percentage of the fluid can be recovered to provide better process economies.

How Operating Conditions Affect Thermal Stability

Simply selecting a fluid with an operating temperature range that matches the process temperature range is not enough. Consideration needs to be given to conditions that can cause temporary fluid overheating—such as equipment malfunctions or temperature control fluctuations. In addition, localized fluid overheating is present in nearly all heat-transfer systems, although it often goes undetected.

Under normal operating conditions, the fluid film temperature at the wall of heater tubes is designed to be between 50 and 75°F (28 and 42°C) above the bulk temperature of the fluid. Flame impingement on heater tubes can raise the fluid temperature by hundreds of degrees and cause accelerated degradation. Low fluid velocity or complete loss of fluid circulation can cause the volume of fluid in the heater tubes to become overheated.

By choosing a fluid with high thermal stability, most problems associated with localized or temporary temperature excursions beyond the upper process temperature limit are prevented. In addition to minimizing problems associated with fluid overheating, a high-stability fluid will result in better efficiency and less frequent fluid replacement than a lower-stability fluid.

How Thermal Stability Is Measured

The basis of all heat-transfer fluid testing is to heat the fluid to a given temperature and analyze samples for physical and chemical changes periodically. Two types of tests are used to determine fluid stability: static ampoule testing and forced-circulation testing.

Static Ampoule Testing In static apoule testing, ampoules or cylindrical steel vessels are filled with small amounts of fluid, blanketed with nitrogen, sealed, and heated in an oven. Each test involves enough ampoules so that multiple samples can be taken throughout the test period. This procedure allows any random data points to be identified quickly.

The ampoules, each containing a precisely measured amount of fluid, are heated in a convection oven and held at a constant temperature. Samples are removed each week of the test and allowed to cool to condense any light vapors in the ampoule. The contents are weighed and analyzed for chemical composition. The inside of the ampoule is examined for signs of deposits.

The percentage degradation is plotted for each week of the test and fitted with a linear regression line for zero or

* Trademark of The Dow Chemical Company.

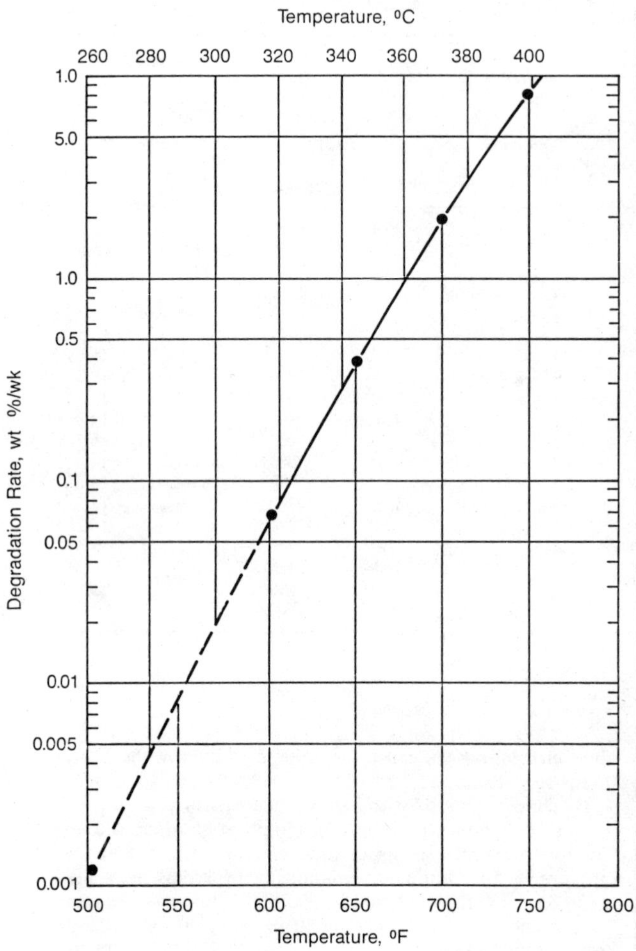

Fig. 4.1 Isothermal degradation rate base curve for fluids used within temperature range of 550 to 750°F (288 to 399°C) (Dowtherm G fluid).

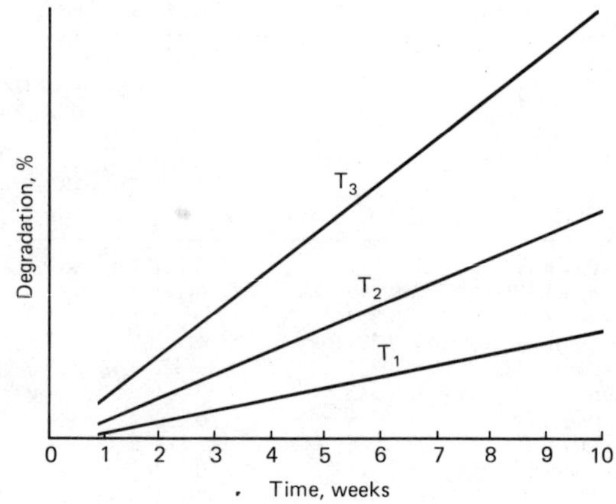

Fig. 4.2 Typical thermal degradation profile of one fluid at three temperatures ($T_1 > T_2 > T_3$).

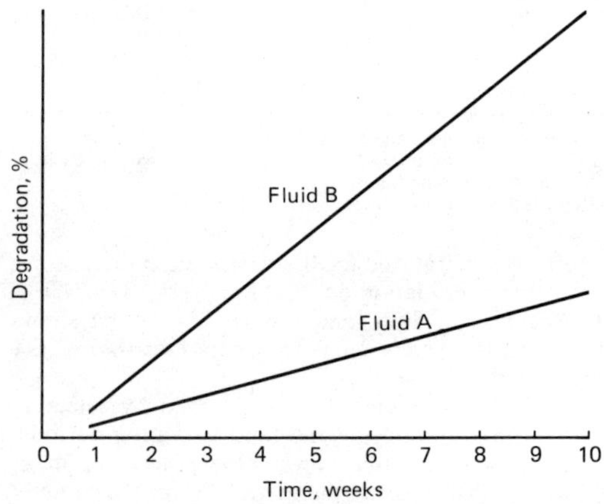

Fig. 4.3 Typical thermal degradation profiles of two fluids at one temperature.

first-order kinetics. The slope of the line is the degradation rate at that temperature. Figure 4.2 shows what the results might look like for 10-week tests at three temperatures. By duplicating the procedure exactly, tests of different fluids can be compared directly. The ranking of fluids by thermal stability is illustrated in Fig. 4.3

While static ampoule testing is helpful for relative fluid stability rankings, the degradation rates obtained tend to be higher than those observed in actual industrial use. The size of the ampoule and the isothermal temperature in the ampoule test are obviously factors in this behavior. In addition, the limited volume of fluid involved in ampoule testing does not permit physical property measurements such as acidity and viscosity.

To obtain these data and to develop a correlation between ampoule and industrial degradation rates, fluids are tested in forced-circulation test loops.

Forced-Circulation Loop Testing In forced-circulation loop testing, fluid is circulated through a small-scale heat-transfer system like the one shown in Fig. 4.4. The test loop has sampling points where fluid can periodically be drawn off and analyzed.

A typical test loop contains 30 to 50 gal (100 to 200 L) of fluid. This permits generous fluid samples throughout the test period so pH and viscosity are easily measured. Solid material (such as degradation products or contaminants) is also more easily detected and analyzed due to the larger volume.

Test Validity Both static ampoule and forced loop tests use time and temperature as the key variables. However, other variables must be carefully monitored to ensure that the resulting data can be correlated to actual use conditions. It is especially important to prevent contamination of test equipment and fluids. Fluid contamination that can be tolerated in full-scale systems can cause disastrously skewed results in the small volume of fluid used in ampoules.

While the fluid volume is significantly greater in forced

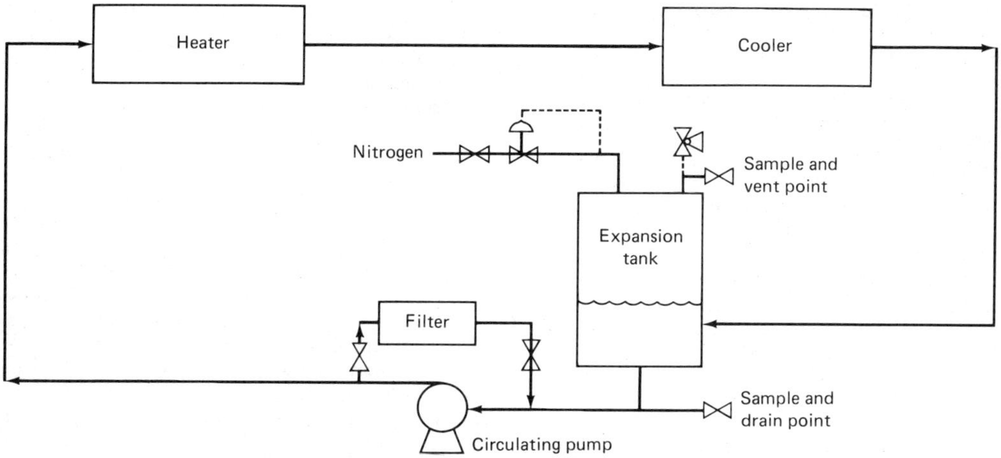

Fig. 4.4 Typical thermal test loop.

loop testing, there is still considerably less fluid than in an industrial system. Contamination of the fluid can still cause disproportionate fluid problems.

Oxygen is the number one enemy of accurate testing. If the system is opened for any reason during the test, the test should be terminated. Proper equipment cleaning between tests is also extremely important to ensure that degradation products from one test do not affect the next test. Even the equipment cleaning chemicals are carefully selected to avoid chlorinated solvents or other materials whose trace residues are incompatible with heat-transfer fluids.

One check made to ensure that contamination has not invalidated the test involves chemical analysis of the fluid before and after the test. When the fluid is first introduced into the system, it is circulated through a sidestream filter. The filter cartridge is saved for comparison with a second filter used to filter the fluid when the test is completed (during the test, no fluid is filtered so that any filterable residue will appear in the weekly samples). An excessive amount of chloride or oxygen compounds in the final fluid residue indicates contamination and invalid test results.

Another method of checking the validation of test results is to look for a correlation between the results of static ampoule testing and forced loop testing. As illustrated in Fig. 4.5, static ampoule tests will give higher degradation rates, but the relative ranking of fluids should remain the same independent of the test method.

Evaluation of System Degradation Rate and Fluid Operating Costs

Within a system it is possible to determine a given fluid's annual degradation with a reasonable degree of accuracy with the following procedure.

First, establish the temperature profile for the system. The temperature profile specifies the temperature of the heat-transfer fluid in various segments of the system and the percentage of the total volume contained in each of those segments. In existing heat-transfer systems there may already be records of the temperature at various points in the system. In new system designs one can estimate temperatures from the heat source and through each user in the process.

Second, use the thermal stability data for the fluids to be compared. The data should include degradation rates for the fluid at several temperatures around the fluid's recommended use range.

For calculating purposes, the proportionality between degradation and temperature is assumed to be constant for fluids discussed here. Only the degree of degradation varies. For example, if fluid *A* is five times more stable than fluid *B* at 550°F, the same relationship will hold true at 600°F. Using the base curve in Fig. 4.1 to calculate an initial degradation rate, it is possible to determine the degradation rates for many fluids simply by scaling the results with a relative stability multiplier factor from Table 4.3.

The final piece of information required is the estimated

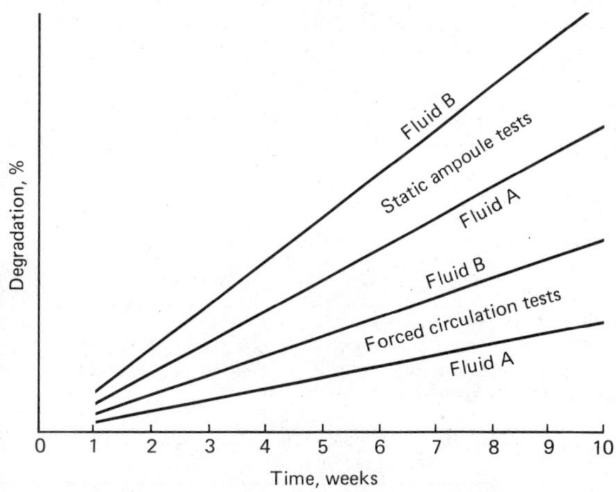

Fig. 4.5 Correlation of static ampoule and forced-circulation tests at temperature *T*.

Table 4.3. Relative Stability of Representative Heat-Transfer Fluids

Fluid description	Trade name	Relative stability*
Mixture of diphenyl (DP) and diphenyl oxide (DPO)	Dowtherm A	0.0024
Mixture of di- and triaryl ethers	Dowtherm G	1.0
Aklylated aromatic	Downtherm J	1.5
Hydrogenated terphenyl	Dowtherm HT	1.6
Aromatic blend	Dowtherm LF	2.2
Petroleum oils	Various	20–50

* Relative stability data generated in forced-circulation unit and ampoule test at 600°F (315.6°C) at The Dow Chemical Company. A low number represents a low degradation rate and thus higher stability.

annual operating time of the system. A system run cyclically throughout the year will naturally have a lower annual degradation percentage than a system that runs continuously.

The system's degradation rate is calculated by summing up the degradation for each volumetric component. The calculation is shown in Table 4.4 for Dowtherm G fluid (stability data taken from Fig. 4.1) and for a typical petroleum oil (stability data taken from Table 4.3).

The total fluid makeup per year can be estimated by adding the degradation losses and the mechanical losses. The latter, of course, depend on the process requirements and the operating practices. Fluid losses through pump seals, valve packings, and other equipment are generally in the range of 5 to 10 percent per year of the total fluid in the system. Obviously, a system properly monitored, maintained, and utilizing a fluid of suitable stability will minimize operating cost and provide designed heat-transfer efficiency.

Effects of Contamination on Thermal Stability

Contamination of an organic heat-transfer fluid can greatly accelerate the degradation and fouling in the system. The principal types of fluid contaminants are now discussed.

Oxygen The most frequently experienced contaminant is oxygen from the air. Above 300°F (149°C) most organic fluids oxidize at a measurable rate. The oxidation products include low boiling and insoluble products. These products can increase the fluid viscosity, foul the system, and contribute to corrosion. Inert gas blanketing is recommended to prevent oxygen contamination in the expansion tank.

Oxidizing Agents Besides oxygen, any strong oxidizing agent will have a significant effect on a heat-transfer system. In practice, the source of these materials is cross contamination from the process or residuals from chemical cleaning procedures.

When system cleanup is required, cleaning chemicals and inorganic salts must be carefully flushed from the system. The presence of these materials and trace amounts of water can cause excessive corrosion and fouling of the system at operating temperatures.

Other Organic Materials Most organic process contaminants are not thermally stable when exposed for an extended time to the temperatures associated with a heat-transfer system. Thus, the heat-transfer system should be monitored routinely to detect this potential problem. Some contaminants can accelerate the degradation of the heat-transfer fluid while others foul the system by their own cracking and polymerization modes.

4.3 ENGINEERING EVALUATION

After having narrowed the choice of heat-transfer fluids by reviewing physical property and thermal stability data, an engineering evaluation is made prior to the final fluid selection. The basic calculations made are to determine the heat-transfer film coefficient and the fluid flow pressure drop for each respective fluid.

Relative Film Coefficients

Many studies have been made to define correlations between the film coefficient h_f and the fluid's physical properties. The Sieder and Tate correlation[1] gives reliable coefficients for organic fluids. It is expressed by

$$h_f = 0.023\,\frac{k}{d}\left(\frac{dv\rho}{\mu}\right)^{0.8}\left(\frac{C_p\mu}{k}\right)^{1/3}\left(\frac{\mu}{\mu_w}\right)^{0.14}$$

Table 4.4. Evaluation of System Fluid Degradation for Dowtherm G Fluid and a Petroleum Oil*

			Degradation rate, wt%/week			
			Dowtherm G fluid		Petroleum oil	
Fluid location	Fluid temperature, °F/°C	Volume, %	Rate @ temperature	System rate	Rate @ temperature	System rate
Film in heater	650/343	1.0	0.3960	0.0040	7.920	0.079
Heater discharge	600/316	34.0	0.0636	0.0237	1.390	0.473
Return loop	550/288	34.0	0.0103	0.0000	0.206	0.070
Expansion tank	450/232	31.0	0.0000	0.0000	0.002	0.001
Total degradation per week, wt%				0.0312		0.623
Total degradation per year, wt%				1.62		32.4

* Degradation for Dowtherm G fluid derived from isothermal test data as shown in Fig. 4.1. Degradation of petroleum oil from Table 4.3 ranges from 20 to 40 greater than that of Dowtherm G fluid. Degradation calculations are based on the lowest value for the oil. Yearly degradation rates are based on 52 weeks of operation.

where C_p = specific heat
d = diameter
k = thermal conductivity
v = fluid velocity
w = wall temperature
μ = viscosity
ρ = density

Using the physical properties from Table 4.1, the film coefficient has been calculated for each fluid and tabulated in Table 4.2. This relative study is for 600°F (315.5°C) and for fluid velocities of 5 and 7 ft/s (1.52 and 2.13 m/s). The optimum velocity, in any given case, will be an economic balance between cost of power and cost of heating surface. The two velocities selected are within the range of typical heat exchangers.

Special Heat-Transfer Considerations for Low Process Temperatures

In applications where low-temperature capabilities are directly required by the process, such as for cooling exothermic reactions in batch procession, there is one additional precaution to be taken in regard to the heat-transfer fluid. The heat-transfer efficiency will drop off sharply when the fluid dynamics of the system switch from turbulent to laminar flow. This situation is likely to occur as a fluid with high viscosity is cooled to near the fluid's pumpability limit. Because of laminar fluid flow and the resulting loss of heat-transfer efficiency, it is not economical to design a chilled loop to operate near a heat-transfer fluid's lowest pumpable temperature.

In order to ensure turbulent flow and maintain heat-transfer efficiency, the fluid selected should have a low viscosity at the minimum required process temperature. The design fluid velocity must yield a Reynolds number Re greater than 3000, where $\mathrm{Re} = dv\rho/\mu$.

Fluid Pressure Drop

The design pressure drop, in an organic heat-transfer system, is based on process restrictions or use of standard practices. Fluid transfer lines are frequently sized by assuming a reasonable velocity.[2] In process exchangers the velocity is dictated by process requirements and economics. The typical fluid velocities used in these two cases range between 3 and 10 ft/s (0.91 and 3.05 m/s). For the relative pressure drop data listed in Table 4.2, standard calculations[3] were made using the selected velocities of 3.0 and 7.0 ft/s (0.91 and 2.13 m/s).

4.4 OVERALL CONSIDERATIONS FOR FLUID SELECTION

In selecting an organic heat-transfer fluid for a high-temperature process, the operating and ambient temperature requirements are first defined. If moderate temperatures are required [below 500°F (260°C)], then multiple fluids are available. When low-temperature pumpability or high-temperature requirements are needed, then the fluid selection is restricted.

After the system temperatures are defined, a review by product can be made. Thermal stability is recognized as the most critical factor in fluid selection. If the fluid cannot pass this screening, then it need not be evaluated for engineering properties, flammability, toxicity, and cost.

REFERENCES

1. E. N. Sieder and G. E. Tate, *Ind. Eng. Chem.*, **28**, 1429 (1936).
2. R. H. Perry and C. H. Chilton, *Chemical Engineer's Handbook*, 5th ed., McGraw-Hill, New York, 1973, pp. 5–31.
3. "Flow of Fluids," Tech. Paper 410, Crane Company, New York, 1969.

Chapter **5**

Molten Salts

JACK D. RICKLY
Olin Research Center

Heat-transfer fluids known as molten salts are solid chemical salts which, when heated beyond melting, have a wide liquid temperature range and other properties useful for transporting and storing heat.

The most common mixture has been in use since the 1930s.[1] A blend of potassium nitrate, sodium nitrite, and sodium nitrate (53/40/7 percent by weight, respectively), it melts at 288°F (415 K) and can be used up to 900°F (755 K). This mixture is the eutectic or lowest melting formulation of these salts. It is best known as Hitec, a trademarked product of Coastal Chemical. Hitec was originated by Du Pont and recently owned by Olin. This and similar compositions are called heat transfer salts (HTS). The HTS designation is also used with other formulations, even those chemically incompatible with Hitec. Nonetheless, HTS will refer to the Hitec-type formulation in the remainder of this chapter.

Drawsalt, 60/40 percent by weight sodium nitrate and potassium nitrate, can be used if the lower melting point of HTS can be sacrificed to obtain higher thermal stability and lower cost. It melts from 432 to 473°F (495 to 518 K) and can be used to 1050°F (843 K). Drawsalt has had extensive testing for solar power applications.[2] The more expensive eutectic (46/54 percent by weight) melts at 432°F (495 K). Often also referred to as drawsalt, it will be referred to here as eutectic drawsalt.

Different salts are used in other applications, but nitrate and nitrite salts so dominate thermal transfer and storage that this chapter deals with them exclusively. Janz and others discuss salts outside the scope defined here.[3,4]

Warning Listings of eutectic salt mixtures, even for thermal transfer or energy engineering, may include mixtures which are explosive. No mixture of oxidizing salts with reducing salts, liquid or powdered metals, or organic materials should be used for thermal transfer media. Especially avoid mixtures of nitrate or nitrite salts (both are oxidizing) with ammonium salts, with organic materials such as urea, or with cyanide, cyanate, thiocyanide, thiocyanate, sulfide, polysulfide, sulfite, or thiosulfate salts, or with sulfur.

The advantages of molten salts are high heat capacity and other good physical properties, excellent thermal stability, low cost, good materials compatibility, negligible vapor pressure or flash point problems, decades of industrial experience, and multiple supply sources. The option of using a water dilution system for HTS is another advantage.[5] Available from American Hydrotherm, New York, it retains the HTS upper temperature limit, but allows system cooling with a liquid medium (of changing composition) to much lower temperatures, such as 100°F (311 K).

Disadvantages include high melting points in the anhydrous systems, a tendency for the melting point of HTS to drift upward with slow oxidation, the advisability of using a nitrogen blanket on HTS to retard deterioration, the propensity of molten salts to leak around seals, and an increase in corrosive aggressiveness if the salt is contaminated or deteriorates. In contrast to HTS, drawsalt is not used with a nitrogen blanket since it has a stable melting range.

The information in this chapter is believed to be accurate. However, the responsibility for a successful and safe molten salt system remains that of the designers, engineers, contractors, and operators. Good design and engineering practices never permit reliance on one source for important information. No license is granted or implied to practice anything patented that is discussed in this section.

5.1 PHYSICAL PROPERTIES

Physical property information includes melting behavior, viscosity, density, heat capacity, and thermal conductivity. Vapor pressure, as usually defined, is negligible. The salt does attempt to equilibrate with some components of its cover gas and can generate significant pressures of oxygen and/or nitrogen oxides under thermal or chemical degradation.[6]

The relative amounts of nitrates and nitrites in an operating system can vary considerably over time. The physical properties that depend significantly on the nitrite content are melting behavior and density.

The melting behavior bears explanation. Eutectics have a melting point, which is the temperature at which the mixture melts completely. Noneutectic mixtures have a melting range. At the lower temperature, or solidus, melting begins. For compositions near the eutectic, the most melting occurs at the solidus. At the liquidus, or higher temperature of the range, the mixture becomes entirely liquid. The liquidus is the key temperature for thermal transfer considerations. Note that as HTS degrades, it changes from eutectic to noneutectic.

Equations which relate physical properties to temperature and composition are given in Table 5.1.

5.2 CHEMICAL PROPERTIES

The major reaction occurring in molten nitrate-nitrite salt mixtures is the drive toward equilibrium with oxygen in the cover gas,

$$NO_2^- + \tfrac{1}{2}O_{2(g)} \rightleftharpoons NO_3^-$$

The equilibrium concentration of nitrite varies as the square root of the oxygen partial pressure, and ranges from tens of parts per million to a few tenths of a percent for atmospheric oxygen partial pressure throughout the temperature range of use (to perhaps a percent near the upper range of drawsalt).[13] This reaction is rather benign since both nitrate and nitrite are soluble in the melt, noncorrosive, and stable.

The consequence of this reaction for drawsalt is a slight decrease in the liquidus due to a slight increase in the nitrite

Table 5.1. Physical Properties of Molten HTS and Drawsalt

	HTS		Drawsalt	
	Physical property	Reference	Physical property	Reference
Liquidus, °F (±5°F)*	$446.8 - 6.0769C + 0.18341C^2 - 0.0033042C^3$	3*b*, pp. 786–795	473 ± 1	8
Volume change on melting $\Delta V_{fus}/V$, %	4.0 ± 0.5	7	4.6 ± 0.5	3*a*, pp. 396–411
Heat of fusion ΔH_{fus}, Btu/lb	≈35	1	49	8
Heat capacity C_p, Btu/lb · °F	0.37 ± 0.03	1	0.38 ± 0.02	†
Thermal conductivity k, Btu/ft · h · °F	0.29 ± 0.12	9	0.31 ± 0.07	3*a*, pp. 396–411
Viscosity η, cP (±5%)‡*	$\exp[-1.404 + 1747.7/(T + 116.8)]$			
Surface tension γ, lb/ft (±1%)*	$0.0092 - 2.4 \times 10^{-6} T$	§	$0.0108 - 2.4 \times 10^{-6} T$	3*a*, pp. 396–411
Density ρ, lb/ft³*	$131 - 0.0256T$	11	$129 - 9.6 \times 10^{-6} T^2$	¶

* C is nitrite as % by weight $NaNO_2$; $0\% < C < 40\%$; T is in °F.

† Determined to 800°F.[8]

‡ Recent values[10] are lower. Those shown summarize previous HTS[1,11] and recent eutectic drawsalt[12] data. It seems viscosities of the two salts are very close.[12] Viscosities for noneutectic (60/40) drawsalt may be 3 to 5% lower than shown here (Ref. 3*a*, pp. 396–411).

§ Values shown are for eutectic drawsalt, which seem to be very close to HTS values.[12]

¶ ±1% (Ref. 3*a*, pp. 396–411). There are also recent values for the eutectic drawsalt.[12].

CONVERSION FACTORS: 2.326 kJ/kg = 1 Btu/lb; 4.187 kJ/kg · K = 1 Btu/lb · °F; 1.731 J/m · s · K = 1 Btu/ft · h · °F, 1 mPa · s = 1 cP; 68.54 mN/m = 1 lb/ft; 16.018 kg/m³ = 1 lb/ft³ = 0.13368 lb/gal.

SOURCE: Adapted from the references cited.

content. For HTS the equilibrium drives in the other direction (although it will not normally attain equilibrium), converting some of the nitrite to nitrate. This causes a gradual upward drift in liquidus for HTS systems and is the reason that nitrogen is often used to blanket HTS. The nitrogen blanket greatly reduces the partial pressure of oxygen in the cover gas, reducing both the rate of oxidation of nitrite and its final equilibrium concentration.

Although this reaction is slow[13] and oxygen diffusion is rather rapid (Ref. 3*b*, pp. 786–795), the rate of the reaction is limited by the gas-liquid interface surface area.[13] For this reason, designs which cause much bubble entrainment should be avoided, although the sparging of a nitrogen blanket is probably a good idea for HTS.

The most damaging decomposition reactions involve the loss of nitrogen oxides or nitrogen from the salt,

$$2NO_2^- \rightleftharpoons NO_{(g)} + NO_{2(g)} + O^{2-}$$

$$NO_2^- + NO_3^- \rightleftharpoons 2NO_{2(g)} + O^{2-}$$

$$3NO_2^- \rightleftharpoons 2NO_{(g)} + NO_3^- + O^{2-}$$

$$5NO_2^- \rightleftharpoons N_{2(g)} + 3NO_3^- + O^{2-}$$

Nitrogen oxides in the cover gas tend to retard this decomposition, so turning over the cover gas too frequently should be avoided unless the corrosiveness of nitrogen oxides or nitric acid vapor is a problem. Although oxide is almost totally converted to peroxide and superoxide in very pure, dry nitrate melts,[14,15] this does not occur in the presence of water vapor. Sufficient water vapor will buffer the oxide through hydroxide formation,

$$O^{2-} + H_2O_{(g)} \rightleftharpoons 2OH^-$$

This should improve salt stability.[16] Unless special attention is given to drying the cover gas, there should be adequate water vapor in the salt. Where dry nitrogen is used, some steam may at times be added. Too much moisture in the cover gas, however, will react with the nitrogen oxides to form nitric acid vapor. Hydroxide is soluble in the melt but is corrosive and encourages the leaching of chromium and molybdenum from stainless steels to form melt-soluble chromates and molybdates. The latter may slightly destabilize the salt. In addition, hydroxides react readily with carbon dioxide from the atmosphere to produce carbonates.

Analytical tests convert any sodium oxide into the hydroxide since the sample is dissolved in water. Peroxides or superoxides, if present, may also be converted to hydroxide. Samples containing hydroxide or oxide may pick up carbon dioxide from the air, converting some of those impurities to carbonates. Thus levels of carbonate as analyzed may be higher than those actually present in the system. This is one advantage of analyzing for total alkalinity; it does not distinguish hydroxides from carbonates and so is not influenced by their interconversion.

The rate of thermal decomposition of the salt depends upon many factors and is not well understood. Factors known to have a great effect are the hottest surface temperature contacting the melt and the hottest bulk salt temperature. Other factors include the cover gas composition, the ratio of the liquid volume to the liquid-solid and liquid-gas interface surface areas, the containment materials, the impurities, and the degree of agitation.

Although water vapor is fairly benign, carbon dioxide will decompose the salt, forming carbonates,[17]

$$CO_{2(g)} + 2NO_3^- \rightleftharpoons CO_3^{2-} + 2NO_{2(g)} + \tfrac{1}{2}O_{2(g)}$$

Sodium carbonate has limited solubility in the melt at low temperatures and will coat the coldest surfaces (usually the cold-side heat exchangers) if its solubility is exceeded at the coldest skin temperatures. Solubility in the eutectic drawsalt is approximately given by[17]

$$\log_{10} S = 3.1063 - \frac{3500.3}{T + 460} = 3.1063 - \frac{1944.6}{T'}$$

while for HTS,[18]

$$\log_{10} S = 2.6357 - \frac{2742.4}{T + 460} = 2.6357 - \frac{1523.5}{T'}$$

where S is the carbonate solubility expressed as percent by weight of Na_2CO_3 (a common format for lab analyses), and T and T' are the temperature in °F and K, respectively. For example, solubility in drawsalt at 500°F (533 K) is 3.91 percent as Na_2CO_3. By simple solubility theory, the carbonate solubility in noneutectic (60/40) drawsalt could be as low as 60 percent that of the eutectic. If carbonate is over 0.17 percent (as Na_2CO_3) in HTS or 0.80 percent in drawsalt, or if total alkalinity (as NaOH) is over 0.07 percent in HTS or 0.30 percent in drawsalt, it is usually cause for corrective action.

Impurities are present in commercial molten salt mixtures; others appear during operation. Aside from hydroxide and carbonate, chloride and sulfate are of most concern since they greatly affect the corrosiveness of the melt. Both should be less than 0.3 percent by weight when expressed as NaCl or Na_2SO_4.

Of cosmetic concern are aluminosilicate impurities which gel and form a scum or fines. These form by a reaction between impurities after the salts are dissolved or melted. They are usually not a problem and may even be beneficial.

Tests of thermal stability and corrosive aggressiveness do not give precise results, especially if they use commercial salts and construction materials to simulate actual use conditions. The experimental scatter of the test results makes definite conclusions difficult to reach, although sometimes the results are suggestive. Results of such tests have suggested that commercial drawsalt of reasonable quality appears to be superior to otherwise identical mixtures made of purer analytical reagent grade chemicals. This perception is indirectly supported by a patent, including theoretical justification, which claims additives to drawsalt to improve stability and compatibility.[19] Many of the additives claimed are chemically similar to impurities which are already in commercial salts (or which quickly appear when used with common containment materials) in quantities slightly exceeding their solubilities. Certainly any testing of salts or additives in molten salt applications should include a comparison with undoped commercial grade salts, allowing contact with representative materials of construction.

5.3 MATERIAL COMPATIBILITY[20]

HTS or drawsalt operated below 800°F (700 K) is compatible with most materials of construction. However, no contact should be permitted with organic materials, carbon, sulfur, with reactive metals such as aluminum, magnesium, or their alloys, or with powdered iron. Use with cast iron or copper is not recommended. At higher temperatures low-carbon stainless steels are recommended, such as SS 316L or SS 304L. Steels containing molybdenum, such as SS 347L, are somewhat less desirable. The low-carbon designation is critical at temperatures over 850°F (454°C) and in thin performance parts, such as in valves. Verification is important since low- and normal-carbon steels are sometimes interchanged by suppliers. Code restrictions may need special consideration when specifying the low-carbon steels.

For high heat flux at the higher temperatures, high nickel alloys such as Incoloy 800 should be used. Where contamination is sometimes unavoidable, critical surfaces may require more costly materials, such as the Hastelloys. Usually this is not specified initially, but it may be considered after actual lifetime information on the alternatives has been obtained.

For salts which are occasionally contaminated in use by chloride or sulfate, it may be necessary to use the high-nickel alloys even at lower temperatures. Even then, salt contamination should be corrected.

Glass and quartz are susceptible to etching if the salt temperature or the alkalinity is high. Alumina and magnesia (the oxide ceramics) appear to withstand almost the entire gamut of salt conditions.

5.4 DESIGN AND CONSTRUCTION

Steam tracing of pipes, valves, and flanges is the usual method of dealing with the melting point of HTS. For drawsalt, electric heat tracing is needed. With electric heat tracing, it is important to make it almost impossible for molten salt to leak onto the heaters or wiring, even in the smallest amounts. This requires great care, since molten salt can leak through the smallest joints, holes, and seals, and will "creep" up surfaces hot enough to keep it molten. Venting from the system should be directed safely outside since it may contain fine salt particles.

Where it is important to avoid salt leaks, welding should be used for pipe bends and connections. Exposed salt lines and valves should be insulated. At high temperatures, valves may need radiation fins to protect the packing glands. It is important that no leakage occur onto solid fuel or built up carbon or soot. Bearings or rotary seals should not be in contact with the salt;[21] cantilever pumps that permit no contact of the molten salt with the packing gland can be used for circulation. Manually operated steel gate valves and automatic valves can be used if they are installed so that they can drain freely. It is prudent to configure the system so that it can drain completely from top to bottom, via gravity, into the sump in the event of total system failure. A bypass around the cold-side heat exchanger will allow drainage even if the exchanger is plugged and will facilitate a gradual system startup.

Metallic asbestos gaskets, asbestos sheet gaskets, and asbestos packing have all been used successfully with HTS. Organic valve-packing lubricants or excessive pipe dope should be avoided. Graphite-lubricated asbestos is satisfactory if the graphite content is minimal.[20]

Avoid high heat fluxes at heater tubes or hot-side heat exchanger surfaces in systems working near the upper temperature range of the salt or when bringing a system up to its operating temperature. There will be a "burn-in" period during startup of a system. During this period, the initial salt decomposition rate will be relatively high, as will be the formation of nitrogen oxides. Such behavior may also occur in the event of temperature excursions or process stream leaks. In normal operation the hottest surface seen by the salt should be no more than 100°F (56 K) above the nominal rated bulk upper temperature limit.

Worst-case conditions (leaks, loss of circulation, plugging, etc.) should never allow salt to contact a surface hotter than 1150°F (892 K) for HTS or 1250°F (951 K) for drawsalt, as violent reaction with the surface material may result. Since molten salt is a good electrical conductor (about 2 Ω/cm), this possible additional source of inadvertent heating should be considered.

Fire control methods in the vicinity of a molten salt unit can include carbon dioxide and approved dry powder-type extinguishers. Vaporizing liquid (carbon tetrachloride) or water extinguishers (other than fine spray sprinklers or low-velocity fog nozzles) should not be used. Note that even small amounts of water can cause violent steam explosions if they penetrate the surface of molten salt. Clean, dry sand is useful for slagging and diking spills of molten salt.

Additional information on design and construction is available,[21–25] as are experimental engineering data.[26] When reading the literature, bear in mind that design has become more conservative in the last decade or so, especially with regard to temperature limits.

5.5 SALT CHARGING

Salt can be charged molten if delivered that way. More common is to first charge to a portable melter or directly to a sump equipped with immersion heaters or steam coils adequate to melt the salts and bring them up to a high enough temperature to circulate. With HTS, a water dilution system can be used. Small amounts of water can aid in melting the salt, but with some pumps the water vapor may cause pump cavitation until the salt becomes rather dry at about 400°F (479 K). A nitrogen sparge can be used to speed the drying of the melt after "boiling" stops.

The salt heaters should be designed to prevent localized overheating during startup. The salt freezes with volume contraction, so remelting should be done in such a way as to prevent entrapped melting. For this reason, heaters in the sump should penetrate the salt surface to deal with a drain-down and freeze problem. This will also avoid the settling of sludge onto heated surfaces, which impedes the flow of heat and can cause disastrous local overheating.

Since HTS is hygroscopic, old bags or opened drums may become solidly caked. Additional effort should then be planned to deal with breaking the cake. Care must always be used to prevent splashing of the molten salt since the high heat capacity results in extremely severe burns. High-temperature protective gloves as well as face shields and protective clothing and boots should be worn. Clothes are best if natural fiber; shoes are best if leather.

5.6 SALT MAINTENANCE

At a minimum, regular testing of the melting range and total alkalinity should be done, perhaps monthly. Less frequent tests of sulfate, chloride, and any other likely contaminants should also be performed.

Sampling for the analyses requires special care. The sample should be taken molten into a vessel of about the needed sample size (nickel or stainless steel is acceptable). When cool, the entire sample must be ground and mixed since it does not freeze uniformly.

The most common maintenance for HTS is to restore its melting point by replacing some of the salt with a "butting salt" of high nitrite content and the correct sodium-to-potassium ratio. The replacement amount is chosen to bring the nitrite of the final mixture to 40 percent by weight expressed as sodium nitrite. Note that bringing up the nitrite by adding sodium nitrite alone is not satisfactory.

Using this procedure with the usual butting salt (54 percent potassium nitrate and 46 percent sodium nitrite by weight), a substantial portion of the existing salt may need to be replaced. Note that this will dilute contaminants, improving their levels as well. However, if impurities and total alkalinity are low, it may be more cost-effective to replace a smaller fraction of the salt with a more expensive all-nitrite butting mixture of 50 percent potassium nitrite and 50 percent sodium nitrite by weight.

REFERENCES

1. W. E. Kirst, W. M. Nagel, and J. B. Castner, "A New Heat Transfer Medium for High Temperatures," *Am. Inst. Chem. Eng. Trans.*, **36,** 371–394 (1940).
2. R. W. Carling, C. M. Kramer, R. W. Bradshaw, D. A. Nissen, S. H. Goods, R. W. Mar, J. W. Munford, M. M. Karnowsky, R. N. Biefeld, and N. J. Norem, "Molten Nitrate Salt Technology Report," SAND80-8052, NTIS, 1981, 118 pp.
3. G. J. Janz, C. B. Allen, N. P. Bansal, R. M. Murphy, and R. P. T. Tompkins, "Physical Properties Data Compilations Relevant to Energy Storage," pts. II and IV, NSRDS-NBS 61, Superintendent of Documents, U.S. GPO, Washington, DC. (*a*) pt. II, 1979, 442 pp.; (*b*) pt. IV, 1981, 870 pp.
4. G. J. Janz, *Molten Salts Handbook,* Academic Press, New York, 1967.
5. H. F. Payne, D. J. I. Davies, and J. G. Parker, "Heat Transfer Salt Circulating Systems with Water Addition," *Chem. Eng.*, pp. CE188–CE190 (1963).
6. K. H. Stern, "High Temperature Properties and Decomposition of Inorganic Salts; Part 3, Nitrates and Nitrites," *J. Phys. Chem. Ref. Data,* **1,**747–772 (1972).
7. J. P. Petitet, M. Fraiha, R. Tufeu, and B. LeNeindre, "Experimental Determination of the Volume Change of Pure Salts and Salt Mixtures at their Melting Point," *Int. J. Thermophys.*, **3,** 137–155 (1982).
8. D. F. Rogers and G. J. Janz, "Melting-Crystallization and Premelting Properties of $NaNO_3$-KNO_3. Enthalpies and Heat Capacities," *J. Chem. Eng. Data,* **27,** 424–428 (1982).
9. R. Santini, L. Tadrist, J. Pantaloni, and P. Cerisier, "Measurement of Thermal Conductivity of Molten Salts in the Range 100–500°C," *Int. J. Heat Mass Transfer,* **27,** 623–626 (1984).
10. P. G. Gaune, "Viscosity of KNO_3-$NaNO_2$-$NaNO_3$ Mixtures," *J. Chem. Eng. Data,* **27,** 151–153 (1982).
11. "HITEC® Heat Transfer Salt; Properties, Uses, Storage and Handling," Tech. Brochure, E. I. du Pont de Nemours and Co., Wilmington, DE, 1976, 9 pp.
12. D. A. Nissen, "Thermophysical Properties of the Equimolar Mixture $NaNO_3$-KNO_3 from 300 to 600°C," *J. Chem. Eng. Data,* **27,** 269–273 (1982).
13. D. A. Nissen and D. E. Meeker, "Nitrate/Nitrite Chemistry in $NaNO_3$-KNO_3 Melts," *Inorg. Chem.*, **22,** 716–721 (1983).
14. R. W. Mar and C. M. Kramer, "Pressure-Temperature Composition Relationships for Heated Drawsalt Systems," *Solar Energy Mat.*, **5,** 71–79 (1981).
15. P. G. Zambonin and J. Jordan, "Redox Chemistry of the System O_2–O_2^-–O_2^{2-}–O^{2-} in Fused Salts," *J. Am. Chem. Soc.*, **91,** 2225 (1969).
16. G. S. Picard, T. A. Flament, and B. L. Tremillon, "Acidity and Thermochemical Stability of Molten Sodium Nitrate and Nitrite Mixture," *J. Electrochem. Soc.: Electrochem. Sci. Techn.*, **132,** 863–868 (1985).
17. S. H. White and U. M. Twardoch, "A Study of the Interactions of Molten Sodium Nitrate–Potassium Nitrate 50 mol % Mixture with Water Vapor and Carbon Dioxide in the Air," SAND80-8182, NTIS, 1981, 95 pp.
18. J. Alexander, Jr., and S. G. Hindin, "Phase Relations in Heat Transfer Systems," *Ind. Eng. Chem.*, **39,** 1044–1049 (1947).
19. R. J. Walter and B. J. Gerik, "Salt Corrosion Inhibitors," U.S. Patent US 4,534,794, Aug. 13, 1985.
20. "Olin Hitec® Heat Transfer Salt," Tech. Brochure, Olin Corp., Stamford, CT, 1984, 9 pp.
21. P. A. Rottenburg, "Heat Transfer Media for Use at Elevated Temperatures," *Trans. Inst. Chem. Eng.*, **35,** 21–36 (1957).
22. H. P. Voznick and V. W. Uhl, "Molten Salt for Heat Transfer," *Chem. Eng.*, 129–135 (1963).
23. E. G. Bohlmann, "Heat Transfer Salt for High Temperature Steam Generation," ORNL-TM-3777, NTIS, 1972, 42 pp.
24. J. Singh, *Heat Transfer Fluids and Systems for Process and Energy Applications,* Dekker, New York, 1985, chap. 6.
25. W. B. Johnson and W. M. Nagle, "Circulating HTS as a Pilot Plant Heat Transfer Medium," *Ind. Eng. Chem.*, **39,** 971–974 (1947).
26. H. W. Hoffman and S. I. Cohen, "Fused Salt Heat Transfer, Pt III: Forced-Convection Heat Transfer in Circular Tubes Containing the Salt Mixture $NaNO_2$-$NaNO_3$-KNO_3," ORNL-2433, unclassified, NTIS, 1960, 26 pp.

Part 6

Special-Purpose Materials

Chapter 1

Phase-Change Thermal Storage Materials

GEORGE A. LANE

The Dow Chemical Company
Central Research Laboratories
Inorganic Materials and Catalysis Laboratory

When energy use and availability are out of phase, energy storage is required. Thermal energy can be stored as sensible or latent heat. Latent, or phase-change material (PCM), energy storage is based on the heat absorbed or released when a material passes through a reversible phase transition, usually from the solid state to the liquid state.

1.1 ADVANTAGES OF PHASE-CHANGE MATERIALS (PCMs)

Latent-heat storage systems have the advantages, compared with sensible-heat storage systems such as rock beds or water tanks, of greater storage capacity per unit weight or volume, less control complexity, and greater system efficiency.

Latent-heat systems also utilize some sensible-heat storage above and below the PCM melting point. In the following equations, the first term represents the latent heat; the second term, the sensible heat below the melting point; and the third term, the sensible heat above the melting point. In practical PCM systems the latent-heat term predominates.

$$Q = ma_m \Delta h_m + \int_{T_i}^{T_m} mc_p \, dT + \int_{T_m}^{T_2} mc_p \, dT$$

$$= m[a_m \Delta h_m + \bar{c}_{p_s} (T_m - T_i) + \bar{c}_{pl} (T_2 - T_m)]$$

where
Q = quantity of heat stored
m = mass of heat storage medium
a_m = fraction melted
Δh_m = heat of fusion per unit mass
T = temperature
T_i = initial temperature
T_m = melting temperature
T_2 = final temperature
c_p = specific heat (isobaric)
$\bar{c}_{p_s}$ = average specific heat between T_i and T_m
$\bar{c}_{pl}$ = average specific heat between T_m and T_2

As an example, for a PCM melting at 30°C, with density of 1.6 g/cm^3, 188 kJ/kg (45 cal/g) heat of fusion, and 2.1 kJ/kg · K (0.5 cal/g · °C) specific heat operating over the range 20 to 40°C, the latent-heat term is 188 kJ/kg (45 cal/g), the sensible-heat contribution is 42 kJ/kg (10 cal/g) with 21 kJ/kg (5 cal/g) for the frozen PCM and 21 kJ/kg (5 cal/g) for the melted material, and the total storage capacity is 230 kJ/kg (55 cal/g) or 368 kJ/cm^3 (88 cal/cm^3).

In comparison, water, specific heat 4.2 kJ/kg · K (1.0 cal/g · °C) and density 1.0 g/cm^3, operating over this same 20°C range, has a capacity of 84 kJ/kg (20 cal/g), or 84 kJ/cm^3 (20 cal/cm^3). To equal the capacity of the same weight of PCM in the preceding example, water would have to swing from 20 to 75°C or 20 to 108°C to equal the volumetric capacity. A pressurized tank would be required in the latter case. In practice, a water tank about double the size of the PCM store would be used.

The comparison with rock beds is even more favorable. With a specific heat of 0.8 kJ/kg · K (0.2 cal/g · °C) and a density of 2.6 g/cm^3, the rock bed would have a storage capacity of only 17 kJ/kg (4 cal/g) or 42 kJ/cm^3 (10 cal/cm^3) from 20 to 40°C. A bed of the same weight as the PCM store would have to operate over the range 20 to 295°C to achieve equal capacity or over the range 20 to 196°C for volumetric parity. Since this is impractical, one would choose a rock bed about five times the size of the PCM unit.

Another fundamental advantage of PCM storage in systems using solar collectors, either air- or water-cooled, derives from the fact that as the operating temperature of a collector increases, the collection efficiency drops.[1] PCM systems maintain lower collector temperatures with improved efficiency.

When used as the heat source–heat sink with heat pumps, PCMs have another advantage. Since PCM storage is isothermal, a heat pump can be configured to operate at maximum efficiency at the PCM melting point. High values of coefficient of performance (COP) have been obtained for such systems.[2]

1.2 REQUIREMENTS FOR PCMs

Lane has listed the important criteria for selecting a practical PCM:[3]

1. Thermal properties
   - Suitable phase-transition temperature
   - High latent heat of transition
   - Good heat transfer
2. Physical properties
   - Favorable phase equilibria
   - Low vapor pressure
   - Small volume change
   - High density
3. Kinetic properties
   - No supercooling
   - Sufficient crystallization rate
4. Chemical properties
   - Long-term chemical stability
   - Compatibility with materials of construction
   - No toxicity
   - No fire hazard
   - No nuisance factor
5. Economics
   - Abundance
   - Availability
   - Cost-effectiveness

These factors are given different weights, of course, and the most important issues in research and development have been supercooling, phase-equilibrium behavior, materials of construction, and encapsulation technology.

Temperature

In selecting a PCM for a particular application, the operating temperature of the heating or cooling system must be matched to the transition temperature of the storage material. In this regard, PCMs are less versatile than sensible-heat storage materials; no single PCM will meet the diverse requirements of forced-hot-air, hydronic, air conditioning, domestic hot water, heat pump, and other systems. Thus, a family of PCMs with a spectrum of melting temperatures has been developed, so the system designer has a fairly good range of heat storage materials from which to choose.

In matching the system operating temperature with the PCM transition temperature, the PCM melting point must be chosen at a temperature interval above the operating temperature. This interval must be large enough to provide a sufficient temperature gradient for a satisfactory heat withdrawal rate. However, the temperature interval should not be excessive, since heat losses will increase and system efficiency will decrease. An interval of about 5 to 10°C has been employed for many systems.

Stability to Segregation

In the early years of PCM development, most of the salt hydrates studied were of the type that segregate during freezing and melting, with denser material settling to the bottom of the container, thus greatly reducing the heat storage capacity.

This segregation results from the phase equilibrium properties of the material and has been a considerable impediment to PCM heat storage development. Because segregation is critical, PCM phase-equilibrium studies have been very important and will be discussed more fully in a later section.

Density

Among the important physical properties for PCMs are the density and the change in density during phase transition. High density is desirable to allow a smaller size of storage container. However, lower hydrates, with fewer water molecules coordinated to the inorganic salt, are more dense than materials with higher hydration levels, yet they have lower heats of fusion, i.e., less storage capacity. The two factors must be balanced in choosing a storage material.

Density change during phase transition poses problems for PCM heat storage systems. Storage containers must be of sufficient strength to withstand the forces of expansion and contraction. Most salt hydrates are "normal," i.e., expand on melting and shrink on freezing. Water is atypical; it expands as it freezes. While a minimal density change during melting would seem to be desirable, this property is not compatible with other PCM properties of paramount importance.

The melting process consists of disruption of the crystal lattice of a compound, destroying the long-range order of the molecules. The energy needed for this disruption is the fusion energy, which is useful for heat storage. This process is necessarily associated with a change in density as the molecules become more disordered. The density change and the energy absorbed generally rise and fall together, although the relationship has many exceptions.

A notable exception is sodium sulfate decahydrate (Glauber's salt), which has very little density change during phase transition yet a very high heat of transition. The reason is that incongruent melting materials such as sodium sulfate decahydrate do not truly melt but rather decompose to the anhydrous salt and a saturated salt solution, with a combined density near that of the hydrate.

Because of the drawbacks of the segregating variety of PCM, it is usually preferable to choose a nonsegregating salt hydrate and deal with the effects of density change during melting.

Heat Storage Capacity

The heat of transition should be as high as possible, especially on a volumetric basis, to minimize the physical size of the heat store. However, other factors can be more important. For example, because incongruent melting materials lose storage capacity rapidly and are difficult to stabilize, it is always preferable to choose a stable, congruent melting PCM, even if the heat of fusion is somewhat lower.

Highly hydrated salts have large heats of transition because of their water content, but these are the most likely hydrates to form unstable incongruent or semicongruent melting materials. On balance, phase stability during melting and high density are more important than high heat of transition.

Phase-Transition Kinetics

Some salt hydrates that otherwise hold great promise as heat storage PCMs have inherently low crystallization rates, too low to provide a sufficient rate of heat withdrawal in a practical thermal storage system. While some methods of accelerating the crystallization process are known, usually little can be done to convert these salt hydrates into suitable PCMs.

A kinetics problem which has garnered much attention is supercooling. Most salt hydrates drop many degrees below the melting point before crystallization begins. When crystallization does begin, the temperature rises to the melting point until freezing is complete. Supercooling is undesirable for PCMs since it interferes with heat withdrawal. For most salt hydrate PCMs, supercooling has been mitigated by adding nucleators, chemicals to initiate the freezing process.

Chemical Stability and Safety

Besides the phase-segregation instability problems mentioned above, PCMs can suffer from degradation by loss or gain of water of hydration, chemical decomposition, or incompatibility with materials of construction.

To prevent alteration in degree of hydration in salt hydrate PCMs, the encapsulating material must be of sufficient thickness and impermeability to prevent in- or out-migration of water vapor. PCMs, especially organic materials, can also degrade by thermal decomposition, hydrolysis, or oxidation.

Inexpensive metal or plastic encapsulant materials are preferred for containing and protecting the PCM, for example, mild steel, aluminum, or polyethylene. Many organic PCMs swell plastic containers and exude from the surface, so metal encapsulants often are preferred.

While some salt hydrates are inert with mild steel, especially in the absence of oxygen, many others are incompatible with metal containers, so plastics are a popular encapsulant.

Because some PCM containers tend to leak, the PCM must be nontoxic, noncorrosive, nonflammable, and safe for ultimate disposal.

1.3 PHASE-EQUILIBRIUM PROPERTIES

Salt hydrates can be divided into two broad categories with subdivisions on the basis of their phase-equilibrium behavior:

1. Segregating hydrates
   - Semicongruent melting
   - Incongruent melting
2. Nonsegregating hydrates
   - Congruent melting
   - Quasi-congruent melting
   - Congruent isomorphous
   - Eutectic

Since segregation has caused a multitude of problems for PCM systems, and phase-segregation behavior is determined by which category the system falls in, it is important to employ these terms with technical precision.

Unfortunately, among the salt hydrates the segregating type is more prevalent than the nonsegregating variety, so the selection of stable materials is restricted. If a nonsegregating hydrate is available with the correct melting point and other properties, it should be selected in preference to a segregating material.

When a stable, nonsegregating hydrate is not available in the desired temperature range, one must turn to a semicongruent or, worse yet, an incongruent-melting material. These can be used only if stabilization (e.g., mechanical agitation, thickening, or gelling) is employed to combat phase separation.

Congruent Melting

Congruent-melting mixtures are those which, when melting or freezing at equilibrium, have exactly the same composition in the liquid and solid phases. An example is magnesium nitrate hexahydrate, $Mg(NO_3)_2 \cdot 6H_2O$, which melts at 89°C to form a liquid having a composition the same as that of the hexahydrate crystal. When this liquid is frozen, $Mg(NO_3)_2 \cdot 6H_2O$ crystals form. Since no anhydrous or lower hydrate species are formed, no chemical or physical segregation can occur.

Isomorphous Compositions

Crystals of the components of isomorphous mixtures are mutually soluble and form solid solutions. This can happen when lattice parameters of the components are nearly identical and the chemical makeup is similar. Isomorphous mixtures can be either incongruent or congruent melting. In the latter case, segregation does not occur during melting or freezing. An example of congruent melting is the minimum phase-transition point (15°C) composition of calcium chloride hexahydrate and calcium bromide hexahydrate.

Quasi-Congruent Melting

Quasi-congruent melting is a term coined to indicate materials which technically are semicongruent-melting (see below) but melt and freeze as if they were congruent-melting. In this case kinetics, not equilibrium, determines the behavior of the material, and segregation does not occur. These materials are always suspect since segregation could begin after a long period of desirable behavior. However, some materials have been proved effective by long testing and use; an example is magnesium chloride hexahydrate, $MgCl_2 \cdot 6H_2O$, which melts and freezes without chemical separation at 117°C.

Semicongruent Melting

For semicongruent melting materials, the liquid and solid phases in equilibrium during a phase transition are of differing composition because of conversion of the hydrate to a lower-hydrated material through loss of water. This conversion is termed a *peritectic reaction.*

An example of this behavior is disodium phosphate dodecahydrate, $Na_2HPO_4 \cdot 12H_2O$. When this crystalline material is heated to 35.2°C, a peritectic reaction occurs to yield crystalline heptahydrate. This releases 5 mol of water of hydration, dissolving some of the salt to form a saturated solution:

$$Na_2HPO_4 \cdot 12H_2O \longrightarrow (1 - x)\ Na_2HPO_4 \cdot 7H_2O + xNa_2HPO_4(aq)$$

Segregation now can occur because the heptahydrate is denser than the liquid phase and can sink to the bottom of the container. Once the heptahydrate has precipitated, the temperature must increase further to the melting point of $Na_2HPO_4 \cdot 7H_2O$ (48°C) before further melting occurs (another peritectic reaction, losing five additional moles of water). Unless stirred, these segregated phases tend to remain apart, drastically reducing the heat storage capacity as cycles of thawing and freezing continue.

Semicongruent melting PCMs are usually thickened or gelled in an effort to prevent physical separation of the lower hydrate from the liquid-phase saturated solution.

Incongruent Melting

The phase-transition behavior of incongruent melting salt hydrates is the same as that of semicongruent materials, except that the peritectic reaction forms an anhydrous salt rather than a lower hydrate.

Segregation tendencies are increased for this class of material because anhydrous salts have a higher density than do lower hydrates. A notable example of an incongruent-melting PCM is Glauber's salt, or sodium sulfate decahydrate. For this material at 32°C, the peritectic reaction

$$Na_2SO_4 \cdot 10H_2O \longrightarrow (1 - x)\ Na_2SO_4 + xNa_2SO_4(aq)$$

yields anhydrous sodium sulfate and a saturated solution of the salt. Precipitation of the anhydride leads to severe segregation of the PCM, worsened by the fact that aqueous sodium sulfate solutions have an inverse solubility-temperature relationship; that is, as temperature increases, the salt becomes less soluble.

Incongruent-melting PCMs also usually are thickened or gelled to impede the segregation, with varying success. They are more difficult to stabilize than are semicongruent-melting materials.

Eutectics

A *eutectic* is a minimum-melting composition of two or more components, each of which melts and freezes congruently, forming a mixture of the component crystals during crystallization.

Eutectics nearly always melt and freeze without segregation since they freeze to an intimate mixture of crystals, leaving little opportunity for the components to separate. On melting both components liquefy simultaneously, again with separation unlikely.

An example of a eutectic PCM is magnesium chloride hexahydrate-magnesium nitrate hexahydrate, $MgCl_2 \cdot 6H_2O$-$Mg(NO_3)_2 \cdot 6H_2O$, which melts at 58°C.

Some segregating PCM compositions have sometimes been incorrectly called *eutectics,* since they are minimum-melting. Because one of the components undergoes a peritectic reaction during phase transition, however, they should more properly be termed *peritectics*. Examples are the mixtures of Glauber's salt with various chlorides, e.g., $Na_2SO_4 \cdot 10H_2O$-$NaCl$, with phase transition at 18°C.

1.4 THICKENERS AND GELLANTS FOR PCMs

Numerous attempts have been made to stabilize segregating PCMs by thickening or gelling. In many cases success has been poor. In general, thickening has only delayed the segregation processes, and gelling has been more successful when a tight matrix structure has been generated. Proof of stability should be demanded for this type of PCM, including not only thousands of freeze-thaw cycles in the laboratory but also simulations more closely representing field-use conditions.

Mineral Thickeners

The effectiveness of a mineral powder in thickening liquids depends strongly on the size and shape of its particles and the ionic charges on their surfaces. Long, needle-like particles are usually best, and flat, thin platelets are also often effective. Particles approaching a spherical shape are generally least effective. Mineral thickeners act by interfering with liquid flow by forming a network of solid particles. High-aspect-ratio particles and ionic surface charge are helpful in forming and maintaining such networks.

One of the first uses of mineral powder for PCM thickening was the development of attapulgite clay-thickened Glauber's salt by Dr. Maria Telkes.[4] This mineral consists of lathlike particles approximately $1 \times 0.1 \times 0.005$ to $0.010\ \mu m$. This PCM has been studied, tested, and used for more than a decade, yet there is still controversy regarding its stability.

Another mineral-thickened PCM is Chliarolithe, developed in France[5] and available commercially in Europe.[6] This product consists of $CaCl_2 \cdot 6H_2O$, thickened with diatomaceous earth. The inventors studied the thickening efficacy of different diatoms, silica-bearing skeletons of microscopic animals. They found needle-shaped varieties to be best, followed by cylindrical species. Disk-shaped particles were found to be worst.

Polymeric Thickeners and Gellants

Long-chain polymers can be used to thicken liquids if the polymer is soluble. In solution, the chains are swollen and dispersed, acting like long, randomly oriented "ropes" in the liquid, impeding flow of the material.

Many attempts have been made to thicken PCMs with polymers, e.g., cellulosics, even peat moss. Most of these efforts have been unsuccessful because the solubility of polymers decreases as the ionic strength of a solution increases. In concentrated salt solutions, i.e., salt hydrates, few polymers are sufficiently soluble to provide sufficient thickening through increased viscosity.

Polymers with shorter chains have improved water solubility but thicken poorly. They can be used successfully if the polymer chains are cross-linked after mixing, thus extending their length greatly. The result is termed a *hydrogel.* Products based on this technology are available by import from Europe.[7]

Hydrophilic-Lipophilic Gelation

More recently, improved gelled PCMs have been developed which rely on long, thin, cylindrical micelles to thicken the liquid.[8] The micelles are formed by adding surfactants to the melted salt hydrate PCM. Ordinarily, surface-active agents form spherical micelles, but by matching the hydrophilic-lipophilic balance of the surfactant with the ionic charge density of the salt hydrate, cylindrical micelles can be formed.

An advantage of this type of thickening is that the solutions have pseudo-plastic rheology; that is, they thin when sheared and can be pumped or poured readily from one container to another. Filling of PCM capsules is easier, and after filling, the material becomes gelled again.

1.5 NUCLEATION AND CRYSTALLIZATION

The processes of nucleation (i.e., the initiation of crystal formation) and crystallization (i.e., crystal growth) are of great importance in PCM technology. Nearly all salt hydrates supercool before they freeze; that is, when they are cooled, the temperature drops below the equilibrium melting-freezing temperature before freezing begins. Once crystallization has been initiated, the temperature rebounds to the equilibrium melting-freezing temperature and is maintained until freezing is complete.

Supercooling

A well-behaved PCM must display very little supercooling—ideally, none. There has been much effort to eliminate supercooling in salt hydrate PCMs. Supercooling is a natural phenomenon in most materials, and salt hydrates are especially prone to it.

It is important in heat storage systems to reduce supercooling to a minimum since it interferes with the ability to withdraw stored heat. For example, if one attempts to withdraw heat from a PCM store melting at 27°C, using air heat-transfer fluid at 22°C, more than 5°C supercooling will not allow any crystallization and thus no heat withdrawal.

Various mechanical means of providing nucleation have been used. For example, a "cold finger," held below the freezing point, maintains a constant supply of PCM crystals, which act as initiation centers for crystal growth. Most PCMs, however, contain additives to minimize supercooling by acting as initiation sites for crystallization.

Nucleators

Crystal growth begins at tiny centers where nucleation occurs. Without the presence of such sites, pure salt hydrates supercool to 50 to 100°C below their melting points. Few materials display this extent of supercooling since spots on the container wall, tiny particles of dust, or impurities can serve as nucleation centers, often reducing the amount of supercooling to about 10 to 20°C. Also, supercooling tends to be lessened as the container size increases since there is a higher probability of incidental nucleating centers being present. For small, encapsulated PCM masses, supercooling can be a significant problem.

To reduce supercooling to less than a few degrees, it is necessary to use additives that have crystal lattice structure features that initiate crystallization. These features are such that the nucleator crystal lattice is similar to that of the PCM crystal.

These similarities can be divided into three types: isomorphous, isotypic, and epitaxial. Isomorphous and isotypic nucleators are isostructural with the PCM, crystallizing in the same crystal system, with similar lattice dimensions. Generally, good nucleators have dimensions within about 15 percent of those of the PCM. Isomorphous nucleators have the added feature of being related chemically to the PCM, sufficiently similar that the crystals are mutually soluble, forming solid solutions over some composition range.

Epitaxial nucleators do not have the same crystal structure

as the PCM. Presumably, however, there is some crystal lattice feature with affinity for the PCM, so that crystallization is initiated. In retrospect, it is sometimes possible to explain the nucleation activity of an epitaxial nucleator. More often, it remains a mystery. Prediction of epitaxial behavior is rare, confined to simple cases where the distances between crystal planes is a small multiple of those of the PCM. Epitaxial nucleators are most often found by trial-and-error testing of a large number of candidates, usually chosen by intuition.

Isostructural nucleators lend themselves to a more scientific approach. By finding in the literature materials with crystal structures that match that of the PCM, narrowing the list to those compounds that appear practical, and testing the candidates for their effect on supercooling, scientists have discovered many effective nucleators. Simple coincidence of crystal structure, however, is no guarantee of nucleation ability, and more candidates fail than succeed. Many other factors are of critical importance, such as degree of solubility, temperature stability, degree of hydration, and chemical interactions.

A classic study in PCM nucleation was the work of Telkes[9] in discovering that sodium tetraborate decahydrate is a practical nucleator for sodium sulfate decahydrate. Since then (1976) many investigators have discovered new nucleator-PCM compositions. Lane and Rossow have patented a number of nucleation additives for calcium chloride hexahydrate,[10] magnesium nitrate hexahydrate,[11] and magnesium chloride hexahydrate.[12] PCM nucleation has also been studied in the Soviet Union, Canada, the Federal German Republic, and Japan, where Kai et al.[13] have been especially active.

Crystal Growth

In order to be a workable PCM, a salt hydrate, once nucleated, must crystallize at a rate sufficient to release energy fast enough to provide the necessary heat input to the system. Most salt hydrates are adequate in this regard, although there are some notable failures, such as calcium nitrate tetrahydrate.

If crystallization proves to be too slow, such as when the hydrate melt is highly viscous, preventing ions from moving to the crystallizing surface, a viscosity reduction additive might be helpful. Most often, however, the material must be abandoned as a PCM.

1.6 PCM ENCAPSULATION

Nearly all PCMs, including salt hydrate PCMs, must be encapsulated to be of practical use in energy storage. The discrete masses of PCM that are surrounded by encapsulation media can range from a hefty chunk weighing a ton or more, contained in a large tank with heat exchanger, down to particles a few micrometers in diameter.

PCM encapsulation is more complicated than is immediately apparent, since the capsule serves not only as a container for the melted PCM but also as a heat-transfer surface, vapor barrier, and structural member and often as a decorative element. Encapsulants thus have rather stringent requirements, and more than one commercial entry has foundered, not because the PCM was inadequate, but because the package was improperly designed.[14]

Requirements

It is essential that the encapsulant have the required physical properties, the PCM-encapsulant combination be compatible and stable, and the package possess the requisite heat-transfer capabilities.

Metal containers are often used for encapsulation because they seal easily, have good physical strength, possess good heat-transfer properties, and are readily available at a low cost. However, since some salt hydrate-metal combinations tend to corrode, a metal encapsulant should be chosen carefully in such systems. Aluminum is not suitable with most salt hydrates. Usually, mild steel can be used if air is excluded from the interior, since oxygen accelerates the attack of many salts on steel. Some nitrate salts cause stress crack corrosion in mild steel. Stainless steel cannot be used with chloride salts, again because of stress crack corrosion.

Organic PCMs are usually compatible with metal encapsulants; however, some are not. For example, fatty acid PCMs attack both aluminum and steel.

Plastic materials are used most often with salt hydrate PCMs, but embrittlement has been encountered in some cases. Plastic materials frequently encounter compatibility problems, softening and swelling, with organic PCMs.[15]

The most important physical properties of the encapsulant are container strength and flexibility, temperature resistance, ultraviolet (UV) stability, and vapor barrier properties.

Most PCMs undergo a substantial volume increase during melting and shrinkage during freezing. This causes pressure changes in the air space above the PCM in the sealed container, imposing a stress on the encapsulant. Also, localized, unrelieved volume increases frequently occur during melting, causing further stress. Metal containers must be of sufficient thickness and strength to withstand these stresses. Plastic encapsulants accommodate these stresses through a combination of strength and flexibility.

Above about 100 to 140°C, depending on the material, plastics soften and lose strength, so they are seldom used for PCMs melting above approximately 70 to 90°C. Metals become the encapsulants of choice.

Vapor barrier properties are of utmost importance for salt hydrates or any other PCMs that have volatile components. Metal containers have excellent barrier properties. Plastics vary greatly in resistance to vapor passage, and this resistance decreases rapidly as the temperature increases. To contain the water vapor of a salt hydrate, plastic containers usually must be somewhat thicker than strength requirements alone would dictate.

If plastic encapsulants are exposed directly to sunlight in operation, the material must be stable to UV radiation. This usually is accomplished by using a plastic containing a stabilizer such as carbon black.

The encapsulant material serves both as the PCM containment and the heat-transfer surface. Thus it is essential that the container be designed to ensure the needed heat-transfer rate in both the heat storage and heat withdrawal modes.

When air is used as the heat-transfer medium, the heat conductivity of the encapsulant is not critical since the limiting factors in heat transfer are the air film coefficient and the air heat content. With a liquid heat-transfer medium, the lower conductivity of plastics becomes significant.

All factors must be considered in designing for heat transfer, including the effect of solid buildup during heat withdrawal and convection during heat storage.

Bulk Storage in Tanks

When the PCM is contained in a single large mass, as in a tank, the tank itself rarely is the heat-transfer surface; rather, a heat exchanger is immersed in the PCM. While this is analogous to a heat storage water tank, there are important differences. Since a PCM tank stores much more heat in a given volume, it must also contain proportionately more heat-transfer surface to provide a given rate of heat withdrawal. In addition, solid buildup on the heat exchanger further reduces the heat-transfer rate, requiring more surface area.

Whereas water storage tanks are shipped empty to the site, installed, and then filled, PCM tanks are best filled at the factory and shipped full. They must be sufficiently rugged to withstand the additional stresses of this treatment. In spite of the extra complexity and cost of the PCM tank, the space conservation has proved to be well worth the price, particularly in commercial installations.

To increase the heat transfer rate, the technique of direct-contact heat transfer has been applied to bulk tank storage. While this involves increased mechanical complexity, the size of the heat exchanger is minimized.[16]

Macroencapsulation

The size of PCM macrocapsules varies greatly, from plastic pipes several inches in diameter and several feet long to shapes the size of an aspirin tablet. In contrast to the bulk storage tank, the PCM macrocapsule incorporates no moving mechanical equipment and uses its own surface for the heat-transfer process. Macroencapsulated systems can use either liquid or air for heat transfer. Macrocapsules are quite amenable to mass production and shipping.

The encapsulant material can be either plastic or metal, and development of plastic film packages has also been attempted.[17] If they meet the criteria outlined previously, common packaging materials such as bottles, cans, and film packages are recommended for macroencapsulated systems. In some cases, these capsules are designed to be incorporated into building structural materials.

Over the last several years, numerous macroencapsulated PCM products have come on the market. The shape and size of these have varied widely. Some of the earliest products were flat, shallow plastic trays[18] or flat plastic film containers.[19] These packages were designed for use with incongruent melting Glauber's salt, and the designers hoped that the small vertical dimension would decrease segregation and facilitate rehydration.

Congruent melting PCMs were first marketed in plastic pipes several inches in diameter and several feet long, filled with $CaCl_2$-based PCM and hermetically sealed.[20] The same PCM soon was marketed in sealed "tin cans."[21]

A further development of the technology was the introduction of steel or plastic panels filled with $CaCl_2$-based PCM. Some of these panels were designed to be fitted between the studs of a house wall, with glazing, functioning as a solar collector.[22] One such product is fabricated from translucent plastic sheets, allowing its use as a glazing material.[23]

Most of these macroencapsulated products were designed for use in arrays in either direct-gain solar systems or centrally located heat storage boxes. Another class of product is the plastic sphere, filled with PCM and sealed, intended for use in large numbers in heat storage tanks.[24]

Home builders prefer to install modular units rather than build up heat storage devices at the site. An important development, therefore, has been the design of window- or wall-type units incorporating encapsulated PCM.[25] A few are on the market, and there is a need for other innovative designs.

Microencapsulation

Microencapsulated PCM systems are characterized by small, individual particles of PCM, surrounded by encapsulant. In most cases, the encapsulant forms a continuous web throughout the PCM mass. Ideally, the encapsulant matrix functions as the physical containment, moisture barrier, and heat-transfer surface.

Microencapsulation is a difficult technology but is promising because it makes it possible to fabricate structural, functional, or decorative materials directly out of PCM.[26] Few microencapsulated systems are on the market. A cement-encapsulated Glauber's salt product is available, further encapsulated in a stainless-steel pipe.[27]

1.7 AVAILABLE COMMERCIAL PCMs

As stated previously, a selection of PCMs is needed to meet the variety of applications for energy storage. The melting point of the PCM is matched with the needs of the application. This discussion of commercially available PCMs is organized, therefore, by the temperature range.

Cold Storage for Freezing

A long-established technology is the use of PCMs which freeze below the melting point of ice in cooling plates for cold storage applications. The PCMs are usually eutectic salt-ice mixtures, and the compositions are closely guarded by the manufacturer. These mixtures seldom supercool, so there is little need for nucleators. Since the compositions usually are eutectics, they do not segregate.

Ice as a PCM

The use of frozen water to store energy, in this case "coldness," is an ancient practice. In not too distant times, a substantial industry was involved in cutting ice from lakes and rivers in the winter and storing it in "ice houses" for summer sale.

The simplest form of present-day ice storage unit uses a refrigerant coil to freeze water in a tank. Freezing is terminated well before the tank is entirely frozen. The stored coldness is recovered by circulating water into and through the tank.

Another type is the "ice shucker," which consists of refrigerated plates arranged vertically above an ice storage bin. Water is frozen on the plates to a certain thickness and then caused to drop off by a short melting cycle. Chilled water is circulated through the bin to recover the stored coldness as required.

Yet another commercial product uses as a heat exchanger a tightly wound mat of small-diameter plastic tubing, rolled up and positioned in a water tank. Chilled ethylene glycol solution is pumped through the heat exchanger to freeze the water. Coldness is recovered from the tank by circulating the glycol solution in reverse through the heat exchanger.

A disadvantage in the use of ice as a PCM is that it melts at a lower temperature than necessary to provide building cooling and dehumidification. Refrigeration equipment is stressed to expend more energy than ideally necessary.

Coolness Storage

Although PCMs which melt somewhat higher than water (5 to 15°C) are more expensive coolness storage media than ice, the savings in operating at a higher temperature usually gives them an advantage. Materials providing storage at about 5 to 8°C are desired for cooling and dehumidification, while those melting at about 8 to 15°C provide only cooling.

In a typical commercial building, cooling is usually needed in some portions of the building, even in winter. The equipment is similar to the tank with heat-exchange coils described above for ice. The PCM is frozen during off-peak hours, and coolness is withdrawn as needed during the day.

The cooling equipment can be operated in either a full- or partial-storage mode. In the first case, the entire cooling demand is generated during nighttime hours and stored and the refrigeration equipment shut down during peak hours.

There is a choice between two operating strategies in the partial storage mode, and the most economical one can be selected. If load leveling is the objective, the refrigeration equipment is operated around the clock at capacity and storage is used to supply varying demand. If utility demand charges are high, demand limiting may be the objective and the system is operated to reduce the peak load.

The choice of coolness storage PCMs is rather limited, particularly in the low-temperature portion of the range. In Table 1-1, only one PCM is listed in the 5 to 8°C range and only two between 8 and 15°C. The Glauber's salt-based material melting at 8°C is incongruent-melting. If thickened with mineral powder to retard segregation, the long-term stability of this PCM is still not proved. A reportedly stable gelled composition based on a hydrogel is available.

Table 1.1. Commercially Available Phase Change Materials

PCM	Melting point, °C	Heat of fusion, cal/g	Density, liquid, g/cm³
Coolness storage			
$Na_2SO_4 \cdot 10H_2O$-NH_4Cl-KCl	8	29	1.49 gel
PE Glycol	11	24	1.12
$CaCl_2$-$CaBr_2$-$6H_2O$-KBr	12	32	1.78 gel
Heat buffers and sinks			
$CaCl_2$-$CaBr_2$-$6H_2O$	15–34	34	1.78 gel
$Na_2SO_4 \cdot 10H_2O$-NaCl	18	35	1.48 gel
$CaCl_2 \cdot 6H_2O$	27	46	1.56
Moderate temperature			
$Na_2SO_4 \cdot 10H_2O$	32	60	1.48 solid
$CaBr_2 \cdot 6H_2O$	34	28	1.96
Neopentyl glycol	43	31	. . .
$Na_2S_2O_3 \cdot 5H_2O$	48	48	1.67
Paraffin wax	50	64	0.76
Intermediate temperature			
$MgCl_2$-$Mg(NO_3)_2$-$6H_2O$	58	32	1.52
$NaCO_2CH_3 \cdot 3H_2O$	58	54	1.28
$Na_4P_2O_7 \cdot 10H_2O$	70	44	1.80 gel
$Mg(NO_3)_2 \cdot 6H_2O$	89	39	1.55
$NH_4Al(SO_4)_2 \cdot 12H_2O$	95	64	1.65 solid
High temperature			
$MgCl_2 \cdot 6H_2O$	114	40	1.45
Polyethylene	132	48	0.96 solid

CONVERSION FACTOR: 1 cal/g = 4.181 kJ/kg.

Polyethylene glycols of varying chain lengths are available with several melting point ranges, e.g., approximately 11 and 23°C. Since these are organic PCMs, they suffer from low density and low thermal conductivity.

The $CaCl_2 \cdot 6H_2O$-$CaBr_2 \cdot 6H_2O$-KBr composition is an isomorphous minimum melting mixture which is congruent-melting. It is gelled with a hydrophilic-lipophilic system, however, since normal manufacturing tolerances can produce compositions which segregate. Its high density is an advantage.

Heat Buffers and Heat Sinks

Near room temperature (15 to 30°C) there are several applications in which the PCM buffers diurnal swings in temperature, provides heat for spaces kept below ordinary room temperature, or acts as a heat source–heat sink in heat pump systems.

Mixtures of $CaCl_2 \cdot 6H_2O$ and $CaBr_2 \cdot 6H_2O$ are mutually soluble in all proportions and form solid solutions. By adjusting the composition, one can obtain melting points ranging from the minimum melting point of 15°C up to the melting point of $CaBr_2 \cdot 6H_2O$ (34°C). However, only the two extremes are congruent melting. There are available 16 and 20°C compositions, gelled by the hydrophilic-lipophilic technique.

These materials, as well as others based on Glauber's salt, have been tested in systems for heating greenhouses. The objective is to prevent daytime heat buildup by storing excess solar heat and using it to keep the nighttime temperature from dropping into the range where vegetables, berries, and bedding plants are damaged.

The Glauber's salt-based material, $Na_2SO_4 \cdot 10H_2O/NaCl$, melts incongruently at 18°C and segregates poorly. Clay-thickened material is more stable, and a hydrogel-stabilized composition reportedly is stable.

The 27°C congruent-melting PCM based on phase equilibrium modified $CaCl_2 \cdot 6H_2O$ is probably the most widely used PCM, excluding ice. It is used for passive solar panels, attached residential greenhouses, central solar heat storage, and heat pump systems. In addition to its principal use in heat storage panels and other macroencapsulated products, it has proved effective in commercial installations having multiple water-sourced heat pumps, replacing water storage tanks.

Moderate Temperatures

In the 30 to 50°C temperature range, traditionally considered for hot-air heating systems, there are a number of PCMs, but none is completely satisfactory. Glauber's salt, $Na_2SO_4 \cdot 10H_2O$, has been the subject of far more research, development, and testing than any other PCM. It is incongruent-melting, with phase transition at 32°C, and segregates badly. The clay-thickened material is more stable, but studies reveal eventual segregation.[28] The material gelled with polymeric hydrogel is reportedly stable.

Calcium bromide hexahydrate melts congruently at 34°C, and its high density is favorable. However, it is more expensive than chloride or sulfate salt hydrates, and its heat of fusion is lower.

Neopentyl glycol is interesting. It undergoes a solid-to-solid transition at 43°C, rather than melting to a liquid. Presumably, this might allow encapsulation in a lighter-weight package or with encapsulating material of less strength, thus decreasing costs. The PCM itself is quite costly, however. In common with paraffin wax, it suffers the disadvantages previously mentioned for organic PCMs.

Hypo-$Na_2S_2O_3 \cdot 5H_2O$, is semicongruent-melting, with phase transition at 48°C. It is subject to segregation. A stable hydrogel product is available, however. The cost of this PCM is higher than for the chlorides and sulfates.

Intermediate Temperatures

PCMs in the 50 to 100°C temperature range have been proposed for water heating and off-peak electric heater applications.

The eutectic of $MgCl_2 \cdot 6H_2O$ and $Mg(NO_3)_2 \cdot 6H_2O$, which melts at 58°C, is well behaved and should serve as an excellent heat storage material for hydronic heating or domestic hot-water systems.

Sodium acetate trihydrate, $NaCO_2CH_3 \cdot 3H_2O$, is incongruent melting at 58°C. It was used in the nineteenth century as a PCM in France to heat railway coaches. Segregation problems have hindered its applications until recently, when it became available as a hydrogel-stabilized material.

Sodium pyrophosphate decahydrate, $Na_4P_2O_7 \cdot 10H_2O$, has also recently been developed as a hydrogel-thickened material. Its phase transition is at 70°C and is incongruent.

Magnesium nitrate hexahydrate, $Mg(NO_3)_2 \cdot 6H_2O$, melts congruently at 89°C. It is a candidate for off-peak electric heater systems and also for laundry or dishwashing applications.

Ammonium alum, $NH_4Al(SO_4)_2 \cdot 12H_2O$, which melts congruently at 95°C, has been studied for water-heating systems.

High-Temperature PCMs

In the range above 100°C the main application for building heating-cooling is hot-side thermal storage for absorption air conditioning.

At higher temperatures, the use of PCM heat storage for industrial processing and solar electric power applications is an interesting and attractive field. Fused-salt PCMs are available with melting points ranging up to several hundred degrees.[29–31]

Magnesium chloride hexahydrate, $MgCl_2 \cdot 6H_2O$, is a quasi-congruent melting PCM with phase transition at 114°C and does not segregate. It has been proposed for use with off-peak electric heaters and absorption air conditioners.

Cross-linked polyethylene is an interesting and promising material, which is form-stable; i.e., it maintains its shape above the crystalline to amorphous polymer transition at about 132°C. Thus it can be used without encapsulation in most applications. It has been suggested for use with absorption air conditioners, as well as in various industrial applications.

1.8 SUMMARY

At first glance, the development of heat storage PCMs seems deceptively simple. For more than a decade since the first energy crisis, however, a large number of laboratories around the world have applied their skills to the problem. With the application of basic chemical and physical principles, nucleators have been discovered to mitigate supercooling, phase-equilibrium studies have elucidated segregation behavior and in more than one case used to change a segregating material to a stable PCM, salt hydrates have been gelled to prevent gross separation, phase-change heat-transfer technology has been applied to the systems, and workable encapsulation systems have been designed and tested.

As a result, a family of PCMs has been developed and put on the market, a variety of encapsulated PCM products is commercially available, and modular units for housing construction are being designed.

REFERENCES

1. H. C. Hottel and B. B. Woertz, "The Performance of Flat-Plate Solar-Heat Collectors," *Am. Soc. Mech. Eng. Trans.*, **64,** 91 (1942).
2. D. A. DeNoyer, 'Trent Project," Final Report, Report ERL Project 380080.07A, Consumers Power Company, Lansing, Mich., 1981.
3. G. A. Lane, *Solar Heat Storage: Latent Heat Materials,* vol. I, *Background and Scientific Principles,* CRC Press, Boca Raton, Fla., 1983.
4. M. Telkes, U.S. Patent 3,986,969, 1976.
5. J. D. Silvain, M. Schneider, X. Berger, A. Jaffrin, and L. Bourdeas, European Patent Application 19,573, 1980.
6. Solvay & Cie., "Pour la Stockage Thermique à Basse Tempér-ature Casotherm 281," Paris, 1983.
7. P. J. C. Kent and J. K. R. Page, U.S. Patent 4,209,413, 1980.
8. G. A. Lane, A. S. Teot, and H. E. Rossow, U.S. Patent 4,585,572, 1986.
9. M. Telkes, "Nucleation of Supersaturated Inorganic Salt Solutions," *Ind. Eng. chem.*, **44,** 1308 (1952).
10. G. A. Lane and H. E. Rossow, U.S. Patent 4,272,390, 1981.
11. G. A. Lane and H. E. Rossow, U.S. Patents 4,272,391, 1981; 4,283,298, 1981; 4,272,392, 1981; 4,271,029, 1981; 4,273,666, 1981; 4,329,242, 1982; and 4,402,846, 1983.
12. G. A. Lane and H. E. Rossow, U.S. Patents 4,338,208, 1982 and 4,406,805, 1983.
13. J. Kai, H. Kimura, and K. Mutoh, Japanese Patent Kokai, 51-70193, 1976.
14. G. A. Lane, *Solar Heat Storage: Latent Heat Materials,* vol. II, *Technology,* CRC Press, Boca Raton, Fla., 1986.
15. G. A. Lane, A. C. Kott, G. L. Warner, P. B. Hartwick, and H. E. Rossow, "Macro-Encapsulation of Heat Storage Phase Change Materials for Use in Residential Buildings," Report ORO/5217/8, The Dow Chemical Co., Midland, Mich., 1978.
16. J. D. Wright, M. S. Bohm, and R. S. Barlow, "Oil/Salt Hydrate Direct Contact Heat Exchange Experiments," Report TR-252-1614, Solar Energy Research Institute, Golden, Colo., 1983.
17. G. R. Frysinger, "Life and Stability Testing of Packaged Low-Cost Energy Storage Materials for Use in Residential Buildings," Report ORNL/Sub-7585/1, U.S. Department of Energy, Washington, D.C., 1980.
18. D. Humphries, "Saskatchewan Minerals to Construct Solar Heat Storage Tray Plant," *Toronto Globe,* September 26, 1980.
19. J. Kohler and D. Lewis, "Phase Change Products for Passive Homes," *Solar Age,* **8,** 65 (1983).
20. P. B. Hartwick, "Calcium Chloride Hexahydrate Compact Thermal Energy Storage in a Small Greenhouse," paper presented at Great Lakes Solar Greenhouse Conference, Midland, Mich., 1978.
21. "How to Reduce Your Heat Bills," Texxor Corporation, Omaha, Nebr., 1979.
22. P. J. Moses, M. L. Bennett, J. E. Kosinski, D. E. Snyder, and J. A. Wild, "Application of the Enerphase™ Thermal Energy

Storage Panel," *Proc. 9th Natl. Passive Solar Conf.*, Columbus, Ohio, September 24–26, 1984.
23. A. V. Sedrick, "Translucent Phase Change Material Thermal Storage System," *Proc. Am. Sect. Int. Sol. Energy Soc.*, **5**(2), 980 (1980).
24. J. Patry, "Stockage de l'energie la chaleur latente ou l'avenir au présent," *Equip-Mag,* **817,** 109 (1981).
25. K. P. Zilisch, U.S. Patent 4,446,850, 1984.
26. J. S. Best and W. J. McMillan, U.S. Patent 4,003,426, 1977.
27. B. J. Boardman, U.S. Patent 4,277,357, 1981.
28. S. B. Marks, "Calorimetric Testing of Improved Glauber's Salt Phase Change Thermal Energy Storage Material," *Proc. 3d Int. Conf. Alternative Energy,* Miami, December 1980.
29. G. J. Ganz, C. B. Allen, J. R. Downey, Jr., and R. P. T. Tomkins, "Physical Properties Data Compilations Relative to Energy Storage. I. Molten Salts: Eutectic Data," Report NSRDS-NBS 61, Part I, 1978.
30. G. J. Ganz, C. B. Allen, N. P. Bansal, R. M. Murphy, and R. P. T. Tomkins, "Physical Properties Data Compilations Relative to Energy Storage. I. Molten Salts: Data on Single and Multi-Component Salt Systems," Report NSRDS-NBS 61, Part II, 1979.
31. G. J. Ganz and R. P. T. Tomkins, "Physical Properties Data Compilations Relative to Energy Storage. IV. Molten Salts: Data on Additional Single and Multi-Component Salt Systems," Report NSRDS-NBS 61, Part IV, 1981.

Chapter 2

Fusible Alloys

WILLIAM B. HAMPSHIRE
Tin Research Institute, Inc.

2.1 INTRODUCTION

The fusible alloys are a series of metal combinations characterized primarily by their low melting points, ranging from about 200°C down to room temperature, or even below. Although these metals are soft in comparison to the higher-melting metals, they possess greater strength, ductility, and toughness than do other low-temperature materials. The fusible alloys also possess other valuable metallic characteristics, such as good electrical and thermal conductivity, corrosion resistance, the ability to be remelted and recast, and low vapor pressure.

A unique property of certain fusible alloys is that they expand when they solidify, whereas most materials contract. This expansion facilitates the filling of small details in casting. In other applications the composition may be selected to give near-zero casting contraction, allowing very accurate replication of a surface.

2.2 ALLOY COMPOSITION

The fusible alloys are usually comprised of bismuth, tin, and lead, with possible additions of cadmium, indium, antimony, or gallium. The melting points of some of these are as follows, although the properties of the pure metals are not of primary importance:

Metal	Melting point, °C	Melting point, °F
Lead	327	621
Cadmium	321	610
Bismuth	271	520
Tin	232	449
Indium	156	314

In general, these metals, when combined to produce alloys, form simple eutectic systems. By definition, a certain composition called the *eutectic* composition has a single melting temperature, i.e. a melting point, below that of the pure metals that comprise the alloy. Thus the eutectic behaves like a pure metal in that it changes from completely liquid to completely solid (or vice versa) at a single temperature, its melting point.

Table 2.1 shows the more important fusible alloy compositions. Although the eutectic compositions are important, in many traditional applications it is necessary only to be near the eutectic composition. Therefore, several of the traditional alloys simply represent convenient proportions. For example, Wood's metal contains 4:2:1:1 proportions of bismuth, lead, tin, and cadmium, respectively, and Lipowitz's metal contains 15:8:4:3 parts of these metals. In either case the melting range is no more than 5°F (3°C), which is rather small in practical terms.

In recent years the fusible alloys have been used in more sophisticated applications; thus pressure is mounting for standardization of the compositions. A national standard from the American Society for Testing and Materials (ASTM) will probably concentrate on the true eutectic compositions (which may be more readily formulated with today's improved and faster analytical techniques) and may further expand the applications of these alloys. The Unified Numbering Systems are being used more often and may eventually supplant the current convention, which is to refer to the alloys by melting point, such as the 136°F alloy or the 117°F alloy.

2.3 ALLOY PROPERTIES

Mechanical Properties

Previous applications of the fusible alloys relied more heavily on the melting point than on any strength value, so there was relatively little interest in mechanical properties. This situation was further complicated by the lack of a national standard and the inherent nature of the alloys. This latter factor relates to the fact that room temperature is relatively close to the melting points of the alloys and, consequently, strength values determined at room temperature depend greatly on the rate of testing, as creep effects will have more influence on slower tests. The relative softness of these alloys renders their strength values sensitive to casting and cooling variations, and these alloys tend to age over a period of weeks, thus further complicating strength studies.

With these limitations in mind, Table 2.2 presents some mechanical and physical properties for selected fusible alloys. These values should be considered approximate.

Dimensional Properties

As mentioned above, certain fusible alloys, and specifically those containing more than about 55 percent bismuth, will

Table 2.1. Fusible Alloy Compositions

	Composition, wt %					Eutectic (or solidus) temp.	
Alloy	Sn	Bi	Pb	Cd	Others	°C	°F
Sn-Ag eutectic	96.5	. . .	. . .	. . .	3.5 Ag	221	430
Sn-Zn eutectic	92.0	. . .	. . .	. . .	8.0 Zn	199	390
Sn-Pb eutectic	62.0	. . .	38.0	. . .	. . . . .	183	361
Sn-Cd eutectic	67.0	. . .	. . .	33.0	. . . . .	176	349
293° alloy	51.2	. . .	30.6	18.2	. . . . .	145	293
Bi-Cd eutectic	. . .	60.0	. . .	40.0	. . . . .	144	291
Bi-Sn eutectic	43.0	57.0	. . .	. . .	. . . . .	138	281
266° alloy	40.0	56.0	. . .	. . .	4.0 Zn	130	266
Bi-Pb eutectic	. . .	55.5	44.5	. . .	. . . . .	124	255
In-Sn eutectic	48.0	. . .	. . .	. . .	52.0 In	117	243
217° alloy	25.9	53.9	. . .	20.2	. . . . .	103	217
D'Arcet's metal	25.0	50.0	25.0	. . .	. . . . .	96	205*
Onion's metal	20.0	50.0	30.0	. . .	. . . . .	96	205†
205° alloy	15.5	52.5	32.0	. . .	. . . . .	96	205
198° alloy	. . .	51.7	40.2	8.1	. . . . .	92	198
Lipowitz's metal	13.3	50.0	26.7	10.0	. . . . .	70	158‡
Wood's metal	12.5	50.0	25.0	12.5	. . . . .	70	158§
158° alloy	13.1	49.5	27.3	10.1	. . . . .	70	158
136° alloy	12.0	49.0	18.0	. . .	21.0 In	57	136
117° alloy	8.3	44.7	22.6	5.3	19.1 In	47	117
Ga-Sn eutectic	8.0	. . .	. . .	. . .	92.0 Ga	20	68
63° alloy	12.0	. . .	. . .	. . .	82.0 Ga 6.0 Zn	17	63

*Liquidus 208°F (98°C).
†Liquidus 212°F (100°C).
‡Liquidus 163°F (73°C).
§Liquidus 162°F (72°C).
SOURCE: *Fusible Alloys Containing Tin,* published by the Tin Research Institute, 1967.

Table 2.2. Some Mechanical and Physical Properties of Selected Fusible Alloys

	117° alloy	136° alloy	Lipowitz's metal	Bi-Pb eutectic	Bi-Sn eutectic
Yield temperature, °C	47	58	70	124	138
Density, g/cm³	8.86	8.57	9.38	10.27	8.72
Tensile strength, kg/mm²	37	43	41	44	55
Elongation, percent (slow loading)	1.5	50	200	60	200
Hardness, Brinell no.	12	14	9.2	10.2	22
Maximum load (5 min), kg/mm²	...	...	28	28	62
Safe load (sustained), kg/mm²	...	...	2	2	3
Electrical conductivity, % IACS	3.9	3	4	3	4.5
Specific heat, cal/g · °C	0.035	0.032	0.040	0.03	
Latent heat of fusion, cal/g	...	...	6.67	4.50	
Mean linear coefficient of thermal expansion, per °C	...	...	22×10^{-6}	...	15×10^{-6}

CONVERSION FACTORS: 1.0 cal/g · °C = 1.0 Btu/lb_m · °F; 1.0 cal/g = 1.8 Btu/lb_m; 1.0 g/cm³ = 62.4 lb_m/ft^3; 1.0 kg/mm² = 9.8×10^6 Pa = 1421 lb/in²; °F = 1.8°C + 32.

SOURCE: *Fusible Alloys Containing Tin,* The Tin Research Institute, Columbus, Ohio 1967.

expand when they solidify in direct opposition to normal metallic behavior. Although this is an interesting property, the more important factor for some applications is how the dimensions change with solidification plus cooling to room temperature and subsequent aging over time.

The various fusible alloys may show a variety of behaviors. Some with high bismuth content will grow through all three stages of solidification, cooling, and aging. Others, such as the lead-bismuth eutectic, shrink when solidified, change little during cooling to room temperature, and then grow during aging to exceed the original mold dimensions. The 136°F alloy, the bismuth-indium-lead-tin eutectic, varies by no more than a few one-hundredths of 1 percent of the original dimensions over the course of some weeks. Figure 2.1 summarizes the behavior of some common fusible alloys.

Corrosion Properties

Another material property important in some applications of fusible alloys is their corrosion resistance. In a safety device when an alloy is expected to melt at a specific temperature, corrosive attack may cause premature failure by accelerating rupture or weakening of the alloy or may provide support to the alloy, delaying activation of the device. This second effect can normally be avoided by applying a small, constant tensile force on the alloy, perhaps with a spring.

The metals which make up the fusible alloys generally have good atmospheric corrosion resistance because they form thin, adherent oxide layers on the surface. This is true of tin, lead, and cadmium. Thus fusible alloys containing these metals are expected to perform similarly, although certain air or water contaminants, such as chlorine or sulfur, may cause attack.

The variety of possible environments and possible fusible alloy compositions makes formulating specific guidelines very difficult. Because the various alloying metals tend to form simple eutectics, the phases present in fusible alloys tend to be relatively pure metals. For example, the tin-lead eutectic, widely used as a solder, is a mixture of relatively pure tin and lead phases, with a corrosion resistance roughly related to the reactions of the two separate phases present. In some environments, potable water, for example, the lead phase becomes passivated and further corrosion is not a significant problem. In other cases, however, one phase may be continually attacked in preference to the other, and in particularly aggressive environments, both may be attacked. Galvanic considerations dictate that thorough testing be done.

2.4 APPLICATIONS

Soldering and Sealing

Many electronic components are heat-sensitive, so that even the tin-lead solders melting at 361°F (183°C) are near the upper limit. It is required, therefore, to solder with a fusible alloy, such as the tin-bismuth eutectic, especially if step-

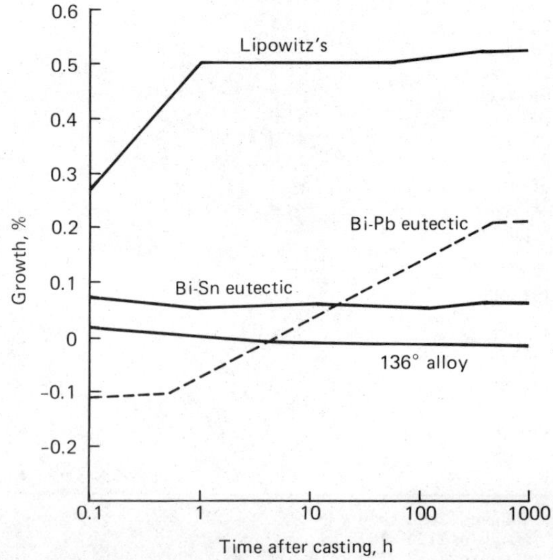

Fig. 2.1 Dimensional changes (shrinkage plus subsequent growth) for some fusible alloys. (*From Fusible Alloys Containing Tin, Tin Research Institute, Columbus, Ohio, 1967.*)

soldering is necessary. Although the fusible alloys are reasonably interchangeable with solders, it should be remembered that they do not wet surfaces quite so readily so it may be necessary to pre-tin the surfaces to be soldered. The mechanical properties of soldered joints made with some fusible alloys are shown in Table 2.2.

Many of the fusible alloys, especially Wood's metal and the indium-containing alloys, will wet glass and other vitreous materials and therefore can be used for glass-to-metal or glass-to-glass seals. Another sealing application uses molten fusible alloy to provide a gastight seal for industrial furnaces.

Safety Devices

One of the older applications of fusible alloys is as a link to activate sprinklers, alarms, or fire doors when the ambient temperature rises above the alloy melting point. Wood's metal, for example, forms the link in the 160°F (71°C) sprinkler head in common use. This activation temperature is sufficiently higher than normal ambient temperatures to ensure proper operation.

In many applications the fusible alloy is used as a solder to hold components in place rather than a massive fusible alloy link. This generally gives a stronger, more reliable type of linkage. Other forms of safety link include boiler plugs that melt when the liquid level falls too low and fuel shutoff devices that prevent burner fuel from feeding a runaway fire.

Pattern Making

Because of the dimensional properties reviewed above, the fusible alloys are widely used for making patterns and for molds, dies, and cores for plastics and rubber fabrications. The tin-bismuth eutectic shows near-zero volume change and allows making metal patterns from, say, wood patterns, without the complicated shrinkage allowances that would be necessary with an alloy that contracts. In fact, the wood patterns themselves can be given a much longer life by metallizing them using an immersion under pressure in a fusible alloy bath.

Complicated interior cavities in plastics forming can be produced by using fusible alloys which are then melted out for recovery. Obviously such cores must be used for materials which melt below the melting point of the fusible alloy. Such is not the case for fusible alloys as mold materials where there is the opportunity for water-cooling the molds.

Fusible alloys are also used in electroforming, where a metallic item of complex design is fabricated by electroplating on a pattern. When a fusible alloy is used as the pattern, separation of the article from the pattern is eased by melting out the pattern.

Other Uses

Fusible alloys, such as Wood's metals, are used for tube bending and similar metal-forming operations where temporary internal support is required. In such applications it is important to coat the metal to be fabricated with thin mineral oil to facilitate separation and cool the cast fusible alloy as quickly as possible to produce a fine structure with good ductility. Fabrication of the metal-fusible alloy assembly should be done at a slow, steady speed.

There are a variety of other metal fabrication uses for fusible alloys where parts and tools must be held in place temporarily, or even for temporary repairs. Unusual shapes that might be difficult to handle can be made more regular by filling in with fusible alloy. Where an accurately placed hole must be drilled in a very hard material (e.g., permanent magnets), an oversized hole can be drilled, filled with a fusible alloy, and redrilled at the precise location required.

One important area of application derives from the fact that, although the fusible alloys have low melting points, their boiling points are relatively high. The resulting low vapor pressure allows them to be held molten with very low metal losses. These metal baths are used for various heat treatment procedures and as coolants for reactors. Overheating may lead to excessive oxidation, but, in general, a wide range of temperatures can be accommodated.

SUGGESTIONS FOR FURTHER READING

1. C. J. Evans, "Fusible Alloys for Industry," in *Tin and Its Uses,* International Tin Research Institute, Uxbridge, Middlesex, U.K., No. 133, 10–14 (1982).
2. C. J. Evans, "Fusible Alloys Containing Tin," in *Tin and Its Uses,* International Tin Research Institute, Uxbridge, Middlesex, U.K., No. 87, 3–7 (1981).
3. J. A. Gonicberg and M. L. Ritch, *Principles of Centrifugal Rubber Mold Casting,* A. J. Oster Co., Providence, R.I., 1980.
4. O. J. Seeds, "Low Melting Alloys," in *Properties and Selection of Metals, Metals Handbook,* vol. 1, 8th ed., American Society for Metals, Metals Park, Ohio, 1961, pp. 863–864.
5. *Fusible Alloys Containing Tin,* International Tin Research Institute, Uxbridge, Middlesex, U.K., No. 175 (1967).

Chapter 3

Low-Expansion Materials

ERNEST G. WOLFF
Mechanical Engineering Department
Oregon State University

Sufficient heat or coolness will cause any material to grow or shrink. For applications requiring dimensional stability with tolerances on the order of microinches, micrometers, or (optical) wavelengths, a design engineer must determine the expected temperatures and review the coefficients of thermal expansion (CTE) of candidate low-expansion materials. The average linear CTE may be defined as

$$\alpha = \frac{\Delta L / L_0}{\Delta T} \tag{3.1}$$

where L_0 is the initial length of the sample. The mean volumetric CTE for an isotropic material is approximately equal to

$$\beta \cong 3\alpha[1 + (T_2 - T_1)\alpha] \tag{3.2}$$

Low-expansion materials are defined as those with a linear CTE of less than $1.8 \times 10^{-6}\,°C^{-1}$ at room temperature. Low CTEs can be obtained with selected metals and alloys, glasses and ceramics, composites, and their combinations. Figure 3.1 illustrates thermal strain behavior versus temperature for typical low-expansion materials. Handbook data[1] are limited for this class of materials, especially at temperatures below ambient. Table 3.1 gives general design data at room temperature for the materials discussed.

3.1 METALS AND ALLOYS

In 1896 Charles E. Guillaume discovered that the CTE of iron-nickel alloys was a minimum at about 36 percent nickel, and the "invariable" dimensions of this alloy led to its trade

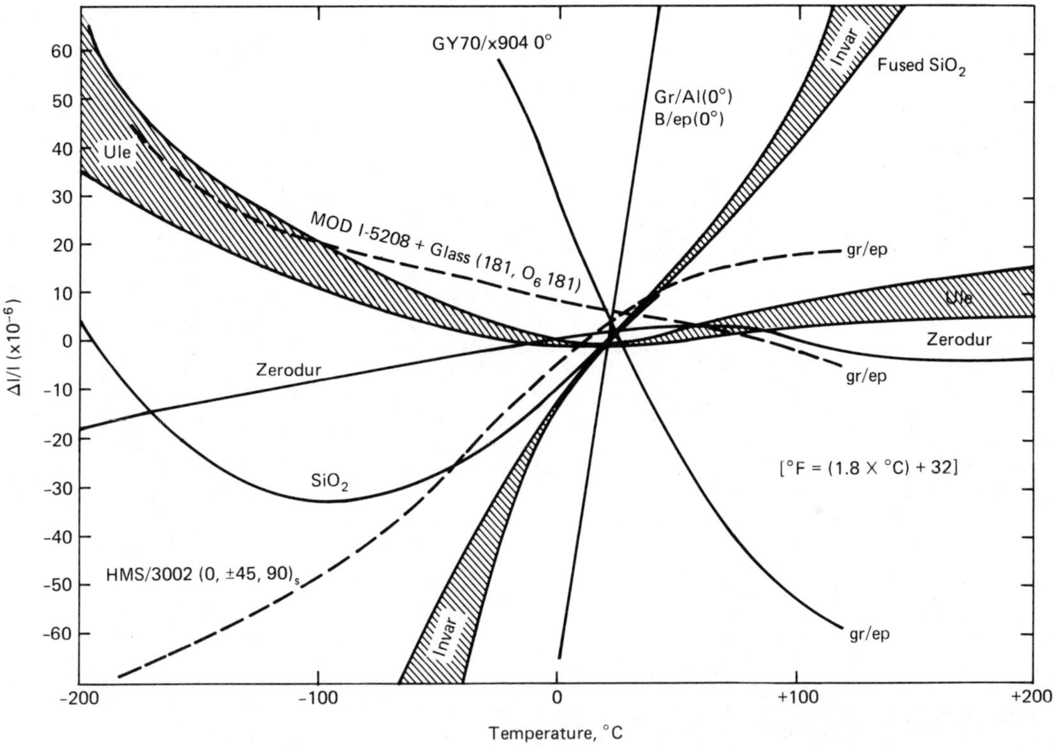

Fig. 3.1 Thermal strain for various low-expansion materials. The curves for graphite-epoxy (gr/ep) apply to the zero-degree direction and are also described in terms of fiber-matrix type and laminate ply orientations. B/ep (0°) refers to unidirectionally reinforced boron-epoxy.

Table 3.1. Relative Properties of Low-Expansion Materials

Material	Density δ, lb/in³ × 10³	Modulus of elasticity E, lb/in² × 10^{-6}	Poisson's ratio ν, × 10²	Specific heat C_p, Btu/lb·°F × 10²	Thermal conductivity K, Btu·in/h·in²·°F × 10²
Invar	291	21		29	50
Silica	79	11	14	7	22
ULE	80	10	17	6	18
Zerodur	91	13	24	8	20
CER-VIT	90	13	25	18	8
LAS	90	13	26	20	8
Cordierite	97	8	12	20	16
NZP	114	15	(20)*	20	(10)
Hf, Ta-W-O	280	20	(20)	15	(80)
PBT film	55	15	. . .	30	2
Gr-Mg (0°)	66	34	30	24	560
Gr-Gl (0°)	72	45	(20)	15	230
Gr-Ep (iso)	60	43	27	21	160
Gr-C (3D)	50	13	4	16	800

*( ) Denotes estimates

CONVERSION FACTORS: 1 g/cm³ = 0.036127 × lb/in³; 1 GPa = 0.145 × 10⁶ lb/in²; 1 Btu/lb·°F = 100 cal/g·°C; 1 gm·cal·cm/cm²·s·°C = 20.17 Btu·in/h·in²·°F = 4.187 W/cm·°C.

name, Invar. Today, *Invar* means a large class of alloys with this basic composition but modified to achieve other desirable properties, such as machinability (with selenium), high strength (with titanium), low change of elasticity with temperature (with chromium and molybdenum), and corrosion resistance (with cobalt and chromium). Applications for Invar alloys include laser housings, geodetic tapes, and microwave components.[2] Methods of fabrication include melting, casting, and hot-pressing of atomized, prealloyed powders. The CTE can be modified to be zero and even negative. Invar sheet, decarburized to about 0.01 percent carbon, heated to 845°C, water-quenched, heated to 95°C, and air-cooled, showed a CTE of 0.01×10^{-6} °C^{-1} over the range 5 to 30°C.[3] Further decarburization and thermomechanical treatments could lower this value or stretch the applicable temperature range. However, treatments which induce low CTEs (such as water quenching) may be incompatible with long-term dimensional stability, which generally requires slow cooling.[2,4]

Super Invar (nominally Fe-30Ni-6Co) was studied in the 1930s.[5,6] In addition to exceptionally low thermal expansion, this alloy has been identified as one of the two best materials for temporal stability (along with Zerodur, a Schott Glass Optical Co. glass-ceramic).[7] Super Invar can exhibit both low positive and low negative values of the CTE.[8] The $\Delta L/L$ curves tend to be discontinuous below 500°C because of the paramagnetic-to-ferromagnetic (A2) inflection temperature in the region 300 to 500°C and the gamma + alpha-to-alpha phase transition (Ar3), which usually occurs below ambient temperatures but can also start as high as 100°C. The CTE and the actual Ar3 temperature are sensitive to plastic deformation,[9] heat treatment, alloying additions, and trace impurities.[10]

3.2 GLASSES AND CERAMICS

Unlike Invar-type alloy research and development, efforts to develop near-zero CTE glasses, glass-ceramics, and ceramics are both current and increasing, reflecting the need for lightweight optical components. Fused-silica-based glasses are represented by Homosil (W. C. Heraeus-Schott Co.), Vitreosil (Thermal Syndicate Ltd.), Type 104 (General Electric), and code no. 7943 (Corning Glass Co.). Fused silica is available in many forms, including the Lockheed development LI-900, an exceptionally low thermal conductivity, 99.7 percent pure amorphous silica fiber-based material used for the Space Shuttle insulating tiles. The National Bureau of Standards (NBS) provides dense fused silica as the standard for low-expansion materials (NBS-SRM-739).

ULE (Corning ultra-low-expansion glass code no. 7971) is a synthetic amorphous, isotropic, single-phase, transparent silica glass with 7 percent titanium dioxide. Attempts by the NBS to adapt ULE glass to a lower CTE-level standard (about 3×10^{-8} °C^{-1} compared to 5×10^{-7} °C^{-1} for silica near room temperature) have been hindered by the difficulty in homogenizing the TiO_2 addition. Related ULE mirror uniformity problems have led to nondestructive methods for measuring CTE variations.[13] ULE mirrors are less sensitive to thermal cycling (than Zerodur[11]) up to temperatures of at least 350°C.[14]

Glass-ceramics include Zerodur, Cer-Vit (Owens-Illinois), and the Russian Astro-Sitall.[12] Compositions are typically based on the lithia-alumina-silica (LAS) system, which contains several low-expansion (crystalline) phases. From the basic formulation of Li_2O-Al_2O_3-$nSiO_2$, we identify β-eucryptite ($n = 2$), β-spodumene ($n = 4$), lithium orthoclase ($n = 6$), and petalite ($n = 8$). Substantial regions in this ternary phase system have negative CTEs, especially near the β-eucryptite and β-spodumene compositions.[15–17] Zerodur contains 70 to 75 percent by weight of crystalline β-eucryptite with a negative CTE and a vitreous β-spodumene phase with a positive CTE. The constituents are melted, refined, homogenized, and then formed by application of conventional glass-making methods. The CTE is controlled by heat treatments, but this also means that there are limits to the reheating processes that may be needed to form complex shapes—since they affect nucleation and crystallization in the microstructure.[18] β-Eucryptite can be processed below 800°C and combined with silica to give a near-zero CTE material.[19] Corning glass code no. 9617 or Pyroceram has a fine-grained crystalline β-spodumene structure and can be fabricated into sheets by sagging or vacuum forming. β-K (Centerflex Ceramics Corp.) has a similar composition and a near-zero CTE. LAS ceramics can be combined with potash, boron, and most alkaline earths to form low-expansion glazes.[20]

Cordierite ($2MgO$-$2Al_2O_3$-$5SiO_2$) is another class of glass-ceramic whose already low CTE can be tailored, e.g., by substituting other oxides such as GeO_2 for the silica.[21–23] Without additions it is difficult to obtain a reproducible CTE because of the anisotropy of the basic cordierite crystal and the likelihood of microcracking in sintered parts.[24] Cordierite is of special interest as a substrate for catalysts which control automotive exhausts because of its stiffness, high-temperature stability, and chemical inertness. The tendency to microcrack (as a result of anisotropic crystals) is avoided by hot-pressing to a moderate bulk density followed by surface densification techniques such as glazing, laser processing, and flame spraying.[22]

Negative CTE phases are found in the $CaTi_4P_6O_{24}$ (CTP) and $NaZr_2P_3O_{12}$ (NZP) systems. A subfamily of the latter, namely, $Na_{1+x}Zr_2P_{3-x}Si_xO_{12}$ (where x lies between 0 and 3) is called the *Nasicon family*. The CTP group can be modified with ionic substitutions to give a controllable CTE in the range -30 to $+60 \times 10^{-6}$ °C^{-1}.[25] Diphasic compositions such as CTP − 30% Nb_2O_5 or NZP + 20% $Zn_3(PO_4)_2$ have near-zero CTEs in the range 0 to 100°C. These materials may be synthesized by solid-state reactions using oxides as precursors, followed by cold-pressing and sintering.

Oxide ceramics with both negative and low-positive CTEs include phases in the Ta-W-O, Nb-Hf-O, Hf-W-Ta-O, Hf-Ti-O, Ta-V-O, Al-Ti-O, and La-O systems.[26–29] Hafnia-titania compounds are used as flame tube liners, largely because of the excellent thermal shock resistance associated with a low CTE. Low and even negative CTEs in oxides such as NZP, V_2O_5, and Nb_2O_5 may be explained in terms of microcracking.[30,31] Nonoxide ceramics with low CTEs, especially at low temperatures, include Si_3N_4, BN, and BP.

3.3 POLYMERS

There are few low-expansion polymeric materials. A recent development is poly(*para*-phenylene benzobisthiazole) or (PBT), a member of a new class of high-modulus, high-strength, thermally stable, aromatic heterocyclic polymers.[32,33] It can be fabricated in the form of thin ribbons or films, and the properties depend on the degree of oriented crystallinity induced by heat tratment. Ordered polymers have been under development by the U.S. Air Force for over 10 years, and PBT is the most highly developed ordered polymer. The most striking aspect of PBT is the stiffness, which is two to three times that of Kevlar. The axial CTE of PBT fibers and the tailorable "fiber" direction in the film can be as low as -10×10^{-6} °C^{-1}. It is highly anisotropic, with the CTE being positive in the transverse direction. Studies are also in progress on molecular composites, where rod-like PBT is embedded in a thermoplastic matrix such as nylon.[34]

3.4 COMPOSITES

Metal Matrix

The axial stiffness of graphite fibers approaches the stiffness of the *a* direction in graphite single crystals, about 145 × 10^6 lb/in^2 (1000 GPa). At the same time, the axial CTE of the fiber reaches -1.62×10^{-6} °C^{-1}.[35] Together, the stiffness and the negative CTE can be used to design any value of CTE above about -1×10^{-6} °C^{-1} in some direction in a relatively low stiffness (and positive CTE) matrix, either metal, plastic, or ceramic. Magnesium, with its Young's modulus of 6.5 × 10^6 lb/in^2 (44.8 Gpa), is a primary metallic candidate for a zero CTE with unidirectional fibers.[36] However, the internal stresses generated in the material during thermal cycling as a result of the thermal expansion mismatch will ultimately induce plastic flow in the matrix to give thermal strain hysteresis. As long as the matrix deforms elastically and there is a perfect, adherent fiber-matrix interface, the Turner equation may be used to predict the composite CTE in the fiber direction:

$$\text{CTE (comp)} = \frac{(\text{CTE}_f)(E_f)(V_f) + (\text{CTE}_m)(E_m)(1 - V_f)}{(E_f)(V_f) + (E_m)(1 - V_f)} \tag{3.3}$$

where *E* and *V* are the Young's modulus and volume fraction, respectively, and the subscripts *f* and *m* refer to the fiber and the matrix. Increasing the matrix stiffness, such as through alloying or heat treatment, will tend to reduce thermal strain hysteresis on cycling but will also make it more difficult to maintain a near-zero CTE. Angle-ply laminated plates can be made to show a near-zero CTE along the angle bisecting the fiber orientations even though the matrix is stiffer (e.g., with aluminum); however, this will raise the level of residual stresses.

Ceramic Matrix

The negative axial CTE of graphite fibers coupled with a cross-ply orientation in a glass matrix can give very low in-plane CTE values, depending on the fiber volume fraction.[37] Ceramics with isotropic negative CTE values can also be tailored toward a zero CTE with the addition of positive CTE reinforcing fibers or particulates. In a study aimed at mirror applications,[38] 2 and 4 weight percent CuO was added to Li-Al-Si- and Zn-Al-Si-based glasses and melted and either cast or drawn. The latter system crystallized at 625 to 725°C to give very low CTE values—about -0.14×10^{-6} °C^{-1} for 0 to 300°C. Heating was used to generate CuO or Cu films, depending on the atmosphere, on the surface via diffusion. Glass inhomogeneity and film discontinuities were two problems encountered, but the self-generated metallic surface showed good fiber-substrate adhesion. Other systems studied include Cu, Ni, and steels with glass,[39,40] Cu-TiO_2,[41] and Nb_2O_5 with Ni and Super Invar.[42] Powder metallurgy techniques have been used to fabricate copper and Super Invar dispersions in Ta-W-O, Ta-V-O, and La_2O_3 matrices.[27] Figure 3.2 demonstrates the type of control of the CTE (near-ambient) that could be achieved with a three-phase mixture prepared by solubilization-gelation methods. The dotted curve gives the CTE of a glass-crystalline two-phase mixture of β-eucryptite assuming the fully crystalline phase has a CTE of -8×10^{-6} °C^{-1}.[43] As a result of anisotropy, 100 percent crystalline β-eucryptite would easily develop microcracks and have low strength. The glass phase can be partly crystallized by heat treatment in the 700 to 1000°C range. By taking a 50/50 mixture of glass and crystals and adding a metallic dispersoid, it can be seen that about 55 percent of Super Invar will give a zero CTE and a minimum thermal expansion mismatch. Aluminum would give a higher thermal conductivity with a high CTE mismatch; titanium is a good compromise.

Plastic Matrix

The low-inplane CTE values obtainable with graphite-reinforced epoxy and other polymeric matrices are well known. The CTE can be readily predicted with standard laminate codes now increasingly available for calculators[44] and personal computers.[45] The approach is to derive the constitutive relations for laminates from basic ply or lamina properties such as inplane longitudinal and transverse stiffnesses, Poisson's ratio, and the shear stiffness. Figure 3.3 shows that a zero CTE in a single direction is possible by varying the ply angle θ in a symmetrical (about the midplane) layup of +θ or −θ layers. A [0, ±θ, 0] layup, however, may be preferable because of the low sensitivity of the CTE to misorientation of the ply angle, an advantage in fabrication. Fiber-rein-

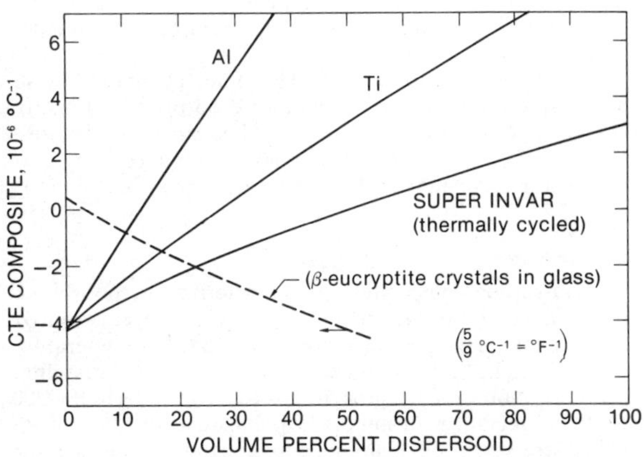

Fig. 3.2 Variation of predicted CTE with volume percent of metallic dispersoids in a 50/50 mixture of β-eucryptite glass and crystals.

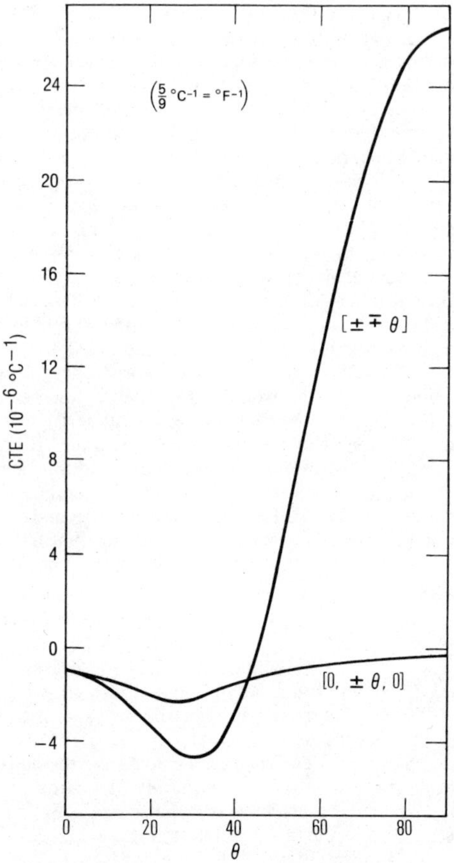

Fig. 3.3 Variation of CTE in the zero-degree direction with ply angle for four-ply laminates of a high-modulus graphite-epoxy matrix composite. θ is the variable ply angle relative to an arbitrary zero-degree direction in the laminate plane.

forced plastics are not low-expansion materials in the through-thickness direction—an important factor in the design of end fittings and joints.

Deviations from predicted CTE values commonly occur as a result of moisture absorption or desorption and microcracking.[46] Coatings such as metallic films can minimize moisture effects, while "preconditioning" the composite by low-temperature cycling stabilizes the microcracks and thereby the CTE.

Carbon Matrix

Carbon-carbon composites are characterized by poorly adhering fibers, porosity, and a variety of carbonaceous matrix constituents. The general shape of the $\Delta L/L$–temperature curve is similar to the *a* direction for single-crystal graphite. The transition from negative to positive CTE falls in the 200 to 400°C region, with minor deviations due to relief of residual stresses, graphitization on heating, and changes in the microstructure such as microcracking and debonding.[47] Felted and five-dimensional weave graphite-carbon composites tend to have negative CTEs below 300°C,[41] but positive values can also be obtained at ambient temperature.[48] Near-zero CTE in all directions may be achieved with metallic coatings on the fibers and complex three-dimensional weave patterns. (Boron nitride may be a future complement to carbon-carbon systems because the CTE in the *a* direction of the hexagonal form is also negative.)

3.5 STRUCTURES

Combinations of materials to form a macrocomposite can lead to low-expansion components and structures. The concept of a slightly negative CTE graphite-epoxy tube with positive CTE end fittings (e.g., Al, Ti, Invar, or molded graphite cloth in an epoxy) is common for large space structure design concepts.[49] Tubes with high thickness-diameter ratios require more complex analysis and/or iterative processes in the design cycle. A low angle-ply layup was found to give axial CTE values of $0 \pm 9 \times 10^{-8}$ °C^{-1}.[50] Cladding a negative CTE (graphite/epoxy) rod with a positive CTE metal (Al, steel) permits fine-tuning of the CTE (by machining or etching) after fabrication.[51] Hybrid composites such as combinations of graphite and glass fibers in a polymeric matrix can be used to achieve low CTEs. These compounds can be produced inexpensively by pultrusion methods for truss structures and antenna and optics supports.[52]

3.6 MEASUREMENT OF EXPANSION

The measurement of the CTE of low-expansion materials has been reviewed.[1,2,41,53] In general, techniques such as laser interferometry are required to provide the necessary thermal strain resolution. Auxiliary methods, such as x-ray diffraction, are recommended when microcracking is suspected. Sample preparation steps, such as machining, may induce residual stresses which affect the CTE. Edge/end[54,55] and size effects also suggest that a component or structure be tested in its entirety. Ultimately, CTE measurements should be tied to the laser wavelength standard of length.

3.7 SUMMARY

There are a wide variety of materials with tailorable low-thermal-expansion characteristics. CTE values in the 10^{-7} to 10^{-6} °C^{-1} range are commonplace; materials with CTE in the 10^{-8} to 10^{-7} °C^{-1} range are plentiful, but the CTEs are somewhat more difficult to achieve reproducibly and measure reliably. CTE values of less than 10^{-8} °C^{-1} are rare, and auxiliary effects leading to dimensional instability require attention, such as the relief of internal stresses. It can be expected that many of the materials mentioned here will be further improved in terms of low thermal expansion over a wide temperature range.

REFERENCES

1. Y. S. Touloukian, R. K. Kirby, R. E. Taylor, and T. Y. R. Lee, "Thermal Expansion," in *Thermophysical Properties of Matter,* vol. 12, *Metallic Elements and Alloys,* 1975; vol. 13, *Nonmetallic Solids,* 1977, IFI/Plenum Press, New York.
2. C. W. Marschall and R. E. Maringer, *Dimensional Instability,* Pergamon Press, Oxford, 1977.
3. C. E. Marschall, *SAMPE Symposium,* vol. 21, 1976, p. 261.

4. C. S. Susskind, D. L. Dull, S. A. Eselun, and E. G. Wolff, *Proc. 2d Int. Conf. on Mechanical Behavior of Materials, ICM II,* Amer. Soc. of Metals, Metals Park, Ohio, 1976, pp. 1893–1897.
5. H. Scott, *Trans. AIME,* vol. **89,** 506 (1930).
6. H. Masumoto, *323rd Sci. Rep., Tohoku Univ.*, vol. **23,** 265 (1934).
7. J. W. Berthold, S. F. Jacobs, and M. A. Norton, *Appl. Optics,* **15,** 1898 (1976).
8. A. I. Zakharov, A. M. Perelpelkina, and A. N. Shiryaeva, *Metallovedenii Termicheskaya Obrabtka Metallov,* No. 6, 62 (1974).
9. A. R. Rosenfield and B. L. Averbach, *J. Appl. Phys.,* **27,** 51 (1979).
10. P. Hidnert and R. K. Kirby, *J. Res. NBS,* paper no. 2602, vol. **27,** 51 (1979).
11. Schott Optical Glass Company Information. Brochure, no. 1, pp. 15–19 (1985).
12. Izmeritel'naya Teknika no. 1, pp. 36–37 (1980).
13. H. E. Hagy and W. D. Shirkey, *Appl. Optics,* **14,** 2099 (1975).
14. H. E. Bennett, J. J. Shaffer, and D. K. Burge, *Appl. Optics,* **23,** 2734–2737, 2852–2853 (1984).
15. W. Ostertag, G. R. Fischer, and J. P. Williams, *J. Am. Ceram. Soc.,* **51,** 651 (1968).
16. E. J. Smoke, *J. Am. Ceram. Soc.,* **34,** 87–90 (1951).
17. M. F. Hochella and G. E. Brown, *J. Am. Ceram. Soc.,* **69,** 13–18 (1986).
18. O. Linding, *SPIE Proceedings,* vol. **400** (*New Optical Materials*), 78–80 (1983).
19. G. Orcel and L. Hench, *Proc. 10th Annual Conf. Am. Ceram. Soc.,* January 19–24, 1986.
20. E. F. Connor and R. A. Eppler, *Bull. Am. Ceram. Soc.,* **52,** 180–184 (1973).
21. I. M. Lachman, R. D. Bayley, and R. M. Lewis, *Bull. Am. Ceram. Soc.,* **60,** 202 (1981).
22. Y. Mehrotra, *Proc. SPIE Conf.,* San Diego, Calif., August 1984.
23. D. L. Evans, G. R. Fisher, J. E. Geiger, and F. W. Martin, *J. Am. Ceram. Soc.,* **36,** 629–634 (1980).
24. Y. Hirose, H. Doi, and O. Kamgaito, *J. Mat. Sci. Lett.,* **3,** 153–155 (1984).
25. D. K. Agrawal and R. Roy, *J. Mat. Sci.,* **20,** 4617–4623 (1985).
26. S. R. Skaggs, Los Alamos Report LA-UR-77-1307, 1977.
27. N. Saka and J. Boustani, MIT, private communication, 1980; also J. Boustani, M.E. (Eng.) thesis, MIT, 1981.
28. C. E. Holcombe, *Bull. Am. Ceram. Soc.,* **59,** 1219 (1980).
29. R. Ruh, G. W. Hollenberg, E. G. Charles, and V. A. Patel, *J. Am. Ceram. Soc.,* **59,** 495–499 (1976).
30. O. R. Biswas, S. Chandrateya, and J. A. Pask, *Bull. Am. Ceram. Soc.,* **58,** 792 (1979).
31. T. Oota and I. Yamai, *J. Am. Ceram. Soc.,* **69,** 1–6 (1986).
32. J. R. Minter, K. Shimamura, and E. L. Thomas, *J. Mat. Sci.,* **16,** 3309–3318 (1981).
33. J. A. Odell, A. Keller, E. D. Atkins, and M. J. Miles, *J. Mat. Sci.,* **16,** 3309–3318 (1981).
34. W. F. Hwang, D. R. Wiff, and C. Verschoore, *Polym. Eng. Sci.,* **23,** 789–791 (1983).
35. E. G. Wolff, *J. Comp. Mat.,* **21,** 81 (November 1987).
36. E. G. Wolff, B. K. Min, and M. H. Kural, *J. Mat. Sci.,* **20,** 1141–1149 (1985).
37. K. M. Prewo, J. J. Brennan, and G. K. Layden, *Bull. Am. Ceram. Soc.,* **65,** 305–313, 322 (1986).
38. P. P. Piroz and G. Dube, AFML-TR-78-139 (1978).
39. B. R. Powell, G. E. Youngblood, D. P. H. Hasselman, and L. D. Bentsen, *J. Am. Ceram. Soc.,* **63,** 581 (1980).
40. L. N. Yagupol'skaya, S. V. Ivanova, E. V. Lysenko, and N. I. Shcherban, *Poroshkovaya Met.,* **11,** 54–58 (1976).
41. K. G. Kumar, C. Pavithran, and P. K. Rohatgi, *J. Mat. Sci.,* **24,** 1299 (1979).
42. E. G. Wolff, in *Thermal Expansion 8,* T. H. Hahn (ed.), Plenum Press, New York, 1984, pp. 211–225.
43. H. F. Gillery and E. A. Bush, *J. Am. Ceram. Soc.,* **42,** 175–177 (1959).
44. K. T. Knoll, in *Advances in Composite Materials,* A. R. Bunsell (eds.), "TI-59 Programs for Analysis of Advanced Composite Materials," *Proceedings ICCM 3,* Paris, August 1980, pp. 826–839.
45. D. L. Reed, General Dynamics/Fort Worth Report no. FZM-5494 (1970). The SQ5 Point Stress Analysis program is available for the IMB PC from H. D. Neubert & Associates, Inc., 6051 Prado Street, Anaheim Hills, Calif., 92807. See also "Think Composites," Box 581, Dayton, Ohio, 45419 for general composites software.
46. S. A. Eselun, H. D. Neubert, and E. G. Wolff, SAMPE Symp., **24,** 1299 (1979).
47. J. Jortner, *Proc. 25th Biennial Conf. on Carbon,* Univ. Paris, June 1981.
48. R. G. Naum, C. K. Jun, and P. T. B. Schaffer, in *Thermal Expansion—1971,* M. G. Graham and H. E. Hagy (eds.), American Institute of Physics, New York, 1971, p. 279.
49. B. Abt, K. Grunewald, and M. Schneerman, *SAMPE Symp.,* **24,** 1333 (1979).
50. R. J. Carter and A. J. Brzezinski, *SAMPE J.,* **22** (2), 17–21 (1986).
51. P. M. Bluck and R. R. Johnson, *SAMPE Symp.,* **29,** 1115 (1984).
52. T. J. Dunn, A. J. Cwiertney, W. L. Freeman, and R. Johnson, NASA-CR-160558 (also NASA Technical Brief), 1980 p. 323.
53. E. G. Wolff, *SAMPE Tech. Conf. Ser.,* **9,** 57–72 (1977).
54. J. Jortner and F. I. Clayton, *JANAF Rocket Nozzle Mat. Rep.,* **1** (5), (1979).
55. H. M. Kural and A. M. Ellison, *SAMPE J.* **16** (5), 20 (1980).

Chapter **4**

Refractories

J. P. HUTCHINSON
Harbison-Walker Refractories
Division of Dresser Industries, Inc.

F. H. WALTHER
Garber Research Center
Harbison-Walker Refractories
Division of Dresser Industries, Inc.

4.1 APPLICATIONS AND TYPES OF REFRACTORIES

Refractories are defined as nonmetallic materials of construction (e.g., brick, concretes) used to build or line furnaces operated at high temperatures. The primary requirement is that they retain their physical stability and chemical identity at the high temperatures. However, this is rarely the sole requirement for satisfactory service. Refractories must usually also be able to withstand thermal stress resulting from expansion and/or contraction; loading from the weight of the furnace parts and furnace contents; mechanical wear or abrasion; and chemical attack by furnace contents, slag, or furnace atmosphere.

Applications of Refractories

Refractories are used in a wide variety of applications, many of which are listed below. One or more pieces of equipment in each process included in the following list will be made of or contain refractories, and often the equipment is unique to the process:

Applications of Refractories

Aluminum processing	Glass plant
Boilers	Heating and heat-treating
Carbon baking	Furnaces
Casting metals	Incinerators
Cement production	Lead production
Ceramic kilns	Lime kilns
Coke ovens	Nickel smelting
Copper processing	Petroleum and petrochemicals
Cupolas	Pulp and paper production
Electric furnaces	Sodium silicate production
Ferroalloy production	Steel plant furnaces
Fluid-solids reactors	Zinc production

Kinds of Refractory

Refractories with a broad range of physical, chemical, and thermal properties are produced to meet the needs of diversified industrial applications. Most refractories are classified on the basis of chemical composition or physical properties. The main classes are silica, alumina-silica, high alumina, basic, and insulating. These, plus various special refractories, provide the capability to accommodate acid, neutral, and basic slags, as well as fumes and dusts generated in industrial environments.

Raw Materials Used

Fireclay refractories or "firebrick" are made of naturally occurring clays and kaolins, a portion of which usually requires precalcining before forming into brick and subsequent firing in periodic or continuous kilns. The quality of the brick obtained will depend in large measure on the quality of clay or kaolin used, and it does vary considerably depending on the chemistry of the particular deposit employed.

High-alumina refractories are produced from mixtures of fireclay and naturally occurring high-alumina minerals such as bauxite, kyanite, and sillimanite for classes containing up to 70 percent alumina. For alumina content of 70 percent or more, bauxites and chemical-process alumina are the primary ingredients. While refractoriness generally increases with alumina content, performance characteristics can vary sharply according to the accessory oxide (oxides other than Al_2O_3 and SiO_2) content. Those constructed of high-purity alumina are really a different type of brick than those made of natural ingredients. High-alumina brick refractories are formed and fired in about the same way as are fireclay brick.

Silica brick refractories are manufactured from either quartzite or quartz pebble rock, and most of the rock is mined from open pits or quarries and the minerals are widely distributed throughout the United States. The practice of beneficiating the raw materials by washing is well established. The mineral changes which occur during the firing of silica brick are substantial and cause the brick to expand. The firing process is therefore carried out on a careful burning schedule.

Basic refractories are manufactured from magnesite, dolomite, and magnesite in combination with natural graphite or other carbon sources, chrome ore, or magnesite and chrome ore mixtures.

Magnesite is the mineral $MgCO_3$, while actually magnesia (MgO) is used in basic refractories. Previously the magnesia was made by burning naturally occurring magnesite rock at high temperatures. Today most of the magnesia is made from magnesium hydroxide which is derived from seawater or well brines.

The principal source of chrome used in the manufacture of basic refractories is a chrome spinel mineral composed of about 30 to 50 percent Cr_2O_3 and the balance, of FeO, MgO, Al_2O_3, and SiO_2. The mineralogy of chrome-containing refractories is complex and varies with the source of the chrome ore. Important chrome deposits are found in the Philippines, Cuba, and South Africa.

Magnesite-chrome or chrome-magnesite brick is formed from mixtures of magnesite and chrome ore that are fired at high temperatures. The most important and widely used class in this category is the *direct-bonded* magnesite-chrome brick. These are fired at extremely high temperatures to develop a direct bonding between the magnesite and chrome ore particles. This results in a brick with a good balance of physical properties, including excellent hot strength and good resistance to thermal shock and chemical attack by slags.

The addition of carbon to a magnesite refractory originally stemmed from the observation that carbon is not easily wetted by slag. Thus carbon can prevent slag from entering the brick and eroding the magnesite. The earliest magnesite-carbon brick were pitch-bonded products that contained about 5 percent carbon. This carbon content was sufficient only to fill the pore structure of the brick. More recently, brick with carbon contents from 8 to 30 percent have been introduced. The premium-quality products contain flake graphite additions. The resulting brick possess good oxidation resistance and high thermal conductivities. Some magnesite-carbon brick contain additions of aluminum, magnesium, and silicon powders to improve hot strength and oxidation resistance. The principal applications for these brick are basic oxygen furnaces, electric arc furnaces, and steel ladle slag lines.

Pure dolomite is the mineral $MgCO_3$-$CaCO_3$, although in practice the proportion of $MgCO_3$ may vary from the ideal 1:1 ratio and the mineral will contain impurities. The raw dolomite stone is calcined to form dense $MgO \cdot CaO$ grains. These grains are used in making burned dolomite brick as well as unburned tar or resin-bonded brick. Although subject to deterioration by hydration, these classes of refractories have developed a niche in applications such as rotary cement kilns, argon-oxygen decarburization furnaces, and steel ladles.

Silicon carbide is an entirely synthetic mineral prepared by heating silica and carbon at high temperatures. Since refractories produced from it can oxidize at high temperatures, they may be fired in kilns with nitrogen atmospheres or may contain an additive which protects them against oxidation.

Zircon is a naturally occurring mineral found in Australia and elsewhere often in the form of beach sand. Since the mineral is not massive, it cannot be crushed to the same particle grading used in other refractories. Zircon brick are difficult to press into shapes and have a characteristic smooth surface texture.

Refractory Brick

A general outline of successive stages in the production of refractory brick is shown below. Both partly processed materials and finished products may be obtained at each stage.

Production step	Processed materials and finished products
Crushing, grinding, screening, mixing	Granular materials, hot- and cold-setting mortars
Water and plasticizing agents are added to batches which will be formed; bonding agents are used with plastics, castables, ramming mixes	Castables, ramming mixes (see discussion of monolithic refractories)
Forming	Plastic refractory
Drying or curing	Chemically bonded brick (unburned)
Firing	Fired refractory brick of all types
Cutting or grinding to size	Insulating fire brick, refractory shapes which must meet rigid size tolerances

Refractory brick are manufactured in a great number of standard sizes as well as in a variety of special shapes.

Table 4.1 shows some physical properties for the more common types of industrial brick. In addition to these types, industrial refractories are made of essentially pure carbon, zirconia (ZrO_2), and large additions of high purity Cr_2O_3. Fusion-cast refractories are made by melting ingredients in an electric furnace and pouring into molds. The main types are zirconia-alumina-silica, alumina, and basic. These special products have low permeability and show excellent resistance to corrosion by fluid slag and molten glass. The main use is for glass-making furnaces.

Insulating Firebrick

Refractory insulations include fiber and mineral wool shapes, insulating rotary kiln liners, lightweight silica brick used over glass tank crowns, and a series of fireclay–high-alumina products which are the utility materials. The properties of the latter group are shown in Table 4.2. They are used mostly as backup to hard brick walls in the manner shown in Fig. 4.3 but may be the primary material of construction in services which do not involve heavy loading or mechanical abuse. Their low iron content is useful in special-atmosphere (reducing) environments such as annealing furnaces.

Monolithic Refractories

Monolithic refractories are special mixes of unformed refractory materials that are cast, rammed, gunned, or troweled into place, forming relatively joint free linings. Most of them contain some sort of binder to impart strength or "set" after they are placed in the field. The binders are sometimes organic but more often are mineral in nature (sodium silicate, phosphate in various forms, various hydraulic cements, etc.).

Table 4.1. Physical Properties of Refractory Brick

Type of brick	Density, lb/ft^3	Apparent porosity, %	Cold crushing strength, lb/in^2	Modulus of rupture, lb/in^2
Fireclay				
Superduty	144–148	11.0–14.0	1800–3000	700–1000
High-duty	132–136	15.0–19.0	4000–6000	1500–2200
Low-duty	130–136	10–25	2000–6000	1800–2500
High-alumina				
60% class	156–160	12–16	7000–10,000	2300–3300
70% class	157–161	15–19	6000–9000	1700–2400
85% class	176–181	18–22	8000–13,000	1600–2400
90% class	181–185	14–18	9000–14,000	2500–3000
Corundum class	185–190	18–22	7000–10,000	2500–3500
Silica (superduty)	111–115	20–24	4000–6000	600–1000
Basic				
Magnesite, fired	177–181	15.5–19.0	5000–8000	2600–3400
Magnesite-chrome, fired	175–179	17.0–22.0	4000–7000	600–800
Magnesite-chrome, unburned	185–191	...	3000–5000	800–1500
Chrome, fired	195–200	15.0–19.0	5000–8000	2500–3400
Chrome-magnesite, fired	189–194	19.0–22.0	3500–4500	1900–2300
Chrome-magnesite, unburned	200–205	...	4000–6000	800–1500
Magnesite-carbon	170–192	9.0–13.0		1000–2500
Dolomite	165–192	5.0–20.0	1500–3500	500–2500
Fused cast magnesite-chrome	205–245	1.0–15.0	High	High
Silicon carbide	160–166	13.0–17.0	9000–12,000	3000–5000
Zircon	225–232	19.5–23.5	7000–11,000	2300–3300

CONVERSION FACTORS: lb/ft^3 × 16.02 = kg/m^3; lb/in^2 × 6.895 = kPa.

Table 4.2. Properties of Insulating Firebrick

	Class					
	20	23	26	28	30	33
Maximum service temperature, °F	2000	2300	2600	2800	3000	3300
Density, lb/ft^3	29	31	48	51	58	82
Cold crushing strength, lb/in^2	110	145	170	220	275	1000
Alumina content (Al_2O_3), %	39	39	46	46	64	77
Iron content (Fe_2O_3), %	0.7	0.6	1.0	1.4	0.6	0.4
Thermal conductivity, Btu · in/h · ft^2 · °F						
500°F	0.8	0.9	1.7	1.7	2.1	6.0
1000°F	1.0	1.1	1.9	2.0	2.4	5.9
1500°F	1.2	1.3	2.2	2.3	2.8	6.3
2000°F	...	1.6	2.7	2.9	3.4	7.3
2400°F	...	...	...	...	4.0	8.4

CONVERSION FACTORS: lb/ft^3 × 16.02 = kg/m^3; lb/in^2 × 6.895 = kPa; Btu · in/h · ft^2 · °F × 0.144 = W/m · K.

The primary advantages of monolithic refractories are ease of (1) inventorying and (2) installation in the field. Under certain conditions they may provide better insulation, lower permeability, and better resistance to thermal stress than their fired-brick counterparts.

There are dozens of proprietary compositions on the marketplace, including mortars, both dense and insulating types of castable or concrete, ramming mix, gunning mix, plastics (moldables), and dry granular products.

Table 4.3 compares several types of refractory monoliths with their fired-brick counterparts. It can be seen that while the monoliths have the same maximum service limit and generally the same chemistry as the brick, they are characteristically less dense and weaker than the brick. This means that they typically will not perform as well as brick under heavy loading or heavy abrasion.

However, it should be noted that a new class of highly engineered monolithic refractory has recently become available which has excellent abrasion resistance, good resistance to thermal shock, and unusually high hot strength. These products have physical properties approaching fired brick and are replacing brick in many applications. Castables in this new class have a cement content of 10 percent or less and a low water requirement.

4.2 PROPERTIES OF REFRACTORIES

Physical

The significant physical properties of a refractory depend on its mineral makeup and how it was manufactured. Unfired refractories consist of a mixture of particles of various sizes; the coarser particles may be ¼ in or more in diameter, while the finer particles may be 200-mesh (about 75 μm) or less.

After the refractory has been fired or heated, the fines react to form a ceramic bond between the larger particles. The properties of the ceramic bond, or matrix, are particularly dependent on composition, manner of forming, and temperature of heating.

The physical properties of refractories most readily determined at ambient temperatures (see Table 4.1) are as follows:

1. Bulk density is the ratio of weight or mass to volume, usually expressed in pounds per cubic foot. Bulk density has obvious practical significance in determining quantities of materials to be purchased or what sorts of supporting structures may be required in building a furnace. It also determines the ability of a refractory with a given composition to store heat.

2. Apparent porosity (volume percentage of open or inconnected pores) indicates the ability of a refractory to resist

Table 4.3. Comparison of Typical Monolithic Refractories to Brick

	Superduty fireclay		70% alumina		Zircon	
	Brick	Plastic (moldable)	Brick	Castable* (concrete)	Brick	Ramming mix
Maximum service temperature, °F	3000	3000	3200	3200	3000	3000
Pounds per cubic foot†	146	138	159	146	228	203
Modulus of rupture, lb/in^2‡	700–1000	75–100	1700–2400	400–800	2300–3300	1300–2300
Chemical analysis, %						
Al_2O_3	41.9	45.6	69.2	70.8		
ZrO_2	...	...	...	...	66.0	64.7
SiO_2	53.2	49.8	26.2	23.3	32.3	32.8
Others	4.9	4.6	4.6	5.9	1.7	2.5

*Castables of the new low-cement class will have density and strength very close to those of the fired brick.
†lb/ft^3 × 16.02 = kg/m^3.
‡lb/in^2 × 6.895 = kPa.

penetration by slags and fluxes or permeation by furnace gases. Porosity also influences thermal flow. High-porosity brick will be more insulating in character but more subject to penetration and permeation.

3. The room temperature strength of refractories is measured for both compression (cold crushing strength) and tension (modulus of rupture or cross-breaking strength). As seen in Table 4.1, refractories characteristically have crushing strength several times greater than cross-breaking strength.

Room-temperature strength is indicative of how well refractories were manufactured, will handle and ship, and will withstand impact or abrasion at relatively low temperatures. However, room-temperature strength provides little or no indication of strength at operating temperatures. This property must be determined by other testing at high temperatures. These hotter tests are discussed in the section on thermal properties.

Some other physical properties sometimes determined at ambient temperatures are permeability, apparent specific gravity, abrasion resistance, electrical resistivity, modulus of elasticity, heat capacity, and emissivity.

Chemical

Only a few chemical elements having high-melting-point compounds—usually oxides—are found in the earth's crust in sufficient quantity to be useful in manufacturing industrial refractories. These include silicon, aluminum, magnesium, calcium, chromium, zirconium, and carbon. However, many of these are found in deposits of sufficient purity to enable direct use or use after minor beneficiation. These include clays (SiO_2, Al_2O_3), bauxite (AlO_3), quartzite (SiO_2), magnesite ($MgCO_3$), chrome ore (Cr_2O_3, MgO, Al_2O_3, FeO), and zircon (ZrO_2 SiO_2). Carbon is obtained from naturally occurring graphite or manufactured coke. In some cases refractory oxides are obtained by chemical processing such as MgO from chloride brines, Al_2O_3 from bauxite, ZrO_2 from zircon, and Cr_2O_3 from chrome ores.

Chemical analysis of refractories serves as a basis for classification and as a guide for predicting their behavior in service. In industrial manufacture the main value of chemical analysis is in quality control of raw materials and finished product. In addition to their primary components, most refractories contain accessory oxides which can have a great effect on service performance. These impurities may result in the formation of low-melting compounds in an otherwise very refractory composition; hence considerable research has been directed at reducing or eliminating them.

Chemical analysis can also be helpful in selecting refractories for a particular service. For example, basic refractories (magnesite, dolomite, and magnesite-chrome ore mixtures) should generally be used in contact with basic slags and acid refractories (silicon, fireclay, zircon) in contact with acid slags. Chrome, high-alumina, and carbon refractories usually resist both but may have other limitations. Refractories with high iron content should be avoided where a furnace atmosphere is strongly reducing. Certain refractories cannot be used together at high temperatures, e.g. fireclay and magnesia.

Table 4.4 shows typical chemical analysis for the common industrial refractories of this presentation. It will be seen that relatively few of them consist of more than 90 percent of any single component and that some, particularly in the basic category, are rather complex chemically.

Of the three types shown in Table 4.4*a* (fireclay, high-alumina, silica), alkalies constitute the most serious impurity or accessory, and, generally, the less the better. However, others such as TiO_2, Fe_2O_3, CaO, and MgO should also be low to obtain best performance characteristics. The lime (CaO) content of silica brick is an exception. It is deliberately added during manufacture and actually improves service performance.

For the basic refractories listed in Table 4.4*b*, the MgO, Cr_2O_3, and Al_2O_3 may be seen as the primary refractory components with SiO_2, Fe_2O_3, and CaO the accessory oxides. However, basic refractories often have a spinel mineral structure which is tolerant of considerable iron content, thus explaining the high "others" totals in most cases.

The special refractory brick listed in Table 4.4*c* are essentially single-component systems. Zircon is naturally-occurring zirconium silicate and is refractory in that form. Silicon carbide will contain a small amount of SiO_2 as a result of oxidation during manufacture but it is not harmful.

Alumina-carbon and magnesia-carbon brick are special refractories not shown in Table 4.4*c*.

Thermal Effects

When refractories are hot, their linear dimensions and volumes are greater than when they are cold as a result of their reversible thermal expansions. With some refractories, additional volume change, usually not reversible, results from mineral inversions and transformations and sometimes from other causes, such as the secondary expansion of some refractory clays and the overfiring of others.

Within a single brick in a furnace wall or arch, there is commonly a temperature drop of many hundreds of degrees between the hot face and the cold face. Consequently, differences in the amounts of thermal expansion in different sections of the brick will cause internal stresses to develop. In a refractory structure, pressure inherent in the furnace construction, such as arch stresses, generally will be increased by thermal expansion of the brickwork. Unequal volume changes resulting from rapid fluctuations in temperature may cause the development of higher stresses within the refractories than they are capable of withstanding.

Maximum Service Temperature

An important consideration in selecting refractory materials is their maximum service temperature. In Table 4.3 several types of refractory brick are contrasted with comparable grades of refractory monoliths, and "maximum service temperature" is an item shown. This property is not simple and must be determined by conducting a number of tests which then must be cross checked. When numbers are presented in the manner exhibited in Table 4.3, it has to be assumed that temperature alone is the major factor to be handled in service and that there is no heavy loading, abrasion, slagging, etc. Following are some of the most common values considered in determining maximum service temperature:

Melting Point The terms *melting point* and *fusion temperature* refer to the transition of a solid material into the liquid form under the influence of heat. Commercial refractories rarely have true melting points. When heated, they typically

Table 4.4. Chemical Analyses of Common Industrial Refractories

(*a*) Refractory Brick of the Alumina-Silica Types

Type of brick	Alumina (Al_2O_3), %	Silica (SiO_2), %	Alkalies (Na_2O, K_2O, Li_2O), %	Others (TiO_2, Fe_2O_3, CaO, MgO), %
Fireclay				
Superduty	41.9	53.2	1.2	3.7
High-duty	41.0	51.6	2.0	5.3
Low-duty	30	62	3.7	4.3
High-alumina				
60% class	58.0	38.0	0.1	3.9
70% class	69.2	26.2	0.2	4.4
85% class	86.5	8.9	0.1	4.5
90% class	88.5	11.0	0.2	0.3
Corundum class	99.2	0.4	0.2	0.2
Silica				
Conventional	0.52	95.9	0.1	3.4
Superduty	0.2	96.5	0.04	3.4

(*b*) Basic Brick

Type of brick	Magnesia (MgO), %	Chrome (Cr_2O_3), %	Alumina (Al_2O_3), %	Others (SiO_2, Fe_2O_3), %
Magnesite, fired	98.2	...	0.2	1.6
Magnesite-chrome, fired	55.7	4.1	3.3	36.9
Magnesite-chrome, unburned	72.1	8.8	9.7	9.5
Chrome, fired	18.4	30.4	32.1	19.1
Chrome-magnesite, fired	36.1	24.8	22.3	16.8
Chrome-magnesite, unburned	31.5	26.6	23.4	18.5

(*c*) Special Refractory Brick

Type of brick	Silica (SiO_2), %	Zirconia (ZrO_2), %	Carbide (SiC), %	Others (Al_2O_3, TiO_2, Fe_2O_3), %
Zircon	32.3	66.0	...	1.7
Silicon carbide	5.6	...	90.8	3.6

soften progressively over a temperature range within which the refractory contains both crystalline and liquid phases.

Some refractories such as silica (SiO_2), corundum (Al_2O_3), and magnesite (MgO) consist of essentially pure oxides in crystalline form and, therefore, have a relatively narrow range of melting. Others, such as fireclay and many high-alumina products fuse and soften gradually over a relatively wide range of temperatures.

Pyrometric Cone Equivalent A standard method of evaluating the high-temperature softening behavior of fireclay and some high-alumina refractories is the determination of their *pyrometric cone equivalent* (PCE). This method is described in detail in the *Manual of A.S.T.M. Standards on Refractory Materials*.

A ground sample of the material to be tested is molded into the form of "test cones" and mounted in a ceramic plaque at a slight angle with the vertical, with a series of standard pyrometric cones having known time-temperature softening values. The plaque is heated at a definite rate until the test cones soften and bend. The number of the standard cone whose tip touches the plaque at the same time as the tip of a test cone is reported as the PCE value of the test cone.

The PCE does not indicate a definite melting point or fusion point because the test is not a measurement of temperature but is merely a comparison of thermal behavior in terms of standard cones.

The PCE test is widely used in evaluating the refractory quality of raw fireclays and the quality control of fireclays while being mined and of fireclay products during manufacture but is not of much value otherwise.

Permanent Dimensional Changes In the firing of refractory brick, permanent changes in dimensions generally occur, altering the individual brick from the mold size to the fired size. It is desirable that these changes be completed in the firing in order to prevent excessive expansion or shrinkage later in service. However, the changes in dimensions and in volume require time as well as suitable temperature for their completion. As firing proceeds, the rate of change in dimensions gradually diminishes to such an extent that it is rarely feasible to attain complete expansion or contraction during the firing process. In service, therefore, if the furnace temperature is sufficiently high and maintained for a sufficient length of time, there will be an additional change in dimensions, generally slight and of a permanent character. Excessive changes in dimensions in service are objectionable because of their harmful effects on the stability of the furnace.

A reheat test is commonly used to measure the dimensional changes in refractory brick. However, the results of a short-time reheat test may not provide a sufficient basis for predicting the long-term volume changes which a refractory may undergo in service at a different temperature and under the different conditions.

In the standard ASTM Reheat Test, Method C113, brick are placed in a furnace and gradually heated to a predetermined temperature which is held constant for 5 h. Afterward cooling measurements are made to determine both linear and volume changes.

In the reheat test at 2550°F (1400°C), high-duty fireclay brick generally show a slight linear contraction of up to 1.5 percent. In the test at 2910°F (1600°C), superduty fireclay brick usually shrink up to 1.0 percent in length. Some fireclay brick expand slightly in the reheat test. However, the expansion generally does not occur if the brick are subjected to a slight pressure.

For high-alumina brick of the 50 to 80 percent alumina classes, the temperature of the reheat is usually 2910°F (1600°C). Some brick show a linear shrinkage of up to 2.0 to 2.5 percent; others, an expansion of up to 3.5 percent or more. For brick of the 90 percent alumina class, corundum brick, and mullite brick, the temperature of the test may be 3140°F (1725°C) and the linear change is usually within the range of minus 0.5 percent to plus 1.5 percent.

In the reheat test at 2910°F (1600°C), most basic brick show a slight linear shrinkage, varying from a few tenths of a percent or less for some types of brick, up to 2.5 percent for others. Chemically bonded chrome-magnesite brick have a slight linear expansion in the test, generally amounting to less than 1 percent. High-fired chrome magnesite brick usually have little or no shrinkage in the reheat test at 2910°F (1600°C) and a shrinkage of only some tenths of a percent at 3000°F (1650°C).

The Load Test The standard laboratory tests for refractories include one for the determination of the behavior of brick under load at high temperatures. In this test, two 9-in (22.9-cm) straight brick are set on end in a furnace of specified design and a vertical load of 25 lb/in^2 (175 kPa) is applied. In the standard procedure, ASTM Method C16, the temperature is gradually raised in accordance with a definitely prescribed schedule. For fireclay and high-alumina brick, the maximum temperature is held for 1½ h, after which the furnace is allowed to cool by radiation to 1830°F (1000°C) or lower before the load is removed. The amount of subsidence is reported as percentage of the original length. The normal temperature of the hold for high-duty fireclay brick is 2460°F (1350°C); for superduty fireclay brick, 2640°F (1450°C). However, recent improvements in the high-temperature strength of certain types of high-duty fireclay brick, especially checker brick, have led to their being tested at 2640°F. For most classes of high-alumina brick, the top temperature of testing is 2640°F (1450°C); for those of the 90 percent alumina class and mullite and corundum brick, it is 2900°F (1595°C) and 3200°F (1760°C).

Results of such testing on fireclay and high-alumina brick are shown on the upper portion of Table 4.5.

The lower portion of Table 4.5 shows results for silica, basic, and silicon carbide brick. It will be seen that test results are reported as "withstands load to" rather "subsidence after heating to." This reflects a marked difference in the way the latter types behave under the combined forces of heat and mechanical stress. They typically retain their integrity until a particular temperature is reached, after which failure occurs abruptly rather than showing gradual subsidence over time as with fireclay and high-alumina brick. Test procedures are different for those brick which are simply heated on a rising curve until failure occurs.

Load testing provides very significant data relative to maximum service temperatures in the field. More recently, the preferred test is the ASTM method for measuring the thermal expansion and creep of refractories under load (C832). Commonly referred to as a "creep test," the test is more sophisticated than the conventional ASTM load test. Rather than measuring the subsidence of a sample *after* testing, it provides a continuous monitoring of this sample's dimensional changes *during* the test.

Thermal Expansion Allowance

In the construction of a furnace, proper consideration should be given to the thermal expansion of the refractories and

Table 4.5. Use Limit as Indicated by 25 lb/in^2 Load Testing

Type of brick	Results of load testing
Fireclay	
Superduty	1.0–3.0% subsidence after heating at 2640°F
High-duty	Withstands load to <2640°F
Low-duty	Withstands load to <2640°F
High-alumina	
60% class	0.1–0.5% subsidence after heating at 2640°F
70% class	0.4–1.0% subsidence after heating at 2640°F
85% class	0.2–0.8% subsidence after heating at 2640°F
90% class	0.0–0.4% subsidence after heating at 3200°F
Corundum class	0.1–1.0% subsidence after heating at 2900°F
Silica (superduty)	Withstands load to 3060°F
Basic	
Magnesite, fired	Withstands load to >3200°F
Magnesite-chrome, fired	Withstands load to 2700°F
Magnesite-chrome, unburned	Withstands load to 2950°F
Chrome, fired	Withstands load to 2800°F
Chrome-magnesite, fired	Withstands load to 3020°F
Chrome-magnesite, unburned	Withstands load to 3020°F
Silicon carbide	Withstands load to 2800°F
Zircon	0.1–0.8% subsidence after heating at 2900°F

CONVERSION FACTOR: °C = (°F − 32)/1.8.

allowance for expansion made as needed. Usually, the vertical expansion in walls is provided by allowing them to move freely upward. Horizontal expansion of walls is provided by means of vertical expansion joints. Compensation for transverse expansion of furnace roofs and bottoms under certain conditions may be accomplished by release of tie rods.

The provision for expansion must be adequate, as insufficient allowance in some cases will cause pinch spalling. However, an excessive allowance is undesirable as it may cause structural weakness, as well as excessive leakage of gases or air infiltration.

Customary maximum allowances for thermal expansion of refractories used at high temperatures are given in Table 4.6. The figures presented in Table 4.6 are based on an assumed furnace temperature of 2600°F or higher. The amount of expansion allowance for any particular case is usually fixed by previous experience. Frequently it may be lower than the figures given in Table 4.6 because of lower operating temperatures, the bedding-in of joints, shrinkage of mortar in the joints, thermal expansion of furnace binding, stretching or letting out of tie rods, or other causes.

In order to provide most advantageously for the thermal expansion of the refractories, it is frequently necessary to estimate the expansion of the steel binding of the furnaces. As shown in Fig. 4.1, the expansion of steel over its usable temperature range is about three times that of fireclay brick.

Expansion Joints in Walls

The methods of allowing for thermal expansion of refractory brick differ considerably, depending on such factors as the dimensions of the wall, the thermal expansion of the refractories, the amount and kind of insulation, the kind of bonding mortar, and the operating conditions.

With high or long walls, it is especially important that sufficient provision for expansion be made. The thickness of the wall and the amount of insulation determine the temperature drop from the hot face to the cold face and the corresponding difference in amount of thermal expansion. With most refractories, the expansion at the hot face is considerably greater than in the cooler parts of the wall. Silica brick actually attain nearly their full expansion at any point in a wall where the temperature is 1000°F or higher and about two-thirds of their expansion at any point where the temperature has reached 500°F.

The horizontal expansion joints through a wall which has a thickness of two or more brick may be staggered. In thin walls, expansion joints extending through from the inner to the outer face cannot always be avoided. For such cases, leakage of gas can be reduced to a minimum by packing the outside ends of open joints with a compressible inorganic packing material or by covering them temporarily with a loose course of brick.

When the mortar used is the heat-setting type, the vertical expansion joints may be staggered as desired. However, if the brick are laid with an air-setting mortar, the expansion joints will not function properly unless those in successive courses are in vertical alignment. Brick which are required to move with respect to each other, as the material expands and contracts on heating and cooling, should not be bonded together with an air-setting mortar.

Individual preferences generally govern the choice of the kinds of expansion joint. The types illustrated in Fig. 4.2 are in common use and are regarded as satisfactory.

It should be noted that there cannot be open expansion joints in contact with molten metal, slag, glass, or other components. In the case of a plane vertical or horizontal surface in such contact, allowance can be made at the ends beyond the point of liquid contact. Curved surfaces can expand vertically and/or horizontally away from the molten charge—into some sort of compressible backup lining. Monolithic linings are often useful in such circumstances.

For monolithic refractory constructions, expansion allowances are not critical unless the installation is particularly massive. When joints are required, allowances should be somewhat less than for brick of a similar composition.

Table 4.6. Thermal Expansion of Refractory Brick in High Temperature Service

Type of brick	Approximate expansion, per foot, in	Approximate expansion, per meter, cm
Fireclay		
Superduty, High-duty, Low-duty	3/32	0.76
High-alumina		
60% class, 70% class, 85% class	3/32	0.76
90% class	1/8	1.04
Corundum class	5/32	1.30
Silica	5/32	1.30
Basic		
Magnesite, fired	1/4	2.08
Magnesite-chrome, fired	1/4	2.08
Magnesite-chrome, unburned	3/16	1.55
Chrome, fired	1/8	1.04
Chrome-magnesite, fired	1/8	1.04
Chrome-magnesite, unburned	5/32	1.30
Silicon carbide	3/32	0.76
Zircon	3/32	0.76

Spalling

Spalling may be defined as the loss of fragments or spalls from the face of a refractory brick or structure through cracking and rupture, with exposure of inner portions of the original refractory mass. In various types of furnace operation, the service conditions are necessarily such as to expose the brickwork to spalling influences. The resistance of refractory brick to spalling is enhanced by proper design of the shapes, optimum sizing of the grains, and close control of each step in manufacture. For a given refractory, the amount of spalling which occurs in service can be reduced to a minimum by proper design, construction, and operation of the furnaces.

Spalling of refractory brick is of three general types: thermal, mechanical, and structural.

Thermal Spalling

This type of spalling is caused by stresses resulting from unequal rates of expansion or contraction in different parts of the brick and usually is associated with rapid changes in temperature. The brick of greatest resistance to thermal

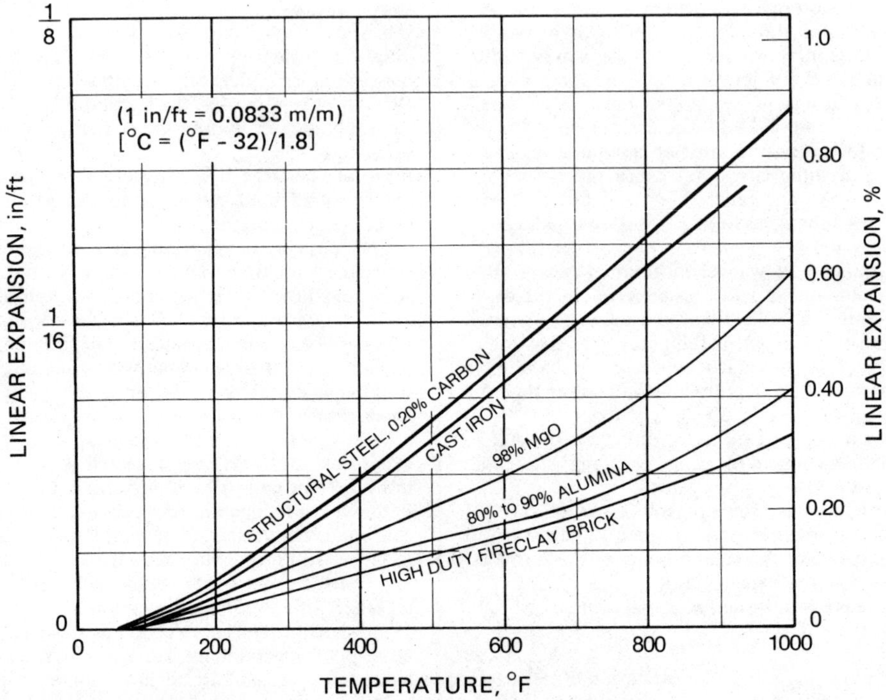

Fig. 4.1 Thermal expansion of various materials used in furnace construction.

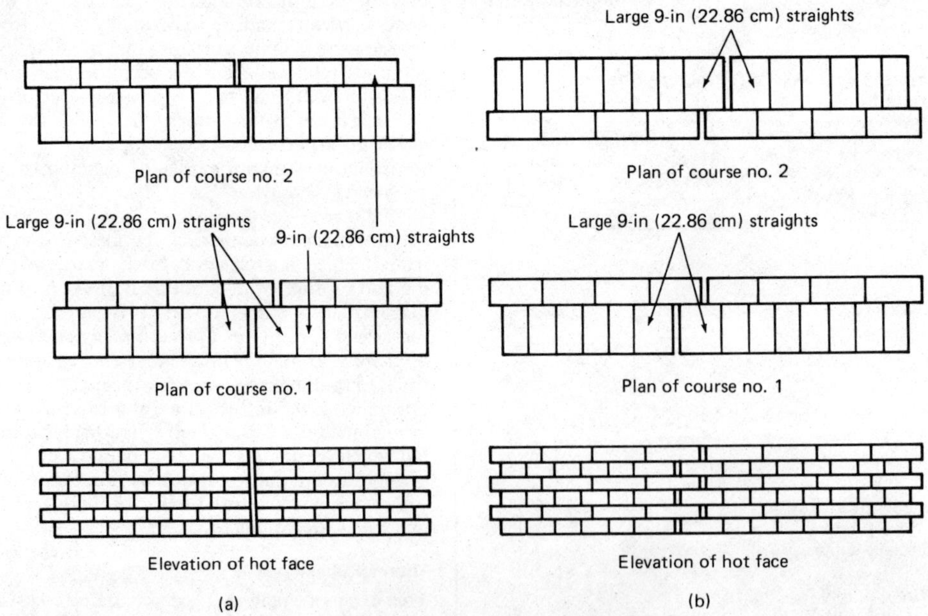

Fig. 4.2 Expansion joints. (*a*) Straight expansion joint in brick wall 13½ in (34.3 cm) thick, all-header construction; (*b*) staggered expansion joint in brick wall 13½ in (34.3 cm) thick, alternate header and stretcher construction.

spalling are those having the lowest average thermal expansion, with freedom from a high rate of thermal expansion through any narrow temperature range, high tensile strength, and a texture conducive to flexibility and relief of stress. With other factors equal, the temperature gradients and the stresses which cause spalling are least in brick having the highest thermal conductivity. A refractory of a given mineral composition will usually have the maximum resistance to spalling when the extensibility, the ratio of strength to modulus of elasticity s/E is a maximum.

Fireclay and high-alumina brick have fair to excellent resistance to spalling depending on composition, grain sizing, porosity, and degree of firing during manufacture.

Silica brick generally are quite sensitive to thermal changes below 1200°F (650°C) and quite resistant above that temperature because of an abrupt crystalline inversion of the primary mineral phase.

Basic brick in general have less resistance to spalling than do fireclay or high-alumina brick, while silicon carbide and zircon are better in terms of spall resistance.

Mechanical Spalling

The spalling of a refractory brick caused by stresses resulting from impact or pressure is known as *mechanical spalling*. Shattering or spalling of brickwork may result from such influences as the rapid drying of wet brickwork; inadequate provision for thermal expansion; or pinching of hot ends of brick, especially in furnace arches.

The brick which are strongest and toughest at operating temperatures have the greatest resistance to mechanical spalling.

"Pinch" spalling is frequently observed in sprung arches because the hot ends of the brick expand more than the cold ends. This condition is aggravated by rapid heating because the furnace structure cannot adjust rapidly to the expansion forces. The differential expansion may cause a concentration of arch stresses on relatively small bearing surfaces at the inner ends of the brick. Insulation decreases the temperature gradient through a roof arch and tends to reduce pinch spalling somewhat.

Structural Spalling

Sructural spalling is defined as the spalling of a refractory unit caused by stresses resulting from differential changes in the structure of the unit. The word *structural,* as used in this connection, refers not to the furnace or the furnace lining assembly, but the texture of the individual brick units in the lining.

Contributing to structural spalling are changes which occur in the texture and composition of the hotter portions of the brick through the action of heat, the absorption of slags or fluxes, and the reaction of slag with the refractory. These alterations may result in the formation of zones in the brick which differ in mineral content and chemical and physical properties. Following the development of a zoned structure, the hot ends of the brick may spall or fall off for various reasons, including differences in thermal expansion or contraction between adjacent zones, increased sensitivity to rapid temperature change, or the presence of shrinkage or expansion cracks.

Monolithic refractories, not having been fired prior to installation, are more resistant to thermal spalling than are their fired-brick counterparts, at least until they have become vitrified or slagged in service.

Being lower in strength than fired brick, however, monolithic refractories may be more subject to mechanical spalling.

Thermal Conductivity

When a furnace is heated, thermal energy flows into the refractory structure, causing a temperature difference to develop between inside and outside surfaces of walls and arches. Part of this thermal energy is stored in the refractory structure and its foundation, and part flows through the walls, roofs, and hearths and is lost to the outside air through radiation and convection. The amount of heat which so escapes is often of considerable importance in the economy of the process. In estimating the quantities of heat flowing through the parts of a furnace, use is made of a property known as the *thermal conductivity* or *K value* of the material involved in the construction. The K value differs not only for different materials, but usually also for the same material at different temperatures (see Table 4.7).

The main factors influencing thermal conductivity are the chemical and mineral composition of the refractory, the amount of amorphous material (glass or liquid), and the porosity and temperature of the refractory.

For any particular type of refractory, the proportion of pore space is unquestionably the most important factor affecting the amount of heat which will flow through it at a given temperature. Within the temperature range covered, the thermal conductivity decreases with increasing porosity.

At atmospheric temperatures, the thermal conductivity of a glass is considerably lower than that of crystalline material of the same composition. With rising temperatures, the conductivity of glasses tends to increase, while that of crystalline material tends to decrease. However, in refractory bodies consisting of crystal aggregates with a limited amount of glass, this effect is often obscured by other factors. The conductivity of a refractory in service at high temperature is changed somewhat either by an increase in the amount of glass or liquid it contains or by devitrification of the glass.

High thermal conductivity is desirable for refractories used in constructions requiring the transfer of heat through brickwork, as in retorts, muffles, by-product coke-oven walls, and recuperators. In most types of furnaces, however, low thermal conductivity is desirable for heat conservation but is usually less important than other properties of the refractories.

Heat Flow

The rate of heat flow in refractory constructions can be calculated only approximately for several reasons, including the following:

1. The thermal conductivity values, on which the computations are based, are seldom known with a high degree of accuracy.
2. The temperatures of the hotter and the cooler faces of the refractory wall or roof are rarely known with exactness.
3. The conductivity of a refractory may be altered in service by mineral changes in the refractory, vitrification of the refractory, or absorption of slags or other materials.
4. The thickness of a furnace lining may not remain constant during operation. The lining may become thinner be-

Table 4.7. *K* Values for Refractory Brick at Various Mean Temperatures, Btu · in/ft² · h · °F

Type of brick	Mean temperature, °F 600	1200	1500	1800	2200	2600
Fireclay						
Superduty	9.8	9.9	10.2	10.5	11.4	12.8
High-duty	8.1	8.3	8.5	8.7	9.2	10.0
High-alumina						
60% class	13.0	12.9	13.1	13.3	14.1	15.7
70% class	15.6	14.4	14.4	14.3	14.6	14.9
85% class	18.4	16.2	16.9	17.5	19.6	22.9
90% class	21.9	18.5	17.6	17.6	17.9	18.8
Corundum class	34.6	22.6	20.7	18.7	17.9	18.3
Silica (superduty)	9.0	10.1	11.0	11.8	13.5	16.1
Basic						
Magnesite, fired	73.2	43.9	36.1	31.7	30.5	32.3
Magnesite-chrome, fired	17.9	15.2	14.7	14.2	14.7	16.2
Magnesite-chrome, unburned	18.4	17.2	15.7	14.5	14.7	16.0
Chrome, fired	15.2	15.0	15.1	14.5	13.3	12.9
Chrome-magnesite, fired	11.8	12.3	12.4	12.5	12.5	13.6
Magnesite-carbon	150.0	115.0	106.0	100.0	95.0	90.0
Silicon carbide	121.2	112.0	107.5	103.0	97.0	94.6
Zircon	22.4	16.5	16.6	16.6	16.8	17.9

CONVERSION FACTOR: 1 Btu · in/ft² · h · °F = 0.144 W/m · K.

cause of wear and transmit more heat or thicker because of the formation of a solid coating (as in rotary cement kilns) and transmit less heat.

5. The rate of heat flow is influenced by the pressure of the furnace gases and the permeability of the refractory. A positive pressure tends to force hot gases out through the walls and a negative pressure, to draw cold outside air into the furnace.

6. The rate of heat flow may be influenced by the thickness of the joints and the character of the bonding mortar.

Other important factors which affect the amount of heat flowing through refractory constructions include the emissivity of the refractory or metal furnace shell, the kinds of gases within the furnace, and external convection currents and other air currents.

Effects of Gases

In refractory structures built of insulating firebrick or other very porous refractories, the type of furnace atmosphere can have a very appreciable effect on heat loss through the refractory walls. This is especially true of atmospheres with a high content of hydrogen.

Most protective atmospheres, including dissociated ammonia, exothermic gases, and endothermic gases, contain appreciable percentages of hydrogen. The conductivity of the hydrogen must be taken into consideration in calculating the heat losses through the refractory walls of controlled-atmosphere, heat-treating furnaces.

The thermal conductivity of hydrogen is approximately seven times as great as that of air. Consequently, the presence of hydrogen gas in the pores of a porous refractory product increases the rate of heat flow through the refractory. With an atmosphere of 100 percent hydrogen, the heat flow through an insulating firebrick or other refractory with similar porosity would be two to two and one-half times as great

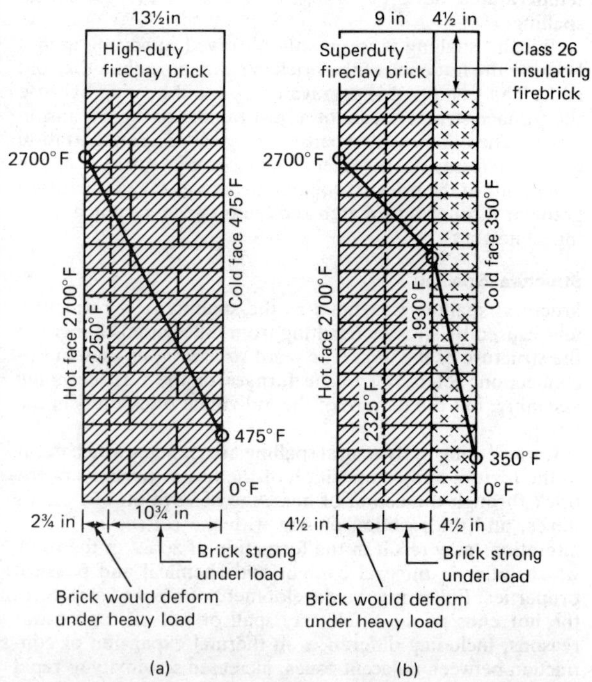

Fig. 4.3 Temperature gradients through furnace walls. (*a*) Uninsulated fireclay brick wall; (*b*) insulated fireclay brick wall. [Conversion factors: 1 in = 2.54 cm; °C = (°F − 32)/1.8.]

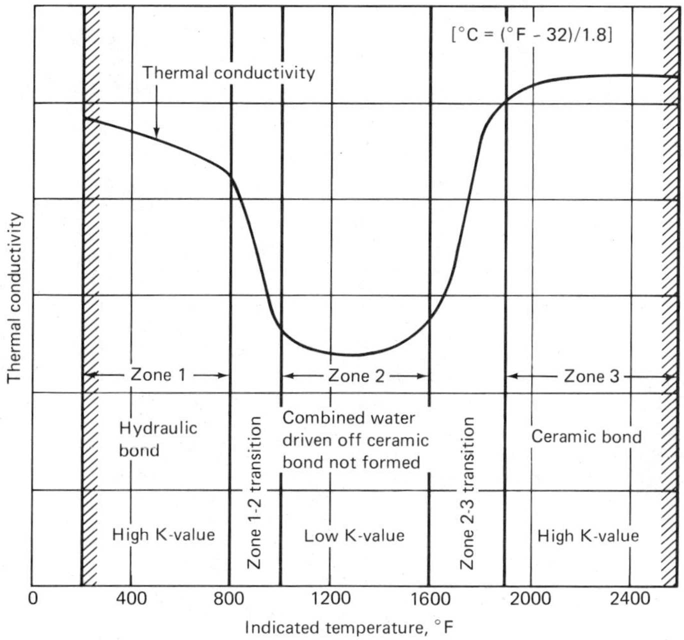

Fig. 4.4 Variations in thermal conductivity through a castable wall heated from one side only.

as in an atmosphere of air. At lower concentrations of hydrogen, the increase will be less and the relationship is almost directly proportional to the percentage of hydrogen present.

As compared with the conductivity of air, that of nitrogen is 100 percent, carbon dioxide about 59 percent, and carbon monoxide about 97 percent.

Figure 4.3 shows the thermal gradients through (*a*) an uninsulated, one-component refractory wall and (*b*) an insulated two-component refractory wall. Addition of more insulation will result in a shallower temperature gradient in the working lining of a vessel and allow deeper penetration of metal and slag into the lining.

Figure 4.4 shows the rather complicated situation relative to thermal conductivity that can occur in a refractory castable wall during service. In calculating heat flow in this case, it is necessary to know the *K* value for the product after a given thermal history.

Figure 4.5 presents a graph useful for calculating the heat-transfer properties of refractories. It indicates approximate heat loss in relation to outside face temperatures.

4.3 SELECTION OF REFRACTORIES

Factors Governing Selection

Economy in the use of refractories is governed by the selection of the kinds best suited for the intended purpose. A careful study of furnace design and operating conditions will greatly aid in the selection.

From the multiplicity of factors listed below, it may appear

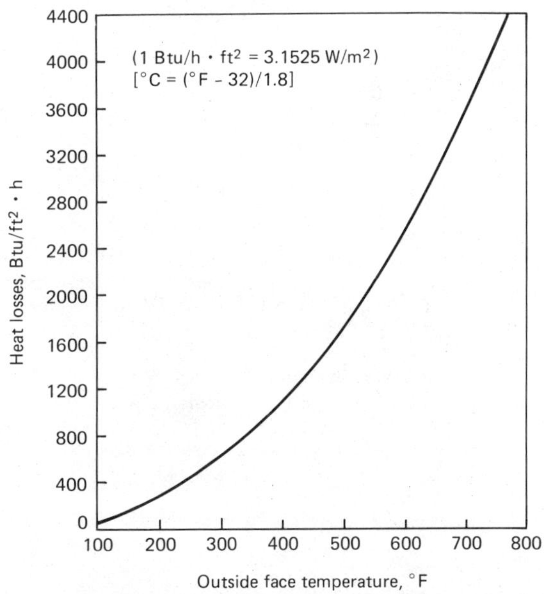

Fig. 4.5 Heat loss from vertical wall to still air at 70°F through radiation and convection.

that the proper choice of the most suitable material would be exceedingly difficult. Sometimes this is true. However, there are usually data on hand from previous experience under similar conditions.

Moreover, the best refractory selection often depends on a few requirements so important that the other factors will play a minor role. In some cases refractoriness, maximum service temperature, alone will be the deciding factor, and in others high refractoriness will have to be coupled with resistance to thermal shock. In still other cases resistance to metals, slags, or disintegration by reducing gases may be the governing factor. In some cases high insulating value is desirable, but in others high thermal conductivity may be needed. Major factors need to be identified.

1. Operating factors
   - Function of the furnace
   - Nature of material being processed
   - Rate and continuity of operation
   - Range and rapidity of temperature changes
   - Chemical attack by metals, slags, ash, etc.
   - Fluidity of molten metal or slag
   - Velocity of furnace gases
   - Abrasion from contained solids or gases
   - Impact from charging
   - Erosion by molten furnace contents
   - Impinging flames or hot spots
2. Furnace design and construction
   - Type of furnace
   - Design and dimensions of walls and arches
   - Loads imposed on the lining
   - Conditions of heating (one or more sides)
   - Amount of insulation
   - Air- or water-cooling
   - Type of refractory construction (brick or monoliths)
   - Methods of bonding or support
   - Provision for thermal expansion
   - Mechanics of any moving furnace parts
3. Factors related to the refractories themselves
   a. Properties at room temperature (brick only)
      - Workmanship and physical strength
      - Density, porosity, permeability
      - Chemical and mineral composition
      - Uniformity
      - Size and design

   b. Properties at high temperatures (all types)
      - Refractoriness or maximum service temperature
      - Reversible thermal expansion
      - Resistance to thermal shock
      - Resistance to chemical attack
      - Resistance to mechanical impact or stress
      - Resistance to abrasion or erosion
      - Permeability to gases or liquids
      - Volume stability (bloating or shrinkage)
      - Resistance to gases and fumes
      - Thermal conductivity
      - Heat capacity
      - Electrical resistance

   c. Economic factors
      - Delivered cost
      - Cost of installation (brick vs. monoliths)
      - Special shapes or forming required
      - Service life

Special Factors for Monolithic Refractories

1. *Estimating coverage:* Instead of making brick counts, monolithic construction requirements are determined by calculating the volume to be filled and the pounds of monolith required to fill that volume. Monolithic material requirements per cubic foot (or kilograms per cubic meter) are typically shown on industrial product data sheets.

2. *Anchoring:* Previously, monolithic refractories were used because they are quicker and easier to install than prefired brick shapes. However, in many cases monolithic refractory installations require anchors to ensure that the material remains securely in place. Further, the anchors must be affixed to the furnace shell or some sort of supporting structure by bolting or welding. There are, therefore, some extra expenses involved in monolithic refractory construction.

Anchors are made from carbon steel or stainless wire, rod, or castings as well as prefired refractory shapes. Refractory manufacturers will usually supply any anchors needed as well as estimates of requirements. Figure 4.6 shows a typical furnace construction employing both metal and ceramic anchors.

3. *Initial heat-up:* A controlled heat-up schedule is required for most monolithic refractories, particularly castables or concretes bonded with hydraulic setting calcium aluminate cements. Heat will remove the free or mechanical water as steam at 212°F (100°C). Water of hydration will be driven off between 400 and 1600°F (205 to 872°C). Initial heating rates should raise the temperature about 50°F (10°C) per hour. Holds (one h per inch of thickness) should be incorporated at 400 to 500°F (205 to 260°C), 1000 to 1200°F (540 to 650°C), and the maximum preheat temperature.

Ordering Refractories

Refractory technology has become increasingly specialized in recent years; thus it is often necessary to understand an application area well before selecting refractories. Technical representatives of the larger refractory suppliers have been trained to do this and are available for consultation.

When ordering standard brick sizes and shapes or monolithic refractories, it is not necessary to send the suppliers furnace drawings. A complete description and correct dimensions for the shapes required will usually suffice. When ordering shapes designed to fit a circular lining, provide both the inside and outside diameters of the lining.

Orders for brick should include enough mortar material of the right kind to lay the brick. Information on these points is again available from the refractory supplier.

4.4 FURNACE DESIGN AND CONSTRUCTION

Sound furnace design and construction—whether carried out by experienced masonry specialists working on a high-production furnace or by a contractor building a municipal incinerator—begins with a few fundamental ideas required to produce satisfactory performance.

Furnace Foundations

The foundation must function at the temperature produced by the furnace. For many industrial furnaces, contractors build foundations of concrete, consisting of a crushed-stone aggregate, sand, and binder of hydrated Portland cement.

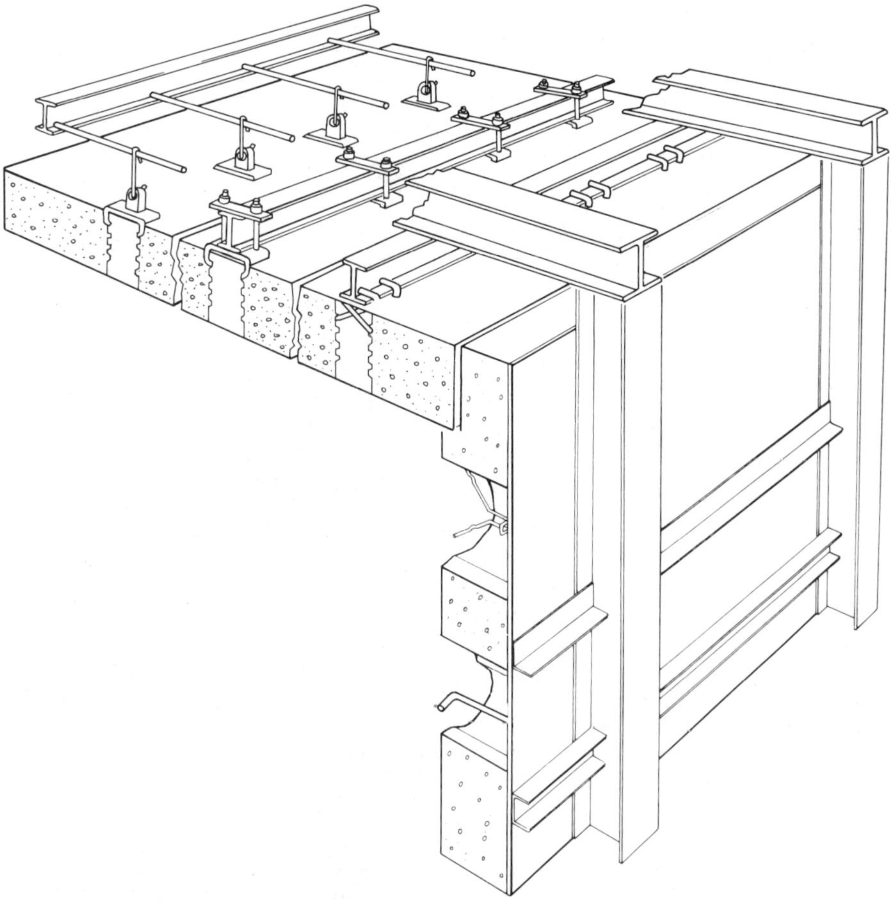

Fig. 4.6 Typical castable furnace construction showing ceramic anchor systems in the roof and metallic anchors in the sidewall.

Under normal conditions, Portland cement concrete has been safely used for furnace foundations of up to 700°F (371°C). When the temperature reaches 900°F (483°C), dehydration of the cement reaches a point where the concrete retains little mechanical strength.

Ordinary aggregates include quartz pebbles, silica gravel, crushed silica rock, and/or crushed limestone. An aggregate of silica rock in any form will expand sufficiently at temperatures up to 1000°F (538°C) to set up stresses in the concrete and weaken the foundation. Limestone or dolomite rock in an aggregate will calcine at somewhat higher temperatures and weaken considerably at temperatures above 700°F (371°C). Calcined fireclay, in sizes of up to 1 in (2.54 cm), can be substituted for conventional aggregate. Its thermal expansion is low, and it will not shrink at the highest temperature to which concrete is normally subjected.

As mentioned, however, Portland cement should not be used above 900°F (483°C). When this temperature may be exceeded in service, it is simpler and safer to use a prepared refractory castable which contains both heat resistant aggregate and cement phases.

Ventilation

High-temperature furnace operation also calls for ventilation in the lower courses of brickwork or the upper part of the foundation. Good furnace design often requires placement of the furnace on plates, girders, or low brick piers to allow air circulation under the vessel. Sometimes cross-flues are formed in the top of the concrete foundation. At others, pipes 3 in (7.62 cm) in diameter or larger are embedded in the foundation.

Course Configurations

Courses of brick laid in a wall so that the lengths parallel the face of the wall are called *stretchers*. Brick whose length runs at right angles to the face are headers. In soldier courses the brick stand on end, and in rowlock courses they lie on one edge.

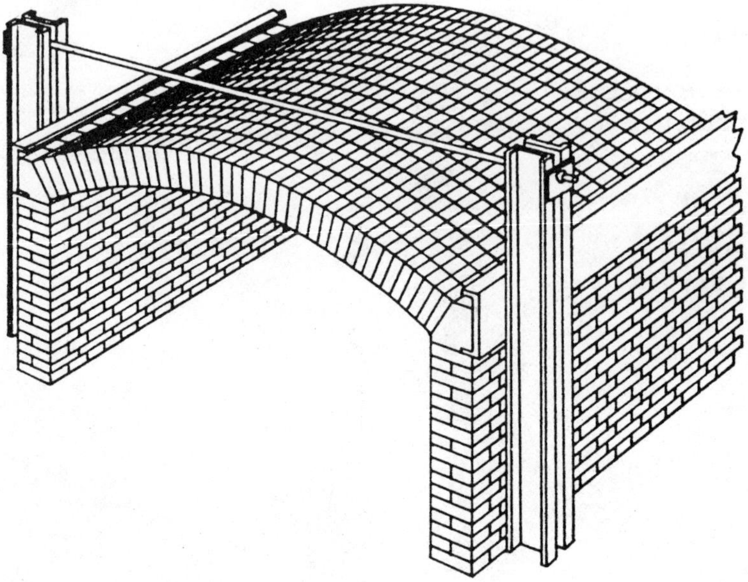

Fig. 4.7 Ring arch roof construction.

Header courses tend to spall less than do stretchers at the hot face of a furnace wall because they expose a smaller area to the high temperature. However, stretcher courses expose fewer joints than do headers, which is advantageous in applications where joints tend to wear more rapidly than brick.

Bonding, tightening construction through combinations of headers and stretchers and offsetting vertical joints, strengthens and stabilizes furnace walls. The type of bond selected for any particular furnace will depend on the design of the furnace, thickness of the walls, the need for gastight construction, severity of operating conditions, and the need for easy maintenance.

In any case, the wall should be bonded to permit transfer of the loads to the cooler part of the wall when the inner, hotter portion loses its ability to carry them. Walls must be designed to carry structural loads at high temperatures.

Alternate header and stretcher courses probably provide the most common arrangement for standard industrial furnaces. Large [9-in (22.9-cm)] brick break joints start ends of walls and turn corners. This is the construction shown in the sidewalls of the furnace in Fig. 4.7.

Courses mostly consisting of headers are often used advantageously in 9- and 13½-in (22.9- and 34.3-cm) walls subject to high temperatures, heavy loads, and slag attack. This construction is usually preferred for basic brick walls. This bond provides stability and easy replacement, but most expansion joints pass entirely through 9-in (22.9-cm) walls.

In composite wall construction consisting of two or more kinds of brick in inner and outer courses, the courses are sometimes tied together. Usually, the more refractory brick goes into the interlocking courses.

However, when brick have marked differences in rates of thermal expansion, the backup courses should not be tied to the inner courses. This is especially true when the temperature gradient through the wall makes a significant difference in total expansion.

Brick Counts

The number of refractory straights required to build a simple wall can be determined from Table 4.8. Multiplying brick per square foot from the appropriate row, depending on wall thickness, by the area of the wall will provide the brick count. For walls not of simple rectangular shapes, determine the volume in cubic feet and multiply by the number of brick per cubic foot.

Table 4.8. Number of Refractory Straights Required per Square Foot of Wall or Floor

Wall or floor thickness, in	Brick dimensions, in	
	9 × 4½ × 2½	9 × 4½ × 3
2½	3.6	
3	. . .	3.6
4½	6.4	5.3
5	7.2	
6	. . .	7.2
7½	10.8	8.9
9	12.8	10.7
13½	19.2	16.0
18	25.6	21.4
27	38.4	32.1
Number of brick per cubic foot	17.0	14.2

Wall Thickness Versus Height

Wall thickness must bear some relation to height and unsupported length. In straight walls not supported, a 4½-in (11.4-cm) thickness will carry heights of up to 3 ft (0.91 m); a 9-in (22.9-cm) thickness, 3 to 8 ft (0.91 to 2.44 m); a 13½-in (34.3-cm) thickness, 8 to 12 ft (2.44 to 3.66 m); and 18-in (45.7-cm) thickness, walls higher than 12 ft (3.66 m).

Walls with unsupported length more than one and one-half times their height should be somewhat thicker, and thermal spalling conditions may indicate additional thickness. Walls of cylindrical furnaces and stacks, with adequate backing, may be somewhat thinner for a given height than straight walls.

Cylindrical Construction

Cylindrical walls, arches, and domes are built with brick tapered to turn circles. Arch brick slope from edge to edge so that the length of the brick parallels the furnace wall like a stretcher while the wedge tapers from end to end so that it faces into the furnace like a header. A 9 × 4½ × 2½-in (22.9 × 11.4 × 6.4-cm) arch makes a 4½-inch (11.4-cm) lining, while the same wedge shape makes a 9-in (22.9-cm) lining. In basic brick, key shapes may also taper along the edges. Combinations of these shapes taper in two dimensions to turn domes.

The wall shown in Fig. 4.8 is built of "circle" brick, which, unlike wedges, arches, or keys, have rounded inside and outside chords. Figure 4.8 shows the circle brick being set on a course of brick laid on the flat, which is a common method of placing floor brick.

Monolithic Construction

In many cases, brick sizes and shapes or the type of bond will be chosen to minimize the number of joints in the lining. Monoliths—not without construction or thermal expansion joints—present the fewest joints and opportunities for penetration by metal, slag, or furnace gases.

Ramming mixes or castable refractory materials are often used to fill places where brick would be cut to fit. For example, ramming mixes or dry refractory materials can be used to protect the toe of the skewback when building arches. On many installations the irregular space between electric furnace roof brick and electrode ports is filled with ramming mix or castable refractories.

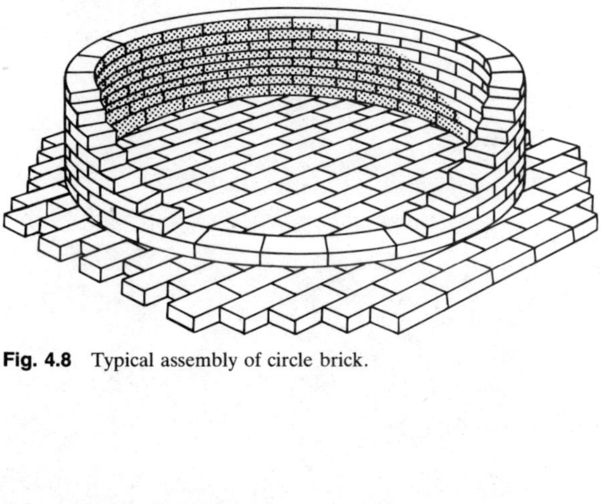

Fig. 4.8 Typical assembly of circle brick.

Joint Thickness

Brickwork laid with heat-setting mortars should have thin joints, either dipped or poured. The brick should be rubbed or tapped into place to produce as much brick-to-brick contact as possible. Joints made with an air-setting mortar generally can be made somewhat thicker, but such joints should be completely filled.

Thermal Expansion Allowance

In furnace construction, proper allowance must be made for thermal expansion. Usually, vertical expansion allowances permit walls to move freely upward, and horizontal expansion allowances appear at joints in the brick. The subject is covered in some detail in the earlier discussion of thermal expansion.

Furnace Hearths

The construction of furnace hearths presents special problems. Some furnace bottoms must withstand impact and abrasion from a charge of scrap metal. Many must resist penetration by metal or slag accompanied by corrosion or erosion.

Liquid pressure may tend to float out hearth brick. When this is the case, invert arch or invert dome bottom construction may be very helpful.

Many furnace hearths are built of refractory brick, usually seated on a monolithic refractory bed. Others have subbottoms constructed of brick with working hearths composed of monolithic refractories such as dead-burned magnesite or a ramming mix.

Arch Construction

Arches form the roofs of most furnaces, combustion chambers, and flues, providing the standard solution to the problem of spanning the high-temperature process with refractories. Sometimes arches span wall openings, and sometimes they carry the weight of walls or checkerwork. Most arches are built of brick, but monolithic materials are gaining popularity throughout the industry.

In a true arch, the design of the whole determines the shape of each block or structural unit. Theoretically, each joint is a small piece of the radius of the circle of which the arch is a segment. Each end of the arch rests on a skewback. The arch becomes self-supporting after all of the pieces fit into place, but it must be supported until the final, center shapes—the keys—fit into place. When it is complete, the arch "springs" from the sloping faces of the skewback shapes. The skewbacks cut off the arc of the circle on the outside radii of the arc.

The sprung arch exerts—on the skewbacks—a downward vertical force and an outward horizontal force, a distribution of its weight. The vertical force may be carried by steel beams or by the furnace walls or a combination of walls and steel buttresses. The horizontal thrust of a roof arch travels through the skewbacks to a steel supporting system known as the *binding* and composed of beams and tie rods. The tie rods, usually above the furnace, link one side of the furnace to the other, and balance opposing forces one against the other.

In traditional furnace design, the binding consists of:

1. Horizontal buttress beams running lengthwise of the furnace in contact with the skewbacks where possible

2. Vertical beams, "buckstays," spaced at intervals along the furnace walls and usually set in concrete

3. Horizontal tie rods, I-beams, or channels extending across the furnace above the roof to connect the upper ends of opposite buckstays

In ring arches, each course of brick forms a separate ring running across the roof and the joints are continuous across the roof. Ring arches require somewhat less labor for initial construction. Cold repairs are easier to make and offer better resistance to spalling. However, ring arches require support at the end of the furnace to forestall outward displacement.

Figure 4.7 shows ring arch construction and also some details on the placement of skewbacks, beams, and buckstays.

In bonded construction, all joints are broken and the rings interbond for a stronger construction. Bonded roofs are better adapted for hot repairs but demand more skill of the brick masons and more uniformity in the brick.

Sometimes furnace designers strengthen ring arch roofs by using a longer brick in every third or fourth ring. This construction creates ribs across the roof, and the ribs remain strong when the thinner parts of the roof wear away. The strength of the ribs helps when roofs must be patched.

In suspended arches, a steel superstructure helps to carry the weight of the roof—otherwise distributed through the arch to the walls and binding. Suspended arches are often used with dense, heavy basic brick.

In suspended construction, the refractories carry smaller loads and therefore suffer less deformation at high temperatures than they otherwise might. Furthermore, it is easier to make allowance for thermal expansion to avoid thermal stresses and pinch spalling with suspended construction.

Monolithic arches—cast-over forms much like those required for brick arches—have replaced brick in some applications. Skews may be eliminated in monolithic arches, but design considerations remain the same as those for brick arches. However, a monolithic arch should provide all of the advantages of monolithic construction, including reduced cost and downtime.

4.5 THERMAL-MECHANICAL ANALYSIS OF REFRACTORY STRUCTURES

In the earlier section on thermal effects on refractories, it will be noted that factors such as *thermal strress, stability under hot load, thermal expansion, thermal spalling,* and *thermal conductivity* (or *heat flow*) were all discussed in terms of simplified steady-state situations—as if furnaces and high temperature vessels do not experience change in service.

In the section on heat flow, there was a brief description of the differences existing during heat-up of a furnace and after heat-up is completed and steady state has been reached. Heat-up is just one example of a non-steady-state situation. When furnaces are filled and emptied, ladles are tipped to pour, or linings wear thinner in service, there are also non-steady-state conditions.

In the section on heat flow, it was also noted that calculations made to cover non-steady-state conditions are quite complicated and beyond the scope of the present discussion. This is also true of stress calculations.

However, demand for non-steady-state calculations has been increasing, and recently, with the assistance of computers, it has become possible to study certain thermal effects in dynamic settings.[5]

REFERENCES

1. F. H. Norton, *Refractories,* 4th ed., McGraw-Hill, New York, 1968.
2. C. B. Clark and J. S. McDowell, "Refractories," in *Industrial Minerals and Rocks,* American Institute of Mining and Metallurgical Engineers, Calif., 1961.
3. Harbison-Walker Refractories Company, *Modern Refractory Practice,* William Feather Co., Cleveland, 1961.
4. Harbison-Walker Refractories Company, *Handbook of Refractory Practice* (published internally), 1980.
5. C. A. Schacht, "Improved Analytical Procedures and Material Properties for Predicting the Thermomechanical Behavior of Refractory Linings of Teeming Ladles," *Ceramic Engineering and Science Proceedings*, American Ceramic Society, vol. 7 (Jan.–Feb. 1986), pp. 209–219.

Chapter 5

Thermally Conductive Plastics

RAYMOND E. MESLOH
Battelle Memorial Institute

Generally, one considers plastics as excellent insulators, both electrically and thermally—and they are. In fact, for many applications plastics provide a distinct advantage over metals, e.g., in areas of appearance, cost, weight, formability, maintenance, and electrical properties. However, plastics could fulfill many applications if they were slightly more thermally conductive.

The disadvantages of the low thermal conductivity of plastics become apparent in both product characteristics and manufacturing processes, especially in the more sensitive electronics applications. In plastic applications which have inadequate performance as a result of heat buildup, thermal distortion, and heat degradation, a solution is available in the form of thermally conductive metal-polymer composites.

5.1 MODIFIER MATERIAL CONSIDERATIONS

Fillers and reinforcements have been used to enhance the characteristics of pure plastic resins. Cost, strength, stiffness, heat deflection temperature, and coefficient of friction are just a few of the properties altered. The thermal conductivity of a resin can be modified as well. This is accomplished by choosing a filler which has a conductivity higher than that of the base resin. Metal and ceramic-glass-based particulates have been used, with particle size and shape ranging from the low-aspect-ratio chunky blocks and spheres to the higher-aspect-ratio platelets, flakes, and fibers. Considerable effort by many investigators[1] has been given to developing methods for increasing the conductivity of plastic in this and other manners.

The four qualities of a filler, important in its choice, are the aspect ratio, chemical compatibility with the resin, inherent properties (tensile stength, elastic modulus, thermal conductivity, and density), and processibility. Additives to plastics which are used in large concentrations, up to 50 percent, and which give little property enhancement are termed *fillers*. Fillers, in general, have low aspect ratios, are mineral in nature (and thus compatible with most resins), and are readily processible because of their powder or small granular shape. Their strength characteristics play a minor role in the composite performance since the small size greatly reduces the opportunity for bonding with the polymer and subsequent reinforcement.

Once the aspect ratio of the inorganic component exceeds 20:1, property improvement begins to escalate. Depending on the particle surface chemistry, reinforcement can occur at an aspect ratio as low as 20:1, or, with no compatibility, require an aspect ratio of as high as 100:1.

Chemical compatibility has two aspects, stability with and adhesion to the resin matrix. A resin and filler system can be tailored to increase compatibility by altering the resin backbone, the metal filler alloy, or the surface chemistry through a coating process.

A compatible system takes advantage of the reinforcing effect of the filler, so the inherent strength of the filler particle becomes important. Not all chemically identical products exhibit equal mechanical or thermal properties since their strengths are also a function of how the products are made.

Fillers of identical mechanical properties can exhibit different levels of processibility. This is often due to chemical compatibility with the resin or the surface roughness of the particles. Discriminatory evaluation usually demands actual on-line testing.

General Conductivity Modifiers

The thermal conductivity of most plastics falls in the range of 0.05 to 0.2 Btu/h · ft² · °F/ft (0.086 to 0.346 W/m · °C). Except for plastic foams [which go as low as 0.01 Btu/h · ft² · °F/ft (0.0173 W/m · °C) as a result of the void-gas spaces], the resins are at the lowest end of the range of conductivities for solid materials at room temperature.

Plastics can be changed into partially conductive materials or "*semi-insulators*" with the addition of certain modifiers. In fact, any material having a conductivity greater than the plastic can be used as a conductivity enhancer. However, considering other factors and composite properties, not all materials would be desirable or even usable. Some common materials and forms for filling plastics are:

Alumina	Powder
Aluminum	Fiber, flake, powder
Asbestos	Fibers
Brass	Powder
Calcium carbonate	Powder
Glass	Fiber, flake, powder, mats
Graphite	Fiber, flake, powder
Kaolin	Powder
Mica	Flakes
Magnesium oxide	Powder
Molybdenum disulfide	Powder
PTFE	Powder
Stainless steel	Fibers
Talc	Powder

However, within each family of fillers—the minerals, the glasses, and the metals—the variations in conductivity have only a small effect on the thermal conductivity of the composite. Glass, while usually added as a reinforcing agent to improve tensile strength and modulus properties of a composite, also increases the thermal conductivity. The normal range of conductivity for base polymers is 0.1 to 0.2 Btu/h · ft² · °F/ft (0.173 to 0.346 W/m · °C); 20 to 30 percent glass-reinforced versions increase conductivity to 0.15 to 0.25 Btu/h · ft² · °F/ft (0.26 to 0.432 W/m · °C). Talc-filled versions, even if chemically coupled to improve mechanical properties, may show virtually no change in thermal conductivity. Metal powders, such as iron and aluminum, can increase heat transfer through a resin. However, such materials must be used at high loadings, often over 50 percent by weight, because of their low aspect ratio and high density.

The step beyond the powder fillers and modified resins is those resins whose property profiles are changed dramatically by the incorporation of high-aspect-ratio conductive modifiers. This family of particulates includes metalized glass fibers, graphite fibers, stainless-steel fibers, and quick-quenched aluminum fibers and flakes.

High-Aspect-Ratio Materials

Fibers and flakes are characterized by high aspect ratios (ratio of length to diameter or thickness) which aid in thermal conduction as well as reinforcement of the resin. Flakes tend to orient parallel to the flow of the resin and form layered parts. This gives a high content of filler in minimal thickness. Fibers also orient, but their ends give rise to high stress concentrations leading to warpage and dimensional change due to stress relief in service. Spheres are lower in stress concentration effects but are also lower in conductivity enhancement because of their low aspect ratio.

In order to be used effectively, these reinforcing fillers must be amenable to the process intended. The initial high aspect ratio of a long brittle material is quickly reduced in the tortuous high-shear environment of extrusion and injection molding, while other particles, short to begin with, give up the efficiency of greater aspect ratio. In all cases efficiency of these modifiers is dependent on the aspect ratio of the particle in the final molded part.

Graphite fibers are being used with greater frequency in high-performance applications, usually for their reinforcing properties, although a slight improvement in thermal conductivity is a bonus in some applications. Metalized coatings do give an appreciable increase in conductivity.

Glass fibers metalized by vacuum deposition or dip coating with aluminum or nickel have been tested for electrical and thermal property enhancement. These products, like the carbon fibers, must be processed as little as possible to avoid particle breakup in use. When employed in sheet molding compound (SMC) and some bulk molding compound (BMC) compounds, sufficient aspect ratio remains to give better thermal and electrical conductivity than that seen in injection molding applications.

For levels of improvement beyond that which is obtained with glass, graphite, and mineral fillers, it is necessary to limit the composite development to the metal fillers. These typically do not give the high reinforcement values possible with long glass or graphite fibers but do enhance thermal conductivity to an even greater extent.

Metallic Conductivity Enhancers

The use of metal particulates as additives is not new to the polymer industry. Fine aluminum and copper powders, ball-milled until a flat platelet form is achieved, are commonly blended into the plastic at levels of 1 to 5 percent to obtain a metallic appearance. However, no increase in thermal conductivity will be seen.

Aluminum, iron, and ferrite powders are added to plastic resins to enhance or develop specific properties. A 60 percent by weight loading of a ferrite will give the compound magnetic properties. A loading of 40 percent aluminum powder will increase the heat conductivity of a formulation nearly as much as a 70 percent loading of bronze or steel powder. Even at these high levels, the increase in conductivity is limited because of the low aspect ratio of the spherical powder.

To achieve the high aspect ratio needed for property increase, the metal must be either drawn into a wire or formed into a flake. In both cases, it is necessary to reduce the thickness of the particle to a range below 0.002 in (50 μm). This serves the dual purpose of improving processibility and allows a high aspect ratio to be achieved without unmanageable particle length.

The aspect ratio of metal fibers can be varied by changing the thickness or the length. Although efficiency increases in direct proportion to the increase in aspect ratio, processibility tends to decrease. Commonly used aluminum fibers are 0.003 in (80 μm) thick and 0.125 in (3 mm) long. Steel fibers of this thickness could destroy a plastic injection molding machine, so the commonly used products range from 0.0005 to 0.002 in (12 to 50 μm), with L/D ratios of 1000.

Metallic flake and fiber fillers[2,3] for plastics are now made by a unique rapid solidification process called *melt spinning*. When compounded into a typical resin system, these flakes cause an increase in flexural modulus, in electrical conductivity, and a significant increase in thermal conductivity. Equal loadings of aluminum powder and flake in two specimens of phenolic resin produced ratios of 2/1 and 8.4/1 increase in thermal conductivity, respectively.

5.2 THERMAL AND MECHANICAL PROPERTIES

In the development of a thermally conductive composite, the choice of plastic resin is based on chemical resistance, heat deflection temperature, and processibility. The choice of flake or other filler is based on the degree of conductivity needed. Because the conductivity is a function of the aspect ratio, the mixing process must not cause a significant degradation of the flake. Loadings of 35 volume percent, about 50 to 55 weight percent, have been achieved, and in some cases higher loadings are possible. Because the effect of a filler on a resin is based on its volume loading, not weight loading, the same effective volume loading of flake in several polymers will be several different weight percents.

Presented in Table 5.1 are typical thermal and mechanical properties of modified plastics with high-aspect-ratio metallic fillers.

The data in Table 5.1 assume that good adhesion exists between the resin and the filler. If not, and a thin gas layer exists around the filler particles and is prevalent throughout the composite, the thermal conductivity of the composite could actually decrease. Often the lack of good adhesion will become a problem at high loading values even though it is

Table 5.1. Thermally Conductive Plastic Properties

Resin base	Specific gravity	Tensile strength, ksi	Composite approximate cost, $/lb	Fill rate, wt% material	Thermal conductivity*	
					Filled	Unfilled
Aluminum flake						
ABS	1.54	3.3	3.50	40	0.51	0.10
ABS/PC	1.54	5.2	2.85	40	0.54	0.10
PC	1.54	6.4	3.60	40	0.57	0.11
Nylon 6,6	1.48	9.5	3.35	40	0.48	0.14
PBT	1.63	7.8	3.30	40	0.56	0.10
Stainless-steel fiber						
ABS	1.33	5.3	5.00	7	0.11	
PC	1.27	10.0	5.00	7	0.13	
Nipan fiber						
ABS	1.20	13.0	12.00	20	0.33	
PC	1.40	15.0	12.75	20	0.46	
Nylon 6,6	1.38	19.5	15.75	30	0.58	
Glass						
ABS	...	14.5	...	30	0.13	
PBT	...	22.0	...	40	0.29	
PC	...	18.5	...	30	0.21	
Polyethylene	...	10.0	...	30	0.22	0.17
Polystyrene	...	13.5	...	30	0.17	0.12

*Measured through the plane, Btu/h · ft² · °F/ft × 1.73 = W/m · °C.

SOURCE: The properties listed were adopted from published data of the following companies: ComAlloy International, Nashville, Tenn.; LNP Corp., Malvern, Pa.; Mobay Chemical Corp., Pittsburgh, Pa.; Wilson Fiberfil International, Evansville, Ind. The approximate costs of the filled composites were obtained through verbal communications with employees of these companies.

not experienced at lower ones. This is related to processing parameters, poor wetting factors, and voids between closely packed particles not being filled with resin.

The geometric shape of the filler has a very pronounced effect on the thermal conductivity of the resultant composite. Three-, two-, and one-dimensional shapes—spheres, flakes, and fibers—can be added alone or in combination to obtain various properties. The most prominent change is the increase in both electrical and thermal conductivity. An upper limit of the aspect ratio is on the order of 30 to 40. For higher aspect ratios there is a tendency for the fibers and other high-aspect-ratio fillers to bend and clump, defeating the effect of their initial higher aspect ratio on thermal conductivity. Such behavior will also degrade mechanical properties.

Thermal Property Estimation

The thermal conductivity of filled plastics can be determined either through theoretical estimation or by direct experimental measurement by standard tests. The experimental method is more accurate if care is taken in setting up the test, but generally more time-consuming than the theoretical estimation technique. Theoretical prediction is best applied in the extension of test data from one plastic to another plastic with the same filler. The following are parameters to be considered in a theoretical model:

1. Filler conductivity
2. Shape and aspect ratio
3. Resin conductivity
4. Volume loading
5. Packing and orientation
6. Bonding

In addition, it is possible to estimate the thermal conductivities for various shaped fillers, although with less confidence. There are two other main properties of the modified resin which change as a function of the modifier and its volume fraction—density and specific heat. These two quantities are required in some methods of determining thermal conductivity and in solving transient heat-transfer problems.

The thermal conductivity of the particulate filler is very important in its effect on the composite thermal conductivity. However, when the filler conductivity reaches a "breakover" point, continued increases have little benefit. For metal fillers whose thermal conductivities are appreciably higher (100 to 2000 times) than that of the base resins, variations in thermal conductivity of the alloy have a surprisingly small effect on the conductivity of the composite. In comparing conductivity K of copper, $K = 226$ Btu/h · ft² · °F/ft (391 W/m · °C) to a filler of carbon steel $K = 42$, (72.6 W/m · °C) there is a change in conductivity of 500 percent. However, a change of only 5 to 6 percent in the relative conductivity factor B [see Eq. (5.1)] is produced. Thus, for metal modifiers, their thermal conductivities are not as important as their other properties, e.g., corrosion resistance, ease of forming and manufacture, density, bonding, and aspect ratio.

Rather than alter the composition of the metal filler, a change to a base resin with a higher thermal conductivity should be considered. This could produce a much greater increase in the overall conductivity of the composite plastic.

Thermal conductivity of the filled plastic varies parabolically with percent of loading by volume. The upper limit of fill percentage depends greatly on particle shape and ability to process, as well as the point of degradation of other desirable properties with loading. For high-aspect-ratio modifiers such as flake or fiber loading volumes of up to 30 to 35 percent are achievable, while with spheres a maximum of 50 percent can be obtained with little difficulty.

Thermal Conductivity—Theoretical

Theoretical quantification of the thermal conductivity of a filled plastic serves two purposes, the estimation of the thermal conductivity and the identification of the variables which influence the thermal conductivity. Knowledge of the influential variables permits the composite to be tailored to meet given thermal as well as electrical and mechanical requirements.

Progelhof et al.,[4] in their review of prediction methods for the thermal conductivity of composites, state that the Lewis and Nielsen[5,6] correlation for two phase systems best fits ex-

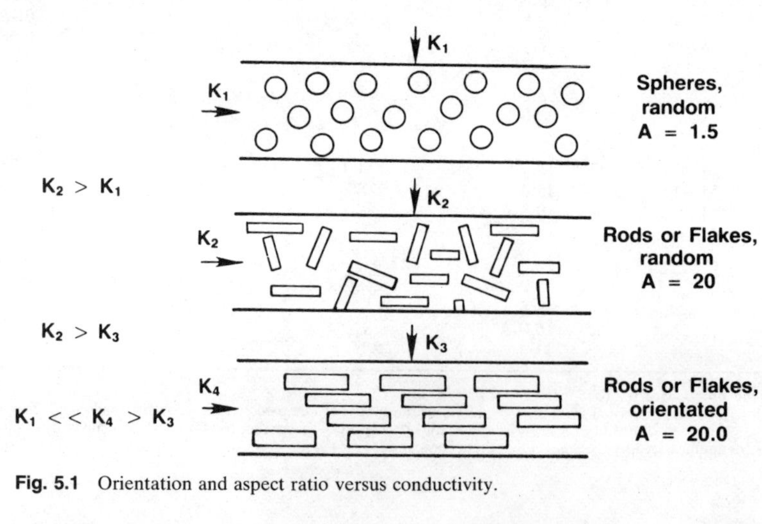

Fig. 5.1 Orientation and aspect ratio versus conductivity.

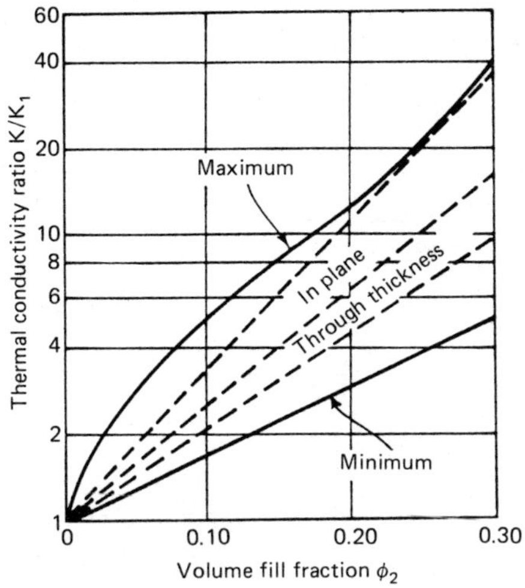

Fig. 5.2 Thermal conductivity ratios as a function of fill fraction for aluminum-flake-filled resins.

perimental data for the range of conductive flake, fiber, and spherical fillers being investigated here. This was also verified by Bigg.[7] The equations are

$$K = K_1 \frac{1 + AB\phi_2}{1 - B\psi\phi_2} \tag{5.1}$$

where

$A = k_e - 1$

$B = (K_2/K_1 - 1)/(K_2/K_1 + A)$

$\psi = 1 + \phi_2 (1 - \phi_m)/\phi_m^2$

K = thermal conductivity composite, Btu/h · ft² · °F/ft (W/m · °C)

K_1 = thermal conductivity base resin, Btu/h · ft² · °F/ft (W/m · °C)

K_2 = thermal conductivity modifier, Btu/h · ft² · °F/ft (W/m · °C)

A = shape constant, $A = f(L/D)$ or $f(L/t)$, nondimensional (nd)

$f(L/D)$ or $f(L/t)$ = function of length to diameter or length to thickness

k_e = Einstein coefficient, nd

B = relative conductivity factor, nd

ψ = packing factor, nd

ϕ_2 = volume loading fraction, nd

ϕ_m = maximum packing fraction, nd

Detailed derivation of this equation and determination of the variables, especially the Einstein coefficient and, thus A, and the maximum packing fraction for a given filler are described by Nielsen[6] and Bigg.[7] As mentioned previously for modifiers with conductivities above a given value, the two main factors which enhance composite conductivities are volume loading and aspect ratio. The effects of aspect ratio and filler orientation on conductivity are shown in Fig. 5.1. A layering tendency is relatively easy to obtain through processing control in molding, but uniform packing is greatly dependent on the part geometry and flow during molding and is more difficult to achieve. Generally for thin cross sections the inplane conductivity of the part is the same in all directions but quite different from the through-thickness conductivity. Thus a minimum of two conductivity determinations must be made to describe the thermal conductivity of the composite. The theoretical ranges for aluminum-flake-filled resins are shown in Fig. 5.2.

Specific Heat, Density, and Volume Percentage

Theoretically, the specific heat for the composite c is given by

$$c = \frac{1}{(1 - W)/c_1 + W/c_2}$$

where c_1 and c_2 are the specific heats of the resin and filler, respectively, and W is the weight fraction of composite.

The densities of a filled plastic may also be calculated given the percentage by weight W (in fractional form) of the filler desired:

$$\gamma = \frac{1}{(1 - W)/\gamma_1 + W/\gamma_2}$$

where γ, γ_1, and γ_2 are the density of composite, base resin, and filler, respectively, and W represents the weight fraction of filler.

Likewise, the percentage by volume in fractional form of the filler in the composite can be calculated:

$$\phi = \frac{\gamma - \gamma_1}{\gamma_2 - \gamma_1}$$

The use of these formulas assumes complete wetting of the filler by the resin and no voids or entrapped air bubbles.

Mechanical Properties—Estimates

In most cases the effects on mechanical properties of filling various resins with metallic or other fillers and reinforcements must be determined experimentally. However, as shown in Table 5.2, the effects can be estimated in terms of improvements +, degradation −, or no to little change 0.

Table 5.2. Mechanical Property Effects of Fillers

		Fibers		
	Powders ($A \sim 1$)	Short ($A < 10$)	Long ($A > 10$)	Flakes ($A > 10$)
Tensile strength	0	0	+	−
Ultimate strength	0	0	+	−
Elastic modulus	0	0	+	+
Flexural modulus	0	0	+	+
Impact	−	−	+	−
Heat-deflection temperature	+	+	+	+
Fatigue	−	−		−
Creep	+	+	+	
Thermal conductivity	0	0	+	+

REFERENCES

1. R. B. Seymour (ed.), *Polymer Science and Technology,* vol. 15, *Conductive Polymers,* Plenum Press, New York, 1981.
2. Y. V. Murty and F. R. Mollard (eds.), *Continuous Casting of Small Cross Sections,* The Metallurgical Society of AIME, Warrendale, Pa. 1981.
3. "Transmet Opens New Rapid-Solidification Facility," *Metals Powder Report,* vol. 37, No. 8, pp. 427–428, Shrewsbury, U.K., August 1982.
4. R. C. Progelhof, J. L. Throne, and R. R. Teutsch, Methods for predicting the thermal conductivity of composite systems: A review, *Polym. Eng. Sci.,* **16** (9), 615–625 (September 1976).
5. T. Lewis and L. Nielsen, Dynamic mechanical properties of particulate-filled polymers, *J. Appl. Polym. Sci.,* **14,** 1449–1471 (1970).
6. L. E. Nielsen, The thermal and electrical conductivity of two phase systems, *Ind. Eng. Chem. Fund.,* **13** (1), 12–20 (1974).
7. D. M. Bigg, Conductive polymeric compositions, *Polym. Eng. Sci.,* **17** (12), 842–847 (December 1977).

Chapter **6**

Bimetallic Materials

ANTHONY J. IZBICKI

Hood & Company, Inc.

6.1 DEFINITION AND PRINCIPLE OF OPERATION

Bimetallic materials, also called *thermostat metals,* are engineering materials generating mechanical output analog to thermal input. They are materials composed of two or more metallic layers of differing coefficient of thermal expansion such that the radius of curvature of the composite changes with temperature change.

Thermostat metals of two components (one of higher coefficient of thermal expansion and one of lower coefficient of thermal expansion) and three components (with a center layer to provide controlled electrical resistivity) are the most common configurations. The principle of operation of two- and three-component systems is that of mechanical relief of thermally induced stress. When the component material of higher coefficient of thermal expansion (i.e., one with a high rate of dimension change with respect to change in temperature) is bonded to a component of lower coefficient of thermal expansion (i.e., one with a low rate of dimension change with respect to change in temperature), the component with the higher coefficient cannot change dimension with temperature as it would in a free state.

The restraint of dimension change caused by the bonded condition results in internal stress loading of the composite. The internal stress loading of the composite is mechanically relieved by change of shape. For the special case defined by plane bond interfaces between the components of different coefficients of thermal expansion with composite length dimension >> width dimension > thickness dimension, the shape change is by change of radius of curvature (deflection). The plane bonded case defined is used for most applications. Other configurations are not of general interest and will not be discussed.

6.2 FLEXIVITY

The primary measure of thermostat metal deflection with temperature change is flexivity as defined by ASTM standard

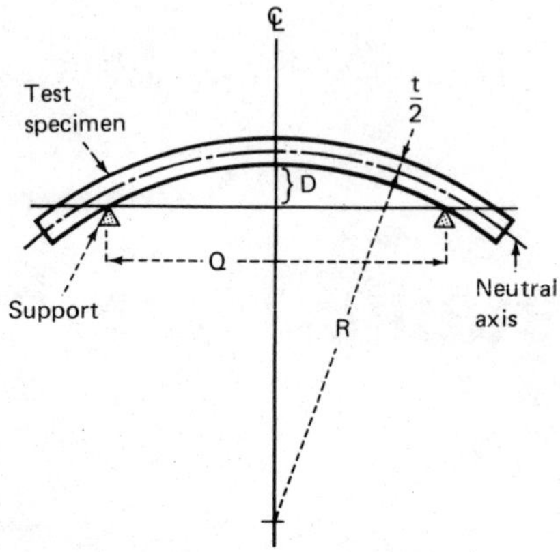

Fig. 6.1 Flexivity definition geometry.

B106, Test Method for Flexivity of Thermostat Metals (see Fig. 6.1).

Flexivity as defined is a bulk property intrinsic to a given alloy combination in a given temperature range. The parameters of the composite combination that determine flexivity are coefficient of thermal expansion, the volume percent of each component, and the elastic modulus of each component. It should be noted that coefficient of thermal expansion is a function of temperature and, therefore, flexivity is also a function of temperature. Design equations for thermostat metals are all derivable special cases of Eq. (6.1) defining flexivity.

Flexivity F, the change of curvature of the longitudinal centerline of the specimen per unit temperature change for unit thickness, is given by the equation

$$F = t\,\frac{(1/R_2) - (1/R_1)}{T_2 - T_1} \tag{6.1}$$

To determine the flexivity between any two temperatures T_1 and T_2, it is necessary to measure the curvatures $1/R_1$ and $1/R_2$ at temperatures T_1 and T_2, respectively. To find the curvature at either temperature, measure the distance D. Then the curvature is as follows:

$$\frac{1}{R} = \frac{8D}{Q^2 + 4Dt + 4D^2} \tag{6.2}$$

where R = radius of curvature of longitudinal centerline of specimen, in or mm
t = thickness of test specimen, in or mm
Q = distance between support points, in or mm
D = (for point supports) perpendicular distance between longitudinal centerline of lower surface of specimen midway between point supports and straight line joining support points, in or mm

6.3 OTHER IMPORTANT PARAMETERS

Electrical resistivity of thermostat metals is the parallel-circuit value of the individual components. The value for two-

Table 6.1. Nominal Composition of ASTM Standard Thermostat Metals in Weight Percent

Thermostat metal designation	Elements											
	High-expansion component							Intermediate component		Low-expansion component		
	Ni	Cr	Mn	Cu	Fe	Al	C	Ni	Mn	Ni	Fe	Co
TM1	22	3	...	...	75	...	...	...	...	36	64	...
TM2	10	...	72	18	...	...	...	...	...	36	64	...
TM3	25	8.5	...	...	66.5	...	...	...	...	42	58	...
TM4	25	8.5	...	...	66.5	...	...	...	...	45	55	...
TM5	25	8.5	...	...	66.5	...	...	...	...	50	50	...
TM6	22	3	...	...	75	...	...	...	...	40	60	...
TM7	14	...	9.5	...	71.5	5	...	...	...	36	64	...
TM8	10	...	72	18	...	...	...	...	...	36	64	...
TM9	22	3	...	...	75	...	...	100	...	36	64	...
TM10	22	3	...	75	...	...	...	100	...	36	64	...
TM11	22	3	...	...	75	...	...	100	...	36	64	...
TM12	22	3	...	...	75	...	...	100	...	36	64	...
TM13	22	3	...	...	75	...	...	100	...	36	64	...
TM14	22	3	...	...	75	...	...	100	...	36	64	...
TM15	22	3	...	...	75	...	...	100	...	36	64	...
TM16	22	3	...	...	75	...	...	100	...	36	64	...
TM17	22	3	...	...	75	...	...	100	...	36	64	...
TM18	19.4	2.25	...	...	78.3	...	0.5	...	...	42	58	...
TM19	19.4	2.25	...	...	78.3	...	0.5	...	...	39	61	...
TM20	18	11.5	...	...	70.5	...	...	...	...	36	64	...
TM21	18	11.5	...	...	70.5	...	...	...	...	42	58	...
TM22	100	...	...	...	...	...	...	...	...	36	64	...
TM23	10	...	72	18	...	...	...	...	...	42	58	...

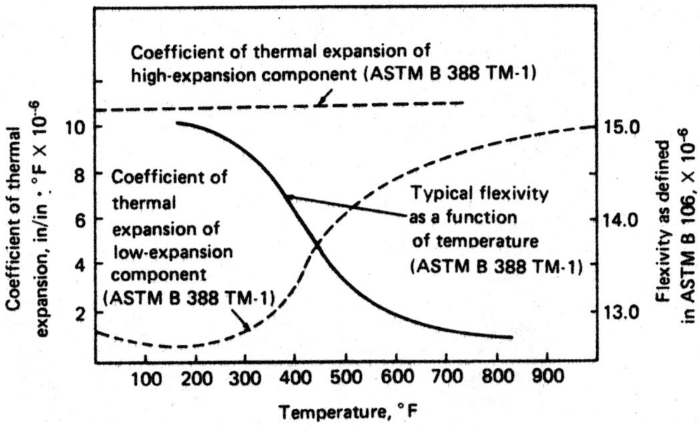

Fig. 6.2 Example of flexivity as a function of temperature.

Table 6.2. Properties of Thermostat Metals (Inch-Pound Units)*

Properties, units†	ASTM type TM1	TM2	TM3	TM4	TM5	TM6	TM7	TM8
Maximum sensitivity temperature range, °F	0–300	0–400	200–600	250–700	300–850	100–550	0–300	0–400
Maximum recommended temperature, °F	1000	500	1000	1000	1000	1000	900	500
Flexivity × 10^{-6}								
50–200°F	15.0	21.5	10.4	8.4	6.3	12.9	14.5	15.9
100–300°F	14.6	21.1	10.6	8.6	6.4	12.9	14.5	15.9
Electrical resistivity at 75°F, Ω_{CMF}	475	675	440	400	350	440	560	850
Modulus of elasticity, lb/in² × 10^6	25	20	25	25	25.5	25	24	19.5

	TM9	TM10	TM11	TM12	TM13	TM14	TM15	TM16
Maximum sensitivity temperature range, °F	0–300	0–300	0–300	0–300	0–300	0–300	0–300	0–300
Maximum recommended temperature, °F	900	900	900	900	900	900	900	900
Flexivity × 10^{-6}								
50–200°F	11.2	12.4	13.5	13.8	14.0	14.7	14.9	15.0
100–300°F	10.5	12.7	13.3	13.7	14.0	14.3	14.4	14.5
Electrical resistivity at 75°F, Ω_{CMF}	100	125	150	175	200	250	300	350
Modulus of elasticity, lb/in² × 10^6	26	26	26	25.5	25.5	25.5	25	25

	TM17	TM18	TM19	TM20	TM21	TM22	TM23
Maximum sensitivity temperature range, °F	0–300	200–600	150–450	0–300	200–600	0–300	200–600
Maximum recommended temperature, °F	900	900	900	900	900	900	500
Flexivity × 10^{-6}							
50–200°F	15.4	12.0	14.3	13.9	10.6	10.2	18.0
100–300°F	14.6	12.0	14.1	13.9	10.9	10.2	18.0
Electrical resistivity at 75°F, Ω_{CMF}	400	420	456	472	430	95	565
Modulus of elasticity, lb/in² × 10^6	25	25	25	25	25	26	20

* Refer to ASTM standard B388 for complete tables of properties and tolerances. Specific heat of 0.12 Btu/lb · °F and density of 0.29 lb/in³ may be used for approximate calculations.

† U.S. Customary System (inches, pounds, ounces, and degrees Fahrenheit) is used throughout this chapter. Flexivity does not readily convert to metric values because of the difference in definition between ASTM and DIN standards.

component systems is determined by the electrical resistivity of the high expansion and the low-expansion components. Three-component systems are tunable to meet predetermined electrical resistivity values by selection of the third-layer electrical resistivity and its volume percent of the composite. Thermostat metal electrical resistivity is determined by ASTM standard B63.

The modulus of elasticity applicable to thermostat metals is determined by ASTM standard B223 (cantilever beam method). This measure is representative of the majority of practical application configurations for estimating mechanical spring rate and force developed.

The maximum sensitivity range is the temperature range of maximum thermal response. It is useful for selection of thermostat metals applicable to a design temperature range. Detailed flexivity as a function of temperature curves is available for final design selection.

6.4 THERMOSTAT METAL TYPES

ASTM standard B388, Standard Specification for Thermostat Metal Sheet and Strip, describes 23 industry standard products. Other types with a variety of flexivity, electrical resistivity, and mechanical properties are commonly available. Table 6.1 and Table 6.2 present the properties of the industry standard types and are representative of the range of properties available.

Table 6.3. Selected Design Formulas

Cantilever beam at free end	
$D = \dfrac{0.53F\,\Delta T\,L^2}{t}$	free deflection
$P = \dfrac{2.12FE\,\Delta T\,bt^2}{L}$	force developed by restraint of deflection
$P = \dfrac{4EDbt^3}{L^3}$	mechanical stiffness
Simple beam at center	
$D = \dfrac{0.1325F\,\Delta T\,L^2}{t}$	free deflection
$P = \dfrac{8.48FE\,\Delta T\,bt^2}{L}$	force developed by restraint of deflection
$P = \dfrac{64\,EDbt^3}{L^3}$	mechanical stiffness
Spiral coil	
$A = \dfrac{64F\,\Delta T\,L}{t}$	angular deflection
$P = \dfrac{1.485FE\,\Delta T\,bt^2}{r}$	force developed by restraint of deflection
$P = \dfrac{0.0232EAbt^3}{L\,r}$	mechanical stiffness

Symbols: D = deflection in inches for beams
A = angular degrees of rotation for coils
ΔT = temperature change
F = flexivity in range of temperature change
E = elastic modulus in pounds per square inch
L = free deflection length in inches
P = load or force in ounces
r = radius of spiral coil in inches
b = width of strip in inches
t = thickness of strip in inches

6.5 DESIGN

Thermostat metal applications generally apply mechanical output analog to thermal input for sensing and control. Thermal input is commonly ambient condition, heating elements or direct heating from electrical current.

The most frequently encountered configurations are simple beam, cantilever beam, and spiral coil. The general formulas for design are given in Table 6.3.

The formulas given in Table 6.3 are useful for cases in which length is much greater than width and deflection is small enough to approximate change of curvature as linear deflection. Thermostat metal design involves compromise (see Table 6.4). Consideration of flexivity as a function of temperature within the design range is critical to performance, especially at the high-temperature end, where flexivity of many commonly used types falls off rapidly.

When significant force and deflection are required simultaneously, the available temperature change should be divided in half. Deflection can then be estimated by means of the unrestrained thermal deflection formula, and force can be estimated with respect to the force developed by complete restraint of deflection formula, using one-half of the available temperature change in each formula.

Thermostat metal applications involve many configurations in addition to those presented in Table 6.3. "Snap-acting" devices in blade or disk form are, for example, a special engineering problem beyond the scope of a general-interest discussion.[1–3]

6.6 ADDITIONAL CONSIDERATIONS

Stress relief (commonly referred to as *heat treat*) is an important consideration. Fabrication of strip, parts, and assemblies results in residual stress loading at each step. To stabilize deflection with temperature change (thermostat metal deflects as a mechanical relief of thermally induced

Table 6.4. Thermostat Metals Design Compromises

Deflection with temperature change varies
Inversely with the thickness
Directly with the square of the length
Not affected by changes in width within length >> width constraint
Force exerted with temperature change varies
Directly with the square of the thickness
Inversely with length
Directly with width within length >> width constraint
Maximum load varies
Directly with the square of the thickness
Inversely with length
Directly with width within length >> width constraint

Table 6.5. ASTM Standards for Thermostat Metals

B63	Resistivity of Metallically Conducting Resistance and Contact Materials
B70	Change of Resistance with Temperature of Metallic Materials for Electrical Heating
B106	Flexivity of Thermostat Metals
B223	Modulus of Elasticity of Thermostat Metals (Cantilever Beam Method)
B362	Mechanical Torque Rate of Spiral Coils of Thermostat Metal
B388	Thermostat Metal Sheet and Strip
B389	Thermal Deflection Rate of Spiral and Helical Coils of Thermostat Metal
B478	Cross Curvature of Thermostat Metals
B753	Thermostat Component Alloys
C351	Mean Specific Heat of Thermal Insulation
E92	Vickers Hardness of Metallic Materials

stress), parts or assemblies are stress-relieved prior to use. The stress-relief condition should be determined for each application. Typical conditions are 1-h soak at a temperature between 500 and 1000°F. If the maximum operating temperature is within this range, 100°F above the maximum temperature is recommended.

Corrosion resistance of the majority of thermostat metals (consisting of Ni-Cr-Fe and Ni-Fe alloys) is adequate for most operating environments. When additional corrosion resistance is needed or fabrication methods warrant, thermostat metals can be plated. Plating type and thickness should be selected for minimum effect on thermal and electrical characteristics. An exception worth noting is the high-expansion alloy specified for ASTM standards B388 TM-2, TM-8, and TM-23, consisting of 72% Mn, 18% Cu, and 10% Ni. This alloy has poor corrosion resistance, is susceptible to stress corrosion cracking, and is difficult to plate. The alloy is widely used, however, for its advantages of high coefficient of thermal expansion and high electrical resistivity.

6.7 SUMMARY

Thermostat metals are widely used for sensing and control. They retain many advantages over solid-state devices in that no external power supply is needed for function, the response is intrinsic to a small volume of metal, and the functional element is tolerant of abuse and overload conditions.

The utility of thermostat metals involves many diverse areas with specific and unique requirements. Although many alloy combinations are standard, each device application requires custom engineering and production. The engineered product nature of thermostat metals suggests the need for close cooperation between producer and user engineering departments.

REFERENCES

1. M. Kutz , *Temperature Control,* Krieger, Huntington, N.Y., 1975.
2. W. H. Wittrick, D. M. Myers, and W. R. Blunden, "Stability of a Bimetallic Disk," *Quart. J. Mech. Appl. Math.,* **6,** 15–31 (1953).
3. A. M. Wahl, "Analysis of the Valverde Thermostat," *J. Appl. Mech.,* **66,** 183–189 (1944).

Chapter 7

Heat-Transfer Compounds

W. C. BACHMAN
Dylon Ceramic Technologies Company

E. L. PIPER
Dylon Industries, Inc.

7.1 INTRODUCTION

Many industrial operations must control the temperature of pipelines and vessels to prevent undesirable effects such as freezing, corrosion, condensation, and viscosity change. For this purpose, the process lines are "traced" with oil- or steam-heated lines or with electrical heaters. To increase the area of thermal contact between these tracers and the process lines, heat-transfer compounds are used. The result is a major reduction in the cost of heat tracing as well as a significant savings in energy costs.

In this chapter, some of the applications served by heat-conductive compounds are cited, leading to a discussion of the composition and properties of these materials. Analytical approaches to designing heat-tracing systems are also discussed.

7.2 APPLICATIONS

If the lowest anticipated ambient temperature is below the freezing point of a liquid being moved in a pipe or stored in a vessel, heat tracings are used. Examples of industrial materials which often require the use of traced lines are phosphoric acid, molten sulfur, glacial acetic acid, benzoic acid, cresol, naphthalene, phthalic anhydride, sorbitol, and paraxylene.

The mobility of highly viscous liquids is increased greatly through the use of heat tracing. Crude oils, polymeric materials, waxes, bitumen and tar residues, and caustic soda liquor are examples of thick fluids for which heat-traced process pipes are needed to facilitate transfer at reasonably fast rates.

Heat tracing may also be required for gases whose dew

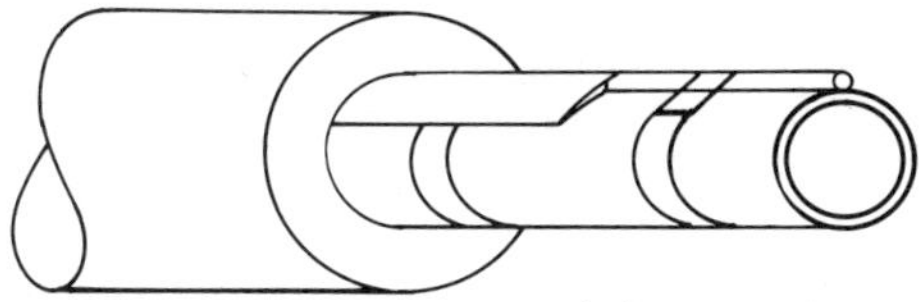

Fig. 7.1 Steam-traced process line using trowelable heat-transfer compound.

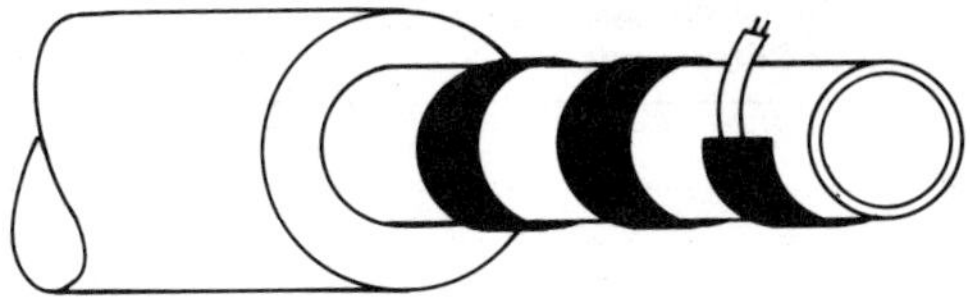

Fig. 7.3 Electrically traced process line using heat-transfer compound in tape form.

points are above ambient temperature in order to prevent undesirable condensation. For example, fuel gas in oil refineries which is contaminated by a liquid phase is likely to cause difficulties at the burners. Similarly, natural gas containing moisture must be moved through traced lines to prevent freezing at critical locations, such as control valves. Compressor suction lines are also heated to prevent damage to the seals and vanes due to the condensation of corrosive gases such as water vapor and H_2S.

7.3 COMPOSITION AND PROPERTIES

The key ingredient in heat-transfer compounds is graphite, a highly conductive ceramic available in powdered form as either artificial graphite or natural flake. Occasionally, carbon blacks are also used as fillers. The binder systems used are designed to provide a pliant, moldable consistency so that maximum contact area can be realized between the process line and the tracer. These binders may be organic or inorganic depending on the intended application. Figures 7.1, 7.2, and 7.3 show three schematic representations of different usages for heat-transfer compounds. In all three tracer designs, the process-tracer system is surrounded by pipe insulation.

In Fig. 7.1, a steam line is shown banded to a process pipe. In this case, the area through which heat can be conducted from the tracer to the process is greatly increased by the use of a puttylike combination of graphite and binder which may be applied with a trowel or by pumping. Heat-transfer compounds are available which can be hard-setting or can remain plastic, depending on the application.

In the hard-setting heat-transfer compounds, the binder system is normally a water-extendable inorganic and dries on standing in air. Typically, the drying time required ranges from 1 to 5 days depending on the thickness of the compound in cross section. The recommended curing procedure is to air-dry for 24 h, followed by a 4- to 12-h cure at about 200°F (93.3°C). An advantage of the inorganic hardening compound is its ability to withstand temperatures of up to 1800 °F (982°C). A disadvantage is the tendency of the binder system to deteriorate when exposed to moisture for extended time periods.

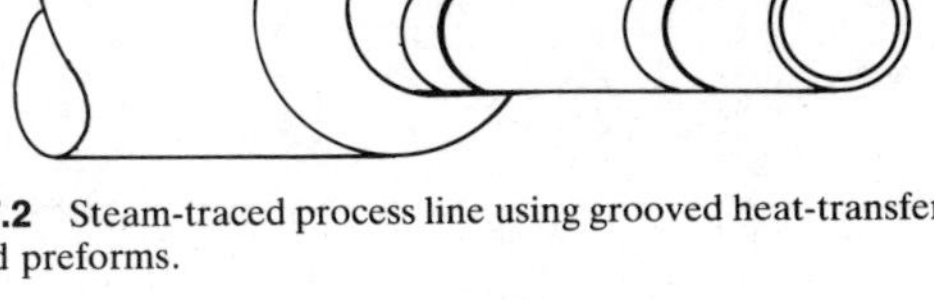

Fig. 7.2 Steam-traced process line using grooved heat-transfer compound preforms.

Epoxy-bonded graphite compounds are also hard-setting and are available in one- or two-component systems. At room temperature, the epoxies chemically set to a gel in 3 to 5 h, achieve handling strength overnight, and reach maximum strength in approximately 2 weeks. Preferably, the chemically setting materials should be allowed to remain at room temperature for about 4 h and then heated to 300°F (149°C) for an additional 4-h period. These resinous compounds are highly resistant to water, acids, and salts. Typical applications include the tracing of off-shore oil lines, solar heating systems, and cryogenic lines.

Nonhardening heat-transfer compounds are bonded by an organic material which limits the upper usage temperature to approximately 400°F. These materials are applied to areas where periodic removal of the compound is anticipated for such operations as thermocouple installation or repair, maintenance work on valves and instruments, and use of cartridge heaters. Since no drying time is needed, these materials are ideally suited for emergency repairs on areas where the heat-tracing compound has been damaged or chipped off for any reason. The nonhardening compounds can be brought to operating temperatures immediately.

Some typical properties of trowelable heat-transfer compounds are given in Table 7.1.

The thermal conductivity of the nonhardening organic compounds is typically 0.2 to 0.3 Btu/h · ft · °F (0.35 to 0.52 W/m · K). The thermal conductivity of the resin-based hardening compounds is in the range of 0.6 to 0.8 Btu/h · ft · °F (1.03 to 1.40 W/m · K). The inorganic hardening compounds generally have higher thermal conductivity.

Table 7.1. Typical Properties of Trowelable Heat-Transfer Compounds

	Hardening		Nonhardening, organic
	Inorganic	Resin	
Density, lb/gal	13–14	12–13	10–13
Consistency, NLGI no.*	1	5	1
Flash point, COC °F*	None	460	415
Maximum usage temperature, °F	1800	425	400
Minimum usage temperature, °F	−250	−65	−200
Minimum application temperature, °F	40	40	40

CONVERSION FACTORS: 1.0 lb/gal = 119.8 kg/m^3; °C = (°F − 32)/1.8.
*NLGI = National Lubricating Grease Institute; COC = Cleveland open-cup method.

Table 7.2. Typical Properties of Preformed Heat-Transfer Compounds

	Grooved tape	Plain tape
Cross section, in	2 × ½	2 × ⅛
Unit weight, lb/ft	0.3	0.15
Consistency, NLGI no.	6+	6+
Flash point, COC °F	>400	>400
Maximum usage temperature, °F	>400	>400
Minimum usage temperature, °F	−200	−200
Minimum application temperature, °F	40	40
Thermal conductivity, Btu/h · ft · °F	0.5	0.5

CONVERSION FACTORS: 1.0 lb/ft = 1.49 kg/m; 1.0 in = 2.54 cm; 1.0 Btu/h · ft · °F = 1.73 W/m · K; °C = (°F − 32)/1.8.

Heat-transfer compounds are also available in extruded, tapelike preforms which facilitate a fast, convenient means for line tracing. The cross sections of these tapes can be simple rectangles, for example, 2 in (5.08 cm) in width by ⅛ to ½ in (0.32 to 1.27 cm) thick or can contain a centrally located groove for the tracer line. Typically, the tapes are available in 3-ft (0.91-m) lengths. Figure 7.2 depicts this form of heat-transfer compound, showing the tracer nested in the grooved preform which is positioned on the process pipe. The entire system is banded together and then enclosed with pipe insulation. Figure 7.3 illustrates the use of extruded flat tapes with spirally wound electrical tracing.

As in trowelable forms of heat-transfer compounds, graphite is the common filler material in preformed compounds. A nonhardening polymeric binder is used to provide a tacky, pliant consistency which is strongly adherent to the process and tracer lines. The tack of the preformed strips aids in installation as a temporary means for securing the tracer-process system prior to banding. To facilitate handling, the strips or tapes are normally separated by a protective release paper which is readily removed immediately preceding application. Since heat-conductive materials, such as graphite, are also electrically conductive, caution must be exercised when heat-transfer compounds are used on electrically traced process lines. Either the electrical heating lines must be well insulated or a less conductive form of heat transfer compound must be used, e.g., one in which carbon black is substituted for graphite.

Some of the properties of preformed heat-transfer compounds are listed in Table 7.2.

7.4 SYSTEMS DESIGN

To avoid over- or underdesigning of the tracing requirements for a given system, a calculation of the minimum tracing needs is advisable. Because of the large number of variables involved in this calculation and the complexity of an exact solution, simplifying assumptions are often made.

Bertram et al.[1] have treated the problem of a spirally traced process pipe using heat-conductive compounds. Beginning with a consideration of the heat flow through a differential element of this system, they have derived an expression for the total heat transferred by the tracer, either to the process fluid or lost to the surroundings. Typical calculations are given for ensuring a minimum temperature for a stagnant fluid and maintaining a minimum process pipe wall temperature to prevent condensation. The problem of heat tracing the bottom of a stainless-steel vessel is also treated.

Many users[2] of heat-tracing compounds as well as suppliers[3,6] of these materials have developed computer programs for solving problems dealing with the design of heat-traced equipment. In addition to the speed with which the calculations can be performed, the convenience of storing engineering data relating to heat-transfer coefficients, thermal conductivities, and saturated steam properties make computers attractive in solving tracer design problems. Blackwell[2] has described a computer program which calculates the surface temperature of the insulated pipe and the heat transferred per unit length of pipe and estimates the number of linear tracers required, both with and without the use of heat-transfer compounds. A derivation of the equations used in this program may be found in Kern[4] and Kohli.[5] The first equation describes the rate of heat loss per unit length as follows:

$$Q = \frac{\pi h_a d_o}{12} (T_s - T_{\text{amb}}) \tag{7.1}$$

Another equation permits the calculation of the outside surface temperature of the insulation as

$$T_s = \frac{T_a + XT_{\text{amb}}}{X + 1} \tag{7.2}$$

Table 7.3. Values of h_a for Pipes in Still Air

Nominal pipe diameter, in	$T_s - T_{\text{amb}}$, °F												
	0	10	20	30	40	50	100	150	200	250	300	400	500
½	1.66	1.73	1.80	1.87	1.94	2.01	2.37	2.72	3.08	3.43	3.79	4.50	5.21
1	1.64	1.71	1.78	1.84	1.91	1.98	2.32	2.66	3.01	3.55	3.69	4.37	5.06
2	1.54	1.60	1.67	1.74	1.81	1.87	2.21	2.54	2.88	3.21	3.55	4.21	4.88
4	1.45	1.52	1.58	1.65	1.71	1.78	2.11	2.43	2.76	3.08	3.41	4.06	4.72
8	1.37	1.43	1.49	1.56	1.62	1.68	2.00	2.32	2.64	2.95	3.27	3.91	4.54
12	1.31	1.37	1.44	1.50	1.56	1.63	1.95	2.26	2.58	2.90	3.22	3.85	4.49
24	1.25	1.31	1.37	1.43	1.49	1.56	1.87	2.18	2.49	2.80	3.11	3.74	4.36

CONVERSION FACTORS: h_a = Btu/h · ft² · °F based on still air (1.0 Btu/h · ft² · °F = 5.678 W/m² · K) (1.0 in = 2.54 cm).

Table 7.4. Correction Factor for h_a at Different Wind Velocities

Wind velocity, mi/h	$T_s - T_{amb}$, °F												
	0	10	20	30	40	50	100	150	200	250	300	400	500
2.5	1.50	1.50	1.49	1.49	1.49	1.48	1.46	1.45	1.43	1.41	1.39	1.36	1.32
5.0	1.79	1.79	1.78	1.78	1.77	1.77	1.74	1.72	1.69	1.66	1.64	1.59	1.53
10.0	2.25	2.24	2.24	2.23	2.22	2.21	2.17	2.13	2.09	2.05	2.01	1.93	1.85
15.0	2.62	2.61	2.60	2.59	2.58	2.57	2.52	2.47	2.42	2.37	2.32	2.22	2.12
20.0	2.90	2.89	2.88	2.87	2.86	2.85	2.79	2.73	2.67	2.61	2.56	2.44	2.32
25.0	3.13	3.11	3.10	3.09	3.08	3.06	3.00	2.94	2.88	2.82	2.76	2.63	2.51
30.0	3.29	3.28	3.27	3.26	3.24	3.23	3.17	3.11	3.05	2.99	2.92	2.80	2.68
35.0	3.42	3.41	3.40	3.39	3.38	3.37	3.31	3.26	3.20	3.15	3.09	2.98	2.87

CONVERSION FACTOR: 1.0 mi/h = 0.447 m/s.

where

$$X = \frac{h_a d_o \ln(d_o/d_i)}{24k_i} \tag{7.3}$$

a dimensionless parameter describing the size of the insulation, its thermal conductivity, and film heat-transfer coefficient. Typical values of h_a can be determined from Tables 7.3 and 7.4.

The number of tracers needed without and with heat-transfer compound, respectively, is then given by

$$N_{woc} = \frac{Q}{C_{woc}(T_{st} - T_p)} \tag{7.4}$$

and

$$N_{wc} = \frac{Q}{C_{wc}(T_{st} - T_p)} \tag{7.5}$$

where N is the number of parallel running tracers (Figs. 7.1 and 7.2) or the length of tracer per unit length of pipe for wound tracers (Fig. 7.3).

Typical experimentally determined thermal conductance values for tracers with (C_{wc}) and without (C_{woc}) heat-transfer compound are given in Table 7.5. Typical calculated tracer lengths with heat-transfer compound are about one-tenth of those without.

Table 7.5. Thermal Conductance Values for Tracer Lines in Btu/h · ft · °F

Tube size, in	C_{woc}	C_{wc}
0.375 OD	0.295	3.44
0.500 OD	0.393	4.58
0.625 OD	0.490	5.73

CONVERSION FACTORS: 1.0 in. = 2.54 cm; 1.0 Btu/h · ft · °F = 1.73 W/m · °C.

NOMENCLATURE

C_{woc} = thermal conductance, tracer to pipe, without heat-transfer compound, Btu/h · °F · ft of pipe

C_{wc} = thermal conductance, tracer to pipe, with compound, Btu/h · °F · ft of pipe

d_i = inside diameter of insulation, in

d_o = outside diameter of insulation, in

h_a = film heat-transfer coefficient to air (corrected for wind), Btu/h · °F · ft^2

k_i = thermal conductivity of insulation, Btu/h · ft · °F

N_{wc} = number of tracers required with heat-transfer compound

N_{woc} = number of tracers required without heat-transfer compound

T_a = average temperature of pipe and tracer, °F

T_{amb} = air temperature, °F

T_{st} = average temperature of hot medium, °F

T_p = temperature in pipe, °F

T_s = outside surface temperature of insulation, °F

REFERENCES

1. C. G. Bertram, V. J. Desai, and E. Interess, "Designing Steam Tracing," *Chem. Eng.*, April 3, 1972, p. 74.
2. W. W. Blackwell, "Estimate Heat-Tracing Requirements for Pipelines," *Chem. Eng.*, Sept. 6, 1982, p. 115.
3. W. C. Bachman, *Engineering Guide to Pipe Heat Tracing,* Dylon Industries Inc., Berea, Ohio.
4. D. Q. Kern, *Process Heat Transfer,* McGraw-Hill, New York, 1950, p. 18.
5. I. P. Kohli, "Steam Tracing of Pipelines," *Chem. Eng.*, March 26, 1979, p. 157.
6. Thermon Manufacturing Co., *Steam Tracing of Pipes,* Catalog T-100-M, June 1984.

Part 7

Heat Exchangers and Related Devices

Chapter **1**

Heat-Exchanger Technology

GRAHAM WALKER
University of Calgary

1.1 INTRODUCTION

Heat exchangers are devices designed to enhance or facilitate the flow of heat. Countless examples are found in everyday life, for every living thing is equipped in some way with a type of heat exchanger. Mammals have complicated heat exchangers. The primary unit is the lungs, which not only cool the body by saturating the expelled air with water vapor but also serve as complex mass exchangers taking oxygen from air and carbon dioxide from the blood. The skin acts as a supplementary heat exchanger, changing its character to promote or inhibit the transfer of heat from the body depending on the temperature, humidity, and velocity of the air.

Fortunately, industrial heat exchangers are much less complicated, although the trappings of technological civilization make them no less vital. Synthetic heat exchangers are found in all facets of our lives. Boiling water or frying an egg requires a heat exchanger. Refrigerators operate on a vapor compression cycle using two heat exchangers, one to cool the freezer compartment and the other to transfer to the air heat from the freezer compartment plus the work required to drive the system. Automobiles are equipped with a heat exchanger (misnamed the "radiator"), and all electrical and electronic equipment must be provided with heat exchangers for cooling.

Building heating and cooling systems involve the use of heat exchangers. Electricity is generated in baseload electric power stations that depend on heat exchangers for generating and condensing the steam used to drive the turbine-powered alternators. Oil refineries and chemical processing plants use many different heat exchangers. Food processing, baking, brewing, mixing, and freezing all involve the use of heat exchangers of one sort or another.

It is evident that heat exchangers exist in great numbers, widely distributed throughout industry and commerce. They are found in an enormous range of heat transfer capacities. The smallest [>2 Btu/h (>1 W)] are included in miniature cryocoolers for infrared thermal imaging, heat-seeking missile guidance, or superconducting electronic applications. The largest [<6 × 10^9 Btu/h (<2 GW)] are the boilers, condensers, and condenser cooling water air coolers in base-load electric power stations.

This vast range of capacities embraces a great diversity of types, shapes, and arrangements of heat exchangers. The majority serve to facilitate the energy transfer between two fluid streams at different temperature levels. Other cases involve solid-mass heating or cooling systems such as the convective cooling fins found on electrical apparatuses dissipating the internal resistance heat generated. In food processing, sides of beef are cooled for refrigerated preservation and meat or bread are heated for cooking.

1.2 CLASSIFICATION OF HEAT EXCHANGERS

Heat exchangers used to transfer heat between two or more streams of fluid may be classified broadly into two groups: regenerative and recuperative.

Regenerative Heat Exchangers

A regenerative heat exchanger has a single set of flow channels through a relatively massive porous matrix. The hot and cold fluids pass alternately through the matrix. When the hot fluid is passing, called the "hot blow," heat is transferred from the fluid to heat the matrix. Later when cold fluid is passing, called the "cold blow," heat is transferred from the matrix to the fluid and the matrix cools. When continuous flow of fluid is required, a duplicate matrix must be provided with quick-acting valves to switch the flow periodically. The matrices will then experience the cold blow and the hot blow successively, periodically, and alternately. In other cases a single matrix is used and the flow is switched cyclically by moving the matrix or the flow headers controlling the fluid flow to and from the matrix.

Figure 1.1*a* shows some of the possible variations of regenerative heat exchangers. Regenerators are presently used in only a few specialized applications, principally exhaust air preheaters in large electric power stations, in steel plants, in gas turbines, and in industrial gas liquefaction processing. There is a family of regenerative machinery—Stirling, Ericsson, Vuillemier, and Gifford-McMahon cycle engines—

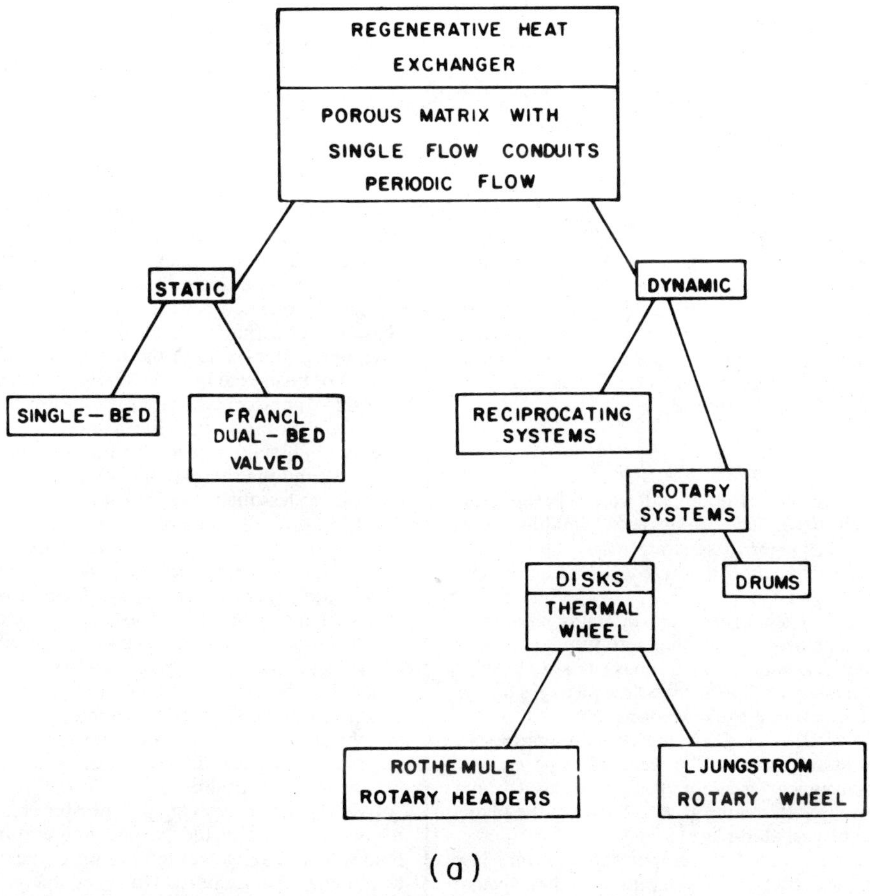

Fig. 1.1 Design variants and classification of regenerative (*a*) and recuperative (*b*) heat exchangers.

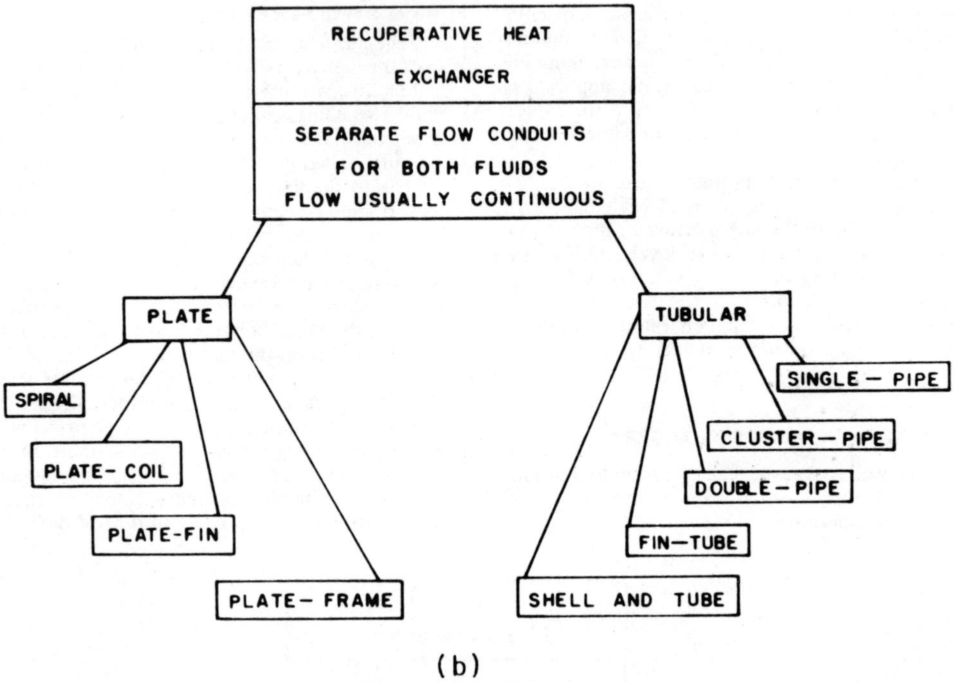

Fig. 1.1 (*Continued*)

all of which embody regenerative heat exchangers. Presently recuperative exchangers are preferred for the majority of cases, but further applications of regenerative heat exchangers appear likely as attempts are made in the future to utilize low-temperature thermal effluent streams. For these moderate-temperature applications, regenerative heat exchangers are attractive because the matrix can be made cheaply from gravel aggregate, paper and plastic honeycomb, or any finely divided material.

Recuperative Heat Exchangers

A recuperative exchanger is equipped with separate flow conduits for each fluid. The fluids flow simultaneously through the heat exchanger in separate paths, and heat is transferred from the hot fluid to the cold fluid across the walls of the flow section.

Recuperative heat exchangers exist in many variations (Fig. 1.1*b*) but can be broadly classified into tubular or plate types. Tubular heat exchangers have tubes (mostly, but not always, circular in cross section) for the flow passages of one or both of the fluids. Plate exchangers have stacks of deformed plates in which the fluid flow passages are impressed. The stacks of plates may be brazed or welded in a permanent assembly (plate-fin types) or simply compacted in a frame by compression bolts with sealing gaskets between the plates (plate-and-frame or gasketed-plate).

Heat exchangers for very high temperatures or very low temperatures require special consideration. Other special cases arise with corrosive fluids, thick viscous liquids, solid-liquid slurries, or multicomponent vapors condensing at different temperature levels.

Exceptionally high heat transfer may be attained in systems where the fluids boil or condense and the situation is further complicated because the fluids assume complex two-phase patterns that are difficult to analyze and predict.

To newcomers, one of the most surprising and disturbing aspects of heat-exchanger technology is the lack of precision in design or performance prediction. For new and unfamiliar systems, it is rarely possible to predict performance closer than ± 20 percent. In use, the internal surfaces can become "fouled" and the thermal performance progressively can deteriorate to less than half the "as-new" performance. In such cases the ease of cleaning the surfaces to restore the performance is an important selection criterion.

Vibration is another area of concern, particularly in large, highly rated systems with high-velocity flows. The vibration may arise out of the fluid motion, because of pressure pulses in the fluids, or may be structureborne vibration, resulting from mechanical vibration originating outside the exchanger. When the frequency of the vibration corresponds to the natural frequency or related harmonics of some part or assembly, the amplitude of vibration may build to dangerous levels, resulting in physical damage to the structure or leakage between the fluid streams.

Every process involving the transfer of heat from one fluid to another must include a net temperature drop. The hot fluid will cool to a lower temperature; the cool fluid will heat to a higher temperature. However, the *maximum* temperature attained by the *cool* fluid will always be *less* than the *maximum* temperature of the *hot* fluid. Similarly, the *minimum* temperature of the hot fluid will always be greater than the *minimum* temperature of the cool fluid. There must always be a temperature difference between the fluids necessary to induce the transfer of energy from the hot fluid to the cold fluid. This is the imperative requirement of the second law of thermodynamics.

Similarly, energy will always be expended in causing fluids to flow through the exchanger. This becomes manifest in the pressure difference at inlet and outlet of the two fluids. Frictional degradation of pressure energy during the flow causes the fluid pressure at outlet to be less than it is at inlet.

Thermal performance is generally related to the frictional pressure drop. Improvements in thermal performance will likely be accompanied by increased pressure drop. With high-density fluids (liquids), the friction power expenditure is generally small relative to the rate of heat transfer. With low-density fluids (gases), it is easy to effect an improvement in the heat-transfer rate that requires an equivalent or greater expenditure of mechanical energy in the form of compressor or fanwork. Such improvement in heat transfer is costly, for in most systems mechanical energy is "worth" up to 10 times its equivalent in low-grade thermal energy. The design and use of a heat exchanger always requires careful consideration of both the heat-transfer rates attained and the mechanical power necessary to achieve them.

Careful selection of the proper materials to use for the fluids to be handled is another vital aspect of heat-exchanger technology. Corrosion, ultimately leading to failure of the heat exchanger (to separate and contain the fluids), occurs in many forms that must be recognized and sufficiently understood to provide a sound basis for logical and economic choices of materials.

Many heat exchangers serve the double duty of transferring thermal energy from one fluid to another and acting, also, as a containment vessel for fluids under pressure. Some exchangers consist of two such pressure vessels intricately combined.

The separate but related functions of the heat exchanger are recognized by performing the design in two stages: thermal design and mechanical design. *Thermal design* is concerned with those aspects of heat transfer and hydraulics associated with the provision of adequate surface area and flow guidance to ensure that the exchanger is capable of the required thermal duty with an acceptable fluid pumping power. *Mechanical design* is concerned with all matters necessary to ensure the integrity and durability of the exchanger as a pressure vessel. All heat exchangers and pressure vessels with internal pressures greater than 15 lb/in^2 (1 atm) must be designed to conform to the Boiler and Pressure Vessel Code formulated by the American Society of Mechanical Engineers (ASME). This code has been incorporated by reference into the safety regulations of all the regulatory authorities in North America. All countries have similar regulations, some, but not all, of which are as stringent as the ASME Boiler and Pressure Vessel Code.

The two activities of thermal and mechanical design are sometimes carried out simultaneously by the same designer. More frequently, however, there are specialized groups concerned only with one or the other activity. This is advantageous in that designers can develop a deeper awareness of the fundamentals and be aware of recent developments in their specialties. However, mechanical and thermal designs may proceed independently of each other, which can lead to unnecessary constraints, expense, or delays as a result of a breakdown in communication between the two groups.

1.3 TUBULAR EXCHANGERS

Tubular exchangers are used in great numbers, far more than any other type of exchanger. They are produced in the widest variety of sizes and styles, ranging from the tiny units used in miniature cryocoolers to the giant installations containing thousands of tubes used as condensers in baseload power stations.

Tubular exchangers are widely used because the technology is well established for manufacturing precision metal tubes capable of containing high pressures in a variety of materials. There is virtually no limit to the range of pressures and temperatures that can be accommodated. Furthermore, the holes for tubes in the *tubesheets*, the endplates to which the tubes are attached, are easily made by drilling. This facility for manufacturing established the use of tubular exchangers and led to the growth of major manufacturers and the development of industry standards for manufacture. Tubular exchangers were the first choice in new developments because of their widespread use, their relatively low cost, their familiarity to users, and the many years of satisfactory service routinely obtained.

Classification by Use

Tubular exchangers can be broadly classified by the character of the fluids they are used to handle.

Liquid-Gas This is by far the biggest use of tubular heat exchangers, typically air-cooling of a hot effluent. In terms of annual production, automobile radiators are by far the leaders of the field, tens of millions per year. Similar devices are necessary for all the other water-cooled engines used in trucks, locomotives, and other mobile equipment and stationary power. A liquid is pumped through the tubes with very high rates of convective heat transfer. The air, in crossflow over the tubes, may be in the forced or free convective mode. Heat-transfer coefficients on the air side will be lower than those on the liquid-side heat transfer. Fins will probably be used on the outside (air side) of the tubes to compensate.

Liquid-Liquid Another very wide use, typically cooling water on one side, is used to cool a hot effluent stream. Both fluids are pumped through the exchanger so that the principal mode of heat transfer is forced convection. The relatively high density of liquids results in very high rates of heat transfer, so there is little incentive in conventional situations to use fins or other devices to enhance the heat transfer.

Gas-Gas This type of exchanger is found in the exhaust gas-air preheating recuperators of gas turbine systems, steel furnaces, and cryogenic gas liquefaction systems. In many cases one gas is compressed so the density is high while the other is at low pressure with a low density. Normally the high-density fluid flows inside the tubes. Internal and external fins are sometimes provided to enhance the heat transfer.

Condensers These may be liquid- (water-) or gas- (air-) cooled. Alternatively, the heat from condensing streams may be used for heating another processing fluid. Normally the condensing fluid is routed:

1. Outside the tubes with liquid cooling, i.e., a water-

cooled steam condenser. Fins are not necessary either inside or outside the tubes.

2. Inside the tubes with gas coolings, i.e., an air-cooled Freon condenser heat pump or refrigerating system. Fins are normally provided on the outside of the tubes. The tubes may be flattened to obtain a better surface area-flow cross-section ratio.

Evaporators-Boilers This important group of tubular heat exchanger can be subdivided into two classes:

1. *Fired Systems:* These involve the products of combustion of fossil fuels (coal, oil, natural gas) at very high temperatures but ambient pressure (and hence low density) and generate steam under pressure. Normally the water-steam is contained inside the tubes (water tube boiler), and pin fins, studs, or other forms of extended surface may be applied to the tubes to enhance the heat transfer. In smaller systems the combustion products may pass through the tubes (fire tube boilers). Locomotive boilers and small low-pressure boilers (Lancashire boilers) are of this type.

2. *Unfired systems:* This embraces a great variety of different boiler-evaporator heat exchangers extending over a broad temperature range from high-temperature nuclear and steam generators to very-low-temperature cryogenic gasifiers for liquid natural gas evaporation. Many chemical and food processing applications involve the use of steam or electric heating to evaporate solvents, concentrate solutions, distill liquors, or dehydrate compounds.

Types of Tubular Exchangers

Single-Pipe Heat Exchanger The simplest conceivable tubular exchanger is the single bare tube through which fluid passes. Heat is exchanged between it and the environment surrounding the tube. Many building heating systems are of this type. The perimeter heating pipe is installed horizontally at ground level, and hot water or steam is circulated through the tube. Air flows in crossflow by natural convection over the pipe and heats the room. The heat transfer from the bare pipe may be enhanced by incorporating fins on the external surface. It is commonplace nowadays to find finned-perimeter heating tubes consisting of rectangular aluminum fins mounted on a small-bore thin wall copper pipe. The fin is squeezed onto the pipe during manufacture but is not otherwise secured on the pipe. Routine handling and differential thermal expansion with repeated heating and cooling of the pipe in service can loosen the fins with the result that they contribute essentially nothing to the heat-transfer process. If fins are necessary, it is vital that there be good thermal continuity between the metal and the pipe. Use of the same metal for the pipe wall and the fins and joining with soldered, bronzed, or conducting epoxy joint resolves many problems of this nature.

Clustered-Pipe Heat Exchanger A development of the single-pipe heat exchanger is shown in Fig. 1.2*a*. Two or more tubes are joined by a thermally conducting medium, so that

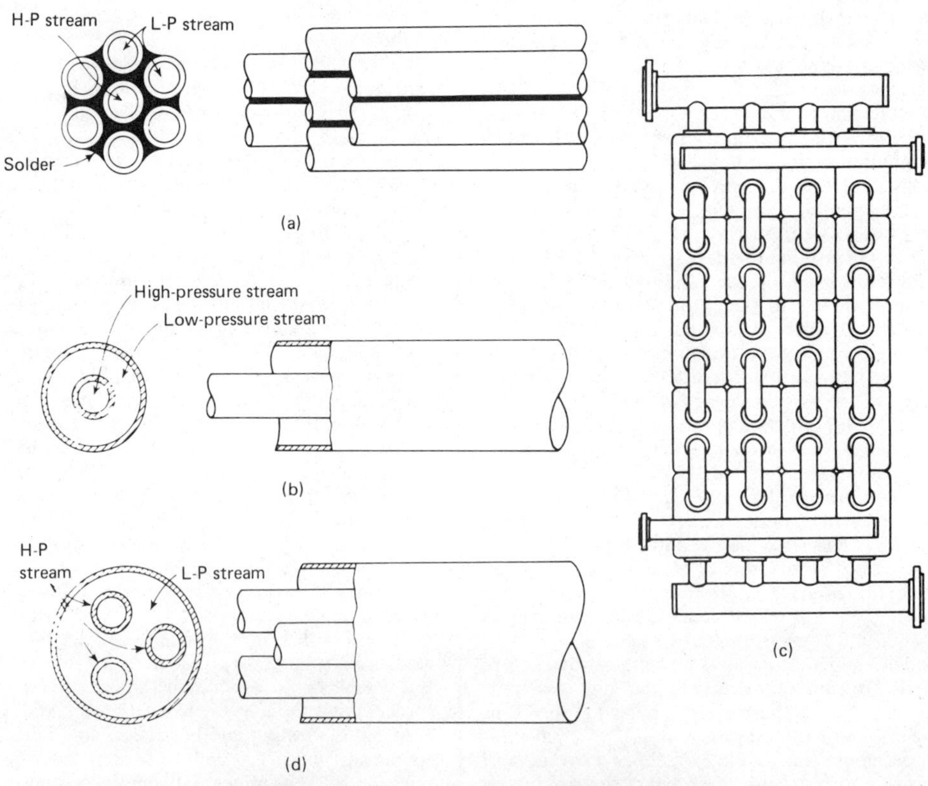

Fig. 1.2 Simple tubular exchangers. (*a*) Cluster-type tubular exchanger; (*b*) double-pipe heat exchanger; (*c*) multiple stack of double-pipe heat exchangers; (*d*) multiple-tube heat exchanger.

heat is transferred between fluids flowing in the tubes. Sometimes a cluster of tubes is arranged around a central core tube. High-density fluid passes through the core tube. The return stream of the low-density fluid passes through the multiple tubes arranged around the core tube. This construction is favored in some small cryogenic counterflow heat exchangers. The conducting medium connecting the tubes is deposited by electroplating and is periodically discontinuous to reduce axial conduction along the tubes.

Double-Pipe Heat Exchanger A conceptually similar system is the double-pipe heat exchanger shown in Fig. 1.2*b*. This consists simply of a central tube contained within a larger tube and was at one time widely used for small-capacity systems. It was relatively cheap and easy to manufacture and offered great flexibility since it could be made in any length and the tubes were selected from a wide range of sizes and materials available. It was customary to operate with the high-pressure and high-density or corrosive fluid in the small inner tube with the less demanding fluid (water) in the outer tube. This minimized the requirement for thick-walled, expensive alloy materials to contain the corrosive, high-pressure, and high-temperature fluid.

Stacks of double-pipe heat exchangers were commonly found (Fig. 1.2*c*) in either parallel or series arrangements. The standard fittings used for double-pipe heat exchangers are no longer so widely available; as a consequence, the practice of fabricating double-pipe heat exchangers has been largely superseded by custom or standard off-the-shelf shell-and-tube exchangers.

Tube-and-Shell Exchangers Sometimes, to increase capacity or reduce the length required, more than one internal tube is incorporated within the outer-tube enclosure as shown in Fig. 1.2*d*.

The most common form of multitubular heat exchanger is the *tube-and-shell* exchanger shown in Fig. 1.3*a*. It is widely used for liquid-liquid heat transfer, liquid-condensing vapor applications, and liquid-evaporating vapor systems.

The great number of tube-and-shell heat exchangers in use

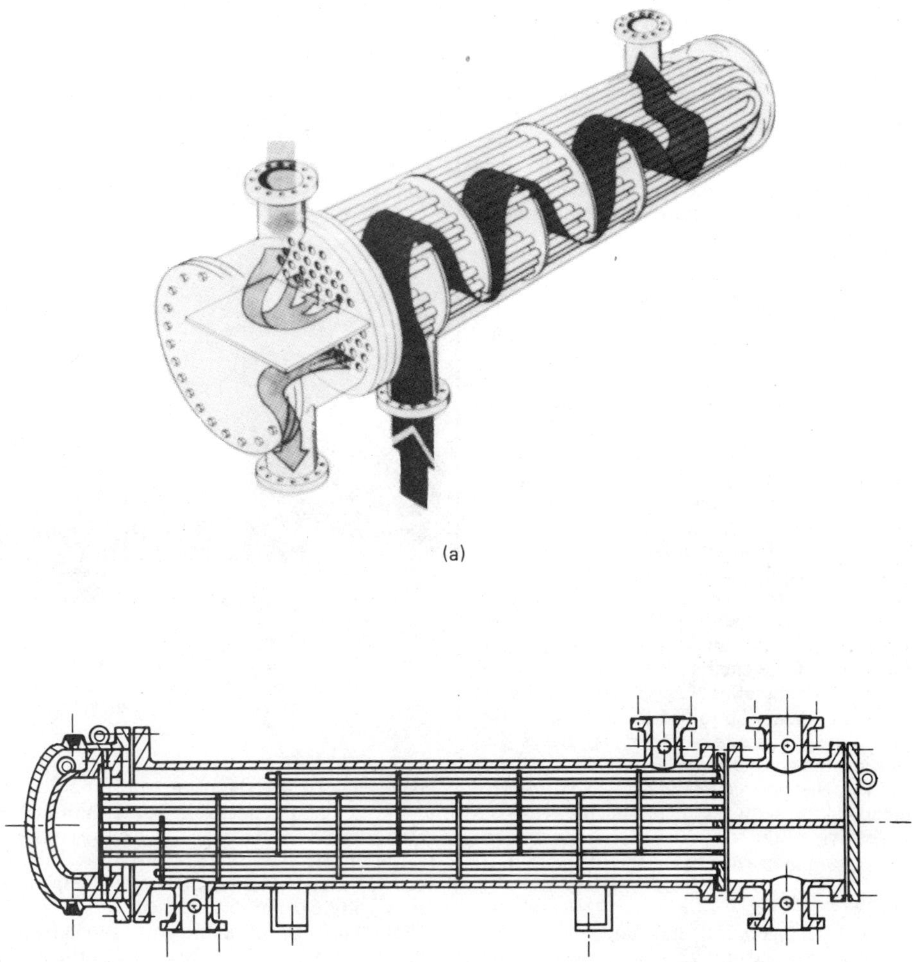

(a)

(b)

Fig. 1.3 (*a*) Tube-and-shell heat exchanger; (*b*) type AES TEMA tubular exchanger.

has led to the emergence of industry standards which have been widely adopted. The best-known standards for the high-class custom made tubular heat exchangers used in the chemical and process industries are the Tubular Exchanger Manufacturers' Association (TEMA) Standards. These include a basic nomenclature and classification scheme for tube-and-shell heat exchangers.

The constituent parts of a tube-and-shell exchanger comprise four principal subassemblies: (1) front end, (2) rear end, (3) tube bundle, and (4) shell. These subassemblies exist in a variety of types and can be arranged in different combinations. The standard terminology recommended by TEMA identifies the various parts and components. A typical combination of subassemblies is shown in Fig. 1.3*b*, designated in TEMA nomenclature as a

> . . . split-ring floating heat exchanger with removable channel and cover, two tube passes, single pass shell, 23¼ in. inside diameter with tubes 16 ft. long. The TEMA designation should be: SIZE 23-192 TYPE AES.

The type of exchanger is designated by a three-part alphabetical combination characterizing the stationary front head, the shell, and the rear head, i.e., Type AES. The complete designation also includes the size of the exchanger. This is a combination of two numbers indicating (1) the shell-and-tube bundle diameter and (2) the tube length. The nominal diameter is the inside diameter of the shell in inches rounded off to the nearest integer. The nominal length is the tube length in inches. For U-tubes, the length is taken as the straight length from the end of the tube to the tangent of the bend.

The type AES heat exchanger shown in Fig. 1.3*b* is of relatively complicated (and therefore costly) construction which can be completely dismantled to facilitate cleaning both the inside and outside tube surfaces. The floating rear head arrangement accommodates any credible differential movement of the tube assembly without restraint. Other less costly combinations are possible; the differences between the various types are principally design or construction features incorporated to facilitate cleaning the inside or outside tube surfaces and to accommodate the differential thermal expansion arising from large temperature differences in the fluids. The possible variations are virtually endless.

Internal leakage from one fluid stream to another through the packing around the floating header sheet may sometimes cause difficulties with the type of construction shown in Fig. 1.3*b*. This can be avoided by other forms of construction so that any leakage of either fluid through the packed joints appears at the exterior of the shell and is readily detected with no possibility of contaminating the other stream. Packing glands designed to contain up to 600 lb/in^2 (4.09×10^6 N/m^2) pressure can be designed, but there is always the possibility of a leak to the surroundings. An external packed head is to be avoided when the shell fluid is toxic, corrosive, expensive, or inflammable. With very high tube pressures or situations where leakage of the tube fluid is unacceptable, a double-tubesheet type of construction is sometimes used.

Extended-Surface Tubes

Most shell and tube exchangers are fabricated with "bare" tubes, but sometimes extended-surface tubes are used. The extended fin surface may be added to a bare tube by welding or some form of mechanical rolling. Alternatively, the tube may be extruded to produce an integral continuous fin.

Finned tubes for shell-and-tube exchangers sometimes have the longitudinal fins shown in Fig. 1.4*a* rather than the radial fins favored for crossflow air heaters and coolers. A typical two-tube-pass, single-shell-pass exchanger with longitudinal fins is shown in Fig. 1.4*b*.

Finned-tube exchangers are particularly suitable in applications where there are large differences in the heat-transfer

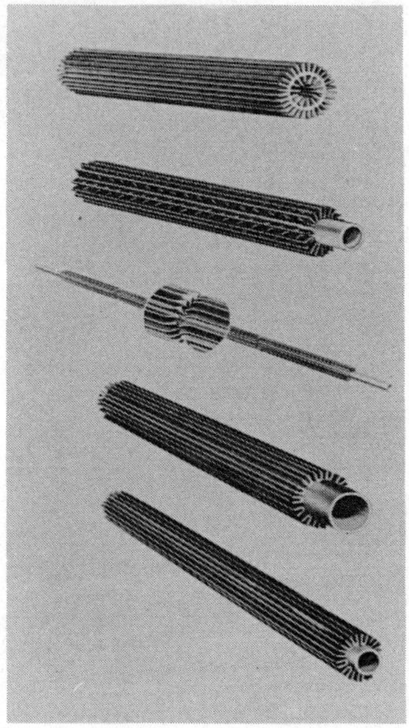

(a)

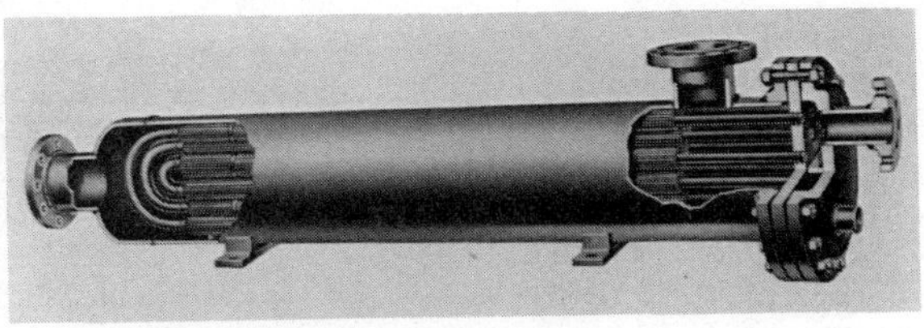

(b)

Fig. 1.4 Longitudinal finned tube heaters. (*a*) Longitudinal finned tubes; (*b*) two-tube-pass, single-shell-pass-line heater with longitudinal finned tubes. (*Courtesy Brown Fintube Co., Tulsa, Okla.*)

coefficients of the shell-side and tube-side fluids. One example is steam condensing inside the tubes with very high rates of heat transfer to heat a thick viscous oil moving rather slowly on the shell side. In these cases the additional area for heat transfer on the shell side compensates for the difference in heat-transfer coefficient.

Internally finned tubes are occasionally used with the fins attached to the tube by welding or bronzing. Alternatively, the tube may be extruded with internal and external fins integral with the tube such as the double-fluted tube configuration shown in Fig. 1.5.

Fig. 1.5 Extended-surface tube, double-fluted tube configuration.

Tube-and-Fin Heat Exchangers

Many applications for heat exchangers involve the transfer of heat between a liquid or vapor and a gas. In most cases the gas is atmospheric air used for cooling the liquid-vapor stream or itself being heated or cooled for air conditioning. The liquid varies according to the application—water, refrigerant, acids, oils, or alcohols. Condensing vapors include steam, refrigerants, and hydrocarbons.

Tube-and-fin heat exchangers are well suited to this duty. They consist simply of a bank of finned tubes, perhaps a single row but customarily several rows deep with manifold or U-tube connections at the ends of the tube sections. A typical example is shown in Fig. 1.6*a*. The liquids flow through the tubes, and air flows in crossflow over the tubes.

The heat-transfer coefficient for liquid flow in the tubes is orders of magnitude greater than for airflow over the tubes. To compensate, the area for heat transfer on the air side is increased by the provision of fins.

In cases where space and volume are not at a premium, fin design is guided largely by economic considerations. Many air-cooled systems utilize aluminum fins wrapped or extruded over the base tube.

Applications A familiar application of tube-and-fin heat exchangers is the automotive "radiator." Another familiar application is the air-cooled tube-and-fin heat exchanger found at the rear of domestic refrigerators.

The same system is used in larger-scale refrigerating and air conditioning plants and in heat pumps where refrigerant is either condensed or evaporated and air is either heated or cooled depending on the use. Similarly, hot-water or steam heating and chilled-water or brine cooling systems are used to heat or cool air for comfort systems in building planes, ships, vehicles, etc. Air coolers are used for lubricant oil cooling systems on aircraft, ships, vehicles, and stationary gas turbine or diesel engine power generators.

In much larger sizes, air coolers are used in petroleum refineries, in the process industries, and in all manner of commercial enterprises for dissipating waste heat to the atmosphere. Condensers for steam turbines are sometimes direct-air-cooled.

Natural and Forced Convection The tubes in the tube bank may be arranged horizontally to establish natural-convective air motion, but the low-heat-transfer coefficients characteristic of natural-convection systems demand the exchanger be of great size. The alternative is a forced-convection system with the airflow driven by fans over the tubes. This allows use of a smaller exchanger with consequent savings in capital cost and reduces the need for tubes to be mounted horizontally.

The drive fan can be mounted upstream or downstream of the tube bank (Fig. 1.6*b*). When the fan is mounted upstream, the arrangement is called a *forced-draft system* and when mounted downstream, an *induced-draft system*. Fans

(a)

(b)

Fig. 1.6 Tube-and-fin heat exchangers. (*a*) Tube-and-fin heat exchanger for air conditioning system; (*b*) induced-flow tube-and-fin heat exchanger for air conditioning. (*Courtesy Custom Coils Ltd., U.K.*)

are almost invariably of the axial flow type with large-volume, low-pressure characteristics and selected to impart an air velocity of about 3 to 16 ft/s (1 to 5 m/s) through the tube bundle. They usually have four to six blades and a diameter somewhat less than the width or length of the tube bank.

Construction Air-cooled heat exchangers can be generally categorized into two types: (1) single tube pass and (2) multitube pass. In the single-tube-pass type (Fig. 1.7*a*) the finned tubes are located at the tube ends in header tanks or manifolds. The tube ends are secured by tube rolling in heavy-walled manifolds or by brazing and soldering in header tanks for less arduous duties.

In multitube-pass heat exchangers (Fig. 1.7*b*) the tubes are interconnected at the ends by a U connection, allowing the liquid in the tubes to make two or more traverses of the tube bank. This type of exchanger is widely favored for the condensing and evaporating systems in steam heating or refrigerating and air conditioning.

(a)

(b)

Fig. 1.7 Air-cooled heat exchangers. (*a*) Single-tube-pass air cooler; (*b*) multitube-pass air-cooled heat exchanger. (*Courtesy Brown Fintube Co., Tulsa, Okla.*)

The tubes used in tube-and-fin heat exchangers range from small-diameter [0.2-in (5-mm)] tubes found in compact aircraft oil coolers or air conditioning systems to the larger-diameter [2-in (5-cm)] tubes found in viscous oil and flue gas service. Standard tube lengths range up to 40 ft (12 m) with tube bundles of up to 12 ft (4 m) wide and as many as 30 tubes deep. Fin heights vary up to 0.75 in (2 cm) with up to 10 fins per inch (4 fins per cm) of tube length. The ratio of extended surface area to bare tube area ranges from 6 to 20.

Aluminum is an economical fin material for temperatures up to 750°F (400°C) on the liquid side. For low-temperature work up to 250°F (120°C), the aluminum fin may be mechanically wrapped under tension to the roughened surface of the base tube. This provides an economical tube suitable for use in a low-vibration situation. For higher temperatures, the fin may be wrapped under tension in a groove machined in the base-tube wall or the fins may be produced by a high-pressure cold-extrusion process. In corrosive conditions, a base tube other than aluminum may be used.

For the smaller air conditioning and steam heating coils and automotive radiators, thin-walled copper and brass tubing with soldered or brazed joints are widely used. The fins are a flat helically coiled copper or aluminum strip or wire coil wound into the tube and secured by dip brazing or soldering.

Tank Heaters and Coolers

Tubular exchangers are often used for heating and cooling the contents of tanks. The contents may require heating to reduce the viscosity and facilitate pumping, prevent solidification of waxes or other solids, or melt the contents prior to processing. The contents may require cooling to prevent the buildup of excessive vapor pressure to slow down a reaction or to induce the selective deposition of solids. In batch processing operations, heating or cooling the contents is frequently involved for evaporation, concentration, etc.

There are, in general, two types of tank heating systems: (1) gross contents heating and (2) suction heating. For gross heating, a tubular coil is located in the tank or walls of a thermally insulated tank. Internal tank coils are made in a wide variety of configurations: flat spirals, helical coils, hairpin tube bundles, and ring headers. Because of natural convection, heating coils tend to be located at the bottom of the tank and cooling coils at the top.

For suction heating, a hairpin tube bundle is located at exit from the tank. The heating process is preferential and confined almost entirely to fluid leaving the tank rather than the remaining contents of the tank. The tube bundle may be actually inserted in the tank or incorporated in the suction line leaving the tank. The latter form, called a *line heater,* may be used in any application where heating of pipeline contents is required.

Fired Heaters

Many tubular exchangers in a wide variety of types and sizes are available where the heating process is caused by the hot products of combustion of gas, oil, or coal. These systems are used for raising steam, superheating steam, and heating oils and vapors generally and are called *fired heaters.*

Evaporators

The process of evaporation is widely used in the chemical and refining industries. There are large numbers of different types of evaporators, and most incorporate some form of tubular heat exchanger.

In the majority of cases the heating medium is steam located in the tubes with evaporating fluid passing over the tube bundle in the shell. Sometimes a sequence of an evaporator system in series, called a *multiple-effect evaporator,* is used to gain economy in steam consumption by sequential decrease in the steam temperature and pressure from one evaporator to the next.

Scraped-Surface Exchangers

Scraped-surface heat exchangers are used for processes likely to result in the substantial deposition of solids on the exchanger surface. Such processes include the precipitation of crystals and waxes in chilling applications. Use of a scraped-surface exchanger prevents the accumulation of significant buildup of the solids.

Scraped-surface heat exchangers are most commonly of double-pipe construction with the process fluid in the inner pipe and the cooling or heating fluid (steam or water) in the annulus. A rotating element is contained within the tube and is equipped with spring-loaded blades scraping the inside tube walls.

Nonmetallic Tubular Exchangers

Most heat exchangers of all types are fabricated from metals. Metals are available in a wide range of types and forms with different strength characteristics. They are easily worked and readily joined by welding, brazing, or soldering. Economic considerations dictate the use of metal of lowest cost that is compatible with the fluids to be used in the heat exchanger. The most significant characteristic of a metal from the aspect of fluid compatibility is its resistance to corrosion.

Corrosive liquids and vapors such as hydrochloric, sulfuric, and phosphoric acids require use of exotic metals such as titanium, tantalum, zirconium, Hastelloy, and other alloys. Certain fluids are so corrosive that no metal alloys can be found with suitable corrosion resistance within the bounds of economic feasibility. The alternative is to use nonmetallic systems.

Plastic Materials Plastic materials, particularly polytetrafluoroethylene (PTFE) (Teflon or Fluon), are particularly favored because of their exceptional resistance to chemical attack or degradation. Very effective heat exchangers are constructed from clusters of long, thin flexible tubes of PTFE secured at the tube ends to headers of the same material.

Graphite Materials Graphite is also widely used for heat exchangers with highly corrosive fluids. Graphite has a high resistance to corrosion, is a good conductor of heat, and has a low coefficient of thermal expansion and thus is resistant to thermal shock. It is readily machinable but cannot be welded and has low tensile strength. Equipment design is therefore based on loading the graphite in compression. The temperatures and pressure in graphite exchangers are generally limited.

The difficulty of low tensile strength can be overcome by ingenious design whereby the graphite components are subject to compressive rather than tensile stress. One approach is to use metal tie rods in tension to maintain graphite components in compression loading even when an internal pressure is applied. Another approach is to enclose the graphite within a steel shell. Installation of the close-fitting graphite components is accomplished at elevated temperatures so that on cooling the metal shell contracts and preloads the graphite components in compression.

For higher-pressure applications, the cross-bore technique is often used wherein rows of parallel holes are drilled in solid blocks of impervious graphite. Two sets of holes are drilled, perhaps of different diameter, at mutually perpendicular axes, one set for the corrosive tube-side fluid and the other for the noncorrosive shell-side fluid. Incorporation of this cross-bore technique in tube-and-shell or cubic exchangers is illustrated in Fig. 1.8.

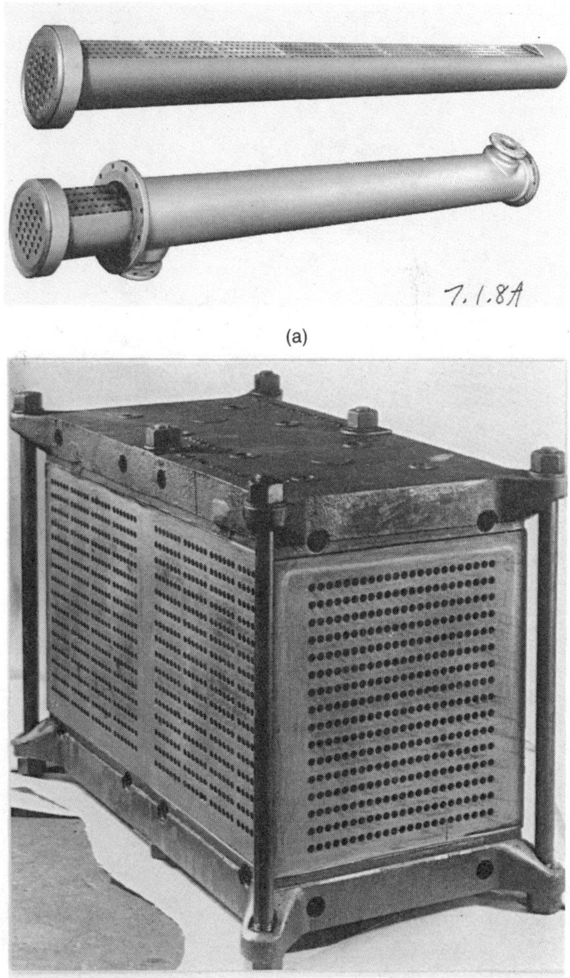

(a)

(b)

Fig. 1.8 Graphite heat exchangers. (*a*) Cross-bore graphite tube-and-shell heat exchanger; (*b*) cross-bore graphite "cubic" heat exchanger. (*Courtesy Carborundum Co., Solon, Ohio.*)

Glass Materials Glass is another material used for heat exchangers in special situations. It is particularly attractive for use with acids and other corrosive fluids. Glass is corrosion-resistant to virtually all fluids. It is also favored in pharmaceutical and food processing where purity of the product is essential. If the process contacts only glass and Teflon, the possibilities for contamination are greatly reduced.

The processing of chlorinated and brominated organic fluids is often facilitated in impervious borosilicate glass vessels to overcome the problem of permeation existing with many other materials.

Glass heat exchangers are commonly found in the two types shown in Fig. 1.9. One is a typical conventional tube-and-shell heat exchanger in which the tubes are constructed of glass with PTFE tubesheets and a PTFE threaded fitting securing each tube in the tubesheet. The shell may also be of glass if both the shell-side fluid and the tube-side fluid are

(a)

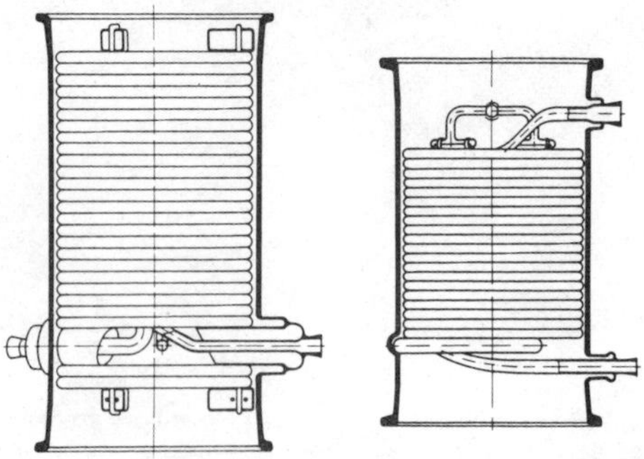

(b)

Fig. 1.9 Glass heat exchangers. (*a*) Tube-and-shell heat exchanger with tubesheet and tube seals; (*b*) glass coil condenser and boiler. (*Courtesy Corning Glass Co.*)

corrosive. In such cases the permissible operating pressures and temperatures are quite limited. Higher pressures and temperatures are possible with carbon steel shells and for corrosive fluids, the shells may be lined with rubber or Teflon.

Figure 1.9*b* shows glass coil condenser and coil boiler elements. In one case the coolant, water, circulates in the coil. In the other case steam or other heating fluid circulates in the coil. The elements can be stacked one upon the other in a vertical column as part of a circulating loop. In recent years, large glass tube exchangers for heat recovery from contaminated air-exhaust streams have been introduced.

Noncircular Tubes

Normally the tubes used in tubular heat exchangers are circular in cross section. There are a few exceptions to this general rule in which the tubes are flattened to an elliptic shape or rectangular form with semicircular ends.

One example of the use of noncircular tubes is the lamella heat exchanger produced by the Swedish Alfa-Laval Company (Fig. 1.10). The lamella heat exchanger is considered a variant of the conventional tube-and-shell exchanger and designed specifically for longitudinal countercurrent "tube-side" flow of both tube and shell fluids.

Heat Pipes

The elements of a heat-pipe heat exchanger are shown in Fig. 1.11*a*. An element consists simply of a metal tube closed at both ends and is about a quarter full of liquid. One end of the tube is heated and the other end is cooled. Liquid in the tube evaporates at the heated end and condenses at the cooled end. There is, therefore, a flow of vapor from the heated to the cooled end and a flow of condensate liquid back to the cooled end.

With a heat pipe, very large amounts of energy can be transferred over relatively long distances (several meters) with only a degree or two of temperature difference between the heated and cooled ends. The effect is the same as would be attained with a solid metal bar having a superlative thermal conductivity, orders of magnitude greater than the best thermal conductors known—pure silver, copper, and aluminum.

Figure 1.11*b* shows a multielement heat-pipe heat exchanger suitable for waste heat recovery in an inlet air preheater or similar application. There is a very large literature about heat pipes. Reay has provided a good summary of information about design, application, and availability with an extensive bibliography.[6]

1.4 PLATE HEAT EXCHANGERS

Plate heat exchangers are less widely used than tubular heat exchangers but offer advantages to make them the system of choice in many cases.

Plate heat exchangers can be classified in four principal groups:

1. Plate-and-frame or gasketed-plate heat exchangers used as an alternative to tube-and-shell exchangers for low- and medium-pressure liquid-liquid heat-transfer applications.

2. Spiral heat exchangers used as an alternative to tube-and-shell exchangers where low maintenance is required,

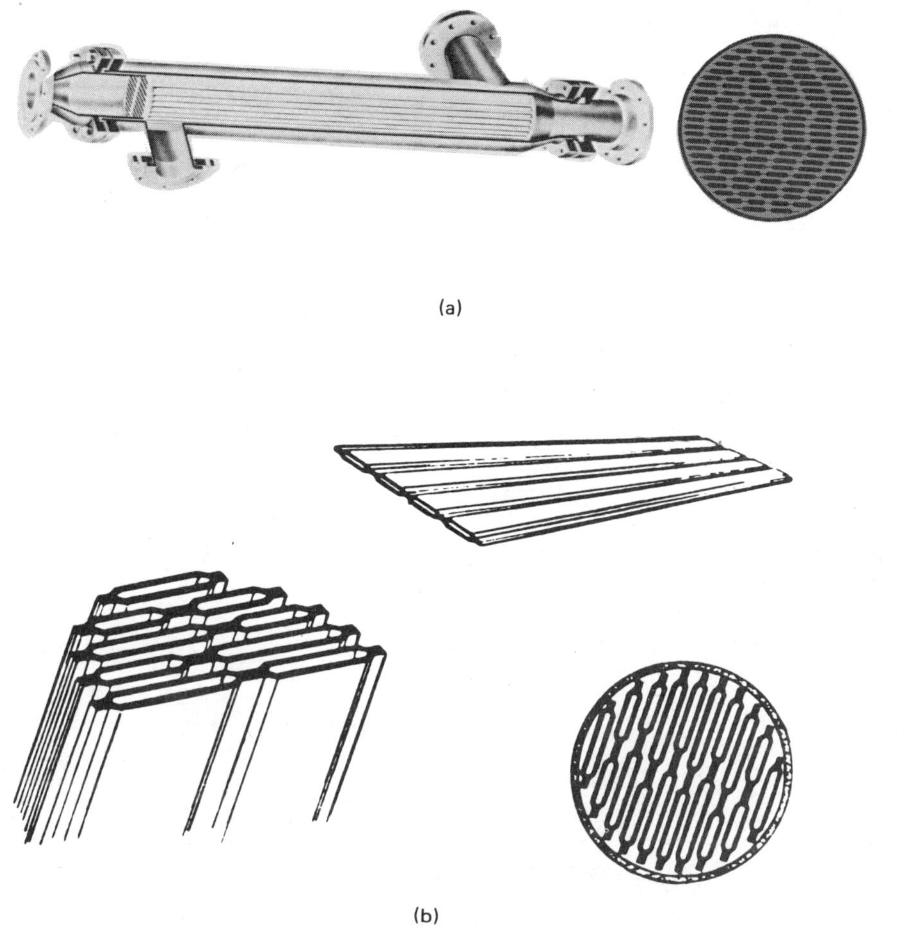

Fig. 1.10 Lamella heat exchanger. (*a*) Cross section of lamella bundles in shell; (*b*) section of lamella heat exchanger. (*Courtesy Alfa-Laval.*)

particularly with fluids tending to sludge or containing solids in suspension.

3. Plate-coil heat exchangers made from previously embossed plates to form a conduit or coil for liquids coupled by fins.

4. Plate-fin heat exchangers. A stack of die-formed corrugated plates welded or brazed to provide compact heat-exchanger surfaces for various gas-gas exchange processes. Commonly used in cryogenic applications and exhaust-inlet air preheating waste heat-recovery systems for buildings, gas turbines, etc.

Plate-and-Frame or Gasketed-Plate Heat Exchangers

Construction Plate-and-frame or gasketed-plate heat exchangers are made of an assembly of the pressed metal plates (Fig. 1.12). These are aligned on a frame and secured between covers by tie bars as shown in Fig. 1.13. Gaskets set in the grooves around the periphery of each plate contain the fluids and direct the flow distribution. Ports for the inlet and outlet of hot and cold fluids are punched or blanked into the corners of each plate. When the plates are assembled, the ports are aligned to form the distribution headers for flow through the plate pack.

Fluid flow occurs in the fine passages between alternate plates. Except at the ends, there are hot fluids flowing on both sides of the two plates forming the conduit in which the cold fluids flow and vice versa. The distribution of fluids to alternate plates is effected by selectively removing part of the circumferential gasket separating the plates as shown at the left-hand ports in the specimen plates shown in Fig. 1.12. Intermixing of the two fluids in the event of gasket failure is prevented by the provision of leakage grooves in the gasket near the headers.

Plates can be made of any metal that can be die-formed and can be virtually any shape. Plate heat exchangers are manufactured in a very wide range of sizes, including large units (Fig. 1.14).

Advantages of Plate-and-Frame Heat Exchangers The principal advantage of plate-and-frame heat exchangers is the

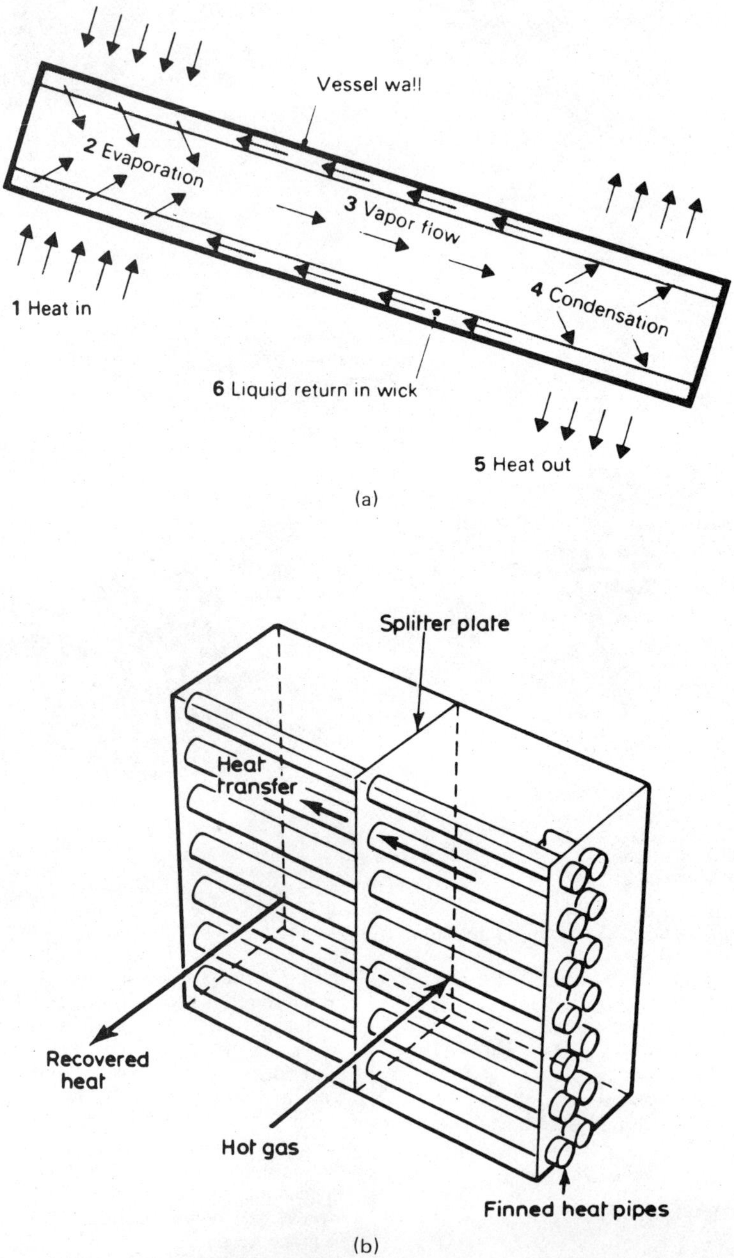

Fig. 1.11 (*a*) Heat-pipe heat exchanger; (*b*) arrangement of multielement heat-pipe heat exchanger for waste heat recovery. (*After Reay.*[6])

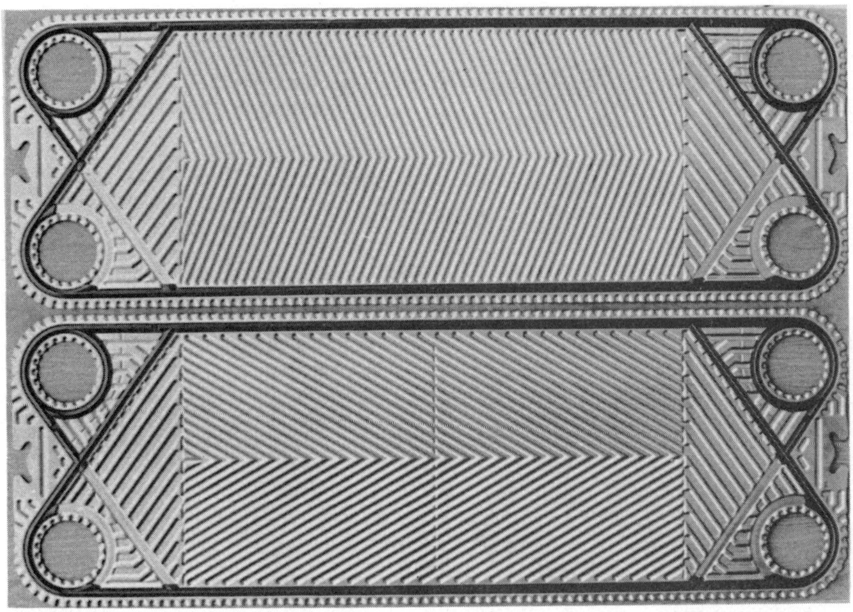

Fig. 1.12 Die-formed pressed metal plastes. (*Courtesy Alfa-Laval.*)

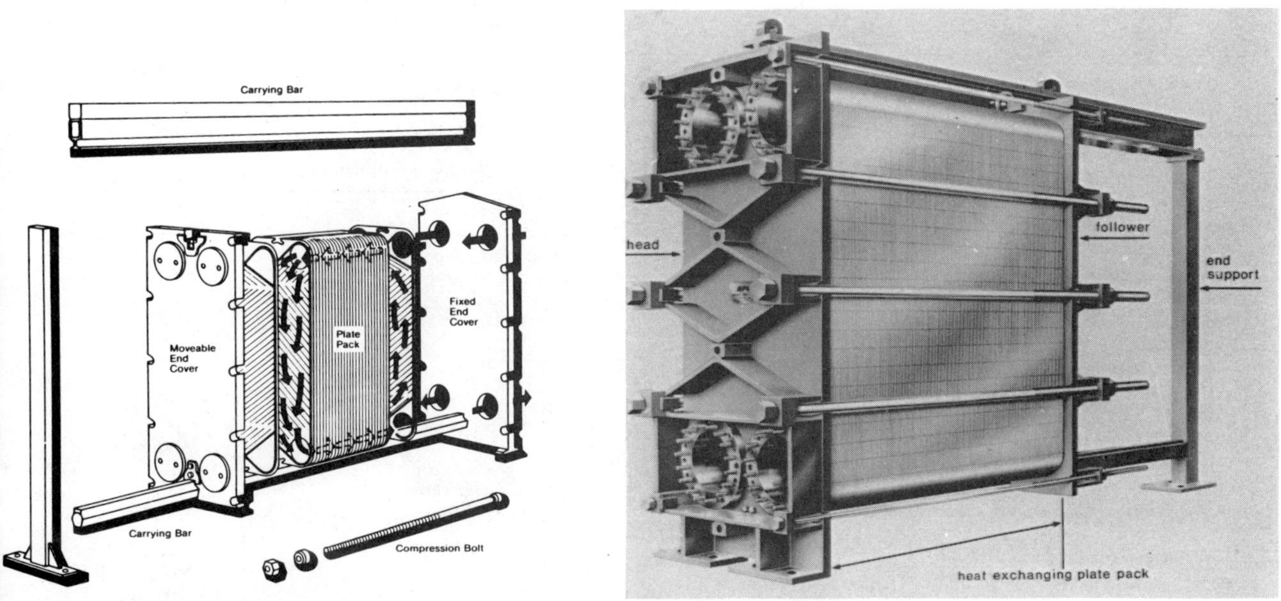

Fig. 1.13 Construction elements of plate-and-frame heat exchanger. (*Courtesy APV.*)

Fig. 1.14 Large plate-and-frame heat exchanger. (*Courtesy Alfa-Laval and APV.*)

accessibility of the heat-exchange surface. The follower may be rolled back along the frame and the plates separated for inspection and cleaning.

Other advantages claimed for plate-and-frame heat exchangers compared with tube and shell exchangers are:

1. Only one-third to one-half the surface area for heat transfer is required so that savings in cost can be realized. The reduced surface area is a result of the relatively high turbulence achieved in the corrugated flow paths between the plates and increased exchanger effectiveness because of true countercurrent flow and because of reduced fouling (one-tenth to one-fourth that of shell-and-tube exchangers) as a result of the turbulent fluid action and polished surface finish of the plates. The close spacing between adjacent plates is equivalent to the use of very-small-diameter tubes with consequent high heat-transfer coefficients.

2. The reduced surface areas required permit substantial savings in the space required for plate-and-frame heat exchangers. Little clearance space around the exchanger is required to permit dismantling for cleaning compared with the exchanger length required to permit withdrawing a tube bundle.

3. The improved exchanger effectiveness resulting from true countercurrent flow results in an increased temperature spread and allows a closer approach to ideality, a particularly significant feature in heat-recovery systems where the temperature potential is limited.

4. The thermal rating of the plate-and-frame exchanger can be readily increased or decreased by varying the number of plates. This is a most attractive feature for systems likely to involve substantial future changes in load.

5. The increased effectiveness of the plate-and-frame heat exchanger increases the temperature rise of the cooling water so reducing the necessary flow with consequent savings for piping, valves, pumps, and operating cost.

6. The internal void volume of a plate-and-frame heat exchanger is only 10 to 20 percent of the equivalent tube-and-shell exchanger. Less process media are involved so that better process control and faster response times can be achieved.

7. The reduced surface requirement and volume of fluid contained combine to reduce the gross mass of the plate-and-frame exchanger to approximately one-sixth of the equivalent shell-and-tube exchanger. Further savings result from lower shipping, installation, and foundation costs.

Disadvantages of Plate-and-Frame Heat Exchangers The only disadvantage of plate-and-frame exchangers is the large number of surfaces that must be sealed by gaskets. Elastomeric materials are normally used for most applications, but the maximum temperatures and pressures at which these can be used is limited to about 400 lb/in^2 (2.7 MPa) and 300°F (150°C). Compressed-asbestos-fiber gaskets may be used up to 600°F (350°C), but the pressures for which they are suitable is limited to about 240 lb/in^2 (1.8 MPa).

Comparison with Shell-and-Tube Exchangers A good technical review of plate-and-frame heat exchangers was presented by Marriott[5] including the comparison of plate-and-frame and tube-and-shell heat exchangers reproduced in Table 1.1. Marriott projects a cost for the exchangers in

Table 1.1. Comparison between Plate and Tubular Heat Exchanger for Typical Water-Water Duty

	Hot side	Cold side
Flow, m^3/h	50	50
Temperature in/out, °C	80/40	20/60
Results of Calculation for Both Types		
	Plate	Tubular
Heat-transfer area, m^2	25	85
Overall heat-transfer coefficient (clean)	5200 kcal/ m^2 · °C · h	1750 kcal/ ft^2 · °C · h
Fouling factor, m^2 · °C · h/ kcal	0.00006	0.0001
Overall heat transfer, kcal/m^2 · °C · h	3960	1500
Pass system (hot/cold)	1/1	8/baffled
Channels or tubes per pass	40	56
Plate or tube size	0.32 m^2	17/20 mm × 3 m
Calculated pressure drop, meters–water gage (mwg)	4	6
J, mwg/number of transfer units	2	3
Pumping power, hp	1.1	1.65
Weight (empty), kg	615	2400
Weight (full), kg	720	3100
Overall size (including cleaning space), m	1.5 × 0.7 × 1.4	7.0 × 0.7 × 0.7
Floor area required, m^2	1	5

SOURCE: After Marriott.[5]

Table 1.2. Comparison of Areas and Pressure Drops for Shell-and-Tube (S and T) and Plate Heat Exchangers (PHEs)

	Flow rate, kg/h			Areas, m², and pressure drops, kN/m²					
				S and T design			PHE design		
Case	Fluid A	Fluid B	Temperature duty, °C	m²	kN/m² A	kN/m² B	m²	kN/m² A	kN/m² B
1	6080 hydro carbon	24,265 water	160 A 49 / 49 B 33.5	28	2.9	21	8.6	2	15
2	392,725 water	351,350 seawater	41.7 A 28 / 37.3 B 22	1,125	131	131	337.5	75	60
3	26,820 gasoline	90,900 water	105 A 35 / 35 B 26	83.6	21.4	30	26.2	10	30
4	7955 solvent	8820 water	60 A 40 / 35 B 26	170	20	30	41.4	20	25
5	67,570 desalter effluent	214,775 seawater	105.5 A 38 / 41 B 20	139.5	75	55	35.3	30	70

SOURCE: After Fuller.[1]

stainless steel and for a design pressure of 142 lb/in² of \$2700 to \$4000 for the plate-and-frame heat exchanger and \$5500 to \$6000 for the tube-and-shell (fixed-tubesheet) heat exchanger (1970 U.S. dollars). In mild steel the tubular unit would have cost about \$3500, about the same as the plate-and-frame unit in stainless steel.

Marriott suggested that the plate-and-frame heat exchanger was well suited to the common duty of cooling a process stream of fresh or treated noncorrosive, nonfouling water by means of a cheap, readily available, highly corrosive, brackish water supply. Titanium was normally used for construction, as this was the *"only material capable of withstanding indefinitely the action of brackish water from any source, coast, estuary, river, canal, lake or bore hole."*

More recently Fuller[1] has provided further confirmation of the reduced area and pressure drop requirements of plate heat exchangers compared with shell-and-tube exchangers. Table 1.2 shows area and pressure drop requirements of plate and shell-and-tube exchangers for five different liquid-liquid heat-exchange duties. In all cases the area requirement for the plate heat exchanger is one third to one-quarter that of the equivalent tube-and-shell exchanger. Fuller also included the cost data reproduced in Fig. 1.15. This shows the approximate cost in pounds sterling per square meter of heat-transfer area as a function of the exchanger size. Four characteristic curves are given, one for an all-stainless-steel plate heat exchanger (curve 4) and three for shell-and-tube heat exchangers with all stainless steel (curve 1), with stainless tubes and mild steel shell (curve 2), and with mild steel tubes and shell (curve 3). The curve is somewhat misleading unless it is recalled that the necessary area for the plate heat exchanger is very much less than that for the tube-and-shell. Superficial inspection of Fig. 1.15 shows that *area for area,* the stainless-steel plate heat exchanger is substantially more

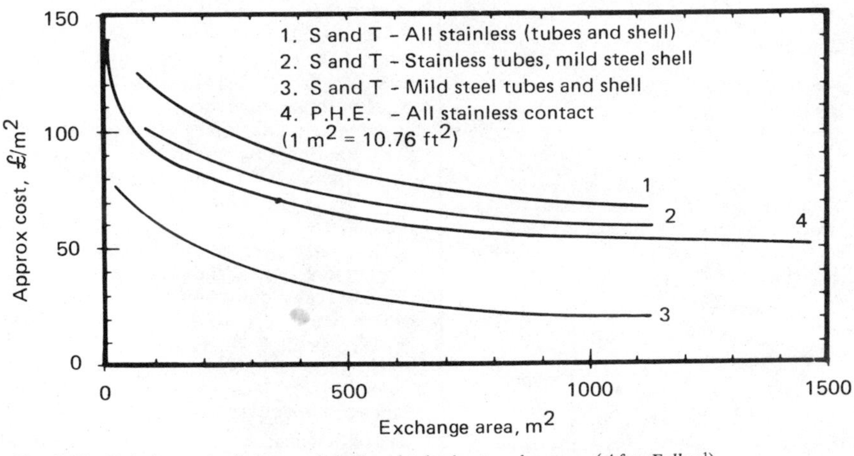

Fig. 1.15 Relative cost of plate and shell-and-tube heat exchangers. (*After Fuller.*[1])

expensive than the mild steel tube-and-shell. However, on a *duty-for-duty* basis the cost differential is greatly reduced and may often be reversed.

Spiral Plate Heat Exchangers

The spiral plate heat exchanger was developed to recover waste energy from contaminated water effluents in pulp mills, and its use has been extended to other process applications. Spiral heat exchangers are available as a general replacement for tube-and-shell exchangers in low- to moderate-pressure service but are particularly favored for difficult situations involving highly viscous liquids or dense slurries.

Construction The elementary construction of a spiral heat exchanger is illustrated in Fig. 1.16. A pair of long strips is rolled around a split mandrel in scroll or spiral fashion. The strips are spaced a uniform distance apart by means of studs welded to the sheets; for very viscous liquids and dense slurries, however, the spacer studs are not included. One side of the strip is curved so as to contact the adjacent straight strip. This seam is welded so as to seal the spiral channel. Alternatively in some cases, spacer bars are welded into each spiral channel at the sides as illustrated in Fig. 1.16*b*. Only one side of each channel is closed in these ways, and the closure is applied to alternate sides. The open sides of the two channels are closed by gasketed cover plates clipped to rigid flanges around the periphery of the cylindrical chamber enclosing the spiral elements as shown in the assembled unit in Fig. 1.17. This construction precludes the possibility of

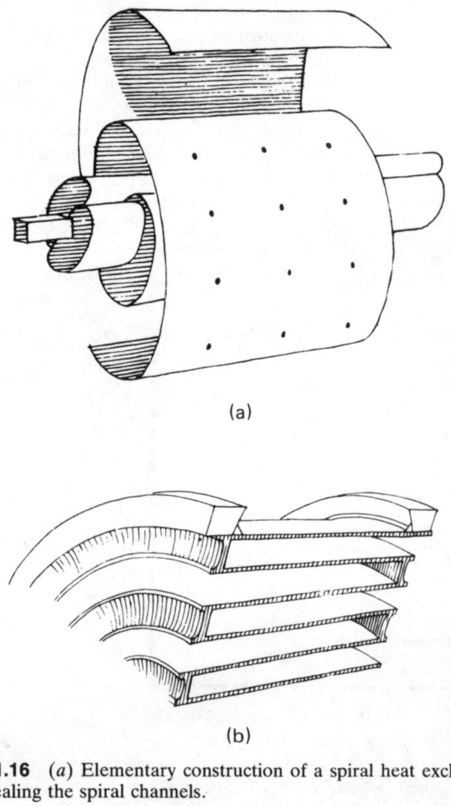

(a)

(b)

Fig. 1.16 (*a*) Elementary construction of a spiral heat exchanger; (*b*) sealing the spiral channels.

Fig. 1.17 Assembled spiral heat exchanger. (*Courtesy Alfa-Laval.*)

leakage between the streams yet allows ready accessibility to both channels for inspection, maintenance, or cleaning.

Advantages of Spiral Plate Heat Exchangers Flow characteristics in the spiral conduits correspond to flow on the tube side of a shell-and-tube exchanger rather than the shell side. Therefore, with both fluids in tube-side flow, the spiral heat exchanger attains a higher overall heat-transfer coefficient. The rectangular flow sections can be dimensioned for optimum flow conditions for there is no requirement for the flow section to be the same for both fluids or to even remain constant in passage from inlet to outlet.

The continuously curved flow section contributes to improved overall heat-transfer coefficients in the same way as curved pipe. Furthermore, at velocities resulting in laminar flow in straight-line tube flow, the continuously varying passage realizes turbulent flow.

The true countercurrent flow increases the exchanger effectiveness. In combination with the higher overall heat-transfer coefficient, this permits a reduction of approximately 20 percent in the required heat-transfer area with consequent savings in size, weight, and cost, including reduced transport, foundation, and installation charges.

The scrubbing effect of the fluids in both spiral passages inhibits deposition of sludge and other deposits. Fouling factors for spiral heat exchangers are commonly one-third the value of those recommended for conventional shell-and-tube exchangers. This makes the spiral exchanger preferred for fluids tending to sludge or those containing solids in suspension. Thermal insulation of the spiral exchanger is rarely required; the cool fluid enters at the periphery and flows in the outer passage.

Maintenance requirements tend to be low because the scrubbing action of the fluids inhibits the permanent deposition of scale and other solids. Furthermore, the side covers can be readily removed for inspection or mechanical cleaning. The single flow passage of each fluid facilitates cleaning by chemical means without removing the end covers.

Panel-Coil Heat Exchangers

The basic construction of these heat exchangers is two embossed metal plates welded together (Fig. 1.18). They form

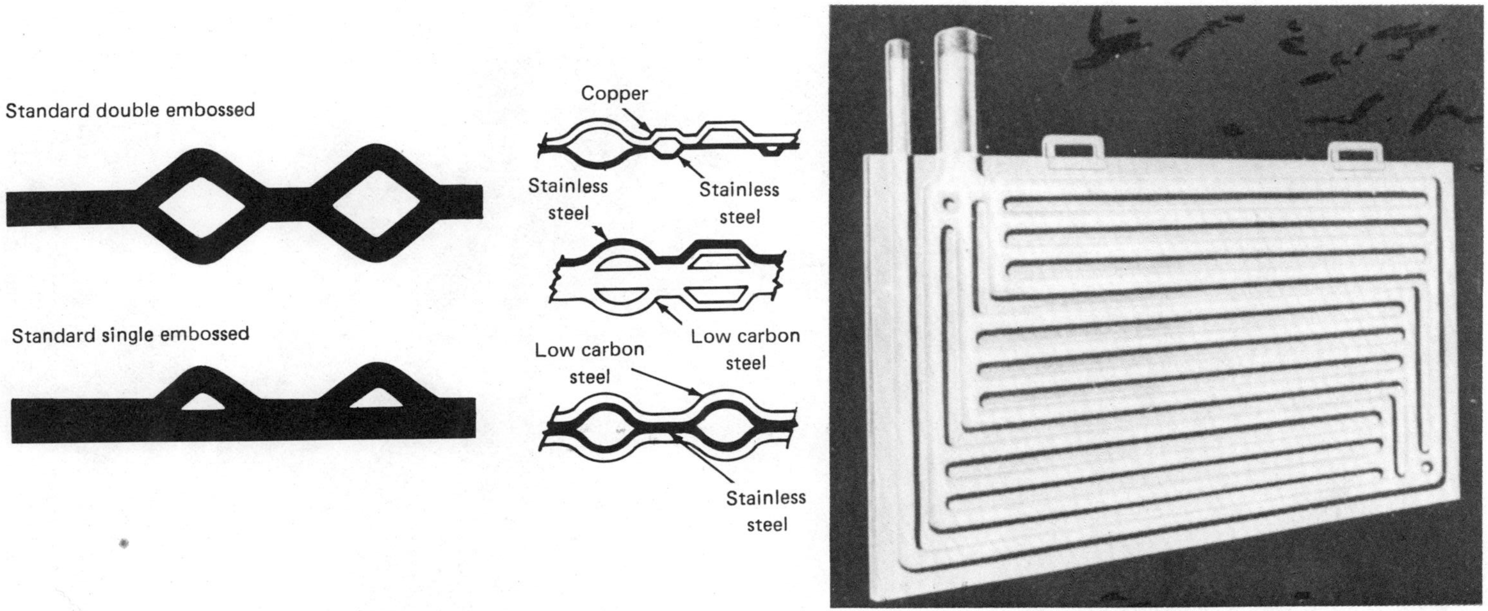

Fig. 1.18 Basic construction of a panel-coil heat exchanger.

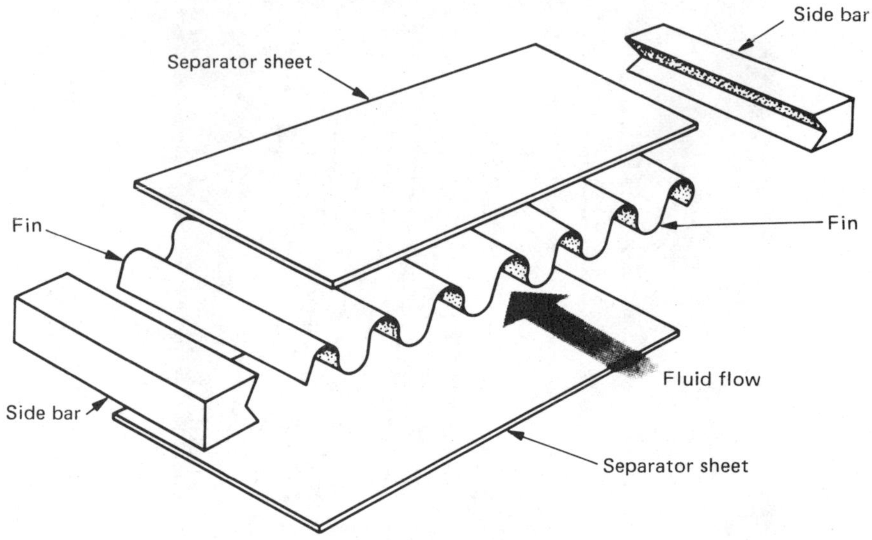

(a)

(b)

Fig. 1.19 (*a*) Construction details of a single-element plate-fin heat exchanger; (*b*) crossflow and counterflow arrangements of plate-fin heat exchangers. (*After Lenfestey.*[3])

a flat panel with a series of flow channels for the heat-transfer medium coupled by relatively thick fins.

In the *double-embossed* type both plates are embossed with identical matching flow channels. In the *single-embossed* type one flat plate and one embossed plate are welded together with a combined panel thickness equal to approximately half the equivalent double-embossed panel. Of course, the section area of the flow channel in the single-embossed design is also half that of the double-embossed type.

There is little restriction on the patterns that can be embossed into the plate; thus flow channels of virtually any size, shape, and distribution can be readily obtained. The flow sections can be a single continuous coil or distributed in a more complicated fashion.

Panels can be fabricated in virtually any materials that can be welded or otherwise joined. Metals are used almost invariably. Carbon steel plates are most economical where conditions permit, but many other combinations are routinely produced for corrosive fluids or other special service.

After fabrication of the basic panel, it can be rolled or pressed into a variety of different shapes or configurations. It can be simply mounted adjacent to a wall as part of the perimeter room heating system; it can be hung in a tank for liquid heating; it can be incorporated as part of the structure of a vessel; or it can be clamped on the inside or outside of a vessel. Multiple units can be ganged together in endless variety for air or liquid heating. Panels can be made in flat pancake form or as cones with wide or shallow angles.

Panel-coil heat exchangers are remarkably versatile, inexpensive, and applicable in a surprisingly wide range of industrial uses. They can be produced in virtually any flexible material available in sheet form and in a variety of thicknesses to withstand any design pressure.

Plate-Fin Heat Exchangers

The elements of a plate-fin heat exchanger consist of a crimped or corrugated die-formed fin plate sandwiched between flat metal separator plates (Fig. 1.19). Side bars are located along the extremities of the fin sections. The heat exchanger is formed by a stack of such elements welded or dip-brazed to form a rigid matrix.

Stacks of elements can be assembled and joined in virtually any size and shape. They are completed by welding on flow distributing manifolds or headers; see Fig. 1.20 for a large plate-fin heat exchanger in the course of manufacture. Several such units may be eventually incorporated into very large heat exchangers as shown in Fig. 1.21.

A very wide variety of fin plates have been used. Typical elements are shown in Fig. 1.22. In some cases a wavy corrugation is favored to promote turbulence and enhance the heat transfer. In other cases the fins are sometimes louvered or dimpled, again to enhance the heat transfer and promote turbulence. Measures such as those described above are effective in enhancing heat transfer but are invariably accompanied by increased fluid frictional effects manifest in an increased pressure drop in the core.

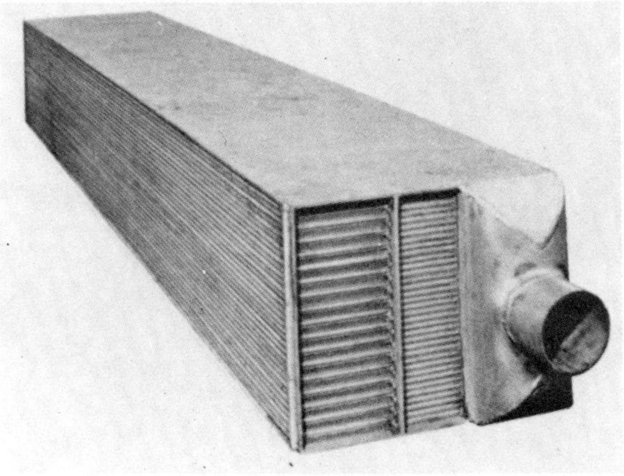

Fig. 1.20 Large plate-fin heat exchanger in course of manufacture with one flow header attached. (*After Lenfestey.*[3])

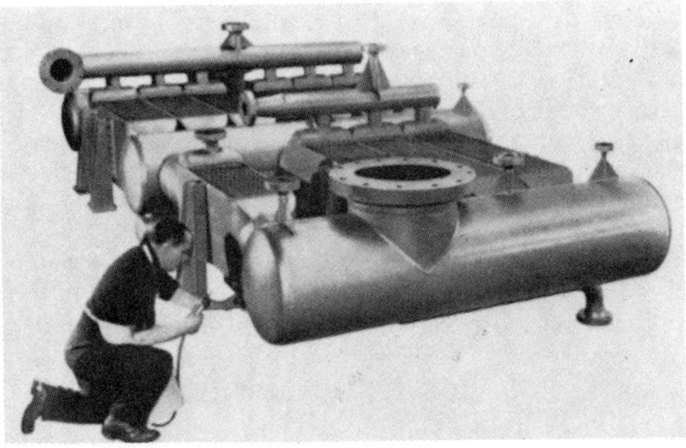

Fig. 1.21 Large complicated plate-fin heat exchanger in course of manufacture. (*After Lenfestey.*[3])

The alternate corrugated fin plates need not be of the same height; i.e., uniform spacing of the separator plates is not necessary. This is advantageous where the density of the hot and cold fluid streams is different, such as a liquid on one side and a gas on the other. In gas turbine recuperators the hot exhaust products on one side are at atmospheric pressure while the air being preheated is at pressures of 4 to 10 atm. Similarly, in recuperative counterflow plate-fin heat exchangers used in cryogenic systems the return stream will be at a much lower density than the high-pressure inlet stream en route to the expansion engine or valve.

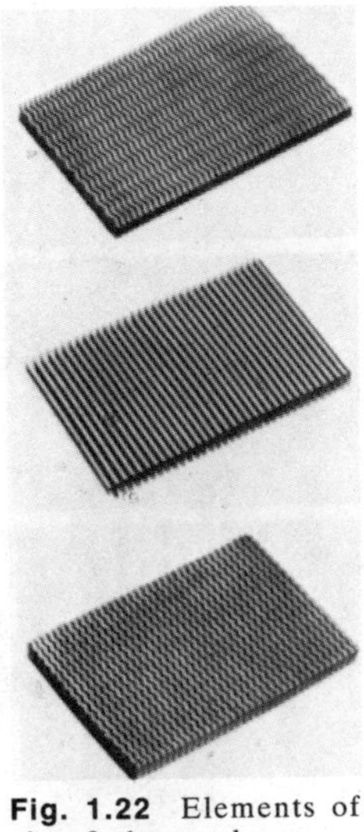

Fig. 1.22 Elements of plate-fin heat exchangers.

Plate-fin heat exchangers used in cryogenic and gas liquefaction plants are generally of aluminum throughout with dip or furnace brazing as the preferred method of construction. Heat exchangers of extraordinarily compact construction can be achieved at moderate cost, small size, and low weight.

Plate-fin heat exchangers are also used for closed-cycle gas turbines. In these applications, temperatures are very high and stainless-steel or nickel alloy sheets are used instead, welded or brazed as required.

Similar systems have been used for open-cycle gas turbines and heat-recovery exchangers in other fired systems, but the hot side soon becomes fouled by the products of combustion, and the small flow passages eventually become blocked. No really effective, low-cost, and convenient method of cleaning the very compact matrices of plate-fin heat exchangers has yet been developed. Primarily because of the relative cleanliness and purity of fluids in cryogenic and gas liquefaction systems, plate-fin heat exchangers have assumed a dominant position in those applications.

The major development of interest in energy conservation will result in the near future in thermal recovery from relatively low temperature, thermal effluent streams. This will surely provide many opportunities for application of plate-fin heat exchangers. Some will likely be of brazed aluminum construction; however, there also appear to be substantial fields of application for plate-fin exchangers of low-cost plastic, waxed paper, and aluminum foil with epoxied or other low-cost cemented construction. One form of low-cost aluminum heat exchanger for thermal energy recovery systems is shown in Fig. 1.23*a*. It consists of a stack of die-formed metal plates rolled at the seams of adjacent plates to form alternate flow channels with mutually perpendicular axes of flow. The assembled exchanger is shown in Fig. 1.23*b*. It can

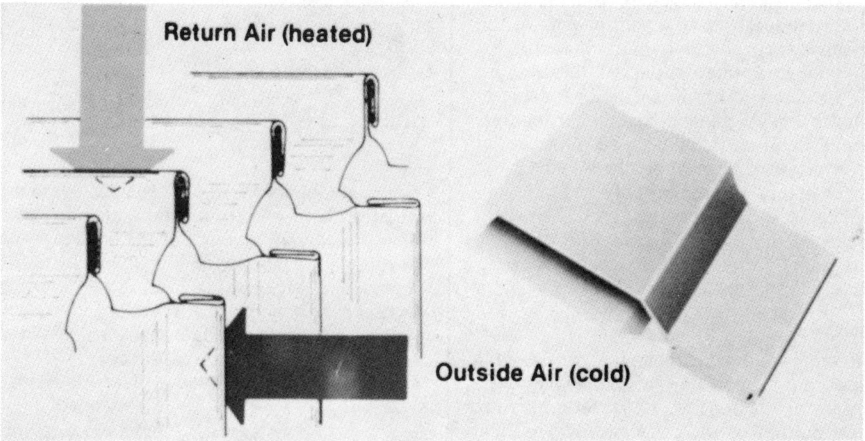

(a)

(b)

Fig. 1.23 (*a*) Low-cost aluminum plate heat exchanger for energy recovery; (*b*) assembled aluminum plate heat exchanger. (*Courtesy ACS, Elk Grove, Ill.*)

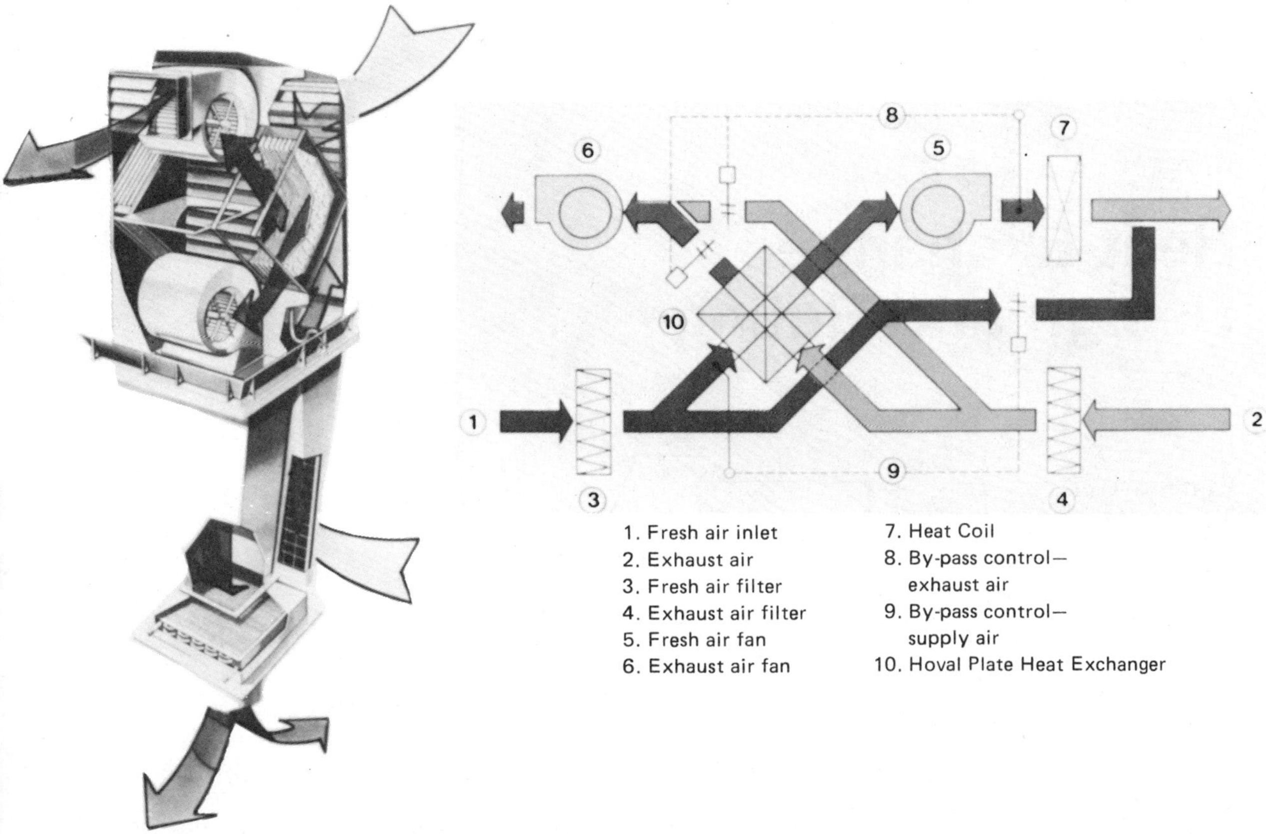

Fig. 1.24 Heat exchanger incorporated in exhaust energy recovery system. (*Courtesy ACS, Elk Grove, Ill.*)

be incorporated in the exhaust-energy-recovery system shown in Fig. 1.24.

The use of plastics for heat exchangers is of current interest, including research work on plate-fin heat exchangers of rigid polyvinyl chloride sheets intended for air conditioning applications. The moderate pressure and temperature environment permitted the use of vacuum-formed plastic sheets so thin that the relatively low thermal conductivity of the plastic was not significant.

For design and comparative evaluation purposes of plate-fin heat exchangers, the best compilation of heat-transfer and friction data is that given by Kays and London.[2]

ACKNOWLEDGMENT

This material was reproduced from part of the notes for short courses on heat exchangers presented by the author in 1980 at the University of California, Los Angeles and at the University of Bath, U.K. The notes were eventually published in revised form as *Industrial Heat Exchangers—The User Guide,* G. Walker (ed.), Hemisphere Publishing Corp., New York, 1983.

REFERENCES

1. R. Fuller, "Plate Heat Exchangers in Sea Water Cooling Systems," *Proceedings of the First Mechanical Engineering Conference, Cooling with Seawater,* Institute of Mechanical Engineering, London, 1979.
2. W. M. Kays and A. L. London, *Compact Heat Exchangers,* 2d ed., McGraw-Hill, New York, 1964.
3. A. G. Lenfestey, "Low Temperature Heat Exchangers," in *Progress in Cryogenics,* Vol. 3, K. Mendelssohn (ed.), Heywood, London, 1961, pp. 23–48.
4. A. G. Lenfestey, "Compact Heat Exchanger Assemblies for Gas Separation Plants," *Proceedings of the Second International Cryogenic Engineering Conference,* Ilfiffe, Guildford, U.K., 1968, pp. 47–49.
5. J. Marriott, "Where, When and How to Use Plate Heat Exchangers," *Chem. Eng.,* **78** (April 5, 1971).
6. D. A. Reay, *Heat Recovery Systems,* E. & F. N. Spon., London (Halstead-Wiley, New York), 1979.

Chapter **2**

Heat-Exchanger Design and Rating

MARTIN GOLLIN
ARCO Chemical Company

2.1 INTRODUCTION

A heat exchanger is a processing unit which performs the thermal conditioning of a solid, fluid, or surface. While heat exchangers are usually designed on the basis of a specified set of "design" operating conditions (heat-duty, fluid flow rates, temperatures, etc.), these conditions are, in fact, a limited characterization of the actual operating situation and requirements.

It is important for the designer to assess the consequences of the failure of a unit to perform its required duty. While the cost of a unit may be measured in thousands of dollars, the cost of a system being inefficient or shut down because of the failure of an exchanger can be much greater. Generally, effort expended during the design stage will be well spent in terms of overall service and reliability.

The material in this chapter aims at describing the factors which should be considered when preparing a heat exchanger design or specification. It is not intended as a treatise on the thermal and mechanical design of heat exchangers. These topics are covered exhaustively in several publications (Refs. 1–5).

2.2 DUTY

Definition

It is of paramount importance that the purpose of the unit be understood and clearly defined. It may simply be a matter of stating that a certain amount of thermal energy must be transferred. However, it is more likely that the duty should be stated in terms of maintaining a fluid or surface temperature at a specified value (within a certain tolerance). The flow rates, temperatures, and physical properties of the fluids then enable a heat duty to be specified.

It is important to define a unit in this way because even a well-designed exchanger will rarely operate as specified because of (1) the inherent inaccuracies in the heat-transfer and pressure-drop correlations, (2) the additional surface which is added to account for fouling over the life of the unit, and (3) the tendency of designers to add surface to be "safe."

In some cases an ability to transfer additional heat can be accommodated by the reduction in temperature driving force that occurs as the outlet temperatures approach the inlet temperatures of the other fluid. In other cases positive control must be employed. In either event, when the purpose of the unit is fully understood, the need for control can be assessed. Little can be done for an undersized unit.

A further complication arises when a single process stream is required to carry out several duties in successive exchangers. In such cases the effects described above can be magnified because any tendency of the first unit in the network to remove too much heat under "clean" conditions results in insufficient heat and temperature differentials to allow the following units to achieve their design duty. The design and control of the system under these conditions requires careful consideration.

Heat Balance Check

Although it would appear to be unnecessary, experience shows that checking specified heat duty with a heat balance performed on each fluid stream saves time and confusion. Errors which often occur are due to the use of the wrong specific heat, latent heat or enthalpies, or confusion between the flow rates of the various fluids under different operating conditions.

2.3 PROCESS CONDITIONS

Fluid(s) involved in heat transfer can be characterized by temperature, pressure, phase, physical properties, toxicity, corrosivity, and fouling tendency. All of these must be considered when assessing the type of unit to be used.

Temperature

The temperature level and profile of each fluid must be considered.

Design Temperature This parameter is important as it indicates whether a material with high-temperature creep resistance or low-temperature impact resistance will be required. The design temperatures will also indicate whether precautions must be taken against freezing (if water is present) and whether thermal expansion differentials could be of importance. These topics will be considered individually later in this chapter. It is important that the designer understand the conditions that set the design temperatures, as these generally refer to upset conditions that could affect the design of the unit in other ways.

Temperature Profiles The temperature profile(s) for the unit reflects the operation of the unit under design conditions. Figures 2.1 and 2.2 show the temperature profiles for two exchangers with single-phase fluids on either side. For the unit shown in Fig. 2.1, the outlet temperature of the hot fluid T_2 is greater than the outlet temperature of the cold fluid t_2, and there is no temperature "cross." For the unit shown in Fig. 2.2, this is not so and there is a temperature cross. The importance of this distinction is that where there is a temperature cross the unit must be designed so that the flow is pure countercurrent, or, if multipass units must be used (usually so that the tube-side velocity will be increased), multiple shells must be used. The number of multiple shells required can be determined using a graphical technique as shown in Fig. 2.2, where the number of shells required is 3.

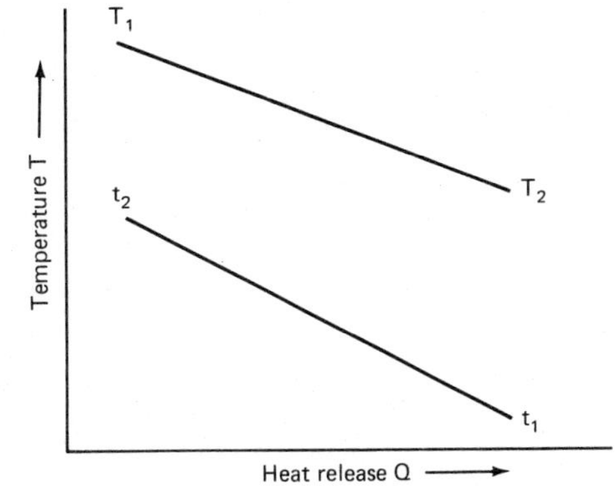

Fig. 2.1 Temperature profile with no temperature cross.

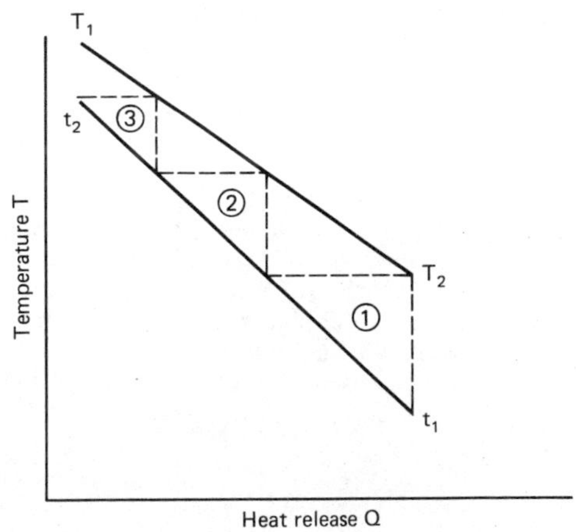

Fig. 2.2 Temperature profile with a temperature cross.

The existence of a temperature cross can lead to a constraint on the type of unit and designs which have lower overall coefficients than would otherwise be possible. If such a situation is determined to exist, the operating conditions should be examined to find a method of removing the cross (e.g., by increasing the water flow to a cooler and thus lowering the cold fluid outlet temperature for the same heat duty).

The temperature profile for a fluid undergoing a phase change (such as condensing) is quite different. Figure 2.3 shows the temperature profiles for a pure component being

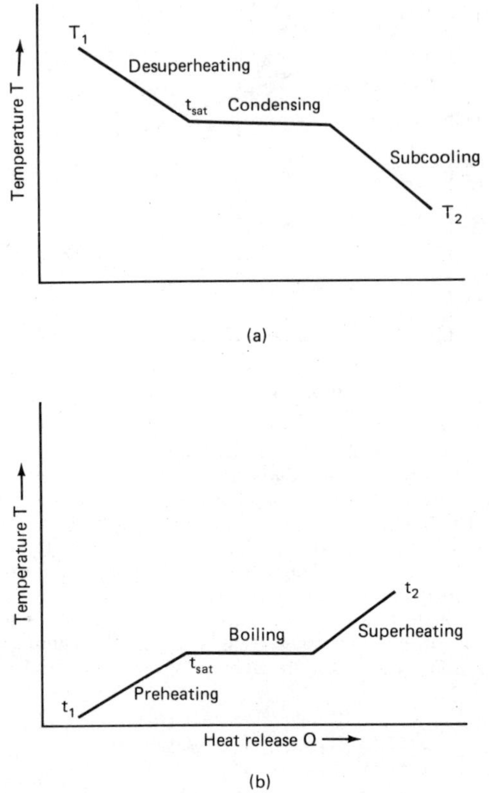

Fig. 2.3 Temperature profile with a pure component phase change. (*a*) Condensing; (*b*) boiling.

(*a*) cooled, condensed, and subcooled and (*b*) heated, vaporized, and superheated. Figure 2.4 shows a multicomponent stream being cooled with a single component condensing out of a noncondensible gas (T_D to T_2) with heat removed by a cooling fluid (t_1 to t_2).

The temperature profile for the pure component system

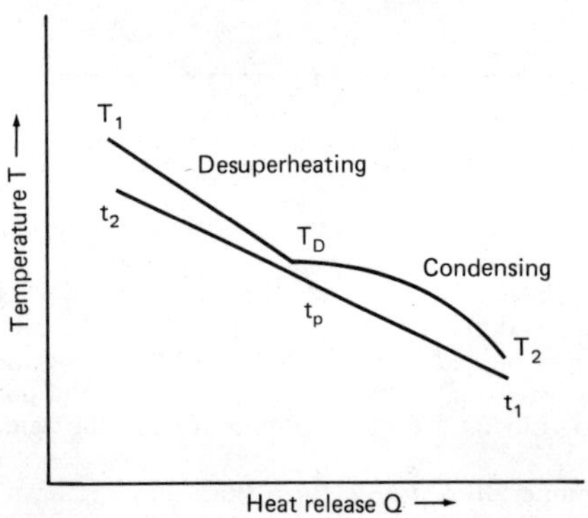

Fig. 2.4 Temperature profile for condensation of a component from a noncondensible gas.

is straightforward, but the system with a noncondensible gas shows the existence of a temperature pinch occurring between the dew point of the hot stream T_D and the corresponding temperature of the cooling fluid t_p. This pinch point reduces the overall temperature driving force for the unit and, as described above, will require either pure countercurrent flow or multiple shells to carry out the duty. If such a situation is found, the possibility of altering the temperature profile of the cooling fluid should be considered.

In most cases, from an initial examination of the temperature profiles, a certain amount of information can be obtained as to the type of unit which will be required.

Temperature Driving Force The effective temperature driving force is a measure of the actual potential for heat transfer which exists at the design conditions. In a pure countercurrent exchanger the effective temperature difference is defined by the log mean temperature difference, ΔT_{LM}, where, referring to Fig. 2.1,

$$\Delta T_{LM} = \frac{(T_1 - t_2) - (T_2 - t_1)}{\log_e [(T_1 - t_2)/(T_2 - t_1)]}$$

This equation is valid if the hot fluid is a condensing single component ($T_1 = T_2$) or the cold fluid is a boiling single component ($t_1 = t_2$).

When the unit does not have pure countercurrent flow, the log-mean temperature difference (LMTD) must be adjusted. There are several ways of doing this, but perhaps the most convenient is to use the *F*-factor method, where the effective temperature difference ΔT_{eff} is given by

$$\Delta T_{eff} = F\, \Delta T_{LM}$$

The *F* factor can be determined analytically for each flow arrangement but is usually presented graphically in terms of the thermal effectiveness *P* and the heat capacity ratio *R* for each flow arrangement where

$$P = \frac{t_2 - t_1}{T_1 - t_1} \qquad R = \frac{T_1 - T_2}{t_2 - t_1} = \frac{WC}{wc}$$

where T_1 = hot-fluid inlet temperature
T_2 = hot-fluid outlet temperature
t_1 = cold-fluid inlet temperature
t_2 = cold-fluid outlet temperature
W = hot-fluid mass flow rate
C = hot-fluid heat capacity
w = cold-fluid mass flow rate
c = cold-fluid heat capacity

There are many published charts for the various flow arrangements possible (Refs. 2, 4, 5), and the reader should consult these. Figures 2.5 to 2.7 show the *F* correction charts for several of the more common arrangements.

Figure 2.5 shows the chart for a 1–2 exchanger, which uses one pass for one fluid and two passes for the other. The F_{min} line indicates the practical limitation on the use of the unit and corresponds to the situation where the two outlet temperatures are the same. As a rule of thumb, the minimum *F* correction factor which should be used is 0.8.

In situations where the heat release curves are nonlinear (see Figs. 2.3 and 2.4), the approach described above is not applicable and a "weighted" temperature difference must be

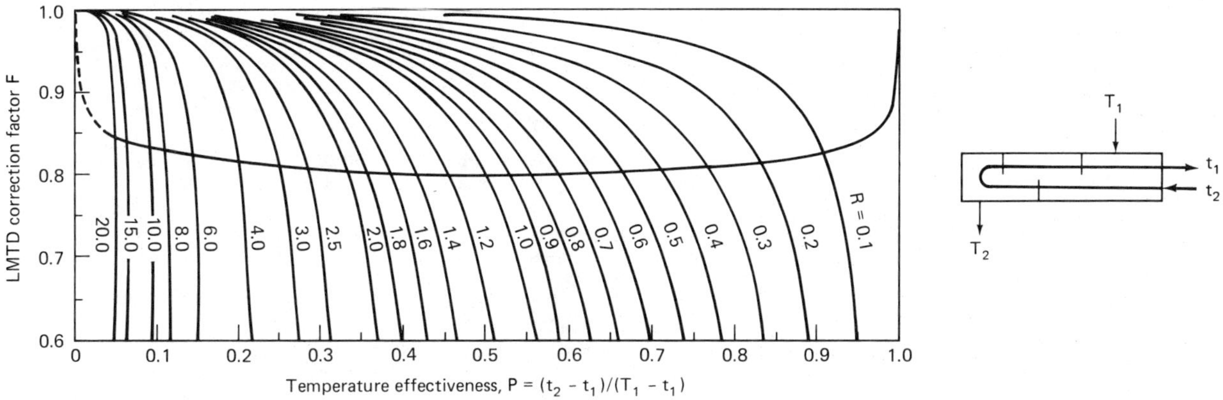

Fig. 2.5 LMTD correction factor F as a function of P and R for the 1–2 exchanger with the shell fluid mixed. (*From Rohsenow et al.,*[5] *reproduced by permission.*)

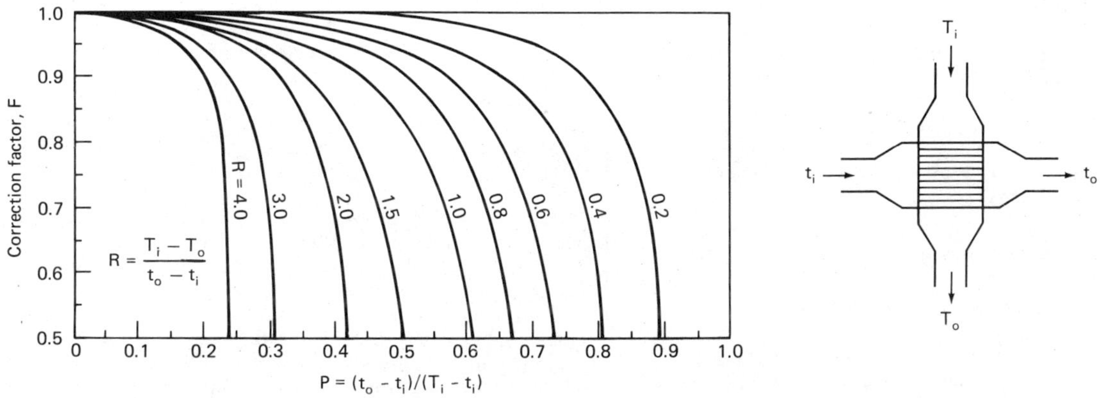

Fig. 2.6 LMTD correction factor F as a function of P and R for a crossflow exchanger with one fluid mixed. (*From Pitts and Sissom,*[6] *reproduced by permission.*)

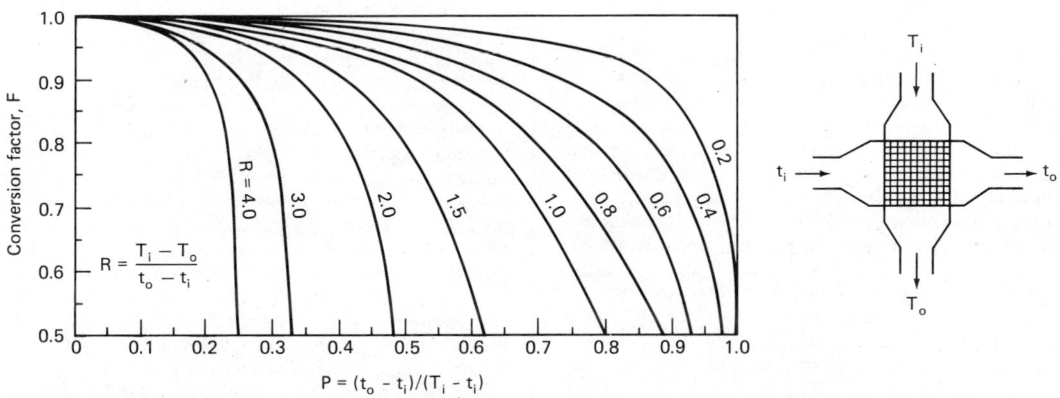

Fig. 2.7 LMTD correction factor F as a function of P and R for a crossflow exchanger with both fluids mixed. (*From Pitts and Sissom,*[6] *reproduced by permission.*)

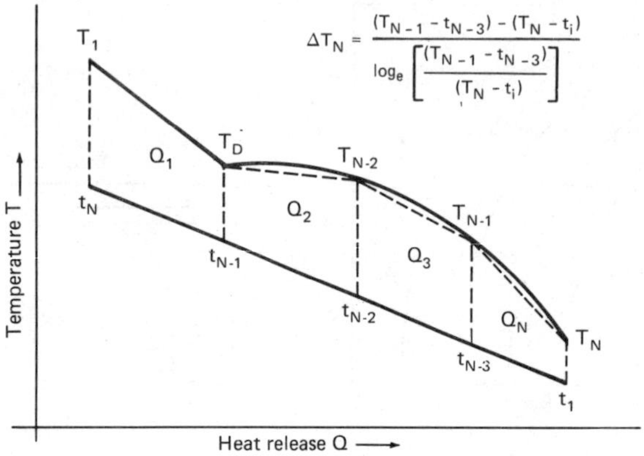

Fig. 2.8 Calculation of weighted LMTD for nonlinear heat-release processes.

determined. The method by which this is done is as follows;

1. The heat-release versus temperature curves for each fluid are constructed (see Fig. 2.8).
2. The complete heat-release curve is approximated by a number of sections in which the heat release curves can be represented by a series of straight lines.
3. The heat duty of each section Q_N is calculated.
4. The LMTD for each section ΔT_N is calculated.
5. The ratio $Q_N/\Delta T_N$ is calculated for each section.
6. The overall LMTD ΔT_{LM} is then calculated from

$$\frac{1}{\Delta T_{LM}} = \frac{1}{Q_{tot}} \sum_{1}^{N} \frac{Q_N}{\Delta T_N}$$

where Q_{tot} = the total heat duty of the unit and Σ_1^N $(Q_N/\Delta T_N)$ is the sum of $Q_N/\Delta T_N$ for sections 1 to n.

The F correction factor can be applied to each section or an estimate made and applied to the final LMTD to obtain the actual ΔT_{eff}. This value of ΔT_{eff} is useful in carrying out the initial analysis of the unit. As the design of condensers is often carried out in "stepwise" manner, however, this procedure may not be directly applicable to the final design calculations.

Pressure*

The pressure level of the fluids has a significant effect on the type of unit selected. Vapor-phase volumetric flow rate is directly affected by pressure, and at low pressures the combination of high volumetric flowrates and low allowable pressure drops can have the effect of requiring a design that maximizes the area available for flow (e.g., crossflow or split flow with multiple nozzles). A low allowable pressure drop usually results in low-heat-transfer coefficients (as convective heat transfer is obtained at the expense of pressure drop). In general, higher heat transfer rates are obtained by placing the low-pressure gas on the outside of tubular surfaces.

Low-pressure condensing services are also usually limited by the small allowable pressure drop, and the use of in-tube vertical downflow condensing in such designs should be considered.

* See also "Pressure Drop" in *Addendum* at the end of the chapter.

At high pressures the vapor phase volumetric flow rates are lower and allowable pressure drops are greater. These lead to more compact units. The higher the pressure, however, the greater will be the required thickness of the pressure containing membranes and the more advantageous it is to place the high-pressure fluid on the tube side. This is due to the thinner tube wall required to resist internal pressure than external pressure.

Above 150 lb/in^2 (1000 kPa) the use of designs such as plate exchangers, spiral exchangers, and other compact types becomes impractical because of the difficulty in pressing the plate thicknesses required at these pressure levels.

Fluid Phase

The phase of the fluid(s) within a unit is an important consideration in the selection of the type of unit to be used.

Liquid-Liquid Liquid-phase fluids are generally the simplest to deal with. The high density and the favorable values of the other transport properties allow high-heat-transfer coefficients to be obtained at relatively low pressure drops.

Many of the compact-type heat exchangers, such as plate or spiral, are ideally suited to relatively low pressure liquid services with small to moderate flows. Caution should be used in selecting these types of units when the fluids are toxic or flammable.

Unless one of the fluids has a very low film coefficient, the use of an enhanced heat-transfer surface on one side will be futile as the rate of heat transfer will be limited by the other surface. Use of enhanced surfaces on both sides will be effective but may be difficult to justify on economic grounds.

Vapor-Vapor Vapor-phase fluids generally achieve lower overall coefficients, must meet stricter pressure drop limits, and have larger inlet and outlet nozzles relative to liquids. All of these factors make the use of plate or spiral heat exchangers less attractive and the use of some type of compact or tubular unit preferable. Many vapor-vapor services are suited to a coil-in-duct-type unit. As described previously, the use of enhanced surfaces is not normally effective when applied to one side only. It is more effective to employ enhanced heat-transfer surfaces on both sides of vapor-vapor exchangers as a result of the low overall coefficients otherwise obtained.

Liquid-Vapor In these units the use of enhanced surfaces to improve the vapor side coefficient in relation to the liquid coefficient should be seriously considered. As enhanced surfaces can most easily be applied to the outside of tubes in the form of fins, it is usually preferable to place the vapor phase on the outside of a tubular unit unless other considerations indicate otherwise.

Condensing Vapors When a condensing vapor is to be handled, it is almost always preferable that the fluid path be downward so as to minimize the possibility of the condensate blocking the vapor path and causing "slugging" or other two-phase flow instabilities. Exceptions to this rule may be dephlegmators and other units where extreme care is taken in calculating the two-phase flow regimes. Where the vapor condensing rate may be affected by the presence of noncondensible gases, provision should be made to vent these from the system at a "highpoint" in the unit.

Boiling Liquids The vaporization of a fluid can be achieved from a pool or utilizing forced flow. With pool boiling, heat is supplied to a liquid which vaporizes and passes out of the unit. For this situation there must be sufficient disengagement area for the liquid droplets which are carried up with the vapor to fall back to the surface of the liquid prior to exit of the vapor from the unit.

The other method used to vaporize a liquid is to use a forced flow (produced by either natural circulation or a pump). In this method only a portion of the fluid entering the unit is vaporized and the vapor and liquid are separated outside the unit.

The flow path of the fluid in vaporizing services should always be upward to minimize two-phase flow instabilities. The choice of which side to place the vaporizing fluid on will probably be determined by other considerations (e.g., pressure or material requirements).

When natural-circulation systems are used, there will be an interaction between the design of the exchanger and the piping layout.

In cases where the phase-equilibrium data are uncertain for boiling liquids, a useful approach is to heat the fluid at a pressure greater than required and then to generate the vapor required by flashing the liquid across an orifice.

Physical Properties

Before undertaking design work, a designer must determine the physical properties of the fluids as accurately as possible. In many cases the accuracy of the physical property determinations is very poor and allowance for this should be made in the design.

There are computer-based physical property databases which can be utilized if literature sources are insufficient.

The process conditions should be carefully noted as the physical properties and phase equilibria are susceptible to variation with pressure and temperature.

Phase Equilibrium

When the fluids to be processed are undergoing a phase change (boiling or condensation), the phase equilibrium must be determined. This relates to the heat-release curve and the vapor-weight-fraction curve.

The heat-release curve is used in determining the overall temperature driving force for the unit, and the vapor-weight-fraction curve is used in determining the two-phase flow regime in various sections of the unit.

It is important to understand that the phase equilibrium is a function of pressure. Thus the pressure drop allocated to the stream must be adhered to, otherwise the dew point and heat release may differ from those specified.

Toxicity

Where the fluids to be processed are toxic, special precautions need to be taken to minimize the potential for leakage. This can involve the use of welded joints, the use of double tubesheets (see Fig. 2.9) where the space between the tubesheets is enclosed and connected to a vent/drain system, or double-walled tubes.

Corrosivity

Where the fluids require the use of a corrosion-resistant material, it is important that the cost of the unit be minimized by reducing the amount of such material.

In general, if only one fluid is corrosive, that fluid should be placed on the tube side of the unit. This principle should be carefully examined in the following two cases.

The first case is when the other fluid is cooling water and

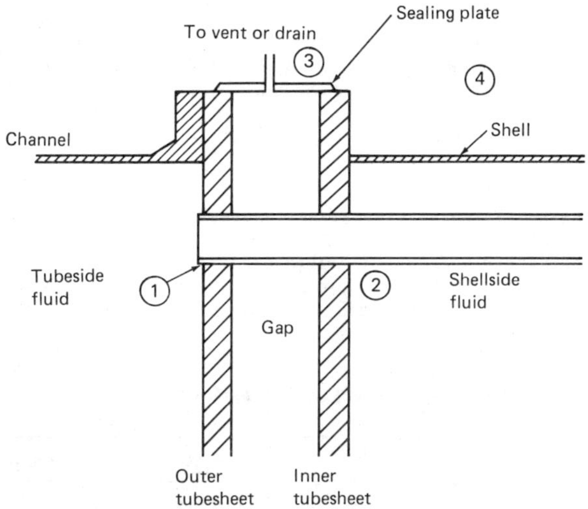

Fig. 2.9 Double-tubesheet design used to minimize leakage of fluids to atmosphere or into another process stream. (1) Tubes expanded or welded to outer tubesheet; (2) tubes expanded into inner tubesheet; (3) drain required at bottom of gap; (4) sealing plate to confine leakage and/or protect unit from debris, snow, rain, etc.

the corrosion-resistant material is stainless steel. The reason for this is the susceptibility of austenitic stainless steels to stress corrosion cracking and other problems associated with impurities in cooling water. The back of the tubesheet of shell-and-tube exchangers consists of many crevices (where the tubes enter) which are ideal sites for the deposition and concentration of impurities which can initiate attack of the stainless steel. The rear of the tubesheet is also an area of low water flow rate (and hence high metal temperature); this is also a condition which will favor the stress corrosion mechanism. Thus, while many exchangers with stainless-steel tubes operate with cooling water on the shell side, alternative arrangements should be carefully considered. Placing the water on the tube side removes nearly all crevices and allows high water flow rates over all surfaces. The geometry of other types of heat exchangers should be carefully examined to ensure that a similar coincidence of crevices and water does not occur.

The second case is when the pressure of the other fluid is high enough to justify placing that fluid on the tube side and using the corrosion-resistant material for both the shell and the tubes. The cost of this can be reduced by using clad plate or weld overlay in place of solid plate or forgings.

Fouling Tendency

The fouling tendency of a fluid is one of the most significant design parameters in terms of both thermal and mechanical design.

In the thermal design process, the fouling factors applied to account for laydown of deposits on the heat-transfer surfaces can be up to 50 percent of the total thermal resistance. Thus the use of a high fouling factor can have a severe economic penalty and significantly alter the performance of the unit under "clean" conditions. The choice of fouling factor is, in many cases, based not on any experimental data but rather on a subjective opinion combining judgment and a desire to use a hidden safety factor. Another very important determination to be made at the start of design work is whether either side of the exchanger will require mechanical cleaning during its lifetime.

Tables 2.1 and 2.2 give a summary of the typical fouling heat-transfer coefficients recommended for various fluids from Ref. 5.

If a fluid has a fouling tendency and no other parameters are involved, it should be placed on the tube side where high velocities and smooth surfaces will minimize the rate of buildup of the fouling layer. If mechanical cleaning is required, it can be carried out by water jets and mechanical tube cleaners.

Table 2.1. Typical Fouling Heat-Transfer Coefficients for Various Fluids

	Btu/h·ft²·°F	W/m²·K
Heat Transfer Fluid Fouling		
Oils		
Fuel oil	200	1136
Transformer oil	1000	5678
Engine lube oil	1000	5678
Quench oil	250	1420
Gases and vapors		
Manufactured gas	100	568
Engine exhaust gas	100	568
Steam (non-oil-bearing)	2000	11360
Exhaust steam (oil-bearing)	1000	5678
Refrigerant vapors (oil-bearing)	500	2840
Compressed air	500	2840
Industrial organic heat transfer media	1000	5680
Liquids		
Refrigerant liquids	1000	5680
Hydraulic fluid	1000	5680
Industrial organic heat transfer media	1000	5680
Molten heat transfer salts	2000	11360
Chemical Processing Fouling		
Gases and vapors		
Acid gas	1000	5680
Solvent vapors	1000	5680
Stable overhead products	1000	5680
Liquids		
MEA and DEA solutions*	500	2840
DEG and TEG solutions†	500	2840
Stable side draw and bottom product	1000	5680
Caustic solutions	500	2840
Vegetable oils	330	1874
Natural Gas-Gasoline Processing Fouling		
Gases and vapors		
Natural gas	1000	5680
Overhead products	1000	5680
Liquids		
Lean oil	500	2840
Rich oil	1000	5680
Natural gasoline and liquefied petroleum gases	1000	5680

* Monethanolamine and diethanolamine solutions.
† Diethylene glycol and triethylene glycol solutions.
SOURCE: From Rohsenow et al.[5] Reproduced by permission.

If other factors require that a fouling fluid be placed on the shell side, the tubes should be placed on a square pitch and the unit designed so as to have a removable tube bundle. The fluid velocity should be as high as possible and the baffle spacing designed so as to minimize the low-flow areas which are inevitable on the shell side of baffled shell-and-tube exchangers. The use of U-tube exchangers, with their single tubesheet, is recommended in such situations. The alternative designs require a floating head with the consequent larger shell diameters and leak-prone seals. When U-tube exchangers are used, care must be taken to ensure that there is no temperature cross.

The use of plate and spiral exchangers with fouling liquids has been reported to be very successful. If design conditions permit, the use of these types of units should be seriously considered for fouling liquids.

2.4 SAFETY AND PROTECTION*

There are several important factors which must be considered in addition to those already described;

FREEZE PROTECTION. In units where low ambient or operating temperatures can occur and a liquid (e.g., steam or condensate) may be trapped and freeze, special care must be taken to ensure that this does not occur. Designs which allow for self-draining should be used, and in extreme cases designs such as the vertical-tube bayonet (where the flow of hot fluid up the center tube aids in the prevention of freeze-up in the outer tube) should be considered.

PRESSURE RELIEF. In certain situations where a very-high-pressure fluid is placed in the tubes and a low-pressure fluid is on the shell side, the unit must be protected against the possibility of tube rupture. It is not normally economical to design the shell-side for the full design pressure of the tube-side fluid so that the unit is protected by a pressure relief system. This may involve bursting disks or other relief devices which may be connected to a vent system if the tube-side fluid is hazardous.

TUBE METAL TEMPERATURES. In certain situations the temperature of one of the fluids is so high that the choice of tube material becomes difficult. In these situations the designer must examine the actual metal temperatures which will exist in the unit, taking into account the temperatures of both fluids and their individual heat-transfer coefficients. For example, in a boiler the boiling coefficient is so much higher than that of the hot gas that the tube metal temperature is only a few degrees hotter than the saturation temperature of the water. Thus a reasonable design metal temperature can be selected. However, it is important that such an analysis be considered when setting the controls for the system as, if the tubes dry out, the metal temperature will soar and failure of the tubes occur.

CODE OF CONSTRUCTION. Certain design codes must be complied within the design of a heat exchanger. A list of those applicable in the United States is given in Table 2.3. A good working knowledge of the important sections of these codes is important in ensuring a safe and reliable unit.

* See also "Differential Thermal Expansion" and "Vibration" in *Addendum* at the end of the chapter.

Table 2.2. Water Fouling Heat-Transfer Coefficients

Temperature of heating medium	≤240°F (115°C)		240–400°F (115–205°C)	
Temperature of water	≤125°F (52°C)		>125°F (52°C)	
	Water velocity, ft/s (m/s)		Water velocity, ft/s (m/s)	
Types of water	3 (1) and less B*(W)†	Over 3 (1) B(W)	3 (1) and less B(W)	Over 3 (1) B(W)
Seawater	2000 (11360)	2000 (11360)	1000 (5680)	1000 (5680)
Brackish water	500 (2840)	1000 (5680)	330 (1870)	500 (2840)
Cooling tower and artificial spray pond				
Treated makeup	1000 (5680)	1000 (5680)	500 (2840)	500 (2840)
Untreated	330 (1870)	330 (1870)	200 (1135)	250 (1420)
City or well water (such as Great Lakes)				
River water	1000 (5680)	1000 (5680)	500 (2840)	500 (2840)
Minimum				
Average	500 (2840)	1000 (5680)	330 (1870)	500 (2840)
Muddy or silty	330 (1870)	500 (2840)	250 (1420)	330 (1870)
Hard (>15 grains/gal)	330 (1870)	500 (2840)	250 (1420)	330 (1870)
Engine jacket	330 (1870)	330 (1870)	200 (1135)	200 (1135)
Distilled or closed-cycle condensate	1000 (5680)	1000 (5680)	1000 (5680)	1000 (5680)
Treated boiler feedwater	2000 (11360)	2000 (11360)	2000 (11360)	2000 (11360)
Boiler blowdown	1000 (5680)	2000 (11360)	1000 (5680)	1000 (5680)
	500 (2840)	500 (2840)	500 (2840)	500 (2840)

* B = Btu/h · ft² · °F.
† W = W/m² · K.
SOURCE: From Rohsenow et al.[5] Reproduced by permission.

2.5 DESIGN

The design of a heat exchanger requires the consideration of the factors outlined above and the use of a design methodology that will allow the selection of the optimum unit. It is worthwhile pointing out, however, that it is unlikely that any two designers will arrive at exactly the same design for a given set of conditions as the process involves many judgments.

STEP 1: *Assess the heat-transfer mechanisms involved.* The heat-transfer mechanisms which can play a part in the design of an exchanger are natural convection, forced convection, pool boiling, forced flow boiling, condensation, radiation, and conduction.

While not all of these mechanisms will be significant in a particular unit (particularly natural convection), they can have effects on the heat-transfer system. For example, heat loss by natural convection and radiation from a piping system may influence the temperature into an exchanger, and if this heat loss is large in comparison to the heat load on the exchanger, it will be an important factor to be considered in the design of the unit.

STEP 2: *Select nozzle size.* It is important that the designer select a suitable size for each of the nozzles in the unit. The nozzle must be of sufficient size to ensure that the pressure drop through the nozzle is not too high while minimizing the cost of the nozzle.

Table 2.4 presents a methodology for determining the minimum inside diameter of a nozzle for various types of fluid entering or leaving the unit. The actual size of nozzle used will depend on the pressure, material, corrosion allowance, and pipe schedule.

Table 2.3. Applicable Codes of Construction

Standards of Tubular Exchanger Manufacturers' Association

API Standard 660. Heat exchangers for general refinery service

Heat Exchanger Institute, Inc.
- Standard for power plant heat exchangers
- Standard for steam surface condensers
- Standard for closed feedwater heaters

ASME Boiler and Pressure Vessel Code
- Section I, Power Boilers
- Section VIII, Division 1 Pressure Vessels

Table 2.4. Calculation of Minimum Nozzle Inside Diameter*

	Velocity head loss through nozzle VH, lb/in²	
Fluid	Shell side	Tube side
Liquid	0.35	0.7
Dry gas	0.35	0.35
Wet vapor	0.17	0.2

* The minimum inside diameter is calculated from:

$$d_{min} = \sqrt[4]{\frac{M^2}{3.54 \times 10^6 \times VH \times \rho}}, \text{ in}$$

where M = mass flow rate of fluid through nozzle, lb/h
ρ = density of fluid, lb/ft³
VH = velocity head loss through nozzle

CONVERSION FACTORS: 1 lb/hr = 0.4536 kg/hr; 1 in = 0.0254 m; 1 lb/ft³ = 16.02 kg/m³; 1 lb/in² = 6.895 kPa.

Table 2.5. Typical Overall Heat-Transfer Coefficients U_o for Shell-and-Tube Exchangers

Fluid 1	Fluid 2	Total Fouling Reistance (1), h · ft² · °F/Btu	U_o (1, 2), Btu/h · ft² · °F
Water (3, 4)	Water	0.0015	250–300
	Gas, about 10 lb/in² gage	0.001	15–20
	Gas, about 100 lb/in² gage	0.001	30–40
	Gas, about 1000 lb/in² gage	0.001	60–100
	Light organic liquids (5)	0.0015	125–175
	Medium organic liquids (6)	0.002	75–125
	Heavy organic liquids (7)	0.0025	40–75
	Very heavy organic liquids (8)	0.004	
	Heating		10–40
	Cooling		5–15
Steam (no noncondensibles)	Gas, about 10 lb/in² gage	0.0005	15–20
	Gas, about 100 lb/in² gage	0.0005	35–45
	Gas, about 1000 lb/in² gage	0.0005	70–110
	Light organic liquids (5)	0.001	135–190
	Medium organic liquids (6)	0.0015	80–135
	Heavy organic liquids (7)	0.002	45–80
	Very heavy organic liquids (8)	0.0035	15–45
	Water	0.001	300–400
Light organic liquids (5)	Light organic liquids (5)	0.002	100–130
	Medium organic liquids (6)	0.0025	70–100
	Heavy organic liquids (7)	0.003	
	Heating		40–75
	Cooling		25–50
Light organic liquids (5)	Very heavy organic liquids	0.004	
	Heating		20–50
	Cooling		5–25
Medium organic liquids (6)	Medium organic liquids (6)	0.003	50–80
	Heavy organic liquids (7)	0.0035	
	Heating		30–50
	Cooling		15–35
	Very heavy organic liquids (8)	0.0045	
	Heating		15–30
	Cooling		5–25
Heavy organic liquids (7)	Heavy organic liquids (7)	0.005	10–30
	Very heavy organic liquids	0.006	5–15
Gas, about 10 lb/in² gage	Gas, about 10 lb/in² gage	0	10–15
	Gas, about 100 lb/in² gage	0	15–20
	Gas, about 1000 lb/in² gage	0	15–25
Gas, about 100 lb/in² gage	Gas, about 100 lb/in² gage	0	20–30
	Gas, about 1000 lb/in² gage	0	25–35
Gas, about 1000 lb/in² gage	Gas, about 1000 lb/in² gage	0	35–60
Water	Condensing light organic vapors, pure component (5, 9)	0.001	150–200
	Condensing medium organic vapors, pure component (6, 9)	0.001	100–150
	Condensing heavy organic vapors, pure component (7, 9)	0.002	75–100

NOTES: (1) The total fouling resistance and the overall heat-transfer coefficient are based on the total outside tube area. (2) Allowable pressure drops on each side are assumed to be about 10 lb/in² except for (a) low-pressure gas and condensing vapor where the pressure drop is assumed to be about 5 percent of the absolute pressure, and (b) heavy organics where the allowable pressure drop is assumed to be about 20 to 30 lb/in². (3) Aqueous solutions give approximately the same coefficients as water. (4) Liquid ammonia gives about the same results as water. (5) "Light organic liquids" include liquids with viscosities less than about 0.5 cp, such as hydrocarbons through C_8, gasoline, light alcohols, and ketones. (6) "Medium organic liquids" include liquids with viscosities between about 0.5 and 1.5 cp, such as kerosene, straw oil, hot gas oil, absorber oil, and light crudes. (7) "Heavy organic liquids" include liquids with viscosities greater than 1.5 cp, but not over 50 cp, such as cold gas oil, lube oils, fuel oils, and heavy and reduced crudes. (8) "Very heavy organic liquids" include tars, asphalts, polymer melts, greases, etc., having viscosities greater than about 50 cp. Estimation of coefficients for these materials is very uncertain. (9) These values may be used for vapor mixtures when the condensing range of the vapor is less than half of the temperature difference between the outlet coolant and the vapor. If the condensing range is greater than this, or if there are significant amounts of noncondensible gas present, the coefficient should be reduced towards the values shown for gas cooling; in these cases, the accuracy of the estimation is very uncertain.

CONVERSION FACTORS: h · ft² · °F/Btu = 0.176 (m² · K)/W; Btu/h · ft² · °F = 5.68 W/m² · K.

SOURCE: From Kakac et al.[7] Reproduced by permission.

The calculation of the nozzle size will allow a check to be made on the proposed piping sizes for the unit.

STEP 3: *Select materials.* The materials of construction are selected on the basis of the fluid property temperatures and design conditions.

STEP 4: *Select type of unit.* Using the parameters described above, the designer should select the type of unit which is to be used. This unit may be a proprietary type, such as a plate, a spiral, or an air cooler, which will be designed by a manufacturer of such equipment. For cases such as this the designer must prepare a specification which clearly defines the heat duty which the exchanger must perform and all other relevant information which could affect the design.

STEP 5: *Determine size and layout parameters.* In practically every situation a size limit will be placed on the unit. This may be due to plot, transportation, or weight considerations.

There may also be a preference as to the positions of the nozzles or even the orientation. For certain units, such as thermosiphon reboilers, there will be a definite requirement to interact with the piping design to ensure that the pipe routing is satisfactory from both hydraulic and stress viewpoints.

In certain cases, particularly where hot fluids are to be handled and/or large lines are involved, the stresses imposed on the unit by the expansion of the piping becomes an important consideration.

STEP 6: *Perform preliminary thermal design.* A preliminary calculation of the required surface area (based on the effective temperature difference, the heat duty, and an estimated overall heat-transfer coefficient) should be made so as to assess the possible size and cost of the unit and whether multiple shells are required.

Table 2.5 gives a selection of overall coefficients for various combinations of fluids and fouling factors. These are approximate, and other conditions, such as pressure drop, temperature profile, and relative fluid flow rates, could cause the actual coefficients to vary from those shown.

Even if a proprietary design is to be used, it is useful to make an assessment of the expected size of the equipment. This will allow a preliminary layout to be made prior to the receipt of the design.

STEP 7: *Perform detailed design.* As stated previously, this chapter is not intended to deal with the detailed design of a heat exchanger. However, it is worthwhile to state that while relatively simple designs can be carried out manually, the majority of thermal and mechanical design is carried out by heat-exchanger vendors using computer methods such as those developed by H.T.R.I., H.T.F.S., and others.* Vendors have also developed their own computerized procedures.

STEP 8: *Check design.* Whether the design has been carried out by an outside company or internally, it must be checked to ensure that it meets specifications. In addition to carrying out checks on the thermal performance, the weight, size, and nozzle locations and orientations should be examined to ensure that they are compatible with the rest of the system.

* Heat Transfer Research, Inc., Alhambra, Cal; Heat Transfer and Fluid Flow Services, A.E.R.E. Harwell, Oxon, United Kingdom.

The materials of construction should be reexamined, and where differential thermal expansion could be of concern, this should be analyzed using the metal temperatures which will actually exist. The worst case metal temperature differentials may occur under startup or transient conditions and an examination of the performance of the unit under such conditions should be carried out.

STEP 9: *Control scheme.* As described previously, the operation of an exchanger may vary widely with time as a result of the effects of fouling and changes in flow. In some services this is unimportant, while in others such a variation will cause a problem with the system in which the exchanger is installed. In these cases a system of control will be required that will be able to accept the changes in performance of the unit and yet still enable it to meet its service requirements.

Various types of control are as follows:

1. Where it is important that the cold fluid not be heated to too high a temperature, the unit can be designed with cocurrent flow. This is a self-controlling system, as any increase in the cold fluid outlet temperature causes a reduction in the temperature driving force and the tendency to overperform is reduced. This type of control obviously requires that there not be a temperature cross. In order that not too high an economic penalty be paid, it is usually only feasible to design such a system where the temperature differences are large.

2. Another system whereby a fluid may be protected from being heated to too high a temperature is by attemporation. In this system the fluid is subjected to secondary cooling at some point in its flow path. The cooling may be carried out in a second exchanger (surface attemporation) or by the addition of fluid at a lower temperature (spray attemporation). This scheme is widely used in the control of the temperature of superheated steam passing to steam turbines.

3. Perhaps the most widespread type of control used is that of bypassing a portion of one of the fluid streams around the unit. This has the effect of limiting the amount of heat available and pinching the outlet temperatures. The bypass is controlled by the movement of a control valve situated in the bypass line. The position of this valve is set by the control system to ensure that the specified parameter (temperature) is met. Two factors must be considered in the design of such a system. First, the pressure drop of the bypass line must be such that a sufficient flow of fluid will pass through with the valve fully open and the remainder of the fluid passing through the exchanger. Second, extreme care must be taken to ensure that the reduction in flow of the bypassed fluid will not lead to any undesirable consequences in the heat exchanger such as an increased tendency to foul at lower flow rates, increased metal temperatures, or fluid maldistribution in units with parallel flow paths.

2.6 RATING OF AN EXISTING DESIGN

As can be seen from the previous discussion, the design of an exchanger involves more than simply an analysis of one operating condition. Operation under other conditions is required, and it is useful to be able to predict the operation of an exchanger under such situations.

There is a relatively simple technique available for manual use, and many of the computer programs available have such a capacity. However, it is important to realize that it is not as accurate as a full design calculation and it makes use of the performance characteristics for one condition extrapolated to other conditions.

CONSTANT-*UA* METHOD. In this method the equation $Q = UA\ \Delta T$ is taken as characteristic of an exchanger at a particular set of design conditions.

It is assumed that (1) the fluids have a constant specific heat (i.e., linear heat-release vs. temperature relationships), (2) the correction factor to the LMTD is constant, and (3) the overall heat-transfer coefficient used in design is applicable.

These assumptions restrict the use of the method to single-phase fluids or pure components undergoing a phase change (boiling or condensing). It also assumes that the physical properties of the fluids remain approximately the same at the new conditions as at the design condition.

Given these assumptions, the only adjustments that can be made are to the overall coefficient based on clean or fouled conditions and to the inside and outside coefficients based on changes in flow.

Generally the tube-side coefficient can be assumed to vary with mass flow rate to the 0.8 power and the outside coefficient with flow rate to the 0.6 power. Thus, if

H = inside heat-transfer coefficient
h = outside heat-transfer coefficient
W = tube-side fluid flow rate
w = shell-side fluid flow rate

then the inside and outside coefficients under the new operating conditions are given by

$$H_{\text{new}} = H_{\text{base}} \left(\frac{W_{\text{new}}}{W_{\text{base}}}\right)^{0.8} \qquad h_{\text{new}} = h_{\text{base}} \left(\frac{w_{\text{new}}}{w_{\text{base}}}\right)^{0.6}$$

Table 2.6. Equations for Predicting Boiler Performance

$$E = \exp\left(\frac{AU}{W_H C_H}\right) \tag{1}$$

$$E = \frac{T_1 - t}{T_2 - t} \quad \text{and} \quad T_2 - t = \frac{T_1 - t}{E} \tag{2}$$

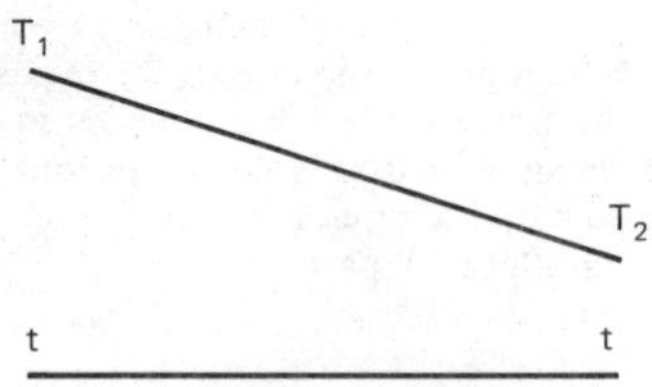

where T_1 = hot-fluid inlet temperature, °F (°C)
T_2 = hot-fluid outlet temperature, °F (°C)
W_H = hot-fluid flow rate, lbs/hr (kg/s)
C_H = specific heat of hot fluid, Btu/lb°F (J/kg · °C)
A = surface area of boiler, ft² (m²)
U = overall heat-transfer coefficient for boiler, Btu/h · ft² · °F (W/m² · °C)

NOTE: Base T_2 on heat absorbed by water, allow for heat losses separately.

Table 2.7. Equations for Predicting Cocurrent Heat-Exchanger Performance

$$m = \frac{W_C C_C}{W_H C_H} = \frac{T_1 - T_2}{t_2 - t_1} = \frac{1}{V}$$

$$V = \frac{W_H C_H}{W_C C_C} = \frac{t_2 - t_1}{T_1 - T_2} = \frac{1}{m}$$

$$E = \frac{T_1 - t_1}{T_2 - t_2}$$

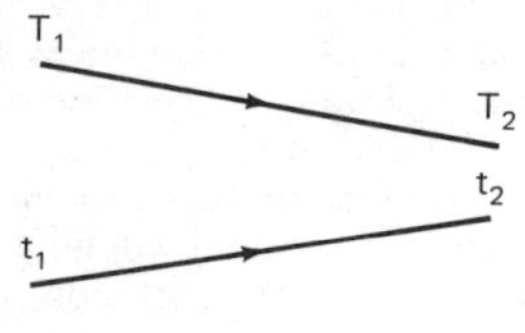

	Temperature rise of cold fluid, $t_2 - t_1$
$E = \exp\left[\frac{AU(m+1)}{W_C \times C_C}\right]$	$(T_1 - t_1)\frac{E-1}{E(m+1)}$
	$(T_2 - t_1)\frac{E-1}{E+m}$
$E = \exp\left[\frac{AU(V+1)}{W_C C_C V}\right]$	$(T_1 - t_2)\frac{E-1}{(Em)+1}$
	$(T_2 - t_2)\frac{E-1}{m+1}$
	Temperature drop of hot fluid, $T_1 - T_2$
$E = \exp\left[\frac{AU(m+1)}{W_H C_H m}\right]$	$(T_1 - t_1)\frac{E-1}{E(V+1)}$
	$(T_2 - t_1)\frac{E-1}{(EV)+1}$
$E = \exp\left[\frac{AU(V+1)}{W_H C_H}\right]$	$(T_1 - t_2)\frac{E-1}{E+V}$
	$(T_2 - t_2)\frac{E-1}{V+1}$

where T_1 = hot-fluid inlet temperature, °F(°C)
T_2 = hot-fluid outlet temperature, $°F_2$(°C)
t_1 = cold-fluid inlet temperature, $°F_2$(°C)
t_2 = cold-fluid outlet temperature, $°F_2$(°C)
W_H = hot-fluid mass flow rate, lbs/h (kg/s)
C_H = hot-fluid specific heat, btu/lb · °F (J/kg · °C)
W_C = cold-fluid flow rate, lbs/h (kg/s)
C_C = cold-fluid specific heat, Btu/lb · $°F_2$ (J/kg · °C)
A = surface area of exchanger, ft² (m²)
U = overall heat-transfer coefficient for exchanger, Btu/h · ft² · °F (W/m² · °C)

NOTES: Base T_2 in the above formula on the heat absorbed by the cold fluid and allow for heat losses separately. The LMTD and the hot gas outlet temperature shown on the specification sheet should allow for heat losses.

Table 2.8. Equations for Predicting Countercurrent Heat-Exchanger Performance

$$m = \frac{W_C C_C}{W_H C_H} = \frac{T_1 - T_2}{t_2 - t_1} = \frac{1}{V}$$

$$V = \frac{W_H C_H}{W_C C_C} = \frac{t_2 - t_1}{T_1 - T_2} = \frac{1}{m};$$

$$X = \frac{T_1 - t_2}{T_2 - t_1}$$

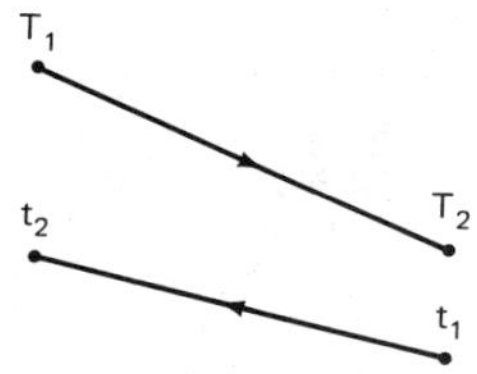

	Temperature rise of cold fluid, $T_1 - T_2$
$m > 1,\ X > 1.1$	$(T_1 - t_1)\dfrac{(X-1)}{(Xm) - 1}$
$X = \exp\left[\dfrac{AU(m-1)}{W_C C_C}\right]$	$(T_2 - t_1)\dfrac{(X-1)}{(m-1)}$
	$(T_1 - t_2)\dfrac{(X-1)}{X(m-1)}$
$m < 1,\ X < 0.91$	$(T_1 - t_1)\dfrac{X^{-1} - 1}{X^{-1} - m}$
$X^{-1} = \exp\left[\dfrac{AU(V-1)}{W_C C_C V}\right]$	$(T_2 - t_1)\dfrac{X^{-1} - 1}{X^{-1}(1 - m)}$
	$(T_1 - t_2)\dfrac{X^{-1} - 1}{1 - m}$
$m = 1,\ 0.91 < X < 1.1$	$\dfrac{T_1 - t_1}{X + 0.5(1 + m)}$
$X = \dfrac{W_C C_C}{AU}$	$\dfrac{T_2 - t_1}{X + 0.5(1 - m)}$
	$\dfrac{T_1 - t_2}{X + 0.5(m - 1)}$

	Temperature drop of hot fluid, $T_1 - T_2$
$V < 1,\ X > 1.1$	$(T_1 - t_1)\dfrac{(X-1)}{(X-V)}$
$X = \exp\left[\dfrac{AU(1-V)}{W_H C_H}\right]$	$(T_2 - t_1)\dfrac{(X-1)}{(1-V)}$
	$(T_1 - t_2)\dfrac{(X-1)}{X(1-V)}$

Table 2.8. Equations for Predicting Countercurrent Heat-Exchanger Performance (*Continued*)

	Temperature drop of hot fluid, $T_1 - T_2$
$V > 1,\ X < 0.91$	$(T_1 - t_1)\dfrac{1 - X}{V - X}$
$X^{-1} = \exp\left[\dfrac{AU(V-1)}{W_H C_H}\right]$	$(T_2 - t_1)\dfrac{1 - X}{V - 1}$
	$(T_1 - t_2)\dfrac{1 - X}{X(V - 1)}$
$V = 1,\ 0.91 < X < 1.1$	$\dfrac{(T_1 - t_1)m}{X + 0.5(1 + m)}$
$X = \dfrac{W_H C_H m}{AU}$	$\dfrac{(T_2 - t_1)m}{X + 0.5(1 - m)}$
	$\dfrac{(T_1 - t_2)m}{X + 0.5(m - 1)}$

where T_1 = hot-fluid inlet temperature, °F (°C)
T_2 = hot-fluid outlet temperature, °F (°C)
t_1 = cold-fluid inlet temperature, °F (°C)
t_2 = cold-fluid outlet temperature, °F (°C)
W_H = hot-fluid mass flow rate, lb/h (kg/s)
C_H = hot-fluid specific heat, btu/lb · °F (J/kg · °C)
W_C = cold-fluid fow rate, lb/h (kg/s)
C_C = cold-fluid specific heat, Btu/lb · °F (J/kg · °C)
A = surface area of exchanger, ft² (m²)
U = overall heat-transfer coefficient for exchanger, Btu/h · ft² · °F (W/m²·°C)

NOTES: (1) For crossflow substitute the term AUF in the above formula for AU, where F is the LMTD correction factor, $F = 1.0$ for more than three passes. (2) Base T_2 in the above formula on the heat absorbed by the cold fluid and allow for heat losses separately. The LMTD and hot gas outlet temperature should allow for heat losses.

If finned tubes are used or radiation, condensation, or boiling heat transfer is occurring, other adjustments to the individual coefficients must be made.

The new overall heat-transfer coefficient is then calculated by

$$\frac{1}{U} = \frac{1}{H}\frac{d_o}{d_i} + r_i\frac{d_o}{d_i} + r_w + r_o + \frac{1}{h}$$

where d_o = outside tube diameter
d_i = inside tube diameter
r_w = wall resistance
r_i = inside fouling resistance
r_o = outside fouling resistance

The *UA* method allows the determination of any two terminal temperatures given the other two and can, with little effort, be incorporated into a computer program or used with a programmable calculator. Tables 2.6 to 2.8 show the relevant equations for exchangers with boiling or condensation, cocurrent flow, and countercurrent flow.

The calculation method is as follows;

1. The flow rates of the fluids under the new conditions are calculated.
2. The modified inside and outside heat transfer coefficients are calculated.
3. The fouling factors to be used are selected.
4. The overall heat-transfer coefficient under the new operating conditions is calculated.
5. The product *UA* (where *A* is the effective area of heat transfer for the exchanger) is calculated.
6. The value of $W_H C_H$ (where W_H is the mass flow rate of the hot fluid and C_H, the specific heat capacity of the hot fluid) is calculated.
7. The value of $W_C C_C$ (where W_C is the mass flow rate of the cold fluid and C_C, the specific heat capacity of the cold fluid) is calculated.
8. The temperature correction factor *F* to the LMTD is determined (initially this will be the value under design conditions, but an iterative calculation will allow a determination to be made under the new conditions) and applied to the *UA* term, which now becomes *UAF*.
9. For the different types of units, the calculation method will vary slightly
   a. *For boilers,* the method is straightforward (see Table 2.6) with the value of *E* being calculated and then the value of the unknown temperature from Eq. (2) in Table 2.6.
   b. *For cocurrent flow* (see Table 2.7), the value of *m* is calculated, then the value of *E*, then the value of either $(t_2 - t_1)$ or $(T_1 - T_2)$, and finally the value of the remaining unknown temperature.
   c. *For countercurrent flow* (see Table 2.8), the value of *m* is calculated, then the value of *X*, then the value of either $(T_1 - T_2)$ or $(t_2 - t_1)$ using the equation indicated by the values of *X* and *m*, and finally the value of the final unknown temperature.

It is possible to use this method to calculate the performance of units with nonlinear heat-release curves. This may be done by treating the unit as several units in series, each with a linear heat-release curve and iterating to find the overall temperature profile. This is a tedious matter if a large number of sections are required.

ADDENDUM

Process Conditions

PRESSURE DROP. In many designs, the overall coefficient which can be achieved is limited by the pressure drop allowed through the unit for one or both of the fluids. In such cases, the basis for setting this value should be examined to determine whether it is possible to increase the pressure drop and reduce the size of the exchanger. In a system where a process stream carries out several duties in series, the total stream pressure drop should be allocated so as to reduce the total cost of the exchanger system.

Safety and Protection

DIFFERENTIAL THERMAL EXPANSION. When a large differential in thermal expansion can exist within an exchanger (such as between shell metal and tube metal), consideration should be given to utilizing a design which will minimize the resultant stresses. Such designs would include floating tubesheets or U-tubes. Large temperature differentials alone need not be the cause of this problem; the use of, for example, a carbon steel for one component and a stainless steel for another component can also result in problems due to the difference in the coefficient of thermal expansion of the two materials.

VIBRATION. When a fluid flows over tubes or through a passage, vibration is possible. This vibration can cause damage and/or increased pressure drop. Vibration analysis is a complex issue beyond the scope of this chapter, but should be considered during design. This is particularly true for vapor phase fluids.

REFERENCES

1. J. P. Gupta, *Fundementals of Heat Exchanger and Pressure Vessel Technology,* Hemisphere, New York, 1986.
2. E. V. Schlunder (ed.), *Heat Exchanger Design Handbook,* 6 vols., Hemisphere, New York, 1986.
3. K. P. Singh and A. I. Soler, *Mechanical Design of Heat Exchangers and Pressure Vessel Components,* Arcturus, Cherry Hill, N.J., 1984.
4. N. P. Cheremisinoff (ed.), *Handbook of Heat and Mass Transfer,* Gulf, Houston, Tex., 1986.
5. W. M. Rohsenow, J. P. Hartnett, and E. N. Ganić (eds.), *Handbook of Heat Transfer Applications,* McGraw-Hill, New York, 1985.
6. D. R. Pitts and L. E. Sissom, *Theory and Problems of Heat Transfer,* McGraw-Hill, New York, 1977.
7. S. Kakac, A. E. Bergles, and F. Mayinger (eds.), *Heat Exchangers—Thermal-Hydraulic Fundamentals and Design,* Hemisphere, New York, 1980.

Chapter 3

High-Performance Heat-Transfer Surfaces

JAY A. KOHLER

KEITH E. STARNER

York International Corporation

This chapter is a review of some of the heat-transfer surfaces and mechanisms which have been applied to enhance heat-exchanger performance.

3.1 SINGLE-PHASE FLOW

General Principles

Single-phase flow can be classified as internal flow or external flow. Although most of the methods for enhancing internal flow are based on flow inside circular tubes, the same methods can be applied to noncircular shapes. External flow includes flow over plates and over tubes.

Single-phase flow can be further classified as turbulent or laminar. In either turbulent or laminar flow, the principal mechanisms of enhancement are the use of secondary heat-transfer surface and the disruption of the unenhanced fluid velocity and temperature profile.

Turbulent Flow Table 3.1 presents correlations for internal single-phase, turbulent flow of several types of roughened surface tubes including those shown in Fig. 3.1. As a point of reference, equations for internal flow in smooth tubes are also presented in Table 3.1. Because of the wide variety of geometries for external flow and the corresponding wide variation in flow behavior, there are no comparable equations for external flow over smooth or enhanced surfaces.

Turbulent flow exhibits a low-velocity flow region immediately adjacent to the wall with velocity approaching zero at the wall. Most of the thermal resistance occurs in this low-velocity region, which is known as the *laminar sublayer*. The principal objective for turbulent flow enhancement is to reduce the thickness of this laminar sublayer.

Table 3.1. Correlations for Internal Turbulent Flow

Tube type	Nusselt number	Internal area	Friction factor	Reference
Smooth	$\text{Nu} = 0.027\ \text{Re}^{0.8}\ \text{Pr}^{0.333}\ (\mu/\mu_w)^{0.14}$	. . .	. . .	1
Smooth	$\text{Nu} = \dfrac{(f/2)\ \text{Re Pr}}{1 + 12.7\sqrt{f/2}\ (\text{Pr}^{0.667} - 1)}$	. . .	. . .	2
Spiral fins	$\text{Nu} = 0.023\ \text{Pr}^{0.4}\ \text{Re}^{0.8}\left(\dfrac{A_x}{A_{fc}}\right)^{0.10} \times \left(\dfrac{A_n}{A_a}\right)^{0.5} (\sec\alpha)^3$	Actual	$f = 0.046\left(\dfrac{A_x}{A_{fn}}\right)^{0.5} (\cos\alpha)^{0.5}/\text{Re}^{0.20}$	3
Various (see Fig. 3.1)	$\dfrac{\text{Nu}_e}{\text{Nu}_s} = \left\{1 + \left[2.64\ \text{Re}^{0.036}\,(e/d)^{0.212} \times \left(\dfrac{p}{d}\right)^{-0.21}\left(\dfrac{\alpha}{90}\right)^{0.29} \text{Pr}^{-0.024}\right]^7\right\}^{1/7}$	Nominal	$\dfrac{f_e}{f_s}$ = function of $\left(\dfrac{p}{d}, \dfrac{\alpha}{90}, \dfrac{e}{d}, \text{Re}, \beta, n\right)$ n = number of sharp corners facing the flow	11
Corrugated	$\text{Nu} = \dfrac{\text{Re Pr}\sqrt{f/2}}{\beta\sqrt{\text{Pr}}\ (\text{Re}\sqrt{f/2})^{0.127} + \gamma}$ γ, β = functions of $\dfrac{e}{d}, \dfrac{p}{d}$	Nominal	$\sqrt{\dfrac{f}{2}} = \dfrac{-1}{2.46\ \ln\,(r + 7/\text{Re})^m}$ r, m = functions of $\left(\dfrac{e}{d}\right)$	4
Spiral fins	$\text{Nu} = \dfrac{\text{Re Pr}\sqrt{f/2}}{\beta\sqrt{\text{Pr}}\ (\text{Re}\sqrt{f/2})^{0.136} + \gamma}$ β = function of $\left(\dfrac{e}{p}, \dfrac{e}{d}\right)$ γ = function of $\left(\dfrac{p}{d}\right)$	Nominal	$\sqrt{\dfrac{f}{2}} = \dfrac{-1}{2.46\ \ln\,(r + 7/\text{Re})^m}$ r, m = functions of $\left(\dfrac{e}{l}, \dfrac{p}{d}\right)$ l = lead of ridge	4

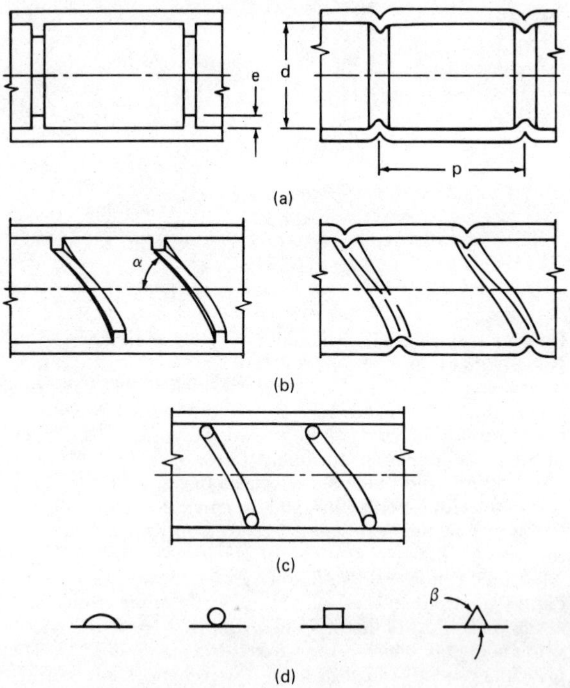

Fig. 3.1 Internal roughness.[11] (*a*) Transverse ribs; (*b*) helical ribs; (*c*) wire coils; (*d*) profile shapes.

For fully developed flow in a smooth tube, the dimensionless laminar sublayer thickness is

$$y^+ = \frac{y\sqrt{g_c\tau_o/\rho}}{\nu} = 5 \tag{3.1}$$

Using the shear stress τ_o for a smooth tube, the ratio of laminar sublayer thickness y to diameter d is

$$\frac{y}{d} = 25\ \text{Re}^{-0.875} \tag{3.2}$$

For example, if Re = 20000 and d = 1.0 in (25 mm), the laminar sublayer thickness is 0.004 in. (0.10 mm). To be effective in providing enhancement, any roughness must extend through the laminar sublayer. For wire coil inserts in tubes, Edwards and Sheriff[5] recommend that the wire height exceed twice the laminar sublayer thickness. The enhancement will increase as the wire penetrates into the buffer layer which is defined as $5 < y^+ < 30$, or six times the laminar sublayer thickness. However, the friction factor increases much faster than the heat-transfer coefficient when enhancements protrude into the buffer layer. Figure 3.2 presents the performance of a ridge-roughened tube.

In addition to roughness as a means for disturbing the boundary layer, methods for providing a swirl flow also can reduce the boundary-layer resistance. Spiral inserts and protrusions may cause a spiral flow in tubes, producing secondary flow patterns and increasing the velocity near the wall.

Several types of external flow enhancements take advantage of higher coefficients in the developing flow region im-

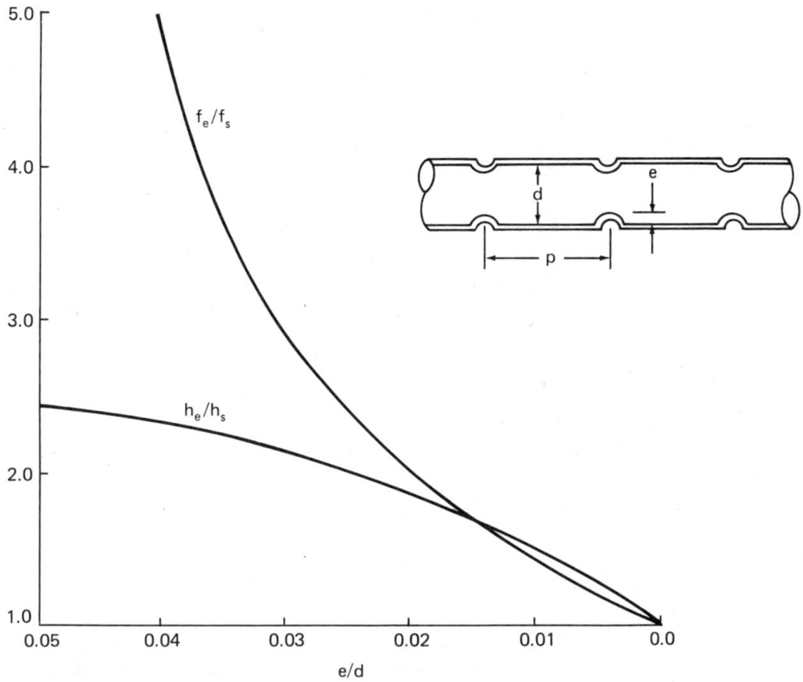

Fig. 3.2 Effect of ridge height on coefficient enhancement with Re = 39,800 and p/d = 0.496.[8]

mediately downstream of a leading edge. In this region the laminar sublayer increases from zero thickness at the leading edge. When repeated interruptions are provided, such as the discontinuities in the surface shown in Fig. 3.6, a higher average heat-transfer coefficient results.

Laminar Flow In laminar flow, thermal energy transfer is accomplished only by conduction, as opposed to turbulent flow, where both conduction and bulk fluid mixing contribute to energy transfer. In fully developed flow, the fluid velocity and temperature vary across the entire flow channel width so that the thermal resistance is not just in the region near the wall, as in turbulent flow. As a result, small-scale surface roughness is not effective in enhancing heat transfer. Most laminar flow surface enhancements employ some method of swirling the flow or creating turbulence. For example, Uttarwar and Raja Rao[10] observed a transition from laminar to turbulent flow at a lower Reynolds number in a wire coil-insert tube than in a smooth one. The use of secondary surface also is very common. If, however, the secondary surface creates small flow channels, as with closely spaced fins, performance can be adversely affected.

Surface temperature variation is one important factor affecting the heat-transfer coefficient. Also important are the length of the heat-transfer device, the velocity and temperature profiles at the device entrance, and natural-convection effects.

Correlations for internal laminar flow in several types of tubes are presented in Table 3.2. Because of the variety of factors affecting laminar flow heat transfer, testing of a new surface in its proposed application is highly recommended.

Internal Flow

Internal Fins Tubes having internal fins as an integral part of the base tube are available from a variety of manufacturers. Fins can be either longitudinal or have a spiral. The critical parameters are shown in Fig. 3.1.

TURBULENT FLOW. Carnovos[3] tested 21 different geometries in a turbulent flow of air and developed the correlation shown in Table 3.1. A more universal form for turbulent flow developed by Withers[4] for multiple lead fins is also given in Table 3.1. The parameters β, γ, r, and m, which are defined in Ref. 4, are a function of the fin geometry. Table 3.3 shows the enhancement levels for various pitches and numbers of leads.

LAMINAR FLOW. Internal fins, when used for laminar flow, are typically larger than for turbulent flow. Fins which constitute 10 to 20 percent of the diameter are common. Fins should be spaced far enough apart to prevent narrow passages between them which would restrict flow. Marner and Bergles[6] reported on tests of an internally finned tube with constant wall temperature. The 1-in (25-mm)-diameter tube had 16 fins which were 0.082 in (2.1 mm) high and had 84 percent more surface than a smooth tube. The correlations they developed are shown in Table 3.2. Marner and Bergles reported an increase in Nusselt number of 200 and 300 percent over a plain tube during heating. During cooling the Nu increase was marginal because of restricted flow between the fins. The isothermal friction factor was 70 to 240 percent over that for plain tubes. Watkinson et al.[7] tested spiral and longitudinal fin tubes with constant wall temperature; the

Table 3.2. Correlations for Internal Laminar Flow

Tube type	Nusselt number	Internal area	Friction factor	Reference
Smooth	$Nu = 1.86\left[Re\,Pr\left(\frac{d}{L}\right)\right]^{1/3}(\mu/\mu_w)^{0.14}$ (constant wall temperature)	. . .	$\frac{16}{Re}$	1 6
Twisted tape	$Nu = 1.322\left[\pi\,Re\,Pr\left(\frac{d}{4L}\right)\right]^{0.458}\left(\frac{\mu}{\mu_w}\right)^{0.14}$	Nominal	$\frac{54}{Re}$ Isothermal	6
Internal fin	$Nu = 0.253\left[\pi\,Re\,Pr\left(\frac{d}{4L}\right)\right]^{0.722}\left(\frac{\mu}{\mu_w}\right)^{0.14}$ (heating) $Nu = 1.138\left[\pi\,Re\,Pr\left(\frac{d}{4L}\right)\right]^{0.427}\left(\frac{\mu}{\mu_w}\right)^{0.14}$ (cooling)	Nominal	$\frac{27}{Re}$ "	6
Spiral fins	$\frac{Nu_e}{Nu_s} = 10.3\left(\frac{b}{p}\right)^{0.5}\left(\frac{d}{d_e}\right)^{0.4} \times \left(\frac{A_a}{A_n}\right)Re^{-0.07}$	Nominal	$\frac{16.4(d_e/d)^{1.4}}{Re}$	7
Straight fins	$\frac{Nu_e}{Nu_s} = 1.31\,n^{-0.5}\left(\frac{d}{d_e}\right)^{0.21} \times \frac{A_a}{A_n}Re^{0.13}$	Nominal	"	"

NOTE: Correlations valid for test geometries only.

developed correlations are given in Table 3.2. At Re 500 the heat-transfer coefficient increase over the smooth bore tube ranged from 8 to 224 percent, with most observations being between 20 and 60 percent. The friction factor increase was from 18 to 131 percent, with most observations being between 20 and 80 percent.

Corrugated and Repeated-Ridge Tubes Corrugated tubes are constructed with circumferential or spiral indentations as shown in Figs. 3.1 and 3.3. Corrugated tubes are relatively easy to manufacture and, as a result, can be relatively low in cost.

Withers[4] developed a turbulent flow correlation for single-lead corrugated tubes as shown in Table 3.1. The factors β, γ, r, and m, which are defined in Ref. 4, are a function of the corrugation geometry. Table 3.4 shows the performance for some typical cases with different ridge heights.

Table 3.3. Typical Effect of Number and Pitch p of Internal Fins on Turbulent Flow Enhancement (Pr = 5)

				Re = 10000		Re = 50000	
No. of leads	e/d	e/p	p/d	h_e/h_s	f_e/f_s	h_e/h_s	f_e/f_s
6	0.03	0.15	0.2	2.51	2.89	2.55	3.36
6	0.03	0.06	0.5	1.72	1.88	1.63	1.88
6	0.03	0.038	0.8	1.59	1.45	1.51	1.45
12	0.03	0.15	0.2	2.23	2.21	2.00	1.93
12	0.03	0.06	0.5	1.48	1.34	1.53	1.34
12	0.03	0.038	0.8	1.20	1.21	1.39	1.21

SOURCE: Withers.[4]

Transverse, or repeated-ridge, tubes are shown in Fig. 3.1. The principal parameters are ridge height e and pitch p. Kalinin and Yarkho[8] investigated repeated-ridge tubes for turbulent and transitional flow. Figure 3.2 shows the effect of ridge height and pitch on heat transfer and friction-factor enhancement.

Inserts Removable inserts of many types have been used in tubes to improve heat transfer. Most common are ones

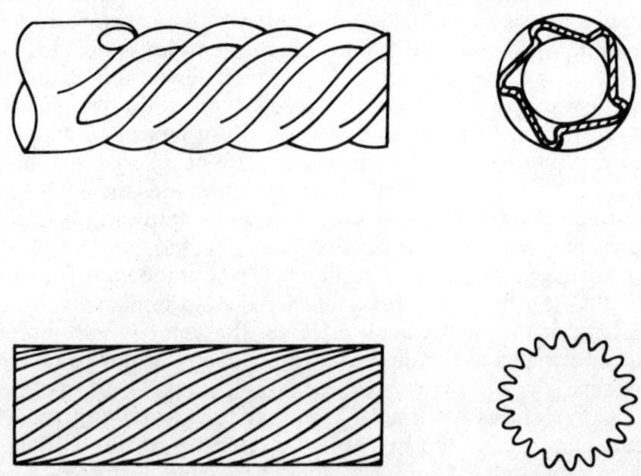

Fig. 3.3 Spirally fluted tubes. (*Turbotec and GA Technologies.*)

Table 3.4. Typical Effect of Ridge Height of Corrugations on Internal Turbulent Flow Enhancement (Pr = 5)

			Re = 10000		Re = 50000	
e/d	e/p	p/d	h_e/h_s	f_e/f_s	h_e/h_s	f_e/f_s
0.01	0.025	0.4	1.40	1.24	1.36	1.24
0.02	0.05	0.4	1.60	1.77	1.54	1.77
0.03	0.10	0.4	2.28	4.23	2.22	4.45

SOURCE: Withers.[4]

which are spiral or swirl devices and impart spiral motion to the fluid. They act as flow disturbers and do not contribute additional area since the thermal contact with the tube is generally poor. However, the additional surface does contribute to frictional loss. Most inserts are continuous and extend the entire tube length. They must be secured in place to ensure that they do not move under the force of the fluid. This can be done by providing a friction fit or by mechanically fastening them to the heat exchanger. Tension-wound wires (Fig. 3.1) can be used.

Inserts can be successfully used in laminar flow since the insert can provide bulk flow disturbance. Marner and Bergles[6] tested a twisted tape (Fig. 3.4) with constant wall temperature. They found a Nu increase of 50 to 125 percent over a plain tube for cooling and heating. The isothermal friction factor was 70 to 240 percent higher. Van Rooyen and Kroger[9] tested internally finned tubing with twisted tapes using SAE 10 and 20 oils. The use of a tape improved the internally finned tube performance, with the tightest tape pitch providing the greatest heat transfer. Uttarwar and Raja Rao[10] tested seven different wire coil inserts in oil with a Prandtl number between 300 and 675 and Re between 30 and 700. On the basis of a constant pumping power comparison, the wire with e/d of 0.08 and p/e of 5.0 showed an improvement of 350 percent. They also concluded that the net improvement in heat transfer was much better at Re below 200 than at higher flow rates and that the net improvement for laminar flow is higher than that for turbulent flow.

Miscellaneous Enhancement Means A general correlation was developed by Ravigururajan and Bergles[11] for turbulent flow in internally ribbed tubes. Included in their study were the various types of tubes and the profile shapes shown in Fig. 3.1. Data were taken from 17 previous studies to develop the correlation shown in Table 3.1.

Considerable experimental work has been performed on intentionally roughened surfaces. Referred to as sand-grain-roughness, testing included various-sized granular surfaces. Despite good performance potential, it has received little commercial use because of the difficulty of applying it economically with controlled roughness size and good adhesion.

External Flow

Flow Perpendicular to Tube Banks For externally low finned tubes in shell-and-tube heat exchangers, Bell[12] recommends use of the results of Briggs et al.[13] as applied to a shell-and-tube design method such as the Delaware method.[12] At Re above 1000, the heat-transfer coefficient is

Fig. 3.4 Twisted-tape insert.[6]

the same as for plain tubes, whereas at lower Re, the finned-tube coefficient becomes slightly less than that of the plain tube. Bell reports uncertainty in the friction factor and recommends a value of 1.5 times that for a plain tube.

For heat exchange to gases, such as in large air-cooled heat exchangers, tubes with relatively high fins are used. Weierman[14] categorized the various tube types by the method of bonding the fin to the tube and reviewed application limits for each type. Performance has been summarized by Rabas and Eckels[15] and Weierman[16] for some specific geometries.

Several augmented designs have been used to improve the performance or cost-effectiveness of solid-fin geometries. Among the geometries used are solid fins with slots or punched-out sections, wire loops brazed to the outside of the tube, and sharp spikes which are gouged out of the base tube.[17] Eckels and Rabas[18] report on tests of pin or segmented fins and one called the *K-Y fin*.

Axial Flow Along Tubes Applications involving flow axial to tubes include (1) flow in the annulus between concentric tubes, as in coaxial heat exchangers, and (2) shell-and-tube exchangers where flow is primarily axial. Circumferentially finned tubes, as shown in Fig. 3.9, can be used to improve performance for turbulent flow. They are not useful for laminar flow because of the lack of fluid motion between the fins. Knudsen and Katz[19] tested several geometries using flow in the annulus between an externally finned tube and the inside of a plain tube. Their tests with water produced increased heat-transfer coefficients of 25 to 100 percent beyond the increase as a result of the added area.

Flow Over Parallel Plates Tube-and-plate heat exchangers (see Fig. 3.5) are made of continuous parallel plates which serve as fins. Various modifications have been made to the fin to enhance performance. The most common is the use of wavy or corrugated fins (Fig. 3.5). The level of enhancement depends primarily on the height and pitch of the corrugation.

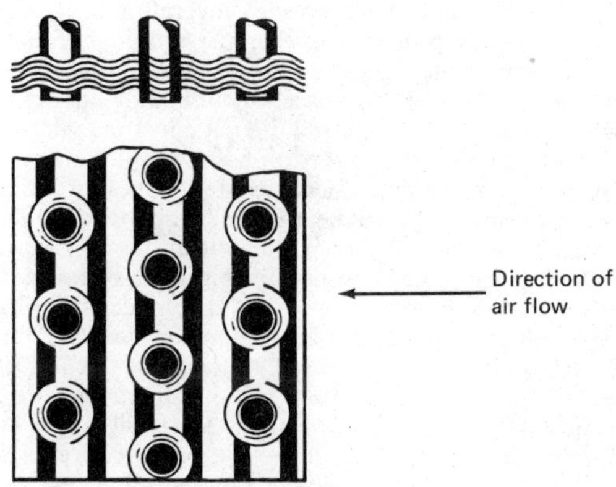

Fig. 3.5 Wavy fin for tube-and-plate heat exchanger.

A more recent method is the use of louvered or slit fins (Fig. 3.6).

Nakayama[21] summarized previous tests by Sparrow and Hajiloo[22] and Cur and Sparrow[23] for offset fins of compact heat exchangers. Nakayama reports that the enhancement is due to the shedding of wakes from upstream segments.

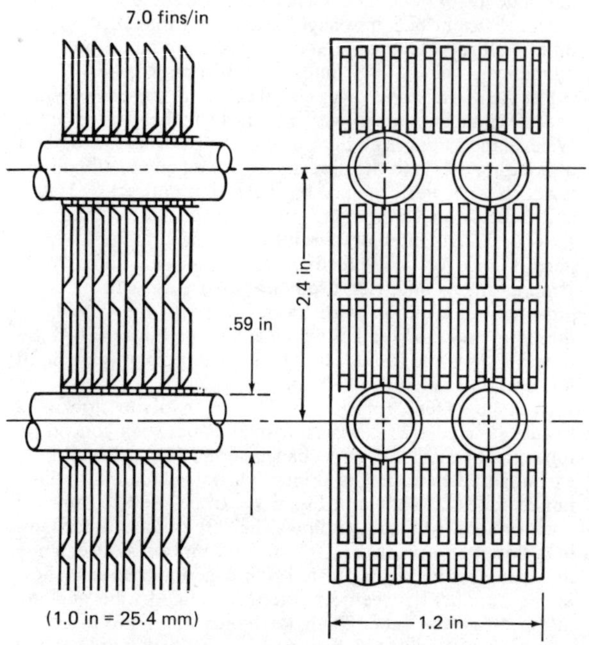

Fig. 3.6 Slotted tube-and-plate fin of Forgó.[20]

3.2 CHANGE OF PHASE

The transfer mechanisms that are important in the change-of-phase process are gas or liquid convection with mass transfer at the gas-liquid interface, conduction through thin liquid films, and surface nucleation.

High-performance surfaces are designed to utilize and enhance mechanisms which will yield overall improvement to the heat-transfer process. For example, although a moderate amount of nucleate surface boiling enhancement can be very effective in a relatively quiescent pool boiling environment, such enhancement may have little net effect in a high-velocity, convection-dominated, two-phase flow stream.

In most phase-change heat-transfer processes the dominant resistance to flow of heat is conduction through or convection to a liquid layer on a heat-transfer surface. High coefficients can be achieved with thin films of low-conductivity liquid if film resistance is the only effective resistance present. The heat-transfer coefficient for static conduction through a film of liquid is equal to the thermal conductivity divided by the thickness. For a 0.001-in film, the heat-transfer coefficients for water, R-11, and R-12 (refrigerants 11 and 12) are 3450, 625, and 525 Btu/h · ft^2 · °F, respectively. Actual heat-transfer coefficients would often be higher because of fluid motion. Thus, maintaining a thin film can be an effective means of promoting high performance. Thin films can be achieved during two-phase heat transfer by use of mechanical means, gravity, surface tension forces, shear flow, or bulk liquid displacement by vapor bubbles.

Condensation Enhancement

Heat-transfer enhancement is possible on either the shell side or tube side of a condenser. Efforts to utilize the higher performance of dropwise condensation have concentrated mainly on steam condensation and appear limited in potential because of the inevitable surface contamination problems and the increased thermal resistance of nonwetting coatings. The focus here will be on filmwise condensation, which is the predominant mode experienced in practical applications.

Shell-Side Condensation *Shell-side condensation* implies condensation on the outside of tubes or surfaces in direct contact with a vapor space. The performance of the film condensation process is by conduction through liquid films and is described in the classical works of Nusselt.[24] Figure 3.7 shows film thicknesses for a vertical surface as a function of surface length and mass flux to the surface. Surface length is a major deterrent to good heat transfer in gravity controlled liquid films on vertical surfaces because of increasing film thickness. Enhancement of the heat-transfer coefficient involves reducing the liquid film thickness or modifying the flow pattern of the liquid layer to promote better heat transfer. Figure 3.8 shows techniques such as fluted tubes and vertical fins and wires, some of which are loosely attached, which are used to reduce film thickness. Such devices utilize surface tension to pull the condensate into crevices and thus make the film thinner over the major part of the surface. Various investigators have been able to use selected surface roughness to promote heat transfer by causing turbulence in the condensate boundary layer.[25] An increase of a factor of nearly 2 was observed.

Because of the strong effect of length it is not appropriate to compare the heat-transfer performance of a vertical surface with an enhancement device to heat-transfer performance on a long vertical tube without enhancement. For design purposes, a horizontal tube with a much shorter gravity flow length would be the logical unenhanced reference. For example, a vertical 4-ft (1.2-m)-long fluted tube produced R-11 heat-transfer coefficients three times higher than that of a smooth horizontal tube.[26]

One of the most successful forms of condensation enhancement has been the use of finned tubing in the horizontal orientation. Use of short radial fins on a horizontal tube (Fig. 3.9) can increase the heat-transfer coefficient in R-12 by a factor of nearly 2. Combining that with the additional heat-transfer area will yield a net increase of a factor of 5 over a plain 0.75-in (19-mm)-diameter tube. Finned tubes of this type having 11 to 40 fins per inch (430 to 1570 fins per meter) typically are not used with higher-surface-tension fluids, such as water, since liquid holdup between the fins will result in flooding of the surface. Rudy and Webb[27] have shown that condensate flow over the horizontal finned tube is dominated by surface tension rather than by gravity.[28] A factor of 3 increase in the condensing coefficient of ammonia was obtained by use of spaced wire wrap on a horizontal tube—showing the importance of surface-tension forces.[29] Application of finned tubing in a shell-and-tube heat exchanger

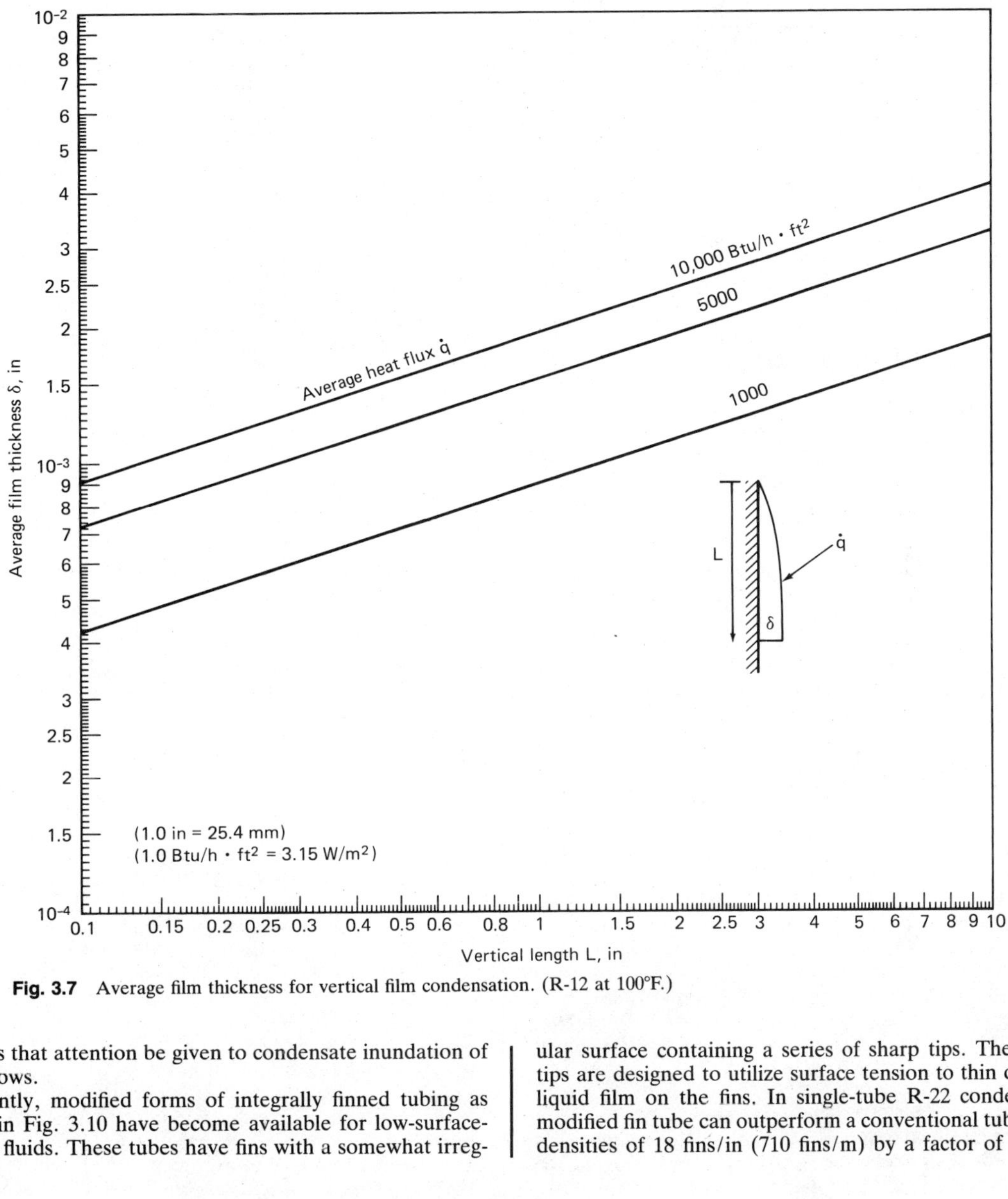

Fig. 3.7 Average film thickness for vertical film condensation. (R-12 at 100°F.)

requires that attention be given to condensate inundation of lower rows.

Recently, modified forms of integrally finned tubing as shown in Fig. 3.10 have become available for low-surface-tension fluids. These tubes have fins with a somewhat irregular surface containing a series of sharp tips. These sharp tips are designed to utilize surface tension to thin down the liquid film on the fins. In single-tube R-22 condensing, a modified fin tube can outperform a conventional tube having densities of 18 fins/in (710 fins/m) by a factor of 3 on the

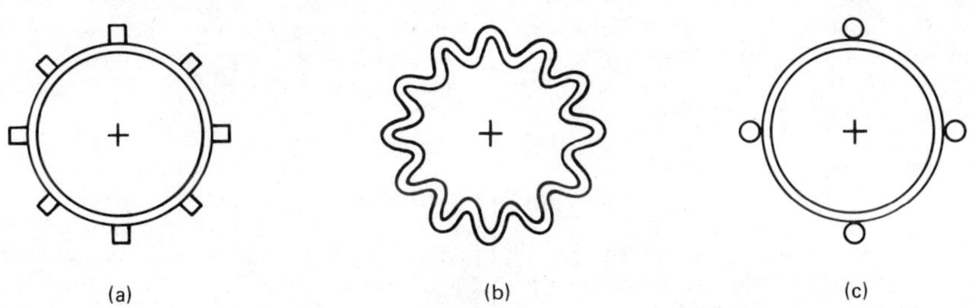

Fig. 3.8 Vertical condensation enhancement devices. (*a*) Fins; (*b*) flutes; (*c*) wires.

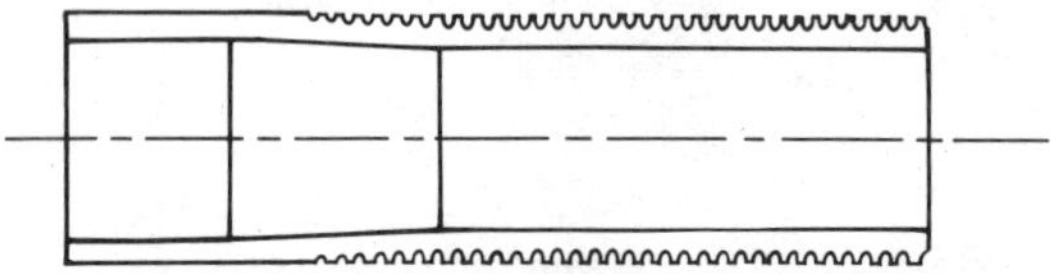

Fig. 3.9 Cross section of an externally finned tube.

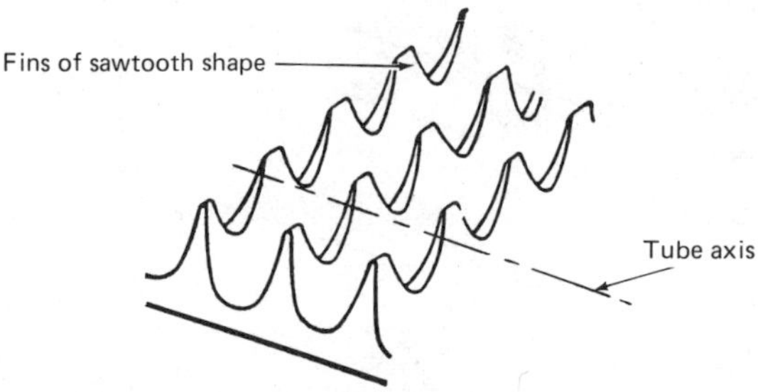

Fig. 3.10 Modified annular fins for condensation enhancement.[30]

condensing side.[30] Such surfaces, however, are very sensitive to liquid loading from other tubes.[27]

In-Tube Condensation With a combination of gas and liquid flowing inside a tube during condensation, the augmentation devices mentioned for single-phase heat transfer have been found to be effective. Specifically, twisted-tape inserts and internal fins have received attention.[31] These devices serve mainly to add extra area or to disturb the boundary layer of the liquid film that flows along the tube wall in an annular fashion. Tube-side improvements of a factor of 2 with R-113 have been observed for internal fin tubes at the expense of a minimal increase in pressure drop. Twisted tapes can give a 30 percent heat-transfer increase but with more than 200 percent increase in pressure drop.

Most recently a new generation of tubes with fine, closely spaced helical internal fins has found application for fluorocarbon condensers.[21] The enhancement mechanism appears to be a combination of added area, thin-film production, and surface roughness. For a typical refrigerant condenser application, condensing performance improvements of 100 percent and greater are reported with minimal pressure drop increase.

Evaporation Enhancement

As with condensation, it is possible to enhance evaporation on either the shell side or tube side of a heat exchanger. The mechanisms important for enhancement are thin-film evaporation, convective boiling, and nucleate boiling. In many cases, combined mechanisms will be found responsible for enhancement results.

Thin-Film Vaporization As with condensation, the challenge is to produce a thin film to minimize film thermal resistance. For efficient operation, complete surface coverage is also required. Thomas[32] discusses a variety of techniques applicable to top-fed vertical plates and tubes. The fluted vertical tube seems to be the most successful thin-film vaporization device. The flutes serve to thin down the film on the crests of the surface. This thinning mechanism is explained by Thomas as a combination of Marangoni instabilities (surface-tension-induced flows), rollwaves, and roping effects, all of which serve to wet the crests of the flutes intermittently in a thin film to prevent dry spots from forming. Such mechanisms are also apparently at work to produce very thin films in controlled experiments on short plain surface vertical tubes.[32] Experiments with water produced evaporation coefficients up to six times higher than that predicted from the combination of calculated falling film Nusselt thickness[24] and conduction through a laminar layer.

Nucleation Enhancement The nucleation process from gas-filled surface cavities of conical and enhanced reentrant shapes is shown in Fig. 3.11.[33] The equation describing a force balance across a bubble with the temperature difference

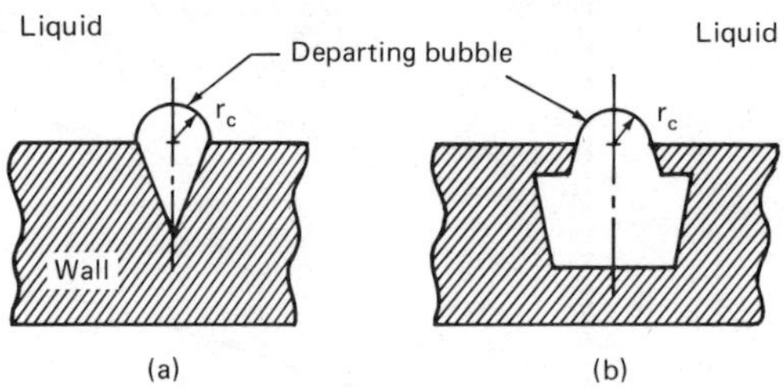

Fig. 3.11 Nucleation from (*a*) conical and (*b*) reentrant cavities.

ΔT_{sat} required to sustain this pressure difference for bubble growth is

$$\Delta T_{sat} \simeq \frac{2\sigma RT^2}{h_{fg}\, p r_c} \tag{3.3}$$

At a saturation temperature of 35°F a cavity radius r_c of 0.001 in requires a wall superheat of only 0.1 and 0.2°F for refrigerant 22 and ammonia, respectively. For refrigerant 11, the required superheat increases to 1.6°F. The advantage of the reentrant cavity is that, if properly sized, the shape of the cavity mouth prevents liquid floodback and thus the cavity is a stable source of gas for repeated nucleation. The conical cavity, as might be found on the surface of a commercial finish metal surface, tends to deactivate itself as a result of capillary action drawing liquid into the cavity. The optimum reentrant cavity size has the largest possible radius to provide minimum required superheat but is not so large that liquid surface tension cannot prevent floodback. If floodback does occur, reactivation of the cavity must take place from even smaller conical cavities within the larger reentrant ones and at much higher temperature differences. Subsurface visual studies[34] resulted in the conclusion that thin film evaporation within a reentrant cavity is the principal heat-transfer mechanism in enhanced surface boiling. Vapor within the cavity drives out all but a thin-surface film of liquid on the cavity walls which then produces high heat-transfer coefficients.

The evolution of enhanced surface geometries for nucleate boiling is discussed by Webb.[35] Many types of surfaces have evolved from the concept of the reentrant cavity. They cover the range from modified deep surface scratches to machined interconnected tunnels and porous metal coatings. Some examples of enhanced surface construction are shown in Fig. 3.12. It is typical to find boiling heat-transfer enhancement of a factor of 2 to 5 with many surface types and a factor of

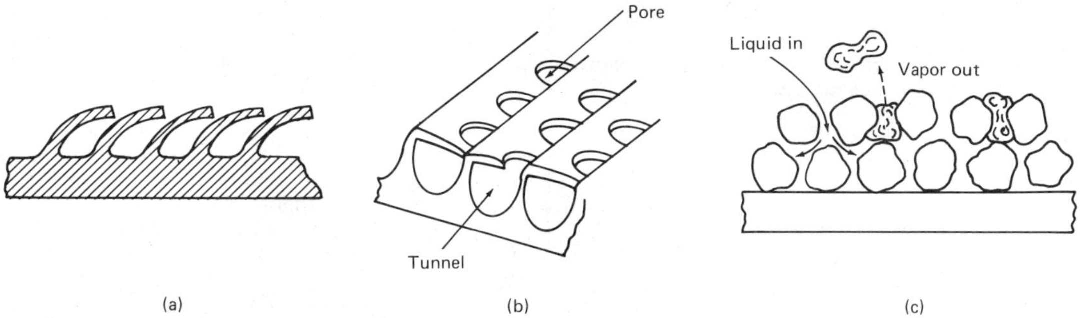

Fig. 3.12 Enhanced nucleation surfaces. (*a*) Bent fins;[36] (*b*) modified fins;[37] (*c*) porous surface.[38]

up to 10 with porous coatings.[38] Commercial application of a porous boiling surface is described by Starner and Cromis.[39]

Extended surfaces in the form of short fins have been found to be effective for use in boiling outside of tubes.[40] In addition to the added surface, the boiling coefficient per unit area of surface is also enhanced relative to bare tubes. Figure 3.13 shows the relative performance of finned and porous coated tubes versus plain tubing in pool boiling of R-12.

It has been found that the use of nonwetting coatings on a surface can significantly increase boiling performance.[41] It is theorized that the coating works in conjunction with naturally occurring surface cavities to produce enhancement. Teflon spots have been found effective with boiling water on metal surfaces. Having the nonwetting coating only in local spots appears more effective than a complete surface coating. Enhancements of a factor of 5 to 10 for commercially smooth and pitted surfaces with water have been reported.[42]

Convective Boiling—Shell Side If significant convection is present on the shell side of a tube bundle, it is possible to have all three mechanisms of thin-film wetting, nucleation,

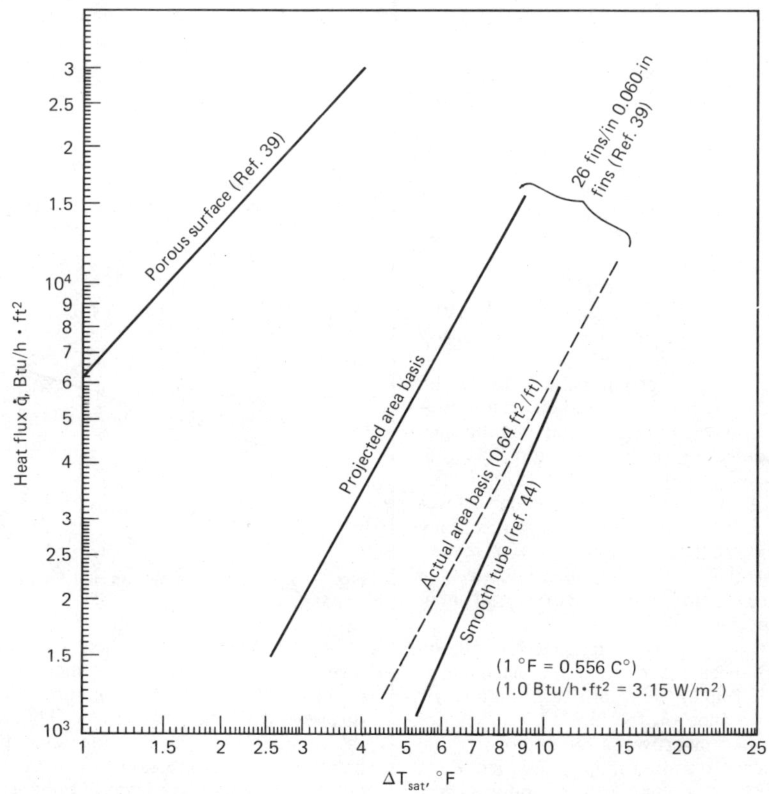

Fig. 3.13 Comparison of various boiling tube surfaces. (R-12 pool boiling at 40°F; 0.75-in-diameter tube.)

and convection present.[43] The convection contribution will tend to play a larger part for high-surface-tension fluids operated at low surface superheat where nucleation is more difficult. Figure 3.14 shows the importance of convection in shell-side boiling and illustrates how control of mass flow rate over the tubes can lead to boiling enhancement.[44] Frictional or gravitational pressure drop will be accompanied by a saturation temperature drop which can be significant in some applications. Performance enhancement often can be accomplished by devising ways, such as sprays, to wet the tubes throughout the bundle without using a large inventory of liquid.

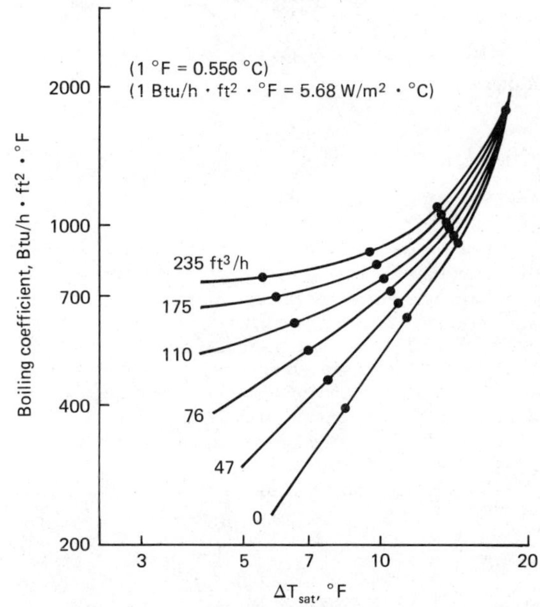

Fig. 3.14 Effect of vapor injection on coefficients in R-12 shell-side boiling.[44]

The external integral fin concept has been found to be especially effective in convective dominated shell-side boiling. The fins aid nucleation and are very effective in the two-phase flow stream, provided fin spacing is not too close. Fin densities of from 19 to 40 fins per inch (750 to 1570 fins per meter) are typical. Application of enhanced boiling surfaces to convective bundles will give different results depending on the relative magnitude of the boiling and convective contributions. This is illustrated in Ref. 45, where single-tube and bundle performance for three types of surfaces in shell-side boiling are compared.

Convective Boiling—Tube Side In-tube convective boiling can be improved by enhancement of any of the mechanisms mentioned for shell-side boiling. As a practical matter, the surface configurations possible on the inside of tubing are limited. Lavin and Young[46] investigated the use of internal ridges, axial and helical fins, and an internal divider. Each flow regime of nucleate boiling, annular flow, and mist flow showed enhancement compared to a plain tube. Other possible enhancement devices are similar to those covered under sensible-heat transfer. The delicate balance of pressure drop increase and heat-transfer enhancement must be observed as increased pressure drop with evaporating fluids results in a boiling-temperature penalty.

Tubes with small, closely spaced helical fins mentioned for in-tube condensation also have been found to be effective for evaporation.[21] Some typical applications show a 100 percent increase in heat transfer with less than 10 percent increase in pressure drop. Such surfaces appear to be effective over the full range of application from nucleate boiling to superheating.

3.3 METHODS OF TESTING AND EVALUATION

Testing

A common method for determining turbulent-flow, single-phase coefficients from heat-exchanger tests is the Wilson method.[47] In this method, the coefficient on one side is held constant while the coefficient to be determined is varied by making a series of tests at different flow rates. The primary requirement is that the equation describing the coefficient to be measured be of the form

$$\mathrm{Nu} = C_1 \, \mathrm{Re}^{C_2} \, \mathrm{Pr}^{C_3} \tag{3.4}$$

and that the constants C_2 and C_3 are known in advance. The objective of the test is to determine C_1. As shown below, the total resistance R_T is composed of a constant resistance and a variable resistance which is the one to be determined.

$$R_T = R_C + R_V = R_C + \frac{1}{C_1}\left(\frac{1}{k \, \mathrm{Re}^{C_2} \, \mathrm{Pr}^{C_3}/d}\right) \tag{3.5}$$

This equation is that of a straight line with slope of $1/C_1$ and an intercept of R_C. A straight line through the test points will then provide the unknown C_1 (Fig. 3.15).

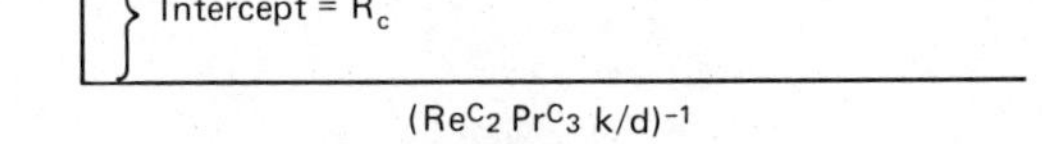

Fig. 3.15 Example of Wilson plot for turbulent flow.

Localized variations in heat transfer may affect the performance of a particular enhanced surface. In some applications, flow maldistribution prevents uniform conditions throughout the heat exchanger. This can be exaggerated by the unique pressure drop characteristics of the surface enhancement. In cases where enhancement is due to two-phase flow mechanisms, the level of enhancement throughout the exchanger is likely to vary as a result of local quality and void fraction variations. Leong and Cornwell,[48] for example, demonstrated how significant variations in boiling coefficient occur throughout a kettle reboiler. Measurement of local heat-transfer coefficients in heat exchangers can provide important insights into the optimal use of enhanced surfaces. Measurement of the performance of individual components in an exchanger or the simulation of the expected variation in conditions with a small, controlled test apparatus can provide useful information.

Evaluation

Evaluation of augmented surfaces should be oriented toward meeting specific objectives. These objectives can be reduction in size, weight, or cost. Alternatively, the improvements can be in the heat-exchanger performance characteristics such as increased capacity, reduction in temperature difference, or reduction in fluid pressure losses.

When considering prospective augmentations for a given application, it is important to focus on the dominant heat-transfer resistance in order to provide the greatest improvement to the overall coefficient. An obvious but often overlooked point is the need to identify the relative values of all resistances. If fouling or contact resistances are involved, the relative value of these must be considered. Frequently, the use of enhanced surfaces must be justified economically. Thus the additional cost, if any, of such surfaces must be established.

Enhancement for Same Heat-Exchanger Performance In single-phase flow, fluid pressure drop frequently will be a limiting factor. If an existing surface is being replaced with one having higher friction factor and heat-transfer characteristics, a larger flow area might be used to reduce the fluid velocity and pressure drop. In heat exchangers where the spacing between surfaces can be varied, as in tube-and-plate exchangers, the spacing can be increased to control the pressure drop.

For internal tube flow in shell-and-tube exchangers, or any other exchanger which has flow passages of a fixed cross-sectional area, the fluid travel length can be optimized for a given surface. For multiple pass exchangers, the travel length is equal to the tube length times the number of passes, so that the exchanger length can be controlled by adjusting the number of passes.

When cost reduction is the primary objective, it is useful to compare surfaces on the basis of cost per unit of heat-transfer. In addition to savings in surface cost, there often are additional savings due to size, weight, or manufacturing cost when less surface is needed.

Enhancement for Different Heat-Exchanger Performance High-performance surfaces often are used to improve overall performance. Some of the various methods for evaluating the value of the performance improvement versus any increases in exchanger cost are as follows:

1. The incremental cost can be determined and compared to other methods of achieving improved performance.
2. Costs can be compared to the ultimate value of performance improvements to the user.
3. Where energy savings result from power reduction, the value of heat-transfer enhancements can be evaluated using payback period or life-cycle cost analysis.

Performance Evaluation Criteria Various criteria have been developed for evaluating the performance of enhanced surfaces in single-phase flow. When pressure drop is the limiting factor, evaluations may be made simply in terms of the theoretical pumping power:

$$P = \Delta p\, W \tag{3.6}$$

Bergles et al.[49] have developed a method of evaluating heat-transfer enhancements which is based on eight criteria. In this evaluation method, the specific criterion applying to a problem depends on the nature of the "fixed" design parameters and the design objectives. Methods for evaluating gas-to-gas heat exchangers have been cited by Webb.[50]

Fouling An important consideration in whether high-performance surfaces should be used is the effect which fouling will have on the heat-exchanger performance. Unfortunately, this is an area of considerable uncertainty, and no specific recommendations can be made. Many surfaces have been successfully used in fouling environments with no greater fouling accumulation than with unenhanced surfaces. On the other hand, in some heavily fouling environments, the surface could be fouled to the extent that the surface is rendered ineffective. Surfaces with very fine structures are more likely to be made ineffective than large structured surfaces. In general, the same factors which influence fouling on unenhanced surfaces will influence fouling of high-performance surfaces.

It is very useful to consider how an allowance for fouling in design (fouling factor) can influence the overall prospect for reduction of heat-exchanger surface through use of enhancement. In Ref. 51 it is shown that as a result of increased heat flux, a heat exchanger utilizing enhanced surface is usually influenced the most in surface usage for a given fouling factor. This is the result of the temperature drop across the fouling film, which is equal to the product of fouling factor and local heat flux. Theoretically, this situation can be greatly enhanced if the local heat flux on the fouling side is reduced by the addition of secondary surface such as fins.

NOMENCLATURE

A, A_a = actual heat-transfer area
A_{fc} = open-core free-flow area
A_{fn} = nominal flow area = $\pi(d/2)^2$
A_n = nominal heat-transfer area = $\pi d L$
A_x = actual flow area
b = interfin spacing
C_1, C_2, C_3 = constants
d = hydraulic diameter
d_e = equivalent diameter
d_f = diameter over fins
e = roughness height
f = friction factor = $\tau_o\, 2g_c/\rho V^2$
g_c = gravitational constant
h = heat-transfer coefficient
h_{fg} = heat of vaporization
k = thermal conductivity
L = tube length
m = operand in friction factor correlation[4]
n = number of fins in cross section

P = pumping power
p = pitch between ridges, absolute pressure [Eq. (3.3)]
Δp = pressure drop
$\dot{q}$ = heat transfer per unit area
R = universal gas constant
R_c = constant portion of thermal resistance
R_t = total thermal resistance
R_v = variable portion of thermal resistance
r = operand in friction factor correlation[4]
r_c = cavity radius
T = absolute temperature
ΔT_{sat} = saturation temperature minus wall temperature
W = mass flow rate
y = boundary-layer thickness

Dimensionless Variables

Nu = Nusselt number = hd/k
Pr = Prandtl number = $c_p \mu/k$
Re = Reynolds number = $\rho Vd/\mu$
y^+ = dimensionless boundary-layer thickness

Greek Letters

α = angle of spiral relative to tube axis
β = fin profile angle in Fig. 3.1, operand in heat-transfer correlation[4]
γ = operand in heat-transfer correlation[4]
δ = liquid film thickness
μ = dynamic viscosity
ν = kinematic viscosity
ρ = density
σ = surface tension
τ_o = shear stress = $f\rho V^2/2g_c$

Subscripts

e = enhanced
s = smooth or unenhanced

REFERENCES

Single-Phase Heat Transfer

1. E. N. Seider and C. E. Tate, "Heat Transfer and Pressure Drop of Liquids in Tubes," *Ind. Eng. Chem.*, **28,** 1429 (1936).
2. B. S. Petukhov, "Heat Transfer and Friction in Turbulent Pipe Flow with Variable Physical Properties," *Adv. Heat Transf.*, **8,** (1964).
3. T. C. Carnavos, "Cooling Air in Turbulent Flow with Internally Finned Tubes," *Heat Transf. Eng.*, **1** (2), 41 (1979).
4. J. G. Withers, "Tubeside Heat Transfer and Pressure Drop for Tubes Having Helical Internal Ridging with Turbulent/Transitional Flow of Single-Phase Fluid," *Heat Transf. Eng.*, **2** (1, 2), 48(Part 1), 43(Part 2) (1980).
5. F. J. Edwards and N. Sheriff, "International Developments in Heat Transfer," Conference on International Developments in Heat Transfer, ASME, 1961, pp. 415–425.
6. W. J. Marner and A. E. Bergles, "Augmentation of Highly Viscous Laminar Tubeside Heat Transfer by Means of a Twisted Tape Insert and an Internally Finned Tube," *Advances in Enhanced Heat Transfer,* ASME, 1985.
7. A. P. Watkinson, D. L. Miletti, and G. R. Kubanek, "Heat Transfer and Pressure Drop of Internally Finned Tubes in Laminar Oil Flow," *National Heat Transfer Conference,* ASME, San Francisco, 1975.
8. E. K. Kalinin and S. A. Yarkho, "Augmentation of Heat Transfer in Gases and Liquids Flowing in Tubes," *Heat Transf. Sov. Res.*, **4** (2), 79 (1972).
9. R. S. Van Rooyen and D. G. Kroger, "Laminar Flow Heat Transfer in Internally Finned Tubes with Twisted-Tape Inserts," *Sixth International Heat Transfer Conf.*, Toronto, 1978.
10. S. B. Uttarwar and M. Raja Rao, "Augmentation of Laminar Flow Heat Transfer in Tubes by Means of Wire Coil Inserts," *ASME J. Heat Transf.*, **107,** 930 (1985).
11. T. S. Ravigururajan, and A. E. Bergles, "General Correlations for Pressure Drop and Heat Transfer for Single-Phase Turbulent Flow in Internally Ribbed Tubes," *Conf. Augmentation of Heat Transfer in Energy Systems,* ASME, Miami Beach, 1985.
12. K. J. Bell, "Delaware Method for Shell Side Design," in *Heat Exchangers: Thermal-Hydraulic Fundamentals and Design,* S. Kakac, A. E. Bergles, F. Mayinger (eds.), Hemisphere, New York, 1981.
13. D. E. Briggs, D. L. Katz, and E. H. Young, "How to Design Finned-Tube Heat Exchangers," *Chem. Eng. Progr.*, **59** (11), 49 (1963).
14. C. Weierman, "Finned Tubes Can Lower Heat-Transfer Costs," *Oil Gas J.*, **73** (44), 64 (1975).
15. T. J. Rabas and P. W. Eckels, "Heat Transfer and Pressure Drop Performance of Segmented Extended Surface Tube Bundles," *Fifteenth National Heat Transfer Conf.*, San Francisco, 1975.
16. C. Weierman, "Correlations Ease the Selection of Finned Tubes," *Oil Gas J.*, **74** (36), 94 (1976).
17. R. L. Webb, "Air-Side Heat Transfer in Finned Tube Heat Exchangers," *Heat Transf. Eng.*, **1** (3), 33 (1980).
18. P. W. Eckels and T. J. Rabas, "Heat Transfer and Pressure Drop Performance of Finned Tube Bundles," *ASME J. Heat Transf.* **107,** 205 (1985).
19. J. G. Knudsen and D. L. Katz, "Heat Transfer and Pressure Drop in Annuli," *Chem. Eng. Progr.* **46,** 490 (1950).
20. L. Forgo, "Some Extra-High Capacity Heat Exchangers of Special Design," in *Heat Exchangers: Design and Theory Sourcebook,* N. H. Afgan and E. U. Schlunder (eds.), Hemisphere, New York, 1974.
21. W. Nakayama, "Enhancement of Heat Transfer," Seventh International Heat Transfer Conf., Munich, 1982.
22. E. M. Sparrow and A. Hajiloo, "Measurements of Heat Transfer and Pressure Drop for an Array of Staggered Plates Aligned Parallel to an Air Flow," *ASME J. Heat Transf.* **102,** 426 (1980).
23. N. Cur and E. M. Sparrow, "Experiments on Heat Transfer and Pressure Drop for a Pair of Colinear, Interrupted Plates Aligned with the Flow," *Int. J. Heat Mass Transf.* **21,** 1069 (1978).

Two-Phase Heat Transfer

24. W. Nusselt, "Die Oberflachenkondensation des Wasserdampfes," *Z. Ver. Deutsch. Ing.*, **60,** 541 (1916).
25. R. L. Webb, "The Use of Enhanced Surface Geometries in Condensers: An Overview," in *Power Condenser Heat Transfer Technology,* P. J. Marto and R. H. Nunn (eds.), Hemisphere, New York, 1981.
26. R. W. Murphy and J. W. Michel, "Enhancement of Refrigerant Condensation," *ASHRAE Trans.* **90** (Part 1B), 72 (1984).
27. T. M. Rudy and R. L. Webb, "Theoretical Model for Condensation on Horizontal Integral-fin Tubes," *Heat Transfer—Seattle* 1983, AIChE Symposium, Vol. 79, No. 225, AIChE, New York.
28. K. O. Beatty, Jr. and D. L. Katz, "Condensation of Vapors on Outside of Finned Tubes," *Chem. Eng. Progr.*, **44** (1), 55 (January 1948).
29. A. Thomas, J. J. Loreny, D. A. Hilles, D. T. Young, and N. F. Sather, "Performance Tests of the 1 MW Shell and Tube Exchangers for OTEC," *Proceedings 6th OTEC Conference, U.S. Dept. Energy,* June 19–22, 1979.

30. K. Takahashi, W. Nakayama, T. Senshu, and H. Yoshida, "Heat Transfer Analysis of Shell-and-Tube Condensers with Shell-Side Enhancement," *ASHRAE Trans.* **90** (Part IB), 60 (1984).
31. M. Luu and A. E. Bergles, "Enhancement of Horizontal In-Tube Condensation of R-113," *ASHRAE Trans.* **86** (Part 1), 293 (1980).
32. D. G. Thomas, "Prospects for Further Improvement in Enhanced Heat Transfer Surfaces," *Desalination,* **12,** 189 (1973).
33. P. Griffith and J. D. Wallis, "The Role of Surface Conditions in Nucleate Boiling," *Chem. Eng. Progr. Symp. Ser.*, **56** (49), 49 (1960).
34. J. Arshad and J. R. Thome, "Enhanced Boiling Surfaces—Heat Transfer Mechanism and Mixture Boiling," *ASME/JSME Joint Thermal Engineering Conference Proceedings, Honolulu,* Vol. 1, March 1983, pp. 191–197.
35. R. L. Webb, "The Evolution of Enhanced Surface Geometries for Nucleate Boiling," *Heat Transf. Eng.*, **2** (3, 4), 46 (January–June 1981).
36. R. L. Webb, "Heat Transfer Surface Having a High Boiling Heat Transfer Coefficient," U.S. Patent 3,696,861, October 10, 1972.
37. K. Fujie, W. Nakayama, H. Kuwahara, and K. Kakizaki, "Heat Transfer Wall for Boiling Liquids," U.S. Patent 4,060,125, November 29, 1977.
38. R. M. Milton, "Heat Exchange System," U.S. Patent 3,384,154, May 21, 1968.
39. K. E. Starner, and R. A. Cromis, "Energy Savings Using High Flux Evaporator Surface in Centrifugal Chillers," *ASHRAE J.*, **19,** 24 (December 1977).
40. D. F. Katz, J. E., Myers, E. H. Young, and G. Balekjian, "Boiling Outside Finned Tubes," *Petroleum Refiner,* **34** (2), 19 (February 1955).
41. R. J. Hummel, "Means for Increasing the Heat Transfer Coefficient between a Wall and Boiling Liquid," U.S. Patent 3,207,209, September 21, 1965.
42. R. K. Young and R. F. Hummel, "Improved Nucleate Boiling Heat Transfer," *Chem. Eng. Progr.*, **60** (7), 53 (1964).
43. K. Nakajima, "Boiling Heat Transfer Outside Horizontal Multitube Bundles," *Heat Transf. Jpn. Res.*, **7** (2), 1 (1978).
44. D. B. Robinson and D. L. Katz, "Effect of Vapor Agitation on Boiling Coefficients," *Chem. Engr. Progr.*, **47** (6), (June 1951).
45. S. Yilmaz, J. W. Palen, and J. Taborek, "Final Report, Evaluation of High Efficiency Heat Transfer Techniques," U.S. Department of Energy, DOE/ET/13328, Contract no. DE-AC03-78ET13328, December, 1981.
46. J. G. Lavin and E. H. Young, "Heat Transfer to Evaporating Refrigerants in Two-Phase Flow," *AIChE J.*, **11,** 1124 (November 1965).

Testing and Evaluation

47. E. E. Wilson, "Basis for Rational Design of Heat Transfer Apparatus," *Trans. Am. Soc. Mech. Eng.*, **37,** 47 (1915).
48. L. S. Leong and K. Cornwell, "Heat Transfer Coefficients in a Reboiler Bundle," *Chem. Eng.*, (343), 219 (1979).
49. A. E. Bergles, A. R. Blumenkrantz, and J. Taborek, "Performance Evaluation Criterion for Enhanced Heat Transfer Surfaces," Fifth International Heat Transfer Conference, 1974.
50. R. L. Webb, in *Heat Transfer Design and Data Book,* E. U. Schlunder (ed.), Hemisphere, New York, 1983.
51. K. E. Starner, "Effect of Fouling Factors on Heat Exchanger Design," *ASHRAE J.*, **18,** 39 (May 1976).

Chapter 4

Heat Pipes

WALTER B. BIENERT
Dynatherm Corporation

4.1 INTRODUCTION

Heat pipes are a special class of heat-transfer device available in many geometries, including pipes, plates, chambers, annuli, and various bent structures. In many ways they resemble solid conductors, but their effective conductivity can exceed that of the best metallic conductors by several orders of magnitude. The high conductance is achieved by evaporation and condensation of an internal, two-phase working fluid which is circulated by a built-in capillary pump. This heat-transport mechanism is very effective but also imposes several restrictions on the operation and application of heat pipes: (1) the capillary pump has only a finite capacity; hence heat pipes have a maximum heat-transport capability beyond which they cease to function; (2) heat pipes are sensitive to orientation; (3) a given heat pipe is limited to a finite temperature range for optimum operation; and (4) since the heat pipe is partially filled with a fluid, it must not be punctured, and there are limits regarding exposure to extreme temperatures or pressures.

Most applications of heat pipes utilize their high effective conductance. Examples are electronic packages where high local heat rates must be safely dissipated or spacecraft radiators where other cooling mechanisms are not available.

Heat pipes are nearly isothermal over their entire surface; this feature is often utilized where uniform temperatures are important (in industrial processes and calibration systems). In commercial heat exchangers, high conductance and redundancy are desired features: instead of using a single fluid loop, banks of heat pipes, each a self-contained heat-transfer device, provide the redundancy. Finally, heat pipes can be used to control temperatures and heat flow rates. These special variable-conductance heat pipes (VCHPs) are used frequently in spacecraft temperature control.

This chapter discusses the fundamentals of heat pipes in sufficient detail to provide an understanding of their operation, recognize their limitations, evaluate their function as heat-transfer devices, and perform a preliminary design analysis. For more comprehensive treatment of the subject, the reader is referred to Refs. 1 to 5.

4.2 PRINCIPLE OF OPERATION

The principle of operation is shown in Fig. 4.1. The hermetically sealed heat-pipe envelope, often a tube or a pipe, is evacuated and then partially filled with a suitable fluid. In the absence of heat sources or heat sinks, the liquid and vapor phases of the fluid are in thermal equilibrium. When heat is applied to any section of the heat pipe (evaporator), vaporization occurs locally, raising the pressure in the vapor phase. The increased pressure causes a flow of vapor to the unheated, cooler sections of the pipe where the excess vapor condenses (condenser). The cycle is completed by returning the liquid to the evaporator. Heat is transported in the form of latent heat of the working fluid. In practical heat pipes, the evaporator and the condenser are often separated by a section with no heat transfer across the external boundary (adiabatic section).

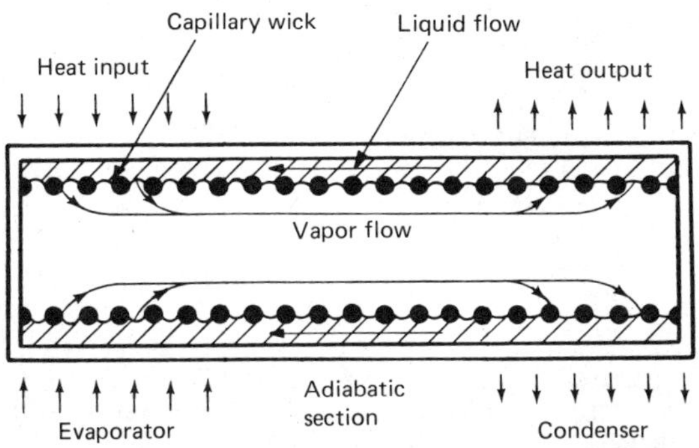

Fig. 4.1 Schematic of heat pipe.

Most heat pipes utilize capillary pumping to compensate for the hydrodynamic losses in the liquid and the vapor. In some applications, however, gravity or centrifugal forces are substituted for the capillary pump, or at least contribute to the pumping.

Capillary pumping is a surface effect; it requires the presence of a structure with very small pores, typically 10^{-3} to 10^{-2} in (0.025 to 0.25 mm). Such a structure is commonly referred to as a "wick." When a wick is saturated with a fluid, the interface between the liquid and vapor is curved and can sustain a finite pressure difference. Fluids that wet the wicking material (all heat-pipe fluids do) have an interface that is concave on the vapor side. In this case, the pressure in the vapor exceeds that of the liquid by an amount Δp_c:

$$\Delta p_c = \frac{2\sigma}{r} \tag{4.1}$$

where σ represents the surface tension and r the radius of curvature of the interface. The radius may vary from nearly infinite (flooded wick) to a minimum value r_c that is equal to the radius of the pores of the wick. Thus the maximum capillary pumping is given by

$$\Delta p_{c,\max} = \frac{2\sigma}{r_c} \tag{4.2}$$

When a thermal load is applied to the heat pipe, the radius of curvature varies along its length, the smallest radius occurring at the evaporator.

4.3 HEAT-TRANSPORT CAPABILITY

The amount of heat that can be transported is determined by a balance between the available capillary pumping and the hydrodynamic losses in the liquid and, to a lesser extent, in the vapor. The following equations apply to a "wicked" heat pipe, in the absence of gravity and with laminar flow in liquid and vapor.

Single Evaporator and Condenser at Opposite Ends

Heat is applied uniformly along the evaporator and removed uniformly along the condenser. The maximum axial heat flow rate is given by

$$Q_{\max} = \frac{2K_w A_w N_l F_l}{r_c L_{\text{eff}}} \tag{4.3}$$

where K_w = permeability of the wick
A_w = its cross-sectional area
r_c = radius of its pores

The parameter N_l combines several properties of the liquid phase of the working fluid:

$$N_l = \frac{\sigma \rho_l \lambda}{\mu_l} \tag{4.4}$$

where σ = surface tension
λ = latent heat
ρ_l = density of liquid
μ_l = dynamic viscosity of liquid

The ratio F_l of the viscous pressure loss in the liquid to the total pressure loss is

$$F_l = \frac{\Delta p_l}{\Delta p_l + \Delta p_v} = \left(1 + \frac{32 K_w A_w \nu_v}{D_{h,v}^2 A_v \nu_l}\right)^{-1} \tag{4.5}$$

where ν_v, ν_l = kinematic viscosities of vapor and liquid
$D_{h,v}$ = hydraulic diameter of the vapor channel
A_v = cross-sectional area of channel

The effective length L_{eff} is defined by

$$L_{\text{eff}} = \tfrac{1}{2}L_e + L_a + \tfrac{1}{2}L_c \tag{4.6}$$

where L_e, L_a, and L_c are the lengths of evaporator, adiabatic section, and condenser, respectively.

QL Approach

Heat pipes with nonuniform heat input and output or with multiple evaporators and condensers are best analyzed using the QL approach. The QL requirement is defined by

$$(QL)_{req} = \int_0^L Q(x)\, dx \tag{4.7}$$

where $Q(x)$ is the local, axial heat flow rate; it can be calculated from the specified distribution of heat sources and heat sinks.

For a heat pipe with uniform heat input and output along a single evaporator and condenser, the QL requirement becomes

$$(QL)_{req} = Q_{max} L_{eff} \tag{4.8}$$

The QL requirement must be less than the QL capability. The latter is defined by

$$(QL)_{cap} = \frac{2K_w A_w N_l F_l}{r_c} \tag{4.9}$$

The QL capability is independent of heat-pipe length and the lengths of the evaporator and the condenser. It is completely determined by the wick, the cross section of the pipe, and the working fluid. Heat pipes are frequently specified in terms of QL capability. The combination of Eqs. (4.8) and (4.9) implies that, for a given heat-pipe design, the product of Q_{max} and L_{eff} is fixed; for example, doubling the effective length results in halving the maximum heat load that the pipe can carry.

Segmented Wicks

It is sometimes advantageous to segment the wick along the length of the pipe in order to achieve a higher transport capability. Secondary or distributor wicks at the evaporator are another form of segmentation. In the general case of N segments, the transport capability of the Nth segment is given by

$$Q_{max} = \frac{2K_w A_w N_l F_l}{L_{eff}} \left(\frac{1}{r_b} - \frac{1}{r_a} \right) \tag{4.10}$$

where K_w and A_w are the wick parameters for the Nth segment L_{eff} is its effective length which is calculated from Eq. (4.6) and r_a and r_b are the capillary radii at the beginning and end of each segment. Continuity of the pressure in the liquid requires that

$$r_b \text{ of } N\text{th segment} = r_a \text{ of } (N + 1)\text{th segment}$$

Segmented wicks are best evaluated by iteration: a trial value for Q_{max} is selected, and the required capillary radii are calculated at the beginning and end of each segment. The process is repeated for increasing Q_{max} until one of the capillary radii becomes smaller than the minimum radius which that particular wick can support.

Application of Heat-Transfer Models

The preceding models give adequate results for the majority of heat pipes; however, they break down if the vapor losses are significant. The parameter F_l, which is a measure of the relative importance of vapor losses, should always be close to unity; it should never be less than 0.5. If the vapor flow is turbulent (the liquid is always laminar) or if inertial effects cannot be ignored, different models must be applied. Vapor effects are usually important in low-pressure heat pipes, in particular those using liquid metals as working fluids. Vapor-dominated heat pipes are treated in Refs. 1 and 2.

4.4 EFFECTS OF GRAVITY

Depending on the orientation of the heat pipe, gravity either aids or impedes the flow of the liquid (it has virtually no effect on the vapor flow). If the evaporator is located above the condenser, the heat-transport capability is reduced; likewise, elevation of the condenser increases it. The influence of gravity is accounted for by multiplying the transport capability [Eq. (4.3), (4.9), or (4.10)] by a factor $(1 + \eta)$, where η is defined as

$$\eta = \frac{r_c(h - D \cos \alpha)}{2H} \tag{4.11}$$

where r_c is the pore radius of the wick. The parameter H is the wicking height factor which is a property of the working fluid. It is defined as

$$H = \frac{\sigma}{\rho_l g} \tag{4.12}$$

By convention, the elevation (tilt) of the heat-pipe axis h is considered negative if the evaporator is located above the condenser; D is the interior diameter of the pipe and α the tilt angle (always greater than zero).

Figure 4.2 shows the effects of elevation on the heat-transport capability. At a tilt equal to the "static" wicking height, the wick can maintain itself filled but cannot support any heat transport. Graphs such as Fig. 4.2 are used to extrapolate the performance of heat pipes to zero gravity. Actual test results at zero tilt often yield misleading data because excess fluid in the pipe can create a free puddle which bypasses the wick. This results in an apparent higher capability than can be expected in zero gravity where puddles are ineffective. Use of Eq. (4.11) for positive elevations (evaporator below condenser) is not recommended since the results will always be affected by puddle flow.

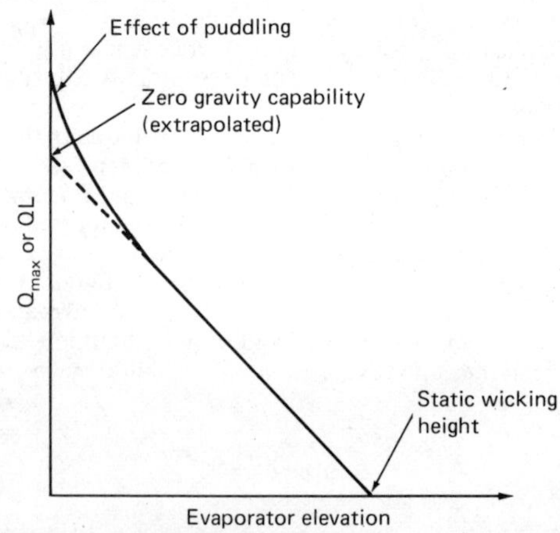

Fig. 4.2 Heat-transport capability versus tilt.

4.5 WICKS

The primary function of the wick is to pump the liquid from the condenser to the evaporator. Capillary pumping is inversely proportional to the size of the pores in the wick; typical pores range from 10^{-3} to 10^{-1} in (0.025 to 2.5 mm). In comparison to gravity, capillary forces are relatively weak. For example, a 100-mesh screen with a capillary radius of 0.005 in (0.127 mm) has the following lifting capability:

Fluid	Temperature, °F	Wicking height, in
Nitrogen	−321	0.68
Ammonia	32	2.54
Water	212	3.84
Sodium	1500	10.00

The second function of the wick is to provide a flow path for the liquid. The wick parameter that determines the flow losses is its permeability K_w. Two types of wicks are in use: homogeneous and composite. In the former type, the permeability is approximately proportional to the square of the capillary radius. This dependence of the permeability on the capillary radius requires a design compromise: strong pumping (small pores) is inherent to large viscous losses. In composites wicks, the permeability and the pumping radius are unrelated. The arterial wick is an example of a composite wick. The advantage of independent optimization of capillary pumping and permeability is partially offset by the fact that composite wicks need to be primed in order to develop their full capability. Figure 4.3 shows practical configurations of both wick types.

Homogeneous Wicks

The most widely used homogeneous wicks are porous structures fabricated from wire mesh screen and axial grooves in the wall of the heat pipe.

Screens Fabrication techniques for screen wicks include spot welding, sintering, and press fitting. The capillary pumping radius is given by

$$r = \frac{w + d}{2} = \frac{1}{2M} \tag{4.13}$$

where d = wire diameter
w = linear opening of mesh
M = mesh number, in^{-1}

The permeabilities of common screen wicks are given in Table 4.1. The values in the table are calculated ones and are reliable to within only plus or minus 25 percent since manufacturing techniques can affect the actual permeabilities substantially.

Table 4.1. Permeability of Common Screens

Mesh number, in^{-1}	Permeability, ft^2
20	8.4×10^{-8}
30	2.9×10^{-8}
50	8.3×10^{-9}
100	2.1×10^{-9}
200	5.0×10^{-10}

Capillary Tubes The pumping radius of a circular, capil-

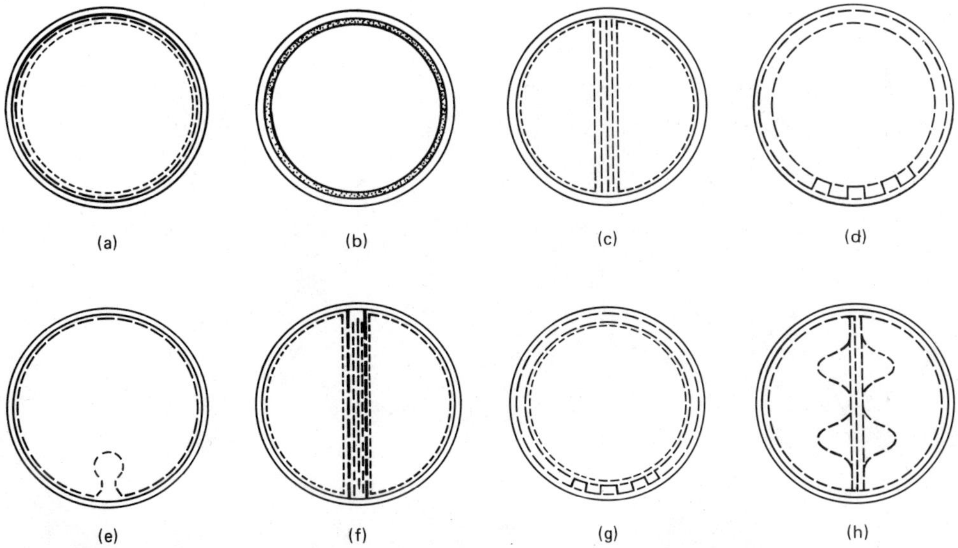

Fig. 4.3 Typical wick configurations. (*a*) Circumferential screen; (*b*) sintered powder wick; (*c*) slab with circumferential distributor wick; (*d*) axial grooves; (*e*) pedestal artery; (*f*) composite slab (coarse screen on inside, fine screen on outside); (*g*) screen-covered grooves; (*h*) slab with auxiliary arteries.

lary tube is one-half its diameter ($r_c = D/2$), and the permeability K_w is given by $K_w = D^2/32$.

Axial Grooves Axial grooves are fabricated by extruding or forging the wall material; they are readily available only in aluminum and copper. Typical grooved heat pipes are 0.25 to 1 in (6.4 to 25 mm) in diameter and contain 15 to 50 axial grooves. The grooves are 0.04 to 0.06 in deep (1 to 1.5 mm), and the groove opening ranges from 0.01 to 0.03 in (0.25 to 0.75 mm).

The pumping radius of an axial groove is equal to the groove opening ($r_c = w$). The permeability is given by

$$K_w = \frac{\beta D_h^2}{32} \tag{4.14}$$

where D_h is the hydraulic diameter of a groove ($D_h = 4 \times$ area/wetted perimeter) and β is a shape factor, which ranges from ⅔ to 1. The equations for the heat-transport capability in Sec. 4.3 must be modified for axially grooved heat pipes: the change in capillary radius along the length of the pipe causes the liquid area and the permeability to vary. In order to account for this variation, an integration must be performed along the length of the heat pipe. The analytical method is described in Ref. 6. As a first approximation, Eq. (4.3) or (4.9) may be used, provided average wick properties are employed: at the evaporator, the capillary radius is assumed to be at its minimum value, and the area and permeability are calculated accordingly; at the condenser, the grooves are assumed to be completely filled. The transport capability of several axially grooved, aluminum ammonia heat pipes is given in Fig. 4.4.

Composite Wicks

Arteries The most widely used composite wick consists of one or several arteries that are formed from wire-mesh screen. Typical arteries have diameters of ⅛ in (3 mm) and are fabricated from 200-mesh screen. If the artery is fully primed (completely filled), the screen from which the artery is fabricated provides the capillary pumping. However, in the unprimed condition (partially filled or containing vapor or gas bubbles), the capillary pumping is reduced to that of the diameter of the artery itself. The two cases are summarized below:

	Wicking parameters of sample artery, ⅛-in (3-mm) diameter, 200-mesh screen	
	Primed	Unprimed
Pumping radius, ft	2.1×10^{-4}	5.2×10^{-3}
Permeability, ft²	3.4×10^{-6}	3.4×10^{-6}
K_w/r_c, ft	1.6×10^{-2}	6.5×10^{-4}

The transport capability of the arterial wick is proportional to K_w/r_c. If the artery is unprimed, its capability is reduced by a factor of 25 (for the preceding sample case). The need for priming makes arterial heat pipes difficult to use and also somewhat unreliable: priming must be accomplished under reduced load, and arteries have a tendency to deprime if traces of noncondensable gas are present in the working fluid.

Composite Porous Wicks These wicks are fabricated from porous materials (e.g., screens) of different pore sizes. Examples are multiple layers of screens in which the outer layer (the one exposed to the vapor) has a finer mesh size than do the inner ones. For these wicks, the capillary pumping radius is determined by the finer, outer screen. The product

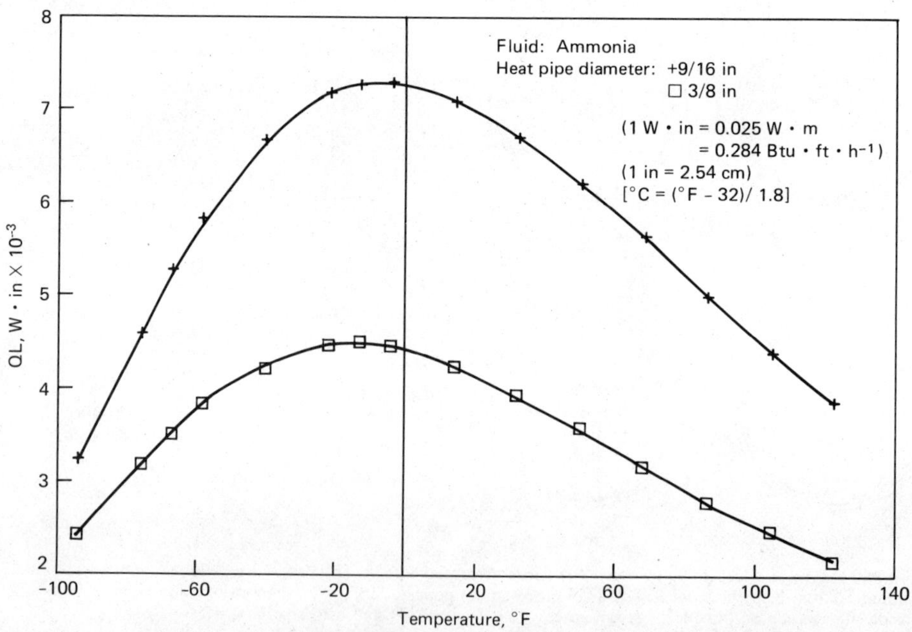

Fig. 4.4 Heat-transport capability QL of axially grooved heat pipes.

of permeability and wick area is the sum of the products of the individual wicks. Composite porous wicks are not as sensitive to depriming as arterial wicks are since they have multiple interconnected flow paths.

Secondary Wicks

Secondary wicks are normally used in the evaporator and, to a lesser extent, in the condenser. The purpose of the secondary wick in the evaporator is to transport the liquid from the main axial wick (e.g., an artery or a slab; see Fig. 4.3) to the heat input surfaces. Examples of such secondary wicks are bridges and circumferential screens and grooves. The performance can be evaluated using the model for segmented wicks [Eq. (4.10)]. Care must be exercised in selecting the proper values for the wick area A_w and the effective length L_{eff}.

4.6 FLUID SELECTION

Several criteria must be considered in selecting a working fluid for a heat pipe:

Temperature	Vapor pressure
Transport properties	Compatibility
Handling	

Temperature

A heat pipe cannot function below the freezing point or above the critical temperature of its working fluid. As a first selection criterion for a fluid, the specified operating temperature range must fall between these two thermodynamic limits. For some fluids (e.g., water) it is also desirable to avoid exposure to temperatures below freezing points for that fluid.

Vapor Pressure

As a rule of thumb, the vapor pressure at the lowest operating temperature should be no less than 0.1 atm, and at the highest operating temperature, it should not exceed 10 atm. Below 0.1 atm, vapor effects become a significant contributor to the hydrodynamic losses. Heat pipes can be operated successfully in this regime, but the simple models described here break down. Operation below 0.01 atm is generally unfeasible because molecular flow is encountered.

Vapor pressures above 10 atm have no detrimental effect on the function of the heat pipe; however, safe containment of the fluid by the heat pipe envelope becomes a problem.

Transport Properties

The physical properties of the working fluid that affect the heat-transport capability are as follows:

Property	Units
Surface tension	lb_f/ft (N/m)
Heat of vaporization	Btu/lb_m (J/kg)
Density (liquid and vapor)	lb_m/ft^3 (kg/m^3)
Dynamic viscosity (liquid and vapor)	$lb_f \cdot s \cdot ft^{-2}$ ($N \cdot s \cdot m^{-2}$)

The hydrodynamic models given in Secs. 4.3 and 4.4 utilize three parameters that are groups of the preceding fundamental properties:

Grouped Parameters	Units
Liquid transport factor N_l	$Btu \cdot h^{-1} \cdot ft^{-2}$ ($W \cdot m^{-2}$)
Wicking height factor H	ft^2 (m^2)
Ratio of kinematic viscosities ν_v/ν_l	None

Table 4.2 gives the three parameters and the vapor pressure for six common heat-pipe working fluids.

Compatibility

A major factor in the selection of a working fluid is its compatibility with other materials in the heat-pipe system. Critical to the performance of a heat pipe is the noncondensable

Table 4.2. Fluid Properties

T, K	P, N/m²	N_l, W/m²	H, m²	ν_v/ν_l
		Nitrogen		
65	1.74E + 04	7.68E + 09	1.38E − 06	16.1
70	3.86E + 04	8.28E + 09	1.28E − 06	10.3
75	7.61E + 04	8.66E + 09	1.17E − 06	6.9
80	1.37E + 05	8.68E + 09	1.06E − 06	4.9
85	2.29E + 05	8.29E + 09	9.51E − 07	3.7
90	3.60E + 05	7.55E + 09	8.46E − 07	2.9
95	5.40E + 05	6.59E + 09	7.42E − 07	2.3
100	7.77E + 05	5.50E + 09	6.36E − 07	1.8
105	1.08E + 06	4.34E + 09	5.28E − 07	1.5
110	1.47E + 06	3.17E + 09	4.15E − 07	1.2
115	1.94E + 06	2.00E + 09	2.99E − 07	1.0
120	2.51E + 06	9.15E + 08	1.74E − 07	0.9
		Ammonia		
200	8.67E + 03	1.09E + 11	5.97E − 06	150.4
220	3.39E + 04	1.07E + 11	5.62E − 06	53.2
240	1.03E + 05	1.12E + 11	5.06E − 06	24.8
260	2.56E + 05	1.16E + 11	4.49E − 06	14.1
280	5.51E + 05	1.13E + 11	3.94E − 06	9.1
300	1.06E + 06	1.00E + 11	3.39E − 06	6.1
320	1.87E + 06	8.07E + 10	2.79E − 06	4.2
340	3.08E + 06	5.87E + 10	2.14E − 06	3.0
360	4.80E + 06	3.71E + 10	1.48E − 06	2.2
380	7.15E + 06	1.67E + 10	8.46E − 07	1.7
400	1.03E + 07	1.45E + 09	2.11E − 07	1.2
405	1.13E + 07	2.20E + 07	8.06E − 09	1.0
		Freon-11		
300	1.14E + 05	1.17E + 10	1.25E − 06	6.0
310	1.59E + 05	1.13E + 10	1.18E − 06	4.9
320	2.17E + 05	1.09E + 10	1.10E − 06	4.1
330	2.89E + 05	1.04E + 10	1.03E − 06	3.4
340	3.79E + 05	9.78E + 09	9.52E − 07	2.9
350	4.87E + 05	9.06E + 09	8.77E − 07	2.5
360	6.18E + 05	8.25E + 09	8.03E − 07	2.1
370	7.74E + 05	7.37E + 09	7.32E − 07	1.8
380	9.57E + 05	6.46E + 09	6.64E − 07	1.6
390	1.17E + 06	5.56E + 09	5.95E − 07	1.4
400	1.42E + 06	4.70E + 09	5.23E − 07	1.2
410	1.70E + 06	3.89E + 09	4.37E − 07	1.1

Table 4.2. Fluid Properties (*Continued*)

T, K	P, N/m²	N_l, W/m²	H, m²	ν_v/ν_l
		Methanol		
280	5.86E + 03	3.16E + 10	3.03E − 06	125.7
300	1.77E + 04	3.82E + 10	2.87E − 06	61.2
320	4.58E + 04	4.46E + 10	2.70E − 06	34.0
340	1.06E + 06	4.95E + 10	2.54E − 06	20.4
360	2.21E + 05	5.25E + 10	2.37E − 06	13.0
380	4.27E + 05	5.35E + 10	2.17E − 06	8.8
400	7.66E + 05	5.26E + 10	1.95E − 06	6.4
420	1.28E + 06	4.94E + 10	1.68E − 06	4.9
440	2.03E + 06	4.27E + 10	1.38E − 06	3.9
460	3.05E + 06	3.19E + 10	1.06E + 06	2.9
480	4.43E + 06	1.68E + 10	7.06E − 07	2.0
500	6.42E + 06	4.16E + 09	2.95E − 07	1.1
		Water		
280	9.94E + 02	1.31E + 11	7.61E − 06	752.8
300	3.52E + 03	2.03E + 11	7.36E − 06	427.6
320	1.05E + 04	2.81E + 11	7.08E − 06	224.6
340	2.71E + 04	3.57E + 11	6.77E − 06	145.2
360	6.21E + 04	4.23E + 11	6.46E − 06	90.5
380	1.29E + 05	4.73E + 11	6.14E − 06	59.3
400	2.45E + 05	5.06E + 11	5.82E − 06	40.7
420	4.36E + 05	5.20E + 11	5.48E − 06	29.0
440	7.32E + 05	5.16E + 11	5.12E − 06	21.2
460	1.17E + 06	4.95E + 11	4.74E − 06	15.8
480	1.78E + 06	4.57E + 11	4.33E − 06	11.9
500	2.64E + 06	4.06E + 11	3.89E − 06	9.0
		Sodium		
800	9.46E + 02	2.24E + 12	1.82E − 05	19520
850	2.28E + 03	2.25E + 12	1.79E − 05	9228
900	5.01E + 03	2.26E + 12	1.76E − 05	4746
950	1.02E + 04	2.25E + 12	1.72E − 05	2615
1000	1.95E + 04	2.24E + 12	1.69E − 05	1527
1050	3.52E + 04	2.22E + 12	1.66E − 05	939
1100	6.04E + 04	2.19E + 12	1.62E − 05	604
1150	9.86E + 04	2.15E + 12	1.58E − 05	406
1200	1.54E + 05	2.09E + 12	1.55E − 05	283
1250	2.29E + 05	2.02E + 12	1.51E − 05	205
1300	3.27E + 05	1.93E + 12	1.47E − 05	153
1350	4.50E + 05	1.84E + 12	1.43E − 05	118

NOTES: T = temperature (K), T(°F) = 1.8T(K) − 460; P = vapor pressure (N/m²), multiply by 1.45 × 10^{-4} to obtain lb/in² absolute; N_l = liquid transport factor (W/m²), multiply by 0.3172 to obtain Btu · h^{-1} · ft^{-2}; H = wicking height factor (m²), multiply by 10.76 to obtain ft²; ν_v/ν_l = ratio of kinematic viscosities.

gas that may be generated because of chemical reactions. This gas will collect in the condensing regions and cause condenser blockage. The length of this "gas slug" is inversely proportional to the vapor pressure and thus strongly dependent on the operating temperature. It is common practice to perform gas tests well below the normal operating temperature; this procedure accentuates the length of the slug and facilitates the measurement. For example, aluminum-ammonia heat pipes intended for use at 77°F (25°C) are normally tested at −40°F (−40°C). Gas generation is usually strongest during the first 10 to 30 days of operation. If absence of gas is very important, an initial burn-in period lasting several weeks is recommended. At the end of that period, the gas should be removed and only then should the heat pipe be sealed. In high-temperature heat pipes, corrosion and erosion of the container and wick can also result in damages. Solid precipitates can accumulate in the evaporator. This leads to increased resistance to fluid flow which results in lowering of the heat transport capability. Table 4.3 lists the results of compatibility tests of various material-fluid combinations.

Table 4.3. Compatibility Data of Various Fluids and Materials

	Aluminum	Stainless steel	Carbon steel	Copper	Nickel
Acetone	C	C	. . .	C	C
Ammonia	C	C	C	I	C
Dowtherm-A	. . .	C	. . .	C	
Freon-11	C	C			
Lithium	. . .	I	. . .	. . .	I
Mercury	. . .	I	C		
Methanol	I	C	C	C	
Potassium	. . .	C	. . .	. . .	C
Sodium	. . .	C			C
Water	I	I	I	C	C

NOTES: C = compatible, I = incompatible, blank = no data.

Handling

Practical considerations may exclude some fluids from particular applications in spite of otherwise favorable properties. A brief discussion of these considerations follows.

Flammability Many fluids are flammable. Although heat pipes are hermetically sealed, flammability may be a concern if large numbers of pipes are installed (e.g., heat-pipe banks in heat exchangers) or if the pipes are used in very confined spaces (e.g., manned spacecraft).

Toxicity The same comments apply as for flammability. In addition, processing during fabrication may pose dangers to personnel. Very few fluids fall into this category.

Stability Some fluids decompose at elevated temperature without apparent corrosion. An example is Dowtherm A, which would otherwise be a useful fluid in the 500 to 750°F (260 to 400°C) range, where few other fluids are available. The same applies to the glycols which cover the same temperature range.

Danger Liquid metals are usually considered dangerous materials. One of these is mercury, which is not flammable but has an extremely toxic vapor. Mercury vapor could escape unnoticed from a leaking heat pipe if it is operated above the atmospheric boiling point [674°F (357°C)]. Alkali metals are widely used in heat pipes because of their excellent transport and heat-transfer properties. They are the only choice above 900°F (500°C). Because of the danger involved, handling of these metals should be left to an experienced heat-pipe manufacturer.

Cost The cost of the working fluid is normally a negligible contributor to the overall cost of a heat pipe.

4.7 HEAT-PIPE CONDUCTANCE

The conductance, or its inverse, the resistance, is the most important parameter to the design engineer who wants to incorporate a heat pipe into a heat-transfer system. Unlike the heat-transport capability which defines the limit of operation, the overall conductance is the one parameter that is used in heat-transfer calculations. The conductance of a heat pipe is essentially independent of the length; it is completely defined by the geometries of the evaporator and the condenser.

The complete thermal model of a heat pipe is shown in Fig. 4.5. Since the model involves a series of individual resistances, it is more convenient to evaluate the overall resistance than the conductance. The overall resistance is given by

$$R = R_{\text{ext},e} + R_{\text{wall},e} + R_{\text{wick},e} + R_{\text{evap}} + R_v + R_{\text{cond}} + R_{\text{wick},c} + R_{\text{wall},c} + R_{\text{ext},c} \quad (4.15)$$

The first four terms in Eq. (4.15) are resistances associated with the evaporator and the last four terms, with the condenser. The external resistances $R_{\text{ext},e}$ and $R_{\text{ext},c}$ are determined by the installation and are not part of the heat pipe; however, these resistances are often significant and must be included in the overall network. Resistances R_{evap} and R_{cond} are thermal resistances associated with the kinetics of evaporation and condensation: they are very small and may be neglected. The vapor resistance R_v is negligible for all low-, ambient-, and moderate-temperature heat pipes; it is the only resistance that is length-dependent.

A simplified resistance model which excludes R_v, R_{evap}, R_{cond} and the external resistances is sufficient for most analyses. It consists of only four terms as given by

$$R = R_{\text{wall},e} + R_{\text{wick},e} + R_{\text{wick},c} + R_{\text{wall},c} \quad (4.16)$$

In terms of conductances, this model can be written as follows:

$$\frac{1}{G} = \frac{1}{G_{\text{wall},e}} + \frac{1}{G_{\text{wick},e}} + \frac{1}{G_{\text{wick},c}} + \frac{1}{G_{\text{wall},c}} \quad (4.17)$$

The wall conductances are given by the standard equation for a cylinder:

$$G_{\text{wall}} = \frac{2\pi\kappa L}{\ln (D_{\text{ext}}/D_{\text{int}})} \quad (4.18)$$

where κ = thermal conductivity of the wall material
L = length of evaporator or condenser
$D_{\text{ext}}, D_{\text{int}}$ = external and internal pipe diameters

Most heat pipes have sufficiently thin walls to permit the use of the following simple equation:

$$G_{\text{wall}} = Ah_{\text{wall}} \quad (4.19)$$

where A is the evaporator or condenser area and h_{wall} is an equivalent heat transfer coefficient of the wall. It is given by

$$h_{\text{wall}} = \frac{\kappa}{t} \quad (4.20)$$

where t is the wall thickness and κ its thermal conductivity. The error of using Eqs. (4.19) and (4.20) is less than 10% for wall-to-diameter ratios of up to 0.1.

The wick conductances G_{wick} are defined by

$$G_{\text{wick}} = Ah_{\text{wick}} \quad (4.21)$$

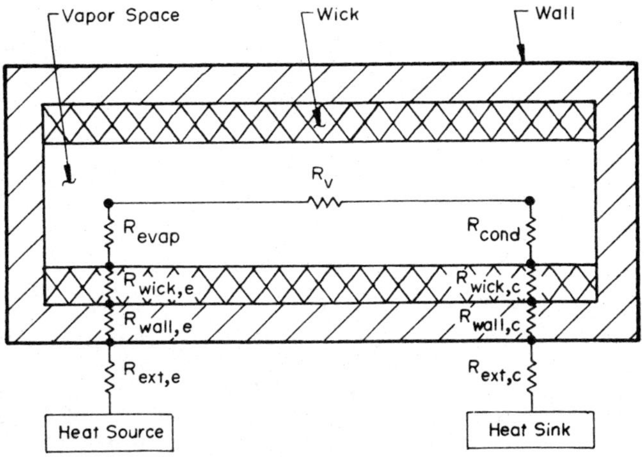

Fig. 4.5 Thermal model of a heat pipe.

where A is the evaporator or condenser area and h_{wick} is the heat-transfer coefficient associated with the wick. In most heat pipes, the resistance associated with the wick predominates among all other resistances. Existing data, while extensive, are often unreliable. Table 4.4 gives recommended values for several wicks and fluids that are commonly used in low- and medium-temperature heat pipes. These values apply to the evaporator; the equivalent condenser coefficients are usually about 25 percent higher.

Nonuniform Heat Input

In many applications of the heat pipe, heat is applied or removed nonuniformly. Examples are one-sided heat input/output, very short heating/cooling zones, and localized two-dimensional heat sources. In all the above cases, evaporation or condensation occurs not only under the physical area of the heat source or sink but also in adjacent areas which are conductively coupled to the source or sink. Design equations for three common cases are given below.

One-Sided Heat Input If heat is applied or removed over one-half the circumference of a tubular heat pipe, the con-

Table 4.4. Evaporator Heat-Transfer Coefficients

Type of wick	Fluid	Heat-transfer coefficient, Btu/h · °F · ft²
Single layer of sintered-copper screen	Water	2500
	Alcohols	1500–2000
	Freons	800
Multilayer of stainless-steel screen [1/16 in thick] spot-welded or press-fit	Water	1000
	Alcohols	500
	Freons	200
Grooves	Water	1800
	Ammonia	1500–2000
	Methane	200–300
	Freons	100–200

CONVERSION FACTOR: 1 Btu/h · °F · ft² = 5.674 W/m² · K.

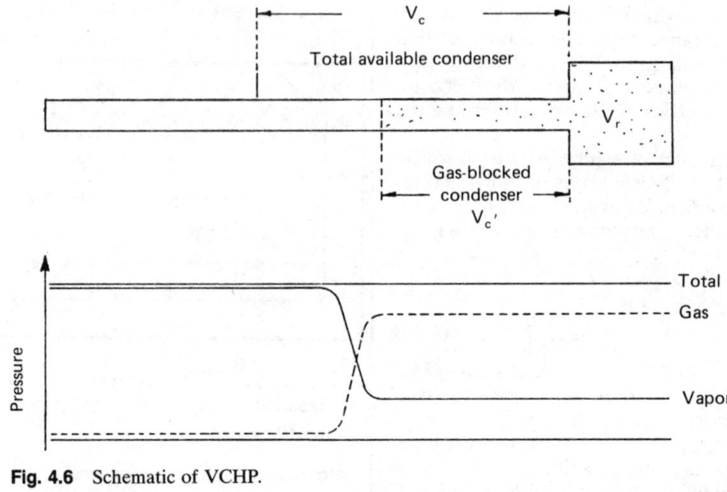

Fig. 4.6 Schematic of VCHP.

tribution of the other half is accounted for by multiplying the actual input/output area by a factor F:

$$F = 1 + \frac{\tanh m}{m} \tag{4.22}$$

with

$$m = \frac{\pi}{4}\sqrt{\frac{hD^2}{\kappa t}} \tag{4.23}$$

where h = heat-transfer coefficient of the wick
D = pipe diameter
κ = wall conductivity
t = wall thickness

Short Evaporators or Condensers Assume that heat is applied or removed over a very short length l of the heat pipe. The ratio of effective-to-actual length is given by

$$R = 1 + 2\sqrt{\frac{\kappa t}{hl^2}} \tag{4.24}$$

where l is the actual length of the heat input/output zone; κ, t, and h were defined with Eq. (4.23).

Localized Two-Dimensional Heat Sources Small, circular heat input zones are sometimes encountered with flat heat pipes. For this case, the ratio of effective-to-actual heat-source diameter can be approximated by

$$R = 1 + 1.97z - 0.10z^2 \tag{4.25}$$

where $z = \kappa t/hd^2$ and d is the actual diameter of the heat source; κ, t, and h were defined previously. Equation (4.25) is valid for $z < 2.5$.

4.8 VARIABLE-CONDUCTANCE HEAT PIPES

The primary application of variable-conductance heat pipes (VCHPs) is to control the temperature of a heat source when the heat load varies or the temperature of the heat sink is not constant. Several techniques are available to achieve variable conductance; the most common one involves the addition of a noncondensing gas to the working fluid. The theory of VCHPs is relatively complex.

Figure 4.6 shows the schematic of a gas-loaded VCHP. A fixed amount of gas has been added to the working fluid and a reservoir is located at the end of the condenser. The vapor sweeps the noncondensing gas toward the condenser. Some of the gas is trapped in the reservoir, and the remainder blocks part of the condenser. The location of the interface between vapor and gas is strongly affected by the vapor pressure of the working fluid. For instance, if the vapor pressure increases, the noncondensable gas is compressed and the condenser blockage reduced: this is equivalent to increasing heat-pipe conductance which, in turn, tends to lower the vapor pressure. The result is a nearly stable heat-pipe temperature.

A simple VCHP provides adequate control, provided the sink temperature does not vary much. For example, an ammonia VCHP can maintain the vapor temperature at 86 ± 3°F (30 ± 1.5°C) while the heat load varies from 0 to 100 percent, if the sink is held constant at 32°F (0°C). However, if the sink temperature varies over a wide range, other, more sophisticated VCHPs must be employed. These are discussed in detail in Refs. 3 and 4.

REFERENCES

1. P. Dunn and D. A. Reay, *Heat Pipes*, Pergamon, Oxford, U.K., 1976.
2. S. Q. Chi, *Heat Pipe Theory and Practice*, McGraw-Hill, New York, 1976.
3. B. D. Marcus, *Theory and Design of Variable Conductance Heat Pipes*, NASA CR 2018, TRW Systems Group, Redondo Beach, Calif. 1972.
4. E. A. Skrabek, *Heat Pipe Design Handbook*, Dynatherm Corporation, Cockeysville, Md., 1972, Contract NAS9-11927, NTIS no. N74-22569.
5. P. J. Brennan and E. J. Kroliczek, *Heat Pipe Design Handbook*, B&K Engineering, Inc., Towson, Md., 1977, Contract NAS5-23406.
6. H. Jen, *Summary Report for Axially Grooved Heat Pipe Study*, B&K Engineering, Inc., Towson, Md., 1977, Contract NAS5-22562.

Chapter 5

Air- and Liquid-Cooled Heat Sinks

VICTOR LORENZETTI

EG&G Wakefield Engineering, Inc.

5.1 INTRODUCTION

Electrically powered components generally produce more heat than they can dissipate unassisted. The main function of a heat sink is to protect the electrical device from the heat it produces as a by-product of normal operation. If not removed, this heat will cause the device to exceed its safe operating temperature. In this circumstance, device performance, life, and reliability are tremendously reduced.

Transistor devices (and other semiconductors) contain a silicon (or germanium) crystal (or chip) through which varying amounts of current pass. The junction where the current control takes place is also where the power dissipation is the greatest. The junction temperature, which is equivalent to the overall chip temperature, becomes the critical requirement for design control. A generally accepted fact is that, for every 10°C reduction of device junction temperature, operating life doubles. The design objective is to hold the junction temperature below the maximum specified by the device manufacturer.

Most air-cooled heat sinks are manufactured as "stampings," as parts made from "extrusions," or as custom-designed heat sinks. Commercially available heat sinks are offered in many stamped and extruded shapes; a typical design of each is shown in Fig. 5.1*a* and 5.1*b*.

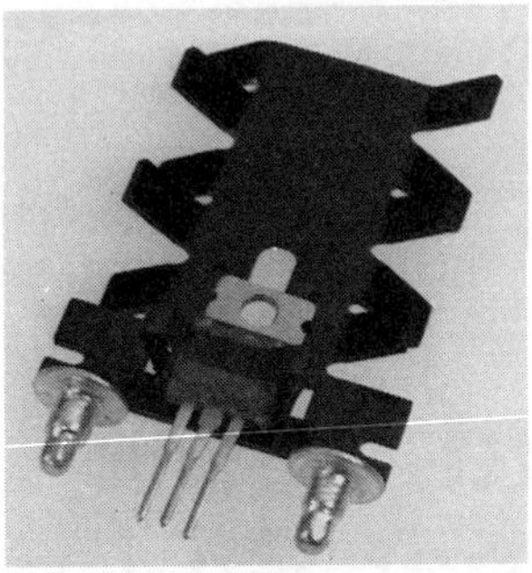

(a)

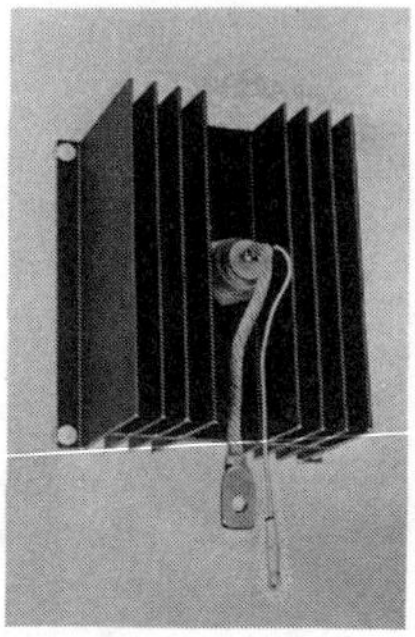

(b)

Fig. 5.1 Examples of commercial heat sinks. (*a*) Stamped heat sink; (*b*) extruded heat sink.[1a]

Manufacturers of heat sinks provide operating performance data to aid in the selection process. This process requires determination of the desired functional requirements and then matching this need to dissipation capability. Small stamped heat sinks rarely rate above power levels of 10 W, while extruded types can dissipate many hundreds of watts, depending on size, type, and operating conditions. Thermal transmission effects involved in removing heat from a powered device to the heat sink and then to the surroundings requires one or more of the following heat-transfer processes:

- Conduction
- Radiation
- Convection

The methods for dissipating heat considered here are natural convection in air, radiation, forced convection with air, and forced convection with liquids.

A generalized characterization of heat-sink size and performance is shown in Fig. 5.2. This figure illustrates the volume required for heat sinks for a range of thermal resistances for both natural- and forced-convection applications. The natural-convection curve is based on data where the applied heat loading was more or less localized rather than being spread uniformly. Typically, to reduce the thermal resistance in natural convection by 50 percent, the heat-sink volume must be approximately quadrupled.

The selection of a heat sink to maintain a semiconductor at a desired operating temperature requires knowledge of:

1. The available volume of space to be occupied
2. The maximum allowable device junction temperature
3. The power dissipated by the device
4. The device configuration
5. Ambient conditions (temperature, airflow)

For most applications, the designer should know the limit of the heat-sink surface temperature, as well as the ambient conditions. With a given set of design constraints, the required thermal resistance can be calculated as

$$\theta_{sa} = \frac{T_s - T_a}{Q} \tag{5.1}$$

where T_s = heat sink temperature, °C
T_a = ambient temperature, °C
Q = input power, W
θ_{sa} = thermal resistance, sink to ambient, °C/W

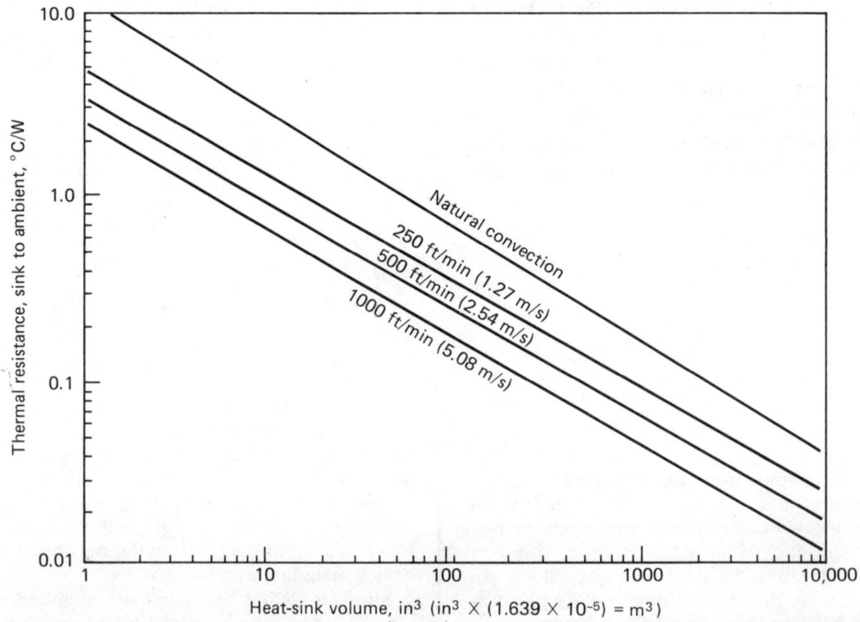

Fig. 5.2 Relationship between thermal resistance and heat-sink volume.[1a] The natural-convection curve is based on 50°C rise of air temperature.

5.2 HEAT-SINK PERFORMANCE AS A FUNCTION OF POWER DISSIPATION

Junction temperature is a function of the sum of the thermal resistances between the junction and the ambient air, the amount of heat being dissipated, and the ambient-air temperature level.

Figure 5.3 is a simplified drawing of a semiconductor mounted on a heat sink. Three thermal resistances are readily identified as shown in Fig. 5.3.

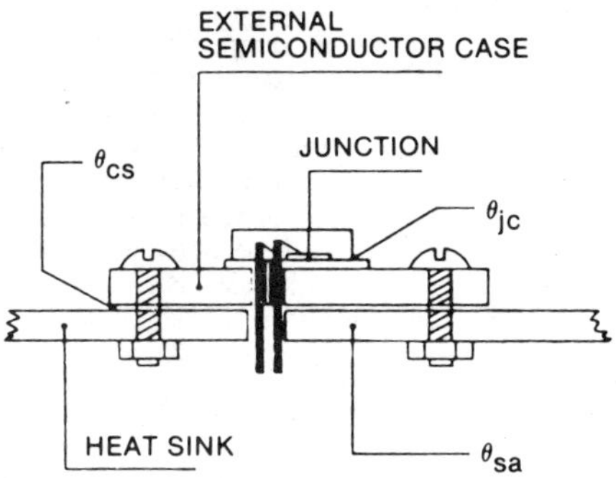

Fig. 5.3 Location of thermal resistances.[1a]

The total resistance θ_{tot} is the sum of the individual resistances:

$$\theta_{tot} = \theta_{jc} + \theta_{cs} + \theta_{sa} \tag{5.2}$$

where θ_{jc} = thermal resistance between the semiconductor junction and its external case, °C/W
θ_{cs} = thermal resistance from case to heat sink, °C/W
θ_{sa} = thermal resistance from heat-sink surface to ambient air, °C/W

The thermal resistance from heat-sink surface to ambient air θ_{sa} is a function of the convection heat-transfer coefficient h and the surface area A of the heat sink as shown in the following formula:

$$\theta_{sa} = \frac{1}{hA} \tag{5.3}$$

The heat-transfer coefficient h for both natural and forced convection is a complex function and cannot be easily generalized.

With a semiconductor or any powered modular device mounted on a heat sink, the relationship between junction temperature rise above ambient temperature and power dissipation is given by

$$Q = \frac{T_j - T_a}{\theta_{jc} + \theta_{cs} + \theta_{sa}} \tag{5.4}$$

where Q = power dissipated, W
T_j = junction temperature, °C
T_a = ambient air temperature, °C

and θ_{jc}, θ_{cs}, and θ_{sa} are as described previously. In most applications, values for all these parameters are known or can be found except that for the thermal resistance from heat-sink surface to air θ_{sa}. The value of this parameter then becomes the basis for sink selection.

Data for heat-sink selection with forced convection are generally presented in terms of θ_{sa}. For natural convection, data are presented in terms of ΔT_{sa} (temperature difference between sink and air). This results in the following simplified form:

$$\Delta T_{sa} = (T_j - T_a) - Q(\theta_{jc} + \theta_{cs}) \tag{5.5}$$

It eliminates the requirement of multiplying θ_{sa} by Q to obtain the maximum allowable ΔT_{sa} and thus gives that parameter for direct comparison with data as presented on most heat-sink manufacturers' graphs.

EXAMPLE Assume that a semiconductor with a TO-3 case must be operated so that its junction temperature will not exceed 125°C when it is dissipating 10 Watts to ambient air at a temperature of 50°C. The value of θ_{jc} for this device established by the manufacturer is 1.5°C/W, and θ_{cs} is estimated to be 0.09°C/W because a thermal joint compound is being used. For this example, the joint compound is assumed to have an average thickness of 0.0015 in (0.0381 mm); a later section gives alternatives and a calculation procedure for this interface. The thermal resistance schematic of this example is shown in Fig. 5.4. Solving Eq. (5.4) for the unknown value of θ_{sa} (thermal resistance from sink to air) produces the following equation:

$$\theta_{sa} = \frac{T_j - T_a}{Q} - (\theta_{jc} + \theta_{cs}) \tag{5.6}$$

Inserting known values, we obtain

$$\theta_{sa} = \frac{125 - 50}{10} - (1.5 + 0.09) = 5.9\text{°C/W}$$

This is the largest value of θ_{sa} that can be used. Heat sinks providing smaller values of θ_{sa} are also acceptable since the resulting junction temperature will be less than the 125°C specified. With this value of θ_{sa} we could then proceed to evaluate the various heat sinks and locate one which will provide this performance. Manufacturers of stamped heat sinks provide data for low-power applications, and selections can be made readily. If an extrusion is required, however, most manufacturers present data based on a 3-in vertical height. A later section explains the relationship of performance as a function of extrusion length (height when oriented vertically).

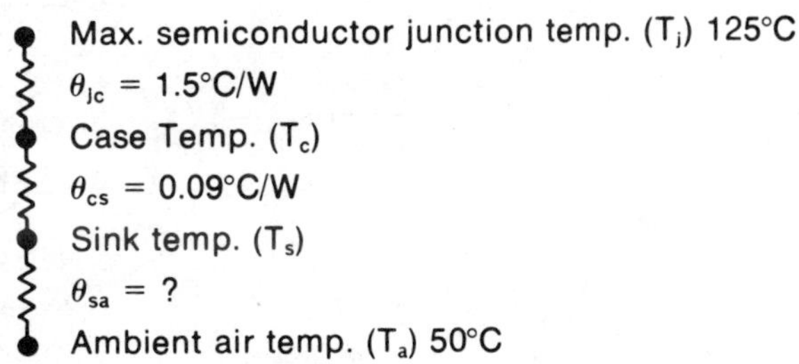

Fig. 5.4 Thermal-resistance schematic.

5.3 DESIGN ANALYSIS FOR A HEAT SINK IN NATURAL CONVECTION

When a powered heat-generating component is mounted to any heat sink, the heat will move from the point of contact through the heat sink and then to the surroundings. For this to happen, the surroundings must always be at a lower temperature than the heat sink. Where natural convection is concerned, the process in equation form can be expressed as

$$\theta_{sa} = \frac{\rho l_1}{A_1} + \frac{1}{h_c A_2 + h_r A_3} \tag{5.7}$$

The first group of terms represents the conduction component of θ_{sa} where A_1 is the average cross-sectional area available for heat flow by conduction, l_1 is the average path length that the heat must take, and ρ is the thermal resistivity of the heat-sink material. The second group of terms in the equation represent the convection component of the thermal resistance of the heat sink where h_c is the convection heat-transfer coefficient and A_2 is the heat-sink area exposed to the ambient surroundings. The third group of terms represents the radiation component where h_r is a simplified expression and can be called the *radiational heat-transfer coefficient.* Finally, A_3 represents the effective area available for radiational heat transfer. In many cases of heat-sink design A_3 is significantly less than A_2.

The conduction component $\rho l_1 / A_1$ above is not easily used because it is difficult to assess the share of power distributed through a given segment of the heat sink. By making a few assumptions, the design procedure can be simplified, leading to a determination of a more useful requirement—that of calculating the expected average temperature of the heat-sink surface. Once an average temperature difference and the resulting heat-transfer coefficient have been determined, the overall equation for thermal resistance including radiation can be rewritten as:

$$\theta_{sa} = \left[\frac{1}{h_c A_c + h_r A_r}\right] (1.897) \tag{5.8}$$

where θ_{sa} = thermal resistance, sink to ambient, °C/W
h_c = convection heat-transfer coefficient, Btu/h · ft² · °F (W/m² · °C)
h_r = radiation heat-transfer coefficient, Btu/h · ft² · °F (W/m² · °C)
A_c = convection surface area, ft² (m²)
A_r = radiation surface area, ft² (m²)

(*Note:* The constant, 1.897, is used to convert to °C/W with British units only.) When evaluating the heat flow path through a stamped or extruded heat-sink design, the temperature gradient through the thickness is usually small and can be ignored.

For a stamped sheet metal or extruded heat-sink design, the heat flow from where the device is mounted will spread radially outward to the edges (Fig. 5.5). With a given or planned size in mind, one can determine the distances to several locations and calculate temperatures as shown in the following procedure. For an extrusion, it is advisable to determine the average temperature in the flat back plane separately; the temperature gradient in the fins can be accommodated by the use of fin effectiveness. It is to be noted that the effectiveness normally is applied to the finned surface only; the exposed area at the base between the fins is considered prime surface area and is assumed to be 100 percent efficient. Since all calculations are considered approximate, it will simplify matters to consider all surface area to be equally effective and to assume fin effectiveness η for typical commercial heat sink as follows:

Use η = 0.80 for forced convection
Use η = 0.90 for natural convection

Procedure for Temperature Averaging of Fin

If we visualize the heat sink as a fin of a given thickness extending from the device location, a model for analysis purposes can be prepared as in Fig. 5.6.

For a given design (Fig. 5.6), we determine:

T_s = base temperature (device mounting location), °F (°F = 1.8°C + 32)
T_a = ambient temperature (known or assumed), °F (°C)
t = fin thickness (if not constant thickness assume average), ft (1 ft = 0.3048 m)
L = fin length (length of conduction path from device location to furthest dimension), ft
k = material thermal conductivity (for aluminum use a value 100), Btu · ft/h · ft² · °F [(Btu · ft/h ·ft² · °F) × (1.731) = W/m · °C]
h = film coefficient; Btu/h · ft² · °F, for natural convection [Eq. (5.9)]; for forced convection [Eq. (5.10)] [(Btu/h · ft² · °F) × (5.68) = W/m² · °C]

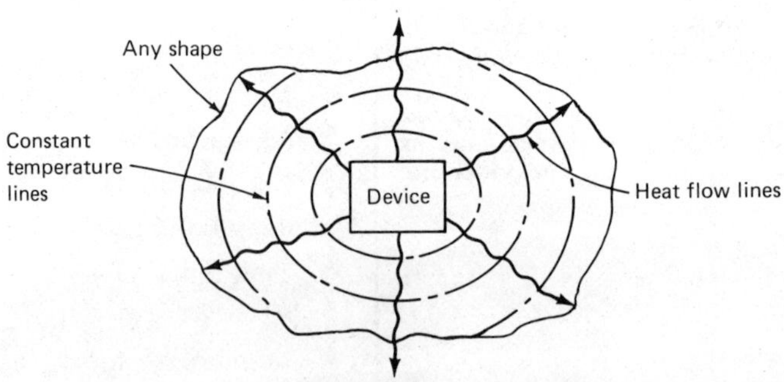

Fig. 5.5 Example of temperature flow lines.

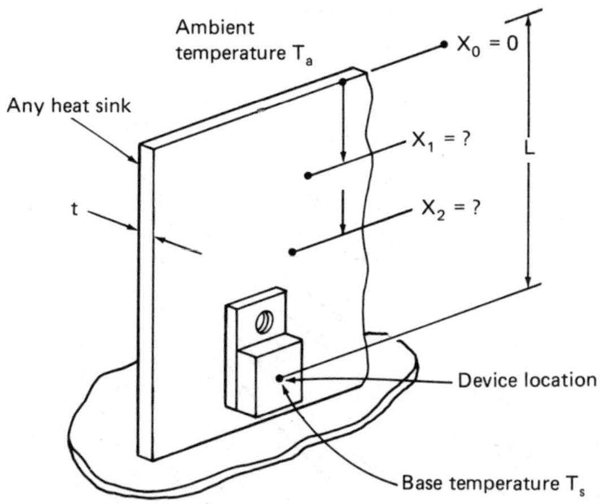

Fig. 5.6 Heat-sink model for temperature averaging.

X = assumed distance from fin tip to desired location X_0, X_1, X_2, ft (m)

T_X = temperature at location X, °F (°C)

Determine temperature differential base to ambient:

$$\Delta T_{sa} = T_s - T_a$$

Determine fin effectiveness parameter:

$$M = \left(\frac{2h}{kt}\right)^{1/2}$$

Determine temperature rise above ambient at any distance X^2:

$$\Delta T_{Xa} = \frac{\Delta T_{sa} \cosh MX}{\cosh ML} \quad \text{°F (°C)}$$

Then fin temperature at any location X is

$$T_X = \Delta T_{Xa} + T_a, \quad \text{°F (°C)}$$

By assigning various values to X at selected locations from the fin edge toward the heat source, gradient temperatures for determining an average can be established. Compare the calculated average temperature to the assumed one and, if the two are not reasonably close, repeat the procedure until results are acceptably comparable.

5.4 DETERMINING HEAT-TRANSFER COEFFICIENTS—PLANE SURFACES

Natural Convection

For a vertical plane surface in natural convection laminar flow conditions simplified for air at sea level, the film coefficient is expressed as[3]

$$h_c = 0.29 \left(\frac{\Delta T}{L_v}\right)^{1/4} \quad \frac{\text{Btu}}{\text{h} \cdot \text{ft}^2 \cdot \text{°F}}$$
$$= 1.41 \left(\frac{\Delta T}{L_v}\right)^{1/4} \quad \frac{\text{W}}{\text{m}^2 \cdot \text{°C}} \tag{5.9}$$

where h_c = heat-transfer coefficient, Btu/h · ft² · °F (W/m² · °C)

ΔT = assumed average surface to ambient temperature difference, °F (°C)

L_v = vertical height of surface, ft (m)

Forced Convection

For a plane surface in forced convection simplified for air at sea level, the film coefficient is expressed as[4];

$$h_c = 0.305 \sqrt{\frac{V}{L_c}} \quad \text{Btu/h} \cdot \text{ft}^2 \cdot \text{°F}$$
$$= 3.86 \sqrt{\frac{V}{L_c}} \quad (\text{W/m}^2 \cdot \text{°C}) \tag{5.10}$$

where h_c = heat-transfer coefficient, Btu/h · ft² · °F (W/m² · °C)

V = linear air velocity, ft/min (m/s)

L_c = length of fin parallel to airflow, in (m)

Determining the Radiative Coefficient

When the heat-sink temperature is considerably above the ambient temperature, there is the potential for a significant radiational cooling effect. In forced-air systems, convection effects are high compared to radiation, which can be ignored. For natural convection, however, the convection heat-transfer coefficient is sufficiently modest that radiation can be useful. Published catalog performances of heat sinks include radiation.

A quick determination of the radiative heat-transfer coefficient can be made by using the graph as shown in Fig. 5.7.[5] This graph is based on an emissivity ϵ of 0.90. For emissivities other than 0.90, the coefficient value can be modified by the following correction:[5]

$$h_r^* = \frac{\epsilon_s}{0.90} h_r \tag{5.11}$$

where h_r^* = corrected radiation coefficient, Btu/h · ft² · °F (W/m² · °C)

ϵ_s = emissivity in question

h_r = surface radiation coefficient determined from Fig. 5.7, Btu/h · ft² · °F (W/m² · °C)

5.5 DETERMINING THE EFFECTIVE RADIATIVE SURFACE AREA

The effective surface area of simple geometries can be calculated as the area that actually "sees" the surroundings. For this simplified condition, it is the total external surface and can be used for both convective A_c and radiative A_r heat-transfer surface areas. As the shape of the heat sink acquires more complex folded geometry, much of the actual surface does not "see" the surroundings, and the effective surface is subject to reduction because of this view factor. For the sheet metal form of heat sink, the designer must calculate or estimate the effective radiation surface area.

When an extruded design is being considered as the heat sink (Fig. 5.8), the effective area for radiation is considerably different from the actual surface area. This is particularly true when the ratio of fin height to fin spacing exceeds 4 to

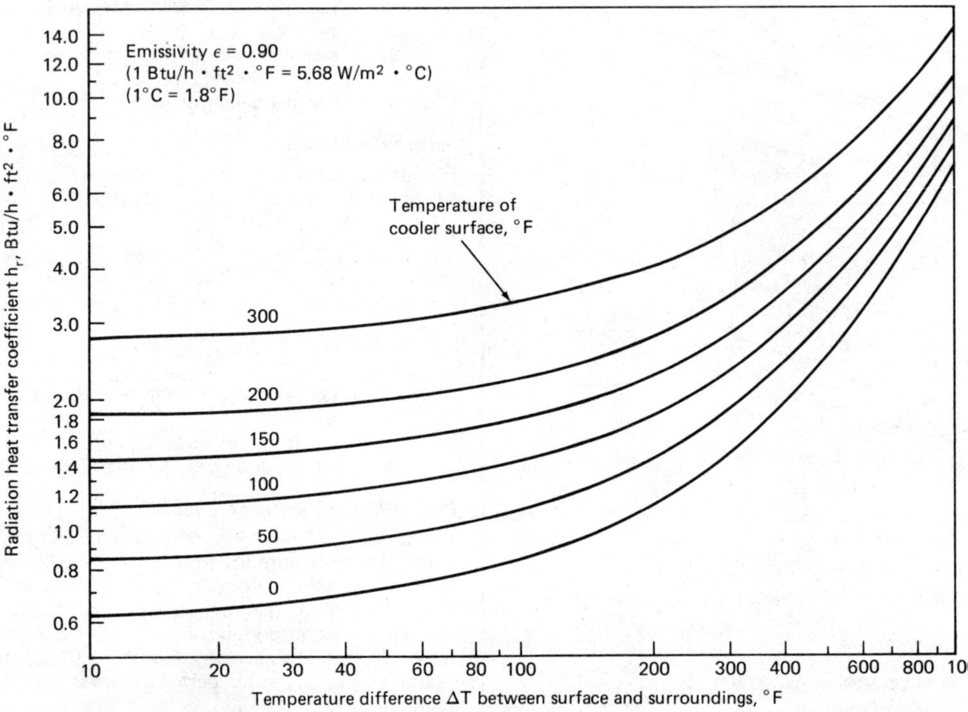

Fig. 5.7 Relationship between radiative coefficient and surface temperature difference. (*Reprinted with permission from Genium Publishing Corporation, 1145 Catalyn Street, Schenectady, N.Y., Heat Transfer Book, Section 505.2.*)

1. In such cases, the approximate radiational area A_r can be calculated as:†

$$A_r = 2H(W + L) + WL \tag{5.12}$$

Configuration constraints can be a dominating factor. Air-cooled sinks are in many cases shaped to utilize available space for effective packaging with adjacent components and very often are used as part of the enclosure. For odd shapes, the area is defined by the "string" perimeter, that is, the perimeter formed by wrapping an imaginary string snugly around the shape as shown in Fig. 5.9, where the approximate radiational area A_r would be

$$A_r = L(2S + W_2) + H(W_1 + W_2) \tag{5.13}$$

5.6 DETERMINING HEAT-TRANSFER COEFFICIENTS—PARALLEL SURFACES

Natural-Convection Coefficient.

For parallel-surface natural-convection laminar conditions and air at moderate temperatures, the approximate heat-transfer coefficient and the resulting thermal performance of an existing (or planned) heat-sink design can be calculated.

† *Note:* With sinks which are attached to an enclosure, the flat back side is not included as it does not "see" ambient conditions. Equation (5.12) is for enclosure mounting.

Using the extrusion model shown in Fig. 5.10, the following input parameters can be determined:

S_f = average space width between fins, in (m)
H_f = fin height from base, in (m)
ΔT_{fa} = assumed average surface to ambient temperature difference, °F (°C)
L_f = sink vertical length, in (m)
h_c = heat-transfer coefficient, Btu/h · ft² · °F (W/m² · °C)

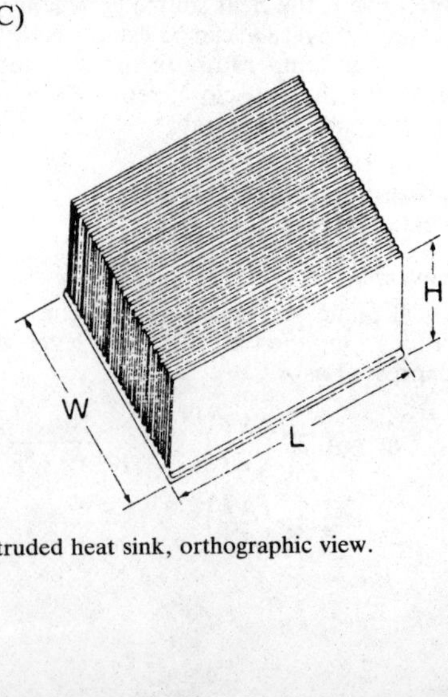

Fig. 5.8 Extruded heat sink, orthographic view.

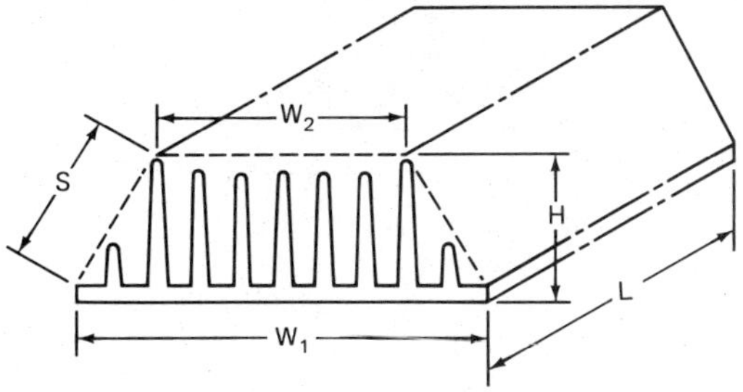

Fig. 5.9 Example of string outline for irregular shapes.

It is noted that this method is usually used for ratios of H_f/S_f from approximately 5 to 10 for moderate fin temperatures and is a simplified derivative of Ref. 5. For other fin ratios, this method is still useful for establishing an approximate level of performance.

1. Calculate the effective heat-transfer coefficient, h_∞ that would exist for a single isolated vertical surface:

$$h_\infty = 0.156 \left(\frac{\Delta T_{fa}}{L_f}\right)^{1/4} \quad \frac{\text{Btu}}{\text{h} \cdot \text{ft}^2 \cdot {}^\circ\text{F}}$$

$$= 1.41 \left(\frac{\Delta T_{fa}}{L_f}\right)^{1/4} \quad \frac{\text{W}}{\text{m}^2 \cdot {}^\circ\text{C}}$$

2. Calculate a dimensionless fin spacing parameter F:

$$F = 5.53 S_f \left(\frac{\Delta T_{fa}}{L_f}\right)^{1/4}$$

where units are degrees Fahrenheit and inches, or

$$F = 100.7 S_f \left(\frac{\Delta T_{fa}}{L_f}\right)^{1/4}$$

where units are degrees Celsius and meters.

3. Determine heat-transfer coefficient correction factor

$$\frac{h_c}{h_\infty} = \frac{1.68 F^3}{24}\left[1 - \exp\left(\frac{-24}{1.68 F^3}\right)\right]$$

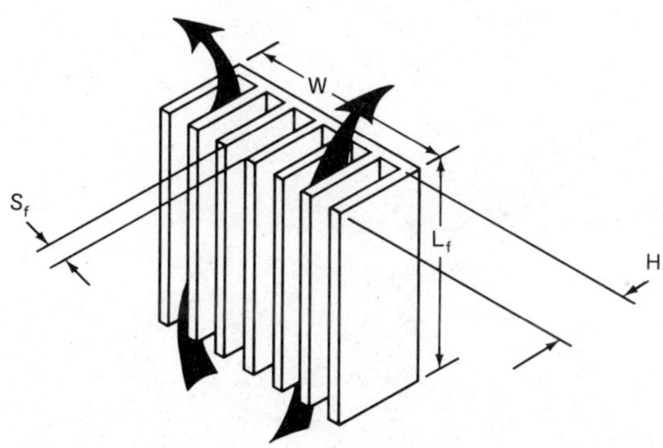

Fig. 5.10 Heat-sink dimensional model (vertical mounting).

4. Apply correction factor

$$h_c = \left(\frac{h_c}{h_\infty}\right) h_\infty \qquad \frac{\text{Btu}}{\text{h} \cdot \text{ft}^2 \cdot {}^\circ\text{F}} \left(\frac{\text{W}}{\text{m}^2 \cdot {}^\circ\text{C}}\right)$$

Forced-Convection Coefficient

A detailed analysis for this type of problem normally requires determining values of parameters such as Reynolds and Prandtl numbers. For initial design purposes, a reasonable determination for the heat-transfer coefficient can be made by using Eq. (5.10). The airflow velocity must be known or assumed, and the heat transfer area can be extracted from a manufacturer's catalog. Once these quantities are determined, we can calculate the thermal resistance for an existing or planned heat-sink design by using Eq. (5.3), repeated here for convenience:

$$\theta_{sa} = \frac{1}{hA}$$

It is to be remembered that when forced convection is in use, radiation generally is not an important factor and is usually ignored.

EXAMPLE Let us assume that we wish to determine the thermal resistance of a large commercially available extrusion, no. 4559[1a] 28 in long. Other assumed conditions are moderate air temperature and air velocity provided at 700 ft/min. From Eq. (5.10):

$$h_c = 0.305 \sqrt{\frac{V}{L}} = 0.305 \sqrt{\frac{700}{28}} = 1.525 \text{ Btu/h} \cdot \text{ft}^2 \cdot {}^\circ\text{F}$$

Heat-transfer area (from manufacturer) is indicated at 87.1 in^2/in. Since this extrusion will be mounted so that its back plane will not be effective, a correction is made by subtracting the extrusion width that is given as 10.78 in. Therefore, 87.1 − 10.8 = 76.3 in^2/in, so that the effective surface area is 76.3 in^2/in × 28 in = 2136.4 in^2 = 14.84 ft^2. Then

$$\theta_{sa} = \frac{1}{hA} = \frac{1}{1.525(14.84)} = 0.0442 \text{ h} \cdot {}^\circ\text{F/Btu}$$

Correcting to conventional thermal resistance units, we obtain

$$\theta_{sa} = (0.0442)(1.897) = 0.084 {}^\circ\text{C/W}$$

As an alternative to the calculation procedure above, a graph for determining forced-convection thermal resistance directly is shown in Fig. 5.13.

5.7 THERMAL RESISTANCE—GRAPHICAL SELECTION METHOD

Natural-Convection Selection

Most extrusion manufacturers present performance data for a 3-in (7.62-cm) vertical height only. To make use of these data and to offer a suggested method for other length extrusions, the graph shown in Fig. 5.11 was developed as a guide.[1a] It should be recognized that a graphical solution that will cover all operating conditions is only approximate because of differences in fin spacing, thermal loading, length, and the associated variations of average temperature differences. Once the preferred extrusion is selected, Fig. 5.11 will assist in establishing the approximate length necessary to obtain a desired thermal resistance.

EXAMPLE Suppose that it is desired to use an extrusion with a

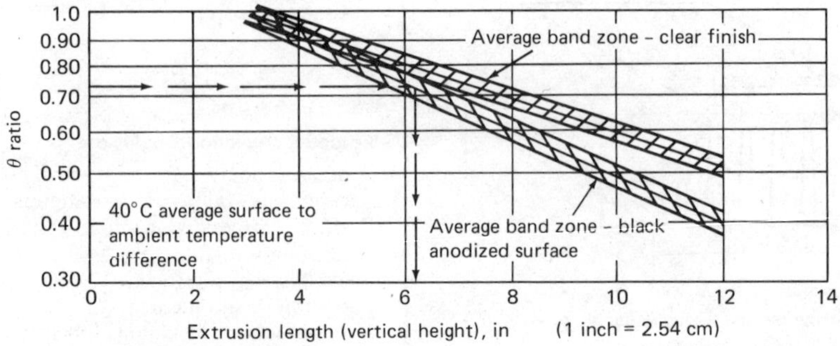

Fig. 5.11 Natural convection length selector. (*Reprinted with permission from EG&G Wakefield Engineering, Standard Products Catalog.*)

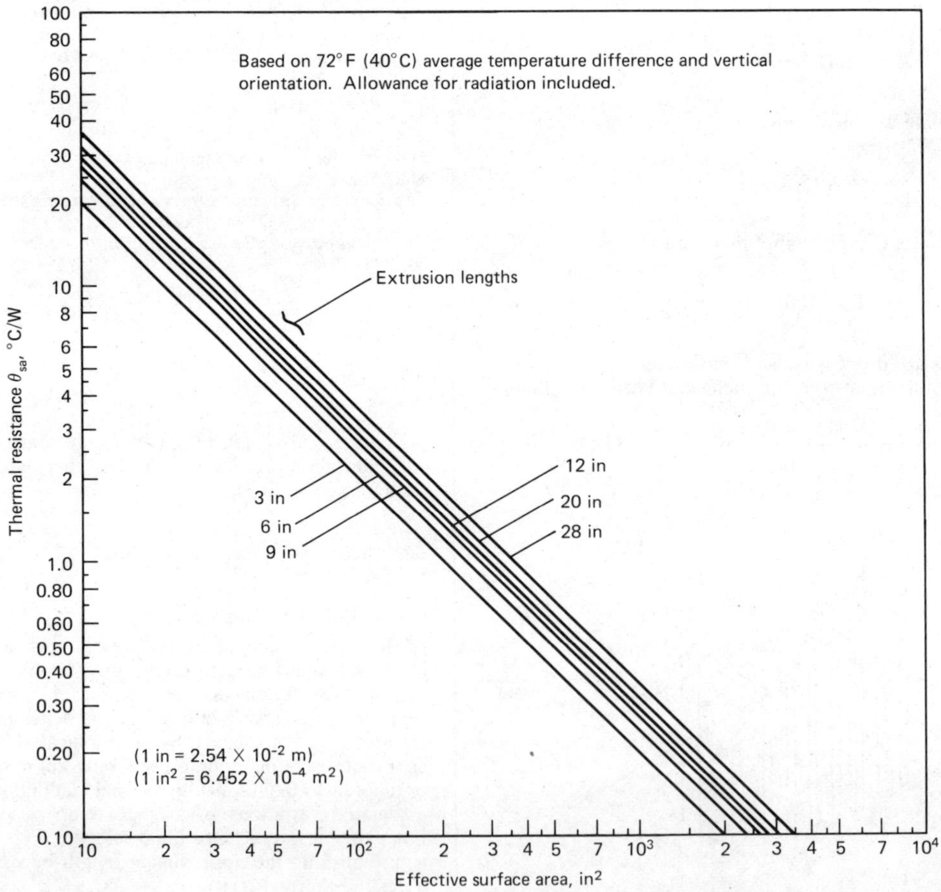

Fig. 5.12 Natural-convection-resistance chart. (*Printed with permission from EG&G Wakefield Engineering.*)

black anodized finish having a thermal resistance of 2.8°C/W. From the manufacturer's data we find the thermal resistance to be 3.9°C/W for a 3-in length. The θ ratio is defined as

$$\theta \text{ ratio } = \frac{\theta \text{ desired}}{\theta \text{ for a 3-in length}}$$

The θ ratio for our example is 2.8/3.9 = 0.72. Reading horizontally across at a θ ratio of 0.72, we find the approximate center of the appropriate shaded zone and read vertically down to determine that the length required is 6.25 in. Any length greater than this would also be acceptable at the user's option. It is important to note that the selected values are based on the assumption that the extrusion length is mounted in the vertical position only and includes the effects of surface finish as indicated.

When it becomes necessary to estimate the natural-convection thermal resistance of any existing extruded heat sink, a graph such as that in Fig. 5.12[1a] can yield a quick approximate answer. In its development, an average surface to ambient-temperature difference of 40°C was selected. An allowance for radiation has been included. It is assumed that fin spacing is reasonably adequate for fluid mobility. It is not possible to account for all the possible variations in a simplified form, but this graph should be a useful guide in determining the approximate thermal resistance for a given size.

EXAMPLE Let us assume that we wish to evaluate the thermal resistance of an extrusion 8 in long with a surface area given as 102.7 in²/in. The total surface area is then 102.7 (in²/in) × 8 in = 821.6 in². Finding the surface area on the abscissa, read in Fig. 5.12 vertically up to an estimated position between the 6- and 9-in lines; then left to read θ_{sa} of 0.32°C/W.

Forced-Convection Selection

The forced-convection performance of extruded heat sinks determined by the same method as in the example above can be estimated from a graph such as that in Fig. 5.13.[1a] On the basis of theory and experience, together with some basic assumptions, this graph should yield reasonable and useful information. The graph is based on an assumed velocity of 700 ft/min (3.55 m/s) and should be understood to be a useful approximation of predicted thermal performance. As in the natural convection section, it is again assumed that device quantity and location add no unusual heat-distribution effects.

When forced convection is the cooling method, it is im-

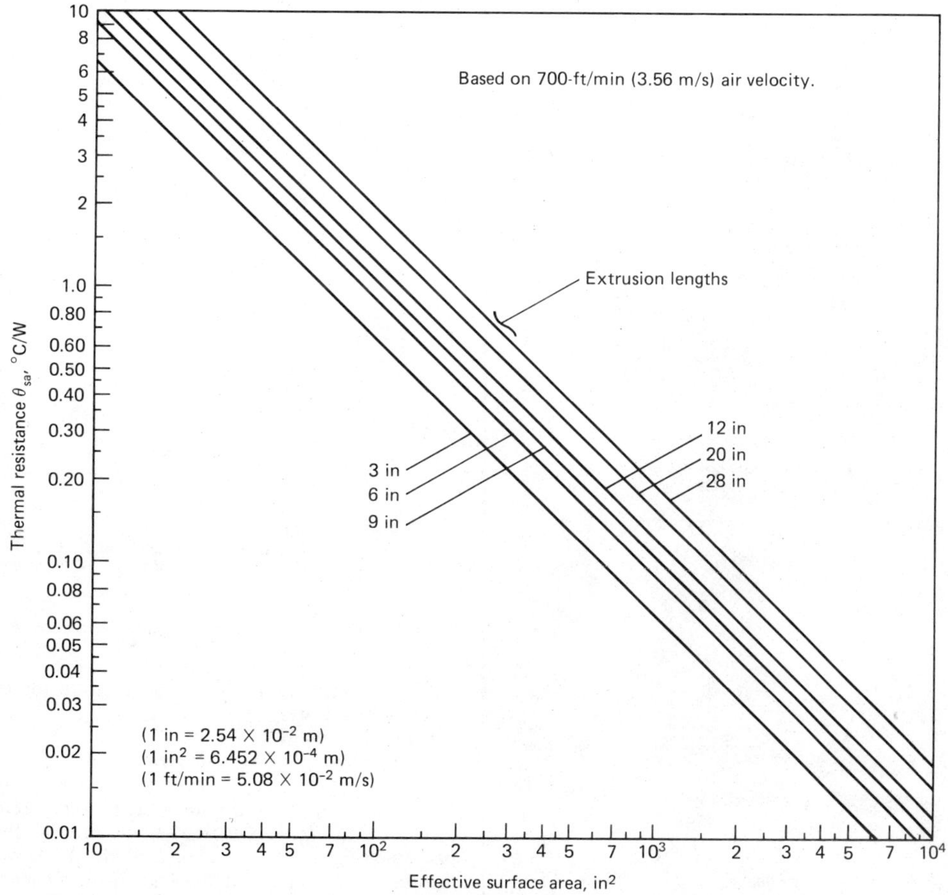

Fig. 5.13 Forced-convection-resistance chart. (*Reprinted with permission from EG&G Wakefield Engineering, Standard Products Catalog.*)

portant to assess the temperature change of the air. When making these calculations, one must determine whether the operating conditions prevent the air from rising above effective functional limits. As air moves through the spaces within the extrusion, it accepts the heat produced by the input wattage, causing a rise in temperature along its length. It is imperative that the outlet temperature of the air never exceed the desired maximum surface temperature of the heat sink and, in fact, remain as far below this temperature as is practical.

If the temperature rise of the air is not small in comparison to the temperature difference between the junction and the ambient, the logarithmic mean temperature difference between the junction and the air passing through the heat sink should be used to design the heat sink.

5.8 LIQUID-COOLED HEAT SINKS

Liquid cooling, although avoided wherever possible, can dissipate greater amounts of heat than can air cooling. Liquid-cooled heat sinks (sometimes called *cold plates* or *chill plates*) can be used where one or more of the following conditions exist or are desirable:

- When power levels exceed 10 to 15 W/in^2 (1.5 to 2.3 W/cm^2) of contact area
- When thermal resistance must be lower than that attainable with air
- When packaging problems require efficient space management
- When surface temperatures must be kept uniform
- When cabinet noise is a consideration
- When operating conditions are environmentally dirty or hazardous

Liquid cold plates are available from manufacturers as standard product or can be custom designed as shown in Fig. 5.14*a* and 5.14*b*. By design, these cold plates are configured as one- or two-sided plate surfaces and lend themselves for component mounting. Base material is provided for threaded holes and inserts. Many combinations of machining detail can be provided. Manufacturers of liquid cold plates present performance data in graphical form such as shown in Fig. 5.15.[1b] Most systems in operation use water as the coolant; however, other liquids can be used to satisfy different system or performance objectives.

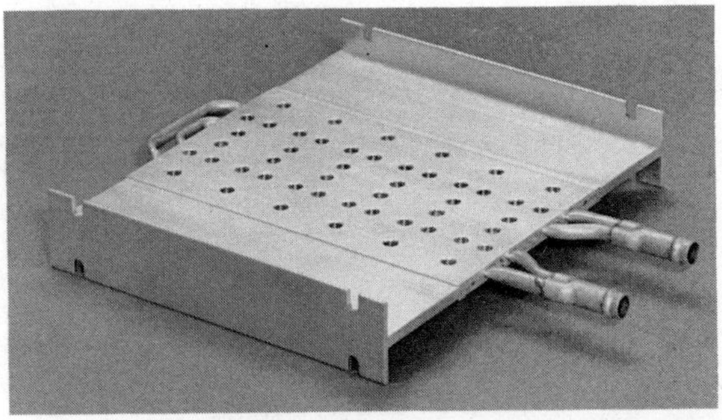

(a)

(b)

Fig. 5.14 Examples of liquid cold plates. (*a*) Standard cold plate; (*b*) custom cold plate.[1b]

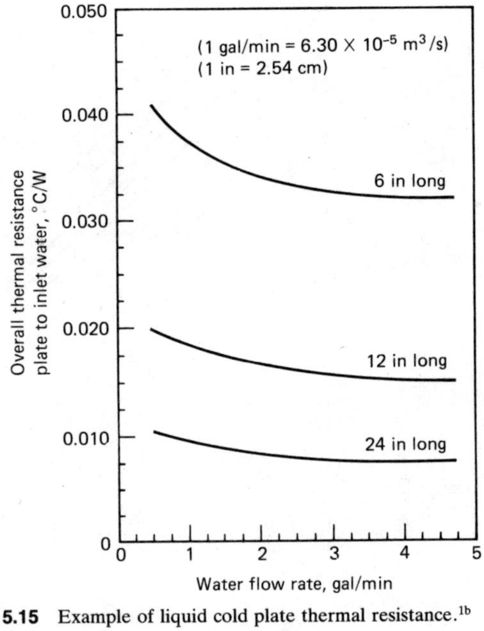

Fig. 5.15 Example of liquid cold plate thermal resistance.[1b]

When liquid cooling has been selected, the two types of systems in use are:

- Closed system
- Open system

The closed system is designed to circulate the same liquid constantly using a fixed volume as in Fig. 5.16*a*. The open system, as in Fig. 5.16*b*, flows liquid through the cold plate once and the effluent is then discarded. As one would expect, there are advantages and disadvantages for each of the methods.

Open- and Closed-System Considerations.

Some of the advantages and disadvantages are:

Advantages	Disadvantages
Closed System	
Water quality controllable	Operating components: heat exchanger, reservoir, pump, etc. required
Flow rate uniform	Maintenance increased
Internal cleanliness maintained	Overall reliability reduced
System temperature more consistent	Initial and operating costs higher
Constant water supply not required	Operating temperature set by ambient conditions
Corrosion potential minimal	
Constant drain not required	
Open System	
Operating components not required	Cost of water
Maintenance reduced	Uncertainty of water chemistry
Electrical power not required for components	Higher corrosion potential
Chilled water could be supplied if available	System temperature can vary seasonally
	Water flow rate variation
	Water supply uncertainty
	Water waste treatment must comply with local and/or federal codes

The greatest advantage of liquid cooling is its capacity for large amounts of heat dissipation. If the performance shown in Fig. 5.15 and a flow rate of 2 gal/min (1.26×10^{-4} m^3/s) with a cold plate 12 in (0.304 m) long are assumed, an estimate of a thermal resistance θ_{pw} of 0.017°C/W can be made. Taking the reciprocal of 0.017°C/W (58.8 W/°C) allows a better perspective on the allowable power levels for various temperature rises. Using 58.8 W/°C (now a conductance), the allowable power levels have been calculated and shown in Table 5.1.

The contacting surface area for the cold plate used in this example is 7.75×12 in = 93 in^2 (600 cm^2) and at the rise of 30°C gives us a watt density of 1764/93 = 19 W/in^2 (2.90 W/cm^2). With large amounts of power to be dissipated, the use of liquids and the resulting economy of required space for packaging becomes obvious and possibly attractive for many difficult thermal management problems. The higher heat capacity of water results in a very small rise in temperature even with a substantial power input to the cold plate surface.

Cold-Plate Size Selection Procedure

The thermal resistance between the cold plate and the water can be regarded as the sum of four resistances: (1) the resistance due to plate metal conductance, (2) the contact resistance at the tube outer diameter (O.D.), (3) the resistance of the tube wall, and (4) the convective resistance between the tube inner diameter (I.D.) and the moving fluid. Two of these, the contact resistance and the tube wall resistance, are generally a small percentage of the total and are considered negligible.

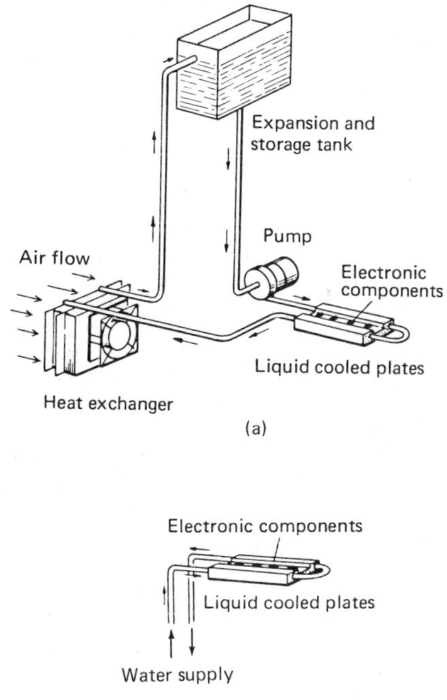

Fig. 5.16 Examples of liquid cooling systems. (*a*) Closed system; (*b*) open system.

With a heat-generating component mounted on a cold plate, the relationship between junction temperature rise above the inlet water temperature and power dissipation is given by

$$Q = \frac{t_j - t_w}{\theta_{jc} + \theta_{cp} + \theta_{pw}} \tag{5.14}$$

Table 5.1. Example of Temperature Rise and Power Level Based on Fig. 5.15

Plate temperature rise above inlet water, °C	Allowable power level, W
10	588
20	1176
30	1764
40	2352

NOTE: Data based on 12-in (30.48-cm) length and 2-gal/min (1.26×10^{-4}-m^3/s) flow.

where Q = power dissipated, W
t_j = junction temperature, °F (°C)
t_w = inlet water temperature, °F (°C)
θ_{jc} = thermal resistance from junction to semiconductor case, °F/W (°C/W)
θ_{cp} = thermal resistance from case to cold plate, °F/W (°C/W)
θ_{pw} = thermal resistance from cold plate to coolant water, °F/W (°C/W)

Values for θ_{jc} are rated by the device manufacturer and are not controllable; θ_{cp} is a somewhat controllable variable and can be calculated by the method described in this chapter. For standard products, the thermal resistance from cold plate to inlet water θ_{pw} is generally determined by the manufacturer for various flow rates as in Fig. 5.15.[1b]

For many applications, the mounting surface temperature is specified rather than the device junction temperature. In these cases, the required cold-plate thermal resistance is expressed as

$$\theta_{pw} = \frac{t_p - t_w}{Q} \tag{5.15}$$

where t_p is the cold-plate surface temperature in Celsius.

Operating performance (Fig. 5.15) resistances are based on inlet water temperatures because for many applications the temperature rise is small and considered negligible. In unusual cases where the heat load may be very high or the water flow very low, it may be necessary to account for water temperature rise in the cold plate. Most published data are based on water as the liquid. For other liquids, the performance can be considerably different from water and should be considered and evaluated early in the design phase.

Performance Analysis for Liquid Cold Plates

Using the design and construction features of a commercial liquid cold plate,[1b] and allowing for the negligible losses as previously described, we express thermal resistance as

$$\theta_{pw} = \left(\frac{X}{kA_1} + \frac{1}{hA_2}\right) C_1 \quad \text{°C/W} \tag{5.16}$$

where X = average length of heat-flow path, ft (m)
k = metal thermal conductivity, Btu/h · ft · °F (W/m · °C)
A_1 = average cross section of heat path, ft² (m²)
h = liquid film coefficient, Btu/h · ft² · °F (W/m² · °C)
A_2 = effective liquid side area, ft² (m²)
C_1 = conversion constant = 1.897 for British units = 1.0 for SI units

Accurate analysis of a typical cold plate requires an involved and detailed heat-transfer analysis, and it is very difficult to characterize a simplified calculation procedure to cover all designs and their heat-loading effects. For this reason, manufacturers of this type of cold plate determine performance by actual testing, and then publish the data in graphical form.

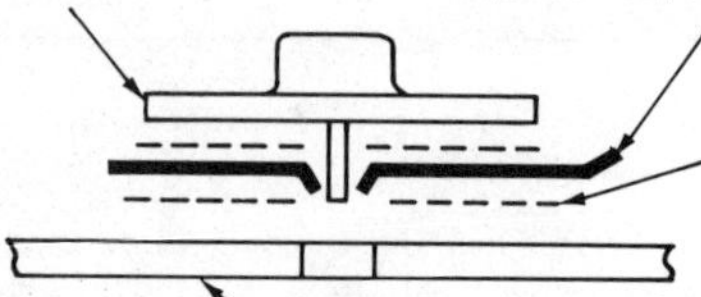

Fig. 5.17 Example of transistor and interface material application.[1a]

5.9 IMPROVING THERMAL PERFORMANCE

As mentioned earlier in this chapter, the contact resistance θ_{cs}, between a device and its heat sink is controllable but must be calculated with respect to the design being considered. When planning for this interface attachment, it is important to remember that all contacting surfaces have some kind of mechanical finish and are rarely perfectly flat over the total contacting surface area. Surface finish and flatness effects can result in a small solid-to-solid contact area and will create a significant thermal resistance. This high resistance results from the fact that all the void areas contain air which has a very high thermal resistivity of 1200°C · in/W (30.48°C · m/W). Therefore, almost any method of attaching a semiconductor device to its heat sink is better than a simple dry mount.

Device attachment must consider whether there is a need for electrical isolation from the heat sink. If no isolation is required, the optimum resistance can be attained by using a thermal compound. If electrical isolation is required, several options can be considered:

- Resilient pad material (no thermal compound required)
- Beryllium oxide wafer with thermal compound
- Mica wafer with thermal compound
- Coated aluminum wafer with thermal compound

Table 5.2. Thermal Resistivity Values for Common Materials

Material	Thermal resistivity, ρ, °C · in/W
Still air	1200
Mylar film	236
Silicone grease	204
Mica	66
Fiberglass-silicone resilient material	56
Silicone or oil-based thermal compound	56
Thermally conductive dielectric epoxy	48
Anodized coating	5.6
Alumina	1.2
Steel, carbon	0.84
Beryllium oxide	0.32
Aluminum	0.19
Copper	0.10

CONVERSION FACTOR: (°C · in/W) × 0.0254 = (°C · m/W).
SOURCE: Reprinted with permission from EG&G Wakefield Engineering, *Standard Products Catalog*.

(a) (b) (c)

Fig. 5.18 Examples of custom-designed heat sinks. (*a*) Mechanically attached fins with heat pipes included; (*b*) convoluted fin assembly; (*c*) fabricated fin epoxy bonded assembly.[1b]

The thermal resistance from device case to heat sink θ_{cs} can be calculated by

$$\theta_{cs} = \frac{\rho t}{A} \tag{5.17}$$

where ρ = thermal resistivity, °C · in/W (°C · m/W)
t = average layer thickness, in (m)
A = contact area, in^2 (m^2)

Figure 5.17 shows a typical arrangement of mounting a semiconductor device to a heat sink. To aid in the calculation procedure, a tabulation of thermal resistivity values for commonly used materials is given in Table 5.2.[1a]

The considerations for the joint design between a device and its heat sink should not be minimized. These should include the life expectancy, the frequency of device removal, and the operating environment.

Example (Multiple Layer, Fig. 5.17) Assume a semiconductor device in a TO-3 style case is to be mounted on a heat sink with a dielectric coated insulating wafer. The actual layers and the individual resistances in that interface would be:

Thickness, in	Layer	Thermal resistance θ, °C/W
0.001	Thermal compound (assumed)	0.056
0.0035	Thermally conductive dielectric epoxy	0.168
0.025	Aluminum	0.005
0.0035	Thermally conductive dielectric epoxy	0.168
0.001	Thermal compound (assumed)	0.056
	Total θ_{cs}	0.453

5.10 CUSTOM-DESIGNED HEAT SINKS

Although standard air- and liquid-cooled cold plates and heat sinks are commercially available, there is a great demand by equipment designers for unusual solutions to unique cooling and packaging problems. To satisfy these requirements, many unconventional and innovative design concepts available are for those with special cooling problems. Some of these concepts are illustrated in Fig. 5.18*a* to 5.18*c*.[1b]

Figure 5.18*a* illustrates a passive design utilizing heat pipes. Heat from a modular device is effectively transported through heat pipes to the finned section and dissipated by natural convection to the surroundings. The unique design of Fig. 5.18*b* is an all-copper convoluted fin heat sink used to cool the hot footprint of a traveling-wave tube (TWT). The large heat sink in Fig. 5.18*c*, not extrudable, was fabricated from mechanically formed fins and is an alternative in packaging of multiple devices of a power supply.

REFERENCES

1. EG&G Wakefield Engineering, Wakefield, Massachusetts, Components Division: (a) *Standard Products Catalog,* 1985; (b) *Active Cooling Products and Capabilities Catalog,* 1985.
2. D. Q. Kern and A. D. Kraus, *Extended Surface Heat Transfer,* McGraw-Hill, New York, 1972, p. 89, Eq. (2.9).
3. W. H. McAdams, *Heat Transmission,* 3d ed., McGraw-Hill, New York, p. 173, Eq. (7-5b).
4. International Rectifier, *Heat Exchanger Data Book AN-701,* p. 6. (Modified for English units.)
5. Genium Publishing Corporation, *Heat Transfer Book,* Secs. 504.3 and 505.2, 1145 Catalyn Street, Schenectady, N.Y.

Part 8

Electric Powered Heating

Chapter **1**

Electric Resistance Heating-Element Materials

ROBERT WATSON
The Kanthal Corporation

1.1 INTRODUCTION AND DEFINITIONS

Resistance heating materials are used for a wide variety of applications ranging from defroster elements in refrigerators to furnace heating elements operated at temperatures in excess of 3000°F (1650°C). In general, four factors characterize resistance heating materials—resistivity, maximum temperature, temperature coefficient of resistance, and oxidation resistance. This chapter reviews the generally available materials and lists the critical properties. The materials are grouped into oxidation-resistant and non-oxidation-resistant materials, with subgrouping into metals, ceramic-metals, and ceramics.

The selection of resistance heating materials must be based on optimization of several key characteristics of the available materials. In order to assure a basic understanding of the terminology of resistance heating materials, the following definitions are provided.

1. *Resistivity:* This characterizes the ability of a material to inhibit electrical current flow in the presence of an applied voltage. The common units of resistivity are in ohms per square mil-foot or ohms per circular mil-foot (ohms-square

millimeter per meter). Resistivity is independent of shape or length. Resistance is generally either an exact value (ohms) or given in a value per unit length (ohms per foot). By knowing the resistivity of a material, one can calculate the resistance (in ohms per length) from the following formula:

Round wire:

$$\Omega/\text{ft} = \frac{\text{resistivity, } \Omega/\text{cir mil} \cdot \text{ft}}{(\text{wire diameter}/1000)^2}$$

Rectangular forms:

$$\Omega/\text{ft} = \frac{\text{resistivity, in } \Omega/\text{mil}^2 \cdot \text{ft}}{\text{thickness}/1000 \times \text{width}/1000}$$

where wire diameter, thickness, and width are in units of inches.

2. *Temperature coefficient of resistance:* This is a correction factor for changes in the resistivity of a material with respect to temperature. A material that has an increase in resistivity with increasing temperature has a positive temperature coefficient, while one that has a resistivity decrease with increasing temperature has a negative temperature coefficient. Most materials have a positive temperature coefficient. Materials for high-temperature heating [above 400°F (200°C)] will have a temperature coefficient (of resistance) (TC) given as a correction factor which is to be multiplied times the base (room temperature) resistivity value. Sometimes the resistivity change is given in units of ohms per ohm-°C in parts per million (ppm). In most cases this ppm figure is used for materials applied at low temperatures and will have a very small change in resistivity (such as a wire-wound resistor).

3. *Maximum temperature:* The maximum temperature of a material is the generally accepted temperature that should not be exceeded by the material while in use. It should be noted that this maximum temperature does not refer to the temperature of a chamber, furnace, or device containing the resistance element but the element itself. In some cases it may be possible to operate above this maximum temperature, but the life that may be expected will be decreased exponentially, and an extremely short life may result.

1.2 ENVIRONMENTAL CONSIDERATIONS

Most materials used for resistance heating fall into one of two categories: oxidation-resistant materials or non-oxidation-resistant materials. While the oxidation-resistant materials are developed primarily for use in air, they can often be used in other gases or vacuum as well. The non-oxidation-resistant materials generally require a protective atmosphere (hydrogen, nitrogen, etc.) or a vacuum environment.

1.3 OXIDATION-RESISTANT MATERIALS

Oxidation-resistant materials fall into three general classifications: metals, ceramic-metals (cermets), and ceramics.

Metallic Oxidation-Resistant Materials

The most widely used resistance heating materials are metallic. These materials fall into three general categories: nickel-chrome, iron-chrome-aluminum, and precious metals. The nickel-chrome materials were developed in the United States in the early 1900s and have been used as the traditional metallic resistance heating material in the United States. Table 1.1 lists the typical characteristics of four nickel-chrome alloys.

In the 1930s a second group of heat-resistant alloys was developed, the iron-chrome-aluminum alloys. These materials were developed in Sweden and are used in lieu of nickel-chrome alloys in many applications in Europe and the Far East. Use in the United States has not been as widespread because of the prior design experience with nickel chrome. The iron-chrome-aluminum alloys in the past have been primarily limited to applications which require some of the unique advantages of the iron-chrome-aluminum materials; however, more conversions to this alloy are seen as a result of recent improvements. Table 1.2 lists the properties of several iron-chrome-aluminum alloys.

In order to operate in an oxidizing environment (or air), the heating element material must be able to form an oxide layer that protects the material from further oxidation. If the material is unable to form a stable, protective oxide surface (such as aluminum exposed to weather), it will be continuously oxidized until failure occurs. Accordingly, one key property for these oxidation-resistant materials is the ability to form a stable oxide. With the nickel-chrome materials, the oxide that forms at elevated temperatures is a chromium oxide (CrO_2). This oxide is generally a dark green, blackish color. While this oxide does adhere to the base material reasonably well, it does become relatively thick over time and has a tendency to spall or flake off, exposing more of the base material. The oxide that forms on the iron-chrome-aluminum materials is an aluminum oxide [alumina (Al_2O_3)]. The aluminum oxide is a grayish color and has the advantage that it more tightly adheres to the base material and thus has less tendency to spall.

Table 1.1. Nickel-Chrome Alloys

Designation	Composition	Resistivity, Ω/cir mil · ft	Maximum temperature in air, °F (°C)
ASTM* A Grade	80%Ni-20%Cr	655	2195 (1200)
	70%Ni-30%Cr	705	2280 (1250)
ASTM C Grade	60%Ni-15%Cr-25%Fe	668	2055 (1125)
ASTM D Grade	35%Ni-20%Cr-45%Fe	626	2000 (1100)

* American Society For Testing and Materials, Philadelphia, Pa.

Table 1.2. Iron-Chrome-Aluminum Alloys[1]

Composition	Resistivity, Ω/cir mil · Ft	Maximum temperature in air, °F (°C)
5.5%Al-22%Cr-balance Fe	872	2550 (1400)
5.0%Al-22%Cr-balance Fe + additions	837	2550 (1400)
5.0%Al-22%Cr-balance Fe	837	2425 (1330)
4.5%Al-22%Cr-balance Fe	812	2335 (1280)
5.0%Al-15%Cr-balance Fe	752	1920 (1020)

The potential advantages of aluminum oxide are the generally higher temperature capability and the normally longer life of the iron-chrome-aluminum alloys. Table 1.3 shows the relative lifetimes of several materials. These data have been generated from ASTM tests (ASTM standards B76-81 and B78-81) or modification of these procedures. In this test, the heating element material is heated to a specified temperature and then allowed to cool to room temperature. The cycle is then repeated until the sample fails. While there is no direct correlation between this test and how the element will perform in an actual application, the results can be useful on a relative basis. The values in Table 1.3 are the percentages of the life expected for the 5.5%Al-22%Cr-balance Fe alloy at 2190°F (1200°C). A value of 42 for ASTM A-grade material at 2190°F (1200°C) indicates that it will have 42 percent of the life of the 5.5%Al-22%Cr-balance Fe material at 2190°F (1200°C). The chart can be used for comparing different alloys and to evaluate what change in life might be expected from changes in the element temperature.

Other considerations for the oxide properties are sometimes important in the alloy selection. First, if it is important to preserve purity of the ware during heating, it may be preferable to use the iron-chrome-aluminum alloys because its oxide will tend to spall less and be less apt to spot or discolor the ware. This is an extremely important factor in choosing heating elements for use in glassware or porcelain furnaces. The type of oxide formed also has an effect on the element temperature. Since the chromium oxide has a higher emissivity than the aluminum oxide, a nickel-chrome element will run slightly cooler than the iron-chrome-aluminum material at equivalent surface loading.

The resistance change with respect to temperature is an important design consideration. It must be realized that, with some materials, the change in resistance is so great that if the element were to be designed with respect to room-temperature properties only, the power may be insufficient at operating temperatures. Table 1.4 lists the typical temperature coefficient of resistance for the nickel-chrome and iron-chrome-aluminum materials.

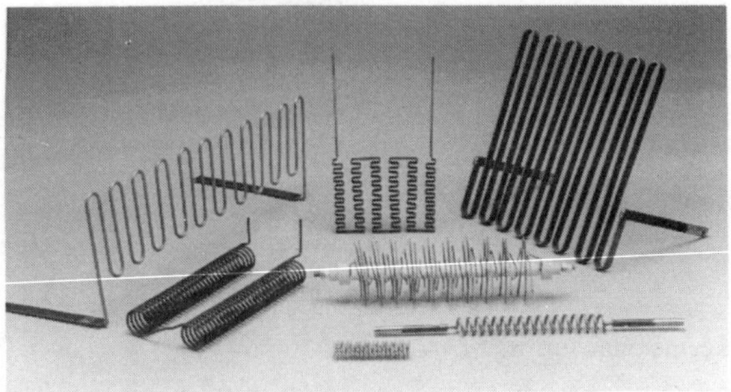

Fig. 1.1 Typical metallic heating-element configuration. (*Courtesy of Kanthal Furnace Products, Hallstahammar, Sweden.*)

In general, either type of material can be formed by traditional metalworking methods. It is important that the element terminations be made of a material that is compatible with the heating section. It is usually recommended that nickel-chrome materials be joined to nickel-chrome terminals, while iron-chrome-aluminum materials be joined to iron-chrome-aluminum terminals.

Both nickel-chrome and iron-chrome-aluminum materials are formed into a wide variety of shapes, forms, and configurations. In some cases the configuration will determine which alloy is chosen. As the iron-chrome-aluminum alloys are iron-based, they will typically exhibit a lower hot strength than nickel-chrome materials. This is generally important in air heaters where the heating element may be required to support itself. If the element temperature exceeds 1475°F (800°C), the required and the actual strength should be compared. Figure 1.1 shows typical heating-element forms.

The majority of metallic heating elements are formed into helically wound coils, wrapped onto a form, or formed into sinuous loop elements. The heating wire is often combined with other materials to form heaters.

Table 1.3 Relative Life* of Iron-Chrome-Aluminum and Nickel-Chrome Alloys in Air[1]

Alloy	Temperature, °F (°C) 2010 (1100)	2190 (1200)	2370 (1300)
Nickel-chrome			
ASTM A	236	42	NR†
ASTM C	238	44	NR
ASTM D	74	20	NR
Iron-chrome-aluminum			
5.5%Al-22%Cr	516	100	34
5%Al-22%Cr + additions	849	132	35
4.5%Al-22%Cr	407	96	24

* Relative life figures, with 5.5%Al-22%Cr-balance Fe 100 percent.
† Not recomended.

Table 1.4. Temperature Coefficients of Resistance for Nickel-Chrome and Iron-Chrome-Aluminum Alloys[1,2]

	Nickel-Chrome Alloys				Iron-Chrome-Aluminum Alloys				
Temperature, °F (°C)	ASTM A	70%-Ni-30%Cr	ASTM C	ASTM D	5.5%Al-22%Cr	5.0%Al-22%Cr	5.0%Al-22%Cr +	4.5%Al-22%Cr	5.0%Al-15%Cr
68 (20)	1.00	1.00	1.00	1.00	1.00	1.00	1.00	1.00	1.00
212 (100)	1.01	1.01	1.02	1.03	1.00	1.00	1.00	1.00	1.01
392 (200)	1.02	1.02	1.04	1.06	1.00	1.01	1.01	1.01	1.02
572 (300)	1.03	1.04	1.05	1.10	1.00	1.01	1.01	1.01	1.03
752 (400)	1.04	1.05	1.07	1.13	1.01	1.02	1.02	1.02	1.04
932 (500)	1.05	1.06	1.08	1.15	1.01	1.03	1.03	1.03	1.05
1112 (600)	1.04	1.06	1.09	1.17	1.02	1.04	1.04	1.05	1.08
1292 (700)	1.04	1.05	1.09	1.19	1.02	1.04	1.04	1.06	1.09
1472 (800)	1.04	1.05	1.10	1.21	1.03	1.05	1.05	1.06	1.10
1652 (900)	1.04	1.05	1.11	1.22	1.03	1.05	1.05	1.07	1.11
1832 (1000)	1.05	1.05	1.11	1.24	1.04	1.06	1.06	1.07	1.11
2012 (1100)	1.06	1.06	1.12	1.24	1.04	1.06	1.06	1.07	...
2192 (1200)	1.07	1.07	...	...	1.04	1.06	1.06	1.08	...
2372 (1300)	...	...	...	...	1.04	1.06	1.06	1.08	...
2552 (1400)	...	...	...	...	1.04	...	1.07	...	...

Other Metallic Materials

Platinum is a noble metal that is sometimes used as a heating element at high temperatures. Unlike the previously mentioned metallic materials, it does not form a protective oxide coating. It can be used in oxidizing environments at high temperatures because it is extremely stable and nonreactive with oxygen.

Platinum can be used up to 2912°F (1600°C) in oxidizing environments and is often alloyed with other materials to extend its useful temperature range. For example, a Pt-20% Rh alloy can be used up to 3092°F (1700°C). Platinum is generally not recommended for use in vacuum or hydrogen because of possible reactions from contamination that may be reduced to a base metal.

Platinum and platinum alloys have relatively high temperature coefficients of resistance. Table 1.5 lists the resistivity of platinum and platinum alloys at two temperatures.

A number of other lower-temperature materials are also used for special heating purposes. One example is the use of a thin foil of aluminum, stainless, or brass alloy that is sandwiched between two flexible layers to form a foil heater. Foil elements ordinarily have a low power dissipation but have the ability to apply the power over a large area, with the desired distribution. In addition, since foil elements are low in mass, they can heat rapidly to temperatures of 752°F (400°C).

Ceramic Metals

Ceramic-metals, or cermets, are a class of materials that exhibit properties of both ceramics and metals. One of the most interesting for resistance heating elements is molybdenum disilicide. One property for which this material has been widely known is its high oxidation resistance, one of the highest of any electrically conductive material. Unfortunately, the ceramic properties of the material are that it is extremely brittle and cannot be directly fabricated into usable heating-element forms. In the 1950s a powder metallurgical method was developed that allowed the material to be formed into U-shaped heating elements that were usable to element temperatures of 3092°F (1700°C). A second quality has subsequently been developed that allows element temperatures of up to 3272°F (1800°C).

The key to the oxidation resistance of this material is the formation of a silicon dioxide layer. Just as in the metallic materials, this oxide layer acts as a barrier to further oxidation and allows the element to survive at the high temperatures. This oxide forms whenever the element temperature is above 1830°F (1000°C) in air. As with all materials that form such protective oxides, the optimum life can be obtained by heating the elements in air to form a good oxide layer before actual use in the application. With molybdenum disilicide heating elements, any material that will react with the silica layer will reduce the maximum operating temperature. Some of the worst offenders to the silica layer are members of the halogen family. These materials (e.g., chlorine or fluorine) can prevent the formation of a stable oxide layer, causing the element to continuously loose the oxide layer or preventing it from forming.

These elements are widely used in laboratory and industrial furnaces. While they were originally developed for high temperatures, their use at lower temperatures [1472 to 2192°F (800 to 1200°C)] is expanding. The low-temperature use of these elements is due to their good corrosion resistance to most environments. The corrosion resistance can result in very long life at these lower temperatures. In addition to the use of this material in furnaces, improved fabrication techniques are allowing the use of this material in lower-temperature applications where rapid heating, high element temperature, or high power concentrations are beneficial.

Table 1.5. Platinum and Platinum Alloys

Composition	Resistivity, Ω/cir mil · ft At 68°F (20°C)	Resistivity, Ω/cir mil · ft At 2732°F (1500°C)	Maximum temperature in air, °F (°C)
Pt	65	320	2910 (1600)
Pt-10%Rh	115	320	3000 (1650)
Pt-20%Rh	125	350	3092 (1700)

Molybdenum disilicide is traditionally manufactured in three hot-zone diameters, 0.118 in (3 mm), 0.236 in (6 mm), and 0.354 in (9 mm). In order to terminate the hot zone, it is welded to molybdenum disilicide material having twice the diameter of the hot zone. The terminal can then pass outside the furnace insulation where a lower-temperature low-resistance connection can be made. Figure 1.2 shows some typical element configurations.

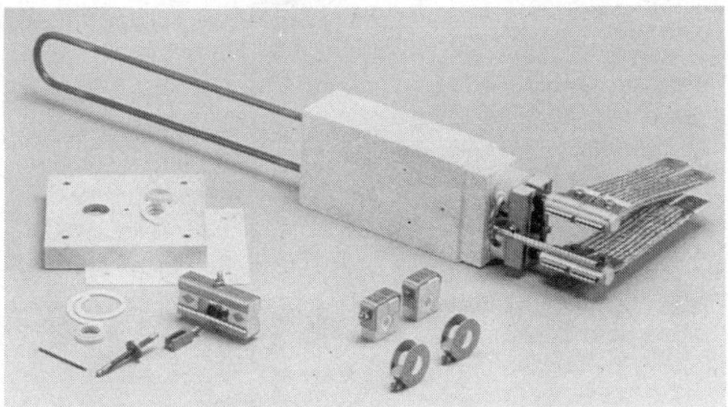

Fig. 1.2 Typical molybdenum disilicide heating-unit assembly. (*Courtesy of Kanthal Furnace Products, Hallstahammar, Sweden.*)

The resistivity of molybdenum disilicide is strongly positive, increasing approximately 13 times from room temperature to 3000°F (1650°C). Table 1.6 lists the resistivity and the resistance in ohms per foot of both the standard and improved grades of molybdenum disilicide at various temperatures.

Because of the highly positive temperature coefficient, special consideration has to be given to starting these elements. While the element itself can tolerate the thermal shock, the current inrush is too high for power supplies sized for use at operating temperatures. Two alternatives have been used to avoid the inrush problem. First is the use of three voltage steps in starting up the elements. This voltage stepping can be accomplished by switching either a multitapped transformer or the element connections. The most common method of dealing with the current inrush is the use of phase-angle

Table 1.6. Molybdenum Disilicide Resistivity[3]

Temperature		Resistivity, Ω/cir mil · ft	
°F	°C	Standard material	3000°F (1800°C) material
68	20	180	168
212	100	265	235
392	200	367	325
572	300	481	427
752	400	602	536
932	500	721	650
1112	600	842	770
1292	700	963	890
1472	800	1089	1010
1652	900	1221	1136
1832	1000	1347	1264
2012	1100	1475	1396
2192	1200	1612	1534
2372	1300	1752	1672
2552	1400	1905	1803
2732	1500	2058	1956
2912	1600	2211	2109
3092	1700	2381	2245
3272	1800		2398

fired silicon controlled rectifiers (SCRs or thyristors). These solid-state devices can limit the current by turning off when the current exceeds a preset value.

Because of the rapid heating capability of the molybdenum disilicide heating elements, they are well suited for use in modern ceramic-fiber-lined furnaces. There has been some concern that the molybdenum disilicide was not suited for cycling furnaces. The reason for breakage during cycling of a furnace can, in most cases, be traced to inadequate clearances around the feedthrough refractory. As the refractory expands and contracts, it is important that no mechanical stress be placed on the elements.

Ceramic Materials

Silicon carbide is a synthetic ceramic that was originally developed as an abrasive. Heating elements were made from this material in the United States prior to the 1940s. Silicon carbide elements are produced by forming grains of silicon carbide into bars which are fired to bond or sinter the grains of silicon carbide together. After firing the bars, reduced-resistance terminals (cold ends) are developed by either "welding" a low-resistance silicon carbide connector to the ends or impregnating the ends of the bar with silicon metal in order to reduce the resistance.

When silicon carbide heating elements are heated in air, an oxidation process begins that will slowly increase the resistance of the bar. Initially the oxidation process is slow, but the oxidation rate later continues to climb at an ever-increasing rate. Because of this increase of resistance with time (aging), special considerations must be given to the power supplies used with silicon carbide heating elements. In most cases it is recommended that a power supply with the capacity of twice the nominal operating voltage of new elements be used.

The resistivity of silicon carbide is so low that a transformer will be required in most cases. Additionally, the resistivity of silicon carbide first decreases (negative temperature coefficient of resistance) and then increases (positive coefficient of resistance). Figure 1.3 is a typical resistance curve.

One advantage of silicon carbide heating elements is strength at elevated temperatures in compression. The strength is sufficient to allow elements to be suspended with supports only at the ends.

Silicon carbide elements are provided in several forms. The most common is the straight rod-type element, with diameters ranging from 0.375 in (9.5 mm) to 2.125 in (54 mm). The maximum length increases with increasing diameter, with lengths of over 10 ft (3 m) possible with the largest diameters.

In addition to the straight elements, some manufacturers join straight lengths together at one end to make U elements. These U elements are generally used when connections are available at only one end.

Another type of silicon carbide element is the spiral-cut

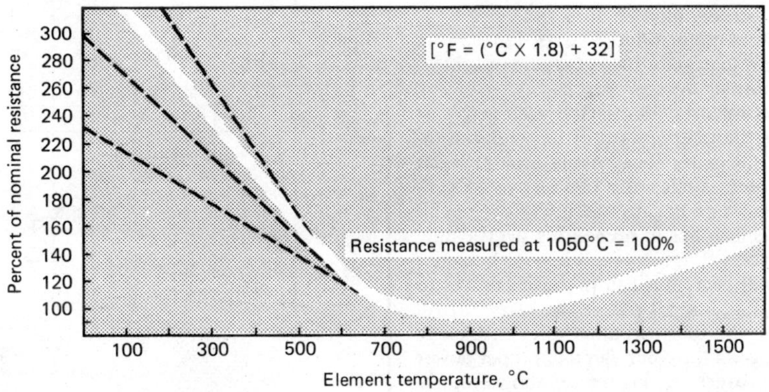

Fig. 1.3 Resistance characteristic of silicon carbide heating elements. (*Courtesy of Kanthal Limited, Perth, Scotland, U.K.*)

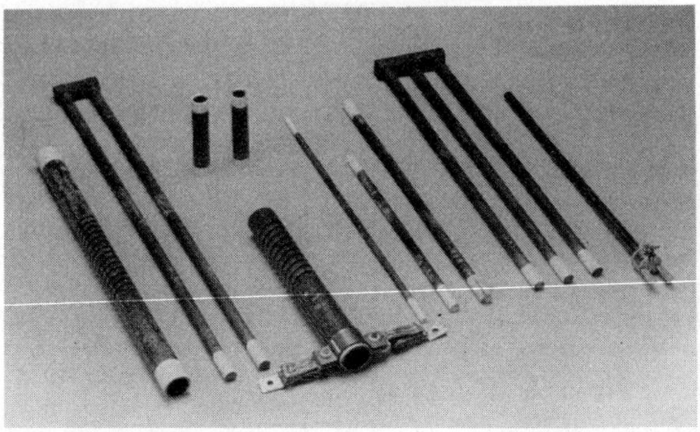

Fig. 1.4 Typical silicon carbide heating-element configurations. (*Courtesy of Kanthal Furnace Products, Hallstahammar, Sweden.*)

heating element. These elements can be used with the electrical connections on opposite ends or with both connections at the same end. Several common types of silicon carbide heating elements are shown in Fig. 1.4.

Lanthanum chromite is an oxide compound that can be used at furnace temperatures of up to 3300°F (1815°C). These elements are rod-type with enlarged terminal sections for electrical connections on each end. The life that can be expected from these elements is strongly temperature dependent. Life is about 500 h at 3300°F (1815°C) furnace temperature and over 2500 h at 3100°F (1700°C). The resistivity of lanthanum chromite decreases rapidly up to a temperature of about 1200°F (650°C), above which the resistivity decreases very slowly with increasing temperature. The elements do age; however, the aging effect is very small (less than 10 percent).

One disadvantage with this type of element is that, at high temperatures, Cr_2O_3 is evaporated from the element. In order to prevent contamination of the material being heated, it is recommended that the elements be separated from the material by a shield of pure alumina ceramic. Lanthanum chromite can be used in oxidizing atmospheres, or atmospheres with at least 0.1 atm of oxygen partial pressure. If these elements are used in other environments (hydrogen, nitrogen, argon, or vacuum), they should be shielded from the atmosphere by nonporous, high-alumina ceramics.

Despite the drawbacks mentioned above, lanthanum chromite is one of the few materials that is capable of achieving a 3272°F (1800°C) furnace temperature.

Zirconia is a ceramic material that is not conductive at room temperatures. As the temperature of the material increases the resistivity of the material decreases until, at about 1472°F (800°C), it becomes sufficiently conductive to generate heat. The zirconia has a negative temperature coefficient, which can cause difficulties in preventing overheating the element. With a fixed voltage, as the element becomes hotter, the resistance will decrease, causing more power to be generated. With this additional power the element will increase in temperature and the resistance will continue to fall. Without careful design of the controller and proper power supply, the element can overheat. Figure 1.5 shows the resistance curve for a typical zirconia heating-element system.

The principle of the zirconia heating element has been known since the 1800s; in fact, a patent was issued in the early 1900s for a furnace utilizing the zirconia heater. While the zirconia element has the potential of being used at temperatures above 3632°F (2000°C), its use has been limited because of the large number of technical problems associated with the furnace system. These problems range from the material compatibility of the furnace chamber at 3632°F (2000°C) to problems in starting and controlling the element. Recent advances have been achieved with this technology as increasing interest has developed in higher-temperature material processing.

Use of Oxidation-Resistant Materials in Other Environments

As indicated previously, the oxidation-resistant materials are most often gaged by their performance in air or other oxidizing environments. In many heating applications it can be desirable to use these materials in other environments. Table 1.7 lists the maximum recommended temperatures for the materials reviewed above.

Approximate Relative Costs

As a general rule, the initial cost of an element increases with increasing maximum temperature capability. When the initial cost is examined, the total lifetime of the element

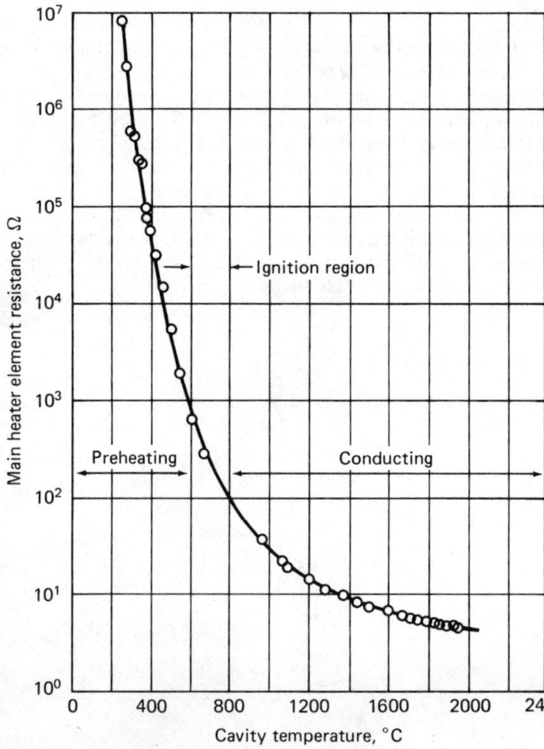

Fig. 1.5 Resistance curve for zirconia heating-element system. (*Courtesy of Artcor, Inc., Costa Mesa, Calif.*)

Table 1.7. Maximum-Use Temperatures (in °C) of Oxidation-Resistant Materials in Other Atmospheres[1-4]

Atmosphere	Nickel-chrome alloys				Iron-chrome aluminum				
	ASTM A	70%Ni-30%Cr	ASTM C	ASTM D	5.5%Al-22%Cr	5.0%Al-22%Cr	4.5%Al-22%Cr	$MoSi_2$ standard	SiC
Air	1200	1250	1125	1100	1400	1400	1280	1700	1600
Nitrogen	1200	1250	1125	1100	1200	1200	1050	1600	1350
Hydrogen									
(15°C dewpoint)	1200	1250	1125	1100	1350	1350	1250	1460	1200
Hydrogen (dry)	1200	1250	1125	1100	1350	1350	1250	1350	1200
Exothermic									
(15%H_2-70%N_2-10%CO_2-5%CO)	<900	1200	1125	1100	1250	1250	1150	1600	1400
Endothermic									
(20%CO-40%H_2-40N_2)	1100	1200	1100	1050	1200	1200	1100	1400	1250
Vacuum									
10^{-3} torr	1100	1100	1000	950	1200	1250	1100	1100	1100
10^{-5} torr	1000	1000	900	850	1050	1100	900	900	900
CO_2								1600	1600

CONVERSION FACTOR: °F = 1.8 × °C + 32.

Table 1.8. Properties of Molybdenum, Tantalum, and Tungsten[5,6]

Property	Molybdenum	Tantalum	Tungsten
Melting point, °C	2610	3030	3390
Maximum-use temperature, °C	1900	2200	2500
Resistivity at 20°C, Ω/cir mil · ft	32	75	34
Resistivity at 1000°C, Ω/cir mil · ft	192	325	198
Resistivity at 1500°C, Ω/cir mil · ft	289	433	301
Resistivity at 2000°C, Ω/cir mil · ft	379	523	397
Resistivity at 3000°C, Ω/cir mil · ft	...	...	619
Formability	Fair	Poor	Good

CONVERSION FACTOR: °F = 1.8 × °C + 32.

should also be reviewed so that lifetime cost can be compared. While a metallic heating element would be initially the least expensive for a furnace operating at 2372°F, (1300°C), the true operating cost for a molybdenum disilicide or silicon carbide element may be less expensive. The lifetime cost is one reason why the molybdenum disilicide is being used more frequently in lower-temperature [below 2000°F (1100°C)] furnaces with protective atmospheres.

Another consideration for element selection would be the necessity of a special or more costly power supply. The type of power supply will have a bearing on the initial cost of the heating system.

1.4 NON-OXIDATION-RESISTANT MATERIALS

The non-oxidation-resistant materials fall into two categories: metallic and carbon or graphite. Unlike the oxidation-resistant materials, these heating elements cannot be heated to high temperatures in air. If they are heated in air above a critical temperature, they are rapidly oxidized and soon fail.

Metallic Materials

Three commonly used non-oxidation-resistant metallic heating element materials are molybdenum, tungsten, and tantalum. These materials exhibit some similar properties in that they all have strongly positive temperature coefficients; have very high melting temperatures; are more difficult than traditional metallic materials to form, shape, or join; and require operation in either vacuum or a protective atmosphere to prevent oxidation. Table 1.8 lists the physical data for these materials.

Both molybdenum and tungsten can be used in almost any atmosphere that protects it from oxidizing. Tantalum, however, must be used only in gases that do not contain oxygen, nitrogen, hydrogen, or carbon. Tantalum will form nitrides in nitrogen, hydrides in hydrogen, and carbides in carbon.

In using these materials, it is important to give special consideration to the supports that are in contact with the heating elements. While all three can be used with high-purity alumina (Al_2O_3) up to 3450°F (1900°C), the temperature must be reduced if the elements are supported on lower-grade materials. If the elements are supported on mullite (65%Al_2O_3-35%SiO_2), the maximum temperature must be reduced to about 3100°F (1700°C). On standard firebrick the maximum temperature must be reduced even further to approximately 2200°F (1200°C).

Nonmetallic Materials

Graphite elements are used in many applications requiring high temperatures in vacuum or nonoxidizing atmospheres. In air, graphite will begin to oxidize at about 750°F (400°C); however, in nitrogen, hydrogen, or rare gases it can be used at temperatures of up to 5400°F (3000°C). In a vacuum, elements are limited to about 4000°F (2200°C). At the highest temperatures graphite does not melt, but rather will sublime. The sublimation process accelerates with increasing temperatures resulting in decreased life.

The resistivity of graphite is relatively high. While the resistivity may vary depending on the grade or quality of

Table 1.9. Maximum-Use Temperature (in °C) of Non-Oxidation-Resistant Materials in Other Atmospheres[5-7]

Atmosphere	Molybdenum	Tantalum	Tungsten	Graphite
Air	450–600	500	500–700	400
Nitrogen	1600	500	1350	2200
Hydrogen (15%°C dewpoint)	1400	450	1400	
Hydrogen (dry)	1900	450	2500	2200
Exothermic (15%H_2-70%N_2-10%CO_2-5%CO)	1200	450	1300	
Endothermic (20%CO-40%H_2-40%N_2)	1100	450	1000	
Vacuum				
10^{-2} torr	1700	...*	2000	
10^{-4} torr	1800	2200	2400	2200

* Becomes brittle as a result of action of gas residues (either oxygen or nitrogen).
CONVERSION FACTOR: °F = 1.8 × °C + 32.

graphite, it will, in general, be approximately 4800 $\Omega/mil^2 \cdot ft$. The temperature coefficient of graphite is negative at low temperature. At about 1450 to 1830°F (800 to 1000°C) the temperature coefficient becomes positive; the resistivity at 2190 to 2550°F (1200 to 1400°C) is the same as at room temperature. Some of the key factors that make graphite an excellent heating element material are the following:

Low coefficient of thermal expansion
High strength at higher temperatures
Resistance to thermal shock
Resistance to chemical attack
High thermal conductivity
Low vapor pressure [3.61×10^{-11} torr at 1830°F (1000°C)]
Ability to be machined
High thermal emissivity (about 85 percent the emissivity of a blackbody)[7]

Non-Oxidation-Resistant Materials in Other Environments

As with the oxidation-resistant materials, the non-oxidation-resistant materials are often used in environments other than vacuum. Table 1.9 lists the typical maximum element temperatures allowable for various environments.

1.5 CONCLUSION

Many different types of heating-element material are available. In choosing which material to use, two important factors are the anticipated maximum element temperature and the compatibility of the element with the environment. In order to provide the proper type of power control, it is necessary to consider how the resistivity will change with respect to the element temperature.

Once these factors are considered, one or two candidate element materials can be selected. It will then be necessary to size the heating element for the power required. These calculations are reviewed in subsequent chapters. On the basis of these calculations, it can then be determined whether the operating criteria are exceeded. If the results indicate that the element is operating far below the maximum temperature possible, it may be cost-effective to choose a material that has lower maximum temperature capability or redesign the element so that it will operate closer to its performance limit.

REFERENCES

1. *Kanthal Resistance Heating Handbook for Industrial Furnaces,* Kanthal Furnace Products, Hallstahammar, Sweden, 1986.
2. *The Nikrothal Handbook,* AB Kanthal, Hallstahammar, Sweden, 1969.
3. *Kanthal Super Handbook,* Kanthal Furnace Products, Hallstahammar, Sweden, 1984.
4. *Kanthal Silicon Carbide Electric Heating Elements,* Kanthal Limited, Perth, Scotland.
5. R. Kieffer and F. Benesovsky, "Metallic Heating Element Materials for High Temperature Furnaces," *Metallurgia,* **58** (347) (September 1958).
6. *Custom Fabricated Parts of Tungsten Molybdenum and Their Alloys,* GTE Products Corporation, Towanda, Pa.
7. *Carbon and Graphite Products,* Airco Carbon Products Department, St. Marys, Pa.
8. *1986 Annual Book of ASTM Standards,* Vol. 03.04, American Society of Testing and Materials, Easton, Pa., 1986.

Chapter 2

Characteristics and Application of Electric Resistance Heaters

JEFFREY P. MELLY
Watlow Electric Manufacturing Company

This chapter summarizes the characteristics and application of electric resistance heaters. For each commonly used heater type there is a description of construction; typical applications; temperature, watt density, and voltage capabilities; thermal response; efficiency*; and heater life considerations where appropriate.

A method for determining heater needs for given applications is also presented. This method includes defining the application problem; determining energy requirements; reviewing application factors; selecting heater types, sizes, and quantities; and evaluating control methods.

* *Efficiency* in this chapter is defined as the percentage of the generated heat energy that is transferred to the object to be heated. A low efficiency indicates that much of the energy generated is lost to the surroundings.

2.1 DESCRIPTION AND CHARACTERISTICS OF VARIOUS HEATERS

The characteristics of the various commonly used heater types are described in the following paragraphs and are summarized in Table 2.1.

Cartridge Heaters

CONSTRUCTION. Cartridge heaters are cylindrical in shape and consist of a resistance element wound around a ceramic core. The wound core is concentrically placed within a metal sheath tube and surrounded with granular magnesium oxide material. This electrically insulates the resistance element from the tube and provides a medium for heat transfer to the sheath tube. Electrical terminations are made at one end. Higher-performance, longer-lived cartridge heaters are *swaged* (a rotary hammer compacting process) to improve internal heat transfer and electrical insulating properties. Swaging densifies the granular insulation material. Cartridge heaters are widely applied and are manufactured in a wide range of lengths, diameters, heat flux-to-length variations, wattages, lead arrangements, and internal sensor and control mechanisms. (See Fig. 2.1.)

TYPICAL APPLICATIONS. The following equipment uses cartridge heaters: plastic molds, runners, and gating systems; platens; barrels on extrusion equipment; packaging seal bars, platens, hot melt glue systems, labeling and marking systems; medical laboratory equipment, including chemistry analyzers, chromatography equipment, ovens, autoclaves, and kidney dialysis systems; photographic processing equipment; environmental rooms; and food-warming equipment and steamers. Cartridge heaters are capable of both high-temperature and high-watt-density operation, but care must be taken to apply cartridge heaters at compatible watt density and sheath temperature ratings.

TEMPERATURE RANGE. Up to 1600°F (870°C) sheath temperature is recommended.

WATT DENSITY RANGE. Up to 400 W/in^2 (62 W/cm^2) and higher under special conditions is recommended. As a rule, a cartridge heater can operate at 1600°F (870°C) at 40 W/in^2 (6.2 W/cm^2).

VOLTAGE. Typically 120 and 240 V is used, with 480 V and higher requiring special design or sizing considerations.

THERMAL RESPONSE. Cartridge heaters depend on conductive heat transfer in most applications, and good conduction requires close fit (hole ID–heater OD) between the heater and the work to be heated. As fit looseness increases, watt density must be decreased accordingly. Watt densities of liquid-immersion cartridge heaters vary according to liquid viscosity and heat capacity, and care must be taken if film boiling conditions (departure from nucleate boiling) might be possible.

EFFICIENCY. Cartridge heaters are 100 percent efficient when immersed.

HEATER LIFE. Operating at 1600°F (870°C) internal temperature with modest cycling and with the proper hole fit, a properly designed and constructed cartridge heater will provide approximately a one-year life.

Flexible Heaters

CONSTRUCTION. Flexible heaters consist of a pattern-wound resistance wire or chemically machined resistance foil bonded between two layers of a flexible insulating material. The standard insulating material is fabric-reinforced silicone rubber. Other insulating materials used include Kapton,* reinforced Neoprene, Mylar,* and mica. The heaters range between 0.007 and 0.055 in thick, depending on the insulating material and resistance element used. Flexible heaters can be vulcanized, adhesive-attached, or clamped to the surface requiring heat. Flexible heaters can be fabricated and shaped to fit to curved or angular surfaces. Integral thermostats, control sensors, varying surface watt densities, insulation systems, holes, cutouts, notches, and clamping devices are all possible for flexible heaters. (See Fig. 2.2.)

TYPICAL APPLICATIONS. Freeze and condensation protection of equipment and materials exposed to weather; food-warming equipment; medical equipment, including blood chemical analysis and kidney dialysis equipment; heating and preheating on engines, rocket motors, spacecraft, and satellites; and heating of materials for forming and for copying and printing applications are typical applications.

* Registered trademarks of the Du Pont Company.

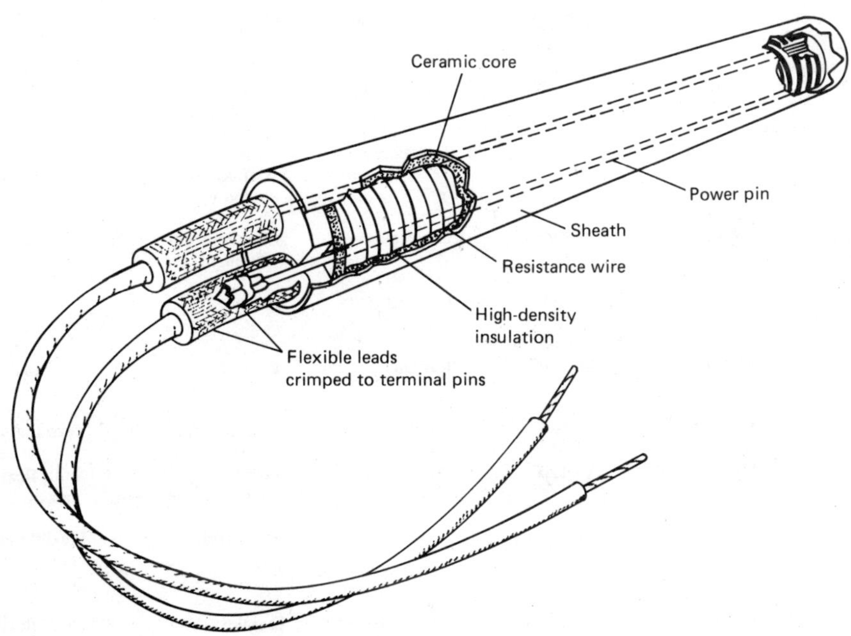

Fig. 2.1 Typical swaged cartridge heater.

Table 2.1. Heater Selection Matrix

Heater description	Watt density, W/in²	Application temperature: Range, °F	Application temperature: Transfer mode	Sheath material	Availability of typical lead configurations
1. Cartridge heater (swaged)	Up to 400	To 1000 To 1600	Conductive Conductive	Stainless steel Incoloy	Stranded, insulated Lead wires, ceramic beads, over pins above 500°F at lead end
2. Flexible	Up to 75 Up to 60 Up to 15 Up to 12 Over 100	To 500 To 400 To 250 To 225 To 1100	Conductive Conductive Conductive Conductive Conductive	Silicone rubber Kapton Neoprene Mylar Mica	Insulated flexible Insulated flexible
3. Ceramic fiber	Up to 30	To 2200	Radiant	Molded ceramic fiber	Formable high-temperature strips, flexible wire
4. Cast-in	Up to 100	To 700	Conductive	Aluminum, bronze	Basic tubular terminations
5. Tubular	Up to 150	To 350 To 750 To 1200 To 1600	Conductive Conductive Conductive Conductive	Copper Steel Stainless steel Incoloy	Pins, screw lugs, ceramic-metal seal, threaded stud, spade overmolded, and straight flexible
6. Flat tubular	Up to 60	To 1200 To 1400	Conductive Conductive	Stainless steel Incoloy	Terminal pins, lead wires
7. Finned tubular	Up to 80	To 1200	Convection	Incoloy/ Stainless steel	Terminal, enclosure, NEMA† classes
8. Immersion tubular assembly	Up to 60	To 500	Convection	Copper, steel, stainless steel, Incoloy	Terminal, enclosure, NEMA classes
9. Circulation tubular assembly	Up to 60	To 500	Convection	Copper, steel, Incoloy	Terminal, enclosure, NEMA classes
10. Duct tubular assembly	Up to 40	To 800	Convection	Incoloy	Terminal, enclosure, NEMA classes
11. Radiant tubular	Up to 50	To 1600	Radiant	Incoloy	Terminal box, enclosure
12. Channel band	Up to 40	To 650	Conductive	Aluminum, brass	Terminal box, studs
13. Ceramic insulated band	Up to 45	To 1200	Conductive	Zinc steel, stainless steel	Stud, flexible, and terminal boxes
14. Mica band and strip	Up to 55	To 900	Conductive	Zinc steel, stainless steel	Flexible leads with braid or hose, studs, screws, and terminal boxes
15. Mineral-insulated band and strip	Up to 150	To 1400	Conductive	Stainless steel	Stud, flexible, and terminal boxes

Table 2.1. (*Continued*)

Mounting and installation techniques	Formability and flexibility	Durability	Corrosive resistance	Efficiency
Inserted into a tight-fitting hole	Rigid	Very good	Good for the Incoloy sheath; other sheaths available; moisture and hydrocarbons can effect lead end	Very good; all energy used by work
Adhesive, clamping, bonding, fasteners, self-gripping	Flexible and in various shapes; mica is less flexible	Fair; abrasion or contact with sharp objects can pierce the insulation	Fair to good, depending on insulation and the agent	Good, some losses; reduce with insulation
Clamping straps, interlocking edges, pins/washers, external frames with edge support, cementing	Rigid but cuttable with simple hand tools	Fair mechanical strength; excellent thermal shock resistance	Fair to good; element wire can be affected by furnace atmosphere	Very good in closed furnaces; good as a high-temperature radiation source
Bolting or clamping	Rigid but machinable	Very good	Good; moisture and hydrocarbons can affect lead ends	Moderate; back-side losses do occur
Bulkhead fittings, locator rings, brackets, standoffs, clamping on to collars	Semirigid; field forming is possible	Good	Dependent on sheath material selection as various materials perform best in particular applications	Moderate to good; round shape in clamping applications adds to losses; good in convection and immersion
Same as tubular	Semirigid, factory-formed	Good	Same as tubular	Moderate to good; flat side is good in conduction
Attached to structure inside housing	Factory-assembled	Good	Brazing material is sometimes poor	Good
Over-the-side, screw-plug, flange into tank mounting	Factory-assembled	Good	Same as tubular	Good
Heater assembly bolted to tank	Factory-assembled	Good	Same as tubular	Good
Flanged housing affixes to duct	Factory-assembled	Good	Same as tubular	Good
Attached to reflector assembly	Factory-assembled	Good	Same as tubular	Fair (because of scatter loss)
Clamped onto the parts	Factory-formed	Good	Typical for aluminum	Good
Clamping, banding clamp tabs	Rigid and factory-formed	Good	Moisture and hydrocarbons can be a problem	Good
Clamping, banding, Wedge-Lok, clamp tabs	Rigid and factory-formed	Moderate to good as leads at exit are somewhat vulnerable	Moisture and hydrocarbons can find entry to the heater	Good when well-clamped and shrouded or insulated
Clamping, banding, clamp bars	Rigid and factory-formed	Very good	Moisture and hydrocarbons can be a problem at the terminal	Good; back-side and internal insulation cuts losses

Table 2.1. (*Continued*)

Heater description	Watt density, W/in²*	Application temperature: Range, °F	Application temperature: Transfer mode	Sheath material	Availability of typical lead configurations
16. Mineral-insulated cable heater	Up to 30	To 1200	Conductive	Stainless steel, Inconel	Flexible leads protected by hose or braid
17. Flexible heating cable, constant-wattage	Up to 10	To 250	Conductive	Plastic	Flexible
18. Flexible heating cable, self-regulating	Up to 10	To 275	Conductive	Plastic	Flexible
19. Elements on ceramic substrate	20–200	To 1800	Conductive, convective, radiant	Ceramic	Pins, flexible strip
20. Silicon carbide, cermets	Up to 70	To 3100	Radiant	None, exposed	Aluminized rod with braided bussing
21. Graphite (requires nonoxidizing atmosphere)	Up to 1500	To 5500	Radiant	None, exposed	Custom-made
22. Open-coil and ribbon	Up to 100	To 2500	Convective, radiant	None, exposed	Studs, pins
23. Radiant-heating panels	Up to 24	800–1100	Radiant	High-emissivity stainless steel	Terminal box, raceway, and stud terminals
24. Quartz radiant heaters	Up to 40	To 1800	Radiant	Quartz tube	Studs

* Allowable watt density is highly dependent on the specific application details. Watt density may be increased under certain operating conditions.
† National Electrical Manufacturers' Association.
CONVERSION FACTORS: To convert from W/in² to W/cm², multiply by 0.16 in²/cm². °C = (°F − 32)/1.8.

TEMPERATURE AND WATT DENSITY RANGE BY INSULATING MATERIAL. The following ranges are recommended: silicone rubber insulation 80 to 500°F (27 to 260°C), up to 75 W/in² (11.6 W/cm²); Kapton insulation up to 400°F (204°C), up to 60 W/in² (9.3 W/cm²); Neoprene insulation up to 250°F (121°C), up to 15 W/in² (2.3 W/cm²); Mylar insulation up to 225°F (107°C), up to 12 W/in² (1.9 W/cm²); and Mica insulation up to 1100°F (593°C), up to 100 W/in² (15.5 W/cm²). Care must be taken to select the watt density for a specific operating temperature. As watt density increases toward the maximum, heater operating temperature must decrease.

THERMAL RESPONSE. Flexible heaters have a somewhat slower heat-up time because the flexible insulating materials are resistant to heat flow, and the generally used lower watt densities are conducive to lower heat fluxes. They are capable of applying heat evenly over large areas.

EFFICIENCY. An efficiency of 90 to 95 percent is possible when the heater is bonded to the part surface. The efficiency of flexible heaters can be enhanced by adding back-side insulation of foam or batting and by using insulation thicknesses that are thinner on the work side of the resistance element. Another means of maximizing heater life is to avoid the detrimental effects of contamination that can attack leads and permeate the internal structure of the heater. Contaminants can degrade the insulation or resistance element. Protecting against contamination by electrolytic liquids, gases, and organic compounds will avoid premature failure. Fol-

Table 2.1. (*Continued*)

Mounting and installation techniques	Formability and flexibility	Durability	Corrosive resistance	Efficiency
Clamping, coiling, inserting in a hole or groove	Semirigid, easily bent or formed	Moderate; can be crushed or ruptured	Good; moisture and hydrocarbons can find entry through lead ends	Moderate, keeping close contact to the work is sometimes difficult
Wrapping, tie-wrap banding	Flexible, easily wrapped	Fair, abrasion or contact with sharp objects can pierce insulation	Fair to good depending on the corrosive agent and jacketing material used	Good
Wrapping, tie-wrap banding	Flexible, easily wrapped	Fair, abrasion or contact with sharp objects can pierce insulation	Fair to good depending on the corrosive agent and jacketing material used	Good
Clamping, free-standing, inserted in hole	Extremely rigid	Thin section heaters are very brittle	Good, outer ceramic covering is impervious to most chemicals	Good
Bulkhead fittings, ceramic locator rings, standoffs, clamping blocks	Extremely rigid	Highly susceptible to mechanical damage and thermal shock	Fair to excellent depending on corrosive agent and particular cermet used	Good
Mechanical support is required above 2500°F	Rigid	Easily damaged; must be protected	Poor, must be operated in controlled atmospheres	Good
Supported by ceramic insulators	Rigid	Susceptible to mechanical damage, must be protected	Fair to good; element wire can be affected by atmosphere	Good
Studs or brackets	N/A	Good; emitter of stainless steel is protected	Moisture and hydrocarbons falling into a panel can be a problem	Good when placed close to the work
With reflectors and supporting structures	Rigid	Susceptible to mechanical damage, must be protected	Good	Good

lowing the manufacturer's application guide on temperature versus watt density and fit, clamping, and attaching will promote good life.

Ceramic-Fiber Heaters

CONSTRUCTION. Ceramic-fiber heaters consist of a resistance wire embedded in the face of a molded, rigidized ceramic fiber body which provides both support and thermal insulation. The fiber and element are molded together to the desired shape in a liquid-based process. Drying provides strength, rigidity, and dimensional stability. The molded ceramic-fiber heater has excellent thermal insulation properties and a low thermal mass, both of which improve efficiency and processing time in furnaces, crucibles, and similar equipment. The resistance wire may be either fully embedded or partially exposed on the heated surface, and the element configuration is usually sinuated (formed in one plane) or coiled (helically wound). The resistance wire itself is usually a special high-temperature alloy of iron-chrome-aluminum that develops its own protective oxide coating to retard the effects of destructive wire oxidation over time. Surface coatings of various materials can be added to the heated surface. These coatings enhance durability and can be applied to increase radiant heat transfer. Ceramic-fiber heaters are available in flat-panel, semicylindrical, and full-cylinder shapes. Some manufacturers design and build custom shapes to suit particular applications. (See Fig. 2.3.)

TYPICAL APPLICATIONS. Electric furnaces, including heat-treating, aluminum holding and melting, semiconductor diffusion conveyor, brazing, and glass and fiber-optic annealing and coating furnaces; analytical instrumentation, including oxygen analyzers, total carbon analysis, and gas chromatography; medical-dental furnaces; high-temperature process heating, such as wire coating and annealing, ceramics man-

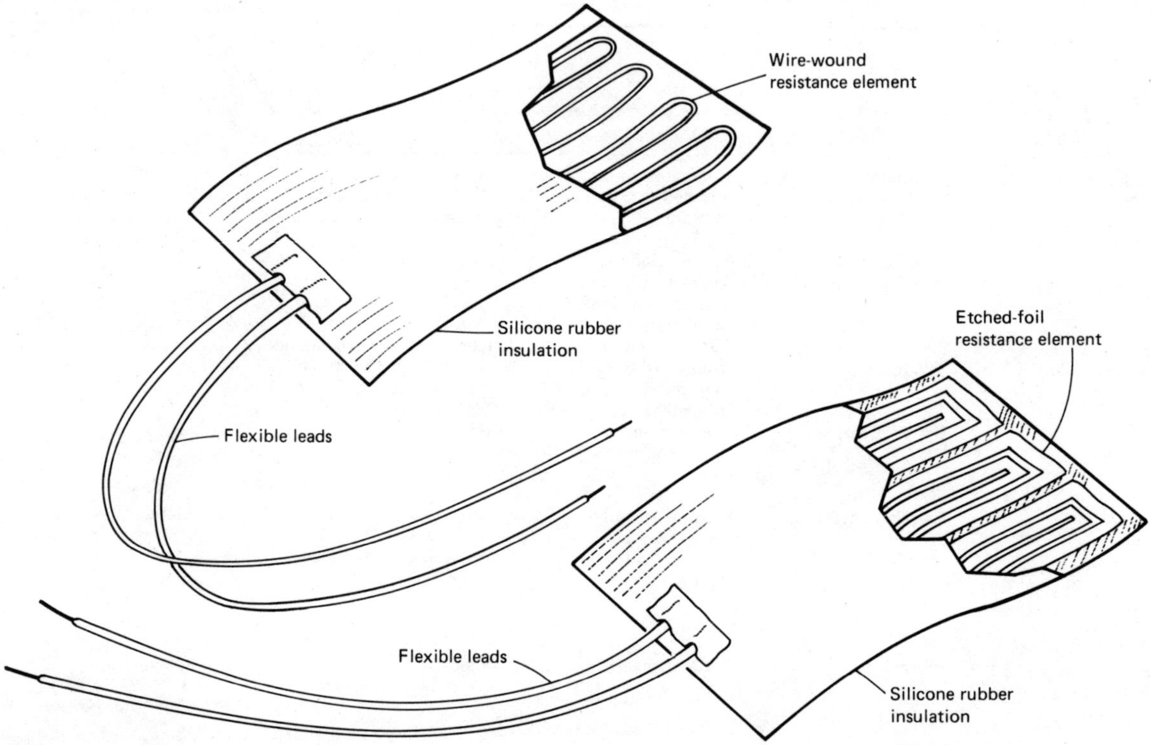

Fig. 2.2 Typical flexible heater.

ufacturing, composite material manufacturing, and petrochemical processing; and furnace retrofitting are typical applications.

TEMPERATURE RANGE. The recommended range is from 1000°F (538°C) up to 2200°F (1204°C).

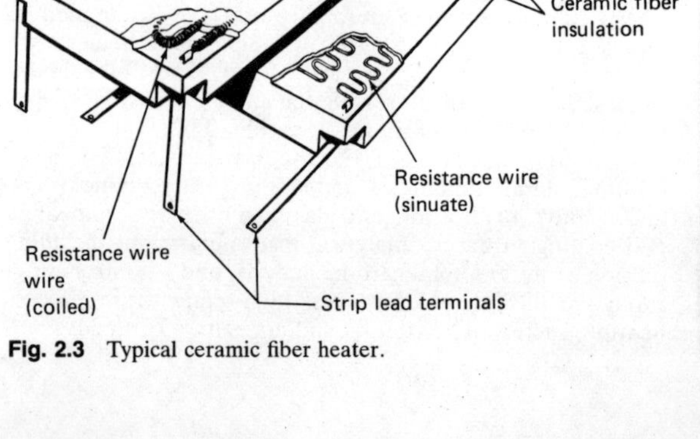

Fig. 2.3 Typical ceramic fiber heater.

WATT DENSITY RANGE. Up to 30 W/in^2 (4.6 W/cm^2) is recommended.

THERMAL RESPONSE. Ceramic-fiber heaters reach rated surface temperature in 2 to 5 min under full-rated power input. Chambers reach operating temperature in 5 to 15 min depending on the size of the chamber. The principal mode of heat transfer is by radiation, and heat transfer to the work and the heating enclosure starts immediately on element heat-up.

EFFICIENCY. Standard 2-in (5-cm)-thick heaters are essentially unidirectional heaters, with 90 to 95 percent heat transfer through the hot-face. The built-in ceramic-fiber material thickness behind the element retards back-side losses, but additional layers of ceramic-fiber blanket are sometimes used to further reduce furnace losses.

HEATER LIFE. While depending on operating temperature and element design, a minimum life is expected to be 2000 h at an element temperature of 2300 to 2400°F (1260 to 1316°C).

Cast-in Heaters

CONSTRUCTION. Cast-in heaters consist of a sheathed heater, usually a tubular heater, that has been cast into a metal shape using permanent mold or sand cast techniques. Open tubes and inserts used for mounting control devices or fluid cooling can be cast into the part at the same time. The most common shapes of cast-in heaters are flat rectangles

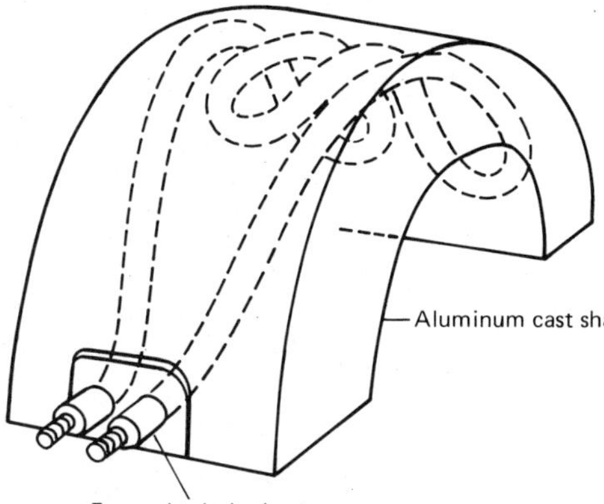

Fig. 2.4 Typical cast-in heater.

and semicylinders, but virtually any shape capable of being cast can be designed with a cast-in heater element.

Cast-in heaters are cast using aluminum, bronze, or magnesium; aluminum is most common. Cast-in heaters can be machined to enhance conductive heat transfer from the surface, facilitate mounting, and obtain close size tolerances. Coatings such as anodizing, paint, and plastics or fluoroplastics can be applied to the cast-in heater surfaces. (See Fig. 2.4.)

TYPICAL APPLICATIONS. Commercial food equipment such as grills, warmers, and fryers; plastic equipment platens and barrel heaters; packaging equipment and glue melting; equipment requiring heated flat surfaces, such as T-shirt decal and paper presses; and medical equipment and freeze protection are typical applications.

TEMPERATURE RANGE. Up to 700°F (371°C) for aluminum alloys, and possibly higher depending on metal alloy, is recommended.

WATT DENSITY RANGE. Up to 60 W/in^2 (9.3 W/cm^2) on the casting surface is recommended.

THERMAL RESPONSE. The thermal response time of a cast-in heater is slower than that in other methods of heating because the heater must heat the thermal mass of the casting prior to heating the work. The advantages of cast-in heaters are the uniformity of the heat distribution within the casting, ruggedness, and customized cast shapes.

EFFICIENCY. Efficiency is approximately 85 percent. Thermal efficiency can be enhanced by ensuring that there is intimate contact between the casting and the material to be heated. Insulating or shrouding the casting faces will also increase efficiency.

HEATER LIFE. The service life of cast-in heaters is generally many years. The intimate contact between the heating element and the metal keeps the resistance wire running at a temperature close to the metal casting, thereby avoiding the life-shortening wire oxidation effects caused by high operating temperature.

Tubular Heaters

CONSTRUCTION. Tubular heaters typically contain a single, helically wound, and centered resistance wire in a metal tube surrounded by compacted magnesium oxide insulating material and fabricated round in cross sections. Electrical terminations are usually made at both ends. Commonly used metal tube sheath materials include copper, steel, and various stainless steels. Each sheath material satisfies application requirements based on such factors as cost, maximum temperature capabilities, and corrosive nature of the material to be heated. Tubular heaters are rigid but can be formed into a variety of shapes. They can be fabricated in various standard diameters of ¼ to ¾ in and in lengths of up to 15 ft.

Tubular heaters can be manufactured with a wide variety of modifications, attachments, and special configurations. These include numerous terminals, terminations and lead arrangements, special terminal housings and overmolds to protect against moisture and explosion, unheated terminal end lengths, locator rings and brackets for mounting, and bulkhead and double-ended terminal construction. Flat or nonround cross sections can be fabricated using multiple elements for better heat transfer and for polyphase connection. (See Fig. 2.5.)

TYPICAL APPLICATIONS. Tubular heating elements are often used to construct other heater assemblies. These include bulkhead, plug, over-the-side immersion heaters, circulation heaters, forced-convection driers and duct heaters, cast-in heaters, extruded-channel heaters, and radiant heating assemblies. Tubular heating elements are used in food-warming and cooking equipment, furnaces, ovens and driers, film-processing equipment, plastics and packaging equipment, bulk-material heating, and freeze and condensation protection.

TEMPERATURE RANGE. The following ranges are recommended:

Up to 1600°F (871°C) (Incoloy sheath)
Up to 1200°F (649°C) (stainless-steel sheath)
Up to 750°F (399°C) (steel sheath)
Up to 350°F (177°C) (copper sheath)

WATT DENSITY RANGE. The recommended range is 2 to 150 W/in^2 (0.3 to 23.2 W/cm^2) depending on the nature of the application.

THERMAL RESPONSE. Similar in time to a cartridge heater of equal watt density in the same type of application.

EFFICIENCY. As with other heaters in immersion applications, tubular heaters impart nearly 100 percent of their energy into the medium. The round cross section of the tubular heater makes it less efficient than flat heaters in clamp-on applications because of the limited surface contact of the round element.

HEATER LIFE. Most tubular applications are designed for a 5-year operational life. The industry literature defines allowable watt densities and temperatures that are more conservative than some of the other heater types.

Flat Tubular Heaters

CONSTRUCTION. A flat tubular heater has an oval cross section in which the two opposite sides have been flattened. Depending on its cross-section dimensions, a flat tubular heater can contain multiple resistance elements. (See Fig. 2.6.)

TYPICAL APPLICATIONS. Flat tubular heaters are used in immersion applications, deep-fat fryers, and steam kettles and to clamp on metal tank sides. Their primary advantage

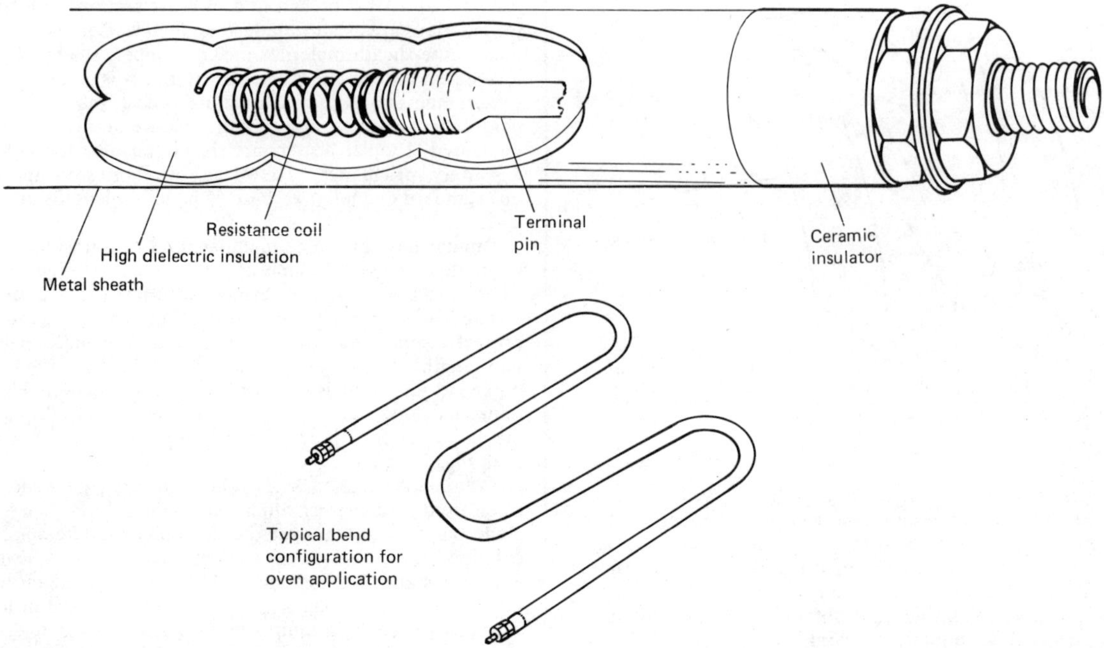

Fig. 2.5 Typical tubular heater.

is a larger surface area as compared to a round tubular heater. This allows better heat transfer to liquids and flat plates.

TEMPERATURE RANGE. The maximum recommended temperature is 1400°F (760°C).

WATT-DENSITY RANGE. Up to 150 W/in² (23.2 W/cm²), depending on application, is recommended.

THERMAL RESPONSE. This is better than that with round tubular elements because of the smaller cross-sectional area.

EFFICIENCY. Efficiency is approximately 80 percent. Heat transfer is better than that with round tubular elements for clamp-on applications because of flat surface construction.

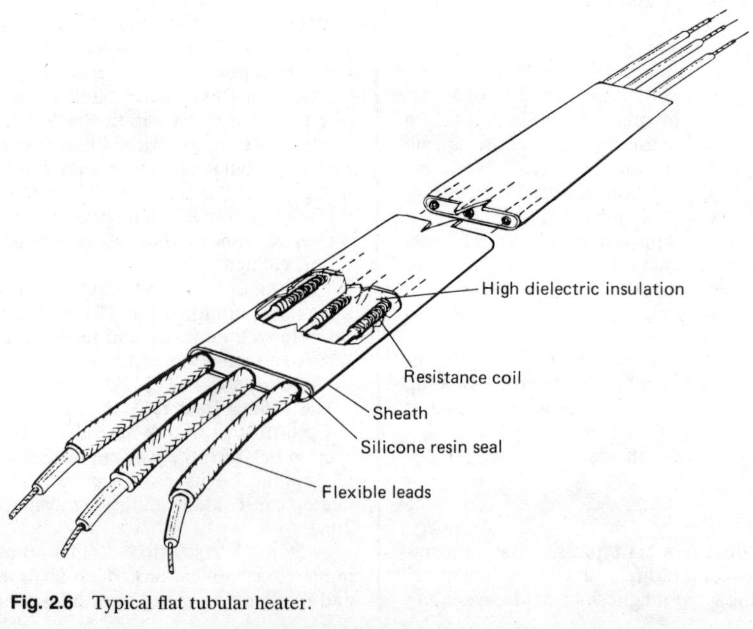

Fig. 2.6 Typical flat tubular heater.

Size and formability allow designs with more kilowattage in specified volume than that achieved with round tubulars.

HEATER LIFE. Heater life is the same as that for round tubulars.

Finned Tubular Heaters

CONSTRUCTION. Finned tubular heaters consist of a tubular heating element with a series of fins attached to the heating element's steel sheath to increase the surface area. (See Fig. 2.7.)

TYPICAL APPLICATIONS. Process gas and air applications, including drying and gas heating, are common.

TEMPERATURE RANGE. Up to 1200°F (649°C) with a stainless-steel sheath, or as limited by the fin-attaching braze material, is recommended.

WATT-DENSITY RANGE. Up to 80 W/in^2 (12.4 W/cm^2) is recommended.

THERMAL RESPONSE. This is better than that obtained with nonfinned units, allowing increased watt density and lower operation temperatures.

EFFICIENCY. Efficiency is 100 percent when gas is heated.

Immersion Heaters—Tubular

CONSTRUCTION. Immersion heaters consist of hairpin or straight tubular heating elements brazed, welded, or mechanically fastened into a plug, fitting, or flange to be plumbed into a fluid flow system for fluid heating. A terminal enclosure exists usually on or extended beyond the flange or plug. (See Fig. 2.8.)

TYPICAL APPLICATIONS. Tubular immersion heaters are used in applications requiring the heating of fluids or highly viscous materials.

TEMPERATURE RANGE. The recommended range is up to 500°F (260°C) as dictated by the physical characteristics of the various fluids and gases.

WATT DENSITY RANGE. Up to 60 W/in^2 (9.3 W/cm^2) is recommended.

THERMAL RESPONSE. This is fully dependent on the wattage input, the fluid, and the system design.

EFFICIENCY. The efficiency is 100 percent from element to liquid. The overall system efficiency depends on the amount of insulation around the vessel and piping network.

Circulation Heaters—Tubular

CONSTRUCTION. A circulation heater is a heater assembly consisting of a flange or screw plug immersion heating assembly fabricated as a part of a vessel with fluid inlet and outlet openings. (See Fig. 2.9.)

TYPICAL APPLICATIONS. Tubular circulation heaters are used for heating fluids in closed systems.

TEMPERATURE RANGE. The recommended range is −100 to 500°F (−73 to 260°C).

WATT DENSITY RANGE. The recommended range is 2 to 60 W/in^2 (0.3 to 9.3 W/cm^2).

THERMAL RESPONSE. This is fully dependent on wattage input, the fluid, and the system design.

EFFICIENCY. Element heat input is nearly 100 percent, but there will be some heat loss through the vessel and piping depending on their insulation.

Duct Heaters—Tubular

CONSTRUCTION. Duct heaters consist of arrangements of tubular heating elements, usually hairpin-shaped, mounted in a frame, all for insertion into a duct of moving air or gas. The heaters project outward in a cluster at right angles to the flow, thus allowing air to flow through the elements.

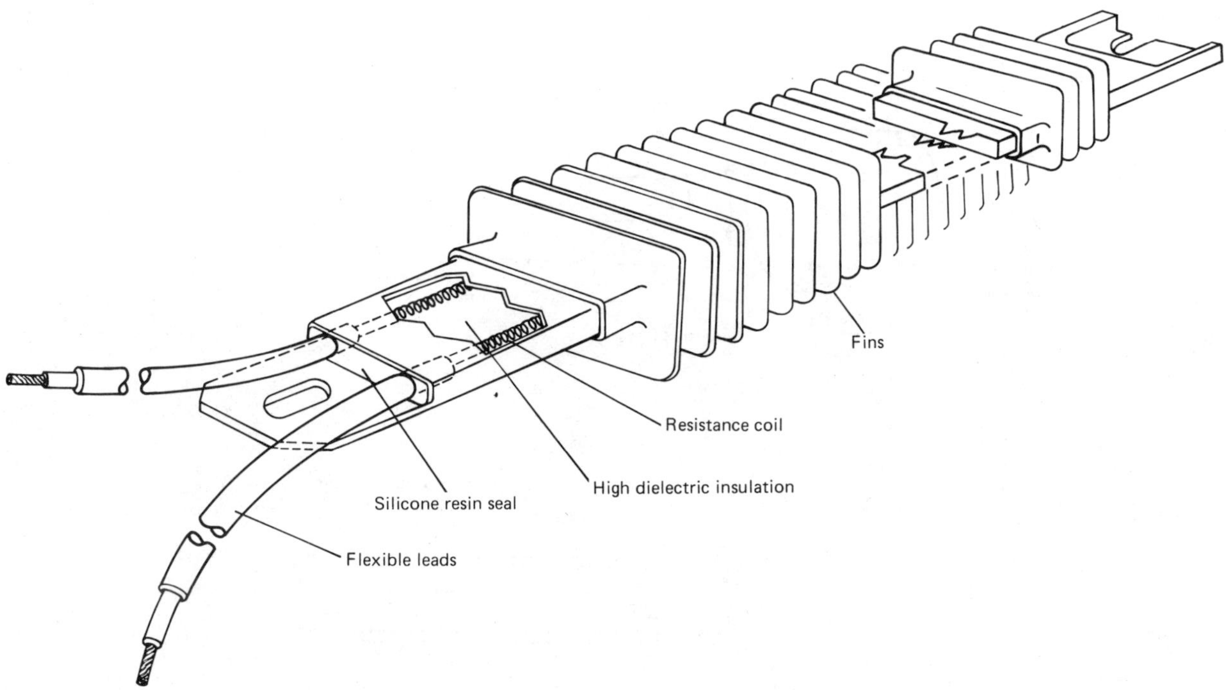

Fig. 2.7 Typical finned tubular heater.

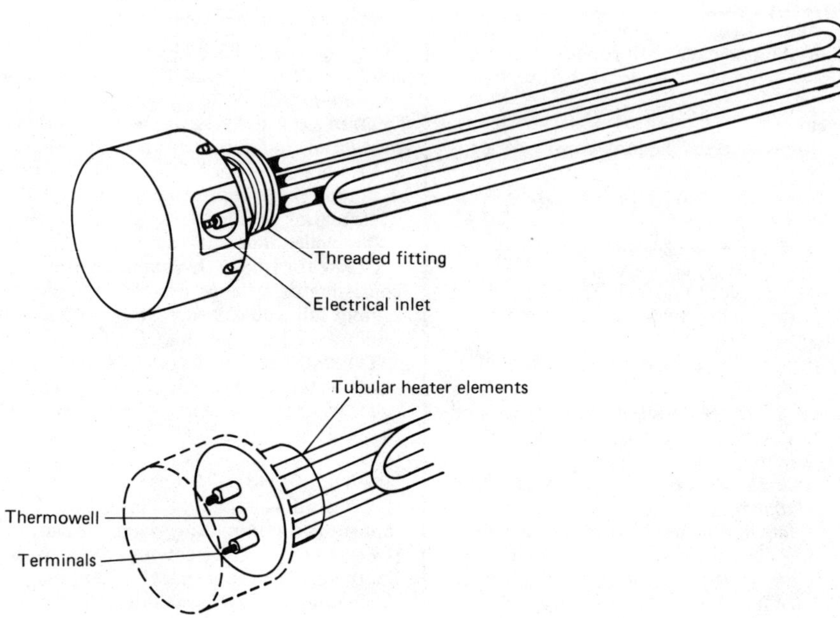

Fig. 2.8 Typical immersion heater—tubular.

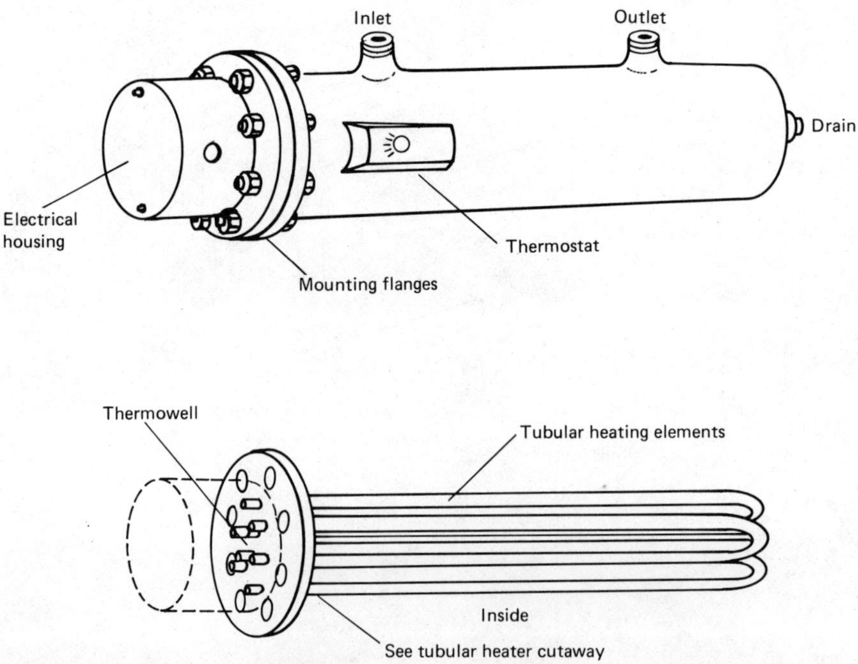

Fig. 2.9 Typical circulation heater—tubular.

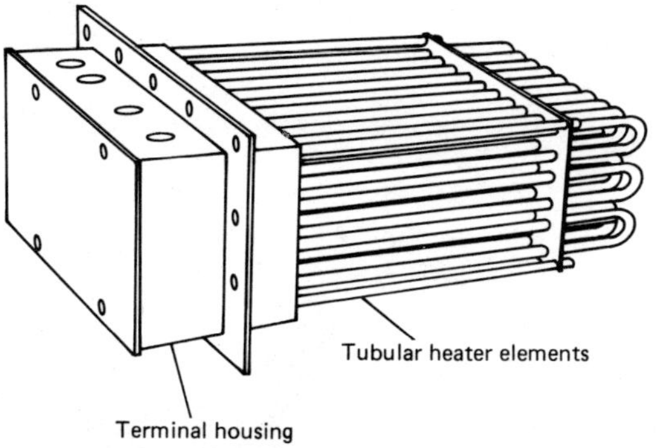

Fig. 2.10 Typical duct heater—tubular.

There is a terminal housing exterior to the duct for electrical connections. (See Fig. 2.10.)

TYPICAL APPLICATIONS. Tubular duct heaters are used for heating air or other gases in a duct system, process drying, or heating.

TEMPERATURE RANGE. Up to 800°F (427°C) is recommended.

WATT DENSITY RANGE. The recommended range is 2 to 40 W/in^2 (0.3 to 6.2 W/cm^2).

THERMAL RESPONSE. This depends on watt density, mass flow rates, the heat capacity of the gas, and the heating-element configuration.

EFFICIENCY. Although 100 percent of the heat energy is available to the gas for heating, insulation of the duct and plenum system to maintain high efficiency should always be considered.

Radiant Heaters—Tubular

CONSTRUCTION. Tubular radiant heaters consist of tubular heating elements mounted in front of a polished aluminum reflector, all within a mounting structure. (See Fig. 2.11.)

TYPICAL APPLICATIONS. Comfort heating; process drying; plastic forming; commercial food equipment; thermal-responsive adhesives, epoxies, and paints are typical applications.

TEMPERATURE RANGE. A tubular sheath temperature of 1200 to 1600°F (649 to 871°C) is recommended.

WATT DENSITY RANGE. The recommended range is 30 to 50 W/in^2 (4.6 to 7.7 W/cm^2) on tubular sheath.

THERMAL RESPONSE. It normally takes 3 to 5 min to stabilize sheath temperature.

EFFICIENCY. Most radiant systems direct 50 to 60 percent

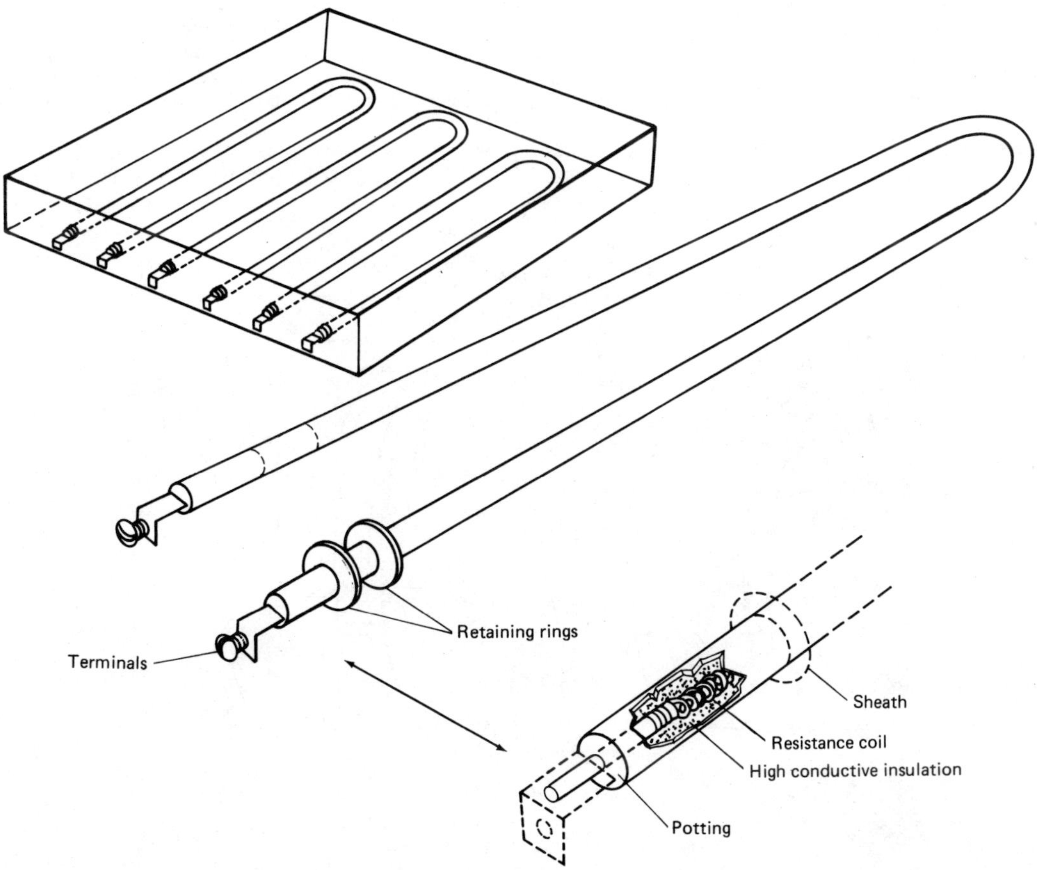

Fig. 2.11 Typical radiant heater—tubular.

of the radiant energy toward the work; the balance is lost to the surroundings. This efficiency decreases when the reflector becomes oxidized or dirty. Heat received by the work decreases as a function of the heater distance from the work.

Channel Band Heaters—Tubular

CONSTRUCTION. These heaters consist of an aluminum extrusion with formed, continuous channels. A tubular heating element is placed in the channels. These channel band heater extrusions are formed in circular segments which, with the heater, are held together by a metal clamping strap. This provides an uninsulated cover for the exposed heating element and supports a terminal or location device. The extrusion channels can also accommodate tubing for cooling or fluid heating. (See Fig. 2.12.)

TYPICAL APPLICATIONS. Tubular channel band heaters are used for heating of barrels, plastic injection molding, extruder and blow molding equipment, and heating other cylindrical surfaces.

TEMPERATURE RANGE. Up to 650°F (343°C) is recommended.

WATT DENSITY RANGE. Up to 40 W/in^2 (6.2 W/cm^2) on the inside surface is recommended.

THERMAL RESPONSE. Thermal response of channel band heaters is moderately fast if the tubular heating element is held in close contact within the aluminum channel; a loose heater element retards heat transfer and slows response.

EFFICIENCY. The efficiency of channel band heaters is generally 75 to 80 percent. Efficiency can be improved with good thermal contact between the heater surfaces and the work. Shrouding or insulating will decrease back-side heat losses and improve the total wattage absorbed by the work.

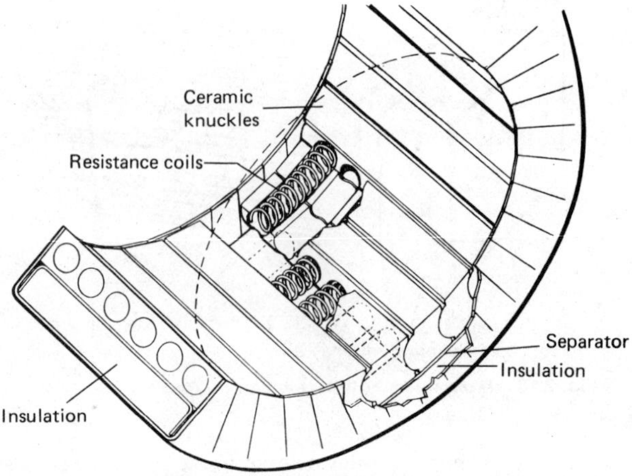

Fig. 2.13 Typical ceramic-insulated band heater.

Ceramic-Insulated Band Heaters

CONSTRUCTION. Ceramic band heaters ("knuckle" heaters) consist of a coiled resistance wire strung through loose but interlocking ceramic insulators, forming a heating assembly that can be wrapped around cylindrical surfaces. The coil insulator assembly and a backing layer of fiber insulation are placed within a protective metal housing. The housing is formed and has serrated edges to provide for curving flexibility. The outer housing is split to provide for opening and wrapping. In some constructions an inner metal structure is built in to support additional back-side insulation. In all con-

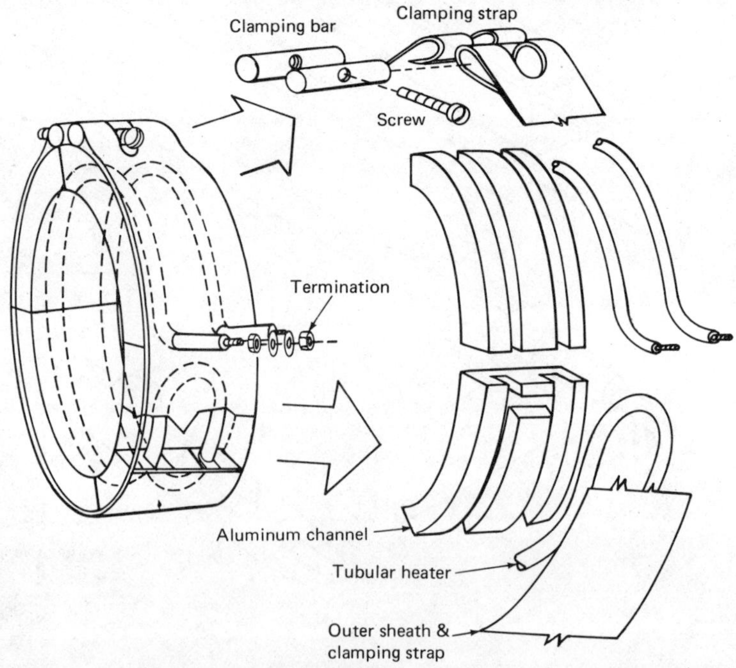

Fig. 2.12 Typical channel band heater—tubular.

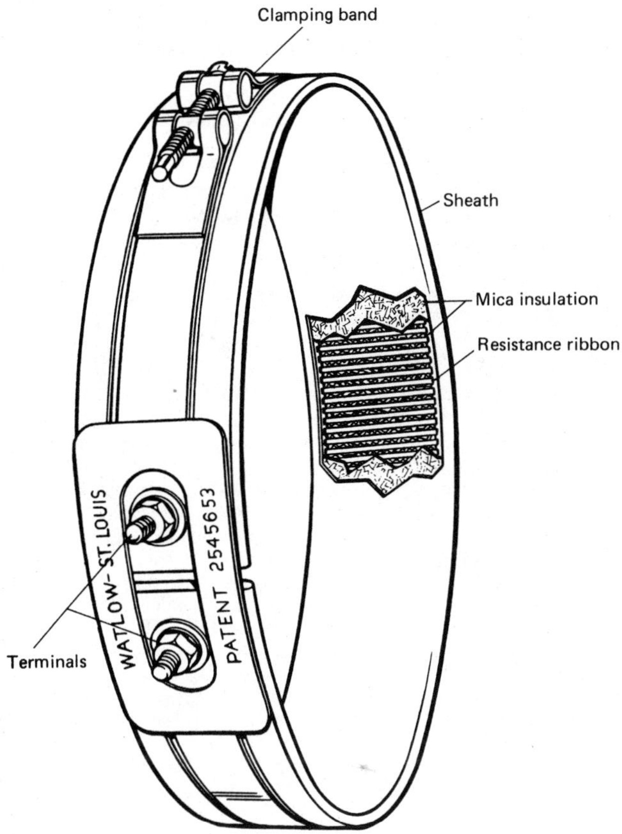

Fig. 2.14 Typical mica band heater.

structions the outer shroud provides a clamping mechanism. (See Fig. 2.13.)

TYPICAL APPLICATIONS. Ceramic band heaters are used for barrel heating on plastic injection molding, extruder, and blow molding equipment.

TEMPERATURE RANGE. Up to 1200°F (649°C) is recommended.

WATT DENSITY RANGE. Up to 45 W/in^2 (7.0 W/cm^2) on the work surface is recommended.

THERMAL RESPONSE. Ceramic-insulated band heaters are slower to respond and more difficult to control because of their large thermal mass. Because of the heat capacity of the mass of ceramic, heat-up and cool-down are slower.

EFFICIENCY. The efficiency of ceramic band heaters with an additional insulating shroud should be between 75 and 85 percent.

Mica Band and Strip Heaters

CONSTRUCTION. Mica band and strip heaters consist of a resistance wire or ribbon wound around a piece of mica, which is placed between two additional pieces of mica to provide electrical insulation; the mica and element assembly is typically placed in a metal sheath that serves as a protective cover. The resulting heater assembly can be shaped or curved, usually into a curved semicylindrical shape to become a band heater, or left flat to be a strip heater. Terminals or lead wires usually project perpendicular to the major surface. Holes, cutouts, and mounting hardware can be provided. A clamping method is provided. (See Figs. 2.14 and 2.15.)

TYPICAL APPLICATIONS. The following applications are typical: for band heaters—plastic injection molding and extruder equipment, round dies, packaging equipment, and pipe heating; for strip heaters—dies and molds, food warming, freeze and moisture protection, packaging equipment and glue melting, storage tank, general industrial process equipment heating, hoppers, storage bins, and large equipment surfaces.

TEMPERATURE RANGE. Up to 900°F (482°C) with steel sheath is recommended.

WATT DENSITY RANGE. The recommended wattage is up to 55 W/in^2 (8.5 W/cm^2) for band heaters and up to 40 W/

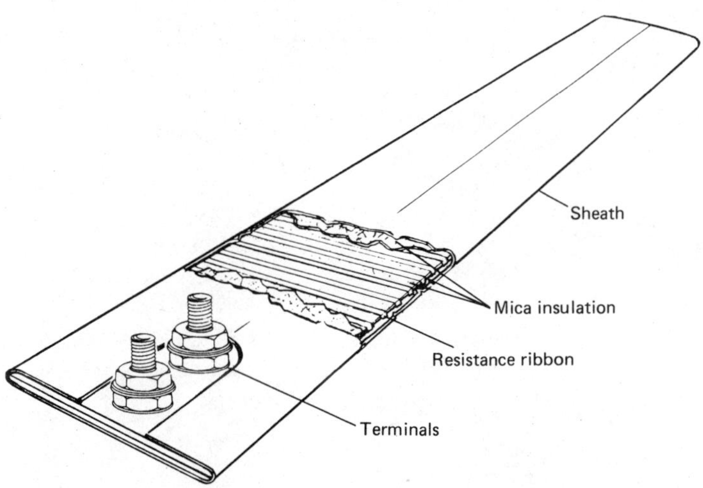

Fig. 2.15 Typical mica strip heater.

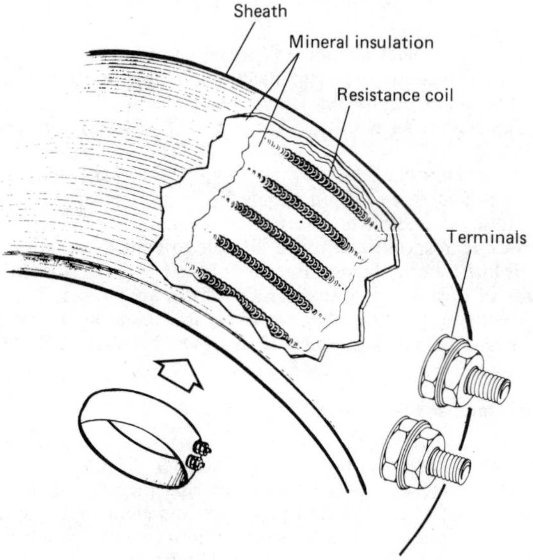

Fig. 2.16 Typical mineral-insulated band heater.

in^2 (6.2 W/cm^2) for strip heaters. The difference exists because of the superior methods of clamping band heaters to the work.

THERMAL RESPONSE. Thermal response of mica-insulated strip and band heaters is good. The mica insulation is thin, allowing very rapid response.

EFFICIENCY. High efficiencies (75 to 85 percent) can be achieved with mica-insulated heaters by using good clamping techniques and insulating or shrouding the back side of these heaters to limit losses.

Mineral-Insulated Band and Strip Heaters

CONSTRUCTION. A mineral-insulated strip or band heater is constructed of a resistance element embedded in a pliable or flowable metal oxide insulating material, all surrounded by a metal sheath. These heaters fall into two categories of construction: thick heaters using a resistance coil in a flowable insulation powder and the thin mineral-insulated strip or band heater, which uses a small coiled wire or sinuate wire placed between a thin insulating system. Various terminal and lead arrangements are possible; the typical construction consists of terminal studs positioned perpendicular to the major surface. (See Figs. 2.16 to 2.18.)

TYPICAL APPLICATIONS. Mineral-insulated band and strip

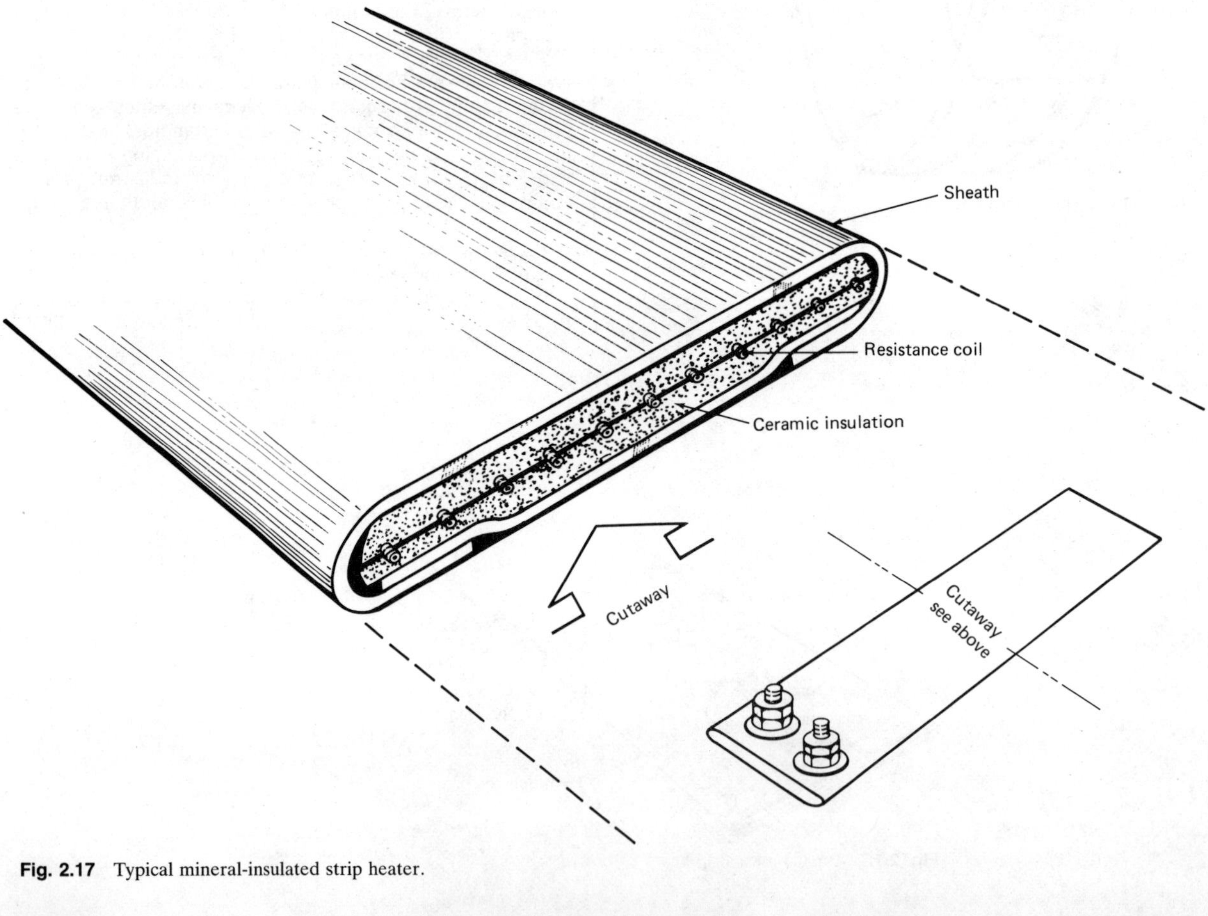

Fig. 2.17 Typical mineral-insulated strip heater.

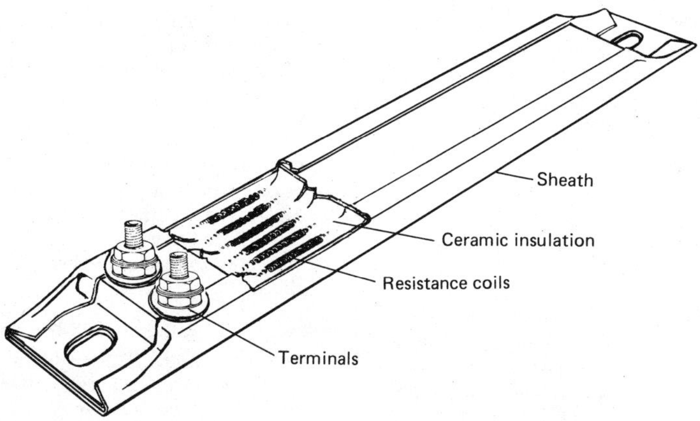

Fig. 2.18 Typical mineral-insulated thick-strip heater.

heaters are used for plastics processing, ovens and furnaces, crucibles, freeze protection, and air and gas duct heaters.

TEMPERATURE RANGE. Up to 900°F (482°C) is recommended for thick construction and up to 1400°F (217°C), for thin construction.

WATT DENSITY RANGE. Up to 30 W/in^2 (4.6 W/cm^2) is recommended for thick construction and up to 150 W/in^2 (23.2 W/cm^2), for thin construction.

THERMAL RESPONSE. The thin-construction heater provides good thermal response and is more responsive than knuckle (ceramic band) heaters. The thick-construction heaters are somewhat less responsive because of greater mass.

EFFICIENCY. The efficiency of mineral-insulated strip and band heaters is quite high for the thin construction that contains its own back-side insulation. With good clamping, these heaters can be 80 to 90 percent efficient when shrouded. The thick construction is somewhat less efficient and requires shrouding and insulation to achieve maximum efficiency.

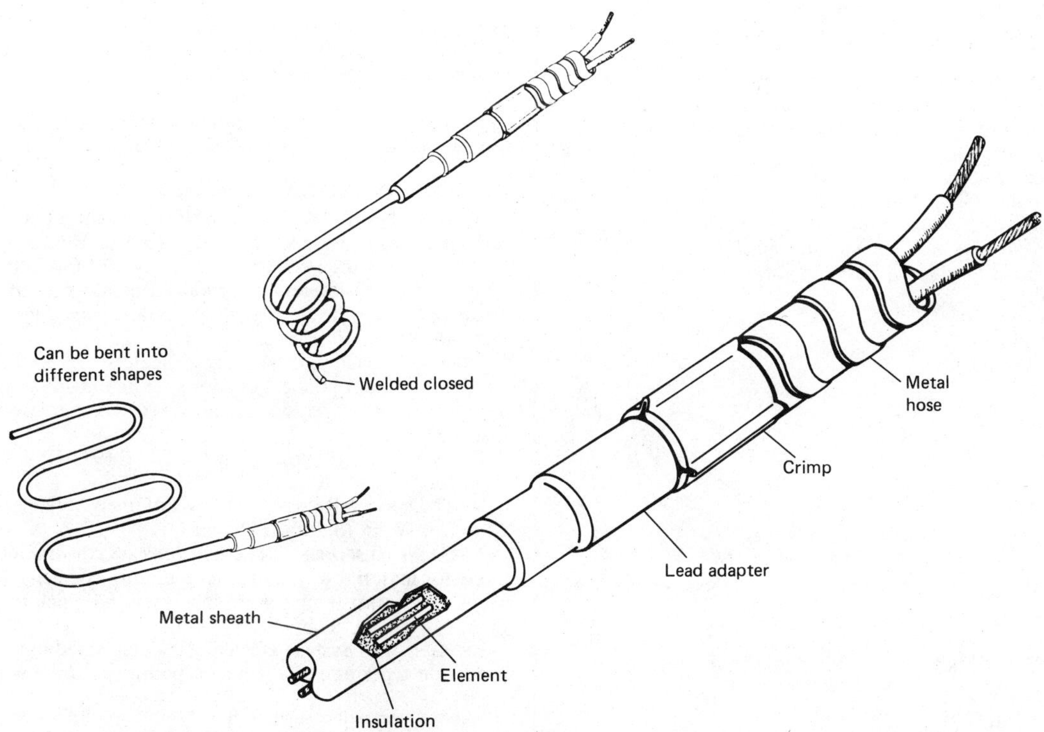

Fig. 2.19 Typical mineral-insulated cable heater.

Mineral-Insulated Cable Heaters

CONSTRUCTION. A cable heater is made by encasing resistance wires within a ceramic mineral insulation, all within a metal tube. The assembly is mechanically reduced in diameter, usually by drawing or swaging. The ceramic insulation, usually magnesium oxide, provides electrical insulation and a heat-transfer medium. Cable heaters are fabricated mostly in diameters of ¼ in or less and, as such, are bendable and formable for conformal applications. Construction can accommodate unheated lead sections and internal thermocouples. Square and rectangular cross sections can be constructed. (See Fig. 2.19.)

TYPICAL APPLICATIONS. Mineral-insulated cable heaters are used in plastic processing machinery nozzle and sprue bushing heating; process heat tracing for gas and liquid heating within pipes or vessels; packaging, sealing, and hot glue application equipment; industrial heating and drying equipment; small ovens and furnaces; analytical instrumentation; photographic processing equipment; and lead bonding and soldering equipment.

TEMPERATURE RANGE. Up to 1200°F (649°C) is the recommended sheath temperature.

WATT DENSITY RANGE. Up to 30 W/in^2 (4.6 W/cm^2) on the sheath is recommended.

THERMAL RESPONSE. Cable heaters respond quickly on being energized. The lower thermal mass allows more rapid heat-up than tubular or cartridge heaters.

EFFICIENCY. The efficiency of cable heaters depends on their ability to maintain contact with the material being heated. Forces of thermal expansion and the flexibility of the heater make clamping somewhat more difficult. Efficiencies of 60 percent are typical for clamped heaters. Insulation around the cable heater can improve efficiency, if used carefully.

Flexible Heating Cable—Constant-Wattage

CONSTRUCTION. Constant-wattage heating cable is constructed of a series of resistors connected in parallel in modules or zones. These resistors are wound around a pair of flexible conductors to ensure alternate pole contact at intervals down the length. This system keeps wattage output nearly equal from zone to zone regardless of length. The resistor system is covered with a plastic outer jacket, and sometimes a metal braid is added to provide abrasion resistance. (See Fig. 2.20.)

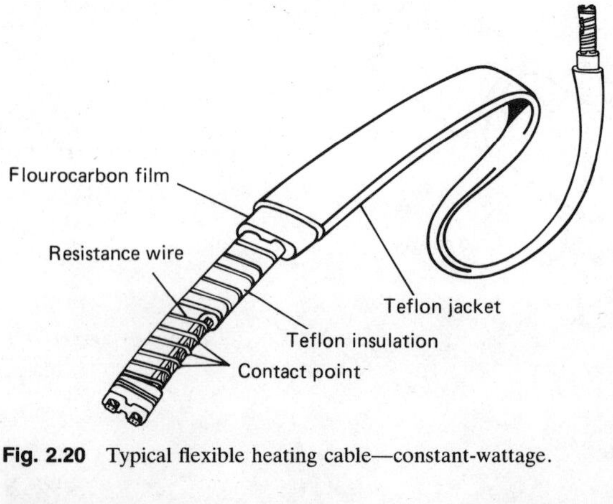

Fig. 2.20 Typical flexible heating cable—constant-wattage.

TYPICAL APPLICATIONS. Constant-wattage heating cable is used mostly in heat tracing and freeze protection.

TEMPERATURE RANGE. Up to 250°F (121°C) is recommended.

WATT DENSITY RANGE. Up to 10 watts per lineal foot (32.8 W/m) is recommended.

THERMAL RESPONSE. Thermal response is relatively slow because of the low wattage.

EFFICIENCY. Efficiencies of 50 percent for uninsulated applications are typical. Constant-wattage flexible cable should be insulated to reduce losses and maximize the heat transfer to the work.

Flexible Heating Cable—Self-Regulating

CONSTRUCTION. Self-regulating heating cable is composed of two or more parallel, flexible conductors separated by a semiconducting polymerized heating element extruded over and between the conductors. An insulating jacket is extruded over the element conductor system. Self-regulating heating cable conducts electricity and produces heat within the element polymer up to a determined temperature at which point the polymer resistance becomes high and the heat output is reduced until the element temperature falls again. These materials with large increases in resistance as temperature increases are called *positive temperature coefficient (PTC) materials.** (See Fig. 2.21.)

TYPICAL APPLICATIONS. Self-regulating heating cable is used in heat-trace systems, protecting piping and tanks from freezing and maintaining temperature of material within.

TEMPERATURE RANGE. Up to 275°F (135°C) is recommended.

WATT DENSITY RANGE. The recommended density is 10 W per lineal foot of cable (32.8 W/m).

THERMAL RESPONSE. Self-regulating heating cable is relatively slow to reach its cutoff temperature.

EFFICIENCY. Typical efficiencies of 50 percent for uninsulated applications. Self-limiting heating cable should be used with insulation to maximize the heat input to the work.

Element-on-Ceramic Substrate Heaters

CONSTRUCTION. A typical heater consists of a screen-printed metal resistance material placed on a ceramic substrate, which is then covered with a layer of insulating material and then sintered to produce an integral structure. These heaters are usually small in size and can be fabricated in a wide variety of shapes. (See Fig. 2.22.)

TYPICAL APPLICATIONS. These heaters are used in scientific instrument applications for spot heating of apparatus or complex processes where the fused coating resists corrosive attack.

TEMPERATURE RANGE. Up to 1800°F (982°C) is recommended.

WATT DENSITY RANGE. This is extremely variable—from 20 to 200 W/in^2 (3.1 to 31 W/cm^2).

THERMAL RESPONSE. These heaters are constructed from a resistor material with a high temperature coefficient. They provide very fast temperature increases and quick response time.

EFFICIENCY. System efficiency is determined entirely by insulation types and heat-transfer pathways to the work.

* Low-power, self-regulating PTC heaters constructed of barium titanate ceramic elements mounted on a copper base and coated with silicone rubber are used in the temperature control of electronics.

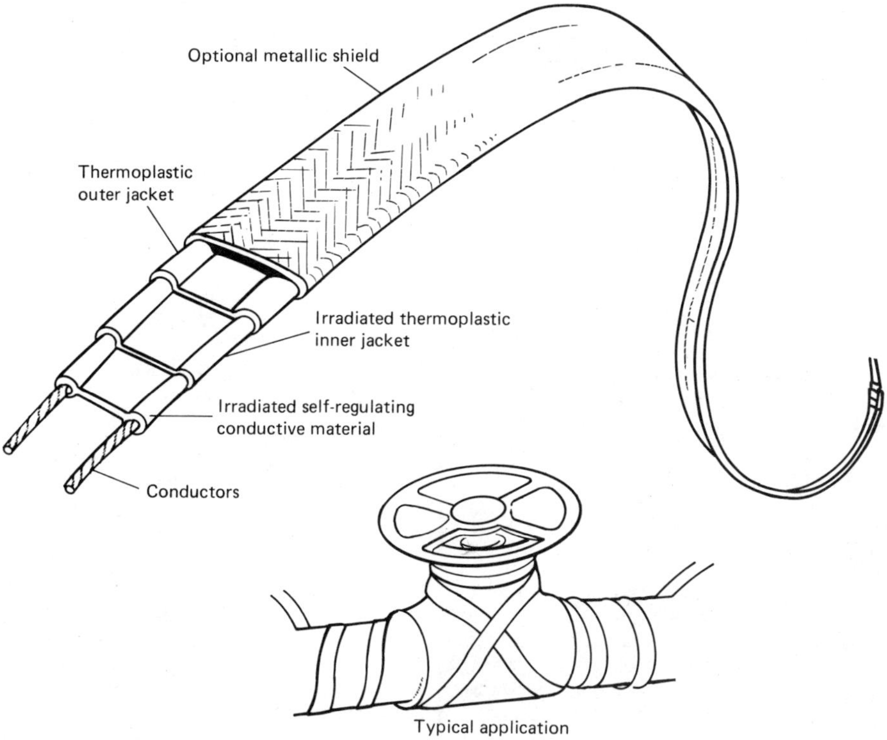

Fig. 2.21 Typical flexible heating cable—self-regulating.

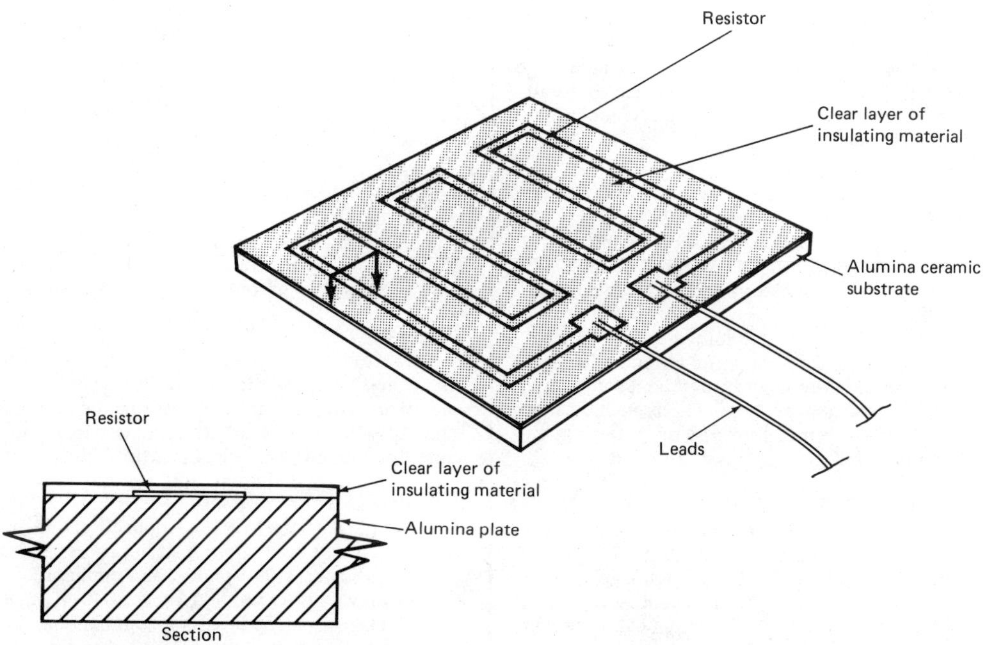

Fig. 2.22 Typical element-on-ceramic substrate heater.

Silicon Carbide, Cermet Heaters

CONSTRUCTION. Silicon carbide (SiC) elements are typically fabricated into rod or tubular straight lengths in different cross-sectional areas and dimensions. Their length contains a central heated zone and a terminal zone at each end. The terminal zone is treated to reduce resistance and operate at a lower temperature to accommodate terminal connections. Other cermet (*cer*amic-*met*al) materials are also used as heating elements. The most notable, molybdenum disilicide ($MoSi_2$), is a pressed powder that is then sintered and formed into hairpin-shaped heating elements. During operation, a glass (SiO_2) forms over the element surface. (See Fig. 2.23.)

TYPICAL APPLICATIONS. Silicon carbide and other cermet elements run exposed and are high-temperature radiant sources typically used above 1800°F (982°C). Applications include glass melting and holding, ceramic firing, ore calcining, heat treating, brazing, sintering, and primary metal processing.

TEMPERATURE RANGE. Up to 3100°F (1704°C) is recommended.

WATT DENSITY RANGE. Up to 70 W/in^2 (10.8 W/cm^2), depending on the heated chamber temperature and the ambient atmosphere, is recommended.

THERMAL RESPONSE. The thermal response of both silicon carbide and other cermet elements is moderate to slow since the resistance increase with temperature during each heat-up is very large (SiC, 150 percent; $MoSi_2$, 500 percent). Both elements usually require low voltage multitap transformers.

EFFICIENCY. Typical efficiency equals 100 percent. Absorption of the available heat output depends on the design of the containing chamber and the magnitude of system losses.

HEATER LIFE. Used properly with continuous operation, which is the most advantageous condition for these heaters, silicon carbide elements can last well beyond 2000 h of operation. It should be noted that silicon carbide elements age with time in service, which causes the resistance to greatly increase and the wattage to decrease. Aging is compensated for by increasing the supply voltage.

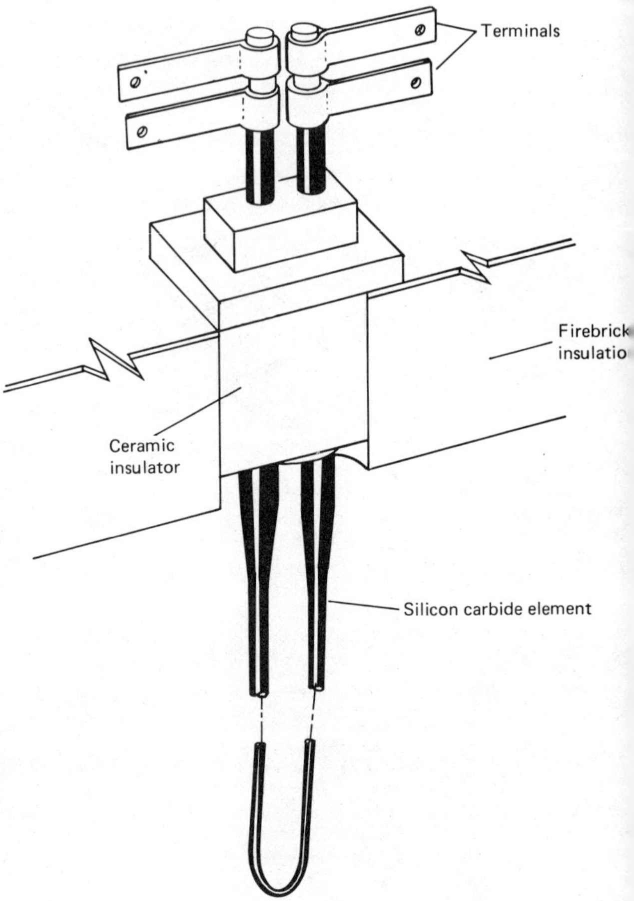

Fig. 2.23 Typical silicon carbide, cermet heater.

Graphite Heaters

CONSTRUCTION. Graphite elements are available in rod, tape, and machined solid forms. The exposed elements are terminated to graphite bussing strips and mechanically supported. (See Fig. 2.24.)

TYPICAL APPLICATIONS. A graphite element runs exposed and must operate in a vacuum or a nonoxidizing atmosphere. It is a very-high-temperature radiant source used at 2500°F (1371°C) and above. Applications include laboratory furnaces, refractory metal processing, sintering, calcining, and ceramic firing.

TEMPERATURE RANGE. Up to 5500°F (3038°C) is recommended.

WATT DENSITY RANGE. This is consistent with the element and furnace design—up to 1500 W/in^2 (232 W/cm^2).

THERMAL RESPONSE. The thermal response of graphite elements is moderate to slow as the resistance-versus-temperature variation requires a controlled application of electrical power.

EFFICIENCY. Overall system efficiency depends on insulation and system losses.

HEATER LIFE. Used properly with continuous operation, graphite elements can last in excess of 2000 h. The atmosphere requirements are either vacuum or nonoxidizing systems. The cost of graphite vacuum furnaces is high.

Open-Coil and Ribbon Heaters

CONSTRUCTION. Open-coil heating elements usually take the form of an assembly and are constructed without a metal sheath or insulation material covering the resistance element. The exposed elements are a resistance wire or ribbon, usually coiled or looped back and forth through supporting insulators. The exposed element and insulators are mounted into a support structure or housing. (See Figs. 2.25 and 2.26.)

TYPICAL APPLICATIONS. Applications include convective and radiant heating, industrial process air and gas heating, large furnaces, hot plates, and appliances.

TEMPERATURE RANGE. Up to 2500°F (1371°C), depending on the resistance alloy, is recommended.

WATT DENSITY RANGE. Denoted in terms of total wat-

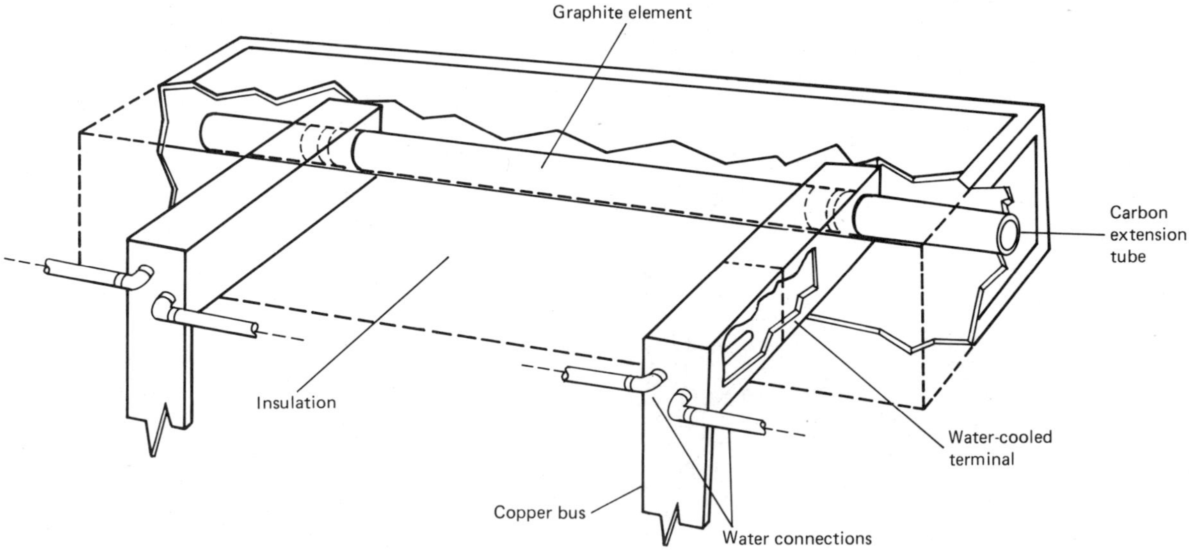

Fig. 2.24 Typical graphite heater.

tage output divided by total area for the system, up to 100 W/in^2 (15.5 W/cm^2) is the recommended density.

THERMAL RESPONSE. Thermal response is rapid.

EFFICIENCY. Total system efficiency depends on the proper use of insulation and the limiting of losses.

Radiant-Heating Panels

CONSTRUCTION. Radiant-heating panel construction techniques vary, but all have in common a flat, rectangular emissive face behind which exists a resistance heating element. One common construction uses resistance elements embedded in a ceramic fiber insulating structure. The other construction uses individual emitter strips, fabricated of metal, holding a mica-insulated resistance element. (See Fig. 2.27.)

TYPICAL APPLICATIONS. Radiant-heating panels are used for plastic sheet processing, drying, plastic thermoforming, and textile processing.

TEMPERATURE RANGE. Surface heating to 800 to 1100°F (427 to 593°C) yields best results.

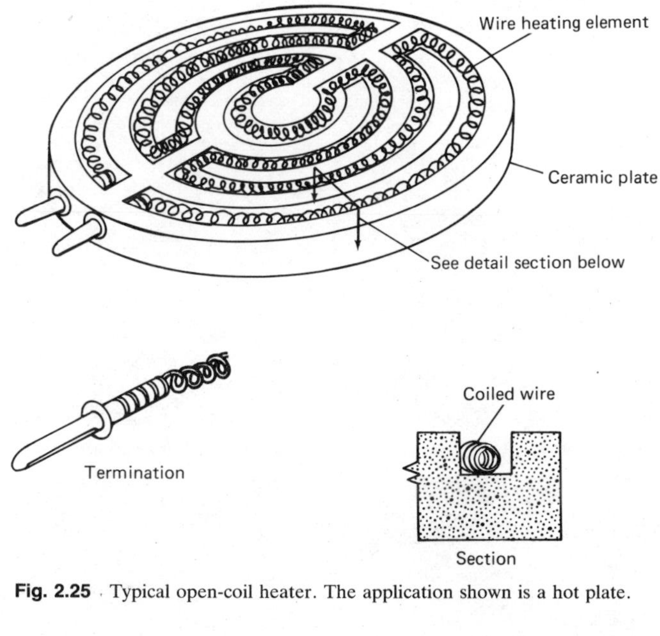

Fig. 2.25 Typical open-coil heater. The application shown is a hot plate.

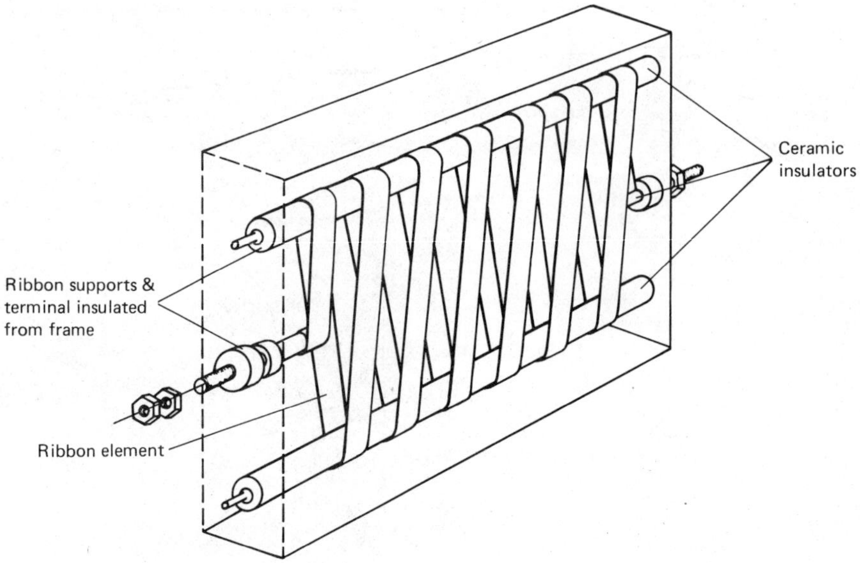

Fig. 2.26 Typical ribbon heater.

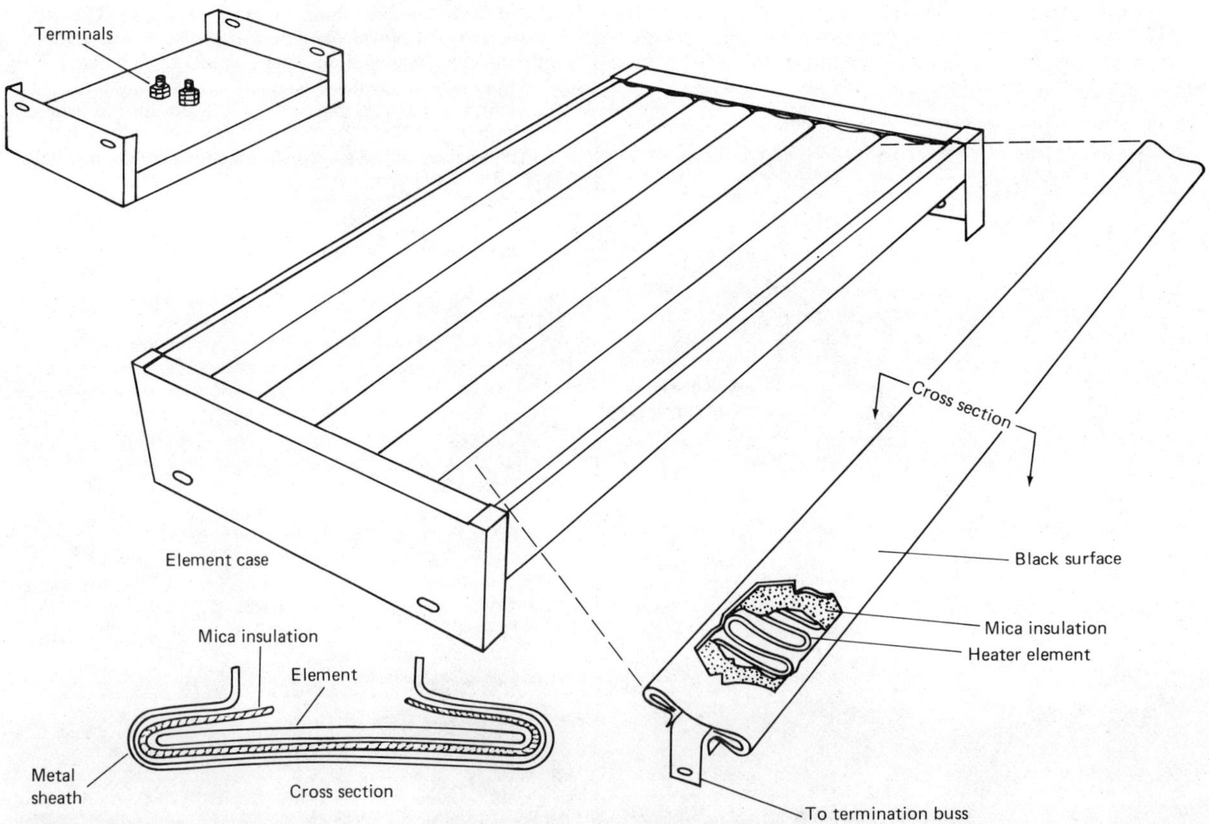

Fig. 2.27 Typical radiant heating panel.

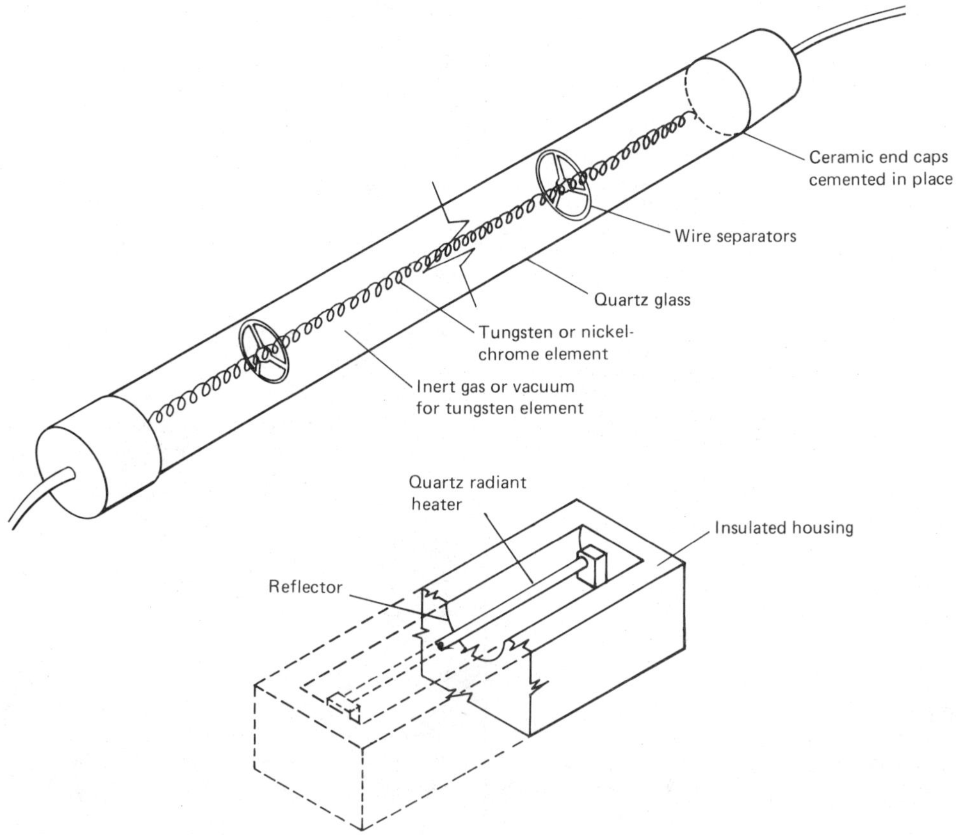

Fig. 2.28 Typical quartz radiant heater.

WATT DENSITY RANGE. Up to 24 W/in^2 (3.7 W/cm^2) is recommended.

THERMAL RESPONSE. Thermal response is moderate as the system has some thermal mass, especially the ceramic-fiber-based construction.

EFFICIENCY. The efficiency of radiant panels is better than that obtained from reflector-based radiant sources because a panel gives wide area coverage and can be brought closer to the work; evenness of heat is good.

Quartz Radiant Heaters

Construction Quartz heaters consist of a coiled high-temperature resistance wire alloy 80%Ni-20%Cr or pure tungsten, housed inside a quartz tube. The quartz tubing is terminated with ceramic insulating caps and a mounting system that allows the quartz tube to expand and contract thermally. A quartz heater is typically backed by a reflector to direct the radiant energy output. Quartz is used because it is transparent to infrared radiation and is capable of high operating temperatures. (See Fig. 2.28.)

TYPICAL APPLICATIONS. Shrink packaging and lamination, thermoforming, fusing plastics, vulcanizing rubber, metal heating and melting, drying operations, space heating, sterilization, and industrial process warming operations are typical applications.

TEMPERATURE RANGE. The maximum recommended temperature for heating elements may be as high as 1800°F (982°C).

WATT DENSITY RANGE. Up to 40 W/in^2 (6.2 W/cm^2), based on tube surface area, is recommended.

THERMAL RESPONSE. Quartz heaters are capable of generating their full heat-capacity output quickly.

EFFICIENCY. The efficiency of the radiant quartz heater is greatly affected by the ability to focus or concentrate the radiant output on the work and the absorption of the work itself. Typical efficiency is 45 to 90 percent. Radiant systems must work at high temperatures to be effective and are less efficient overall than systems where heat transfer is by conduction.

2.2 ELECTRIC RESISTANCE HEATER APPLICATION

Introduction

This section is devoted to the application of electric heating elements in various environments and is organized according to a step-by-step method that is modified to handle specific application problems. First, the method is defined; then, each step in the method is described. The approach uses a static heat-transfer analysis in an effort to clearly state the application solution. First-level approximations have been used throughout and then supplemented with safety factors. The use of more detailed thermal analysis is encouraged if the application warrants a more exact solution.

TRANSIENT VERSUS STEADY STATE. Because of the complex nature of transient heat flow calculations, only a steady-state approach is utilized. Transient cases are handled with a simplified technique that defines the starting and finishing points but does not allow for analysis of points in between. Again, the use of more detailed thermal analysis is encouraged when the application warrants a more exact solution.

Method for Heater Selection

A first step in solving electrical heating problems is the determination of required heat. The heat requirement must be converted to electrical power and the most practical heater then selected for the job. Whether the problem is heating gases, liquids, or solids, the approach to determining the power requirement is the same. All electrical heating problems involve the following steps:

- I. Definition of the problem; a brief statement and sketch of the known requirements and parameters concerning the problem
- II. Calculation of heat energy required (in kilowatthours or British thermal units)
  - A. System losses under operating conditions
    1. Conduction
    2. Convection
    3. Radiation
  - B. Heat energy required for process start-up
  - C. Heat energy required to maintain process
- III. Conversion from heat energy to power (kilowatthours to kilowatts) based on start-up heat and start-up time
- IV. Review of application factors to evaluate the operating environment, life requirements, mechanical considerations, and operating costs
- V. Selection of the type, size, and number of heaters, while considering factors such as efficiency and total system cost
- VI. Consideration of the temperature-control method

Defining the Problem

The heating problem must be clearly stated, specifically defining the operating parameters:

- Minimum start and maximum finish temperatures expected
- Maximum flow rate of heated material(s)
- Required start heat-up time and process cycle times
- Weights and dimensions of both heated media and containing vessels
- Effects of insulation and its thermal properties
- Electrical requirements such as voltage, control methods, and electrical limitations
- Other requirements:
  1. Safety considerations
  2. Heater life requirements
  3. Efficiency and cost factors

In view of these items, an assessment must be made as to whether enough information is available for problem solution. As a minimum, the thermal properties of both the process materials and the containing vessel will be required.

Energy Calculations

The heat energy (in kilowatthours or British Thermal units) required to satisfy the system needs will depend on:

1. Heat energy required for start-up (in kilowatthours or British Thermal units)
2. Heat energy required to maintain operation at the desired temperature (in kilowatthours or British Thermal units). The electrical heater powers required (in kilowatts) for start-up and operation will be the heat energy (in kilowatthours) for each case divided by the required start-up or working cycle time. The required power (in kilowatts) rating of the heater will be the greater of these two values times factors that account for efficiency of the heater and the uncertainty in the heat requirement estimate. Calculation of the absorbed thermal energy should consider both the sensible-heat energy requirements calculated using the specific heat capacity of the material and the latent-heat requirements (or phase change) calculated using the latent heat of fusion or vaporization. All materials absorbing heat, including the heated media, work being processed, containers, racks, belts, etc., should be considered.

Heat losses occur in systems in three ways: by conduction, by convection, and by radiation. These values should be determined individually and summed for total losses. During start-up of a system, the losses are zero and rise to 100 percent (of steady state) at process temperature. A good approximation of actual losses (during the heat-up period) will be obtained by calculating the rate of heat loss (in British Thermal units or kilowatthours) at the process temperatures and then multiplying the result by two-thirds. The following equations can be used to estimate the required power. Both the start-up power and the operating power must be analyzed before heater selection. Start-up power is given by

$$P_S = \left[\frac{Q_A + Q_C}{t_s} + \frac{2}{3}(\dot{Q}_L)\right]\frac{(1 + \text{SF})}{E/100}$$

where Q_A = heat absorbed by materials during heat-up, Btu or kWh
Q_C = latent heat absorbed during heat-up, Btu or kWh
$\dot{Q}_L$ = conduction, convection, radiation losses, Btu/h or kW
SF = safety factor
t_s = start-up time required, h
E = efficiency of heater, %

and operating power is given by

$$P_O = \left[\frac{Q_B + Q_D}{t_c} + \dot{Q}_L\right]\left(\frac{1 + \text{SF}}{E/100}\right)$$

where Q_B = heat absorbed by processed materials in working cycle, Btu or kWh
Q_D = latent heat absorbed by materials heated in working cycle, Btu or kWh
$\dot{Q}_L$ = conduction, convection, radiation losses, Btu/h or kW
SF = safety factor
t_c = cycle time required, h
E = efficiency of heater, %

A safety factor of varying size should always be added to allow for unknown or unexpected conditions. The size of the safety factor is dependent on the accuracy of the wattage calculation. Heaters should always be sized for a higher value than the calculated figure. A factor of 10 percent (i.e., 0.10) is adequate for small systems that are closely calculated; 20 percent additional wattage is more common. Values of 20 to 35 percent should be considered for systems with large convection losses (doors opening) or large radiation losses with uninsulated edges. The efficiency of the heater depends on a number of variables. (See following discussion of efficiency.) Note that heat losses from the heater itself are included in heater efficiency and should not be included in the term $\dot{Q}_L$.

The start-up and operating watts are displayed in a graphic format in Fig. 2.29 to allow visualization of the various energy component parts. The area of each block represents the amount of energy required. The Q_A and Q_C areas will remain constant such that a change in start-up time will require a corresponding change in power required. With the use of this graphic aid, the following evaluations are possible:

1. Compare start-up watts to operating watts.
2. Evaluate the effects of lengthening start-up time such that start-up wattage equals the operating wattage (possibly use a timer to start the system before the shift begins).
3. Recognize that more production capacity exists than is being utilized. (A short start-up time requirement needs more wattage than the processing wattage.)
4. Identify energy usage and redesign or add insulation to reduce wattage requirements.

After the entire system has been considered, start-up time, production capacity, and insulating methods should be reevaluated. The addition of thermal insulation may be cost-justified by energy savings, reduction of number of heaters, or longer heater life. The selection of heaters will then be made utilizing the most efficient system.

Review of Application Factors

Other factors must be considered before selection of a particular heater type. These include environmental factors, heater life, operating costs, mechanical requirement, and efficiency.

Operating Temperatures Heat produced must be dissipated. The temperature of a heater will rise until the losses equal the input. Operating temperatures usually indicate the temperature of the working process. Actually, the temperature of the heater is most important in choosing a particular heater type for an application. Three temperatures should be considered in the application problem: (1) the starting temperature for the process, (2) the maximum process temperature, and (3) the temperature of the heater when the process is at maximum temperature. The heater temperature will be greater than the maximum process temperature. This increased temperature is dependent on the type of heater, the nature of the heat-transfer path between the heater and the work, and the insulating system around the heater. The heater temperature can be considerably higher than the final working temperature of the process. This circumstance can be very detrimental to heater life.

Efficiency All electrical resistance heaters convert 100 percent of the electrical energy into thermal energy. (In the case of quartz tube heaters, some energy is shorter-wavelength visible light.) However, all thermal energy does not get to the heated parts. Different types of heaters exhibit specific efficiencies under similar conditions. These efficien-

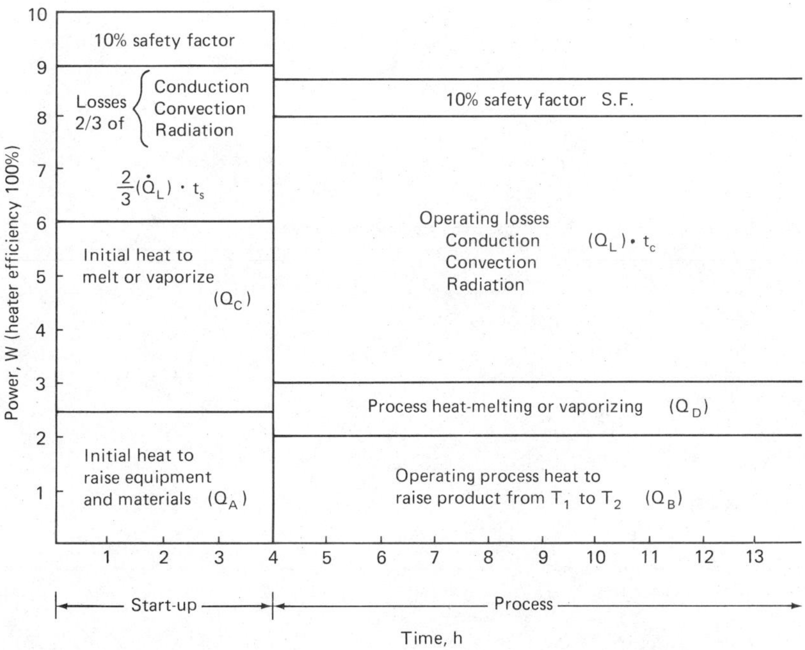

Fig. 2.29 Power evaluation diagram. (*Reprinted courtesy of Watlow Electric Mfg. Co.*)

cies are approximated in Table 2.2. These efficiencies must be accounted for in the overall wattage calculations.

Watt-Density Limitations (Watts per Square Inch) Every type of electric heater has specific temperature and mounting considerations that establish maximum allowable watt density ratings. These considerations are a function of heater surface temperature. The maximum operating watt density is based on the assumption that heater life must meet a given minimum requirement. Factors considered in establishing these ratings are:

1. Element operating temperature
2. Temperature limits of insulating materials
3. Quality of conduction path (clamping, hole fit)
4. In immersion applications:
   a. Boiling characteristic and vapor pressure of liquid
   b. Fluid viscosity and film heat-transfer coefficients
   c. Corrosion rate of sheath materials
   d. Temperature of fluid decomposition

In conjunction with desired life, watt density is used to calculate both the required number of heaters and the size of the heaters. Determining the size and quantity of heaters involves establishing the total power requirements and the maximum allowable watt density (utilizing manufacturer supplied guidelines).

Mechanical Considerations Installation and removal of heaters must be carefully considered at the time of equipment design. Installation of heaters is generally not a problem since the area is relatively clean and cool and often partially disassembled. Removal of heaters, however, is sometimes attempted while equipment is still warm. There may also be some level of contamination present that makes heater removal difficult.

Electric heaters are generally susceptible to mechanical damage (especially the leads) and must be protected as much as possible. The mechanical durability is rated for each heater type described in Table 2.1. Depending on the possibility of mechanical damage and the susceptibility of the heater to damage, extra protection may be required. For these reasons, full access must be provided (in the design process) for ease of heater replacement. This is usually done with shrouds or guards over the heaters. These guards also serve a secondary purpose in that they may minimize convective heat losses from the back of heaters and increase efficiency of the system.

In all applications where the heater must be attached to a surface, it is extremely important to maintain as intimate a contact as possible to aid heat transfer. Heaters mounted on the exterior of the part should have clamping bands or bolts to facilitate this contact, and heaters inserted in holes should have hole fits as tight as practically possible. Provision for removal of failed heaters can be accomplished by through-drilling holes in blocks or utilizing split-plate constructions.

Environment Contaminants are the primary cause of shortened heater life. Decomposed oils and plastics (hydrocarbons in general), conductive pastes used as antiseize materials, and molten metals and metal vapors can all create situations that affect heater life. Some heater constructions are sealed more securely against contaminants than others. In analyzing applications, all possible contaminants must be listed in order to be able to fully evaluate the proposed heater.

EXAMPLE Heat is required to maintain molten zinc in the passageways of a zinc die-casting machine. The possible contaminants for this application are (1) molten zinc metal, (2) hydraulic oils, (3) high-temperature antiseize materials, and (4) moisture—especially if die cooling is aided by water circulation. All of these factors indicate that a highly sealed heater construction is required.

The corrosiveness of the materials heated or the materials that will contact the heater must also be taken into consideration. Even if a heater is completely sealed, the choice of the external sheath material is very important to heater life.

Explosive environments require that the heater be completely isolated from potentially dangerous areas. This is accomplished by inserting the heater in protective wells and routing the wiring through sealed passageways out of the hazardous area. Very close fusing is required in these critical applications to minimize the magnitude of potential failures.

Heater Life Requirements Maximum heater life is achieved by minimizing the heater temperature and the temperature variation caused by cycling the heater on and off. The higher the temperature, the shorter the service life.

Mineral insulated heaters using traditional (nickel-chromium) alloys for resistance elements are subject to the life-limiting factor of wire oxidation. The winding wire oxidizes

Table 2.2. Typical Efficiency* of Electric Resistance Heaters

Heater type	Immersed or fitted into a hole	Tightly clamped or bonded	Radiant (noncontact): Without extra insulation	Radiant (noncontact): With insulation
Mica strip or band	. . .	80+	. . .	. . .
Tubular	100	70+	45	90
Flexible	100	90+	. . .	. . .
Swaged cartridge	100	70+	45	90
Radiant panel	. . .	. . .	45	90
Mineral-insulated cable	100	60+	45	90
Mineral-insulated strip or band	. . .	80+	. . .	. . .
Ceramic fiber	. . .	—	45	90

* The efficiencies listed are only approximate and are highly dependent on specific application details. In general,

$$\text{Efficiency} = \frac{\text{watts absorbed by work}}{\text{watts input}} \times 100 = \text{percent}$$

at a rate dependent on the element temperature. If the element temperature is known, it is possible to project a heater life as shown in the following table (for mineral insulated cartridge, tubular, cable, strip, and band heaters):

Internal element temperature, °F (°C)	Approximate life, h
1500 (815)	7000
1600 (871)	2000
1700 (927)	667
1800 (982)	260
1900 (1038)	80
2000 (1093)	40
2100 (1149)	16

Generally, application charts and operating recommendations of manufacturers use maximum 1600°F (871°C) internal temperature to ensure expected heater life of greater than 1 year. Heaters utilizing lower temperature insulating materials (silicone rubber and mica) have life-limiting factors associated with excessive thermal cycling and operating above the temperature limits of the insulation. Excessive thermal cycling occurs whenever the temperature of the heating element undergoes a large excursion between the ON and OFF cycles. This may be caused by any of the following:

1. Very high watt density
2. "Long" ON and OFF cycles
3. Poor conduction path between heater and work
4. Forced cooling (quench)

What constitutes an excessive thermal cycle varies according to the heater type. A total temperature excursion of more than 25 percent of the desired temperature is a good yardstick definition.

Insulation temperature limits are accounted for in Table 2.1. Maximum temperature limits are approximately

Silicone rubber	550°F (288°C)
Mica	1200°F (649°C)
Magnesium oxide	1800°F (982°C)
Ceramic fiber	2200°F (1204°C)

Lead Considerations It is not only important to consider the type of electric heater and placement and wattage requirements but also necessary to consider the types of electrical leads used and the methods by which they exit the heated area. General considerations in selecting various lead types are:

1. Current rating
2. Temperature of lead area
3. Contamination in the lead area
4. Flexibility requirements
5. Abrasion-resistance requirements
6. Relative cost

The design should provide for ease of removal of the heater from the equipment. The use of terminal blocks for electrical connectors is recommended wherever possible. Table 2.3 is a composite listing of the various attributes of lead wire conductors and insulating systems.

Temperatures listed indicate actual physical operating limits of various wire types. Wires are sometimes rated by agencies (Canadian Standards Association, Underwriters' Laboratories, Verband Deutscher Elektrotechniker) for operating at much lower temperatures. In this case, the rating agency temperature limit is the maximum level at which this wire has been tested. If agency approvals are required, do not exceed the established temperature limits.

When selecting wiring for electric heater circuits, consider that the wiring may be operating at a temperature above room ambient. A temperature above ambient may be the result of conducted heat from heater terminals, radiation from heated surfaces, conductor self-heating, or high ambient temperatures. In high-temperature areas, wiring should employ high-temperature insulation and conductors. Outside the heated zone, conventional wiring methods and materials are adequate.

Electrical Considerations Electric resistance elements generate power (watts) when a voltage is applied. The standard relationships between voltage, resistance, current, and

Table 2.3. Electrical Power Lead Characteristics

Lead types	Maximum lead area temperature °F	Contamination resistance	Flexibility	Abrasion resistance	Relative cost
		Lead Protection Methods			
Metal overbraid	. . .	Average	Good	Excellent	Moderate
Flexible conduit	. . .	Good	Average	Excellent	Moderate
		Lead Insulation Types			
Ceramic beads over solid nickel	1200	Poor	Poor	Average	High
Asbestos with silicone-impregnated glass braid	1000	Poor	Good	Average	Low
Glass with silicone-impregnated glass braid	750	Poor	Good	Average	Low
Silicone rubber	500	Good	Good	Poor	Low
TFE, FEP, PTFE	500	Excellent	Good	Good	Low
Thermoplastic (PVC)	150	Good	Good	Good	Low

CONVERSION FACTOR: °C = (°F − 32)/1.8.

power as expressed by Ohm's law and Joule's law apply to electric resistance heaters. Figure 2.30 displays these basic equations and their various combinations. These relationships are applicable as described in the following.

VOLTAGE. All ac waveforms are converted to their effective dc values. The voltage used is the true root-mean-square (rms) voltage. The rms value can be easily determined by measuring the voltage across the heater terminals with a voltmeter that reads true rms. Most meters today will measure true rms. Even the nonstandard "clipped, clamped, or phase-fired" waveforms that solid-state controllers create can be measured and expressed as true rms voltages.

CURRENT. The current is also measured as an rms value that represents the exact dc equivalent current. The equations shown in Fig. 2.30 apply equally to both dc and ac (rms).

RESISTANCE. Since most materials change resistance as the temperature is changed, it is important to know the temperature at which the resistance is specified. This will be at either room temperature (for quality assurance and manufacturing purposes) or the operating temperature (for system design purposes). Most wattage calculations use the resistance at operating temperature. The complex heater impedance is almost entirely resistive. For all except the largest heater systems, the power factor can be considered 1.0. Large systems usually will exhibit a slight capacitive component.

THREE PHASE. Three-phase heater applications can be evaluated using standard electrical engineering relationships. These are shown for both *delta* and *wye* connections in Fig. 2.31. Only balanced configurations are shown.

TESTING. Another important consideration is electrical testing. In addition to the basic resistance measurements, the insulation systems (when they exist) are tested for weaknesses that may cause premature failure.

The insulation resistance is measured between the element and the sheath by applying a dc voltage (usually 500 V), measuring the small current that flows, and converting ($R = V/I$) to insulation resistance. This test measures the bulk insulation resistance properties.

The insulation dielectric strength (breakdown voltage) is usually measured by applying a high ac voltage across the insulation layer and measuring the current leakage. By compensating for the capacitive and inductive effects, the leakage current due to the resistive component can be determined. Usually, if the heater passes a dielectric strength of 1000 V plus twice the rated voltage (for 1 min), the heater is considered acceptable (i.e., a 240-V heater must pass a 1480-V test with less than 2-mA leakage). This test measures element spacing and insulation integrity.

Since the two tests are conducted under different conditions, the results are not directly comparable; however, results of both tests must be considered in evaluating heater integrity.

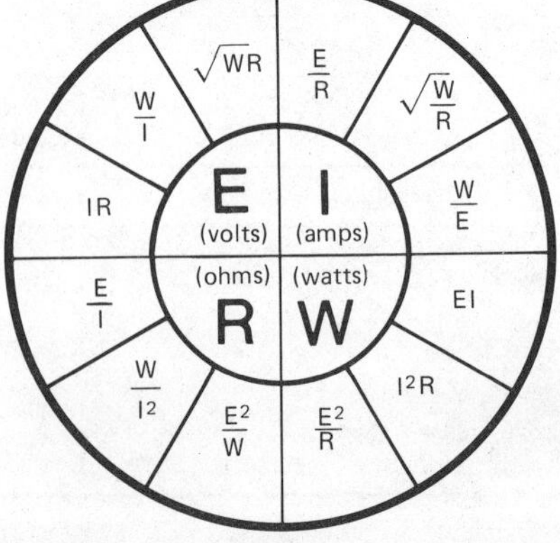

Fig. 2.30 Summary diagram of Ohm's and Joule's laws.

Selection of Heater Type, Size, and Quantity

After the required wattage has been calculated and various heater attributes considered, the number of possible heater types should be narrowed considerably. Several additional factors must be considered before final heater type selection.

Initial Installation Cost Specific installation costs must be considered for each type of heater:

Application type	Installation cost	Unit efficiency, %
Immersion	Medium	High (100)
Block with a hole	High	High (100)
Tightly clamped or bonded	Medium	Medium (60–98)
Radiant applications	Low to high	Low (45–90)

Compromises must be weighed—the higher installation cost cartridge heater is more efficient, completely protected, and enclosed. The cartridge heater requires accurate drilling and possibly reaming for securely fitting holes to provide good performance. A strip heater requires a smooth mounting surface with drilled and tapped mounting holes. Unit efficiency is adequate; however, a back-side shroud or insulation boosts efficiency to 90 percent. A radiant panel mounts on external brackets and offers noncontact heating but is relatively inefficient. The unit efficiency can be greatly enhanced by providing insulating shrouds to eliminate edge losses; however, this adds cost to the system.

Replacement Cost The cost of a new heater, lost production time, removal, and installation of the new heater must be considered. Generally, these costs are actually greater than expected. Heater life must be such that replacement can be scheduled and planned during off-peak production times to avoid lost production capacity.

Heater Life Expectancy and Reliability Heater life is based on several factors:

1. Operating temperature
2. Mechanical damage
3. Corrosion
4. Contamination
5. ON/OFF cycling

While attention to environmental and physical factors can eliminate shortened heater life, the operating temperature limitation must be provided for early in the design phase. By utilizing the maximum safe watt density, expected life will result. If longer life is required, larger size or additional heaters may be added—reducing the operating watt density and increasing life.

DEFINITIONS for both WYE and DELTA (Balanced Loads)

V_P = phase voltage
V_L = line voltage
I_P = phase current
I_L = line current
$R = R_1 = R_2 = R_3$ = resistance of each branch (equal resistances);
$W_{delta} = 3W_{wye}$ = heating power

THREE-PHASE WYE

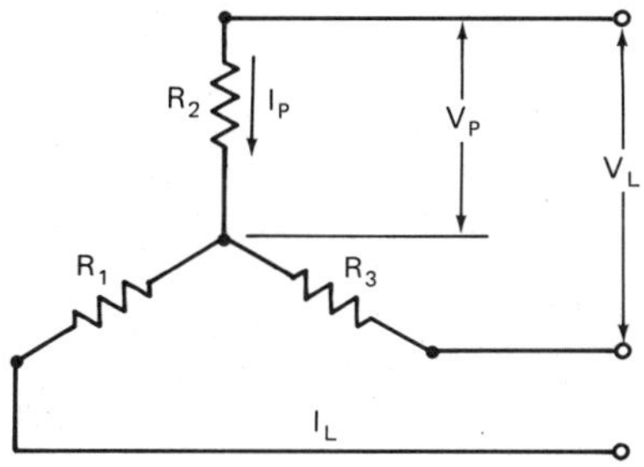

EQUATIONS FOR WYE ONLY

$$I_P = I_L$$
$$V_P = V_L/1.73$$
$$W_{wye} = V_L^2/R = 3(V_P^2)/R = 1.73\ V_L I_L$$

THREE-PHASE DELTA

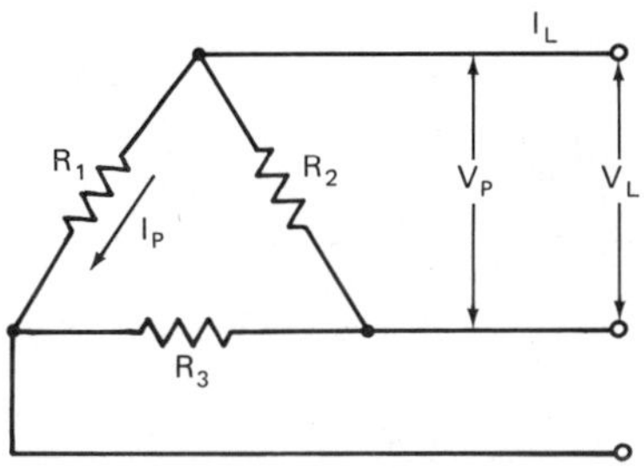

EQUATIONS FOR DELTA ONLY

$$V_P = V_L$$
$$I_P = I_L/1.73$$
$$W_{delta} = 3(V_L^2)/R = 3V_P I_p = 1.73 V_L I_L$$

Fig. 2.31 Balanced three-phase relationships for delta and wye circuit configurations.

Operating Cost The total system operating cost is a composite of two factors. It is usually best to examine cost on an annual basis:

1. Total cost of energy (kWh) × ($/kWh)
2. Replacement cost

EXAMPLE Two barrels on different plastic extruders are operated for 40 h/week. Five 1000-W mica band heaters are used on each extruder. Each extruder has a shroud surrounding the heaters, but one extruder also has extra insulation placed inside the shroud. The heaters are controlled to maintain a constant barrel operating temperature. Expected heater life is 1 year for the uninsulated application.

Costs		
Heater cost		$12 each
Maintenance labor rate		$20/h
Lost production time rate		$50/h
Energy cost		$0.07/kWh
	Uninsulated	Insulated
Actual power usage, kWh		
	4.06	2.38

Annual Energy Cost		
2080 h × 4.06 kW/h × $0.07/kWh	$591	
2082 h × 2.38 kW/h × $0.07/kWh		$346
Replacement Cost		
5 heaters at $12 each	60	60
4 h of labor to install × $20/h	80	80
4 h of lost production time × $50/h	200	200
Total	$931/yr	$686/yr

The cost of operation is much less when shrouds are used. Also, with the reduced wattage in case 2 (and subsequently effective lower watt density), it is probable that the heaters can operate longer than the originally estimated 1 year. However, the initial cost will be higher for case 2. The payback on insulating shrouds must be considered with respect to the extended life and reduction in equipment downtime.

Safety The performance of an electric heater must be taken into consideration if a heater failure could cause serious problems. If a heater is used in a medical application where there are life-and-death considerations, careful fusing, attention to current leakage, and stringent electrical characteristics must be designed into the product. If a heater is used in an industrial environment where hazards exist, attention to enclosures around the heaters is necessary to prevent possible heater failures from becoming dangerous to personnel or the plant.

Heater Type Selection With an understanding of the relationships between operating costs, replacement costs, and heater life, the rational selection of a heater type is possible. If detailed estimates have been made as described herein, then the type selection may be obvious. If detailed estimates are not possible, then selection from Table 2.1 should narrow the scope to two or three possibilities—all of which should work.

After selecting a heater type, the calculated wattage should be reevaluated to ensure that appropriate system safety factors and heater efficiencies have been applied. With the heater type, total wattage, and watt density (based on life-vs.-watt density considerations) defined, the size and quantity of heaters can be computed.

STEP 1. Calculate the minimum total area of heaters required:

$$\text{Minimum heated area} = \frac{\text{total watts}}{\text{maximum allowable watts per unit area}}$$

STEP 2. On the basis of mechanical requirements, life requirements, and all other variables, determine an initial heater size and quantity and test for feasibility.

$$\text{Actual heater area (no. of heaters)} \times \text{(area of each heater)} \geq \text{minimum heated area}$$

STEP 3. Revise the quantity and size of heaters until the above relationship is satisfied.

Evaluation of Proper Control Methods

Many different methods have been used to control heat, including the following: bimetallic-metallic elements, capillary-bulb thermostats, thermocouples, resistance tempera-

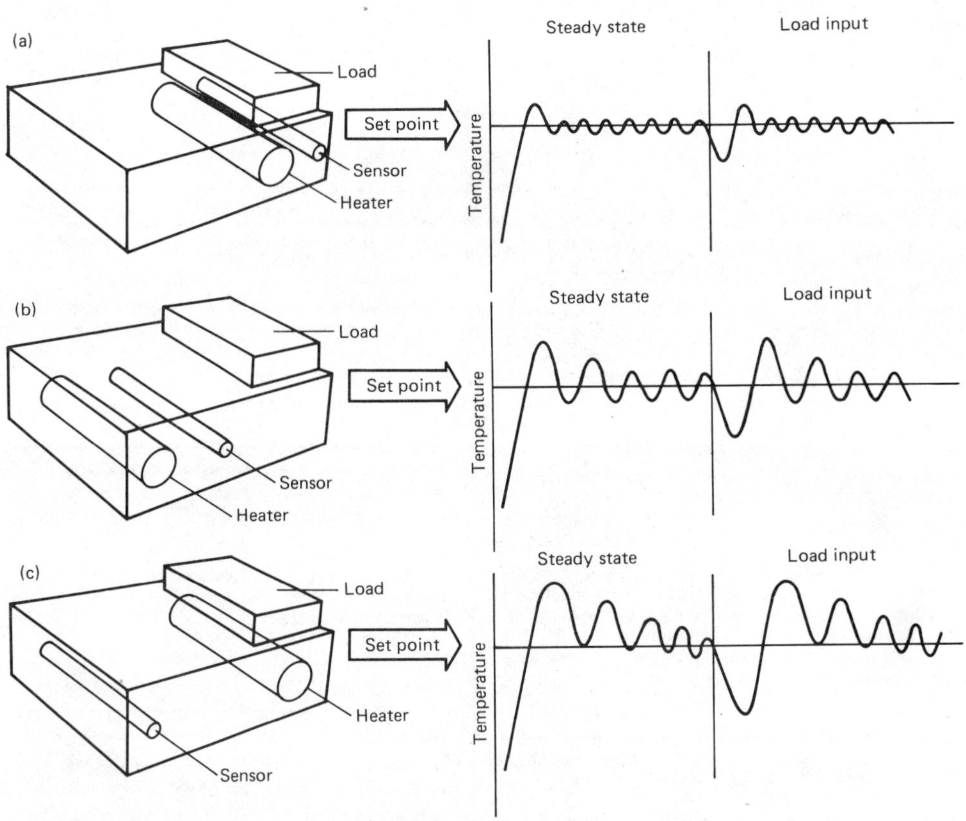

Fig. 2.32 Thermal response in three systems with different placement of sensors, heaters, and work. (*a*) *Best*—the heat source is close to the work and the sensor is close to both the heat source and the work. This short heat-conduction path minimizes thermal lag. (*b*) *Marginal*—the heat source is distant from the work and the sensor is located between the source and the work. The longer heat conduction path increases thermal lag of the system, but being located midway, the sensor can respond to work or heater changes without excessive lag. (*c*) *Poor*—the heat source is close to the work and the sensor is distant from both the heat source and the work. The sensor is too far from the heat-conduction path to respond to temperature changes without excessive lag. The sensor is also located too far from the workload.

ture devices (RTDs), thermistors (negative temperature coefficient), and (positive temperature coefficient) (PTC) devices. Many heater types are available with internal sensors. In all control situations for thermal systems there are four major elements:

1. *The load:* This is the material to be heated and is generally associated with a cycle time or a time rate of passage over or through the heated part.
2. *The heater:* The electrical heating device.
3. *The heat-transfer medium:* This is the block of material or the liquid that transfers the heat energy to the load.
4. *The sensor:* The sensing device that controls the heating element to maintain a specific temperature for the load.

With these four elements in mind, consideration must be given to the placement of these various items with respect to each other as indicated by Fig. 2.32.

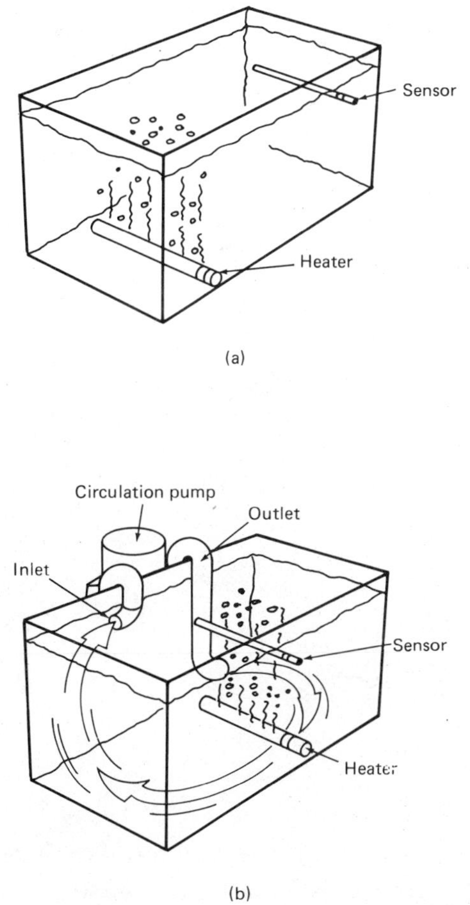

Fig. 2.33 Sensor placement in convection applications. (*a*) *Poor liquid heating control*—The convection heat path between the heater and the sensor is too long, creating excessive lag and temperature overshoot. (*b*) *Best liquid heating control*—the shorter convection heat path between the heater and the sensor and a circulation pump to mix materials and reduce heat gradients, minimizing thermal lag and temperature overshoot.

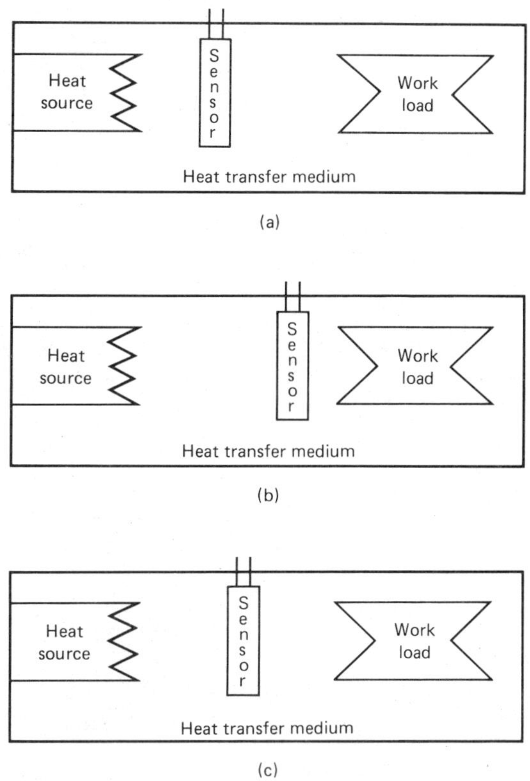

Fig. 2.34 Sensor location as a function of static/dynamic thermal loading. (*a*) Sensor near the heat source; (*b*) sensor near the workload; (*c*) sensor halfway between.

Conduction, Convection and Radiation Applications The various types of applications can be categorized with regard to the primary mode of heat transfer—conduction, convection, or radiation. A conduction system might consist of the heating element in direct contact with the heat-transfer medium which, in turn, heats the load. Shown in Fig. 2.32 are three different conduction systems using different temperature sensor locations.

The radiation and convection heat-transfer applications, in general, encompass both liquid- and gas-type applications. In these cases, placement of the sensing element is critical to stable performance. The examples shown in Fig. 2.33 are for liquid tank heating. As shown, the sensing element must be located relatively close to the heating element in order to sense possible overheating of elements or the medium in which they are immersed.

Sensor Location Proper placement of the sensor in relation to the workoad and heat source can compensate for various types of energy demands from the workload. Proper sensor placement can limit the effects of thermal lags in the heat-transfer process. The heater controller can respond to the temperature change feedback only from the sensor. Thus sensor location will influence the ability of the controller to regulate the temperature about the desired set point.

When the heat demand is static (i.e., steady), placing the sensor closer to the *heat source* (Fig. 2.34*a*) will keep the heat fairly constant throughout the process. In this type of system the distance between the heat source and the sensor is small (minimal thermal lag); therefore, the heater will cycle frequently reducing the potential for overshoot and undershoot at the workload. With the sensor placed at or near the heat source, it can quickly sense temperature changes, thus maintaining tight control. This placement is suggested for soaking or radiant panel applications.

When the heat demand is dynamic (i.e., time-varying), placing the sensor closer to the *workload* (Fig. 2.34*b*) will allow the sensor to "see" the temperature changes (in the workload) faster and the controller to take the appropriate output action more quickly. In this type of system the distance between the heat source and the sensor is notable (significant thermal lag); therefore, the heater cycles will be longer, causing a wider swing between the maximum (overshoot) and minimum (undershoot) temperatures at the workload. In this situation, it is also possible for the heater to exceed the recommended operating temperature. It is recommended that the electronic controller selected in this situation include the features (anticipation and offset ability) to compensate for these conditions. With the sensor at or near the workload, it can quickly sense heat absorption or loss, making it an ideal placement for dip tanks or deep fat fryers.

When the heat demand fluctuates between static and dynamic, placing the sensor "halfway" between the heat source and the workload (Fig. 2.34*c*), will divide the heat-transfer lag times equally. Because the system can produce some overshoot and/or undershoot, it is recommended that the electronic controller selected for this situation include the features (anticipation and offset ability) to compensate for these conditions. This sensor location is most practical in most thermal systems.

Chapter 3

Dielectric and Microwave Heating

PAUL MOFFITT

J. A. Callanan Company

Although radio-frequency (RF) dielectric heating has been used in industry since World War II, it is best known today in the form of the microwave oven that has become so commonplace in home kitchens.

The advantages of dielectric heating stem from the fact that the heat is generated directly within the work material itself. This allows for very precise control of the amount of heat and where it is generated. Because of the inherently shorter thermal time constants involved it tends to be much faster than processes that require transfer from a heated medium outside the workpiece. RF heating also facilitates the internal heating of materials which are poor thermal conductors.

3.1 TYPICAL RF DIELECTRIC HEATING APPLICATIONS

Dielectric heating has been found to be advantageous in the following main areas:

- Plastics heating, welding, and processing
- Food processing
- Wood manufacturing
- Adhesive drying and curing
- Ceramics
- Rubber processing
- Medical (diathermy, hyperthermia, and cauterizing)

Textile and paper drying
Chemical processing

Dielectric heating remains a growing field with several new applications being investigated each year.

3.2 THEORY OF DIELECTRIC HEATING

Dielectric heating occurs when an insulating (dielectric) material is subjected to a rapidly changing (1-MHz or higher frequency) electric field. Although there are several mechanisms by which this heating may occur, the most common is by what is called *rotational* or *orientation polarization*. If the molecules of the dielectric possess a natural electric dipole moment (i.e., by nature they are polar), then it is possible by rotation of the dipole moment to transfer energy from the electric field to the molecule. The transferred energy manifests itself as heat. Dielectrics which have molecules that are nonpolar are normally not good candidates for dielectric heating.

To develop a simple equation that shows the effect of the variables involved in the heating process, the work material can be modeled as a dielectric between two conductive electrodes connected to an RF generator as shown in Fig. 3.1. This is easily recognizable as a capacitor with C representing the capacitance contributed by the dielectric.

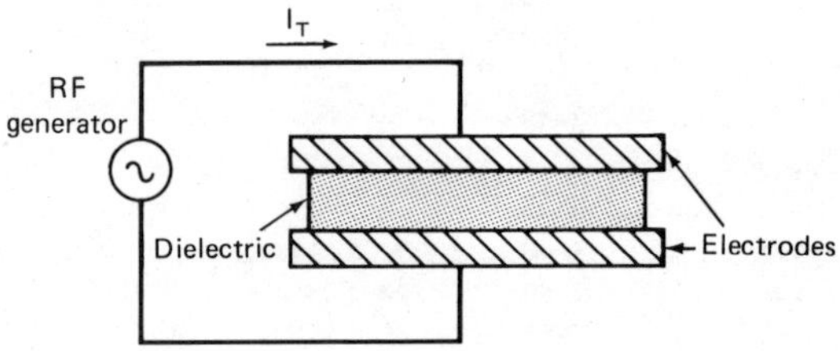

Fig. 3.1 Simplified model of dielectric heating process.

The energy generated as heat can be shown as a resistance R_{loss} connected in parallel with the capacitor (Fig. 3.2). It must be understood that R_{loss} is not a dc resistance that can be measured with a standard dc ohmmeter. The bulk resistance of a dielectric is by definition very high. The resistance R_{loss} is the equivalent loss resistance that appears when the dielectric is placed in a rapidly changing electric field. The current through the dielectric is complex and can be written as

$$I_T = I_{loss} + jI_C \tag{3.1}$$

where I_T = total current through work material
I_{loss} = loss current through work material
I_C = capacitive current through work material

The capacitive current I_C is represented in vector form in Fig. 3.3. A perfect resistor would have a vector only along the real or 0° axis, while a perfect capacitor would have all its current along the imaginary or 90° axis. The ratio between the reactive and the resistive currents can be represented as

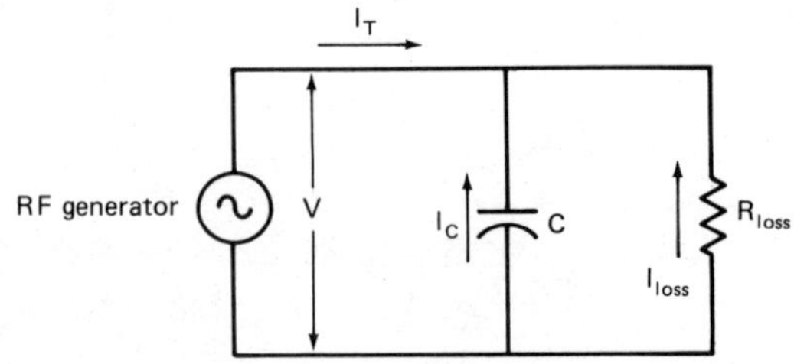

Fig. 3.2 Equivalent circuit model of Fig. 3.1.

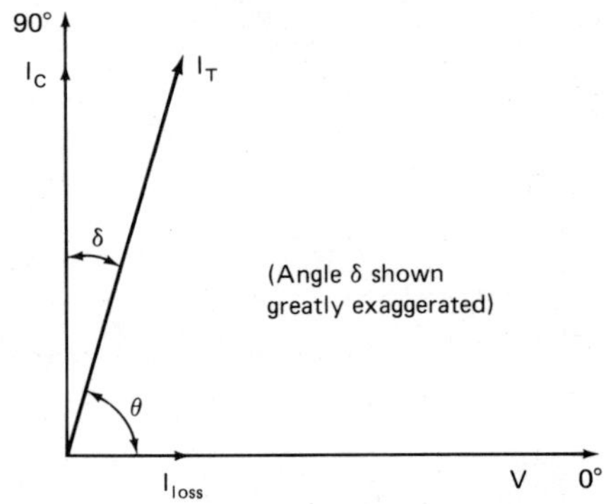

Fig. 3.3 Vector diagram of current through dielectric.

the tangent of angle δ. At a given frequency of applied voltage this can be written as $\tan \delta = 1/(2\pi f C R_{loss})$, where f is the frequency. The capacitance of a dielectric between two flat plates can be calculated with good accuracy as

$$C = \epsilon_0 \frac{kA}{d} \tag{3.2}$$

where C = capacitance, F
ϵ_0 = permittivity constant = 2.248×10^{-13} F/in (8.854×10^{-12} F/m)
k = dielectric constant of the dielectric
A = surface area of the dielectric, in² (m²)
d = thickness of the dielectric, in (m)

The power dissipated in the dielectric can be written as

$$P = \frac{V^2}{R_{loss}} = 5.56V^2 fk \tan \delta \frac{A}{d} \times 10^{-11} \tag{3.3}$$

for SI units; for USCS units this would be

$$P = 1.41V^2 fk (\tan \delta) \frac{A}{d} \times 10^{-12} \tag{3.4}$$

where P = power dissipated, W
V = voltage across dielectric
A = area between electrodes, in² (m²)
d = thickness of dielectric, in (m)

This equation is often expressed in the form of power dissipated per unit volume:

$$P_v = 1.41E^2fk\ (\tan\delta) \times 10^{-12} \qquad (3.5)$$

where P_v is the power density in watts per cubic inch and E is the electric field strength in volts per inch. A more complete formula which takes into account the possible nonuniformity of the electric field is

$$P_v = 1.41fk\ (\tan\delta) \int_V (\mathbf{E}\cdot\mathbf{E}^*)\ dV \times 10^{-12} \qquad (3.6)$$

where * indicates the complex conjugate.

The amount of power needed for a given temperature rise of the work material can be calculated as

$$P = \frac{(T_2 - T_1)mC_p}{t1055}\ \text{W} \qquad (3.7)$$

where T_1 = beginning temperature, °F
T_2 = desired final temperature, °F
C_p = specific heat of the material, Btu/lb · °F
m = weight of the material, lb
t = time, s

Although these equations are valid and accurate for the simplified model presented here, the fact that there are very few published tables of k and tan δ makes them useful mostly for reference value.

3.3 FACTORS INVOLVED IN DIELECTRIC HEATING

Dissipation Factor

Probably the most important factor in dielectric heating is tan δ, or the *dissipation factor,* as it is sometimes called. This factor is high in materials with polar molecules and quite low in materials with nonpolar molecules. Generally, dielectrics which lend themselves to dielectric heating have a dissipation factor of 0.01 or greater. The dissipation factor of some common dielectrics is given in Table 3.1.

The dissipation factor varies with frequency (Fig. 3.4*b*), temperature (Fig. 3.4*d*), humidity, and in some cases applied voltage. Most of these variations can be explained by noting that the dipole electric moment of the molecule has a transition time. When the period of the excitation voltage is equal to the transition time of the molecules, the dissipation factor is at its maximum. The transition time is related to the viscosity of the material, which is proportional to its temperature. In RF heating processes that involve a phase change of the dielectric, such as from solid to liquid, it is important to take into account that the dielectric constant and dissipation factor will also change from one phase to the other.

Although the tan δ of many materials has been measured, the customary measurement frequency is usually 60 Hz, 1 kHz, or 1 MHz, all below the common frequencies used in RF heating. Another term that may be seen in tables is the power factor of a material. This is the cosine of the angle θ in Fig. 3.3 and for most materials will have a numerical value

Table 3.1. Dielectric Constants and Dissipation Factors for Selected Material

Material composition	Dielectric constant at			Dissipation factor at		
	1 MHz	100 MHz	3000 MHz	1 MHz	100 MHz	3000 MHz
Acetal	3.7	. . .	. . .	0.005	. . .	. . .
Beeswax, yellow	2.53	2.45	2.39	0.0092	0.0090	0.0075
Butyl rubber	2.90	2.82	2.75	0.0010	0.0010	0.0009
Douglas fir	1.93	1.88	1.82	0.026	0.033	0.027
Epoxy resin (Araldite CN-501)	3.62	3.35	3.09	0.019	0.034	0.027
Ethyl alcohol (absolute)	24.5	23.7	6.5	0.090	0.062	0.250
Ice (from pure distilled water)	4.15	3.45	3.20	0.12	0.035	0.0009
Mahogany	2.25	2.07	1.88	0.025	0.032	0.025
Neoprene rubber	6.26	4.5	4.00	0.038	0.090	0.034
Nylon 6	4.0	. . .	. . .	0.01	. . .	. . .
Organic polysulfide, fillers	110	30	16	0.39	0.28	0.22
Paper (typical)	2.99	2.77	2.70	0.038	0.066	0.056
Phenol-formaldehyde (Bakelite BM120)	4.36	3.95	3.70	0.0280	0.0380	0.0438
Phenol-formaldehyde, 50% paper laminate	4.60	4.04	3.57	0.034	0.057	0.060
Polycarbonate	3.0	. . .	. . .	0.010	. . .	. . .
Polyethylene	2.26	2.26	2.26	<0.002	0.002	0.00031
Polypropylene	2.5	. . .	. . .	<0.005	. . .	. . .
Polystyrene	2.56	2.56	2.55	<0.00007	<0.0001	0.00033
Polytetrafluoroethylene (Teflon)	2.1	2.1	2.1	<0.0002	<0.0002	0.00015
Polyvinyl formal	2.92	2.80	2.76	0.019	0.013	0.0113
100% polyvinyl chloride	2.88	2.85	2.84	0.0160	0.0081	0.0055
Polyvinyl choride, flexible	5.0	. . .	. . .	0.1	. . .	. . .
Silicone-rubber compound	3.20	3.16	3.13	0.0030	0.0032	0.0097
Sodium chloride (fresh crystals)	5.90	. . .	. . .	<0.0002	. . .	. . .
Water (distilled)	78.2	78	76.7	0.040	0.005	0.157

SOURCE: Adapted from A. R. Von Hipple (ed.), *Dielectric Materials and Applications,* Wiley, New York, 1954; *Plastics,* 8th ed. D.A.T.A., San Diego, Calif., 1986.

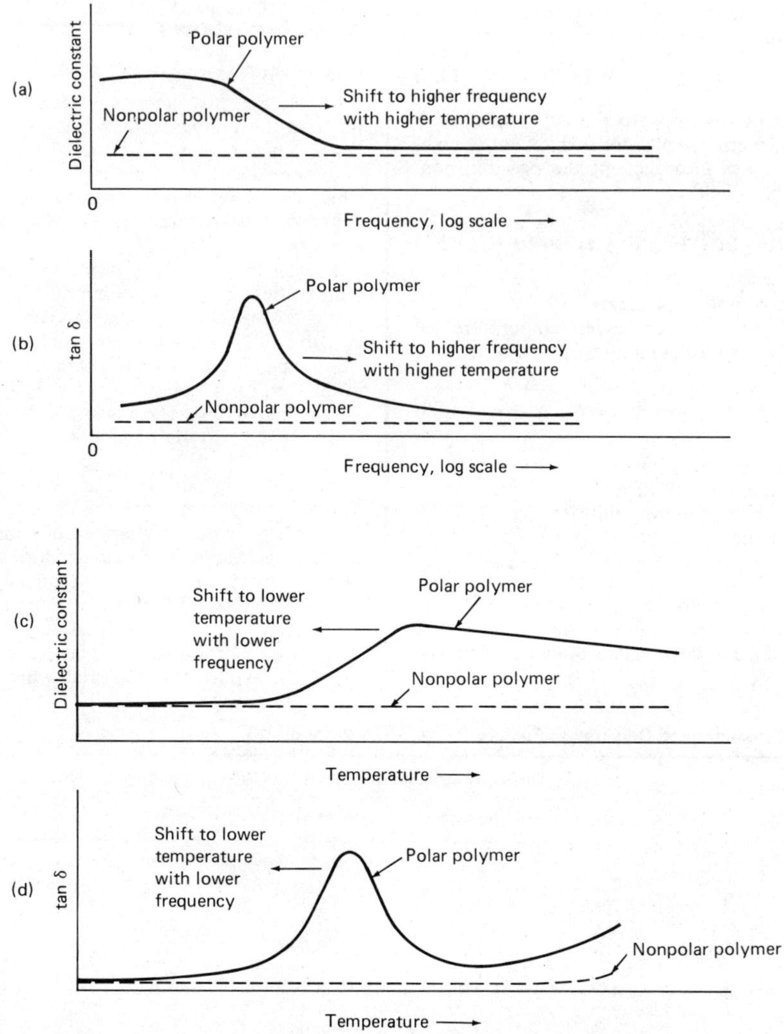

Fig. 3.4 Variation of dielectric constant and tan δ with respect to frequency and temperature. (*Adapted from Ref. 1.*)

very close to that of the dissipation factor of the material. Because of the difficulty in obtaining exact values, it is best to evaluate a sample of any new material at the frequency of interest.

Dielectric Constant

The dielectric constant also varies with frequency (Fig. 3.4*a*), temperature (Fig. 3.4*c*), humidity, and in some cases applied voltage. Although it is not as important as tan δ in determining heatability of a dielectric, it becomes very important when heating dielectrics made of several different types of material. For stacked layers of differing materials the RF potential voltage across the layers divides in inverse relation to the capacitance of the layer. (See Fig. 3.5 for a two-layer example.) If we wish to calculate the voltage potential across

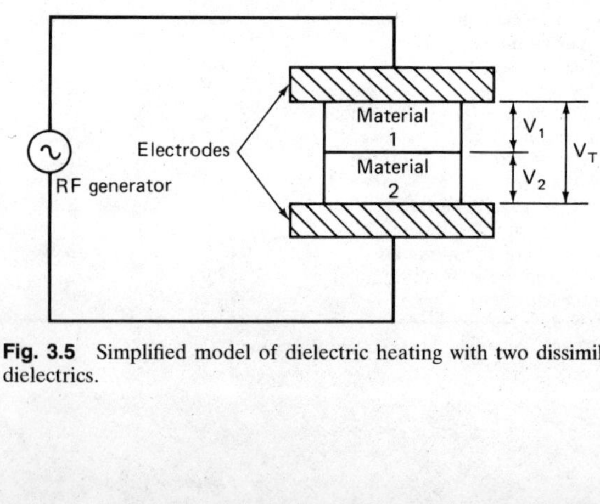

Fig. 3.5 Simplified model of dielectric heating with two dissimilar dielectrics.

the jth material, we can use a variation of the voltage divider formula from electrical engineering, assuming that the total area of all materials is the same:

$$V_j = V_T \frac{k_j}{d_j} \frac{1}{\sum_{i=1}^{i=n} (k_i/d_i)} \tag{3.8}$$

where k_i = dielectric constant of ith material
d_i = thickness of ith material
V_T = total voltage across all dielectrics

This equation assumes that the effect of the dissipation factor of each material is negligible, which is true in most cases.

Voltage

Although it may seem that the most efficient way to increase power loss within the dielectric would be to increase the voltage across the dielectric, a practical limit is reached when the dielectric undergoes voltage breakdown. In most situations this means ruining the workpiece and possibly damaging the application electrodes. Voltage breakdown of a dielectric is difficult to predict since it takes on the form of avalanche conduction; that is, current increases linearly with increasing applied voltage until the breakdown voltage is reached, and then increases dramatically at the breakdown voltage. The amount of stress that the dielectric is subjected to is difficult to calculate exactly as it is dependent on the geometry of the workpiece and applicator electrodes, as well as the electrical characteristics of the workpiece. Since the voltage gradient at the edge of a razor edge or needlepoint is extremely high in value, these shapes should be avoided in designing electrodes for use in dielectric heaters. Another situation which may cause voltage breakdown is a discontinuity caused by going from a material with a very high dielectric constant to one with a low dielectric constant. It should be kept in mind that air at atmospheric pressure breaks down at 74 V/mil (29,100 V/cm) or less, which is much less than most dielectrics, and will often be the limiting factor in electrode design. Many microwave heating processes take place in a partial vacuum in which the air surrounding the dielectric may break down at an even lower voltage gradient (see Fig. 3.6).

At higher frequencies or with larger workpieces the voltage distribution along the piece may become a problem.

Frequency

The frequency used is determined by the application, with each frequency range having its own unique set of advantages and disadvantages. Although any frequency RF could be used for dielectric heating, the following industrial, scientific, and medical (ISM) frequencies established by the Federal Communications Commission (FCC) are usually chosen as the frequency of operation in the United States.

A manufacturer is allowed unlimited radiation on these frequencies. Operating any equipment on the frequencies of 490 to 510 kHz, 2170 to 2194 kHz, 8354 to 8374 kHz, 121.4 to 121.6 MHz, 156.7 to 156.9 MHz, and 242.8 to 243.2 MHz is prohibited, since these are frequency bands that have been set aside for safety, search, and rescue communications.

Because of their wide bandwidth limits the frequencies 27.12, 915, and 2450 MHz are almost always used. It is possible to use other frequencies, but radiation from the equipment should be carefully monitored to prevent it from interfering with radio communications equipment.

For plastic welding, and many drying applications, 27.12 MHz is the preferred frequency. As can be seen from Table 3.2, an electrode or die can become rather large at this frequency before the designer must worry about uneven voltage distribution due to wavelength. A higher frequency would weld or dry faster for a given voltage, but each setup would require more extensive engineering development to ensure even voltage distribution and minimum radiation.

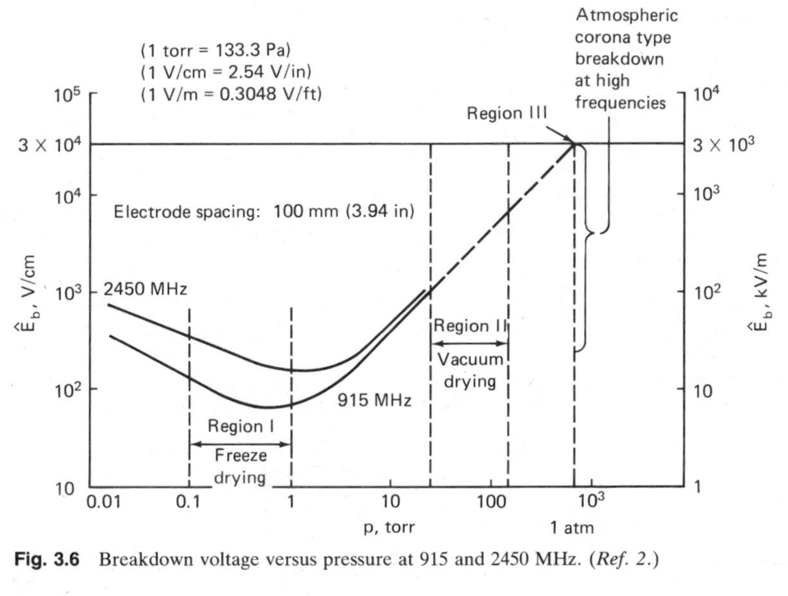

Fig. 3.6 Breakdown voltage versus pressure at 915 and 2450 MHz. (*Ref. 2.*)

Table 3.2. FCC ISM Frequencies

ISM frequency	Tolerance	Wavelength
6.78 MHz	±15.0 kHz	145 ft (44.4 m)
13.56 MHz	±7.0 kHz	72.6 ft (22.1 m)
27.12 MHz	±163.0 kHz	36.3 ft (11.1 m)
40.68 MHz	±20.0 kHz	24.2 ft (7.37 m)
915 MHz	±13.0 MHz	1.08 ft (32.8 cm)
2,450 MHz	±50.0 MHz	4.82 in (12.2 cm)
5,800 MHz	±75.0 MHz	2.03 in (5.17 cm)
24,125 MHz	±125.0 MHz	0.49 in (1.24 cm)
61.25 GHz	±250.0 MHz	0.19 in (4.99 cm)
122.50 GHz	±500.0 MHz	0.09 in (0.24 cm)
245.00 GHz	±1.0 GHz	0.05 in (0.12 cm)

NOTE: Europe also uses 433.92 MHz ± 870 kHz.

By convention, 915 and 2450 MHz are termed *microwave frequencies* and are the preferred frequencies for many drying and bulk or volumetric heating processes. This is because a much lower voltage potential may be used to achieve the same amount of heating. The frequency 2450 MHz is especially attractive at the lower power ranges because of the availability of inexpensive magnetron generator tubes from the consumer microwave market.

Power and Time

The power dissipated within the work material is normally not constant with respect to time, as it is dependent on all the variables already mentioned, which will vary during the heating process. As such, the power will need to be integrated over time to gain a good estimate of the amount of heating being done. The amount of power available from the generator often is limited for economic, voltage breakdown problems, or other reasons. The tradeoff is an increase in the amount of time required to perform the process.

Heat Loss

The dielectric is actually a heat generator, and caloric energy will normally flow from the material, instead of the other way around as in normal heating processes. With thin materials the amount of energy lost may be so great as to prevent the accomplishment of the desired temperature rise. One technique to reduce the heat flow is to use other dielectrics, called *buffer material,* between the work material and the electrodes to thermally insulate between them. Another technique is to heat the surrounding ambient or the electrodes by conventional heating methods.

3.4 DESCRIPTION OF EQUIPMENT

RF generators are normally sized by output power, ranging from a few watts to hundreds of kilowatts in size, and output frequency. At the lower frequencies (1 to 200 MHz) the generators are normally a separate unit from the electrodes, coupled to the electrodes via transmission lines or short straps. As the power output and/or frequency increases, it is more common to physically render the generators and electrodes as an integral system to reduce losses in the feed system or transmission line. At the higher frequencies (915 MHz and above) hollow ducts called *waveguides* are commonly used to connect the generator and the applicator assembly.

Figure 3.7 is a block diagram of a typical dielectric heater. The heart of the generator is a power oscillator. Because of the power level and frequency involved the active component of the oscillator is a power vacuum tube. The high-voltage power supply output ranges from 1000 V for a very small generator to 15,000 V or more for a very large generator.

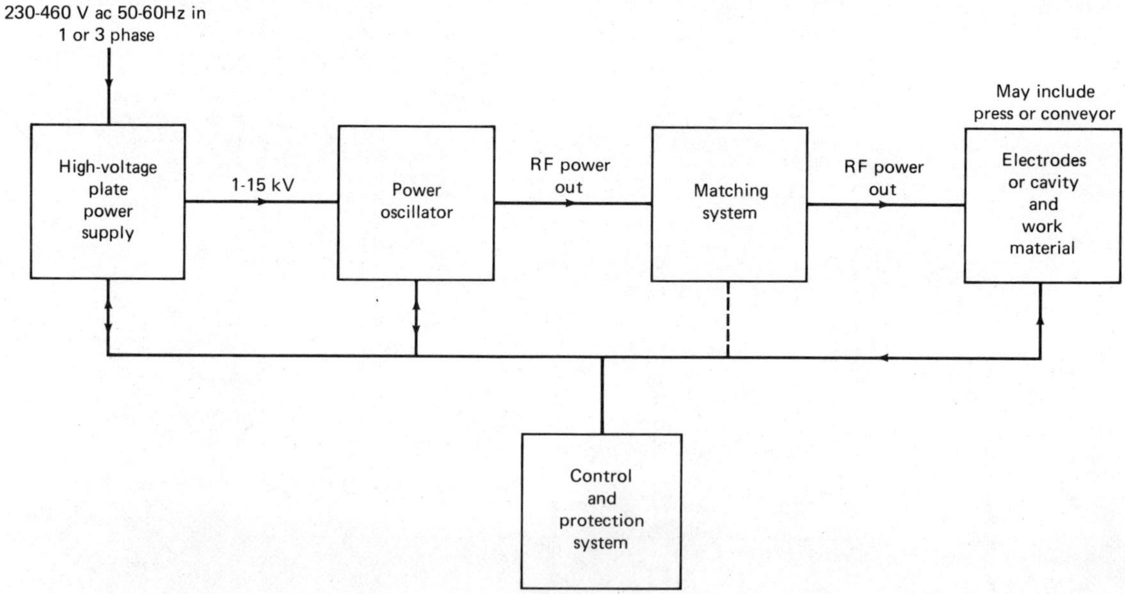

Fig. 3.7 Block diagram of typical dielectric and microwave heating.

The matching system is responsible for transforming the impedance of the load to an impedance that the power oscillator can transfer power into. It is important to note that this load impedance, made up of a combination of the electrodes or applicators and the work material, may change dramatically during the heat cycle. This is because the dielectric constant tan δ and possibly the dimensions of the work material itself may be changing. If the load is not properly matched to the output of the oscillator, damage to the tube is possible. For this reason special, extra rugged tubes and/or automatic tuning of the matching network may be employed. Coupling the power into the work material is discussed further in Sec. 3.5.

The control system energizes the cooling system, tube filament, and RF energy at the proper time in the heating cycle. Additionally, the control system may be called on to control the press, conveyor system, or other auxiliary equipment involved in the heating process. Another job of the control system is to protect the expensive components against fault conditions such as excessive current draw, excessive power dissipation, or high-voltage breakdowns, called *arcs,* between the electrodes.

Until recently the output frequency of dielectric heaters used in the high-frequency range (3 to 30 MHz) was whatever happened to be convenient for the particular application. It is now the common practice to use generators that are frequency-stabilized to operate within one of the ISM frequency bands. This eliminates many of the legal problems that resulted from interference as there is no restriction on RF energy radiated within these bands.

For a typical microwave heater, the control system and high-voltage power supply are much the same as in a high-frequency heater, but the power oscillator tube is almost always a special tube called a *magnetron.* Other than the tube, the major differences between microwave and high-frequency equipment lie in the matching network and the way the RF energy is applied to the work material.

The process of generating the RF energy involves a certain amount of waste power. Generally 40 to 60 percent of the power used by the generator from the power line is dissipated inside the generator as heat.

Fig. 3.8 A 10-kW plastics welder operating on 27.12 MHz. (*Courtesy of J. A. Callanan Company.*)

In 1986 the purchase price of generators operating in the high-frequency range was about \$1.00 per watt. At the lower powers this ratio will be higher because of the disproportionate cost of the control system with respect to the RF section. Generators operating in the microwave range can be purchased for \$2.00 to \$5.00 per watt. The life of the generator can be quite long; 20 to 30 years is not unusual with proper maintenance. The power tube normally has a lifetime of about 5000 h or more of continuous operation and is designed to be easily replaceable.

Figure 3.8 shows a 10-kW plastics welder frequency stabilized to operate in the 27.12-MHz band.

3.5 APPLICATION INFORMATION AND EXAMPLES

Coupling the RF Energy to the Work Material

Figure 3.9 shows several methods of applying energy to the work material. The methods illustrated in Fig. 3.9*a, b,* and *d* are the methods of choice for almost all high-frequency dielectric heating. The amount of airspace above and below the work material in the method of Fig. 3.9*b* may be varied to control the amount of voltage developed across the work material but should be kept small enough that voltage breakdown of the air from corona does not become a problem. The methods of Fig. 3.9*e* and *f* are quite common in the microwave industry and are not used at the lower frequencies because the minimum cavity dimension required to resonate properly at any frequency is one-half the wavelength of that frequency, making the cavity very large for anything below 915 MHz. The methods of Fig. 3.9*g* and *h* are two of a class of applicators called *waveguide applicators;* Fig. 3.9*g* is useful for heating thin films, and Fig. 3.9*h* is used for heating filament-type materials.

Because the wavelength of the RF energy at microwave frequencies may be shorter than the size of the work material itself, there is a tendency to develop hot and cold spots within the material to be heated corresponding to the voltage maximums and minimums which exist in the RF field. See Fig. 3.10 for an example of the electric field within one type of microwave cavity. It is possible for more than one voltage distribution pattern or mode to exist in the same cavity. By careful design it is possible to excite several of these modes at the same time. This is called *multimode operation* and allows the work material to be more evenly heated as the number voltage maxima and minima are greater than those normally existing in single-mode operation. Some manufacturers use metal paddles in front of the cavity feed point to shift from one pattern or mode of operation to another pattern or mode. The same result can be had by moving the work material itself through the use of turntables or conveyor belts. In the larger-microwave-cavity heaters it is common practice to use multiple magnetrons and feeds into the cavity, which tends to "fill in" the voltage maxima and minima.

Ensuring that the energy sent to the work material is absorbed as heat in the work material can often be a problem. It is not unusual for the currents circulating in the oscillator and feed components and/or cavity walls to reach hundreds or even thousands of amperes in magnitude. Without careful

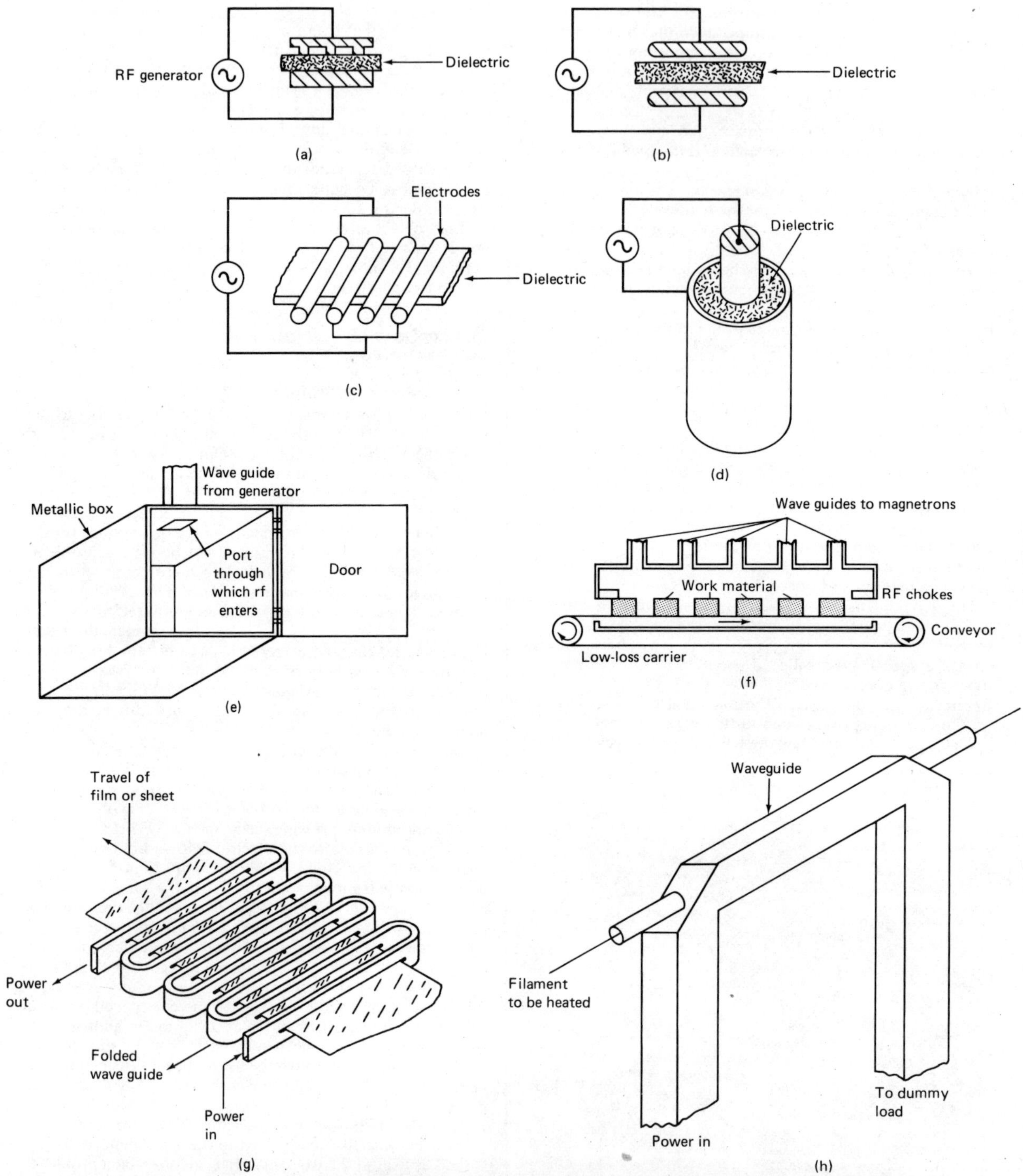

Fig. 3.9 Methods of applying RF energy to the dielectric. (*a*) Contact electrodes; (*b*) noncontact electrodes; (*c*) stray field; (*d*) coaxial; (*e*) cavity; (*f*) multiple-feed cavity; (*g*) folded waveguide (meander); (*h*) straight waveguide.

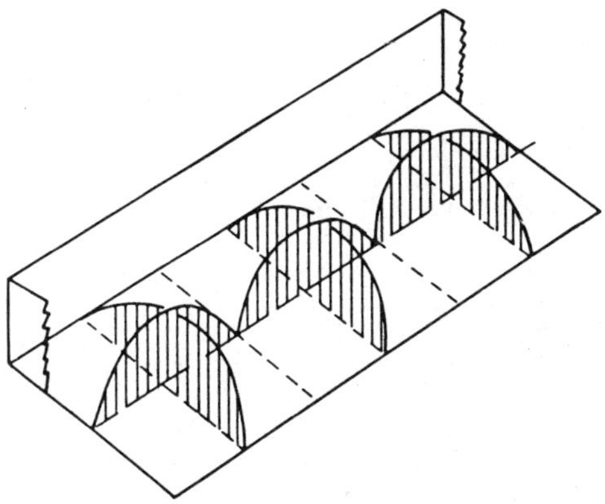

Fig. 3.10 Electric field potential within simple microwave cavity or waveguide (TE_{103} mode). (*Ref. 2.*)

design, this can cause considerable I^2R losses and even damage in the feed components and oscillator section.

At microwave frequencies any energy not absorbed in the work material tends to be reflected back to the magnetron, possibly damaging it and the waveguide feed. The most common method used to prevent this is the ferrite circulator, which allows power to travel only in one direction through it (shown schematically in Fig. 3.11). Any power coming from the magnetron that is not absorbed in the work material is "circulated" to a "dummy" load, where it is absorbed. The dummy load is often circulating water, thus making it easy to carry away the waste heat. The objective is to make sure that as much of the energy as possible goes into the work material and not the dummy load. The percentage of energy coming from the magnetron coupled into the work material is called the *coupling efficiency* and in well-designed systems will be 60 to 80 percent.

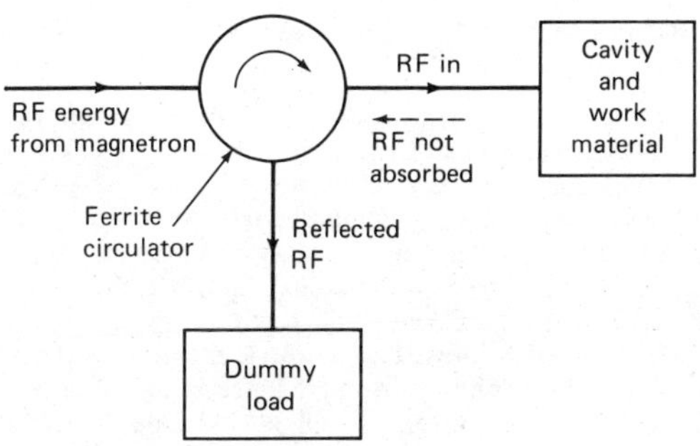

Fig. 3.11 Schematic representation of the operation of a ferrite circulator.

Mixed Dielectrics

For a process that may require sections made up of different materials such as those shown in Fig. 3.12*a*, the dielectrics can be modeled as capacitors connected in parallel, shown in Fig. 3.12*b*. Note that the voltage V across the sections will be dependent on the total equivalent impedance of all the parallel sections. In most cases the two materials will heat at different rates, resulting from differing dielectric constants, dissipation factors, specific heats, and other variables. This may or may not be a problem, depending on what the designer wants to accomplish.

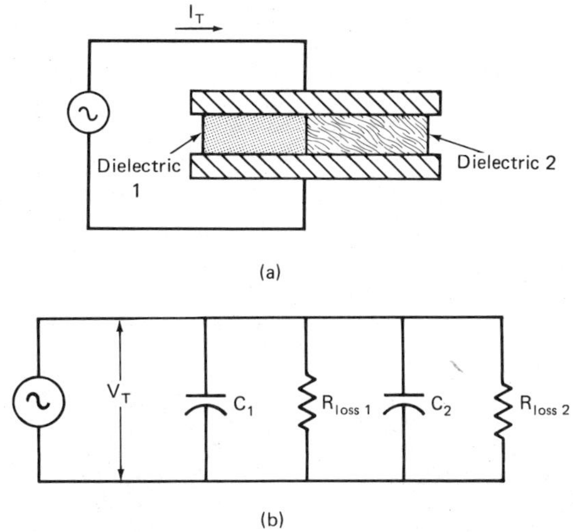

Fig. 3.12 Heating two dissimilar dielectrics in parallel. (*a*) Physical model; (*b*) electrical model.

Another case where mixed dielectrics may be used is in heating a dielectric which does not respond to RF heating. A catalyst dielectric is sometimes mixed in with the original dielectric with the catalyst dielectric acting as a built-in heating element.

Measurement of Power and Temperature

The most accurate way to measure the amount of power going to the electrodes or applicator system is by the use of a directional coupler. This device can be placed in line with a coaxial transmission line or waveguide and is sensitive to the power flowing in a single direction, either toward or reflected back from the workload. Directional couplers are commonly used with microwave heaters to give an indication of how well the applicator system is coupled to the waveguide. Subtracting the reflected power from the forward power will give the amount of power dissipated in the workload (disregarding any power lost in the applicator or electrodes).

At the lower frequencies, it is possible to obtain a relative indication of power by measuring the voltage at the electrodes or current going to the electrodes. This will not give a true indication of power as the impedance of the workload will change as the heating cycle progresses. A similiar technique can be used with some microwave applicators by using probes to measure the strength of the E or H fields at selected points near the wall of the cavity. It is also possible to obtain a relative indication of power dissipated in the workload by monitoring the plate or cathode current of the oscillator tube.

All the standard methods of temperature measurement can be employed in RF and microwave heating, but the presence of the RF field presents additional difficulties for any device that uses metallic conductors, as the conductors cause distortion of the field and may carry RF energy back to any electronics involved in the measurement instrumentation, causing interference or damage. If the change in the voltage distribution is not of major importance, the interference can often be eliminated by using shielding and filters.

The use of optical infrared pyrometers, both noncontact and fiber-optic pyrometers, is becoming increasingly popular in both RF and microwave heating because of their inherent greater immunity to RF interference.

Automatic Power Control

Because the amount of power delivered to the work material may vary considerably during the heating cycle, it often is desirable to control the power delivery by some automatic means. Whether the control process is open- or closed-loop will depend on what the designer wishes to accomplish. The two main methods of control are controlling the amount of power the generator generates and controlling the amount of power transferred to the work material by adjusting the matching system.

The easiest way of controlling the amount of power generated is by controlling the anode voltage of the power oscillator tube. This is usually done by controlling the voltage at the primary source of the high-voltage transformer, through the use of switchable taps on the primary, thyristor controllers, or using variable-reactance systems. With some magnetrons it is possible to control the output power of the magnetron by controlling the strength of the magnetic field that is a necessary part of magnetron operation.

Controlling the power delivered to the load by varying the matching system usually involves some sort of mechanical drive system with the inherent wear problems involved with moving parts. On the positive side, automatic control of the matching system usually allows for shorter heating cycles as the maximum amount of power can be transferred during the whole cycle.

Use of Auxiliary Heating

It is common practice to use more conventional heating systems in conjunction with RF dielectric heating. One reason is to prevent heat generated with the work material from escaping to the cooler ambient at the surface, ensuring even heat distribution or, in some instances, a more rapid temperature rise within the center of the material being heated. Most drying applications use RF heating to drive moisture to the surface where its evaporation is aided by the use of heated air. An additional use of conventional heat is to heat the material to a temperature at which its dissipation factor is sufficiently high to employ RF heating.

Penetration Depth

At microwave frequencies the penetration depth of the RF becomes a factor. At these frequencies the amount of heating done decreases at an exponentional rate from the surface toward the center. The formula for calculating the penetration depth is

$$D_p = \frac{\lambda}{2\pi} \frac{1}{[2k(\sqrt{1 + \tan^2\delta} - 1)]^{1/2}} \tag{3.9}$$

where D_p is the penetration depth in meters where energy is equal to 0.368 of surface value, and λ is the wavelength of RF energy.

Examples

Let us calculate what is needed to make a 2-in^2 weld in vinyl 0.024 in (0.0609 cm) thick with a typical tan δ of 0.1 and a dielectric constant of 3.5. The measured breakdown voltage of vinyl is 375 V/mil (1.48×10^5 V/cm), corresponding to a total voltage of 9000 V across the 24-mil (0.0609 cm) thickness. This value applies only if the material is placed between two very large, flat electrodes. In practice, the die normally has a fairly sharp discontinuity at its corner that will cause the electric field at that point to approach the voltage breakdown value when the potential difference between the electrodes is only about 2000 V rms. We will choose a 1000-V, 27.12-MHz voltage between the electrodes to give us a safety margin. The power fed to the work material is then 1117 W. If the specific heat of the vinyl is 0.4, the time calculated to raise the plastic to 350°F (176.6°C) is about 0.25 s. In this situation the calculated is very far from the actual situation because the model is overly simplified and does not include the amount of heat lost into the welding electrodes, which is very significant. The measured time to make this weld is closer to 2 s. As the thickness of the work material increases, the actual results come closer to the calculated results.

At the higher frequency (915 MHz), voltage breakdown is normally not so much of a problem, so the amount of power needed for the desired temperature rise is the only calculation to be made. A common application is tempering meat, that is, raising the temperature of frozen meat to just below the point of thawing for further processing. The dielectric that is being heated is mostly frozen water, which has a specific heat of 1.0 and dissipation factor of 0.0009. To raise the temperature of 50 lb (22.7 kg) of meat from 0 to 28°F (−77.8 to −2.22°C) in 5 min would take 5 kW, according to Eq. (3.7), assuming a coupling efficiency of 100 percent. Another rule of thumb is 50 to 60 Btu per pound (1.16×10^5 to 1.39×10^5 J/kg) of frozen meat.

3.6 GOVERNMENT REGULATIONS AND SAFETY CONSIDERATIONS

Because RF heaters are a possible source of harmful interference to communications equipment such as two-way radio, broadcast stations, and microwave repeaters, their use is regulated by the government. In the United States the FCC is the pertinent agency, and the rules governing the use of industrial, scientific, and medical equipment can be found in Part 18 of the FCC rules and regulations. In Europe the post office of each individual country usually has the job of setting radiation limits, although there is an attempt being made to have one international set of rules through the International Special Committee on Radio Interference (CISPR).

The radiation limits of Part 18 of the FCC rules can be summarized as they apply to industrial RF heaters by saying that below 5275 MHz all radiation must be below 10 μV/m at a distance of 1 mi from the equipment. If the equipment in question is operating in one of the established ISM frequencies, the field strength limits apply only to energy emitted outside of that particular ISM band, which would consist of harmonics of the fundamental frequency operating within the ISM band and possible intermodulation products.

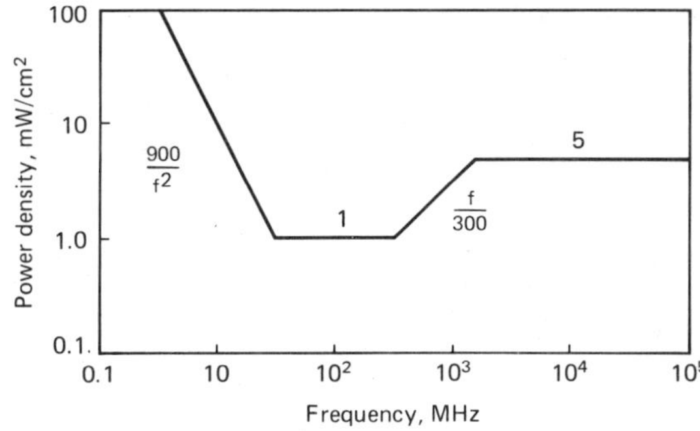

Frequency range, MHz	E^2, V^2/m^2	H^2, A^2/m^2	Power density, mW/cm^2
0.3–3	400,000	2.5	100
3–30	4,000 $(900/f^2)$	0.025 $(900/f^2)$	$900/f^2$
30–300	4,000	0.025	1.0
300–1,500	4,000 $(f/300)$	0.025 $(f/300)$	$f/300$
1,500–100,000	20,000	0.125	5.0

Fig. 3.13 ANSI C95.1-1982 power-density limits; RF protection guide. (*Note: f* is the frequency in megahertz. Based on average SAR limit of 0.40 W/kg in exposed tissue.)

RF heating equipment that operates within one of the ISM frequency bands previously could be FCC prototype-certified. This means that the radiation from the equipment had been measured by the manufacturer and the data sent to the FCC, who certified that the equipment did indeed comply with its rules. The purchaser was thus relieved from the legal burden of proving compliance. A more recent procedure, called *verification,* allows the manufacturer to simply make the measurements and keep them on file, without the necessity of sending the data to the FCC.

If the equipment cannot meet the limits, some sort of shielding can be implemented to reduce the radiation to below the limit. If the equipment itself cannot be shielded sufficiently to meet the limits, it would be necessary to enclose it inside a completely shielded room, often called a *screen room*.

Biological side effects to the people who operate RF heating equipment are also possible. Limited research on the effects of RF energy on humans has been done to date, so this area remains controversial. The Occupational Safety and Health Administration (OSHA) regulation Part 1910.97 states that no one should be exposed to power-density levels greater than 10 mW/cm^2, averaged over a 6-min period. A widely accepted recomended standard is the Americal National Standards Institute (ANSI) C95.1-1982, which takes into account the fact that the biological hazard to humans is dependent on the frequency of the RF energy. A graph of the standard is shown in Fig. 3.13.

Electric shocks or burns from improper operation of the equipment are also possible. The voltages used in the oscillator section of the heater are extremely lethal. Touching the electrodes or applicators or placing a hand inside a microwave cavity while RF energy is present will also cause severe burns.

REFERENCES

1. D. G. Fink (ed.), *Standard Handbook for Electrical Engineers,* 11th ed., Sec. 4, McGraw-Hill, New York, 1978.
2. A. C. Metaxas and R. J. Meredith, *Industrial Microwave Heating,* Peter Peregrinas, London, 1983.

SUGGESTIONS FOR FURTHER READING

ANSI C95.1-1982, American National Standards Institute, New York, 1982.

Dittrich, H. F. and H. Stanley, *Tubes for RF Heating,* in Amperex (N.V. Philips' Gloeilampenfadrieken ©), Einhoven, The Netherlands, 1971.

FCC Rules and Regulations, Vol. II, Part 18, Washington, D.C., 1985.

Heatsealing and High Frequency Welding of Plastics, H. P. Zade, Temple Press, London, 1959.

The Journal of Microwave Power, published quarterly by the International Microwave Power Institute, Clifton, Va.

Microwave Power in Industry, EPRIEM, 3645 Electrical Research Institute, Palo Alto, Calif., 1985.

Plastics, 8th ed. D.A.T.A., Inc., San Diego, Calif., 1986.

Chapter **4**

Induction Heating

N. V. ROSS
Ajax Magnethermic

Induction heating is a method of heating electrically conducting materials with alternating current electric power. Alternating current electric power is used in a copper coil to create an alternating magnetic field. This alternating magnetic field is used to induce alternating electrical voltages and current in a workpiece that is closely coupled to the coil. These alternating currents generate electric resistance losses and heat the workpiece.

The application of induction is predominantly in the processing, forming, treating, and fabrication of metals. However, there are many other specialty applications that take advantage of the particular characteristic and design possibilities of induction heating. Induction heating systems range in size from 200-MW systems in steel-forming mills to small systems, on the order of 1 kW, that are used in heat treating small components and in heat forming of a variety of materials.

4.1 CHARACTERISTICS OF INDUCTION HEATING

Induction heating can be used for precise and repeatable heating of electrical conducting objects. There can be important advantages over heating with fuels and heating by ordinary electrical resistance heating.

Compared to heating with fossil fuels, induction heating is generally considered to be clean, rapid, and well suited to automatic control. Because of the relatively rapid heating rates, nontransference of heat through the workpiece surface, and the ability to control the atmosphere of the workpieces, it is possible to reduce or eliminate problems with scaling, hot and cold spots, and other surface effects, such as decarburization, that are encountered with traditional heating furnaces. In heating large steel billets, scale losses have been reduced to 0.5 percent from the 3 to 4 percent experienced with conventional furnaces.

It is also often claimed that induction heating allows greater production flexibility since downtime and startup times are substantially reduced. Additionally, induction heat may require less floor space and require relatively limited operator attention. Because of the compactness of the equipment, closer ties between processing stations are possible. Inherent advantages of induction heating are that the electrical components are not subject to mechanical wear or wastage and that the workpiece handling components are not subject to high temperature.

Use of induction heating may also result in energy conservation. The efficiency in converting electric power to heat in the workpiece is on the order of 50 to 85 percent compared to efficiencies in the range of 40 percent or less for high-temperature fuel-fired furnaces. This efficiency advantage alone can largely offset the higher cost of electric power. Added to this is the elimination of startup, standby, and shutdown heat losses. Also, the flexibility of induction heating has allowed the better retention of residual heat from upstream metal-processing steps. In continuous casting operations, for example, it has been possible to heat slabs to rolling temperatures as they are fed directly from the casting operation to the rolling mill.

Induction heating may have an environmental advantage to the plant operator in that no combustion products are produced and heat radiation into the work area about the furnace is reduced compared to conventional fuel-fired equipment.

A final important characteristic of induction heating is the ability to deliver heat into metal pieces (or the metal components of assemblies of composites) without direct contact with electric heating elements or the heating medium. This characteristic can be used to advantage in various assembly and forming operations involving both metallic and nonmetallic materials.

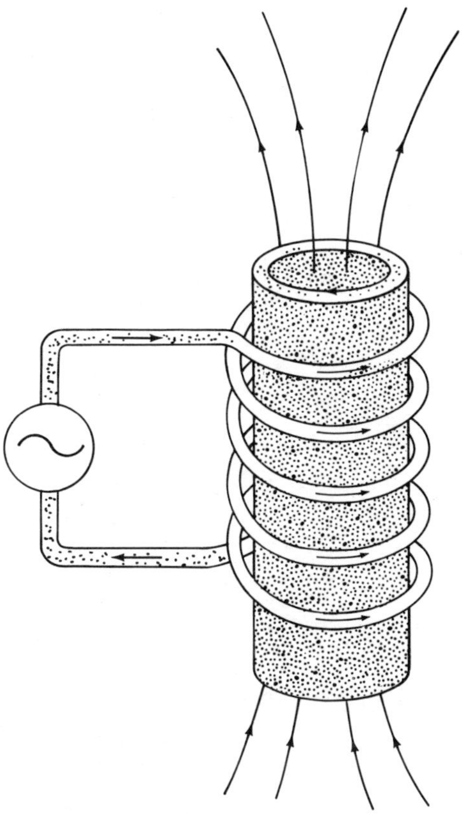

Fig. 4.1 Induction heating coil.

4.2 FUNDAMENTAL RELATIONSHIPS OF INDUCTION HEATING

If an alternating current flows through a coil, a magnetic field is produced that varies with the amount of current. The field is concentrated inside the coil. If an electrically conductive load is placed inside the coil (Fig. 4.1), eddy currents will be induced inside the load. The eddy currents will flow in a direction opposite to the current flow in the coil. These induced currents produce a magnetic field in the opposite direction of the field produced by the current in the coil; they prevent the field from penetrating to the center of the load. The eddy currents are therefore concentrated at the surface and decrease towards the center. This is often referred to as skin effect and is illustrated by Fig. 4.2.

The depth where the current density drops to a value of 37 percent of the surface value is called the penetration depth. All of the current under the curve can be thought of as being contained within this depth. This simplifying assumption is useful in calculating the resistance of the current path in the load. Since the load has resistance to the current flow, heat will be generated. The amount of heat generated is a function of the product of resistance and the current squared. The result of squaring the current is illustrated by Fig. 4.3. The

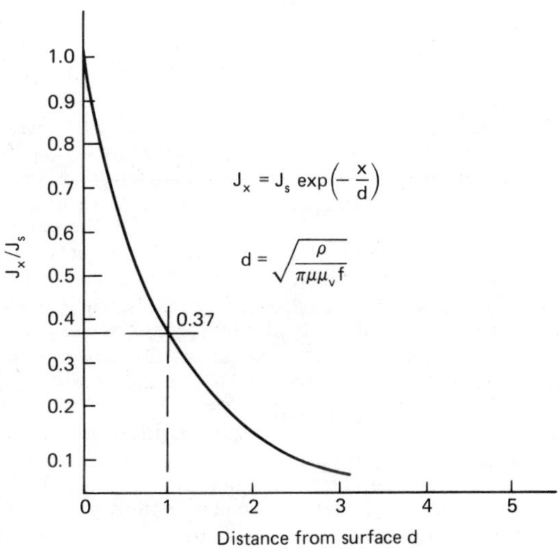

Fig. 4.2 Current density profile.

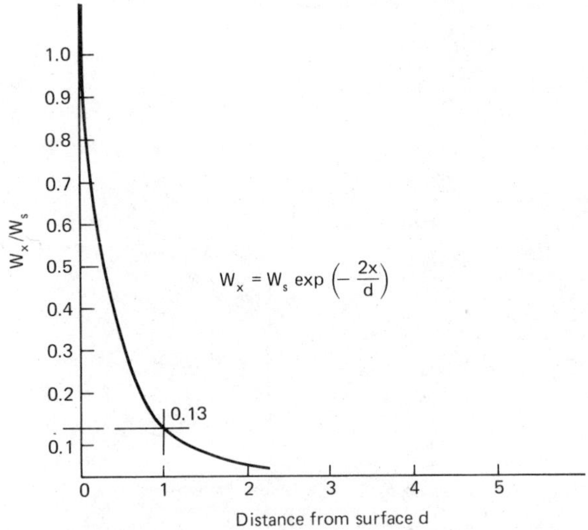

Fig. 4.3 Power density profile.

curve is much steeper than that in Fig. 4.2 and, at the depth of penetration, the heating effect has fallen to 13 percent of that at the surface.

The depth of penetration is one of the most important factors in any discussion concerning induction heating. It is usually calculated first since so much depends upon it. The general formula for depth of penetration is given by

$$d = \sqrt{\frac{\rho}{\pi \mu \mu_v f}} \tag{4.1}$$

The depth of penetration is a function of three variables, two of which are related to the load. The variables are the electrical resistivity of the load ρ, the relative permeability of load μ, and the frequency f of the alternating current in the coil. The magnetic permeability of vacuum (μ_v) is a constant equal to 3.19×10^{-8} H/in ($4\pi \times 10^{-7}$ H/m).

One of the major reasons for calculating the depth of penetration is to determine how much current will flow within the load of a given size. If the load is very large compared to the depth of penetration, maximum current will flow. If the load is very small compared to the depth of penetration, very little current will flow within the load. The amount of heat generated is related to the current squared. It is therefore imperative to have a large current flow in the load. By way of illustration, refer to Fig. 4.4*a*, *b*, and *c* which depicts the behavior of rectangular loads of three thicknesses relative to the depth of penetration. The current flow and resulting heat generation is shown for the different size loads.

A rule of thumb for design is that, for reasonable efficiency, the ratio of load thickness to depth of penetration should be greater than three. One can control this ratio by the correct choice of penetration depth.

It can be shown that the relative permeability is a function of power density and frequency. Therefore, it is possible to choose a penetration depth by selecting the frequency and power density such that for a given load thickness the ratio is greater than three. The depth of penetration versus frequency is plotted for nonmagnetic loads at various resistivities in Fig. 4.5. The depth of penetration versus power density for magnetic carbon steel at various frequencies is plotted in Fig. 4.6.

Induction coil efficiency is calculated by the ratio of the equivalent resistance of the load to the total resistance of the equivalent resistance of the load and the resistance of the coil.

$$\eta = \frac{R_{eqv}}{R_{eqv} + R_{coil}} \tag{4.2}$$

The equivalent resistance of the load incorporates a correction factor K_2 that takes into account the skin effect on the ac resistance. The factor K_2 is a function of the ratio of load dimensions and depth of penetration and depends upon load shape. For cylindrical loads, K_2 is plotted against a/d in Fig. 4.7, where a is the load diameter.

Equation (4.2) can be presented in the following form for cylindrical loads of diameter a and cylindrical coils of inside diameter b made of copper:

$$\eta = \frac{1}{(0.001/\sqrt{\mu\rho})\,(b/a)(1/K_2) + 1} \tag{4.3}$$

where μ = relative permeability of load
ρ = resistivity of load, $\Omega \cdot$ in ($1.0\ \Omega \cdot \text{in} = 2.54\ \Omega \cdot$ cm)
b = inside coil diameter
a = outside load diameter
K_2 = correction factor (Fig. 4.7)

Equation (4.3) is plotted in Fig. 4.8 for nonmagnetic loads, where $\mu = 1.0$. It is seen that as the resistivity increases, the coil efficiency increases. It is lowest for heating copper, $\rho = 1.0\ \mu\Omega \cdot$ in ($2.54\ \mu\Omega \cdot$ cm), and highest when heating graphite, $\rho = 500\ \mu\Omega \cdot$ in ($1270\ \mu\Omega \cdot$ cm). The relatively high electrical resistance and permeability of steel below the Curie point allow for efficiencies in the range of 70 to 90 percent, depending on the coil design. Efficiencies in the range of 50 to 80 percent can be achieved when heating above the Curie point for stainless steel (where the magnetic permeability falls to unity), and for brass, bronze, titanium, and aluminum. Efficiencies are lowest when heating copper and are typically less than 50 percent.

Radiation and convection heat losses from the load can reduce efficiency when heating times are long and the load temperature is high.

The appropriate power density, in terms of rate of heat input per unit surface area of the load, depends on the geometric characteristics and physical properties of the load and the limitations and objectives in heating the load. In bulk or through heating, the maximum acceptable surface to centerline temperature difference may constrain the power density. In other applications, the power density will be dictated by other thermal response characteristics of the load. The strength and frequency of the produced magnetic field are the variables in coil design that are used to establish the desired power density at the load surface.

Generally, induction heating systems function with power densities in the range of 10 to 5000 kW/ft^2 (100 to 50,000 kW/m^2).

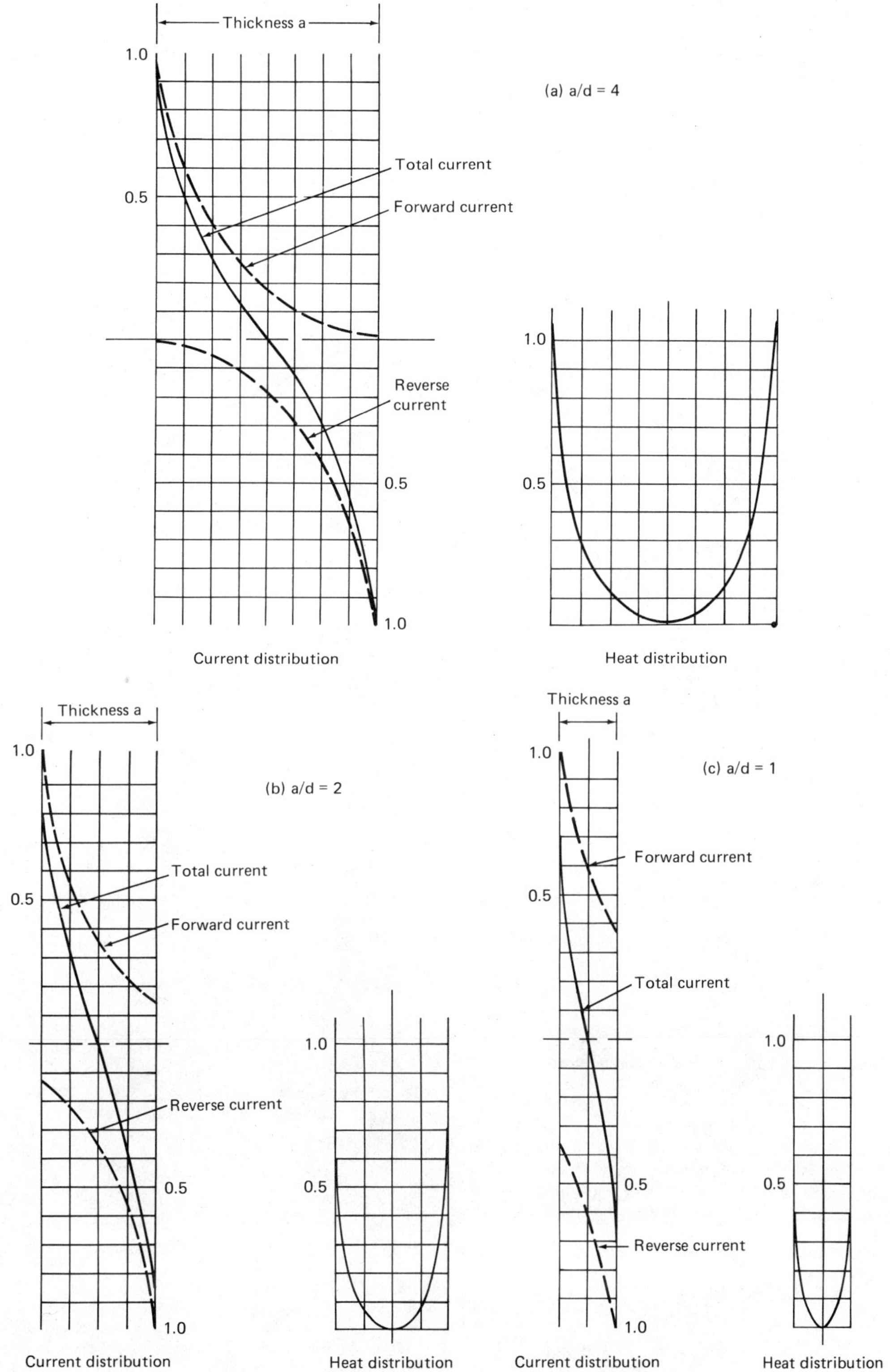

Fig. 4.4 Current and heat distributions for different thickness to depth of penetration ratios.

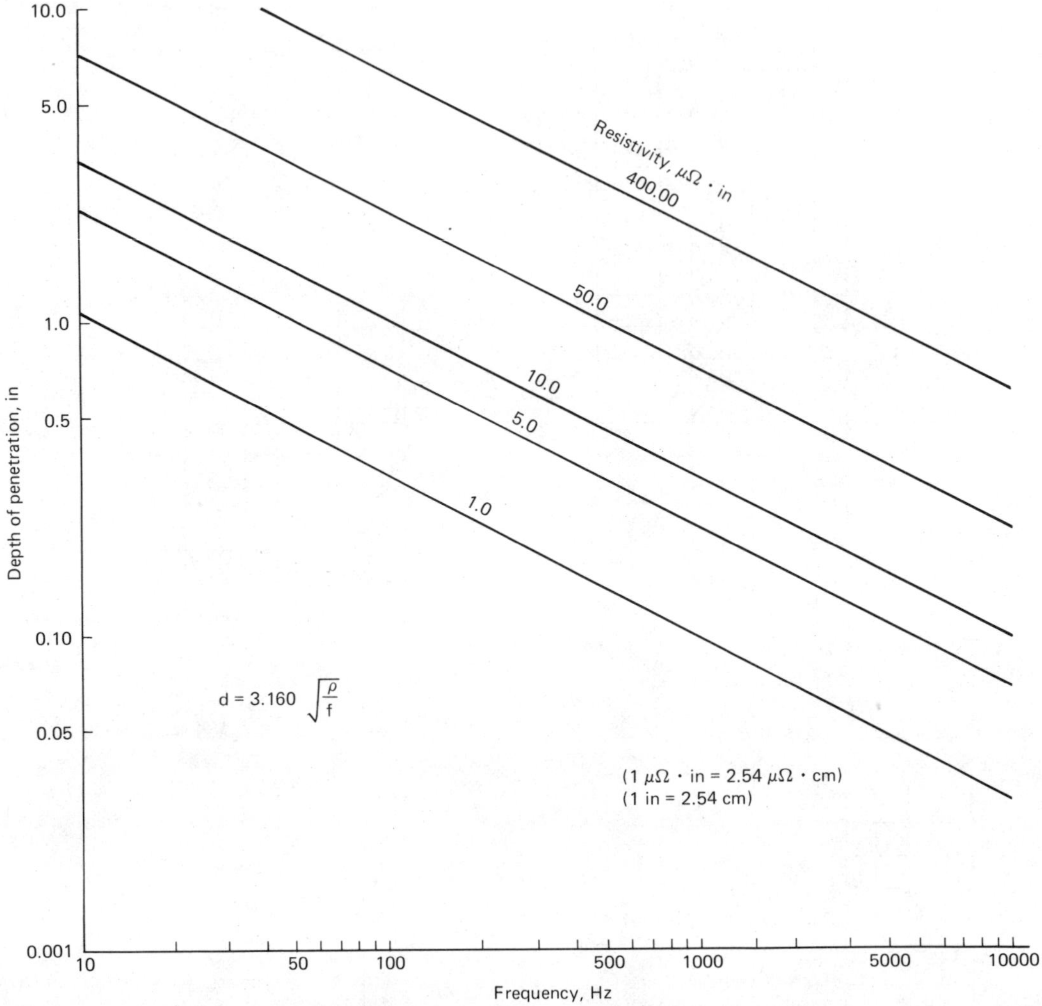

Fig. 4.5 Depth of penetration for nonmagnetic loads.

4.3 APPLICATIONS OF INDUCTION HEATING

Some of the major applications of induction heating are given in Table 4.1. Other applications include the annealing and stress relieving of metal parts, the heating of the internal surfaces of plastics-forming equipment, the bonding of plastics through the induction heating of iron-based plastic filler material, heating of silicon for processing, the heating of plastic encapsulated metal parts, bonding and unbonding of rubber or plastic and metal, and the heating of metal parts for shrink fits.

4.4 INDUCTION HEATING SYSTEM DESIGNS

Coils

The design of an induction heating coil is dependent on

1. The kind of material to be heated
2. The shape and dimensions of the load
3. The required temperature rise (amount and time)
4. The desired production rate

These parameters dictate the required power density and frequency. The major coil design parameters of power, voltage, current, and number of turns are determined by analyzing the electromagnetic system formed by the power supply, the coil, and the workpiece. Analysis and experience guide the final electrical design of a coil.

The better the electromagnetic coupling between the coil and the workpiece, the better the system efficiency. Best efficiencies are usually achieved when the coil forms a helix about the workpiece. Nonetheless, a wide variety of coil configurations, including internal, single loop, hairpin, pancake, split, and contoured coils, are successfully used.

The electromagnetic coupling between the coil and the workpiece is limited by mechanical design factors. Consid-

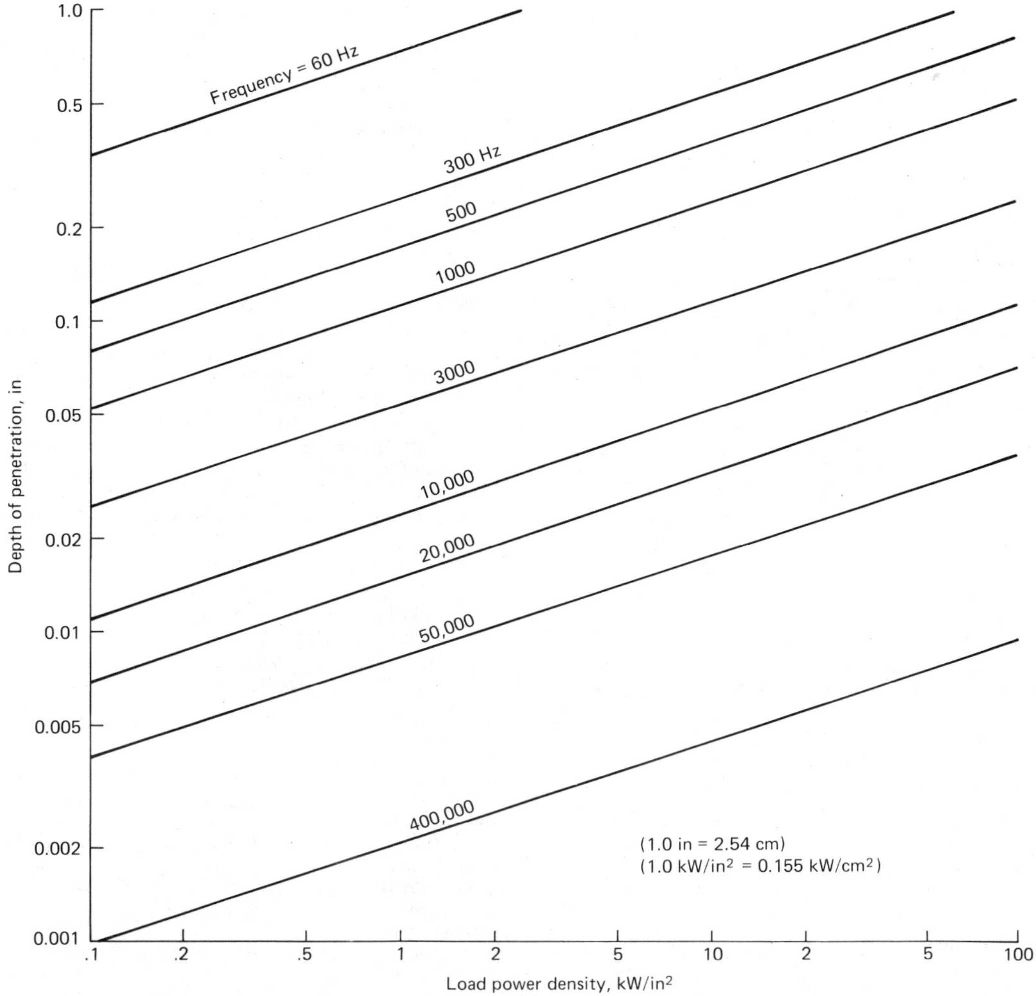

Fig. 4.6 Depth of penetration for carbon steel at 1300°F (704°C) as a function of power density and frequency.

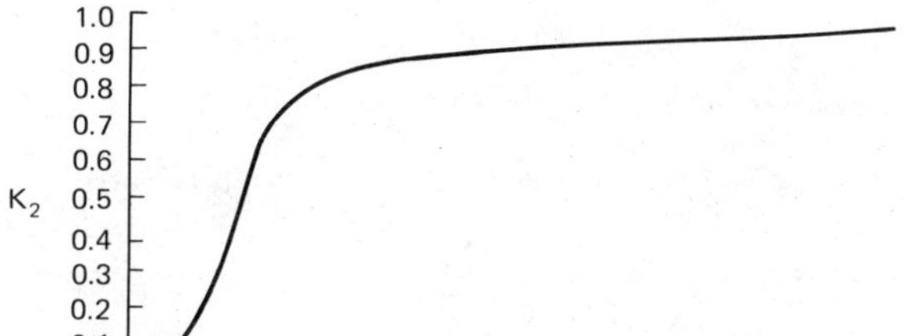

Fig. 4.7 Coefficient of correction K_2 for load resistance.

eration must be given to the thermal and electrical insulation requirements, the structural support of the coil and the workpiece, and the means of handling the workpieces for loading and unloading from the coil.

The allowable or desired temperature profile in the workpiece determines the appropriate power density or rate of energy input to the workpiece per unit surface area [see Eq. (4.4)]. If the desired production rate requires a total energy input greater than is possible with the heating of a single workpiece, coils are designed to accommodate several workpieces. Individual workpieces are moved through the coil at sufficient rate to achieve the desired production rate and the coil is of sufficient length to yield an acceptable rate of energy input to each workpiece. The same principles apply in the design of coils for heating rods and tubing.

In some cases, the close coupling of the coil and the workpiece is facilitated by the relative movement of the coil and

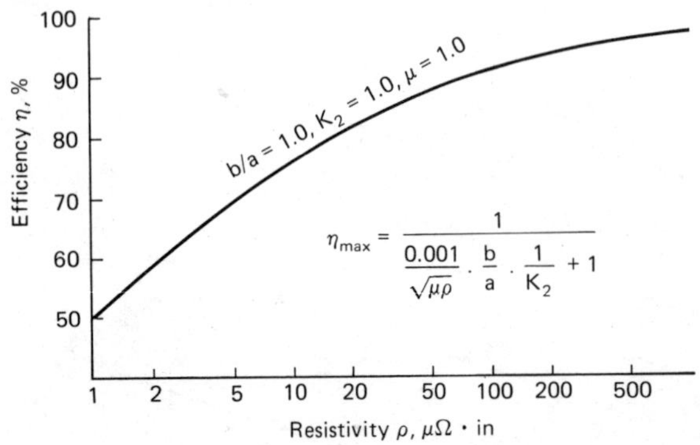

Fig. 4.8 Maximum coil efficiency as a function of load resistivity. [Coil is made from copper; resistivity units are ohm-inch (1 Ω · = 2.54 Ω · cm); load is nonmagnetic (μ = 1.0); b = coil inside diameter; a = load diameter; K_2 = coefficient for load resistance.]

workpiece. For example, spheres may be rotated within a circular coil to allow for heating over the entire surface of the sphere.

Coils are always made from copper and are usually hollow tubes with water cooling. Various coatings are used to protect the coils from the high temperature of the workpiece and to provide electrical insulation. These coatings include cements, fiberglass, and ceramics. Thermal insulation and refractory linings between the coil and the workpiece are used to protect the coil and its insulation. With coils operating at line frequency, it is possible to use stainless steel liners between the workpiece and the refractory lining.

Power Supplies

Induction-heating, alternating-current, power supply systems are classified by the frequency of the current supplied to the coil. These systems can be classified as line-frequency systems, motor-alternator systems, solid-state systems, and radio-frequency systems.

Line-frequency systems operate at 50 or 60 Hz available from the power grid. These are the lowest cost systems and are used for bulk heating of large billets because of the large depth of penetration. The lack of a need for frequency conversion equipment is the major economic advantage of this class of system. However, power factors are relatively low and capacitors are used for power factor correction. The low frequency of line-frequency systems results in the most uniform heating. Systems of over 200 MW capacity have been installed.

Motor-alternator systems and solid-state frequency converter systems are used in the medium frequency range (about 100 Hz to 50 kHz). Motor-alternator systems have been increasingly replaced by solid-state systems but maintain advantages of proven reliability and easier maintenance. Solid-state systems employ silicon controlled rectifiers (SCR) or thyristors. These devices function as fast-acting switches to produce the required coil frequency. The advantages of solid-

Table 4.1. Applications of Induction Heating

Application	Justification	Important characteristics	Size range	Frequency range
Bulk heating of billets and slabs for forming, forging, annealing, melting	Reduced scale, operating flexibility, reduced equipment size, short heating time	Heating within about 0.5 in (13 mm) of surface; heat rate limited by conduction into interior of slab or billet; after transient temperature rise, temperature profile constant through heatup; soak time needed for temperature to equalize	1 kW to 200 MW	50/60 Hz [over 6 in (15 cm) dia.] to 450 kHz [0.25 in (0.6 cm) dia.]
Surface hardening	Rapid heating of surface only possible, metallurgical advantages over conventional through-hardening, easily automated, energy efficient	Used with water, brine, polymer, or air quench; hardening depths from 0.01–0.25 in (0.25–6 mm); heating times may be less than 1 s; scanning coils and ferromagnetic flux concentrators used	1 kW to 4 MW	3–450 kHz
Soldering, brazing, welding	Fast and direct heating	Wide variety of coil designs used, preplaced solder or brazing material often used, tube butt and seam welding widely used	1–500 kW	Mostly 400–500 kHz
Vessel heating	High-power input possible	Air- or water-cooled coil wraps around vessel	1–1000 kW	60 Hz to 5 kHz
Melting	Controlled atmosphere, speed, operating costs, automatic stirring	Coreless furnace melts charge inside refractory vessel that is surrounded by induction coil, channel furnace heats molten metal in side loop that is connected to main vessel, temperatures to over 5000°F (2760°C)	Up to 20 MW	50/60 Hz to 10 kHz

state systems are high efficiency, variable frequency control for load following, and rapid startup.

Radio-frequency systems, about 100 kHz and up, allow for shallow depths of penetration and high power density. The internal components of these systems typically include a high-voltage ac transformer, a high-voltage ac-to-dc converter using rectifier diodes, a high-voltage, self-excited, radio-frequency oscillator using an industrial triode, and a load-matching transformer. Radio-frequency induction heating systems up to about 1 MW have been developed for high-power applications such as tube welding.

4.5 COIL SIZING FOR THROUGH HEATING OF CYLINDRICAL LOADS

The size of the coil needed for the through heating of a cylindrical electrical conductor can be established from the following relationships. The load may be in the form of a stationary billet that is heated in a single coil that encompasses the entire billet, a steady progression of billets moving through a coil, or a continuous rod moving through a coil.

First, the maximum allowable power density must be determined. In this type of heating, there occurs an initial transient temperature profile in the load. However, during most of the heating time, the temperature rises uniformly in the load as depicted in Fig. 4.9. The allowed power density is determined from the allowed surface to centerline temperature difference. This temperature difference is given by

$$\Delta\theta_{sc} = \frac{a}{4\lambda}(QW_g - W_r) \tag{4.4}$$

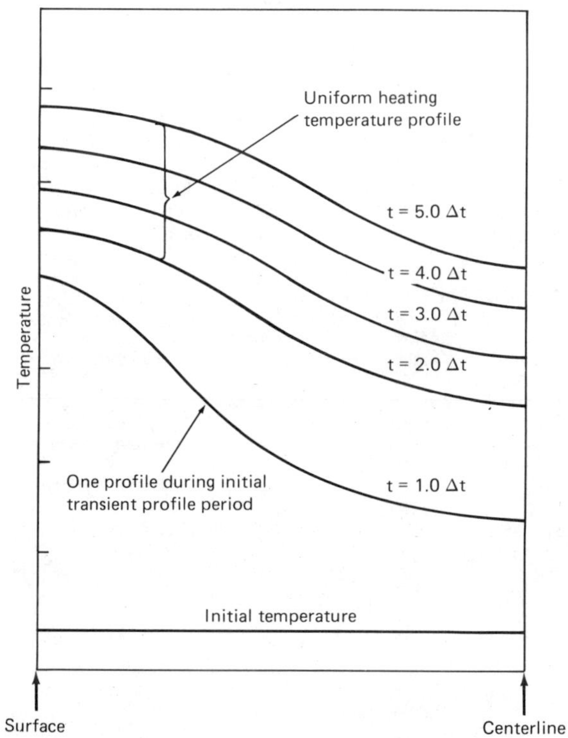

Fig. 4.9 Temperature response of load with induction through heating.

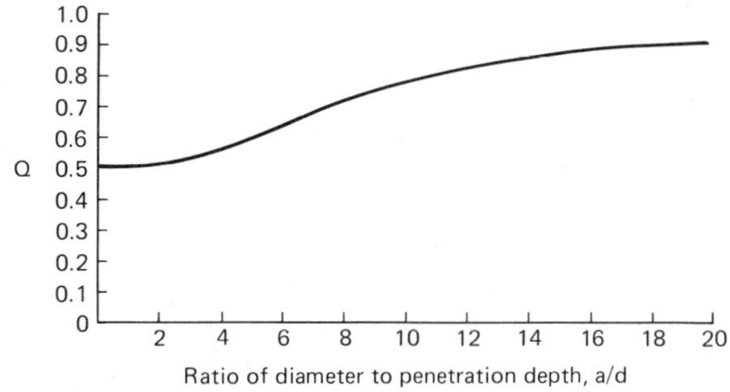

Fig. 4.10 Coefficient Q for calculating final temperature difference between surface and centerline of round load.

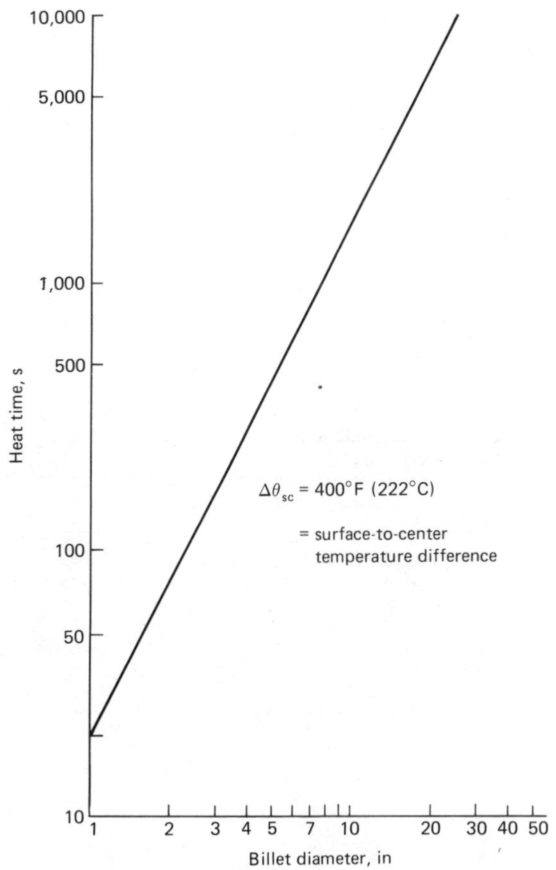

Fig. 4.11 Time for heating carbon steel billets from 70 to 2250°F (21.1 to 1232°C). (1.0 in = 2.54 cm.)

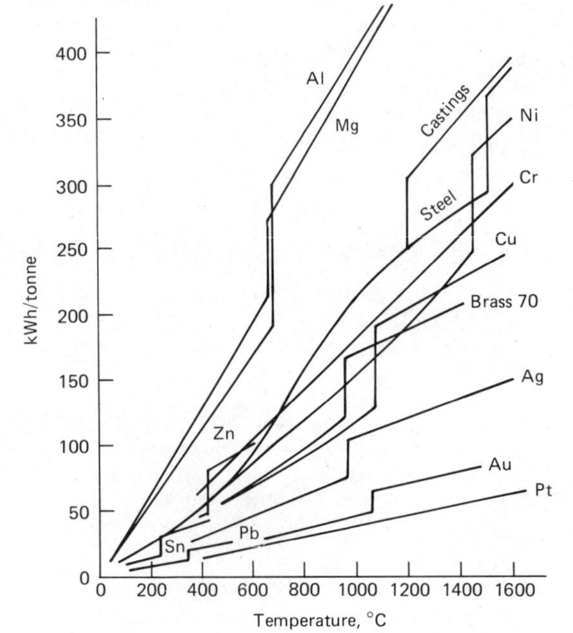

Fig. 4.12 Energy content of various metals. (°F = 1.8°C + 32; 1 kWh/tonne = 4.46 × 10^{-4} kWh/lb.)

where a = diameter of load
λ = thermal conductivity of load
Q = function of a/d (Fig. 4.10)
W_g = energy flux into load
W_r = radiation heat flux loss from load
$\Delta\theta_{sc}$ = surface-to-center temperature difference

Table 4.2. Typical Efficiency of Induction Heating Coils

Material	Final temperature °F (°C)	Efficiency
Carbon steel	2282 (1250)	0.75
Carbon steel	1292 (700)	0.80
Stainless steel	2282 (1250)	0.70
Brass	1472 (800)	0.60
Copper	1652 (900)	0.50
Aluminum	932 (500)	0.60

Next, the heating time can be calculated from

$$t = Q\frac{a^2 c\delta}{16\lambda}\left(\frac{\Delta\theta_m}{\Delta\theta_{sc}}\right) \tag{4.5}$$

where t = heat time
δ = density
$\Delta\theta_m$ = temperature rise
c = specific heat capacity

and when the radiant heat loss from the load is small. When the radiant heat loss is significant, a more accurate relationship is

$$t = \frac{Q(a^2/16)(c\delta/\lambda)\,\Delta\theta_m}{\Delta\theta_{sc} + (a/4\lambda)(1 - Q)W_r} \tag{4.6}$$

When radiant losses are significant or when the physical properties of the load vary significantly with temperature, the heating time is best calculated in a stepwise fashion for a series of temperature rises that yield the total desired temperature rise.

For a coil encompassing an entire stationary billet, the production rate of the coil (mass heated per unit time) is simply the billet mass divided by the heating time. For continuous heating of billets or rods, the length of the coil is dependent on the desired production rate as given by

$$L = \left[\frac{\text{production rate}}{(\pi/4)a^2\delta}\right] t \tag{4.7}$$

where L is the length of the coil. Figure 4.11 presents typical times for heating carbon steel to rolling temperature.

The total electrical energy required to operate the induction heating system is determined by multiplying the production rate by the required heat energy as given in Fig. 4.12 and dividing by the coil efficiency. Table 4.2 may be used to estimate coil efficiency. If radiant heat losses from the load are important, the calculated required electrical power must be increased by the factor $W_g/(W_g - W_r)$.

BIBLIOGRAPHY

Brown, G. H., C. N. Hoyler and R. A. Bierwirth, *Radio Frequency Heating,* Van Nostrand, New York, 1947.

Cable, J. W., *Induction and Dielectric Heating,* Reinhold, New York, 1954.

Davies, J. and P. Simpson, *Induction Heating Handbook,* McGraw-Hill, London, 1979.

Lozinskii, M. G., *Industrial Applications of Induction Heating,* Pergamon, Oxford, 1969.

May, E., *Industrial High Frequency Electric Power,* Chapman and Hall, London, 1949.

Semiatin, S. L. and D. E. Stutz, *Induction Heat Treatment of Steel,* American Society of Metals, Metals Park, Ohio, 1985.

Stansel, N. R., *Induction Heating,* McGraw-Hill, New York, 1949.

Chapter 5
Plasma Heating

S. L. CAMACHO
Plasma Energy Corporation

5.1 PLASMA HEATING AND THE PLASMA TORCH

Plasma heating is heating by electricity. The plasma torch is the hardware used in the heating process. Both alternating-current (ac) and direct-current (dc) power are used in plasma heating.

Plasma heating is accomplished by initiating and maintaining a gaseous electrical current conductor (the plasma arc column) and using the electrical resistance of the arc column to convert electricity into heat. The initiation and maintenance of the plasma arc column is the function of the plasma torch.

The plasma torch may be used to heat and melt solids, heat and vaporize solids and liquids, or heat gases and raise their enthalpies. The temperatures that are achievable by plasma heaters are the highest sustainable temperatures known in industrial heating.

Several configurations of plasma torches are available. The most common type is the *arc plasma torch* (APT) with electrodes made of copper or refractory metals. A plasma torch without electrodes, called *induction plasma torch* (IPT), is available for specific heating applications.

A dc-operated APT is more stable and therefore is more commonly used; ac-operated APT and IPT are operated at 50 to 60 Hz (cycles per second) or at several kilohertz.

Induction plasma torches are operated at kilohertz frequencies. The heat is induced in a flowing gas by high-frequency oscillations without electrodes in the IPT.

Arc plasma torches incorporate electrode(s) in their design and are manufactured in two basic types: transferred and nontransferred. Both types are equally effective in converting electricity into heat. A convertible APT design incorporates into one shroud assembly the choice of selecting to operate in either transferred arc mode or nontransferred arc mode. Physical exchange of the collimator and electrode is required to effect the conversion from transferred to nontransferred mode.

Figure 5.1 illustrates the transferred arc, nontransferred arc, and induction plasma torches. Note that both the transferred arc and nontransferred arc torches require positive and negative arc attachment points. Each attachment point represents an interface between the gaseous plasma arc column and a solid electrode. The IPT in Fig. 5.1*c* does not require electrodes. The nontransferred type of torch uses metallic rear and front electrodes as arc attachment points. The transferred type of torch uses the rear electrode for the internal attachment point but depends on the workpiece to anchor the external attachment point and complete the electrical current-conducting path.

The most common APT design is the gas vortex-stabilized torch as shown in Fig. 5.1*a* and *b*. The plasma gas is introduced tangentially through the vortex generator between the rear electrode and the collimator or the front electrode, depending on the APT type. Part of the injected gas rotates and moves rearward to the rear electrode and determines the position of the arc attachment at the point where the rotating gas changes direction and moves forward to the collimator or front electrode. The remaining part of the injected gas rotates and moves forward, mixes with the gas from the rear electrode, and exits through the collimator or front electrode. The rotating gas or vortex forms a low-pressure central core on the common axis of the electrodes within which the plasma column is stabilized.

The plasma arc column may be initiated and maintained in almost any gas medium (e.g., air, oxygen, nitrogen, carbon monoxide) in a wide range of pressures from soft vacuum to 20 atm and in the range of temperatures from 1000 to 13,500°F (540 to 7500°C).

The electrodes of the APT are water-cooled in order to retard the rate of metal losses through vaporization at the arc attachment points. Excessive metal vaporization of the electrode material leads to premature electrode failure and process disruption. A water-cooled shroud integrates and protects the major components of the plasma torch.

The water-cooling of plasma torch components results in energy losses. The rear electrode, collimator, front electrode, and shroud absorb heat from the plasma column and from the furnace environment. Collectively, these losses range from a low of 8% for the transferred torch to a high of 30% for the nontransferred torch.

5.2 PLASMA TORCH ELECTRODES

The electrode(s) of the APT is (are) the only consumable part of the torch. The electrode is fabricated from different

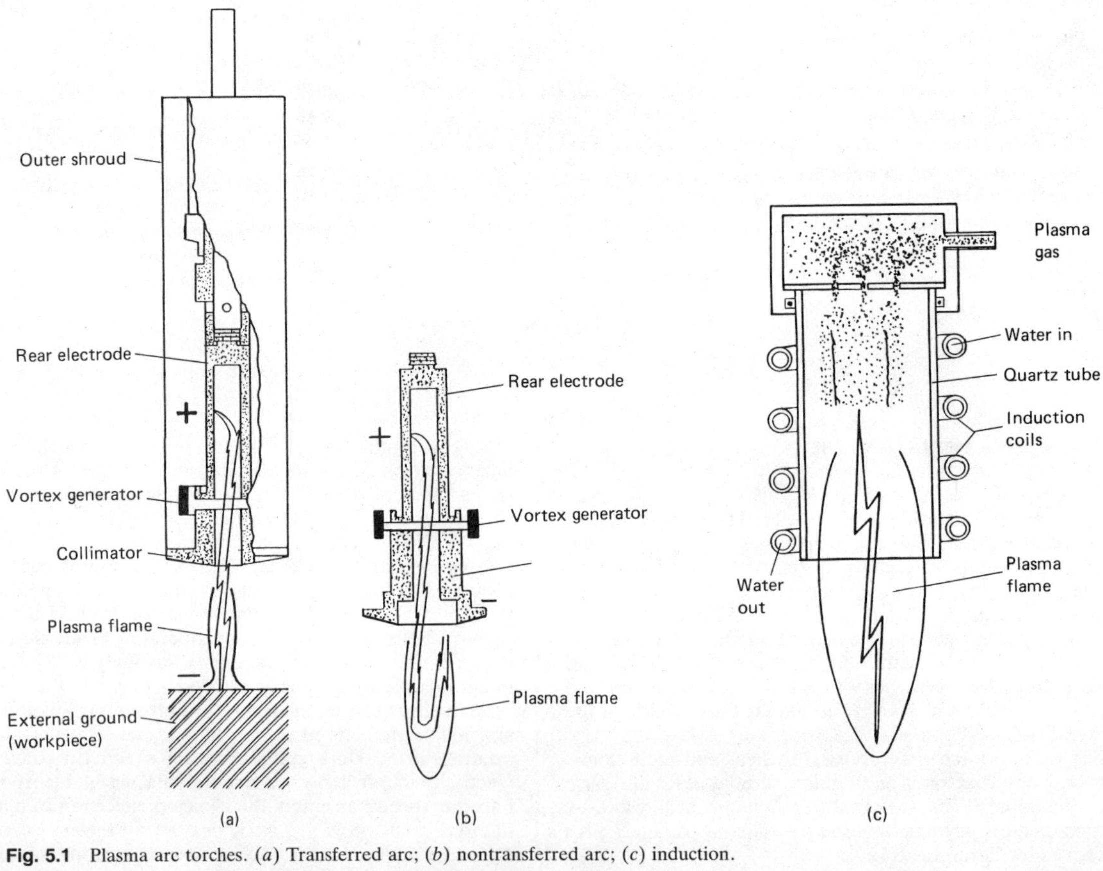

Fig. 5.1 Plasma arc torches. (*a*) Transferred arc; (*b*) nontransferred arc; (*c*) induction.

materials, such as copper alloys, steel, tungsten, and graphite. The electrode configuration varies from the pointed-tungsten-rod shape of the cutting and welding torches to the hollow tubular copper electrode shown in Fig. 5.1*a* and 5.1*b*.

The useful life of the electrode is determined by the material used to fabricate it and by the arc current. Material considerations include factors such as the physical and thermal properties of the base metal, the type and amount of alloying elements, the casting method, and heat treatment. The arc current density at the plasma-electrode interface is the other consideration.

Copper is the preferred electrode material because of its high thermal conductivity and diffusivity. The copper may be alloyed with zirconium or chromium to improve its strength and oxidation resistance at high temperatures, or it may be alloyed with silver to control metal losses through oxidation. Alloying with silver is especially beneficial when the process requires pure oxygen as the plasma gas. These copper alloys are readily available.

The current density at the arc attachment points influences the useful life of the electrode in a significant way. High current densities yield short electrode life because the electrode is eroded by the high-current arc. The current density is difficult to measure, so the absolute current level is used to determine the useful life of the electrode.

A point of reference on electrode life is as follows: a 1-in-internal-diameter electrode made of copper with 2 percent zirconium alloy may be operated at approximately 4500 A with air as plasma gas and yield a useful electrode life (200 h or more). The electrode life may be increased to several thousand hours by changing the given parameters.

5.3 THE INDUCTION PLASMA TORCH

The plasma state may be generated and maintained without electrodes or attachment points. Induction plasma is generated by the electrodeless method of induction with high-frequency alternating current. This method of generating and maintaining the plasma state in the IPT is uniquely different from the method employed by the APT.

Figure 5.1*c* illustrates the basic configuration of the IPT. The inductive coupling between the flowing-gas medium and the induction coil heats a portion of the flowing gas to plasma temperatures. Efficient gas mixing of the plasma column and surrounding gas generates a volume of high-temperature gas that is used as a heat source free of electrode contamination, an advantage of the IPT.

For certain applications the IPT is capable of slightly higher plasma temperatures because of the absence of arc attachment points in this type of torch. In the APT the arc attachment points are anchored to relatively cold surfaces that extract energy from the plasma column and cool the column.

The power level of the IPT is limited to about 300 kW by the need for high-frequency power and a special containment vessel. Frequencies in the kilohertz range are required to effect the proper electromagnetic coupling between the induction coil and the gas volume. The containment vessel must be transparent to the magnetic field of the induction coil; otherwise, the heating will be induced within the containment vessel and not within the gas volume. Metallic containment vessels cannot be used with IPT.

5.4 THE PLASMA HEATING SYSTEM

The major components of a plasma heating system (PHS) include the plasma torch, power supply, water-gas system, and control panel. The power supply conditions the electricity to match the voltage and current requirements of the plasma torch. The water-gas system conditions and monitors the flow rates of water coolant and plasma gas to the torch. The control panel integrates the safety lockouts to protect the PHS and provides for visual and recorded data acquisition of the PHS operating parameters.

The power supply uses the latest in solid-state electronic components. Three-phase ac power of 480 or 4160 V is converted into dc power of 500 to 1500 V by thyristors for the APTs. A choke coil is added to the dc side to provide stability to the plasma torch when operating near the maximum output of the power supply. Stabilization is needed to counter the negative current characteristics of the plasma state which predicates that the arc voltage will decrease slightly (about 0.25 to 1.5 V/A) when arc current is increased. The plasma column will extinguish randomly without the choke coil.

Plasma torches are available in power levels from 60 to 6000 kW. The torch diameters range from 2 in (5 cm) for the small torches to 12 in (25 to 30 cm) for the large ones. Multiple torches are used to satisfy higher heating requirements. The torch length is determined by the furnace or reactor design but is in the range of 20 to 100 in (50 to 250 cm). The plasma torches are cooled with high-pressure water. The water is deionized in some applications. The range of water flow rates is 10 to 100 gal/min at inlet pressures of about 200 lb/in^2 gage. The range of gas flow rates is 3 to 100 standard ft^3/min at inlet pressures of 40 to 100 lb/in^2 gage.

Starting the APT requires a plasma starter, a low-power source which is normally included in the power-supply enclosure. High-frequency (HF) and high-voltage (HV) starters are the most common types that are used to start APTs.

The HF starter initiates a sustained high-frequency spark in the gap between two components of the torch, usually the rear and front electrodes. The electric spark in the gap ionizes the gas and makes the gas conduct electric current from the main power source.

The HV starter initiates a single pulse of high voltage that ionizes the gas in the gap and provides a conducting path for the high current of the main power source.

The IPT is started by simply energizing the induction coil with HF oscillations.

5.5 METHODS OF PLASMA HEATING

Heating with the APT is by high-temperature gas convection, radiation, and conduction. The plasma arc column is generated at about 21,500°F (12,000°C), and the high-temperature arc column is used to heat the surrounding gas to an average bulk temperature of about 7200°F (4000°C). These temperatures have been measured by the absolute intensity measurement technique for the 4300-Å argon atomic line using a JY-HR 1000 Monochrometer. The enthalpy of the plasma-heated gas may range from 100 to 400 Btu/standard ft^3 (3500 to 14,500 kJ/normal m^3).

The heat is delivered to the workpiece from the transferred arc torch by the plasma-heated gas and is generated in the workpiece by the resistance of the arc attachment point. The heat energy from the nontransferred arc torch is delivered by the plasma-heated gas only. The IPT delivers its heat energy in a similar way to the nontransferred arc torch.

The range of gas velocities is 650 to 2000 ft/s (200 to 600 m/s) in the vicinity of the plasma torch collimator or front electrode.

About 60 to 65 percent of the heat energy that reaches the workpiece or feed material is absorbed when heating with a plasma torch. The remainder of the energy is lost to the walls of the furnace and lost as enthalpy of the effluent gas. The temperature of the plasma-heated gas effluent is higher than the effluent from oxygen-fuel burners, making the recycling of effluent gas energy a more efficient process.

The power output of the APT is varied by changing the arc current or the arc voltage or both. The bias to the thyristors is altered to change the arc current. The length of the transferred arc column or the gas flow rate of the plasma gas is altered to change the arc voltage. The power output of the IPT is changed by varying the frequency and changing the electromagnetic coupling between the coil and the plasma gas. Power change is easy to do, and the response of the APT and IPT are immediate. This rapid response feature makes plasma heating adaptable to programming for automated heating.

Plasma Heating in Refractory-Lined Furnaces

The APT is used in this method of applying plasma heating. The torch is operated in the transferred arc mode to the molten metal or slag at the bottom of the furnace as shown in Fig. 5.2.

The plasma torch electrode is connected to the positive lead, and the ground is connected to the negative lead of the power supply. A good seal is maintained around the water-cooled shroud of the plasma torch to prevent inward entrainment of ambient air and to prevent gas effluent from leaking around the torch.

The feed material, introduced from the top of the furnace, makes its way downward by gravity to the plasma-heated zone of melting. The gas from the plasma torch, plus whatever other gases are evolved from the feed material, flows upward through the feed and exits from the furnace. The enthalpy of the exiting gas preheats the incoming feed material. The molten metal product and molten slag are tapped from different heights on the furnace.

Plasma Heating in Water-Cooled Furnace

When metal purity and control of metal evaporation are important process parameters, the use of a water-cooled furnace (Fig. 5.3) is recommended. This furnace features a water-cooled hearth and mold inside a pressure vessel. The water-cooled furnace permits operation at the desired pressure and temperature to minimize the evaporation losses of alloying elements. This method of plasma heating by an APT is ideally suited for melting scrap of titanium alloy.

Plasma melting of alloyed titanium scraps preserves the aluminum and vanadium alloys by operating the plasma furnace above the vaporization pressure of these elements. Compare plasma melting with electron beam (EB) melting, where the alloying elements are selectively vaporized by the high-vacuum requirement of the EB gun that is lower than the vaporization pressure of the alloying elements.

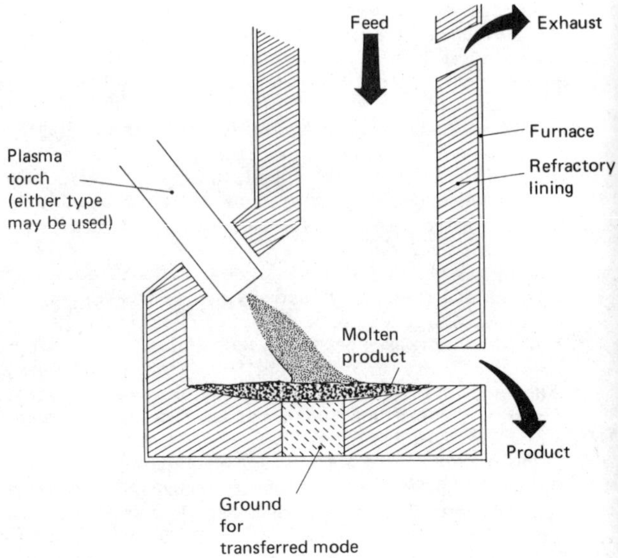

Fig. 5.2 Plasma heating in refractory-lined furnace.

Plasma Heating in In-Flight Reactors

The in-flight heating approach (Fig. 5.4) uses an APT and is recommended for the production of metal powders and advanced ceramic powders. This method of plasma heating is employed when the feed material is required to be in contact with the plasma flame in order to effect certain changes in the material structure, shape, or composition.

In-flight plasma heating is employed primarily to spheroidize random-shaped particles of metal and ceramics or to calcine fine powders. The technique has been employed in the direct reduction of ores of iron, ferrochrome, ferromanganese, ferronickel, etc. The IPT may be used in in-flight plasma heating.

Plasma Heating of Gases

The heating of gases by plasma is relatively simple. The nontransferred APT (Fig. 5.1*b*) and the IPT (Fig. 5.1*c*) are recommended for heating gases.

The gas to be heated is introduced into the plasma torch as the plasma gas and heated directly by the plasma column. Alternatively, the gas may be heated indirectly by introducing the gas downstream of the plasma torch, if the type of gas or the flow rate is not compatible with the requirement of the plasma torch for a stable arc. The direct heating approach is used when high gas enthalpy is required. The indirect heating approach is used when relatively low enthalpy is preferred.

5.6 ADVANTAGES OF PLASMA HEATING

The salient advantages of plasma heating are described as follows:

1. *Massless heat:* The method of releasing heat energy

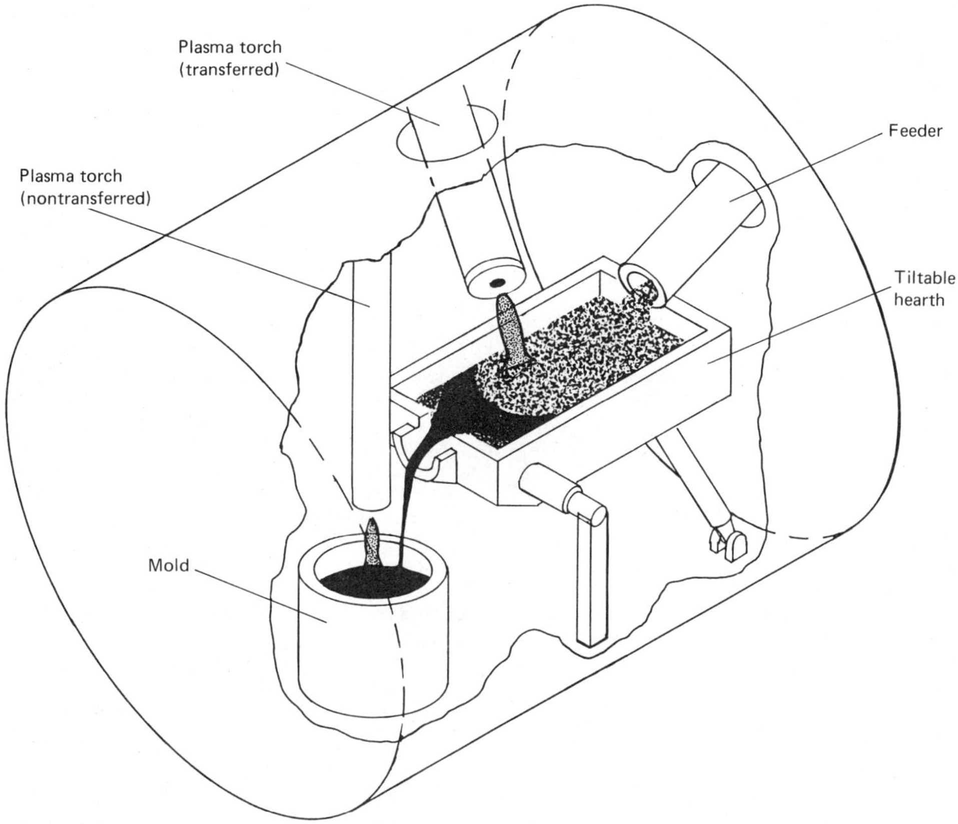

Fig. 5.3 Plasma heating in water-cooled furnace (controlled atmosphere and pressure).

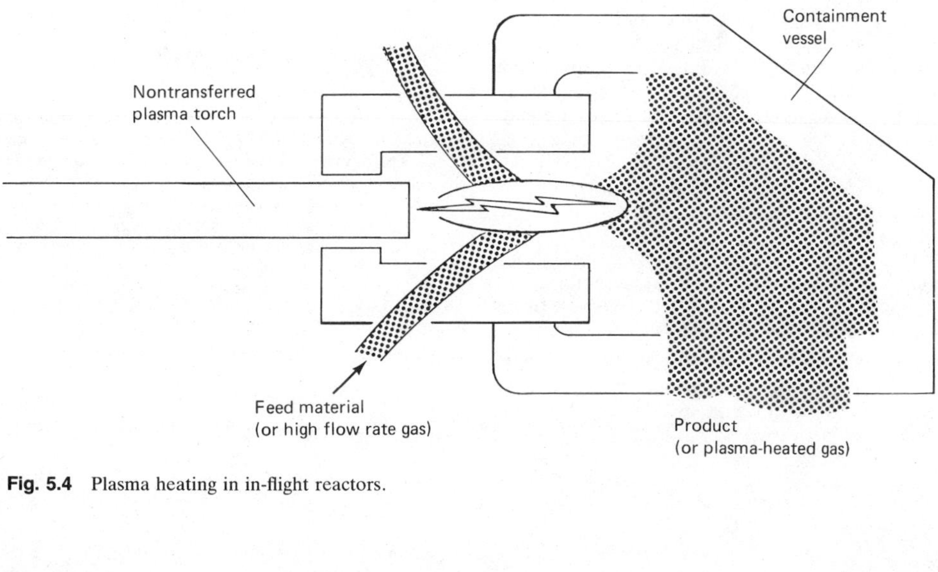

Fig. 5.4 Plasma heating in in-flight reactors.

with virtually no mass simplifies processing and control. Process effluent will be much less than the process effluent from oxygen-fuel heaters.

2. *Controlled furnace atmosphere:* Because the plasma torch is compatible with almost any gas (e.g., reducing, oxidizing, neutral, inert gases) the furnace atmosphere can be selected or tailored to match process requirements. Oxygen, air, argon, carbon monoxide, hydrogen, methane, nitrogen, and helium are typical gases used in the plasma torch.

3. *High temperatures:* At high temperatures heat-transfer rates and reaction kinetics are rapid and complete. The plasma torch in practical applications routinely achieves high temperatures [7000 to 13,500°F (3900 to 7500°C)]. These temperatures are not achievable by fossil-fuel burners. [Oxygen-fuel burners may be able to achieve a maximum temperature of 5000°F (2750°C); air-fuel burners attain lower temperatures because of the nitrogen dilution.]

4. *Smaller furnace-reactor size:* The plasma release of heat does not involve large quantities of gas; thus the furnace or reactor volume and the size of the furnace or reactor openings can be relatively small.

5. *Long-term availability of electric energy assured:* Electricity from nuclear fission reactions is plentiful and is predicted to remain so. Fusion reactions when harnessed will also be used to produce electricity. The assured future availability of electrical power is among the driving forces to apply plasma heating in industrial processes.

5.7 APPLICATIONS OF PLASMA HEATING

Plasma heating is employed in a variety of different applications. Small tungsten-tip plasma torches of 60 kW or less are used in plasma cutting, plasma welding, and plasma deposition. A plasma deposition torch is used to deposit metal on metal, metal on refractory, refractory on metal, refractory on refractory, etc. The tungsten electrode of these small APTs is protected from oxidation by the use of argon as plasma gas. In most cases these torches are hand-held, small-sized devices that operate at low voltages (20 to 60 V) and moderate currents (≤1000 A).

Larger APTs rated at 150 to 6000 kW are employed in industrial processes to melt scraps of steel and titanium; to smelt ores of iron, chromium, manganese, etc.; to provide a heat source for blast furnaces and cupolas; to augment heat production in electric arc furnaces and blast furnaces; to provide a heat source for temperature control of transfer ladles and tundishes in continuous steel casting operations; and to pyrolyze hydrocarbon materials such as oil, coal, peat, plastic, and rubber. These torches use a hollow tubular electrode (Fig. 5.1*a* and *b*) and operate in the ranges of 400 to 1200 V and 400 to 5000 A.

New applications of plasma torches are emerging. Plasma torches are used as heat sources in the production of powder metals, high-temperature ceramics, and refractories; to recover precious metals from spent catalysts and electronic scrap; and to generate very reactive gas species for use in the destruction of dioxins and other hazardous chemicals. Because of the ability of the plasma torch to generate high temperatures at low gas flow rates, the torch is used to pyrolyze hazardous wastes such as polychlorinated biphenyl (PCB)-laden oil and other hydrocarbon wastes.

The plasma torch is being used to dispose of municipal, commercial, and industrial wastes. The pyrolysis of waste materials by plasma heating leads to a final disposal of the wastes by converting them into harmless gases and vitrified slag. In some cases the high-temperature pyrolysis produces selected metal vapors which are condensed and recovered for recycling.

5.8 ECONOMICS OF PLASMA HEATING

The energy source for plasma heating is electrical, which is among the more expensive energy sources. An energy unit of electricity costs about four times more than the same energy unit of gas, oil, or coal. This energy comparison is often referred to as the *energy cost ratio* (ECR). Even with an ECR of 4, plasma heating can compete with other heating methods in selected applications. In some applications (ceramics, powder metallurgy, pyrolysis, etc.) plasma heating is the only applicable heating technology.

When taken collectively, the advantages of plasma heating can offset the energy cost disadvantage. The energy consumption of a process may be reduced as a result of the higher process efficiencies gained from high temperature, low mass, and smaller furnace or reactor. The following example points out the potential process efficiency gains that may be realized by employing plasma heating.

EXAMPLE A small, closed, rotary-hearth furnace and an open-type slot furnace have process efficiencies of 25 and 15 percent, respectively, using natural gas as fuel. These furnaces consumed about 3 million British thermal units of natural gas per ton of steel heated. It is not practical to equip these small forging furnaces with waste-heat recuperators.

A cost analysis was made to heat the same steel by plasma heaters to the same forging temperatures in a reduced-size furnace. The analysis shows that the furnace will require about 350 kWh per ton of plasma energy to heat the steel. The process efficiency of the plasma-heated furnace will be 65 percent.

The price of natural gas at the selected site was \$3.50 per million British thermal units, and the price of electricity was about 4 cents per kilowatthour. The gas flow rate in the rotary hearth was about 38,000 ft^3/ton of steel heated, and the gas flow rate in the plasma furnace was calculated to be 200 ft^3/ton.

The estimated energy cost for plasma heating was \$14 per ton, compared with \$10.50 per ton for the cost of heating with natural gas. The difference in energy cost could be compensated for by other operating improvements with plasma heating. Scaling losses, labor and maintenance costs, and scrap losses will be reduced.

The capital cost of the plasma heating system was about three times the capital cost of the gas heating systems. However, the lower cost of the reduced-size furnace and the reduced gas volume handling equipment made up part of the capital cost differential.

When all cost factors were considered, the plasma heating approach netted a cost saving of about 30 percent of the cost of the natural gas operation. The payoff period was less than 2 years.

Chapter 6

Laser Heating

STEVEN A. LLEWELLYN
Llewellyn Associates, Inc.

6.1 INTRODUCTION

For many years infrared light sources have been widely applied as both diffuse and concentrated heating tools in a wide range of industries. The invention of the laser in 1960 at Hughes Research laboratory introduced a device capable of delivering light in a form not previously available.[1]

The term *laser* has become a most familiar acronym for light amplification by stimulated emission of radiation. A laser produces light from stimulation of a *gain medium* with energy being introduced to this medium by several different methods. This energy releases light particles—photons—which are amplified in a resonant, mirrored cavity. A light beam that is highly collimated, coherent, and virtually monochromatic is extracted from the cavity by using a partial reflector as one of the cavity mirrors. The unique properties of laser-generated light have made this device an increasingly important source of energy in the 1980s.

The class of lasers of interest in thermal design is known as the material processing laser. Material processing lasers are normally defined as lasers used in applications where a change of physical state of a material is achieved by interaction with a laser beam. Even within this definition, the versatility of laser light allows it to be applied to several processes—welding, hole drilling, cutting, and heat treatment of both metals and nonmetals, as well as raising the temperature of a surface.

6.2 LASERS IN MATERIAL PROCESSING

There are two types of lasers in common use in industrial laser processing—the carbon dioxide laser and the neodynium laser. Some CO_2 lasers can produce output powers of up to 25 kW, but the majority of today's material-processing gas laser installations are units providing 25 W to 2 kW. The solid-state neodynium laser is available in several forms. The neodynium atoms can be suspended in a glass host "rod." They can also be added to the more popular yttrium aluminum garnet host—the so-called YAG laser. It is the neodynium that is the "gain" medium in the YAG laser. Material processing YAG lasers are currently available in power outputs of up to 400 W. In some applications (e.g., drilling) the ability of the solid-state laser to deliver high-power pulses is more valuable than its average power output capability. These different lasers are used for all the processes described in Fig 6.1. However, the wavelength differences—10.6 μm for CO_2 (far infrared) and 1.6 μm (near infrared) for neodynium—make many applications a better task for one or the other unit.

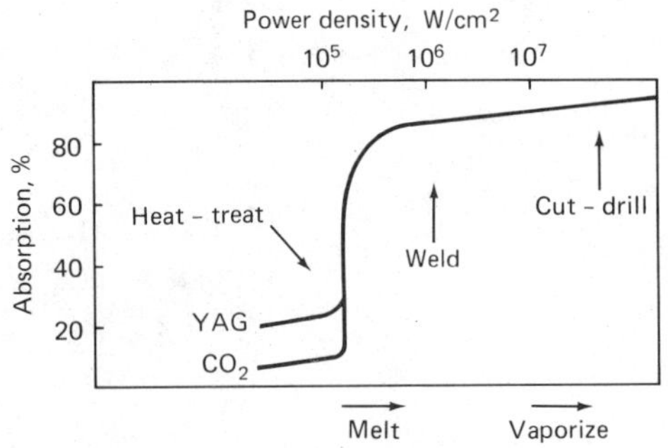

Fig. 6.1 Absorption characteristics and process domains for typical metal as a function of incident power density.

These CO_2 and YAG lasers emit in the infrared portion of the spectrum—a familiar heat source to a wide area of industrial and even domestic users. However, there are very substantial differences between conventional infrared heat sources and laser light. The coherence and monochromaticity of laser-generated light allows the beams generated by material processing lasers to be focused to very high brilliances and thus to produce very high power densities. This high brilliance helps to overcome one major problem of lasers used for industrial processes—reflectivity of the light by the workpiece. Some materials are so reflective to a particular laser light wavelength that no significant work can be done until the power available is able to overcome the reflections (e.g., copper or gold and the CO_2 laser).[2] It is the high power density produced by a focused laser beam that allows light to weld, drill, and cut metals and nonmetals.

Beam power densities in the range of 10^5 W/cm^2 and higher (see Fig. 6.1) provide the energy to vaporize the material to a vapor-plasma in which the beam can travel and produce welding penetrations previously allowed only by electron beams. The addition of a coaxial gas jet creates a high-power-density cutting tool that can cut metals and nonmetals at high speed with comparatively low heat inputs because of the high brilliances achieved by the focused laser beam.[3]

From Fig. 6.1 it can be seen that increasing power density from the 10^5 W/cm^2 region delivers energy in a form where the majority of the power delivered to the workpiece is released into vaporizing the material, provided the energy is absorbed into that material. This is possible with metals such as steel or tungsten and nonmetals such as paper and rubber. It is in this power-density region that hole drilling and cutting with lasers are carried out. Of more general interest is the region below 10^5 W/cm^2, where lasers can be used as heat sources and where they have the ability to heat materials without vaporizing the surface of the workpiece.

6.3 THE LASER AS A HEAT SOURCE

The physics of the light-amplification phenomenon demands that about 50 percent of the energy delivered to the laser cavity be retained within the laser to continue the process. Including the other energy conversion inefficiencies involved, lasers are not very efficient heat sources. Comparing electrical power supplied to the laser with laser light output, a 1-kW CO_2 laser is about 10 to 15 percent efficient and has a capital cost of approximately $100 per watt. A typical YAG laser is 3 to 5 percent efficient, and its capital cost is about twice as much per watt as the CO_2 unit. Therefore, in evaluation of the laser as a heat source in comparison with more conventional (and more efficient) heating methods, there must be a major reason for considering such an inefficient and relatively capital intensive method of generating heat. The precision and control that are possible with laser light generally provide that justification.

Laser energy, because of the way it is generated and the ways in which it can be manipulated, is a highly controllable form of energy. Lasers can produce continuous power or can be pulsed at frequencies up to 15 kHz with some low-power devices. Even high-power (400 to 600 W) units can produce pulse lengths of 40 to 50 μs, and this capability allows the beam to deliver very small, accurate quantities of energy to a workpiece. When this controllability of power is combined with high-speed positioners that direct the laser beam, a tool is created that can produce heat where it can be used effectively to carry out precise heating. The laser cannot efficiently be used to heat a complete component to a particular temperature, but it can be used to produce isolated heated areas on a larger part and deliver that power in such a way that other temperature-sensitive areas are not heated appreciably.

In some applications this precision is supplemented by other advantages, but in almost every case where a laser is used in a low-power-density application as a heat source, it is justified on the basis of controllability.

6.4 APPLICATION EXAMPLES

Heat-Treating Metals

Early lasers had less capability to produce high power densities than their modern counterparts. Accordingly, low-power-density applications such as heat treatment, particularly transformation hardening, were explored by applications engineers in the early 1970s. While the installation of lasers for this work has not met the early predictions, there are numerous examples of this successful process in the industrial workplace.

An example of the use of a laser for local precise heating is shown in Fig. 6.2.[4] The cast-iron component shown is from the steering assembly of an automobile. Millions of these units have been laser-hardened since 1974. The tracks shown inside the bore of the housing are made using a 1.2-kW carbon dioxide laser with a moving optical assembly that directs the beam to a programmed path. These tracks are positioned to produce hardened areas in the bore where local high surface loading previously produced early part failures.

Laser beams generated by current medium and high-power lasers, when focused, produce a power density that is sufficiently high to cause surface melting of metals. To make the

Fig. 6.2 Laser-hardened steering-gear housing. (*Lumonics Material Processing Corp.*)

heat-treatment process work, the beam is either defocused or oscillated by a moving mirror or integrated by a multi-surfaced mirror. This enlarging of the beam spot area lowers the power density and increases the coverage of the beam. However, it also usually takes the beam into a region where the losses to reflectivity from a shiny metal surface are so high that not enough energy can be coupled to the workpiece to make the process efficient. Therefore, most laser heat treatment of metals is done with an absorbent coating applied to the metal prior to heating. Coatings vary from sprayed or dipped black paint to more complex carbon-based proprietary coatings. Some dark-colored commercial coatings such as black-oxide also seem to work well.[5] Most of the coating is evaporated away from the treated areas and is easily removed from untreated zones.

Laser Soldering

Both CO_2 and neodymium lasers have been used for many years in soldering applications.[6] Lasers cannot compete with conventional processes in the soldering of a large number of components to a circuit board. However, the soldering of selected, heat-sensitive components to a section of a large circuit or to a ceramic substrate is both feasible and used in production situations. For soldering operations the laser beam is used with optics designed to produce large beam spots (typically 1.5-mm diameter) which produce power densities outside the vaporization range. Components are generally precoated with solder, and a correct process description would be *laser reflow-soldering.*

Care must be taken with the positioning of the laser energy, particularly with beam direction. Solder already present can act as a focusing mirror, and thermal damage to other adjacent components can occur. Shielding may be employed to overcome this problem or assembly schedules modified to allow the soldering to be done before the parts likely to be damaged are fitted. To allow the laser to solder several components in one operation, the beam may be split into two independent beams and directed over the workpiece using computer numerical control mirror motion.

6.5 SUMMARY

The two applications for lasers as heat sources that have been described demonstrate the inherent capability of the laser to act as a high-speed, precise, selective heating tool. For applications where such a tool can be applied effectively, today's industrial laser (proved in the production shop) can offer flexibility and return on investment to the manufacturing engineer.

REFERENCES

1. T. H. Maiman, "Stimulated Optical Radiation in Ruby," *Nature,* **187,** 493–494 (1960).
2. D. M. Roessler, "An Introduction to the Laser Processing of Materials," in *The Industrial Laser Annual Handbook,* M. Levitt and P. A. Belforte (eds.), PennWell, Tulsa, Okla., 1986, pp. 16–30.
3. S. A. Llewellyn, "Laser Cutting—State of the Art," *The Laser vs. the Electron Beam,* Reno, Nev., November 1985 (Bakish Material Corp., Englewood, N.J.).
4. W. W. Duley, *Laser Processing and Analysis of Materials,* Plenum, New York, 1983.
5. S. A. Tower, "Transformation Hardening with Lasers," *Manufact. Eng.,* **84,** 145–147 (August 1984).
6. D. U. Chang, "Experimental Investigation of Laser Beam Soldering," *Welding J.,* **65,** 33–41 (October 1986).

Part 9
Refrigeration

Chapter 1

Vapor Compression Refrigeration

HARRY J. SAUER, JR.
RONALD H. HOWELL
University of Missouri, Rolla

1.1 PRINCIPLES OF OPERATION

The Vapor-Compression Cycle

Continuous refrigeration can be provided by several different processes. In the vast majority of applications, the vapor-compression system, often termed the simple compression cycle or the reversed Rankine cycle, is used to obtain the refrigeration. The basic single-stage refrigeration system consists of an evaporator, a compressor, a condenser, and a refrigerant flow-control device.

Figure 1.1 illustrates schematically the basic vapor-compression cycle. Here cold low-pressure liquid and vapor refrigerant enters the evaporator and boils, absorbing heat from some other substance, such as air or water. The refrigerant vapor leaves the evaporator as a cool low-pressure gas and flows to the compressor, where its pressure and temperature are increased. It is discharged from the compressor as a hot high-pressure gas into the condenser, where it is condensed into a liquid. The cooling fluid, air or water, must be at a lower temperature than the refrigerant gas. The high-pressure liquid flows from the condenser through the expansion valve to the evaporator. The expansion valve meters the liquid flow and reduces the high-pressure liquid to a cold low-pressure liquid and vapor as it reenters the evaporator.

The first law of thermodynamics applied to the complete system requires that

$$Q_L + W = Q_H \qquad \text{Btu/h}$$

The rate of heat absorbed at the low temperature Q_L, called the capacity of the unit, is sometimes indicated as tons of refrigeration, where

$$\begin{aligned} 1 \text{ ton of refrigeration} &= 12{,}000 \text{ Btu/h} \\ &= 200 \text{ Btu/min} \\ &= 3.514 \text{ kW} \\ &= 4.712 \text{ hp} \end{aligned}$$

An important performance parameter for refrigeration systems is the coefficient of performance (COP). In principle, the COP is a measure of how efficiently the system provides the desired cooling. Symbolically,

$$\text{COP} = Q_L/W$$

The vapor-compression refrigeration cycle can be illustrated on a pressure-enthalpy diagram of the refrigerant, as shown in Fig. 1.2.

Subcooled liquid, at point A, decreases in pressure as it goes through the expansion or metering valve. Some of the liquid flashes into vapor and cools the fluid entering the evaporator at point B. As the refrigerant passes from point B to C, the remaining liquid receives heat and changes from a liquid to a vapor, but it does not increase in pressure. However, enthalpy does increase. Superheating normally occurs between point C, where the vapor passes the saturation curve, and point D. As the vapor passes through the compressor, point D to E, its temperature, pressure, and enthalpy increase significantly because of the work input for compression. Line EF indicates that the vapor is desuperheated within the condenser before it attains a saturated condition and begins to condense. Line FG represents the change from saturated vapor to saturated liquid within the condenser. Line GA represents subcooling before the liquid flows through the expansion or metering device.

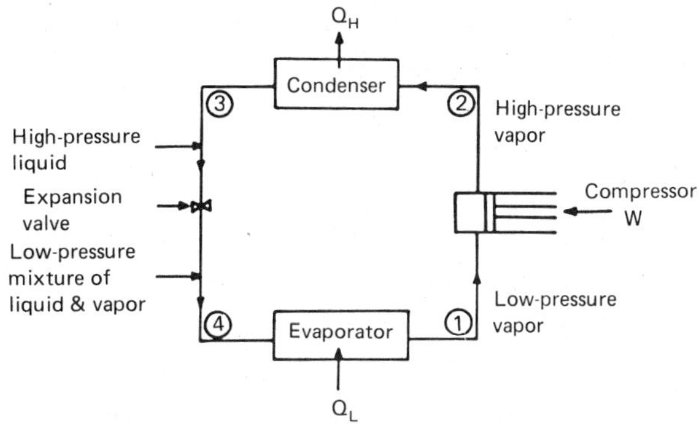

Fig. 1.1 Basic vapor-compression refrigeration cycle. (*Reprinted by permission of John Wiley & Sons, Inc. from H. J. Sauer, Jr., and R. H. Howell, Heat Pump Systems. Copyright © 1983, John Wiley & Sons, Inc.*)

The first law of thermodynamics can be applied to each component of the system individually.

Compressor Figure 1.3 shows the mass and ideal energy flows for a compressor. The rate of energy inflow must equal the rate of energy outflow during steady-state operation. Hence,

$$mh_1 + W = mh_2 \qquad \text{or} \qquad W = m(h_2 - h_1)$$

where m is the rate of refrigerant flow, and h_1 and h_2 are the enthalpies of the refrigerant at the inlet and the outlet of the compressor, respectively.

Condenser Figure 1.4 shows the mass and energy flows for a condenser, where

$$mh_2 = mh_3 + Q_R \qquad \text{or} \qquad Q_R = m(h_2 - h_3)$$

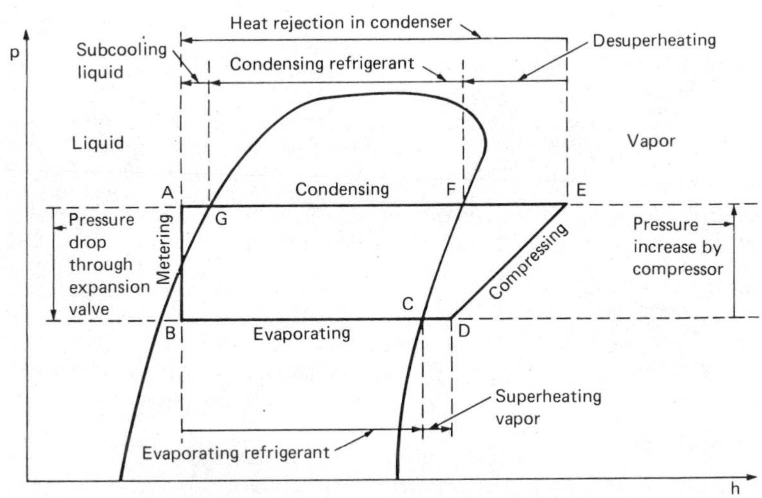

Fig. 1.2 Ideal pressure-enthalpy diagram for a refrigeration cycle. (*Reprinted by permission of PWS Publishers, Inc., from D. C. Look, Jr., and H. J. Sauer, Jr., Engineering Thermodynamics.*)

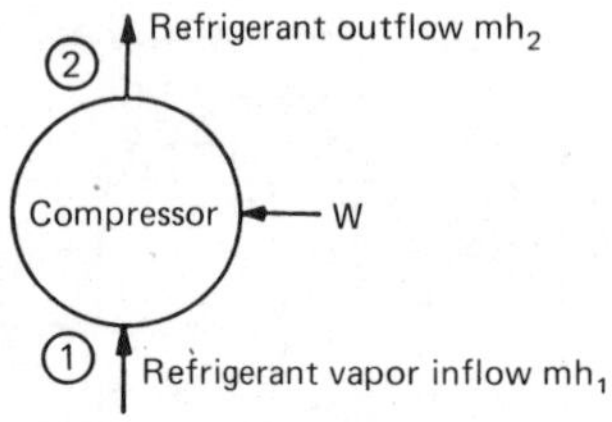

Fig. 1.3 Compressor flows.

The heat leaving the condensing refrigerant Q_R must equal the heat energy Q absorbed by the fluid or body that receives it. The heat balance for a cooling fluid is

$$Q_R = m_{\text{fluid}} c_p (t_{\text{out}} - t_{\text{in}})_{\text{fluid}}$$

where m_{fluid} = flow of cooling fluid through condenser
t_{out} = temperature of cooling fluid leaving condenser
t_{in} = temperature of cooling fluid entering condenser
c_p = specific heat of cooling fluid

Expansion Device This device is essentially a flow restrictor such as a small valve with a seat opening or a long length of small-bore tubing. Ideally there is no heat transfer and no work so that

$$mh_3 = mh_4$$

Dividing both sides by m gives

$$h_3 = h_4$$

Evaporator Figure 1.5 shows the ideal mass and energy flows for an evaporator, where

$$mh_4 + Q_A = mh_1 \quad \text{or} \quad Q_A = m(h_1 - h_4)$$

For the case where the evaporator cools a fluid,

$$Q_A = m_{\text{fluid}} c_{\text{fluid}} (t_{\text{in}} - t_{\text{out}})_{\text{fluid}}$$

where m_{fluid} = flow rate of fluid being cooled
c_{fluid} = specific heat of fluid flowing through evaporator
t_{in} = temperature of fluid being cooled entering evaporator
t_{out} = temperature of fluid being cooled leaving evaporator

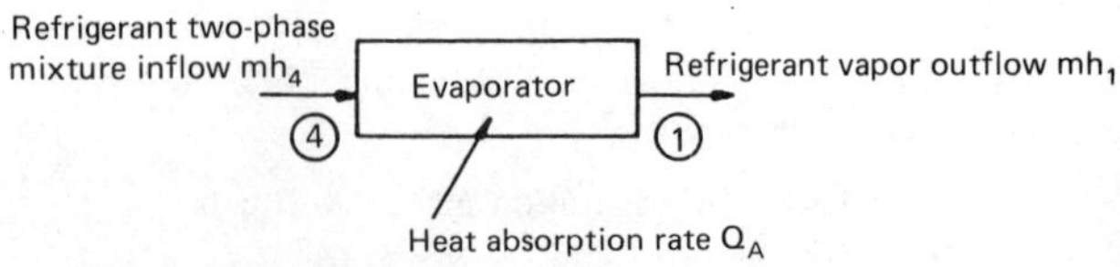

Fig. 1.5 Evaporator flows.

The thermodynamics of refrigeration systems is discussed in detail in Ref. 1.

Refrigerants

Refrigerants are the fluids circulated in refrigeration systems. They absorb heat in the evaporator and reject heat in the condenser.

Ammonia and sulfur dioxide were among the first refrigerants in vapor-compression systems. Today the major refrigerants are the halogenated hydrocarbons, also called chlorofluorocarbons (CFC), which are marketed under trade names such as Freon and Genetron. Most of the early refrigerants were thermodynamically suitable. Some of them, such as methyl chloride, sulfur dioxide, carbon dioxide, ammonia, and some hydrocarbons, became commercially important. However, all of these refrigerants suffered from one or more properties that restricted their general use. Some are toxic or flammable, and sometimes both. Others are lacking in chemical stability or operate at very high pressures. In the 1930s the fluorocarbon refrigerants were developed and introduced into the refrigeration and air-conditioning industry. R-11 and R-12 were among the first and others followed.

The refrigerant system has been greatly expanded since the 1930s by the American Society of Heating, Refrigerating and Air-Conditioning Engineers and now includes inorganic, hydrocarbon, azeotropic, and mixed refrigerants. A description and explanation of the refrigerant numbers is given in Ref. 1.

Both tabular and graphic data for thermodynamic and thermophysical properties of refrigerants are listed in Ref. 1. Figure 1.6 is an example of the graphic data for refrigerant-12.

Table 1.1 shows the theoretical calculated performance of a number of refrigerants for the U.S. standard cycle of 5°F (−15°C) evaporation and 86°F (30°C) condensation. The table can be used to compare the properties of various refrigerants, but actual operating conditions will be somewhat different from the theoretical calculated data. The results in Tables 1.2 and 1.3 illustrate the effect of different evaporating and condensing temperatures on the theoretical capacity and

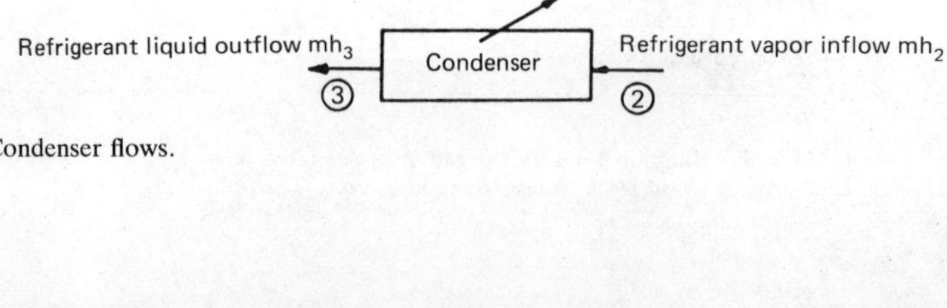

Fig. 1.4 Condenser flows.

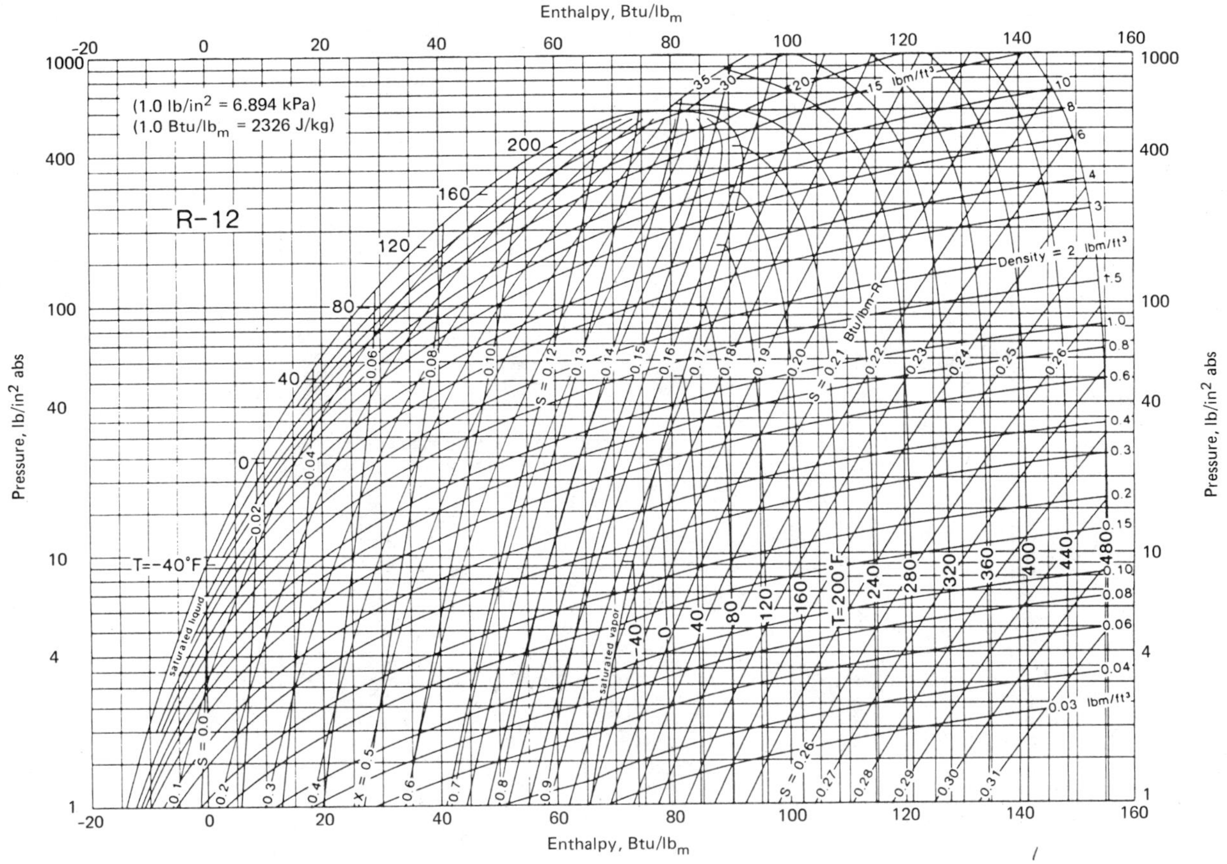

Fig. 1.6 Pressure-enthalpy diagram for refrigerant-12. (*Reprinted by permission from ASHRAE Handbook 1985 Fundamentals.*)

horsepower required for compression of refrigerants R-12 and R-22.

The choice of refrigerant for an application often depends on characteristics not directly related to its ability to remove heat. Such characteristics are flammability, toxicity, density, viscosity, condensing and evaporating pressure, and availability. As a rule, the choice of a refrigerant is a compromise between conflicting desirable properties. For example, the pressure in the evaporator should be as high as possible to avoid air leakage into the evaporator while, at the same time, a low condensing pressure is desirable so that dangerous pressure conditions do not prevail.

Refrigerant-12 is the refrigerant most commonly used with reciprocating compressors in domestic refrigerators and freezers, commercial and industrial air-conditioners, and some automotive refrigeration systems. Refrigerant-22 competes with refrigerant-12 in smaller air-conditioning units and is competitive with ammonia in industrial low-temperature systems where toxicity is of concern. The greatest use of ammonia is for large industrial systems and low-temperature applications. Refrigerant-11 and refrigerant-113, which have much lower densities, are the most widely used refrigerants in centrifugal compressor systems. The advantages of refrigerant-12 over refrigerant-22 are lower cost, head pressure, and compressor discharge temperature, and less tendency to leak. An advantage of refrigerant-22 is a lower volume flow per unit of refrigeration and, hence, a smaller compressor for a given refrigeration capacity.

1.2 SYSTEM COMPONENTS

One of the major tasks in the design of a refrigeration system is the selection of the components: compressor, condenser, evaporator, refrigerant flow control device, and controls.

The characteristics of each of the components are affected by the other components, and the system resulting from all of the components must perform properly at design conditions as well as at part load operation. Most refrigeration systems have varying operating conditions of both the heat source and the heat sink media, and this results in nondesign operating conditions. Selection of the optimum components for a particular system requires a familiarity with the effects

Table 1.1. Comparative Refrigerant Performance per Ton, Based on 5°F Evaporation and 86°F Condensation*

Refrigerant No.	Refrigerant Name	Evaporator pressure, lb/in² gage	Condensing pressure lb/in² gage	Compression ratio	Net refrigerating effect, Btu/lb	Refrigerant circulated, lb/min	Liquid circulated, in³/min	Specific volume of suction gas, ft³/lb	Compressor displacement, ft³/min	Horse-power, hp	Coefficient of performance	Compressor discharge temperature, °F
170	Ethane	221.3	661.1	2.86	58.6	3.41	342.9	0.53	1.82	1.953	2.41	122
744A	Nitrous oxide	294.3	922.3	3.03	85.2	2.35	71.2	0.28	0.66	1.310	3.60	
744	Carbon dioxide	317.5	1031.0	3.15	55.5	3.62	167.1	0.27	0.96	1.840	2.56	151
13B1	Bromotrifluoromethane	63.2	247.1	3.36	29.3	6.86	123.8	0.38	2.63	1.030	4.25	124
1270	Propylene	37.0	167.0	3.51	173.0	1.1	61.5	2.61	3.03	1.046	4.51	108
290	Propane	27.2	140.5	3.70	121.0	1.65	94.0	2.48	4.09	1.030	4.58	97
502	22/115 azeotrope	36.0	175.1	3.75	45.7	4.38	99.4	0.82	3.61	1.079	4.37	99
22	Chlorodifluoromethane	28.2	158.2	4.03	70.0	2.86	67.4	1.24	3.55	1.011	4.66	128
115	Chloropentafluoroethane	24.0	135.8	3.89	29.1	6.88	151	0.77	5.30	1.17	4.02	86
717	Ammonia	19.6	154.5	4.94	474.4	0.422	19.6	8.15	3.44	0.989	4.76	210
500	12/152a azeotrope	16.4	112.9	4.12	60.6	3.30	80.3	1.50	4.95	1.01	4.65	105
12	Dichlorodifluoromethane	11.8	93.3	4.08	50.0	4.00	85.6	1.46	5.83	1.002	4.70	101
40	Methyl chloride	6.5	80.0	4.48	150.2	1.33	40.9	4.47	5.95	0.962	4.90	172
600a	Isobutane	3.3†	44.8	4.54	111.5	1.79	91.0	6.41	11.50	1.083	4.36	80
764	Sulfur dioxide	5.9†	51.8	5.63	141.4	1.41	26.6	6.42	9.09	0.968	4.87	191
630	Methylamine	9.9†	46.8	6.13	304.0	0.66	28.2	15.54	10.23	0.978	4.81	
600	Butane	13.2†	26.9	5.07	128.6	1.56	75.9	9.98	15.52	0.953	4.95	88
114	Dichlorotetrafluoroethane	16.1†	22.0	5.42	43.1	4.64	89.2	4.34	20.14	1.049	4.49	86
21	Dichlorofluoromethane	19.2†	16.5	5.96	89.4	2.24	45.7	9.13	20.43	0.941	5.01	142
160	Ethyl chloride	20.5†	12.4	5.83	142.3	1.45	45.8	17.06	24.82	0.906	5.21	106
631	Ethylamine	23.1†	10.0	7.40	225.5	0.89	349.0	32.32	38.67	0.855	5.52	
11	Trichlorofluoromethane	23.9†	3.5	6.19	66.8	2.99	56.6	12.21	36.54	0.938	5.03	111
611	Methyl formate	26.3†	1.6†	7.74	189.2	1.06	29.9	48.25	51.00			
610	Ethyl ether	26.9†	4.9†	8.20	126.3	1.58	62.9	35.00	55.40	0.822	5.74	
30	Methylene chloride	27.6†	9.5†	8.60	134.6	1.49	30.9	49.90	74.30	0.963	4.90	205
113	Trichlorotrifluoroethane	27.9†	13.9†	8.02	53.7	3.73	66.5	27.38	102.03	0.973	4.84	86
1130	Dichloroethylene	28.3†	15.8†	8.42	114.3	1.75	38.3	63.60	111.20	0.973	4.83	
1120	Trichloroethylene	29.6†	26.2†	11.65	91.7	2.18	41.6	229.40	502.00	0.980	4.82	

* Saturated suction vapor except for refrigerants-113, -114, and -115. In these cases, enough suction superheat was assumed to give saturated discharge vapor.

† Inches Hg vacuum.

CONVERSION FACTORS: lb/in² gage = 6.894 kPa gage; 1 Btu/lb = 2326 J/kg; 1 lb/min = 0.00756 kg/s; 1 in³/min = 2.73×10^{-7} m³/s; 1 ft³/lb = 0.062 m³/kg; 1 ft³/min = 4.72×10^{-4} m³/s; 1 hp = 0.7457 kW; °C = (°F − 32)/1.8.

SOURCE: *ASHRAE Handbook 1985 Fundamentals*.

Table 1.2. Effect of Temperature on Capacity,* Btu/h

Evaporation temperature, °F	Condensing temperature, °F			
	60	80	100	120
	Refrigerant-12			
−20	5240	4790	4320	3841
0	8280	7590	6890	6160
20	12590	11580	10550	9470
40	18510	17080	15620	14090
	Refrigerant-22			
−20	8672	7987	7279	6536
0	13518	12479	11402	10277
20	20303	18782	17203	15553
40	29553	27386	25139	22789

* Constant compressor displacement 4 ft^3/m (1.89×10^{-3} m^3/s).
CONVERSION FACTORS: 1 Btu/h = 0.2931 W; °C = (°F − 32)/1.8.
SOURCE: *ASHRAE Handbook 1985: Fundamentals*.

of changing operating conditions on the performance of each component.

Compressor

Types The types of compressors commonly used in refrigeration systems are reciprocating, rotary vane or screw, and centrifugal compressors. Reciprocating compressors are used more than any other type for systems in the general range of 0.5 to 100 tons.

The most widely used reciprocating compressor is the halocarbon compressor, which is manufactured in three design types: open, semihermetic or bolted hermetic, and welded-shell hermetic. Ammonia compressors are available only in the open design because of the incompatibility of ammonia with motor materials.

Open-type compressors are those where the crankshaft extends through a seal in the crankcase to accommodate an external drive.

Table 1.3. Effect of Temperature on Theoretical Horsepower per Ton

Evaporation temperature, °F	Condensing temperature, °F			
	60	80	100	120
	Refrigerant-12			
−20	1.03	1.36	1.76	2.24
0	0.70	0.99	1.33	1.73
20	0.43	0.68	0.97	1.31
40	0.20	0.42	0.67	0.95
	Refrigerant-22			
−20	1.03	1.37	1.77	2.24
0	0.71	1.01	1.34	1.74
20	0.44	0.69	0.98	1.32
40	0.20	0.42	0.67	0.97

CONVERSION FACTORS: 1 hp/ton = 0.212 kW_e/kW_r; °C = (°F − 32)/1.8.
SOURCE: *ASHRAE Handbook 1985: Fundamentals*.

Hermetic compressors are those where the motor and the compressor are contained inside the same pressure vessel, with the motor shaft directly connected to the compressor crankshaft and the motor in direct contact with the refrigerant. Figure 1.7 is a schematic drawing of a hermetic reciprocating compressor.

A semihermetic (bolted, accessible, or serviceable) compressor is of bolted construction so that it can be field-repaired.

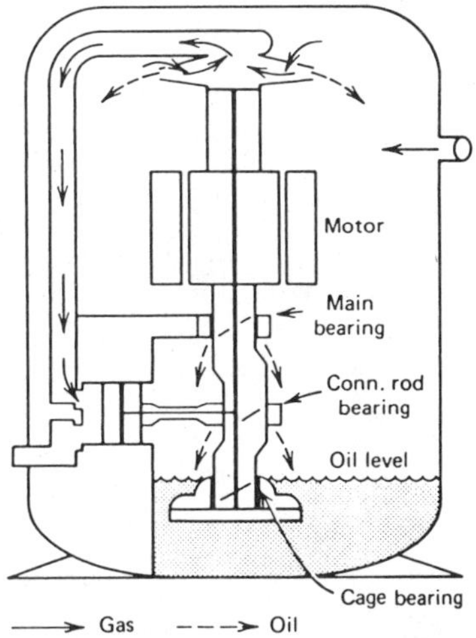

Fig. 1.7 Schematic of hermetic compressor. (*Reprinted by permission of John Wiley & Sons, Inc., from H. J. Sauer, Jr., and R. H. Howell, Heat Pump Systems. Copyright © 1983, John Wiley & Sons, Inc.*)

A welded-shell (sealed) hermetic compressor is one in which the motor compressor is mounted within a steel shell which is weld-sealed. One of the important operating characteristics of reciprocating compressors is the effect of the clearance volume (i.e., the volume occupied by the refrigerant within the compressor that is not discharged by the moving member). This effect is illustrated in the piston-type compressor by considering the clearance volume between the piston and the cylinder head when the piston is in the top dead-center position. The clearance gas remaining in this space after the compressed gas is discharged from the cylinder reexpands to a larger volume as the piston recedes and the pressure falls to the inlet pressure (see Fig. 1.8). As a result, the volume of refrigerant discharged from the compressor is less than the volume displaced by the piston. This effect is expressed by the volumetric efficiency η_v,

$$\eta_v = \frac{m_a}{m_i}$$

where m_a is the actual mass of new gas entering the compressor per stroke and m_i is the theoretical mass of gas represented by the displacement volume and determined at the pressure and temperature at the compressor inlet.

When the effect of clearance alone is considered, the resulting expression is termed clearance volumetric efficiency.

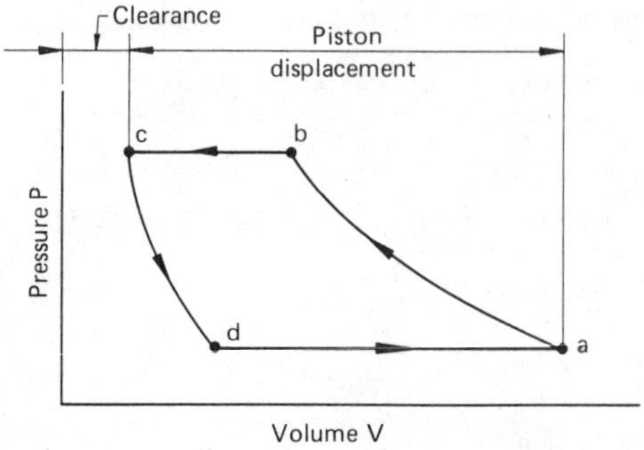

Fig. 1.8 Pressure-volume diagram for reciprocating compressor. (*Reprinted by permission from ASHRAE Handbook 1985: Fundamentals.*)

For the simple cycle shown in Fig. 1.8, the volumetric efficiency becomes

$$\eta_{vc} = \frac{V_a - V_d}{V_a - V_c} = 1 + C - C\left(\frac{v_1}{v_2}\right)$$

where C = clearance ratio = $V_c/(V_a - V_c)$

v_1 = specific refrigerant volume at beginning of compression

v_2 = specific refrigerant volume at end of compression

η_{vc} = clearance volumetric efficiency

Volumetric efficiency is a measure of how well the piston displacement (size) of the compressor is utilized in moving refrigerant vapor through the cycle. Each refrigerant will have a variable volumetric efficiency in the cycle due to changes in v_1. In addition, heat transfer and cycle and component pressure drops will affect the actual volumetric efficiency. The actual volumetric efficiency of a compressor is best obtained by actual laboratory measurements of the amount of refrigerant compressed and delivered to the condenser. The magnitude of the difference between actual and clearance volumetric efficiency is illustrated in Fig. 1.9.

A compressor normally used for air-conditioning will have a clearance volume of about 5 percent. If the compressor has low clearance volume, for example, 2.5 percent, then it is more suitable for low-temperature operation. Such a compressor will provide, for example, about 15 percent greater refrigerating capacity at an evaporator temperature of 0°F (−17.8°C) and a condenser temperature of 110°F (43.3°C). However, this compressor has somewhat more power demand under maximum cooling load conditions than one of medium clearance.

Rotary compressors are characterized by their circular, or rotary, motion instead of reciprocating motion. Most are direct-drive devices. The rotary compressor is also characterized by high volumetric efficiency since small clearance volumes are possible, and a correspondingly low reexpansion loss.

The helical rotary or screw compressor belongs to the broad class of positive displacement-type compressors. Several differing geometries are currently used in screw compressors. Screw compressors used for refrigeration and air-conditioning applications are available in two distinct types: single screw and twin screw. Both are typically used in the oil-injected mode where significant quantities of oil are used to cool, seal, and lubricate the compressor.

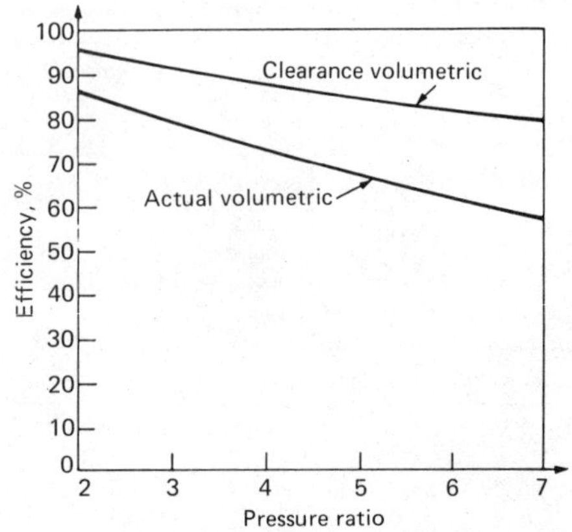

Fig. 1.9 Actual and clearance volumetric efficiencies. (*Reprinted by permission of PWS Publishers, Inc., from D. C. Look, Jr., and H. J. Sauer, Jr., Engineering Thermodynamics.*)

Larger rotary, blade-type compressors are used in low-temperature refrigeration as high-volume, low-stage, or booster compressors. These compressors are applied at saturated suction conditions ranging from −125 to −5°F (−87.2 to −20.6°C) with refrigerants-12, -22, and -717. These compressors are compact and lightweight, and designed for use in industrial processing plants, cold-storage plants, freezing plants, chemical processing plants, or for any application requiring a low-temperature system. Full unloading for starting and capacity control from 100 percent down to 20 percent with proportional power reduction is available for this type of compressor. The unloading is accomplished by using valves in both end covers.

Centrifugal compressors, or turbocompressors, are characterized by a continuous exchange of angular momentum between a rotating mechanical element and steadily flowing fluid. Centrifugal units have greater volumetric capacity, size for size, than do positive displacement machines. For effective momentum exchange, their rotative speeds must be higher, but little vibration or wear results because of the steadiness of the motion and the absence of contacting parts. Centrifugal compressors are used in a wide variety of refrigeration and air-conditioning applications. As many as eight or nine stages can be installed in a single casing.

Compressor Performance An example of the presentation of capacity data is given in Fig. 1.10. This set of curves is for a four-cylinder semihermetic compressor with a 2.38-in (60.3-mm) bore, a 1.75-in (44.4-mm) stroke, and a speed of 1720 r/min operating with refrigerant-22.

In order to be meaningful, compressor curves should be labeled with at least the following information:

Identification
Degrees of subcooling
Compressor speed
Type of refrigerant

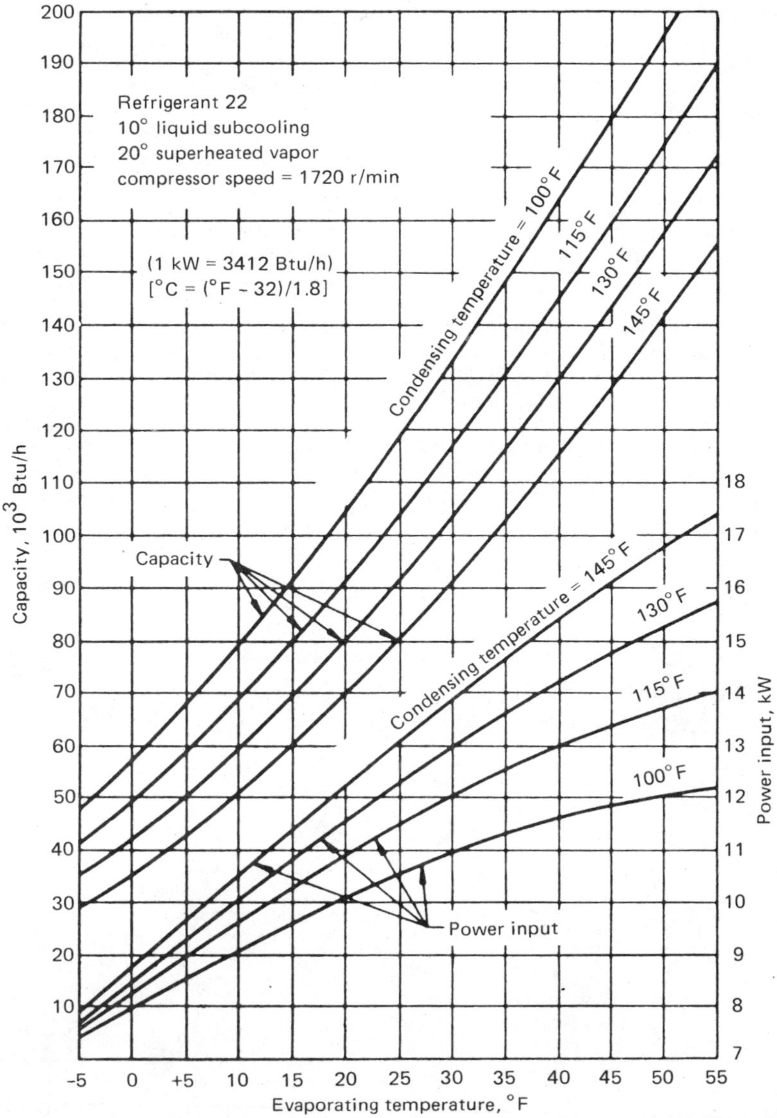

Fig. 1.10 Typical performance curves for reciprocating compressor. (*Reprinted by permission from ASHRAE Handbook 1975 Equipment.*)

Suction gas superheat
Compressor ambient conditions
Maximum power or maximum operating conditions
Minimum operating conditions at fully loaded and fully unloaded operation

With a constant discharge pressure, a positive displacement compressor has a gas pumping rate almost directly proportional to the absolute pressure at the compressor suction. There is some variation because of compression ratio change, but this difference is usually small unless the compression ratio is great enough to result in a lower volumetric efficiency. At compression ratios of 5 or less, the effect of a small change in the compression ratio is not very significant.

Changes in the gas temperature at the suction of the compressor will have a small effect on the capacity for a few degrees change in suction temperatures above −50°F (−46°C).

The relationship between refrigeration capacity and pressure at the compressor suction is useful in judging the effect of suction line pressure loss or lower evaporating pressure.

If the evaporating pressure is 75 lb/in² abs (517 kPa abs), a 2-lb/in² (14-kPa) loss in the suction line will lower the compressor capacity by about 2/75 (14/517), or 2.7 percent. If the evaporating pressure is 20 lb/in² abs (138 kPa abs), a 2-lb/in² (14-kPa) reduction will result in about a 10 percent capacity reduction.

Small increases in compressor discharge pressure, with no increase in condensing temperature (such as results from discharge line loss), have very little effect on the compressor capacity. The principal result is an increase in power. An increase in discharge pressure (the result of higher condensing temperature causing a higher temperature of the liquid to the expansion valve) reduces the capacity because it reduces the heat removal capacity of the refrigerant handled.

Compressor Capacity Control To control compressor capacity, hot-gas bypass or cylinder unloading may be used. The preferred method, cylinder unloading, uses a mechanism which automatically holds open some cylinder suction valves when less than full capacity is required. Cylinder unloading provides unloaded starting, requires less power, has a wider capacity range, and is quiet in operation. Multistep cylinder unloading reduces compressor capacity in predetermined steps, thereby maintaining evaporator-compressor balance.

No-load starting can be done with a manual or an automatic valve in a bypass line between the hot-gas and the suction lines at the compressor. To prevent overheating, this valve should be open only during the starting period, and then close after the compressor is up to full speed and full voltage is applied to the motor terminals.

Where full unloading is required for capacity control, hot-gas bypass arrangements can be used in ways that will not overheat the compressor. In using these arrangements, hot gas should not be bypassed until after the last cylinder unloading step. A hot-gas bypass should (1) give acceptable regulation throughout the range of loads, (2) not cause excessive superheating of the suction gas, (3) not cause any liquid carryover to the compressor, and (4) maintain an oil return to the compressor.

Four of the more common methods of using hot-gas bypass are illustrated in Fig. 1.11. Figure 1.11*a* shows the simplest type of hot-gas bypass. It can dangerously overheat the compressor if used for long periods of time. Figure 1.11*b* shows

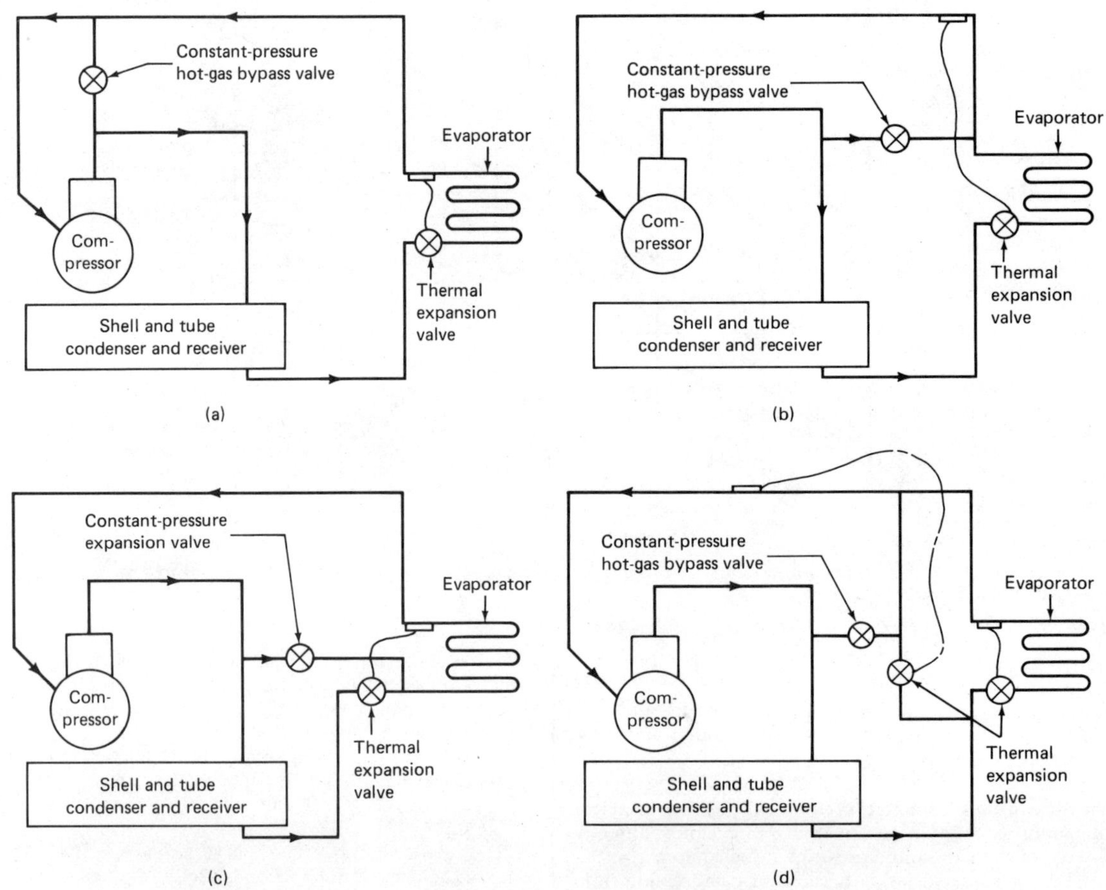

Fig. 1.11 Hot-gas bypass arrangements. (*Reprinted by permission from ASHRAE Handbook 1984 Systems.*)

the use of a hot-gas bypass to the exit of the evaporator. The thermostatic expansion valve bulb should be placed at least 5 ft (1.5 m) downstream from the point of entrance of the bypass to assure good mixing.

In Fig. 1.11*d* the hot-gas bypass enters after the evaporator thermostatic expansion valve bulb. Another thermostatic expansion valve supplies liquid directly to the bypass line for desuperheating purposes. It is always important to install the hot-gas bypass far enough back in the system to maintain sufficient gas velocities in suction risers and other components to assure oil return at any evaporator loading. Figure 1.11*c* shows the most satisfactory hot-gas bypass arrangement. Here the bypass is connected into the low side between the expansion valve and the entrance to the evaporator. If a distributor is used, the gas enters between the expansion valve and the distributor. Refrigerant distributors are commercially available with side inlet connections that can be used for a hot-gas bypass. The pressure drop through the distributor tubes must be evaluated to determine how much gas bypassing can be effected. This arrangement provides good oil return.

Compressor Pressure Controls Most compressors include a low-pressure control, and some also have a high-pressure control, usually mounted on the compressor. A small tube or capillary is connected to a port on the service valve so that it can sense pressure under operating conditions. Both controls may be combined into one control called a dual pressure control. The low-pressure control is in the circuit primarily to protect against loss of refrigerant charge. If the suction pressure goes below the control setting for any reason, it will trip, shutting off the compressor. It is an automatic reset control, that is, it will reset itself when the suction pressure rises above the setting.

The high-pressure control is in the circuit to protect against failure of the condenser fan motor or any other condition that causes the head pressure to increase to an unsafe point. It usually has a manual reset feature, which means that once it trips, it must be reset manually before the compressor can be operated again.

To protect against oil pressure failure, an oil pressure failure switch is commonly used. It measures the differential between the suction pressure and the oil pressure. The switch employs two pressure-sensing bellows; one is connected to the suction side of the compressor, the other to the discharge side of the oil pump. Suction pressure is exerted on one bellows and oil pump discharge pressure on the other bellows. The pressure differential switch is provided with both cut-in and cut-out points.

Evaporator

Types The evaporator transfers heat from the fluid or body being cooled to the boiling refrigerant. Various types of evaporators are commercially available for cooling and dehumidifying air or chilling water or antifreeze solutions.

Direct-expansion finned-tube and shell-and-tube evaporators are the types used in most refrigeration systems (see Fig. 1.12). The finned-tube heat exchanger, also known as an extended surface coil or simply a coil, consists of rows of tubing through which the refrigerant flows and over which the air flows. The shell-and-tube evaporator has a metal shell through which water or brine flows over the refrigerant tubes, which are within the shell and through which the refrigerant flows.

In the flooded cooler, the refrigerant is vaporized on the outside of bare or augmented surface tubes that are submerged in evaporating liquid refrigerant within a closed shell. The cooled liquid flows through these tubes, which may be straight, U-shaped, or coiled.

Evaporator Performance Data The proper choice and sizing of evaporators is probably the most important factor in designing an efficient and economic refrigeration system and in many ways the most difficult. The complex geometries of these heat exchangers, together with wide variations in operating conditions, make precise calculation of sizes from basic physical principles impracticable, and various empirical factors have to be applied to heat-transfer relationships. Although such factors may be well established for available heat exchanger geometries, details are often trade secrets.

Whether the heat exchanger is selected as an off-the-shelf item or designed for the particular application, the following factors need to be taken into consideration:

1. Heat-transfer requirements
2. Cost
3. Physical size
4. Pressure-drop characteristics

Heat-transfer capacity information given in manufacturers' catalogs is usually based on actual laboratory tests. To simplify coil selection, data are usually presented in the form of tables or charts.

Condenser

Types The condenser in a refrigerating system removes from the compressed refrigerant gas the energy gained during compression and the heat absorbed by the refrigerant in the evaporator. The refrigerant is converted back into the liquid phase at the higher condenser pressure and is then available for reexpansion back into the evaporator. The common forms of condensers may be broadly classified on the basis of the cooling medium as (1) water-cooled, (2) air-cooled, and (3) evaporative (air and water) cooled.

Figure 1.13*a* and *b* shows simple sketches of air- and water-cooled condensers. Because the refrigerant vapor gives up heat to the air moving across the condenser surface, it is condensed inside the tubes of the air-cooled condenser. Water-cooled condensers have water flowing inside the tubes, and thus the refrigerant condenses inside the shell, but outside the tubes.

Coils in air-cooled condensers are commonly constructed of copper, aluminum, or steel tubes, ranging from ¼ to ¾ in (6.3 to 19 mm) in diameter. Copper, the most expensive material, is easy to use in manufacturing and requires no protection against corrosion. Aluminum requires exact manufacturing methods, and special protection must be provided if aluminum-to-copper joints are made. Steel tubing is also used, but weather protection must be provided.

Fins are installed on the outside to improve the air-side heat transfer. Most fins are made of aluminum, but copper and steel are also used. The most common forms are plate fins individually fastened to the tube, or a fin spirally wound onto the tube. Other forms, such as plain tube-fin extrusions or tube extrusions with accordion-type fins are also used. The number of fins varies from 4 to 30 per inch (160 to 1180 per meter).

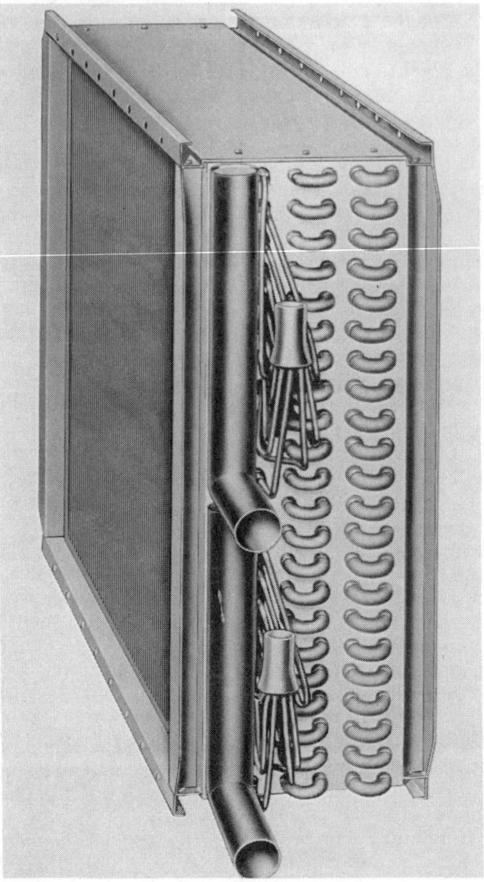

(a)

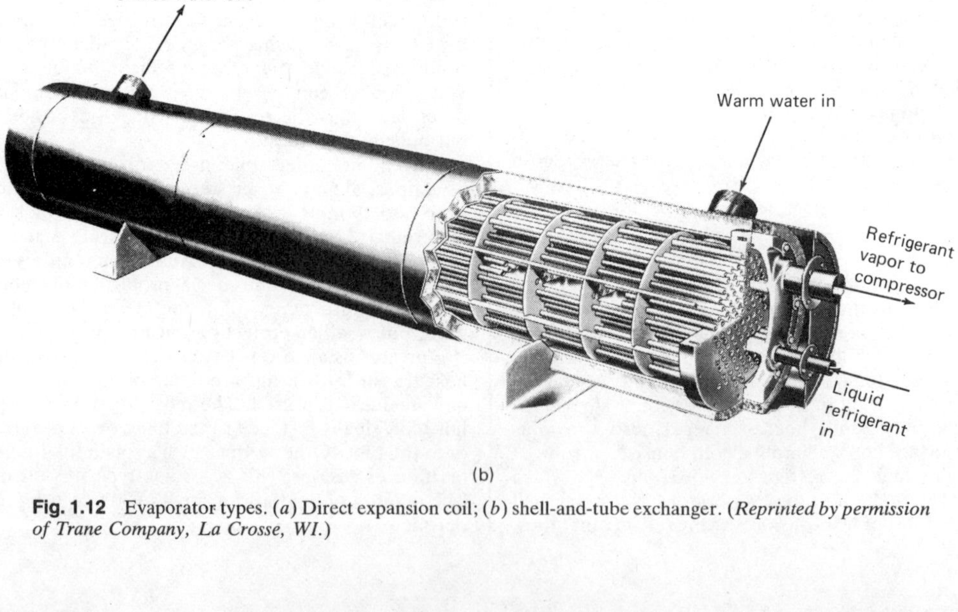

(b)

Fig. 1.12 Evaporator types. (*a*) Direct expansion coil; (*b*) shell-and-tube exchanger. (*Reprinted by permission of Trane Company, La Crosse, WI.*)

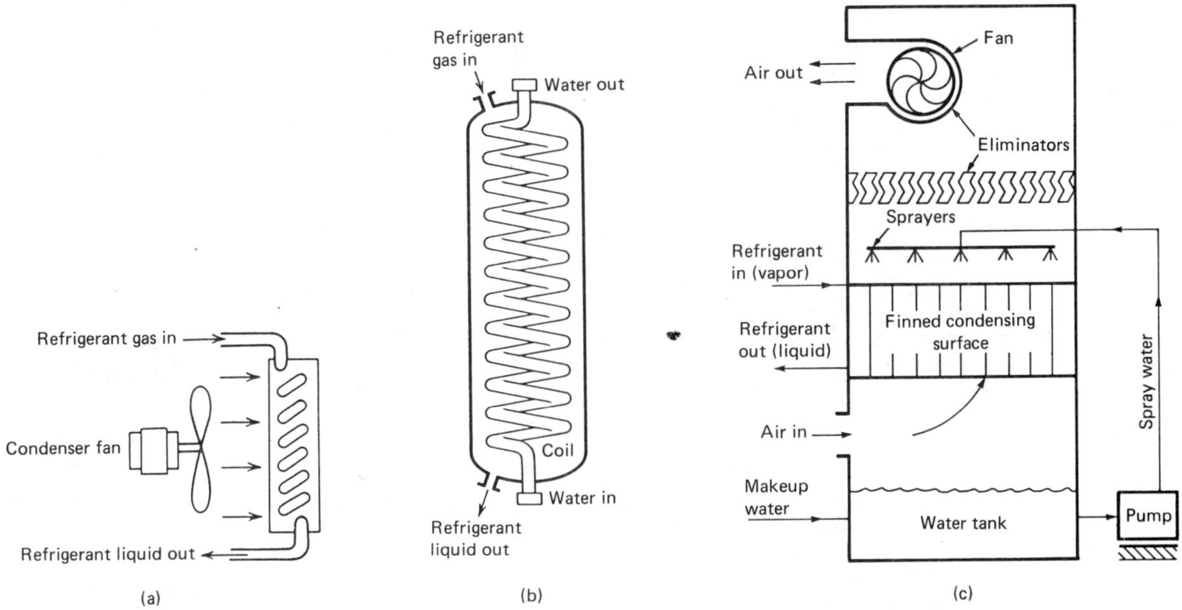

Fig. 1.13 Condenser types. (*a*) Air-cooled; (*b*) water-cooled; (*c*) evaporative. (*Reprinted by permission of John Wiley & Sons, Inc., from H. J. Sauer, Jr., and R. H. Howell, Heat Pump Systems. Copyright © 1983, John Wiley & Sons, Inc.*)

An evaporative condenser is a device that has a coil in which the refrigerant is condensed and a means to supply air and water over the external surface of the coil. Heat is transferred from the condensing refrigerant inside the coil to the coil's external wetted surface and then into the moving air stream, principally by evaporation. Figure 1.13*c* shows one type of evaporative condenser.

Performance and Selection The selection of a condenser depends on the cooling load, the refrigerant used, the source and temperature of the available cooling fluid, the quantity of coolant being circulated, the condenser location, the required operating pressures, and maintenance considerations.

Theoretical values of the heat rejection rate per unit of refrigeration produced in the evaporator can be found in Ref. 3. In actual practice, the heat removed is 5 to 10 percent higher than the theoretical values because of inherent (and irreversible) losses during compression.

The heat-transfer process in an air-cooled condenser has three main phases, (1) desuperheating, (2) condensing, and (3) subcooling. Figure 1.14 shows the changes in temperature of refrigerant-12 passing through the condenser and the corresponding temperature change of the cooling air as it passes through the coil. Desuperheating, condensing, and subcooling zones will vary 5 to 10 percent, depending on the temperature of the entering gas and the exiting liquid; however, Fig. 1.14 is typical for most of the commonly used refrigerants. Condensing takes place in approximately 85 percent of the condenser area at a substantially constant temperature. The drop in condensing temperature indicated in Fig. 1.14 is due to the friction loss through the condenser coil.

A condenser must be able to reject the compressor suction capacity plus a large part of the work input to the compressor. The following equation should be used to determine the total heat of rejection (THR)

$$\text{Condenser capacity (MBH)}^* = \text{compressor capacity (tons)} \times 12 + \text{absorbed power (bhp)} \times 2.5$$

The compressor capacity and the absorbed power should be based on the compressor manufacturer's data at the highest suction pressure anticipated. With the condenser capacity requirements determined, rating tables for a particular type of condenser may be used to select a condenser of the appropriate size.

In many cases, condenser ratings are based on the following data:

Fouling factor, 0.0005 h · ft^2 · °F/Btu (8.8×10^{-5} m^2 · K/W).

Condensing temperature, 105°F (40.6°C).

Performance with city water assumes 75°F (23.9°C) inlet water and 95°F (35°C) outlet water.

Performance with tower water assumes 85°F (29.4°C) inlet water and 95°F (35°C) outlet water.

Water pressure drop does not include any external fittings or valves.

For conditions other than those given, the capacity can be estimated as follows. At a given flow (gal/min) a 10 percent increase in initial temperature difference will give approximately an 8 percent increase in capacity. At a given flow

* 1 MBH = 1000 Btu/h = 0.293 kW; 1 bhp = 2545 Btu/h = 0.7457 kW.

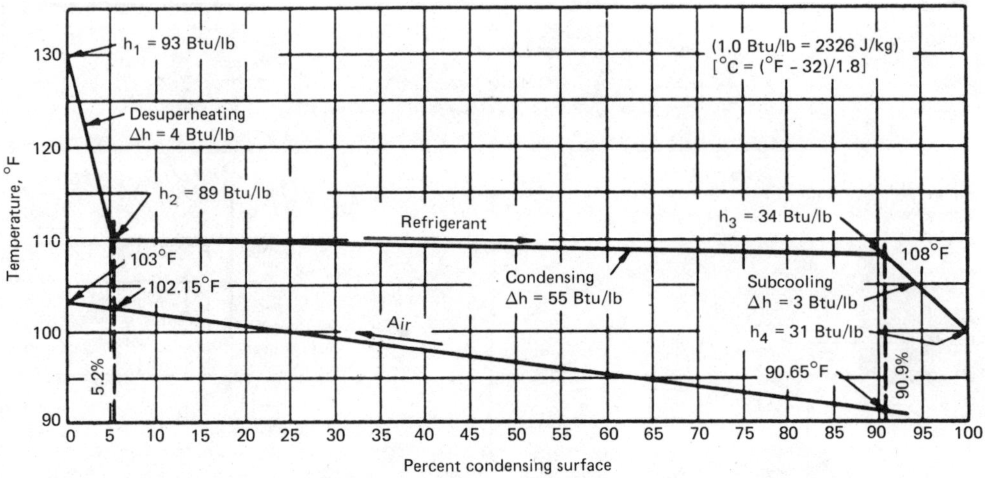

Fig. 1.14 Typical temperature and enthalpy changes in an air-cooled R-12 condenser. (*Reprinted by permission of Trane Company, La Crosse, WI.*)

(gal/min) a 10 percent decrease in initial temperature difference will give approximately a 9 percent decrease in capacity. Initial temperature difference equals condensing temperature minus inlet water temperature.

Condenser Control At part load operation of the refrigeration system or when the cooling medium for the condenser changes in temperature, control of the condenser cooling rate is necessary. If this cooling rate is not controlled, the condensing refrigerant temperature can become too high and make operation inefficient; or it can become too low and cause operational problems in the refrigeration cycle.

If the condenser is air-cooled, usually the flow rate of air to the condenser is varied or cooling air is recirculated. Air flow rate may be achieved by fan motor cycling, fan motor speed control, or throttling with an air flow damper. Partially flooding the condenser with liquid refrigerant can also be used to reduce the effective condensing surface but requires special condenser circuiting or a separate vessel as a receiver.

When the condenser is the evaporative type, control is generally accomplished by varying the air flow rate, partially recirculating the cooling air, or varying the amount of spray water onto the condenser surface. Specific methods used for condensing pressure control with evaporative condensers are (1) cycling the spray pump motor, (2) cycling both fan and spray motors, (3) throttling the spray water, (4) bypassing air, (5) throttling air via dampers, and (6) combinations of these methods.

With water-cooled condensers, head pressure controls are used for maintaining the condensing pressure and conserving water. For cooling tower applications, head pressure controls are used only where it is necessary to maintain condensing temperatures above a minimum. For water-cooled condensers, two types of controls regulate condenser water flow—a water-regulating valve or a control circuit using a water pump. In cooling tower applications, a simple bypass can also be used to maintain the condensing temperature, using a manual or an automatic valve which responds to head pressure change.

Expansion (Metering) Device

Performance Requirements Control of refrigerant flow is a necessary function in any refrigeration system. The expansion device is a refrigerant flow restriction at which neither work nor any significant amount of heat transfer occurs. Therefore, there is no change in refrigerant enthalpy.

The expansion valve controls the flow rate of refrigerant liquid between condenser and evaporator in order to match the evaporation rate. The expansion device must allow the highest flow rate required at the lowest pressure difference available and must also operate with the available pressure difference at the lowest refrigerant rate used. Pressure difference available is equal to the condenser liquid pressure at the outlet, minus the pressure at the evaporator inlet, plus or minus the static head of the refrigerant liquid column, minus other possible pressure losses that result from flow resistance or a pressure difference required to open pilot-operated valves. Pressure loss or gain values, except for liquid column static and valve opening pressures, vary with the refrigerant flow rate. Losses directly related to the flow rate include the liquid line losses due to solenoid valves, manual valves, filters, and the refrigerant distributor.

All of the losses are normally determined based on liquid entering the expansion valve. This requires that the temperature of the liquid in the line be no higher than that corresponding to saturation at the local pressure. This sometimes requires subcooling of this liquid. Liquid receivers should be avoided since they result in little or no subcooling. It is important that the system be designed so that none of the refrigerant entering the expansion valve is in vapor form. A separate subcooler may be desirable, especially if the liquid has passed through a receiver.

Capillary Tubes A capillary tube, as shown in Fig. 1.15, is a small-bore tube that provides a restriction between the outlet of the condenser and the inlet of the evaporator, reducing the pressure. The capillary tube produces a refrigerant pressure drop proportional to the square of the fluid

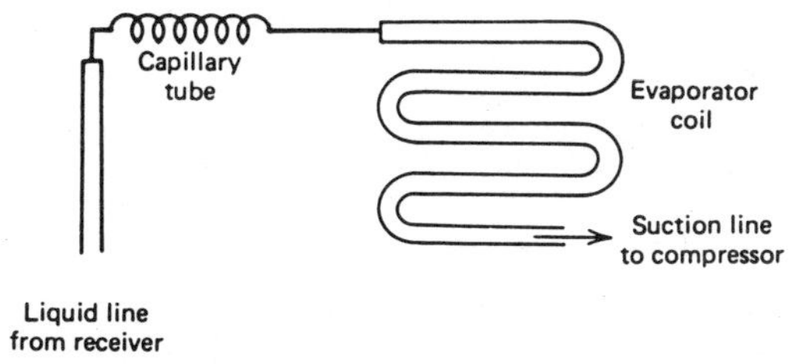

Fig. 1.15 Capillary tube. (*Reprinted by permission of John Wiley & Sons, Inc., from H. J. Sauer, Jr., and R. H. Howell, Heat Pump Systems. Copyright © 1983, John Wiley & Sons, Inc.*)

velocity, which increases as the specific volume of the refrigerant increases. It is sometimes soldered to the outer surface of the suction line for heat exchange (subcooling) purposes.

The initial selection of a capillary tube for a given compressor rating may be determined by using Fig. 1.16. The curves in Fig. 1.16 apply to typical operating conditions and no distinction is made between units which do and do not have capillary tube heat exchangers.

The flow characteristics of a capillary tube can be adjusted independently by varying either the bore or the length. Therefore, it is feasible to select first the most convenient length and then (within certain limits) select a bore to give the required flow. An alternative procedure is to select first a standard bore and then adjust the length as necessary.

More precise techniques for the sizing or selection of capillary tubes are given in Ref. 3. Precision capillary tubing is available in bulk as well as in complete assemblies.

Thermostatic Expansion Valve The thermostatic expansion valve has a moving plug which controls the area available for flow. This valve has three major functions:

1. *Throttling action:* The thermostatic expansion valve separates the high-pressure and low-pressure sides of the system. This pressure difference, between the condenser and the evaporator, permits throttling the liquid from the condenser down to the evaporator pressure.

2. *Modulating action:* The valve also feeds liquid refrigerant into the evaporator at the proper rate at all times. If too much liquid refrigerant enters the evaporator, not all of it will change to vapor and there is danger of passing liquid into the compressor, possibly damaging the compressor valves. If there is not enough refrigerant going to the evaporator, there will not be enough liquid to absorb the heat at the evaporator, reducing the system capacity. By modulating the flow, the valve will maintain the proper amount of liquid in the evaporator.

3. *Controlling action:* The valve must also respond to load changes at the evaporator. As the load of the evaporator increases, the valve moves toward the wide-open position, reducing the pressure drop across the valve port and allowing more refrigerant to flow. This makes more refrigerant available for the compressor to pump. As the load decreases, the valve closes, reducing the amount of refrigerant available to the compressor.

The basic operation of the valve is affected by three forces, as illustrated in Fig. 1.17.

1. Bulb pressure (temperature at the exit of the evaporator) acting on one side of the diaphragm, which tends to open the valve (P_1)

2. Evaporator or suction pressure action on the opposite side of the diaphragm, which tends to close the valve (P_2)

3. Spring pressure on the evaporator side of the diaphragm, which also tends to close the valve (P_3)

When the pressures above and below the diaphragm are equal, the valve is said to be in equilibrium, that is, it will pass the correct amount of refrigerant required by the conditions of the system. The remote bulb senses the leaving temperature of the refrigerant gas exiting the evaporator.

The pressure exerted by the sensing bulb is offset by the adjustable spring at a pressure that assures that the leaving vapor is superheated and contains no liquid. In residential air-conditioning systems, the spring is set to require 6 to 15°F (3.33 to 8.33°C) of superheat, with an average of about 10°F (5.55°C). As the system load changes, the suction temperature and the pressures of the inlet and the outlet of the evaporator will change, but the difference in temperature or superheat should remain the same.

If the load on the evaporator increases, the refrigerant passing through the evaporator will pick up more heat, becoming superheated, and will be at a higher temperature than the refrigerant at the inlet of the evaporator. The expansion valve bulb at the exit of the evaporator will sense this and exert additional pressure on the diaphragm, opening the valve and allowing more refrigerant to flow. Liquid refrigerant at the entrance to the evaporator will evaporate at a faster rate, raising the evaporator pressure and bringing the forces again in balance at the new setting. If there is a very light load on the evaporator, the liquid admitted to the evaporator will not vaporize completely until it almost reaches the outlet of the evaporator. Therefore, its temperature and pressure will be very close to the temperature and pressure at which it entered. This reduction in temperature will be sensed by the bulb, which will exert less pressure on the diaphragm. The valve will tend to close, reducing the flow of liquid.

In all cases, the valve adjusts to changes in load by changing the refrigerant flow, which then matches the pumping rate of the compressor so that the system is always in balance and maintains its basic capacity. Generally the thermostatic expansion valve will maintain a fully active evaporator under all load conditions. The evaporating temperature will vary with the load condition.

If there are no unusual pressure drops in the evaporator [normal design pressure drop is 2 lb/in^2 (14 kPa)], the inlet gas pressure is reasonably close to the exit gas pressure. In this case, the valve will operate properly sensing pressure at the evaporator inlet, point P_2 in Fig. 1.17.

If the pressure drop across the evaporator exceeds 2 lb/in^2, an excessive amount of inlet gas pressure offsets the pressure generated at the remote sensing bulb. For example, a 6-lb/in^2 (41.4-kPa) pressure drop may require an additional 6°F (3.33°C) of superheat to offset the pressure on the diaphragm. This can result in an evaporator starved of liquid. To correct this problem, an external equalizer is tapped into the suction line at a point downstream from the sensing bulb and connected to the underside of the valve diaphragm, as shown in Fig. 1.18. In this way, the effective pressure at the bulb is fed to the valve, and superheat is maintained at a normal level. As a result, the evaporator receives the required amount of refrigerant regardless of the pressure drop through the evaporator.

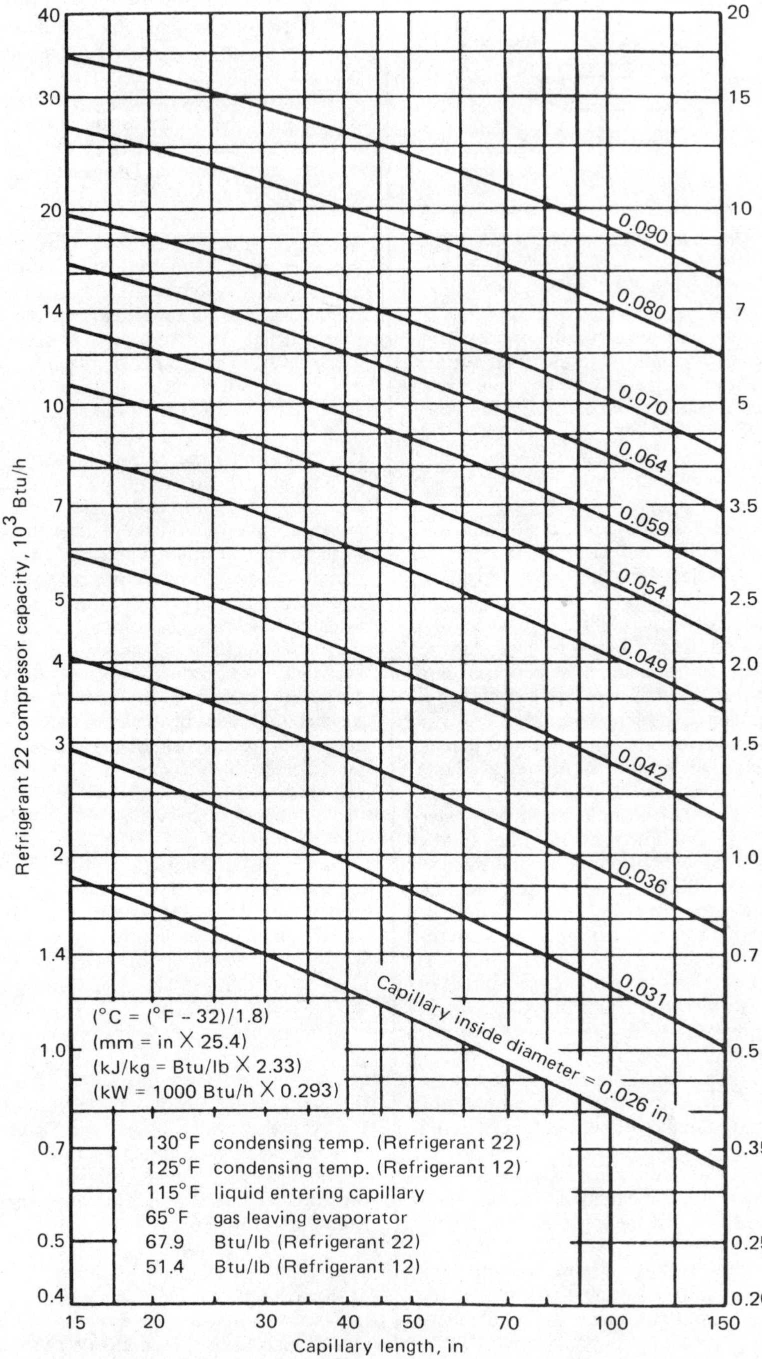

Fig. 1.16 Capillary tube selection chart. (*Reprinted by permission from ASHRAE Handbook 1983 Equipment.*)

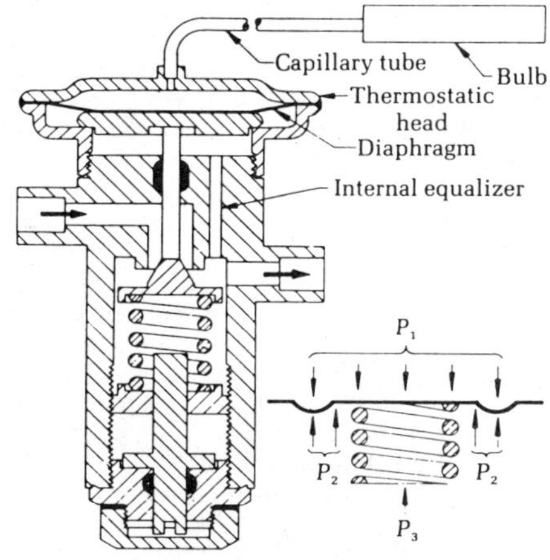

P_1—Thermostatic element's vapor pressure
P_2—Evaporator pressure
P_3—Pressure equivalent of the superheat spring force

Fig. 1.17 Thermostatic expansion valve. (*Reprinted by permission from ASHRAE Handbook 1983 Equipment.*)

Some expansion valves will have a series of tubes at the exit so that they may feed more than one row of an evaporator. These are called distributors and are designed to feed the several rows of a multicircuited evaporator simultaneously.

Thermostatic expansion valve capacity ratings for refrigerants R-12, R-22, R-500, and R-502 are normally based on subcooled, vapor-free, 100°F (38°C) liquid refrigerant entering the expansion valve. They are also rated with a maximum superheat change of 7°F (3.89°C). Ammonia (refrigerant-717) ratings are based on subcooled, vapor-free, 86°F (30°C) liquid refrigerant entering the expansion valve and a maximum superheat change of 7°F (3.89°C).

Steps to be taken in selecting the correct thermostatic expansion valve include the following.

1. Determine the pressure drop across the valve.
   a. Subtract the evaporating pressure from the condensing pressure.
   b. Then subtract all other pressure losses to obtain the net pressure drop across the valve.
2. Select a valve from the capacity tables.
   a. Consider the design evaporating temperature.
   b. Consider the available pressure drop (from 1 above).
   c. The valve capacity should equal or slightly exceed the tonnage rating of the system.
3. Consider the liquid temperature of the refrigerant entering the valve. Apply liquid temperature correction factors if necessary.
4. Find the nominal capacity of the valve from the capacity tables.
5. Decide whether an external equalizer is to be used.

Constant-Pressure Expansion Valve The constant-pressure expansion valve operates using the evaporator or valve-discharge pressure. It controls the mass flow rate of liquid refrigerant entering the evaporator and maintains a constant evaporator pressure. Figure 1.19 is a schematic of a constant-pressure expansion valve which operates in a manner similar to the thermostatic expansion valve, except that it controls flow based on the evaporator pressure. This valve controls the evaporation of the liquid refrigerant in the evaporator at a constant pressure and temperature.

Constant-pressure expansion valves automatically vary the flow of liquid refrigerant to the evaporator and balance the compressor pumping capacity. The low-side pressure can be adjusted by turning the pressure adjustment screw.

The constant-pressure expansion valve should be selected based on the required system capacity and for the lowest expected pressure drop across the valve. It should have an adjustable pressure range to provide the required evaporator pressure. Normally the constant-pressure expansion valve is suitable only for constant-load applications. If used on a variable-load application, this valve will tend to starve the evaporator at the high load and overfeed it at low loads,

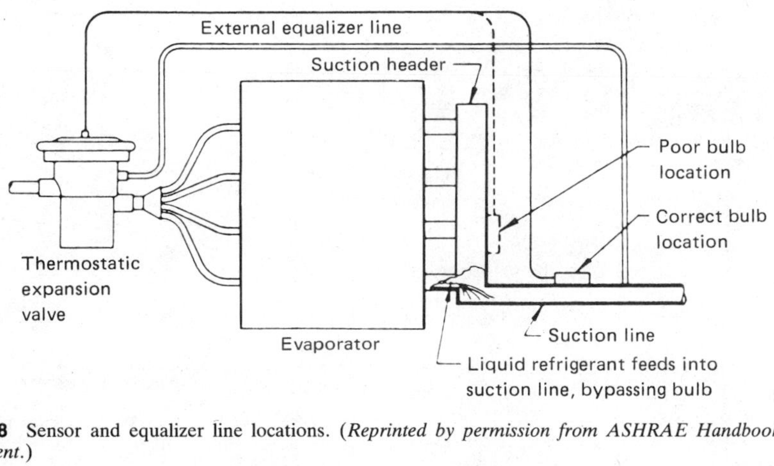

Fig. 1.18 Sensor and equalizer line locations. (*Reprinted by permission from ASHRAE Handbook 1983 Equipment.*)

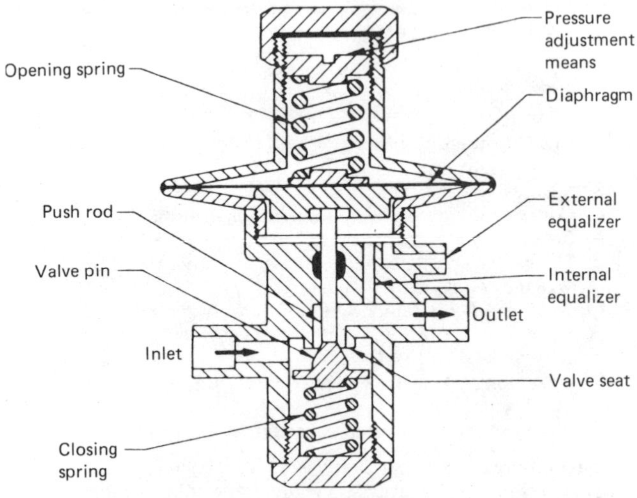

Fig. 1.19 Constant-pressure expansion valve. Valve is used with either internal or external equalizer, but not both. (*Reprinted by permission from ASHRAE Handbook 1983 Equipment.*)

with possible compressor damage resulting. Suitable applications for constant-pressure expansion valves are drink and food dispensers, water coolers, ice cream freezers, and self-contained room air-conditioning units.

High-Side Float Valve A high-side float valve controls the flow rate of refrigerant liquid entering the evaporator such that it equals the rate at which the refrigerant gas is pumped from the evaporator by the compressor. The valve keeps a constant level of liquid in the housing, which is connected to the high-pressure side of the system.

A refrigeration system in which a high-side float valve is used consists ordinarily of a single evaporator, compressor, and condenser. One of the disadvantages of the high-side float valve system is that the amount of system refrigerant charge is critical. The use of an excessive amount of system charge will cause floodback, while an insufficient amount of system charge will cause a reduction in system capacity.

Low-Side Float Valve The low-side float valve carries out the same functions as the high-side float valve, but the float is on the low-pressure side of the system.

For the low-side float valve system the quantity of refrigerant in the system is not critical. The low-side float valve can be used in multiple evaporator systems in which some of the evaporators may be controlled by other low-side float valves and some by thermostatic expansion valves.

1.3 SYSTEM PERFORMANCE

One of the principal design aspects of a refrigeration system, whether factory-built or field-assembled, is the selection of the components, including compressor, condenser, evaporator, refrigerant flow control device, motor, and controls. The characteristics of each of the components have an interrelationship to the other components, and the system formed by combining each component must perform properly at both design conditions and all other conditions that may be expected during operation.

The first step in the design process is to become familiar with the effects of changing the operating conditions on the performance of each component. The next step is to study the performance of the entire system. To make this type of evaluation, the individual performance characteristics of the compressor, condenser, expansion device, and evaporator must be combined. This section outlines a technique for analyzing the overall performance of the complete system.

The operating conditions of a system are determined by solving sets of simultaneous equations which represent the performance of all components. The mathematical representation of the performance of a component such as the compressor, for example, can be complex. Thus, it is simpler to solve the simultaneous equations by graphic means. The expected performance of each component can be obtained from catalog data and plotted as a graph either directly or after suitable calculations.

In the design of a refrigeration system, one of the most important considerations is that of establishing the proper relationship or "balance" between the vaporizing and the condensing sections of the system. It is important to recognize that whenever an evaporator and a condensing unit are connected together in a common system, a condition of balance is established between the two such that the rate of vaporization is equal to the rate of condensation. Since all the components in a refrigeration system are connected together in series, the time-averaged refrigerant flow rate through all components is the same during steady-state operation.

The flow characteristics of the expansion valve have an important effect on the operation of the entire system. The expansion device is expected to deliver the proper quantity of refrigerant to balance the compressor flow rate at all suction and condensing conditions. Proper selection of an expansion device is a first step.

The next step in analyzing the performance of the complete vapor-compression system is to study the behavior of the compressor and the condenser operating together. These two components combine to form the condensing unit. The performance of a condensing unit can be found by superimposing the condenser performance characteristic on compressor capacity curves such as those given in Fig. 1.10. The result is shown in Fig. 1.20.

At any operating point of the condensing unit, identical

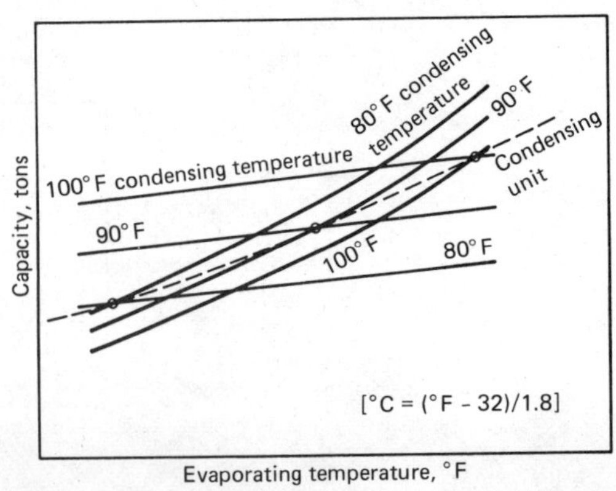

Fig. 1.20 Condensing unit performance as a function of evaporating temperature. (*Reprinted by permission of John Wiley & Sons, Inc., from H. J. Sauer, Jr., and R. H. Howell, Heat Pump Systems. Copyright © 1983, John Wiley & Sons, Inc.*)

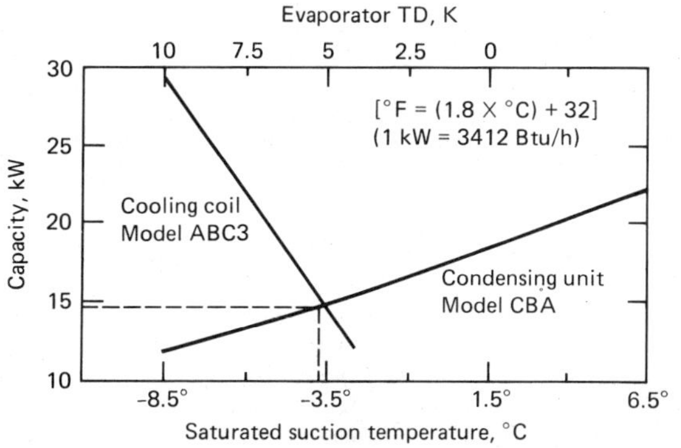

Fig. 1.21 Graphic analysis of system balance. (*Reprinted by permission of John Wiley & Sons, Inc., from H. J. Sauer, Jr., and R. H. Howell, Heat Pump Systems. Copyright © 1983, John Wiley & Sons, Inc.*)

values of the condensing temperature, evaporating temperature, and number of tons of cooling capacity must exist for both the compressor and the condenser. Operating points occur at the intersections of the common condensing temperatures. The line connecting the operating points represents the performance of the condensing unit. Note that as the evaporator temperature is reduced, the condensing temperature is also reduced.

Once the combined performance of the compressor and the condenser has been found, attention is shifted to the evaporator performance. For any particular evaporator and condensing unit, the point of system balance can be determined graphically by plotting the evaporator capacity and the condensing unit capacity on the same graph. Using data taken from the manufacturer's rating tables, the condensing unit capacity is plotted as a function of suction temperature, whereas the evaporator capacity is plotted as a function of the evaporator temperature differential TD. An example of this graphic analysis is shown in Fig. 1.21.

A more complete set of performance data for a particular combination of evaporator and condensing unit is shown in Fig. 1.22. In this case the evaporator is an air-cooling coil.

To demonstrate the analysis of a system, assume that an air-cooling coil has been selected in order to satisfy an air-conditioning load of 250,000 Btu/h (73.3 kW) when air at 65°F$_{wb}$ (18.3°C) enters and the evaporating temperature is about 48°F (8.9°C). Coil capacity lines for various entering wet-bulb temperatures are also given in Fig. 1.22. These lines have been drawn about 3 lb/in^2 (20 kPa) to the left of the evaporating temperature to allow for the pressure loss between the evaporator outlet and the compressor inlet. It is necessary that this 250,000 Btu/h (73.3 kW) be removed by the condensing unit when air is entering at 95°F (35°C). The condensing-unit capacity curves in Fig. 1.22 are for a unit that meets this requirement. Note that the 95°F condensing-unit line intersects the 65°C (18.3°C) coil line at about 260,000 Btu/h (76.2 kW) and just under 46°F (7.8°C) at the 3 lb/in^2 (20 kPa) higher evaporator pressure. At design conditions, the performance is approximately what is needed.

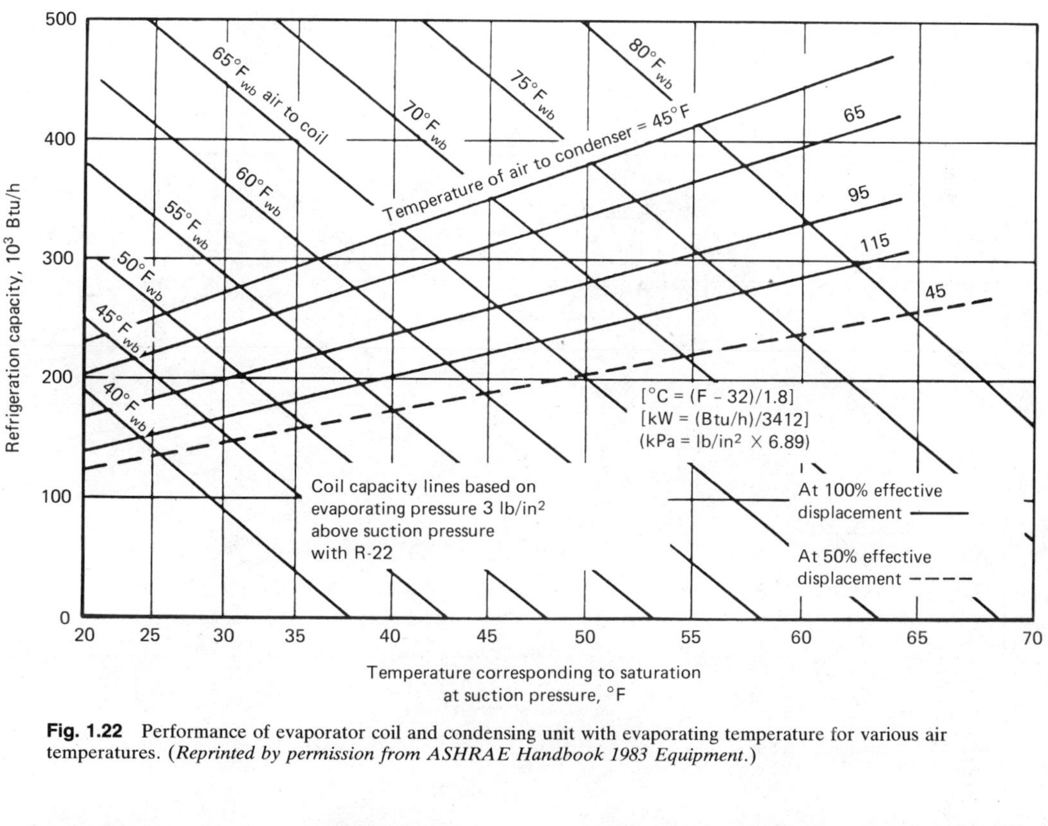

Fig. 1.22 Performance of evaporator coil and condensing unit with evaporating temperature for various air temperatures. (*Reprinted by permission from ASHRAE Handbook 1983 Equipment.*)

When the condenser and the compressor are to be selected separately or the performance is not available for all of the operating conditions expected or to be considered, separate compressor and condenser performance curves need to be plotted. Figure 1.23 is an example of a set of such curves. Compressor performance requires two curves for each suction saturation temperature: the solid line indicating the heat from the evaporator and the broken line indicating the heat to the condenser. The difference between the two represents the energy put into the refrigerant by the compressor. The larger the difference in temperatures, the greater will be the heat energy added by the compressor.

Since both the heat picked up at the evaporator and the energy input to the compressor must be rejected by the condenser, the intersections of the broken lines with the condenser performance lines are the only intersections of any significance. If the suction temperature is 40°F (4.4°C) and the temperature of air to the condenser is 105°F (40.6°C), the intersection of the broken line for 40°F (4.4°C) with the condenser line for 105°F (40.6°C), as shown in Fig. 1.23, is at about 129°F (53.9°C) condensing temperature and 290,000 Btu/h (85.0 kW) heat rejected. Moving vertically down to the solid line for 40°F (4.4°C) the heat removed from the evaporator at this condition is about 213,000 Btu/h (62.4 kW).

The power required by the compressor is an important consideration. Using the values of condensing and evaporating temperatures found from methods outlined above, the power required can be found directly from pertinent equipment data.

1.4 REFRIGERANT PIPING

The performance of a refrigeration system depends on good piping selection. Hot-gas lines carry the hot refrigerant gas from the compressor to the condenser, while suction lines carry cool refrigerant gas from the evaporator to the compressor. The liquid line carries liquid refrigerant from the condenser to the evaporator. Proper sizing of these lines considers the pressure drop and velocity for the hot-gas and suction lines and the pressure drop only for liquid lines.

The suction and hot-gas lines must carry gas, which has a greater volume per unit mass than liquid. Thus, they must have larger diameters than the liquid line. The suction line must be appropriately insulated because it will be colder than the ambient air and it will have a tendency to collect condensate from the air on its surface and also gain heat from the ambient air.

Another pipe will connect the compressor discharge valve to the inlet side of the condenser. Since the compressor and the condenser are often mounted in the same cabinet (condensing unit), this will be a relatively short line, called the hot-gas or discharge line. It will be smaller than the suction line because the gas is compressed at a higher pressure into a smaller volume. It will be hotter than the other lines because of compression.

The liquid line will connect the outlet of the condenser, which is usually at its lowest point, to the inlet side of the expansion device. If the liquid line is exposed to high ambient temperatures, it should also be insulated.

Horizontal suction lines for reciprocating compressors

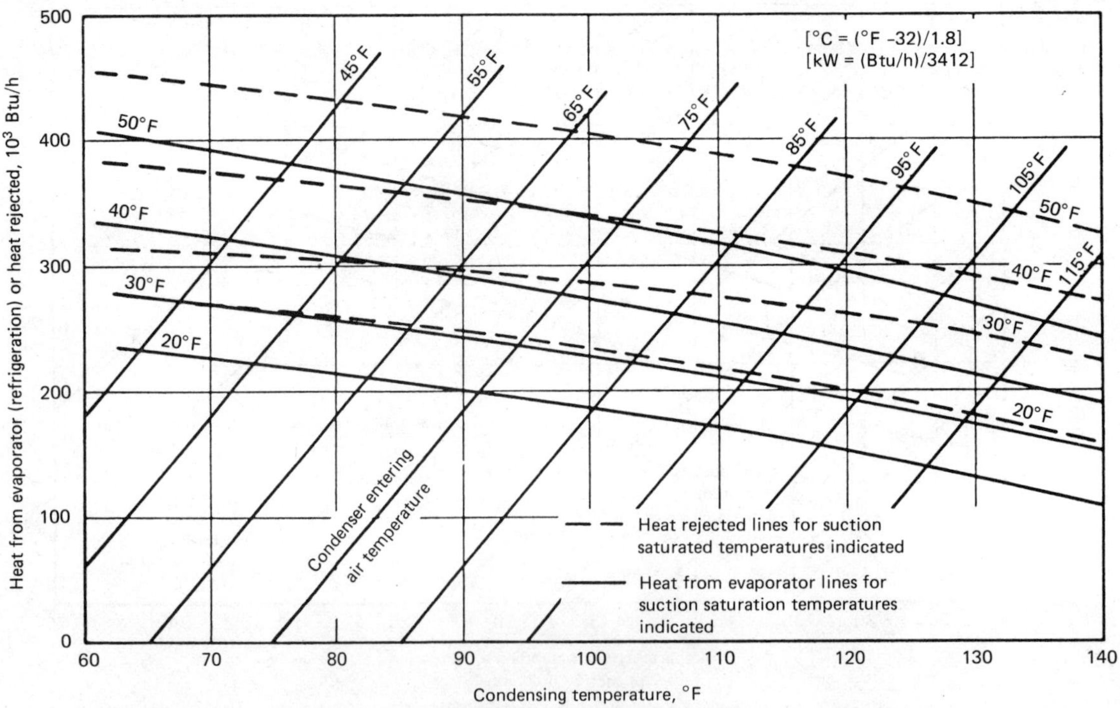

Fig. 1.23 Performance of air-cooled condensing unit with condensing temperature. (*Reprinted by permission from ASHRAE Handbook 1983 Equipment.*)

should be pitched toward the compressor [approximately ⅛ in (3.2 mm) for each 10 ft (3 m) of run] to assure good oil return to the compressor. Since a certain amount of oil will always circulate within the system, deep bends or traps are to be avoided in order for the oil to return to the compressor.

If the evaporator is 10 ft (3 m) or more below the compressor, then an oil trap as shown in Fig. 1.24 must be provided in the suction line near the evaporator. An oil trap is

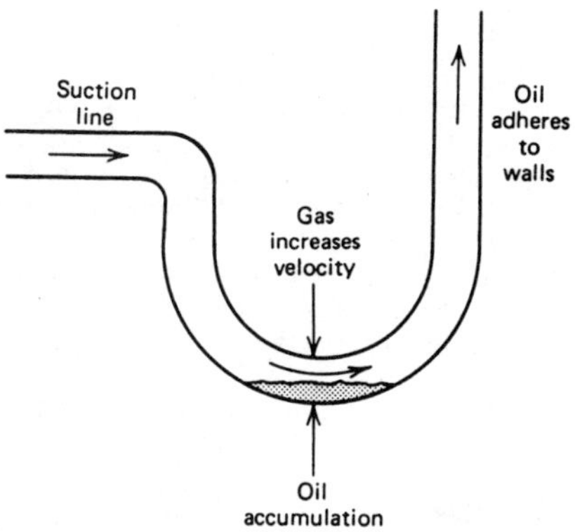

Fig. 1.24 Oil trap for oil return to compressor. (*Reprinted by permission of John Wiley & Sons, Inc., from H. J. Sauer, Jr., and R. H. Howell, Heat Pump Systems. Copyright © 1983, John Wiley & Sons, Inc.*)

needed because the oil returning to the compressor has only the velocity of the return gas to carry it upward against the force of gravity. As it rises, some of it tends to adhere to the walls of the pipe and run back down, collecting at the lowest point. With a trap, the effective tube area is reduced, increasing the velocity of the return gas at this point. The refrigerant gas will reach sufficient velocity to entrain additional oil out of the trap and carry it back to the compressor, eliminating the possibility of oil collecting in the evaporator.

Pressure Drop Considerations

The available tables for line sizing[3] are based on tons of refrigeration capacity for a line and provide the friction drop per 100 ft (30.48 m) of equivalent pipe length corresponding to 1°F (0.56°C) change in saturation temperature. Pressure drops are given in degrees because this method is convenient and is generally accepted throughout the industry.

Liquid lines are normally designed so that the pressure drop is no larger than that corresponding to about a 1 or 2°F (0.56 or 1.1°C) change in saturation temperature. Examples of some liquid pressure losses for a change of 1°F (0.56°C) saturation change at 100°F (37.8°C) condensing pressure are approximately as follows.

Refrigerant	lb/in²	kPa
R-12	1.8	12.4
R-22	2.9	20.0
R-50	23.1	21.4

The pressure drops that occur in the liquid line are caused by accessories such as solenoid valves, strainer driers, and hand valves, as well as by the actual pipe and fittings.

The pressure drop occurring in the suction line reduces the system capacity because it requires the compressor to operate at a lower suction pressure in order to maintain the desired evaporating temperature. Suction lines are usually sized so that friction causes a pressure drop no greater than the equivalent of about a 2°F (1.1°C) change in saturation temperature.

The equivalent pressure loss at 40°F (4.4°C) saturated suction temperature is given approximately by the following data.

	Suction loss		Pressure loss	
Refrigerant	°F	°C	lb/in²	kPa
R-22	2	1.1	1.81	12.47
R-22	2	1.1	2.94	20.26
R-502	2	1.1	3.14	21.63

At suction temperatures lower than 40°F (4.4°C) the pressure drop equivalent for a given degree change decreases. This means that low-temperature lines should be sized for a very low pressure drop.

It is important to minimize the pressure loss in hot-gas lines because such losses increase the required compressor horsepower per ton of refrigeration and decrease the compressor capacity. It is usual practice not to exceed a pressure drop corresponding to a 2°F (1.1°C) change in saturation temperature at that pressure.

Refrigerant Line Velocities

Practical design velocities in refrigerant lines are based on pressure drop, noise, and oil entrainment. The following table gives recommendations for various refrigerants at design load.

	Practical gas line velocities for various refrigerants, ft/min	
Refrigerant	Suction line	Discharge line
R-12, R-22, R-502	1200–4000	2000–3500
R-717 (ammonia)	500–5000	3000–6000

CONVERSION FACTOR: 1 ft/min = 0.3048 m/min.

The higher gas velocities are sometimes encountered in relatively short suction lines on comfort air-conditioning or other applications where the operating time is only 2000 to 4000 h/yr and low initial cost of the system may be more significant than low operating cost. Industrial or commercial refrigeration applications, where equipment may run almost continuously, should be designed with refrigerant velocities at the lower end of the given velocity range.

Materials

Depending on system sizes and methods of installation, steel or copper lines can be used. However, it is very important that the material used be compatible with the refrigerant in the system.

Copper and copper-bearing materials are normally attacked by ammonia when moisture is present and are not used in ammonia piping systems. Iron and steel piping, fittings, and valves are suitable for ammonia gas and liquid. Systems with refrigerants R-12, R-22, and R-502 normally use copper.

1.5 REFRIGERATION ACCESSORIES

Liquid-Suction Heat Exchanger

Very often liquid-suction heat exchangers are used to subcool the liquid refrigerant and superheat the suction gas. Several types of liquid-suction heat exchangers are available. The simplest form of a heat exchanger is obtained by soldering the suction and liquid lines together in a counterflow pattern, and then insulating the lines as a unit. Shell-and-coil or shell-and-tube heat exchangers are also used.

Filter Drier

All refrigeration units should have a filter drier installed to clean out any dirt or foreign material and remove water that might clog the expansion device. There are several different types of filter driers, but most combine an activated alumina desiccant (drying agent) with a molecular sieve. Also, a metal screen will be used at the entrance to trap solids. Filters are mounted in the liquid line, usually near the condensing unit. Whenever the compressor must be changed because of a burnout, the liquid line drier should also be changed. It is also good practice to change the drier on an older system that has been opened or is suspected of containing moisture. Replaceable element filter driers are preferred for large systems, since the drying element can be replaced without breaking any refrigerant connections.

Moisture indicators can be installed in refrigerant liquid lines in order to provide an indication of when the drier cartridge should be changed.

Strainer

Strainers are used in liquid and suction lines in order to protect automatic valves and the compressor from any foreign material such as pipe and welding scale, rust, and metal chips.

Filter driers perform the dual function of refrigerant drying and filtering out particles far smaller than those trapped by mesh strainers. No other strainer is needed in the liquid line if a good filter drier is used.

Liquid-Moisture Indicator

All refrigeration systems should have a means to check for proper refrigerant charge. The common devices which are used are a liquid line sight glass, a liquid level test cock on the condenser or receiver, and an external gage glass with equalizing connections and shutoff valves. A properly installed sight glass will show bubbling when there is insufficient charge. The preferred location for liquid indicators is in the liquid line as close as possible to the receiver outlet, or the condenser outlet if no receiver is used.

When sight glasses are installed near the evaporator, often a complete liquid condition at the sight glass cannot be obtained because of pressure drop in the liquid line. Subcooling will eliminate the vapor formation.

Sight glasses with double ports and seal caps, which provide added protection against leakage, are preferred. Moisture-liquid indicators large enough to be installed directly in the liquid line serve the dual purpose of liquid line sight glass and moisture indicator. Moisture indicators use chemicals which change color in the presence of water.

Receiver/Accumulator

Nearly all large systems have a refrigerant receiver which is used to store temporary excess quantities of refrigerant when parts of the system are pumped down for repairs, evaporators are defrosted, or temperature changes cause an accumulation of refrigerant. The receiver should be large enough to hold the entire system charge.

A surge drum is used on the suction side of almost all flooded evaporators to prevent liquid flow to the compressor. The exceptions are shell-and-tube coolers and similar shell-type evaporators, which provide ample surge space above the liquid level or contain eliminators to separate gas and liquid.

Safety Relief Devices

It is essential for pressure vessels used in refrigerant systems to be protected from rupture due to pressures in excess of the vessel design pressures. The Safety Code for Mechanical Refrigeration (ANSI/ASHRAE 15-1978) outlines requirements for relief valves used on pressure vessels for refrigeration. Also, Section VIII, Division I, of the ASME Boiler and Pressure Vessel Code requires various relief devices to prevent the operating or standby pressure from exceeding design pressures.

Relief valves are most commonly pop-type designs, which abruptly open when the inlet pressure exceeds the outlet pressure by the valve setting pressure, as illustrated in Fig. 1.25. Seat design is such that once lift begins, the resulting increased seat area causes the valve to pop wide open against the force of the setting spring.

Two devices performing similar safety relief operations are the fusible plug and the rupture member, which are also illustrated in Fig. 1.25. The plug contains a fusible member that melts at a specific temperature, which corresponds to the safe pressure of the refrigerant. This device is limited in application to pressure vessels with internal gross volumes of 3 ft^3 (0.085 m^3) or less and internal diameters of 6 in (152.4 mm) or less. The rupture member device contains a disk designed to rupture at a predetermined pressure.

Evaporator Pressure Regulator

The evaporator pressure regulator (back pressure regulator) controls the evaporator pressure. It is used in the evaporator outlet or suction line wherever low limit control of the evaporator pressure or temperature is required.

Evaporator pressure regulators are used on finned-coil evaporators where frosting needs to be prevented or evaporator pressure must be maintained constant and higher than the suction pressure in order to prevent dehumidification. These regulators are also used on water and brine chillers in order to prevent freeze-up under low-load conditions.

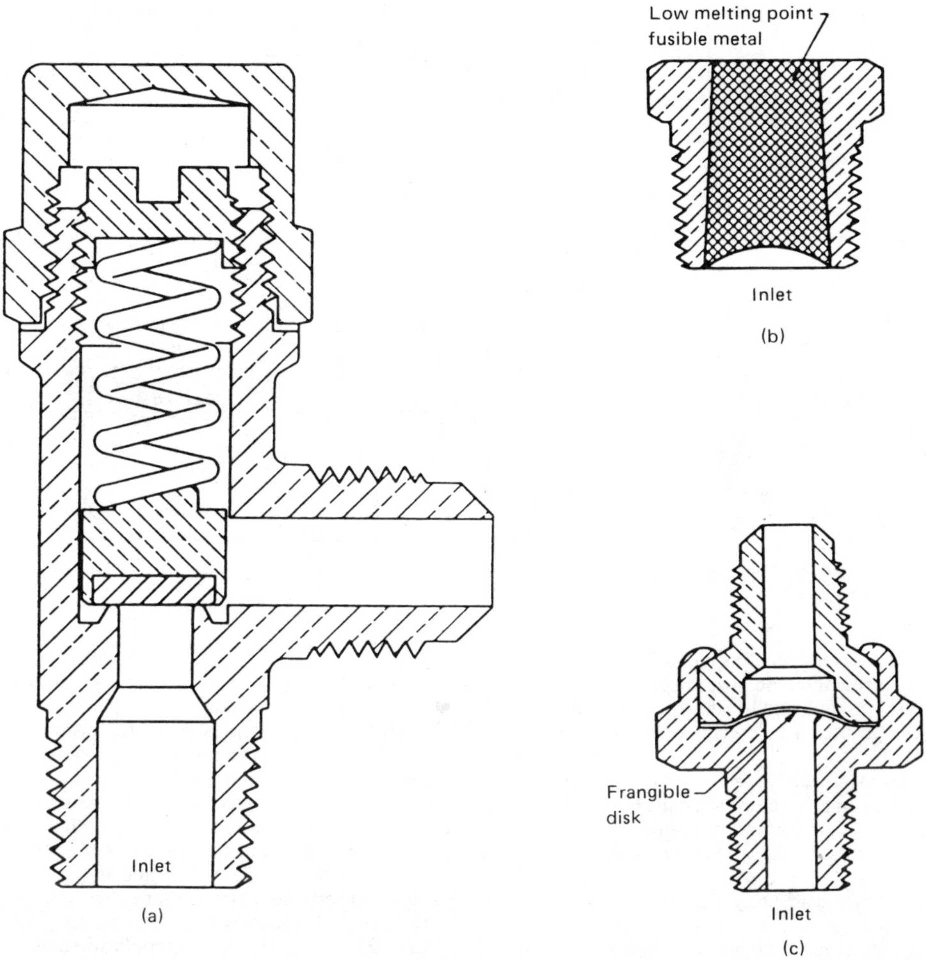

Fig. 1.25 Safety relief devices. (*a*) Pop-type safety relief valve; (*b*) fusible plug; (*c*) rupture disk. (*Reprinted by permission from ASHRAE Handbook 1983 Equipment.*)

Suction Pressure Regulator

The suction pressure regulator limits the compressor suction pressure, the regulator outlet pressure, to a maximum value. This type of regulator is used in the suction line at the compressor on any refrigeration installation in which the liquid expansion valve cannot limit the suction pressure and compressor motor overload would otherwise exist.

Direct-acting suction pressure regulating valves sense only the outlet or downstream pressure. The suction pressure regulator should be selected for the required capacity and system refrigerant, the required regulator inlet pressure, and the lowest practical pressure drop across the regulator in order to minimize any loss in system capacity.

Condenser Pressure Regulator

Condenser pressure regulating valves and condenser bypass valves are used for controlling cold weather operation of air-cooled condensers and are built and operate similarly to evaporator pressure regulators and suction pressure regulators. These valves respond to compressor discharge pressure or condenser inlet or outlet pressure, depending on the location of the valve in the condenser circuit.

1.6 MULTISTAGE SYSTEMS

There are refrigeration systems that have more than one stage of compression, and these are defined as multistage systems. In general, there are two types of multistage systems, compound and cascade.

At high compression ratios, the refrigerant gas and the lubricating oil can reach temperatures that are high enough to cause the oil and the refrigerant to break down. This can be a problem in reciprocating compressors, but it is generally not a problem with rotary screw compressors, where cool oil or liquid refrigerant is injected directly into the compressor to cool the gas stream. Another problem that occurs at high

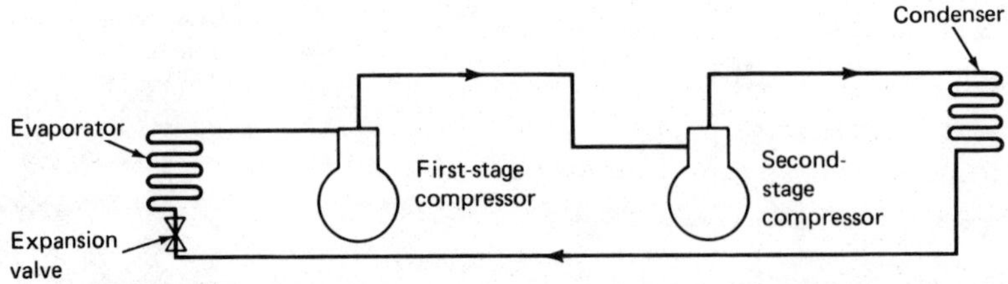

Fig. 1.26 Basic elements of two-stage compound system. (*Reprinted by permission from ASHRAE Handbook 1984 Systems.*)

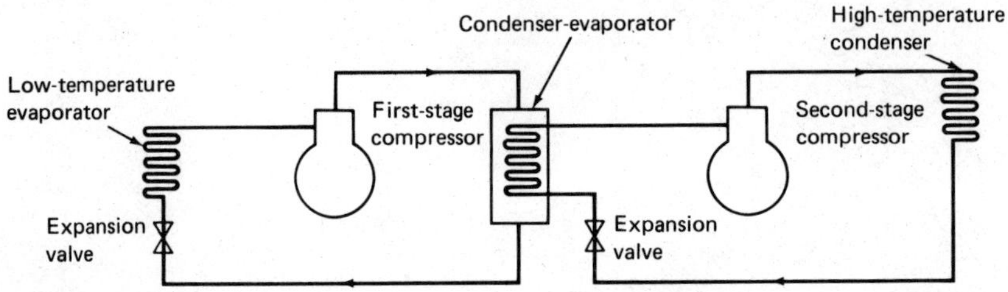

Fig. 1.27 Basic elements of a cascade system. (*Reprinted by permission from ASHRAE Handbook 1984 Systems.*)

pressure ratios across reciprocating compressors is the severe decrease in compressor volumetric efficiency. This occurs because reexpansion and small leakages become significant at a high discharge pressure, causing an increase in power demand, a loss in refrigerant flow rate, and consequently a loss in refrigeration capacity.

For these reasons, manufacturers usually limit the pressure ratio of multicylinder machines to approximately 9:1 and rotary screw compressors to approximately 19:1. When the overall system pressure ratio approaches or exceeds these limits, the pressure ratio affecting the individual compressors must be reduced.

Many refrigeration installations have refrigeration loads that require widely varying temperatures throughout the plant. Multistaging enables the selection of evaporating temperatures for these loads that correspond to the intermediate suction pressure that can be handled most efficiently when the high-stage compression equipment is at or near this equivalent suction pressure. In such systems, the upper stage or stages are used for the high- and intermediate-temperature loads, and the low-stage boosters are used for the lower-temperature loads.

A compound system is a multistage application in which compressors are interconnected in series using the same refrigerant. Figure 1.26 illustrates a simple outline of a two-stage compound system. In addition to the components shown, compound systems are likely to contain additional components such as desuperheaters and liquid coolers.

A cascade system is a multistage application in which two separate refrigerants are interconnected so that one provides the means of heat rejection for the other. Figure 1.27 shows a simple cascade system.

The cascade system is most often used where the pressure ratio of the condenser and the evaporator is greater than can be handled with commercial compressors or heat exchangers or where the service conditions make a cascade system more economical by possibly reducing both the displacement of the lower stage and the line sizes. Using a cascade system eliminates the problem of balancing the oil in the crankcases between separate stage compressors. Although cascade systems require an additional temperature difference between the condenser of the low-temperature cascade and the evaporator of the high-temperature cascade, these systems usually do not require desuperheaters and subcoolers.

Each refrigerant circuit in the cascade system may be either a single-stage or a compound system.

REFERENCES

1. *ASHRAE Handbook 1985 Fundamentals,* American Society of Heating, Refrigerating and Air-Conditioning Engineers, Atlanta, GA, 1985.
2. *ASHRAE Handbook and Product Directory, 1979, Equipment,* American Society of Heating, Refrigerating and Air-Conditioning Engineers, Atlanta, GA, 1979.
3. *ASHRAE Handbook 1983 Equipment.* American Society of Heating, Refrigerating and Air-Conditioning Engineers, Atlanta, GA, 1983.

Chapter 2

Thermoelectric Refrigeration

RAYMOND MARLOW
Marlow Industries, Inc.

2.1 INTRODUCTION

Thermoelectric refrigeration is a solid-state method of achieving a cooling or heating effect. The cooling phenomenon was discovered by Peltier in 1834. The application of this cooling effect remained minimal until the development of semiconductor materials. With the advent of semiconductor materials came the capability for a wide variety of practical thermoelectric refrigeration applications.

2.2 GENERAL THEORY OF OPERATION

Thermoelectric refrigeration is achieved when a direct current is passed through one or more pairs of *n*- and *p*-type semiconductor materials.[1] Figure 2.1 is a diagram of a single pair consisting of *n*- and *p*-type semiconductor materials. In the cooling mode, direct current passes from the *n*- to the *p*-type semiconductor material. The temperature T_C of the interconnecting conductor decreases and heat is absorbed from the environment. This heat absorption from the environment (cooling) occurs when electrons pass from a low energy level in the *p*-type material through the interconnecting conductor to a higher energy level in the *n*-type material. The absorbed heat is transferred through the semiconductor materials by electron transport to the other end of the junction T_H and liberated as the electrons return to a lower energy level in the *p*-type material. This phenomenon is called the Peltier effect.

A second phenomenon is also important in thermoelectric refrigeration. When a temperature differential is established

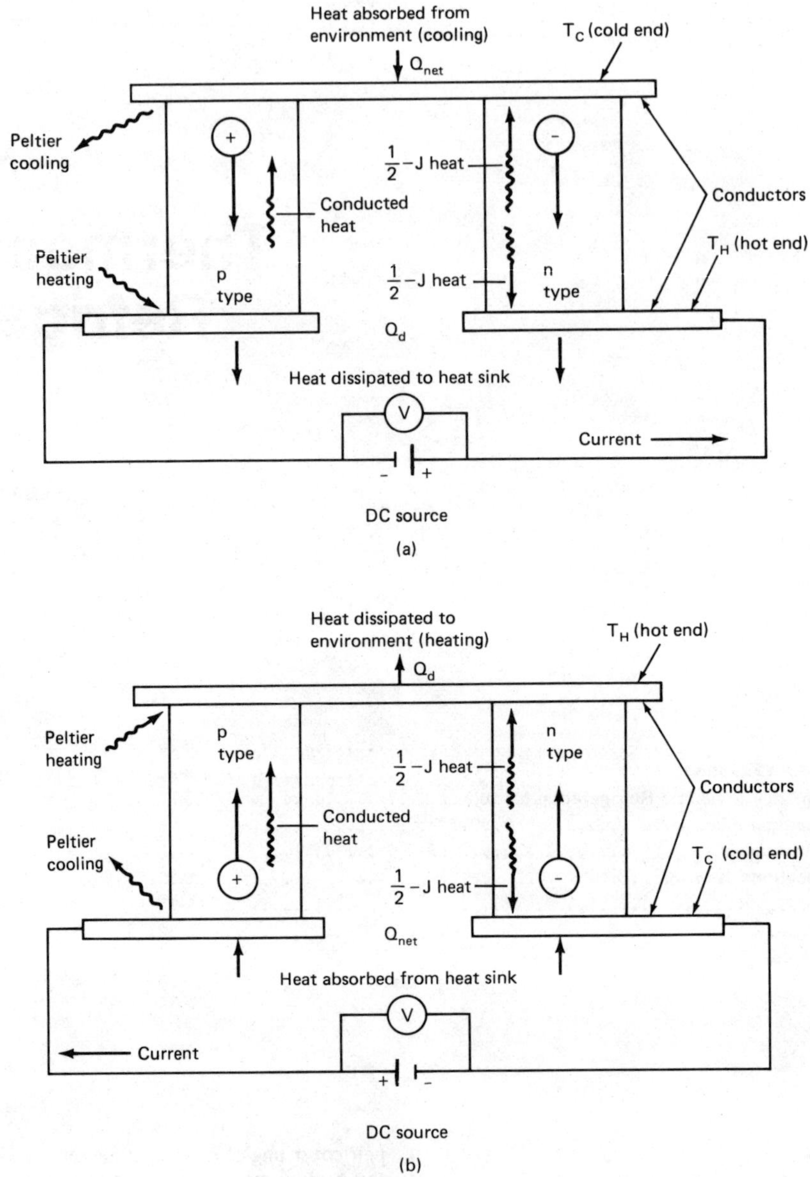

Fig. 2.1 Schematic of thermoelectric module operation. (*a*) Cooling mode; (*b*) heating mode.

between the hot and cold ends of the semiconductor material, a voltage is generated. This voltage is called the Seebeck voltage, and it is directly proportional to the temperature differential. The constant of proportionality is referred to as the Seebeck coefficient.

The Peltier effect is controlled by the Peltier coefficient, defined as the product of the Seebeck coefficient of the semiconductor material and the absolute temperature. The Peltier coefficient relates to a cooling effect as current passes from the *n*-type material to the *p*-type material, and a heating effect when current passes from the *p*-type material to an *n*-type material, as shown in Fig. 2.1. Reversing the direction of the current reverses the temperature of the hot and cold ends.

Ideally, the amount of heat absorbed at the cold end and the heat dissipated at the hot end are dependent on the product of the Peltier coefficient and the current flowing through the semiconductor material. Practically the net amount of heat absorbed at the cold end due to the Peltier effect is reduced by two sources, conducted heat and Joule heat. Due to the temperature differential between the cold and hot ends of the semiconductor material, heat will be conducted through the semiconductor material from the hot to the cold end. As the current is increased, the temperature differential, and thus the conducted heat, increases because the Peltier cooling effect increases. However, the other loss, Joule heat, is proportional to the square of the current and, therefore, eventually becomes the dominant factor. At any given current, thermal equilibrium is established at the cold end when the Peltier effect at the cold end is equal to the sum of the conducted heat plus one-half of the Joule heat. The other half of the Joule heat goes to the hot end.

As the current continues to increase and Joule heating becomes the dominating factor, a point is reached where additional current will result in less net cooling. The current at which no further cooling can be achieved is the maximum current I_{max}. Maximum voltage V_{max} and maximum temperature differential ΔT_{max} will also occur for any given heat load at the maximum current.

The net heat dissipated at the hot end is the sum of the net heat absorbed at the cold end plus the applied electric power. The coefficient of performance (COP) used to define the cooling "efficiency" is defined as the net heat absorbed at the cold end divided by the applied electric power.

The properties of semiconductor materials—electrical resistivity, thermal conductivity, and Seebeck coefficient—that define their maximum cooling capabilities are temperature-dependent. This is why simple formulas are not dependable for thermoelectric refrigeration design. For any one semiconductor material there is a temperature range over which that particular material has the best performance. The most widely used thermoelectric material for refrigeration in the temperature range of −184 to 446°F (−120 to 230°C) is a pseudo-binary alloy, $(Bi, Sb)_2(Te, Se)_3$, commonly referred to as bismuth telluride.

The refrigeration capability of a semiconductor material is dependent on a combined effect of the material's Seebeck voltage, electrical resistivity, and thermal conductivity over the operational temperature range between the cold and hot ends. The Seebeck coefficient squared divided by the product of electrical resistivity and thermal conductivity is called the figure of merit. Each of the *n*- and *p*-type semiconductor material properties varies as a function of temperature, and therefore the figure of merit for each material is temperature-dependent. It can be shown that the maximum temperature differential that can be achieved by a single pair of *n*- and *p*-type material is directly proportional to the "temperature averaged" figure of merit of each semiconductor material. Therefore, maximizing the figure of merit is the major objective in the selection and optimization of thermoelectric materials. The figure of merit of the semiconductor material limits the temperature differential, whereas the length-to-area ratio of each *n*- and *p*-type semiconductor material defines the heat pumping capacity.

2.3 DESCRIPTION OF A THERMOELECTRIC REFRIGERATION MODULE

More than one pair of semiconductors are usually assembled together to form a thermoelectric module. Within the module each semiconductor is called a pellet, and a pair of pellets is called a thermocouple. Thermoelectric modules, as shown in Fig. 2.2, are typically classified as single-stage modules or multistage modules.

Single-Stage Module

A single-stage module as shown in Fig. 2.2*a* consists of several thermocouples connected thermally in parallel and electrically in series to increase the operating voltage of the module. These thermocouples are interconnected with good electric conductors such as copper. The conductors must be electrically isolated from the device being cooled; otherwise the module will be electrically short-circuited to the surface being cooled. However, the electrical isolation material must also be a thermally conductive material to minimize the temperature difference between the conductor and the device being cooled. The module shown in Fig. 2.2*a* has a ceramic plate on the top and bottom surfaces of the module. Alumina ceramics typically provide the electrical isolation and thermal conductance that satisfy this requirement.

Thermoelectric modules vary in size from a single pair of pellets that may be as small as 0.06 in (1.5 mm) square up to groups of pairs that may be 2.0 in (51 mm) square. The limit in making smaller modules is related to the assembly of smaller pellets and the mechanical integrity of the pellets.

A limiting factor in the use of single-stage modules is related to the figure of merit of the semiconductor material. Regardless of the amount of power applied to a single-stage module, the coldest temperature that can at present be reached with bismuth telluride with a hot-side temperature of 80°F (27°C) and no thermal load is approximately −29°F (−34°C) in still air and −53°F (−47°C) in a vacuum.

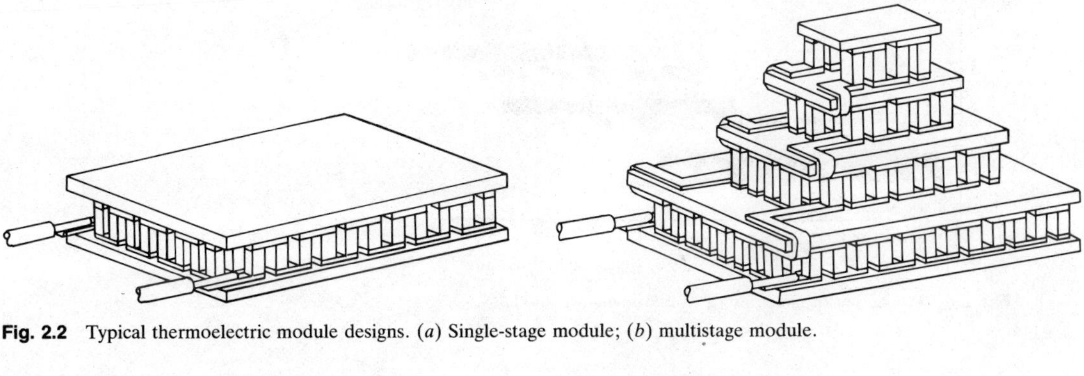

Fig. 2.2 Typical thermoelectric module designs. (*a*) Single-stage module; (*b*) multistage module.

Multistage Module

When the desired temperature differential cannot be obtained with a single-stage module, a multistage module is required. A typical four-stage module is shown in Fig. 2.2*b*. Multistage modules are essentially single-stage modules stacked in a vertical array. Typically a multistage module is pyramid shaped because the lower stage must pump the heat dissipated by the upper stages in addition to the active heat load on the top stage. Therefore, there are always more pellets in the lower stage than in the upper stages when all of the pellets are connected in series.

As additional stages are added to a module, colder temperatures can be achieved. A practical limit in the number of stages is presently between eight and ten stages. With an eight-stage module, a temperature of −218°F (−139°C) on the cold end has been achieved when the hot-end temperature was 127°F (53°C).[2]

2.4 OPERATIONAL CHARACTERISTICS

The basic operational characteristics of a thermoelectric module include the cold- and hot-side temperature, heat pumping capacity at the cold side, heat dissipated at the hot side, input current, and voltage. The acceptable range of each of these characteristics must be established by the design engineer in order to select a commercial thermoelectric module. Considerations in the final selection are typically cost, reliability, and performance.

Number of Stages

The first consideration in the selection of a module is to determine the number of stages required to obtain the desired temperature differential defined by the desired cold- and hot-end temperatures. The temperature differential achievable for various modules depends on the atmosphere in which the module operates. A greater temperature differential can be reached when the module is operating in a vacuum compared with operation in dry air, nitrogen, or other gases because the passive heat load is less in a vacuum. Figure 2.3 shows the practical cold-side temperature range achievable with present thermoelectric materials for various numbers of stages in dry nitrogen and vacuum. Figure 2.3 also shows an overlap in the number of stages of a module relative to the desired temperature differential. This is an important design consideration since, in general, the cost of a module is influenced by the number of stages. As the number of stages increases, the cost of the module increases. For minimum cost, the modules with the fewest number of stages would normally be selected.

The design of multistage modules is complex. Typically multistage modules are custom designed for optimum performance in a specific application. Computer programs to assist in the design of complex multistage modules have been

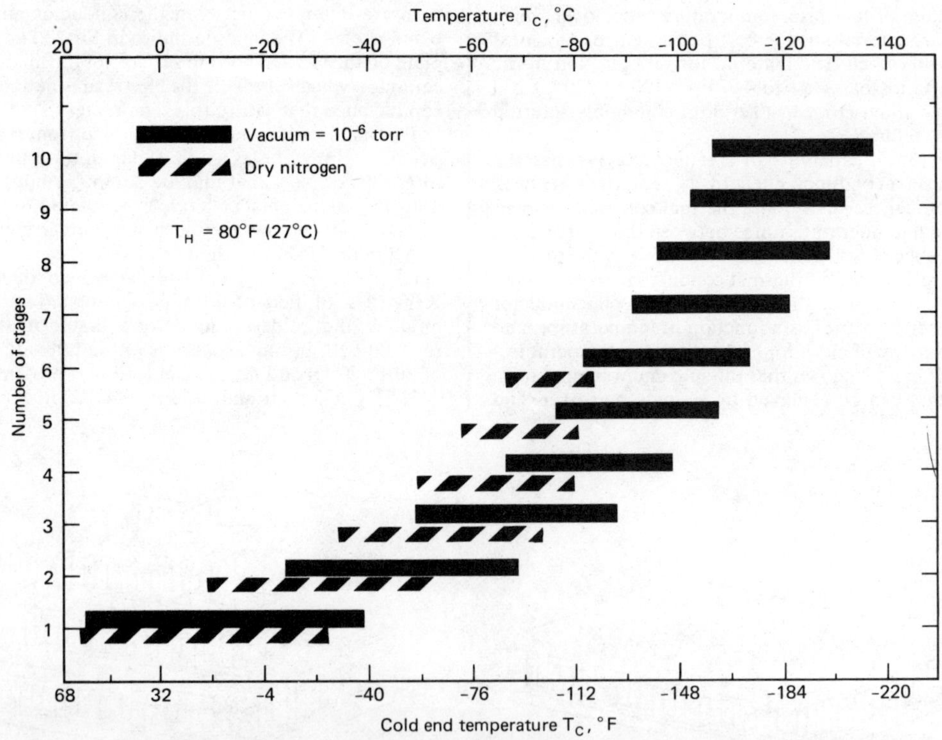

Fig. 2.3 Typical performance range of thermoelectric modules.

developed and applied by thermoelectric module manufacturers.

Temperature Differential and Heat Load

Manufacturers of thermoelectric modules provide tables that define the maximum performance capabilities of various modules. A typical format is shown in Table 2.1. The heat pumping capacities of various modules cover a wide range. The maximum heat pumping capacity is defined as the amount of heat required to suppress the temperature differential to zero. By dividing the maximum temperature differential that can be achieved with no heat load by the maximum heat pumping capacity with no temperature differential, the slope of the temperature differential versus heat load line can be defined. The actual performance is such that the temperature differential suppression is linear with the heat load.

Performance Curves

After one or more candidate modules have been selected based on data similar to those in Table 2.1, more detailed performance characteristics need to be considered. Typical module performance curves are illustrated in Fig. 2.4. One curve shows the effect of cold-side temperature T_C as a function of current I when the net heat load is zero ($Q_{net} = 0$). As discussed in the next paragraph, the condition of zero net heat load actually includes several parasitic heat loads that are present during the testing of a module. For single-stage modules the tests are normally conducted in room air conditions; for multistage modules they are typically conducted in a vacuum.

The relationship between the cold-side temperature and the net heat pumping capacity when operating at maximum current is also given in Fig. 2.4. As the heat load is increased, the temperature differential that can be achieved is decreased. The net heat pumped Q_{net}, defined in the performance curve, is the net heat load of the device being cooled. If the device being cooled is a small electrical component, the net heat load consists of the active Joule heating I^2R within the component and the heat conducted through the lead wires. The parasitic heat loads attributed to thermal radiation and convection on the lateral surfaces of the module are accounted for in the zero net heat load condition. Radiation and convection heat loads will need to be determined when the device being cooled has large surfaces.

The third curve in Fig. 2.4 relates the input voltage to the input current. As the current increases, the voltage increases. The current is not linearly proportional to the voltage due to the temperature dependence of the material properties.

Design Options

There are three usual objectives in the design and application of thermoelectric modules, (1) maximum heat pumping, (2) maximum COP, and (3) maximum speed of response.

Maximum heat pumping requires the module to operate at the current and voltage that pumps the maximum amount of heat over the specified temperature differential. The currents and voltages listed in Table 2.1 apply to operation at the current for maximum heat pumping.

Table 2.1. Summary Performance and Geometry of Typical Thermoelectric Cooling Modules

	ΔT_{max} when $Q = 0$ and $T_H = 80.6°F$ (27°C)							Dimensions				
	Vacuum, 10^{-6} torr		Dry N_2, 760 mm Hg					Base ceramic		Top ceramic		
Module	°F	(°C)	°F	(°C)	Q_{max}, W	I_{max}, A	V_{max}, V	Width, in (mm)	Length, in (mm)	Width, in (mm)	Length, in (mm)	Height, in (mm)
1	121	(67)	110	(61)	0.52	1.0	0.8	0.16 (3.96)	0.16 (3.96)	0.16 (3.96)	0.16 (3.96)	0.10 (2.41)
2	122	(68)	115	(64)	2.45	2.0	2.0	0.26 (6.60)	0.26 (6.60)	0.26 (6.60)	0.26 (6.60)	0.09 (2.20)
3	122	(68)	115	(64)	55.4	6.0	15.2	1.50 (38.1)	1.50 (38.1)	1.50 (38.1)	1.50 (38.1)	0.17 (4.40)
4	162	(90)	151	(84)	1.04	1.5	2.1	0.26 (6.60)	0.26 (6.60)	0.16 (3.96)	0.16 (3.96)	0.15 (3.80)
5	193	(107)	176	(98)	6.29	4.7	7.4	0.86 (21.72)	1.11 (28.27)	0.34 (8.61)	0.51 (12.98)	0.44 (11.10)
6	200	(111)	157	(87)	0.85	1.1	6.9	0.52 (13.26)	0.68 (17.22)	0.16 (4.06)	0.31 (7.98)	0.35 (8.80)
7	238	(132)	194	(83)	0.66	3.8	6.5	0.86 (21.72)	1.11 (28.27)	0.20 (5.08)	0.20 (5.08)	0.80 (21.10)

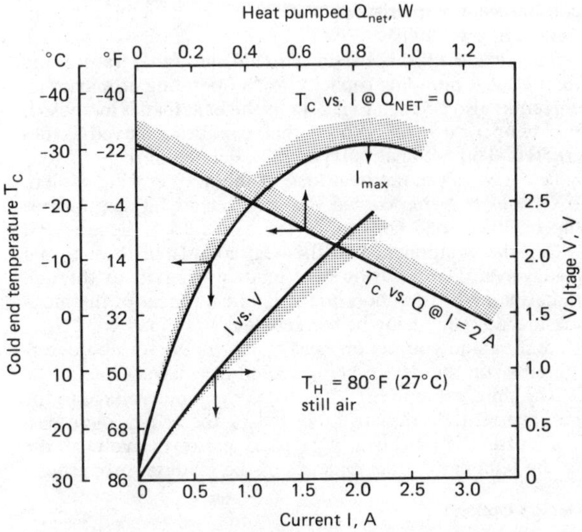

Fig. 2.4 Thermal and electrical performance of typical thermoelectric modules.

The current for operation at the maximum COP is generally less than the current for maximum heat pumping. For example, when operating at the current for achieving maximum COP at a hot-end temperature of 80°F (27°C), the maximum COP is about 0.2 for a 104°F (40°C) temperature differential. When the temperature differential is 72°F (22°C), the maximum COP is about 1.0.

The design of a module for maximum speed of response is more complex. Consideration must not only be given to the thermal and electrical characteristics of the module, but also to the thermal mass of the module and the device being cooled. In addition, the characteristics of the heat sink must be considered in the overall design. Temperature differentials of 144°F (80°C) from a hot-side temperature of 140°F (60°C) have been achieved in 2 s for properly designed modules with minimized heat loads. Typical single-stage or multistage modules will achieve the maximum cold-end temperature within 60 to 90 s after power is applied. This response time applies to the cold end of the module when there is no added mass on the cold end and the hot end is attached to a heat sink capable of absorbing the heat with a minimum rise in temperature.

A major feature of a thermoelectric module is the ability to stabilize the temperature of a device when the desired control temperature is within approximately ±9°F (5°C) of the ambient temperature. A conventional heater has the ability to heat a device above the ambient temperature but does not have positive action capability when the ambient exceeds the control temperature. Separate heating and cooling systems may be eliminated by using a thermoelectric module.

Mechanical Dimensions

The dimensional characteristics of typical modules are shown in Table 2.1. Typically the top and bottom surface areas of a single-stage module have the same dimensions except for ledges for electrical connections on some modules. The cold-stage surface area of a multistage module is typically smaller than the hot-stage surface area.

The flatness of the cold and hot surfaces of a module, the parallelism between the cold and hot surfaces, and the tolerances on the height of the module are important mechanical characteristics of the modules. These tolerances become increasingly important to overall performance when groups of modules are sandwiched between two heat exchangers.

Electrical Characteristics

Table 2.1 shows the maximum direct current together with the voltage that will achieve the maximum temperature differential in the cooling mode. For maximum performance the ripple component of the direct current should not exceed 10 percent.

Depending on the direction and the amount of current, thermoelectric modules have the ability to cool, heat, or stabilize temperatures regardless of whether or not the cold surface is above or below the ambient temperature. By applying the correct amount of current at the desired polarity, the temperature of a device can be stabilized as the ambient temperature oscillates. Temperature stability depends on the quality of the temperature controller, thermal contact, thermal mass, and heat load capacity. Tolerances of ±0.18 to ±0.54°F (±0.1 to ±0.3°C) at a stabilized temperature are reasonable.

The module with the lowest current would normally be selected because it makes the power supply temperature controllers less expensive for a given output power. However, the total power to the module is the major factor. Low current modules will, therefore, require higher voltages to maintain the same power input. High-voltage modules require a larger number of pellets, thereby increasing the cost. Therefore, when selecting a module, a tradeoff is made between module cost and power supply or temperature controller cost, or both.

Each bismuth telluride thermocouple operating at the current for maximum temperature differential has approximately a 0.12-V drop when the hot end is at 80°F (27°C). This means that approximately 10 thermocouples are required per volt of input power for maximum cooling. Even more thermocouples will be required for maximum COP.

2.5 PRACTICAL CONSIDERATIONS

Thermal

The practical heat load for thermoelectric modules ranges from a few milliwatts to tens of watts. This practical limitation is related to the competitiveness of other refrigeration methods in both cost and efficiency. Conventional mechanical refrigeration systems are normally not competitive with thermoelectric refrigeration when the heat loads are under 25 W. Thermoelectric refrigeration can be used to pump heat loads much greater than 25 W, but these are usually for very specialized applications.

The lowest practically achievable temperature is about −148°F (−100°C). Since the efficiency of a thermoelectric module (single-stage or multistage) decreases as the temperature decreases, the heat pumping capacity is typically limited to only a few milliwatts at those cold temperatures.

In heating applications, the practical limit is about 176°F (80°C). This temperature limit is imposed primarily by the manufacturing techniques used to assemble the thermoelectric modules. Although typical thermoelectric modules are assembled at soldering temperatures of 280°F (138°C), they are not capable of operating at these higher temperatures for extended periods of time without degrading in performance. The gradual degradation in performance with time is due to contaminates diffusing into the thermoelectric material resulting in changing cooling properties. It should be noted that less input power is required to heat than to cool.

Environmental

Because thermoelectric modules are solid-state and have no moving parts, high reliability is an inherent feature. However, the reliability may be affected by environmental extremes.

Moisture With power applied to the module, excessive moisture within the module can cause galvanic corrosion at the thermoelements, solders, and conductors. The moisture can also provide an electric short-circuit within the module, which would reduce the module's performance. Generally, a moisture barrier, gasket, or dry atmosphere is required to maintain reliability.

High Temperature Continual operation at high temperatures accelerates thermal diffusion of metal ions into the pellets, thus decreasing the performance of the module. In addition, high temperature can cause the solder to migrate into the pellets, which results in a damaged joint, thereby reducing performance. Typically, this is not a problem below 176°F (80°C). However, the temperature at which these factors begin to affect the performance of specific modules should be defined to the design engineer by the module manufacturer.

Thermal Cycling Rapid temperature cycling from cool to hot can also affect the performance of the module over time. As temperatures change, the module will expand or contract. This action stresses the pellets in the module. Although one major feature of thermoelectric modules is the ability to heat and cool, consideration should be given to minimizing rapid power cycling.

Mechanical Stress Mechanical stresses applied during shock or vibration are not a major failure mode if the modules are properly mounted. In general, the module is strongest in the compression mode. The most likely failure mode of the solder joints is in the shear direction, which is parallel to the top or bottom surfaces of the module. Most modules will withstand shock levels of 1000g over a frequency range of 10 to 2000 Hz and random vibration of about 65g rms from an acceleration power density of 0.01 to 5.0g^2 Hz over a frequency range of 50 to 20,000 Hz.

Vacuum A vacuum is the most reliable environment in which to use a module. Thermoelectric modules are not susceptible to failure as the result of any level of vacuum. When in a vacuum, the effective heat pumping capacity of the module is increased, thereby providing colder operating temperatures as compared to operating in air for the same input power.

Radiation No form of radiation has been noted in the literature that has induced failure of thermoelectric modules. Several tests have demonstrated that high neutron bombardment and gamma radiation do not affect performance.

Installation

Mounting The module may be pretinned and soldered to a heat sink. Usually the hot end is soldered to the heat sink and the cold end is greased with a nonsetting thermal compound for good heat transfer. Soldering the top or bottom to a heat sink can lead to mechanical breaking if the thermal expansion mismatch between the module and the attached heat sink is large. This is especially true if the hot end and the cold end are both mechanically constrained between the heat sinks.

A good thermal epoxy is another way of bonding or attaching the module. The same failure modes as for soldering also apply to epoxy.

Another way of mounting a module is between two clamped plates. Extreme caution should be used for small or multistage modules. Improperly applied compression loads can cause very large shear loads, which will fracture the pellets.

The reason individual modules are not larger than about 2.0 in (51 mm) square is due to mechanical considerations. The modules tend to bow much like a thermostatic bimetallic element. This bowing between the hot and cold ends of the module is due to the fact that one end is contracting and the other end is remaining constant or expanding. The stress induced by these expansion coefficients tends to strain the individual pellets on the outer edge of the module. When the distance across the module increases beyond about 2.0 in (51 mm), the strain is too great for the pellets and they crack or break. For this reason, groups of modules are assembled onto heat exchangers rather than mounting one large module to the heat exchangers.

Handling The module is a rigid assembly sandwiched between ceramic plates. The lead wires are attached to stiff lead conductors. Excessive torque on the lead wires will result in pulling the wires loose or snapping the lead conductor loose from the circuit. The ceramic plates on the top and bottom of the module may be cracked or broken by mishandling, which increases the probability of electric short circuits.

Table 2.2. Applications of Thermoelectric Modules

Electronic components	Small refrigerators	Instruments	Special
Infrared detectors	Boats	Dew point sensor	Dehumidifiers
Laser diodes	Mobile homes	Cold bath	Water coolers
Charge-coupled devices	Offices	0°C reference source	Air-conditioning
Blackbody reference source	Hotels	Microtome	Display cabinets
Voltage reference source	Portable picnic	Blood coagulator	Cream coolers

Heat Sinking Adequate heat sinking is required to dissipate the heat load and power of the module without an excessive hot-end temperature rise. Failures have occurred where the solder joints on the hot side of the pellets were melted because of inadequate heat transfer to the heat sinks.

2.6 RANGE OF APPLICATIONS

The application of thermoelectric modules ranges from the cooling of electronic components such as infrared detectors with heat loads in the milliwatts and temperatures of −171°F (−113°C) to special applications where 5 tons of air-conditioning is achieved in a railroad passenger car with a temperature difference of 11°F (6°C).[3] Table 2.2 shows a list of typical applications of thermoelectric refrigeration.

2.7 SUMMARY

Thermoelectric refrigeration is a highly reliable and practical method of cooling electronic components and small-volume refrigerators from a cost and efficiency viewpoint.

When heat loads are less than 25 W and temperature requirements are not below −148°F (−100°C) or above 176°F (80°C), thermoelectric refrigeration is practical for a wide variety of applications.

REFERENCES

1. H. J. Goldsmid, *Electronic Refrigeration,* Pion, London, 1986.
2. R. Buist, J. Fenton, and D. Tuomi, "Low Temperature Thermoelectric Cooler for 145 K Operation," Night Vision Laboratory, VA, Sept. 1971.
3. J. G. Stockholm, "Prototype Thermoelectric Air Conditioning of a Passenger Rail Coach," presented at the 4th International Conf. on Thermoelectric Energy Conversion, TX, Mar. 1982.

Chapter 3
Heat-Driven Refrigeration

DONALD K. MILLER
York International Corporation

3.1 DEFINITIONS

Absorbent: A solution that absorbs refrigerant at a relatively high temperature to create a vapor pressure in an evaporator corresponding to a low refrigerant temperature. The weight concentration of refrigerant in the absorbent increases.

Absorber: A heat and mass transfer device that permits simultaneous heat and mass transfer from refrigerant to absorbent at a relatively high temperature with attendant heat of condensation, heat of solution, and some subcooling. The heat is removed by a coolant, such as cooling tower water, air, or a process fluid.

Condenser: A heat exchanger that condenses refrigerant vapor to a liquid, removing superheat from the vapor, latent heat during condensation, and sometimes sensible heat from the liquid at a relatively high temperature. The heat is removed by a coolant such as cooling tower water, air, or a process fluid.

Coefficient of performance (COP): A number ratio of beneficial cooling compared to the amount of heat input to the generator.

Desorber: See *Generator.*

Design working pressure (DWP): Pressure required by prevailing code or the specifications.

Evaporator: A heat exchanger that transfers heat at a

relatively low temperature from a pumped fluid (that is used for cooling elsewhere) to a liquid refrigerant that evaporates at a lower temperature to form a vapor.

Fouling. The forming of a thermal resistive film on a heat-transfer surface internally or externally, which reduces the overall rate of heat transfer.

Generator: The heat exchange device for boiling a solution of refrigerant and absorbent to produce evaporation at a high temperature; called reboiler or desorber in ammonia absorption systems. The weight concentration of refrigerant in the absorbent decreases.

Heat sink: The fluid medium or coolant that removes heat from the thermal cycle at an appropriate temperature level for the absorbers, condensers, and perhaps other heat exchangers.

Heat source: The fluid medium that furnishes thermal energy to a heat exchange device in the thermal cycle.

Intercooler: A device for cooling refrigerant liquid to a lower temperature in order to increase the thermal efficiency of a refrigeration system.

ppm: Parts per million, usually by weight.

Reboiler: See *generator.*

Reflux: Relatively cool liquid refrigerant or absorbent solution recycled to the top of a distillation tower to increase the refrigerant content of the vapor mixture rising from the top of the tower.

Refrigerant: A volatile fluid that increases in heat content by virtue of evaporation with latent heat addition or by virtue of superheating of the vapor and which releases heat upon condensation, absorption, or sensible cooling.

Suction line superheater: A heat exchanger to transfer heat from refrigerant liquid to refrigerant vapor at the evaporator in order to improve thermal efficiency in an ammonia absorption refrigeration system.

Subcooling: The process of heat removal from a liquid by virtue of sensible heat exchange in a device without lowering the vapor pressure of the fluid.

Tons: The capacity of a refrigeration unit, system, or device, with each ton equivalent to a rate of heat transfer of 12,000 Btu/h (3.52 kW).

Trimming of charge: The addition or deletion of a refrigerant or absorbent to establish the proper inventory of each in the unit or system.

Wetting agent: A substance that reduces surface tension to cause complete wetting of the heat-transfer surface and thus enhance the heat-transfer rate.

3.2 EQUIPMENT FOR ABSORPTION MACHINES

Two basic types of industrial heat-driven refrigeration units or systems are discussed in this chapter, the lithium bromide absorption unit and the ammonia absorption system. Both types use an absorbent and a refrigerant. A refrigeration load increases the concentration of the refrigerant in the absorbent to form a refrigerant-rich solution. Heat is used in a reboiler to boil the refrigerant-rich solution and remove the accumulated additional refrigerant. The refrigerant-poor solution is then recycled in the unit or system to the absorber, where it again absorbs the refrigerant that had been removed by boiling. Other heat-driven systems, such as steam-jet and organic Rankine cycle systems, are not widely used in industry, hence are not discussed in this chapter.

Configuration Concepts

Lithium Bromide Absorption Units The configuration of the lithium bromide absorption unit is primarily as a prepackaged standard-size unit, semihermetically tight at the point of manufacture. This is dictated by the necessity to limit the extent of piping and the requirement to leak-test with a mass spectrometer for high-vacuum service. The designed fluid flow distributions and internal pressure drop restrictions are usually critical and must be closely controlled.

Each internal part must be free of any residual oil or dirt. The internal surfaces of the shells and many of the internal parts and inside pipe surfaces are washed with a suitable corrosion inhibitor and then dried before assembly.

All shells are usually horizontal, and during operation, regeneration of the solution is accomplished by a simple boiling process with no distillation towers required.

Ammonia Absorption System The configuration and fabrication concept for ammonia absorption systems differs markedly from that for lithium bromide units. There is considerable flexibility for the design to accommodate any configuration deemed essential to meet the demands of a specific application, custom-designed and tailored to specific requirements.

Internal flow distributions and pressure drop restrictions are much less severe than those for lithium bromide units. Shells are fabricated using conventional heat exchanger techniques.

Components

Lithium Bromide Absorption Unit (Fig. 3.1) Components consist of an absorber and an evaporator, with each containing heat exchange tubing, usually in the lower shell; a generator and a condenser, with each containing heat exchange tubing, usually in the upper shell; a second generator in a separate shell for a two-stage unit; one or more solution

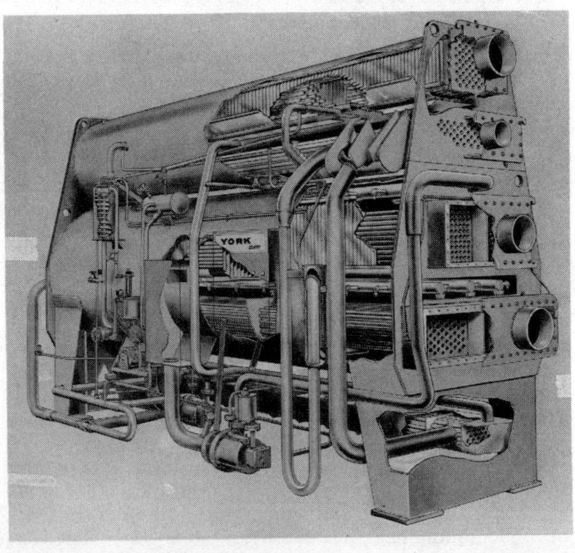

Fig. 3.1 Single-stage lithium bromide absorption unit. (*By permission of York International Corporation.*)

heat exchangers suitably tubed; one or more solution pumps and a refrigerant pump; spray nozzle banks for the evaporator and the absorber; an intermittent or a continuous purge system using a vacuum pump or jet purge device plus a purge gas collection vessel which may contain a water-cooled coil over which solution is cascaded to effect more efficient purging with removal of water vapor; necessary internal baffles, eliminators, drain pans, orifices, and piping; external piping, gaskets, valves, heads, and supports; and electric, electronic, or pneumatic controls as appropriate.

Ammonia Absorption System (Fig. 3.2) Components consist of absorbers and condensers (liquid or air-cooled); solution heat exchangers as required; remote evaporators or coolers; pumps of vertical or horizontal configuration using mechanical seals and couplings with almost any type of driver (such as electric motor, steam turbine, or gas engine); one or more fractional distillation towers with trays or suitable packing to effect intimate contact of gravity falling solution and countercurrent rising vapors; reboilers containing heat exchange surfaces; solution tanks and refrigerant receivers; suction line to liquid refrigerant heat exchangers; intercoolers of flash, shell-and-tube, or shell-and-coil configuration; piping and valves; and suitable controls of electric, electronic, or pneumatic types.

Fig. 3.2 Ammonia absorption refrigeration system, coffee freeze-drying at −67°F (−55°C). (*By permission of Borsig GmbH, Berlin.*)

Materials of Construction

Lithium Bromide Absorption Unit Evaporator tubes are usually finned copper or copper nickel; absorber tubes are usually plain surface copper or copper nickel; condenser tubes are usually plain surface copper or copper nickel; generator tubes are placed in shallow bundles and are usually finned surface copper nickel; solution heat exchanger tubes are usually mild steel; shells are mild steel; internal baffles are mild or stainless steel; internal eliminators and drain pans are stainless steel; piping is mild steel, stainless steel, copper, or copper nickel; pumps have either impregnated cast iron or cast steel housings, bronze impellers, stainless shafts, stainless liners for hermetic motors, carbon and ceramic seals, and suitable O rings and gaskets; other gaskets are of reinforced neoprene or Teflon; purge pumps are single or two-stage suitable for high-vacuum evacuation of inert gases saturated with water vapor.

Ammonia Absorption System Materials are usually steel. Copper or copper nickel are not suitable for contact with ammonia.

Operating and Design Working Pressures

Lithium Bromide Absorption Units These units use water as a refrigerant. The DWP of the shells is generally limited to 15 lb/in^2 (103.4 kPa) gage, and the units are code exempt on the shell side. For chemical or petrochemical applications, the DWP can be raised to 75 lb/in^2 (516.8 kPa) gage, with code construction and welding required. The tube side of standard units is generally 150 lb/in^2 (1033.5 kPa) gage, although units have been designed for up to 500 lb/in^2 (3445 kPa) gage.

Ammonia Absorption System These systems use ammonia as the refrigerant. The low-pressure-side shells are usually designed for 174 lb/in^2 (1200 kPa) gage DWP and the high-pressure-side shells for 250 lb/in^2 (1722.5 kPa) gage and sometimes 300 lb/in^2 (2607 kPa) gage.

Installation Practices

Lithium Bromide Absorption Unit These units are generally shipped in one piece, although the larger sizes may be shipped in two or three pieces. They are quiet in the absence of cavitation and are usually mounted indoors on simple foundations or on floors that are level. Corrugated vibration pads may be used on upper-floor installations.

Outdoor installations require careful attention to the freeze-up potential for refrigerant and small water lines as well as the crystallization potential for solution lines and pumps.

Solution charging is accomplished from drums of inhibited solution and preferably a source of distilled water.

Ammonia Absorption System These installations are typically encountered in chemical plants. Most installations are outdoors and there is little concern for freeze-up of the solution or the refrigerant. Simple foundations and structural steel are used and the shells are properly leveled for any gravity distribution requirements. The systems operate quietly, so there is no concern with noise or vibration.

Insulation for evaporators, intercoolers, and low-temperature lines is required. The towers, heat exchangers, and reboilers may be insulated to conserve energy.

Leak detection is accomplished with nitrogen or air pressurization and soap solution testing followed by evacuation, then ammonia pressurization and wet litmus paper, hydrochloric acid vapor, or burning sulfur taper testing. The systems are usually charged from tankcars or truck tanks of refrigerant-grade ammonia and preferably distilled water. A solution of inhibitor (0.2 percent commercial sodium dichromate, by weight of the total charge) is injected into the system.

Practical Size Limits

Lithium Bromide Absorption Unit The maximum practical nominal size of prefabricated lithium bromide units is generally limited to 1400-ton (4928-kW) capacity.

Ammonia Absorption System There really is no limit for the maximum size or capacity of a field-constructed custom-made ammonia absorption system. The maximum size system reported to date is 6285 tons (22,123 kW); however, this is merely governed by the marketplace.

Thermal Cycle Difference

Lithium Bromide Absorption Unit Water is used as the refrigerant, and generally a 58 to 65 percent by weight lithium bromide solution is used to absorb the refrigerant. A pump sprays recirculated solution over tubes of the absorber where heat of condensation and solution are removed. Some subcooling of the solution is also accomplished. This maintains a low-pressure condition to produce refrigeration at the evaporator in one step at only one temperature level. The solution strongest in refrigerant is pumped through a heat exchanger to the generator, where a heat source such as steam, hot water, or another hot fluid in the tubes removes the absorbed refrigerant by a simple boiling process. Only the refrigerant is distilled, and it is then condensed on condenser tubes.

In two-stage units, the distilled refrigerant from the first-stage generator is used as a steam heat source for a lower-temperature generator so that the solution is concentrated in two steps. Then there are also two steps of heat exchange for the solutions. This increases the efficiency for the cycle.

Water, which is circulated first to the tubes of the absorber and then to the tubes of the condenser, removes heat of absorption and condensation, respectively. Parallel flow may also be used.

Ammonia Absorption System[1–3] Ammonia is used as the refrigerant and an ammonia-water solution as the absorbent. Ammonia refrigerant evaporated at low temperature is absorbed in a relatively weak solution of ammonia and water, where it condenses and is absorbed in the solution to form a strong ammonia solution. The heat of condensation and solution is removed by a coolant, such as cooling tower water or ambient air, at a relatively high temperature.

The strong solution is sometimes pumped through heat exchangers to multiple stages of absorption for multiple-step refrigeration at successively higher temperatures, whereupon the solution is further strengthened before it is pumped to the distillation tower. This improves efficiency. For single-stage absorption, solution is pumped through the exchangers directly to a distillation tower. In the tower, the solution contacts hot rising vapors of ammonia and water boiled off the reboiler at the base of the tower. This then is a fractional distillation process since both the water and the ammonia refrigerant are volatile. Near the top of the tower, the rising vapor mixture contacts relatively cooler condensed liquid of almost 100 percent ammonia, called reflux.

The high-purity ammonia vapor off the top of the tower is condensed by the cooling action of cooling tower water, well water, or ambient air. In some cases, the vapor off the top of the tower is cooled and condensed by other streams of fluids available within the plant. The hot weak ammonia solution leaving the base of the tower and reboiler is passed through successive stages of heat exchange to perhaps several stages of absorption.

The cooling water from the condenser usually is next fed in series to the absorbers, although the reverse direction has been used depending on the design concept for a given application. Water flow can also be in parallel to the condensers and absorbers.

There are also designs of ammonia systems where distillation is accomplished in two stages with two separate towers. In such a system, the vapors off the top of the lower-stage tower contain a high concentration of water and are reabsorbed in an absorber. The efficiency of such a system is less than for one where there is only one distillation step; however, the advantage is that lower-temperature heat sources can be utilized.

The weight concentration of ammonia in the solutions is usually from 5 to 50 percent (by weight) with no danger of freeze-up or crystallization from 25 to −100°F (−3.9 to −73.3°C). Because of slight amounts of water carryover in the ammonia condensate at the condenser, measured in parts per million, this water finds its way to the evaporators. Therefore, a small sample is bled or pumped continuously or periodically from evaporators to an absorber.

Temperature Limits of Fluids

Lithium Bromide Absorption Unit The maximum temperature of 90/10 copper nickel surfaces contacting lithium bromide solutions (to avoid excessive corrosion) is about 320°F (160°C). In the shallow flooded generator, it is necessary to use skip tubing and roll the tubes at the tube supports or have some other suitable anchoring or isolation device in order to avoid fretting corrosion at the tube supports.

At the other extreme of temperature, 65 percent by weight lithium bromide in solution crystallizes at 105°F (40.6°C) and 60 percent crystallizes at about 60°F (15.6°C).

Ammonia Absorption System Ammonia will begin to break down into its elements, nitrogen and hydrogen, at a temperature of 570°F (298.9°C). A lower temperature, below 350°F (176.7°C), is deemed prudent to preclude excessive corrosion products in the presence of mild contaminants such as air, even with a corrosion inhibitor present.

Controls and Instrumentation

Lithium Bromide Absorption Unit Combinations of electric, electronic, and pneumatic controls are used as appropriate. Typical controls include proportional heat input controls, freeze-limit thermostat, refrigerant and solution level controls, cooling water bypass valve, and thermally protected motors.

Ammonia Absorption System Controls are applied as necessary to evaporators, intercoolers, absorbers, solution tanks, reboilers, distillation towers, and condensers. Commonly used controls include ratio flow controls, pressure and temperature controls, level controls, proportional heat input controls, and hot-gas bypass.

3.3 ENERGY SOURCES

Primary energy sources normally used are gas or oil, which are easily modulated to a prime energy-driven reboiler. Examples of *secondary energy sources* are steam, recirculated hot water, hot oil, hot kerosene, and hot brines.

Waste streams could be exhaust steam from a steam turbine, exhaust gases from a gas turbine, stack gases from a power station or furnace, steam or hot water from a diesel engine jacket, or almost any stream containing heat energy that might previously have been thrown away when extensive energy conservation measures were not exercised. Obviously, the energy stream must be at a sufficient temperature level to drive the system. *Process streams* could be any chemical or petrochemical fluids that are presently used for heat exchange within a process, but which could be more efficiently used in a comprehensive energy balance to produce refrigeration.

3.4 APPLICATIONS AND LIMITS

Lithium Bromide Absorption Unit The minimum effluent chilled fluid temperature for a fully loaded lithium bromide absorption unit is at least 39 to 40°F (3.9 to 4.4°C), with a minimum of 36°F (2.2°C) refrigerant temperature. These units were initially designed for the express purpose of cooling water for the air-conditioning of buildings. There have been many other applications,[4] such as indirect cooling of wort in breweries; indirect heat exchange to cool air for industrial processing; cooling of lean oil while using hot kerosene or hot gasoline as the heat source; cooling of hydrocarbon liquids while condensing hot mixed hydrocarbon gases as the heat source; cooling of water-based brines using exhaust steam from steam turbines and hot water from diesel engine jackets.

The minimum heat source temperature that can be used is a function of the available cooling fluid temperature for the absorber and condenser and the coefficients of heat transfer that can be developed. In general the heat source must be at least as high as 180°F (82.2°C). For higher than 250°F (121.1°C) heat sources, there can be special precautions in the design of the generator and the heating fluid circuit.

Application to hydrocarbon fluid cooling and using hot gasolines or condensing hydrocarbon mixtures as the heat source derates the unit so that its capacity is 35 to 50 percent of nominal tons (kW) when using the usual 85 to 90°F (29.4 to 32.2°C) cooling water. In addition, for such applications heavy fouling factors further derate the unit capacities and several alterations of the unit must be accomplished to meet applicable codes.

Ammonia Absorption System These systems have considerable flexibility and can accommodate a wide variety of applications with very few limitations on attainable low temperatures or restrictions on the upper limit of heat source temperatures. The minimum practical effluent chilled fluid temperature for an ammonia absorption system evaporator is perhaps as low as −75°F (−59.4°C). Furthermore, multitemperature levels are attainable with the systems designed for optimum efficiency that match the available heat source and heat sink temperatures as well as the quantities of heat and heat rejection provided.

A major manufacturer of these systems[5,6] lists applications such as gas cooling at 26.6°F (−3°C) using a steam heat source; process cooling at −22 to −49°F (−30 to −45°C) using hot air, hot water, and steam for heat; freeze drying at −54.4°F (−48°C) and −67°F (−55°C) using heavy fuel oil and thermo oil as prime energy heat sources; and ice cream manufacture at three temperature levels, with one as low as −49°F (−45°C), using heavy fuel oil for prime energy heat source.

These plants are operated automatically, unattended except for data logging and minimal maintenance. Systems have operated for 35 years and longer.

3.5 THERMAL PERFORMANCE AND UTILITIES

The typical thermal performance and utilities for lithium bromide absorption units and ammonia absorption systems are given in Tables 3.1 and 3.2.

3.6 ADVANTAGES AND DISADVANTAGES

Advantages

Lithium Bromide Absorption Unit These units utilize low electric power inputs; preengineered standard units can be modified for a special application; units are preassembled at the factory and require a relatively small space for installation; they have low-pressure operation and a low DWP of shells; the refrigerant is nontoxic and nonhazardous; there is no leakage or loss of refrigerant to atmosphere; they are readily placed indoors on upper floors.

Ammonia Absorption System These systems are custom-designed to match exactly a wide range of loads and temperature levels for the evaporators with multitemperature level capability; the refrigerant cycle can be combined with centrifugal compressor equipment (combination systems); they are readily designed for water-cooled or air-cooled absorbers and condensers; they are adaptable to almost any

Table 3.1. Performance Data for Lithium Bromide Absorption Units

Basis: Maximum capacity for single-stage absorption units,[7,8] chilled water flow 2400 gal/min per 10^3 tons with 10°F range; cooling water (coolant) flow 3600 gal/min per 10^3 tons; power for solution and refrigerant pumps 9 kW per 10^3 tons.

	Chilled water temperature, °F		
	50	45	40
COP			
85°F coolant	0.713	0.690	
75°F coolant	0.753	0.729	0.704
Minimum heating fluid temperature out, °F			
85°F coolant	226	236	
75°F coolant	202	212	223

NOTE: L/s = gal/min × 0.0631; kW = tons × 3.52; °C = (°F − 32)/1.8. The COP of two-stage units is 60 to 70% greater and the heat source at least 100°F (55.6°C) higher.

SOURCE: York International Corp., York, PA.

Table 3.2. Performance Data for Ammonia Absorption Systems

Basis: Single-stage absorption; single-stage desorption; cooling water (coolant) supply temperature 77°F with 14 to 18°F range; size of system 1000 tons.

	Data at evaporator temperature, °F		
	+14	−22	−58
COP	0.61–0.54	0.50–0.45	0.34–0.31
Minimum heating fluid temperature out, °F	230–265	285–320	345–385
Generator heat, 10^6 Btu/h	20–22	24–27	35–39
Coolant rate, gal/min	3600–4700	4000–5300	5400–6900
Power for system pumps, kW	60–90	70–110	90–130
	% Change per °F Coolant Temperature Change		
Heat input	0.4–0.6	0.4–0.6	1.6–2.2
Coolant flow	0.3	0.3	1.1–1.7
Pump power	1–2	1–3	2–3
Heating fluid temperature out	2–2.5	2–2.5	2.5–3

NOTE: °C = (°F − 32)/1.8; °C range = °F range/1.8; kW = tons × 3.52; MW = 10^6 Btu/h × 0.293; L/s = gal/min × 0.0631. For evaporative condenser, use lower heating temperatures, reduce coolant flow to 60% of above values, and increase electric power to 180% of above values (fans included).

SOURCE: W. Malewski, Borsig GmbH, Berlin.

available heat source of sufficient temperature level; they use conventional heat exchanger and chemical plant technology with steel tubing, shells, and piping; they have positive pressures above −28°F (−33.3°C) evaporator temperature; purging is simple; they use a low-cost refrigerant which is also biodegradable; there is no oil contamination of the refrigerant (except in combination systems); simple foundations and structures are used with all outdoor installation; there is seldom any concern with solution or refrigerant freeze-up; they are not subject to refrigerant liquid carryover damage; the electric energy requirement is only 5 to 10 percent of that for mechanical refrigeration; they are ideal for handling peak loads greatly in excess of nominal plant capacity, using ammonia and weak-solution storage tanks; semiskilled personnel can operate them; the noise level is less than 65 dBa measured at 1-m distance; and maintenance and repair costs are less than 50 percent of those for mechanical refrigeration.

Disadvantages

Lithium Bromide Absorption Unit These units operate continuously at high vacuum, and any inward leakage of air subjects internal parts to potential corrosion; the solutions are susceptible to crystallization, and refrigerant lines are susceptible to freeze-up; the minimum evaporator temperature is 36°F (2.2°C), and only one level of evaporator temperature is available; the criticality of pressure drop for refrigerant vapor and the necessity for a prepackaged concept restricts the maximum practical size of units; they are not readily adapted to air-cooled construction in industrial sizes; the solution charge is expensive; a more complex control of solution chemistry is required; no remote evaporators or booster compressor combinations are feasible; more expensive copper and copper nickel heat exchange tubing is required (steel tubing is only permitted in the absence of vapor-liquid interfaces); more complex purging systems and procedures are required; charging and withdrawal of solution for maintenance is cumbersome; a unit is highly susceptible to corrosion should disassembly for maintenance be necessary; and installation outdoors requires special measures for shutdown and standby protection.

Ammonia Absorption System High shell DWPs are required; ammonia is semihazardous, making indoor installations less desirable; there is greater consumption of space; more engineering and relatively greater field labor are required for installation; higher flow rates of cooling water or higher temperature ranges and, therefore, larger cooling towers or heat rejection devices are required.

3.7 MAINTENANCE REQUIREMENTS

Lithium Bromide Absorption Unit The best policy for the maintenance of a lithium bromide absorption unit is not to break into the unit solution side unless absolutely essential. This avoids the possibility of corrosion and contamination from ambient substances. Generally, maintenance involves checking for leak tightness, heat exchanger fouling, seal and valve diaphragm inspections, and solution and refrigerant testing (annual).

With an effective preventive maintenance program, these units can experience a life of at least 16 years. Some units are still performing after 25 years.

Ammonia Absorption System Maintenance is considerably simplified. With system designs that have positive pressure in the system at all times, the only effect of leaks is the loss of refrigerant and perhaps some water. Generally, maintenance involves trimming of the charge to the proper levels; checking the concentrations of corrosion inhibitors; purging when necessary; checking for heat exchanger fouling; inspection of pumps, seals, and valves; cleaning of strainers and upkeep of painted surfaces.

The service life of ammonia absorption equipment with reasonable preventive maintenance can easily reach 50 years.

3.8 COSTS, WEIGHTS, AND DIMENSIONS

Lithium Bromide Absorption Unit The 1986 budget costs for a standard single-stage lithium bromide absorption unit are \$200 per ton (\$57 per kW) at 1200 tons (4224 kW) to \$285 per ton (\$81 per kW) at 400 tons (1408 kW). Budget costs for special applications with other fluids that require capacity derating and code construction of shells can reach three to four times these values. Weights for a single-stage standard unit are 72 lb/ton (9.3 kg/kW) at 1200 tons (4224 kW) to 70 lb/ton (9 kg/kW) at 400 tons (1408 kW). Weights for special fluid units that are capacity derated and have code construction of shells are about twice these values.

Dimensions of length/width/height are 27/08/14 ft (08/02/04 m) at 1200 tons (4224 kW) to 22/06/10 ft (07/02/03 m) at 400 tons (1408 kW).

Ammonia Absorption System The 1986 budget costs for ammonia absorption systems range from \$1600 to \$3300 per ton (\$455 to \$938 per kW) at 1200 tons (4224 kW) for evaporator temperatures of +14 to −58°F (−10 to −50°C). They also range from $2300 to $4000 per ton ($653 to $1136 per kW) at 400 tons (1408 kW) for the same temperature ranges. Costs do not include steel structures, evaporators, erection, or installation. Erection, installation, and startup costs range from 25 to 35 percent additional.

Dimensions of length/width/height range from 36/43/40 to 53/46/40 ft (11/13/12 to 16/14/12 m) at 1200 tons (4224 kW) for the above temperature ranges. At 400 tons (1408 kW) the dimensions of length/width/height range from 26/36/33 to 33/43/33 ft (08/11/10 to 10/13/10 m) for the same temperature ranges.

REFERENCES

1. "Absorption Refrigeration Cycles," in *ASHRAE Handbook 1985 Fundamentals,* American Society for Heating, Refrigeration and Air-Conditioning Engineers, Atlanta, GA, chap. 1.0, pt. III.
2. C. L. Rescorla and D. K. Miller, "Ammonia-Absorption Refrigeration Systems," in *Plant Engineering Handbook,* 2d ed., McGraw-Hill, New York, 1959, p. 41.
3. M. Bogart, *Ammonia Absorption Refrigeration in Industrial Processes,* Gulf Publ. Co., Houston, TX, 1981.
4. J. Ishida, "Recent Applications of Large (LiBr) Absorption Units in Japan," presented at the ASHRAE Symposium, paper PH-79-3.
5. G. Holldorff, "Revisions up Absorption Refrigeration Efficiency," *Hydrocarbon Process.,* July 1979.
6. G. Holldorff and W. Malewski, "Cogeneration for the Simultaneous Supply of Power, Heat and Refrigeration," *Transactions of the 1986 Citrus Engineering Conf.,* vol. 32, Mar 27, 1986, pp. 1–22.
7. "Standard for Lithium Bromide Absorption Water Chilling Packages," Air-Conditioning and Refrigeration Institute, ARI Std. 560-82.
8. D. K. Miller, "The Performance of Water Cooled Lithium Bromide Absorption Units for Solar Energy Applications," *Heating/Piping/Air-Cond.,* Jan. 1976.

Chapter 4

Expendable Coolants

ERIC C. GUYER
Editor in Chief

4.1 EXPENDABLE COOLANTS VERSUS POWERED REFRIGERATION

Materials that are used to absorb heat and then are discharged to the environment are referred to as expendable coolants. Expendable coolants are used to supply cooling at temperatures that are below and above the ambient temperature. Generally, expendable coolants are most efficient when heat is absorbed in the change of phase of the coolant (i.e., liquid-to-vapor, solid-to-liquid, and solid-to-vapor).

Expendable coolants are used as a practical and economic alternative to powered refrigeration systems in a variety of circumstances. This includes cases when the rate of cooling (in British thermal units per hour or in watts) is large but is of short duration and when the rate of cooling is low but is of extended duration. An example of the former is the current use of ice banks for emergency cooling of the atmosphere within nuclear power plant containment buildings. An example of the latter is the use of liquid nitrogen for cooling infrared radiation sensors.

Factors involved in the choice between expendable coolant system or a power refrigeration system include

1. Weight and size limitations
2. Power availability
3. Supply and maintenance requirements
4. Lifetime
5. Response characteristics
6. Reliability and safety
7. Cost

Depending on the cooling duty (i.e., total energy, power, and temperature), the weight and size of expendable coolant systems may be greater or less than that of powered refrigeration systems. Expendable coolants can have a particular advantage when electric power is limited or unavailable. Expendable coolants do need to be replenished, involving a material and labor cost, and this factor trades against the relatively low maintenance that is characteristic of many powered refrigeration systems. Expendable coolant systems that require coolant storage at a temperature below ambient, such as liquid nitrogen or ice, will lose total cooling capacity over time due to heat gains from the environment. Relatively rapid startup and response to varying cooling load are possible with expendable coolant systems. Total comparative costs should include initial system hardware costs, maintenance costs, and utility costs (power or coolants). Even in cases where there is no onsite electric or heat utility, an

Table 4.1 Heat Absorption Performance of Expendable Coolants at Atmospheric Pressure

Note: Many of the materials listed present important physical and/or chemical hazards.

Substance	Liquid density, lb/ft³	Bp,* °F	Heat of vaporization, Btu/lb	Specific Heat, Btu/lb · °F				Heat absorption								
								Btu/lb					Btu/ft³			
				Bp	200°F	450°F	1500°F	Bp to 0°F	Bp to 100°F	Bp to 200°F	Bp to 450°F	Bp to 1500°F	Bp to 0°F	Bp to 100°F	Bp to 200°F	Bp to 450°F
Hydrogen, H_2	4.37	−423	194	3.42	3.45	3.47	3.62	1,645	1,990	2,335	3,200	6,966	7,190	8,696	10,200	13,980
Water, H_2O	62.4	212	970	0.45	0.45	0.47	0.52	. . .	228†	328†	1,419†	1,944†	. . .	14,230†	20,470†	88,550†
Helium, He	7.62	−452	11	1.24	1.24	1.24	1.24	573	697	821	1,132	2,437	4,360	5,310	6,250	8,620
Ammonia, NH_3	40.6	−28	589	0.48	0.53	0.59	0.81	602	651	703	844	1,572	24,440	26,430	28,550	34,250
Hydrogen fluoride, HF	62.0	67	650	0.34	0.34	0.35	0.38	. . .	661	695	782	1,165	. . .	40,982	43,000	48,500
Methane, CH_4	25.9	−258	248	0.50	0.58	0.71	1.09	382	438	495	675	1,638	9,890	11,340	12,800	17,500
Methanol, CH_3OH	48.1	148	472	0.39	0.39	0.46	. . .	. . .	. . .	492	600	. . .	. . .	. . .	23,650	28,850
Ethanol, CH_3CH_2OH	48.1	173	367	0.41	0.41	0.45	. . .	. . .	. . .	377	486	. . .	. . .	. . .	18,100	23,350
Ethane, C_2H_6	34.1	−127	210	0.33	0.50	0.63	1.0	259	299	345	486	1,290	8,830	10,200	11,750	16,600
Ethylene, C_2H_4	35.3	−155	208	0.33	0.45	0.55	0.83	264	303	345	474	1,168	9,320	10,700	12,200	16,700
Carbon dioxide, CO_2‡	62.0	−110	252	0.20	0.22	0.24	0.30	274	295	317	377	656	16,990	18,290	19,650	23,400
Carbon monoxide, CO	49.3	−313	91	0.25	0.25	0.26	0.29	171	196	219	283	577	8,430	9,660	10,790	13,950
Nitrogen, N_2	50.2	−320	86	0.25	0.25	0.25	0.28	166	191	216	279	569	8,330	9,590	10,820	14,000
Oxygen, O_2	71.2	−297	92	0.22	0.22	0.23	0.26	157	179	201	260	525	11,180	12,740	14,330	18,540
Neon, Ne	75.2	−411	37	0.25	0.25	0.25	0.25	137	162	187	249	507	10,300	12,180	14,080	18,700
Argon, A	88.0	−302	68	0.13	0.13	0.13	0.13	104	117	130	161	292	9,150	10,300	11,450	14,190
Krypton, Kr	162.3	−241	47	0.13	0.13	0.13	0.13	77	90	103	135	268	12,500	14,600	16,750	21,910

* Boiling point at atmospheric pressure.

† Value given is heat absorption for ice at 0°F to indicated temperature with volumetric capacity calculated from the liquid density. For ice at 32°F, subtract 15 Btu/lb or 958 Btu/ft³. For water at 70°F, subtract 182 Btu/lb or 11,350 Btu/ft³.

‡ Solid carbon dioxide assumed cubed and occupying 65% of total volume. Sublimes with evaporation.

CONVERSION FACTORS: °C = (°F − 32)/1.8, 1 Btu/lb · °F = 4187 J/kg · K, 1 Btu/lb = 2326 J/kg, 1 Btu/ft³ = 37.257 J/m³.

SOURCE: Adapted from F. E. Schroeder, E. E. Towe, P. H. Lake, and R. H. Wunderman, "Study of Equipment Cooling Systems," Boeing Airplane Co., WADC TR-59-253, Nov. 1959.

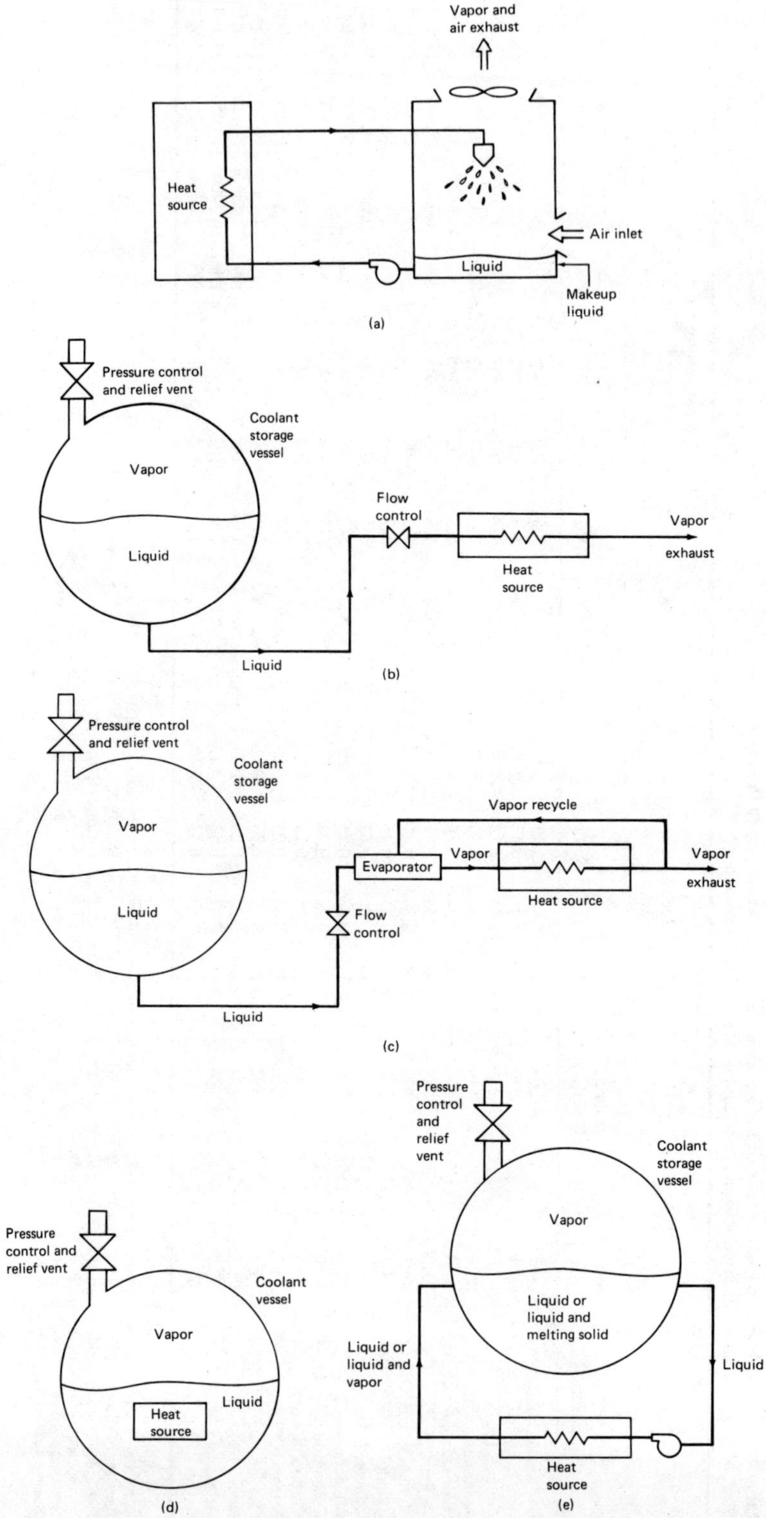

Fig. 4.1 System configurations for expendable coolant systems. (*a*) Evaporative air cooling; (*b*) liquid-wetted heat source; (*c*) vapor-cooled heat source; (*d*) immersed heat source; (*e*) liquid-cooled heat source.

expendable coolant system should not be employed unless it is superior to a powered refrigeration system with its own fuel or electric power storage unit.

4.2 SELECTION OF EXPENDABLE COOLANT

The selection of an expendable coolant for a specific application should be based on a consideration of the required cooling temperature, the heat absorption and heat transfer characteristics of the candidate coolants, and the practical usage factors associated with each of the different coolants.

Table 4.1 provides the heat absorption characteristics of a number of materials that have potential application as expendable coolants. Some general comments with regard to the materials listed are as follows:

Water (Ice, Chilled Water, Ambient Water) Water is by far the most widely used expendable coolant. It is the best choice for systems involving cooling above 32°F (0°C), except for special circumstances.

Ammonia Ammonia provides high cooling capacity per unit mass and volume while having an atmospheric pressure boiling point that is lower than the freezing point of water. However, the toxicity of the exhaust may be a problem. The vapor pressure at room temperature is approximately 130 lb/in^2 (896 kPa), making long term room temperature storage a practical possibility.

Alcohols Alcohols are suitable for cooling at moderate temperatures above ambient. Flammability may be a concern. The possibility exists for use as first a coolant and then as a fuel.

LNG and Liquid Hydrogen, Ethane, and Ethylene These materials can be considered when small amounts of low-temperature or cryogenic refrigeration are needed along with a fuel supply.

Liquid Cryogens (He, Ar, Kr, Ne, O_2, N_2) Liquid nitrogen is the most commonly used and least expensive low-temperature expendable coolant. Liquid inert gases provide the lowest temperature capability, but the latent heats are generally low. All liquid cryogens must be maintained at low temperature to avoid high storage vessel pressure. This is normally accomplished by venting vapor that is the result of liquid evaporation due to vessel heat gain.

Carbon Dioxide Use of solid carbon dioxide (dry ice) has the advantage in that there is no need to deal with a liquid phase at atmospheric pressure. This accounts for its popularity as an expendable coolant in shipping containers. Cooling can be supplied at a much lower temperature than with water ice. Liquid carbon dioxide can be stored at room temperature, but the vapor pressure is approximately 1000 lb/in^2 (6894 kPa).

Halogenated Hydrocarbon Refrigerants (R-12, R-11, R-502, etc.) These materials offer a wide selection of operating temperatures at near atmospheric pressure but should be avoided as expendable coolants because of current concern about the effect of these materials on the environment. They are also relatively expensive compared to the alternatives. Because of reasonable vapor pressure at ambient temperature, it is possible to maintain indefinite storage in a pressure vessel of practical design. See Chap. 1 for physical properties.

4.3 SYSTEM CONFIGURATION

A number of configurations for cooling systems using expendable coolants are shown in Fig. 4.1. Figure 4.1*a* depicts the configuration normally used when water is used as the expendable coolant in an evaporative air cooling system. Water temperatures approaching the wet bulb temperature of the air (and the dew point temperature with more complex configurations) can be attained with water supplied to the system at ambient temperature.

A common design for all vaporizing expendable coolants is shown in Fig. 4.1*b*. The heat source is cooled at the boiling temperature of the coolant provided that the liquid coolant adequately wets the surface of the heat exchanger. Potential problems are dry areas and liquid carryover out the exhaust. Note that when the exhaust temperature is lower than the ambient temperature, it may be desirable to use the residual cooling capacity of the exhaust vapor for some useful purpose, such as reducing the heat gain to the liquid storage vessel. Within the limits of the mechanical design of the system, a valve at the vapor exhaust can be used to increase heat exchanger pressure and thus increase cooling temperature.

The system shown in Fig. 4.1*c* is applicable if the use of direct liquid cooling as in Fig. 4.1*b* is difficult or undesirable. In such a system a large fraction of the exhaust vapor is recirculated and is recooled by direct-contact evaporation from the liquid phase. Because of the temperature rise of the vapor in cooling the heat source, temperature uniformity may be more difficult to establish. Also, heat-transfer coefficients will be lower, resulting in larger required heat-transfer area or larger operating temperature difference.

The configuration shown in Fig. 4.1*d* may be the simplest for many applications. In this case, the heat source is immersed in the expendable coolant liquid. The potential for maintaining uniform temperature at the heat source is good. Evaporated liquid is typically vented through a pressure control valve. Exhausted vapor may be used to reduce heat gain to the storage vessel from the environment.

The configuration of Fig. 4.1*e* has application to water cooling systems that are chilled by ice and liquid evaporating systems where it is desirable to use the liquid as the medium to cool the heat source.

Part 10
Combustion Heat from Clean Fuels

Chapter **1**

Combustion Heat from Clean Fuels

RICHARD J. REED
North American Manufacturing Co.

1.1 INTRODUCTION

Designing and building a heat source into a piece of equipment necessitates a study of the economics of first costs and operating costs. The final segment of this chapter compares costs for combustion with those for electricity, while the first objective of this chapter is to provide the designer with data and know-how relative to various combustion options.

Applications

Applications covered in this section include process heating for domestic, commercial, and industrial use—curing, distilling, drying, heating for forming, heat treating (crystalline restructuring), melting, polymerizing, and preheating. Some examples of equipment and processes are:

- Boilers
- Die-cast machines
- Die, ladle, mold, and platen heaters
- Food preservation and preparation
- Glass fire polishing
- Heaters for liquids such as water, thermal fluids, molten salts, and molten metals
- Plastic and rubber molding and extrusion
- Soldering equipment

Clean Fuels

Clean fuels could include all fuels if it were possible to justify the cost of cleanup of the fuels or their products of combustion. Fuels with heavier molecules (solids and liquids) tend to have higher carbon-hydrogen ratios and therefore tend to smoke or soot if not carefully controlled.

The focus of this chapter is on small and compact heat-producing devices involving combustion of natural gas, liquefied petroleum (LP) gas, and clean gaseous fuels derived from petroleum, coal, and refuse. Kerosene and distillate fuel oils (Nos. 1 and 2) are usually considered to be clean burning fuels, but the scope of this chapter is for sizes that are generally below the economically feasible range for these fuels, except on emergency standby bases. Some of the data and discussion in this chapter applies to these light liquid fuels as well as to gases.

Natural Gas The primary component of natural gas is methane, CH_4, with some ethane, C_2H_6, and lesser amounts of nitrogen, oxygen, and carbon dioxide. Table 1.1 lists analyses and properties of selected natural gases and other clean gases.

LP Gas The primary component of LP gas is propane, C_3H_8, with small amounts of propylene, butane, and butylene. It is generally wise to warn users of small, low-temperature combustion equipment to beware of LP gas containing more than 1 or 2 percent of the latter three components because they may cause smoking and sooting.

LP gas must be vaporized to burn in gas burners. This may necessitate installation of a vaporizer if consumption rate exceeds about 100,000 Btu/h (30 kW), especially if the liquid fuel storage tank is in a cold climate. LP gas can be used as an alternative or standby replacement for curtailed natural gas, but it is necessary to readjust the air-fuel ratio.

Derived Fuels Derived fuels should be scrubbed before use in small-combustion equipment and used with caution because their properties may fluctuate with time.

1.2 FUEL PROPERTIES

The designer of heating equipment needs to understand the fuel-handling and safety limitations of fuels specified for use and must know about the heat-releasing capabilities of the fuels.

Properties Affecting Handling and Safety

Most properties that make burner starting easier and make burners more reliable also, unfortunately, make the fuel more hazardous to handle and transport.

Table 1.1a. Analyses of Typical Gaseous Fuels

	Analysis % by volume, mol %								
Type of gas	CH_4	C_2H_6	C_3H_8	C_4H_{10}	CO	H_2	CO_2	O_2	N_2
Acetylene, commercial				(97.1% C_2H_2, 2.5% C_3H_6O)				0.084	0.28
Butane, commercial, natural gas	...	...	6.0	70.7n-23.3iso-					
Butane, commercial, refinery gas	...	...	5.0	50.1n-16.5iso-			(28.3% C_4H_8)		
Digester (sewage), Bergen, N.J.	59.0	0.05	...	...	...	...	39.4	0.16	0.57
Digester (sewage), Decatur, Ill.	68.0	...	...	...	...	2.0	22.0	...	6.0
Landfill, Cagistrio, 1981	53.4		(other hydrocarbons 0.17)		0.005	0.005	34.3	0.05	6.2
Natural, Alaska	99.6	...	...	...	...	...	...	...	0.4
Natural, Algeria LNG, Canvey	87.20	8.61	2.74	1.07	...	...	...	...	0.36
Natural, Gaz de Lacq	97.38	2.17	0.10	0.05	...	...	...	...	0.30
Natural, Groningen, Netherlands	81.20	2.90	0.36	0.14	...	...	0.87	...	14.40
Natural, Kuwait, Burgan	86.7	8.5	1.7	0.5n-0.2iso-	...	...	1.8	...	0.6
Natural, Libya LNG	70.0	15.0	10.0	3.5	...	...	...	...	0.90
Natural, North Sea, Bacton	93.63	3.25	0.69	0.27	...	...	0.13	...	1.78
Natural, Birmingham, Ala.	90.0	5.0	...	...	...	...	...	...	5.0
Natural, East Ohio	94.1	3.01	0.42	0.28	...	0.01	0.71	0.01	1.41
Natural, Kansas City, Mo.	84.1	6.7	...	...	...	...	0.8	...	8.4
Natural, Pittsburgh, Pa.	83.4	15.8	...	...	...	...	...	...	0.8
Producer, Koppers-Totzek*,‡			(0.28 H_2S)		58.7	32.9	7.0	...	1.1
Producer, Lurgi,† commercial	5.8		(6.6 H_2O, 0.1 H_2S)		14.1	20.9	12.5	...	40.0
Producer, Lurgi,† subbituminous	10.2		(0.4 C_mH_n, 0.3 H_2S)		17.1	40.2	31.4		
Producer, IGT,† lignite	28.2		2.1 C_mH_n, 0.4 H_2S)		19.5	24.5	24.6	...	0.6
Producer, BCR,† West Ky.	15.6		(1.4 H_2S)		44.0	24.4	14.0	...	0.6
Producer, W-G, bituminous†	2.7	...	...	...	28.6	15.0	3.4	0.0	50.3
Producer, Winkler†	1	...	...	...	10	12	22	...	55
Propane, commercial, natural gas	...	2.2	97.3	0.5					
Propane, commercial, refinery gas	...	2.0	72.9	0.8			(24.3% C_3H_6)		
Sasol, South Africa	28.0	...	...	...	22.0	48.0	...	1.0	
SNG, no methanation	79.3	...	...	...	1.2	19.0	0.5		

* O_2-blown.
† Air-blown.
‡ Illinois coal.
SOURCE: *Combustion Handbook,* North American Mfg. Co., Cleveland, Ohio 44105, 1986.

Table 1.1*b*. Properties of Gaseous Fuels from Table 1.1*a*

Type of gas	Gas gravity	Calorific value at STP: Btu/ft³ Gross	Btu/ft³ Net	kcal/m³ Gross	kcal/m³ Net	Gross Btu/ft³ of std. air	Gross kcal/m³ of std. air
Acetylene, commercial	0.92	1492	1444	13,280	12,850	122.9	1094
Butane, commercial, natural gas	1.98	3219	2991	28,640	26,615	105.1	935.1
Butane, commercial, refinery gas	1.97	3170	2955	28,200	26,289	105.6	939.9
Digester (sewage), Bergen, N.J.	0.93	598	544	5,322	4,841	106.1	944.2
Digester (sewage), Decatur, Ill.	0.77	695	632	6,186	5,621	106.2	944.7
Landfill, Cagistrio, 1981	0.88	540	495	4,804	4,404	105	934
Natural, Alaska	0.56	1009	917	8,976	8,160	106.0	942.7
Natural, Algeria LNG, Canvey	0.64	1139	1039	10,132	9,245	105.8	941.3
Natural, Gaz de Lacq	0.57	1029	936	9,154	8,326	105.9	942.5
Natural, Groningen, Netherlands	0.64	887	807	7,894	7,184	105.9	942.3
Natural, Kuwait, Burgan	0.64	1094	997	9,730	8,873	105.8	941.5
Natural, Libya LNG	0.78	1339	1227	11,913	10,919	105.6	940.0
Natural, North Sea, Bacton	0.59	1032	939	9,181	8,356	105.9	942.2
Natural, Birmingham, Ala.	0.60	1000	910	8,897	8,096	105.9	942.4
Natural, East Ohio	0.59	1027	935	9,140	8,318	105.9	942.3
Natural, Kansas City, Mo.	0.63	970	883	8,630	7,858	105.9	942.1
Natural, Pittsburgh, Pa.	0.63	1123	1025	9,995	9,116	105.8	941.4
Producer, Koppers-Totzek*,‡	0.71	290	275	2,581	2,447	131.3	1168
Producer, Lurgi,† commercial	0.80	173	158	1,538	1,405	123.6	1100
Producer, Lurgi,† subbituminous	0.74	304	275	2,705	2,447	122	1087
Producer, IGT,† lignite	0.79	499	457	4,442	4,066	112.8	1003
Producer, BCR,† West Ky.	0.77	388	362	3,465	3,221	120.4	1071
Producer, W-G, bituminous†	0.84	168	159	1,499	1,415	129.6	1153
Producer, Winkler†	0.98	117	111	1,041	988	188.3	1676
Propane, commercial, natural gas	1.52	2499	2316	22,236	20,603	105.1	935.1
Propane, commercial, refinery gas	1.50	2462	2289	21,902	20,367	106.0	942.7
Sasol, South Africa	0.41	511	463	4,544	4,117	118.7	1056
SNG, no methanation	0.47	869	787	7,731	7,003	107.8	958.7

* O_2-blown.
† Air-blown.
‡ Illinois coal.
SOURCE: *Combustion Handbook,* North American Mfg. Co., Cleveland, Ohio 44105, 1986.

Gas Gravity Gas gravity (see Table 1.1*b*) affects the flow and mixing characteristics of a gas. It is also called *specific gravity relative to air,* and is evaluated as

$$\text{Gas gravity} = \frac{\text{density or specific weight of gas}}{\text{density or specific weight of stp air}}$$

where stp (standard temperature and pressure, 60°F and 14.696 lb/in² abs) air has a specific weight of 0.0763 lb/ft³.

Gas gravity greater than 1.0 (heavier than air) means that if there is a leak, the gas will spread across the floor and fill low places. (Workers have been asphyxiated in access pits near propane leaks.) A spark will cause a fire to flash along the gas-air interface.

Gas gravity slightly less than 1.0 (e.g., natural gas, about 0.6) means that the gas will mix easily with air. If this mixture accidentally accumulates in a confined space, a spark will cause an explosion.

Gas gravity much less than 1.0 (e.g., hydrogen, 0.07) means that the gas may leak through joints that are considered tight for other gases.

Minimum Ignition Energy The concept of minimum ignition energy is most easily understood if evaluated as minimum ignition temperature, which is tabulated in Table 1.2. This represents the lowest energy level at which combustion of a given fuel can start, whether spontaneously or by spark, glow plug, or compression (autoignition).

Flammability Limits Flammability limits (Table 1.2) give the lowest and highest percentages of a gas in a gas-air mixture that will burn. Beyond these lower and upper flammability limits (formerly called *inflammability limits*), the mixture is either too lean or too rich to burn at the specified temperature and pressure. With increases in temperature, these limits broaden.

Natural gases have relatively narrow flammable ranges (5 to 15 percent), whereas derived fuels containing hydrogen and carbon monoxide have ranges as much as seven times as wide. This means that the derived fuels may be easier to

Table 1.1c. Properties of Gaseous Fuels from Table 1.1a and b

Type of gas	Wobbe index	Vol. air / Vol. fuel	Stoichiometric products of combustion: $\%CO_2$ dry*	$\%H_2O$ wet	$\%N_2$ wet	Total vol. / Vol. fuel	Flame temperature,† °F
Acetylene, commercial	1559	12.14	17.4	8.3	75.8	12.66	3966
Butane, commercial, natural gas	2287	30.6	14.0	14.9	73.2	33.10	3543
Butane, commercial, refinery gas	2261	30.0	14.3	14.4	73.4	32.34	3565
Digester, (Sewage), Bergen, N.J.	619	5.64	18.1	17.8	67.3	6.63	3285
Digester, (Sewage), Decatur, Ill.	791.5	6.55	14.7	18.4	69.7	7.52	3368
Landfill, Cagistrio, 1981	581.7	5.16	17.6	17.7	67.9	6.10	3276
Natural, Alaska	1352	9.52	11.7	18.9	71.6	10.52	3472
Natural, Algeria, LNG, Canvey	1423	10.76	12.1	18.3	71.9	11.85	3483
Natural, Gaz de Lacq	1365	9.71	11.7	18.8	71.6	10.72	3474
Natural, Groningen, Netherlands	1107	8.38	11.7	18.4	72.0	9.40	3446
Natural, Kuwait, Burgan	1364	10.33	12.2	18.3	71.7	10.40	3476
Natural, Libya, LNG	1520	12.68	12.5	17.4	72.2	13.90	3497
Natural, North Sea, Bacton	1345	9.74	11.8	18.7	71.7	10.77	3473
Natural, Birmingham, Ala.	1291	9.44	11.7	18.6	71.8	10.47	3468
Natural, East Ohio	1336	9.70	11.9	18.7	71.7	10.72	3472
Natural, Kansas City, Mo.	1222	9.16	11.8	18.5	71.9	10.19	3461
Natural, Pittsburgh, Pa.	1446	10.62	12.0	18.3	71.9	11.70	3474
Producer, Koppers-Totzek	343	2.21	27.2	12.1	63.9	2.75	3605
Producer, Lurgi, commercial	192.8	1.40	17.7	17.6	67.7	2.22	2977
Producer, Lurgi, subbituminous	354	2.49	23.4	19.6	61.5	3.20	3317
Producer, IGT,‡ lignite	562	4.43	18.7	17.5	67.0	5.24	3406
Producer, BCR,‡ West Ky.	444	3.23	22.3	14.7	66.0	3.88	3514
Producer, W-G, bituminous	183.6	1.30	18.5	9.8	73.5	2.08	3167
Producer, Winkler	118.2	0.62	24.1	9.3	68.9	1.51	3016
Propane, commercial, natural gas	2029	23.8	13.7	15.5	73.0	25.77	3532
Propane, commercial, refinery gas	2008	23.2	14.0	14.9	73.2	25.10	3560
Sasol, South Africa	794.4	4.30	12.8	21.0	68.8	4.94	3584
SNG, no methanation	1264	8.06	11.3	19.8	71.1	8.96	3485

* Ultimate.
† Theoretical (calculated) flame temperatures, in air, dissociation considered. Although these temperatures are lower than those reported in the literature, they are all computed on the same basis; so they offer a comparison of the relative flame temperatures of various fuels.
‡ With 100% oxygen.
SOURCE: *Combustion Handbook,* North American Mfg. Co., Cleveland, Ohio 44105, 1986.

light and maintain stable, but that leaks may be more prone to cause explosions or uncontrolled fires.

Properties Affecting Heat Release

The composition, or chemical analysis (Table 1.1*a*), and gas gravity affect heat release indirectly, the latter because it influences flow rates through control and burner orifices.

Calorific Value Calorific value (Table 1.1*b*) refers to the chemical energy released by the combustion of a fuel. The *heating value* is related, and is measured by a calorimeter in which a sample of the fuel is burned and the heat release absorbed by a surrounding water jacket. Standards for this measurement (ASTM standards D1826 and D2382) require that the temperature of the final products of combustion (POC) be at or corrected to 60°F. This is identified as the *gross heating value* or *higher heating value*. When modified by subtracting the latent heat of condensation of the water vapor in the POC, it is termed *net heating value* or *lower heating value*. Fuel-burning efficiencies are usually calculated as follows:

$$\text{Efficiency} = \frac{\text{useful output}}{(\text{heating value}) \times (\text{fuel gas flow rate})} \times 100\%$$

In the United States, the gross heating value is usually used in the denominator, but many other parts of the world use net heating values. Except for the recent development of condensing heat exchangers, the net heating value is a more logical approach when comparing the performance of various fuels.

Available Heat This is the amount of heat that is actually available to do useful heating and to satisfy wall losses, opening losses, and conveyor losses from the combustion chamber and furnace. (See Fig. 1-1.) A formula for available heat is

$$\text{Net heat} = \text{gross heat} - (\text{moisture loss})$$

where "moisture loss" is the latent heat in water vapor in the flue gases

Table 1.2. Combustion Characteristics of Fuels*

Fuel	Minimum ignition temperature, °F (°C)		Calculated flame temperature,† °F (°C)		Flammability limits % fuel gas by volume‡				Maximum flame velocity, ft/s (m/s)		% theoretical air for maximum flame velocity
					In air		In O_2				
	In air	In O_2	In air	In O_2	Lower	Upper	Lower	Upper	In air	In O_2	
Acetylene, C_2H_2	581 (305)	. . .	4770 (2632)	5630 (3110)	2.5	81.0	. . .	. . .	8.75 (2.67)	37.0 (11.3)	83
Butane, commercial	896 (480)	. . .	3583 (1973)	. . .	1.86	8.41	. . .	. . .	2.85 (0.87)		
Butane, n-C_4H_{10}	761 (405)	541 (283)	3583 (1973)	. . .	1.86	8.41	1.8	49	1.3 (0.40)	11.6 (3.55)	97
Carbon monoxide, CO	1128 (609)	1090 (588)	3542 (1950)	. . .	12.5	74.2	16	94	1.7 (0.52)	. . .	55
Carbureted water gas	. . .	. . .	3700 (2038)	5050 (2788)	6.4	37.7	. . .	. . .	2.15 (0.66)	. . .	90
Ethane, C_2H_6	882 (472)	. . .	3540 (1949)	. . .	3.0	12.5	3	66	1.56 (0.48)	. . .	98
Gasoline	536 (280)	. . .	. . .	. . .	1.4	7.6					
Hydrogen, H_2	1062 (572)	1040 (560)	4010 (2045)	5385 (2974)	4.0	74.2	4	94	9.3 (2.83)	38.5 (11.7)	57
Hydrogen sulfide, H_2S	558 (292)	. . .	. . .	. . .	4.3	45.5					
Methane, CH_4	1170 (632)	1033 (556)	3484 (1918)	. . .	5.0	15.0	5.1	61	1.48 (0.45)	14.76 (4.50)	90
Methanol, CH_3OH	725 (385)	. . .	3460 (1904)	. . .	6.7	36.0	. . .	. . .	. . .	1.6 (0.49)	
Natural gas	. . .	. . .	3525 (1941)	4790 (2643)	4.3	15.0	. . .	. . .	1.00 (0.30)	15.2 (4.63)	100
Producer gas	. . .	. . .	3010 (1654)	. . .	17.0	73.7	. . .	. . .	0.85 (0.26)	. . .	90
Propane, C_3H_8	919 (493)	874 (468)	3573 (1967)	5130 (2832)	2.1	10.1	2.3	55	1.52 (0.46)	12.2 (3.72)	94
Propane, commercial	932 (500)	. . .	3573 (1967)	. . .	2.37	9.50	. . .	. . .	2.78 (0.85)		
Propylene, C_3H_6	. . .	. . .	. . .	5240 (2893)	2.4	10.3	2.1	53			
Town gas (Br. coal)	700 (370)	. . .	3710 (2045)	. . .	4.8	31.0					

* For combustion with air at standard temperature and pressure. Unless otherwise noted, data are from *North American Combustion Handbook,* North American Mfg. Co., Cleveland, Ohio 44105, 1986.

† Flame temperatures are adiabatic—calculated for stoichiometric ratio, dissociation considered.

‡ In a fuel-air mix. Example: for methane, LEL = 5% or 95 volumes air/5 volumes gas = 19.1 air-gas ratio. Stoichiometric ratio is 9.53:1, so excess air is 19 − 9.53 = 9.47 ft^3 air/ft^3 gas or 9.47/9.53 × 100 = 99.4% excess air.

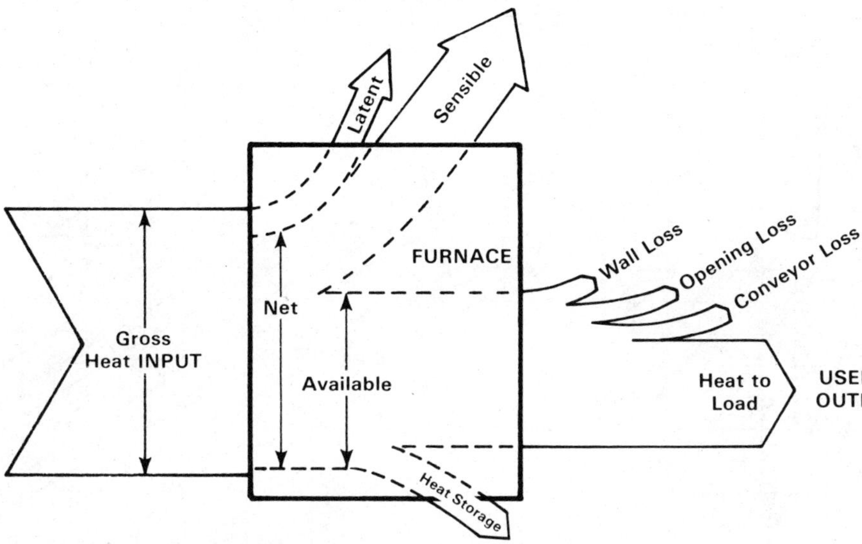

Fig. 1.1 Sankey diagram for a furnace, oven, kiln, boiler, heater, or incinerator. The widths of the arrows are approximate quantitative indicators of where the input energy goes.

$$\text{Available heat} = (\text{net heat}) - [\text{dry (sensible) flue-gas loss}]$$
$$= (\text{gross heat}) - (\text{total flue-gas loss})$$

These values can be expressed in any consistent set of units, such as British thermal units per unit of fuel, per unit of time, or per unit of processed product. Available heat is very often expressed as a percentage of gross heat, as plotted in Fig. 1.2*a* to *d* and tabulated in Table 1.3. This is the *best possible efficiency* for the selected flue-gas exit temperature, assuming zero wall, opening, and conveyor losses.

Example 1.1 A die-casting machine will include a small molten aluminum holding chamber, the flue-gas exit temperature of which is estimated to be 1600°F. The required available heat input is 100,000 Btu/h. Assuming 60°F combustion air and 10 percent excess air, find the required heat input, fuel capacity, and air capacity.

Enter Fig. 1.2*a* at 1600°F on the bottom scale, following the vertical line to the 60°F, 10 percent excess air curve. Move horizontally left and read 54 percent available. Dividing the required 100,000 Btu/h by 0.54 (54/100 percent), gives 185,200 Btu/h required gross input.

Selecting Birmingham natural gas as typical, Table 1.1*b* lists 1000 Btu/ft^3 and Table 1.1*c* lists 9.44 ft^3 air required per cubic foot of gas. Therefore, the required fuel capacity will be 185,200 Btu/h divided by 1000 Btu/ft^3, or 185.2 ft^3/h of gas. The required air capacity will be 185.2 ft^3/h gas × 9.44 ft^3 air/ft^3 gas = 1748 ft^3/h air.

Example 1.2 An integral heater for heat-transfer liquid for a plastic processing machine is predicted to have a stack gas temperature of 500°F when fired at 100 percent of rating with natural gas with 2% oxygen (dry basis) in the exhaust gas. At full rating the load is 4000 lb/h to be raised 200°F. The fluid has a specific heat of 0.7. Determine the required available heat input, the fuel consumption, and the required combustion air input.

The available heat required will be 4000 lb/h × 200° × 0.7 Btu/lb · °F = 560,000 Btu/h. From Fig. 1.2*b*, at 500°F and 2% oxygen, read 81.1 percent available, and note that 2% oxygen corresponds to 9.5 percent excess air. The required gross heat input will therefore be

$$\frac{(560{,}000 \text{ available Btu/h})(100 \text{ gross Btu})}{81.1 \text{ available Btu}} = 691{,}000 \text{ gross Btu/h}$$

Table 1.1*b* shows 1000 gross Btu/ft^3 for a typical natural gas, so the fuel required will be 691,000 Btu/h divided by 1000 Btu/ft^3 = 691 ft^3/h of gas. Table 1.1*c* lists 9.44 ft^3 of air required per cubic foot of fuel gas, but this case will use 9.5 percent in excess of that, or 9.44 × 1.095 = 10.34 ft^3 of air per ft^3 of gas; so the required comburent will be 691 ft^3/h gas × 10.34 air/ft^3 gas = 7145 ft^3/h air.

Reaction Intensity This is an indicator of compactness possible for a combustion chamber. Although compactness may be desirable, a compact combustion chamber will probably require more expensive containment materials, have undesirable emissions, and operate at a higher temperature level with consequent lower fuel efficiency.

Reaction time at normal pressure is about 0.001 s for a premixed industrial flame and is inversely proportional to combustion intensity (Table 1.4).

Substitutability Substitutability of one fuel for another makes it easier to live with fuel supply shortages and price fluctuations. It is preferable that such interchanging of fuels be done without major changes in combustion chambers, burners, piping, or controls. Some considerations are (1) equal heat input rate; (2) fluid handling capability of flues, burners, piping, valves, and regulators; (3) burner stability; (4) heat release pattern; and (5) furnace atmosphere.

Gas-for-gas substitution often involves some readjustment to maintain the same heat input and to maintain the correct air-fuel ratio. This is because (1) the heating value H of the substitute gas may be different and (2) a different gas gravity G will cause a different gas flow rate through valves, regulators, and burner orifices. The *Wobbe index* is a measure of substitutability on both these counts:

$$\text{Wobbe index} = \frac{H_o}{G_o^{1/2}} = \frac{H_s}{G_s^{1/2}}$$

where the subscripts o and s stand for original gas and substitute gas, respectively.

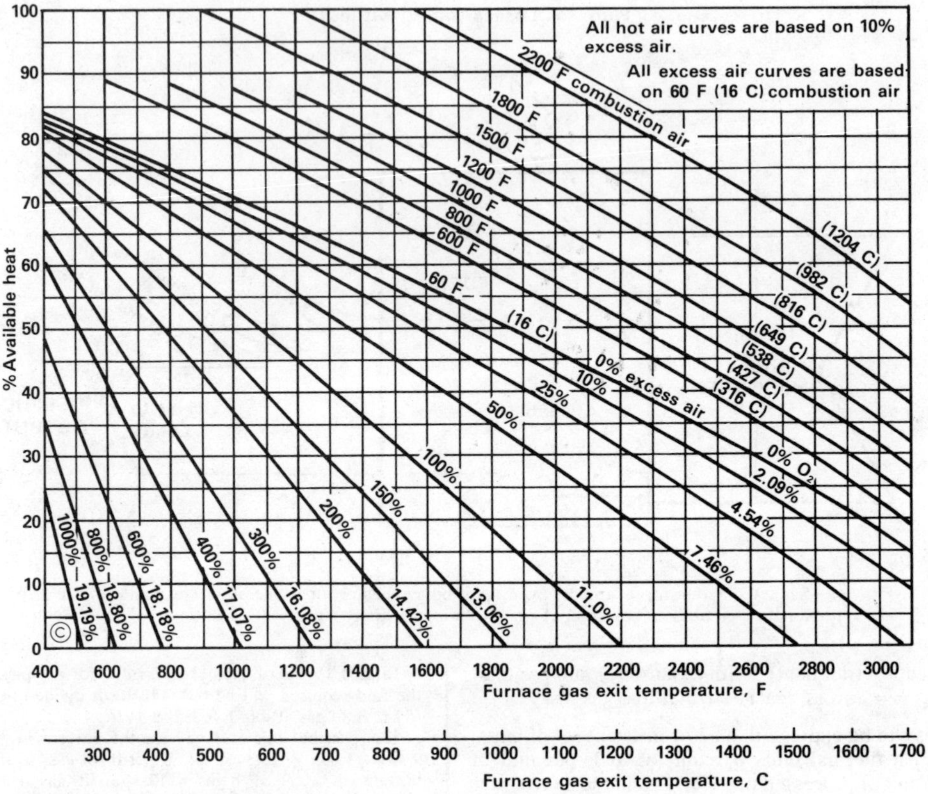

Fig. 1.2a Available-heat graph for a typical natural gas. See also Fig. 1.2b to d and Table 1.3. See Examples 1.1 and 1.2. (*Courtesy of North American Mfg. Co.*)

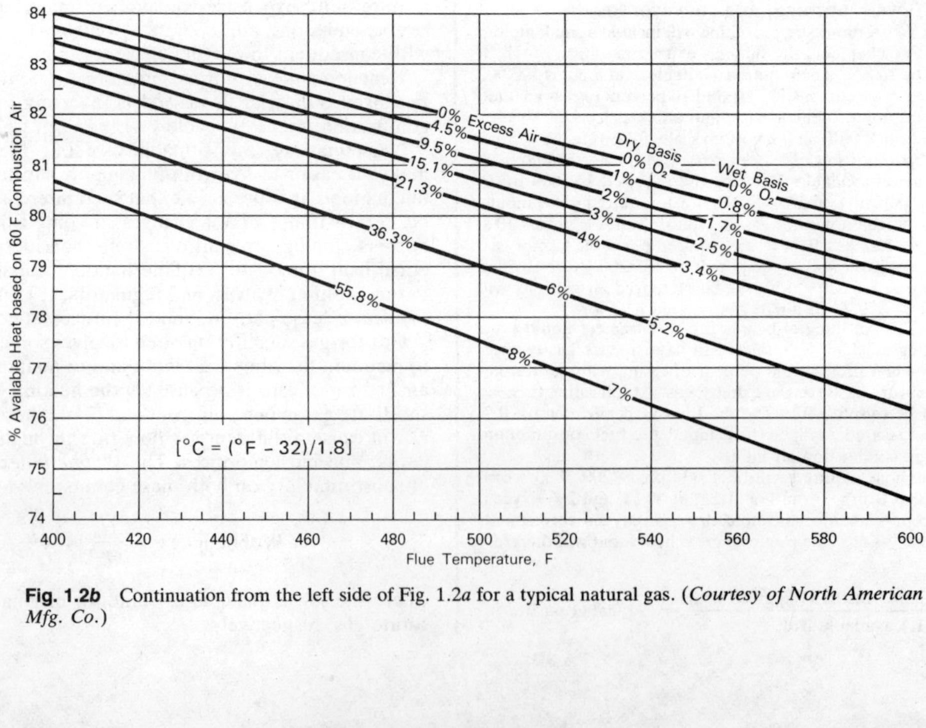

Fig. 1.2b Continuation from the left side of Fig. 1.2a for a typical natural gas. (*Courtesy of North American Mfg. Co.*)

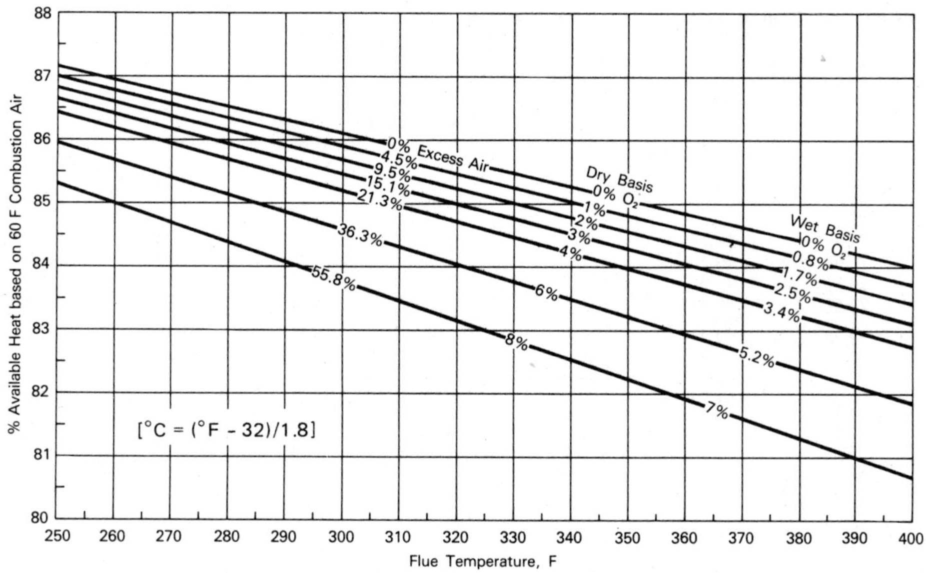

Fig. 1.2c Continuation from the left side of Fig. 1.2*b*. This range should not be used unless equipment is protected from water and acid condensation. (*Courtesy of North American Mfg. Co.*)

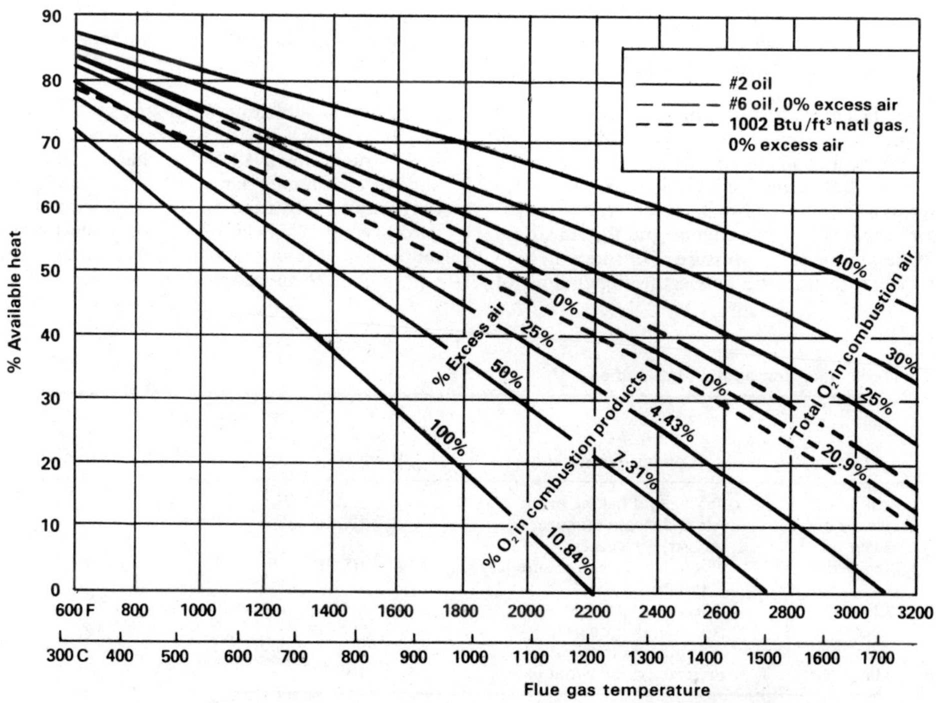

Fig. 1.2*d* Available-heat graph for cases involving oxygen enrichment of the combustion air. Note that this is for distillate no. 2 fuel oil, with a short dashed line indicating a correction for natural gas. (*Courtesy of North American Mfg. Co.*)

Table 1.3. Percent Available Heat for Commercial Propane, with 60°F air

Flue temperature °F	Percent excess air							
	0	5	10	25	50	100	150	200
300	87.80	87.59	87.38	86.75	85.69	83.59	81.48	79.37
400	85.72	85.42	85.12	84.22	82.72	79.71	76.71	73.70
500	83.61	83.22	82.83	81.65	79.70	75.78	71.87	67.96
600	81.46	80.98	80.50	79.05	76.63	71.80	66.97	62.15
700	79.29	78.71	78.13	76.41	73.53	67.77	62.01	56.26
800	77.08	76.41	75.74	73.73	70.38	63.69	56.99	50.30
900	74.84	74.07	73.31	71.02	67.20	59.56	51.91	44.27
1000	72.57	71.71	70.85	68.27	63.97	55.37	46.78	38.18
1100	70.28	69.32	68.36	65.49	60.71	51.15	41.58	32.02
1200	67.95	66.90	65.84	62.68	57.41	46.87	36.33	25.79
1300	65.60	64.45	63.30	59.84	54.08	42.55	31.03	19.50
1400	63.22	61.97	60.72	56.96	50.71	38.19	25.67	13.16
1500	60.82	59.47	58.11	54.06	47.30	33.78	20.26	6.75
1600	58.39	56.93	55.48	51.12	43.86	29.33	14.80	0.28
1700	55.93	54.38	52.82	48.16	40.39	24.84	9.30	...
1800	53.45	51.79	50.14	45.16	36.88	20.31	3.74	...
1900	50.94	49.18	47.42	42.14	33.34	15.73	...	...
2000	48.40	46.54	44.68	39.08	29.76	11.11	...	...
2100	45.84	43.87	41.90	35.99	26.15	6.45	...	...
2200	43.24	41.17	39.09	32.87	22.49	1.75	...	...

CONVERSION FACTOR: °C = (°F − 32)/1.8.

The second column in Table 1.1*c* lists Wobbe index values for a number of gaseous fuels. If the substitute gas has the same Wobbe index as the original gas, no change of valve settings will be necessary to maintain the same input after the fuel change.

If an installation has many heating units for which it would be a great nuisance to readjust many burners when natural gas is curtailed often, a propane-air mixing station may be installed to serve a whole group of heating units. Table 1.5 illustrates how this is used to achieve the same Wobbe index.

All the above assumes that the original and the substitute gas will be supplied at the same pressure. For safety in handling, a gas-air mixture must be above the upper limit of flammability, i.e., too rich to burn until it is further diluted with the normal amount of combustion air at the burners. A gallon of liquid commercial propane vaporizes to 36 ft^3 of gas; commercial butane, 31.5 ft^3, at standard temperature and pressure.

Oil-for-gas substitution is quite easy for single-port burners of 250,000 Btu/h capacity or greater because nozzle-mix dual-fuel burners are available in that range. For smaller sizes and for continuous flame and other small-port burners, the only way to substitute oil for gas is with an external vaporizing system. This can be justified only for systems totaling more than 1×10^6 Btu/h, with frequent curtailments, and with a good technical and maintenance staff.

Table 1.4. Combustion Intensities

Fuel	Combustion situation	Combustion intensity	
		Btu/h · ft^3 · atm	kcal/h · m^3 · atm
All	Drying and baking ovens	10^2	10^3
Heavy oil	Air and steam atomizer	10^4–10^5	10^5–10^6
Light oil	Industrial furnaces	10^6	10^7
Light oil	Gas turbines with pressure jet burners	10^6–10^7	10^7–10^8
Liquid	Rocket	10^8	10^9
Gas	Nozzle-mix burners* (turbulent diffusion)	10^4–10^6	10^5–10^7
Gas	Premixed,† industrial flames	10^6–10^7	10^7–10^8

* *Furnace* volume used in intensity evaluation.
† *Flame* volume used in intensity evaluation.
SOURCE: R. H. Essenhigh, "An Introduction to Stirred Reactor Theory Applied to Design of Combustion Chambers," in *Combustion Technology,* H. B. Palmer and J. M. Beer (eds.), Academic, New York, 1974, pp. 389–391.

Table 1.5. Wobbe Index Illustration

	Natural gas	Commercial propane	Air	Mix = 58.6% propane + 41.4% air
H, Btu/ft^3	1025	2615	0.0	1532
G	0.592	1.552	1.0	1.324
Wobbe	1332	2099	0.0	1332

CONVERSION FACTOR: 1 Btu/ft^3 = 37,258 J/m^3.

1.3 BURNER CONFIGURATIONS AND OPERATING CHARACTERISTICS

Premix Burners

Premix burners are usually simply nozzles and manifolds for a gas-air mixture that has been prepared in a mixer somewhere upstream. They may have more thorough mixing than nozzle-mix burners, but their principal use is in systems where it is necessary to spread a small amount of heat over a wide area—as across the width of a piece of cloth being singed. Only a single mixture line needs to be piped to each flame location, whereas, a nozzle-mix arrangement would require separate air and gas lines to each.

A great variety of heat-release patterns and concentrations are possible by design of the manifold and the ports (nozzles) thereon. Early models were simply drilled pipes, but they had poor stability, lacking adequate flame holding configurations. Most low-capacity designs rely on the cooling effect of the metal nozzles to position the flame. Higher-capacity versions use baffles and bluff bodies to position the flames by velocity control.

For a given burner design, increasing the mixture pressure causes the length and, to a lesser extent, the width of each flame envelope to increase. A greater proportion of primary air shortens the flame. The stability of the array of flames depends on the size and spacing of the ports. For natural-gas burners, a typical maximum port size is ⅛-in diameter or $\frac{1}{16}$-in slot. It is recommended that the gross input rate for natural gas be in the range from 12,000 to 20,000 Btu/h · in^2 of port opening. (This is five to eight times the laminar flame velocity.) Fuels containing hydrogen can go higher. Spacing must be close enough to assure reliable lighting for the anticipated mixture pressure. Enclosed burners do not require as close spacing as those in the open. Locating ports too close together may interfere with adequate aeration by the secondary airstream.

All premix burners have a rather narrow range of stable firing rates and air-fuel ratios. They may be subject to flashing back, blowing off, or yellow tipping. Flashback, burning back in the manifold at the mixing point, limits the low fire rate. It may occur if the mixture pressure drops below 0.25 in of water column with natural gas or below 0.40 in of water column with faster-burning gases. Blowoff or lifting of the flame away from the nozzle (and perhaps being extinguished) limits the high fire rate. Yellow tips occur with poor mixing or localized rich mixtures.

Situations that may cause the problems described above are cross-drafts, variable mixture pressure, nozzle overheating, mixture overheating, turndown to too low an input, and turnup to too high an input. Figure 1.3 shows the above limits for a specific configuration and suggests the usual design range.

Atmospheric Burners These burners are like those used on domestic ranges and many types of domestic heaters. They use very low gas pressure to inspirate one-third to three-fourths of the required combustion air through a simple venturi. Input intensity and turndown range are very limited. Most rely on natural draft to supply the balance of the required combustion air (secondary air) around the burner nozzles, with the result that the air-fuel ratio is uncontrolled, usually with much excess air.

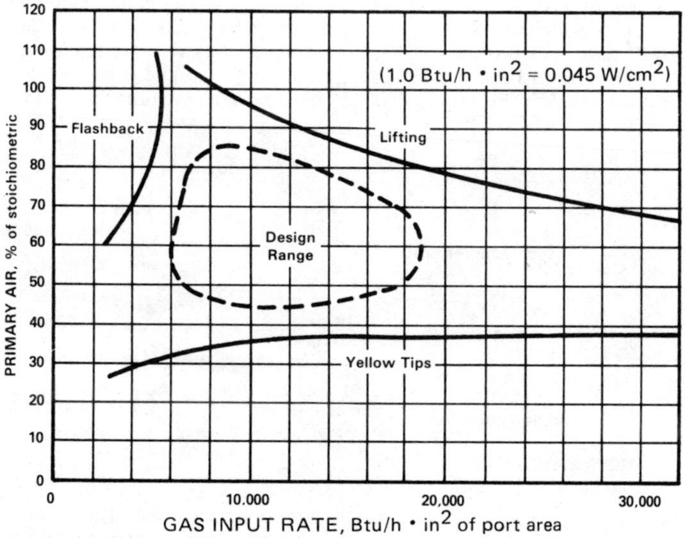

Fig. 1.3 A typical premix burner limitation.

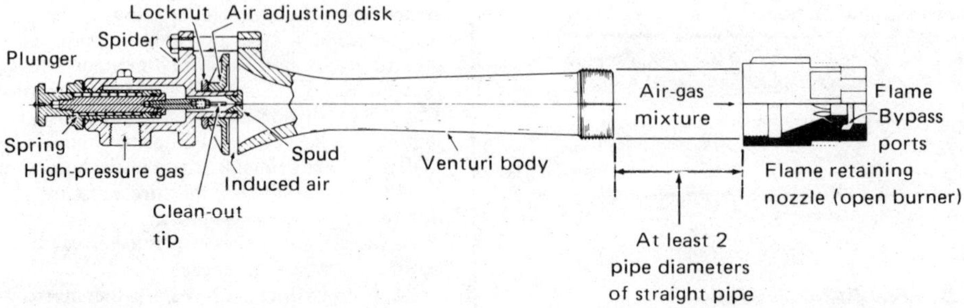

Fig. 1.4 Inspirator (gas-jet) mixer. High-velocity gas flows through the spud, entraining and mixing with air induced in proportion to the gas flow. The venturi converts mixture velocity to static pressure. Air-gas ratio is adjusted by positioning the air disk. The spring-loaded spud-cleaning plunger is used only with dirty gases. This type of mixer is often used with burners such as shown in Fig. 1.5.

Because the low gas pressure of atmospheric burners generates an even lower mixture pressure, the velocity of the POC is so low that convection heat transfer varies with the amount of "draft," or secondary airflow. An induced-draft fan can improve the convection heat transfer. The flames are usually nonluminous, so radiation is low. Radiation can be enhanced by flame impingement on or attachment to some refractory or metal material that can, in turn, radiate to the load. A ceramic fiber matrix burner utilizing this principle is described later in this chapter.

High-Pressure Inspirator Mixer-Burner Combinations These are an industrial, heavy-duty extension of the atmospheric burner principle. (See Fig. 1.4.) These enter a fuel consumption range where their poor fuel efficiency (because of inadequate air-fuel-ratio control) usually causes users to opt for "power" burners, discussed below. Inspirator mixers rely on the velocity energy of a 5 to 30 (or higher) lb/in^2 gage gas jet to induce the necessary air (10 times the gas volume for natural gas, 24 times for propane).

Power Premixer-Burner Combinations Aspirator mixers in which blower air induces the flow of gas are used in these systems. This arrangement assures a more steady air-fuel ratio throughout the turndown range and provides higher mixture pressures than atmospheric mixer-burner arrangements with less gas pressure than required by inspirators. (See Fig. 1.5*a* and *b*.) A total burner area 2.5 times the mixer throat area usually gives good performance.

Nozzle-Mix Burners

Nozzle-mix burners are more versatile than premix burners in terms of variety of flame shapes, width of air-fuel ratio, extent of turndown range, and dual-fuel adaptability. Nozzle-mix burners have displaced premix, large-port burners of the conventional feather-shaped flame. There are many nozzle-mix configurations designed to enhance radiation by using the burner's own refractory tile or adjacent wall as the radiator.

High-Velocity Burners High-velocity burners such as that shown in Fig. 1.6 enhance convection heat transfer and might be considered for jet-impingement heating. They incorporate partial combustion (which expands the gases) upstream of a final restrictive nozzle, thus generating a high-momentum stream of products of combustion. The final nozzle may be refractory or alloy.

Air and Fume Heating Burners These burners (Fig. 1.7*a* and *b*) utilize all-metal configurations that promote mixing of the burner POC with the passing air or fume stream. These are widely used in recirculating ovens and in fume incinerators. Because the primary load to be heated is air, excess air does not constitute a wasteful use of fuel. Air-fuel-ratio

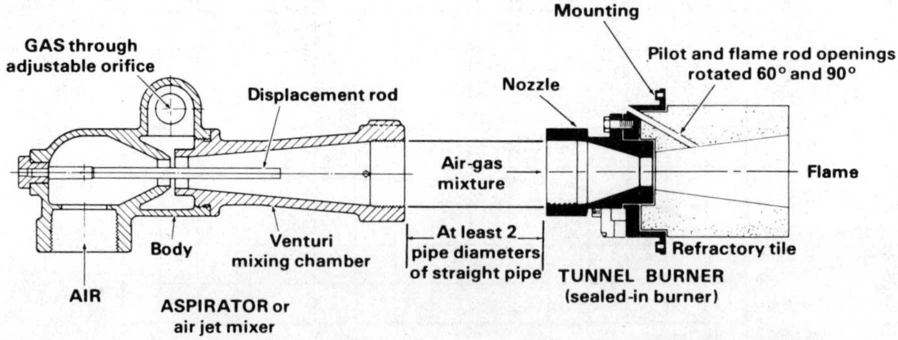

Fig. 1.5a Aspirator (air-jet) mixer. Blower air enters at lower left, usually at about 0.5 lb/in^2 gage. Gas is pulled into the airstream from an atmospheric regulator through a V-port orifice used for adjusting air-gas ratio.

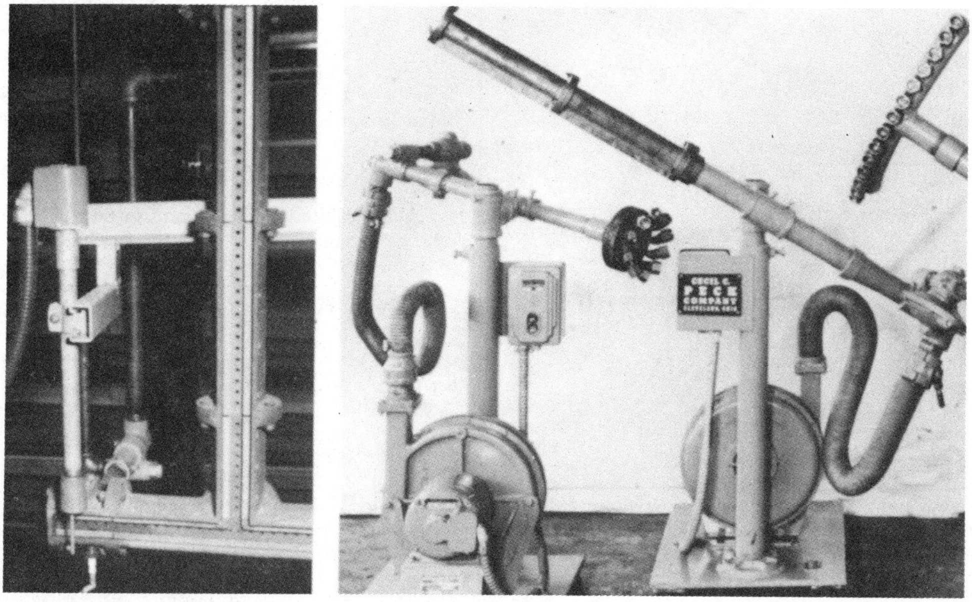

Fig. 1.5*b* Power premix gas burner systems—including blower, aspirator mixer, atmospheric regulator, valve, and a variety of burners. Continuous flame burners, with two extra rows of ports under stainless baffles, are more stable than drilled pipe or ribbon burners.

control and careful mixing design are necessary to prevent quenching of the flames before combustion is complete.

Radiant Tube Burners

Radiant tube burners fire through alloy or conducting ceramic (silicon carbide) tubes used to increase heat-transfer area for indirect heating. As the name implies, the original purpose was to provide a radiant-heat source in the form of a row of glowing tubes while protecting the product load from contact with the POC. They are now also used for convection-type indirect-air heaters. For pipes 3 in and smaller, premix burners are usually used, often with pull-through eductors. For larger pipe sizes, nozzle-mix burners are used, and energy-saving regenerator arrangements are available to preheat the incoming cold combustion air with the outgoing hot POC.

Immersion Burners

Immersion burners are used to heat liquids. The most common method is to fire a burner inside a metal tube immersed in a vat of liquid. Because the heat-transfer coefficient h is so much better on the liquid outside than the gas inside, some means should be provided to enhance POC velocity. Swirl or high-velocity burners are more effective than corebusters or internal longitudinal fins, both of which foul easily, reduce capacity, and require extra draft.

Inserting radiant premix fiber-matrix burners (Fig. 1.8) into an immersion tube gives nearly uniform heat transfer along the tube length. These fiber-matrix burners provide high total heat transfer for a given maximum liquid film temperature because of the high percentage of radiant output.

Gas flames have been immersed beneath the surface of liquids, such as water solutions, but they require a great deal of careful design. Turndown range is usually very limited. Chemical reaction between the products of combustion and the liquid may alter liquid composition or prevent thorough combustion.

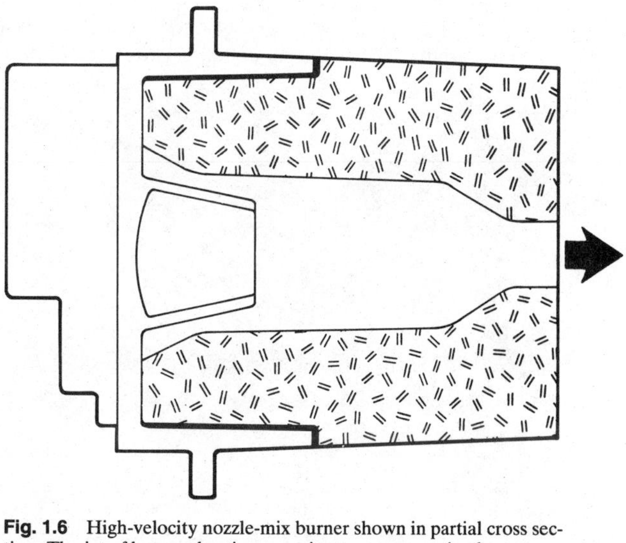

Fig. 1.6 High-velocity nozzle-mix burner shown in partial cross section. The jet of hot combustion gases increases convection heat transfer and induces strong recirculation within a furnace.

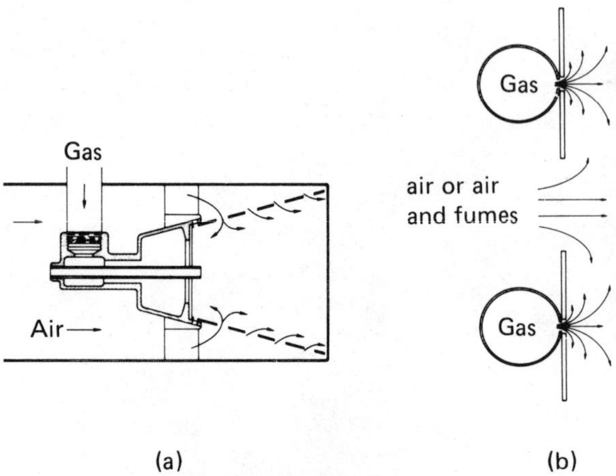

Fig. 1.7 (*a*) In-duct burner. This kind of burner is often installed in the ducts of recirculating ovens, such as shown in Fig. 1.11*b*. (*b*) Fume incinerating grid. The perforated stainless-steel plate attached to rows of raw-gas manifolds would be fitted all the way across the fume duct so that no fume could bypass the grid of ports and flames.

1.4 COMBUSTION CONTROL

Air-Fuel-Ratio Control

The stoichiometric or ideal volumetric ratio of air to fuel is listed for a number of the gaseous fuels shown in Table 1.1. Operating with less than the ideal air-fuel ratio will result in waste from unburned fuel, may emit aldehydes or even carbon monoxide, and may form smoke or foul heat-transfer surfaces with soot and coke. Operating with more than the ideal air-fuel ratio (excess air) wastes fuel by heating an unnecessary load of air, but this is less harmful than fuel-rich operation, so most heaters are adjusted to operate with a minimal (5 to 15 percent) amount of excess air.

Extreme deviation from the ideal air-fuel ratio can present a safety hazard and/or result in pollution, loss of reliability,

Fig. 1.8 Fiber-matrix burner for an immersion tube. The three parts shown in a vertical position would be attached in series to the end of the horizontal segment, making a 6½-in diameter × 16-ft long radiant heat source that operates within a 10½-in-diameter tube. This burner is rated at 2,400,000 Btu/hr.

and reduced heating effectiveness because of lowered flame temperature.

Air-fuel ratio is often initially set by the appearance, sound, and smell of the flame and POC. It should then be set more accurately by using a flue-gas analyzer and adjusting for the maximum percent carbon dioxide CO_2 or minimum percent oxygen O_2 without smoking or showing yellow flame tips. Figure 1.9 shows how flue gas analysis varies with air-fuel ratio.

For atmospheric burners and inspirator mixers, air-fuel ratio can be adjusted by using the air shutters. For aspirator premixers and nozzle-mix burners, there should be a fine-tuning valve (adjustable limiting orifice) in the fuel line near each burner for setting air-fuel ratio. In the units shown in Fig. 1.5, a multiple-fluted V-port plug, advanced by a fine thread, is built into the gas connection of the aspirator mixers for this purpose.

All air-fuel-ratio settings are vulnerable to changeable fuel and air-supply pressures, so line pressure regulators must be installed upstream.

For burners that are operated at a variety of input rates (modulated), an automatic air-fuel-ratio control is essential to maintain efficiency, cleanliness, reliability, and high flame temperature at all firing rates. Low-cost systems use linkage arrangements to modulate fuel and air valves together. The most widely used air-fuel-ratio control system is pressure-balanced, using special regulators to make the air pressure "track" the fuel pressure or the fuel pressure to track the air pressure. Sophisticated systems actually meter the fuel flow and airflow and use digital controllers to compare the flows and adjust accordingly. The *oxygen trim control* system uses a continuous oxygen sensor in the flue-gas stream to bias the air-fuel-ratio control system.

Combustion Chamber Pressure Control

Combustion chamber pressure control always makes air-fuel-ratio control work better because the downstream (furnace) pressure may influence flow of fuel and air into the combustion chamber.

Combustion chambers should be of tight construction to prevent air in-leakage caused by the negative draft produced by the hot column of gases in the chamber and stack. Such air leakage often "fools" a flue-gas analyzer, resulting in a faulty air-fuel-ratio setting. The air leaking in is rarely mixed well enough with the fuel to participate in the combustion reaction, but it does absorb some heat, thereby reducing efficiency.

In-leaking cold air may also cause localized chilling of part of the load in direct-fired heaters or part of the heat exchanger in indirect-fired units.

Some form of automatic damper control (furnace pressure control) is desirable on any modulated heater to minimize the high cost of cold-air infiltration.

Input Control

Automatic temperature control (or pressure control, in the case of a boiler) may be as simple as a bimetal sensor actuating a solenoid-operated fuel valve, or as complex as a computer-directed program responding to several cascaded sensors and/or timers.

It is important to emphasize that combustion quality considerations must be part of the control logic and that ramp

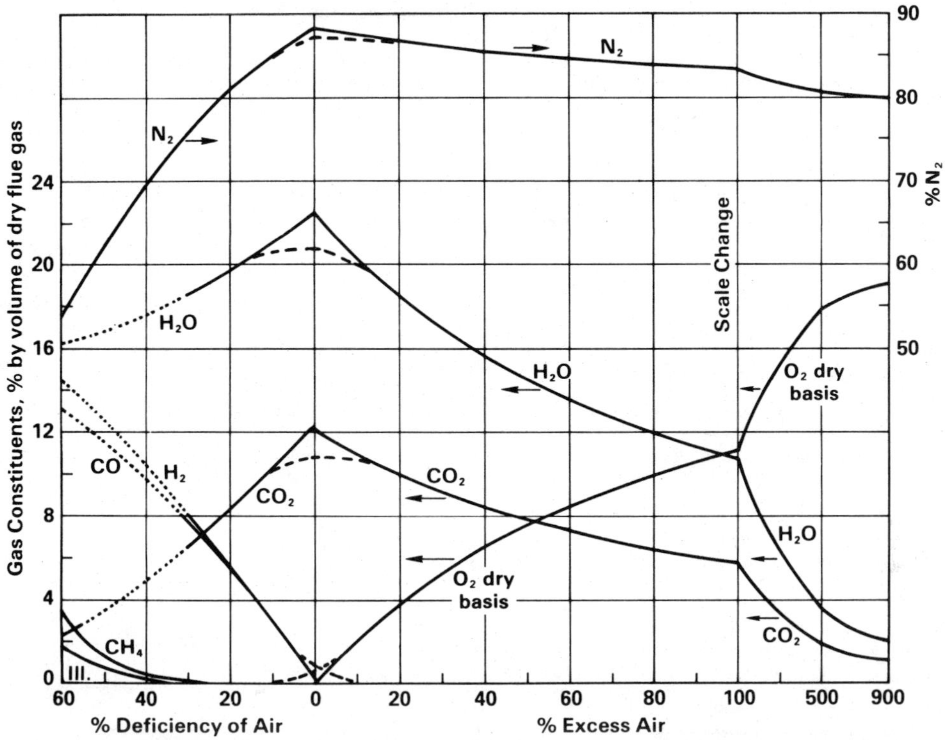

Fig. 1.9 Effect of air-fuel ratio on flue-gas analysis of methane combustion products. The percent CO_2 will be higher and percent H_2O lower for propane and liquid fuels, but most other variables shown will be about the same for all fossil fuels. The dashed lines indicate variations that may result from poor mixing or flame quenching.

rates must be within reason for practical combustion operation and nonharmful heat-transfer effect on the equipment and the load.

Improperly programmed combustion can lead to explosions and fuel-fed fires, so it is imperative that safety aspects be incorporated into programmed controls. There are many governmental and insurance regulations covering fuel-fired equipment. The following are a few general suggestions:

1. Purge combustion chambers with air, steam, or inert gas before start-up and after shutdown. Accumulation of unburned fuel is a major cause of explosions.
2. Cleanliness is essential. Plugged air filters, for example, can cause burners to go rich, then out.
3. If the same flame-supervising device monitors both pilot and main flame, the pilot must be programmed out (interrupted) after the main flame is lighted.
4. Redundancy in fuel shutoff valves is desirable, as is a leak test system.
5. Operating instructions and nameplates must emphasize the need for regular maintenance and checking of shutoff valves and interlocks.
6. Installation instructions should include a clearly marked fuel shutoff station at a safe but visible distance from any heater. This makes it possible to discontinue a fuel-fed fire after an explosion.
7. Fuel lines and regulators should be protected from damage by impact, wear, vibration, or heat.
8. Large airtight observation ports encourage awareness of the combustion process and its maintenance needs.

1.5 DIRECT HEATING

If the products of combustion (nitrogen, carbon dioxide, and water vapor) make direct contact with the surface(s) or fluid to be heated, the process is termed *direct heating*. By contrast, in *indirect heating,* the products of combustion transfer their heat through some separating wall to another fluid (as in a heat exchanger). (Figure 1.14*a* to *c* illustrates direct heating; Fig. 1.14*d* to *h*, indirect heating.)

Conventional Convection Heating

Conventional convection heating rates can be evaluated by the usual heat-transfer formula

$$Q = hA\,\Delta T$$

where Q = heat flow rate
h = convection film coefficient
A = exposed surface area
ΔT = log-mean temperature difference (LMDT) between the POC stream and the surface or fluid being heated

The h values are dependent primarily on Reynolds number, the ratio of momentum forces to viscous forces, which determines the ability of the main stream to reduce the thickness of the stagnant, insulating fluid boundary layer. Reynolds number is dimensionless: density × velocity × dimension/viscosity. Of these, velocity is a predominant design variable, as seen in Table 1.6.

Jet-Impingement Heating

Jet-impingement heating utilizes flame or hot gases in direct contact with a load to be heated. It is most often applied in fast heating of small pieces. Examples are surface hardening

Table 1.6. Convection Coefficients for Turbulent Flow h_c in Btu/ft² · h · °F

Air, natural convection	1–5	
Air or superheated steam, forced convection	5–50	(see also tables below)
Oil*, forced convection	10–300*	
Water, forced convection	50–2000	(see also tables below)
Water, boiling	500–5000	
Steam, condensing	1000–10000	

Flow configuration	Air or products of combustion at 200°F mean film temperature† and <1 lb/in² gage						Water at 150° mean film temperature				
Across a single pipe	Nominal pipe size, in	Velocity, ft/s 1	3	8	20	50	Nominal pipe size, in	Velocity, ft/s 1	3	8	20
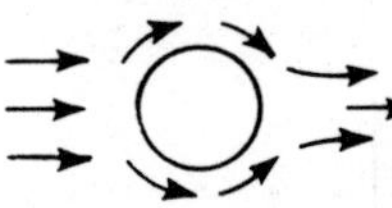	½	1.9	3.6	6.4	11.0	18.6	½	530	1070	2360	4940
	1	1.6	3.0	5.3	9.1	15.4	1	450	980	2160	4530
	2	1.2	2.4	4.2	7.1	12.0	2	360	870	1930	. . .
	4	0.9	1.8	3.2	5.4	9.2	4	310	770	. . .	. . .
	8	0.7	1.4	2.4	4.1	7.0	8	280	. . .	. . .	. . .
Across a bank of tubes‡	Actual tube OD, in	Velocity,§ ft/s 1	3	8	20	50	Actual tube OD, in	Velocity,§ ft/s 1	3	8	20
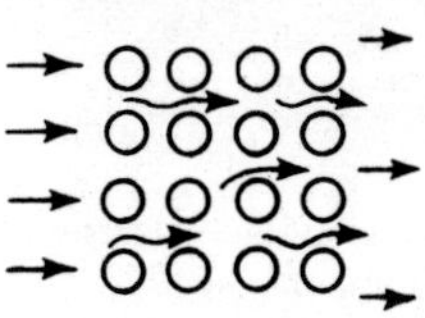	⅜	2.6	5.2	9.3	16.2	28.1	⅜	950	1850	3330	5780
	¾	2.0	3.9	7.1	12.3	21.3	¾	720	1400	2530	4380
	1½	1.5	2.9	5.3	9.3	16.1	1½	550	1060	1910	3320
	3	1.1	2.2	4.0	7.1	12.2	3	410	800	1450	2510

Parallel to a flat surface	Velocity, ft/s 3	8	20	50
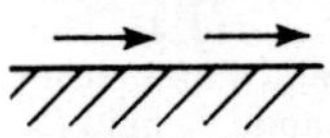	0.9	1.8	3.7	7.4

Inside a pipe	Nominal pipe size, in	Velocity, ft/s 3	8	20	50	125	350	Nominal pipe size, in	Velocity, ft/s 1	3	8	20
	1	. . .	2.43	5.06	10.5	21.9	50.1	½	420	1020	2240	4650
	2	0.97	2.12	4.42	9.20	19.1	43.7	1	380	920	2010	4190
	4	0.85	1.86	3.86	8.04	16.7	38.2	2	330	800	1760	3660
	8	0.74	1.62	3.37	7.01	14.5	33.3	4	290	700	1330	3200
	16	0.65	1.42	2.95	6.14	12.7	29.2	8	250	610	1340	2790

* For 150 SSU fuel oil, $h_c \approx 1.3 \times$ ft/s; for 2500 SSU fuel oil, $h_c \approx 9 \times$ ft/s.

† For other mean air film temperatures, use the following multipliers:

T, °F ⟶	100	200	600	1000	1400	1800
Factor ⟶	1.09	1.0	0.76	0.66	0.56	0.50

‡ h_c may be one-quarter to one-third higher for triangular pitch arrangement: 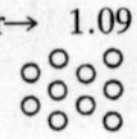

§ Average velocity in space between tubes.

NOTE: 1 ft/s = 0.3048 m/s; 1 Btu/(h · ft² · °F) = 5.678 W/(m² · K).

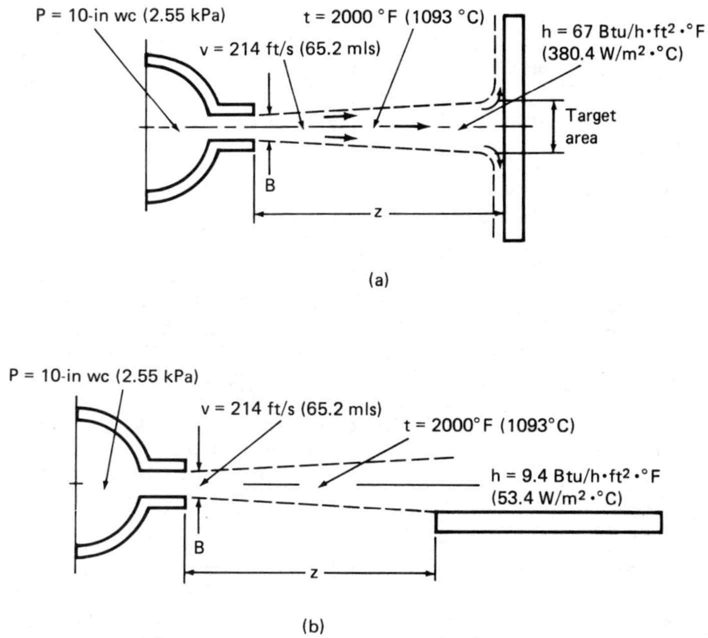

Fig. 1.10 (*a*) Jet-impingement heating at a 90° angle results in a local heat-transfer coefficient within the target area of 67 Btu/h · ft² · °F for a pressure drop of 10 in of water column and $Z/B = 8$. (*b*) Parallel-flow heating, at a 0° angle, results in a local heat-transfer coefficient of 9.4 Btu/h · ft² · °F but diminishes as the velocity degrades.

of gears and engine valves; fire polishing of glassware; and localized heating for brazing, soldering, hot punching, and forming. All of these involve intentionally heating only part of the load. Entire pieces can be uniformly heated by impingement heating using moving jet arrays or by conveyor movement of pieces past arrays of overlapping jet streams.

The critical variables in impingement heating are velocity and nozzle-surface configuration. Figure 1.10*a* and *b* gives example heat-transfer coefficients for the two extremes of perpendicular and parallel flow. Within the target area of the 90° case, h is typically six times that for a 0° case.

The centerline velocity of a cold (stp) jet issuing from a manifold nozzle is

$$V = 67.6\ \Delta P^{1/2}$$

where V is velocity in feet per second and P is pressure drop in inches of water column. The velocity dissipates both laterally and longitudinally.

The Z/B ratio is an important parameter: Z is the distance from the jet nozzle to the surface being heated, and B is the nozzle dimension—the diameter in the case of round nozzles; the width for slots. Optimum heat-transfer coefficients are usually obtained with Z/B in the range of 7 to 8. Although h will be less with $Z/B < 7$, the ΔT will be higher. This advantage is used where very localized heating is desired, but short Z/B configurations result in an undesirable striping if larger areas are to be heated (Fig. 1.11).

The data of Fig. 1.10*a* and *b* are for jets without combustion. The thermal expansion and turbulent reaction of combustion will increase the velocities and the width of the jets, resulting in more prompt merging of adjacent jets. A lack of data on these effects forces equipment designers to resort to trial and error procedures for accomplishing uniform heating.

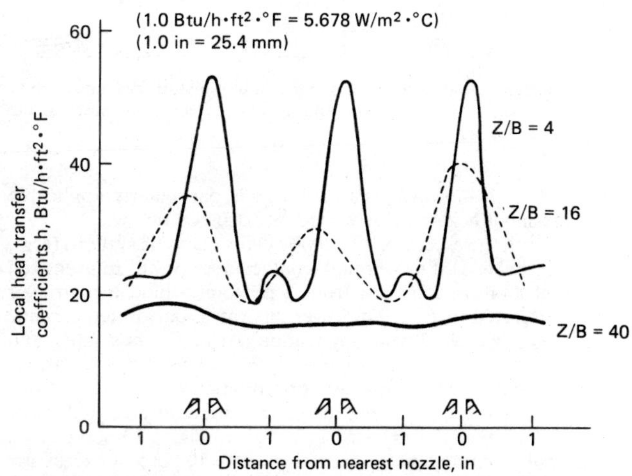

Fig. 1.11 Arrays of jets merge into a continuous "sheet" of hot gas or flame as distance from the nozzles increases. Temperature also dissipates with distance, making the uniformity even better with increasing distance.

Radiation Heating*

Radiant heating is often used in direct-fired situations requiring a uniform flux over an extensive surface area. Low-temperature examples are textile and paint drying. Radiant-combustion systems have wide use in high-temperature direct-fired systems and in a range of indirect-fired applications.

Radiant heating depends on surface radiation to achieve a high component of radiant flux to the product. This can be accomplished in one of the following ways:

1. *Convention heating:* The radiating surface is the lining of a conventional furnace, relatively large compared with the flame, heated by a combination of gas radiation, luminous radiation, and low-velocity convection.

2. *External-contact heating:* The radiation surface is the "face" of a "radiant burner" heated by high-velocity impinging convection, or surface combustion, as in Fig. 1.12. Some configurations allow the POC to flow across an adjacent wall, thus enhancing wall radiation.

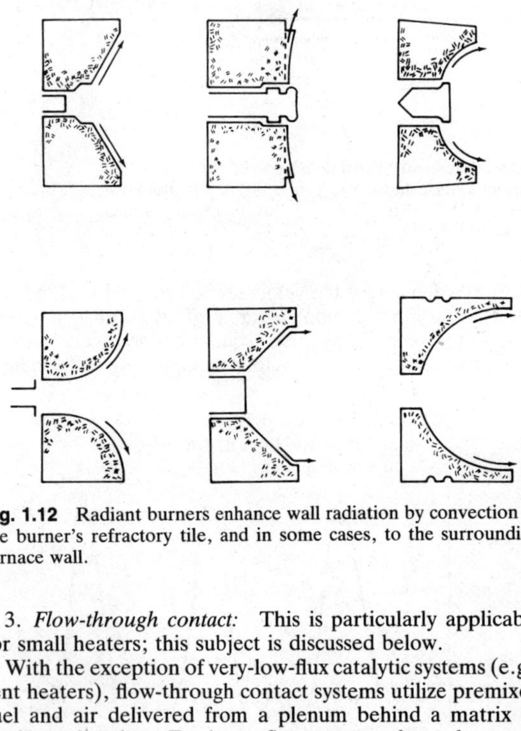

Fig. 1.12 Radiant burners enhance wall radiation by convection to the burner's refractory tile, and in some cases, to the surrounding furnace wall.

3. *Flow-through contact:* This is particularly applicable for small heaters; this subject is discussed below.

With the exception of very-low-flux catalytic systems (e.g., tent heaters), flow-through contact systems utilize premixed fuel and air delivered from a plenum behind a matrix or perforated surface. For lower flux rates, perforated ceramic tiles have found wide application. At higher flux rates, and for greater fuel turndown and preheat variability, various ceramic porous media are appropriate.

Ideally, the gases exit the surface zone fully burned at the surface temperature. In such a case, Fig. 1.13 serves as a design nomograph for natural gas with 10 percent excess air.

EXAMPLE 1.3 An ideal premix flow through a burner fired at a net heating rate of 90,000 Btu/h with an emissivity-view factor product of 0.8 has a preheated air stream at 500°F and is heating a product at 1200°F (sink temperature). In Figure 1.13, at *A* (90,000/0.8 = 112,500), proceed vertically to *B* on the 1200°F sink temperature curve, then horizontally (to the right in *this* case) to the base preheat curve at *C*, then parallel to the appropriate sink temperature curve (1200°F) to the correct 500°F preheat curve at *D*. The surface temperature (2100°F) is on the top abscissa at *D* and the percent of the net fuel flux input radiated from the burner surface (52 percent) is read on the ordinate at *D*.

As the input to a flow-through radiant burner increases, the flame will begin to detach and the radiant fraction will be greatly reduced. Ultimately, the burner may perform as an attached blue flame burner, or the flame may detach totally. The ability to maintain radiant surface combustion may be enhanced by catalytic materials.

Flow-through radiant burners can provide highly uniform heat fluxes. They also provide almost immediate radiant efficiencies in excess of 50 percent without the need for large volumes of radiative gas. Other benefits are very low noise and low pollutant emissions (typically 10 ppm NO_x).

The current temperature limit for the radiant surfaces is about 2200°F. Structural design must assure that the perforated or porous surface materials will not crack, shrink, or separate, because that would result in a short circuit and loss of surface temperature uniformity and could lead to explosive flashback into the plenum.

The uniform flux and high radiant efficiency have encouraged use of flow-through radiant-heating devices in an array of important applications. The perforated tile type, which can be fed with inspirated low-pressure gas, has found broad use in domestic heaters and some commercial warming applications. Porous ceramic block systems have been used in high-temperature heat-treating systems. Porous fiber matrix systems are used in textile drying, commercial and residential appliances, and industrial boiler and process applications. An example of a segmented industrial process burner can be seen in Fig. 1.8.

1.6 INDIRECT HEATING

Indirect heating avoids contact between the source of heat (flame and hot products of combustion) and the receiver of heat (workpiece, product, load, or process fluid to be heated). Some sort of conductive dividing wall separates the source and receiver of heat. This "heat exchanger" may take many forms—double-pipe, shell-and-tube, plate-type, and finned heat-transfer surfaces, or simply a tank or drum surrounded by a furnace.

Indirect heating invariably costs more in fuel and equipment; so direct heating is preferred unless either:

1. The quality of the end product being heated would be affected by some chemical or metallurgical reaction with the POC (CO_2, H_2O, N_2, and maybe SO_x or NO_x).

2. Combustible or hazardous gases would be released during heating of the product (e.g., solvent vapors in some drying or curing processes.

3. The product requires containment because of its fluid (gas or liquid) state.

Some heating equipment has used indirect heating in the past to achieve even heating of the product, but that is no longer necessary. It is certainly much more economical to use direct heating, even if extra heat-spreading components are required.

* By Dr. Robert M. Kendall, President, Alzeta Corporation, Santa Clara, Calif.

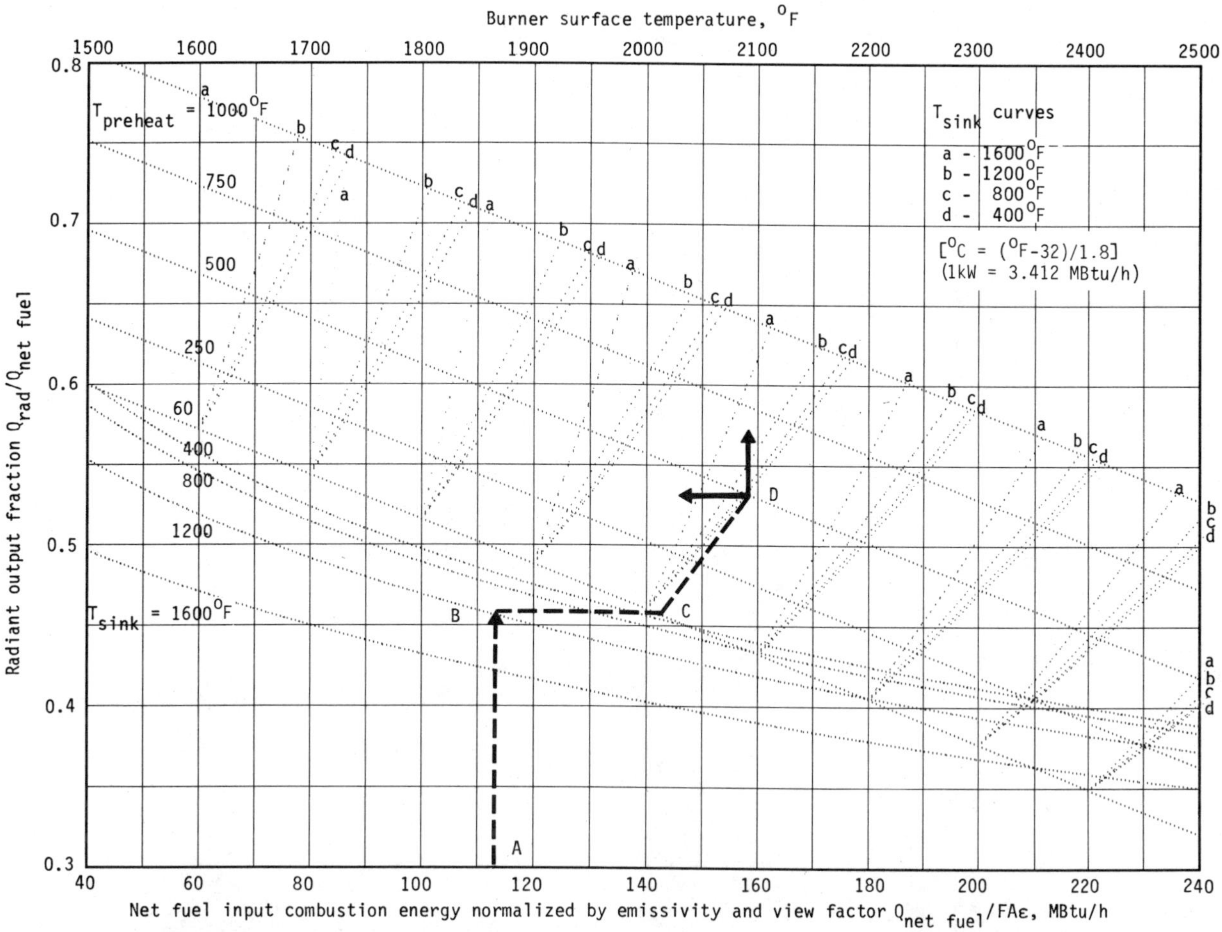

Fig. 1.13 Nomograph for determination of operating variables with a ceramic fiber (flow-through type) radiant heater, using natural gas with 10 percent excess air. (See Example 1.3.)

Heat-Transfer Media

Heat-transfer media used for indirect heating may be gas, vapor, liquid, or solid . . . most commonly air, steam, water, or fluidized bed. Table 1.7 lists properties of water and some other heat-transfer liquids. For more details on thermal fluids, refer to Part 5, Chap. 4 (by Seifert) on organic heat-transfer fluids.

Heat-Transfer Coefficients

Figure 1.14 gives schematic* representations of several forms of direct (Fig. 1.14*a* to *c*) and indirect (Fig. 1.14*d* to *h*) heating with "round-number" comparisons of overall coefficient of heat transfer U. All of these involve some radiation, and radiant burners may increase U, but the usual low temperature level (<1200°F) minimizes the radiation component. The U values listed are typical for conventional convection.

Conventional convection can be enhanced by (1) higher fluid velocities (see Table 1.6) (baffling is one way to increase velocity), (2) addition of extra heat-transfer surface (e.g., fins), (3) raising the temperature differential (more intense input rate, but this usually raises fuel cost), or (4) bombarding the surface with hot solids (fluidized bed).

Comparative Fuel Costs

The resistance to heat flow is always greater for indirect heating than for direct heating, so the heat flow will be less with the same products of combustion temperature, or the POC gas will have to be hotter to achieve the same heat flow.

The usual way to compensate for the added resistance of an indirect heating configuration is to increase the fuel input to raise the combustion chamber and POC temperature. This usually shortens the life of the equipment, increases the fuel consumption, and reduces the fuel efficiency. The fuel saving

* Not to scale. Flames shown may represent dozens of premix flames on one or several burner manifolds, or they may represent one or more larger nozzle-mix burners.

Table 1.7. Properties of Some Heat-Transfer Liquids (See Part 5 for more complete data)

	Maximum temperature °F	Pour point, °F	Melt point, °F	Specific gravity	Specific heat, Btu/lb · °F		Thermal conductivity Btu ft/ft² · hr · °F		Viscosity, cp	
					At 70°F	At ( )°F	At 70°F	At ( )°F	At 70°F	At ( )°F
Ethylene glycol	325–400	. . .	−60	1.12	0.625	0.710 (300)	0.167	. . .	20.0	1.0 (300)
Polyalkylene glycols	400–565	0 to −45	. . .	0.99–1.08	0.44	0.64 (500)	0.099–0.121	0.095–0.099 (300)	90–320	1.5–1.85 (500)
Water	450	. . .	32	0.8*	1.00	0.66 (600)	0.349	0.78 (600)	0.98	0.81 (600)
Aromatic base hydrocarbons	425–850	−18 to −80	. . .†	0.89–1.13	0.37–0.46	0.63–0.72 (600)	0.071–0.083	0.063 (600)	6.2–387	0.29–0.42 (600)
Triaryldimethane	500	−31	. . .	1.03	0.36	0.60 (572)	0.17	0.14 (600)	32	0.37 (600)
Mineral oils	525–680	15–25	. . .	0.85–0.94	0.38–0.48	0.62–0.72 (600)	0.070–0.078	0.059–0.065 (600)	190 to 10 000	0.25–1.0 (600)
Mercury	600	. . .	−38	12.86*	0.033	0.0325 (850)	4.8	8.1 (600)	. . .	1.0 (400)
Diphenyl-diphenyl oxides	700–805	. . .	12–22	1.06	0.379	0.591 (700)	0.081	0.057 (700)	4.5	0.30 (700)
Salts (HTS)	800	. . .	300	1.83‡	. . .	0.37	. . .	0.42 (600)	. . .	2.0 (600)
Sodium	1500	. . .	208	0.89‡	0.253	0.32 (1000)	82.08	37 (1000)	. . .	0.3 (600)
NaK(78 wt% K)	1500	. . .	10	0.81‡	. . .	0.20 (1000)	. . .	15 (1000)	. . .	0.2 (600)

* At 600°F compared to 62.35 lb/ft³ water at 60°F.
† Except Therminol 88 melts at 293°F.
‡ At 500°F compared to 62.35 lb/ft³ water at 60°F.
CONVERSION FACTORS: °C = (°F − 32)/1.8; 1 Btu/h · ft · °F = 1.731 W/m · K; 1 Btu/lb_m · °F = 4187 J/kg · K; 1 cp = 1×10^{-3} N · s/m² = 2.419 lb_m/ft · h.
SOURCES: J. L. Boyen, *Practical Heat Recovery,* Wiley, New York, 1975, pp. 102–105.
P. L. Geiringer, *Handbook of Heat Transfer Media,* Reinhold, New York, 1962.

achievable by substituting direct for indirect heating is illustrated in the following example.

EXAMPLE 1.4 A product with surface area A_2 of 1.2 ft² is indirectly heated by a natural-gas POC stream that flues at 1400°F (at the start of heating cycle). The surface separating the product from the POC, $A_{1_i} = A_{1_o}$, has an area of 5 ft² (case *e*, Fig. 1.14). If this were changed to direct heating (case *a*), how much could the POC temperature be lowered while keeping the same heat flow?

In case *e*

$$Q_e = \frac{\Delta T}{1/UA}$$

$$= \frac{1400 - 60}{(1/3 \cdot A_{1_o}) + (1/6 \cdot A_{1_i}) + (1/3 \cdot A_2)}$$

$$= \frac{1340}{[1/3(5)] + [1/6(5)] + [1/3(1.2)]} = 3547 \text{ Btu/h}$$

If case *a* is to have the same heat flow, $Q_a = 3547 = (h_c + h_r)(A_2)(\Delta T) = (3)(1.2)(\Delta T)$, then the temperature difference between the POC and the product need be only 3547/3.6 = 985°, so the chamber temperature need be only 985 + 60 = 1045°F. Using judgment, estimate the POC temperature at 1150°F in case *a*, and then read the percent available heat from Fig. 1.2*a* at a typical 25 percent excess air, as follows:

Case	ΔT, °F	Furnace T, °F	Flue T, °F	Percent available	Percent efficiency
Indirect, *e*	1340	. . .	1400	55	52
Direct, *a*	985	1045	1150	63	60

(a) Heater.

(b) Recirculating air heater, oven.

(c) Fluidized bed.

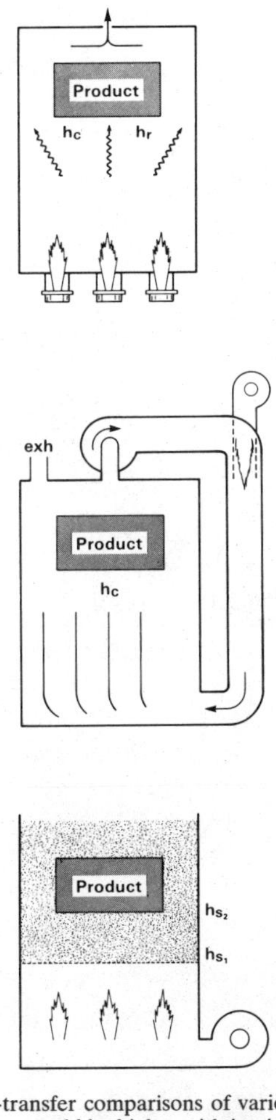

(a) $\dfrac{Q}{\Delta T} = UA = (h_c + h_r) \times A_2$

typically $= 3A_2$

(b) $\dfrac{Q}{\Delta T} = UA = h_{c_2} \times A_2$

typically $= 3A_2$

(c) $\dfrac{Q}{\Delta T} = UA = h_{s_2} \times A_2$

Fig. 1.14 (*a*) to (*c*) Heat-transfer comparisons of various direct heating systems. U and h values are based on conventional convection...would be higher with impingement. Actual values may range from 0.5 to 5 times those shown, where Q = heat flow, Btu/h, or kW; ΔT = temperature difference, °F, or °C; U = overall heat-transfer coefficient Btu/h·ft²·°F, or kW/°C·m²; h = film coefficient, or radiation equivalent, same units as U; A = heat-transfer area, ft² or m²; subscript 2 refers to the product surface; subscript 1 refers to the heat exchanger, *i* for inside surface, *o* for outside surface; subscript *r* refers to radiation; subscript *c* refers to convection (including all not labeled *r*); subscripts *g*, *v*, *l*, *s* refer to gas, vapor, liquid, or solid.

(d) Box oven, air or prepared atmosphere.

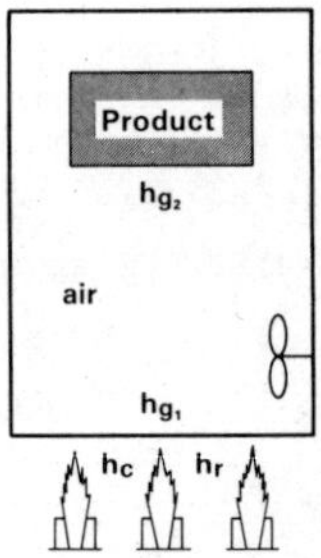

$$\frac{Q}{\Delta T} = UA = \frac{1}{\dfrac{1}{(h_c + h_r) \times A_{1_o}} + \dfrac{1}{h_1 \times A_{1_i}} + \dfrac{1}{h_2 \times A_2}}$$

$$\text{typically} = \frac{1}{\dfrac{1}{3A_{1_o}} + \dfrac{1}{3A_{1_i}} + \dfrac{1}{3A_2}}$$

(e) Air or prepared atmosphere gas circulated through (or over) tubes.

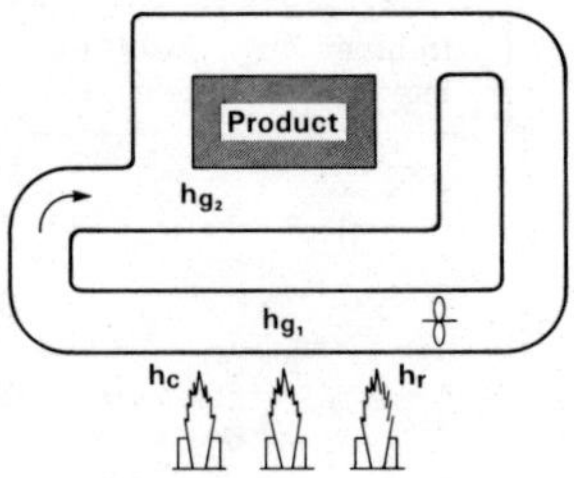

$$= \frac{1}{\dfrac{1}{3A_{1_o}} + \dfrac{1}{6A_{1_i}} + \dfrac{1}{3A_2}}$$

(f) Water (or heat transfer liquid) heater.

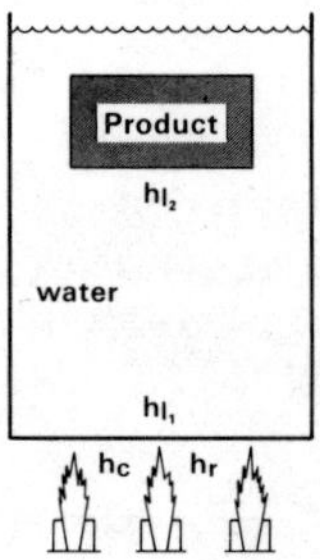

$$= \frac{1}{\dfrac{1}{3A_{1_o}} + \dfrac{1}{200A_{1_i}} + \dfrac{1}{100A_2}}$$

(g) Water (or heat transfer liquid) circulated through (or over) tubes.

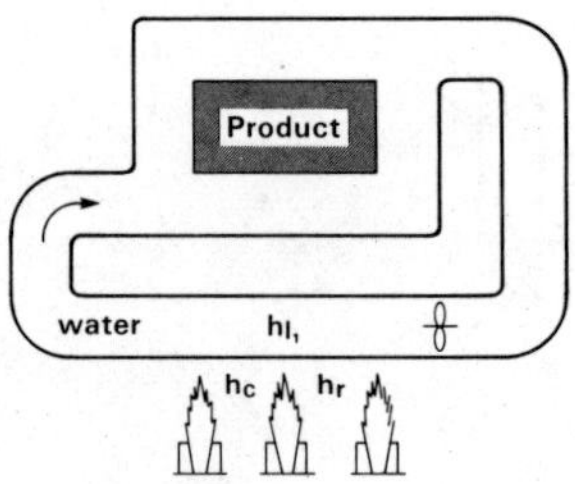

$$= \frac{1}{\dfrac{1}{3A_{1_o}} + \dfrac{1}{500A_{1_i}} + \dfrac{1}{100A_2}}$$

(h) Steam, vapor heater.

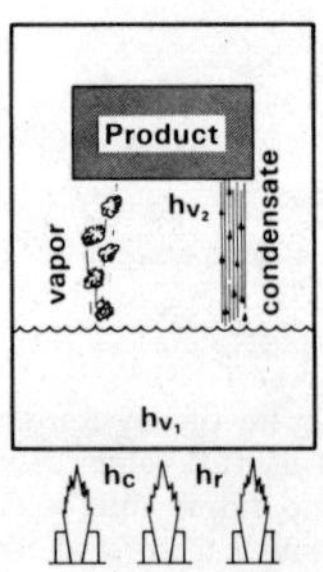

$$= \frac{1}{\dfrac{1}{3A_{1_o}} + \dfrac{1}{1000A_{1_i}} + \dfrac{1}{1000A_2}}$$

Fig. 1.14 (*d*) to (*h*) Heat-transfer comparisons of various indirect heating systems. See caption for Fig. 1.14*a* to *c*.

Assume 3 percent miscellaneous losses; thus percent efficiency will be 3 percent less than percent available. The percent fuel saved by using direct firing (case *a*) instead of indirect firing (case *e*) would be

$$100\left(1 - \frac{\text{eff } a}{\text{eff } e}\right) = 100\left(1 - \frac{52}{60}\right) = 13.3\%$$

Comparative First Costs

Because direct heating eliminates the heat exchanger, it almost always involves less capital investment. For reasonable equipment life, it is often necessary to use better materials of construction in indirect-fired equipment. Even if the same materials were used in each, the size and complexity of the indirect heating equipment may make its first cost higher.

Other Pros and Cons

The primary advantages of indirect heating are elimination of product contamination and possible improved product cleanliness (especially if oil-fired or propane-fired direct heaters happen to get out of adjustment). In most cases, safety, reliability, and pollution emission will be about the same.

Control response will be much slower with indirect heating. If product feed rates vacillate, a simple temperature sensor on the product may be insufficient. A temperature sensor in the intermediate heating medium will react more promptly but may not give as accurate control of product temperature. A cascaded control also tied into feed rate will usually be most effective. In addition, it is usually desirable to have overtemperature cutouts for product and equipment protection in the event of product flow stoppage. These should be actuated by sensors on both product and the heat exchanger. The latter also protect the heat exchanger or dividing wall from damage or shortened life.

1.7 EVALUATING FEASIBILITY—GAS FIRING VERSUS ELECTRICAL HEATING

Operating Costs

In most parts of the world that have both natural gas and electricity available, the bottom line operating cost is usually less with gas, even after correcting for its lower percent available heat. In 1985 the average ratio of electric- to natural-gas costs in the United States was 4:1. That would indicate that any gas-fired equipment having better than 25 percent available heat will have a lower operating cost than the equivalent electrically heated unit. In numerous cases, LP gas (propane) and light fuel oils are also cheaper.

When designing equipment for industrial or commercial installations, steam heating should be considered because steam lines are almost always available in such cases and steam can be generated from the most economical fuel available, changing quite easily as supply and price factors vacillate. The excellent heat transfer possible with steam heating, and the simplicity of control, make this option low in first cost as well as operating cost.

If fear of product contamination by direct contact with products of combustion eliminates the possibility of using a direct fuel-fired heater, a direct electric heater stands a good chance of having lower operating cost than an indirect fuel-fired heater of the same size and product heating capacity, because the indirect heater will have a higher throw-away POC temperature than a direct one. If, however, the product must be surrounded by an inert gas, an indirect heater may be required even with electric heating, and the cost benefit will return to fuel-firing. (Direct electric heating with air as the transfer medium may be damaging to some products.)

The Sankey diagram (Fig. 1.1) shows the heat loss due to throw-away POC in a fuel-fired heater. The magnitude of this "flow heat loss" can be predicted, and the available heat charts and tables (Fig. 1.2*a* to *d* and Table 1.3) make it easy to determine how much heat is left (available) after the flow heat loss is subtracted from the net heating value of the fuel.

Theoretically, electrically heated equipment has no flue heat loss, but if there is any fluid flow into or out of the heating chamber, electric heaters will also suffer such an "unavailable" or flow heat loss. The flow heat loss due to cold-air infiltration through leaks, as a result of the "chimney effect" of a hot chamber, has been known to reduce efficiency of some heaters by 5 to 15 percent.

Induction heating is usually applicable only to long production runs of one size and shape of product. Its copper or copper alloy electrodes do not actually become hot, but they are subjected to radiated and convected heat from the load that they are heating; thus they must usually be water-cooled. The Sankey diagram for induction heating, therefore, has an additional flow heat loss of warmed water going to drain. The nature of induction heating processes is such that many are poorly shrouded from radiation and convection heat loss to the surroundings.

If the heating process is for the purpose of mass transfer (drying or curing), some carrier flow (usually air) is necessary to remove the vapor or off-gas, so there will be a heat loss to the exiting flow.

The miscellaneous losses on the Sankey diagram, other than the flow heat loss, are usually no more than 1 to 5 percent if good engineering (including seals and insulation) have been incorporated in the design of the heating unit.

After evaluating the percent efficiency for electric heating and fuel-firing, Fig. 1.15*a* to *d* can be used to compare actual operating costs per million British thermal unit of *useful heat*. The vertical scales of all four figures are the same, making it easy to compare costs of electricity and various fuels.

Some fuel-fired systems require blowers or fuel pumps. A check should be made to see if their operating costs will be significant.

EXAMPLE 1.5 A recirculating-air heater will use convection to achieve superior uniformity in drying a labyrinth of complex shapes. A fraction of the total circulated air, equal to the makeup (vapor carrier) air, will be exhausted at 800°F. (From steam tables, the heat content of the water vapor exhausted at 800°F and atmospheric pressure is 1433 Btu/lb.) A part of the vapor carrier air will also serve to supply oxygen for combustion, if fired with natural gas. Estimated "fuel" prices are $0.04 per kilowatthour for electricity or $4.40 per thousand cubic feet of natural gas. Annual output will be 600,000 lb of product having 0.25 specific heat. It is to have its water content reduced by 5 percent of its final weight. Compare annual "fuel" costs with electricity and natural gas.

The annual "useful heat" required will be $600{,}000 \times 0.25 \times (800 - 60) + 600{,}000 \times 0.05 \times 1433 = 154 \times 10^6$ Btu per year.

Assuming about 10 percent excess air, Fig. 1.2*a* shows 74 percent available with natural gas. Estimate that wall losses, conveyor losses, and opening losses will lower the actual efficiency to about 70 percent. An electrically heated unit will have *all* the same losses in this case.

Entering the bottom scale of Fig. 1.15*a* at 4 cents per kilowatt hour, and turning left at 70 percent efficiency, read $8.20 per million useful British thermal unit. The annual operating costs with electricity will therefore be $\$8.20 \times 154 \times 10^6$ Btu/yr = $1263. Entering the bottom scale of Fig. 1.15*b* at $4.40 per thousand cubic feet of natural gas,

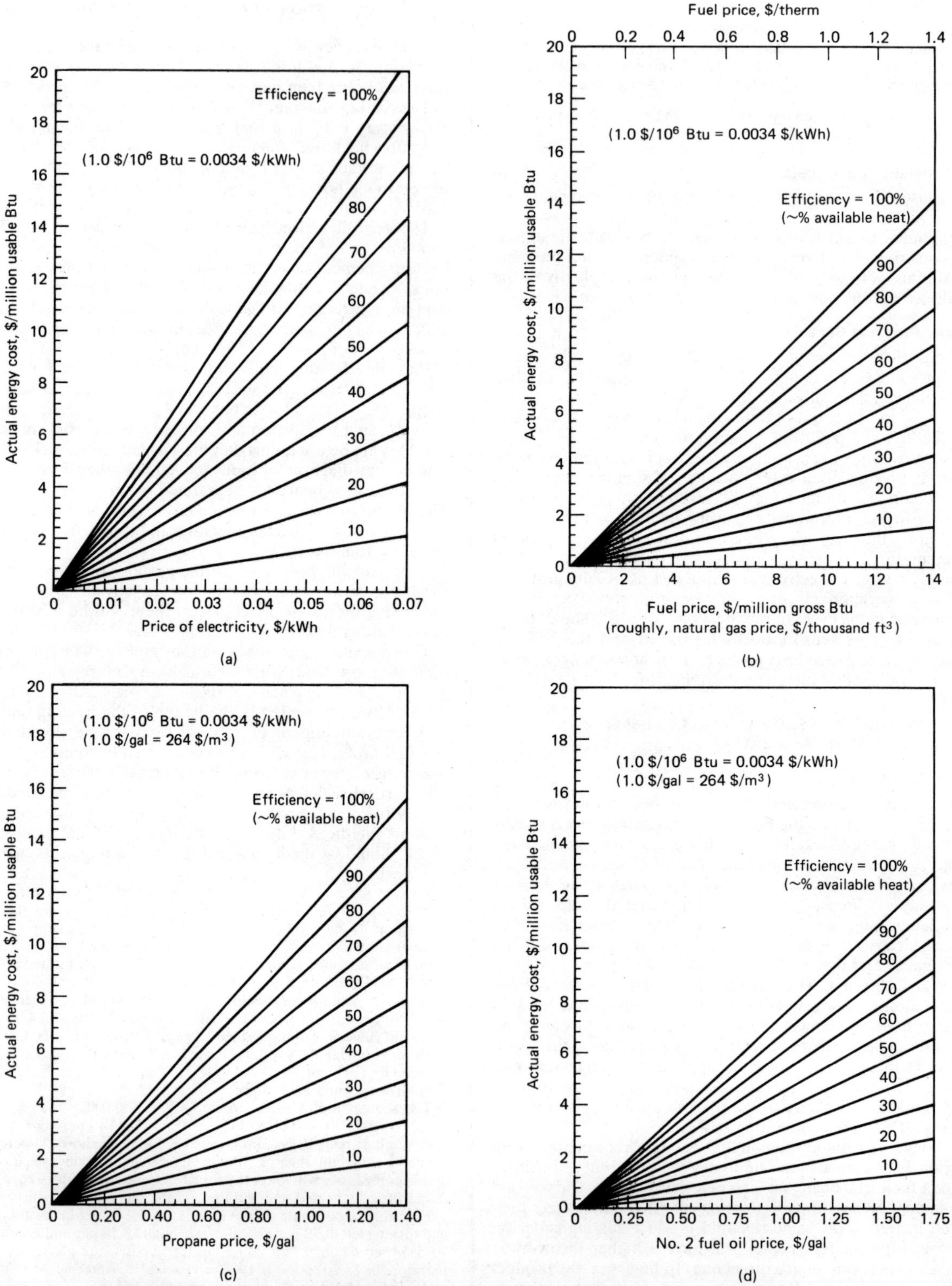

Fig. 1.15 Comparative actual energy costs, for (*a*) electricity, (*b*) natural gas, (*c*) propane, and (*d*) no. 2 fuel oil; with corrections for efficiency (available heat). See Examples 1.5 and 1.6.

and turning left at 70 percent, read 3.30 per million useful British thermal unit. The annual operating cost with natural gas will therefore be $3.30 × 154 = $508, so the annual saving with gas will be $1263 − 508 = $755.

EXAMPLE 1.6 A crucible will be used to melt a die-casting metal and superheat the liquid metal to 1020°F. If electrically heated, radiation will be used and there will be no flow heat loss. If fuel-fired, the exit POC temperature might be 1050°F. The furnace will be well sealed and insulated, so all other losses will be only 1 percent. If electricity will be available at $0.025 per kilowatthour, how low must fuel prices be to obtain competitive operating costs?

Entering Figure 1.2*a* at 1050°F and assuming 10 percent excess air, read 68 percent available heat. This, minus 1 percent other losses, will result in 67 percent efficiency with natural gas. Similarly, efficiencies will be about 69 percent with no. 2 fuel oil and propane. The electric furnace efficiency will be 99 percent.

In Fig. 1.15*a*, the actual electrical energy cost will be $7.3 per million useful British thermal units. Reading horizontally across, this compares to $10.80 for natural gas, $0.94 per gallon for propane, and $1.45 per gallon for no. 2 fuel oil. These figures are based on 1050°F exit POC temperature, which will occur only at the end of the melting cycle, so average efficiencies will be better than those calculated.

First Costs, Maintenance, and Equipment Life

First costs, maintenance, and equipment life must be considered together in the user's total cost of heating equipment. An electric resistance system may be so simple compared to a gas burner system that its first cost will be less. However, to maintain its claim to near 100 percent efficiency, an electric element chamber must have neither air leaks nor flow through it. This means that the element temperature will rise to whatever level necessary to dissipate its energy by radiation and natural convection. Some efforts to make electric heating chambers compact have resulted in very high design heat-flux densities on the element surfaces and consequent high element temperatures. These necessitate extra care in thermal as well as electrical insulation. Life of the element is shortened exponentially as its operating temperature rises.

To heat molds and die blocks, the seemingly simple and convenient solution is to insert several "electrodes" into drilled holes in the dies or molds. An "electrode" is an electric resistance encased in a metal sheath. The replacement costs of this practice are so high that there is a trend toward substituting pumped thermal fluids from adjacent fuel-fired heaters.

If an electric resistance element is heating a heat-transfer liquid, high heat density may cause chemical breakdown of the liquid. That, in turn, may result in charlike deposits on the surface of the element, insulating it and raising its temperature to self-destruction.

Fabricated metal gas burner manifolds and ports offer more flexibility in design and are usually cheaper than cast metals, but they will not last as long.

Controllability

Small, low-cost units usually have only ON/OFF control, which sacrifices some temperature uniformity. On larger and more sophisticated heating equipment, modulating input and air-fuel-ratio control can provide excellent temperature uniformity with respect to time and product dimension. Such controls on fuel-fired installations are usually less expensive than the silicon controlled rectifier (SCR) controls necessary to accomplish the same effect with electric heating.

Hazards

Death or health damage from asphyxiation or electric shock is rare where equipment has been tested and certified by testing agencies such as Underwriters' Laboratory, Canadian Standards Association, and American Gas Association.

Most fuel-fired equipment is connected to a stack that has some natural draft and will carry away unburned fuel or carbon monoxide. Carbon monoxide is generated only when there is a malfunction or major maladjustment of a burner system. It is usually preceded by unpleasant aldehyde odors that serve as a warning of trouble. Blockage or insufficiency of flues or stacks can be a serious problem. Installation instructions of fuel-fired equipment must emphasize the importance of checking stack capacity and condition before heating equipment is connected. Similarly, grounding of electric heating equipment is a vital requirement.

Part 11

Cryogenic Fluids and Design Practice

Chapter 1

Cryogenic Fluids and Design Practice

RALPH C. LONGSWORTH
YAIR LAHAV
APD Cryogenics, Inc.

1.1 PROPERTIES OF COMMON CRYOGENS

Physical Properties and Characteristics

The cryogenic temperature region is generally considered to be below 200°R (111 K), and cryogens are defined as substances whose normal boiling temperature is less than 200°R (111 K). The most common cryogens are helium, hydrogen, neon, nitrogen, air, argon, oxygen and methane. Their properties are summarized in Table 1.1*a* to *c* based primarily on data from Ref. 1. Other cryogens which are not listed are carbon monoxide, 147.8°R (82.1 K); fluorine, 42.4°R (23.6 K); and helium 3, 5.75°R (3.2 K). Table 1.1*a* presents data on the cryogens at the triple point, normal boiling point, and critical point. It also contains saturation temperature data at several low pressures in order to permit determination of the temperatures at which these gases freeze out on cryopumping surfaces or in heat exchangers. Table 1.1*b* presents data for the liquid phase of each of the cryogens at saturation conditions from temperatures near the triple point to temperatures near the critical point. Table 1.1*c* presents data for the cryogens in the gas phase from the normal boiling point temperature to room temperature. Values of viscosity μ and thermal conductivity k at low pressure plot as straight lines on log paper; i.e., they can be expressed by an equation in the form $\mu = aT^b$.

Helium Liquid helium is readily available in industrial areas worldwide and is commonly used for research at cryogenic temperatures and for cooling superconducting devices. It can

Table 1.1*a*. Properties of Cryogens—General

	Helium, He	Hydrogen, H_2†	Neon, Ne	Nitrogen, N_2	Air‡	Argon, Ar	Oxygen, O_2	Methane, CH_4
Molecular weight	4.00	2.02	20.18	28.02	28.96	39.95	32.0	16.04
Normal boiling-point temperature, °R	7.54	36.68	49.0	139.2	141.8	157.0	162.4	201.1
ρ liquid, lb_m/ft^3	7.798	4.418	75.35	50.46	54.56	86.98	71.27	26.5
ρ vapor, lb_m/ft^3	1.04	0.0831	0.596	0.287	0.280	0.363	0.279	0.114
h_{fg}, Btu/lb_m	8.98	192.6	37.07	85.86	88.26	70.19	91.62	219.4
Critical temperature, °R	9.47	59.9	80.0	227.1	238.4	272.1	278.6	343.1
Critical pressure, lb/in² abs	33	188	395	491	547	705	737	670
Triple-point temperature, °R	3.91§	25.2	44.3	113.8	. . .	150.8	97.9	163.2
Triple-point pressure, lb/in² abs	0.736§	1.044	6.27	1.82	. . .	9.99	0.021	1.69
ρ solid at 37°R, lb_m/ft^3	. . .	4.98@TP	90.1	64.1	. . .	109.4	. . .	32.3
h_{sf}, Btu/lb_m	. . .	25.0	7.1	11.1	. . .	12.7	6.0	25.2
Saturation temperature, °R								
10^{-8} torr	. . .	6.68	13.2	42.7	. . .	48.2	50.8	57.6
10^{-4} torr	. . .	9.68	18.1	56.5	. . .	64.6	66.1	78.3
1 torr	2.28	17.2	28.4	75.1	. . .	86.8	97.4	121.1
100 torr	. . .	27.2	39.8	114.1	. . .	132.1	134.1	165.1
100 lb/in² abs	. . .	. . .	. . .	176.6	. . .	197.6	. . .	. . .
Liquid at NBP								
C_p, Btu/lb_m°F	1.070	2.312	0.437	0.490	0.468	0.271	0.405	0.824
μ, lb_m/ft hr	0.00862	0.0319	0.314	0.382	0.405	0.610	0.460	0.285
k, Btu/h · ft · °F	0.01572	0.0685	0.0653	0.0807	0.0815	0.0712	0.0875	0.1116
σ_L, lb_f/ft	6.37×10^{-6}	1.32×10^{-4}	3.30×10^{-4}	6.00×10^{-4}	7.11×10^{-4}	7.59×10^{-4}	9.04×10^{-4}	9.66×10^{-4}

* The symbols used in Table 1.1*a* to 1.1*c* and conversions to SI units are as follows: ρ = density, $lb_m/ft^3 \times 16.02$ = kg/m³; h_{fg} = heat of vaporization, Btu/lb_m × 2326 = J/kg; h_{sf} = heat of fusion, Btu/lb_m × 2326 = J/kg; c_p = specific heat at constant pressure. Btu/lb_m · °F × 4186.8 J/kg · K; p = pressure, lb/in² abs × 6895 = Pa; pressure (vacuum, absolute), torr = 1 mm Hg, torr × 133.3 = Pa; μ = viscosity, lb_m/ft · h × 41,338 = Pa · s; k = thermal conductivity, Btu/h · ft · F × 1.731 = W/m · K; σ = surface tension, lb_f/ft × 14.59 = N/m; temperature, °R × 1/1.8 = K, 0°F = 459.67°R, 0°C = 273.15 K.

† Normal H_2.

‡ Composition of dry air, mole percent: 78.09 N_2, 20.95 O_2, 0.93 Ar, 0.03 CO_2.

§ λ point, transition to superfluid phase.

be used to provide cooling over the temperature range of 2°R (1.1 K) to room temperature and is relatively safe because it is inert. Gaseous helium is used in all the low-temperature closed-cycle refrigerators. Helium is unique in that it remains a liquid at absolute zero temperature and 1 atm pressure, but it can be solidified at 25 atm and less than 2°R (1.1 K). At 3.91°R (2.17 K) helium goes through a phase transition which is referred to as the *lambda* (λ) point because of the shape of the specific heat curve. Below 3.91°R (2.17 K), liquid helium is described as a superfluid because of several unusual physical properties which one needs to understand when working in this temperature region. Despite

Table 1.1*b*. Properties of Saturated Liquids

Temperature, °R	p_{sat}, lb/in² abs	ρ_f, lb_m/ft^3	c_p, Btu/lb_m · °R	μ, lb_m/ft · h	k, Btu/h · ft · °F	h_{fg}, Btu/lb_m	N_{Pr}
			Liquid Helium 4				
2	0.00618	9.05	0.10	. . .	. . .	8.86	
3	0.01458	9.07	0.501	. . .	. . .	9.85	
4	0.825	9.15	0.880	0.00718	0.01029	9.80	0.614
5	2.48	8.92	0.557	0.00912	0.01175	10.1	0.433
6	5.60	8.65	0.688	0.00900	0.01326	10.1	0.467
7	10.65	8.14	0.901	0.00878	0.01479	9.60	0.535
8	18.12	7.53	1.26	0.00853	0.01632	8.30	0.658
9	28.6	6.30	2.75	0.00833	0.01785	5.16	1.28

Table 1.1*b*. (*Continued*)

Temperature, °R	p_{sat}, lb/in² abs	ρ_f, lb$_m$/ft³	c_p, Btu/lb$_m$ · °R	μ, lb$_m$/ft · h	k, Btu/h · ft · °F	h_{fg}, Btu/lb$_m$	N_{Pr}
			Liquid Hydrogen (*para*-Hydrogen)				
26	1.493	4.64	1.700	0.0554	0.0606	195.5	1.55
30	4.210	4.613	1.856	0.0445	0.0636	193.6	1.30
35	11.482	4.475	2.216	0.0340	0.0673	190.7	1.12
40	25.153	4.255	2.577	0.0283	0.0710	185.7	1.03
45	47.64	4.025	2.991	0.0257	0.0748	175.3	1.03
50	81.88	3.710	3.760	0.0232	0.0785	155.7	1.11
55	130.4	3.187	5.706	0.0207	0.0822	116.5	1.44
			Liquid Neon				
45	7.40	77.6	0.426	0.387	0.0673	38.4	2.45
50	17.99	74.6	0.438	0.293	0.0649	36.8	1.98
55	37.12	71.2	0.457	0.225	0.0643	34.6	1.60
60	68.08	67.6	0.485	0.180	0.0634	32.1	1.38
65	114.2	63.4	0.526	0.147	. . .	29.2	
70	179.6	58.1	0.582	0.122	. . .	25.0	
75	268.6	51.6	0.684	0.096	. . .	18.8	
			Liquid Nitrogen				
115	2.07	54.2	0.476	0.707	0.0944	91.4	4.30
120	3.36	53.5	0.478	0.594	0.0898	90.3	3.16
160	47.4	46.6	0.505	0.256	0.0694	78.2	1.86
180	113.1	42.9	0.538	0.206	0.0578	69.8	1.92
200	227.6	38.3	0.617	0.181	0.0463	57.2	2.41
220	406.3	30.7	0.717	0.170	0.0324	34.0	3.77
225	466.5	26.8	. . .	. . .	0.0283	21.1	
			Liquid Argon				
155	12.94	88.03	0.2703	0.634	0.0718	69.8	2.39
160	17.37	86.91	0.2731	0.578	0.0705	69.1	2.24
170	29.74	84.66	0.2803	0.486	0.0674	67.1	2.02
180	47.95	82.42	0.2879	0.416	0.0638	65.0	1.88
190	73.54	79.8	0.2948	0.370	0.0600	62.4	1.82
200	108.0	77.1	0.3026	0.331	0.0558	59.6	1.80
210	153.0	74.4	0.3138	0.298	0.0519	56.5	1.80
220	210.0	71.4	0.3344	0.266	0.0479	53.3	1.86
230	280.4	68.1	0.3692	0.235	0.0438	48.7	1.98
240	365	64.8	0.448	0.201	0.0398	43.2	2.26
250	445	60.8	. . .	. . .	. . .	36.4	
260	584	53.9	. . .	. . .	. . .	26.4	
			Liquid Oxygen				
100	0.035	81.3	0.401	1.86	0.1090	106.0	6.84
120	0.52	78.2	0.398	0.992	0.1016	100.0	3.89
140	3.34	75.1	0.402	0.641	0.0941	96.6	2.74
160	13.03	71.4	0.405	0.472	0.0866	92.2	2.21
180	36.84	67.3	0.409	0.368	0.0785	87.4	1.92
200	85.6	64.0	0.417	0.302	0.0696	80.6	1.81
220	169.2	59.4	0.438	0.273	0.0612	73.2	1.96
240	299.3	54.2	0.476	0.252	0.0528	62.0	2.27
260	489.1	47.6	0.556	0.239	0.0418	46.3	3.18
275	660	37.5	. . .	0.227	0.0312	20.2	

SOURCE: Reprinted by permission from Ref. 6.

Table 1.1c. Properties of Gases at 1 Atm

Temperature, °R	ρ, lb_m/ft^3	c_p, $Btu/lb_m \cdot °R$	μ, $lb_m/ft \cdot h$	k, $Btu/h \cdot ft \cdot °F$	N_{Pr}
			Gaseous Helium		
7.6	1.042	1.63	0.00309	0.0056	0.90
10	0.643	1.45	0.00370	0.0064	0.84
20	0.281	1.283	0.00578	0.0102	0.73
50	0.1095	1.248	0.01047	0.0186	0.71
100	0.0546	1.243	0.0164	0.0292	0.70
150	0.0366	1.241	0.0213	0.0383	0.69
200	0.0272	1.240	0.0258	0.0464	0.688
250	0.0219	1.240	0.0296	0.0533	0.688
300	0.01822	1.240	0.0334	0.0602	0.688
350	0.01561	1.240	0.0368	0.0663	0.688
400	0.01369	1.240	0.0402	0.0725	0.688
450	0.01216	1.240	0.0434	0.0780	0.688
500	0.01093	1.240	0.0463	0.0835	0.688
540	0.01012	1.240	0.0486	0.0876	0.689
			Gaseous Hydrogen (Normal Hydrogen)		
36.7	0.0830	2.85	0.00254	0.0090	0.805
50	0.0576	2.63	0.00354	0.0124	0.751
100	0.0278	2.51	0.00655	0.0228	0.721
150	0.01841	2.62	0.00934	0.0343	0.713
200	0.01380	2.75	0.01094	0.0423	0.711
250	0.01102	2.95	0.01280	0.0525	0.717
300	0.00920	3.10	0.01457	0.0626	0.720
350	0.00788	3.21	0.01607	0.0716	0.720
400	0.00690	3.30	0.0177	0.0814	0.718
450	0.00613	3.36	0.0192	0.0901	0.716
500	0.00552	3.40	0.0206	0.0985	0.712
540	0.00511	3.42	0.0217	0.1050	0.707
			Gaseous Neon		
48.8	0.599	0.280	0.0110	0.0057	0.540
75	0.371	0.257	0.0168	0.0074	0.583
100	0.279	0.250	0.0219	0.0089	0.616
150	0.183	0.246	0.0318	0.0117	0.669
200	0.1382	0.246	0.0381	0.0139	0.674
250	0.1106	0.246	0.0441	0.0162	0.671
300	0.0919	0.246	0.0503	0.0186	0.666
350	0.0790	0.246	0.0560	0.0208	0.663
400	0.0691	0.246	0.0622	0.0231	0.663
450	0.0614	0.246	0.0675	0.0250	0.663
500	0.0550	0.246	0.0726	0.0270	0.663
540	0.0510	0.246	0.0768	0.0285	0.663
			Gaseous Nitrogen		
139.3	0.288	0.259	0.0134	0.00415	0.846
150	0.265	0.258	0.0143	0.00449	0.822
200	0.1947	0.253	0.0186	0.00592	0.795
250	0.1545	0.251	0.0228	0.00736	0.777
300	0.1280	0.250	0.0268	0.00877	0.764
350	0.1098	0.249	0.0306	0.01013	0.752
400	0.0960	0.249	0.0342	0.01148	0.743
450	0.0853	0.249	0.0376	0.01280	0.732
500	0.0767	0.249	0.0409	0.01405	0.725
540	0.0711	0.249	0.0432	0.01507	0.715

Table 1.1c. (*Continued*)

Temperature, °R	ρ, lb_m/ft^3	c_p, $Btu/lb_m \cdot °R$	μ, $lb_m/ft \cdot h$	k, $Btu/h \cdot ft \cdot °F$	N_{Pr}
			Gaseous Argon		
157.1	0.3495	0.1271	0.0200	0.00350	0.726
200	0.2783	0.1268	0.0241	0.00442	0.691
250	0.2212	0.1262	0.0286	0.00545	0.663
300	0.1837	0.1257	0.0328	0.00635	0.648
350	0.1571	0.1252	0.0378	0.00740	0.640
400	0.1372	0.1250	0.0423	0.00831	0.636
450	0.1219	0.1247	0.0468	0.00918	0.636
500	0.1097	0.1247	0.0513	0.01004	0.636
540	0.1014	0.1247	0.0548	0.01072	0.636
			Gaseous Oxygen		
162.3	0.273	0.220	0.0169	0.00470	0.793
200	0.222	0.220	0.0207	0.00582	0.783
250	0.1767	0.220	0.0258	0.00734	0.770
300	0.1468	0.219	0.0304	0.00879	0.757
350	0.1254	0.219	0.0349	0.01024	0.746
400	0.1102	0.219	0.0390	0.01167	0.730
450	0.0974	0.219	0.0430	0.01303	0.720
500	0.0876	0.219	0.0469	0.01440	0.715
540	0.0810	0.220	0.0500	0.01546	0.712
			Gaseous Methane		
201.1	0.114	0.513	0.0108	0.0068	0.815
540	0.0402	0.533	0.0270	0.0198	0.727

SOURCE: Reprinted by permission from Ref. 6.

the fact that 70 Btu (73.8 kJ) of heat will vaporize 1 ft^3 (0.283 m^3) of liquid helium, it is not difficult to provide thermal shielding to reduce heat losses to less than 1 Btu/h (0.3 W) in small systems and less than 20 Btu/h (6 W) in large trailers.

Hydrogen Hydrogen is used as a rocket fuel and in a number of commercial applications. It is transported primarily as a liquid for reasons of economy or for use as a liquid. Mixtures of hydrogen in air in concentrations of 4.0 to 75 percent H_2 by volume are flammable. Thus a great deal of consideration has to be given to safety when working with hydrogen.

Hydrogen can exist in two molecular forms, one with the two protons spinning in the same direction, *para*-hydrogen, p-H_2, and the other with the two protons spinning in opposite directions, *ortho*-hydrogen, o-H_2. At room temperature hydrogen has normal or equilibrium concentrations of 75 percent o-H_2 and 25 percent p-H_2 by volume. At its normal boiling point, normal hydrogen is 99.8 percent p-H_2. If normal H_2 at room temperature is cooled and liquefied in the absence of a catalyst, it will evaporate at the rate of about 1 percent per hour as the heat of conversion, 302.4 Btu/lb, is released. When liquid hydrogen is produced commercially, the o-H_2 is converted to p-H_2 in stages using an ion or nickel-based catalyst.

Neon Neon constitutes 0.0012 percent of air by volume and is produced in small quantities as a by-product in some large air-separation plants. Because of its relatively high cost and small usage, it is not shipped as a liquid. Since it is a noble gas, it is safe and easily handled. There are few applications that require cooling near its normal boiling temperature.

Nitrogen Nitrogen in liquid form is the most commonly used cryogen for general experimental purposes because it is readily available, relatively inexpensive, and relatively safe to handle. It is produced along with liquid oxygen in large air-separation plants. Both of these cryogens are widely distributed in liquid form in bulk quantities.

By connecting a vacuum pump to a bath of liquid nitrogen, temperatures down to the nitrogen freezing point of 114°R (63.3 K) can be reached. Alternatively, if the pressure above the liquid is controlled at elevated pressure, temperatures up to the critical point of 227°R (126 K) can be maintained.

Argon Argon is present in air (0.934 percent by volume) and is available in large quantities in liquid or gaseous form as a product of many large air separation plants. Liquid argon is inert, nontoxic, clear, and colorless. It is comparable to liquid nitrogen in ease and safety of use but is slightly more expensive.

Oxygen Oxygen in liquid and gaseous form requires special system design, operation, and maintenance to be used safely. Oxygen boils at a higher temperature than nitrogen. Liquid air which starts out with about 78 percent N_2 and 21 percent O_2 becomes oxygen-rich as the N_2 evaporates preferentially and its boiling temperature increases. Liquid air should be treated with the same respect and precautions as liquid oxygen.

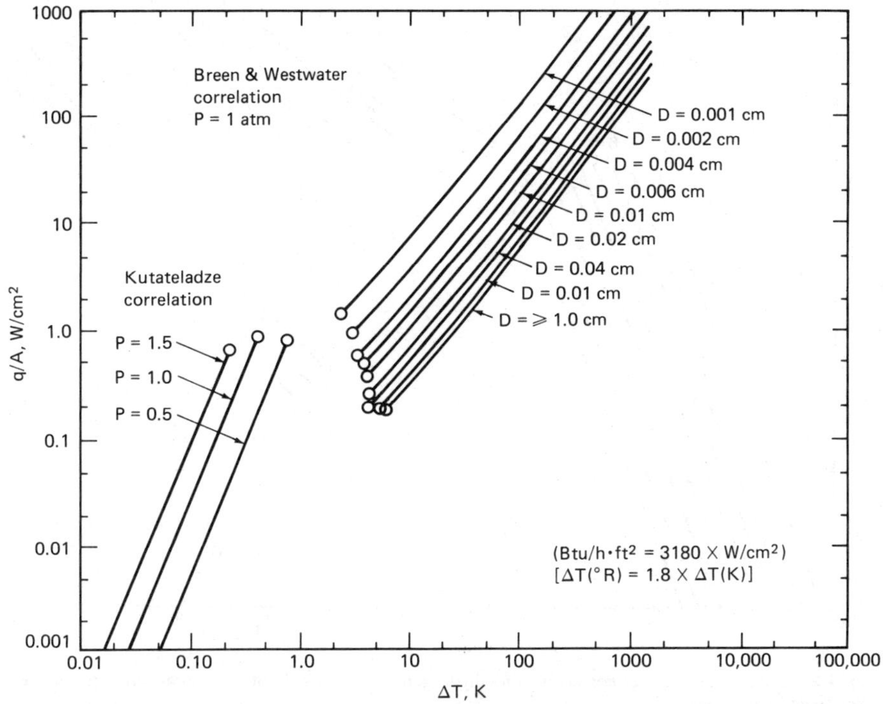

Fig. 1.1 Boiling-heat-transfer correlations for helium on the outside of horizontal tubes of diameter D. (*Reprinted by permission.*[2])

Methane Methane is the main constituent of natural gas and is presently being liquefied and transported in large quantities by ship from remote areas to industrialized regions of the world for use as a fuel. It is not readily available in liquid form and requires special treatment to be used safely.

Boiling and Condensing Heat Transfer

Pool-boiling heat-transfer correlations for helium, hydrogen, and nitrogen are presented in Figs. 1.1 to 1.3. These are based on the Kutateladze correlation in the nucleate boiling region and the Breen and Westwater correlation for film boiling[2]. Experimental data for smooth surfaces in liquid helium are in good agreement with the general correlation. However, other experimental studies[3] with different cryogens, surface conditions, geometry, or orientation show deviations by as much as a factor of 10 from the values presented.

Condensing heat transfer for vertical surfaces is predicted[4] using any consistent units by the relation

$$h = 0.943 \left[\frac{g\rho(\rho - \rho_v)k^3 h_{fg}}{L\mu\,\Delta T} \right]^{1/4} \tag{1.1}$$

where ρ = liquid density
k = liquid thermal conductivity
h_{fg} = latent heat of vaporization
μ = liquid viscosity
ρ_v = gas density
L = length
ΔT = temperature difference between the gas and the surface
g = gravitational acceleration

Table 1.2 lists values of condensing heat-transfer coefficients h using this equation. Experimental data[5] for $L = 1$ in (2.54 cm) with helium give a value of $h = 114$ Btu/h · ft² · °R (0.065 W/cm² · K) over the ΔT range 0.02 to 2°R (0.011 to 1.1 K), and with hydrogen a value of $h = 175$ Btu/h · ft² · °R (0.1 W/cm² · K) over the ΔT range 0.5 to 9°R (0.3 to 5 K). These values are independent of ΔT in the range tested[5] and are slightly lower than predicted by the general correlation. Nitrogen is reported[5] to follow the general equation better than hydrogen or helium.

1.2 REFRIGERATION

Refrigerator System Selections

When designing a system to operate at cryogenic temperatures, it is customary to consider the use of liquid nitrogen or liquid helium as refrigerants. Both are readily available in most areas and can be handled rather easily and safely if a few precautions are observed (see section on safety). Liquid helium costs about 10 times as much as liquid nitrogen by volume and has less refrigeration available but boils at a much lower temperature.

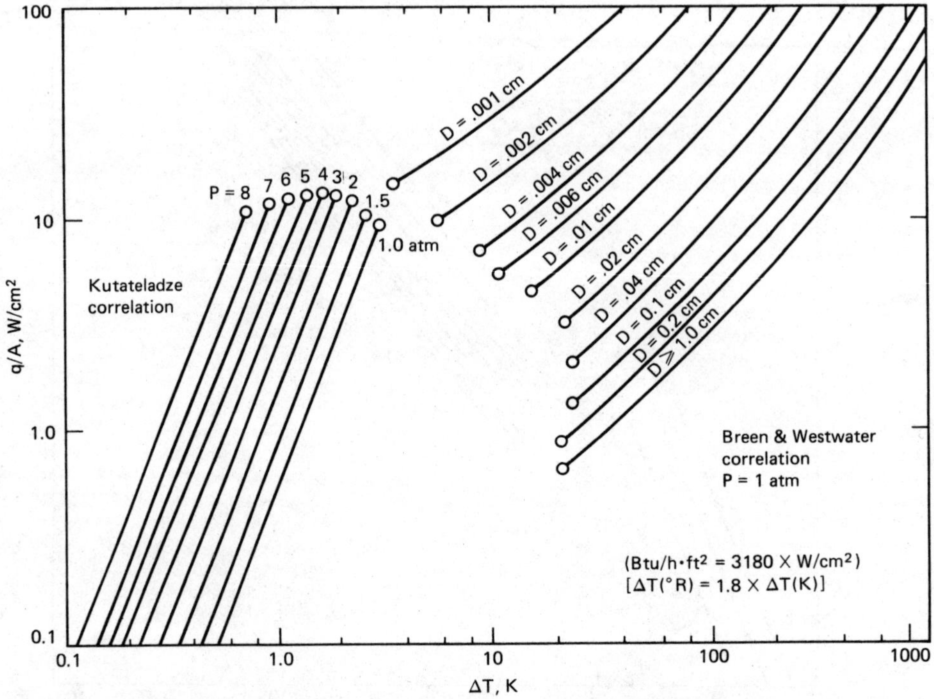

Fig. 1.2 Boiling-heat-transfer correlations for hydrogen on the outside of horizontal tubes of diameter D. (*Reprinted by permission.*[2])

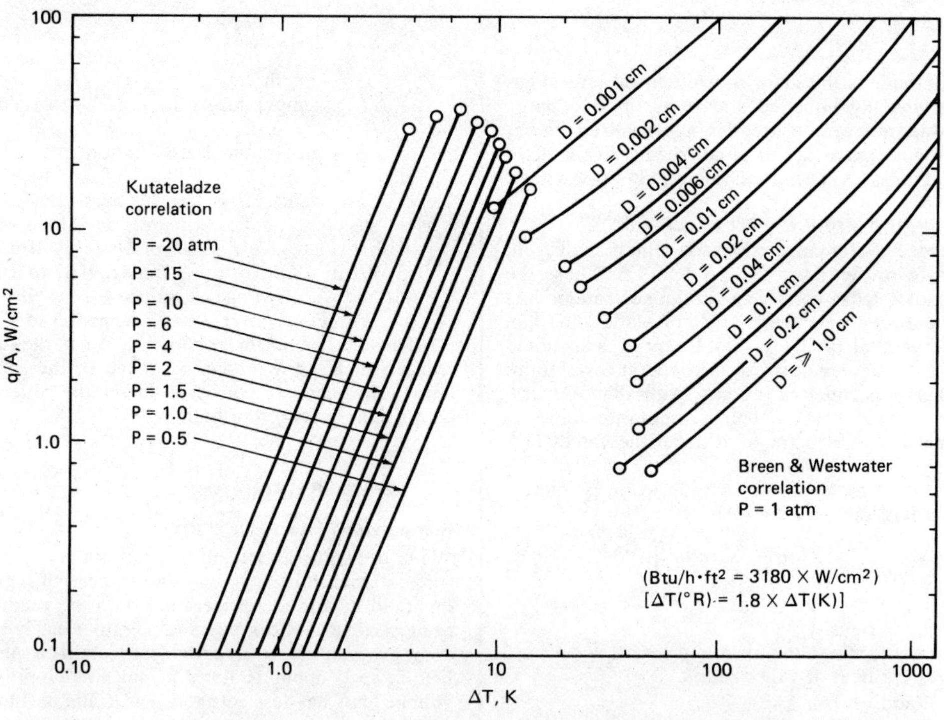

Fig. 1.3 Boiling-heat-transfer correlations for nitrogen on the outside of horizontal tubes of diameter D. (*Reprinted by permission.*[2])

Table 1.2. Condensing Heat-Transfer Coefficient for Vertical Surfaces at Atmospheric Pressures*

Fluid	Helium	Hydrogen	Neon	Nitrogen	Oxygen	Argon
Btu/h · ft^2 · °R	91.1	331	501	558	674	567
W/cm^2 · K	0.052	0.188	0.285	0.317	0.383	0.322

* L = 1 ft (0.305 m), ΔT = 1°R (0.556 K).

If a system is to be used for more than about 3000 h, it is usually more economical to use a closed-cycle refrigerator rather than liquid cryogens. In some cases, the object being cooled can be immersed in liquid helium or nitrogen, and the refrigerator can be used to recondense the boil-off vapor.

Figure 1.4 shows the refrigeration capacity range versus temperature for the types of cryogenic refrigerator that are commercially available. Figure 1.5 shows the most common cryogenic refrigerators in schematic form. All of these refrigerators use helium as the working fluid and have compressors that are separate from the expander. The gas used in cryogenic refrigerators must be clean, i.e., less than 10-ppm contaminants, in order to avoid freeze-out in the cold regions. However, oil removal techniques have been developed that permit the use of oil-lubricated compressors which are very reliable.

Reference 6 is a good general reference on cryogenic technology and contains a section on refrigeration.

Gifford-McMahon Cycle

By far the most commonly used cryogenic refrigerator is the Gifford-McMahon (GM) type. Commercially available GM-type refrigerators are designed around the use of air-conditioning-type compressors to compress helium and operate with supply pressures in the range of 300 lb/in^2 gage (2.17 mPa) and return pressures in the range of 100 lb/in^2 gage (0.79 mPa). Long life is achieved in the expander by operating at relatively low speed (60 to 150 r/min) and having small pressure loading on the main piston seal rings. Most expanders have a regenerator-type heat exchanger packed inside a plastic piston. Refrigeration is produced by having a valve mechanism at the warm end admit high-pressure gas

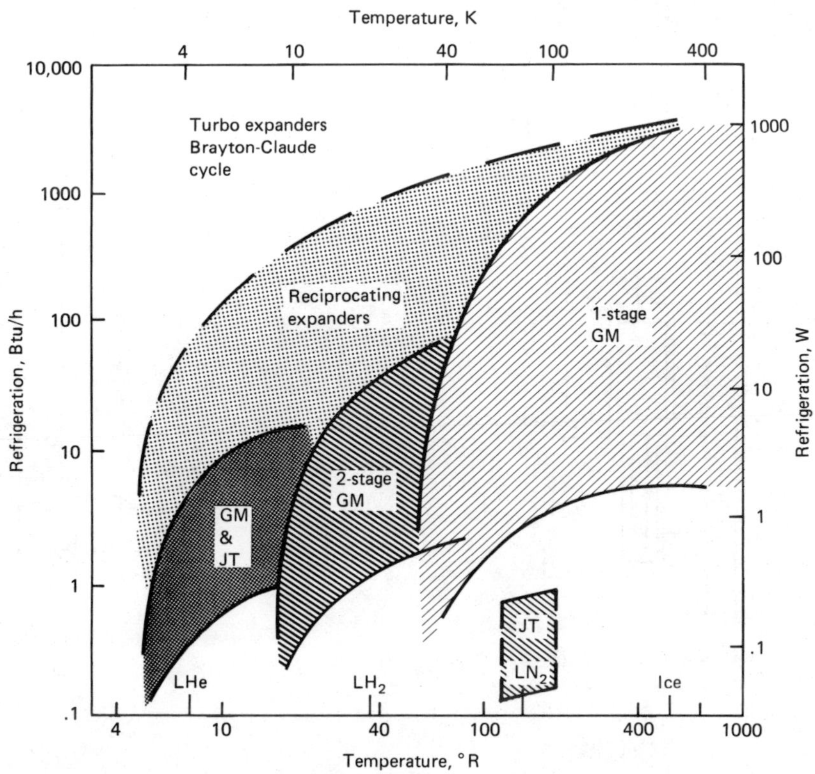

Fig. 1.4 Refrigerator types as a function of temperature and refrigeration. [*Reprinted by permission from R. C. Longsworth and W. A. Steyert, IEEE Trans. Electron. Dev.,* **ED-34,** *5(1987).*]

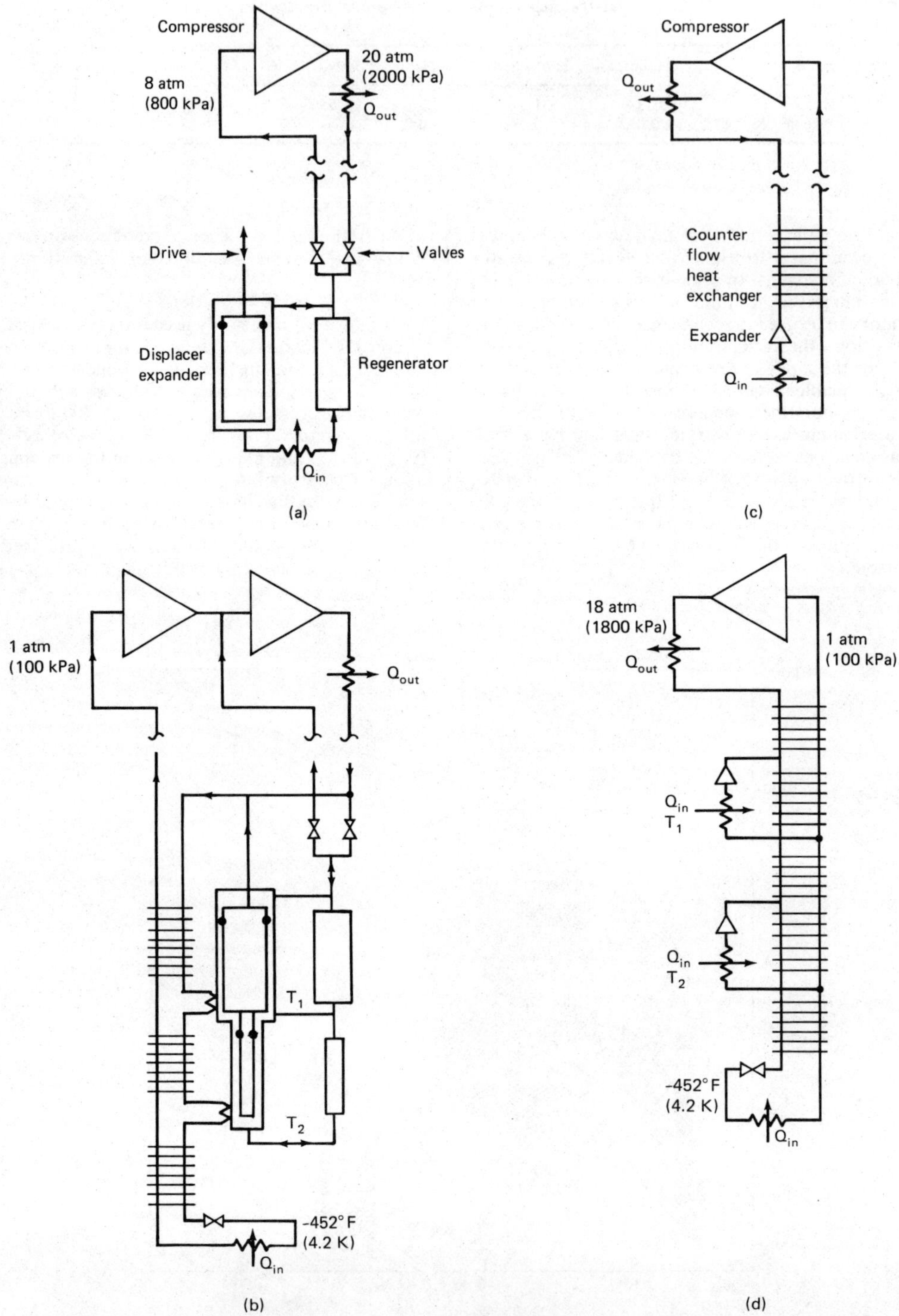

Fig. 1.5 Common cryogenic refrigerator process schematics. (*a*) Single-stage GM cycle; (*b*) two-stage GM plus JT; (*c*) reverse Brayton cycle; (*d*) two expander Claude (or Collins) cycle.

while the cold expansion volume is increasing, then venting the gas to low pressure while the cold volume is decreasing. Most units must run for 20 to 40 min with no load to reach minimum temperature, which is about 54°R (30 K) for single-stage expanders and 16°R (9 K) for two-stage expanders.

Some single-stage units are in use to recondense nitrogen, but two-stage units are in more common use as cryopumps and for general use as a convenient source of low-temperature refrigeration for research applications. A photograph of a typical two-stage unit which has a rated capacity of 7 Btu/h at 36°R (2 W at 20 K) plus 34 Btu/h at 139°R (10 W at 77 K) is shown in Fig. 1.6. Such a unit typically operates in a vacuum and cools small devices that are attached to the cold heat stations.

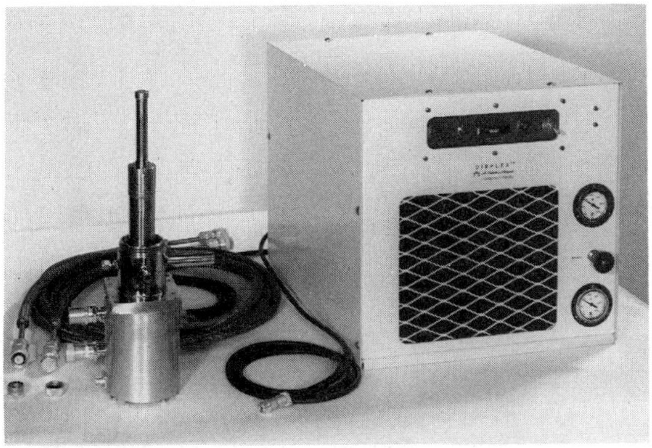

Fig. 1.6 Two-stage GM-type refrigerator. (*Courtesy of APD Cryogenics, Inc.*)

The most common method of producing a few watts of refrigeration at temperatures near the normal boiling point of liquid helium 7.6°R (4.2 K) is to add a counterflow heat exchanger to a two-stage GM-type expander and have a simple orifice to throttle the pressure from about 300 lb/in^2 gage (2.17 MPa) to 0 lb/in^2 gage (101 kPa). The isenthalpic expansion of helium that has been precooled to about 27°R (15 K) (Fig. 1.5*b*) produces refrigeration at 7.6°R (4.2 K) due to the nonideal properties of the gas. This is referred to as *Joule-Thomson* (JT) cooling. An additional stage of compression is required to bring the return gas pressure from about 0 lb/in^2 gage (101 kPa) to about 100 lb/in^2 gage (0.79 MPa). A variable JT valve enables the temperature to be set in the range of ~6.3 to 9.0°R (3.5 to 5.0 K). An active temperature controller is needed for setting higher temperatures.

Reverse Brayton Cycle

When large amounts of refrigeration are required, the reverse Brayton cycle is commonly used. When a JT cold stage is added to the reverse Brayton cycle, it is referred to as a *Claude cycle unit,* and a variation with two expansion stages which is used to liquefy helium is referred to as the *Collins cycle.* The building blocks of these units are usually oil-flooded screw compressors, finned-tube heat exchangers, and reciprocating expansion engines in the smaller sizes and centrifugal compressors, aluminum plate-fin heat exchangers, and gas- or oil-bearing turboexpanders in the larger sizes. With the exception of several small-capacity helium liquefiers, there are few standard refrigerators of this type. Unlike the GM-type refrigerators, these units circulate the working fluid and thus enable the transport of cold gas or liquid over long distances.

Joule-Thomson Cycle

Miniature JT heat exchangers about the size of a short pencil produce about 1 Btu/h (0.3 W) of refrigeration at 140°R (77 K) with N_2 or 155°R (86 K) with Ar supplied from commercial gas cylinders or special small gas bottles which may be charged to 6000 lb/in^2 (141 MPa). Systems of this type, which weigh about 3 lb (1.4 kg) and run for 2 to 3 h, are used in a number of military applications. Small closed-cycle Stirling cycle coolers are also widely used in military applications, but their short life limits their commercial use.

Other Cycles

Small cryocoolers operating on the reverse Stirling cycle are the most common closed-cycle cryogenic refrigerators in use in military applications. They consist of a motor driving a compressor piston and an expander with a displacer that is very similar to a GM expander which operates at the same speed as the compressor but about 90° out of phase. The expander can be driven by the same motor as the compressor piston, or it can be split from the compressor by a short interconnecting gas line in which case the expander is pneumatically actuated.

Instead of achieving a pressure cycle by a motor-driven compressor piston, a displacer with a regenerator can be used to displace and transfer gas from an ambient-temperature volume to a high-temperature volume at about 1100°F (594°C). A pressure ratio of about 1.8:1 is produced in this cycle. This thermal compressor is connected to the same type of expander as used in the Stirling cycle cooler to produce low-temperature refrigeration. Heat is thus the primary energy input. This cycle is called the *Vuilleumier cycle.* Small refrigerators operating on this cycle are being developed for spacecraft applications because the gas is transferred within the refrigerator at essentially the same pressure throughout. The seals and bearings are thus lightly loaded and long life can be achieved. Using thermal energy to power the unit can minimize the refrigeration system weight on a spacecraft.

Cryogenic Refrigerator Efficiency

Frequently, in order to evaluate the feasibility of a given process, one is interested primarily in the power consumption or efficiency of producing refrigeration at a rate Q and temperature T_c relative to ambient temperature T_a. Figure 1.7[7] presents the results of a survey of cryogenic refrigerator efficiency η versus refrigeration rate. Using these data, the input power P can be calculated using the formula

$$P = \frac{Q(T_a - T_c)}{\eta T_c} \tag{1.2}$$

If refrigeration is produced at more than one temperature level, the input power for each level is computed and the power summed.

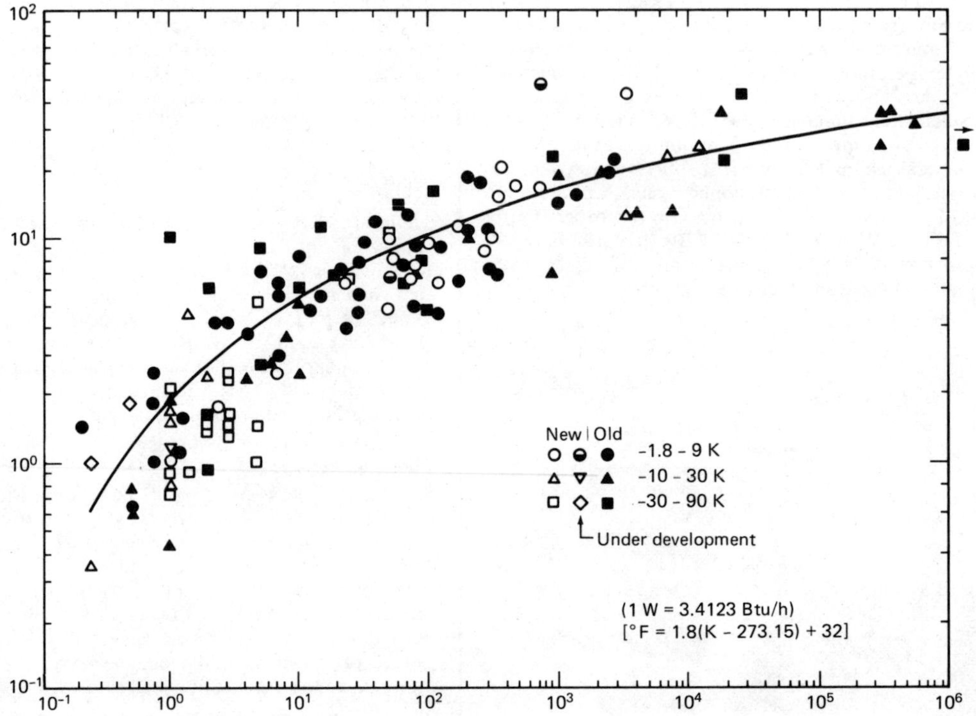

Fig. 1.7 Efficiency of low-temperature refrigerators and liquefiers as a function of refrigeration capacity.[7]

The power required to cool a gas stream which has constant specific heat from T_{c1} to T_{c2} is given by

$$P = \frac{Q}{\eta}\left[\frac{T_a}{(T_{c1} - T_{c2})/\ln (T_{c1}/T_{c2})} - 1\right] \quad (1.3)$$

Table 1.3 lists the ideal power required to produce refrigeration at the normal boiling temperature of common cryogens and the ideal power required to produce 61 in^3 (1 L) of liquid starting at 80°F (300 K).

Table 1.3. Ideal Power Required to Produce Refrigeration at the Normal Boiling Point of Common Cryogens and Power Required to Liquefy Them

Fluid	T, °R (K)	Refrigeration, W/W	Liquefaction, W · h/L
Helium	7.6 (4.2)	70.4	236
Hydrogen	36.7 (20.4)	13.7	278
Neon	49.0 (27.2)	10.1	447
Nitrogen	139.2 (77.4)	2.88	173
Argon	157.0 (87.3)	2.44	185
Oxygen	162.4 (90.2)	2.33	195
Methane	201.1 (111.7)	1.69	129

CONVERSION FACTORS: 1 W = 3.413 Btu/h, 28.32 L = 1 ft^3.

1.3 STRUCTURAL MATERIALS FOR CRYOGENIC SERVICE

Commonly Used Materials

A relatively small number of different materials are commonly used in the design of cryogenic equipment.

Stainless Steel Type 304 stainless steel is commonly used as a general structural material because it has good strength and toughness properties, relatively low thermal conductivity, and good machining and welding characteristics. Also, it can be cleaned for low vacuum outgassing rates and low radiation emissivity and is widely available in sheet and tube form at a moderate price. Carbon steels are not used as structural members because they become very brittle at cryogenic temperatures.

Copper Type OFHC copper is used to transfer heat because it has the highest thermal conductivity of any readily available material and is easy to work. The thermal conductivity is strongly dependent on impurities, so it is essential to specify the grade when ordering copper. OFHC grade has values that are about three times higher at 40°R (22 K) than at room temperature.

Aluminum Type 6061 aluminum is widely used in industrial cryogenic systems for heat exchangers, vacuum shells, and piping. It combines good thermal conductivity and strength while being light and cost-effective. Difficulties in welding, brazing, and soldering aluminum limit its use in small-scale applications.

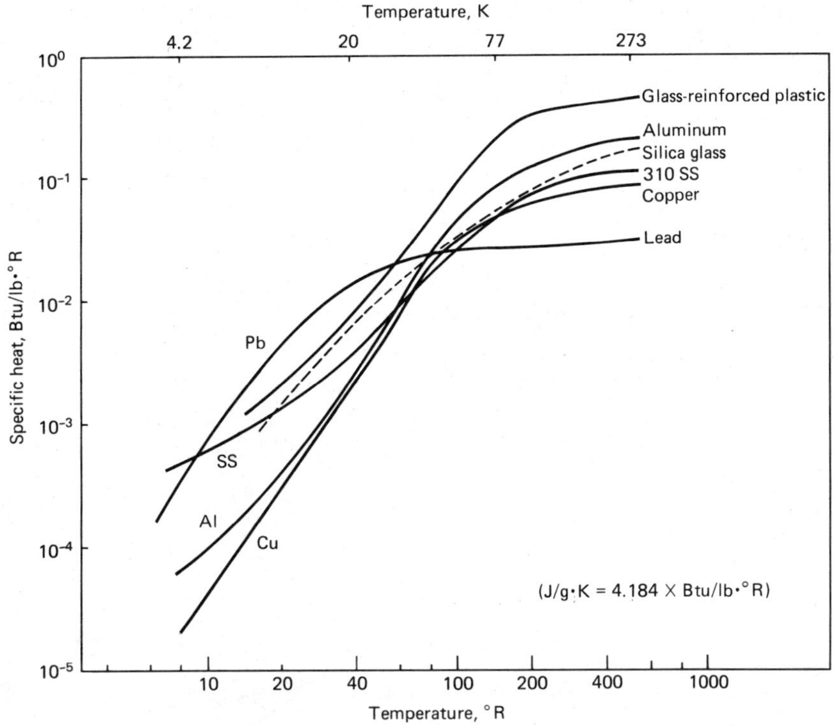

Fig. 1.8 Specific heat of selected materials at low temperatures.

Glass Silica glass is commonly used in glass dewars and as windows on systems where optical access is necessary. It exhibits low thermal conductivity, is transparent, and is easy to fabricate, but it is brittle. Glass dewars have very good vacuum retention characteristics unless they are in a helium-rich atmosphere, in which case helium diffuses through the glass and degrades the insulation effectiveness of the vacuum.

Glass-Reinforced Plastic Glass-reinforced plastic (GRP) Grade G10 is widely used as a support member in systems, under either tension or compression because it has a very high ratio of strength to thermal conductivity. Properties in the direction of the glass fiber, *w* (warp) direction, are significantly different from the properties in the *n* (normal) direction.

Other Materials Materials that are less frequently used include Teflon,* both pure and filled, which remains somewhat flexible and is used as a seal; sapphire, which is unique in having moderately high thermal conductivity while being a good electrical insulator; brass, because it is easy to machine; beryllium-copper, which is effective for springs; and several plastics, including nylon and Delrin.

Properties of Structural Materials

Specific Heat Figure 1.8 shows specific[1,8] heat-versus-temperature data for the commonly used materials. Specific heats for all types of structural material drop to very low values at low temperatures. Data for lead is shown in Fig. 1.8 because it has relatively high specific heat in the 10 to 50°R (5 to 30 K) range and is commonly used for the second-stage regenerator in GM and Stirling cycle cryocoolers.

Enthalpy Table 1.4 lists the enthalpy of selected materials[1,8] with 0°R as the reference temperature. The enthalpies of stainless steel and GRP are based on a limited number of specific heat data point[8]; thus the accuracy is estimated to be +1 to 5 percent. Enthalpy data are used to determine how much heat has to be removed from a system in order to cool it down.

Thermal Conductivity Figure 1.9 presents thermal conductivity data for selected materials versus temperature.[1,6,8,9] High-purity metals have very high thermal conductivities in the liquid hydrogen temperature region, while alloys, plastics, and glass become better insulators as they become cold. Because of the large changes in thermal conductivity as the temperature is decreased, it is necessary to use the thermal conductivity integral to calculate heat flow from one temperature level to another. These data are presented in Table 1.5.

For pure metals, the same mechanism that is responsible for transporting heat is responsible for transporting electricity. Thus there is a well-defined relationship between thermal conductivity and electrical resistivity (Wiedemann-Franz law). This relationship is used to evaluate the thermal conductivity of metals and their purity as well as the thermal resistance of a joint by measuring the electrical resistivity at room temperature and in liquid nitrogen. The ratio will be

*Trademark of E. I. DuPont de Nemours, Inc.

Table 1.4. Enthalpy of Selected Solids at Low Temperature, Btu/lb (Reference Temperature 0°R)

Temperature, °R (K)	Stainless steels 305 to 310*	Copper OFHC	Aluminum 6063	Glass silica	GRP G-10*
540 (300)	40.6	34.2	73.3	50.6	167
360 (200)	20.5	18.2	36.5	22.8	85
180 (100)	4.56	4.56	7.64	4.98	18
144 (80)	2.53	2.59	4.03	2.94	9.9
108 (60)	1.08	1.11	1.57	1.47	3.9
72 (40)	0.293	0.26	0.32	0.51	0.91
36 (20)	0.044	0.015	0.021	0.061	0.10

* Accuracy of stainless-steel and GRP data is estimated to be ±5 percent.
CONVERSION FACTOR: J/g = Btu/lb × 2.326.

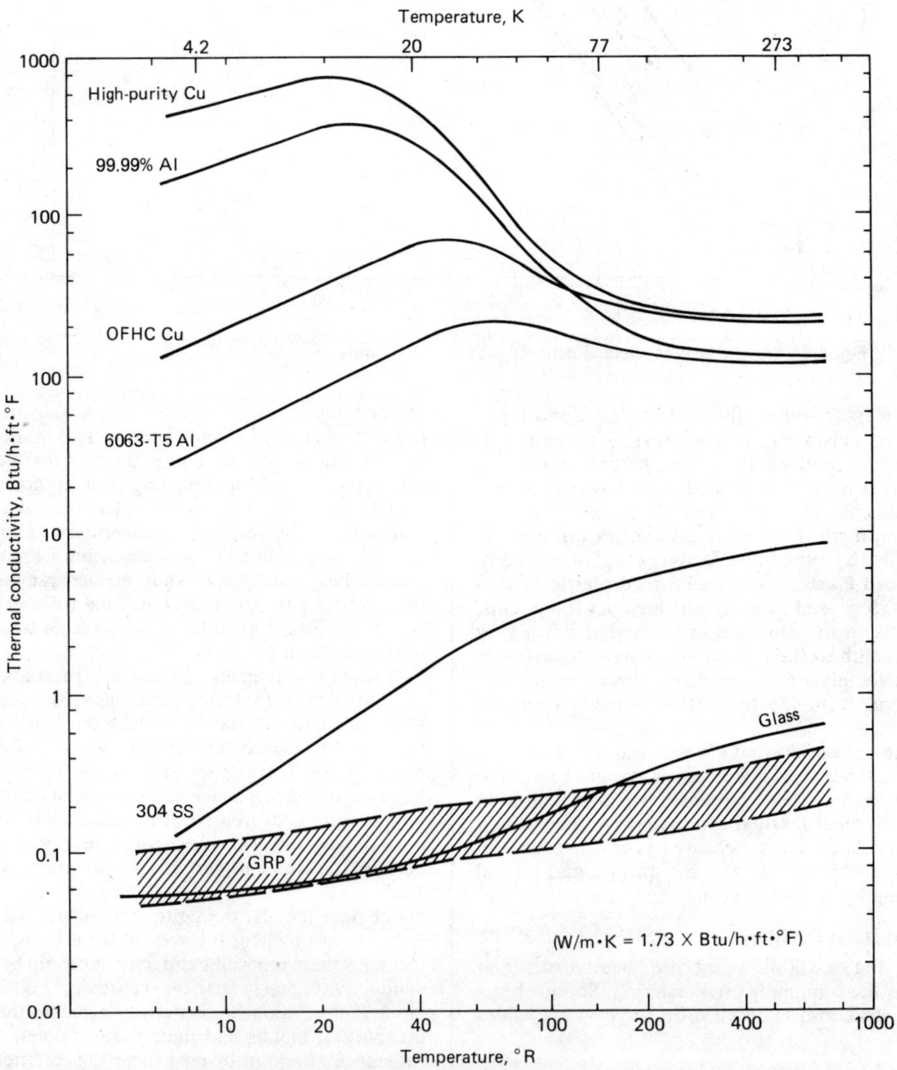

Fig. 1.9 Thermal conductivity versus temperature for selected materials.

Table 1.5. Thermal Conductivity Integrals of Selected Materials at Low Temperature

Temperature, °R (K)	Stainless steel 304	Copper OFHC	Aluminum 6063	Glass silica	GRP G-10 (*w*)	GRP G-10 (*n*)
540 (300)	265	13,165	5,309	17.2	13.6	9.6
360 (200)	144	9,700	3,577	8.92	2.0	5.1
180 (100)	45.7	6,063	1,827	2.53	2.4	1.7
144 (80)	30.2	5,249	1,446	1.68	1.7	1.2
108 (60)	17.1	4,296	1,013	1.00	1.0	0.79
72 (40)	7.14	2,928	537	0.51	0.51	0.43
36 (20)	1.41	953	143	0.17	0.17	0.14

$$\int k(T)\,dT = \frac{qL}{A} \quad \text{Btu/h} \cdot \text{in} \quad \text{(reference Temperature 0°R)}$$

where $k(T)$ = thermal conductivity as a function of temperature
L = length, in
A = area, in^2
q = heat flow, Btu/h

CONVERSION FACTOR: W/cm = 0.115 × Btu/h · in.

the same as the thermal conductivity ratio, and the thermal resistance will be proportional to the electrical resistance.

Thermal Expansion Most materials have significant contraction between room temperature and 150°R (83 K), but below that there is little further change. Table 1.6 presents thermal expansion data as a percentage change in length relative to room temperature for selected materials.[1,6,8] Glass-to-metal transitions are possible at low temperatures by adding other oxides and impurities to glass to change its thermal expansion to match that of alloys of FeNi (Invar). Some of these have essentially no change in dimension as they are cooled.

1.4 INSULATION

Types and General Characteristics

Large industrial-sized equipment and storage tanks containing liquid nitrogen or warmer cryogens are usually insulated with a fibrous material that is filled with dry nitrogen. Since air can liquefy at liquid nitrogen temperature and moisture can freeze at much higher temperature, it is imperative that the outer container for this type of insulation be vaportight. In some instances, small air leaks are compensated by continually purging the insulation with dry nitrogen.

For containment of lower-temperature components, vacuum insulation is used. For gas-filled insulations, conduction is usually the dominant heat-transfer mechanism, while for vacuum insulation, radiation is dominant apart from conduction in the structural supports.

Fibrous Insulation

Fibrous and powder insulations commonly used include fiberglass, rock wool, powdered cork, and perlite. Foam insulations, such as polyurethane, are sometimes used, but they contract significantly and tend to open leakage paths. When insulations are filled with dry nitrogen, heat conduction through the nitrogen tends to dominate, and they all have the same apparent thermal conductivity within a factor of 2, with fiberglass, in particular, being better than average. When insulations are evacuated, radiation becomes the dominant heat-transfer mechanism and the apparent thermal conductivity can vary by a factor of 5. Evacuated fiberglass is one of the poorer insulations. Table 1.7 lists apparent thermal conductivities for several insulations. Reference 6 is rec-

Table 1.6. Thermal Expansion of Selected Materials at Low Temperature

Temperature °R (K)	Stainless steel 304	Copper OFHC	Aluminum 6063	Glass silica	GRP G-10 (W)	GRP G-10 (N)
360 (200)	0.140	0.149	0.201	0.06	0.108	0.346
180 (100)	0.262	0.283	0.370	0.09	0.197	0.603
0 (0)	0.292	0.326	0.415	0.11	0.241	0.706

$$\frac{L_a - L_t}{L_a} - \%$$

where L_a = length of ambient temperature, 527°R (293 K)
L_t = length cold, $L_t < L_a$

Table 1.7. Apparent Thermal Conductivity of Typical Cryogenic Insulations between 540 and 140°R

Insulation	N_2 pressure, torr	Density	Thermal conductivity, Btu/h · ft · °R
Fiberglass	>100	6.9 lb/ft³	1.4×10^{-2}
Fiberglass	$<10^{-2}$	3.1 lb/ft³	1.0×10^{-3}
Aluminized Mylar	$<10^{-4}$	90 layers/in	2.4×10^{-5}

CONVERSION FACTOR: W/m · k = 1.73 × Btu/h · ft · °F.

ommended for additional information concerning the insulation of cryogenic equipment.

Vacuum Insulation

As a general rule, it is necessary to have a vacuum (residual gas pressure) of less than 1×10^{-4} torr in order to eliminate gaseous conduction heat transfer in insulations. Radiation then becomes dominant; and, in simple systems with only a few radiation shields, the emissivity ϵ of the surfaces becomes very important. Radiation shields are either active or passive. Active shields are usually used in liquid hydrogen and helium systems. They are shields that are either cooled with liquid nitrogen, hydrogen or helium boil-off vapors, or an external source of refrigeration. Passive shields are layers of metal foil or aluminized Mylar that "float" at a temperature governed by the heat-flux and temperature-boundary conditions.

The equation for radiant-heat transfer between parallel plates is useful in many cryogenic applications for computing the heat transfer q between surfaces at temperatures T_h and T_c (°R) with N radiation shields having a net emissivity E

$$q = \sigma(T_h^4 - T_c^4)\frac{E}{1 + N} \tag{1.4}$$

where the Stefan-Boltzmann constant σ has the value 1.712×10^{-9} Btu/h · ft² · °R⁴ (5.67×10^{-12} W/cm² · K⁴). For the simple case of no shields, the net emissivity is

$$\frac{1}{E} = \frac{1}{\epsilon_1} + \frac{1}{\epsilon_2} - 1 \tag{1.5}$$

or for the case of concentric spheres

$$\frac{1}{E} = \frac{1}{\epsilon_1} + \frac{A_1}{A_2}\left(\frac{1}{\epsilon_2} - 1\right) \tag{1.6}$$

The emissivity of surfaces is dependent on the temperatures of the hot and cold surfaces and the composition and condition of the surfaces. When radiation shields are interposed, the conduction between the layers of shielding may become dominant and actual experimental data have to be used to predict the heat transfer.

Table 1.8 lists some values of emissivity that have been measured for surfaces commonly found in cryogenic systems. Table 1.8 shows the general range of emissivities for various materials and points out that emissivities decrease as the temperature decreases, decrease as metal purity increases, decrease as the surface becomes smoother, and increase as the surface becomes dirty. Aluminum is the preferred material for radiation shielding because of its very low emissivity and because the oxide layer that protects it from degradation is transparent to infrared radiation. It has been found that parallel surfaces at liquid helium and liquid nitrogen temperature which are covered with aluminum tape (3M Company No. 425) have a net heat flux of 0.0030 Btu/h · ft² (12.4 mW/m²). The smooth layer of high-purity aluminum in this tape has slightly lower emissivity than the values listed in Table 1.8. Figure 1.10 shows values of net emissivity for a variety of cold surfaces that have been tested in which the warm surface is painted black.[10] Some of the surfaces listed include two layers of superinsulation wrapped on the cold surface.

Table 1.9 lists rated heat flux and net emissivities of 0.00025-in-thick aluminized polyester as a function of density. Increasing the density increases the conduction through the layers. Aluminum thickness is about 300 Å. Significant variations occur from batch to batch because of variations in thickness and metal purity.

Table 1.10 lists values of heat flux that were measured for various numbers of layers of aluminized mylar between a surface at 139°R (77 K) (aluminum covered) and one at 7°R (3.9 K) (painted black). These data show that there is an optimum number of layers beyond which the heat flux begins to increase. It is interesting that the minimum value measured, 0.0040 Btu/h · ft² (0.0126 W/m²) is no better than simply coating both surfaces with aluminum tape.

Multiple layers of aluminized Mylar or polyester with a high vacuum is one form of *superinsulation*. Other forms are aluminum foil with fiberglass paper or nylon net as a spacer. The residual gas that is trapped between the foil or Mylar sheets reduces the effectiveness of these superinsulations until it is pumped out—a process that may take several days or weeks, depending on the number of layers and their density. Superinsulation must be kept very clean while it is installed. Such things as oil from fingerprints seriously degrade the emissivity. A very efficient insulation system on a liquid helium container will typically have about 10 layers of superinsulation around the helium vessel, a liquid-nitrogen-cooled shield, and about 50 layers of superinsulation around the liquid nitrogen shield. Research equipment that has to be evacuated quickly will typically have one or two actively cooled shields and no superinsulation.

Table 1.8. Emissivities of Surfaces at Cryogenic Temperatures for Diffuse Thermal Radiation from 150°R (83 K)

Surface	Temperature, °R 7	139	540
Aluminum			
High-purity annealed	0.011	0.018	0.03
Commercial plate	0.018	0.03	0.05–0.08
Copper			
High-purity		0.008	
ETP	0.006	0.015	0.02
Commercial sheet	0.054	0.075	
Stainless steel			
304, polished		0.03	0.05
Commercial sheet	0.078	0.15	

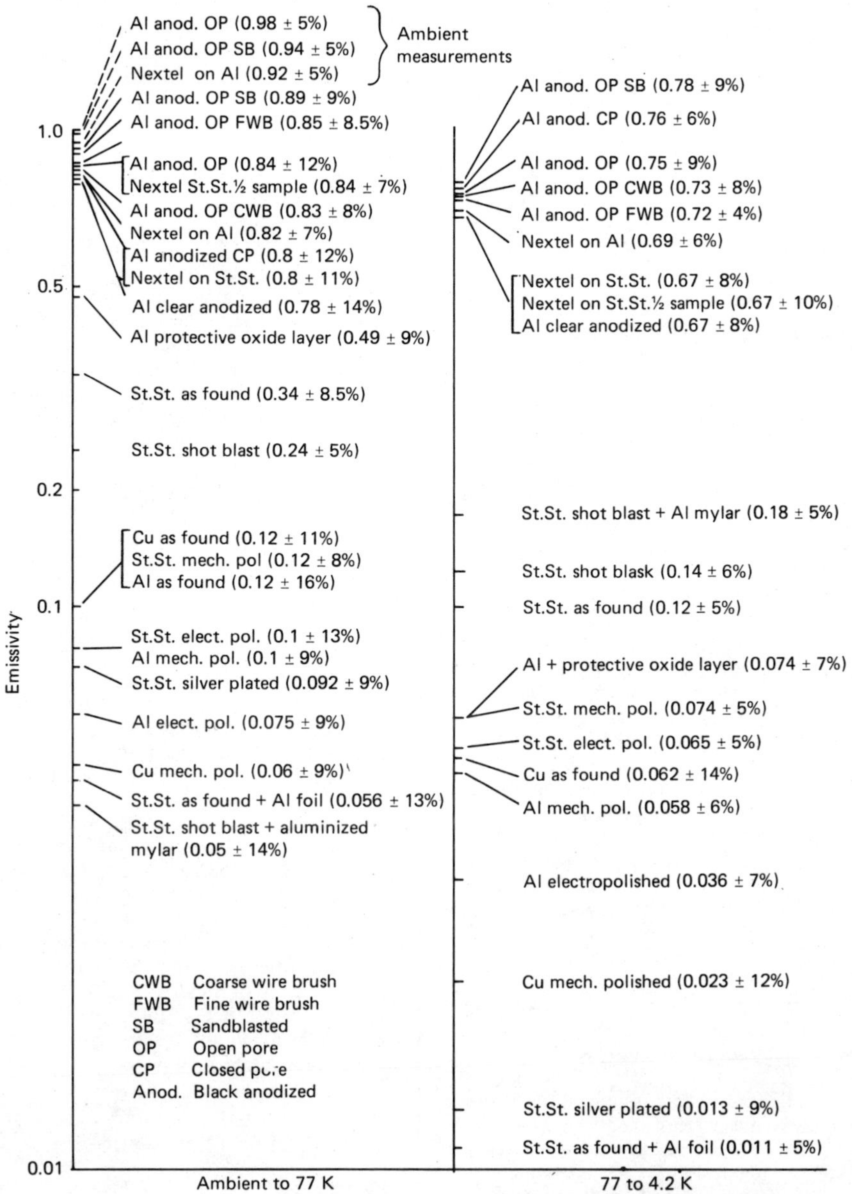

Fig. 1.10 Emissivities of selected metallic surfaces at cryogenic temperatures. (*Reprinted by permission.*[10])

Vacuum Requirements

Familiarity with vacuum technology is almost a prerequisite for working at cryogenic temperatures. O'Hanlon[11] is a recommended reference.

A good vacuum is essential for most cryogenic systems in order to minimize heat losses. The thermal conductivity of gases is proportional to the product of their density and the mean free path of the gas molecules between collisions. Since these are nearly inversely proportional to each other for air (and other cryogens) over a wide range of pressures at normal ambient temperature, the thermal conductivity is nearly constant over a wide range of pressures. As the pressure is reduced, however, a point is reached where the mean free path is greater than the distance between the cold surface and the room-temperature container, and the thermal conductivity decreases proportionally with the pressure. This transition occurs at a pressure of about 1 torr (1 mmHg, 133

Table 1.9. Rated Heat Flux and Net Emissivity of Aluminized Polyester between 540 and 139°R (300 and 88 K)

Layers, in	Heat flux for 50 layers, Btu/h · ft²	Net emissivity per layer
40	0.12	0.042
80	0.20	0.070
120	0.28	0.098

CONVERSION FACTOR: W/m² = 3.152 × Btu/h · ft².

Pa) and when the pressure has been reduced below 1×10^{-4} torr, heat losses by conduction through the gas can usually be neglected.

In order to establish an insulating vacuum, it is necessary to have a vacuum pump that can effectively remove gases at pressures below 1×10^{-4} torr and have sufficient displacement to achieve a low pressure in a reasonable time. Assuming that a pump with sufficient displacement (pumping speed) and minimum pressure capability is being used, it is characteristic when evacuating a vessel to have the pressure drop quickly and steadily at first, then have the pressure drop very slowly until it reaches equilibrium. Initially, the pump is removing bulk air from the chamber; and then begins to remove (1) gases that desorb from the surfaces (typically water), (2) gases that permeate through elastomer seals, (3) hydrogen that diffuses from steel, and (4) gases that enter the chamber through leaks. Steady-state pressure is reached when gases permeating and leaking into the chamber are removed at the same rate they are entering.

When a cryogenic system is cooled down, the vacuum usually improves as gases are cryopumped on the cold surfaces. Water has a vapor pressure of 10^{-12} torr at 213°R (118 K). Small quantities of nitrogen and oxygen can be adsorbed by charcoal at 140°R (77 K), and hydrogen can be adsorbed effectively by charcoal at 36°R (20 K) at a rate about 0.25 lb of hydrogen per pound of charcoal. Charcoal can be bonded to a cold surface with many epoxies or retained in a copper screen basket. Synthetic zeolites are also effective adsorbents of trace gases at cryogenic temperatures but have to be heated above 300°F to remove H_2O while charcoal can be effectively cleaned up by subjecting it to vacuum pressure at room temperature.

It is typical for a leaktight cryogenic system with elastomer O rings and reasonably clean surfaces to operate in the 10^{-6}-torr region. Pressures of 10^{-8} torr are possible, especially if the system is heated to about 160°F (345 K) while being evacuated for a period of about 12 h. In order to get below 10^{-8} torr, it is usually necessary to use metal gaskets and eliminate all organic materials from the system so that it can be heated or baked out about 400°F (480 K) under a higher vacuum for about 12 h before being cooled down. Electropolished stainless steel is preferred for the vacuum housing.

Table 1.10. Heat Flux from 139°F to 7°R (77 to 4 K) for Various Numbers of Layers of Aluminized Mylar

	Heat flux, Btu/h · ft²				
Number of layers	2	4	8	30	65
Heat flux, Btu/h · ft² (× 10^{-3})	5.4	4.7	4.3	4.0	4.4

CONVERSION FACTOR: W/m² = 3.152 Btu/h · ft².

Vacuum Pumps

Mechanical Mechanical pumps are used to get the bulk of the air out of a vacuum system. The most common type of small pump, up to 15 ft³/min (7 L/s), has a rotary eccentric lobe sliding vane with oil lubrication. Single-stage units have a minimum pressure of about 1×10^{-2} torr, and two-stage units can operate to about 2×10^{-4} torr. Larger mechanical vacuum pumps have two rotating impellers with figure-eight-shaped lobes that mesh and are referred to as *Root's blowers*.

Diffusion Pumps Diffusion pumps have a boiler in their bases that vaporizes oil which flows down through nozzles in a stack over the boiler. The high-velocity oil molecules entrain lighter gas molecules and carry them to the base of the pump where the oil is condensed and the gas molecules removed with a mechanical pump. A pressure of less than 0.01 torr is required of the mechanical pump.

Diffusion pumps operate in the range of 1×10^{-8} to 5×10^{-3} torr. Oils which have low vapor pressure are used; nevertheless, oil does tend to migrate back into the vacuum chamber. If a chamber must be kept clean, a liquid-nitrogen-cooled trap or equivalent is required.

Turbomolecular Pumps

As the name implies, a *turbomolecular pump* has a high-speed rotating shaft with a series of rotating and stationary blades that compress gas from the 10^{-8} torr region to a pressure of about 1×10^{-3} torr which is maintained by a mechanical pump.

Cryopumps Cryopumps use small two-stage cryogenic refrigerators to cool extended-surface panels. An inlet louver cooled to about 108°R (60 K) freezes out water and hydrocarbons; behind it is a panel at about 27°R (15 K) that freezes out nitrogen, oxygen, and argon. Charcoal bonded to the back side of the colder panel adsorbs neon, hydrogen, and small quantities of helium. Standard cryopumps have a minimum pressure of about 1×10^{-10} torr and can pump air at pressures of up to 1×10^{-3} torr. Their thermal inertia enables them to almost instantaneously pump a chamber of several cubic feet (or cubic meters) from 1 torr to $<1 \times 10^{-3}$ torr.

1.5 CRYOGENIC INSTRUMENTATION

Thermometry

For most cryogenic applications, diode, thermocouple, and resistance thermometry are the primary means of measuring temperature. Gas, acoustic, capacitance, paramagnetic, or similar kinds of thermometers are specialized instruments used primarily in low-temperature research work. Vapor-pressure thermometry is also used within limited temperature ranges.

Diodes At constant forward current, the voltage drop of a semiconductor diode increases with decreasing temperature. The voltage-versus-temperature function is nonlinear, and since it changes from specimen to specimen, each diode thermometer must be calibrated individually. They are usually connected in the four-wire configuration with a 10-μA forward current on two leads, and voltage is read out on the other two.

Silicon and gallium-arsenic (or GaAlAs, GaAsP) diodes are the most commonly used cryogenic temperature sensors. Their range of operation spans from 2°R (1 K) to well above room temperature. Figure 1.11 shows typical voltage-versus-

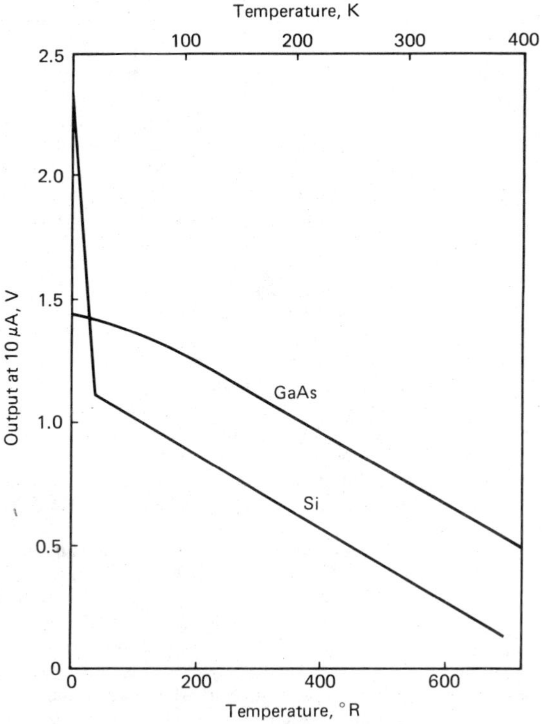

Fig. 1.11 Diode temperature sensor output at 10 μA.

temperature characteristics for commercially available sensors. At the 10-μA forward current, the sensors exhibit a typical 1.5 mV/°R (2.7 mV/K) sensitivity above 70°R (50 K) and as much as 28 mV/°R (50 mV/K) for silicon around 7°F (4 K). Silicon diodes are considered superior to GaAs sensors in terms of lifetime, reproducibility, noise, and sensitivity. Gallium-arsenic diodes are affected less by high magnetic fields than silicon diodes and are thus more frequently used in applications such as monitoring temperatures near superconducting magnets.

Diode thermometers are usually packaged so as to avoid direct contact with their housing, and the primary path of heat flow between the environment and the diode itself is through the electric leads even if the housing is filled with helium. Good thermal contact between the leads and the object being measured is especially important.

A special aspect of diode thermometry is that these sensors rectify any ac noise picked up by the wires connecting them to the measuring electronics and superpose a parasitic dc voltage on the signal which falsifies the temperature reading. To eliminate this effect, either the leads must be properly shielded or a shunt capacitor can be placed in parallel with the diode (if the sampling rate and conversion time of the measuring electronics permits this). The capacitor must be put as close as possible to the diode in order to minimize the area of the ac loop.

Thermocouples Thermocouples are the simplest cryogenic temperature transducers. Their advantages are the lack of need for power supply, absence of self-heating, fast response, and relatively low cost. Disadvantages are the poor sensitivity, 6 to 11 μV/°R (10 to 20 μV/K), the requirement for a reference junction, and the danger of decalibration on mechanical strain. Long leads can introduce errors, and the wires should run without interruption from the junction to the readout meters.

Copper-Constantan (type T) or iron-doped gold Chromel versus Chromel are the commonly used thermocouples for the measurement of low temperatures (Fig. 1.12). The for-

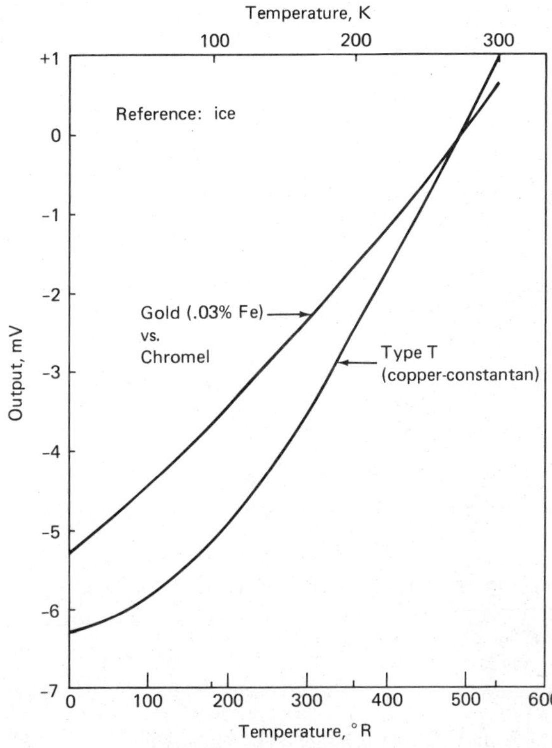

Fig. 1.12 Thermocouple temperature sensor output.

mer is suitable for general thermometry purposes above liquid nitrogen temperature, while the latter pair is used at liquid helium temperatures and above. Copper-Constantan thermocouples follow a common calibration curve; Au(Fe)-Chromel sensors require calibration for each batch of wire made from a given melt.

Providing an appropriate reference and connections free from parasitic emf are of paramount importance at cryogenic temperatures.

Resistance Thermometers Resistance temperature detectors (RTDs) are very stable and accurate temperature sensors. They utilize the temperature-dependence of resistivity of conductors or semiconductors. Platinum RTDs are the most frequently used resistance thermometers in industrial cryogenics above liquid hydrogen temperature. Germanium and carbon glass resistance thermometers are used for temperatures below that of liquid nitrogen (Fig. 1.13). The resistance is measured either in the four-wire configuration or by a bridge circuit.

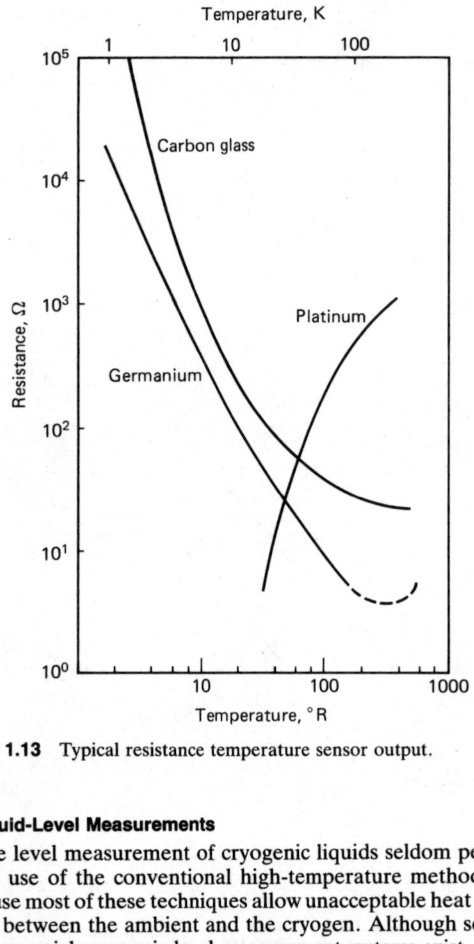

Fig. 1.13 Typical resistance temperature sensor output.

Liquid-Level Measurements

The level measurement of cryogenic liquids seldom permits the use of the conventional high-temperature methods because most of these techniques allow unacceptable heat transfer between the ambient and the cryogen. Although several commercial cryogenic level measurement systems exist, based on various principles, there is no "best" configuration. The design engineer must decide case by case which method or equipment is appropriate for the given application in terms of system geometry, accessibility, maximum permissible heat loss, and resolution. The more common methods that are used include (1) differential pressure between the top and bottom of a tank, (2) weight of the tank, (3) capacitance of two parallel vertical wires, and (4) diodes used at discrete points which sense whether they are in liquid or vapor.

Pressure, Vacuum, and Flow Measurements

Since these measurements are almost invariably implemented at the room-temperature section of a cryogenic system, standard techniques not specific to low temperatures are generally applicable.

1.6 SAFETY

It is good practice to do a safety review on new designs and to establish safety procedures for operating and servicing equipment. The following potential risks must be considered in order to provide a safe working environment. References 12 and 13 are good sources of additional information.

Asphyxiation The common cryogens listed in Table 1.1 are nontoxic. However, with the exception of air and oxygen, they are asphyxiants. When cryogens spill or are vented into a room, they displace some of the air. When O_2 concentration is reduced from 21 to 14 percent, breath and pulse rates increase, and thinking and coordination are diminished. Lower concentrations are dangerous and should be avoided. One would assume that since N_2 is a major constituent of air, an increased concentration is less of a problem than for other gases. The opposite, however, is true in that increased N_2 concentration decreases the blood's ability to transport O_2 disproportionately; thus 16 percent O_2 is the minimum concentration recommended when the balance is N_2. Asphyxiation due to nitrogen is probably the most common cause of death in cryogen-related accidents.

Pressure The volume of cryogens at atmospheric conditions is more than 700 times the volume of their liquids. It is thus imperative that adequate vents and pressure-relief devices be used to ensure that liquids are never trapped as they warm up. One serious problem that can occur is to have moisture from the air freeze out in the neck of a storage dewar and plug it. To avoid this, storage containers should have vents that do not allow air to back-diffuse into the neck tube. They should also be equipped with a secondary vent tube and pressure-relief device.

Loss of Vacuum Pressure-relief valves on liquid storage containers are usually sized on the assumption that there is a sudden loss of insulating vacuum and air rushes in and condenses on the cryogen container, causing a rapid venting of the cryogen. Large cryogen storage containers should be vented outside the building to avoid asphyxiation.

Combustion Hydrogen in air is flammable in concentrations from 4 to 75 percent by volume. It can detonate in concentrations from 18 to 59 percent with very little energy required (as low as 0.02 mJ) and at pressures as low as 1.5 lb/in^2 abs (10 kPa). Methane is flammable in air in concentrations from 5 to 15 percent by volume. It can detonate in concentrations in the range 6.3 to 11.5 percent with little energy required (0.30 mJ) and at pressures as low as 1 lb/in^2 abs (7 kPa).

Oxygen, gaseous and liquid, and liquid air should be handled in extremely clean systems, free of hydrocarbons, in order to avoid combustion and explosions. Metals (e.g., 304 stainless steel) that are compatible with oxygen must be used.

Cryocondensation of Air Air which condenses on very cold surfaces, such as uninsulated liquid nitrogen transfer lines, becomes rich in oxygen as the more volatile nitrogen evaporates first. If this collects on combustible material, it can be very explosive.

Freeze Burns Cold liquids and gases can cause freeze "burns." On open surfaces, at room temperature, the cryogens bounce around as a result of film boiling, and heat transfer rates are low. If the liquid soaks into clothing or is otherwise trapped, it causes very rapid freezing of the skin. Protective gloves and glasses are recommended as a minimum precaution when transferring cryogens.

1.7 CRYOGENIC DESIGN PRACTICE

This section contains general design recommendations. Reference 14 contains additional information that is useful for designing an experimental apparatus.

Material Compatibility

Beyond the obvious safety restrictions of liquid oxygen compatibility, certain materials or composites are incompatible with the cryogenic environment for structural reasons. The cryogenic use of composites or assemblies whose parts exhibit widely different thermal expansion, materials which undergo transitions between two phases (e.g., iron) or radically different mechanical properties, and certain elastomers (as latex and most rubbers, which develop cracks on repeated thermal cycling) should be avoided. The behavior of electronic components, connections, and circuits may strongly change or simply fail at low temperatures. However, many electronic components perform very well at cryogenic temperatures.

Metal-to-Metal Joints

Materials and geometry permitting, fusion welding is the preferred metal-to-metal joint in cryogenic design, especially where the joint serves as a seal. If fusion welding is impractical or if purely structural joints are required, brazing is permissible. Soft solder is commonly used for making good thermally conducting bonds but never as a sealing technique because of its propensity to leak. Roll bonding and friction welding or explosion bonding are sometimes applied, mainly for joining dissimilar metals, such as aluminum and stainless steel. Compression fittings tend to leak at low temperatures and are not used. All of these joints may develop cracks on thermal cycling. Some of them open only when cold. When joints serve also as seals, they have to be carefully leak checked by using standard leak-detection techniques. Leaking cold joints have to be redone, or if it is impractical, soldered over with indium.

Standard fasteners can be used at cryogenic temperatures provided their thermal contraction is equal or higher than the joined pieces over the entire temperature range. The principle is to have differential thermal contraction cause the joint to seal tighter when it becomes cold.

Metal-to-Nonmetal Joints

The differential thermal contraction renders these joints generally unreliable. Small scale [maximum 2 to 3 in (5 to 8 cm)] workpieces may be joined by adhesives or epoxies. A basic restriction of such designs is that the differential thermal contraction may not induce other than compressive stresses in the joint. Fused graded joints between metals and ceramics or glasses are generally reliable. Cryogenic optical windows using such joints are commercially available.

Nonmetal-to-Nonmetal Joints

Reliable joints can be achieved by adhesive bonding if both the joined pieces and the adhesive have matched thermal expansion coefficients. Good cryogenic joints can be formed between certain thermoplastic polymers by ultrasonic or thermal fusion welding.

Cryogenic Seals

The most frequently used static cold seals, in the order of decreasing reliability, are indium gaskets, metal gaskets, and special elastomer seals. Indium seals are made by trapping indium wire in a close fitting shoulder and compressing it until it extrudes from the joint. Metal gaskets include Teflon-coated spring tubing shaped as O rings, and K cross-section spring material coated with Teflon or gold. Smooth surfaces with matched expansion are required for these seals to work. Standard O rings invariably leak. Silver chloride and liquid seals are used sometimes to seal materials of strongly different thermal expansion coefficient. With the exception of bellows with linear or bending motion, no reliable dynamic seals are available for cryogenic applications.

Thermal Contacts at Low Temperatures

An important aspect in the design of cryogenic heat sinks, thermal switches, and insulating structures is that, in vacuum, the heat transfer between two bodies in contact deteriorates with decreasing temperature. It is standard practice to use a thin sheet of indium [0.005 to 0.010 in (0.5 mm) thick] as a gasket material in a bolted joint that has to transfer heat. Vacuum grease is a poor second choice for heat transfer. Contacts in gases or liquids may exhibit orders of magnitude higher overall conductivity than in vacuum as a result of thermal conduction through the gas in the gap. There is no practical way to make a good mechanical thermal switch in vacuum, but a number of techniques have been developed to make gas-filled switches and thermal diodes.

Gas-filled switches typically consist of closely spaced hot and cold plates in a gastight housing which are separated by a vacuum ($<1 \times 10^{-4}$ torr) for thermal isolation or in which gas such as helium or hydrogen is admitted for gas conduction.

Gas-filled tubes can have high heat fluxes from bottom to top when the top temperature causes the gas in the tube to condense and drip to the bottom where it evaporates. The tube can also be designed to promote convective heat transfer when the bottom end is hotter than the top. In either case, if the bottom end is colder than the top, the gas in the tube stratifies or may even freeze out.

Instrumentation Installation

In systems where no excess refrigeration is available, it is of fundamental importance that the heat generated or conducted by the low-temperature instrumentation be minimized. All the wiring connecting the cold sensors with the ambient electronics must be of low thermal conductivity at cryogenic temperatures, and total cross-sectional area of the wire is to be kept as low as the mechanical integrity permits. Insulated Manganin or beryllium copper are usually preferred as lead wires. Low-thermal-conductivity alloys are also poor electrical conductors, and, in addition, the lead resistance is temperature-dependent. These effects have to be canceled by the measurement configuration (e.g., four-wire resistance measurement method or the use of bridge circuitry).

In cases where thermocouples of copper or gold wire, which are good conductors, must extend from the cold region to the ambient, the minimization of the wire diameter is especially important: 3 to 5-mil (75 to 125-μm)-diameter Teflon-coated wires are routinely used. If resistance heaters are used for temperature control, it is standard practice to use the cryostat body as a common terminal at ground potential and lead only one wire per heater to the ambient.

Leads are normally brought through the vacuum wall using glass-Kovar feedthroughs. Teflon plugs in compression fittings are also used, especially with thermocouples which should not be interrupted.

Whenever it is possible, the wires connecting the cold region with the ambient should be thermally anchored to surfaces which can intercept and dissipate the in-flowing heat. In case of multistage refrigerators, the lead wires are to be anchored to each stage having a temperature that is higher than the temperature to be measured. Above a liquid cryogen bath, the wires can be cooled by the evolving cold vapors. In order to avoid short-circuiting with the ground and each other, lead wires are usually insulated over the full length. Care must be taken to provide adequate heat transfer through the insulation by maximizing the anchoring area. Good thermal anchoring is important from another aspect. If the in-flowing heat is not intercepted, it warms up the sensor and results in higher temperature readings.

Long lead wires tend to pick up electromagnetic noise from the environment. If such a problem is present, the lead wires must be shielded. This usually affects only the room-temperature section since the normally used metal enclosure of the cryogenic section serves as a shield itself. In glass or plastic cryostats, special commercially available miniature coaxial cryogenic cables are to be used in order to minimize the heat flux. Coaxial cryogenic cables are preferred as lead wires for ac circuits operating in a cryogenic environment. Heater circuits should always be operated from dc (preferably floating) power supplies. From both safety and noise standpoints, it is important that metal cryostats, housing instrumentation, be grounded.

Heaters

Heaters are usually constructed from alloys that tend to have fairly constant resistance down to liquid helium temperature. However, there is some change; thus it is recommended that both voltage and current be measured to determine heater power. Heaters should be wrapped or clamped in a material that will contract on cooling in order to maintain a high clamping force measurement. If a heater that is mounted in vacuum is not in adequate thermal contact with the object being heated, it can radiate a significant amount of heat, thus giving a misleading heat input, and burnout.

Example

Figure 1.14 is a schematic diagram of a refrigerated liquid helium cryostat which illustrates a number of the cryogenic

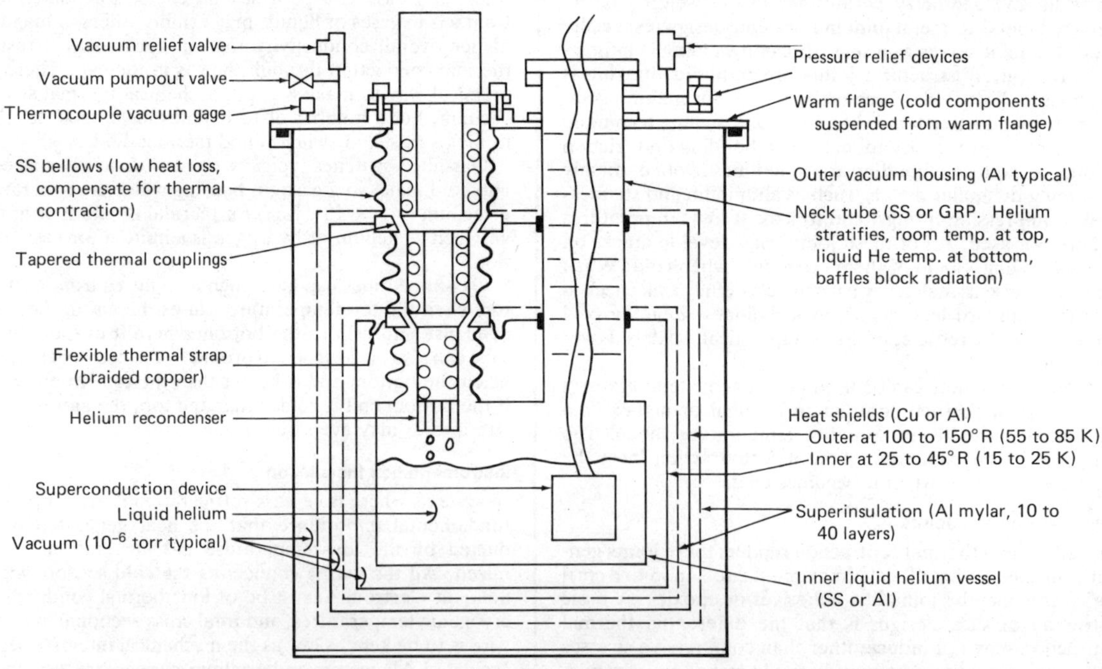

Fig. 1.14 Refrigerated liquid helium cryostat for continuous cooling of a superconducting device. (*Reference sources: U.S. Patents 4,223,540; 4,279,127; 4,484,458; 4,606,201.*)

design features discussed in this chapter. This type of cryostat is being used to cool superconducting magnets for magnetic resonance imaging (MRI) and can also be used to cool superconducting electronic devices. The cryostat consists of an outer vacuum housing and an inner liquid helium vessel with two refrigerated heat shields and superinsulation in between. Two neck tubes are shown, a corrugated stainless-steel bellows sleeve for the refrigerator and a rigid stainless-steel (or GRP) tube for the electrical leads to the superconducting device. The rigid neck tube serves to support the cold structure by suspending it from the warm flange.

Helium gas in the neck tubes stratifies with a temperature gradient from liquid helium temperature at the bottom to room temperature at the top. The refrigerator is a two stage GM unit with a JT heat exchanger that is coiled around the expander cylinder in such a way that the temperature gradient in the refrigerator matches the temperature gradient of the sleeve. The refrigerator plugs into the neck tube, which is flexible enough so that both heat stations can engage. The colder stage is designed to "float" because less heat is being transported across the flexible strap and copper has a very high thermal conductivity in this temperature region. Helium condenses on the cold-end heat exchanger and drips back to the liquid bath. The refrigerator can be removed and serviced with only a small loss of liquid helium.

Baffles are put in the neck tube to block radiant heat. These must be sufficiently flexible to allow helium to vent rapidly if there is a loss of vacuum or the superconducting magnet quenches. A small reseatable relief valve handles normal vent flow during periods when the refrigerator is off, and a larger burst disk vents abnormally high flow rates. A relief valve on the vacuum housing protects the system when it warms up if there has been a leak which caused a large amount of air to be cryopumped on the helium temperature surface.

REFERENCES

1. J. J. Johnson, *A Compendium of the Properties of Materials at Low Temperature, Phase I,* WADD Technical Report 60-56; Part I, "Fluids"; Part 2, "Solids" (available from FSTI), 1960.
2. E. G. Brentari and R. V. Smith, "Nucleate and Film Pool Boiling Design Correlations for O_2, N_2, H_2 and He," *Int. Adv. Cryog. Eng.,* **10** (1965).
3. T. F. Irvine, Jr. and J. P. Hartnett (eds.), *Advances in Heat Transfer,* Academic, New York, 1968, pp. 325–517.
4. W. M. Roshenow and J. P. Hartnett (eds.), *Handbook of Heat Transfer,* Sec. 12, McGraw Hill, New York, 1973.
5. S. Sato and H. Ogata, "Film Condensation Heat Transfer of Helium in a Vertical Tube," *Cryogenics,* **16,** 221, 224 (April 1976).
6. R. Barron, *Cryogenic Systems,* 2d ed., Oxford Univ. Press, New York, 1985.
7. T. R. Strobridge, "Cryogenic Refrigerators, an Updated Survey," NBS Technical Note 655, U.S. Department of Commerce, 1974.
8. R. P. Reed and A. F. Clark (eds.), *Materials at Low Temperatures,* American Society for Metals, Metals Park, Ohio, 1983.
9. R. B. Stewart and J. J. Johnson, *A Compendium of the Properties of Material at Low Temperature, Phase II,* WADD Technical Report 60-56, Part IV, 1961 (available from FSTI).
10. W. Obert, "Emissivity Measurement of Metallic Surfaces Used in Cryogenic Applications," in *Advances in Cryogenic Engineering,* R. Fast (ed.), Vol. 27, Plenum, New York, 1982, pp. 293–300.
11. J. F. O'Hanlon, *A User's Guide to Vacuum Technology,* Wiley, New York, 1980.
12. K. D. Williamson and F. J. Edeskuty (eds.), *Liquid Cryogens,* Vol. 1, *Theory and Equipment,* CRC, Boca Raton, Fla., 1983.
13. B. W. Birmingham, R. H. Kropschot and D. B. Mann (eds.), *Technology of Liquid Helium,* NBS Monograph III, Washington, D.C., 1968.
14. G. K. White, *Experimental Techniques in Low Temperature Physics,* 3d ed., Clarendon, Oxford, 1979.

Part 12

Thermal Sensors

Chapter 1

Fusible and Color-Changing Temperature Indicators

E. KIMMEL
Tempil Division, Big Three Industries, Inc.

1.1 INTRODUCTION

This chapter describes and outlines the area of usefulness of two categories of temperature indicators widely used in industrial and research applications. One of these is based on a physical property of materials, i.e., melting point, and is characterized by high accuracy, coupled with simplicity of use and ease of interpretation. The other is based on the chemical properties of materials and the color changes produced under varying time-temperature conditions.*

1.2 FUSIBLE TEMPERATURE INDICATORS

When a crystalline solid is heated, a temperature will be reached where that solid changes sharply to a liquid. This melting point has a definite, reproducible value that is virtually unaffected by ambient conditions and environment that may cause errors in other temperature-sensing methods. For example, fusible temperature indicators are applicable in induction heating and in the presence of static electricity or ionized air about electrical equipment, where thermoelectrical means of measuring temperature often function erratically.

The Temperature Stick

The most popular of the fusible indicators is a temperature-sensitive stick of calibrated melting point. These chalklike indicators are made in different precalibrated temperature ratings in the range of 100 to 2500°F (38 to 1371°C); each has a temperature-indicating accuracy within about 1 percent of its rating. Functionally, when the workpiece attains the predetermined melting point of the chalk mark, the mark instantly changes from a solid to a liquid phase (liquefies),

* See Part 12, Chap. 10 for discussion of liquid-crystal temperature indicators.

Fig. 1.1 Use of temperature-sensitive stick for railroad car bearing monitoring.

notifying the observer that the workpiece has reached that temperature. Figure 1.1 is a typical application of a temperature stick.

From an interest point of view, the fusible components of sticks with temperature ratings between 100 and 650°F (38 and 343°C) are organic compounds. Those with ratings of 700°F (371°C) and higher are inorganic salts and oxides.

Under certain circumstances, premarking with a stick is not practical. This is often the case when (1) a prolonged heating period is expected, (2) if the surface is highly polished and does not readily accept a stick mark, or (3) the material being marked is one which gradually absorbs the liquid phase of the stick. In such instances, the operator frequently strokes the workpiece with a stick. The point at which the surface achieves the desired temperature is determined by noting when one ceases to make dry marks and begins to leave a liquid smear.

On cooling, a melted mark will not solidify at the exact same temperature at which it melted, so solidification of a melted crayon mark cannot be relied on for temperature indication.

Lacquer

Another form of fusible indicator, the temperature-indicating lacquer, offers flexibility in many applications. This lacquer-type fluid contains a solid material of calibrated melting point suspended in an inert, volatile, nonflammable vehicle. As with stick-type indicators, there are different temperature ratings, covering the range from 100 to 2500°F (38 to 1371°C).

The temperature-indicating accuracy is within ±1 percent.

The lacquer normally has the proper consistency for brushing. If spraying is desired as the mode of application, a compatible thinner can be used to alter the viscosity without impairing the temperature-indicating accuracy.

Phase-changing lacquers are often used when a very smooth or soft surface is to be tested, or in instances where the surface is not readily accessible for application of a chalk mark during the heating process. Within seconds after application, the lacquer dries to a dull matte finish and responds rapidly when the temperature to be indicated is reached. The response delay of a lacquer mark is only a fraction of a second. This time can be reduced to milliseconds by applying a mark of minimal thickness.

On reaching its temperature rating, the lacquer mark will liquefy. On subsequent cooling, however, the fluid will not revert to its original unmelted appearance but, rather, to a glossy or crystalline coating, which is evidence of its having reached the required temperature. Temperature-indicating lacquers, on cooling, will not resolidify at the melting temperature.

The Pellet

The first commercial form of the fusible indicator was the pellet, which continues to be useful in certain applications. Pellets are most frequently employed when extended heating periods are involved, or where the greater bulk of indicator material is necessary. They are also useful when observations must be made from a distance and when airspace temperatures are to be monitored. A typical application is the monitoring of heat zones in furnaces.

Phase-change, temperature-indicating pellets may be in the form of flat tablets, about 7/16 in (1.11 cm) in diameter and 1/8 in (0.32 cm) thick. For special applications, miniature pellets about 1/8 in (0.32 cm) by 1/8 in (0.32 cm) are also available. The range of 100 to 2500°F (38 to 1371°C) covered by the chalk and lacquer type is also covered by the pellets. Common pellet materials are compatible with inert and oxidizing atmospheres. For temperature measurements in hydrogen, carbon monoxide, or other reducing atmospheres, special materials may be required.

Temperature-Indicating Labels

Another variation of the phase-change indicator is the temperature-sensitive label. These self-adhesive-backed monitors consist of one or more heat-sensitive indicators sealed under transparent, heat-resistant windows. The centers of the indicator circles turn from white to black at the temperature ratings shown on the label. The change to black is irreversible, resulting from fusion and absorption of the temperature-sensitive substance into its backing material.

After registering the temperature history of the workpiece, the exposed monitor label can be removed, and affixed to a service report, to remain a part of a permanent record.

Figure 1.2 illustrates the use of a temperature-indicating label for monitoring circuit-board temperatures.

Fig. 1.2 Temperature-sensitive label in circuit-board application.

Advantages of Fusible Indicators

Fusible temperature indicators have at least several advantages over other methods of determining surface temperature. First, the temperature indications obtained are those of the surface being tested. The generally small thermal mass of fusible temperature indicators minimizes the effect of the sensor on the surface temperature that is measured. Heat-transfer effects of the sensor can be minimized.

Second, response to temperature changes is generally good. Because a mark left by a stick or a lacquer is of extremely small mass, it rapidly attains equilibrium with the surface. The transient response is also aided by the fact that there is little conduction of heat away from the surface.

Accurate surface temperature readings can be obtained with fusible indicators with little effort, training, and expense. There are numerous instances, particularly in determining heat distribution in complex systems, in which the relative simplicity of fusible indicators makes surface temperature investigations feasible.

1.3 PYROMETRIC CONES

First developed in Europe and introduced to the ceramic industry in the United States in the late 1880s, pyrometric cones are used mainly to follow the temperature rise and determine the finishing point of ceramic ware being fired in a kiln. These cones are manufactured as slender pyramids 1⅛ (2.86 cm) or 2½ in (6.35 cm) high (base to peak). The cones are composed of raw materials similar to those employed in the manufacture of ceramics. In normal use, one or more pyrometric cones will be placed in the kiln—usually on specially constructed cone holders. In the kiln, the cones will be subject to the same heat treatment as the ceramic ware being fired. The softening and deformation of a specified pyrometric cone, observed through a hole in the kiln, will signal attainment of the desired heat process.

Although attempts have been made to assign definite temperature calibrations to the several available cones, such calibrations do not have precise meaning because the deformation temperature of a particular cone will depend on the rate and duration of heating. A slow rate of heating will point to a lower temperature than a rapid rate—where the difference can be as much as 100 to 200°F (55 to 110°C). Although the time-temperature dependence may be disadvantages for some applications, this is a distinct advantage in the ceramic industry because the physical characteristics of the cone closely parallel the state of the ceramic ware. Many different cones are commercially available, covering the temperature range of 1100 to 3600°F (593 to 1982°C).

1.4 COLOR-CHANGE INDICATORS

The fusible temperature indicators described—stick, lacquer, pellet, and labels—are the most commonly applied forms of single-use temperature indicators. These are phase-change indicators. Some color-changing series of indicators are also worth noting because of their unique application suitabilities.

Table 1.1 Examples of Application of Temperature-Sensitive Material Indicators

Application	Temperature-sensing requirement	Important practical consideration	Temperature-sensitive material applied
Furnace heating of aluminum bronze extrusions prior to forging	Indicate heating to 1500–1700°F (800–900°C)	Many pieces	Temperature-sensitive stick
Monitoring of charge of thermal chemical battery	Alert user to premature activation discharge of battery	Need passive system	Temperature-sensitive label
Monitor railroad freight car journal boxes	Prevent bearing malfunction	Avoid overheating; accidents	Temperature-sensitive stick
Ac motor control	Avoid temperatures above 160°F (70°C)	Inventory record for follow-up	Temperature-sensitive label
Heat-shrinkable cables	Proper temperature control of adhesive	Quality control on job site	Color-changing paint
Pacemaker cardiac assist	Transit and storage record	Protect against excessive heat	Temperature-sensing label
Oven baking of cores	438°F (225°C), 450°F (232°C), 475°F (246°C) temperature range control	Prevent overheating of cores during bonding	Pellets
Manufacture of emergency power systems	Eight temperature indications from 140°F (60°C) to 210°F (99°C) are plotted	Control overheating of units in prototype development	Temperature labels
Postforming of laminate	Heat control from 313°F (156°C) to 325°F (162°C)	Prevent overheating	Phase-changing fusible liquid

Temperature-Sensitive Paints

Temperature-sensitive paints signal attained temperature by a color change when the time-temperature of a surface has reached or exceeded a predetermined value. Information on the temperature distribution on an entire surface may be obtained at a glance. These paints are available in a range of colors and transition temperatures and, because they are oil-resistant, are resistant to oil spray or wash during operation. Temperature indication by color-changing paints is much less accurate than by phase-change indicators; the usual accuracy is within 25 to 50°F (14 to 28°C) but may be as much as 150° (83°C) depending on the duration of heating. Color-changing indicators are used primarily in temperature problems involving relatively long heating periods under equilibrium conditions. They are available in paint form and have calibrations in the range 175 to 1475°F (79 to 802°C).

Steam-Autoclaving Monitors

Steam-autoclaving monitors (steam sterilizer monitors) are available as printing inks for special applications of color-changing indicators. They are adaptations of the color-changing paints previously described. They have been modified for application by existing printing methods and, generally, have pigment systems that are sensitive to humidity as well as to temperatures in the 240- to 275°F (116- to 135°C) range.

Temperature-indicating printing ink provides a dependable steam-autoclave signal useful in the field of sterilization. Under the usual steam-autoclave sterilizing conditions, such ink exhibits an unmistakable transformation in color. It is available for letterpress, offset, flexographic, or silk-screen printing. Viscosity characteristics can be adjusted to meet the needs of all standard sterilization equipment.

1.5 EXAMPLES OF APPLICATIONS

Typical applications of temperature material indicators are presented in Table 1.1. The applications cited generally have required the need for rapid temperature indication without instrument wiring or readout devices.

Chapter **2**

Thermocouple Thermometry

THOMAS KOPACZ

Thermo Electric Company, Inc.

2.1 THERMOCOUPLE USAGE

Thermocouples are the most frequently used type of temperature sensor in industrial applications for several reasons:

1. They provide a degree of accuracy acceptable to most industrial processes.
2. Higher degrees of accuracy may be obtained, when required, by selectively matching the positive and negative thermoelements.
3. Temperature measurements from −453 to 3218°F (−270 to 1820°C) can be measured by selecting the proper type of thermocouple.
4. Properly constructed thermocouples are the most rugged industrial temperature sensors.
5. Because thermocouples can be manufactured in a wide variety of configurations, they can be designed easily to match unique process requirements.
6. Thermocouples can be the most cost-effective temperature measuring sensors for many industrial applications.

2.2 THEORY OF THERMOCOUPLE MEASUREMENT

In its most basic configuration, a thermocouple consists of two wires joined at each end (Fig. 2.1*a*). If the wires are of the same composition, no electric current will flow through the circuit no matter how great the temperature gradient between the hot (T_2) and cold (T_1) junctions. Current will also not flow, even if the wires are of different compositions, if the junction temperatures are identical.

Current will flow continuously through the circuit, however, if two conditions are satisfied: (1) the wires are of different compositions and (2) there is a temperature differential between the junctions (Fig. 2.1*b*). The electromotive force (emf) which induces the current flow is called the *Seebeck emf* in honor of the discoverer of this phenomenon.

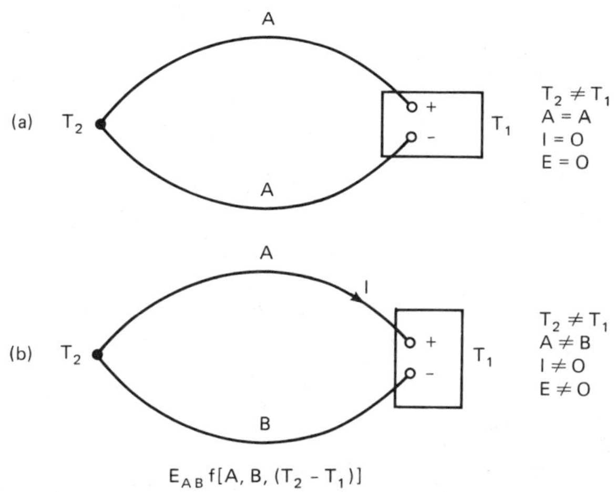

Fig. 2.1 Basic circuitry of thermocouple systems.

The thermal emf of *A* versus *B* is a vector quantity. The material characteristics of the thermoelements *A* and *B* and the temperature difference between the measurement and reference junctions $T_2 - T_1$ determine its magnitude and direction. If (1) the emf-temperature relationship of *A* versus *B* is established, (2) that relationship is a straight line, and (3) the ice point [32°F (0°C)] is set as the reference junction, the temperature of the measurement junction can be determined by measuring the emf of *A* versus *B* as indicated in Fig. 2.2.

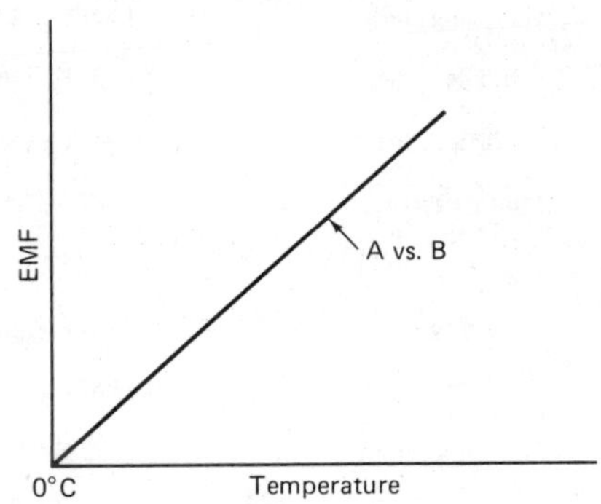

Fig. 2.2 Emf-temperature relationship of *A* versus *B* in a basic thermocouple system.

2.3 LAWS OF THERMOELECTRICS

Homogeneous Circuits

The emf generated in the thermocouple circuit with thermoelements *A* and *B* (Fig. 2.3*a*) is dependent only on the temperature difference between T_2 and T_1. This is true irrespective of the presence of a temperature gradient in the circuit, provided the composition of both thermoelements is homogeneous.

Intermediate Metals

A third metal can be added to a thermocouple circuit without inducing a temperature measurement error if all the junctions of the thermoelements and third metal are the same temperature (Fig. 2.3*b*). However, if there is a temperature difference of 94°F (35°C) between the terminals as shown in Fig. 2.3*c*, a measurement error of approximately 94°F (35°C) will be introduced.

Intermediate Temperatures

The emf generated by a thermocouple with the reference junction at T_1 and the measuring junction at temperature T_2 is equal to the algebraic sum of two thermocouples of the same material: (1) one with the reference junction at T_1 and the measuring junction at temperature T_H and (2) the other with the reference junction at T_H and measuring junction at T_2:

$$E_{AB}\Big|_{T_1}^{T_2} = E_{AB}\Big|_{T_1}^{T_H} + E_{AB}\Big|_{T_H}^{T_2}$$

where $E_{AB}\Big|_{T_1}^{T_2}$ = the emf for a thermoelectric circuit of materials *A* and *B* with reference junction temperature T_1 and measuring temperature T_2

2.4 THERMOCOUPLE STANDARDS

The Instrument Society of America (ISA) and ASTM[1] have standardized eight types of thermocouple and assigned letter designations. Five of these types, J, K, N, T, and E are base-metal thermocouples. The remaining types, R, S and B are noble-metal thermocouples. Table 2.1 shows the typical alloys of the positive and negative thermoelements of these thermocouples, the base compositions, and the range of applications. Types J and K are the most frequently used thermocouples in general industry.

Type J

Type J thermocouples may be employed continuously in either oxidizing or reducing atmospheres up to a temperature of 1400°F (750°C). The positive and negative thermoelements are, respectively, iron and a copper-nickel alloy commonly known as *Constantan*. Thermocouples of this type are used extensively in heat-treating furnaces where the couples are exposed directly to the furnace atmosphere. They also find wide usage in plastics processing and general industry for measurements in the 32 to 1400°F (0 to 760°C) range.

Type K

Type K is one of two base-metal thermocouples capable of measuring temperatures to 2300°F (1250°C). The positive

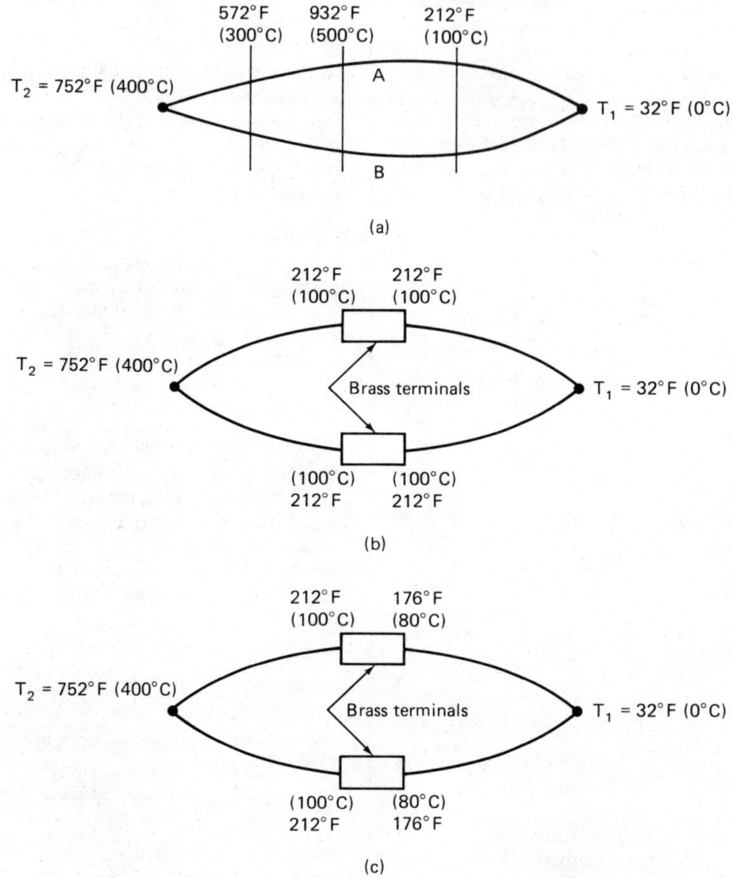

Fig. 2.3 (*a*) Law of homogeneous circuits and (*b*), (*c*) law of intermediate metals.

Table 2.1. IEC-ISA-Designated Thermocouple Alloys

Type	Thermo elements	Typical alloy*	Base composition	Applications: Atmosphere	Applications: Maximum Temperature
J	JP	Iron	Fe	Oxidizing, reducing,	1400°F (760°C)
	JN	Constantan (J)	45%Ni-55%Cu	inert	
K	KP	Chromel	90%Ni-10%Cr	Oxidizing, inert	2300°F (1260°C)
	KN	Alumel	94%Ni-Al,Mn,Fe		
N	NP	Nicrosil	Ni-14%Cr-1.4%Si	Oxidizing, inert	2300°F (1260°C)
	NN	Nisil	Ni-4.4%Si-0.1%Mg		
T	TP	Copper	OFHC Cu	Mild oxidizing,	634°F (370°C)
	TN	Constantan (T)	45%Ni-55%Cu	reducing, inert	
E	EP	Chromel	90%Ni-10%Cr	Oxidizing, reducing,	1598°F (870°C)
		Constantan (T)	45%Ni-55%Cu	inert	
R	RP	Pt-Rh	87%Pt-13%Rh	Oxidizing, inert	2696°F (1480°C)
	RN	Pt	Pt		
S	SP	Pt-Rh	90%Pt-10%Rh	Oxidizing, inert	2696°F (1480°C)
	SN		Pt		
B	BP	Pt-Rh	70%Pt-30%Rh	Oxidizing, inert	3092°F (1700°C)
	BN	Pt-Rh	94%Pt-6%Rh		

* Chromel, Alumel, and Constantan are trade names of Hoskins for KP, KN, and JN-TN. Tophel, Nial, and Cupron are trade names of Cartech for KP, KN, and JN-TN. T1, T2, and Advance are trade names of Driver-Harris for KP, KN, and JN-TN.

thermoelement is Ni-10%Cr alloy; the negative, a Ni-5% (Mn AlSi) alloy. At elevated temperatures, type K couples should be used only in an oxidizing atmosphere. Type N was developed to essentially eliminate the high-temperature drift seen in type K thermocouples. Because of their high-temperature capabilities, type K thermocouples are used extensively to measure temperatures in heat-treating furnaces, jet engines, boiler tubes, and in many applications in chemical and power plants.

Nicrosil-Nisil Type N

This system was designed to exhibit improved oxidation resistance and stability of emf calibration over type K in laboratory and some industrial applications.[2,3] The positive thermoelement is a Ni-14.2%Cr-1.4%Si alloy; the negative, a Ni-4.4%Si-0.1%Mg alloy. In comparison with its type K counterpart, this system does not change on aging in the 930°F (500°C) range and is more oxidation resistant and stable. However, its advantages over type K when used in the form of metal sheathed thermocouples are reported to be minimal.[4] As the emf of Nicrosil-Nisil is about 15 percent lower than that of type K, type K extension wires and instrumentation are not compatible. At present, these systems are used in elevated temperature applications which require only minimal thermocouple calibration changes during operations.

Type T

Type T thermocouples, with copper-positive and Constantan-negative thermoelements, may be installed in both oxidizing and reducing atmospheres up to a maximum temperature of 700°F (350°C). The upper temperature limitation is caused by the poor oxidation resistance in air of copper, the positive thermoelement. Because of the high purity and moisture-resistance advantage of copper (compared with iron in the type J thermocouples), type T units are most commonly used for cryogenic temperature measurements.

Type E

Type E thermocouples have a maximum temperature of 1600°F (900°C) in an oxidizing atmosphere. Their positive thermoelements are made of the same material as that of the type K (Ni-10%Cr) and the negative, Constantan. The emf output of type E thermocouples is larger than that of any other type of metallic thermocouple. This advantage makes it easier to interface with analog-to-digital (A/D) converters; consequently, type E is selected most often when computers or other digital equipment are in service. For this reason, type E sensors are used extensively in power-generation stations.

Type S

Type S thermocouples are designed for continuous use in oxidizing atmospheres up to 2700°F (1450°C). The positive thermoelement is a Pt-10%Rh alloy; the negative, pure platinum. Units of this type should not be installed in reducing atmospheres or in atmospheres containing vapors of lead, zinc, sulfur, phosphorous, or arsenic. Because the high purity of the platinum-negative thermoelement must be maintained, special care must be taken to protect it from contamination by the insulators and the environment. Type S thermocouples find extensive applications in industrial laboratories as a standard for the calibration of base metal thermocouples and other temperature-sensing instruments.

Type R

Type R thermocouples are very similar in design to the type S sensors described previously. The positive thermoelement is a Pt-13%Rh alloy; the negative, platinum. Their thermoelectric power is slightly higher than the type S thermocouple. However, their characteristics, range of applications, and advantages and limitation are the same. Generally speaking, type R thermocouples are used more extensively in industrial environments while type S thermocouples are more often found in laboratories.

Type B

Type B thermocouples may be used in oxidizing or inert atmospheres up to 3100°F (1700°C) on a continuous basis and up to 3200°F (1760°C) intermittently. The positive element is a PT-30%Rh alloy; the negative, a Pt-6%Rh alloy. These alloys are less likely than pure platinum (used for type S and R thermoelements) to be affected by trace impurities from insulators or the environment. Also, under comparable conditions, the type B thermocouple exhibits less grain growth and is more stable than either of the S or R types. Like the S and R, the type B thermocouple should not be used in reducing atmospheres or atmospheres containing metallic and nonmetallic vapors. They are used primarily in glass plants when the maximum operating temperature is near 3100°F (1700°C).

2.5 EMF CHARACTERISTICS OF STANDARDIZED THERMOELEMENTS AND THERMOCOUPLES

The thermoelectromotive force of each thermoelement, as compared with platinum and over a temperature range from 32 to 3100°F (0 to 1700°C), is shown schematically in Fig. 2.4. Three base-metal thermoelements, JP, TP, and KP, have positive thermal emf in comparison with platinum. Their emf is larger than that of the positive, noble-metal thermoelements RP, SP, and BP. The emf of iron is shown as a band because commercial iron is very impure; therefore, the emf difference from one heat to another could be very large. To compensate for this variation, JN thermoelements are melted with slight differences in composition to match the iron thermoelements. Therefore, the emf of JN is also shown as a band even though nominal emf values for JP and JN are available in National Bureau of Standards (NBS) tables.[4] The emf of Pt-TN and Pt-EN, the negative thermoelements for type T and type E thermocouples, are shown as a straight line; different alloys are not required for matching purposes.

The emf values of various ISA-designated thermocouples are shown in Fig. 2.5. The thermoelectric power of base-metal thermocouples is generally much larger than that of the noble-metal thermocouples. However, the maximum operating temperatures of the noble-metal thermocouples are much higher than those of the base-metal types. Among the base-metal thermocouples, types K and N are the only ones

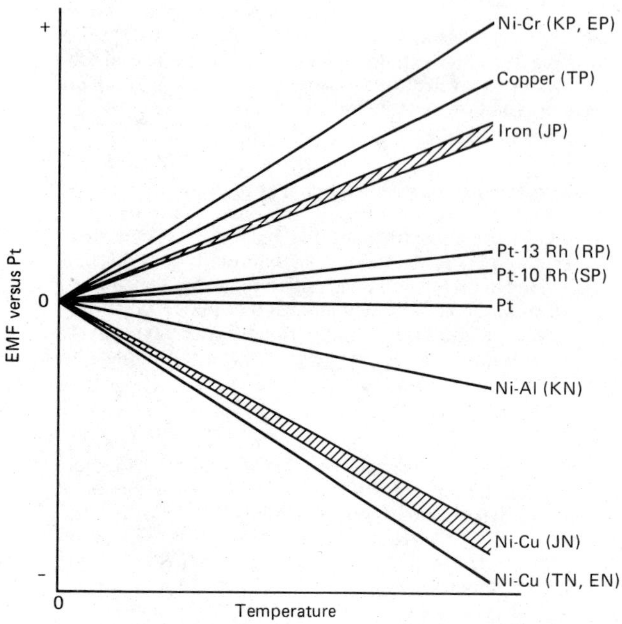

Fig. 2.4 Thermal emf of thermoelements versus platinum.

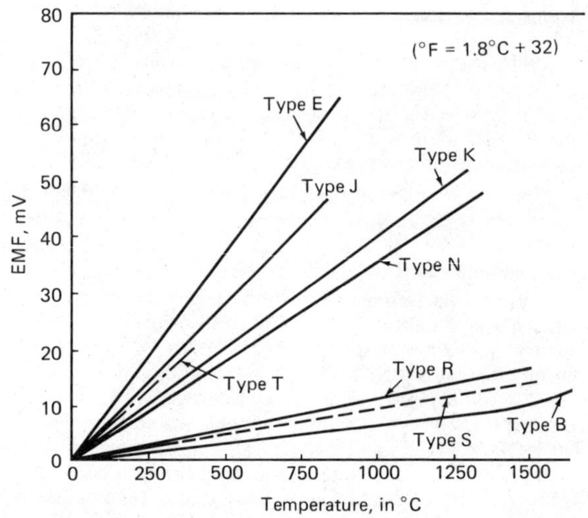

Fig. 2.5 Thermal emf of various ISA-designated thermocouples.

that can be used in continuous operations in the 1600 to 2300°F (900 to 1260°C) range.

2.6 EMF OF NONSTANDARDIZED THERMOCOUPLES

19 Alloy-20 Alloy

This thermocouple was developed for temperature sensing and control at elevated temperatures in hydrogen and other reducing atmospheres. The positive thermoelement is a Ni-19%Mo alloy; the negative, a Ni-1%Co alloy. The emf of this sensor is similar to the type K in the range 32 to 418°F (0 to 200°C). Therefore, type K extension wires and connectors can be used with this thermocouple. However, as its emf is larger than that of the type K at higher temperatures, type K recorders and temperature controllers are not compatible. The 19 Alloy-20 Alloy thermocouple has poorer oxidation resistance than type K and, therefore, should not be used above 1200°F (650°C) in an oxidizing atmosphere. However, it is not susceptible to aging in the 930°F (500°C) range and will remain in calibration in neutral and reducing atmospheres up to 2500°F (1400°C). As a result, type 19 Alloy-20 Alloy thermocouples are widely used in heat treating applications which must be carried out in reducing atmospheres at elevated temperatures.

Platinel I

Platinel I and its sister combination, Platinel II, were designed to approximate the emf characteristics of type K thermocouples. At elevated temperatures, their thermal emf is approximately the same. However, there is a significant deviation in the low-temperature range. Platinel thermocouples are used to measure turbine inlet temperatures in aircraft jet engines because they are stable at high temperatures in air. Like other platinum thermocouples, Platinel I and Platinel II sensors should not be installed in reducing atmospheres or where they might be subject to contamination by sulfur, phosphorous, and silicon.

Table 2.2 Nonstandardized Thermocouples

Thermocouples	Base composition	Application: Atmosphere	Application: Maximum temperature
19 and 20 alloys	82%Ni-18%Mo 99%Ni-1%Co	Reducing	2552°F (1400°C)
Platinel I	83%Pd/14%Pt-3%Au 65%Au-35%Pd	Oxidizing	2372°F (1300°C)
Tungsten-rhenium	W W-26%Re	Reducing	5000°F (2760°C)
Tungsten-rhenium	W-5%Re W-26%Re	Reducing	5000°F (2760°C)

Tungsten-Rhenium

This type of thermocouple was designed for use primarily at elevated temperatures [up to 4200°F (2315°C)] in reducing atmospheres. Two of the more common types are W-3%Re versus W-25%Re and W-5%Re versus W-26%Re. Although these systems are rated for high temperatures, they are usually used below 4200°F (2315°C) because of the temperature limitations of the associated ceramic insulators. The thermocouples are supplied as matched pairs to within ± 1 percent of the actual temperature emf tables. Alternate extension wires are available for each of these combinations. At high temperatures, the factors governing the use of W-Re thermocouples are the type of insulation sheath and protection tubes, the diameter of the thermoelement, and the composition of the atmosphere. It is important to make sure that the insulation resistance (IR) between the sheath and the thermoelements is sufficiently high to eliminate spurious emf, which would cause erroneous temperature readings.

2.7 INITIAL CALIBRATION TOLERANCES

Specifications for standardized thermocouples are shown in Table 2.3. Previously, this thermocouple tolerance table was entitled "Limits of Error," which was misleading because it had no time frame. A thermocouple designed to have a limits of error of ±5 percent as shown in the NBS table may have been misconstrued to be expected to maintain that tolerance over its entire service life. The tolerances stated in the NBS table cannot be expected to be maintained. A thermocouple is manufactured to within a designated tolerance from the NBS table. That calibration will change when the thermocouple is put into service. The magnitude and direction of that change are dependent on temperature, time, depth of immersion, atmosphere, and the environmental conditions to which the thermocouple is exposed. Therefore, the term *initial calibration tolerance* more accurately describes the allowable deviation of the thermocouple from the NBS table when it is new.

2.8 CONSTRUCTION OF THERMOCOUPLES

Metal-Sheathed Thermocouples

Metal-sheathed thermocouples are the most frequently used constructions in industry today. In this design, the thermocouple wires are insulated by a high-purity ceramic insulation, generally magnesium oxide (MgO), and enclosed in a metal sheath (Fig. 2.6). This construction has several advantages: (1) the outside diameter of the sheath may vary according to the needs of the application, typically within the range of 0.010 to 0.50 in OD; (2) the sheath metal may be of any of the austenitic stainless steel (or nickel-based) alloys; (3) the thermocouple unit may be manufactured in long lengths and, properly constructed, can be bent to a radius twice its outside diameter; and (4) it is also extremely rugged.

This construction flexibility allows the user to select the exact combination of thermoelement materials to optimize temperature readings and the best sheath metal composition to match the requirements of the process as to temperature, corrosives, lengths, and outside diameters. Further, because the sheath material is flexible and rugged, installation problems are greatly reduced.

Therefore, it is easy to see why metal-sheathed thermocouples are used for many industrial applications such as

Table 2.3 Initial Calibration Tolerances for Thermocouples (Reference Junction 0°C)

Thermocouple type	Temperature range, °C	Tolerances: Standard (whichever is greater)	Tolerances: Special (whichever is greater)
B	870–1700	±0.5%	
E	0–900	±1.7°C or ±0.5%	±1°C or ±0.4%
J	0–750	±2.2°C or ±0.75%	±1.1°C or ±0.4%
K or N	0–1250	±2.2°C or ±0.75%	±1.1°C or ±0.4%
R or S	0–1450	±1.5°C or ±0.25%	±0.6°C or ±0.1%
T	0–350	±1°C or ±0.75%	±0.5°C or ±0.4%
E*	−200 to 0†	±1.7°C or ± 1%	‡
K*	−200 to 0†	±2.2°C or ±2%	‡
T*	−200 to 0†	±1°C or ±1.5%	‡

* Thermocouples and thermocouple material are normally supplied to meet the tolerances specified in the table for the normal specified range. The same materials, however, may not fall within the cryogenic tolerances in the second section of the table. If materials are required to meet the cryogenic tolerances the supplier should be notified. Tolerances indicated in this table are not necessarily an indication of the accuracy of temperature measurements in use after initial heating of the materials.

† Cryogenic ranges.

‡ Little information is available to justify establishing special tolerances for cryogenic temperatures. Limited experience suggests the following tolerances for types E and T thermocouples:

Type E −200 to 0°C ±1°C or ±0.5% (whichever is greater)
Type T −200 to 0°C ± 0.5°C or ±0.8% (whichever is greater)

These tolerances are given only as guide for discussion between purchaser and supplier. Because of the characteristics of the materials, cryogenic tolerances for type J thermocouples and special cryogenic tolerances for type K thermocouples are not listed.

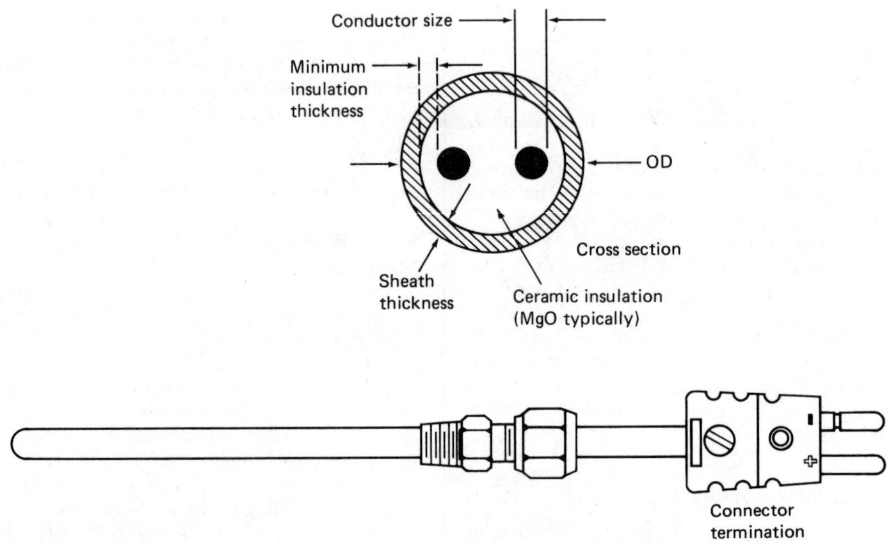

Fig. 2.6 Metal-sheathed thermocouple.

aluminum processing, ovens, kilns and environmental chambers, plastics processing, and stress-relieving of pipe and large vessel welds. The drawing at the bottom of Fig. 2.6 illustrates a typical metal-sheathed thermocouple. They can be supplied in any size, material, and configuration to meet the requirements of specific applications.

Grounded and Ungrounded Metal-Sheathed Thermocouples

Metal-sheathed thermocouples are available in grounded or ungrounded configurations. Figure 2.7*a* illustrates the grounded variety. The thermoelectric elements are welded to the metal sheath becoming an integral part of the weld cap closure. This design features rapid response time and good tip sensitivity. However, the thermoelements are not electrically shielded, and, especially when the wires are small diameter, open circuits at the junction can be caused by repeated temperature cycles. This problem arises because the three alloys (two thermoelements and the metal sheath) usually have different coefficients of thermal expansion. Therefore, they expand and contract at different rates as the temperature cycles stress the welded-wire connection.

Figure 2.7*b* illustrates a typical ungrounded (insulated) thermocouple. The welded thermoelement junction is electrically (and to some extent, thermally) insulated from the weld closure by a mineral insulation, typically magnesium oxide (MgO). Response time is, therefore, longer than that of the grounded type. Typical applications include temperature measurements in conductive solutions and temperature averaging. They are also used wherever circuit isolation is required.

Response Time

The amount of time a thermocouple takes to react to a change in temperature is called *response time*. In many applications, where response time is critical, the type and characteristics of the thermocouple must be carefully considered before a final selection is made. Table 2.4 indicates typical response time for several standard metal-sheathed thermocouples. As the table indicates, response time is affected by

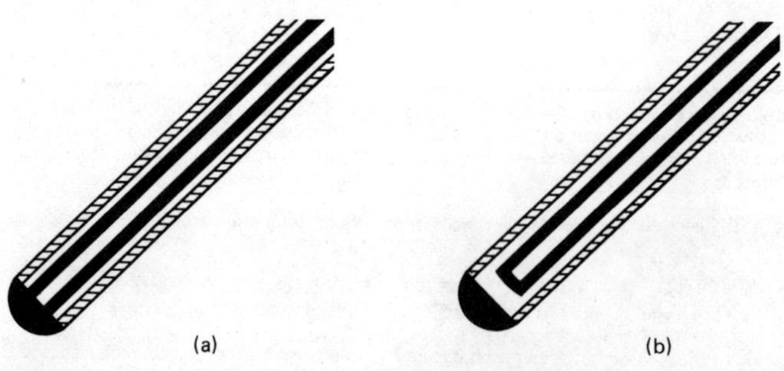

Fig. 2.7 (*a*) Grounded and (*b*) ungrounded measuring junctions.

Table 2.4. Typical Response Time for Standard Metal-Sheathed, Mineral-Insulated Thermocouples

Thermocouple OD, in	Type of measuring junction	Typical response time, s
1/25	Grounded	0.1
1/25	Ungrounded	0.3
1/16	Grounded	0.2
1/16	Ungrounded	0.5
1/8	Grounded	0.7
1/8	Ungrounded	1.3
3/16	Grounded	1.1
3/16	Ungrounded	2.2
1/4	Grounded	2.0
1/4	Ungrounded	4.5
1/4	Exposed-loop	0.1

CONVERSION FACTOR: 1 in = 2.54 cm.

the type of measuring junction and the outside diameter of the metal sheath thermocouple. The temperature system designer must determine the response time requirement of the application and the environment to which the metal sheath will be exposed in order to select the most appropriate size and type of thermocouple unit.

Ceramic-Beaded Thermocouples

Ceramic-beaded thermocouples may be the best solution in certain applications where metal-sheathed thermocouples are not practical or economically feasible. In this design, high-purity alumina insulator beads are strung along the length of the conductor wires and the hot junction is often twisted and welded (Fig. 2.8). Ceramic-beaded thermocouples are able to withstand much higher temperatures than metal-sheathed systems because of the superior temperature resistance of the alumina. Because ceramic beads are relatively fragile, they are often installed inside high-temperature ceramic protection tubes which serve to protect them from both mechanical damage and the atmosphere of the application process.

Fibrous Insulated Thermocouples

The use of fibrous insulated thermocouples is increasing in applications where metal-sheathed and ceramic-beaded thermocouples are consumed by the application process. Their major advantage is that they can be manufactured quickly and at much lower cost than other types. Typical applications are heat-treating, aluminum processing, and bonding of composite materials in autoclaves.

Asbestos and fiberglass insulations were once predominant in elevated temperature, fibrous insulated thermocouples. However, because of the problems associated with asbestos, replacement materials have been developed that do not have its hazardous effects.

Braided ceramic fiber, Cefir* (Fig. 2.9*a*) is one of those materials. Cefir can be applied to high temperature thermocouple conductors such as type K. It offers continuous insulating properties to 2600°F (1427°C) and in many cases, can replace ceramic-beaded insulated thermocouples at lower installed cost.

Synthetic polyamide fiber, B Fibre (Fig. 2.9*b*) is another material which is often used as a direct replacement for asbestos insulation. It has excellent abrasion resistance and a continuous temperature rating of 500°F (260°C). Type Q insulation (Fig. 2.9*c*) is still another material that can replace asbestos in elevated-temperature applications. It has an upper temperature rating of 1200°F (650°C).

These and many other materials can be used to insulate thermocouple wire at elevated temperatures, at lower installed cost, where the application requires high-volume, expendable sensors.

Thermowells

Frequently it is necessary to protect the thermocouple from the environment of the application process. Metal-sheathed and ceramic-beaded thermocouples would quickly fail when subjected to the flows, pressures, or aggressive chemicals of many process industry applications; hence the use of the thermowell.

* Registered trademark of Thermo Electric Co., Inc.

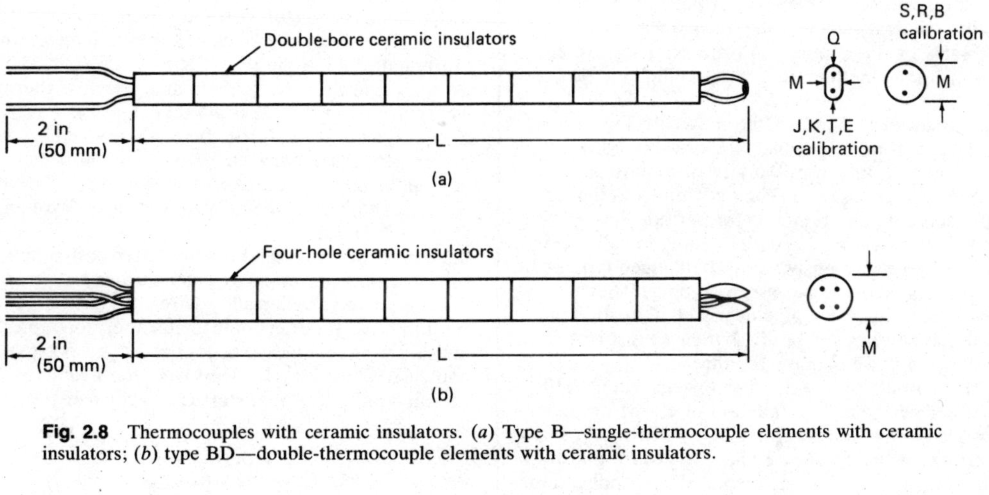

Fig. 2.8 Thermocouples with ceramic insulators. (*a*) Type B—single-thermocouple elements with ceramic insulators; (*b*) type BD—double-thermocouple elements with ceramic insulators.

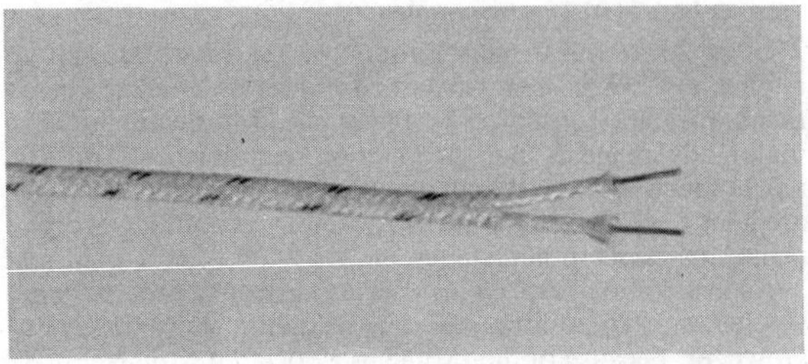

(a)

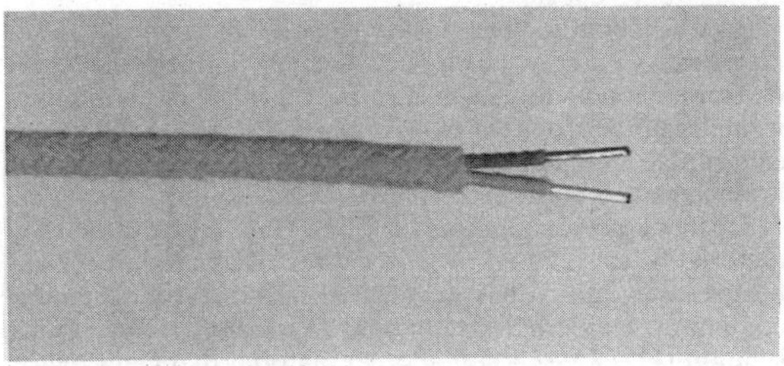

(b)

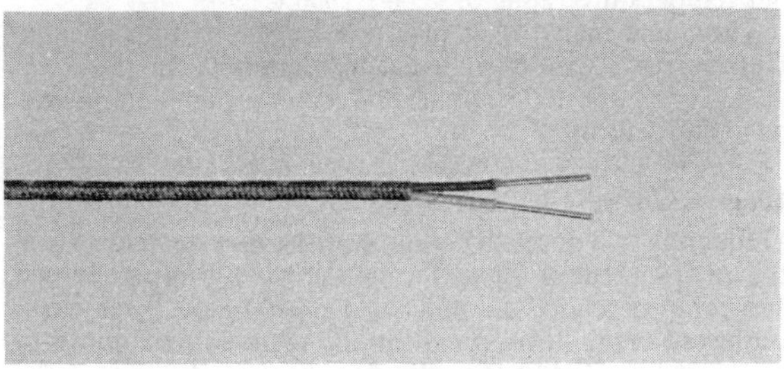

(c)

Fig. 2.9 Fibrous insulated wire. (*a*) Cefir insulated thermocouple wire; (*b*) B Fibre insulated thermocouple wire; (*c*) braided fibrous insulated thermocouple wire.

The typical thermowell is machined from solid bar of various austenitic stainless steel or nickel-based alloys to match the requirements of the application process. They are classified according to the method used to attach them to process equipment.

Flanged Thermowells Flanged thermowells (Fig. 2.10*a*) are attached to a mating flange on a pipe or vessel with bolts. Flanges are available in a wide variety of materials, sizes, and pressure ranges as required. Their advantage is that they can be easily unbolted and removed for periodic inspection.

Threaded Thermowells Threaded thermowells (Fig. 2.10*b*) are usually less expensive than the flanged type and, typically, are applied to low-pressure piping. They screw onto a threadolet welded to a pipe or vessel. Threaded thermowells can also be removed easily, but care must be taken not to strip or otherwise damage the threads.

Weld-in Thermowells Weld-in thermowells (Fig. 2.10*c*) are designed for permanent installation in a pipe or vessel. They are welded in a hole or welding boss. Obviously, since they are permanent, care must be taken to ensure that they will withstand all the adverse elements of the application. Many processes require welded thermowells to monitor temperatures. These must be protected from resonance due to fluid flow or deformation due to viscous media.

Protection Tubes

Protection tubes are typically manufactured from metal pipe or high-temperature ceramic material. They are closed at one end, and a threaded fitting or flange is provided to the other for mounting purposes (see Fig. 2.11*a*). As they do not have the mechanical strength of barstock thermowells, they are often used in low-pressure and flow applications such as stack gas-temperature monitoring.

Ceramic protection tubes (Fig. 2.11*b*) are designed for high-temperature applications where metal tubes could not be used. Materials include alumina, mullite, and cermet (metal-oxide ceramic). Typical applications are glass plant furnace roofs and other systems where elevated-temperature, gaseous atmospheres exist.

2.9 THERMOCOUPLE EXTENSION WIRES

The thermocouple conductors, in wire or cable form, which connect the thermocouple to the temperature sensing or control instrument are known as *extension wires*. The basic circuit diagram of the thermocouple and extension wire system is shown in Fig. 2.12.[5] Thermocouple PN is joined to a temperature controller by extension wire PXNX at head temperature T_H. The hot junction temperature is T_2, and the reference temperature is T_1. The emf E of this thermocouple and extension assembly between T_1 and T_2 is equal to the sum of the thermocouple PN between T_H and T_2 and the emf of the extension wire between T_1 and T_H as expressed in Eq. (*a*) in Fig. 2.12. This equation can be derived from Kirchhoff's law. For this emf to be the same as that when no extension wire is used (Fig. 2.4), the emf of the extension-wire couple must be identical to that of the couple between the range T_1 and T_H. [See Eq. (*c*) in Fig. 2.12.]

As long as Eq. (*c*) in Fig. 2.12 holds, it is not necessary that $E_P = E_{PX}$ and $E_N = E_{NX}$ between the range of T_1 and T_H. If they are the same, the connecting wire or cable is an *extension wire*. However, if $E_P \neq E_{PX}$ and $E_N \neq E_{NX}$ at this range, and Eq. (*c*) in Fig. 2.12 holds true, the connecting wire or cable is known as *alternate* extension wire.

Available extension wires are listed in Table 2.5. Type J, K, N, T, and E thermoelements which meet the emf requirement only in the range 32 to 418°F (0 to 200°C) can be used as extension wire for their respective thermocouples. There are two types of alternate extension wire for the type K thermocouple. One of them, the WX, uses iron as the positive extension wire and Cupronel,* a special nickel-copper alloy, as the negative extension wire. The other one, VX, uses copper as the positive extension wire and a special Cupron,† another nickel-copper alloy, as the negative extension wire. Although both are letter-designated, these two combinations are no longer standardized.

Copper and copper-alloy wires are used as alternate extension wire for the noble metal thermocouples. Types R and S thermocouples use copper as the positive extension and PCLW* or No. 11 Alloy† for the negative. For type B thermocouples, copper versus copper can be employed as

* Trade name of CarTech (Carpenter Technology).

† Trade name of Driver-Harris.

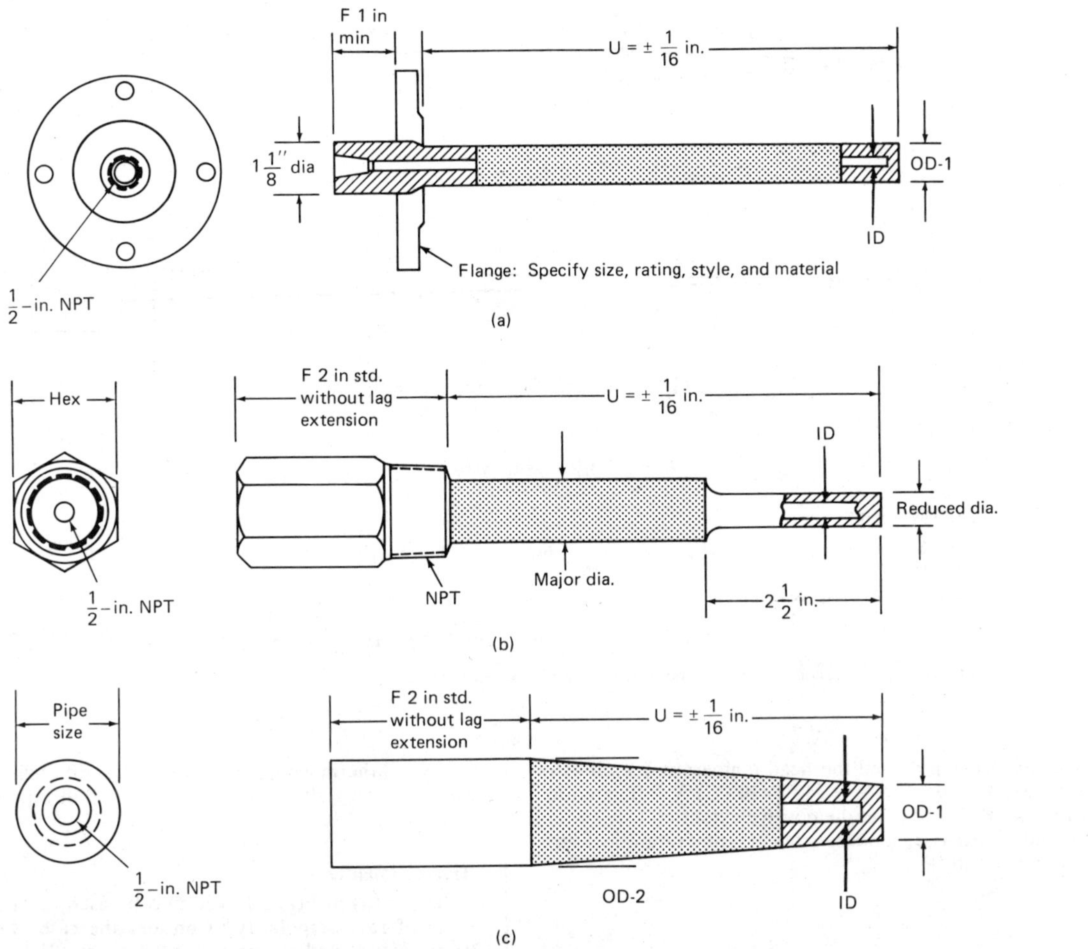

Fig. 2.10 Thermowells made from drilled bar stock. (*a*) Flanged; (*b*) screwed; (*c*) weld-in.

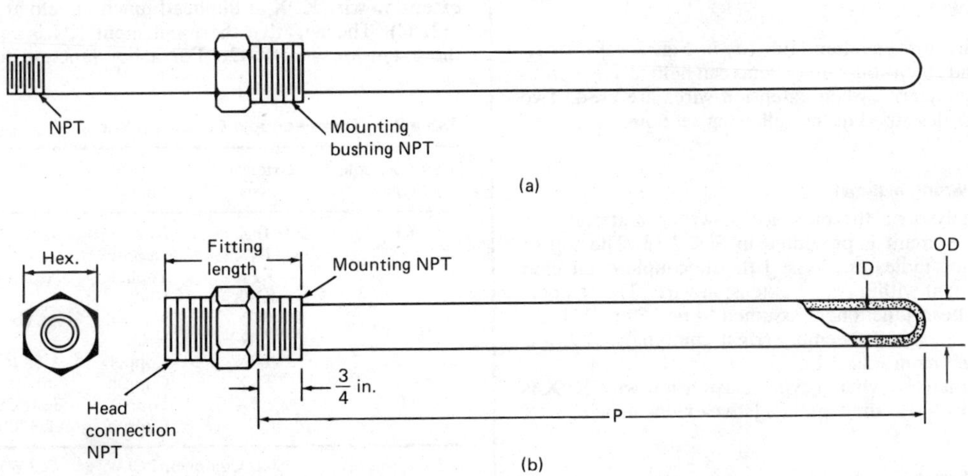

Fig. 2.11 Metal and ceramic protection tubes. (*a*) Metal protection tube with threaded bushing; (*b*) ceramic protection tube with fitting.

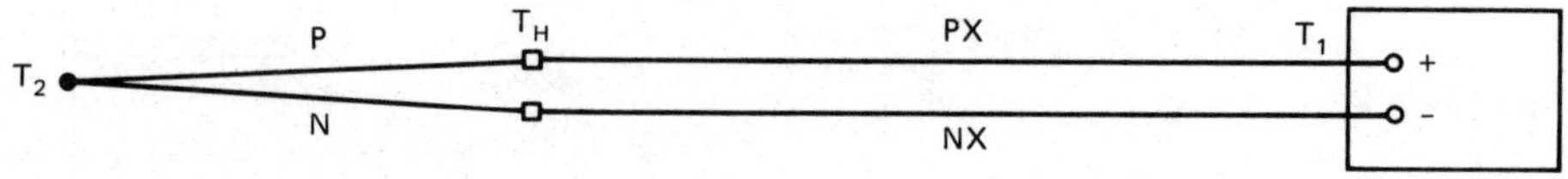

$$E_{\mathrm{ASSEMBLY}}\Big|_{T_1}^{T_2} = (E_P \text{ vs. } E_N)\Big|_{T_H}^{T_2} + (E_{PX} \text{ vs. } E_{NX})\Big|_{T_1}^{T_H} \qquad (a)$$

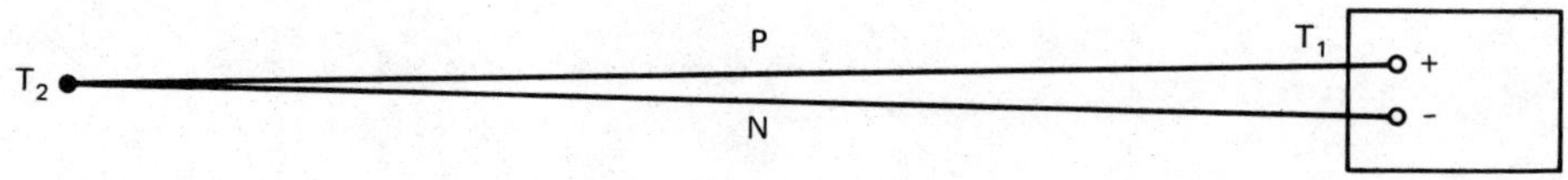

$$E_{PN}\Big|_{T_1}^{T_2} = (E_P \text{ vs. } E_N)\Big|_{T_H}^{T_2} + (E_P \text{ vs. } E_N)\Big|_{T_1}^{T_H} \qquad (b)$$

$$(E_P \text{ vs. } E_N)\Big|_{T_1}^{T_H} = (E_{PX} \text{ vs. } E_{NX})\Big|_{T_1}^{T_H} \qquad (c)$$

EXTENSION WIRE

$$E_P\Big|_{T_1}^{T_H} = E_{PX}\Big|_{T_1}^{T_H} \quad \text{and} \quad E_N\Big|_{T_1}^{T_H} = E_{NX}\Big|_{T_1}^{T_H} \qquad (d)$$

ALTERNATE EXTENSION WIRE

$$E_P\Big|_{T_1}^{T_H} \neq E_{PX}\Big|_{T_1}^{T_H} \quad \text{and} \quad E_N\Big|_{T_1}^{T_H} \neq E_{NX}\Big|_{T_1}^{T_H} \qquad (e)$$

Fig. 2.12 Basic circuit diagram of thermocouple and extension wire.

extension wires provided the head temperature T_H is at or below 212°F (100°C). Above 212°F (100°C) type B extension wire, with PCLW30-6* as the positive extension wire, would be used. Otherwise, serious errors in temperature measurement will occur.

2.10 SOME CIRCUITRY PROBLEMS OF THERMOCOUPLE AND EXTENSION-WIRE SYSTEMS

The basic circuitry equation [Eq. (*a*) in Fig. 2.12] of thermocouple and extension-wire systems can help to solve problems which may arise when extension wires are used. Two examples are described in the following sections.

Presence of Wrong Material

An error analysis on the presence of wrong material in a thermocouple circuit is presented in Fig. 2.13. The top of this illustration indicates a type J thermocouple connected to an instrument with a type J extension wire. The temperature of the head junctions is assumed to be 250°F (121°C). The emf of the type J assembly (right combination) E_{RC} is shown in Eq. (*a*) in Fig. 2.13.

Consider the case when a type K extension wire KPX is inadvertently used as the type J negative extension wire JNX in a type J thermocouple assembly. The emf of this assembly with the wrong alloy present is expressed in Eq. (*b*) in Fig. 2.13.

* Good to N 500°F; at higher temperatures use same wire as thermocouple.

Reverse Polarity

Equation (*a*) in Fig. 2.12 can also be used to calculate the effect of reverse polarity.[6] Consider the case of a type K thermocouple and extension wire assembly. For correct polarity, the positive thermoelement KP is connected to the extension wire KPX at the head junction held at, say, 250°F (121°C). The negative thermoelement KN is connected to the extension wire KNX. For a hot junction temperature

Table 2.5. Thermocouple Extension Wires

Thermocouple type	Extension wire	Typical alloy*	Temperature range
J	JPX	Iron	32 to 392°F (0 to
	JNX	Cupron	200°C)
K	KPX	Tophel	As above
	KNX	Nial	
N	NP	Nicrosil	As above
	NN	Nisil	
T	TPX	Copper	−58 to 212°F (−50
	TNX	Cupron	to 100°C)
E	EPX	Tophel	32 to 392°F (0 to
	ENX	Cupron	200°C)

* Cupron, Tophel, Nial, Cupronel, PCLW, and PCLW30-6 are trade names of Cartech (Carpenter Technology). No. 11 Alloy is a trade name of Driver-Harris.

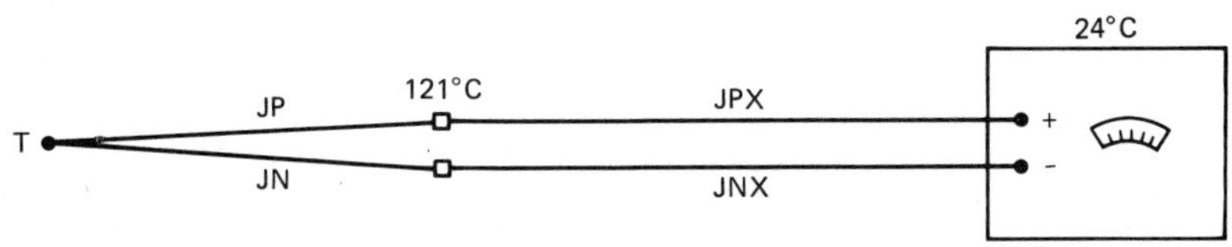

$$E_{RC} = E_{JPvs.JN}\Big|_{121°C}^{T} + E_{JPXvs.JNX}\Big|_{24°C}^{121°C} + E_{JPXvs.JNX}\Big|_{0°C}^{24°C} \qquad (a)$$

WRONG ALLOY (KPX IN PLACE OF JNX)

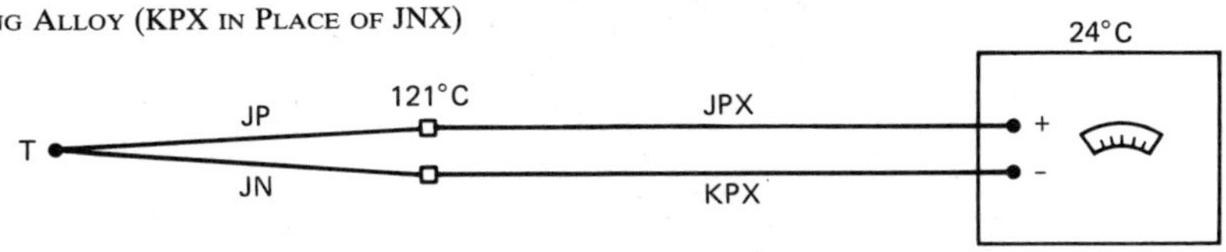

$$E_{WA} = E_{JPvs.JN}\Big|_{121°C}^{T} + E_{JPXvs.KPX}\Big|_{24°C}^{121°C} + E_{JPXvs.JNX}\Big|_{0°C}^{24°C} \qquad (b)$$

TEMPERATURE ERROR
Eq. (*b*)–(*a*)

$$\Delta E = -E_{KPX}\Big|_{24°C}^{121°C} + E_{JNX}\Big|_{24°C}^{121°C} = -2.83 \text{ mV} - 3.49 \text{ mV} = -6.32 \text{ mV} \qquad (c)$$

$$\Delta T = -118°C$$

Fig. 2.13 Error analysis on presence of wrong material in a thermocouple circuit (head temperature assumed to be 121°C).

[1822°F (980°C)], the emf generated at this thermocouple circuit is 40.62 mV.

If the polarity is reversed, i.e., KP is isolated to KNX and KN is joined to KPX, the emf generated by this couple is only 32.68 mV, corresponding to a temperature of 1474°F (787°C). This is nearly 350°F (200°C) lower than the true temperature of the hot junction [1822°F (980°C)].

2.11 GOOD THERMOCOUPLE PRACTICE

Accurate temperature measurement is an important requirement in science and industry. This is particularly true in chemical process industries where tight temperature control is vital. If temperatures are not controlled properly, costly loss of time and materials could be the result. Good thermocouple practice begins with a clear understanding of the basics of thermocouple thermometry and sufficient knowledge about thermocouple material, specifications, extension wires, and construction. Only then is the engineer in a position to select the best thermocouple system suitable to the application process requirements.

Some thermocouple application tips based on experience are as follows:

1. Use premium-grade thermocouple wire for critical requirements.
2. Properly protect thermocouples against physical damage or contamination.
3. Be sure to install the thermocouple in a representative area.
4. Avoid cold working of both the thermocouple and the extension wire.
5. Always be sure of the identity of the thermoelements and the polarity of the connections.
6. Replace thermocouples periodically; the calibration of thermocouples will always change with use.
7. Always isolate thermocouples and extension wires from power lines.

REFERENCES

1. ANSI MC96.1, "Temperature Measurement Thermocouples," Instrument Society of America, Research Triangle Park, N.C., 1982.
2. NBS Monograph 161, *The Nicrosil vs. Nisil Thermocouple Properties and Thermoelectric Reference Data,* National Bureau of Standards, Washington, D.C. 1978.
3. T. P. Wang and C. D. Starr, "Oxidation Resistance and Stability of Nicrosil-Nisil in Air and in Reducing Atmospheres," *Temperature, Its Measurement and Control in Science and Industry,* J. F. Schooley (ed.), Amer. Inst. Physics, 1982.
4. R. L. Anderson, J. D. Lyons, T. G. Kolie, W. H. Christie, and R. Eby, "Decalibration of Sheathed Thermocouples," *Temperature, Its Measurement and Control in Science and Industry,* J. F. Schooley (ed.), Amer. Inst. Physics, 1982.
5. C. D. Starr and T. P. Wang, "Thermocouples and Extension Wires," *Temperature, Its Measurement and Control in Science and Industry,* H. H. Plumb (ed.), Instrument Soc. Amer., Pittsburgh, Pa.
6. T. P. Wang and E. D. Zysk, *ASM Metals Handbook,* Vol. 3, *Thermocouples for Industrial Applications,* American Society for Metals, Metals Park, Ohio, 1980, p. 696.
7. NBS Monograph 125, Supplement 1, *Thermocouple Reference Table,* National Bureau of Standards, Washington, D.C., 1975.
8. Thermo Electric Co., Inc. Catalogues, *Thermocouple Wire, Industrial Thermocouples and RTD's,* and *Miniature/Custom Thermocouples,* Saddle Brook, N.J.

Chapter 3

Resistance Thermometry

BRIAN JONES
Minco Products, Inc.

3.1 INTRODUCTION

Resistance thermometers and thermistors are temperature sensors which change electrical resistance with temperature. The superior sensitivity and stability of these devices are important advantages in low and intermediate temperature ranges. In addition, resistive devices often simplify control and readout electronics.

Resistance thermometers are specified primarily for accuracy and stability from cryogenic levels to the melting points of metals. Resistance thermometers are accurate over a wide temperature range, may be used to sense temperature over a large area, and are highly standardized. The platinum resistance thermometer is the standard temperature instrument used by NBS to interpolate between fixed points in the range 13.81 to 903.90 K (−259.19 to 630.75°C). While normally less stable than resistance thermometers, thermistors offer lower cost and higher sensitivity over more limited ranges. Thermistor sensors can be relatively small and are well suited to product or equipment applications where quantities of use justify the design of special readout circuits.

3.2 RESISTANCE THERMOMETERS

Resistance thermometers may be called *resistance temperature detectors* (RTDs), *platinum resistance thermometers* (PRTs), or *standard platinum resistance thermometers* (SPRTs). These thermometers operate through the principle of electrical resistance changes in pure metal elements.

The traditional sensing element of a resistance thermometer consists of a coil of small-diameter wire wound to a precise resistance value. The most common material is platinum, although nickel, copper, and nickel-iron alloys compete with platinum in many applications.

A relatively recent alternative to the wire-wound RTD substitutes a film of platinum or nickel-iron, which is deposited or screened on a ceramic substrate. Thin-film and thick-film elements allow higher resistances using less metal, thereby lowering cost.

Resistance-Temperature Characteristics

Resistance thermometers exhibit the most linear signal with respect to temperature of any commonly used sensing device.

Small deviations from straight-line response, however, dictate the use of interpolating polynomials to calculate resistance values between fixed temperature points. Platinum follows the modified Callendar-Van Dusen equation over the range −436 to 1292°F (−260 to 630°C):

$$R_t = \begin{cases} R_0\,[1 + At + Bt^2 + Ct^3(t - 100)] & \text{for } t < 0°C \\ R_0\,[1 + At + Bt^2] & \text{for } t \geq 0°C \end{cases}$$

where R_t = resistance at temperature t, °C
R_0 = base resistance at 0°C
A, B, C = constants that describe a given thermometer

The A, B, and C coefficients, individually determined for high-precision laboratory thermometers, are entered into the Callendar-Van Dusen equation to generate computer resistance-versus-temperature tables. Nominal coefficients, used in standard published tables, are:

TCR* (Ω/Ω/°C) = 0.003926 A: 3.9848×10^{-3}
B: -5.870×10^{-7} C: -4.0000×10^{-12}

TCR (Ω/Ω/°C) = 0.003911 A: 3.9692×10^{-3}
B: -5.8495×10^{-7} C: -4.2325×10^{-12}

TCR (Ω/Ω/°C) = 0.003850 A: 3.90802×10^{-3}
B: -5.802×10^{-7} C: -4.27350×10^{-12}

The corresponding equations for copper elements are

$$R_t = \begin{cases} R_1R_0[1 + A_1(t + 200) + B_1(t + 200)^2] \\ R_0[1 + A_2t] \\ R_3R_0[1 + A_3(t - 150) + B_3(t - 150)^2] \end{cases}$$

for t = −200 to −50°C
for t = −50 to 150°C
for t = 150 to 260°C

where R_t = resistance at temperature t, °C
R_0 = base resistance at 0°C
A, B, R_1, R_3 = constants that describe a given thermometer

Nominal values are

TCR (Ω/Ω/°C) = 0.00427 R_1: 1.17058×10^{-1}
A_1: 3.92313×10^{-2} B_1: -7.45044×10^{-6}
A_2: 4.2743×10^{-3} R_3: 1.641145
A_3: 2.62628×10^{-3} B_3: 2.43732×10^{-8}

Note that copper resistance is linear with temperature from −50 to 150°C and nearly linear over other ranges. The equations for nickel-iron are:

$$R_t = \begin{cases} R_0[1 + A_1t + B_1t^2 + C_1t^3] & \text{for } t < 0°C \\ R_0[1 + A_2t + B_2t^2 + C_2t^3] & \text{for } t \geq 0°C \end{cases}$$

* Temperature coefficient of resistance.

where R_t = resistance at temperature t, °C
R_0 = base resistance at 0°C
A, B, C = constants that describe a given thermometer

Nominal values are:

TCR (Ω/Ω/°C) = 0.00518 A_1: 4.68699×10^{-3}
B_1: 8.58992×10^{-6} C_1: 0 A_2: 4.59818×10^{-3}
B_2: 5.89404×10^{-6} C_2: 0

TCR (Ω/Ω/°C) = 0.00527 A_1: 4.63189×10^{-3}
B_1: 6.96196×10^{-6} C_1: -1.72771×10^{-8}
A_2: 4.63189×10^{-3}
B_2: 6.96196×10^{-6} C_2: -5.71203×10^{-9}

Individual manufacturers have developed proprietary curves for nickel elements, working from direct measurements, but there is no simple formula for direct calculation of resistance. Figure 3.1 shows the nonlinear behavior of nickel in comparison to other elements.

Until recently, equations relating resistance to temperature were of interest mainly to researchers and designers of linearizing analog circuits. Now many common instruments such as readouts, data loggers, and controllers contain microprocessors. Interest in direct digital computation of temperature has increased accordingly. A few considerations for the designer or programmer of digital instruments are:

1. Converting resistance to temperature requires finding the root of a polynomial. Instead of root finding, many manufacturers use a lookup table, with resistance-temperature values stored in read-only memory (ROM). Choosing values sufficiently close together, on a 5°C interval, for example, with linear interpolation in between will usually produce acceptable readings.

2. The use of resistance ratios, instead of absolute values, is recommended for measurements of the highest accuracy. A ratio table sets the $R_{0°C}$ value to 1 and uses the A, B, and C coefficients to calculate ratios from this value. The absolute resistance at any given temperature is the product of R_0 and

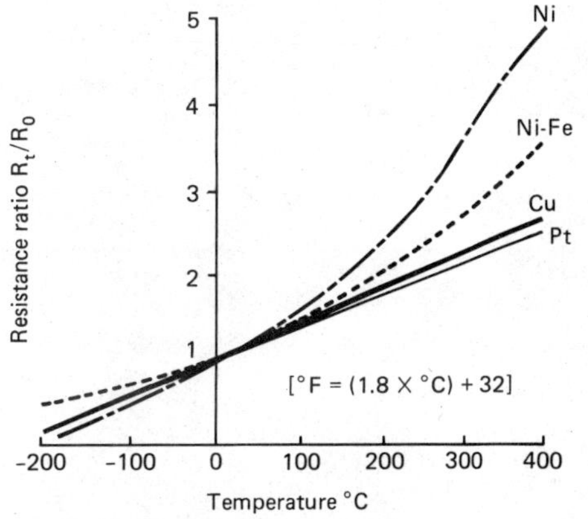

Fig. 3.1 Linearity of metals used in resistance thermometers.

the ratio. Resistance ratios are preferred for their stability. Simple periodic checks of ice-point resistance, with the most recent R_0 used to calculate ratios, will reduce or eliminate the need for complete recalibration and table generation.

Temperature Coefficient of Resistance

The temperature coefficient of resistance (TCR) has many definitions. For resistance thermometers, TCR is normally defined as the average resistance change per degree Celsius over the range 0 to 100°C, divided by $R_{0°C}$:

$$\text{TCR } (\Omega/\Omega/°\text{C}) = \frac{(R_{100°C} - R_{0°C})/R_{0°C}}{100}$$

TCR values for the common elements are, in Ω/Ω/°C

Platinum: $$\frac{(138.50 - 100)/100}{100} = 0.00385$$

or $$\frac{(139.11 - 100)/100}{100} = 0.00391$$

or $$\frac{(139.26 - 100)/100}{100} = 0.003926$$

Copper: $$\frac{(12.897 - 9.035)/9.035}{100} = 0.00427$$

Nickel: $$\frac{(200.64 - 120)/120}{100} = 0.00672$$

Nickel-iron: $$\frac{(917.33 - 604)/604}{100} = 0.00518$$

or $$\frac{(1387.21 - 908.4)/908.4}{100} = 0.00527$$

In one sense, TCR expresses the sensitivity of the resistive wire used in the element, as it defines the average temperature change of a hypothetical 1-Ω thermometer. But specifiers normally use TCR to distinguish between different resistance-temperature curves of the same element material, such as the three curves of platinum shown above. Because all of these curves see widespread use, platinum TCRs must be properly specified to maintain compatibility between thermometers and instruments.

There are four primary curves specified for platinum:

1. *0.003926 Ω/Ω/°C:* The International Practical Temperature Scale (IPTS-68) requires a minimum TCR of 0.003925 for SPRTs. This is achieved only with high-purity wire (99.999% or better) wound in a strain-free configuration. The stresses introduced in manufacturing ordinary industrial models lower TCR.
2. *0.003911 Ω/Ω/°C:* This is sometimes called the "U.S. Industrial Standard." It is lower than laboratory standards as the typical construction of high-temperature ceramic elements imposes strain on platinum wire.
3. *0.00385 Ω/Ω/°C:* This TCR meets the requirements of the international standard IEC 751 (1983), which is essentially identical to the older German standard DIN 43760 (1980). Although the lower TCR indicates less pure platinum, 0.00385 element wire actually costs more because the wire supplier must carefully introduce and control the impurity. The 0.00385 is the highest TCR presently available for thin-film elements.
4. *0.00375 Ω/Ω/°C:* At least one manufacturer now offers thin-film 1000-Ω elements with 0.00375 TCR, intended for low-cost applications.

There are few inherent advantages in specifying any particular TCR over another. Specification of 0.00385, 100-ohm thermometers guarantees the greatest standardization and commercial availability.

Comparison of Element Types

Platinum, because of its wide temperature range and stability, has attained a dominant position in resistance thermometry. Furthermore, advances in element construction have narrowed the price difference between platinum and base-metal thermometers. Nevertheless, nickel, copper, and nickel-iron do have benefits for many applications and should be considered. The primary advantages of the four element types are compared in Table 3.1.

Table 3.1. Comparison of Resistance Thermometer Element Types

Element type	Temperature range, °F (°C)	Resistivity, Ω/cir mil · ft at 68°F (20°C)	Benefits	Base resistance, Ω	TCR, Ω/Ω/°C	Sensitivity, average Ω/°C, 0 to 100°C
Platinum	−436 to 1292 (−260 to 700)	63.8	Greatest range, best stability, good linearity	100 at 0°C 100 at 0°C 100 at 0°C 1000 at 0°C	0.003926 0.00391 0.00385 0.00385	0.392 0.391 0.385 3.850
Copper	−148 to 500 (−100 to 260)	10.7	Best linearity	10 at 25°C	0.00427	0.039
Nickel	−148 to 500 (−100 to 260)	41.5	Low cost, high sensitivity	120 at 0°C	0.00672	0.806
Nickel-iron (70%-30%)	−148 to 400 (−100 to 204)	120.0	Low cost, highest sensitivity	604 at 0°C 1000 at 70°F 2000 at 70°F	0.00518 0.00527 0.00527	3.133 4.788 9.576

Effects of Connection Wire Resistance and Bridge Design

Because an RTD is a resistance-type sensor, any resistance in the extension wires between the RTD and control instrument will add to readings. In some cases one can compensate for this extra resistance with adjustments at the instrument. However, this compensates only when the wires are at a constant temperature since variations in ambient temperature alter copper wire resistance.

Table 3.2 shows resistance values of common copper wire sizes. To approximate error in an uncompensated system, multiply the total length of extension wires by the appropriate value in the table. Then divide by the sensitivity of the RTD element from Table 3.1 to obtain an error figure in degrees Celsius. Measurement error can be significant, especially with small wires or low-sensitivity elements. Fortunately, the use of a three-wire system will reduce errors to a negligible level in most applications.

Figure 3.2 shows a two-wire RTD connected to a typical Wheatstone-bridge circuit. The bridge, a resistive network, translates the RTD's resistance into an electrical signal used by the monitoring or controlling instrument. In this figure, E_s is the supply voltage; E_o is the output voltage; R_1, R_2, and R_3 are fixed resistors; R_t is the sensing element of the resistance thermometer; and L_1 and L_2 are the resistances of the two extension wires.

With a balanced bridge, the voltage drops across the two upper arms, (1) R_1 and (2) $R_t + L_1 + L_2$, are equal and E_o is zero. The fixed resistors R_1, R_2, and R_3 are specified to make the bridge ratios equal at balanced condition:

$$\frac{R_1}{R_2} = \frac{R_t + L_1 + L_2}{R_3}$$

Maximum bridge sensitivity is realized if

$$R_1 = R_2 = R_3 = R_t + L_1 + L_2$$

This sensitivity is desirable, however, only in laboratory systems where the bridge is constantly rebalanced. Typical industrial circuits will have unequal resistors.

The value of R_t at the temperature point of most interest will influence the selection of the bridge arm resistors. Values should also be chosen to limit bridge currents and consequent self-heating in the thermometer. Bridge resistors must be stable and insensitive to ambient temperature variations for best accuracy.

Table 3.2. Resistance of Common Copper Wire Sizes

Wire AWG	Ω/ft at 77°F (25°C)
12	0.0016
14	0.0026
16	0.0041
18	0.0065
20	0.0103
22	0.0165
24	0.0262
26	0.0418
28	0.0666
30	0.1058

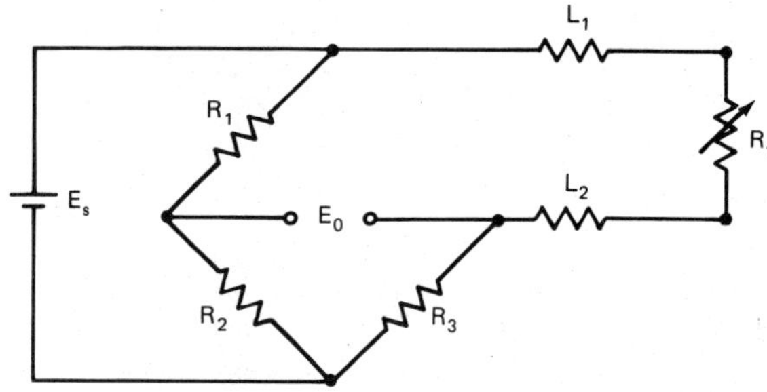

Fig. 3.2 Uncompensated two-wire bridge circuit.

In a two-wire bridge, the wire resistance $L_1 + L_2$ adds directly to readings. If leads are short enough or the sensitivity is high enough, the offset is acceptable. When long extension runs are required between the sensor and the instrument, or sensitivity is low, the specifier should consider a three-wire system. All resistance thermometers with copper elements must have three wires to offset their low sensitivity.

In Fig. 3.3, which shows a three-wire compensating bridge circuit, L_1 and L_2 are now in two separate arms of the bridge, so the measuring currents and voltage drops across them are identical. The third wire L_3 forms part of the output circuit and does not affect the bridge ratios or balance. In fact, no current flows through L_3 when the bridge is in balance.

Proper compensation with the three-wire system depends on these conditions:

1. Wire resistances L_1 and L_2 should be equal. Most manufacturers match wires within ±5 percent. In most cases, therefore, error with a three-wire system is less than 5 percent of the error with a similar two-wire system.
2. Wires should remain bundled together to ensure that ambient temperature changes act equally on all wires.
3. Electronic circuitry connected to E_o should have sufficient input impedance to prevent appreciable current drain through L_3. Normally L_3 acts only as a potential-carrying wire. Any current through it will cause errors when the bridge is out of balance.

Three-wire systems represent a practical compromise between accuracy over distance and the cost of extra signal wires. Although well suited to most industrial areas, they may be affected by electrical noise and contact resistance at the junction points.

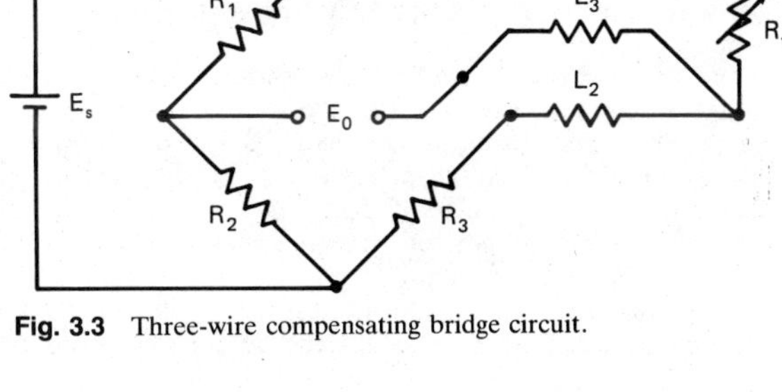

Fig. 3.3 Three-wire compensating bridge circuit.

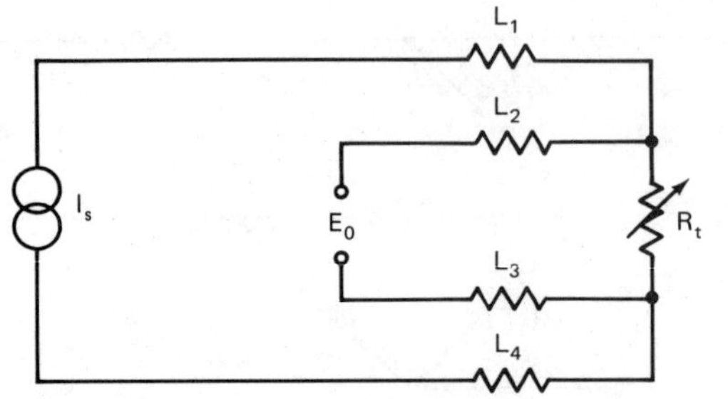

Fig. 3.4 Four-wire system with constant-current source.

Four-wire circuits provide the same resistance compensation as three-wire systems, but also relieve problems with unmatched wires, contact resistances, and thermal electromotive forces (emf's). Thermal emf's are spurious voltages introduced by the thermocouple effect where two dissimilar metals make contact. Laboratory systems, used for the highest-precision measurements, are often variants of the Mueller bridge. The Mueller bridge is basically a switched three-wire system requiring two readings which are averaged to yield a true reading.

The need to throw switches and calculate the mean of readings excludes Mueller bridges from use in automated readout or control. An alternative, still using four-wire RTDs, is the constant-current circuit shown in Fig. 3.4. Here a constant current source drives I_s through L_1 and L_4, and the potential produced by R_t appears across L_2 and L_3. Circuitry connected to E_o must have an impedance sufficiently high to prevent appreciable current flow through L_2 and L_3. In this circuit, the measured potential is independent of wire resistance. Use of an ac current source nullifies emf's.

With the availability of commercial constant-current devices, this type of circuit is becoming increasingly popular. It adapts readily to three-wire sensors as well, with the single wire at one end of the sensing element carrying both current and potential.

Two-Wire Temperature Transmitters

Resistive networks such as the Wheatstone bridge represent a passive solution to the problem of wire resistance. The two-wire temperature transmitter, in contrast, actively amplifies and conditions the RTD signal.

A transmitter, which mounts at or near the RTD location, converts the resistance reading to a current signal proportional to temperature. Unlike voltage or resistance, current must be the same at both ends of a signal loop. This means that temperature signals can be sent thousands of feet over two wires with no loss of accuracy from extension wire resistance or electrical noise.

The standard process control transmitter produces a 4- to 20-mA signal proportional to temperature over a specified range. The signal current also provides power for the transmitter's electronics. Allowable resistance in the signal loop depends on the voltage required by the transmitter at the 20-mA level. Hundreds of ohms—thousands of feet of wire—are seldom a problem.

The disadvantages of transmitters are price—typically about twice that of a resistance thermometer alone—and the need to periodically recalibrate zero and span. On the other hand, cost savings result from the use of inexpensive twisted-pair signal wires over long distances, plus a linear current signal which easily interfaces to voltage input instruments through the use of a load resistor.

Most transmitters mount in a connection head attached to the resistance thermometer or in an instrument rack nearby. Some types of transmitters mount in the same enclosure as the thermometer. This arrangement requires careful attention to the problem of temperature rises induced by heat from the transmitter.

Potential Sources of Error with Resistance Thermometers

Resistance thermometers may produce errors in three ways: (1) the inherent tolerances built into the thermometer, (2) gradients between the thermometer and the medium to be sensed, and (3) errors introduced along the path between the sensor and the readout or control instrument. Some sources of error are electrical; others result from the mechanical construction of the thermometer.

Potential sources of error are described in the paragraphs that follow.

Interchangeability and Conformity *Conformity* specifies the amount the resistance of a thermometer is allowed to deviate from a standard curve (such as the curve produced by the Callendar-Van Dusen equation). Conformity has two components: a tolerance at the reference temperature, usually 0°C, and a tolerance on the slope or TCR. Figure 3.5 shows that a resistance thermometer conforms most closely to its curve at the reference temperature, while the resistance fans out above and below this reference. For example, IEC 751, Class B, requires calibration within 0.12 Ω (0.30°C) at 0°C but allows TCR to deviate from nominal 0.00385 by ±0.000012 Ω/Ω/°C. Thus tolerance spreads to 0.8°C at 100°C, 1.3°C at 200°C, and on up to 3.8°C at 700°C. *Interchangeability* between two thermometers is no more than twice the value of their conformity.

Commercial platinum resistance thermometer elements are available with extremely tight tolerances, to within 0.01 Ω (0.026°C) in some cases. When interchangeability is an overriding consideration, the specifier may consider other means to achieve it. For example, manufacturers may alter their calibration procedures to fix the reference temperature—and tightest tolerance—at a point other than 0°C. As another example, if the difference between two thermometers is more important than absolute temperature, matched pairs—measured to agree within a certain tolerance—may be less expensive than calibrating each thermometer within a small range of nominal.

It is important to note that conformity-interchangeability specifications only denote the relative accuracy of two otherwise identical thermometers mounted side by side in the same environment. They do not include errors acting equally on both thermometers.

Sensitivity The resistance change per degree change in temperature is a function of base resistance and TCR. Although there is no inherent advantage to a higher sensitivity, a larger signal simplifies output electronics and is less susceptible to extension wire effects and electrical noise. In addition, a larger resistance produces the same voltage output with less measuring current, which helps to limit self-heating of the thermometer element.

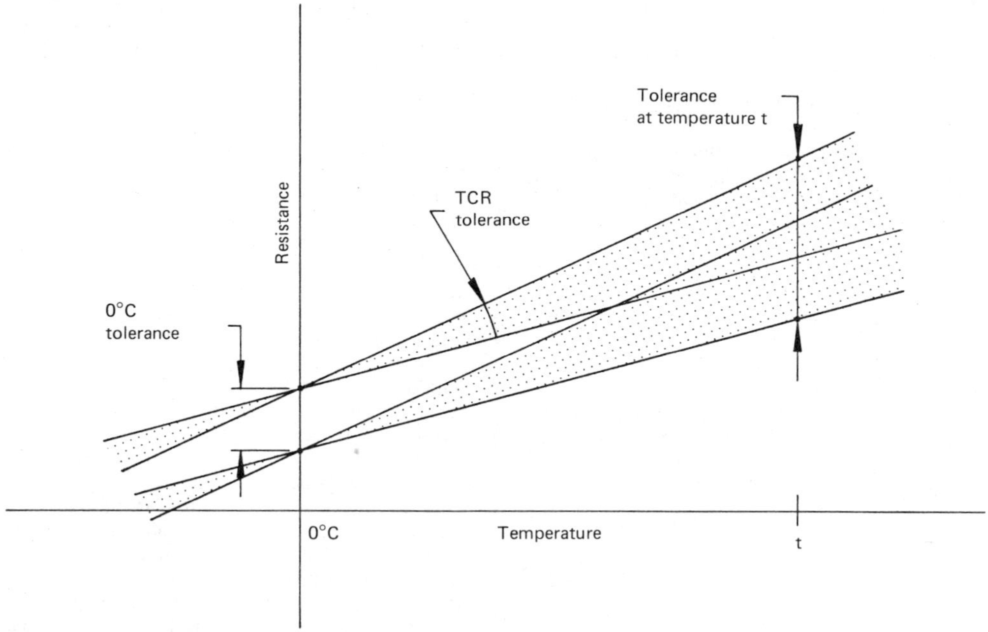

Fig. 3.5 Effects of R_0 and slope tolerance on total tolerance.

Insulation Resistance If the sensing element and extension wires are not completely insulated from the case, a shunting effect occurs in which the case becomes a parallel resistor and lowers readings. In most industrial thermometers, with specified insulation resistances in the 100-MΩ range, error approaches zero. The shunting effect decreases with low-resistance elements, which accounts for the use of 25.5-Ω PRTs in laboratory measurements.

Self-Heating A resistance thermometer is a passive-resistance sensor; it requires a measuring current to produce a useful signal. Because this measuring current heats the element wire above the true ambient temperature, errors will result unless the extra heat is dissipated.

Self-heating is most often expressed in milliwatts per degree Celsius (mW/°C), which is the power in milliwatts ($1000I^2R$) required to raise the thermometer's internal temperature by 1°C. The higher the mW/°C figure, the lower the self-heating.

The generated heat increases with increasing sensing element resistance and with increasing measuring current. The resulting error is inversely proportional to the ability of the thermometer to shed extra heat, which, in turn, depends on thermometer materials, construction, and environment. The worst self-heating occurs when a high resistance is packed into a small body. Thin-film elements, with little surface area to dissipate heat, are an example. Self-heating also depends on the medium in which the thermometer is immersed. Error in still air may be over 100 times greater than in moving water.

Time Constant A time constant indicates the responsiveness of a resistance thermometer to temperature change. A common expression is the time it takes a thermometer to reflect 63.2 percent of a step temperature change in moving water. Response speed depends on the mass of the thermometer and the rate at which heat transfers from the outer surface to the sensing element. A rapid time constant reduces errors in a system subject to rapid temperature changes.

Repeatability The degree of accord between two successive readings with a thermometer is its repeatability. Loss of repeatability results from permanent or temporary changes to the resistance characteristics of the element and may be caused by exposing the thermometer to temperatures at or beyond the endpoints of its specified range. A repeatability test cycles the thermometer between low and high temperatures; any changes to $R_{0°C}$ are noted. A typical repeatability rating for an industrial platinum resistance thermometer is ±0.1°C.

Stability Stability is long-term drift in thermometer readings. A typical specification would limit drift to 0.1°C per year for rated operation. Normal service at points well within the temperature rating typically cause much less drift. Drift is a consequence of the element material (platinum is the most stable), encapsulating materials which could contaminate the element, and mechanical stress placed on the element.

Shock and Vibration Mechanical shock and vibration can alter thermometer readings or cause complete failure. In fact, stability and ruggedness are somewhat exclusive. A laboratory thermometer designed for maximum stability contains an unsupported element which is far too fragile for industrial use.

The elements of most industrial resistance thermometers are fully supported by a bobbin or packing material and therefore stand up quite well to extreme environments. More likely to suffer are extension-wire transition points, which should be properly immobilized. A typical RTD will meet a specification allowing shock of 100g of 8-ms duration and vibration of 10 to 2000 Hz at 20g.

Packaging and Thermal Transfer Sheaths and other structures surrounding resistive elements should maximize heat transfer from the sensed medium, minimize heat transfer from ambients which can alter readings, and provide necessary protection of the element. Proper materials and construction can dramatically improve reading accuracy.

One strategy practicable only with wire-wound resistance thermometers (vs. thermistors, thermocouples, and solid-state devices) is temperature averaging. An element may be wound to average temperature along its entire length, which may range from a few inches to 50 ft.

Resistance Thermometer Types

Examples of commonly available resistance thermometer types, with an emphasis on the design features which take advantage of the benefits listed above and avoid the sources of error, are presented in the following paragraphs.

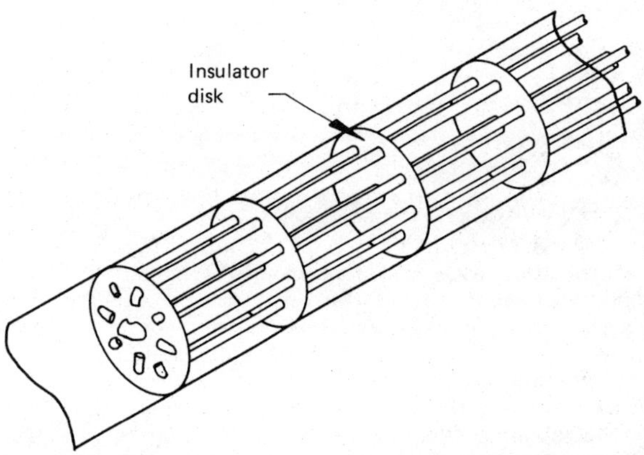

Fig. 3.6 Bird-cage platinum element.

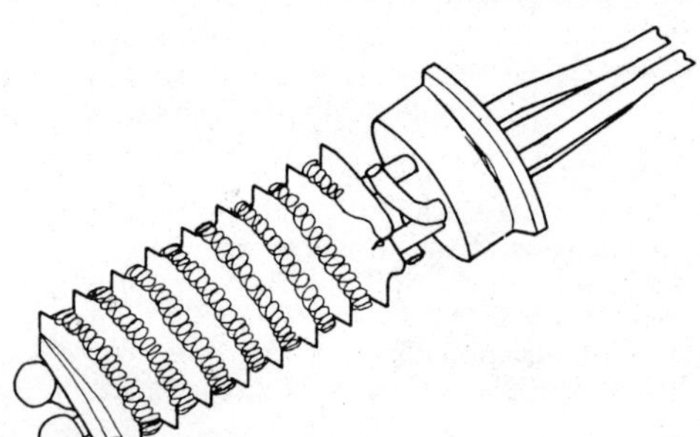

Fig. 3.7 Helical-wound strain-free element.

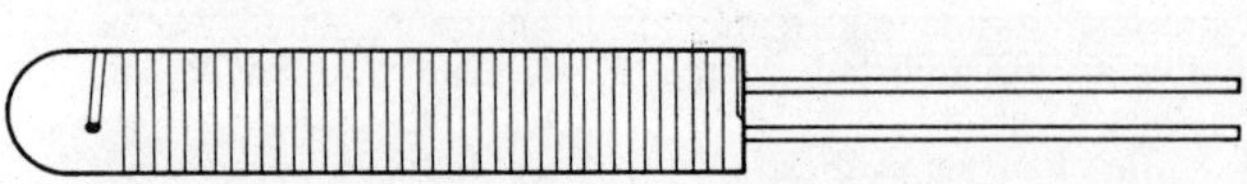

Fig. 3.8 Industrial resistance thermometer element.

Standard Platinum Resistance Thermometers for Laboratory Use The National Bureau of Standards (NBS) specifies the standard platinum resistance thermometer (SPRT) as the standard interpolating instrument used to define temperatures from −259.34 to 630.74°C. For a thermometer to be classified as a primary standard, its R_0 value must change by no more than $4 \times 10^{-6} R_0$ between successive calibrations and its TCR must be greater than 0.003925 Ω/Ω/°C. This requires very-high-purity platinum wound in a nearly strain-free manner.

Figure 3.6 shows a classic "bird-cage" strain-free element, in which the element wire is supported by insulating disks. Figure 3.7 shows a less expensive helical wind. The highest accuracy primary standards encapsulate the element within a glass or quartz sheath, although stainless steel or Inconel may be substituted for secondary, or transfer, standards. The element area must be hermetically sealed to exclude oxidizing agents and helium-filled if used at cryogenic ranges.

The unsupported coils of a strain-free element are quite susceptible to damage from shock and vibration. Even a slight tap on a table can invalidate an expensive calibration.

Industrial Resistance Thermometer Elements Industrial thermometer use demands ruggedness and reliability in addition to accuracy. The need to encapsulate the resistive element for protection from shock and corrosive environments without unduly straining the element has led to a variety of technologies for constructing elements.

Figure 3.8 diagrams a common ceramic element. The element wire, commonly platinum, is wound around a ceramic bobbin, welded to lead wires, and coated with glass. The manufacturer must carefully match materials to prevent thermal expansion strains on the wire.

Higher temperatures are possible with the element shown in Fig. 3.9. Here, a coil of platinum wire is passed through bores in a ceramic tube. The bores are then filled with a ceramic or alumina powder to cushion the coils.

Copper, nickel, and nickel-iron may replace platinum as element materials for lower temperatures, generally below 260°C. Moderate temperatures also permit the use of organic materials in element construction, enabling a wide variety of styles. Resulting benefits include lower cost and faster time response.

Thin-film elements (Fig. 3.10) have arrived within the last 10 years and are now comparable to wire-wound ceramic elements in performance but lower in price. They consist of a flat substrate with a thin film of deposited platinum, laser-trimmed to the proper resistance. Thin-film elements can have high resistances, commonly 1000 Ω. Naked elements exhibit very fast time responses, although they will respond more slowly in probes because of difficulties in transferring heat to the flat element. The user should be careful to limit measuring current, as the small size and high resistance of thin-film elements invite self-heating. Thin-film elements may be more fragile than wire-wound elements.

Industrial RTD Probes The encased probe is the standard resistance thermometer configuration for industrial process control and machinery protection. Most probe cases are of stainless steel or Inconel to withstand high temperatures, although other materials offer advantages at intermediate ranges. For example, the tip-sensitive probe shown in Fig. 3.11 has a copper-alloy tip which conducts heat 20 times better than stainless steel. This design improves thermal con-

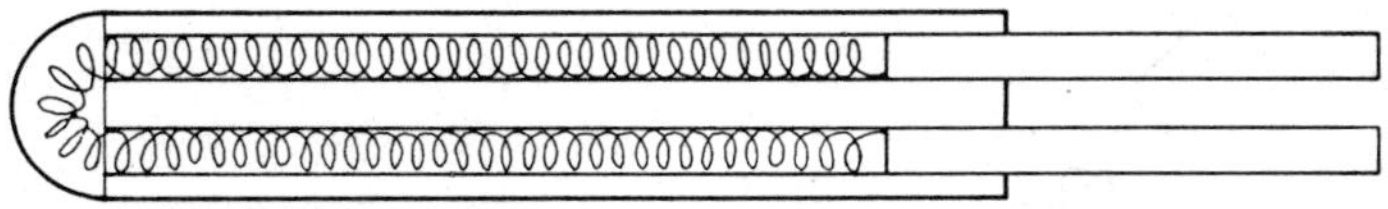

Fig. 3.9 High-temperature element.

tact with sensed surfaces and reduces errors from conduction along the sheath.

Standard probe diameters range from 0.125 to 0.250 in. Smaller probes respond faster when directly immersed, but larger probes are used extensively in standard thermowells. Probe lengths range from a few inches to 10 ft or more.

Figure 3.12 shows the construction of a high-temperature probe. The element fits in the tip, surrounded by high-temperature powder or cement. Extension wires, normally uninsulated, extend back from the element and are encapsulated by powder, cement, or bored ceramic spacers. External wires, often insulated with Teflon or glass braid, are potted with cement at the entry point to seal against moisture.

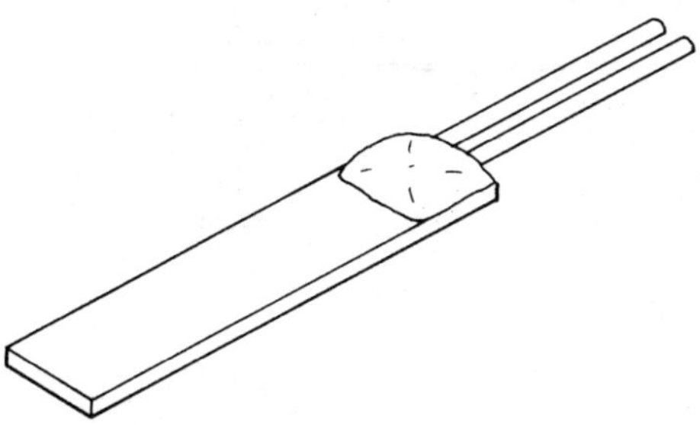

Fig. 3.10 Thin-film element.

The alternative construction shown in Fig. 3.13 places the element, potting, and wire transitions within a module at the tip of the probe. This design allows the user to cut and bend the probe. Temperature is limited to the rating of the external lead-wire insulation: 260°C (500°F) for Teflon and up to 550°C (1022°F) for insulations such as woven mica-glass.

Probe Assemblies A wide variety of mounting fittings and accessories aid probe installation. Selection depends on the nature of the medium being sensed and cost requirements.

Direct immersion of a probe into a process fluid requires a fitting with a pipe thread, which may be adjustable or welded onto the probe. Figure 3.14 shows a typical assembly, with one thread for mounting the probe and another for a connection head. Connection heads provide a transition between probe leads and external signal wires.

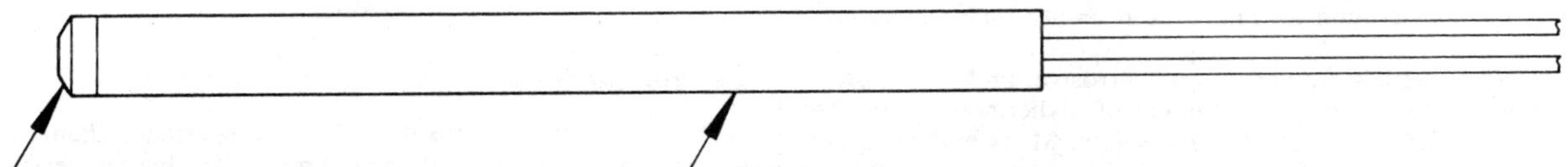

Fig. 3.11 Tip-sensitive probe.

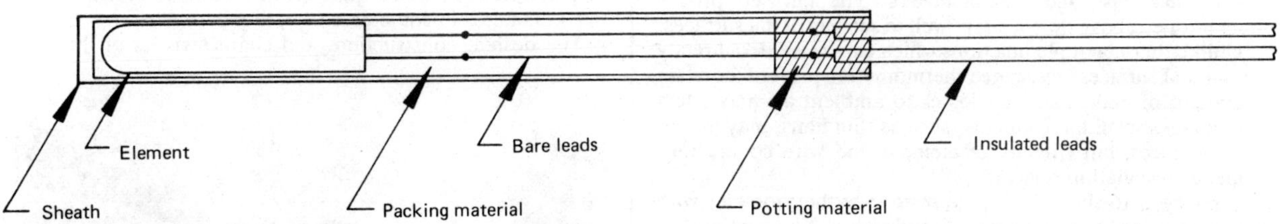

Fig. 3.12 High-temperature probe.

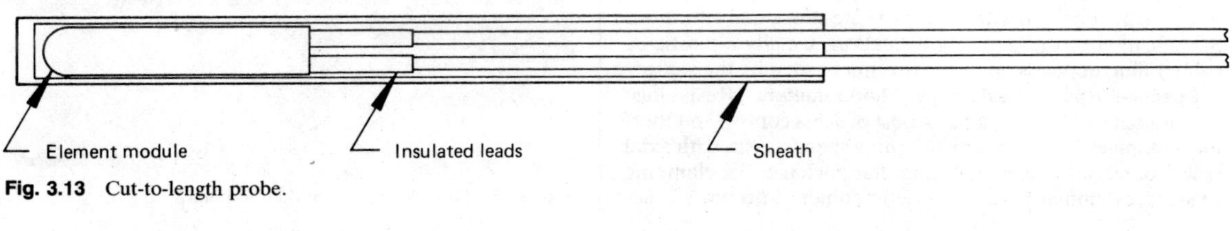

Fig. 3.13 Cut-to-length probe.

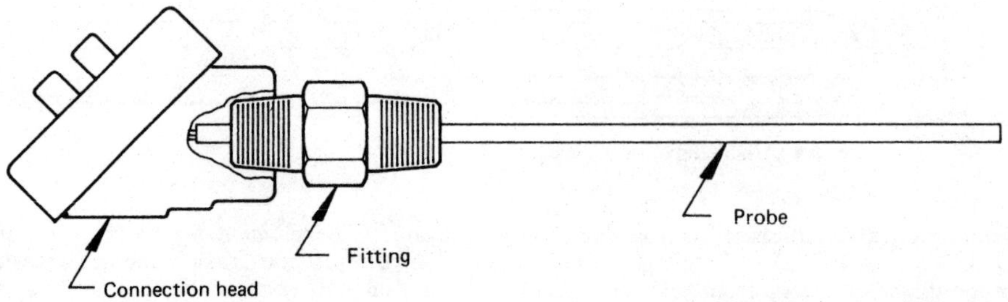

Fig. 3.14 Typical probe assembly.

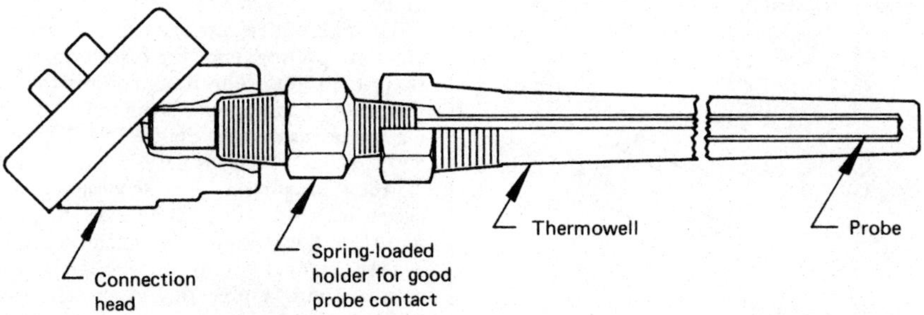

Fig. 3.15 Thermowell assembly.

Mounting in a solid material is best accomplished with a spring-loaded holder, which may be fixed or adjustable. Spring-loading provides good contact of the probe tip and dampens vibration for improved response time and service life.

When liquids are particularly corrosive, under high pressure, or fast-flowing, installation of a thermowell may be necessary for serviceability and safety. Many materials and styles are available to match application requirements. Thermowells drilled from solid bar stock provide the highest pressure ratings, but welded models may cost much less. Figure 3.15 shows a typical thermowell assembly.

Flexible Resistance Thermometers The encased probes mentioned above do not adapt well to sensing of flat surfaces. Unlike thermocouple junctions which can be welded directly to metal surfaces, resistance thermometers present a certain amount of bulk, and heat losses to ambient air may affect readings. Small flat elements, such as thin films, may mount on surfaces, but fragility of element and wire connections makes installation difficult.

Figure 3.16 shows a flexible resistance thermometer with a wire-wound sensing element sandwiched between insulating layers. It conforms closely to sensed surfaces, and has thin insulation. The wire element may be wound to nearly any size to average out temperature gradients, and the flexible construction can withstand extreme shock and vibration. Specification of the covering insulation, usually a plastic or rubber film, depends on the environment seen by the sensor.

Special-Purpose Resistance Thermometers Resistance thermometers readily adapt to most process control and thermal equipment designs. The user may specify cases with axial leads for circuit-board mounting, flat packages for clamping to surfaces, miniature cases for embedment into sensed media, and any metal sheaths and fittings which can be produced by a machine shop. In addition, wire windings may be configured to sense over large areas.

3.3 THERMISTORS

A thermistor operates through electrical resistance changes in semiconductors rather than pure metal. The base material is a mixture of metal oxides pressed into a bead, rod, disk, wafer, or other shape. The bead, with lead wires embedded, is then sintered at high temperatures and coated with epoxy or glass. Beads may be quite small—down to 0.01-in (0.25-mm) diameter in some cases.

The design, construction, and characteristics of thermis-

Fig. 3.16 Flexible resistance thermometer.

tors vary widely among manufacturers. Typical properties are:

1. Thermistors exhibit very large resistance changes, but usually in a direction opposite to resistance thermometers; resistance drops as temperature rises. This is called the *negative temperature coefficient of resistance* (NTC).

2. Base resistances, commonly specified at 25°C, range from thousands to millions of ohms. Thermistor sensitivity dwarfs that of resistance thermometers.

3. Trimming thermistors into a specified resistance tolerance is not easily accomplished. The user must individually calibrate circuits when thermistors are interchanged.

4. Resistance-temperature curves deviate widely from linearity, except over narrow ranges.

5. Thermistors tend to drift more than resistance thermometers, although they stabilize over time.

6. Temperature ranges are moderate, with 300°C (572°F) the common upper limit.

The combination of nonlinearity, high sensitivity, and instability has generally limited thermistors to high-resolution measurements over limited ranges. In such applications, one-point calibration may be sufficient. A classic example is medical thermometry.

Some manufacturers of thermistors offer special models with these characteristics:

1. Positive temperature coefficient (PTC) models, which are used more for current limiting in electronic circuits than for temperature measurement. As current increases through the bead, self-heating drives up resistance dramatically, throttling the current.

2. Thermistors which are closely matched and therefore interchangeable over a specified range.

3. Linearized thermistors or thermistor sets which produce a highly linear output over a limited range.

4. Glass-coated thermistors, specially aged, which may maintain excellent stabilities over moderate temperatures, e.g., ±0.005°C per year over the range 0 to 100°C.

As a general rule, a thermistor acts as a single-purpose sensing device. The designer must relate temperature ranges, resistance-temperature characteristics, and output circuits for each application. The reward is a high degree of resolution and accuracy with a relatively inexpensive system.

Resistance-Temperature Characteristics of Thermistors

In general, the log of the resistance of a thermistor is proportional to the inverse of the temperature of that thermistor. The following equation is valid for ranges up to 100°C in size:

$$\frac{1}{T} = a_0 + a_1 \ln(R_T) + a_3 \ln(R_T)^3$$

where T = absolute temperature, K ($t_{°C}$ + 273.15),
R_T = zero-power resistance at temperature T
a_0, a_1, a_3 = coefficients which describe a given thermistor

This is called the *Steinhart-Hart* equation.

The particular coefficients of the equation depend solely on the composition of the thermistor material and vary widely. There are no published standards to guarantee compatibility, although some popular types may be available from more than one source.

Thermistor Output Circuits

The lack of linearity in thermistor curves has spurred the development of many types of compensating circuits. Two of the simpler ones, to linearize voltage and to linearize resistance, are described below.

Figure 3.17 shows a simple voltage divider circuit. The output voltage is given by

$$\frac{e_o(t)}{e_s} = \frac{1}{1 + sr(t)} = F(t)$$

where $e_o(t)$ = output voltage at temperature t
e_s = source voltage
R = a fixed resistance
$R(t_0)$ = reference, or base, resistance of thermistor
s = $R(t_o)/R$
$r(t)$ = $R(t)/R(t_0)$ (resistance at temperature referenced to base resistance)

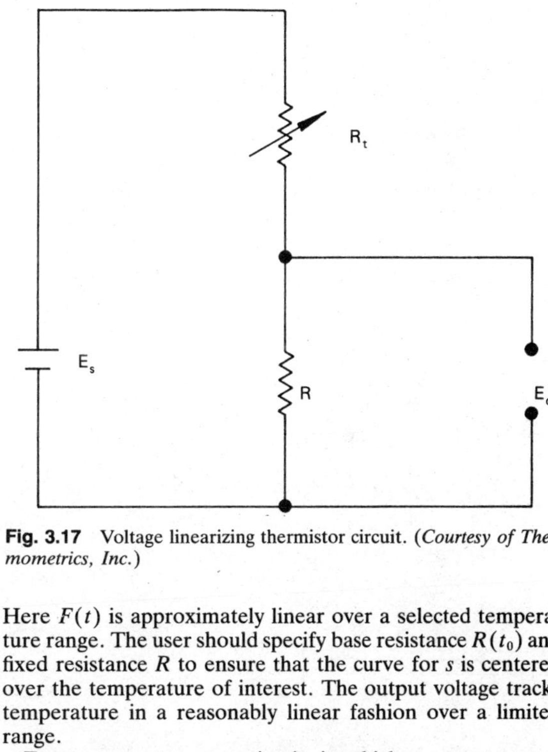

Fig. 3.17 Voltage linearizing thermistor circuit. (*Courtesy of Thermometrics, Inc.*)

Here $F(t)$ is approximately linear over a selected temperature range. The user should specify base resistance $R(t_0)$ and fixed resistance R to ensure that the curve for s is centered over the temperature of interest. The output voltage tracks temperature in a reasonably linear fashion over a limited range.

For a constant-current circuit, in which a current source is driven through the thermistor and voltage drop measured, a linear *resistance* is necessary. An ordinary shunt resistor in parallel with the sensing element (Fig. 3.18) yields this formula:

$$R = R_1[1 - F(t)]$$

where R = overall resistance of circuit
R_1 = a fixed resistance
$F(t)$ = function defined above

Again, response is fairly linear over limited spans.

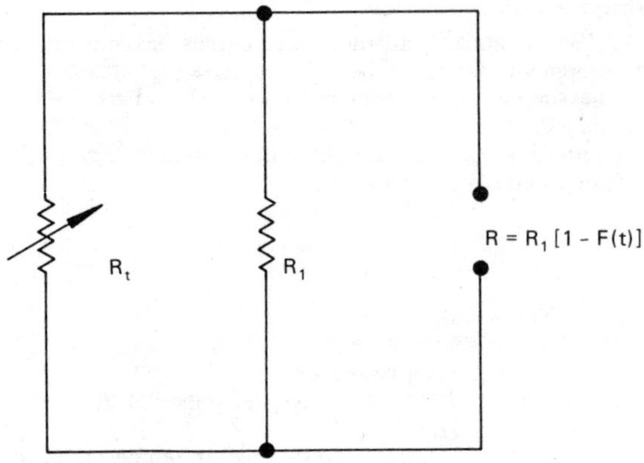

Fig. 3.18 Resistance linearizing thermistor circuit. (*Courtesy of Thermometrics, Inc.*)

Complex circuits, with Wheatstone bridges or other resistive networks, can further linearize thermistor output when necessary. Most employ matched sets of thermistor beads.

Potential Sources of Error with Thermistors

Most of the sources of error listed for resistance thermometers apply as well to thermistors. Thermistors remove one worry—extension wire resistance is normally negligible in relation to their high resistance. Three- or four-wire circuits are rarely required for wire compensation, although they may be needed for linearization with matched sets.

Self-heating, though, can present far more of a problem with thermistors. More heat is generated through the high resistance of thermistors, and there is little surface area for dissipation. The same current through a thermistor and RTD will produce self-heating around 100 times greater in the thermistor. The user should carefully read self-heat specifications and limit current accordingly.

Because of self-heating error, the reference resistance of thermistors is specified at a *zero-power* value. This is an extrapolation from a series of measurements at different sensing currents. It gives the ideal resistance assuming no current and no consequent self-heating.

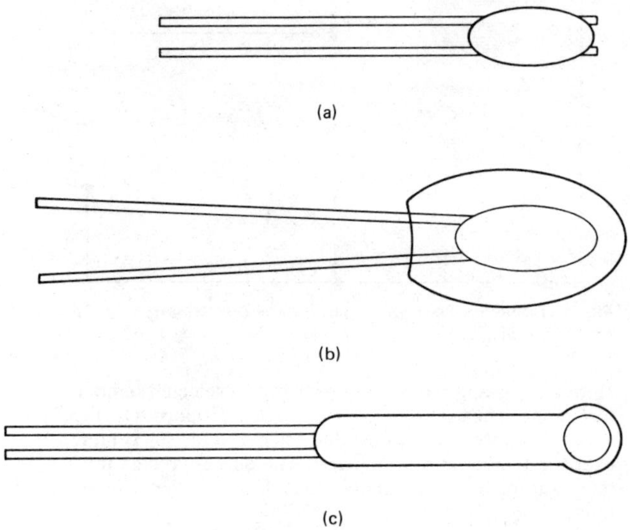

Fig. 3.19 Thermistor types. (*a*) Bare bead; (*b*) glass-encapsulated bead; (*c*) bead with extension leads.

Thermistor Types

Most manufacturers offer beads in the configurations shown in Fig. 3.19: a bare bead, a glass-encapsulated bead, and an assembly with extension wires. These simple elements may be assembled into the same types of cases as RTDs. Bead thermistors are inherently point sensing devices and may fit very small areas, although most environments require some protection for the fragile bead and lead-wire transitions.

Nearly all temperature-sensing applications employ bead-type thermistors. Others, such as rods, chips, or flakes, mount on circuit boards for temperature compensation or current limiting.

ACKNOWLEDGMENTS

The author wishes to thank David Wilson, Chief Engineer, and David Schurr, Product Manager, Minco, for their assistance.

REFERENCES

1. M. Sapoff, "Thermistors for Resistance Thermometry," Chap. 9, *Transducers: Theory and Applications,* J. A. Alloca and A. Stuart (eds.), Reston, Reston, Va., 1984, pp. 184–208.
2. T. W. Kerlin and R. L. Shepard, *Industrial Temperature Measurement,* Instrument Society of America, Research Triangle Park, N.C., 1982, pp. 3–22, 147–197.
3. J. F. Schooley (ed.), *Temperature: Its Measurement and Control in Science and Industry,* Vol. 5, Part 2, American Institute of Physics, New York, 1982.
4. J. T. Riddle, G. T. Furukawa, and H. H. Plumb, *NBS Monograph 126: Platinum Resistance Thermometry,* National Bureau of Standards, Washington, D.C., 1973.
5. T. J. Quinn, *Temperature,* Academic, New York, 1983.

Chapter **4**

Bimetallic Thermometers

PAUL DAVID AuCOIN
Solar-Weston

THOMAS A. PIZZITOLA
Fluor Engineers

4.1 INTRODUCTION

Bimetallic thermometers have been designed to provide accurate and dependable temperature measurements for industrial and laboratory applications. Rugged all-metal construction, exceptional readability, and proven reliability make the bimetallic thermometer a good selection for local temperature indication. They are available in a variety of sizes, types, and ranges.

4.2 THEORY OF OPERATION

A bimetallic thermometer makes use of two physical phenomena: (1) metal changes volume with a change in temperature, and (2) the coefficient of expansion is not the same for all metals. If two metal strips with different coefficients of expansion are bonded together and heated, the resultant strip will tend to bend toward the side of the metal strip with the lower coefficient of expansion. Deflection is directly proportional to temperature change and length of the strip and is inversely proportional to the thickness of the metal.

4.3 TYPES OF ELEMENTS

The most comonly used elements in the bimetallic thermometers are the flat spiral, single helix, and multiple helix as shown in Fig. 4.1. A favorite combination of metals is the

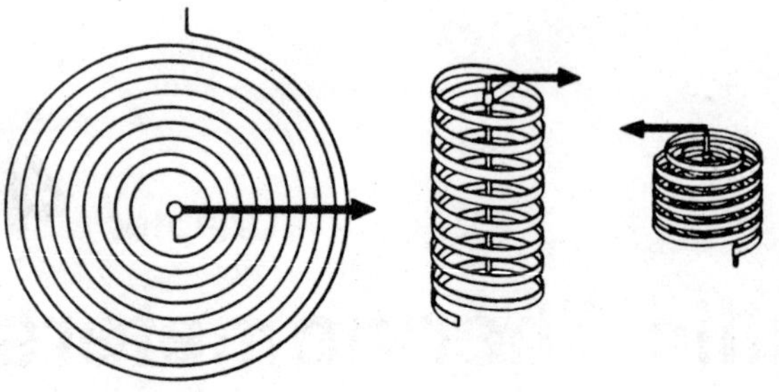

(a) (b) (c)

Fig. 4.1 Principal types of elements used in bimetal thermometers: (*a*) flat spring, (*b*) single helix, (*c*) multiple helix. [*Reprinted with permission from Process Instruments and Controls Handbook, 2d ed., D. M. Considine (ed.), McGraw-Hill, 1974.*]

low-expanding Invar (64%Fe-36%Ni) against a high-expanding nickel-iron alloy with chromium or manganese added. A classical usage of the flat spiral element can be found in the typical low-cost ambient-air temperature thermometer. Most industrial bimetallic thermometers use the single- or multiple-helix elements because of their elongated design which is adaptable to insertion into the bimetallic stem. A single helix moves axially as it winds or unwinds with the temperature change. This requires clearance for a vertical movement of the pointer. The difficulty can be overcome, if desired, by using a multiple element, wound coaxially so as to form coils within coils. This construction is more costly but has an advantage in requiring less immersion depth and fewer parts.

4.4 CONSTRUCTION OF BIMETALLIC THERMOMETERS

Figures 4.2 to 4.4 show the most common design features available by most major manufacturers. Figure 4.5 shows the general construction of a bimetallic thermometer. Most industrial bimetallic thermometers use a helical coil which can

Fig. 4.2 Back-connected bimetallic thermometer—angle form. (*Courtesy of Solar-Weston, Houston, Texas.*)

Fig. 4.3 Bottom-connected bimetallic thermometer—straight form. (*Courtesy of Solar-Weston, Houston, Texas.*)

be designed to fit into a stem more easily than a spiral coil. Stainless steel is the most commonly used material of construction and offers good corrosion resistance. Dial-face diameters are available from 1.0 in (2.54 cm) to 5.0 in (12.7 cm), and stem lengths can vary from as short as 2.0 in (5.08

Fig. 4.4 Universal-angle bimetallic thermometer. The stem can be turned 180° to any desired angle, and the face (dial) may be rotated a full 360°. (*Courtesy of Solar-Weston, Houston, Texas.*)

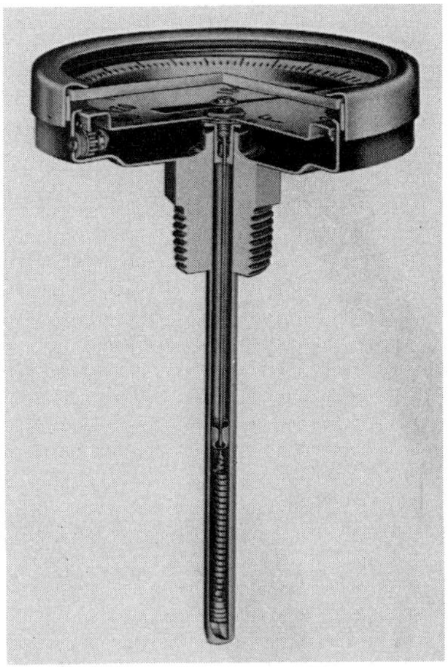

Fig. 4.5 Sectional view of industrial bimetallic thermometer using helical-type bimetallic element. (*Courtesy of Solar-Weston, Houston, Texas.*)

cm) to a maximum of 72.0 in (182.8 cm). Stem mounting in relation to the dial face can be back, bottom, or adjustable (universal) angle as shown in Figs. 4.2 to 4.4.

4.5 CALIBRATION

Calibration of a bimetallic thermometer consists of immersing it in a temperature bath of known temperature and adjusting the length of the bimetallic element to ensure that the pointer will have the proper angular deflection for a given temperature change. This calibration need be done between only two temperatures to achieve linearity over the entire scale.

If the setting of a bimetallic thermometer changes, reset the pointer at only one point on the scale. The thermometer will then be accurate over the entire scale.

If a commercial temperature bath is unavailable, boiling water at 212°F (100°C) or ice at 32°F (0°C) can be utilized so long as the thermometer is provided with an external adjustment provision. (Refer also to ASTM E563, Preparation and Use of Freezing Point Reference Baths.)

4.6 TEMPERATURE RANGES

Bimetallic thermometers are normally available in ranges from −300 to 1000°F (−185 to 538°C) for extreme operating conditions. The practical continuous operating range for a bimetallic thermometer is −100 to 800°F (−73 to 427°C). However, *special* bimetallic thermometers suitable for cryogenic applications and continuous use at 1200°F (649°C) are also available.

4.7 ACCURACY

Industrial bimetallic thermometers will retain accuracy between 0.5 and 1 percent of the scale range at any point on the scale. Stem length selection is critical for accuracy. In liquids, the bimetallic stem should be immersed a total of 2 in (5 cm) into the process medium. In gases, the bimetallic stem should be inserted a total of 4 in (10 cm). This recommendation is especially important with the use of a thermowell.

Most bimetallic thermometers can be calibrated with standards directly traceable to the National Bureau of Standards.

4.8 USE OF THERMOWELLS WITH BIMETALLIC THERMOMETERS

Good piping practice requires the use of a thermowell for installations where pressure may be applied to the bimetallic thermometer stem. Thermowells provide for protection against physical damage as well as corrosive effects of the media flow.

Use of a thermowell also facilitates removal of the thermometer from the line without disturbing the process.

Standard thermowell materials of construction include brass, steel, and stainless steel. Other materials are available. Selection should be based on corrosion-resistance requirements and process pressure.

4.9 SPECIAL APPLICATIONS

Bimetallic thermometers have standard ranges and stem lengths. If the application involves other than standard thermometers (e.g., odd-size connections, liquid-filled dial cases, metric ranges, threads, special temperature ranges, specially marked or calibrated scale plates, odd stem lengths, special materials of construction, finishing, waterproofing, or performance of a military specification), custom thermometers may be obtained from most major manufacturers.

Chapter **5**

Liquid-in-Glass Thermometers*

R. D Thompson†

The liquid-in-glass thermometer consists of a glass envelope, a responsive liquid, and an indicating scale. The envelope is in two parts fused together, a bulb completely filled with the liquid, and a capillary scale section containing the liquid in excess of that required to fill the bulb. The position of the end of the liquid capillary column or index serves, by prior calibration, to indicate the temperature of the bulb. The scale may be marked directly on the capillary tube as in the laboratory or clinical versions, or may be on a separate member mounted alongside the capillary tube as in the domestic and industrial forms.

The liquid-in-glass thermometer reached its zenith as a temperature measuring device near the turn of the twentieth century. Invented over 300 years ago, it came into its own with the international adoption of the Normal Thermometric Scale in 1887. It was used as a standard for the dissemination of the scale from the International Bureau of Weights and Measures to standardizing laboratories throughout the world until the adoption of the International Temperature Scale in 1927. At that time the platinum resistance thermometer replaced the liquid-in-glass thermometer as the primary standard instrument.

Many practical applications have been found for the liquid-in-glass thermometer in addition to its former use as a primary temperature standard. It is employed in scientific work, meteorology, medicine, and industry, and for domestic purposes. It is the most commonly used type of thermometer.

The features that give the liquid-in-glass thermometer an advantage over other thermometers are relatively low cost, simplicity of use, ease of checking for physical damage, and absence of need for auxiliary power. The principal disadvantages are relative fragility, range limitation to 1110°F (600°C), and lack of adaptability to remote reading.

* Reprinted with permission from *Process Instruments and Controls Handbook*, 2d ed., McGraw-Hill, New York, 1974.

† Deceased, formerly Manager, Long Range Planning, Taylor Instrument Co., Process Control Division.

Table 5.1. Temperature Exposure Limits for Various Thermometric Glasses

Glass	Strain point, °F (°C)	Continuous exposure, °F (°C)	Intermittent exposure, °F (°C)
Corning Normal 7560	932 (500)	700 (370)	805* (430)*
Kimble R6	914 (490)	680 (360)	790 (420)
Jena 16 III	923 (495)	690 (365)	795 (425)
Corning Borosilicate 8800	984 (529)	750 (400)	860 (460)
Jena Borosilicate 2954	1018 (548)	790 (420)	900 (480)
Corning 1720	1234 (668)	1005 (540)	1110 (600)
Jena Supremax 2955	1229 (665)	535 (535)	1100 (595)

* Temperature 760°F or 405°C if Corning Standard Thermometer 0041 glass is used for stem.
SOURCE: R. D. Thompson, Recent Developments in Liquid-in-Glass Thermometry, in *Temperature, Its Measurement and Control in Science and Industry*, Vol. 3, Part 1, Van Nostrand Reinhold, New York, 1962, p. 201.

Nearly a hundred years ago cooperative studies were carried out in Germany which involved the scientific glass industry and governmental research laboratories. This work laid the foundations for formulating glasses for thermometers as well as for optical instruments. At the present time there are four basic types of glass used, and the choice is a matter of economics influenced by the range of the thermometer—the higher the range, the higher the cost. Temperature exposure limits as given in Table 5.1 have been established on the basis of thermal stability measurements. Although arbitrary in concept, they may be used as the basis for recommended practice.

Various liquids are used in liquid-in-glass thermometers. Mercury is the choice for higher temperatures or where accuracy is critical. Its advantages are a broad temperature span between its freezing and boiling points, a nearly linear coefficient of expansion, the relative ease of obtaining it in a very pure state, and its nonwetting of glass characteristic. Disadvantages are cost, relatively low coefficient of expansion, and readability characteristics.

For measurements below the freezing point of mercury (ca. −39°C) organic liquids such as toluene, aliphatic hydrocarbons, or organic phosphates are used. Advantages are low freezing points, readability superior to mercury when colored with inert dyes, and lower cost. Disadvantages are lower boiling points, greater tendency to separate in the capillary, and wetting of glass characteristic.

Extension of the range from −39 down to −60°C can also be accomplished by alloying mercury with thallium. The advantages of mercury over organic liquids are not lost, but costs are higher.

Temperature ranges for the common thermometric liquids are given in Table 5.2. The capillary above the liquid is generally filled with a suitable gas. For the better-grade thermometers this is commonly dry nitrogen. Air is often used in the less expensive mercury types as well as in the organic liquid-filled. Electric contact thermometers are generally filled with hydrogen. Maximum registering thermometers are sealed shut in a vacuous condition.

Table 5.2. Temperature Ranges for Common Thermometric Fluids

Fluid	Temperature range, °F (°C)
Mercury	−38–1110 (−39–600)
Mercury alloys	−76–250 (−60–120)
Organic liquids	−328–450 (−200–230)

Styles and designs of liquid-in-glass thermometers are legion, to reflect the multitude of application requirements. Some standardization has been achieved through ASTM, SAMA, and ISO for the laboratory type; through SAMA for the industrial type; and through the U.S. Department of Commerce Voluntary Standards for clinical or fever thermometers.

5.1 LABORATORY THERMOMETERS

Laboratory thermometers are generally built with cylindrical, thin-walled tubing for the bulb fused to the capillary scale tubing. The scale is commonly marked on the scale section after calibration at a number of temperatures which span the range. It may be marked directly on the glass with a stain which is then fired into the glass, or it may be engraved on the surface by an acid etching process. In the latter case the markings are filled with a colored pigment to enhance the readability. Another style, popular in Europe, uses a separate scale, usually of glass, mounted alongside the capillary tube, and the entire scale section is encased in a sealed glass shell. These two styles are illustrated in Fig. 5.1.

Where service conditions are severe, the glass tube may be encased in a metal shell or armor to protect it from physical abuse. This armor may completely shroud the bulb, or the portion near the bulb may be perforated to expose the bulb more effectively to the environment. A section of the armor is cut away to expose the scale section, as shown in Fig. 5.1*c*.

5.2 INDUSTRIAL THERMOMETERS

A variety of thermometers are made in which the tubes are mounted in metal housings. By virtue of their field of application, these are called industrial thermometers. Construction details are shown in Fig. 5.2. As depicted, the scale section is framed in the case opening.

Industrial thermometers are classified by tube construction: i.e., straight or angle; and by case size, i.e., 5, 7, 9, 10, or 12 in. They are also classified as to method of exposure of the bulb to the environment: i.e., with the bulb bare, with the bulb encased in a metal chamber, or with this chamber

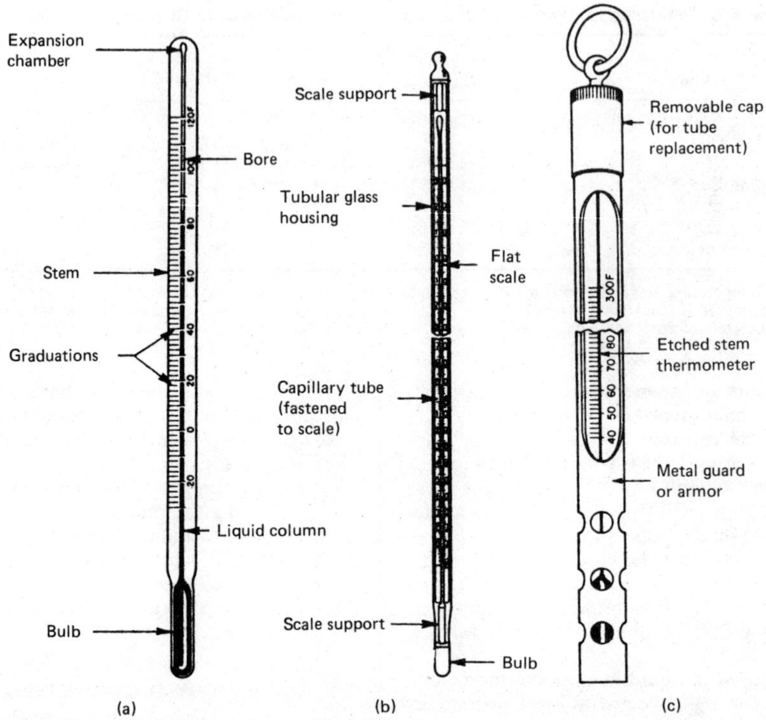

Fig. 5.1 Types of laboratory thermometer. (*a*) Typical common glass thermometer; (*b*) Einchluss, or enclosed-scale, thermometer; (*c*) armored thermometer.

in turn mounted in a well. The metal stem may permit variable immersion at the user's discretion, or it may be provided with a swivel nut, fixed thread, or union connection for fixed-position mounting in the process equipment. The principal dimensional features for these various schemes have been standardized in *SAMA Standard* RC7-1.

The choice of industrial thermometer style is generally governed by the characteristics of the process and the equipment used. Where the process fluid must be confined, insertion in a well is the usual practice, although this automatically adds a time lag.

5.3 PROCESS OR SEMIINDUSTRIAL THERMOMETERS

The separate tube and scale feature of industrial thermometers is used in another class of instruments, commonly called *process* or *semiindustrial thermometers*. Usually the bulb is bare, but it may be protected from mechanical damage by a guard. Typical uses are candy or jelly making, ham boiling, and liquor testing. A variation on this theme involves mounting a cup instead of a guard around the bulb. These so-called cup-case or tank thermometers are used in determining the temperature in storage tanks or vats. The liquid in the cup insulates the bulb so that meaningful readings may be obtained when the temperature of the tank is materially different from ambient.

5.4 SPECIAL DESIGNS OF THERMOMETERS

A variety of special designs of thermometers have evolved over the years. These include maximum registering (used mainly in medicine and meteorology), minimum and maximum-minimum or Sixes type (used in meteorology), sling psychrometer (used in air conditioning and meteorology), Beckmann or differential (used in calorimetry), and electric contact (used for alarm, indication, or control purposes).

5.5 CHARACTERISTICS IN USE

Liquid-in-glass thermometers are designed and calibrated for one of three conditions of use: (1) the bulb and liquid-containing part of the capillary exposed to the temperature to be measured, called *total immersion*, (2) the entire thermometer exposed to these conditions, called *complete immersion*, and (3) the bulb exposed to the temperature to be measured and the scale section exposed to ambient conditions, called *partial immersion*. These conditions are illustrated in Fig. 5.3. If a thermometer calibrated for one set of conditions is used under a different set, corrections for this variance may be determined and applied to the readings. In some cases this correction may amount to a number of degrees and can be very significant, even in work of only moderate accuracy. A very complete discussion of the subject is given in Sec. 5.4 of *NBS Monograph* 90, Calibration of Liquid-in-Glass Thermometers.

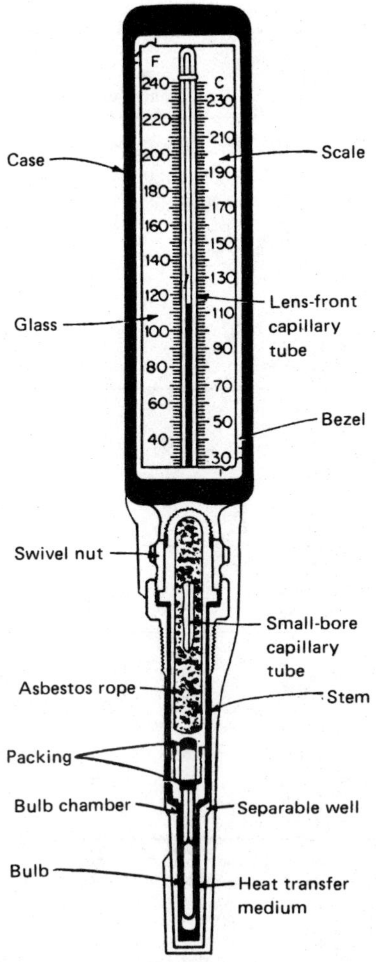

Fig. 5.2 Typical industrial thermometer.

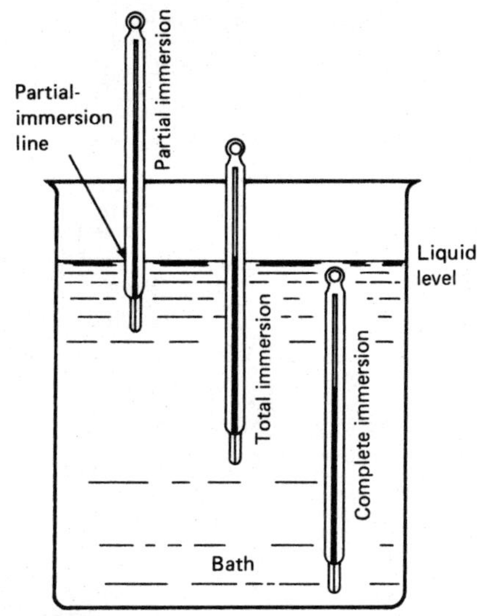

Fig. 5.3 Conditions of use which affect design and calibration of liquid-in-glass thermometers.

Range, sensitivity, and accuracy are interrelated. The range limits for liquid-in-glass thermometers are given in Table 5.2. Sensitivity is a function of the relative volumes of bulb and capillary. The volume of the bulb is limited by the practical considerations of weight of the mercury and strength of the glass. The capillary diameter, on the other hand, is limited by the techniques of tubing production. The net result is that sensitivities rarely smaller than 0.01 to 0.02° are experienced in practice.

Accuracy is influenced by range and sensitivity but is also affected by factors such as thermometer type, quality of work, temperature, and nature of the environment. As with any instrument, accuracy is improved if the thermometer is carefully calibrated and the measured corrections are applied in use. When the highest accuracy is sought, the laboratory type

Table 5.3. Tolerances for Celsius Total-Immersion Mercury Thermometers

Temperature range, °C	Graduation interval, degrees	Tolerance, degrees	Accuracy, degrees	Corrections stated to
Thermometers Graduated under 150°C				
0–150	1.0 or 0.5	0.5	0.1–0.2	0.1
0–150	0.2	0.4	0.02–0.05	0.02
0–100	0.1	0.3	0.01–0.03	0.01
Thermometers Graduated under 300°C				
0–100	1.0 or 0.5	0.5	0.1–0.2	0.1
+100–300		1.0	0.2–0.3	0.1
0–100	0.2	0.4	0.02–0.05	0.02
+100–200		0.5	0.05–0.1	0.02
Thermometers Graduated above 300°C				
0–300	2.0	2.0	0.2–0.5	0.2
+300–500		4.0	0.5–1.0	0.2
0–300	1.0 or 0.5	2.0	0.1–0.5	0.1
+300–500		4.0	0.2–0.5	0.1

Table 5.4. Tolerances for Fahrenheit Total-Immersion Mercury Thermometers

Temperature range, °F	Graduation interval, degrees	Tolerance, degrees	Accuracy, degrees	Corrections stated to
Thermometers Graduated under 300°F				
32–300	2.0	1.0	0.2–0.5	0.2
32–300	1.0 or 0.5	1.0	0.1–0.2	0.1
32–212	0.2 or 0.1	0.5	0.02–0.05	0.02
Thermometers Graduated under 600°F				
32–212	2 or 1	1.0	0.2–0.5	0.2
+212–600		2.0	0.5	0.2
Thermometers Graduated above 600°F				
32–600	5	4.0	0.5–1.0	0.5
+600–950		7.0	1–2.0	0.5
32–600	2 or 1	3.0	0.2–1.0	0.2
+600–950		6.0	0.5–1.0	0.2

Table 5.5. Tolerances for Celsius Partial-Immersion Mercury Thermometers

Temperature range, °C	Graduation interval, degrees*	Tolerance, degrees	Accuracy, degrees†	Corrections stated to
Thermometers not Graduated above 150°C				
0–100	1.0 or 0.5	1.0	0.1–0.3	0.1
0–150	1.0 or 0.5	1.0	0.1–0.5	0.1
Thermometers not Graduated above 300°C				
0–100	1.0	1.0	0.1–0.3	0.1
+100–300	1.0	1.5	0.5–1.0	0.2
Thermometers Graduated above 300°C				
0–300	2.0 or 1.0	2.5	0.5–1.0	0.5
+300–500		5.0	1.0–2.0	0.5

* Partial-immersion thermometers are sometimes graduated in smaller intervals than shown in this table, but this in no way improves the performance of the thermometers, and the listed tolerances and accuracies still apply.

† The accuracies shown are attainable only if emergent stem temperatures are closely known and accounted for.

Table 5.6. Tolerances for Fahrenheit Partial-Immersion Mercury Thermometers

Temperature range, °F	Graduation interval, degrees*	Tolerance, degrees	Accuracy, degrees†	Corrections stated to
Thermometers not Graduated above 300°F				
32–212	2.0 or 1.0	2.0	0.2–0.5	0.2
32–300	2.0 or 1.0	2.0	0.2–1.0	0.2
Thermometers not Graduated above 600°F				
32–212	2.0 or 1.0	2.0	0.2–0.5	0.2
+212–600	2.0 or 1.0	3.0	1.0–2.0	0.5
Thermometers Graduated above 600°F				
32–600	5.0 or 2.0	5.0	1.0–2.0	1.0
+600–950		10.0	2.0–3.0	1.0

* Partial-immersion thermometers are sometimes graduated in smaller intervals than shown in this table, but this in no way improves the performance of the thermometers, and the listed tolerances and accuracies still apply.

† The accuracies shown are attainable only if emergent stem temperatures are closely known and accounted for.

is generally used. The National Bureau of Standards has published tolerances for various ranges and subdivisions to be anticipated in thermometers from quality manufacturers. These data given in Tables 5.3 to 5.6 are reproduced from *NBS Monograph* 90, Calibration of Liquid-in-Glass Thermometers. The tables also list the accuracies attainable under optimum conditions and with the application of the corrections from calibration.

Two characteristics of glass affect the indications of liquid-in-glass thermometers and should be recognized if the ultimate in accuracy is to be attained. The first is a memory effect in which the enlargement of the bulb at a higher temperature persists to some degree when the thermometer is subsequently used at a lower temperature. This effect, called the *temporary depression of the ice point*, is a function of the type of glass. In actual use it may be compensated for by a preconditioning of the thermometer prior to checking for bulb stability. The second effect is experienced at higher temperatures and is a result of aging or secular change in the glass. It is manifest by a contraction of the bulb. The magnitude of this aging change is determined by recalibration at a fixed temperature such as the ice point.

Pressure effects are generally not significant except where fractional degree accuracy is desired. The pressure coefficient of the thermometer can be readily determined by observing the change in reading at constant temperature for a known change in external pressure.

In measurement of temperatures which are changing the lag of the instrument system may be very significant. The lag constant, usually taken as the time to respond through 63 percent of a temperature change, is commonly measured by bringing the thermometer to equilibrium in a well stirred bath at one temperature, transferring it to another bath at a different temperature, and measuring the time for the indication to change by 63 percent of the difference in bath temperatures.

Chapter **6**

Filled-System Thermometers*

G. ROBERT LEAVITT†

Filled-system thermometers are used to provide a remote signal or indication of the temperature of a material for purposes of measurement and/or control. Filled-system thermometers appeared as a method of measurement following the invention of the bourdon spring which is still widely used as an element for converting a pressure or volume change to a usable motion in industrial measurement and control loops. Any reference to *bourdon* in this chapter is intended to also include other measuring elements, such as diaphragms and bellows.

* Reprinted (with adaptation) with permission from D. M. Considine (ed.), *Process Instruments and Controls Handbook,* 3d ed., McGraw-Hill, New York, 1985.

† Formerly Manager, Quality Assurance, Taylor Instrument Co., Rochester, New York.

6.1 PRINCIPLE OF OPERATION

The filled-system thermometer is designed to provide an indication or record of temperature at some distance from the point of measurement. (See Fig. 6.1.) The system usually consists of an element sensitive to temperature (bulb), an element sensitive to pressure or volume changes (bourdon, bellows, or diaphragm), a means of connecting these elements (capillary tubing), and a device for indicating or recording a signal related to the measured temperature. In exceptional cases, the capillary or even the bourdon may be designed to be the temperature-sensitive element, thus eliminating the use of a capillary and/or sensor.

The temperature sensing element contains a fluid which changes its volume or pressure with temperature. The pressure-sensitive element (bourdon) responds to these changes

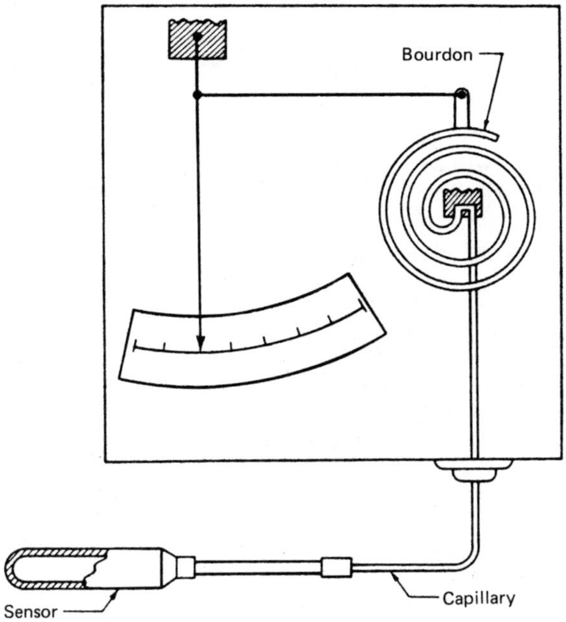

Fig. 6.1 Filled-system thermometer.

by delivering a motion or force to a device which transduces the signal to a usable form. This is commonly a mechanical linkage which drives a pointer or pen, but may be a pneumatic or an electrical device which transmits the temperature signal over long distances. These temperature signals frequently are used for process control purposes.

6.2 RELATIVE ADVANTAGES AND LIMITATIONS

Advantages

The wide use of filled-system thermometers in the process industries may be explained by the following reasons:

1. Fundamental simplicity of the system allows rugged construction, minimizing the possibility of damage or failure in shipment, installation, and use. The amount of upkeep is generally minor.
2. Simplicity of the system allows inexpensive design.
3. As used in the process industries, sensitivity, response time, and accuracy are generally the equal of any other temperature measuring instruments available.
4. The capillary allows considerable separation between the point of measurement and the point of indication. Although capillary lengths of 400 ft (120 m) have been used successfully, it is usually more economical to employ transducers for signal transmission of 100 ft (30 m) or more.
5. The measuring system is self-contained. It does not need auxiliary power unless it is combined with a pneumatic or an electric transmission system.
6. The system can be designed to deliver significant power if necessary to drive indicating or controller mechanisms, including valves.

Limitations

Thermal systems are limited as follows:

1. The bulb size may be too large to fit the available space.
2. The performance characteristics vary considerably with the type of filling fluid, and the user must be certain not to misapply a particular type of system.
3. The maximum temperature is more limited than that in some electrical measuring systems.
4. In case of system failure, the entire unit must be replaced or repaired.
5. Separation of sensing and indicating elements may be limited, depending on other characteristics, such as filling liquid and accuracy requirements.

6.3 CLASSIFICATION OF FILLED-SYSTEM THERMOMETERS

Filled-system thermometers may be separated into two fundamental types: those in which the bourdon responds to (1) volume changes and (2) pressure changes. Those that respond to volume changes are completely filled with a liquid. The liquid expansivity with temperature is greater than that of the sensor metal, the net volume change being communicated to the bourdon. An internal-system pressure change is always associated with the bourdon volume change, but this effect is not of primary importance. The systems that respond to pressure changes are either filled with a gas or partially filled with a volatile liquid. Changes in gas or vapor pressure with changes in sensor temperature are communicated to the bourdon. The bourdon will increase in volume with increase in pressure, but this effect is not of primary importance.

Filled-system thermometers have been classified as follows:

Volumetric Principle	
Liquid-filled (other than Hg)	Class I
Mercury-filled	Class V
Pressure Principle	
Vapor-filled	Class II
Gas filled	Class III

This classification will be used throughout this chapter.

6.4 LIQUID-FILLED THERMAL SYSTEMS

These systems operate on the principle that the temperature signal is generated by and proportional to a volume change in the sensor. They often are referred to as *liquid-expansion* types. They are divided into subclasses to designate the means of providing temperature compensation:

1. *SAMA Class IA* (*Full Compensation*): The compensating means is a second thermal system minus a sensor, or an equivalent means of compensation. (See Fig. 6.2.)
2. *SAMA Class IB* (*Case Compensation*): The compen-

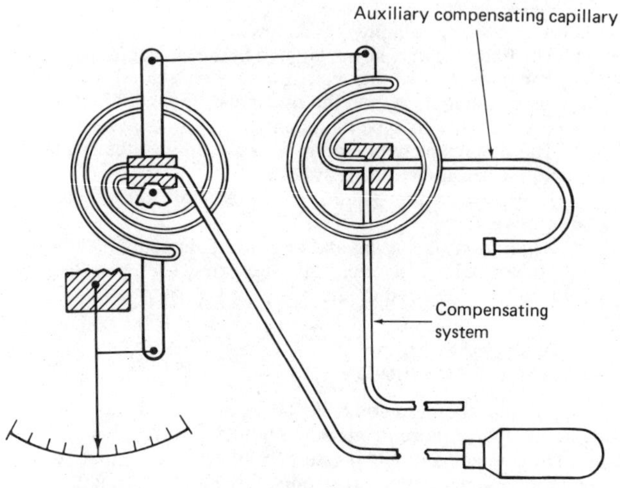

Fig. 6.2 Fully compensated liquid-, mercury-, or gas-filled thermal system (Class IA, IIIA, or VA).

sating means is a thermally responsive system minus the sensor and connecting capillary, or an equivalent means of compensation. (See Fig. 6.3.)

General Design Considerations

A manufacturer of liquid-filled systems generally will provide a single bourdon design to cover all possible temperature spans. The sensors of all Class I or Class V thermal systems from one manufacturer will deliver the same volume change to the bourdon, regardless of range span. Sensor internal volume, therefore, will vary with span, in accordance with Eq. (6.1):

$$\Delta V = V_b \phi R = \text{constant} \tag{6.1}$$

where ΔV = bourdon volume change, cm^3
ϕ = differential expansivity between filling liquid and enclosing sensor, $cm^3/(cm^3)$ (°C)
V_b = sensor volume, cm^3
R = range span, °C

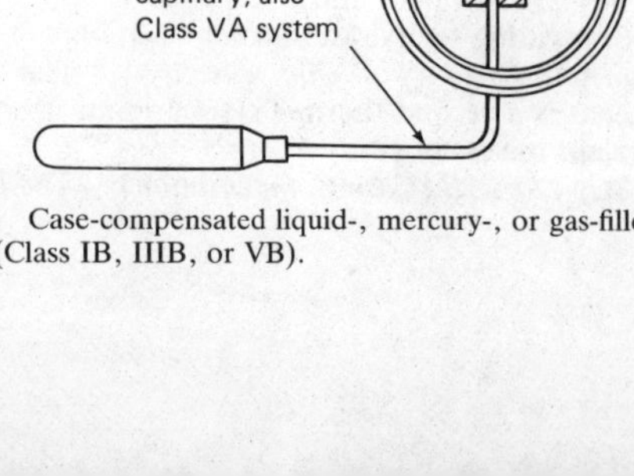

Fig. 6.3 Case-compensated liquid-, mercury-, or gas-filled thermal system (Class IB, IIIB, or VB).

Rearranging Eq. (6.1) yields

$$V_b = \frac{K}{\phi R} = \frac{K'}{R} \tag{6.2}$$

where K and K' are constants. Equation 2 shows that the sensor volume is small for large-range spans, and vice versa. The ΔV utilized by different manufacturers varies considerably but is generally between 0.04 and 0.15 cm^3.

$\phi = 10.8 \times 10^{-4}\ cm^3/(cm^3)$ (°C) for inorganic liquids

$\phi = 13.4 \times 10^{-5}\ cm^3/(cm^3)$ (°C) for mercury

Therefore:

$$V_b = \frac{\Delta V}{\phi R} = \frac{0.04 \text{ to } 0.15}{10.8 \times 10^{-4} R} = \frac{65 \text{ to } 250}{R} \text{ cm}^3 \quad \text{for organic liquids} \tag{6.3}$$

$$V_b = \frac{\Delta V}{\phi R} = \frac{0.04 \text{ to } 0.10}{13.4 \times 10^{-5} R} = \frac{550 \text{ to } 1400}{R} \text{ cm}^3 \quad \text{for mercury} \tag{6.4}$$

Maximum and Minimum Temperatures and Range Spans

The minimum temperature is limited to the freezing point of the organic liquid employed as a filling medium. This is usually between −103 and −328°F (−75 and −200°C), depending on the liquid used.

The maximum temperature to which the liquid is stable is approximately 572°F (300°C). The minimum range span is determined by the maximum bulb size that is practical, generally 22 to 45°F (12 to 25°C), although spans as short as 9°F (5°C) are used. The maximum span sometimes is limited by nonlinearity of expansivity and compressibility of the fluid fill to approximately 270°F (150°), so that standard accuracy may be met with linear dials and charts. Some liquids, however, deliver a linear signal for spans as large as 450°F (250°C). Mercury-filled thermal systems may be used between −39 and +1202°F (−38 and +650°C). Organic liquids freeze at a much lower temperature and are commonly used between −103 and +572°F (−75 and +300°C). Because of the higher expansivity, organic-filled systems are more adaptable to short spans, while the lower compressibility of mercury makes it easier to use on long spans. In either case, the minimum span usually is limited by the largest practical bulb size. (See Fig. 6.4.)

Capillary and Case Compensation

The nonmetallic liquid systems are particularly vulnerable to errors resulting from capillary temperature changes because of the high expansivity of the fill. Only if the capillary volume is very small can these errors be ignored and a Class IB system used. (See Fig. 6.3.)

To neutralize the capillary temperature error an auxiliary thermal system is employed. (See Fig. 6.2.)

The mercury-filled system may be compensated fully by an alternative method. This employs a capillary containing low controlled expansion nickel steel wire which can, by proper selection of dimensions, cancel any net volume changes due to the expansion of the mercury. It should also

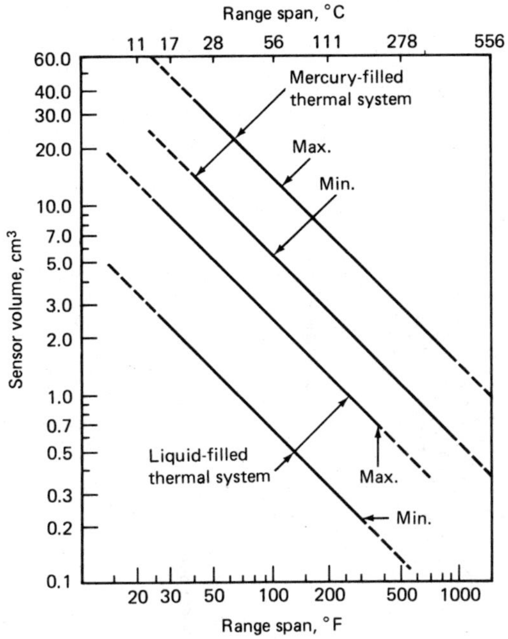

Fig. 6.4 Sensor volume versus range span, Class I and Class V systems. (Maximum and minimum values show extremes among manufacturers.) (Conversion factor: 1.0 in³ = 16.39 cm³.)

be noted that, because of the low expansivity of mercury, Class VB systems can have a capillary up to 50 ft (15 m) with good performance.

A VA system using a compensated capillary, an IB, or a VB often is preferred to a Class IA because of the relative simplicity of construction, lesser vulnerability to abuse, and lower cost.

Since the bourdon volume changes with measured temperature changes, the case compensation can be perfect at only one on scale temperature. In practice, this compensation is adjusted to a specified tolerance with the temperature at midrange. This results in the thermometer being slightly overcompensated at range bottom and slightly undercompensated at range top. Attention must be paid to these details if precise measurements are desired.

Speed of Response

The speed of response of liquid-filled thermometers ranges from 1 to 10 s for 63 percent responses of a bare sensor in well-stirred water. (See Fig. 6.5.) The effect of capillary resistance on speed of response is generally negligible. The low thermal conductivity of the organic fills is somewhat balanced by their smaller bulb size, but for a given diameter mercury systems are faster. To improve response in poor heat-transfer mediums, such as air, sensors often are manufactured from lengths of capillary tubing which may be installed in straight lengths or preformed into relatively tight coils. (See Fig. 6.6.)

Barometric Errors

This effect is negligible in Class I and Class V systems.

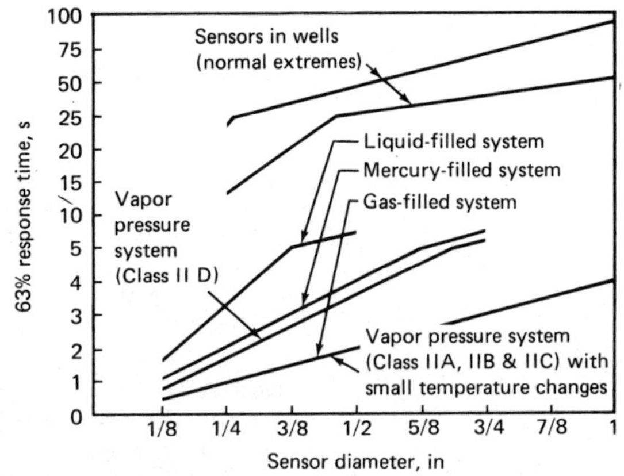

Fig. 6.5 Sensor response versus sensor outside diameter in well-stirred water. (*Note:* ⅛ in = 3.2 mm; ¼ in = 6.4 mm; ⅜ in = 9.5 mm; ½ in = 12.7 mm; ⅝ in = 15.9 mm; ¾ in = 19.1 mm; ⅞ in = 22.2 mm.)

Bulb Elevation Error

Nonmetallic liquid-filled systems are not affected by sensor elevation or depression (relative to the case). There is an effect for mercury-filled systems. If the installed height changes, the instrument should be rezeroed. Functional limits usually are ±30 ft (9 m) unless specially built to permit greater differences in height between bulb and case.

Overrange Protection

Maximum overrange temperature is defined as the maximum temperature to which the bulb of a filled system may be exposed indefinitely without damage to the system. It is usually expressed in percent of span above the upper-scale value.

Classes IB, VA, and VB systems generally are provided with overrange protection of 100 percent of range span, sometimes higher, depending on the design of the bourdon spring or an equivalent element.

On fully compensated systems using two bourdons and a double capillary, especially Class IA, the bourdon expansion capacity is taken up by the potential expansion of liquid in the capillary, and less overrange protection is available.

6.5 VAPOR-PRESSURE THERMAL SYSTEM

This system is defined by SAMA as "a thermal system partially filled with a volatile liquid and operating on the principle of vapor pressure." Four types are employed:

1. *Class IIA*: Designed to operate with the measured temperature above the temperature of the rest of the thermal system. (See Fig. 6.7.)

2. *Class IIB*: Designed to operate with the measured temperature below the temperature of the rest of the thermal system. (See Fig. 6.8.)

3. *Class IIC*: Designed to operate with the measured temperature above and below the temperature of the rest of the thermal system. This type normally requires a larger sensitive portion than Class IIA or IIB. (See Fig. 6.9.)

4. *Class IID*: Designed to operate with the bulb temperature above, below, and at the temperature of the rest

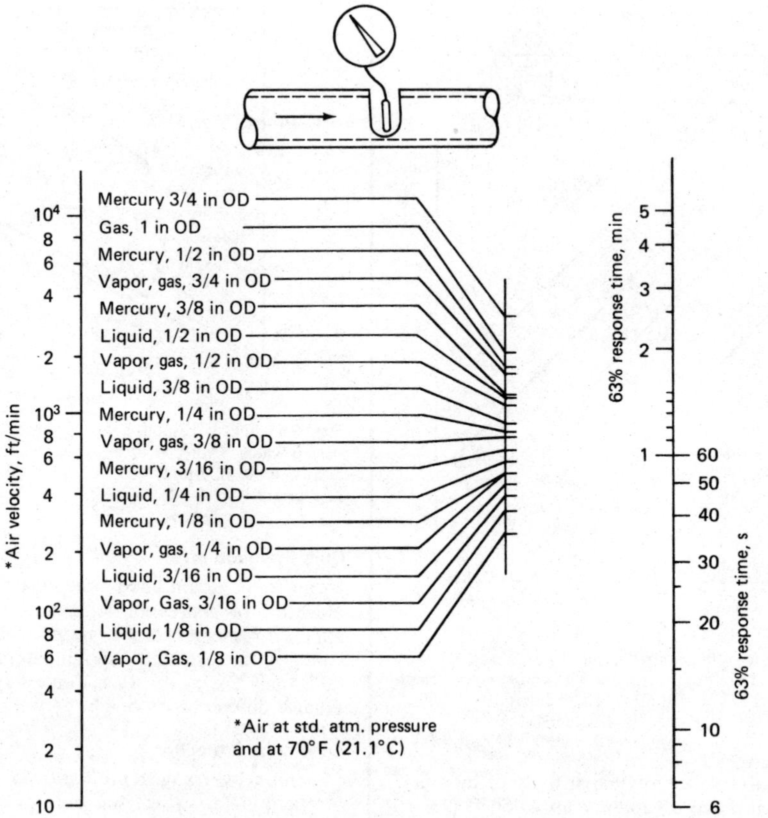

Fig. 6.6 Thermal system response rate in air at various air velocities. (*Note:* 1 ft/min = 0.3 m/min; ⅛ in = 3.2 mm; $\frac{3}{16}$ in = 4.8 mm; ¼ in = 6.4 mm; ⅜ in = 9.5 mm; ½ in = 12.7 mm; ¾ in = 19.1 mm; 1 in = 25.4 mm.)

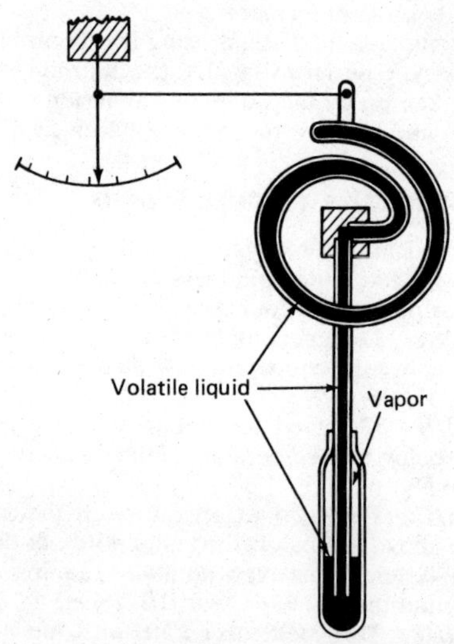

Fig. 6.7 Vapor-pressure thermal system (Class IIA).

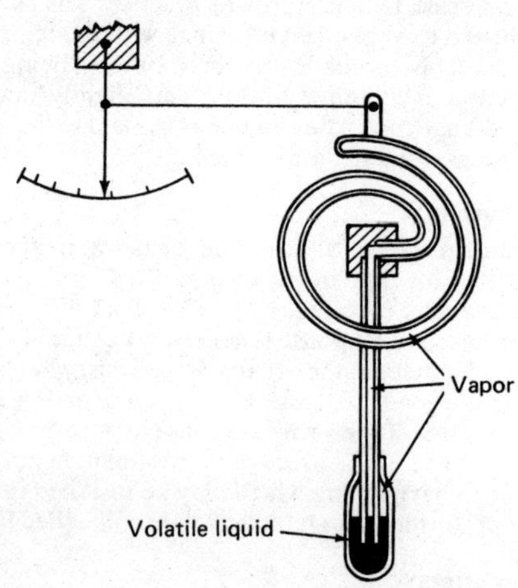

Fig. 6.8 Vapor-pressure thermal system (Class IIB).

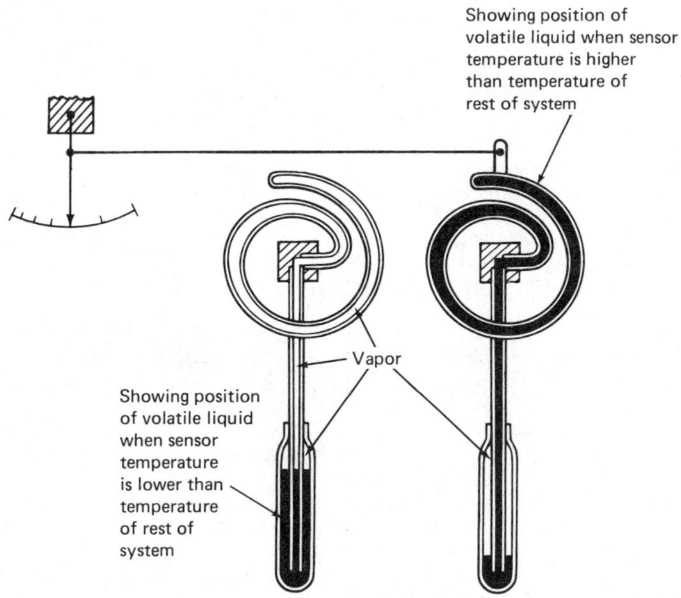

Fig. 6.9 Vapor-pressure thermal system (Class IIC).

of the thermal system. (See Fig. 6.10.) In this type, the volatile liquid is confined to the sensitive portion and a second relatively nonvolatile liquid is used to transmit the vapor pressure to the expansible device.

General Design Considerations

The pressure within the system is the vapor pressure at the interface of the liquid and vapor. It is required that the system be filled so that this interface will always exist within

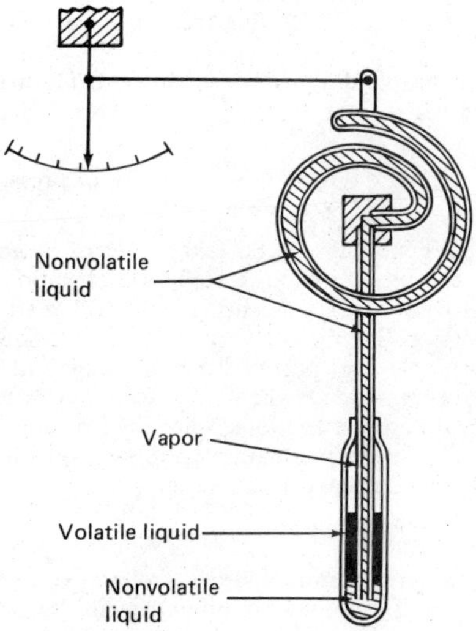

Fig. 6.10 Vapor-pressure thermal system (Class IID).

the sensor. The fill will always be in a liquid state at the coolest part of the system. This combination of requirements and circumstances is responsible for the four basic designs just defined, each best for particular conditions of use.

The Class IIA system, the most common type of vapor-filled thermal system, is designed for the condition that the bulb temperature is always above the temperature of the rest of the system. The bourdon, capillary, and part of the sensor, therefore, always will be liquid-filled. The sensor must be large enough (1) to accommodate the volume change of the fill within the bourdon and capillary due to ambient temperature changes, (2) to accommodate the volume change with temperature of item 1 within the sensor, (3) to accommodate the volume change of the bourdon across the range, and (4) to provide some minimum liquid and vapor space within the sensor. A larger sensor, therefore, may be required for long systems (≧50 to 75 ft) (≧15 to 23 m).

The Class IIB system may have the smallest sensor, since the design does not need to accommodate the effect of items 1 and 3 just given. (See Fig. 6.8.)

The Class IIC system may operate above or below the ambient temperature of the case and capillary, as shown in Fig. 6.9. The bulb volume must accommodate, in addition to the requirements of the Class IIA system, the complete volume of the bourdon and capillary. This system will not give a reliable temperature indication during transfer of the liquid into or out of the capillary and bourdon, which occurs when the sensor temperature crosses the ambient temperature of the bourdon and capillary.

The Class IID system is specifically designed to measure temperature accurately at the ambient temperature of the bourdon or capillary as well as at other temperatures. The actuating volatile liquid is confined to the sensor by a transmitting liquid of low vapor pressure which fills the capillary and bourdon. The means of confinement or trap is generally designed to be large so that actuating liquid getting into the trap cannot enter the capillary under any conditions of use.

The sensor size required is generally somewhat larger than that of Class IIA, since it must accommodate this larger trap and additional volume of transmitting liquid. (See Fig. 6.10.)

The vapor-system indication is a nonlinear function of bulb temperature, the vapor pressure increasing at an increasing rate with rise in the absolute temperature. This is an advantage where the user desires more reading sensitivity toward the top of the range. At least one manufacturer incorporates a varying linkage ratio in order to linearize the pen or pointer motion.

It is desired that the minimum vapor-pressure change across the range be 100 lb/in^2 to minimize barometric pressure effects and to provide adequate bourdon energy to overcome frictional effects. Therefore, various actuating liquids are chosen for different range spans and different temperature levels.

Maximum and Minimum Temperatures

The maximum temperature is limited by the critical point of the liquid employed and by the tendency of most known organic liquids to change chemically at 600°F (316°C) or higher. The minimum temperature is generally limited to approximately −40°F (−40°C) because of loss in reading sensitivity at lower temperatures coupled with the requirement that the bourdon must be able to withstand the vapor pressure of the liquid fill at room (or possibly shipping) temperatures. The minimum scale should be that produced when the fill fluid has a vapor pressure of at least 1.0 atm absolute.

Capillary and Case Compensation

The capillary of Class II systems is insensitive to temperature changes. It is necessary, however, that the capillary temperature of Class IIA systems, when the capillary and bourdon are filled with actuating liquid, not be at a point that exceeds the sensor temperature. The internally contained fluid or vapor within the bourdon is also insensitive to temperature changes qualified as above. However, the modulus of elasticity will decrease with increasing temperature, causing a shift of only 1½ to 2 percent for a 100°F (56°C) case temperature change. In most uses, this effect is not important and no compensation is provided.

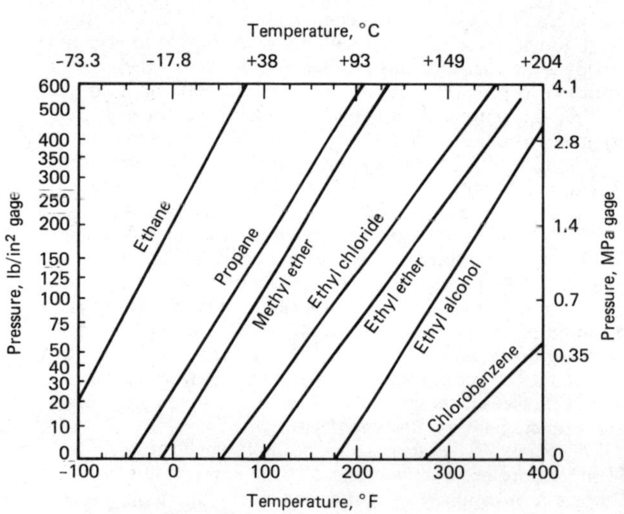

Fig. 6.11 Vapor-pressure-versus-temperature curves for various thermal fills.

Speed of Response

The 63 percent response of Class II systems also varies from 1 to 10 s with the sensor in well-stirred water and a short capillary length. The characteristics affecting response include capillary length and conductivity, viscosity of fluid in the capillary, the size of the system relative to the available heat (especially with Class IIC where reversal occurs), and the heat transfer internally between the bulb and the filling liquid.

Class IID systems are generally slowest because of the sensor size and the viscosity of the fluid.

Barometric Errors

Class II systems are sensitive to barometric changes. They are generally designed to have a minimum of 100 lb/in^2 (690 kPa) across the range so that errors will be kept to ¼ percent of the range.

Bulb Elevation Error

If the bourdon is above the sensor, the pressure within the bourdon equals the vapor pressure in the sensor minus the liquid pressure head along the capillary, or

$$E = \frac{100P_h}{\Delta P}$$

where E = elevation error, % of pressure span
ΔP = thermal fill vapor pressure
P_h = liquid pressure head along capillary, lb/in^2

In the following example, assume:

1. Range is 80 to 200°F. See methyl ether curve of Fig. 6.11.
2. Case is installed above sensor by 50 ft (15 m).

Elevation error will be

$$E = 0.43 \times 50 \times 0.658 \times \frac{100}{329} = 4.3\% \text{ of pressure span}$$

This effect generally can be taken care of by the manufacturer if the user provides installation elevations for the sensor and case so that the instrument will read properly after installation. The Class IIC system will have liquid within the capillary only over part of the range span and is, therefore, not recommended when the sensor and case are at appreciably different elevations, since the instrument cannot read correctly for both situations, i.e., sensor temperature above or below capillary and case temperature ambients.

Overrange Protection

In vapor pressure thermal systems, overrange protection is generally more limited than in other systems because of the increasing rate of vapor-pressure rise with temperature rise. A specific overrange temperature for each range offered should be obtained from the manufacturer. This is generally appreciably less than 100 percent of the range span above the range top. Under some limited conditions, it is possible to fill the system so that all the liquid will be exhausted from the sensor at a sensor temperature above the instrument range, in which case the safe overrange temperature can be increased.

6.6 GAS-FILLED THERMAL SYSTEM: CLASS III

This system is defined by SAMA as "a thermal system filled with a gas and operating on the principle of pressure change with temperature change." The system is usually compensated for ambient temperature effects in one of two ways:

1. *Class IIIA*: With a second thermally responsive system minus the sensor, or an equivalent means of compensation. (See Fig. 6.2.)
2. *Class IIIB*: With compensating means within the case only. (See Fig. 6.3.)

General Design Considerations

The gas-filled system operates on the principle of Charles' law, which states that the absolute pressure of a confined gas is proportional to the absolute temperature. This law is approximated by keeping the capillary and bourdon volume small compared with the bulb volume, thereby preventing an appreciable percentage of the gas mass within the bulb from flowing out as the bulb temperature is increased across the range.

Figure 6.12 shows the system operation in which it is assumed that Charles' law is exactly followed. The manufacturer finds it expedient to keep the angle change across the range a_T–a_B (or a corresponding system pressure change across the range P_T–P_B) constant for all range spans. Curve A for the 0 to 200°F (−18 to 93°C) span is established by filling the system to a particular pressure P_F or angle a_F with the sensor at a particular temperature [say, 32°F (0°C)]. The slope of curve B for a 0 to 400°F (−18 to 204°C) span will

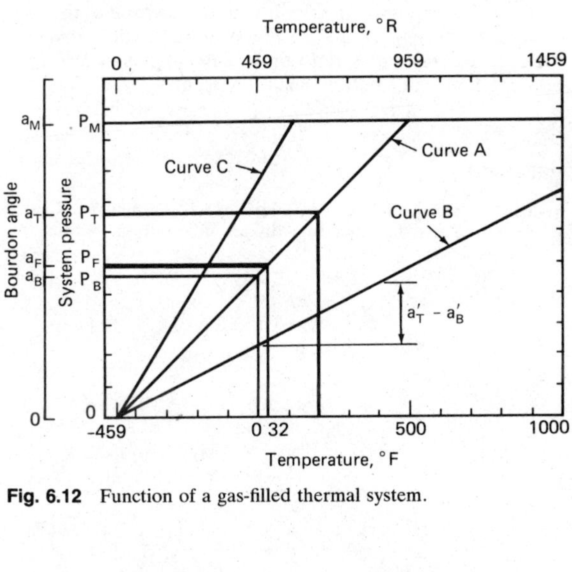

Fig. 6.12 Function of a gas-filled thermal system.

accordingly be one-half that of a 0 to 200°F (93°C) span, i.e., a'_T–a'_B = a_T–a_B. The system may be expressed mathematically by Eq. (6.5):

$$\overbrace{\frac{V_bP_1}{T_{b1}} + \frac{V_cP_1}{T_{c1}} + \frac{V_BP_1}{T_{B1}}}^{\text{Condition 1}} = \overbrace{\frac{V_bP_2}{T_{b2}} + \frac{V_cP_2}{T_{c2}} + \frac{V_BP_2}{T_{B2}}}^{\text{Condition 2}} \tag{6.5}$$

where V_b = volume of bulb, cm³
V_c = volume of capillary, cm³
V_B = volume of bourdon, cm³
T_{b1} = absolute temperature of sensor (condition 1)
T_{b2} = absolute temperature of sensor (condition 2)
T_{c1} = absolute temperature of capillary (condition 1)
T_{c2} = absolute temperature of capillary (condition 2)
T_{B1} = absolute temperature of bourdon (condition 1)
T_{B2} = absolute temperature of bourdon (condition 2)
P_1 = absolute system pressure (condition 1)
P_2 = absolute system pressure (condition 2)

System linearity and sensor and case temperature errors may be calculated from Eq. (6.5). This equation does not include the effect on modulus change of the bourdon. The theory previously described for Class II systems may be utilized to calculate this effect.

Maximum and Minimum Temperature and Range Spans

Gas thermal systems are able to cover the widest range of temperature of any of the filled systems. They are usually limited on the low side by the critical temperature of the gas used, and on the high side by the bulb materials (commonly 5 and 925 K). The maximum span is limited only by the above conditions of use and the nonlinearity due to mass flow from the bulb. The minimum span is limited by the pressure at which the bourdon becomes overstressed.

The gas system lends itself to use with a transducer with biasing springs, making many more ranges, especially with short spans (25 K), available.

Capillary and Case Compensation

Equation (6.6), describing capillary temperature error in Class III systems, is derived from Eq. (6.5),

$$E = \frac{100V_c\,\Delta T_c\,T_b^2}{(V_bT_c + V_cT_b)T_cR} \tag{6.6}$$

where T_c = mean capillary temperature, °R
T_b = absolute temperature of sensor, °R
ΔT_c = capillary temperature change, °F
R = span, °F

It can be observed from the equation that the capillary error increases rapidly as the sensor temperature increases and can be reduced by increasing the sensor volume and/or the span.

Speed of Response

The speed of response of Class III systems is generally fast because the ratio of bulb mass to external surface is more favorable than in other system types. This would be especially true in low-conducting media. (See Fig. 6.6.) The 63

Table 6.1 Comparison of Thermal Systems*

Operating principle	Liquid expansion		Pressure change	
Type	Liquid-filled	Mercury-filled	Vapor pressure	Gas-filled
Class	Class IA and IB	Class VA and VB	Class IIA, B, C, D	Class IIIA and IIIB
Low temperature limit	−125°F (−87°C)	−38°F (−39°C)	−40°F (−40°C)	−450°F (−268°C)
High temperature limit	600°F (316°C)	1200°F (650°C)	600°F (316°F)	1200°F (649°C)
Longest span	350°F (194°C)	1200°F (667°C)	300°F (167°C)	1200°F (667°C)
Shortest span	25°F (14°C)	40°F (22°C)	80°F (44°C)	150°F (83°C)
Typical bulb size, diameter and length, in	⅜ × 5	11⁄16 × 4	11⁄16 × 4	11⁄16 × 4
Dial or chart divisions	Equal	Equal	Nonlinear, larger at range top	Equal
Maximum standard capillary length, ft	200, Class IA 15, Class IB	200, Class VA 50, Class VB	200	200
Capillary temperature compensation	Dual bourdon, Class IA None, Class IB	Compensated capillary, Class VA None, Class VB	None necessary	Generally none
Case temperature compensation	Dual bourdon, Class IA Bimetal strip, Class IB	Bimetal strip or core-type compensator	Usually none	Bimetal strip
Sensor elevation error	Negligible	Generally small	Often large	Negligible
Overrange capability in % of range	0–200%, Class IA 100%, Class IB	100% minimum	Generally small	0–300%, varies with span
Barometric errors	Negligible	Negligible	Usually small	Usually very small
Speed of response	See Fig. 6.5	See Fig. 6.5	See Fig. 6.5	See Fig. 6.5

* These characteristics are to be used only as a guide. Dimensions and functional values will vary with manufacturer.

percent response time is generally between 1 and 4 s in well-stirred water.

Barometric Errors

Class III systems are sensitive to barometric changes. They are generally designed to have a 400 lb/in^2 (2.8 MPa) change across the range, so errors are usually negligible. Special compensation is usually required when they are used with transducers.

Sensor Elevation Error

Since the density of the gas used is very small, the bulb elevation errors are negligible.

Overrange Protection

The maximum temperature for all gas thermometers is that which produces the maximum allowable pressure in the bourdon or the maximum permissible sensor temperature, whichever is lower. (See Fig. 6.12.) It is noted that short spans reach this limiting pressure at a substantially lower temperature.

Averaging Thermal Systems

Class III systems lend themselves uniquely to applications requiring the average temperature of a large enclosure such as an oven or drying chamber. This is so primarily because sensor volume is not critical and capillary-type sensors that can be easily installed will provide adequate volume and minimum sensitivity to bending or uncoiling.

Comparative Performance of Thermal Systems

Table 6.1 is intended as an approximate guide for comparison of the different classes of filled-system thermometers. The values apply principally to bourdon-type systems actuating an indicating pointer or pen. When they are used with transducers which provide biasing spring elements, many of the performance figures such as span, bulb size, response, and overrange may be changed for greater ease of use. Note also that motion response linearity is a direct function of the ratio of sensor internal volume to the volume of the balance of the thermal system. A ratio of at least 40:1 is necessary to produce linearities better than ½ percent.

REFERENCE

1. *SAMA Standard*. RC6-10-1963, Process Measurement and Control Section, Scientific Apparatus Makers' Association, Washington, D.C.

Chapter 7

Radiation Thermometry

ROBERT E. BUCKLEY
Mohawk Associates

GENE D. NUTTER
Instrumentation Systems Center
University of Wisconsin, Madison

7.1 RADIATION THERMOMETRY

We are all familiar with the fact that a heated object radiates heat and that the hotter it is, the more it radiates. The radiant heat originates with the thermal agitation of the atoms and molecules of which the object is composed and is characteristic of all materials at all temperatures above absolute zero. When atoms or molecules within the heated body collide, part of the energy appears in the form of photons, or quanta, of electromagnetic radiation. The number and spectral distribution of the photons depends on the temperature. The relationship between the radiant energy and the temperature of the body is expressed in terms of the Planck radiation distribution function. This allows one to determine the temperature of a heated body by measuring the amount of radiant energy in a known spectral band, from a known area and in a particular direction (Fig. 7.1).

Instruments that perform this function (Fig. 7.2) are known as *radiation thermometers*. Such instruments are employed for noncontact measurements or where other methods are unsuitable and have been widely used in industry for more than 50 years. The scope of radiation thermometry applications has expanded greatly over the past two decades as a result of improved instrumentation. An understanding of certain basic concepts is essential to the correct selection and use of these instruments.

Blackbody Radiation

A body that absorbs all electromagnetic radiation incident on it, reflecting nothing, would appear to be visually black

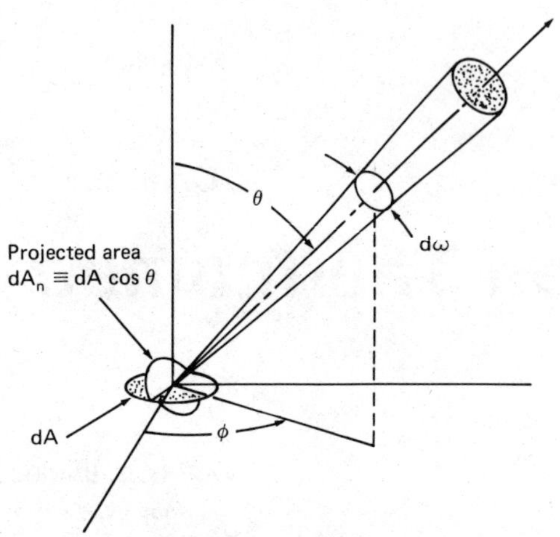

Fig. 7.1 Geometric aspects of spectral radiance, defined to be the derivative of radiance with respect to wavelength.

at room temperature, and is called a *blackbody*. No real material surface is perfectly black, but the concept of a blackbody is of fundamental importance in the theory of radiant heat transfer and forms the basis of radiation thermometry. When such a surface is heated, it radiates in accordance with Planck's radiation law (Fig. 7.3):

$$L_{\lambda b}(\lambda, T) = \frac{c_1}{\lambda^5[\exp(c_2/\lambda T) - 1]} \tag{7.1}$$

where $L_{\lambda b}(\lambda, T)$ = spectral radiance
λ = wavelength of electromagnetic radiation
T = temperature (absolute)

$$c_1 = 2c^2h = 1.191062 \times 10^8 \text{ W} \cdot \mu\text{m}^4/\text{sr} \cdot \text{m}^2 \tag{7.2}$$

$$c_2 = \frac{ch}{k} = 1.438786 \times 10^4 \ \mu\text{m} \cdot \text{K} \tag{7.3}$$

$$c = 2.9979250 \times 10^8 \text{ m/s} \tag{7.4}$$

$$h = 6.626196 \times 10^{-34} \text{ J/s} \tag{7.5}$$

$$k = 1.380622 \times 10^{-23} \text{ J/K} \tag{7.6}$$

Radiation from a blackbody is unaffected by the material composition of the surface, depending only on its temperature. No real material can radiate more than a blackbody at the same temperature.

Radiation from a Material Surface

Emissivity The interior of any opaque solid or liquid absorbs all radiation incident on it, reflecting nothing. It is therefore black and radiates as a blackbody. Absorption and emission take place continuously at every point inside such a body, and a photon travels only a very short distance before it is absorbed. When radiation approaches the outer boundary of such a body, some of it is reflected back into the interior of the body where it is reabsorbed; the remainder passes through the surface where it is emitted. By the conservation of energy, it follows that the reflected component plus the emitted component must add up to the entirety of the radiation, or

$$\rho(\lambda) + \epsilon(\lambda) = 1 \tag{7.7}$$

where ρ = reflectivity
ϵ = emissivity
λ = wavelength in the electromagnetic spectrum

It is clear that a good emitter is a poor reflector and vice versa. Anything, such as oxidation, that affects the reflectivity of a surface must have a complementary effect on its emissivity. Note that this expression implies that the emissivity of a surface can be determined by measuring its reflectivity (e.g., in a spectrophotometer); this is, in fact, a

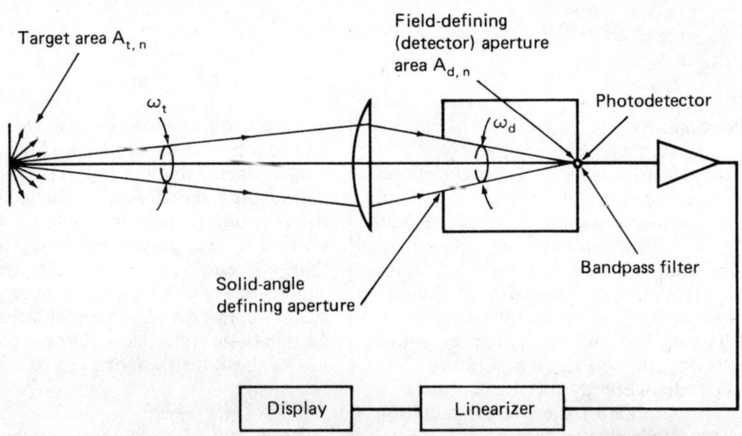

Fig. 7.2 Idealized radiation thermometer. The radiometer output signal is directly proportional to spectral radiance, which is exponentially related to temperature through Planck's law.

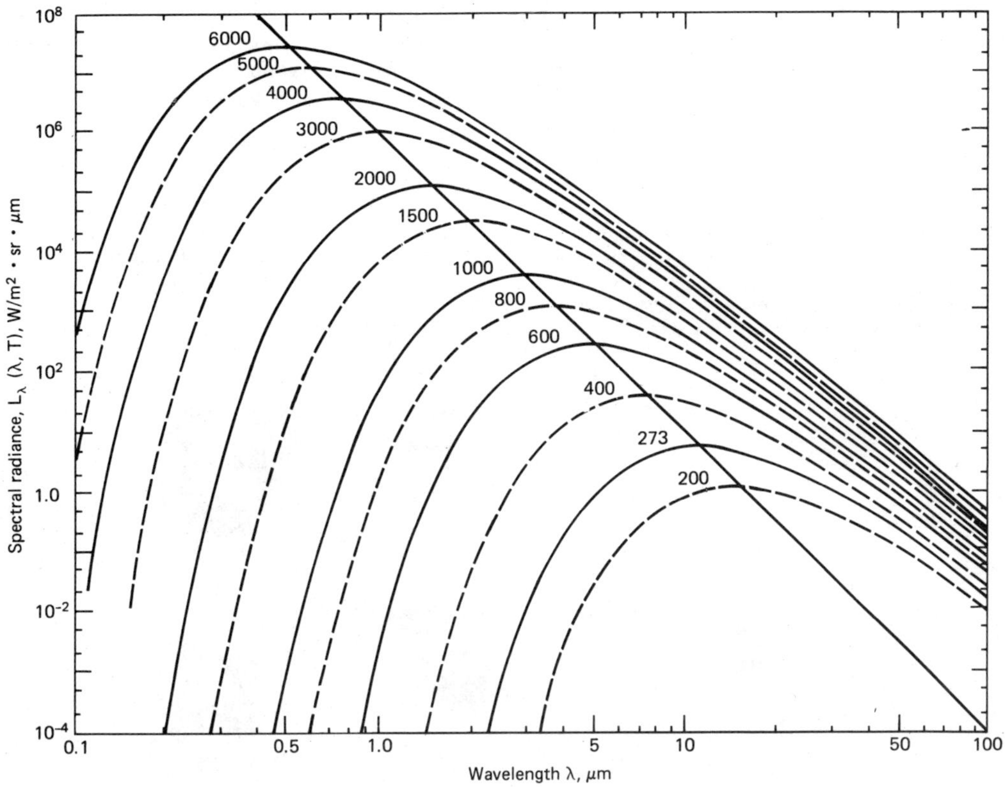

Fig. 7.3 Spectral radiance of a blackbody, from Planck's law. Temperature is shown in kelvins for each curve.

commonly used method of evaluating the emissivity of a specific sample of material.

Cavity Radiation An entirely enclosed cavity in the interior of an opaque isothermal solid material contains blackbody radiation characteristic of the cavity wall temperature. All points on the interior surface emit radiation because of their temperature and also reflect radiation originating from other points on the wall; the sum of the emitted and reflected radiation is *blackbody radiation.*

If a small hole is made in the wall of the cavity, the escaping radiation very closely approximates that of the blackbody radiation in the completely enclosed cavity, and such a cavity can be made to serve as a practical blackbody. Note that the small aperture, as seen from the outside of the cavity, appears to be very black at room temperature, but it will glow more brightly than will a piece of the same material at the same temperature out in the open.

Effect of Surface Roughness A small beam of radiation incident on a perfectly flat surface will reflect only once before leaving the surface, whereas a beam reflecting from a pit in the surface may reflect more than once. Since some energy is absorbed at each reflection, the pitted surface absorbs more energy than the flat surface. By Kirchhoff's law, the emissivity is equal to the absorptivity, so it follows that the emissivity of a rough surface is greater than the emissivity of the smooth surface. Increasing the roughness increases the emissivity. Thus we see that the principal characteristics of emissivity of materials can be understood in terms of the reflectivity.

Atmospheric Effects

The optical medium between the thermally radiating target and the radiation thermometer includes atmospheric gases, vapors, dust, and any windows that might be in the sight path. The atmosphere in most cases is composed of gaseous nitrogen, oxygen, water, and carbon dioxide, with other gases also present to a significant extent in a variety of industrial processes. These gases, vapors, and dust particles absorb, emit, and scatter radiation, and their effects on radiometric measurements range from negligible to severe. By Kirchhoff's law, emission is proportional to absorption. If the concentration and temperature of a vapor-free atmosphere were uniform, it would radiate very nearly as a blackbody in those spectral regions in which it was a strong absorber, with $\rho(\lambda) = 0$; in practice, however, the concentration and temperature of these atmospheric constituents tend to be highly variable with time and position, and vapors are frequently present.

Although the physics of the absorption, emission, and scattering processes is generally understood, the conditions under which it applies are highly variable and seldom well enough known quantitatively to allow for practical corrections to be determined and applied. Once recognized, however, this need not be a serious problem in most cases (gaseous absorption and emission) since there are spectral regions, known as *atmospheric windows,* in which the atmosphere does not absorb and therefore does not emit radiation. These spectral regions are clearly evident in Fig. 7.4. If radiometric measurements are confined to those spectral regions, atmospheric interference—in the absence of particulate scattering—is entirely negligible. When scattering effects are strong, however, as in the case of suspended particulates (dust), steam clouds or sooty flames, the use of an atmospheric window alone does not correct the problem. It is then common practice for the radiometer to view through a tube purged with a clean dry (transparent) gas, using a filter to limit the radiometer passband to an atmospheric window.

Reflected Radiation

Radiation originating from extraneous sources and reflecting from a target surface increases its spectral radiance, $L_{\text{reflected}}$ and causes errors in radiometric temperature measurements:

$$L_\lambda(\lambda)_{\text{observed}} = L_\lambda(\lambda)_{\text{emitted}} + L_\lambda(\lambda)_{\text{reflected}} \qquad (7.8)$$

The magnitude of the reflected spectral radiance depends on the spectral radiance of the extraneous source, the solid angle subtended by it from the target area, and the bidirectional spectral reflectance distribution function of the target. When the magnitude of the reflected radiance is comparable to or larger than that of the emitted radiance, errors in temperature measurement tend to be large. This is most likely to

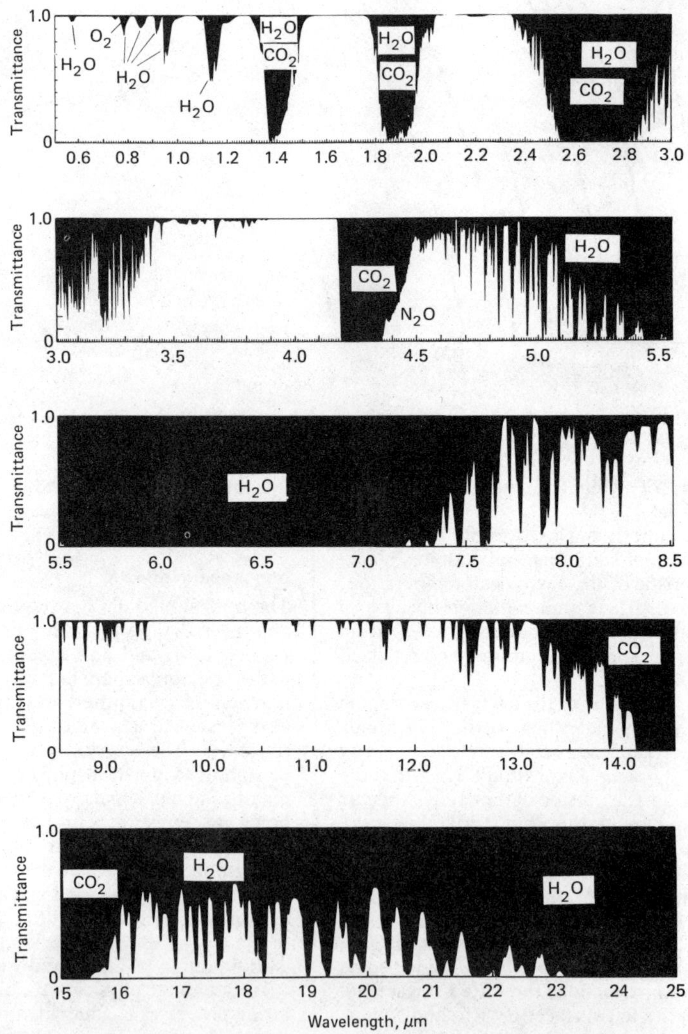

Fig. 7.4 Atmospheric transmission at sea level over a 300-m meter path. (*a*), (*b*), and (*c*) 5.7-mm precipitable water, 26°C; (*d*) 5.7-mm precipitable water, 25°C; (*e*) 2.2-mm precipitable water, 13°C. (*From NRL Report 5453 by H. W. Yates and J. H. Taylor, 1960, adapted by G. D. Nutter.*)

occur when target temperatures are not greatly different from ambient temperature, and the problem intensifies as target temperatures decrease. At target temperatures below about 300 to 400°F (150 to 200°C), reflected radiation is usually the dominant source of error in radiation thermometry.

The target can often be successfully shielded from the extraneous radiation by a suitably placed screen, reducing $L_\lambda(\lambda)_{\text{reflected}}$ to zero. It is also sometimes possible to measure the reflected component separately with the aid of a relatively cool auxiliary reflector of known reflectance located near the target and subjected to the same irradiance and to then subtract that component from the observed spectral radiance [Eq. (7.8)]. The trend in recent years has been toward trying to develop techniques for dealing with reflected radiation, some of which have met with a significant degree of success.

Emissivity Compensation

If the emissivity is known and is essentially independent of wavelength across the radiometer passband, its effects can be compensated for by increasing the radiometer amplifier gain by a factor equal to the reciprocal of the emissivity. When the gain cannot be adjusted, as in the case of a disappearing-filament optical pyrometer or a null-balancing radiometer, a correction can be computed from Wien's law, from which we obtain

$$\frac{1}{T} = \frac{1}{T_{\text{apparent}}} + \frac{\lambda}{c_2} \ln \epsilon(\lambda) \tag{7.9}$$

where λ is the *mean effective wavelength*. When the emissivity changes appreciably across the passband, we must use a different approach that is more direct but more difficult to implement. The radiometer responds with the same signal that it would have if it were viewing a blackbody at the temperature T_{apparent}:

$$\int_0^\infty \tau_o(\lambda)\mathcal{R}(\lambda)\epsilon(\lambda)L_\lambda(\lambda, T)\, d\lambda = \int_0^\infty \tau_o(\lambda)\mathcal{R}(\lambda)L_\lambda(\lambda, T_{\text{apparent}})\, d\lambda \tag{7.10}$$

where $\tau_o(\lambda)$ is the spectral transmittance of the optical system and $\mathcal{R}(\lambda)$ is the spectral responsivity of the photodetector. The true temperature T is determined from the indicated temperature T_{apparent} by numerical integration of the preceding expression. This requires numerical descriptions of $\tau_o(\lambda)$, $\epsilon(\lambda)$, and $\mathcal{R}(\lambda)$. Prior to the advent of computers, this method—although exact in principle—required so much labor that it was seldom used. This no longer is true, although other methods may still be preferable in many cases, particularly for narrow passbands or where fast response is needed.

Severe Ambient Temperature and Dust Environments

It is frequently necessary to use radiation thermometers in severe environments such as those commonly found in industry, where the high temperatures would destroy an unprotected instrument. This is satisfactorily accomplished by the use of air or water flowing through tubes provided in the instrument housing to regulate its temperature. Air or some other dry gas to purge dust from the sight path and lens is also usually supplied through tubes in the instrument housing.

7.2 RADIATION THERMOMETERS

In the following discussion, radiation thermometers are classified as high-temperature or low-temperature devices. High-temperature radiation thermometers are used for temperatures much greater than ambient, typically 500 to 4500°F (250 to 2500°C). Low-temperature radiation thermometers operate from below ambient to 900°F (500°C). A typical range is −40 to 200°F (−40 to 100°C).

High-Temperature Radiation Thermometers

High-temperature radiation thermometers are subdivided into optical pyrometers (disappearing-filament), amplitude-sensing, and ratio or dual-wavelength types.

Optical Pyrometer

For over 50 years, optical pyrometers have been in commercial production and use in the United States and have been manufactured by two companies. These radiation thermometers are manually operated instruments with the human eye as the sensing element.

In the version of the instrument known as the *disappearing-filament optical pyrometer,* the optical system is a low-powered terrestrial telescope viewing the heated "target" area. A lamp filament is located in the image plane and is seen superimposed on the image of the target. A red filter located in the optical path minimizes the dependence of the readings on the eye; in conjunction with the eye, this filter determines the effective wavelength of the instrument. The match is achieved by adjusting the filament current until the filament disappears against the background of the target. The instrument is calibrated by determining the filament current corresponding to the temperature of a blackbody target.

In the other version, the filament current is fixed and the target image is attenuated by variable-extinction wedges. These radiation thermometers are battery-operated, handheld, and quite basic in their operation. The instrument is turned on, often by squeezing a trigger, match is achieved, and the temperature read by the operator.

Since the target must be at a temperature at which luminosity is achieved, the lower temperature is limited to about 1300°F (700°C). The upper temperature is limited in most instruments to about 7200°F (4000°C). Temperatures above about 2500°F (1400°C) are determined by interposing an absorbing glass filter between the target and the filament; this is done to protect the filament, which deteriorates too rapidly at higher temperatures. Although there is no inherent upper limit to the temperature range of this instrument, in practice, temperatures above about 18,000°F (10,000°C) are almost never needed.

Used primarily by the steel industry, the instrument can read temperature to an accuracy of about 0.25 percent of the (absolute) temperature measured under good observing conditions, with repeatability of about 0.1 percent with a single operator. Multiple operators measuring the same target achieve a wider variation in readings.

With this type of radiation thermometer, the target object does not need to fill the field of view of the pyrometer since the eye will still sense the filament brightness match over a portion of the field of view. Accurate measurements of temperature can be made in the presence of moderate amounts of smoke or steam as well.

Amplitude Sensing Radiation Thermometers

Another type of high-temperature radiation thermometer is the electronic type (Fig. 7.5). In this instrument, the sensing transducer or detector converts the incoming radiance into an electrical signal which is read out as a temperature on a meter. Radiance does not vary linearly with temperature. To read out directly in temperature, some form of linearization is required. In simpler instruments using analog meters, linearization is accomplished by shrinking and stretching the scale on the analog meter. In instruments using a digital indicator for readout, an electronic linearizing circuit is used. The range and accuracy of the individual instruments are limited by the design of the linearizing circuit.

In the instrument illustrated in Fig. 7.5, the incoming target radiance is collected by an optical system and directed onto a detector. The detector converts the energy to an electrical signal which is then amplified by the accompanying circuitry and displayed as temperature or provided as an electrical control signal.

In high-temperature electronic radiation thermometers, the radiation from the background which is at ambient temperature constitutes such a small fraction of the energy coming from the target that it can be ignored. Simple instruments therefore can be used quite satisfactorily. Most high-temperature radiation thermometers are designed for permanent installation in remote locations with the indication system located in a control room. Portable instruments that are battery-operated are also available. The fixed-installation radiation thermometers can be equipped with high- and low-temperature alarm relays, sample-and-hold circuits and circuitry to read the highest temperature observed in an interval.

Special radiation thermometers, called *transducers,* provide an output signal which mimics that of a thermocouple for installation into standard 4- to 20-mA control loop. These transducers cover ranges of 500 to 4500°F (250 to 2500°C) with acceptable accuracies as long as the transducer itself is kept at nearly constant temperature. Variations in temperature of the transducer itself, often due to heat load from the target, can materially affect the readings obtained from the devices. Transducers are the lowest first-cost radiometers available.

Also available at low cost is the hand-held radiation thermometer. These simple-appearing pistol-shaped instruments include means for compensating for the instrument ambient-temperature changes but are strongly affected by rapid instrument ambient-temperature variations (e.g., transients). Most hand-held instruments have emissivity adjustments to compensate for graybody emissivity and some have temperature-hold features as well. These radiation thermometers are battery-operated and intended for survey operations rather than temperature recording. In a unit with a temperature range of 900 to 2700°F (500 to 1500°C), short-term accuracy of 2 to 4°F (1 to 2°C) is typical with short-term repeatability of less than 2°F (1°C). Over the long term, both accuracy and repeatability are degraded and are a function of the ambient-temperature compensation method employed.

In the more sophisticated instruments intended for permanent installation, techniques in signal handling are employed to eliminate the gross effects of ambient-temperature changes on the instrument. One technique uses an optical chopper which serves the purpose of converting the dc target signal to an ac signal by alternating the irradiance on the detector between an internal known radiant source and the

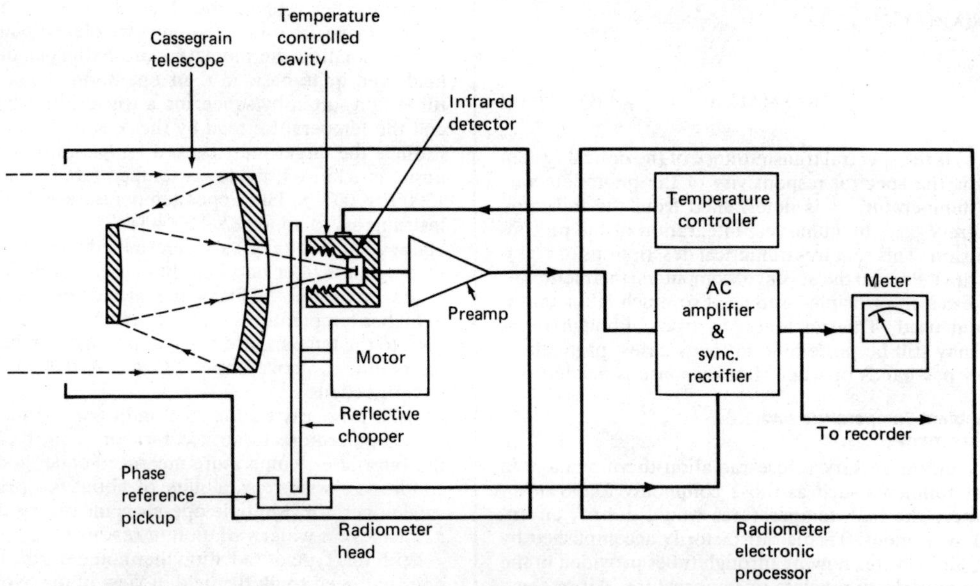

Fig. 7.5 Basic components of a typical radiation thermometer.

target itself. A demodulator which is included in the electronic processing circuitry operates in synchronism with the optical chopper. The demodulator converts the ac signal into a dc signal, maintaining the "sense" of the signal by providing a positive signal when the target is warmer than the reference source and a negative signal when the target radiance is less than the reference, thus reducing or eliminating the dependence of the output signal on the stability of the detector, making it dependent on the stability of a highly stable reference source instead.

The more sophisticated and, therefore, more expensive instruments offer improved accuracy and repeatability over the less expensive instruments. Radiation thermometers that have ranges of 900 to 2700°F (500 to 1500°C), 2200 to 5800°F (1200 to 3200°C), or even higher are offered by several manufacturers. The accuracy is typically 1 percent of span with repeatability of ½ percent of span or better.

Ratio or Dual-Wavelength Radiation Thermometer

The ratio, or dual-wavelength, thermometer is often called a *two-color radiometer.* By making measurements at two wavelengths (both shorter than the peak wavelength of the target radiation), the ratio thermometer determines the slope of the radiance curve and therefore determines the target temperature.

When the slope of the radiance radiation curve rather than the magnitude is measured, the difficulties of emissivity, filling the field of view, and atmospheric attenuation are avoided. This type of instrument is applicable to high-temperature blackbodies or graybodies in smoky or steamy surroundings, or where the target area is partially blocked from view. Two-color radiation thermometers are not recommended for use with selective radiators where the spectral radiance curve is not that of a blackbody or graybody. Such conditions are frequently encountered in practice.

A typical ratio radiation thermometer measures temperature to an accuracy of 9°F at 2700°F (5°C at 1500°C) with a repeatability of 4°F (2°C).

Since two signal channels with either a filtered time-shared detector or two separate detectors are required, the cost of this instrument is greater than that of the instruments previously described.

Low-Temperature Radiation Thermometers

Low-temperature radiation thermometers are sophisticated in design because they must correct for the radiance of the instrument itself as well as measure the target radiance and must incorporate emissivity control measures to allow for target emissivity compensation.

By the nature of Planck's law, low-temperature radiation is long wavelength radiation. For this reason, the low-temperature radiation thermometers are sensitive to the longer wavelengths and are further classified into spectral band and narrowband, each with distinctive applications.

The spectral band radiometers are sensitive in spectral regions where the atmospheric absorption is minimized (i.e., atmospheric windows). Figure 7.4 shows atmospheric absorption and from it, it can be seen that 8 to 14 μm is a long-wavelength window; manufacturers therefore offer instruments sensitive in this region. Instruments are available that measure target temperatures from −75 to 140°F (−60 to 60°C) or 32 to 575°F (0 to 300°C) and have resolutions of less than 0.1°F (0.05°C) over that range and 0.01°F (0.005°C) at the higher temperatures. Such instruments are used in sea surface temperature measurement, crop health determination, and satellite ground truth measurement.

Narrowband instruments are available that are sensitive in the narrow spectral region in which plastics or glasses absorb (and hence radiate). Figure 7.6 provides the spectral transmission characteristics of several kinds of plastic. By collecting energy from a narrow spectral region, the sensitivity of the instrument is reduced but the instrument responds only to the radiant energy from the plastic. At the absorbing wavelengths, the plastic film absorbs and radiates like a blackbody, so reflection and transmission are minimized. An example is that of a radiation thermometer filtered to receive radiation at 3.43 μm in an 0.1-μm-wide band where polyethylene is strongly absorbent (see Fig. 7.6). When using this instrument to view a 0.005-in (0.127-mm)-thick polyethylene film, it is possible to measure the polyethylene film temperature without interference from heaters on the far side of the film.

Instruments of this type, sensitive to 3.43 μm for polyethylene and polypropylene films, or 7.9 μm where polyesters absorb, are commercially available. A typical narrowband instrument of this sort has a range of 120 to 950°F (50 to 500°C) and will offer accuracy of 9°F (5°C) and repeatability of 4°F (2°C).

Specialized Thermometers

In addition to the types of radiation thermometers just discussed are two specialized systems that use fiber optics. The collecting optics can be a simple lens, a mirrored optic, or an optical fiber. Optical fibers have been used for this purpose for more than 15 years and are improving as optical fiber technology improves.

The fiber-optic radiation thermometer has distinct features not always available in other instruments. These features include two-way fiber bundles that permit projection of a light spot on the target for sighting purposes, bifurcated fiber bundles that permit observation of the target with a video camera as well as the detector, and combinations thereof.

The fiber-optic pickup can be extremely small and easily routed to difficult locations where line-of-sight applications are impossible and in its sealed form can be used in high-temperature, high-pressure locations in which other instruments are inapplicable. An example is a flow control for injection molding machines in which the temperature at one or more locations in the plastic flow prior to injection is monitored, providing signals for flow control. In this instance, the range of temperatures measured covers 212 to 750°F (100 to 400°C) with a resolution of 2 to 4°F (1 to 2°C). Since the fiber is exposed to temperatures of 750°F (400°C) and pressures of 30,000 lb/in^2 (2.0 × 10^8 Pa), the uniqueness of this application is evident. It is also possible to measure much higher temperatures when using optical fibers.

The substitution of fiber optics for lens-type optics changes only the "front end" of the instrument. All the options and capabilities of other high-temperature radiation thermometers are available.

An innovative contact technique for measuring temperature has evolved within the past 5 years. As with a thermocouple, intimate contact with the surface to be measured

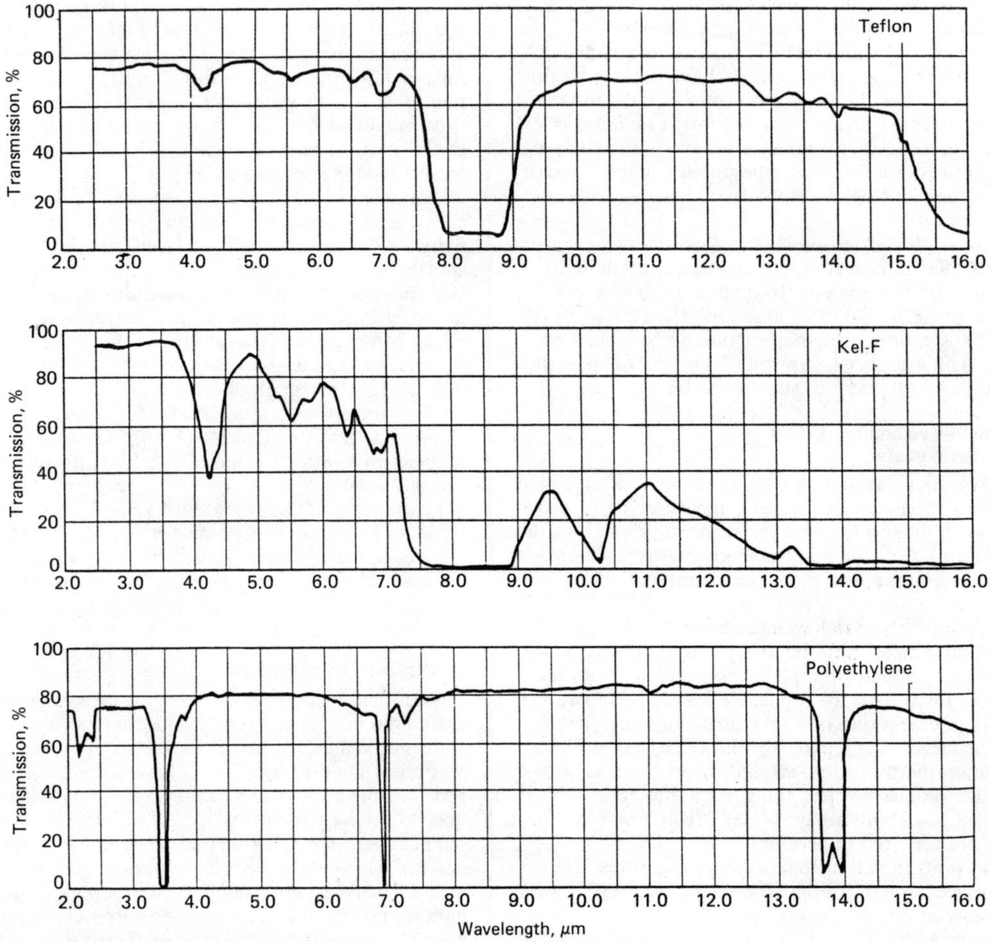

Fig. 7.6 Spectral transmission characteristics of some common plastics. These materials have low emissivity in regions where they transmit well and approximate blackbody emission in the regions of their bands of absorption.

is required. The concept is that an optical fiber can be used to transmit energy at one wavelength to a phosphor at the end of the fiber and to receive energy at another wavelength from that phosphor. In the operation of this instrument, the phosphor is "pumped" by pulses from a filtered xenon light source and the decay time of the phosphor is measured at its "relaxation wavelength." The decay time is a repeatable and sensitive function of the phosphor temperature. Measurements of temperature can be made from −325 to 750°F (−200 to 400°C) with accuracies of 4°F (2°C) without calibration and accuracies of 1 to 2°F (0.5 to 1°C) with spot calibration. The resolution is 0.5 to 1°F (0.25 to 0.5°C) over the entire range. In a medical application in which the fiber is inserted into a vein, the temperature of the blood at the phosphor sensor can be measured without interference from electromagnetic radiation used during therapeutic treatment. Other applications include plasma deposition, ion-beam implantation, and other operations carried out in vacuum systems or strong radiation fields.

7.3 INSTRUMENT SELECTION AND OPERATION

Much of the behavior characteristics of narrowband radiation thermometers can be understood by expressing thermally emitted spectral radiance in terms of a power of the temperature:

$$L(\lambda) \propto T^n \tag{7.11}$$

Where Wien's law is a suitable approximation to the Planck function it can be shown that

$$n = \frac{c_2}{\lambda T} \tag{7.12}$$

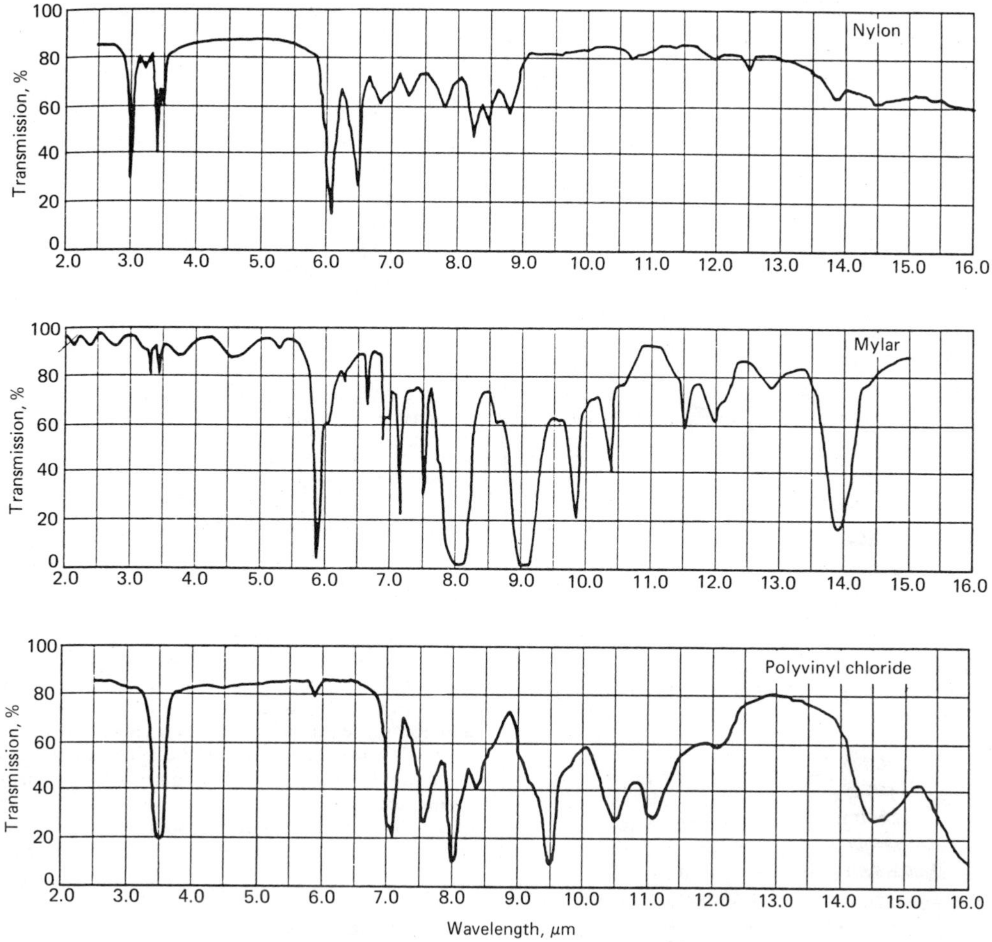

Fig. 7.6 *(Continued)*

It follows that

$$\frac{dT}{T} = \frac{1}{n}\frac{dL_\lambda(\lambda)}{L_\lambda(\lambda)} = \frac{\lambda T}{c}\frac{dL_\lambda(\lambda)}{L_\lambda(\lambda)} \tag{7.13}$$

The typical range of values for n is approximately $5 < n < 20$. The relative error in the measurement of emitted spectral radiance is thus seen to be reduced by a factor of $1/n$ in its effect on the relative error in the temperature; it is apparent that such errors can be minimized by maximizing n.

Corrections for large errors in spectral radiance caused by emissivity can be determined from the graph in Fig. 7.7, for any combination of effective wavelength and temperature.

Spectral Passband

It is seen from Eq. (7.13) that the error is proportional to wavelength, so the wavelength should usually be chosen as short as possible while still having a sufficiently large signal-to-noise ratio. This usually means using the shortest feasible wavelength, as in the case of window or emissivity effects, although there are circumstances (e.g., in the presence of reflected radiation) in which the opposite is true. The spectral band should be narrow enough to fit into an atmospheric window and wide enough to provide sufficient signal.

The choice of spectral passband is often dictated by the presence of strong absorption bands in a given material. This is strikingly true in the case of common plastics, such as polyethylene, which has a strong absorption band centered at 3.43 μm, with a bandwidth typically of only about 0.1 μm. When viewed in that spectral band, polyethylene is opaque, even in very thin [0.001-in (0.025-mm)] sheets; by Kirchhoff's law, it must therefore have an emissivity very close to unity. Radiometers equipped to operate in this band are very successfully used in the plastics industry.

Response Time

Most radiation thermometers are of the type referred to in Fig. 7.2 and have a response time that can be expressed in terms of an exponential constant called the *time constant.*

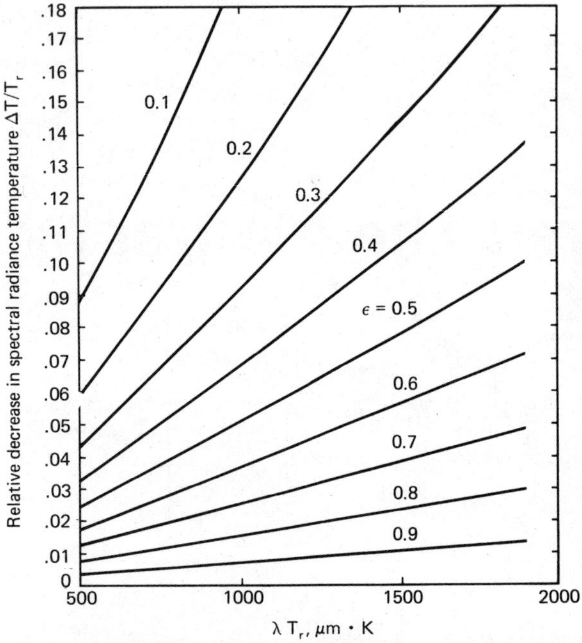

Fig. 7.7 Relative amount by which the observed spectral radiance temperature T_r is reduced by the effect of spectral emissivity, presented as a function of λT_r.

In such cases the error in response to a step function is as illustrated in Fig. 7.8. It can be seen that it takes about five time constants for the radiometer output to come to within 1 percent of its final value. In terms of temperature, the corresponding error can be computed from Eq. (7.13) with $\delta L(\lambda)/L(\lambda) = 1$ percent. The temperature error is thus seen to depend on the time constant, the wavelength, the temperature, and the magnitude of the transient.

Filling the Field of View

The calibration of a radiometer assumes that the field of view is completely filled by a uniformly radiant target. If only a fraction of the field of view is filled, the instrument will indicate too low a temperature by an amount that can be determined from Fig. 7.7 where ϵ is the fraction of the field that is filled.

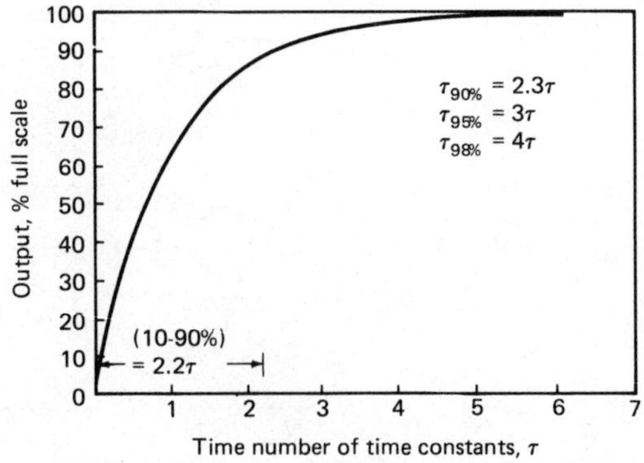

Fig. 7.8 Various definitions of response time in terms of the exponential time constant.

Most radiation thermometers use fixed focus. As long as the uniformly radiant target area extends across the entire field, it can be anywhere in the field defining cone.

7.4 SUMMARY

Radiation thermometry is a technology that is approaching maturity. Advances in photodetectors, optics, and electronics over the past two decades provide a wide range of spectral bands from which to choose and a variety of measurement and control features. Most problems in noncontact temperature measurement can now be successfully addressed by radiation thermometry. The primary areas in which improvements are needed are in situ determination of spectral emissivity and in the measurement of and compensation for reflected radiation. Progress is being made in both areas.

FURTHER READING

1. D. P. DeWitt and G. D. Nutter (eds.), *Theory and Practice of Radiation Thermometry,* in press.
2. G. D. Nutter, "Radiation Thermometry–The Measurement Problem," in J. C. Richmond and D. P. DeWitt (eds.), *Applications of Radiation Thermometry,* ASTM standard STP 895, Philadelphia, 1985.
3. G. R. Peacock, "Radiation Thermometry," in *Process Instruments and Controls Handbook,* 3d ed., D. M. Considine (ed.), McGraw-Hill, New York, 1985, pp. 2.81–2.118.

Chapter 8

Infrared Thermography

ROBERT P. MADDING
Inframetrics Inc.

8.1 INTRODUCTION

Thermography is the process of generating a thermogram by using an infrared imaging system, usually with some means of temperature calibration.[1] A *thermogram* is a photograph or a two-dimensional record of an image which maps the apparent temperature of the scene as sensed by an infrared imaging system.[1] A thermal imaging radiometer is a device which produces thermograms and measures quantitatively the thermal radiation within a definite waveband incident upon it.

Thermograms can be portrayed in black and white or color. Figure 8.1 shows a normal photograph and a black-and-white thermogram. Color thermograms delineate temperature variations quite well at the expense of spatial detail. Black and white thermograms delineate spatial variations quite well at the expense of temperature detail. Modern thermal imaging radiometers display both color and black and white as well as providing absolute temperatures.

The purpose of this chapter is to identify performance criteria for selection of a thermal imaging radiometer, discuss techniques on how to use them effectively, and introduce some of their applications.

8.2 THERMAL IMAGING RADIOMETERS

Thermal imaging radiometers not only display an image of the heat patterns radiating from surfaces but are also designed to produce temperature information.[2–4] This is a big step since none of these instruments measures temperature directly but must infer temperature from the measured radiant energy. The great advantage of thermal imaging radiometers over noncontact thermometers is that they allow rapid assessment of a situation via thermal patterns and show

(a)

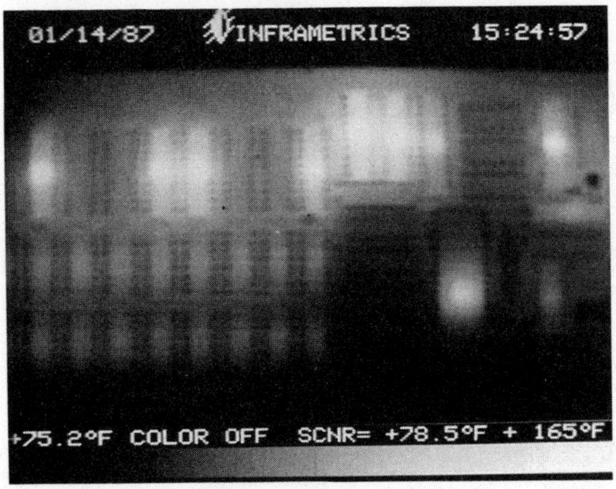

(b)

Fig. 8.1 (*a*) Normal photograph and (*b*) thermogram of printed-circuit board. (*Hal Appleton, Appleton-Kingston Associates.*)

the operator exactly what is being measured. They also have a much faster response time (nanoseconds vs. milliseconds) than do noncontact thermometers. A thermal imaging radiometer performs some million measurements per second. Pattern recognition by the operator facilitates real-time or postprocessing analysis of the right area at the right time. With video recording and computer processing, tremendous amounts of thermal data can be archived, accessed, and analyzed. The major disadvantage of thermal imaging radiometers compared to noncontact thermometers is their cost which is typically 20 to 30 times higher.

Many types of noncontact temperature measurement instruments are available today. Nearly all rely on the same basic physical phenomenon: All real-world objects radiate energy; the amount of energy radiated increases with increasing temperature. The radiated energy is distributed over a band of wavelengths in the electromagnetic spectrum (Fig. 8.2). The spectrum spans from low-frequency radio waves to microwaves, then to the infrared, visible-light, and x-ray wavelengths. The distribution for a specific object temperature is the Planck function.[4,5] The peak of this curve moves toward shorter wavelengths as the object temperature increases (Fig. 8.3). The Planck distribution is modulated by the efficiency of the object as a radiator. This efficiency, called *emittance,*[5] is 100 percent for a perfect radiator, called a *blackbody*. Real objects have emittances that are less than one and which can vary with wavelength.

Objects at or near room temperature have spectral energy distributions that peak in the middle infrared region, near 10 μm. A sufficient amount of energy is radiated to allow detection at great distances by a sensitive instrument. The consistency of the relationship between object temperature and radiated energy allows a calibrated instrument to perform highly accurate noncontact temperature measurements.

High-performance thermal imaging radiometers differ from conventional television cameras in several ways. One class of system operates with one or a few individual detector signals scanned mechanically across the screen rather than with the large photosensitive surface typical of a TV camera. These detectors must be cooled to cryogenic temperatures. Another class of system, the pyroelectric vidicon, is similar to a TV camera, employing a large, uncooled, thermally sensitive surface. However, because of the nature of the pyroelectric effect, these systems are difficult to quantify. The Schottky barrier is also a large surface obviating mechanical scanners. It has not yet been developed into a viable thermal imaging radiometer. Consequently, the discussion here is limited to mechanical scanning systems.

To perform accurate temperature measurements, thermal imaging radiometers incorporate a blackbody reference source that is viewed periodically by the detector for calibration. They use lenses and windows that transmit infrared wavelengths of electromagnetic energy; materials such as germanium, zinc sulfide, and silicon are typical. Optical filters can tailor the spectral response of thermal imaging radiometers to optimize measurement of (or transmission through) specific materials including gases, plastics, and flames. Low-pass, high-pass, bandpass, reject, and attenuating optical filters are available.

While thermal imaging radiometers employ many types of scanners, the specific type is seldom apparent to the operator. Nearly all systems incorporate scan conversion circuitry to create a TV-compatible output. The referencing technique may also be obscure since electronic circuits perform recalibration during the invisible blanking portion of the TV image. Specific information on system design and specifications can be obtained from the manufacturers and in Refs 4, 6, and 7.

8.3 SYSTEM PERFORMANCE

The performance of a thermal imaging system is measured in a way that shows the total quantity of useful output information in a unit of time. This includes a combination of the thermal sensitivity or equivalent random noise level, the scan speed or update rate of the scanner, and the image resolution or number of independent measurement data points within the image. Other considerations include system dynamic range, spectral band sensitivity, calibration, and accuracy. Features such as weight, power consumption, dis-

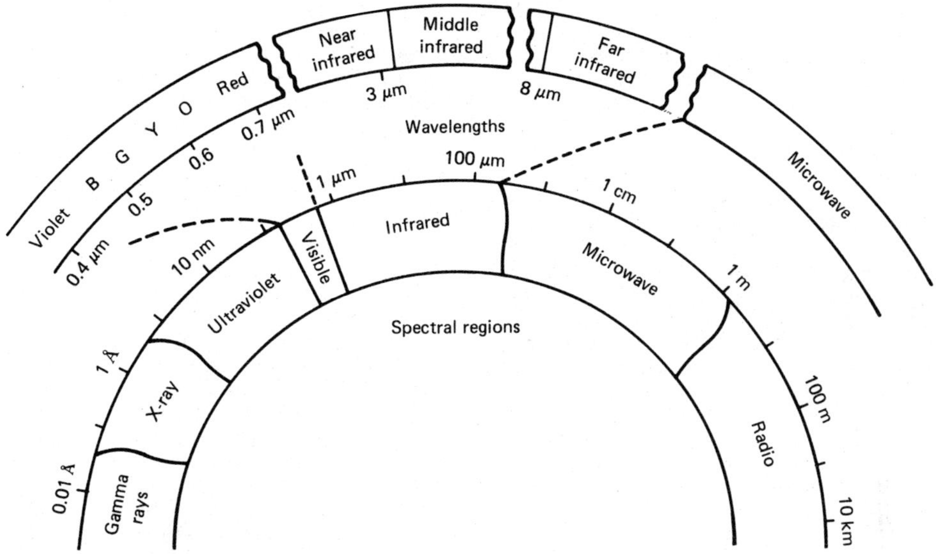

Fig. 8.2 The electromagnetic spectrum.

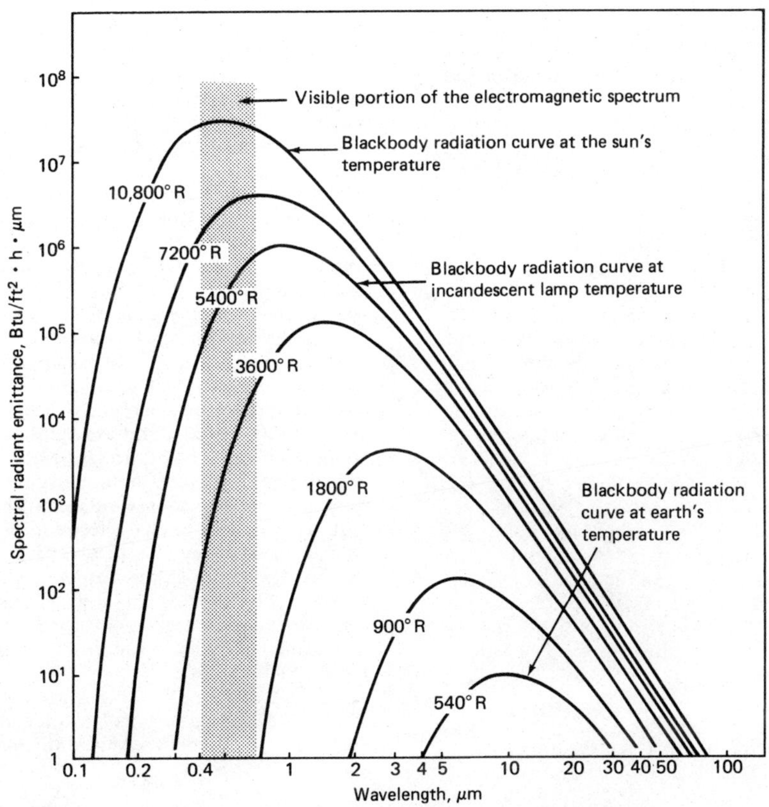

Fig. 8.3 Blackbody radiation curves for various temperatures. (Conversion factors: 1.0 Btu/ · ft^2 · μm = 3.1525 W/m^2 · μm; K = °R/1.8.)

play, and user-friendliness follow these parameters and can be readily evaluated by the user.

An imaging system intended for night vision use, without measurement capability, will generally be specified by a comprehensive parameter known as *minimum resolvable temperature difference* (MRTD).[2] MRTD is measured by determining the minimum thermal contrast needed for a human observer to identify a series of vertical bars using the system under test. The bars get smaller and closer together, thereby testing the system's image resolution. The thermal contrast (temperature difference between the bars and background) is small, testing the system's thermal sensitivity. The observer's eye performs temporal integration, offering a sensitivity improvement that is a function of the system's scan speed. A low MRTD indicates a system with a high-performance figure of merit, and one with greater effectiveness in aiding the user to locate and identify small, low-contrast thermal targets.

A thermal imaging radiometer is a measurement system in addition to a viewer, and the performance is generally specified in a manner that is more useful in measurement applications. Three independent parameters—thermal sensitivity, scan speed, and spatial resolution—replace the combined MRTD for thermal imaging radiometer specification. Thermal sensitivity is specified by noise-equivalent temperature difference (NETD)[8] defining the equivalent system noise level. Scan speed is the frame update rate of the scanner. Spatial resolution is most usefully specified by a continuous slit response function (SRF) curve. The combination of the three yields an indication of the overall performance of a system. Tradeoffs among the three parameters may be made by a system designer for optimization, but it is the *total* that is most difficult to increase. Finally, these parameters must be incorporated into a quantitative system whose accuracy should be specified.

Thermal Sensitivity

Thermal energy is given off by objects in individual bursts known as *photons,* much the same as visible light. The timing between these bursts is random. When very small samples of an object's radiation are made, the random emission of photons is visible as a variation in intensity, or thermal noise. Noise in the thermal image appears the same as the "snow" that is seen when a television camera operates in extremely low light. Both are caused by random noise of photons from either the scene or the light sensitive element in the camera.

A thermal imaging system is characterized by an NETD, the temperature difference between two objects that results in a signal equal to the random background noise of the camera. Averaging of multiple image data points reduces this noise level by the square root of the number of samples averaged. Well-designed instruments will average measurement points to reduce this noise. Noise is significant when operating only on the most sensitive ranges, viewing low-thermal-contrast targets.

Scan Speed

The rate at which complete thermal images are updated by a particular mechanical scanner is the *scan speed* of the system. The rate at which television images are updated on the monitor is the frame rate and is defined by television industry standards. Low scan speeds offer low acoustic noise and low cost as advantages but result in an image that distorts severely when the scanner or scene moves. In addition, low frame rates (below about 50 Hz) on non-TV-compatible systems cause an annoying flicker in the display. Extremely low-frame-rate (slow-scan) systems offer high thermal and spatial sensitivity but are very difficult to focus and point.

Spatial Resolution—Slit Response Function and Instantaneous Field of View

Spatial resolution is a measure of the thermal imaging system's ability to detect and accurately measure the temperature of small objects, where "small" is defined relative to the size of the total image.

Resolution in the thermal image is seen as the ability to present a highly detailed picture as opposed to one which appears unfocused, much the same as the difference between a low-resolution computer graphics image and a high-resolution computer-assisted-design (CAD) system. However, resolution is much more important for a measurement system since the accuracy of the temperature measurement at each point in the image depends on the degree of resolution, even beyond that which appears sharp to the eye.

Resolution is normally determined by characteristics not of the electronics, but of the scanner with its optics scanning mechanism and detector. Typically, an infrared system will detect radiation at the scanner using a detector of finite size, thereby limiting the resolution. The electronics subsequently digitizes the image into picture elements, pixels, smaller than the detector samples. In this way, the pixel digitization does not reduce the system performance. One must be careful to realize that a 256- by 256-pixel image does not usually contain nearly as many true spatial resolution elements (detector samples).

The three-dimensional plot (Fig. 8.4) shows the response of a typical electrooptical imaging system to a point source of illumination. As an example, this might show the relative response of a thermal imaging radiometer to a small, very hot object at a great distance. The vertical axis represents temperature, and the x and y axes represent horizontal and vertical dimensions on the television monitor.

Note that the response of the system to the object is less than 100 percent. This is a result of the small object size; as the object becomes larger, the response gradually approaches 100 percent and the measurement becomes more accurate. There is no particular point where the object measurement is 100 percent accurate, but rather there is a continuous curve of response versus object size.

It is convenient to characterize a thermal imaging system by the relative response to a vertical slit target of various widths. The curve of this response is called a *slit response function* (SRF) and is useful in determining the amount of error or required correction factor for a target that can be approximated by one or more slit shapes. Figure 8.5 shows typical SRF curves.

The measured radiance of an object whose approximate SRF is known is a function of the object radiosity and average surroundings radiosity. An approximate equation in temperature form is

$$T_{app} \doteq (\text{SRF})T_{obj} + (1 - \text{SRF})T_{surr}$$

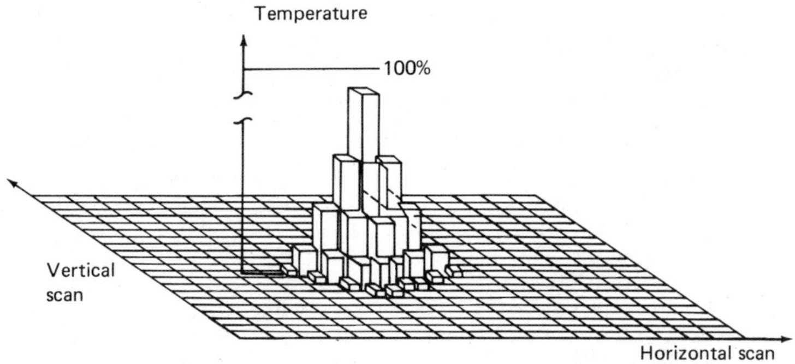

Fig. 8.4 Response of an electro-optical thermal imaging system to a point source of illumination.

The error between apparent and true temperatures is greater when a large difference exists between the object and surroundings temperatures.

The *instantaneous field of view* (IFOV) is defined as the ratio of the detector size to the effective system focal length. Thus a 0.001- by 0.001-in (0.0254- by 0.0254-mm) detector with 0.5-in (12.7-mm) focal length would have a 2-mrad (milliradian) IFOV. The system 50 percent SRF and IFOV should be about the same, with the IFOV being somewhat larger.

Resolution Elements per Image

Regardless of how one specifies resolution, it should be considered in context of the total image as well. A wide field of view (FOV) with good resolution is much more desirable than a narrow FOV with the same resolution. A measure of this is the total number of resolution elements per image. To find the total, divide the horizontal FOV by the resolution element size and multiply the result by the number of infrared (ir) lines per image. For example, a scanner which has a 50 percent SRF of 2 mrad, 20° horizontal FOV, and 200 ir lines per image generates almost 35,000 resolution elements per image. Note that the scanning mechanism determines the number of ir lines per image. The number of TV lines displayed per image, although important, is not the critical specification.

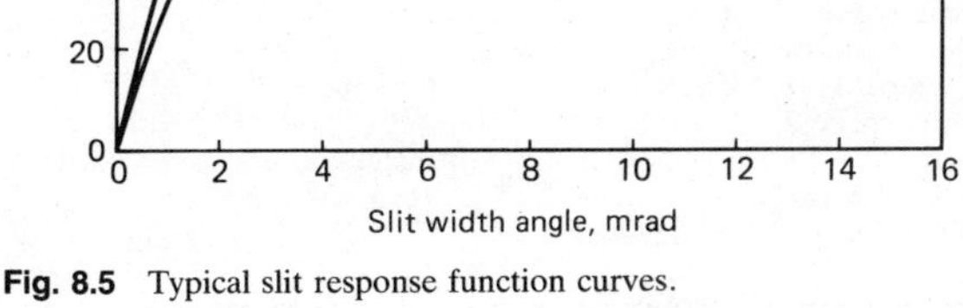

Fig. 8.5 Typical slit response function curves.

Dynamic Range

Dynamic range specifies the ability of a thermal imaging system to resolve fine temperature differences while simultaneously imaging large temperature spans. Dynamic range can be expressed as the number of gray shades, or digital levels of intensity, used to encode the thermal image, much like the number of gradations on a thermometer.

Since many measurements are made after an image has been stored on either videotape, computer disk, or the internal image memory, it is important to be able to accurately measure every object in the image without adjusting the controls. A greater number of levels of encoding allows both goals of fine resolution and wide span to be met simultaneously.

Calibration and Accuracy

Thermal imaging systems measure radiation within a certain spectral band and must convert radiant intensity to temperature by the use of a measured calibration curve. Because of variations in system spectral response, every system has a unique calibration curve. Any modifications to the spectral response, such as the use of optical filters, will create a new curve. Also, any additional optical elements such as telescopes and windows will reduce the transmission of radiation and change the slope of the curve.

The basic calibration curves for modern thermal imaging radiometers are measured at the factory and stored in a read-only memory (ROM) along with any special curves for installed optical filters. Older instruments generally have printed calibration curves which the operator must access to infer temperatures. When additional optics are used, the spectral response is generally sufficiently flat to be insignificant, so that only a transmission factor is required. Some instruments can sense when a particular optic is attached and automatically correct for its inherent transmission loss.

The error in measurement of any system can be divided into a random error (precision) and a fixed error (bias).[9,10] The purpose of instrument calibration is to remove the bias and define the precision. The NETD of a thermal imaging radiometer a measure of its precision. Should manufacturers be able to totally remove bias through calibration, thermal imaging radiometers would be very accurate given typical NETDs of 0.18°F (0.1°C). However, most instruments are calibrated over a wide range of target temperatures, typically several hundred degrees Fahrenheit. Bias is difficult to remove over this broad range. The calibration curve fit to the data points will be imperfect. The quality of the fit depends on the number of calibration data points taken and on the algorithm used.

Thermal imaging radiometers must also endure wide variations in environmental temperatures. Since these instruments are ac coupled devices, they really measure differences. To obtain absolute temperatures, there must be a reference to provide dc restoration. In many instruments the reference is the internal scanner temperature. Thus, if the scanner temperature varies from its value at calibration, the dc restoration level will shift. To provide a versatile instrument, this shift is compensated for electronically. The internal reference is usually measured with a thermistor or other electronic temperature sensor. System accuracy is intimately related to the accuracy of this sensor and the quality of electronic controls to compensate for instrument temperature changes.

Detector sensitivity changes rapidly with detector temperature; thus good stability in detector temperature is mandatory. For years, liquid nitrogen has been used as a detector coolant. As a liquid evaporating under its own vapor pressure, its temperature stability is excellent. Joule-Thompson (JT) cooling is the expansion of a high-pressure gas through a small orifice. Thermoelectric cooling employs the Peltier effect. Its great advantage is that only electricity, no gas or liquid, is used. Another all-electric cooling mechanism is the Stirling engine.

Finally, changes in detector sensitivity, optics transmission, and electronic drift will cause the instrument to go out of calibration over a period of time. Every instrument must periodically be recalibrated according to the manufacturer's recommendations.

To keep instrument costs down, manufacturers will perform a calibration which satisfies the accuracy requirements of the majority of customers. Typical liquid nitrogen cooled instrument accuracy is plus or minus 2 to 5 percent of the temperature reading. In the context of versatility and all the other measurement uncertainties confronting the user, these accuracies are quite reasonable.

Spectral-Band Selection

The earth's atmosphere absorbs radiated energy in the infrared except for two wavelength regions called the *atmospheric windows*.[4,5,8] Typically, the water vapor in the atmosphere absorbs the majority of the infrared energy over much of the spectral band. Many gases absorb as well, but generally at only one or a few narrow spectral lines.

The atmospheric windows allow radiometric measurements with minimal losses. The 8- to 14-μm (long-wave) region is exceptionally free of absorption except with very high atmospheric water content. The 3- to 5-μm (short-wave) region has a relatively high degree of transmission but usually requires compensation when high-accuracy measurements are to be made at path lengths greater than 1 m.

Modern thermal imaging radiometers are available with 8- to 14-μm (long-wave), 3- to 5-μm (short-wave), or 3- to 14-μm (broadband) spectral response. Instruments are also available which detect and image short and long wavebands independently and simultaneously. These arc often called *two-color* systems.

Because of a higher thermal contrast at "earth" temperatures [−4 to about 120°F (−20 to 50°C)] in the long-wave region, greater overall system performance can be achieved. At higher temperatures, the 3- to 5-μm waveband exhibits higher thermal contrast. However, there is always more energy at higher temperatures in a given waveband. Thus, if the 8- to 14-μm waveband is adequate at earth temperatures, it will certainly be adequate at higher temperatures. In fact, filters must often be used to limit the amount of radiant energy impinging on the detector at higher temperatures in both wavebands.

Most common material surfaces have higher emittances in the 8- to 14-μm waveband. Some surfaces have higher emittance in the short-wave region. Often one waveband will be preferred because of the spectral emittance characteristics of the target to be measured.

Sun glint is a far more serious problem in the 3- to 5-μm waveband than in the 8- to 14-μm waveband. There is several hundred times more energy from the sun in the 3- to 5-μm waveband than in the 8- to 14-μm waveband. Thus even a minor reflection of the sun's energy from a surface into the detector can saturate a 3- to 5-μm system while the 8- to 14-μm system remains relatively unaffected. For outdoor work, this requires users of 3- to 5-μm waveband systems to be extremely cautious in their viewing and interpretation of their thermal imagery.

Lower-cost detectors and thermoelectrically cooled detectors are available for the 3- to 5-μm band. The need to correct for atmospheric transmission generally results in less accurate temperature measurements. Broadband systems have the highest sensitivity but are most susceptible to atmospheric losses and variations in atmospheric transmission due to weather conditions.

Specification Summary

Table 8.1 summarizes the key elements to be considered when selecting a thermal imaging radiometer.

8.4 RADIOSITY

Radiosity is defined as the electromagnetic energy coming from a target. As such it is a combination of energy emitted from, reflected from, and transmitted through a target. In

Table 8.1 Key Elements to Selection of a Thermal Imaging Radiometer

Instantaneous field of view	Color image
Field of view	Black-and-white image
Infrared lines/frame	Spectral band
Noise equivalent temperature difference	Type of cooling
Scan speed	Portability
Accuracy	Reliability
Dynamic range	Image-processing compatibility
Direct temperature readout	Technical support
Emittance correction	Cost

equation form, this is

$$R = E_\epsilon + E_\rho + E_\tau \quad (8.1)$$

where E_ϵ = emitted energy $[=\int_0^\infty \epsilon(\lambda)B(\lambda, T_t)d\lambda]$
E_ρ = reflected energy $[=\int_0^\infty \rho(\lambda)B(\lambda, T_b)d\lambda]$
E_τ = transmitted energy $[=\int_0^\infty \tau(\lambda)B(\lambda, T_f)d\lambda]$
ϵ = target emittance
ρ = target reflectance
τ = target transmittance
T_t = absolute target temperature
T_b = absolute temperature of background
T_f = absolute temperature of foreground
$B(\lambda, T)$ = Planck function at wavelength λ and temperature T

Conservation of energy requires

$$\epsilon(\lambda) + \rho(\lambda) + \tau(\lambda) = 1 \quad (8.2)$$

Emittance is the ratio of energy at a given wavelength emitted by a material at temperature T to that of a perfect radiator (blackbody) at the same temperature.[5] *Reflectance* is the fraction of energy reflected by a material at a given wavelength to that incident upon it. *Transmittance* is the fraction of energy transmitted through a target at a given wavelength to that incident upon it. Emissivity, reflectivity, and transmissivity are special cases of emittance, reflectance, and transmittance.[5]

Emissivity applies to a material which is optically polished, flat, and thick enough to be opaque ($\tau = 0$). Few materials exist in the real world which meet these criteria. For opaque materials, the emittance will typically be higher than the handbook values of emissivity. Scratches, bends, and wrinkles all tend to increase the emittance. Also, emittance is spectral in nature.

8.5 GRAYBODY APPROXIMATION

Evaluation of radiosity using Eq. (8.1) is impractical. Spectral values of emittance, etc., are not available; further, thermography instruments average over a finite waveband. Virtually all thermographic radiometric measurements are based on the graybody approximation. A graybody has emittance, reflectance, and transmittance all less than one and constant over all wavelengths of interest. Thus these parameters can be brought outside the integral. In equation form, this is

$$R = \epsilon\int_0^\infty B(\lambda, T_t)\,d\lambda + \rho\int_0^\infty B(\lambda, T_b)\,d\lambda + \tau\int_0^\infty B(\lambda, T_f)\,d\lambda \quad (8.3)$$

Although the resulting integrals can be integrated to provide the Stefan-Boltzmann T^4 law, Eq. (8.3) is intentionally left in the form shown above. In a following section system response will be added, disallowing closed-form integration.

One measures or assumes values for ϵ, ρ, and τ over the instrument response waveband. Caution must be exercised not to assume an emittance value for a waveband other than the system response waveband.

8.6 APPLICATION OF RADIOSITY TO THERMAL IMAGING RADIOMETERS

Thermal imaging radiometers do not measure temperature directly (11). They measure radiosity. The detector output of a thermal imaging radiometer is proportional to the target radiosity sensed by the instrument and is weighted by the optics transmission and detector response. In equation form, employing the graybody approximation:

$$S = \epsilon\int_0^\infty s(\lambda)B(\lambda, T_t)\,d\lambda + \rho\int_0^\infty s(\lambda)B(\lambda, T_b)\,d\lambda + \tau\int_0^\infty s(\lambda)B(\lambda, T_f)d\lambda \quad (8.4)$$

where S is the system output, $s(\lambda)$ is the system response and other terms are defined above. In well-designed radiometers the system output is a linear function of the detector output voltage or current. The system response limits the range of the integral to a waveband, usually 3 to 5 μm or 8 to 14 μm. Detector response is nonlinear with wavelength. The integrals shown above are difficult to perform in the best of cases and are not attempted. Rather, they are written as a temperature dependent radiance value $I(T)$. Thus Eq. (8.4) becomes

$$S = \epsilon I(T_t) + \rho I(T_b) + \tau I(T_f) \quad (8.5)$$

Equation (8.5) is the general radiometric equation employing the graybody approximation. As discussed previously, blackbody sources are used to evaluate S for various temperatures T. For a blackbody, $\epsilon = 1$, $\rho = 0$, and $\tau = 0$. Thus $S = I(T_t)$ for a blackbody. The resulting system calibration is used to infer temperature. Obviously, this can be done directly only with blackbody sources. For real-world targets, the reflected and transmitted energies must be subtracted before target temperatures can be calculated.

8.7 TEMPERATURE MEASUREMENT

Emittance Considerations

Emittance, reflectance, and transmittance must sum to one. Thus they are complementary in the sense that if one is large, the others must be small.

In general, dielectrics have relatively high emittances (0.8 to 0.98). Bright metals have relatively low emittances (0.05 to 0.20). Oxidized metals generally have intermediate emittances (0.4 to 0.7). These are rule-of-thumb values for reasonably flat surfaces. Geometry plays a major role in the emittance of a material. Concave shapes such as recesses and holes can dramatically increase emittance. For example, the thermal image of recessed bolt holes in a hot metal plate will often look "hotter" than the metal surface. This is usually due to the increased emittance caused by the recess. In fact, a recess seven times as deep as its diameter will have an emittance close to one regardless of the material. This emittance contrast can be used advantageously. Table 8.2 gives typical emittances of some common materials. These values are for reasonably flat surfaces. Tables of emissivities for other materials may be found in Refs. 4, 5, 12, and 13.

Many materials, especially metals with thin coatings, can

Table 8.2 Nominal Emittances (Total Normal) of Common Materials

Material	Emittance, 8–12 μm
Aluminum (foil)	0.05
Asphalt pavement	0.93–0.97
Carbon (candle soot)	0.95
Ceramic (e.g. insulators)	0.9
Chipboard	0.9
Concrete	0.9
Copper (polished)	0.05
Copper (oxidized)	0.7–0.8
Gold	0.02
Iron (rusted)	0.65
Paint (26% aluminum)	0.30
Paint (nonmetallic)	0.9–0.98
Plastics (infrared opaque)	0.9–0.95
Powder (Dr. Scholl's Foot Spray)*	0.95
Powder (Gillette Foot Guard Spray)*	0.95
Rubber	0.9–0.95
Silver	0.02
Skin (human)	0.98
Steel (polished)	0.08
Steel (oxidized)	0.80–0.90
Tape (black vinyl electrical)*	0.95
Water	0.98
Wood	0.90–0.95

* Used frequently as a removable emittance reference.

have a dramatic variation of emittance with wavelength in both the 3- to 5- and 8- to 14-μm wavebands. The magnitude of this dependence depends on the material and, the coating and its thickness. Painted metals fall into this category.[5,14] The graybody approximation must be used with caution with these materials. An average value of emittance will give correct results near target temperatures for which it was measured. However, using the graybody approximation for a highly spectral emitter will cause errors at other temperatures due to spectral weighting of Planck's function. The seriousness of the error depends on the spectral nature of the emitter and its temperature.

Emittance varies with viewing angle. The emittance of dielectric materials decreases with increasing viewing angle. The emittance of metals increases with increasing viewing angle.[5] However, if the viewing angle is kept within about 60° of the normal to the target surface, values will remain fairly constant.

Narcissus

Narcissus is the detector seeing its own image. Since detectors in thermal imaging radiometers are cooled to some 200°C below room temperature, detector narcissus appears as a cold spot or spots on a thermal image. Well-designed systems will not exhibit narcissus from their own optics, unless improperly used. However, external optics and reflective targets will often cause narcissus. Flat, polished materials are typical culprits. External windows such as employed on vacuum chambers, semiconductors such as silicon wafers or chips, and metal targets are all capable of producing this annoying effect. The high thermal contrast of the detector to the ambient temperature can still produce a noticeable cold spot, even for highly transmissive window materials such as zinc selenide.

There are four ways to deal with narcissus:

1. Ignore it and make no measurements on or close to the cold spot.
2. Tilt the offending window, optic, or target so that the cold spot is reflected out of the field of view. This can be difficult for microscopic applications where the depth of field is small.
3. For nontransmissive materials, coat with a highly emissive material.
4. Treat the cold spot as a different temperature background. Digital image processing can often be used to subtract the background pixel by pixel, an excellent way to remove narcissus effects.

Distance Considerations

The following measurement techniques all assume an extended source as the target. This means that the spatial measurement resolution of the instrument is small enough to see only the target at the point of measurement. As previously discussed, measurement resolution is typically five times target discrimination resolution (usually quoted as spatial resolution).

Opaque Targets

The system calibration contains values of $I(T_t)$ for various temperatures T_t. Equation (8.5) must be solved for $I(T_t)$ and the system calibration used to calculate the target temperature. Since the transmittance of an opaque target is zero, the result is

$$I(T_t) = \frac{S - (1 - \epsilon)I(T_b)}{\epsilon} \tag{8.6}$$

The instrument measures S. Unknowns are emittance and background energy which is reflected off the target into the detector. Reflectance, calculated using Eq. (8.2), is one minus the emittance. The target emittance can be measured as shown in the following section, or a handbook value can be used.

The energy reflected from the target can be measured or approximated. Some thermographic instruments assume background energy to be that of the ambient environment. They use the internal temperature of the scanner to approximate this radiance value. Caution must be exercised with this approach. Many environments, especially industrial environments, have much higher temperature backgrounds than ambient. Also, a clear sky will be radiometrically as much as 90°F (50°C) colder than ambient air temperature. Thus, sky reflections, common in airborne work, will be calculated much too high with this approach.

It is best to find the blackbody equivalent radiance of the background. There are two simple ways to do this:

1. For smooth targets, the angle of incidence equals the angle of reflection. Estimate the viewing angle between scanner line of sight and the normal at the target surface (θ in Fig. 8.6). Equal to this angle is the angle of incidence ϕ. Point the scanner away from the target at an angle of θ. If the resulting thermal image is reasonably uniform, measure $I(T_b)$. If the thermal image shows gross variations, measure

an average or place a sheet of uniform-temperature material as a background. Use a good emitter such as wood or paper.

2. For convoluted targets, a different approach is necessary. A crumpled and reflattened piece of aluminum foil attached to a piece of rigid material will act as a reasonably good diffuse reflector. Place an appropriately sized piece directly in front of the target. An average measurement of $I(T_b)$ off the foil will give a good value for background energy. Again, if background energies vary greatly, construct a background from high emittance material.

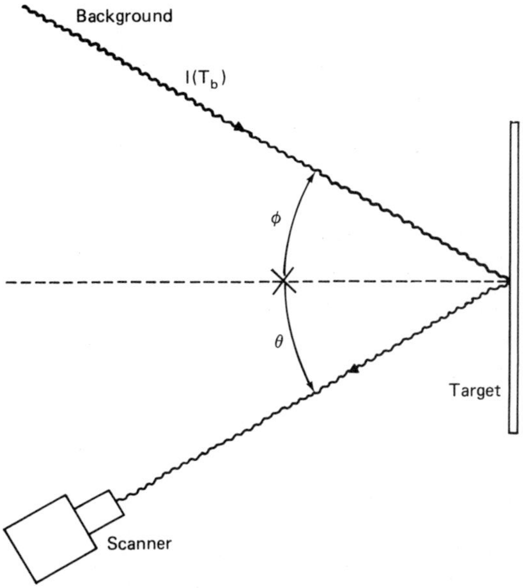

Fig. 8.6 Background energy $I(T_b)$ reflecting off a smooth target.

An artificial background will eventually react to the radiant energies it is screening. Thus it will not remain uniform forever. Use a thermal imaging radiometer to periodically check background uniformity. Emittance varies with viewing angle. Limit viewing angle to within 60° from normal. Units of $I(T_b)$ will be apropos to the instrument used. Some instruments request a temperature and convert this value to $I(T_b)$; in such cases treat the background as a blackbody to obtain the total radiance $I(T_b)$.

When $I(T_t)$ is found, the system calibration can be used to convert this value to a target temperature. Modern instruments will perform the entire calculation for the operator. However, it is important to understand the mathematics necessary to accomplish this feat, especially when dealing with nonopaque targets.

Nonopaque Targets

For targets where the transmittance is not zero, Eq. (8.5) must be used. Two cases are considered: (1) foreground temperature equals background temperature; and (2) foreground temperature does not equal background temperature.

Foreground Temperature Equals Background Temperature In this case Eq. (8.5) simplifies to

$$S = \epsilon I(T_t) + (1 - \epsilon)I(T_b) \tag{8.7}$$

since $\rho + \tau = 1 - \epsilon$ and $I(T_b) = I(T_f)$. The solution to Eq. (8.7) is the same as that to Eq. (8.6). Thus, when background and foreground temperatures are equal, one can use the opaque target calculation and proceed as above.

Background and Foreground Temperatures Unequal In this case, Eq. (8.5) must be solved without simplification:

$$I(T_t) = \frac{[S - \rho I(T_b) - \tau I(T_f)]}{\epsilon} \tag{8.8}$$

Values of emittance, reflectance, and transmittance as well as background and foreground energies must be determined. Background energy was discussed above. Foreground energy can be determined by removing the target and measuring. Once $I(T_t)$ is determined, the target temperature can be found using the system calibration. *Note:* Virtually all thermal imaging radiometers with built-in emittance correction are designed for opaque targets. Thus, the equation to be solved by them must have the form of Eq. (8.6) or (8.7). Do not use automatic emittance correction mode for nonopaque targets where the foreground temperature does not equal the background temperature unless specifically directed by the manufacturer.

8.8 MEASUREMENT OF EMITTANCE, REFLECTANCE, AND TRANSMITTANCE

The reflectance technique and the reference emittance technique are the two common methods of determining emittance and reflectance for opaque objects. In both cases one uses the radiance or level output of the thermal imaging radiometer; temperature need never be determined. The reflectance technique works best on reasonably flat, smooth targets and on transmissive targets. The reference emittance technique is valid only for opaque targets. It works well on rough targets where the reflectance technique is difficult. Transmittance is measured in a manner similar to the reflectance technique.

Reflectance Technique

Reflectance is the ratio of energy reflected from a target to that incident upon it. Thus all that need be done is to measure this ratio. From Eq. (8.5), we see that the emitted and transmitted energies must be taken into account. The simplest way is to measure the difference in reflected energy ratioed to the difference in incident energy. Assuming that the target temperature and the energy transmitted through the target remain constant for the duration of the measurement and are uniform over the area of measurement, the difference measurement will cause them to cancel. If the reflected energies are different, the result is

$$\rho = \frac{\Delta S_{\text{refl}}}{\Delta I_{\text{inc}}} \tag{8.9}$$

where ρ = reflectance
ΔS_{refl} = difference in reflected energy
ΔI_{inc} = difference in incident energy

All that is required, then, is a two-temperature source where each radiosity remains constant over the duration of the measurement. Blackbodies are not necessary. However, the constant radiosity requirement of the sources encourages the use of good emitters.

Figure 8.7 shows the experimental setup. The two sources should be arranged so that they may be viewed directly and their reflection from the target may be viewed by the thermal imaging radiometer. Using the level or radiance mode of the instrument, find the ratio of the two differences indicated. Once reflectance has been obtained, emittance follows simply for opaque objects as one minus the reflectance. For transmissive targets, the transmittance must also be subtracted from one to obtain emittance.

Transmittance Technique

If the two temperature sources are placed behind a transmissive target, one can measure transmittance in a manner similar to the above reflectance measurement. In equation form

$$\tau = \frac{\Delta S_{\text{trans}}}{\Delta I_{\text{inc}}} \tag{8.10}$$

where τ = transmittance
ΔS_{trans} = difference in transmitted energy
ΔI_{inc} = difference in incident energy.

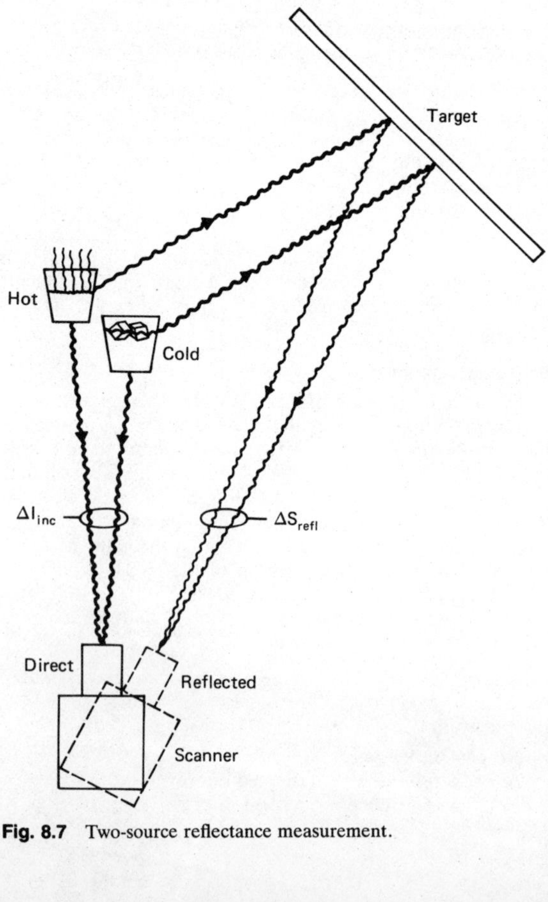

Fig. 8.7 Two-source reflectance measurement.

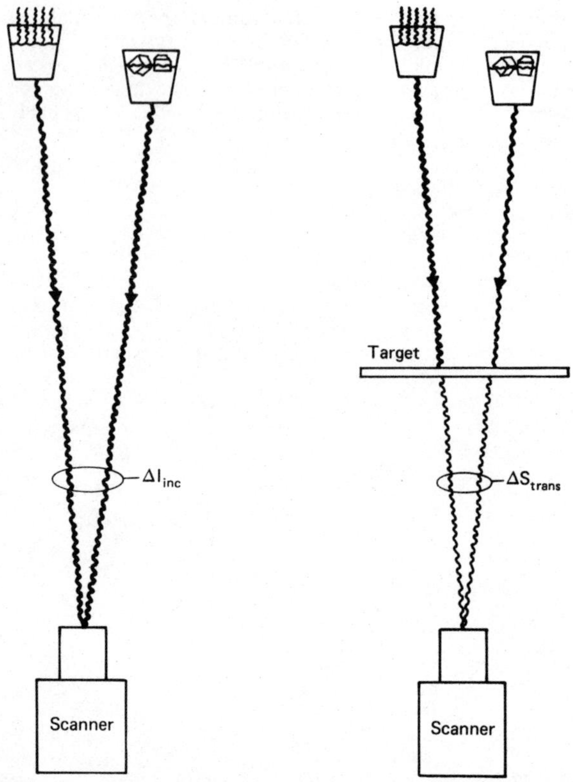

Fig. 8.8 Two-source transmittance measurement.

Figure 8.8 shows the experimental setup. Using the level or radiance mode, measure the radiance difference with the target removed, ΔI_{inc}. Insert the target and measure the difference ΔS_{trans}. Equation (8.10) is valid provided the emitted and reflected energies are constant for the duration of the measurement and uniform over the measurement area.

Reference Emittance Technique

This method works for opaque objects only. A material of known emittance is brought to the same temperature as the target. This is often done by applying the reference emitter to the target (Table 8.2). Of course, the target material must be compatible with the reference emittance material.

In addition to the reference emitter temperature being equal to the target temperature, the background must be uniform. Also, the target temperature must be different from the background temperature (the greater the difference, the better). The equation is

$$\epsilon_t = \epsilon_r \frac{\Delta I_{tb}}{\Delta I_{rb}} \tag{8.11}$$

where ϵ_t = target emittance
ϵ_r = reference emittance
ΔI_{tb} = radiance difference between target and background
ΔI_{rb} = radiance difference between reference and background

As above, the measurement of differences obviates the concern of background contributions. Again no temperatures need be measured. One uses the radiance or level mode. Finding the ratio means that absolute units are not important.

8.9 APPLICATIONS

Thermal imaging radiometers are applied where surface thermal patterns or surface temperatures give clues to a problem which needs to be solved.[4] Often, the problem may occur beneath the surface but through heat-transfer mechanisms will express itself at the surface. Thus one ought to be knowledgeable about thermal imaging radiometers, heat transfer, and the target system. This is frequently a large dose for an individual; many thermography applications are approached as a team effort. Thermography has often been called a "problem-finding tool." As such, further verification is typically required. Spurious inputs have fooled the most diligent thermographers. Experience plays a key role in reducing the intensity of verification but will not replace it. The following applications are briefly described with an eye to the key elements for successful acquisition and interpretation of thermal imaging radiometer data.

Electrical Power Systems

One of the "bread-and-butter" applications of thermography, electrical systems surveys, can save tens of thousands of dollars in preventing electrical equipment outages.[15–17] The principle is straightforward: bad electrical connections are hotter under load than good electrical connections. Since some 80 percent of electrical system failures are due to faulty connections, finding the bad ones goes a long way toward quality maintenance. In addition, overloaded circuits and load imbalances can be found using thermography.

Four factors should be considered to obtain quality data which can be usefully interpreted: target emittance, target distance and size, system load, and solar load.

Target emittance is crucial in electrical systems since much of what is viewed is metallic. To get accurate temperature measurements, the emittance must be known. This is often a rigorous requirement for electrical surveys. The emittance measurement techniques discussed above cannot easily be done in situ because the targets are inaccessible because of their electrical load. However, measurement of typical components brought in from the field is definitely recommended. In lieu of actual measurement, use the rule-of-thumb values in Table 8.2.

As Table 8.2 shows, the variation in emittance is considerable. Assuming an emittance of 1.0 (blackbody equivalent temperature or apparent temperature) will cause an underestimation of the actual target temperature if it is hotter than the background. (A target with emittance less than 1 must be hotter than a blackbody to emit the same energy. So, if the emittance is overestimated, the temperature will be underestimated.)

Often small components are overheated in a small area. All thermal imaging radiometers "see" a spot of finite size as they scan an area. As discussed previously, the resolution element size is typically one-fifth the measurement element size. To assure a good measurement, multiply the IFOV or 50 percent SRF by 5. The inverse of the result will give you a rule-of-thumb multiplier. For example, a system with a 2-mrad IFOV would have a multiplier of 100 $[1/(0.002 \times 5)]$. To use this number, multiply it by the target size (smallest dimension facing the scanner). This will give the maximum distance its temperature can be confidently measured. Thus a 1.625-in- (4.13-cm)-diameter transmission line can be confidently measured at a distance of up to about 162 in or 13 ft.

Conversely, dividing the distance to the target by the preceding number will give the distance over which the temperature will be averaged. For our example, at 10 ft, the system will average over a distance of $^{10}/_{100}$ or 0.1 ft (1.2 in). Most thermal imaging radiometers have a square IFOV; if not, the calculation can be done for both dimensions.

The error introduced by being outside this rule-of-thumb distance depends on the difference between the target temperature and the surroundings temperature (the area around the target). The average of five IFOVs in each direction is a weighted average. The center IFOV has the greatest weight, those next to it less weight, and those at the outer perimeter even less. As shown above, one can compensate for distance effects using an equation. If the target is hotter than its surroundings, being too far away will cause one to estimate the temperature lower than the actual value.

To be on the safe side, telescopic optics are often used in electrical surveys. A 3× telescope will increase the working distance by a factor of 3. So, in the transmission line example above, one can now be 39 ft away and obtain a good measurement. For helicopter surveys of transmission lines, a telescope is essential. There is a tradeoff. The power of the telescope will be limited by the vibration it is subjected to. For inexpensive mounting or hand-held use in a helicopter, a 3× or 4× telescope is about the maximum power one can use.

Electronics Applications

Electronics applications include printed-circuit-board inspection, wire-wrap inspection, location of short-circuited traces in multilayer boards, and power transistor and power hybrid circuit analyses.[18–20] As with electrical system surveys, emittance and spatial resolution play a major role. Usually environment and load can be better controlled, reducing the effect of these variables on the quality of the inspection. Today most inspections fall into research and development or process monitoring categories. Soon automated inspection will be typical; systems are being developed to perform these tasks.

Printed-circuit boards (PCBs) fall into two major categories: conformally coated and uncoated. Conformal coating is done for military boards and other systems exposed to rugged use. Conformal coatings are plastic materials applied to the entire board covering all components. The plastics used are nearly opaque in the 8- to 14-μm waveband and somewhat transmissive in the 3- to 5-μm waveband.[21] The 8- to 14-μm waveband is preferred since uniform emittance is desired. Transmission through the conformal coating results in the emittances of the individual components being seen by a 3- to 5-μm-waveband instrument. Since many transistors are enclosed in metal cans, some integrated circuits are gold-plated, and board traces are tinned with solder, there are dramatic variations in emittance on a typical PCB. Spatial resolution is still quite significant. A system with excellent spatial resolution and wide field of view is desired. Dynamic range is important. Location of a hot component is achieved by pattern recognition of the relatively cool surrounding components. To get spatial detail of these areas which may vary by only a few degrees and still be able to measure the hot component, wide dynamic range is necessary. Six bits (64 levels) is the bare minimum; seven bits (128 levels) or higher is preferred.

One application is to compare the thermal image of a powered-up board under test to that of a standard board. This can be done visually by the operator or automatically by digital image processing through image subtraction. Anomalies in temperature between the test board and the standard board indicate component malfunction. A component may be too hot or too cold (not working). This is a quality control technique.

Design engineers use thermography to indicate design flaws in PCBs. Heat dissipation and power delivered to individual components is often difficult to predict. These studies are performed at the board level and in the cabinet or enclosure housing the entire electronics package. The latter verifies the effectiveness of convective cooling to the various boards and other elements. Since a thermal imaging radiometer cannot "see" through the metal case, the sides are often removed. To preserve the integrity of the convective cooling, a transmissive covering such as polyethylene film is wrapped around the case. Sometimes boards must be removed to map their temperatures. One infers the in situ temperature by extrapolating the observed cooling of the components after the board is removed back to the time when the board was in place. Video recording and subsequent digital image processing are quite useful in this application.

Short circuits in multilayer boards can often be found with thermography when other techniques fail. One knows the short exists, but does not know where it is. By applying current to the short-circuited traces, the path is located by the thermal imaging radiometer showing the temperature rise.

Incorrectly wire wrapped boards will frequently show the mistake thermographically. The culprit wire will often be hotter than the rest.

On the microscopic level, power transistors and hybrid circuits can be examined with a thermal imaging radiometer. Made of silicon (an infrared transmitter) and metals, these low-emittance devices are difficult to measure. Target narcissus is a major concern.

If possible, coat the transistor or hybrid with a high-emittance material such as spray foot powder. Some are conformally coated, allowing direct application of an 8- to 14-μm waveband system. Emittance correction can be done with some systems. A digital image processor is recommended to compensate for target narcissus through image subtraction and emittance correction if the target cannot be coated. If the unpowered image is subtracted from the powered image, the resulting radiosity difference from Eq. (8.5) is

$$\Delta S = \epsilon[I(T_{\text{hi}}) - I(T_{\text{lo}})] \tag{8.12}$$

where ΔS = signal difference at any point
ϵ = emittance at any point
$I(T_{\text{hi}})$ = target radiance at any point with power on
$I(T_{\text{lo}})$ = target radiance at any point with power off

As long as the background does not change, the difference technique will be valid according to Eq. (8.12). When narcissus is present or the background is otherwise nonuniform, the target must not move with respect to the radiometer.

Pattern Recognition Applications

The applications described above rely on self-generated thermal patterns and temperature measurement. In testing of materials for voids or nonuniformities, the thermal signature usually must be created by external application of thermal energy. Also, temperature measurement is not crucial. The resulting thermal pattern is transient with a decay time dependent on the heat capacity and thermal conductance contrast between "good" and "bad" areas, the location of the anomalies, and the thermal loading. Typical applications include detection of:

1. Moisture-damaged insulation in built-up roofing systems, a major maintenance application[22,23]
2. Building defects such as insulation voids in walls[1,24,25,26]
3. Concrete delamination in road surfaces and bridge decks
4. Buried piping such as underground steam line surveys[27]
5. Water absorption in honeycomb structure of aircraft control surfaces[28]
6. Voids in room-temperature vulcanizing (RTV) rubber external protective membrane (EPM) on rocket motor cases[29]
7. Irregularities in thin composites[30]
8. Voids in metals[31,32]

Items 1 to 4 rely heavily on environmental thermal variations to evoke the thermal patterns. In application 1, during the summer months, solar loading and the greater thermal mass of the water soaked insulation combine to produce a temperature contrast between dry and wet areas. The wet areas will be cooler in the day and warmer at night. Heat transfer through the wet insulation produces a higher temperature than the dry areas during the winter months. Application 2 also relies on heat-transfer differences and/or air infiltration and exfiltration during the heating or cooling season. Leaks in buried piping such as underground steam lines will produce a definite surface temperature pattern. Delaminations in road surfaces cause that portion of the highway to be more loosely thermally coupled to the earth. These areas will typically be warmer during the day and cooler at night.

Many modern aircraft employ lightweight, high-strength, honeycomb composite materials for some control surfaces. Over time, tiny surface cracks develop. Water leaks in and creates an imbalance in control surfaces. When the aircraft is at high altitudes, the water freezes. On landing, the higher thermal mass of the water areas causes them to remain cool longer than other areas.

Applications 6 to 8 are generally done under controlled environmental conditions. The EPM on rocket cases is examined after curing when the entire case is warm. The material just above a void will be somewhat thermally decoupled and will cool more quickly. Irregularities in thin composites require application of a uniform heat pulse and subsequent observation of the thermal decay. The flat surface of the material is heated and observed. In the case of void detection in metals, this technique does not work because of the high lateral thermal conductance of metals. Although difficult, the best results have been obtained by applying heat uniformly perpendicular to the metal face and observing the thermal

Table 8.3. Some Thermography Applications

Application	References
Electrical	
Generation	17
Transmission and/or distribution	36
Facility surveys	15,16
Electronic	
Printed-circuit-board inspection	18,19,20
Devices (transistors, hybrids, etc.)	37,38
Roof moisture surveys	22,23
Building diagnostics	1,24,25,26
Composite materials	28,29,30
Steam systems	27
Process monitoring and control	33,34,35
Environmental	39,40

contour lines on the metal face. Since a void restricts heat flow, the temperature just above it will be higher and lower just below causing a "pinch" in the thermal contour lines.

Note that the preceding descriptions are conceptual. The actual details of performing the surveys are much more complex. Refer to specific publications for details.

Applications Summary

From the applications outlined above, it is clear that thermal imaging radiometers are a tool with wide diversity. Table 8.3 summarizes with selected references many applications.

Thermal imaging radiometers provide an image of thermal patterns as well as measurement capability. As such, they give input to both the intuitive and logical aspects of reasoning. Proper application requires a balance of these aspects. With measurement and verification, one can interpret thermal patterns confidently. If one performs temperature measurements with regard to the observed thermal patterns, the true power of the system is realized.

REFERENCES

1. *Applications of Infrared Sensing Devices to the Assessment of Building Heat Loss Characteristics,* ASHRAE Standard, ANSI/ASHRAE 101-1981, The American Society of Heating, Refrigerating, and Air-Conditioning Engineers, Inc., Atlanta, Ga., 1981.
2. J. M. Lloyd, *Thermal Imaging Systems,* Plenum, New York, 1975.
3. G. Gaussorgues, *La Thermographie Infrarouge,* Technique et Documentation Lavoisier, Paris, 1984.
4. R. P. Madding, *Thermographic Instruments and Systems,* University of Wisconsin-Extension Department of Engineering & Applied Science, Madison, Wisc., 1979.
5. W. L. Wolfe, (ed.), *Handbook of Military Infrared Technology,* Office of Naval Research Department of the Navy, Washington, D.C., NTIS No. AD646082, 1965.
6. H. Kaplan, "Thermal Imaging Systems for Measuring Temperature Distribution," *Proceedings ASTM,* 1983.
7. H. Kaplan, *IR Imaging's Latest Crop,* Photonics Spectra, Optical Publishing, Pittsfield, Mass., September 1984.
8. R. D. Hudson, Jr., *Infrared System Engineering,* Wiley, New York, 1969.
9. R. B. Abernethy, et al. and J. W. Thompson, Jr., *Measurement Uncertainty Handbook,* Instrument Society of America, Research Triangle Park, N.C., 1980.
10. E. O. Doebelin, *Measurement Systems: Application and Design,* McGraw-Hill, New York, 1975.
11. R. P. Bruno, G. J. Burrer, and E. A. Saporetti, "Analytical Interpretation of Thermograms Including Digital Processing Techniques," *Proceedings Thermal Infrared Sensing Diagnostics (Thermosense V),* **371,** 230–239, Society of Photo-Optical Instrumentation Engineers (SPIE), Bellingham, Wash., 1982.
12. Y. S. Touloukian, et al., *Thermophysical Properties of Matter,* Plenum, New York, 1972. Vols. 7–9.
13. T. Baumeister, E. A. Avallone, and T. Baumeister, III, *Marks' Standard Handbook for Mechanical Engineers,* 8th ed, McGraw-Hill, New York, 1978.
14. R. P. Madding, "Science behind Thermography," *Thermal Infrared Sensing Diagnostics (Thermosense V) Proceedings,* SPIE, **371,** 2–9, Bellingham, Wash., 1982.
15. G. S. Baird, "Thermographic Aspects of Industrial Predictive Maintenance," *Thermal Infrared Sensing Diagnostics (Thermosense V) Proceedings,* SPIE **371,** 55–59, Bellingham, Wash., 1982.
16. S. A. Seeber, "Infrared Testing: The Least Cost Approach to Effective Electrical Maintenance," *Thermosense VIII, Proceedings,* **581,** 138–146, Bellingham, Wash., 1985.
17. D. C. Stevens, "Monitoring of Generator Stator Core Lamination Repairs Using Thermographic Techniques," *Thermal Infrared Sensing Diagnostics (Thermosense V) Proceedings,* **371,** 38–46, SPIE, Bellingham, Wash., 1982.
18. H. R. Stirn, "Failure Analysis of Multilayer Printed Circuit Boards Using Thermography," *An International Conference on Thermal Infrared Sensing for Diagnostics and Control (Thermosense VI) Proceedings,* **446,** 204–208, SPIE, Bellingham, Wash., 1983.
19. R. W. Sherman, "Infrared Imaging Technique that Comprehensively Verifies Thermal Design of Avionics Equipment," *An International Conference on Thermal Infrared Sensing for Diagnostics and Control (Thermosense VI) Proceedings,* **446,** 164–167, SPIE, Bellingham, Wash., 1983.
20. Boulton, H., "Combining Thermographic, Environmental and Functional Testing for Characterization and Performance Measurement of Electronic Devices and Assemblies," *An International Conference on Thermal Infrared Sensing for Diagnostics and Control (Thermosense VIII) Proceedings,* **581,** 78–84, SPIE, Bellingham, Wash., 1985.
21. H. Kaplan, "The Effect of Conformal Coating on Thermal Infrared Profiles of Printed Circuit Boards," *An International Conference on Thermal Infrared Sensing for Diagnostics and Control (Thermosense VIII), Proceedings,* **780,** 124–129, SPIE, Bellingham, Wash., 1987.
22. W. Tobiasson and C. Korhonen, *Summary of Corps of Engineers Research on Roof Moisture Detection and the Thermal Resistance of Wet Insulation,* Special Report 78-29, Department of the Army, Cold Regions Research and Engineering Laboratory Corps of Engineers, Hanover, N.H., 1978.
23. C. Korhonen, W. Tobiasson, and A. Greatorex, "Comparison of Aerial to On-the-Roof Infrared Moisture Surveys," *An International Conference on Thermal Infrared Sensing for Diagnostics and Control (Thermosense VI) Proceedings,* **446,** 95–105, SPIE, Bellingham, Wash., 1983.
24. P. A. D. Mill, *Thermography—A New Building Science Tool,* Report Series No. 29, Public Works Canada, Ottawa, 1979.
25. J. B. Fang, R. A. Grot, and G. E. Courville, "Heat Loss due to Thermal Bridges in Buildings," *An International Conference on Thermal Infrared Sensing for Diagnostics and Control (Thermosense VI) Proceedings,* **446,** 34–42, SPIE, Bellingham, Wash., 1983.
26. S. J. Marshall, *Infrared Thermography of Buildings, A Bibliography with Abstracts,* Special Report 79-1, United States Army Corps of Engineers, Cold Regions Research and Engineering Laboratory, Hanover, N.H., 1979.
27. H. W. MacKay and G. R. Lawrence, "Application of Airborne Thermal Infrared for Steamline Maintenance," *An International*

Conference on Thermal Infrared Sensing for Diagnostics and Control (*Thermosense VI*) *Proceedings,* **446,** 106–111, SPIE, Bellingham, Wash., 1983.

28. J. Daymond, *Introduction of Thermography as Inspection Tool of Aircraft Composite Control Surfaces Proceedings*, Air Transport Assoc., Non-Destructive Test Forum, New York, 1986.
29. J. B. Chambers, "Thermographic Evaluation of Bond Lines and Material Consistency of Composites," *An International Conference on Thermal Infrared Sensing for Diagnostics and Control* (*Thermosense VII*) *Proceedings,* **520,** 92–95, SPIE, Bellingham, Wash., 1984.
30. Y. L. Hinton, R. J. Shuford, and C. G. Pergantis, "Applications of Real-Time Thermography for Nondestructive Testing of Composites," *An International Conference on Thermal Infrared Sensing for Diagnostics and Control* (*Thermosense VII*) *Proceedings,* **520,** 142–148, SPIE, Bellingham, Wash., 1984.
31. G. L. Orlove, R.P. Madding, C. P. Burger, and B. Raj, "Void Detection in Metals Using Heat Flow Parallel to the Target Surface," *An International Conference on Thermal Infrared Sensing for Diagnostics and Control* (*Thermosense VIII*) *Proceedings,* **582,** 24–26, SPIE, Bellingham, Wash., 1985.
32. C. P. Burger and J. Gryzagoridis, "Transient Field Infrared Video Thermography as a Procedure for Nondestructive Testing," *Proceedings—Measurement Science Conference,* Palo Alto, Calif., January 20–21, 1983.
33. R. P. Madding, "Infrared Sensors and Process Control," *An International Conference on Thermal Infrared Sensing for Diagnostics and Control* (*Thermosense VI*) *Proceedings,* **446,** 9–17, SPIE, Bellingham, Wash., 1983.
34. M. A. Khan, N. H. Madsen, and B. A. Chin, "Infrared Thermography as a Control for the Welding Process," *An International Conference on Thermal Infrared Sensing for Diagnostics and Control* (*Thermosense VI*) *Proceedings,* **446,** 154–163, SPIE, Bellingham, Wash., 1983.
35. T. Bantel, D. Bowman, J. Halase, S. Kenue, R. Krisher, and T. Sippel, "Automated Infrared Inspection of Jet Engine Turbine Blades," *An International Conference on Thermal Infrared Sensing for Diagnostics and Control* (*Thermosense VIII*) *Proceedings,* **581,** 18–23, SPIE, Bellingham, Wash., 1985.
36. R. A. Faccinto, "The Use of Infrared Thermography in Evaluating Substation Transformers," *An International Conference on Thermal Infrared Sensing for Diagnostics and Control* (*Thermosense VIII*) *Proceedings,* **581,** 103–107, SPIE, Bellingham, Wash., 1985.
37. H. Boulton and B. S. Siegal, "Integrating Temperature Test Equipment into Production ATE," *Test & Measurement World,* Interfield, July 1984.
38. H. Boulton, "Testing Hybrids Using Thermography," in *Hybrid Circuit Technology,* Lake, July 1984.
39. R. P. Madding, F. L. Scarpace, and T. Green, III, "Thermal Plumes Along the Wisconsin Shore of Lake Michigan," *Wisc. Acad. Sci. Arts Lett. Trans.,* **LXV,** 86–119, 1977.
40. G. J. Marmer, J. V. Tokar, and R. P. Madding, "Comparison of Thermal Discharges," *Water Resources Bulletin, American Water Resources Assoc.,* **II,** No. 6, 1157–1180, December 1975.

Chapter 9

Humidity Detectors

PIETER R. WIEDERHOLD
General Eastern Instruments Company

9.1 INTRODUCTION

Humidity can be expressed in a variety of different forms: wet-bulb temperature, percent relative humidity, vapor pressure, mixing ratio, dew-frost point, grains per pound, parts per million, and others. These parameters can be measured by a number of different instruments, each capable of accurate measurement under certain conditions and within specific limitations.

This chapter deals with the various types of humidity detectors and the sensors and instrumentation commonly used in industrial automated systems. Only those sensors which are in wide use and for which error analyses have been published by the National Bureau of Standards are discussed.

9.2 DEFINITION OF HUMIDITY

Dry air is a mixture of nitrogen (78 percent) and oxygen (21 percent), with carbon dioxide and other minor constituents making up the remaining 1 percent. The percentage of water gas in air is represented by the humidity.

Dalton's law states that the total pressure P_m exerted by a mixture of gases or vapors is the sum of the pressures of each gas if it were to occupy the same volume by itself. The pressure of each individual gas is called its *partial pressure.* The total pressure of an air-water gas mixture, containing O_2, N_2, and H_2O, is equal to the sum of the partial pressures of each gas:

$$P_m = P_{N_2} + P_{O_2} + P_{H_2O} + \cdots$$

Therefore, the partial pressure of water vapor in air is directly related to measurement of humidity. This vapor pressure varies from 23.2×10^{-6} in of mercury (0.078 Pa) at the −108°F (−77.7°C) frost point of "bone-dry" arctic or industrial dry air to 23.5 in of mercury (79.5 kPa) at the 200°F (93.3°C) dew point of saturated hot air in a product drier.

The ideal humidity instrument would be a linear, wide-range pressure gage, specific to water vapor and employing a primary or fundamental measuring method. Most humidity measurements are made by some secondary instrument which is responsive to a humidity-related phenomenon.

Table 9.1. Humidity Measurement Parameters

Parameter	Description	Units	Typical applications
Wet-bulb temperature	Minimum temperature reached by a wetted thermometer in an airstream	°F or °C	High-temperature driers, air conditioning, meteorology, test chambers
Percent relative humidity	The ratio of the actual vapor pressure to the saturation vapor pressure, with respect to water, at the prevailing dry-bulb temperature	0–100%	Monitoring conditioning rooms, test chambers, pharmaceutical and food packaging
Dew-frost point	Dew point is the temperature to which the air must be cooled to achieve saturation; if the temperature is below 32°F (0°C), it is called the *frost point*	°F or °C	Heat treating, annealing atmospheres, drier control, instrument air monitoring, meteorological-environmental measurements
Volume or mass ratio	Parts per million (ppm) by volume is the ratio of the partial pressure of the water vapor to the partial pressure of the dry carrier gas; ppm by weight is identical to ppm by volume, but the ratio changes according to the molecular weight of the carrier gas	ppm_v and ppm_w	Used primarily to ensure dryness of industrial process gases such as air, nitrogen, oxygen, methane, hydrogen

Humidity parameters most often encountered in scientific and industrial applications are shown in Table 9.1. A psychrometric chart provides a quick means of converting from one humidity format to another since dew point, relative humidity (RH), ambient temperature, and wet-bulb temperature are conveniently related to each other on a single chart.

9.3 HUMIDITY SENSORS

The humidity sensors discussed below represent some of the most commonly used detectors.

Types of Sensors

Primary Sensors Sensors may be divided into primary or fundamental types and secondary types. Fundamental sensors measure the humidity parameter directly based on the humidity parameter definition. Typical fundamental sensors are chilled-mirror dew-point sensors, wet-bulb sensors, and electrolytic sensors.

Fundamental sensors are used as calibration standards. These instruments, especially the chilled-mirror sensor, can have extremely high accuracies [to ±0.2°F (±0.1°C)] and excellent repeatability (virtually no aging and no hysteresis). The chilled-mirror sensor is considered the most accurate commercial humidity sensor, next only to the humidity standards employed at the National Bureau of Standards.

Secondary Sensors Secondary sensors measure a non-humidity parameter, such as resistance, capacitance, impedance, or mechanical expansion of a hair or membrane, which varies with humidity. They are calibrated against a humidity standard so that the relationship between the measured parameter and the actual humidity parameter is exactly known. Instruments using such secondary sensors usually have calibration adjustments.

Wet-Bulb–Dry-Bulb Psychrometer

Psychrometry has long been a popular method for monitoring humidity, primarily because of its simplicity and inherent low cost. A typical industrial psychrometer consists of a pair of matched electrical thermometers, one of which is maintained in a wetted condition.

Water evaporation cools the wetted thermometer, resulting in a measurable difference between it and the ambient, or dry-bulb measurement. When the wet bulb reaches its maximum temperature depression, the humidity is determined by comparing the wet-bulb–dry-bulb temperatures on a psychrometric chart. In a properly designed psychrometer, both sensors are aspirated at an air stream rate between 13 and 33 ft/s (4 and 10 m/s) for proper cooling of the wet bulb, and both are thermally shielded to minimize errors from radiation.

A properly designed and utilized psychrometer, such as the Assman laboratory type, is capable of providing accurate data. However, very few industrial psychrometers meet these criteria and are therefore limited to applications where low cost and moderate accuracy are the underlying requirements.

The psychrometer does have certain inherent advantages:

- Accuracy is superior to all other humidity sensors near saturation. Since the dry-bulb and wet-bulb sensors can be connected differentially, wet-bulb depression can be measured with a minimum of error.
- The simplicity of the device affords easy repair at minimum cost.
- The psychrometer can be used at ambient temperatures above 212°F, (100°C) and the wet bulb measurement is usable up to 212°F (100°C).

Major shortcomings of the psychrometer are:

- Few psychrometers give satisfactory results below 10 percent RH and accuracy is poor below 20 percent RH.
- Wet-bulb measurements at temperatures below 32°F (0°C) are difficult.
- Psychrometers cannot be used in small, closed volumes.

RH Sensors

Commercial RH detectors are secondary devices. They are either electrical sensors operating on the principle of measuring a change of resistance or capacitance or mechanical sensors such as a hair hygrometer.

Resistive Elements These devices are typically electrochemical sensors that offer ruggedness, compactness, and remote electronic readout ability. The three most widely used electronic percent RH sensors are the Dunmore element, the Pope cell, and the polymer sensor.

The *Dunmore sensor* employs a dual-wound inert wire grid on an insulative substrate that is coated with a lithium chloride solution of a controlled concentration. The hygroscopic nature of this salt causes it to take on water vapor from the surrounding atmosphere. The ac resistance of the sensor indicates the prevailing percent RH. Dunmore cells are excellent RH sensors; however, because of the characteristics of lithium chloride, they are usually designed to cover a narrow range of RH, typically about 20 percent.

Wide-range Dunmore sensors can be constructed with a cluster of narrow-range sensors in a common housing, mated with an electrical matching network. This arrangement, however, usually results in a fairly bulky sensor.

The *Pope cell* employs a similar bifilar conductive grid on an insulative substrate. In this case, the substrate is made from polystyrene that has been treated with sulfuric acid, resulting in sulfonation of the longer-chain polystyrene molecules. Because the sulfate radical SO_4 is highly mobile in the presence of hydrogen ions (available from the water molecule in the vapor form), the SO_4^{-2} ions can detach and take on H^+ ions. This alters the surface resistivity of the sensor as a function of the humidity.

In both Dunmore and Pope sensors, the element is arranged in an ac-excited Wheatstone bridge, so that only alternating current flows through the grid. Direct current excitation of either the Dunmore or Pope elements polarizes the sensor, eventually causing loss of calibration.

One significant advantage of the Pope sensor over the Dunmore sensor is that it has a wide range, typically covering 15 to 99 percent RH in a single element. However, the readout of the Pope sensor requires considerable attention since the resistance varies in a nonlinear fashion.

The *polymer sensor* is a macromolecular humidity sensing element composed of an electrode base plate coated with a humidity-sensitive macrosilicone polymer. The resistance of the electrode varies according to the change in relative humidity. The humidity-sensitive material is prepared by polymerizing a solution of quaternary ammonium bases and the reaction of this functional base with a silicone resin. This produces a three-dimensional thermosetting resin characterized by an excellent stability in extreme conditions.

Humidity is measured by the resistance change between the anode and the cathode. When the quaternary ammonium salt contained in the electrode shows ion conductivity caused by the presence of humidity, mobile ions in the film are created. The greater the increase of the humidity in the atmosphere, the greater the ionization becomes, and as a result, the concentration of mobile ions increases.

The Dunmore and Pope sensors are sensitive to contamination. If allowed to become saturated with water, they tend to suffer from "washout," resulting in loss of calibration. Both sensors are subject to errors if contaminated with soluble salts, halogens, or any material that can alter the surface resistivity of the element. Polymer sensors are less sensitive to contaminants because they are bulk-resistivity devices. They also exhibit less hysteresis and aging effects. All three sensors, especially the polymer sensor, respond quickly—in a few seconds.

Capacitive Sensors. In the last two decades several capacitive RH sensors have been developed. The thin-film polymer dielectric type is best known. Its capacitance is dependent on the water absorption in the sensor's dielectric material.

A typical capacitive sensor is a small capacitor with an organic polymer dielectric substrate. The lower electrodes are etched onto a metalized glass plate. The surface is then coated with an active polymer layer using dipping techniques, resulting in a uniform amorphous polymer layer of micron thickness. Most capacitive sensors have a fast response and, like the polymer sensor, are less sensitive to contaminants than the Pope cell and Dunmore cell.

Mechanical Sensors A large number of so-called hair hygrometers are still in use. These hygrometers use sensors that place a human or horse hair in tension and the expansion is a function of ambient relative humidity.

Sensors of this type are of simple and low-cost construction. Attainable accuracies, however, are modest to poor. These sensors also suffer from both aging effects and hysteresis and should be avoided in applications where good accuracy and repeatability are required. One advantage is their relative insensitivity to contaminants and their ability to withstand wetting (condensation) without permanent damage.

Strain-gage sensors have performance characteristics similar to those of the hair-type sensors, although some are more rugged. These sensors tend to be less sensitive to contaminants and are not damaged when accidentally wetted. However, accuracies and repeatability, especially over a wide span, are lower than those for any of the electrical RH sensors.

Dew-Point Sensors

Both fundamental and secondary types of dew point sensors are commercially available. The most commonly used dew point sensors include the following types.

Chilled-Mirror Sensor The condensation-type (chilled-mirror) sensor is one of the most accurate, reliable, and wide-range detectors available for measuring humidity. In the condensation-type hygrometer, a surface is cooled (either thermoelectrically, mechanically, or chemically) until dew or frost begins to condense out. The condensate surface is maintained electronically in a vapor pressure equilibrium with the surrounding gas while surface condensation is detected by optical, electrical, or nuclear techniques (see Fig. 9.1). The surface temperature is then the dew-point temperature by definition.

The largest source of error in a condensation hygrometer stems from the difficulty in measuring condensate surface temperature accurately. Typical industrial versions of the instruments are accurate to ±1°F (±0.55°C) over very wide temperature spans. With proper attention to the condensate surface temperature measuring system, errors can be reduced to less than ±0.4°F (±0.22°C). Condensation-type hygrometers can be made compact by means of solid-state optics and thermoelectric cooling. Condensate detectors and control loops that maintain surface temperatures within ±0.05°F (±0.028°C) of the true dew point are available.

Generally speaking, condensation-type hygrometers work well on clean gases down to −100°F (−73.3°C) frost points. However, below about −40°F (−40°C), response of the instrument is considerably slowed, and susceptibility to errors due to contamination is increased. Also, the task of cooling the surface from high ambients to a low temperature usually requires an ancillary coolant, resulting in additional instrument complexities. Below frost points of −50°F (−45.5°C), instruments such as the electrolytic or aluminum oxide hygrometer are often the better choice.

Another feature of the condensation hygrometer is that if contaminated, it is easily washed and returned to service with no impairment in performance.

Saturated-Salt Dew-Point Sensors The saturated-salt dew-point sensor is widely used for its inherent simplicity, ruggedness, and low cost.

Sensor operation is based on the fact that the vapor pressure of water is reduced in the presence of salt. When water vapor in the air condenses on a soluble salt, it forms a saturated layer on the surface of the salt. This saturated layer has a lower vapor pressure than water vapor in the surrounding air. If the salt is heated, its vapor pressure increases until it matches the water vapor pressure in the surrounding air and the evaporation-condensation process reaches equilibrium. The temperature at which equilibrium is reached is directly related to the dew point.

A saturated salt sensor is constructed with an absorbent fabric bobbin covered with a bifilar winding of inert electrodes and coated with a dilute solution of lithium chloride. Lithium chloride is used as the saturating salt because its hygroscopic nature permits application in relative humidities between 11 and 100 percent.

An alternating current is passed through the winding and salt solution, causing resistive heating. As the bobbin heats, water evaporates into the surrounding air from the diluted LiCl solution. When the bobbin begins to dry out, resistance of the salt solution increases. Increased resistance results in less current passing through the winding. The bobbin cools and water begins to condense, forming a saturated solution on the bobbin surface. Eventually, equilibrium is reached.

Because the sensor operates as a current carrier saturated with Li and Cl ions, the addition of contaminating ions has little effect on its behavior compared to a typical percent RH sensor which operates "starved" of ions and is easily contaminated. A properly designed saturated salt sensor is not easily contaminated. Properly used, a saturated salt sensor has an accuracy of about ±2°F (±1.11°C). Generally speaking, saturated salt sensors are a good choice when a low-cost, rugged, slow-responding, and moderately accurate sensor is needed.

Aluminum Oxide Moisture Sensor An aluminum oxide moisture sensor is a capacitor that is formed by depositing a layer of porous aluminum oxide on a conductive substrate and then coating the oxide with a thin film of gold. The conductive base and the gold layer become the capacitor's electrodes. Water vapor penetrates the gold layer and is absorbed by the porous oxide layer. The number of water molecules absorbed determines the electrical impedance of the capacitor, which is, in turn, a measure of water vapor pressure.

Aluminum oxide humidity instruments are available in a variety of types, from low-cost single-point systems, includ-

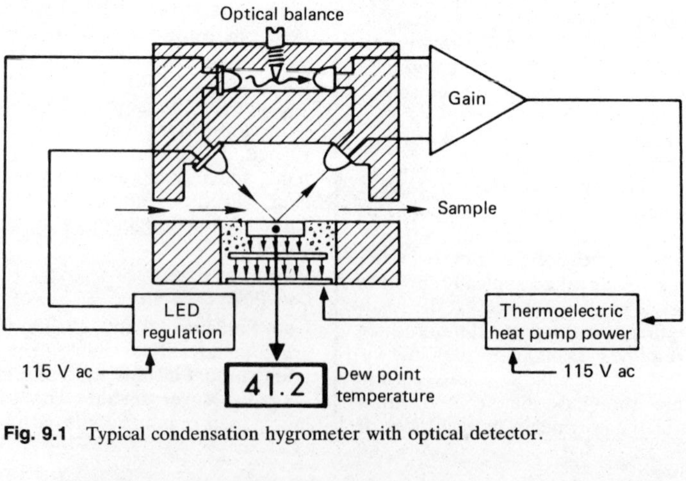

Fig. 9.1 Typical condensation hygrometer with optical detector.

ing portable battery-operated models, to multipoint microprocessor based systems capable of computing and displaying humidity information in different parameters.

Aluminum oxide hygrometers offer lower accuracy than do any of the other dew-point hygrometers mentioned [±4°F (±2.2°C) for the higher dew points to ±8°F (±4.4°C) at low frost points are typical]. They also exhibit aging and hysteresis. Their main advantage is their ability to detect low levels of moisture without need for cooling or special sampling systems.

Dew Cup The simplest instrument for the measurement of dew points is the dew cup. The gas sample is drawn from the furnace or generator across the outside of a polished cup made of chromium-plated copper. The cup is enclosed in a glass container so that moisture can be seen condensing on the cup surface when the dew point is reached. The cup surface is cooled progressively by dropping small pieces of dry ice in acetone (or methanol) until the dew point is reached, as indicated by condensation on the cup surface. The dew cup is most accurate for dew points above the freezing point of water. The use of the dew cup requires a considerable amount of skill and consistency on the part of the operator and it is not recommended for close control.

Fog Chamber The fog chamber is another type of manual instrument that is used throughout industry because of its portability, simplicity, and ability to cover a broad range of dew points without cooling. The fog chamber operates on the principle that a rapidly expanding gas cooled adiabati-

Table 9.2. Comparison of Various Humidity Sensors

Sensor type	Advantages	Disadvantages
Electrical RH	Low power consumption, used in small areas, fast response, high accuracy	Sensitive to contaminants, secondary measurement, hysteresis and aging, narrow operating range, possible condensation near 100% RH, limited accuracy below 10% RH
Mechanical RH	Low cost, low sensitivity to contaminants, condensation tolerant	Limited accuracy; aging and hysteresis
Chilled-mirror dew point	Highest accuracy, NBS traceable, fundamental measurement, laboratory standard, broad dew-point range, broad temperature range, easy to service, fast response, rugged and reliable	Not water-vapor-specific; expensive; cannot be used with SO_2, ammonia, or chlorides; slow response at very low frost points; subject to contamination
Lithium chloride dew point	Simple construction, low cost, high reliability, basic measurement, insensitive to contaminants	Slow response, narrow range, more difficult to service, not suitable below 12% RH, adds heat to measurement area
Aluminum oxide moisture	Small size, used in situ, multiple sensor arrangements broad range (to below −100°F (−73.3°C)	Aging and hysteresis, less accurate, secondary device not uniform, sensitive to contaminants, expensive to replace
Electrolytic cell	Basic measurement more accurate than Al_2O_3, good repeatability, stable, water-vapor-specific	Needs controlled sample flow, sensitive to contaminants, not suitable for in situ use, not available for multisensor arrangements, 2000 ppm maximum, not easy to service, cannot be used with certain gases
Wet-bulb–dry-bulb psychrometer	High accuracy near 100% RH, simple design, low cost, can be used up to 212°F (100°C)	Not accurate below 20% RH lower limit 32°F (0°C), adds moisture to area, needs considerable service

cally will produce a fog only when specific requirements of pressure drop, ambient temperature, and moisture content in the gas sample are satisfied. The atmosphere test sample is drawn into and held under pressure in an observation or fog chamber by a small hand pump. The atmosphere sample is held in the observation chamber for several seconds to stabilize the temperature, after which the quick-opening valve is depressed, releasing the pressure and creating adiabatic cooling, which causes a visible condensation or fog. The procedure is repeated to find the end point—the point at which the fog disappears. The dew point is then determined by referring to a chart based on the initial temperature reading and a pressure ratio reading at the point where the fog disappeared. The method has the same drawback as the dew cup in that it is subject to human interpretation and the skill of the operator. Measurement errors of 9°F (5°C) are not uncommon.

9.4 SUMMARY

The various humidity sensors discussed in this chapter are summarized in Table 9.2.

REFERENCES

1. R. J. List, *Smithsonian Meteorological Tables,* 6th ed, Smithsonian Institution, Washington, D.C., 1979, p. 350.
2. A. Wexler, "Vapor Pressure Formulation for Water in Range 0 to 100°C. A Revision," *J. Res. Nat. Bur. Stand.* (*U.S.*), **80A** (5, 6), 775–785 (September–December 1976).
3. S. Hasegawa and J. W. Little, "The NBS Two-Pressure Humidity Generator, Mark 2," *J. Res. Nat. Bur. Stand.* (*U.S.*) **81A** (1), 81–88 (January–February 1977).
4. J. H. Arnold, "The Theory of the Psychrometer," *Physics,* **4,** 255 (1933).
5. ASHRAE, *Data Book,* 6th ed, American Society of Refrigeration Engineers, New York, 1949, Chap. 4.
6. C. F. Marvin, *Psychrometric Tables,* Publication 235, U.S. Weather Bureau, 1900.
7. R. E. Ruskin, "Principles and Methods of Measuring Humidity in Gases," in *Proceedings of the First International Symposium on Humidity and Moisture Measurement and Control in Science and Industry,* Washington, D.C., Van Nostrand Reinhold, New York, 1963.
8. *Proceedings of the International Symposium on Moisture and Humidity,* Washington, D.C., April 15–18, 1985, Instrument Society of America, Research Triangle Park, N.C.

Chapter **10**

Liquid Crystals

ROCCO A. SAPIENZA
Hallcrest

Although the initial research was performed in 1888 by an Austrian botanist, Freidrich Reinitzer, it was not until the early 1960s that liquid crystals were developed into functional temperature-sensing products for use in space and defense programs. It was during the 1970s that liquid-crystal technology came into general use. Today, liquid crystals can be seen in industrial and consumer products, scientific instrumentation, and digital thermometers.

Liquid crystals are long chains of organic molecules that exhibit the flow properties of a liquid and the optical properties of a crystal. While there are many ordered liquid crystal materials, the most commonly used are part of the cholesterol family and are thus termed *cholesteric liquid crystals.*

When a single cholesteric liquid crystal compound is heated, the crystalline solid will begin to melt. The compound enters an intermediate phase called the *mesophase* or *liquid-crystal phase,* where it exhibits properties of both a liquid and a crystal. While in the mesophase, the compound is capable of scattering white light into its color components. The color of the compound changes as the temperature increases, going from reds, to yellows, to greens, to blues, to violets, and finally becoming transparent as the compound becomes a true liquid. On cooling, the order of color is reversed with violet appearing first. The intensity of the colors produced is dependent on the available light.

Each specific liquid-crystal compound exhibits the liquid-crystal phase at a specific temperature. Combinations of liquid crystals can be made that will exhibit the color changing phenomenon between −4 and 482°F (−20 and 250°C). The temperature range from red to violet can be adjusted so that the change in color will occur in a range as narrow as 2°F (1°C) or as broad as 90°F (50°C).

Liquid-crystal compounds can be used in various shapes, forms, and sizes. They are often used in the form of sheets. Liquid crystals may be made to appear as digits, words, symbols, or graphic designs. For example, digits have been formulated to appear in their respective temperature ranges in a digital thermometer.

Liquid crystals can be calibrated to within ±0.5°F (±0.28°C). Liquid crystals have very little mass and therefore can have a short response time. Liquid crystals are formulated in large batches which produce several hundred thousand sensing units. Within each batch, the degree of precision is generally within ±0.1°F (±0.05°C).

Liquid crystal can be chemically coated and protected from contamination and can be fabricated to be immersible in solutions. Normally, the crystals will maintain accuracy and precision indefinitely without recalibration, as long as they are not subjected to ultraviolet light for extended periods. Liquid crystals are classified as being nontoxic.

The main use of liquid-crystal technology in the consumer product area has been in digital thermometers. In industry, liquid crystals find use in monitoring equipment surface temperatures and in the measurement of air temperature when sensing at many locations is required. A principal advantage of liquid crystals is the capability to measure surface temperature variations over an area.

Chapter **11**

Heat-Flow Transducers

WILSON A. CLAYTON
Hy-Cal Engineering

Heat-flow transducers measure net heat-transfer rate at the transducer sensing surface. They are generally called "calorimeters," which should not be confused with apparatus also called calorimeters used to measure heat content.

11.1 CONTINUOUS TOTAL HEAT-FLOW MEASUREMENT

Total heat-flow rate calorimeters are designed to measure net heat exchange by both convection and radiation in any combination present. Special design constraints must be met to measure conduction heat flow. Transducers are not able to distinguish convection or conduction from radiation heat transfer without special application considerations or modification to suppress one of the heat-transfer mechanisms. The most common modification is the installation of an infrared window over the active sensor, which blocks convection or conduction and changes the transducer into a broad wavelength band radiometer. A highly reflective sensing surface can be used to suppress, but not completely eliminate, sensitivity to radiation heat exchange. Compared to radiometers and pyrometers, sensor technology suitable for measurement of convection or conduction heat transfer is limited, particularly by the requirement for smooth surfaces that avoid disturbances in convection boundary layers.

General purpose heat-flow transducers are suitable for continuous use. Transient calorimeters providing output that must be analyzed in terms of transient response to a heat pulse are not included in this discussion.* General purpose total calorimeters are characterized in terms of steady net heat transfer, with transducer time constant being a factor in selection and application. Useful calorimeters are largely independent of mounting boundary conditions except at low heat source temperatures, requiring only a heat sink pro-

*See Ref. 1 for discussion of transient type calorimetric gages.

vided by fluid cooling or heat sink mass sufficient for the period of heat application. A special case, wide area blanket type heat-flow meters, is discussed separately in Sec. 11.5.

11.2 DIFFERENTIAL THERMOELECTRIC TRANSDUCERS

Total heat-flow rate calorimeters are generally constructed using one of two types of differential thermoelectric sensors. The broadest range of applicability is provided by the circular foil calorimeter and the greatest sensitivity for heat transfer below 60 Btu/ft^2 · s (68 W/cm^2) is provided by the in-depth thermopile calorimeters. Both provide millivolt level outputs driven by temperature gradients between differential thermocouple junctions. Copper-Constantan thermoelectric elements are used except in specialized applications. The depth of the sensing element in the heat-flow direction is kept small so that heat flow is always one dimensional overall through the sensing area into the heat sink body of the transducer. This is a design requirement for independence from mounting boundary conditions that may involve heat loss or gain at the edges of the calorimeter body. Designs that generate output from a temperature difference over a significant depth, greater than 0.02 in (0.5 mm), require a relatively large guard area or installation-dependent calibration that includes effects of usually unstable multidimensional heat flow.

Circular Foil Calorimeters

The circular foil calorimeter is elegant in its simplicity and performance. It was first conceived and demonstrated by Robert Gardon for Corning Glass, and is widely known as the *Gardon gage*.[2] Methods for practical manufacture and calibration were developed by R. A. Whitmore at Hy-Cal Engineering[3] and E. W. Malone at Boeing Co.[4] Figure 11.1 illustrates the thermal and physical cross section of this sensor. Net absorbed heat flow on the foil surface drives a parabolic temperature gradient between the center and the edges of the foil as the heat absorbed in the foil conducts radially to the edge. Direction of heat flow in the heat sink has no effect on this transducer making it independent of its mounting. Output is generated by the differential thermocouple formed through the foil between junctions at the butt welded center wire and edge wire. The preferred construction uses Constantan foil, copper center wire, and copper body, so a copper lead connection anywhere on the copper body replaces the requirement for an edge wire. These materials provide a linear output to 10 mV full scale versus heat flow over a useful operating temperature range to 400°F (200°C) at the foil edge.

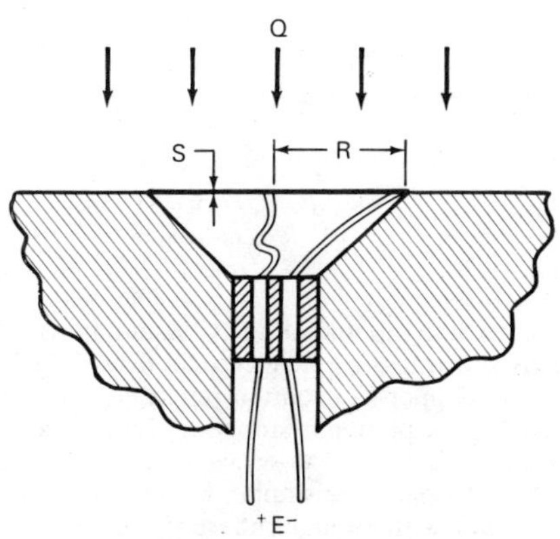

$$E = \frac{QR\epsilon}{4SK}\left[\frac{1}{1 + C\frac{k_w}{k}}\right] = \frac{QR\epsilon}{4S(k + Ck_w)}$$

Fig. 11.1 Circular foil calorimeter. [E = output signal, Q = heat flux, R = radius of foil, S = thickness of foil, ϵ = thermoelectric potential of foil-wire combination, k = thermal conductivity of foil, k_w = thermal conductivity of wire, C = geometry dependent constant (0.025 is typical), $1/[1 + C(k_w/k)]$ = center-wire correction term (often neglected).]

All welded construction provides reproducible, stable transducers. Brazing or soldering is not suitable due to imprecise and unstable location of the effective thermoelectric junctions. Foil thicknesses range from 0.0004 to 0.008 in (0.01 to 0.2 mm) and foil diameters range from 0.04 to 0.2 in (1 to 5 mm) for measurement of heat-flow rates between 3000 and 3 Btu/ft^2 · s (3400 and 3.4 W/cm^2) at 10 mV full-scale output from the smallest, thinnest to the largest, thickest foils. Copper center wire, usually 0.003 in (0.076 mm) diameter, depresses theoretical output by heat conduction down the wire that is proportional to the ratio of the copper wire and Constantan foil conductivities and a coefficient derived from the heat flow geometry within the transducer.[6] This center wire effect is negligible for other foil-wire combinations since copper is the only high-conductivity wire considered.

Circular foil calorimeter output at steady heat-flow rate is driven by ΔT from the relationship in Fig. 11.1, the differential thermoelectric potential ϵ of the foil disk-wire combination and a geometry dependent correction Ck_w for center wire conduction.[7]

The constant C can range from 0.005 for minimum center wire effect to 0.05 for maximum center wire effect, but is fixed for a given design and experimentally averages 0.025 in typical transducers. The nominal center wire correction never exceeded 6 percent for lower conductivity center wires but ranged as high as 28 percent for copper center wires in experiments that compared predicted and measured behavior of circular foil calorimeters constructed from every practical combination of standard thermocouple material used for the foil and wire.[7] Figure 11.2 shows the relative outputs of all the foil-wire combinations evaluated for a median foil temperature 200°F (111°C) above the foil-edge heat sink temperature. Easier comparison of the effects of operating temperature is given by normalizing the results to the lowest body-foil edge temperature and comparing these as plotted in Fig. 11.3.

The Constantan-copper construction (Figs. 11.2 and 11.3) has nominally perfect performance, defined as constant sensitivity ϵ/k, up to 400°F (200°C). Chromel-Constantan provides high sensitivity and good (linear) performance if not heated. The highest sensitivity and best sensitivity stability to the highest operating temperatures is found in a nontraditional foil-wire combination[7] named Fluxfoil. Figure 11.4 shows Fluxfoil experimental data with a superimposed scale of minimum calorimeter full-scale flux levels versus temperature. The minimum flux levels are required for the internal

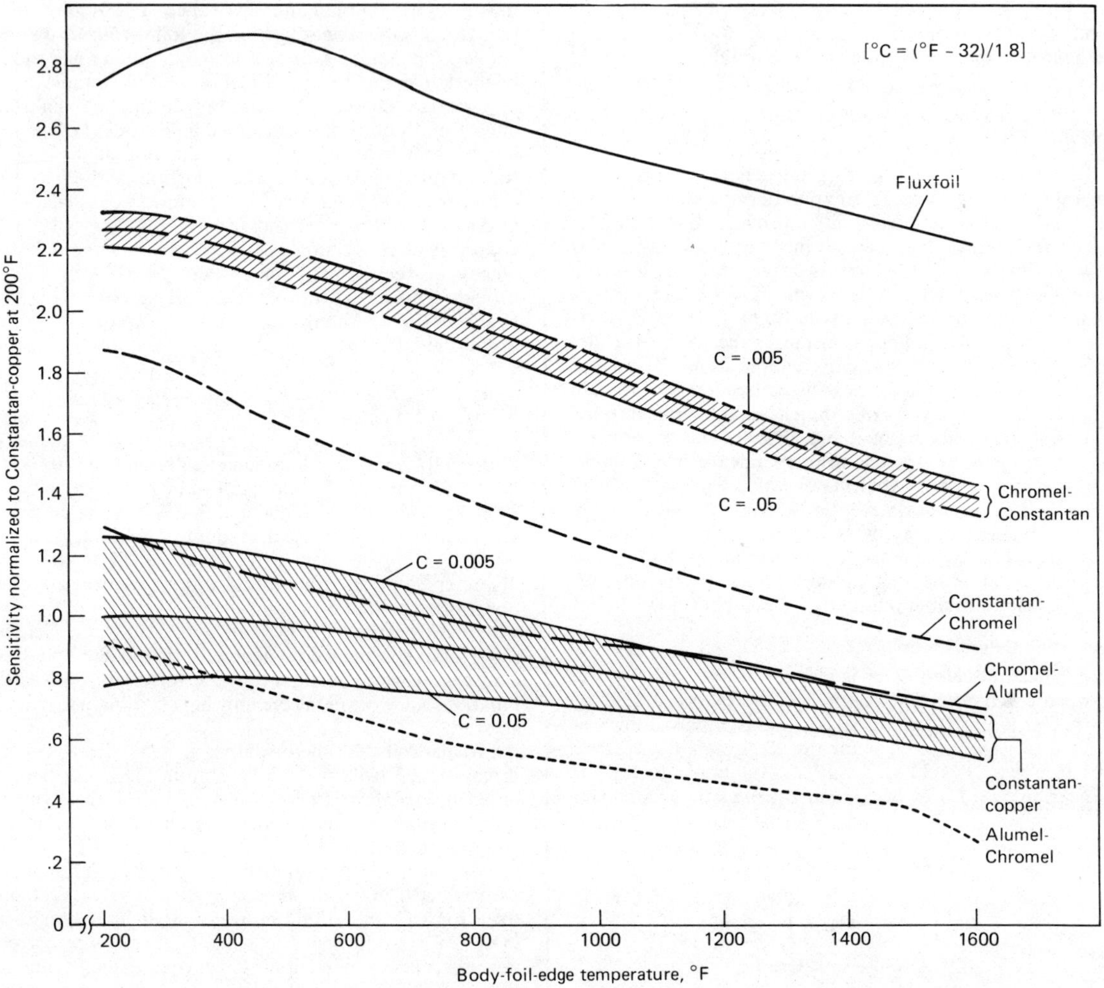

Fig. 11.2 Relative circular foil calorimeter sensitivity. [Base sensitivity, $\epsilon/k = 1.4094\ \mu V \cdot h \cdot ft/Btu$ ($1.4664\ \mu V \cdot m/W$), where $\epsilon = 28.9\ \mu V/°F$ ($52\ \mu V/°C$). Average center-wire effect: $C = 0.025$, except as noted.]

radiation transfer from the back of the foil to the body cavity to remain negligible, otherwise a rapid loss in sensitivity occurs. This radiation transfer effect of high-temperature operation is not included in the relationship in Fig. 11.1, and is a fundamental limitation on minimum net heat-transfer rates that can be measured at high temperatures.

In-Depth Thermopile Calorimeters

In-depth thermopile calorimeters use the thermopile concept for a high-sensitivity measurement of the temperature gradient proportional to heat flow through a small thickness of thermal insulator. Figure 11.5 shows the most common configuration and insulator size. Dense thermopiles are made by winding 0.001-in (0.025-mm) diameter Constantan at a 0.003-in (0.075-mm) pitch. The length of the winding is as required for the number of thermopile junctions desired, but typically less than 0.5 in (12.7 mm). Copper is plated over slightly more than one-half of the Constantan winding width, electrically shorting the high electrical resistance Constantan and forming copper-Constantan junctions on both surfaces where the copper plating is stopped. Hot and cold junctions are automatically formed in series for the thermopiles using this method. Copper-Constantan is the only industrial thermoelectric pair with enough difference in electrical conductivity for application in this method.

Conceptually the output from the configuration in Fig. 11.5 is easily predicted from the expected temperature gradient and the number of thermopile junctions. This is not true in practice due to the actual complexity of heat flow in the composite material formed during assembly of the in-depth thermopile by potting in the calorimeter body with high temperature epoxies. The thermopile must be submerged just below the heat receiving surface of the calorimeter assembly. For all but the lowest heat-flow rates the thermopile must

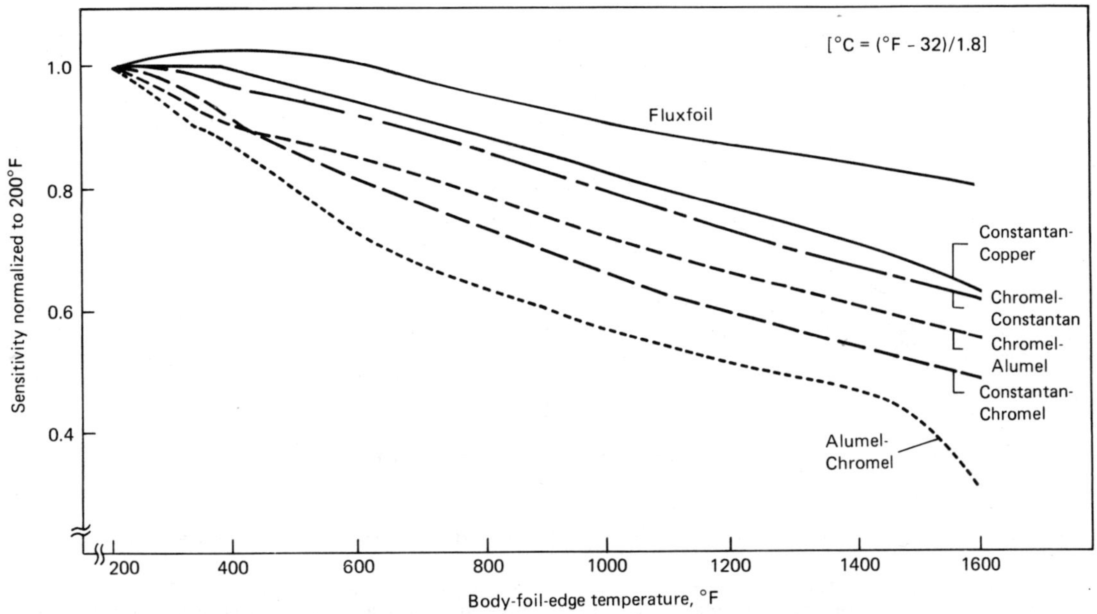

Fig. 11.3 Normalized circular foil calorimeter sensitivity versus temperature. (Average center-wire effect: $C = 0.025$. Foil average temperature 200°F above body.)

be seated in a cavity in the calorimeter body. Heat flow is no longer one-dimensional through the thermopile but all boundary conditions are fixed by the assembly to produce a stable transducer.

All successful combinations of insulating spacer, winding wire, plating, potting materials, and final mounting configurations have been empirically derived to produce calorimeters that have excellent independence of operating temperatures above 32°F (0°C). Temperature correction is required for satisfactory measurements below −58°F (−50°C)

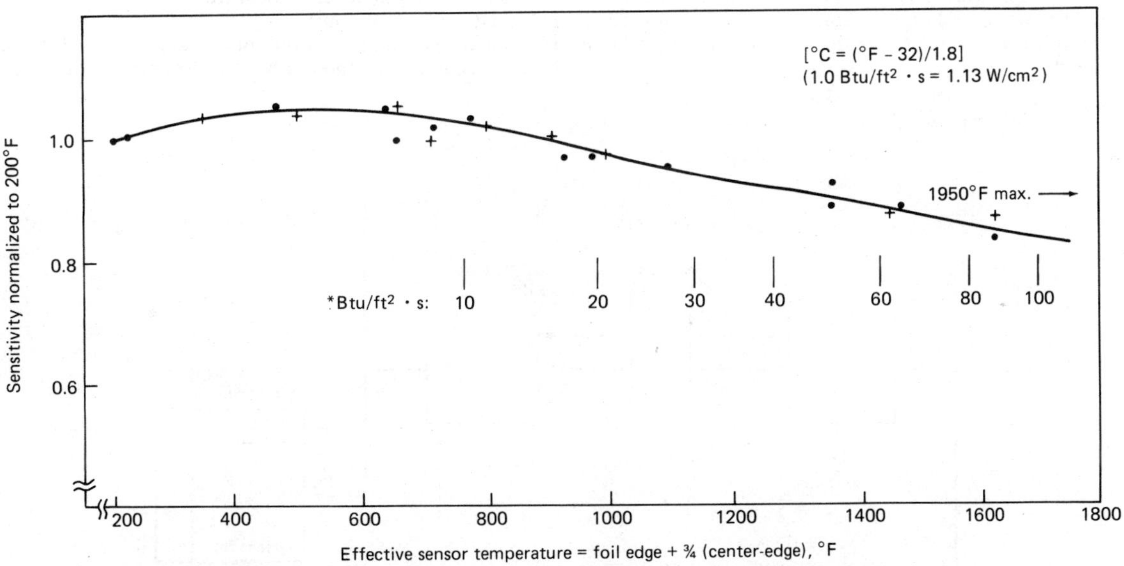

Fig. 11.4 Normalized Fluxfoil calorimeter sensitivity versus temperature. [Base sensitivity, $\epsilon/k = 2.75$ $\mu V \cdot h \cdot ft/Btu$ ($2.86\ \mu V \cdot m/W$). Thermoelectric potential, $\epsilon = 30\ \mu V/°F$ ($54\ \mu V/°C$) nominal. (*Gage 10-mV full-scale limit for negligible internal radiation loss.) (Data for two samples shown.)]

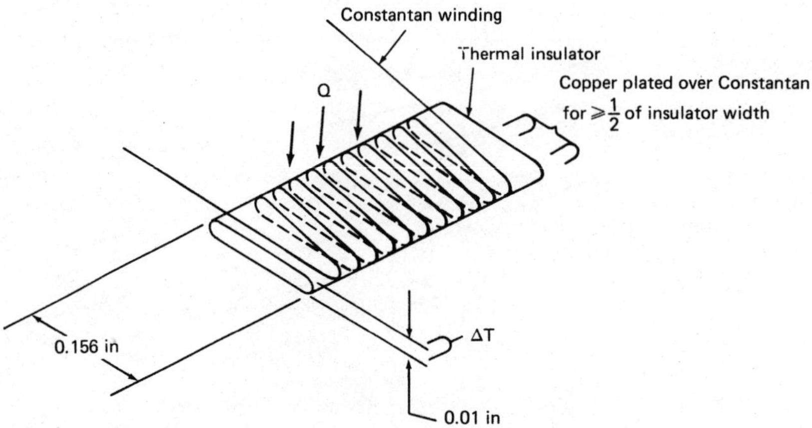

Fig. 11.5 In-depth thermopile with hot junction on top and cold junction on bottom.

for any available in-depth thermopile calorimeter. Low temperature operation is a special concern because only in-depth thermopile calorimeters are sufficiently sensitive for measurement of the low heat-transfer rates associated with low temperatures. Operation down to −328°F (−200°C) is possible. Maximum operating temperature is limited to 500°F (260°C) by the epoxies used in assembly, and then only at less than full rated heat flux level. In-depth thermopiles used for the highest fluxes typically may be only 0.005 in (0.125 mm) in total thickness but will typically experience 9°F (5°C) temperature gradient per Btu/ft^2 · s (W/cm^2) unit of heat transfer. At the maximum available transducer rating of 60 Btu/ft^2 · s (60 W/cm^2), this corresponds to 540°F (300°C) nominal temperature elevation at the hot junctions, hence the cold junction-body temperature may not be allowed to rise more than 50°F (28°C) or the potting materials will be pyrolized.

11.3 HEAT-FLOW TRANSDUCER CALIBRATION

No international standard exists for heat-transfer rate calibration. Practical methods traceable through other measurements have been developed and proven using radiation heat transfer. The universally accepted calibration method was originally patented by Hy-Cal Engineering. Traceability is through traceable optical pyrometer temperature measurement to establish the level of narrow-angle blackbody radiation incident on a transfer standard calorimeter. The transfer standard, calibrated with narrow-angle blackbody radiation, is used with a wide-angle radiation source to establish wide-angle calibrations of a group of working standards covering all levels of heat-transfer rate up to about 300 Btu/ft^2 · s (300 W/cm^2).

Figure 11.6 illustrates the routine application of the wide-angle radiation method of heat-flux calibration. The radiant

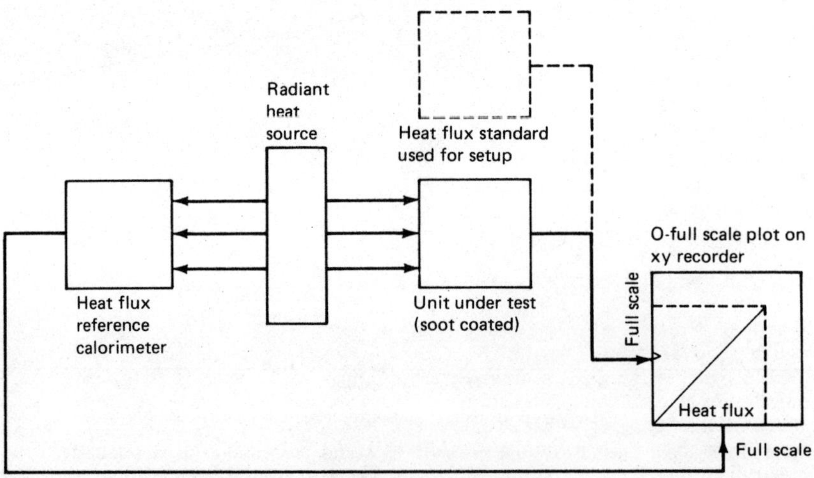

Fig. 11.6 Heat flux calibration schematic.

heat source is a resistively heated graphite plate bathed in inert gas under a water-cooled cover. Viewports are provided to both sides of the plate. Initially a working standard and a "reference" calorimeter are affixed on opposite sides of the source. Outputs are connected to an *xy* recorder as shown. As power is applied to the source, the sensitivity of the heat-flux scale on the *xy* recorder is adjusted to position the output of the working standard on its known slope. The system is then said to be "standardized." Any number of units under test can then be sequentially substituted in the exact position of the working standard and calibrated continuously over the full range by increasing the power through the source starting at zero. The method requires no assumptions providing the calibrations are performed with exact reproduction of the positioning of the working standard when the setup is "standardized." In addition, all exposed surfaces must be uniformly blacked to prevent internal reflection variations.

Most of the uncertainty in this calibration method comes directly from the accuracy with which the original blackbody reference temperature can be determined. A good optical pyrometer may be certified to ½ percent of temperature reading. Radiation level is dependent on temperature to the fourth power, so an initial 2 percent uncertainty results from $\pm½ \times 4 = \pm2$ percent. All other instrumentation transfer errors are relatively minor. The typical rated accuracy of this calibration method is better than ±3 percent of reading. Short-term method repeatability is better than ±½ percent.

Calorimeters for measurements in multiple mechanisms of convection and radiation are always soot blacked and calibrated in terms of absorbed radiation. The permanent coating is applied after calibration and does not impact the calibration sensitivity. Radiometer configurations are not soot coated and are calibrated for incident radiation. This calibration automatically includes the instrument absorbtivity. The standards are always reproducibly sooted.

For the special case of solar instruments, the above source is used with a water filter to obtain very precise and practical calibrations up to solar radiation levels of 100 solar constants (13.6 W/cm^2).

Calibrations beyond levels obtainable with the above method are always extrapolated, which places great importance on well-understood designs and reliable production of linear output transducers.

11.4 TRANSDUCER APPLICATION LIMITATIONS

The descriptions of the operating principles of differential thermoelectric transducers in Sec. 11.2 emphasized behavior over a broad range of operating temperatures. Operating temperature is the second most important application consideration after estimation of the level of heat transfer rate to be measured. In-depth thermopile calorimeters are limited to below 60 Btu/ft^2 · s (68 W/cm^2) and circular foil calorimeters are used only above 3 Btu/ft^2 · s (3.4 W/cm^2) unless low temperature gradient (and low output) is desirable. Normal instruments of both types are limited to about 400°F (200°C) by epoxies and other materials used. In-depth thermopiles can be calibrated for lower-temperature use and circular foils for higher-temperature use within limitations indicated in Fig. 11.4.

After temperature, application constraints are primarily geometric. For convection measurements the sensing surface must be smooth relative to the boundary-layer thickness, and the shape presented must meet measurement parameters. Often these considerations will run into constraints of minimum size, which are about 0.06 in (1.5 mm) thickness on all types *or* exposed diameter on circular foil types. Shape has little constraint otherwise as long as there is some volume to hook-up lead wire. Figure 11.7 suggests widely used practical shapes of circular foil calorimeters. When these transducer types are configured with windows as radiometers (Fig. 11.8) then view factor will be an overriding geometric consideration.

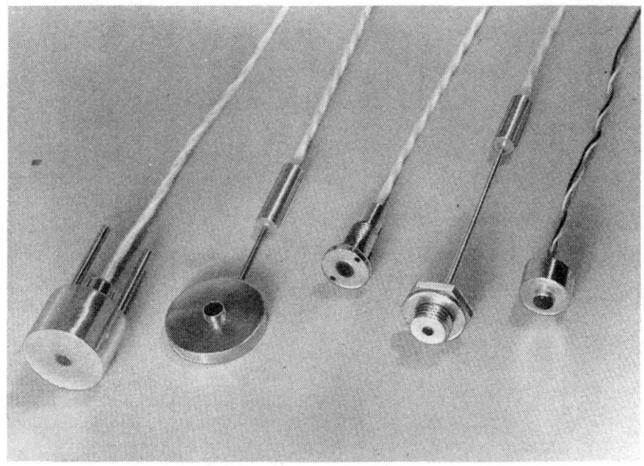

Fig. 11.7 Circular foil calorimeters. Unit at left is water cooled with 1.0-in diameter, a standard in many tests.

None of the calorimeters described here are suitable for transient measurements, so time response may be a limiting consideration. In-depth thermopile or circular-foil calorim-

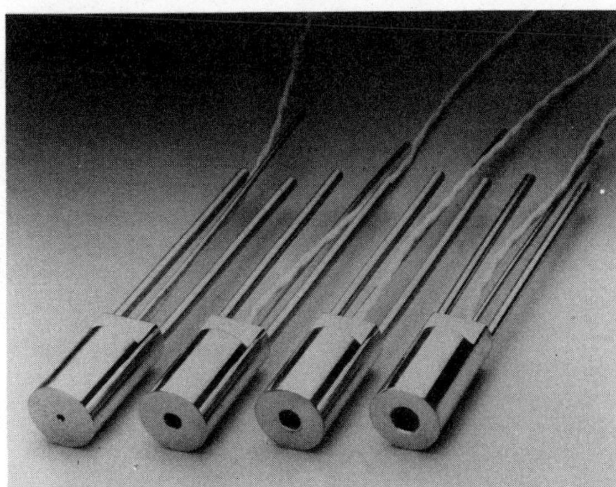

Fig. 11.8 Narrow-angle radiometers. Total view angles 5.5, 11.2, 15, and 30 degrees. In this application only, a thin film thermopile[14] provides fast 10 ms time constant. Water cooled. Gas purge prevents window fouling.

eters can generally not be specified faster than 120- or 15-ms time constants, respectively, with longer time constants required below 5 or 120 Btu/ft^2 · s (5.6 or 136 W/cm^2) heat-flux levels, respectively.

Finally, all of these transducers are designed to generate their rated output at a significant surface temperature rise on the order of 300°F (166°C). This renders these devices generally unsuitable for measurements of transfer coefficients unless they are operated at the low end of their design range to limit temperature gradients. Circular foil calorimeters do offer the advantage that the output is a direct measure of the foil edge to center temperature gradient. The effective temperature is then a mean foil temeprature, usually three-fourths of the edge to center gradient, plus the foil edge or body temperature measured separately. It is generally accepted and experimentally verified that the foil temperature gradient does not cause a significant distortion of the thermal boundary layer.[8]

11.5 HEAT-FLOW METERS

Heat-flow meters are a special case of in-depth thermopile instrument distinguished by the fact that the thermopile is not built into a heat sink to form a complete instrument with fixed internal heat flow. Rather, the thermopile is contained in surrounding material to facilitate its mounting *on* the heat sink included in the measurement. Analysis of this application is extremely difficult due to great dependence on the wide range of boundary conditions encountered. A significant consideration is that these devices cannot always be calibrated by the radiation method described previously, and require a heat-flow stack measurement for calibration. See Refs. 9–13 for recent attempts to understand heat-flow meter measurements.

A typical application of heat-flow meters is in the measurement of heat transfer in various types of building construction. These measurements, of considerable interest worldwide (see References), have proven extremely difficult within demonstrable error lower than about ±15 percent. The measured heat flow is normally lower than undisturbed heat flow without the effects of adding the heat-flow meter. The only predictable exceptions are in cases where the meter is fit into a cavity on the surface which generally reduces and may yield slightly positive errors or where the undisturbed test surface thermal resistance is much lower than the sum of the meter and its contact thermal resistance. The latter case is typical of high surface heat-transfer coefficients where lateral heat flow is relatively unimpeded and the heat-flow direction is from the mounted side.

REFERENCES

1. R. J. Goldstein and H. P. Chiang, "Measurement of Temperature and Heat Transfer," in W. M. Rohsenow, J. P. Hartnett, E. N. Ganić (eds.), *Handbook of Heat Transfer Applications*, 2d ed., McGraw-Hill, New York, 1985.
2. R. Gardon, "An Instrument For The Direct Measurement of Intense Thermal Radiation," *Rev. Sci. Instrum.* **24**(5), 366 (1953).
3. "A Discussion of the Standardized Procedure For Calibrating Heat Flux Transducers," *19th Annu. ISA Conf. Exhib.* (Hy-Cal Engineering Publication TR-209, October 1964).
4. E. W. Malone, "Design and Calibration of Thin-Foil Heat Flux Sensors," *ISA Trans.* **7**, 175–179 (1968).
5. L. W. Woodruff, L. F. Heame, and T. J. Keliher, "Interpretation of Asymptotic Calorimeter Measurements," *AIAA Journal*, **5**(4), 795–797 (1967).
6. N. R. Keltner and M. W. Wildin, "Transient Response of Circular Foil Heat-Flux Gauges to Radiative Fluxes," *Rev. Sci. Instrum.* **46**(9), 1161–1166 (1975).
7. W. A. Clayton, "High Temperature Circular Foil Heat Flux Gauge," *ISA International Instrum. Conf.* (Hy-Cal Engineering Publication TR-417, May 1980).
8. S. A. Striegl and T. E. Diller "The Effect of Entrainment Temperature on Jet Impingement Heat Transfer," *J. Heat Transfer, Trans. ASME*, **106,** 27–33 (1984).
9. G. J. Borell and T. E. Diller, "A Convection Calibration Method for Local Heat Flux Gages," *Natl. Heat Transfer Conf.* (August 1984).
10. Hauser, R. L., "Principles and Performance of *In Situ* Heat Flux Transducers," *ASTM Workshop on Applications of Heat Flow Sensors* (Hauser Laboratories, September 1983).
11. S. N. Flanders, "Confidence in Heat Flux Transducer Measurements," *ASHRAE Trans.* **91,** part 1 (January 1985).
12. H. A. Trethowen, "Measurement Errors With Surface-Mounted Heat Flux Sensors," Building Research Association of New Zealand, 1986.
13. *Heat Flux Measurement—Proc. 2d Workshop IMEKO* (International Measurement Confederation, Budapest, Hungary, 1986; ITC series no. 9, OMIKK-Technoinform, Budapest).
14. R. D. Baxter, Small Size, High-Performance Radiation Thermopile, U.S. Patent 4,111,717 (1978).

Part 13
Automatic Temperature Controls

Chapter **1**

Self-Operated Temperature Controls

D. JOSEPH STADTMILLER
The Partlow Corporation

Certain temperature controllers do not rely on electrical or electronic means of temperature sensing. Generally, these devices are self-operated controls with the energy required to operate the controller obtained from the controlled medium. Similar to electrical-electronic controllers, self-operated temperature controls must have some means of sensing the temperature of the load, specifying the desired set point, and regulating the energy input to the load.

1.1 FILLED-SYSTEM TEMPERATURE CONTROLS

Filled-system thermometers provide the basis for a number of different types of self-operated temperature controls. Filled-system thermometers function on the thermal expansion or vapor-pressure characteristics of various liquid or gaseous media contained in a sensing system. The forces developed as a result of thermal expansion or vapor pressure of the contained medium can be used to produce linear or rotary mechanical movement which, in turn, can operate an energy-flow control device as well as indicate and record temperature.

Generally, the advantages and limitations of filled-system controls parallel those of filled-system thermometers. Advantages are simplicity, ruggedness, minimum upkeep, and self-powering. Sensitivity, response time, and accuracy are comparable to alternatives. Cost can be low. Limitations can include size of sensor, maximum temperature, and failure modes.

Temperature Sensing

Detailed descriptions of the several classes of filled-system thermometers are presented in Part 12, Chap. 6. Table 1.1 is a summary of important characteristics. The classification of filled systems is given by the Scientific Apparatus Makers' Association (SAMA) Standard PMC 6-10-1963. These in-

Table 1.1. Filled-System Operating Comparison

Operating parameter	Type of fill: Mercury	Organic liquid	Vapor	Gas
Low temperature, °F	−38	−100	−40	−400
High temperature, °F	1200	500	600	1200*
Largest span, °F	1200	400	300	1000
Smallest span, °F	25	40†	80	120
Linear vs. nonlinear movement	Linear	Essentially linear	Nonlinear	Essentially linear
Maximum capillary length, ft	200	Class IB, 15; Class IA, 100	200	200
Capillary compensation	Compensating wire within capillary	Class IB, Bimetal; Class IA, Matching System	None	Dependent on span
Accuracy, % span	±1	±1	±1	±1

* Using argon and high-nickel alloys.
† Difficult to produce and may be adversely affected by shipping conditions.
CONVERSION FACTORS: 1 ft = 0.3048 m; °C = (°F − 32)/1.8.

clude Liquid-Filled (Class I) and Mercury-Filled (Class V), which operate on a volumetric expansion principle, and Vapor-Filled (Class II) and Gas-Filled (Class III), which operate on a pressure-change principle.

Organic liquids are used in the liquid-filled type; nitrogen, helium, or argon are commonly used in the gas-filled type; and typical fluids used in the vapor-filled systems are toluene, acetone, ethyl alcohol, methyl ether, and propane. Important considerations in the application of filled-system thermal sensing and control devices are maximum and minimum temperatures and ranges, capillary and case temperature compensation, speed of response, barometric errors, sensor elevation errors, and overrange protection. Part 12, Chapter 6 provides quantitative characterization of the several classes of filled-system thermometers with respect to these application considerations.

Organic liquid-filled systems are most commonly used types, followed by vapor- and gas-filled types. Mercury is a very accurate sensing medium that produces a high-force mechanical movement that is capable of driving very rugged temperature indicators and controllers. Figure 1.1 is an example of a mercury-filled temperature sensor which possesses all the basic functional components common to all filled-system sensors. Pencil and coiled sensors are used. Several variations are illustrated in Fig. 1.2. The best sensor for any application is one that has the maximum outside surface and the minimum internal volume. Coiled sensors generally provide superior response.

Set-Point Types

Each filled-system temperature controller must possess a means for specifying the desired control temperature or set point. This is usually accomplished by positioning some output device (e.g., a switch, valve, or operator) relative to the mechanical movement supplied by the temperature-sensing system.

In situations where the set point is never changed and it is desirable not to have the set-point adjustment easily accessible to personnel, a blind set-point adjustment is desirable. Filled-system controllers can be ordered with the set point adjusted to the desired set-point temperature or can be adjusted in the field. It is most common, however, to see set-point adjustments made on an analog dial with an adjustment knob and pointer. Figure 1.3 shows an example of this type of controller.

Output Types

Filled-system temperature controls can adjust the energy input to the controlled system in a number of ways. The most common type of output is the ON/OFF output that either closes or opens a set of microswitch contacts that will, in turn, energize or deenergize some electrical circuit. This electrical circuit could include a heater or solenoid valve that supplies gas to a gas burner. In either case, either 100 or 0 percent of the available energy is supplied to the load depending on the sensed temperature and the set-point value. Figure 1.4 shows an example of an ON/OFF control that utilizes a microswitch. Set-point adjustments are made by adjusting the switch actuator. As the temperature increases, the element plunger moves the switch actuator until the switch becomes actuated.

Proportional outputs are also used with filled-system temperature controls. The output may be proportional voltage, pneumatic, or electrical resistance as required by the control system.

Control outputs can be in conjunction with temperature indication, transmission, and recording.

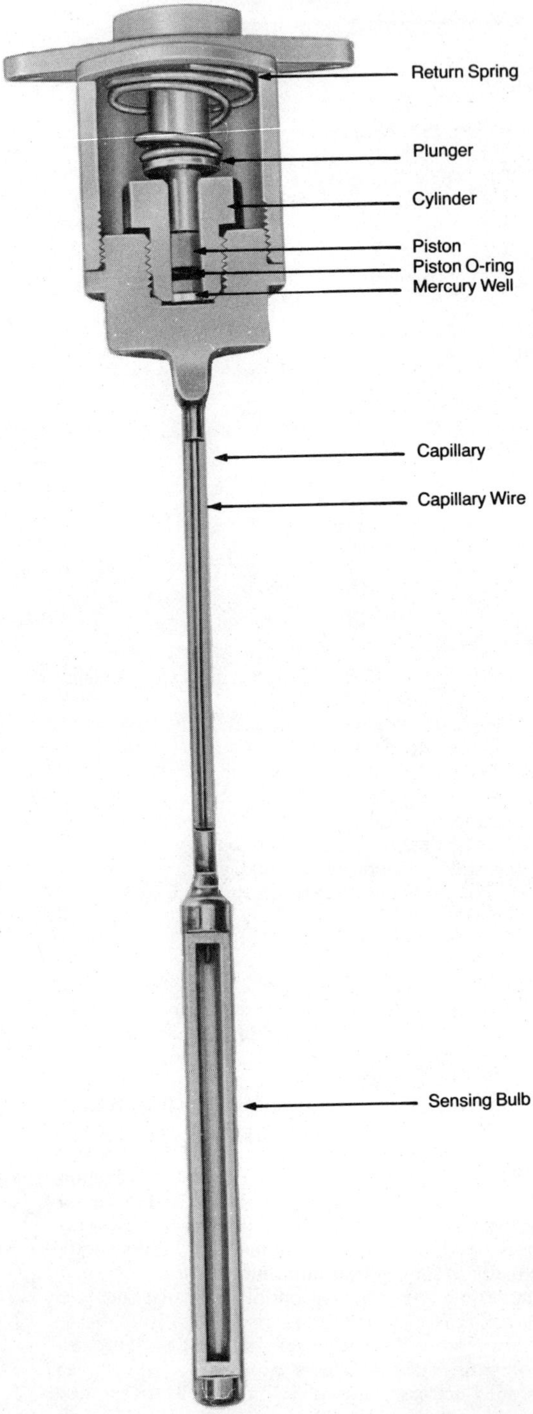

Fig. 1.1 Mercury-filled system. (*Courtesy of the Partlow Corp.*)

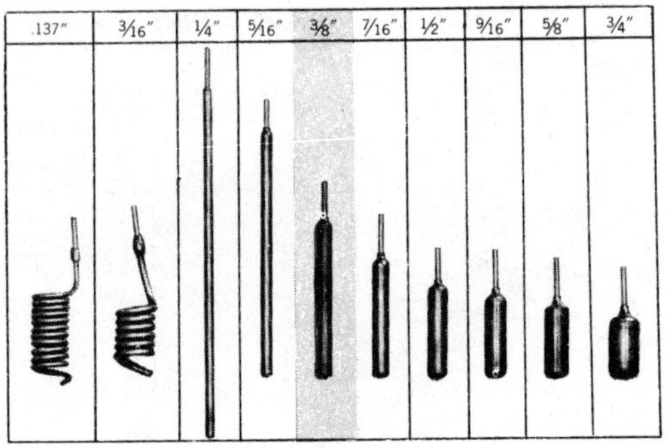

Fig. 1.2 Types of filled-system sensors.

Failure Mode and Overtemperature Protection

All filled-system sensors respond with an increasing mechanical movement as temperature increases. The most common failure mode is for the filled medium to leak out of the sensor, capillary, or the motion or responsive element. When this happens, the temperature controller will indicate an erroneous temperature that is much lower than the actual temperature being sensed. This could be fail-safe operation for cooling applications but not fail-safe for heating applications. Subsequently, all filled-system sensors utilized for heating

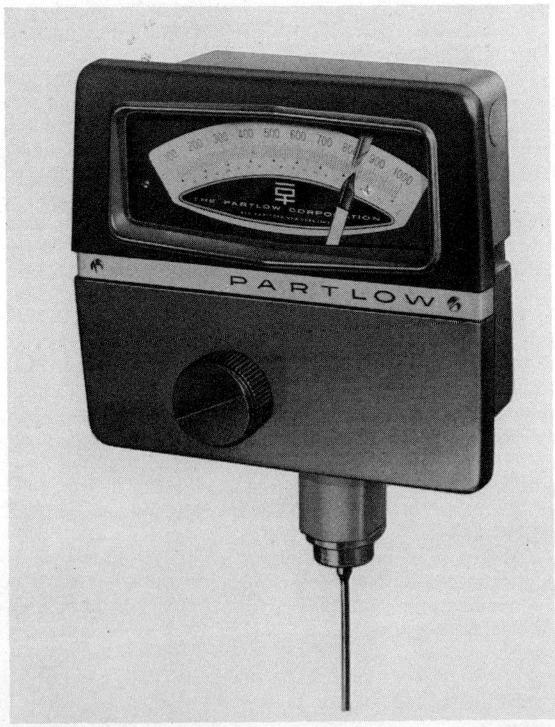

Fig. 1.3 Adjustable set-point filled-system control. (*Courtesy of the Partlow Corp.*)

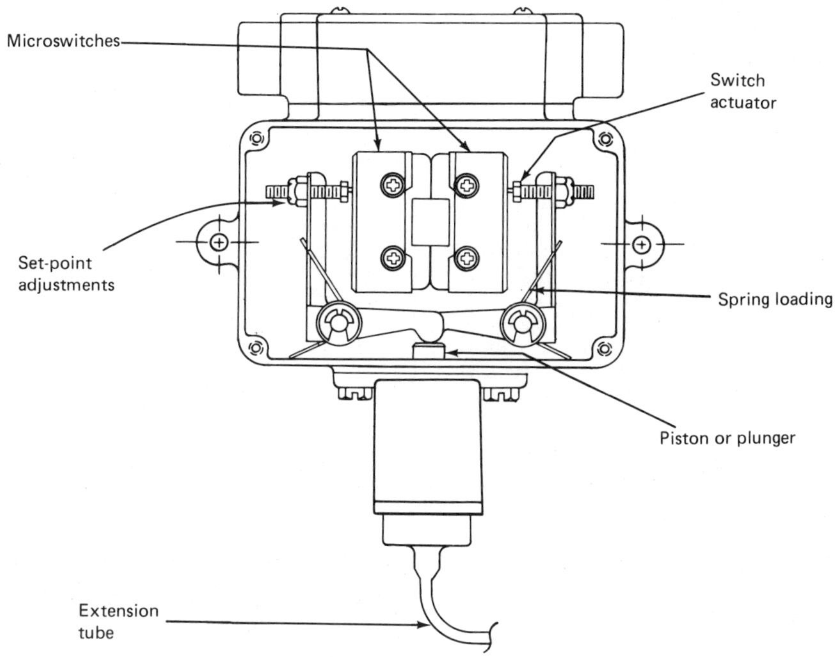

Fig. 1.4 ON/OFF filled-system control.

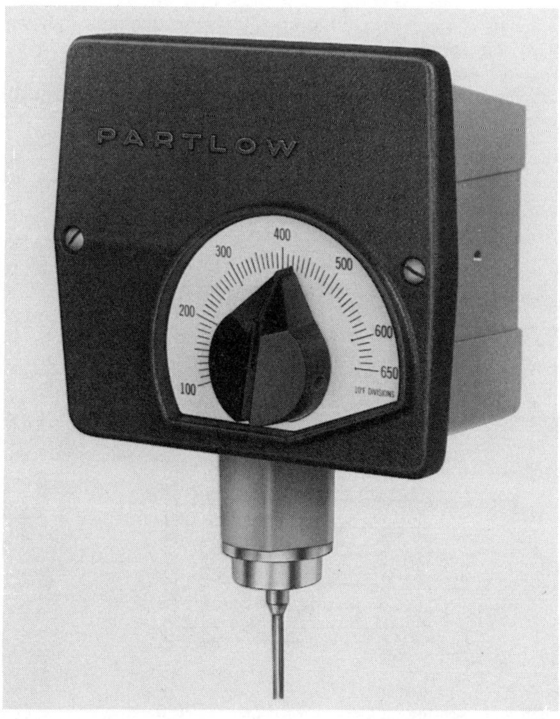

Fig. 1.5 High-limit filled-system device. (*Courtesy of the Partlow Corp.*)

applications should be backed up by an overriding safety device that is totally independent of the primary system.

It is good design practice for the high-limit safety device to incorporate a latched output that must be manually reset to resume normal operation. Also, the set-point adjustment on a safety device should be lockable or inaccessible to the operator.

Filled-system temperature controls should have built-in protection from operating above the maximum operating temperature (i.e., upper-scale value) of the filled-system sensor.

Figure 1.5 is an example of a filled-system high-limit temperature-control device.

1.2 TEMPERATURE REGULATORS

A temperature regulator is a temperature-controlling device that requires no external power and includes within itself a temperature sensor, a temperature reference input, a measuring and thermal actuating element, and the final control element—a valve. Power required for operation is derived from the controlled medium by way of temperature expansion or pressure effects. By the nature of their design, thermal regulators are proportioning controls.

Temperature regulators may be direct-actuated or pilot-actuated and self-contained or remote-sensing. In direct-actuated temperature regulators, the thermal actuating element is directly coupled to the valve plug and provides the force to open and close the valve. Line pressure is used by pilot-actuated regulators to effect valve motion.

In self-contained temperature regulators the entire thermal actuator is contained within the valve body and the actuator serves as the sensing element. The temperature that is controlled is that of the flow exiting the valve. The thermal actuators used in self-contained temperature regulators are bimetallic, liquid-filled, wax-filled, or simple metallic expansion elements. Self-contained temperature regulators are simple and comparatively inexpensive, and the absence of an external valve stem is an advantage with regard to potential for leakage. Common applications are in the regulation of coolant temperature for engines, compressors, and condensers and regulation of the temperature of service water supply. The latter application uses a three-way valve for mixing of hot and cold streams. A self-contained, direct-actuated temperature regulator is shown in Fig. 1.6.

Remote-sensing thermal regulators use a "filled-system" thermal actuator that includes a remotely locatable bulb-type sensor, a capillary connection tube, and an expansion element that powers the valve. Vapor-filled and organic liquid-filled systems (SAMA Classes II and IB; see Part 12, Chap. 6) are most commonly used. Set points are adjustable and the functional temperature range of a regulator is characteristic of the filling fluid. Proportional bandwidths are typically on the order of 10°F (6°C), with applications limited to those under about 500°F (260°C). Figure 1.7 is an illustration of a typical application of a temperature regulator with remote sensing.

Because of the relatively slow response, thermal regulators are best applied in situations having high thermal capacitance.

1.3 BIMETALLIC CONTROLS AND THERMOSTATS

Bimetallic sensors work on the principle of the expansion of metal as temperature increases. These sensors are constructed from two metals that have differing coefficients of thermal expansion. Most often, strips of the two metals are bonded together to form a leaf, coil, helix, or disk. As temperature increases, the bimetal sensor becomes deformed as the two metals expand at their respective and different rates (see Part 6, Chap. 6). This deformity is the mechanical movement that represents the temperature of the bimetal sensor. The movement of bimetal sensors can actuate a pointer to indicate temperature (see Part 12, Chap. 4), perform control functions, or compensate for temperature variations.

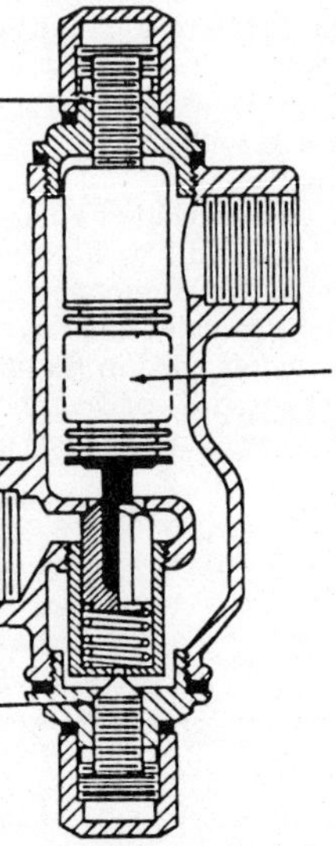

Fig. 1.6 Self-contained temperature regulator. [*From Instrument Engineers' Handbook, rev. ed. Béla G. Lipták (ed.). Copyright 1985 by the editor. Reprinted with permission of the publisher, Chilton Book Co., Radnor, Pa.*]

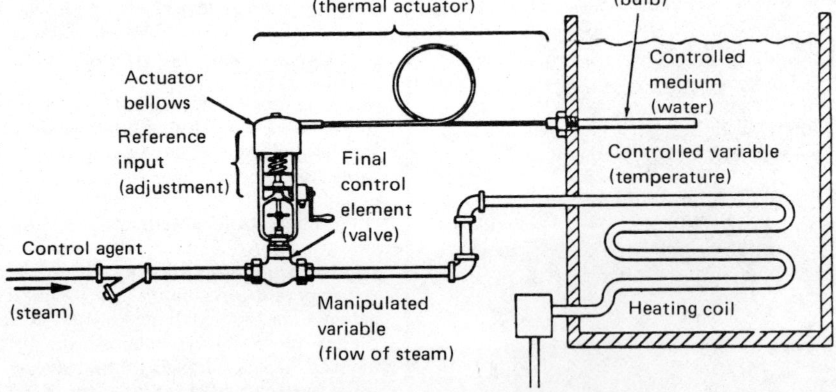

Fig. 1.7 Typical application of temperature regulator with remote sensing element. [*From Instrument Engineers' Handbook, rev. ed. Béla G. Lipták (ed.). Copyright 1985 by the editor. Reprinted with permission of the publisher, Chilton Book Co., Radnor, Pa.*]

Bimetallic sensors have common application in the ON/OFF switching control of electric circuits. Such devices are traditionally referred to as *thermostats*. The operation of the bimetallic sensor is such that one switching function (ON or OFF) is at one fixed temperature, the set point, and the reverse switching function occurs at another temperature that is different from the set point by an amount referred to as the *temperature differential*. Bimetallic thermostats have the advantage of being generally small in size.

Bimetallic elements used in thermostats can be classified as *creep-action* (Fig. 1.8) or *snap-action* (Fig. 1.9). Creep-action elements provide mechanical movements that are proportional to the sensed temperature. The nature of a snap-action element is that it is position-unstable with sufficient temperature change. Temperature differentials (hysteresis) for creep-action thermostats are typically 0.5°F (0.25°C) or higher, while those of snap action are typically 2 to 3°F (1 to 2°C) or higher. The mechanical action of snap-action elements allows larger gaps to be maintained between open electrical contacts, thus reducing electrical arcing and prolonging lifetime. Creep-action elements may have a mechanical adjustment for the set point. The set point of snap-action elements is generally fixed by design. The bimetallic element can itself be an electrical current-carrying component. In such cases, the electrical resistive heating of the element can influence the control operation.

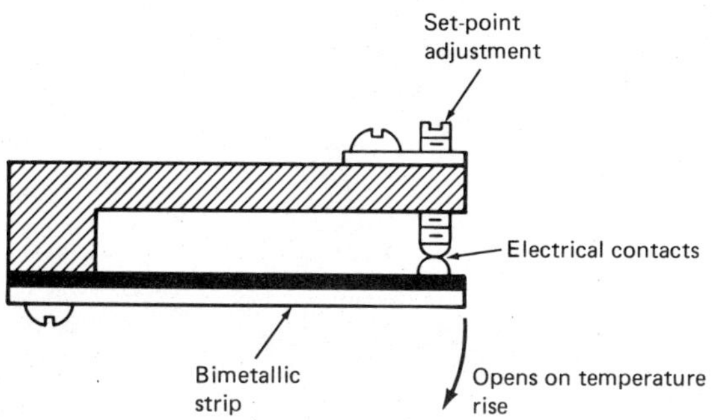

Fig. 1.8 Creep-action bimetallic thermostatic switch.

Bimetallic element thermostats are configured for both surface mounting and fluid immersion. One design uses the outer case of the thermostat as the high-expansion component of the bimetallic element, thereby avoiding the limitations associated with heat transfer into an enclosed bimetallic element.

Bimetallic element thermostats are available with set points from −100 to 1000°F (−73 to 540°C). They can be damaged by overheating. For safety controls, the bimetallic element can be designed to latch in the open-circuit position.

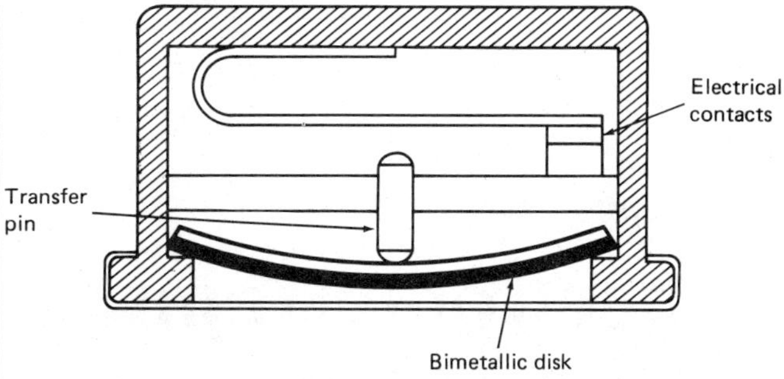

Fig. 1.9 Snap-action bimetallic thermostatic switch.

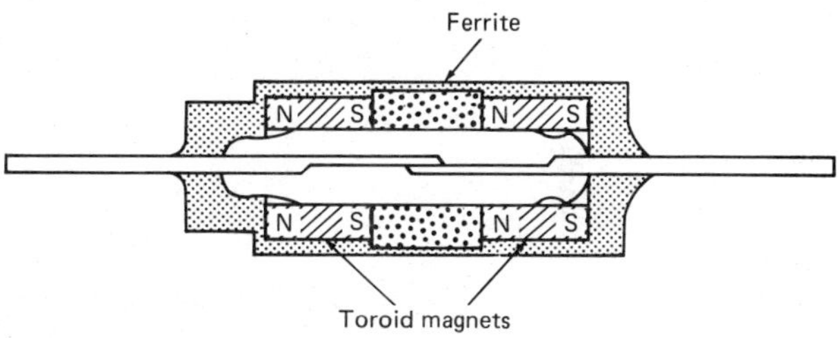

Fig. 1.10 Electroceramic thermostatic switch. (*Courtesy of Midwest Components, Inc.*)

1.4 MERCURY-IN-GLASS CONTROLS

The expansion of mercury in a glass tube is also used to provide thermostatic control. With an increasing temperature, the level of a column of mercury rises to complete a circuit between wires protruding into the glass tube. While the electrical current handling capability is more limited than that of bimetallic element thermostats, operating temperature differentials can be as small as 0.04°F (0.02°C). Applications are in the range of −35 to 400°F (−37 to 204°C). Set points can be fixed or adjustable.

1.5 ELECTROCERAMIC SWITCHES

Thermostatic or temperature-switching control is also achieved with reed switches operated by an electroceramic element. The operation of these devices is based on the phenomenon that certain materials exhibit significant changes in magnetic reluctance when the temperature varies in the vicinity of the Curie temperature of the material. A reed switch is actuated as a result of the changing magnetic field. Commercially available devices have electrical contact ratings on the order of 10 W, operate with about 2- to 3°F (1- to 2°C) differentials, and have fixed set points to 482°F (250°C). A typical configuration of an electroceramic temperature-control switch is shown in Fig. 1.10.

Chapter 2

Electronic Temperature Controllers

D. MITCHELL CARR
Eurotherm Corporation

2.1 INTRODUCTION

An electronic temperature controller uses electrical circuitry to compare the actual temperature, as received from a sensor, to the desired temperature setting and, from this, provides an output signal to either heat or cool the load. The sensor, typically a thermocouple, thermistor, pyrometer, or platinum resistance thermometer, provides an electrical signal corresponding to temperature. The controller can be either analog, using linear circuitry such as amplifiers and filters, or digital, employing microprocessors and other logic circuitry. With few exceptions, both analog and digital controllers use one of several fundamental control algorithms to establish the amount of heating and cooling to be delivered to the load.

2.2 BLOCK DIAGRAM

The block diagram shown in Fig. 2.1 highlights the key elements in a temperature controller. The actual measured temperature must first be converted into a form which can be used by the controller. Since temperature sensors do not provide linear, high-level signals, their output must be conditioned by filters and amplifiers until it is in a form which may be used by the other circuitry in the controller. For a

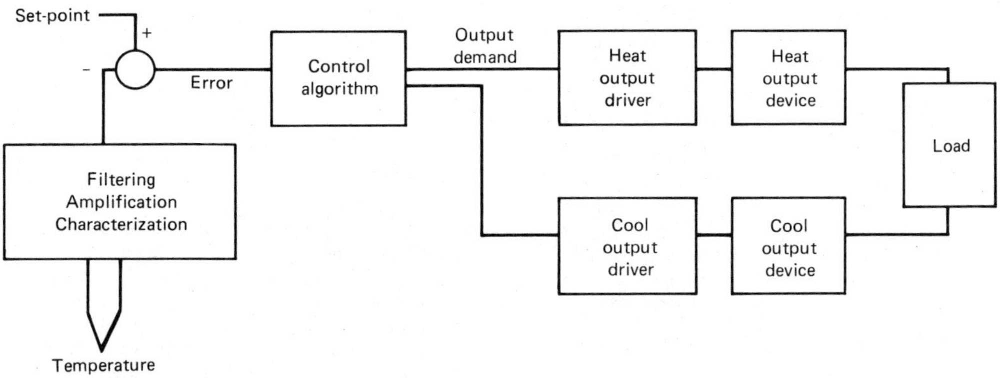

Fig. 2.1 Block diagram of temperature controller.

digital temperature controller, an additional stage is needed which converts the temperature signal into a digital format.

The temperature signal is subtracted from the desired temperature set point to obtain the error signal. The temperature controller is designed to minimize this error signal. The method used by the temperature controller to minimize this error signal is referred to as the *control algorithm.* The control algorithm derives the output demand which is needed to control the temperature. The output demand signal is then supplied to the output circuitry which provides the actual output that heats or cools the load.

2.3 ON/OFF CONTROL

An ON/OFF heating controller applies full heating power to the load if the temperature is below set point and no heating power if the temperature is above set point. In most ON/OFF controllers a hysteresis band is provided to prevent rapid output cycling in cases where the temperature signal is not stable or moves quickly. As shown in Fig. 2.2, with the temperature below set point the output will be fully on in an effort to bring the temperature up to set point. Once the temperature is equal to set point, the output will turn off to prevent any further heating. The temperature will continue to rise for a short time and then begin to fall. The controller output will remain off until the temperature is below set point by an amount equal to the hysteresis band at which time the output will once again turn on.

A controller with an ON/OFF cooling output behaves, with regard to the cooling action, in a way similar to the ON/OFF heating controller. If the temperature is above the cooling set point plus the hysteresis band, the output will be on in an effort to reduce the temperature through cooling action. The hysteresis band for an ON/OFF controller is above the cooling set point and is provided for the same reason as the heating hysteresis band. When the temperature falls to a

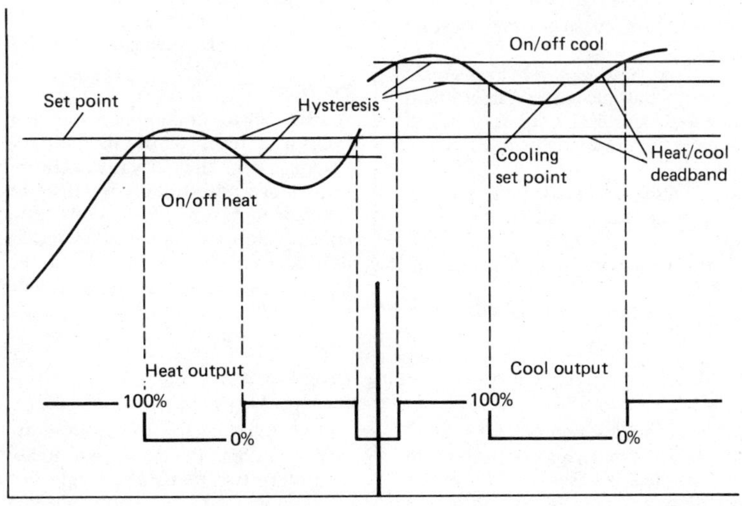

Fig. 2.2 ON/OFF control.

point equal to the cooling set point, the output will turn off. Once the temperature rises to the cooling set point plus the hysteresis, the output will again turn on. Generally, the cooling set point for a heat/cool controller is offset from the heating set point by a heat/cool deadband to prevent cycling between heating and cooling.

ON/OFF controllers are very inexpensive as compared to other types of controllers but, by design, oscillate. For this reason they are not suitable for all applications as the amplitude of the oscillation is at least as large as the hysteresis band setting. In setting up an ON/OFF controller the hysteresis band should be reduced to as small a setting as is allowed. The hysteresis band, however, should not be set so low as to cause excessive cycling of the output.

2.4 PID CONTROL

The mnemonic *PID* refers to *proportional-integral-derivative* control. Most temperature controllers which are not ON/OFF controllers use full PID control or some subset of the algorithm.

Definitions

Proportional Band

- *Proportional band* is usually expressed as a percent of the controller's set-point adjustment range or temperature display range.
- Proportional band is the temperature range around set point in which the controller will increase or decrease the output demand in proportion to the error in an effort to control the temperature. If a heating controller has a span of 800°F (444°C) and the proportional band is set to 10 percent, the output demand will be 100 percent at a point 80° (44.4°C) below set point, 50 percent at a point 40°F (22.2°C) below set point, etc. Error is often expressed in terms of proportional band.
- Proportional band is also expressed as gain. Gain is equal to 100 percent divided by the proportional band (in percent).

Integral Time Constant

- *Integral time constant* is expressed as a time constant usually in minutes or seconds.
- *Integral action* (also referred to as *automatic reset*) is a gradual increase or decrease in output demand as a result of an error between set point and measured temperature. The speed of the output change is determined by the integral time constant.
- Integral time constant is also expressed in *repeats per minute*, which is the inverse of the time constant.

Derivative Time Constant

- *Derivative time constant* is expressed as a time constant usually in minutes or seconds.
- *Derivative action* (also referred to as *rate*) produces additional output demand resulting from a rapid change of input.

Proportional Control

A *proportional-only* controller provides a proportional band adjustment but no other control term. This controller will adjust the output demand gradually based on the error. There will be an error, called *droop*, between set point and actual temperature which will be smaller with smaller proportional band settings. If the proportional band is too small, however, the controller will cause the output demand to oscillate thus causing the temperature to oscillate.

The proportional-only controller has an advantage over the ON/OFF controller in that it can be adjusted so that it does not oscillate. The drawback of this controller is the droop error between actual temperature and set point.

PD Control

A *proportional plus derivative* (PD) controller adds the transient response benefits of a derivative term. Under steady-state conditions there will be a droop error and no contribution from the derivative term. If the temperature changes, however, there will be a change in output demand caused by the derivative term in addition to the ratiometric change caused by the proportional band. A PD controller is often applied to very sluggish loads having a natural period of oscillation longer than 1 h.

Manual Reset

Some PD controllers offer *manual reset* as an option. The manual reset is a fixed output demand level which is added to the output demand from the proportional and derivative terms. The manual reset can be adjusted to eliminate the steady-state droop error, thus allowing the controller to keep the temperature near set point; however, as it is a manual setting, it cannot automatically compensate for changes in ambient conditions. For this reason, the manual reset adjustment may have to be changed frequently to maintain control at set point.

PI Control

A *proportional plus integral* (PI) controller has the advantage of automatically determining the needed steady-state power level. It will control at set point under normal conditions as a result of the integral action provided it has been set up properly. The addition of integral action allows the controller to gradually shift the output demand in an effort to reduce the error. This provides good compensation for changing conditions of load, dissipation, and power availability. PI controllers are often used to control very fast loads, loads having an inherent transport lag, or loads driven by a motor.

PID Control

A PID controller combines the benefits of controlling at set point as a result of the integral action with good transient response from the derivative action. The PID controller does have more adjustments to tune; however, except in some specialized cases, it provides the most stable and accurate control. Note that in many instances a PID controller is tuned as a PI controller simply because the operating personnel do not understand how to adjust the derivative action: this will give good control but will not have the transient response of a properly tuned PID controller.

Cooling Action

For the control of certain loads it is necessary to provide cooling to either bring the temperature down to the set point or to overcome the heat generated by an exothermic process. There are two items of concern in dealing with the cooling channel of a heat/cool controller:

1. The cooling action itself may have to follow a special algorithm to compensate for the characteristics of the cooling

medium. It is very common to pulse water with a solenoid valve to cool a load. If the load is at a temperature above the boiling point, this water will flash to steam and remove a great deal of heat due to the latent heat of vaporization. To handle this drastic cooling action through vaporization, the gain of the system must be very low when the cooling demand is low but high when the water is flowing and not vaporizing. This is accomplished by use of a special output-cycling algorithm as shown in Fig. 2.3 which automatically changes the gain of the output stage as the cooling demand changes by adjusting the cycling characteristic.

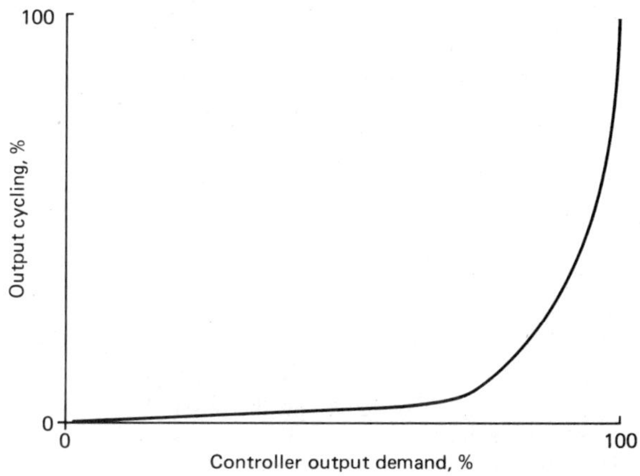

Fig. 2.3 Water-cooling algorithm.

2. The amount of cooling available often does not match the amount of heating available. For this reason, a controller tuned to control in heating will not be tuned optimally for cooling. Most heat/cool controllers provide a *relative cool gain* setting which allows the cooling action to be matched to the heating action. This is accomplished by altering the proportional band when cooling is required. The relative cool gain is typically a multiplier of the proportional band, but this varies between instruments. Some controllers provide a separate set of PID adjustments for cooling.

2.5 CONTROLLER TUNING

Temperature controllers must be tuned properly to work.

Tuning an ON/OFF Controller

For an ON/OFF controller, set the heat and cool hysteresis as small as possible. Be careful that the hysteresis is not too small, however, as this could cause excessive wear on devices such as mechanical contactors. The heat/cool deadband should be at least as large as the cool hysteresis and also large enough to ensure that there is no overlap between heating and cooling.

Closed-Loop Cycling Method

One classical method for tuning proportional, PI, or PID controllers is referred to as the *closed-loop cycling* method. To perform this procedure (Fig. 2.4):

1. Eliminate integral and derivative action.

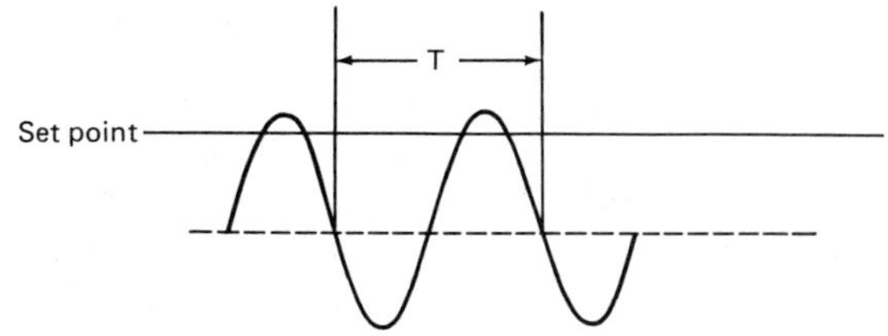

Fig. 2.4 Closed-loop cycling.

2. Reduce the proportional band until the control loop oscillates. (It may be necessary to invoke ON/OFF control.) Measure the period of oscillation T.
3. Increase the proportional band until the process is just stable as denoted by several cycles of oscillation before settling from a perturbation. This value of proportional band P is referred to as the point of *ultimate sensitivity*.
4. Refer to Table 2.1 for the values of proportional band and integral and derivative time constants.
5. Table 2.1 offers values needed to establish quarter-wave damping which may not be suitable for all loads. For adjusting the PID terms to establish alternative response characteristics, refer to Table 2.2 and Fig. 2.5.

Table 2.1. Closed-Loop Cycling Tuning Constants

Controller	Proportional band	Integral time constant	Derivative time constant
Proportional only (P)	$2P$	Not applicable	Not applicable
Proportional and integral (PI)	$2.2P$	$0.8T$	Not applicable
Proportional, integral, and derivative (PID)	$1.67P$	$0.5T$	$0.12T$

Process Reaction Curve Method

Another technique for establishing the correct tuning parameters is the *process reaction curve* method. The procedure to be used for this method is:

1. Place the controller in a manual override mode. (Some controllers offer a manual mode whereby the output can be adjusted directly.)
2. Allow the temperature to stabilize at some low level.

Table 2.2. Varying-Response Tuning Constants

Controller response	Proportional band	Integral time constant	Derivative time constant
Underdamped	P	$0.5T$	$0.125T$
Critically damped	$1.5P$	T	$0.167T$
Overdamped	$2P$	$1.5T$	$0.167T$

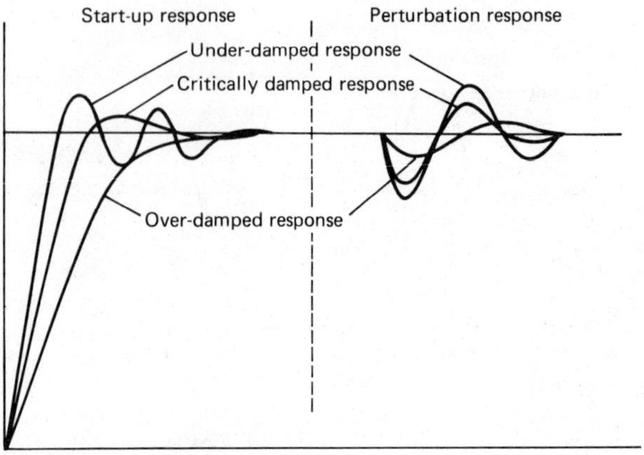

Fig. 2.5 Varying control response.

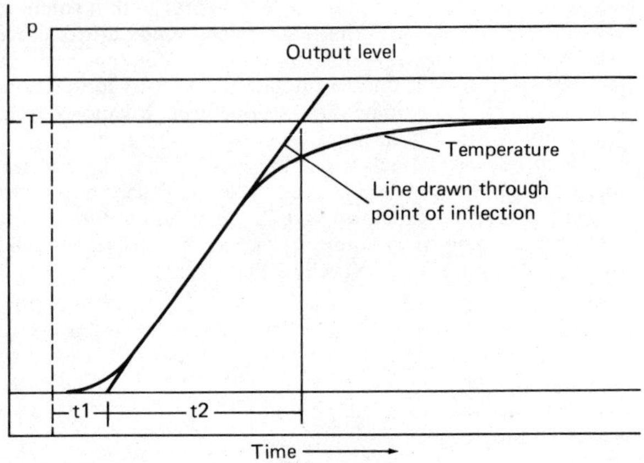

Fig. 2.6 Process reaction curve.

3. Increase the output level by a significant but safe amount, and allow the temperature to stabilize.
4. Record the response to establish a curve or chart recording which resembles the one shown in Fig. 2.6
5. From T = change in temperature expressed as a percent of setpoint range of the instrument and P = change of output used for the evaluation expressed as a percent of the maximum allowable output, calculate $K = T/P$.

Establish the tuning parameters based on the values listed in Table 2.3.

Cooling Action

The instrument must first be set up in heating before the *relative cool gain* parameter may be set. Once the instrument is forced into delivering cooling, observe the temperature to establish whether there is proper control. If the temperature is oscillating, the relative cool gain parameter must be reduced. If the cooling action is sluggish as compared to the heating action, the relative cool gain parameter should be increased.

If the temperature drops drastically as a result of one pulse of cooling, then the amount of cooling medium supplied must be reduced. In the case of water or CO_2 cooling the supply may be reduced by way of a flow controlling valve. Alternatively, it may be possible to reduce the duration of the output ON pulse, thereby reducing the cooling action.

2.6 OUTPUT STAGES

The output stage of a controller converts the control algorithm output demand signal into the physical output which is used to actuate the devices which heat or cool the load. These output stages are sometimes referred to as *drivers*.

Analog Outputs

Many controllers provide a linearly varying analog voltage or current as the output signal, typically 0 to 5 V or 4 to 20 mA. This is then connected to an external driver unit which actuates the heating or cooling device. In establishing this interface, it is sufficient to ensure that the controller output signal is of appropriate magnitude to meet the specifications of the external driver unit.

Cycling Outputs

In the instance of a controller with a cycling output stage, the output signal from the controller will actuate the heating or cooling device by cycling it full ON and full OFF as shown in Fig. 2.7. Varying thc ON time versus the OFF time adjusts

Table 2.3. Process Reaction Curve Tuning Constants

Controller	Proportional band	Integral time constant	Derivative time constant
Proportional only (P)	$\frac{Kt_1}{t_2(1 + t_1/3t_2)}$	Not applicable	Not applicable
Proportional and integral (PI)	$\frac{Kt_1}{t_2(0.9 + t_1/2t_2)}$	$\frac{t_1(30 + 3t_1/t_2)}{9 + 20t_1/t_2}$	Not applicable
Proportional and derivative (PD)	$\frac{Kt_1}{(1.25t_2 + t_1/6)}$	Not applicable	$\frac{t_1(6 - 2t_1/t_2)}{22 + 3t_1/t_2}$
Proportional, integral, and derivative (PID)	$\frac{Kt_1}{t_2(1.3 + t_1/6t_2)}$	$\frac{t_1(32 + 6t_1/t_2)}{13 + 8t_1/t_2}$	$\frac{4t_1}{11 + 2t_1/t_2}$

the output demand level. The definitions relating to the operation of this style of output are:

- ON *time* is the amount of time the output is supplying power.
- OFF *time* is the amount of time the output is not supplying power.
- *Cycle time* is ON time plus OFF time. (Usually stated as the value when the ON time is equal to the OFF time.)
- *Output level* is given by ON/(ON + OFF)
- Minimum ON (OFF) time is the minimum amount of time the output is allowed to be ON (OFF).

The cycle time will be divided into a period of time for which the output is ON and a period of time for which the output is OFF. The output level ratio is determined by the output demand signal from the control algorithm. The minimum ON time should be set to the minimum duration of time needed to successfully engage the output device. For example, if the cooling output minimum ON time is too short, the cooling fan will not rotate. The cycle time should be as short as possible but not be so short as to overstress or damage the equipment.

Phase-Angle Outputs

Certain heaters cannot tolerate being actuated by a cycling output stage and must be controlled using a *phase-angle* output stage. In the case of phase-angle control, a 60-Hz (50-Hz) waveform is continually presented to the heater. Power reduction is accomplished by dividing each half-cycle of this 60-Hz waveform into pieces and only allowing power supply during a small portion of each half-cycle as shown in Fig. 2.8. This type of output stage must be used for actuating transformer primaries, highly inductive loads, and lamps or other heaters such as tungsten or molybdenum disilicide, which undergo a drastic change in electrical resistance as the temperature changes. Phase-angle control is required to prevent blowing fuses as a result of transformer core saturation and also load resistance changes. In addition, it is an excellent method of power modulation for use on very fast loads. Two problems are associated with phase-angle control: it is generally more expensive and generates RF interference.

Output Devices

Many different types of devices can be actuated by the controller output to supply electric power to the load or heater being controlled. Each type has its advantages and disadvantages.

1. *Mechanical contactor*: A set of moving electrical contacts actuated by a coil. Typically actuated by 120 or 240 V ac.
   - *Advantages*: Inexpensive, reliable, may be protected by standard fuses and circuit breakers.
   - *Disadvantages*: Limited life, slow cycling (10- to 60-s cycle time), limited to slow applications, requires high voltage for drive signal.
2. *Mercury contactor*: A column of mercury actuated by a coil and forced up a tube and into contact with a pair of electrodes. Typically actuated by 120 or 240 V ac.
   - *Advantages*: Inexpensive, reliable, may be protected by standard fuses and circuit breakers, faster operation and longer life than mechanical contactors.
   - *Disadvantages*: Limited life, compared to solid-state contactor, medium to slow cycling (3- to 60-s cycle time), limited to medium to slow applications, requires high voltage for drive signal.
3. *Solid-state contactor*: Solid-state power-switching device. Actuated by a dc signal between 3 and 30 V.
   - *Advantages*: Fast cycling (32 ms to 1 s), safe low-level drive signal, inexpensive in the basic form, long life, available from some manufacturers with load failure detection circuitry.
   - *Disadvantages*: Must be protected with special high-speed fuses, high heat dissipation, limited ambient-temperature rating [122 to 158°F (50 to 70°C) maximum].
4. *SCR assemblies*: Solid-state power-switching device. Actuated by SCR gating pulses.
   - *Advantages*: May be used in any firing mode, including phase-angle mode, long life, fast control of high current in specialized applications.
   - *Disadvantages*: Must be protected by special high-speed fuses, high heat dissipation, limited ambient-temperature range [122 to 158°F (50 to 70°C)], special pulse drive signal needed for actuation.

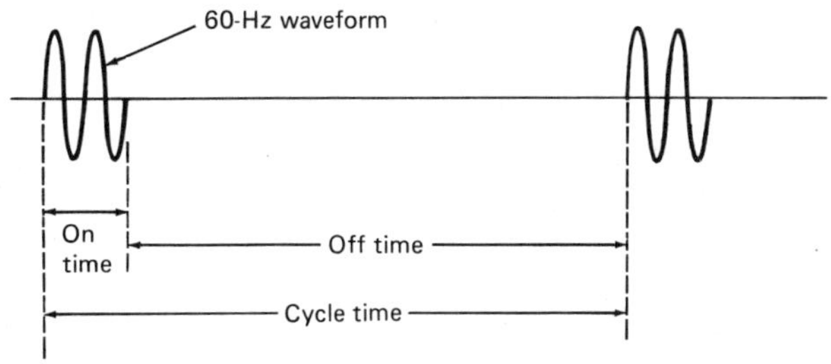

Fig. 2.7 Cycling output.

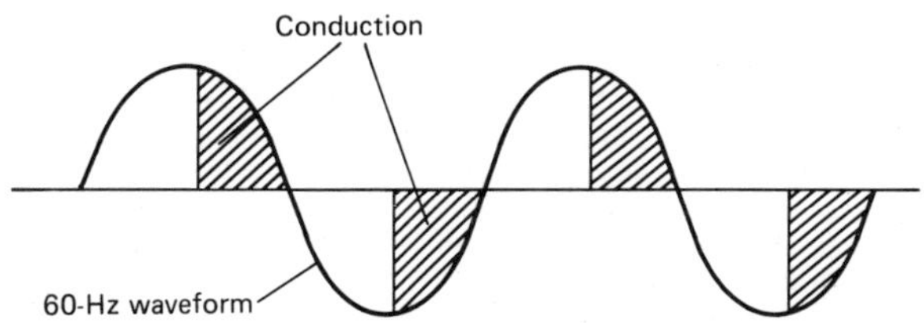

Fig. 2.8 Phase-angle output.

2.7 VALVE POSITIONERS

Some processes use a fluid such as fuel oil, gas, or water as the heating or cooling medium and employ a variable-position motor-actuated valve as the throttling device. In these instances the controller must have a means of positioning the valve. Because of this extra function of valve positioning, the controller has a different set of parameters to set up. The basic elements of the control algorithm are the PID terms; in addition, the controller output stage must be matched to the valve.

Position-Mode Control

A *position-mode* valve positioner uses the output demand signal from the PID algorithm to establish the desired position of the valve. The output stage of the controller measures the position of the valve from a feedback slide wire and pulses the motor open or closed to eliminate the error between desired position and actual position.

Velocity-Mode Control

In the case of a *velocity-mode* valve positioner, the output of a PD controller is used to adjust the velocity of the valve. If there is no error between set point and measured value, there is no velocity. This type of controller does not require the position slide wire as feedback to the controller, but it does mandate the use of motor travel switches in the motor actuator to restrict the movement of the valve. The controller does not have an integral term as this function is accomplished by the valve.

Tuning

Most valve positioners have two parameters in addition to the PID terms which determine the control performance. The *motor travel time* sets the duration of the pulses of power applied to the valve actuator. This should be adjusted to a value equal to the time needed for the valve to travel from the fully closed position to the fully open position. The second parameter, *deadband,* is the amount of error between set point and actual temperature that will be tolerated without causing the valve to move. The deadband parameter, even though it is designed to permit error, prevents excessive movement of the valve and thereby reduces wear. This should be set to the minimum value available which does not cause excessive valve movement.

2.8 ALARMS, INTERLOCKS, AND SHUTDOWN DEVICES

A control system should always be designed in such a way that if a failure occurs, it will have the least amount of consequence. Alarms and interlocks should employ relays and never use Triacs or other solid-state switching devices. The reason for this is that solid-state devices are unreliable when two or more are wired in series according to latching current criteria. Also, when using relays for alarm outputs, the relays should be configured so that they are energized in the safe condition and opening the alarm circuit will cause an alarm condition. By doing this, any failure of power, instrumentation, or wiring will result in an alarm condition.

Sensor Failure Action

Each controller will react to a sensor failure in a predictable manner. There are three distinct actions typical of controllers having an open circuit sensor:

- *Upscale break action*: On failure of the sensor, the input drifts to a high-temperature reading, thus causing heating to shut off and cooling to activate.
- *Downscale break action*: On failure of the sensor, the input drifts to a low-temperature reading, causing full heat to be applied.
- *Instantaneous break action*: The controller immediately detects that the sensor has failed and automatically reverts to the predetermined sensor break power level.

Once the controller establishes that the sensor has genuinely failed to an open-circuit condition, there will be one of the following results:

- Full cooling will be applied.
- Full heating will be applied.
- Both heating and cooling outputs will be disabled.
- A preset output level will be established.
- A preset power level will be established and the controller will be switched into manual override.

Very few controllers can detect the failure of a sensor to a short-circuit condition. In the case of a temperature controller with a thermocouple sensor, it will appear to the controller that the thermocouple is at ambient temperature or the temperature at the location of the short circuit. This could cause the controller to apply 100 percent heat to the load and damage the equipment.

If there is a possibility of damage to equipment when the output is set to full heating for an extended period of time, a *separate* over-temperature alarm unit should be used to shut down power.

In most instances this is the only way of preventing damage from a short circuit sensor.

Alarms and Interlocks

An *interlock* is a form of alarm used to prevent the actuation of a particular device until certain predetermined conditions are met.

Alarms and interlocks are used for two reasons: (1) to notify personnel of a problem and (2) to shut down equipment to prevent damage or injury. Most controllers are available with alarm functions as extra cost options that may be incorporated into the instrument. If the use of the alarm is to be an interlock or as a warning, then this is an appropriate method to be used.

If, however, the alarm is to be used for shutdown purposes to prevent damage or injury, the alarm should be a separate instrument connected to an independent sensor.

In this way, failure of the controller or control thermocouple will in no way affect the performance of the alarm.

2.9 ELECTRONIC CONTROLLER TECHNOLOGY

Electronic temperature controllers are available in two basic formats, analog and digital. Each has its advantages and disadvantages. These must be understood so the correct instrument may be purchased for a particular job.

Analog Controllers

Analog temperature controllers are based on linear-circuit technology employing amplifiers and filters for signal conditioning and control. In some cases these instruments offer digital indications for temperature and set point, but they are still fundamentally analog controllers.

Advantages

- Linear circuitry having infinite resolution with no "stair-step" quantification problems. Allows the analog controller the ability to maintain temperature to a particular point without hunting due to the finite resolution of an analog to digital converter.
- Inherently high RF noise immunity.
- Good reliability at elevated ambient temperatures.
- Capability of controlling very fast processes without regard to access times or analog-to-digital conversion time.
- Easy to modify for special applications.

Disadvantages

- No digital computer interface.
- Control parameter accuracy based on analog compo-

nents having appreciable component tolerances and thermal drift.

- Can be difficult to reconfigure for different input sensors or scale ranges.
- No opportunity for special algorithms such as self-optimizing control.
- Limited range of PID values.

Digital Controllers

Digital temperature controllers are microprocessor-based devices using analog to digital converters to receive the temperature from the sensor. Features for each of these controllers vary with manufacturer and individual controller. The following list of advantages relates to those items which may be offered but are not necessarily available for all instruments.

Advantages

- Digital computer interface for computer supervision and data logging.
- Wide range of digitally reproduceable PID tuning parameters.
- Specialized algorithms such as self-optimizing control.
- Simple to reconfigure for a different sensor or scale range.
- Better user interaction through the use of mnemonics and digital displays.

Disadvantages

- Finite input resolution—could be a problem with high-temperature, high-accuracy applications.
- Finite input update rate—could be a problem with very fast processes.
- More susceptible to RF interference than analog controllers.
- Not easily modified for special applications because of rigid software design.
- More affected [especially very-large-scale integrated-circuit (VLSI) devices such as microprocessors and memory devices] by elevated ambient temperatures than analog circuitry.

Multiloop Temperature-Control Systems

A number of specialized control packages are available which not only control several temperature zones from one instrument but also serve to control other aspects of a process. Packages are available which are capable of complete control of a plastics injection molding machine, a plastics extruder, or a semiconductor diffusion furnace, to name only a few. These instruments incorporate the necessary elements of control for all the temperature zones of the machine as well as the logic functions, alarm strategies, and sequencing operations.

These control packages are designed around two basic formats: scanning systems and distributed systems. A scanning system will use one microprocessor to perform all tasks by scanning each task in sequence. A distributed control package will employ a network of microprocessors, each performing a separate function but in constant communication with a master microprocessor. There are advantages and disadvantages to each.

1. *Scanning systems*
   - *Advantages*:
     - Low cost.
     - Smaller package.
     - All data handling performed by one microprocessor. Thus there is no problem associated with bringing information from one location to another in order to execute a high-speed sequencing task.
   - *Disadvantages*:
     - Less reliable since one failure of a microprocessor shuts down the entire system.
     - Slower because of the large number of tasks being performed by one microprocessor.
     - Reasonable specifications for the PID tuning constants not offered in some commercially available units as the timing of the system is dependent on the number of control loops which are active in the system.
     - Scanner usually based on reed relays which are mechanical devices having a finite life.
2. *Distributed systems*
   - *Advantages*:
     - High reliability. Because of the distribution of tasks to several processing elements, failure of one element does not cause complete shutdown.
     - High speed for each control function since each controller is separate.
     - System program is divided among many microprocessors making the individual programs simpler and more job-specific and thereby more reliable.
     - Typically implemented without the use of mechanical devices for scanning or data acquisition.
   - *Disadvantages*:
     - Higher cost than scanning system.
     - Internal network for communications between microprocessors can slow down the overall execution time.
     - Larger package.

In selecting a complete line control package it is essential that the features be examined for completeness: "Will the package really be capable of performing the entire task?" In addition, the execution time and control accuracy must be examined to ensure that they are adequate for the job.

Appendix

Classified Unit Conversions

Length

1 cm = 10 mm = 0.39370 in = 0.032808 ft = **0.01 m**

1 meter (m)* = 39.370 in = 3.2808 ft = 1.0936 yd
= 1 E+10 angstroms (Å)
= 1 E+6 micrometer (μm)†
= 1000 mm = 100 cm

1 in = 1 E+6 μin = 2.5400 cm = 0.083333 ft = **0.0254 m**

1 ft = 30.480 cm = 12 in = 0.33333 yd = **0.30480 m**

1 mi = 5280 ft = 1.6093 km

Area

1 cm^2 = 0.15500 in^2 = 1.0764 E−3 ft^2 = **0.0001 m^2**

1 m^2 = 1550.0 in^2 = 10.764 ft^2

1 in^2 = 1.2732 E+6 circular mils = 6.4516 cm^2

1 ft^2 = 929.03 cm^2 = 144 in^2 = **0.092903 m^2**

1 mi^2 = 2.7878 E+7 ft^2 = 640.00 acres = 2.5899 km^2

Volume

1 m^3 = 1 E+6 cm^3 = 61,024 in^3
= 1000 (liter) L = 35.315 ft^3

1 L = 1000 cm^3 = 0.26417 gal = **0.001 m^3**

1 in^3 = 16.387 cm^3 = 0.016387 L = 5.7870 E−4 ft^3

1 ft^3 = 28,316 cm^3 = 1728.0 in^3 = 29.922 quart
= 28.316 L = 7.4805 gal = **0.028316 m^3**

Mass

1 kilogram (kg) = 1000 g = 35.274 oz·m = 2.2051 lbm
= 0.06852 slug

1 lbm = 7000 grains = 453.59 g = 0.03108 slug

1 ton = 2000 lbm = **907.18 kg** = 0.90718 metric tons

Density

1 g/cm^3 = **1000 kg/m^3** = 1000 g/L
= 62.427 lbm/ft^3 = 8.345 lbm/gal

Density (*Continued*)

1 lbm/in^3 = 1728.0 lbm/ft^3 = 27.680 g/cm^3

1 lbm/ft^3 = **16.018 kg/m^3** = 0.13368 lbm/gal
= 0.016018 g/cm^3 = 5.7870 E−4 lbm/in^3

Time

1 h = **3600 second (s)** = 60 minute

1 day = **86,400 s** = 1440 min = 24 h

1 year = 8760 h = 365 days

1 solar year = 365.2422 days

Velocity

1 m/s = 196.85 ft/min = 3.2808 ft/s = 2.2369 mi/h

1 ft/s = 0.68182 mi/h = **0.3048 m/s**

1 mi/h = 1.6093 km/h = 1.4667 ft/s

Acceleration

1 m/s^2 = 39.370 in/s^2 = 3.2808 ft/s^2

1*g* (standard free fall) = 32.174 ft/s^2 = **9.8067 m/s^2**

Volume Flow

1 gal/min = 7.4805 ft^3/min
= 0.12468 ft^3/s
= **6.3090 E−5 m^3/s**

1 m^3/min = 1000 L/min = 264.17 gal/min
= 35.315 ft^3/min = 16.667 L/s
= 0.58858 ft^3/s

1 gal/h = 63.090 cm^3/min = 1.0515 cm^3/s

Mass Flow

1 kg/s = 7936.6 lbm/h = 132.28 lbm/min

1 lbm/h = 7.5599 g/min = 0.45359 kg/h

Force

1 newton (N) = 1 E+5 dyn = 0.22481 lbf

1 lbf = 444,822 dyn = 16 ozf = **4.44822** N = 0.45359 kgf

Force per Unit Length

1 lbf/ft = **14.594 N/m** = 0.14594 N/cm

Pressure

1 atm = **101,325 pascal (Pa)** = 2116.2 lbf/ft^2
= 760 mm-Hg at 32°F = 14.696 lbf/in^2
= 33.900 ft-water at 39.2°F = 29.921 in-Hg at 32°F
= 1.0133 bar

1 Pa = 1 N/m^2 = 0.01 mbar = 1.4504 E−4 lbf/in^2
= 1 E−5 bar

1 lbf/in^2 = **6894.7 Pa** = 144 lbf/ft^2
= 51.715 mm-Hg at 32°F = 27.681 in-water at 39.2°F
= 2.3068 ft-water at 39.2°F = 2.0360 in-Hg at 32°F

1 torr = 1 mm-Hg at 32°F

Angle

1 radian (rad) = 63.662 grad = 57.296 degrees (°)

1° = 3600 seconds = 60 minutes = **0.017453 rad**

1 revolution = 400 grad = 360° = **6.2832 rad**

Heat and Energy

1 joule (J) = 1 E+7 erg = 1 N·m = 0.23885 g·cal

1 Btu = **1055.06 J** = 778.17 ft·lbf = 252.00 g·cal

1 kWh = **3.600 E+6 J** = 3412.1 Btu

Heat Flow and Power

1 watt (W) = 3.4121 Btu/h = 1 J/s = 0.73756 ft·lbf/s

1 hp = 2544.4 Btu/h = **745.70 W** = 550 ft·lbf/s

1 boiler hp = 33,471 Btu/h = **9809.5 W**

1 ton of refrigeration = 12,000 Btu/h = **3516.8 W**

Heat Flow per Unit Length

1 W/m = 1.0400 Btu/(h·ft)

1 W/in = 40.946 Btu/(h·ft) = **39.370 W/m**

Heat and Energy Flux

1 Btu/(h·ft^2) = **3.1544 W/m²** = 2.035 E−3 W/in^2

1 W/in^2 = **1550.0 W/m²** = 491.35 Btu/(h·ft^2)
= 3.4121 Btu/(h·in^2) = 0.15500 W/cm^2

Specific Energy, Enthalpy

1 Btu/lbm = **2,326.0 J/kg** = 0.55556 g·cal/g

1 g·cal/g = 1.8 Btu/lbm

Specific Energy per Degree, Specific Heat

1 Btu/(lbm·°F) = **4186.8 J/(kg·K)** = 1 g·cal/(g·°C)

Thermal Conductivity

1 Btu/(h·ft·°F) = 12 Btu·in/(h·ft^2·°F) = **1.7307 W/(m·K)** =
1.4882 kg·cal/(h·m·°C)
= 1.7307 E−2 W/(cm·°C)
= 4.1338 E−3 g·cal/(s·cm·°C)

1 W/(m · K) = 6.9335 Btu·in/(h·ft^2·°F)
= 0.8599 kg·cal/(h·m·°C)
= 0.57779 Btu/(h·ft·°F)
= 2.3885 E−3 g·cal/(s·cm·°C)

Heat Transfer Coefficient

1 Btu/(h·ft^2·°F) = **5.6783 W/(m²·K)**
= 4.8825 kg·cal/(h·m^2·°C)
= 0.48825 g·cal/(h·cm^2·°C)
= 5.6783 E−4 W/(cm^2·°C)
= 1.3563 E−4 g·cal/(s·cm^2·°C)

1 W/(m²·K) = 0.85986 kg·cal/(h·m^2·°C)
= 0.17611 Btu/(h·ft^2·°F)
= 0.085986 g·cal/(h·cm^2·°C)
= 1 E−4 W/(cm^2·°C)
= 2.3885 E−5 g·cal/(s·cm^2·°C)

Thermal Resistivity

1 h·ft·°F/Btu = 241.91 s·cm·°C/g·cal = 57.779 cm·°C/W
= 0.67197 h·m·°C/kg·cal = **0.57778 m·K/W**
= 0.083333 h·ft^2·°F/(Btu·in)

1 m·K/W = 418.68 s·cm·°C/g·cal = 100 cm·°C/W
= 1.7307 h·ft·°F/Btu = 1.1630 h·m·°C/kg·cal
= 0.14423 h·ft^2·°F/(Btu·in)

1 s·cm·°C/g·cal = 0.23884 cm·°C/W
= 4.1338 E−3 h·ft·°F/Btu
= 2.7778 E−3 h·m·°C/kg·cal
= **2.3885 E−3 m·K/W**

Thermal Resistance

1 h·ft^2·°F/Btu = 7373.4 s·cm^2·°C/g·cal
= 1761.1 cm^2·°C/W = 2.0482 h·cm^2·°C/g·cal
= 0.20482 h·m^2·°C/kg·cal = **0.17611 m²·K/W**

1 m²·K/W = 41,868 s·cm^2·°C/g·cal = 10,000 cm^2·°C/W
= 11.63 h·cm^2·°C/g·cal = 5.6783 h·ft^2·°F/Btu
= 1.163 h·m^2·°C/kg·cal

1 s·cm^2·°C/g·cal = 0.23884 cm^2·°C/W
= 2.7778 E−4 h·cm^2·°C/g·cal
= 1.3562 E−4 h·ft^2·°F/Btu
= 2.7778 E−5 h·m^2·°C/kg·cal
= **2.3884 E−5 m²·K/W**

Thermal Diffusivity
1 ft^2/h = 0.092903 m^2/h = 2.7778 E−4 ft^2/s
= **2.5806 E−5 m^2/s**
1 m^2/s = 38,750 ft^2/h = 3600 m^2/h = 10.764 ft^2/s

Temperature
Degrees Celsius ($t_{°C}$) = T_K − 273.15
= ($t_{°F}$ − 32)/1.8
Degrees Kelvin (T_K) = $T_{°R}$/1.8 = $t_{°C}$ + 273.15
Degrees Fahrenheit ($t_{°F}$) = $T_{°R}$ − 459.67
= 1.8 × $t_{°C}$ + 32
Degrees Rankine ($T_{°R}$) = 1.8 × T_K = $t_{°F}$ + 459.67

Viscosity (Kinematic)
1 centistoke = 0.038750 ft^2/h = 0.01 stoke
= 0.0036 m^2/h = 1.0764 E−5 ft^2/s
= **1 E−6 m^2/s**
1 ft^2/h = 25.806 centistoke = 0.25806 stoke
= 0.092903 m^2/h = 2.7778 E−4 ft^2/s
= **2.5806 E−5 m^2/s**
1 m^2/s = 1 E+6 centistoke = 38,750 ft^2/h = 1 E+4 stoke
= 3600 m^2/h = 10.764 ft^2/s

Viscosity (Dynamic)
1 centipoise (cP) = 2.4191 lbm/(h·ft) = 0.01 poise
= **0.001 pascal·s (Pa·s)**
= 6.7197 E−4 lbm/(s·ft)
= 2.0885 E−5 lbf·s/ft^2
= 1.4504 E−7 lbf·s/in^2

Viscosity (Dynamic) (*Continued*)
1 Pa·s = 2419.1 lbm/(h·ft) = 1000 cP = 10 poise
= 0.67197 lbm/(s·ft) = 0.020885 lbf·s/ft^2
= 1.4504 E−4 lbf·s/in^2
1 lbm/(h·ft) = 0.41338 cP = 4.1338 E−3 poise
= **4.1338 E−4 Pa·s**
= 2.7778 E−4 lbm/(s·ft)
= 8.6336 E−6 lbf·s/ft^2
= 5.9955 E−8 lbf·s/in^2

Electrical Resistivity
1 ohm·cm (Ω·cm) = 1 E+9 abohm·cm
= 6.0153 E+6 Ω·circular-mil per ft
= 1 E+6 μΩ·cm = 0.39370 Ω·in
= **0.01 Ω·m**
= 1.1126 E−12 statohm·cm

Form unit conversion factors from equalities between the initial units and the desired units. For example, convert 5 ft to the units of meters:

$$5 \text{ ft} = 5 \text{ ft} \times \frac{1.0 \text{ m}}{3.2808 \text{ ft}} = 1.524 \text{ m}$$

Convert 5 grains to the units of grams:

$$5.0 \text{ grains} = 5.0 \text{ grains} \times \frac{453.59 \text{ g}}{7000 \text{ grains}} = 0.3239 \text{ g}$$

*Preferred SI units are printed in boldface type.
† The exponential notation: E+10 = × 10^{10}; E−4 = × 10^{-4}.
SOURCE: The unit conversions presented are based on ASTM "Standard for Metric Practice," Designation No. E 380 - 85; the unit conversions listed for the British thermal unit and the calorie are based on the "International Table" values as defined by the Fifth International Conference on the Properties of Steam, 1956.

Index